IRRIGATION OF AGRICULTURAL LANDS

AGRONOMY

A Series of Monographs Published by the
AMERICAN SOCIETY OF AGRONOMY

Monographs 1 through 6, published by Academic Press, Inc., should be ordered from:
Academic Press, Inc., 111 Fifth Avenue, New York, New York   10003

Monographs 7 through 13, published by the American Society of Agronomy, should be ordered from: American Society of Agronomy, 677 South Segoe Road, Madison, Wisconsin, USA   53711

IRRIGATION OF
AGRICULTURAL LANDS

Edited by

ROBERT M. HAGAN

PROFESSOR OF WATER SCIENCE, DEPARTMENT OF WATER SCIENCE AND
ENGINEERING, UNIVERSITY OF CALIFORNIA, DAVIS, CALIFORNIA

HOWARD R. HAISE

RESEARCH SOIL SCIENTIST, SOIL AND WATER CONSERVATION RESEARCH DIVISION
AGRICULTURAL RESEARCH SERVICE, US DEPARTMENT OF AGRICULTURE
FORT COLLINS, COLORADO

TALCOTT W. EDMINSTER

ASSOCIATE DIRECTOR, SOIL AND WATER CONSERVATION RESEARCH DIVISION
AGRICULTURAL RESEARCH SERVICE, US DEPARTMENT OF AGRICULTURE
BELTSVILLE, MARYLAND

Managing Editor: R. C. DINAUER

Number 11 in the series
AGRONOMY

American Society of Agronomy, Publisher
Madison, Wisconsin, USA
1967

Second Printing 1969
Third Printing 1974

The American Society of Agronomy, Inc.
677 South Segoe Road, Madison, Wisconsin, USA 53711

Library of Congress Catalog Card Number: 67-17458

Printed in the United States of America

GENERAL FOREWORD

AGRONOMY—An ASA Monograph Series

Several years ago members of the American Society of Agronomy realized the need for comprehensive treatments of specific subject-matter areas in agronomy. A series of monographs entitled "Agronomy" resulted and the first number was published in 1949. The Academic Press, Inc., of New York was the first publisher of the monographs, since the society, a nonprofit organization with no cash reserve, was not initially able to finance the project. In fact, the first six volumes of the series, which were edited by Dr. A. G. Norman, were published by Academic Press, Inc., the source from which they are available today.

During the period 1949–1957 the American Society of Agronomy developed considerably. By 1957 the society operated a headquarters office in Madison, Wisconsin, with a competent editorial staff. Its financial position had improved to the extent that it was able to pursue the monograph project independently, including complete financing and publishing of the series. In recent years this activity of the society has flourished.

Irrigation of Agricultural Lands is the 11th monograph of the Agronomy series. It comes at a time when the science and application of irrigation practices are crucial factors in areas where a delicate balance exists between the supplies of food and fiber and the demands of an exploding population. The subject of irrigation has attracted international attention for many years although scientists of different nations have not always agreed on its underlying principles. The scope of this monograph and the geographical distribution of the authors affirm its importance in serving the needs of scientists throughout the world. The American Society of Agronomy proudly presents this publication for the benefit of mankind.

The eighth number in the series was *Drainage of Agricultural Lands*. Already a supplement to this monograph is in preparation because of recent advances made in the subject and the heavy demand for current information. This publication on irrigation is a timely supplement to the overall subject of water for agriculture.

Two additional Agronomy monographs will be appearing in 1967. Number 12 will be entitled *Soil Acidity and Liming* and number 13 will be entitled *Wheat and Wheat Improvement*. Since soil acidity and liming are important problems affecting crop production in most, if not all, of the world's agricultural nations and since wheat is undoubtedly one of the most important world grain crops, it is felt that these two publications will find wide application and will be of significant educational value.

In 1965, Agronomy monograph 9 on *Methods of Soil Analysis* was printed in two parts: "Part I—Physical and Mineralogical Properties, Including Statistics of Measurement and Sampling" and "Part II—Chemical and Biological Properties." Monograph 10, *Soil Nitrogen* also appeared in late 1965 with a comprehensive treatment of the role of this all important plant nutrient in the soil environment. It might be added that monograph 7, the first published by the American Society of Agronomy, was on the subject of *Oats and Oat Improvement*. Copies of all volumes of the Agronomy monographs beginning with No. 7 through No. 13 may be obtained from the American Society of Agronomy, 677 South Segoe Road, Madison, Wisconsin, 53711.

The fact that the Agronomy monograph series consists of titles primarily in the areas of soil science and crop science should come as no surprise. Members of the American Society of Agronomy are for the most part also members of the Crop Science Society of America and the Soil Science Society of America. The latter societies are outgrowths of the American Society of Agronomy and, in spite of their autonomy and completely separate professional identities, are still closely associated with the founding society. This tri-society association has made it possible for ASA, CSSA, and SSSA to work harmoniously together, to share headquarters office and staff in Madison, and to publish material such as is found in this monograph series in the furtherance of their many mutual professional and scientific objectives.

January, 1967

Matthias Stelly
Executive Secretary-Treasurer
AMERICAN SOCIETY OF AGRONOMY
CROP SCIENCE SOCIETY OF AMERICA
SOIL SCIENCE SOCIETY OF AMERICA

FOREWORD

The application of water to agricultural lands for the purpose of irrigation is one of the alternate uses of this natural resource in many areas. It is essential that water be used effectively and efficiently, whether the supply is limited or excessive.

The practice of irrigation has sometimes been considered to be more of an art than a science. However, present knowledge as revealed in this monograph tends to belie this concept. Brought together in one comprehensive volume by authorities in many professional fields are the principles that form the basis of scientific irrigation—from development of water to its use and reuse. Because of the breadth of material covered and the depth in which it is reviewed, this publication by the American Society of Agronomy will be valuable to all those involved with irrigation, whether in teaching or research, development or practice, or decision making.

The need for a unified reference book for the encouragement and improvement of academic courses in irrigation has been urgent. For despite the growing recognition of its importance, irrigation is now being taught in relatively few institutions and the scope of existing courses varies widely. Where irrigation is taught from the engineering viewpoint, the soil, plant, and agricultural aspects of the subject may not receive sufficient emphasis. Where instruction is given from the viewpoint of the crop and soil scientist, little attention may be given to the engineering aspects of irrigation. This monograph covers the subject from a variety of viewpoints so that it should be useful to instructors and students in all disciplines concerned.

Irrigation of Agricultural Lands should increase the utilization of knowledge now widely scattered in many publications and stimulate new research. It should bring about increased awareness of the information available from other disciplines. It should promote effective cooperation between engineers, soil and crop scientists, and other professional groups whose combined efforts are needed to attain an abundant and permanently successful irrigation agriculture.

The officers and members of the American Society of Agronomy wish to acknowledge the tremendous effort and time expended by the editors, contributing authors, and all others concerned with the preparation of this monograph.

January 1967

ROBERT S. WHITNEY, *President*
American Society of Agronomy

PREFACE

Economic and social development depends upon the achievement of increased agricultural production. This often requires the opening of additional lands to agriculture through new irrigation projects or the improvement of existing irrigation systems and practices to ensure efficient water use and continued productivity. A recent report of the Food and Agriculture Organization of the United Nations suggests that: ". . . improved water management (including irrigation and drainage practices) can probably do more towards increasing food supplies and agricultural income in the irrigated areas of the world than any other agricultural practice." Science and technology in soils, water, plants, and engineering are now sufficiently advanced, if properly implemented, to transform irrigation from an age-old art into a modern science.

Irrigation needs and practices necessarily vary widely. This greatly complicates the planning of new irrigation projects or the operation of existing irrigation systems and irrigated farms. Yet one important benefit of science is the ability to predict what results can be expected in given situations. To make such predictions for situations related to irrigation, the skillful combination of knowledge in such professional fields as climatology, geology, ecology, crop science, soil science, water science and engineering, economics, and other social sciences is essential. The lack of a unifying reference work related to irrigation has made it difficult to locate and utilize the best available knowledge from these many disciplines. There is also an urgent need for a unifying reference work to encourage and improve the teaching of irrigation. Few institutions now offer instruction covering all aspects of irrigation.

Irrigation of Agricultural Lands is designed to provide a comprehensive treatment of the broad field of irrigation. So that the monograph would represent a synthesis of concepts and experiences from many sections of the world, author teams for most chapters were deliberately selected to represent widely separated geographic areas and points of view. Although this monograph emphasizes soil and plant factors involved in planning and operating an irrigation enterprise, chapters are included that summarize knowledge of other factors requiring consideration. The discussion of economics is limited to certain economic principles because their application will vary so greatly among the countries of the world. For similar reasons, legal aspects of irrigation projects have been omitted. Construction details for irrigation works are not included because of the great variety of methods and materials used and the availability of engineering publications, some of which are referenced in this monograph.

This monograph is a comprehensive reference volume which summarizes basic theories, outlines principles, and illustrates applications in practice, and is not a handbook or "how-to-do-it" manual. In some of the 62 chapters, detailed, highly theoretical discussions are presented; in others, theories and principles are illustrated by summarizing irrigation practices found useful in a variety of situations. Such information is of value to both water supply organizations and agriculturalists in the planning, design, and operation of irrigation projects. Detailed bibliographies at the end of each chapter provide references for specialists in each field. This monograph should be useful in any part of the world regardless of the local climate, water supply, soils, and crops.

Although *Irrigation of Agricultural Lands* is published by the American Society of Agronomy, its preparation has involved generous contributions by members of the American Society of Agricultural Engineers, American Society of Civil Engineers, American Society of Plant Physiologists, and several other professional societies. The Editors wish to express their most sincere appreciation to the authors, to the numerous reviewers, and to the officers and staff of the American Society of Agronomy for their contributions, patience, and understanding over the many years involved in completing this monograph.

January, 1967

ROBERT M. HAGAN
HOWARD R. HAISE
TALCOTT W. EDMINSTER

CONTRIBUTORS

Duwayne M. Anderson Soil Physicist, US Army Cold Regions Research and Engineering Laboratory, Hanover, New Hampshire

David E. Angus Senior Research Scientist, Division of Meteorological Physics, Commonwealth Scientific and Industrial Research Organization, Victoria, Australia

Orlin Biddulph Professor of Botany, Department of Botany, Washington State University, Pullman, Washington

James W. Biggar Associate Irrigationist, Department of Water Science and Engineering, University of California, Davis, California

A. Alvin Bishop Head, Department of Agricultural and Irrigation Engineering, Utah State University, Logan, Utah

Milton L. Blanc Research Climatologist and Field Research Coordinator, Environmental Data Service, Environmental Science Services Administration, US Department of Commerce, Tempe, Arizona

Gerard H. Bolt Professor of Soil Physics and Chemistry, State Agricultural University, Wageningen, Netherlands

Robert H. Burgy Professor of Water Science and Civil Engineering, Department of Water Science and Engineering, University of California, Davis, California

Robert B. Campbell Soil Scientist (Physics), Coastal Plains Soil and Water Conservation Research Center, Agricultural Research Service, US Department of Agriculture, Florence, South Carolina (formerly Agronomist, Hawaiian Sugar Planters' Association, Honolulu, Hawaii)

Carl W. Carlson Assistant Director, Soil Management Investigations, Soil and Water Conservation Research Division, Agricultural Research Service, US Department of Agriculture, Beltsville, Maryland

John R. Carreker Research Investigations Leader, Soil and Water Conservation Research Division, Agricultural Research Service, US Department of Agriculture, Athens, Georgia

John W. Cary Soil Scientist, Snake River Conservation Research Center, Soil and Water Conservation Research Division, Agricultural Research Service, US Department of Agriculture, Kimberly, Idaho

Jerald E. Christiansen Professor of Civil and Irrigation Engineering, Water Research Laboratory, College of Engineering, Utah State University, Logan, Utah

Francis E. Clark Chief Microbiologist, Soil and Water Conservation Research Division, Agricultural Research Service, US Department of Agriculture, Fort Collins, Colorado

Arthur T. Corey Professor of Agricultural Engineering, Department of Water Science and Engineering, Colorado State University, Fort Collins, Colorado

Wayne D. Criddle Consulting Engineer, Clyde-Criddle-Woodward, Inc., Salt Lake City, Utah

Herbert B. Currier Professor of Botany and Botanist in the Experiment Station, Department of Botany, University of California, Davis, California

Robert E. Danielson Professor of Agronomy, Department of Agronomy, Colorado State University, Fort Collins, Colorado

John R. Davis Dean, College of Engineering and Architecture, University of Nebraska, Lincoln, Nebraska

Paul R. Day Professor of Soil Physics and Chairman, Department of Soils and Plant Nutrition, University of California, Berkeley, California

William W. Donnan Branch Chief, Southwest Branch, Soil and Water Conservation Research Division, Agricultural Research Service, US Department of Agriculture, Riverside, California

Walton H. Durum Staff Scientist, Quality of Water Branch, Geological Survey, US Department of the Interior, Washington, D.C.

Paul C. Ekern, Jr. Hydrologist and Professor of Soils, Water Resources Research Center, University of Hawaii, Honolulu, Hawaii

Dwight C. Finfrock Agronomist, The Rockefeller Foundation, Bangkok, Thailand

Milton Fireman Soil and Water Specialist, Tipton and Kalmbach, Inc., Denver, Colorado

Joel E. Fletcher Professor of Hydrology, Civil and Irrigation Engineering, Utah Water Reseach Laboratory, Utah State University, Logan, Utah

W. J. Flocker Associate Olericulturist, Department of Vegetable Crops, University of California, Davis, California

David M. Gates Director, Missouri Botanical Garden and Professor of Botany, Department of Botany, Washington University, St. Louis, Missouri (formerly, Professor of Natural History, Institute of Arctic and Alpine Research, University of Colorado, Boulder, Colorado)

Lewis O. Grant Associate Professor of Atmospheric Science, Department of Atmospheric Science, Colorado State University, Fort Collins, Colorado

Niranjan D. Gulhati Formerly Secretary, Ministry of Irrigation and Power, Government of India, and Past-president International Commission of Irrigation and Drainage, New Delhi, India

Jay L. Haddock　Research Soil Scientist, Soil and Water Conservation Research Division, Agricultural Research Service, US Department of Agriculture, Logan, Utah

Robert M. Hagan　Professor of Water Science (Irrigation) and Irrigationist in the Agricultural Experiment Station, Department of Water Science and Engineering, University of California, Davis, California

Howard R. Haise　Research Soil Scientist, Soil and Water Conservation Research Division, Agricultural Research Service, US Department of Agriculture, Fort Collins, Colorado

Warren A. Hall　Professor of Engineering, Water Resources Center, University of California, Los Angeles, California

Ronald J. Hanks　Research Soil Scientist, Soil and Water Conservation Research Division, Agricultural Research Service, US Department of Agriculture, Fort Collins, Colorado

Delbert W. Henderson　Associate Professor, Department of Water Sciences and Engineering, University of California, Davis, California

Archie D. Hess　Chief, Disease Ecology Section, Public Health Service, US Department of Health, Education, and Welfare, Greeley, Colorado

Robert H. Hilgeman　Horticulturist, Citrus Branch Station, University of Arizona, Tempe, Arizona

John W. Holmes　Principal Research Scientist, Division of Soils, Commonwealth Scientific and Industrial Research Organization, Adelaide, S. A., Australia

Marvin Hoover　Extension Cotton Specialist, University of California, Shafter, California

Clyde E. Houston　Extension Irrigation and Drainage Engineer, Department of Water Science and Engineering, University of California, Davis, California

Roger P. Humbert　Western Director, American Potash Institute, Inc., Los Gatos, California

Allen S. Humpherys　Research Agricultural Engineer, Snake River Conservation Research Center, Soil and Water Conservation Research Division, Agricultural Research Service, US Department of Agriculture, Kimberly, Idaho

Theron B. Hutchings　State Soil Scientist, Soil Conservation Service, US Department of Agriculture, Salt Lake City, Utah

Marvin E. Jensen　Research Agricultural Engineer, Snake River Conservation Research Center, Soil and Water Conservation Research Division, Agriculture Research Service, US Department of Agriculture, Kimberly, Idaho

Cornelis Kalisvaart　Senior Scientific Officer and Head of the Agricultural Research Department, Ijsselmeerpolders Development and Settlement Authority, Kampen, Netherlands

Amand N. Kasimatis Extension Viticulturist, Agricultural Extension Service, University of California, Davis, California

Wesley Keller Research Leader, Crops Research Laboratory, Arid Pasture and Range Investigations, Forage and Range Research Branch, Agricultural Research Service, US Department of Agriculture, Logan, Utah

Ludwig L. Kelly Formerly Chief Hydrologist, Soil and Water Conservation Research Division, Agricultural Research Service, US Department of Agriculture, Beltsville, Maryland (now retired)

William D. Kemper Research Soil Scientist (Physics) and Associate Professor, Northern Plains Branch, Soil and Water Conservation Research Division, Agricultural Research Service, US Department of Agriculture, and Department of Agronomy, Colorado State University, Fort Collins, Colorado

E. C. Klostermeyer Entomologist, Irrigated Agriculture Research and Extension Center, Washington State University, Prosser, Washington

Arnold Klute Professor of Soil Physics, Department of Agronomy, University of Illinois, Urbana, Illinois

Paul J. Kramer James B. Duke Professor of Botany, Department of Botany, Duke University, Durham, North Carolina

John N. Landers Irrigation Specialist, IRI Instituto de Pesquisas—USAID, Rio de Janerio, Brazil (formerly Research Assistant, Department of Water Science and Engineering, University of California, Davis, California)

Helmut E. Landsberg Director, Environmental Data Service, Environmental Science Services Administration, US Department of Commerce, Washington, D.C.

Cyril W. Lauritzen Project Supervisor, Soil and Water Conservation Research Division, Agricultural Research Service, US Department of Agriculture, Logan, Utah

John Letey, Jr. Associate Professor of Soil Physics, Department of Soils and Plant Nutrition, University of California, Riverside, California

Robert S. Loomis Associate Professor of Agronomy, Department of Agronomy, University of California, Davis, California

Owen R. Lunt Acting Chairman and Director, Laboratory of Nuclear Medicine and Radiation Biology, University of California, Los Angeles, California (formerly Professor of Plant Nutrition, Department of Botany and Plant Nutrition, University of California, Los Angeles, Calif.)

John R. Magness Formerly Chief, Fruit and Nut Crops Branch, Agricultural Research Service, US Department of Agriculture, Beltsville, Maryland (now retired)

John T. Maletic Soil Scientist and Chief, Land Resources Branch, Office of Chief Engineer, Bureau of Reclamation, US Department of the Interior, Denver, Colorado

Paul D. Marr	Assistant Professor of Geography, Department of Geography, University of California, Davis, California
Stephen J. Mech	Research Agricultural Engineer, Irrigated Agriculture Research and Extension Center, Soil and Water Conservation Research Division, Agricultural Research Service, US Department of Agriculture, Prosser, Washington
J. D. Menzies	Research Microbiologist, US Soils Laboratory, Soil and Water Conservation Research Division, Agricultural Research Service, US Department of Agriculture, Beltsville, Maryland
Yoshiaki Mihara	Chief, Division of Meteorology, National Institute of Agricultural Sciences, Tokyo, Japan
Edward E. Miller	Professor of Physics and Soils, Department of Physics, University of Wisconsin, Madison, Wisconsin
Cleve H. Milligan	Professor of Civil and Agricultural Engineering, Utah State University, Logan, Utah
Dean C. Muckel	Chief, Northwest Branch, Soil and Water Conservation Research Division, Agricultural Research Service, US Department of Agriculture, Boise, Idaho (formerly Research Agricultural Engineer, Reno, Nevada)
Jack T. Musick	Research Agricultural Engineer, Southwestern Great Plains Research Center, Soil and Water Conservation Research Division, Agricultural Research Service, US Department of Agriculture, Bushland, Texas
Francis S. Nakayama	Research Chemist, US Water Conservation Laboratory, Soil and Water Conservation Research Division, Agricultural Research Service, US Department of Agriculture, Phoenix, Arizona
C. E. Nelson	Agronomist, Irrigated Agriculture Research and Extension Center, Washington State University, Prosser, Washington
Donald R. Nielsen	Associate Professor, Department of Water Science and Engineering, University of California, Davis, California
George A. Pavelis	Chief, Water Resources Branch, Natural Resource Economics Division, Economic Research Service, US Department of Agriculture, Washington, D.C.
Howard L. Penman	Rothamsted Experimental Station, Harpenden, Herts, England
Herbert C. Pereira	Director, Agricultural Research Council of Central Africa, Salisbury, Rhodesia
Doyle B. Peters	Research Soil Scientist (Physics) and Associate Professor of Soil Physics, Agricultural Research Service, US Department of Agriculture, and Department of Agronomy, University of Illinois, Urbana, Illinois
Dean F. Peterson, Jr.	Dean, College of Engineering, Utah State University, Logan, Utah

John T. Phelan	Assistant Director, Engineering Division, US Soil Conservation Service, US Department of Agriculture, Washington, D.C.
Marshall B. Rainey	Sanitary Engineer and Project Director, Colorado River Basin Water Quality Control Project, Federal Water Pollution Control Administration, US Department of the Interior, Denver, Colorado
Franklin C. Raney	Associate Professor, Department of Geography, Western Washington State College, Bellingham, Washington
William A. Raney	Chief Soil Physicist, Soil and Water Conservation Research Division, Agricultural Research Service, US Department of Agriculture, Beltsville, Maryland
Ronald C. Reeve	Research Agricultural Engineer, Water Management, Soil and Water Conservation Research Division, Agricultural Research Service, US Department of Agriculture, Columbus, Ohio (formerly, Agricultural Engineer, US Salinity Laboratory, Riverside, California)
Walter Reuther	Chairman, Department of Horticultural Science, University of California, Riverside, California
H. F. Rhoades (deceased)	Professor of Agronomy, Department of Agronomy, University of Nebraska, Lincoln, Nebraska
Sterling J. Richards	Professor of Soil Physics, Department of Soils and Plant Nutrition, University of California, Riverside, California
Herbert S. Riesbol	Assistant to Manager of Hydro Engineering, Bechtel Corporation, San Francisco, California
John S. Robins	Superintendent and Soil Scientist, Irrigated Agriculture Research and Extension Center, Washington State University, Prosser, Washington (formerly Chief, Northwest Branch, Soil and Water Conservation Research Division, ARS, USDA, Boise, Idaho)
August R. Robinson	Research Agricultural Engineer and Director, Snake River Conservation Research Center, Soil and Water Conservation Research Division, Agricultural Research Service, US Department of Agriculture, Kimberly, Idaho
Jack R. Runkles	Professor of Soil Physics, Department of Soil and Crop Sciences, Texas A & M University, College Station, Texas
Leonard Schiff	Research Agricultural Engineer, Soil and Water Conservation Research Division, Agricultural Research Service, US Department of Agriculture, Fresno, California
Richard A. Schleusener	Director, Institute of Atmospheric Sciences, South Dakota School of Mines and Technology, Rapid City, South Dakota
Robert W. Schloemer	Assistant Director, Environmental Data Service, Environmental Science Services Administration, US Department of Commerce, Washington, D.C.

John G. Seeley — Professor of Floriculture and Head, Department of Floriculture and Ornamental Horticulture, New York State College of Agriculture, Cornell University, Ithaca, New York

Aubrey L. Sharp — Formerly Hydrologist, Soil and Water Conservation Research Division, Agricultural Research Service, US Department of Agriculture, Lincoln, Nebraska (now retired)

E. Shmueli — Head, Division of Irrigation, Volcani Institute of Agricultural Research, Rehovot, Israel

Daryl B. Simons — Professor of Civil Engineering and Associate Dean for Research, College of Engineering, Colorado State University, Fort Collins, Colorado

Ralph O. Slatyer — Senior Principal Research Scientist, Commonwealth Scientific and Industrial Research Organization, Canberra, A. C. T., Australia

Dwight D. Smith — Assistant Director, Water Management, Soil and Water Conservation Research Division, Agricultural Research Service, US Department of Agriculture, Beltsville, Maryland

William C. Smith — Assistant Professor Anthropology, Department of Anthropology, University of California, Davis, California

Gilbert G. Stamm — Assistant Commissioner, Bureau of Reclamation, US Department of the Interior, Washington, D.C.

G. Stanhill — Department of Agricultural Meteorology, Volcani Institute of Agricultural Research, Rehovot, Israel

William J. Staple — Soil Physicist, Soil Research Institute, Research Branch, Canada Department of Agriculture, Ottawa, Ontario, Canada

Harry A. Steele — Assistant Director (Planning), Water Resources Council, Washington, D.C. (formerly, Director, Natural Resource Economics Division, Economic Research Service, US Department of Agriculture, Washington, D.C.)

J. R. Stockton (deceased) — Plant Physiologist and Irrigation Specialist, California Agricultural Experiment Station, Shafter, California

Lewis H. Stolzy — Associate Soil Physicist, Department of Soils and Plant Nutrition, University of California, Riverside, California

Lawrence R. Swarner — Agricultural Engineer, Bureau of Reclamation, US Department of the Interior, Boise, Idaho

Champ B. Tanner — Professor of Soil Science, Department of Soil Science, University of Wisconsin, Madison, Wisconsin

Sterling A. Taylor — Professor of Soils and Meteorology and Head, Department of Soils and Meteorology, Utah State University, Logan, Utah

P. W. Terrell — Assistant Chief, Canals Branch, Bureau of Reclamation, US Department of the Interior, Denver, Colorado

Harold E. Thomas	Research Geologist, US Geological Survey, US Department of the Interior, Menlo Park, California
K. Uriu	Associate Pomologist, Department of Pomology, University of California, Davis, California
Yoash Vaadia	Plant Physiologist, Department of Plant Physiology, Negev Institute for Arid Zone Research, Beersheva, Israel
Cornelius H. M. van Bavel	Chief Physicist, US Water Conservation Laboratory, Soil and Water Conservation Research Division, Agricultural Research Service, US Department of Agriculture, Phoenix, Arizona
Frank G. Viets, Jr.	Chief Soil Scientist, Soil and Water Conservation Research Division, Agricultural Research Service, US Department of Agriculture, Fort Collins, Colorado
M. T. Vittum	Head, Department of Vegetable Crops, Cornell University and New York State Agricultural Experiment Station, Geneva, New York
Victor Voth	Specialist in Pomology, South Coast Feld Station, Division of Agricultural Sciences, University of California, Santa Ana, California
Yoav Waisel	Lecturer of Botany, Department of Botany, Tel Aviv University, Tel Aviv, Israel
Lloyd V. Wilcox	Formerly Soil Scientist, US Salinity Laboratory, Soil and Water Conservation Research Division, Agricultural Research Service, US Department of Agriculture, Riverside, California (now retired)
Karl Ernst Witte	Dr. Agr. Dozent, Landwirtschaftliche Fakultät der Universität, Bonn, Germany
Neil P. Woodruff	Agricultural Engineer, Soil and Water Conservation Research Division, Agricultural Research Service, US Department of Agriculture, Manhattan, Kansas

CONTENTS

SECTION IV—SELECTION OF LAND FOR IRRIGATION

SECTION V—SOIL-WATER RELATIONS

SECTION VII—WATER-SOIL-PLANT RELATIONS

37   **Grapes and Berries**
      Part I—Grapes

                                                    A. N. KASIMATIS
      I   Characteristics of the Grapevine   719
      II   Responses to Soil Water Conditions   721
      III   Growth Periods Sensitive to Soil Water Conditions . . .   724
      IV   Effects on Quality   727
      V   Problems Influenced by Irrigation Method or Frequency . .   728
      VI   Summary of Irrigation Recommendations   729

37   **Grapes and Berries**
      Part II—Strawberries

                                                    VICTOR VOTH
      I   Introduction   734
      II   Plant Characteristics   734
      III   Soil Requirements   735
      IV   Irrigation Methods   735
      V   Water Requirements   736
      VI   Irrigating Scheduling   736

38   **Coffee, Tea, Cacao, and Tobacco**

                                                    H. C. PEREIRA
      I   The Irrigation of Coffee   738
      II   The Irrigation of Tea   743
      III   The Irrigation of Cacao   746
      IV   The Irrigation of Tobacco   747

39   **Turfgrass, Flowers, and Other Ornamentals**

                                         O. R. LUNT AND J. G. SEELEY
      I   Introduction   753
      II   Turfgrass   754
      III   Ornamentals   759

SECTION XI—IRRIGATION SYSTEMS

40   **Problems and Procedures in Determining Water Supply**
      Requirements for Irrigation Projects

                                                    G. G. STAMM
      I   Introduction   771
      II   Water Supply   772
      III   Consumptive Use   774
      IV   Leaching Requirement   776
      V   Irrigation Efficiency   776
      VI   Diversion Requirements   778
      VII   Reservoir Operation Studies   783
      VIII Water Shortages   784

SECTION XII—IRRIGATION MANAGEMENT

SECTION XIII—WATER CONSERVATION RELATED TO IRRIGATION

section I

Introduction

1 | Irrigated Agriculture: An Historical Review

N. D. GULHATI[1]
New Delhi, India

WILLIAM CHARLES SMITH
University of California
Davis, California

I. INTRODUCTION

The transition from a hunting and food collecting way of life to one based on agricultural food production affected all aspects of human existence: It was "a cultural revolution by which mankind was able to progress beyond the ecological limitations set by nature to the foraging primitive man, establishing the foundations for the eventual development of civilization" (Armillas, 1961). (*Also, see* Childe, 1946 and 1951; and Braidwood and Willey, 1962.) No less significant than this early food producing revolution is the continuing evolution of agriculture —the development of new systems of agricultural technology, the incorporation of such systems into the framework of society, and their gradual extension to other parts of the world.

Irrigation has played a strategic role in this continuous process of agricultural development. In most of the early civilizations of both hemispheres, as in many nations of today, irrigated agriculture provided, and continues to provide, the agrarian basis of society (Gulhati, 1955). Yet this is not an unmixed blessing. On the one hand, control of water resources permits the establishment of highly productive agricultural practices, and the consequent expansion of human population, in areas where rainfall would be inadequate or unreliable. On the other hand, the operation of a complex irrigation system carries with it certain technological and social imperatives the ignorance of which may lead to disaster. Some of these social imperatives are considered in chapter 2. Other chapters deal with many technical aspects of irrigated agriculture. The present chapter reviews briefly the history of some irrigation developments in southwest Asia and North Africa, southern and eastern Asia, Latin America, and North America.

In the available history of ancient times, there are numerous references to the practice of irrigation from wells, tanks, canals, and directly from the rivers. There are striking examples in many countries of some ancient irrigation works still in service and of others which have been improved with advance in knowledge and techniques and are now working better than when they were originally con-

---

[1] The senior author is primarily responsible for material dealing with irrigation practices in the Old World.

structed several centuries or even millenniums ago. On the other hand, there also exist vestiges of many canals, tanks, and aqueducts, some of them very large, which failed, fell into disrepair, or otherwise went into disuse after operating for a relatively short time. In mainland China, Egypt, India, Pakistan, the Middle East, in countries along the Mediterranean coast, and in parts of the New World, there are several such monuments to colossal human effort, initiative, ingenuity, resourcefulness, and skill (Stamp, 1961; Braidwood and Willey, 1962).

II. SOUTHWEST ASIA AND NORTH AFRICA

More than 5,000 years ago, basin irrigation was introduced in Egypt. Embankments were built along the western bank of the Nile with cross banks to hold floodwaters, and channels were dug to inundate large areas during the high river stage. By the 8th century, about 1.5 million acres were irrigated annually and by the 13th century about 3 million acres, all by basin irrigation. The artesian supply of the western desert in the United Arab Republic is known to have been exploited from ancient times; there are ample signs of developments based on irrigation from this source by wells, 200m deep. About 2,000 such ancient wells, some after repairs and restoration, are still functioning in the oases of the western desert. The system of *kanata* (tunnels), tapping water from the surrounding mountains and leading it by gravity to the lands to be irrigated, is still used in Iran in much the same way as it was 2,500 years ago. In the valleys of the Euphrates and Tigris are remains of two of the largest irrigation canals; one of them is stated to have been 30 to 50 ft deep and about 400 ft wide.

In Egypt, the Delta Barrage was constructed in 1861. The first Aswan Dam was completed in 1902 to provide year-round irrigation waters to the Nile Delta; the Dam was raised in 1912 and again in 1933. Diversion barrages were constructed at Assiut and Zifta in 1902 and at several other places on the Nile in later years. In addition to the net increase in irrigated area in the United Arab Republic since 1900, there has been a large increase in the cropped area, the annual intensity of irrigation in 1960 being 172% (Ministry of Public Works, Cairo, 1950).

III. SOUTHERN AND EASTERN ASIA

Irrigation directly from mountain streams by small diversion canals was practiced in Japan in the Yayoi Era (before 600 B.C.). Equally ancient must be the tiny, picturesque irrigation canals and paddy fields of the Kashmir Valley and the Upper Beas Valley in India. Numerous irrigation canals, big and small, were dug at various times to carry the waters of the Indus system of rivers, during the high water stage, to irrigate lands adjacent to the rivers. Irrigation was developed in the river deltas on the east coast of India and in Southeast Asia. *Atharva-veda,* the ancient Hindu scripture, describes the digging of canals to take water from rivers, symbolizing a river as a cow and a canal as a calf. There is historical evidence of the respective duties of kings and the people in respect to irrigation works having been defined in some parts of India as early as 300 B.C.

Water storage facilities in Japan date from the reign of Emperors Sirjin and Sinjin. In parts of mainland China, Ceylon, and the south Indian plateau, small storages were formed by throwing low embankments across small streams or natural depressions. Many such tanks, built centuries ago, are still in use while others have long been abandoned (Brohier, 1934). In China, early water control measures culminated in the construction of the Imperial Canal in A.D. 700. This Canal, one of the greatest engineering works of all time, is 700 miles long and provides both irrigation and navigation.

Modern large-scale irrigation may be said to have begun in the third decade of the 19th century with the reconstruction of the Cauveri Delta System in southern India, the construction of the two Yamuna Canals in northern India, and the system of deep canals built in the Nile Delta in northern Egypt. It was on these works that the foundation of modern canal irrigation was laid, and lessons in hydraulic engineering were learned at considerable expense. For reasons of economy, canals were built in natural river channels, and these were called on to carry irrigation waters alternately with floodwaters. Depressions were crossed by earthen banks with inadequate provision for the intercepted drainage. Large areas became waterlogged because of seepage from poorly-designed canals and over-irrigation. There were yearly epidemics of malaria. Apart from these, costly mistakes were made in the design and construction of river works; river hydraulics was little understood, and the modern techniques of river control had yet to develop. The canals silted badly and had to be dredged almost every year. But in spite of all these difficulties, these works were worthwhile in the benefits which followed and in providing excellent object lessons and valuable experience which aided and encouraged the developments to follow.

Encouraged by the results of the pioneering works mentioned above, many large irrigation works were undertaken in India in the 19th century. The Ganga Canal, opened in 1854 with a capacity of 8,000 ft^3/sec and picturesque cross-drainage works, is still regarded in India as a classic. Among the other large works of this century are the Godavari and Krishna Canals in south India and the Upper Bari Doab and Sirhind Canals in north India. Toward the close of the century, one of the largest and most successful canals in the world, the Lower Chenab Canal (now in Pakistan), was constructed and brought into operation. Unlike earlier canals, the main object of which was the improvement of then existing cultivation, the cultivators already being in occupation of the land, the Lower Chenab Canal was to serve uncultivated and desert lands belonging to the government, with no resident population other than a few nomads. It was thus necessary, simultaneously with the introduction of irrigation, to colonize the area and settle whole new communities in the areas to be developed. This Canal, about 10,000 ft^3/sec capacity, was a great success, and enormous benefits accrued from it to the new settlers as well as to the Government in terms of water rates and sale proceeds of Government-owned lands. The annual income to the Government alone, within about 20 years, was almost one-half the total capital cost of the canal system. The Khadakvasla Dam, 1 mile long and 107 ft high, was built between 1869 and 1878, in uncoursed rubble masonry, to conserve water for irrigation of lands near Poona in India.

The Indian Irrigation Commission, appointed in 1901, recommended in 1903 a definite line of policy regarding the selection, financing, and maintenance of productive irrigation works. Apart from these self-liquidating or remunerative works, the Commission also examined the desirability of extension of irrigation as a means of protection against famine in areas of insecure and precarious cultivation. Such irrigation, based on expensive storage works, might not be directly remunerative to the Government, but the Commission recommended that the net financial burden which such schemes would impose on the state would not be too high a price to pay for the protection against famine which they would afford. In accordance with the Commission's recommendations, several productive and protective projects were immediately undertaken.

The Triple Canal Project (1905–1915), the first large-scale transbasin diversion (now in Pakistan) comprised two large weirs and one barrage across three different rivers and three large canals, with the principal aim of transferring water from the Jhelum River across the Chenab and Ravi Rivers to irrigate desert areas on the other side of the latter river. Between 1920 and 1930, several large projects were undertaken in India, the largest being the Sukkur Barrage Project (now in Pakistan) with seven canals having a total capacity of 47,000 ft³/sec. (Gulhati, 1955, 1958; Stamp, 1961).

IV. LATIN AMERICA

When European explorers and colonists arrived in the New World, they found the practice of agriculture widespread among the aboriginal inhabitants (Driver, 1961; Steward and Faron, 1959). Small-scale irrigation was practiced in the American Southwest, in some parts of the Caribbean, and in the northern and southern Andes. In the central Andes and in highland Mesoamerica, intensive irrigated agriculture supported dense populations and provided the economic basis upon which were built the complex civilizations of the Aztecs, the Inca, and their predecessors.

Altogether, more than 100 aboriginal cultigens have been identified; of these, the majority were known in Mesoamerica and Peru (Dressler, 1953; Towle, 1961). Recent archaeological excavations in south-central Mexico (MacNeish, 1964; cf. Mangelsdorf et al., 1964), give strong evidence that plant domestication had begun in that area at least as early as 5000 B.C. Settled agricultural villages appeared by 3000 B.C.; around 800 B.C., several hybrid varieties of maize (*Zea mays*) were being grown in irrigated fields. Population increased sharply, ceremonial centers were built, and evidences of trade with other areas of Mesoamerica began to make their appearance. From Mesoamerica the practice of cultivation appears to have spread both northward and southward, reaching the American Southwest by 4000 B.C. and Peru by 2500 B.C. or earlier. It is likely that additional centers of early plant domestication were located in the Andes and Amazon Basin.

By the time of the Spanish Conquest, highly organized and socially stratified societies were established both in the Andes and in Mesoamerica. Rain-fed agriculture alone appears to have been sufficient to support the elaborate cultures of the Olmec and the Maya of lowland Mesoamerica, but in the arid regions of both Mesoamerica and Peru, agricultural technology involved both terracing and extensive irrigation works. For basic references on Mesoamerica and South America, respectively, see Wauchope (1964) and Steward (1946–48). Reviews of recent literature can be found in Braidwood and Willey (1962) and Jennings and Norbeck (1964). The Viru Valley on Peru's arid northern coast may be taken as an example (Willey, 1953). There, the earliest settlements were small, semisubterranean dwellings near the beach, inhabited by a fishing and gathering people who raised some plants but lacked maize. Between 900 and 400 B.C. these settlements were extended upstream into the Valley interior. With maize cultivation came an elaborate and sophisticated ceramic art style (related to that known as "Chavin" throughout the central Andean area) and large, but simple, ceremonial structures. During the next 400 years there was a great population increase. Irrigation probably had its beginnings at this time, and villages were built around pyramid mounds presumed to have ritual significance. Before A.D. 800 the Viru Valley reached a population maximum, with the construction of more concentrated settlements, walled fortifications and an elaborate system of irrigation canals which drew water from a point several miles upriver. By A.D. 1200 a population decline had begun, and the area of settlement was somewhat reduced. This trend, which continued into the Spanish Colonial Period, has been attributed to political and military disturbances (Willey, 1953). However, inadequate drainage and salinization, resulting in partial abandonment of the irrigation system, may have been responsible (Armillas, 1961).

During the Spanish Colonial Period in Latin America, reduction and subsequent expansion of population in many regions was accompanied by associated changes in the level of agricultural production, in cultivation, and in the irrigation of agricultural lands. In this brief survey it is impossible to do more than call attention to these processes of demographic and economic change. Similarly, no more than passing reference can be made to the modern development of irrigated agriculture in Latin America. In Mexico alone, e.g., more than 2,230,000 hectares of land have benefited from irrigation projects carried out since 1926 (Orive Alba, 1960), often in connection with Mexico's land reform (ejido) program. It is clear that the social and political implications of such programs are as important as their technical and economic aspects. Careful study of these and other phases in the agricultural history of Latin America, as of other newly-developing areas of the world, is urgently needed if the contemporary problems of these areas are to be adequately understood.

V.   NORTH AMERICA

Irrigation was practiced on a rather limited, but effective, basis by several aboriginal groups in Southwestern USA. Along the lower Gila and Colorado Rivers, annual flooding created rich alluvial deposits in which crops of maize (*Zea mays* L.), beans (*Phaseolus* spp.), and squash (*Cucurbita* spp.) were raised by the Yuma, Cocopa, Mojave, and Pima Indians (*see* Forde, 1934: Ch. XII, for references). The Hopi Indians of northeastern Arizona, USA, practiced floodwater irrigation in drywashes and constructed small ditch-and-basin systems to utilize the water supplied by natural springs. Similar irrigation arrangements were known to the Zuñi and other Pueblo Indian groups of the Colorado Plateau. The most extensive aboriginal irrigation works in the Southwest are attributed to the "Hohokam," a Pima word meaning "those who have vanished" (Wormington, 1959). These people inhabited the Gila and Salt River drainages of southern Arizona around A.D. 500 to 600. According to Turney (1929), more than 200 miles of ditches and canals were built in the Salt River Valley alone (cf. Haury, 1937). To permit the creation of irrigation works of such magnitude, intervillage cooperation and control must have been required (Haury, 1956). Yet the Hoho-kam did not reach a level of socio-cultural complexity comparable to that of the early civilizations to the south.

The modern history of irrigation agriculture in North America began with the extension of post-Columbian settlement into the arid western regions of the USA. Mediterranean crops (wheat (*Triticum aestivum* L.), barley (*Hordium vulgare* L.), olives (*Olea europaea* L.), and citrus fruits) and irrigation techniques were first introduced by Spanish Americans in the Southwest USA. During the 19th century, Mormons and other Anglo-American settlers adapted these techniques in raising wheat and barley, root crops, and deciduous fruits, as well as providing pasturage for livestock (Thomas, 1920; Logan, 1961).

In the development of irrigation in the American West, as elsewhere, techno-logical innovations became particularly significant (Olin, 1913; Israelson and Hansen, 1962). Small-scale diversion systems were employed in localities where sufficient surface flow was available; similarly, artesian wells were opened to tap underground water supplies. In most areas, however, such resources were lacking or inadequate. The successive development of the wind-driven pump, the gasoline engine pump, mechanical deep-well drilling equipment, and the submerged cen-trifugal pump gave access to increasingly abundant and reliable water supplies. Innovations in water distribution equipment, including portable piping and over-head sprinkler systems, proved advantageous under conditions of porous soil, moderate slope, and high evaporation. Underground tile drainage systems helped in combating salinization and related problems.

The enactment and implementation of relevant legislative measures also did much to promote and direct the development of irrigated agriculture in Western USA. Early efforts in this direction, such as the Desert Land Act of 1877 and the Carey Act of 1894, were designed to stimulate private and state agency par-ticipation. With the Reclamation Act of 1902 (and subsequent amendments), the Federal Government became involved directly in hydrological development (James, 1917; Golze, 1961). As a result, many large-scale enterprises were under-taken which would have been out of reach of state governments or private enter-prise. Nevertheless, such projects typically involved the participation of federal, state, and private agencies. The majority of projects undertaken by the Federal Bureau of Reclamation during the first half of the century were single purpose projects, authorized to perform a particular function such as irrigation, power supply, flood control, etc. Gradually the emphasis was shifted to multiple-purpose projects, combining these and other functions. At the same time, increasing em-phasis was placed on regional (rather than local) development, referring to an entire watershed rather than an area within the watershed.

An outstanding example of this multiple-purpose regional approach to hydro-logical development is the Hoover Dam of the Colorado River Basin Project (Golze, 1961). The intensification of agriculture in California's Imperial Valley and the rapid growth of population in southern California following 1900, posed several related hydrological problems: the need for flood control and irrigation, and the need for hydroelectric power and municipal water supplies throughout the area. It was clear that these problems could be met only through the construction of large-scale water control facilities on the Colorado River. Since the water-use rights of seven states would be involved in such a project, it was first necessary for the states to reach a basic agreement concerning these rights. Such an agreement was formulated in the Colorado River Compact of 1922, which opened the way for field investigations by the Federal Bureau of Reclamation. In 1928, after extensive consideration, the Boulder Canyon Project Act was passed by the US Congress. Pending ratification by six of the seven states, this Act provided for the construction of Hoover Dam and its associated hydroelectric installations and for construction of the All-American Canal to carry water to the Coachella and Imperial Valleys of California. When completed in 1935, this project was the largest of its kind in the world and the first specifically designed to serve multiple functions. In addition to furnishing flood control and irrigation benefits for the lower Colorado River area, the project met a major part of the water and power needs of metropolitan centers along the California coast and provided wildlife refuge areas and recreational facilities. At least as significant as these immediate benefits, however, was the clear demonstration that the solution of major hydrological problems required extensive cooperation between federal, state, and private agencies in the planning and operation of regional water utilization programs.

VI. CONCLUDING REMARKS

By making available the experience of the past, the history of irrigated agriculture provides useful guidance for those concerned with the planning and operation of present and future irrigation systems. Even a brief survey, such as this, is sufficient to indicate the need for better understanding of the interplay of physical factors, technological processes, agricultural practices, and social organizations in the management of water resources.

When a regular water supply was provided for agriculture for the first time in any area, it brought in its wake such vast improvement in agricultural production and assured returns to the farmer that, in many countries, little attention was paid, to begin with, to proper agronomic practices, to soil ameliorative measures, and to the need for drainage. To the people in many lands, accustomed to an uncertain kind of cultivation and a precarious existence, the initial gains from irrigation were such that, for quite some time, little was done to find out if production could be increased still further. But with the developments in agronomic practices, with the use of chemical fertilizers and the proper understanding of soil-water relationships, the full potential of irrigated agriculture, as now realized in some of the advanced countries of the world, is many times more than the initial advantage from irrigation. But in large parts of the world, even today, the production from irrigated agriculture falls far below that obtainable by an optimum combination of irrigation and drainage, appropriate soil reclamation and management measures, and selection of crops best suited to local conditions. These aspects of irrigated agriculture have not received as much attention as the more spectacular engineering feats involved in conserving natural waters and making them usable for irrigation, conveying these waters over long distances, and distributing them equitably among the farmers. Successful irrigation projects involve much more. For a productive and permanently successful irrigated agriculture, attention must be given not only to providing water supply, irrigating efficiently, and draining land as necessary, but also to following sound soil management practices, selecting productive crop varieties, and utilizing all beneficial cultural practices. The history of irrigation clearly points to the need for giving far greater attention to the agricultural phases of irrigation.

Much remains to be learned about the social and cultural settings within which these technological problems are to be confronted. Some scholars have suggested, for example, that in several historical Asian societies, the planning, construction, operation, and maintenance of large-scale irrigation and drainage works tended to generate powerful and oppressive state institutions (Wittfogel 1938; but cf. Adams, 1961). Attempts have been made, with varying degrees of success, to extend this hypothesis to early civilizations of the New World (Steward et al., 1955; cf. Armillas, 1961), and to relate it to the analysis of contemporary political and economic situations (Wittfogel, 1957). In many of its implications this theory is a highly controversial one; nevertheless, it stands as a challenge not only to those social scientists interested in the study of relations between technology and society, but also to legislators, administrators, and technicians concerned with the uses of irrigation in newly-developing areas of the world.

Growing international cooperation should contribute to the solution of many problems of irrigated agriculture. Establishment of the International Commission on Irrigation and Drainage in 1950 has facilitated exchange of information. The International Hydrological Decade, begun in 1965 by the United Nations Educational, Scientific, and Cultural Organization (UNESCO), proposes a sustained and internationally coordinated 10-year program of scientific observations on water resource utilization. This program seeks to accelerate the study of water resources and the worldwide regimen of waters with a view to their rational management in the interest of mankind, to make known the need for hydrological education and research in all countries, and to improve the ability of the less-developed countries to evaluate their resources and use them to the best advantage. One may hope that this and similar programs will serve to stimulate international interest in the complex technical and social problems related to water supply and utilization, and thus contribute to their eventual solution.

LITERATURE CITED

Adams, Robert H. 1961. Early civilizations, subsistence, and environment. p. 269–297. *In* Carl H. Kraeling and Robert M. Adams [ed.] City invincible: A symposium on urbanization and cultural development in the ancient Near East. Univ. Chicago Press, Chicago.

Armillas, Pedro. 1961. Land use in pre-Columbian America. p. 255–276. *In* L. Dudley Stamp [ed.] A history of land use in arid regions. U.N. Educ., Sci., & Cult. Organ. UNESCO Arid Zone Res. 17, Paris.

Braidwood, Robert, and Gordon R. Willey [ed.] 1962. Courses toward urban life. Viking Rund Publications in Anthropology Wenner-Gren Foundation, New York. No. 32. 371 p.

Brohier, R. L. 1934. Ancient irrigation works in Ceylon. The Ceylon Government Press, Ceylon. 381 p.

Childe, V. Gordon. 1946. What happened in history. Pelican Books, New York. 280 p.

Childe, V. Gordon. 1951. Man makes himself. Mentor Books, New York. 242 p.

Dressler, R. L. 1953. The pre-Columbian cultivated plants of Mexico. Botanical Museum Leaflet, Harvard Univ. 16:115–72.

Driver, Harold E. 1961. Indians of North America. Univ. Chicago Press, Chicago. 667 p.

Forde, C. D. 1934. Habitat, economy, and society. Methuen & Co. Ltd. London. 195 p.

Golze, Alfred R. 1961. Reclamation in the United States. The Caxton Printers, Caldwell, Idaho. 486 p.

Gulhati, N. D. 1955. Irrigation in the world: A global review. Int. Comm. Irrig. Drain. New Delhi, India. 226 p.

Gulhati, N. D. Sept. 1958. Worldwide view of irrigation developments. Irrig. Drain. Div. Amer. Soc. Civil Eng., Proc. 14 p.

Haury, E. W. 1937. The Snaketown Canal. p. 50–58. In H. S. Gladwin, E. W. Haury, E. B. Sayles, and N. Gladwin, Excavations at Snaketown. Mater. Cult. Medallion Pap. No. 25. Gila Pueblo, Ariz.

Haury, E. W. 1956. Speculations on prehistoric settlement patterns in the Southwest. 202 p. In Gordon R. Willey [ed.] Prehistoric settlement patterns in the New World. Viking Fund Publications in Anthropology, New York. No. 23.

Israelsen, Orson W., and Vaughn E. Hansen. 1962. Irrigation principles and practices. John Wiley & Sons, New York. 447 p.

James, George W. 1917. Reclaiming the arid West. Dodd, Mead & Co., New York. 411 p.

Jennings, Jesse D., and Edward Norbeck [ed.] 1964. Prehistoric man in the New World. Univ. Chicago Press, Chicago, for William Marsh Rice Univ. 633 p.

Logan, Richard F. 1961. Post-Columbian developments in the arid regions of the United States of America. p. 277–298. In L. Dudley Stamp [ed.] A history of land use in arid regions. UNESCO Arid Zone Res. 17, Paris.

MacNeish, Richard S. 1964. The origins of New World civilization. Sci. Amer. 211:29–37.

Mangelsdorf, Paul C., Richard S. MacNeish, and Gordon R. Willey. 1964. Origins of agriculture in Middle America. In Handbook of Middle American Indians. Robert Wauchope [ed.] Vol. I. Natural environment and early cultures. Robert C. West [ed.] p. 427–445. Univ. Texas Press, Austin.

Ministry of Public Works. 1950. Irrigation and drainage in Egypt. Ministry of Public Works Cairo. 23 p.

Olin, W. H. 1913. American irrigation farming. A. C. McClurg & Co., Chicago. 364 p.

Orive Alba, Adolfo. 1960. La politica de irrigacion en Mexico. Fondo de Cultura Economica, Mexico, D.F. 292 p.

Stamp, L. Dudley [ed.]. 1961. A history of land use in arid regions. UNESCO Arid Zone Res. Ser. 17. 388 p.

Steward, Julian H. 1946–48. The handbook of South American Indians. Bur. Amer. Ethnol. Washington, D.C. Bull. 143. Vol. II, passim.

Steward, Julian, Robert Adams, Donald Collier, Angel Palerm, Karl Wittfogel, and Ralph Beals. 1955. Irrigation civilizations: A comparative study. Pan American Union. Dep. Cult. Affairs. Soc. Sci. Monogr. 1. 78 p.

Steward, Julian, and Louis C. Faron. 1959. Native peoples of South America. McGraw-Hill Book Co., Inc.. New York. 481 p.

Thomas, George. 1920. The development of institutions under irrigation. The Macmillan, New York. 293 p.

Towle, Margaret. 1961. The ethnobotany of pre-Columbian Peru. Viking Fund Publications in Anthology. Wenner-Gren Foundation, New York. No. 30. 80 p.

Turney, Omar A. 1929. Prehistoric irrigation in Arizona. Ariz. Hist. Rev. 2. 163 p.

Wauchope, Robert. 1964. Natural environment and early cultures. Handbook of Middle American Indians. Univ. Texas Press, Austin. 570 p.

Willey, Gordon R. 1953. Prehistoric settlement patterns in the Viru Valley, Peru. Smithsonian Institution. Bur. Amer. Ethnol. Bull. 155. 453 p.

Wittfogel, Karl. 1938. Die Theorie der orientalischen Gesellschaft. Zietschrift fur Sozialforschung 7:90–122. English translation reprint. In Morten H. Fried [ed.] Readings in anthropology. Vol. II. Thomas Y. Crowell Co., New York. 1959.

Wittfogel, Karl. 1957. Oriental despotism: A comparative study of total power. Yale Univ. Press, New Haven, Conn. 556 p.

Wormington, H. M. 1959. Prehistoric Indians of the Southwest. The Denver Museum of Natural History (Denver, Colo.). Pop. Ser. No. 7 (Fourth Printing) 191 p.

2 | The Social Context of Irrigation

PAUL D. MARR

University of California
Davis, California

I. INTRODUCTION

Important irrigation enterprises are found on every continent on the face of the earth. Among the more notable ancient irrigation undertakings were those of Egypt, Iraq, China, Peru, and Mexico. The extensive irrigation works in these areas reflected the technical and administrative ability of the civilizations they served and, in turn, were instrumental in the development of the economy and social life of these areas. The early civilizations in the arid margins and the humid regions of the world, where life depended on irrigated rice, grew in wealth, cultural achievement, population, and power as their irrigation systems were extended and as long as the productivity of the irrigated lands remained high. Salinization of the irrigated soils, silting of canals, and administrative neglect often, though at times almost imperceptibly, reduced the productivity of irrigation systems and adversely affected the nations depending on them.

Today, almost 7,000 years after the beginning of irrigated agriculture in Mesopotamia, irrigation systems continue to be built with increasing technical complexity and in areas never before irrigated. Although most of these systems are being built in countries not as entirely dependent on their irrigation systems as were the early hydraulic civilizations, they are, nevertheless, endeavoring to better the economic conditions, and failures in these irrigation systems may seriously retard a country's economic and social development. The present level of agricultural, engineering, and administrative skills and knowledge of the social and economic milieux of irrigation systems that can be brought to bear on design and operation should make complete failure, or even the partial failure, of a project unlikely. Partial successes or failures of projects, however, do occur. The continued improvement of technical skills and the improvement in our knowledge of the social aspects of projects can improve the likelihood of success and continue to increase the productivity of irrigated lands.

II. INSTITUTIONAL CORRELATES OF IRRIGATION DEVELOPMENT AND DECLINE

In both the Old and the New World where agriculture evolved from dry farming to small-scale irrigation developments and then occasionally into regional irrigation systems, changes occurred in the societies as they developed their irrigation systems (*see* Steward, 1955 and Stamp, 1961). These changes in the economy and culture of a region were often necessary precursors of the further development of complex irrigation systems. The new efficiency of food production associated with irrigation resulted in food surpluses that made possible an increased rural and urban population. In turn, greater wealth was accumulated by the ruling classes which frequently was associated with a cultural florescence such as occurred in Egypt, Mesopotamia, and China.

An example of the historical interrelationship of social institutions and irrigation development was recently reported in work for the Iranian Government. This study is one of a series made as part of a program for restoration of the agricultural economy in the Province of Khuzistan, situated along the southern edge of Mesopotamia at the foot of the Zagros Mountains and on the shores of the Persian Gulf (Adams, 1962). The sequence in the development of irrigation in this province probably began before 5000 B.C. along the upper margins of the valley of the Tigris-Euphrates Rivers where crop production originally depended exclusively on rainfall. Gradually permanent settlement was extended into the drier portions of the valley concurrent with an increasing use of water for irrigating crops. During the following millennium the expanded population was still localized in small communities. After 4000 B.C. a few settlements began to emerge as larger centers, eventually evolved into cities which, in later centuries, became major urban centers. The prominent city in this region for several millennia was Susa, whose massive walls were surrounded by intensively farmed gardens and orchards. Following the conquest by Assyrian invaders during the first half of the first millennium B.C., the area declined. Gradually, under Grecian influence after the conquests of Alexander and then under Sassanian influence, the region developed its agriculture on the basis of a new and boldly-conceived system of weirs and irrigation canals. Under the direction and capable administration of the Sassanian Empire, a regional irrigation system was planned and executed. Ridges were pierced by tunnels for conveying irrigation water, inverted siphons were used for transporting water under periodically flooded streambeds, and kanats were extensively constructed for the first time. Formerly unrelated irrigation works were unified by a system of canals that delivered water across broken topography. The agricultural productivity of Khuzistan was markedly raised through the unprecedented investment of state capital and effective administration of this region-wide system of irrigation works.

The development of the extensive Sassanian irrigation system was paralleled with a directed settlement program for its own citizens and prisoners-of-war. Commercial crops were widely grown, and the cities were known for their handicraft industries, emphasizing the production of fine silks, satins, cottons, and woolens. This new urbanization, the commercial crop production, and the new industries imply a degree of technological progress and a market economy that, for the first time, supplanted the local subsistence economy which formerly prevailed. The market places of Sassanian cities became cosmopolitan centers of trade where Hebrew, Greek, and Syriac were commonly spoken. The milieux that fostered trade and handicraft production also provided a suitable setting for experimenting in manufacturing processes. The arts also flourished, and a university was founded in Jundi Shapur which was noted for its contributions to astronomy, theology, and medicine.

The rich Sassanian legacy which the Arabs acquired in a relatively undestructive campaign began to decline after a brief resurgence. The decline was not reversed until the recent efforts of the Iranian government to restore the province as an important producer of agricultural crops. During 1,300 years of decline that followed, the population greatly decreased, commerce stagnated, and the fine handicraft industries ceased. With the falling into ruin of the fine Sassanian irrigation system the year-round supply of irrigation water ended, and sugarcane (*Saccharum officinarum?*) was no longer produced. The present inhabitants of Khuzistan are again growing sugarcane, but until recently, they were unaware that their province once manufactured sugar and exported it throughout the Abbasid caliphate.

The causes of the decline of Khuzistan's economy was the result of a complex interaction of factors that confronted both the Abbasid caliphate centered at Baghdad as well as Khuzistan in particular. Throughout the eastern caliphate, agriculture and commerce diminished because of a decline in central authority and a consequent rise within the empire of practically independent states that carried on devastating civil wars. Exploitation and over taxation became customary and frequent epidemics decimated the population (Hitti, 1949). Khuzistan felt these forces as did other provinces in the empire. In addition its irrigation works were gradually allowed to fall into disrepair, and the available manpower and capital resources were concentrated in the development of marginal lands. Better lands on the upper plains were being abandoned, because of the neglect of older irrigation works, but new works were constructed to bring water to poorly drained and moderately saline land where large tracts could be obtained for the employment of large gangs of prisoners and slaves. The newly irrigated marginal lands were abandoned within several hundred years. Without the repair of the irrigation works that watered the province's better soils, the agricultural base was shattered, and the area gradually reverted to a subsistence economy.

The history of Khuzistan exemplifies the importance of cultural accomplishments and attitudes in the development and decline of an irrigation system. However, it is doubtful if even the most advanced of the early hydraulic societies could continually cultivate a region that had unfavorable soil and groundwater conditions, when it was clearly not within the technical or physical capabilities of a society to do so. To the north of Khuzistan and between Baghdad and the Zagros Mountains are the Diyala Plains which have been irrigated like Khuzistan for thousands of years. Here, in addition to the problem of soil salinization, was a major problem in the silting of irrigation systems. The ancient means of controlling salinization on the Diyala Plains was probably limited to avoiding over-irrigation (Jacobsen and Adams, 1958). But this practice may lead to an accumulation of salt on the surface because the downward leaching of salt is interrupted and it rises to the surface through capillary action. Drainage practices for lowering the saline water table were unknown but would have been unfeasible in this area of low terrain without modern pumping machinery. Of the three phases of salinization in the area, the earliest, beginning in 2400 B.C. and ending in 1700 B.C., is particularly interesting, because of the length of time that the population of the area continued crop production despite increasing salinization and decreasing crop yields. The first noticeable effect of salinization was a gradual increase in the proportion of salt tolerant barley (*Hordeum vulgare?*) in relation to wheat (*Triticum aestivum?*), and eventually wheat production almost disappeared. This occurred concurrently with a gradual but decisive decline in the absolute yield of barley. By 1700 B.C. the rural population of the plains had dwindled, and the urban centers of Sumeria were reduced to villages. The long struggle with salinization on the Diyala Plains of this period was a major cause of Sumeria's loss of cultural and political power and enabled Babylonia to rise to power in 1700 B.C. (Jacobsen and Adams, 1958). The salinization problem that recurred on the Diyala Plains has been, and continues to be, a major factor limiting the productivity of irrigated lands. It is, however, a tribute to the ability of ancient farmers that they could often continue to practice irrigation in the same area for long periods under difficult soil and groundwater conditions.

The maintenance and desilting of irrigation systems were other major problems facing the ancient hydraulic civilizations. The removal of silt, water plants, and debris to the canal banks was a yearly task which, if neglected, reduced the flow in canals at an increasing rate. The most notable monuments to the efforts of canal maintenance are the mounds of silt along canals used during the Parthian, Sassanian, and Abbasid Periods in Mesopotamia. These mounds, major topographic features found throughout central and northern Iraq, extend for miles and tower above all but the highest mounds built by ancient cities (Jacobsen and Adams, 1958). But more insidious than the silting of canals which can be observed, is the imperceptible rise in the levels of fields from yearly accretions of silt which may gradually cause a reduction in the flow of water from canals. Under these conditions the yearly flood of silt might cause an eventual decline in the productivity of irrigated fields, whereas in areas practicing basin irrigation, the yearly flood of silt brought renewed fertility.

Other indications of the effect of neglect and proper maintenance of irrigation systems come from Egypt. During the late Ptolemaic Period the central government weakened, canal maintenance was neglected, and grain yields declined. During the reign of Augustus, after Egypt had been conquered by Rome in 30 B.C., the Roman army was charged with cleaning and repairing the neglected canals (Westermann, 1917). According to Strabo, the greatest productivity in Ptolemaic times occurred when the recording of the Nile flood on the Memphis nilometer reached 14 cubits (1 cubit = 18 inches), but famine occurred if the nilometer reached only 8 cubits. After these works were repaired and cleaned, readings of 12 cubits resulted in maximum production, and readings of 8 cubits resulted in no hardship (Strabo, ?). Typically, the most critical periods for the effective maintenance of irrigation systems are those of weak central power or of warfare and the following unsettled years which often result in a lack of administrative direction and manpower resources. This is especially serious when the failure in the operation of one section of a complex regional irrigation system endangers the water supply of other sections as well. On occasion, feudalistic landowners did take upon themselves the responsibility of maintaining the irrigation systems without direction from a centralized authority. The latter years of Byzantine Egypt was a time of weak governmental control, but also a time of cultural and economic revival. In this period the estate owners probably filled the gap of direct governmental control either by direction or by default (Hardy, 1931).

Satisfactorily separating cause and effect in the relationship of irrigation to the ancient civilizations is quite impossible, but tendencies do appear, and some tentative statements can be made. It is certain that the great early urban centers of Mesopotamia and Egypt could only be supported by the concentrated and efficiently produced foodstuffs of irrigation systems. These urban centers, and the states or empires associated with them, could only be governed with considerable administrative skill which may have evolved with either the growth of the irrigation system or the increasing complexity of provisioning urban life, ruling cities, and extensive political units. Once the irrigated food producing base was established and satisfactorily maintained, further social developments were not necessarily associated with irrigation. At this point it is likely that irrigation ceased to be a significant variable (Beals, 1955). But unsettled political or military conditions or the gradual delegation by a central authority of the responsibility for maintenance very often led to disaster. While the immediate reasons for the demise of a part or an entire system may have been physical, such as the washing out of weirs or the silting of canals, the knowledge of how to correct these problems persisted at least briefly. The failure to repair or maintain works in such instances was largely the fault of social institutions, rather than the fault of physical forces.

III. SOCIAL AND ECONOMIC BENEFITS OF IRRIGATION DEVELOPMENT

The national purpose for developing, expanding, and maintaining irrigation systems has historically been associated with the desire for national wealth and power. The wealth generated by the irrigation works of the ancient civilizations was concentrated in the ruling class, though at times it was also shared with the religous and merchant classes as well. Today, the expenditures of public funds for irrigation projects are still made with the same historical intent but now include broad social goals which incorporate a balanced concern for the welfare of the nation and for the welfare of the farmer and the region in which he lives. These social goals incorporate the notion of best use or combination of best uses of water and land resources for short and long-term needs. In this context, irrigation developments are increasingly becoming less an end in themselves and more often a part of comprehensive river basin or broad multiple-purpose projects.

A. The Implementation of Irrigation Programs

The first major irrigation legislation in the USA was the result of a mixed social and economic need for Federal assistance in developing the potential irrigation resources of the Western USA. By the time this legislation, the Reclamation Act of 1902, was passed the nation had undergone 25 years of promotional effort by honest developers, dishonest land speculators and "irrigation evangelists." During this period, less costly and less complex stream diversion projects were put into operation, leaving many potential projects undeveloped because of the need for expensive water storage reservoirs, pumping machinery, or lengthy canal works. Also, a lack of knowledge regarding the engineering, agricultural, or financial aspects of irrigation development often led to failure and substantial losses by investors and farmers. It was expected that the direct participation of the Federal Government in irrigation development would reduce the abuse of speculators, insure a scientific approach to land reclamation, and make possible the development of more costly projects. In addition to these goals, which were fulfilled, the provisions of the Reclamation Act were written to assure as wide as possible distribution of land by limiting the amount of land for which a single owner could obtain water rights. Although the Reclamation Act has strong social overtones in offering protection to the unwary settler and assuring a wide distribution of the nation's land resources, it does have the economic requirement that the cost of irrigation projects, and their operational expenses, be repaid to the Federal Government by the benefiting farmers. The largest irrigation works in the Western

USA are Bureau of Reclamation Projects, but water deliveries to about 80% of the irrigated area are made through locally directed irrigation districts. Landowners in these districts finance, construct, and operate the irrigation facilities which at times include expensive and complex water storage and hydroelectric facilities. The successful operation of these enterprises has resulted in local economic growth with a minimum of assistance from state and federal agencies, except in chartering and granting taxing powers to the various districts (Adams, 1952).

Land reclamation in other countries has also been implemented to fulfill a combination of both social and economic goals. In the process of enhancing the economic opportunities of the rural population, other problems are ameliorated—such as slowing the migration from agricultural areas to overcrowded urban areas by providing increased supplies of irrigation water or settlement on newly irrigated lands. In Mexico active governmental concern for irrigation began in the early 1920's, after a decade of revolution and land reform. Works sponsored by the Mexican Government range in size from regional multiple-purpose enterprises, such as the Papaloapan Project, to small village irrigation systems (Orive Alba, 1960). In contrast the United Arab Republic, while having a much longer history of modern irrigation development extending from the middle of the 19th century, has only recently undertaken an irrigation program in the Liberation (Tahrir) Province with the expressed purpose of improving the social and economic condition of the new settlers (Warriner, 1962).

B. The Impact on Local Growth and Stability

The effect of large scale investments in irrigation projects can be seen immediately in the form of newly constructed dams, canals, and irrigated fields, but the maturing of the social and economic complex initiated with the settlement of irrigation farmers often takes years to develop. During this period of regional growth, the development costs of the project and of the initial investments in the individual farms are repaid. Concurrently, substantial investments are typically made in the local farm communities to provide facilities to satisfy the increased rural demand for goods and services. Also, local government must provide new roads, schools, and possibly assist in providing medical and recreational facilities. When the initial period of heavy capital expenditures and growth are achieved, the economy is able to reach a level of stability based on a continuing intensive application of land and water resources (US Dep. Interior, 1949).

The degree to which irrigation projects stimulate and stabilize the social and economic activity of a region can be illustrated in the comparison of partially irrigated and totally dry-farmed areas of the North Central USA. The agricultural production of these two types of areas varies widely through time, but the practice of irrigation cushions the variations in crop production, reducing the possibility of a near complete crop loss. The more diverse pattern of cropping in irrigated areas adds further to the economic stability by enabling the usually diversified irrigation farmer to cope better with individual commodity price fluctuations. The increased agricultural output of irrigated areas provides the base for a greater volume of business activity in local farm trade centers through an increased demand for retail facilities and services and new food processing and shipping facilities. Measurements of the direct impact of production of irrigated agriculture on the local nonfarm economy in terms of the flow of income to labor and capital has been found to equal, approximately, the value of crops produced (US Dep. Interior, 1954). A comparatively higher level of business activity results, in turn, in increased urban land values. The higher urban values along with the higher capital value of irrigated lands results in tax collections that are significantly greater than in areas supported by dry farming. This comparatively favorable tax advantage provides the irrigated area with an alternative,

". . . either in providing the locale with public services they would not have had without the irrigation development, or else there would be a reduction of the tax load to all taxpayers. This in turn would have a great stabilizing influence, both on the public and private economy" (Holje et al., 1956; cf. US Dep. Interior, 1958).

C. The Regional and National Impact of Irrigation

Regional benefits often accruing from irrigation projects include the various ancillary features which in the Western USA are frequently adjuncts to irrigation development. These include facilities for hydroelectric power, flood control, municipal and industrial water supplies, fish and wildlife conservation, and recreation. The construction of any of these facilities alone, would usually entail prohibitive expenses but are feasible when developed as a part of a multiple-purpose water project.

The nation also benefits directly and indirectly from irrigation projects. Many crops can be grown in controlled conditions with irrigation, resulting in the production of choice commodities of a wide variety. The crops are processed and shipped to national markets, and, in return, the nation provides the agricultural sector with equipment and supplies. Services, such as transportation and finance, are stimulated by this volume of trade. The most important benefit to local, state, and federal governments, however, is revenue from taxes. The new wealth created through private and federal investments in the development of water resources is reflected in a continual flow of tax revenues from the project areas and induced economic activity elsewhere in the nation.

In many countries the investments in irrigation projects are important features of development programs. The construction phase provides increased employment, and demand for construction materials results in the expansion or introduction of new industries. The products of new irrigation projects may favorably increase the nation's balance of payments by decreasing a need for imported foodstuffs or by providing new products for export.

The benefits of irrigation projects may also include the provision of efficiently produced food supplies for countries suffering acute population pressures. But even the boldest projects may not be of sufficient scale to offset the increase in population resulting from high birth rates. The population of Egypt, for example, is increasing at a rate of more than 500,000 persons annually and will probably increase even more rapidly. The additional arable land that can be brought into production as a result of the Aswan Dam project is limited to approximately 100,000 acres/year over a 20-year period. In 1959, the estimated man–land ratio for Egypt was almost 4 persons/acre of arable land, one of the highest in the world. Consequently, despite an effort of heroic proportions, the country will continue to increase its density of population per irrigated acre. Here a new project will help stabilize a precarious situation. In this respect ". . . the project will not improve the per capita economic situation in the Egyptian Region as much as it will prevent rapid deterioration" (El Mallakh, 1959).

IV. PROBLEMS OF PROJECT PLANNING AND OPERATION

Consideration of the social and economic aspects of the farmer and his community is becoming an increasingly important part of the planning and operation of irrigation projects in the economically advanced and developing countries. In developing countries a farmer participating in an irrigation scheme may have formerly practiced shifting agriculture. This man requires guidance before he can make a contribution to the irrigation project. In a more economically advanced society a farmer may find himself settled on a project farm which in 20 years may no longer be sufficiently large to provide an acceptable standard of living for his family. In another area a project may result in a greatly increased commercial activity in nearby rural communities, but at the same time result in overburdening public services such as roads, police and fire protection, and the school system. These three situations are all aspects of irrigation project planning which may, or may not, require the attention of project planners, depending on the scale, site, and situation of the project. They do, however, represent factors which can very definitely affect the tangible and qualitative success of irrigation enterprises.

A. Problems of Change in the Agricultural Economy

Planners of irrigation enterprises should attempt to plan with sufficient flexibility to account for the changing economic and social structures of an area. Plans should be made for the possible change from one type of cropping system to another, which might be necessitated by changes in competition or changes in the demand for commodities. Another change occurring at an increasing rate in the Western USA is the urbanization of irrigated lands. Changes of the latter type can be anticipated with some degree of certainty and should be reflected in provisions for supplying increasing amounts of municipal and industrial water supplies.

In some areas, however, the environmental conditions and the institutional constraints on the size of farms do not allow for flexible planning. As an example, a substantial part of the financial difficulties of settlers on one irrigation project in the Western USA arose because the project was originally planned and constructed with the production of hay and feed grain as a basic operation. The project progressed satisfactorily, as long as the farmers found a market for their produce with the cattlemen and sheepmen in the region immediately surrounding the irrigated lands. However, with the construction of good roads and the availability of heavy-duty transport trucks, the farmers lost their market because it was economically advantageous to transport large volumes of hay to the livestock ranches for winter feeding from areas further away where the cost of production was considerably lower. Consequently, the project farmers steadily lost their market for hay and feed grain. The severity of the climate prevented them from attaining crop yields sufficiently high to allow their crops to be sold at prices that could compete with crops from lower elevations where the production per acre was considerably greater. Limitations on the size of farms, combined with climatic factors, have prevented the farm operators from changing to another type of irrigated agriculture, such as on-farm livestock production or dairying.

A more successful example of adapting project plans to economic and policy changes is found in the development of the Columbia Basin Project in central

Washington, USA. The original legislation enacted in 1937, concerning the size of farms, limited holdings to two sizes, 40 acres for single persons and 80 acres for married persons. Before the first unit of the project was opened for settlement it was recommended that the farm size be increased from an average of 55 to 130 irrigable acres, depending on the quality of land (US Dep. of Interior, 1946). In 1957 the acreage limitation on farm size was revised again by the US Congress to conform to standard Bureau of Reclamation acreage limits of 160 acres for single persons and 320 acres for married persons. This last change was largely the consequence of reduced farm income, resulting from acreage controls on important cash crops, such as wheat and sugar beets (*Beta vulgaris*), and a lack of a market or processing facilities for other crops (Franklin et al., 1959). Adjustments were possible in the farm sizes of the Columbia Basin Project because settlement occurred over an unusually long period. If the Project had been completely settled on the basis of the original farm size, the Project would be an agriculturally depressed area today.

B. Planning for Change in the Project Area

The regional impact of irrigation works on the local urban economy varies considerably, depending on the nature of the project and the state of development of the system of cities and towns in, and near, the water delivery area. Projects such as the initial stage of the Central Valley Project in California, USA, which delivered supplemental water to an already highly developed portion of the San Joaquin Valley, had a relatively small initial impact on the cities in its delivery area. The Columbia Basin Project, in contrast, was planned to provide water to approximately 1 million acres of dry-farmed land. The size and total impact of this Project was unique in the history of American water development. Consequently, the Federal Government established study groups to determine the optimum number of new villages and towns for the project area, their most advantageous placement, their characteristics, the need for new agricultural processing industries, and the need for the further extension of existing water, railroad, and highway transportation facilities (US Dep. Interior, 1951). The several committees dealing with these problems were composed of local citizens, representatives of private industry, as well as representatives from a wide range of local, state, and federal agencies. The work of these committees has formed a basis for the subsequent development of the area.

A close degree of liaison between water development agencies in the Western USA and representatives of the locally affected areas is, however, unusual. Typically the concern of water agencies is limited to the construction of water-delivery facilities and interest in the nonfarm economy is limited to the analysis of secondary benefits of a project which comprises one section of a project's feasibility study. Evaluations of this type are made in an aggregate form so that individual local communities do not have a clear view of their future prospects and needs. On the other hand, local land-use planning agencies and other public agencies, such as school and road departments, consider the length of time for the maturation of a project to be greater than they desire to plan for, although future densities of rural population in project areas can be predicted with reasonable accuracy. Closer cooperation between the resource planner and the urban planner could make the period of social and economic adjustment for project areas less of a strain, if the long-range needs could be anticipated with some degree of clarity. The resource agency would benefit by obtaining a deeper insight into the demographic, industrial, and urbanization effects of resource development projects, and the urban and county planner would benefit from having a sounder basis for short and long-range planning (*see* Wurster, 1956).

C. The Human Factor

The success of irrigation enterprises is the product of careful use of a well-designed system by well-trained and equipped farmers. To assure that the investment in physical resources will be managed properly, it is becoming increasingly common to select, train, and provide assistance to project settlers. The selection processes are often rigid, but understandably so, when irrigation projects represent substantial investments in both the irrigation system and in ancillary structures such as homes, schools, and hospitals. The first settlers selected for Egypt's Liberation Province were subject to social, medical, and psychological screening. Of 1,100 applicants with the correct social qualifications, only 132 were admitted to the initial training program after further medical and psychological testing (Warriner, 1962). In the Columbia Basin Project, qualification for purchase of federally-owned farm units included consideration of the character, industry, farm experience, health, citizenship, and capital resources from $5,500 to $8,500 net worth. The final selection was made by the drawing of lots among the qualified (Univ. Washington, Agr. Exp. Sta., 1959). The criteria in these two examples seems harsh but recent investigations of irrigation projects in Wyoming, USA, revealed that the settlement of war veterans, without previous experience or authoritative instruction and supervision in irrigated agriculture, was a major contributing factor to the low economic status of the entire project area.

The degree of assistance given to settlers varies as widely as the criteria for selection. In the USA there may be no special assistance available to new settlers in irrigation projects in addition to those services available to all farmers which include the assistance of farm advisors, cooperative credit, and cooperative marketing. In Mexico's Papaloapan scheme the project provides colonizing families with homes and makes credit available for the purchase of land and farming equipment. Free medical assistance is also provided and is a part of a regional health program which also includes programs for the eradication of malaria and parasitic diseases (Winnie, 1958).

The educational program of schemes such as the Papaloapan Project (Mexico) is especially important. In this Project, as well as in similar schemes in other developing countries, it is not uncommon that colonists were subsistence farmers who lived in almost totally isolated communities. When these farmers become a part of developmental projects they are thrust into a market economy and are expected to become participants in the national economy and culture. To accomplish these ends, the colonists are encouraged to develop new desires and are provided with the means to achieve these desires through the improvement of farm techniques and the commercialization of agriculture. The type of educational program that becomes an integral part of such schemes must be administered with great care and patience. Such changes come as a great shock to colonists from traditional societies and require fundamental cultural adjustments on their part. Changes of this magnitude have taken centuries to accomplish through the historical evolution of more advanced economies. Yet, in the desire for "results" from national investments in irrigation projects, a tolerance for the time element required for cultural change may be lost (Arensberg and Niehoff, 1964; Foster, 1962).

LITERATURE CITED

Adams, Frank. 1952. Community organization for irrigation in the United States. Food Agr. Organ. (FAO) UN Develop. Pap. 19. Rome. 39 p.

Adams, Robert M. 1962. Agriculture and urban life in early southwestern Iran. Science 136:109–122.

Arensberg, Conrad M., and Arthur H. Niehoff. 1964. Introducing social change. Aldine Publ. Co., Chicago. 214 p.

Beals, Ralph L. 1955. Discussion: Symposium on irrigation civilizations. In J. H. Steward et al. [ed.] Irrigation civilizations: A comparative study. Pan American Union. Soc. Sci. Monogr. 1. Washington, D.C. 78 p.

El Mallakh, Ragaei. 1959. Some aspects of the Aswan High Dam Project—Egypt. Land Econ. 35:15–23.

Foster, George M. 1962. Traditional cultures: And the impact of technological change. Harper & Row, New York. 292 p.

Franklin, R. E., W. V. Fuhriman, and B. D. Parrish. 1959. Economic problems and progress of Columbia Basin Project. Washington Agr. Exp. Sta. (Pullman) Bull. 597. 46 p.

Hardy, Edward R., Jr. 1931. The large estates of Byzantine Egypt. Columbia Univ. Press, New York. p. 60.

Hitti, Philip K. 1949. History of the Arabs. 4th ed. Macmillan Co., Ltd., London. Ch. 33. 767 p.

Holje, Helmer C., Roy E. Huggman, and Carl F. Kraenzel. 1956. Indirect benefits of irrigation development. Montana State Coll. Agr. Exp. Sta. Bull. 517. 70 pp.

Jacobsen, Thorkild, and Robert M. Adams. 1958. Salt and silt in ancient Mesopotamian agriculture. Science 128: 1251–1258.

Orive Alba, Adolfo. 1960. La politicia de irrigacion en Mexico. Fondo de Cultura Economica, Mexico D. F. 292 p.

Stamp, L. Dudley [ed.] 1961. A history of land use in arid regions. UNESCO Arid Zone Res. Ser., Vol. 17. Paris. 388 p.

Steward, J. H. [ed.] 1955. Irrigation civilizations: A comparative study. Pan American Union. Soc. Sci. Monogr. 1. Washington, D.C. 78 p.

Strabo,? The Geography of Strabo. (Horace L. Jones, transl. 1932) Book XVIII, chapt. 1, paragraph 3. Vol. 8. Harvard Univ. Press, Cambridge, Mass. p. 13.

US Department of Interior, Bureau of Reclamation. 1946. Farm size and adjustments to topography. Columbia Basin Joint Invest. Probl. 6, and 8. Washington, D.C. 115 p.

US Department of Interior, Bureau of Reclamation. 1949. Opportunities, responsibilities, and needs in irrigation development. Columbia Basin Proj. Develop. Rep. 2. Ephrata, Wash. 40 p.

US Department of Interior, Bureau of Reclamation. 1951. Character and scope. Columbia Basin Joint Invest. Washington, D.C. 29 p.

US Department of Interior, Bureau of Reclamation. Oct. 1954. The growth and contribution of federal reclamation to an expanding economy. Comm. on Interior and Insular Affairs. Print no. 27. US House of Representatives. 83rd Congr. Washington, D.C. 27 p.

US Department of Interior, Bureau of Reclamation. 1958. Evaluations of reclamation accomplishments. Columbia Basin Proj. Washington, 1948–1957, Ephrata, Wash. 32 p.

Warriner, Doreen. 1962. Land reform and development in the Middle East. 2nd ed. Oxford Univ. Press, London. 238 p.

Westermann, W. L. 1917. Aelius Gallus and the reorganization of the irrigation system of Egypt and Agustus. Cl. Philology 12:3.

Winnie, William W., Jr. 1958. The Papaloapan project: An experiment in tropical development. Econ. Geogr. 34: 227–248.

Wurster, Catherine Bauer. 1956. Regional planning: What is it? First Central Valley Reg. Planning Conf., Proc. Central Valley Planning Conf. Comm. Stanislaus Co. Planning Dep. Modesto, Calif. 85 p.

section II

Climatic Environment

II

3

World Climatic Regions in Relation to Irrigation

H. E. LANDSBERG and R. W. SCHLOEMER

Environmental Data Service, Environmental Science Services Administration, Washington, D.C.

I. INTRODUCTION

Irrigation is man's method of partially overcoming deficiencies in the natural pattern of precipitation. These deficiencies may stem from lack of precipitation, unfavorable seasonal distribution for natural plant growth, or introduction of plant species which can take advantage of favorable soil or temperature conditions but require more water than naturally provided. The purpose of irrigation may also be to increase crop yield by supplementing rainfall in normally humid areas.

The character and distribution of total precipitation over an area must be considered from at least two viewpoints, depending on the purpose for which the information is needed. First, the total annual supply and its fluctuations must be examined to determine needs for reservoir storage so that there will be optimal benefits from delayed use. Secondly, the probability distribution of certain values on a seasonal or weekly basis must be considered in order to plan intelligently for the use of the available water supply and for balancing of various user demands (Bradley, 1962). Competitive uses for the total water supply include irrigation, power, navigation, and sanitary use. To these have to be added reservoir evaporation and seepage losses. These competitive uses must then be balanced against the distribution of total precipitation for complete understanding of the water budget (Meigs, 1952b; Papadakis, 1962).

The infinite variations of parameters influencing irrigation needs (Gabites, 1960) cannot be covered briefly, nor are there sufficient observations to depict all variations (Landsberg, 1945; Trewartha, 1961). An attempt will be made to describe the major features of patterns creating the need for irrigation along with some typical examples.

II. GENERAL CIRCULATION AND CAUSES OF ARIDITY

An almost infinite variety of rainfall patterns over the earth is the result of general, areal, and local causes. The primary control is the general circulation of the atmosphere which is the unending motion caused by the unequal heating of various parts of the earth by solar radiation. The primary difference in heating exists between the equator and the poles. Equalizing motions are started and, together with variations introduced by the rotation of the earth, result in three major latitudinally distributed cells of circulation on each hemisphere. Relatively low atmospheric pressure exists near the equator and high pressure in the subtropics, whereas in middle latitudes there are eastward-moving alternating high and low pressure systems. Such large horizontal eddies cause transports of cold air south and warm air north overlaid by the meandering strong zonal current of the high level jet streams (Riehl, 1962). In polar latitudes, shallow high pressure, especially in winter, is overlaid by a strong zonal vortex.

In areas where the circulation patterns cause convergence of air currents and ascending motions, precipitation is likely to form. Conversely in areas of divergence of flow, especially in deep high pressure systems, subsidence of air causes clear skies and lack of rainfall. Seasonally, the contrast of land and sea plays an important role. Winter outflow of dry air from continental high-pressure areas is characterized by lack of precipitation. Summer onshore currents from the oceans may bring copious monsoonal rains in areas under these regimes. Because of the ascending motions, major mountain chains which the normal flow has to cross lead to increased rainfall on the windward side and to decreased precipitation to the lee with the descent of the current.

Locally, orography and minor sources of water such as lakes may introduce fairly substantial differences in precipitation (Gol'tsberg, 1957). These are, however, not of the broad impact as those imposed by the general circulation.

Lack of rainfall is usually characterized by high solar radiation over the earth (Landsberg et al., 1963) (Fig. 3–1). These values are a measure of the energy available at the surface of the earth for utilization by plants if other growth factors are suitable. Of course, much of this energy is used for warming the soil, and for evaporation, or it is lost by radiation back to the atmosphere and space. By the very nature of the general circulation, arid zones are generally the areas of maximum solar radiation. If moisture were available in the atmosphere much of the incoming solar radiation would be intercepted above ground level. In areas where small energies from solar radiation reach ground level, much of the incoming

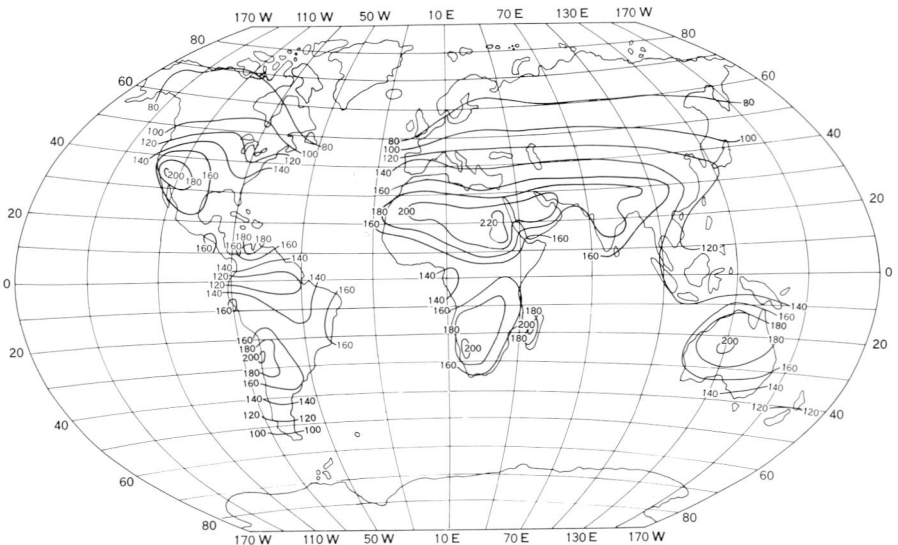

Fig. 3–1. Solar radiation over land areas (after Landsberg et al., 1963). kg cal/cm^2 per year.

energy is dissipated, absorbed, or reflected above the level of the earth by dust or clouds.

III. PRECIPITATION PATTERNS

The extensive drylands of the world generally lie between the latitude parallels of 10° and 35°N and S (Blumenstock, 1958). The semipermanent pressure patterns discussed above, and resultant wind patterns, are distributed in such fashion as to produce easterly winds with an equator-ward component in these belts (trade winds). These areas are immediately north and south of the major tropical convergence zone. The typical areas include Southwest USA, South Central South America, South Africa, North Africa extending into central and southern Asia, and most of western Australia.

Figure 3–2 depicts the major dryland areas of the world after Meigs (1951, 1952a). This chart also shows the principal ocean currents which have a degree of influence on the aridity of bordering land areas. The currents shown are only cool or cold ocean currents. Circulation of air over these cool areas onto the adjacent land areas, where heating takes place rapidly, results in generally desiccating conditions except along the immediate coastal area where fog and dew may contribute to growth of lower types of vegetation.

Another factor contributing to generally arid conditions is distance from source of water supply. Again referring to Fig. 3–2, the aridity increases in generally dry areas as distance from the ocean increases. This characteristic of rainfall patterns is most clearly demonstrated in those areas where there is an offshore prevailing wind rather than an onshore wind.

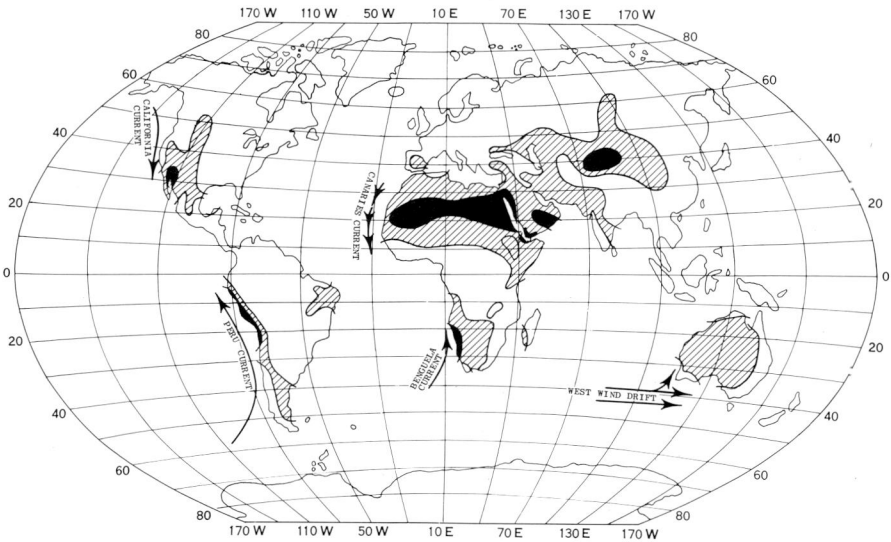

Fig. 3–2. Semiarid and arid areas of the world (after Meigs, 1951 & 1952a) and selected ocean currents (after US Navy Hydrographic Office H.O. 1400 & 1401). Solid color denotes extreme aridity.

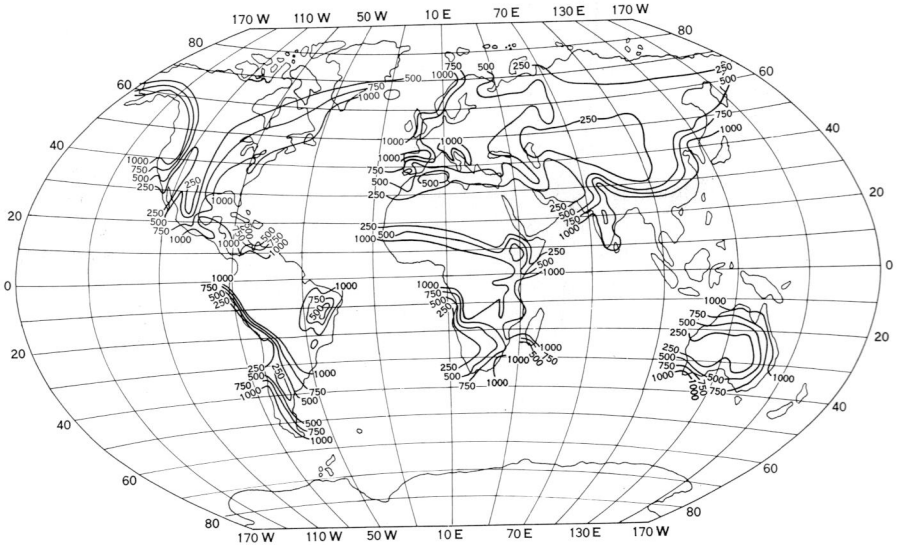

Fig. 3–3. Annual precipitation in areas having < 1,000 cm (40 inches) after Knoch and Schulze, 1952).

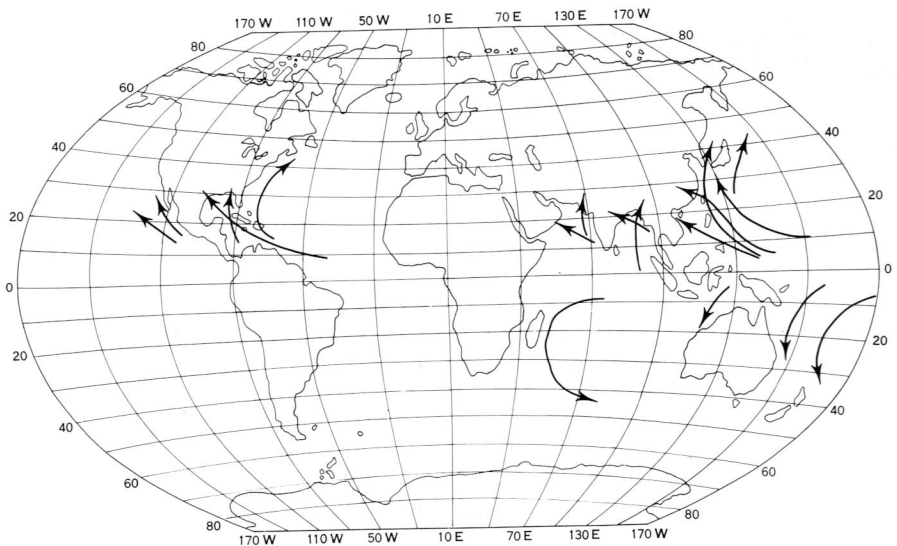

Fig. 3–4. Principal regions of tropical cyclones (after Tannehill, 1956).

If there is a mountain range near a coastal area of onshore winds, there is a resultant increase of precipitation on the windward side and a very rapid decline to arid or semiarid conditions in the lee of the mountains.

These features are all clearly observable on the annual total precipitation map shown in Fig. 3–3 (Knoch and Schultze, 1952). Seasonal distributions on the borders of the arid regions are discussed below. However, note the influences of the

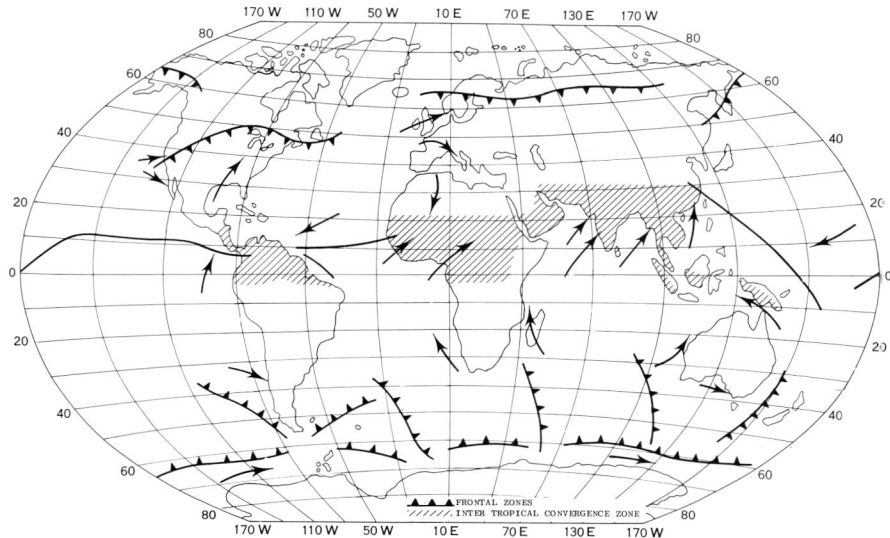

Fig. 3–5. Convergence zones northern summer (after Gentilli, 1958).

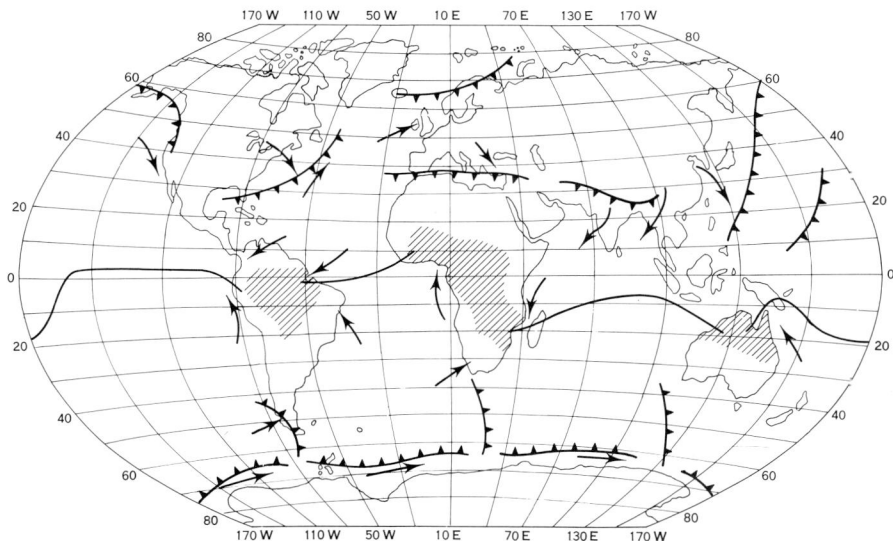

Fig. 3–6. Convergence zones northern winter (after Gentilli, 1958).

predominant triggering mechanisms for rainfall over and above the straight-forward orographic effects discussed above. The tropical cyclones shown in Fig. 3–4 are highly seasonal in character and quite unpredictable in terms of number per year and accompanying rainfall (Tannehill, 1956). Most southeast coastal areas (in the Northern Hemisphere) have recorded their greatest daily and monthly rainfall totals from such storms. These areas may experience semidrouth during part of the critical growing period in years when such storms are absent.

Critical analysis of causes for long-periodic fluctuations and, better still, theoretical concepts for understanding the development and propagation of such storms is an essential task of the future.

Other major rain-producing mechanisms are depicted on Fig. 3–5 and 3–6. The Intertropical Convergence Zone may be crudely described as the zone of meeting near the thermal equator of trade winds of the two hemispheres (Gentilli, 1958). This zone is subject to a seasonal migration lagging somewhat behind the period of high sun, resulting in a distinct dry season (or seasons). Retention of a water supply from the rainy to the dry period is critical since low latitude location coincides with very high evaporation rates from open reservoirs. In these areas, the available water will have to be closely balanced against the factors causing water losses.

The remaining major mechanisms contributing to world precipitation in semiarid areas is the interaction between warm and cold air masses resulting from meridional transport of warm air poleward. The migrant cyclonic storms which form along the major boundaries between warm and cold air transporting moisture from source regions instigate much of the precipitation in the temperate latitudes. The remainder is produced in a very irregular pattern by local thunderstorms, resulting from vertical instability caused by intensive heating. The larger and smaller storms occur at irregular intervals and whole seasons may lack adequate precipitation and suffer from drouth.

As a rather general rule, it can be stated that for land areas the reliability of precipitation from year-to-year is proportional to the total amounts. The smaller the amounts, the greater is the percentage variability of various years (Conrad, 1941). Consequently, areas marginal to the true deserts—or the semiarid zone— show generally the highest risks of moisture deficiencies for agricultural purposes. Here, water conservation has to be developed to the highest degree possible by engineering methods.

IV. THE PATTERN OF WORLD ARIDITY AND WATER SOURCES

A. Permanently Dry Areas

Much of the permanently dry areas of the world can, unfortunately, concern us very little in relation to irrigation. It is true that minor patches of agricultural development can be helped by geological exploration for wells, salt water conversion, technological development of dew collection, and the ever remaining hope for induction of artificial precipitation (see chapter 5). In the latter two cases, high humidity, water condensed in clouds, or water vapor near condensation are a necessary requirement and in those areas defined as permanently dry, even these conditions are generally lacking.

Use of deep wells in these areas has not yet reached the ultimate development. Studies are needed that will relate to replenishment of aquifers locally as well as in areas different from where the aquifer is tapped. Furthermore, the intense evaporation from open reservoirs causes high loss of water. Artificial suppression of evaporation is one of the most hopeful areas of possible artificial interference by man in the hydrologic cycle.

Irrigation is of greatest use in those areas where a remote source of water can provide irrigation water along a river as it passes through an area otherwise devoid or deficient in natural precipitation. The mountain catchment areas serve as a reservoir for irrigated areas in the adjacent foothills or plains. Of greater significance agriculturally are those very large streams such as the Colorado, Rio Grande, and Nile Rivers which have a very large catchment area and then proceed across desert or near desert areas for many miles. The efficacy of reservoirs on these streams is dependent on evaporation and seepage as well as interstate or international agreements on water rights.

B. Seasonally Dry Areas

There are vast areas of the world bordering on the subtropical high pressure belt which receive practically all of the precipitation in one season of the year. Generally these areas lie between 30° and 40° of N and S latitude, most commonly on west coasts. The Mediterranean is the major area of the world which in some schemes of a climatic classification gave the name to a special species of climate. This climatic type occupies a vast area surrounding that sea. Other areas having similar climatic characteristics are southern California, USA, central Chile, South Africa, and western Australia. Wallén (1962) defines the limit of dryland farming (without irrigation) in the Near East as limited by 240 mm/year (10 inches) precipitation and with an interannual variability of 37%.

The typical climatic pattern of these areas is a dry summer and rainy winter. Thus, the reverse of the usual agricultural need is common. Fortunately, many of these areas have a precipitation catchment area within a reasonable distance to supply seasonally needed water supplies for irrigation (Rockies, Andes, Alps, Lebanon, Atlas). These areas are blessed, of course, with high values of summer sunshine and thus lend themselves particularly to specialized agriculture under irrigation systems.

C. Subhumid Areas

The importance of irrigation in the semiarid areas of the world should not completely overshadow the tremendous benefits which can result from supplemental irrigation in those areas which are normally considered to receive sufficient precipitation (Fosberg et al., 1961).

Those areas of the world (southeastern USA and southeastern China are typical) which are subjected to precipitation resulting from monsoonal currents or the interaction of warm and cold air masses normally receive sufficient rainfall. However, it is not unusual to have one or more periods during a crop year when little or no precipitation occurs. Depending upon the stage of plant growth at the time of occurrence, such brief periods of drouth conditions can inflict serious losses in crop production (see chapter 58). The usually abundant rainfall can easily be stored as an emergency reserve in farm ponds or larger reservoirs. The importance of rainfall at various stages of plant growth is now well established so that rainfall probabilities can be used to assess the feasibility of local measures for ensuring an adequate moisture supply at all times.

Stimulation of crop production through supplemental irrigation is a consideration of major importance as the population explosion continues to put pressure on agricultural production requirements. Studies of the frequencies and severities of drouths have shown that these, while as yet unpredictable, follow definite statistical distributions. Hence, they can be included in all plans as calculable risks.

LITERATURE CITED

Blumenstock, David I. 1958. Distribution and characteristics of tropical climates. Pacific Sci. Congr., Proc. 9th. 20:3–24.

Bradley, Charles C. 1962. Human water needs and water use in America. Science 138(3539):489–491.

Conrad, V. 1941. The variability of precipitation. Monthly Weather Rev. 69(1):5–11.

Fosberg, F. R., B. J. Gurnier, and A. W. Kuchler. 1961. Delimitation of the humid tropics. Geogr. Rev. L1(3):333–347.

Gabites, J. F. 1960. How to plan irrigation needs of land. New Zeal. J. Agr. 101(5): 440–441.

Gentilli, J. 1958. The geography of climate. The synoptic world pattern. 2nd rev. ed. The University of Western Australia Press. 172 p.

Gol'tsberg, I. A. 1957. Microclimate and its importance in agriculture. Hydrometeorological Publishing House, Leningrad. (Trans. US Joint Publ. Res. Serv. New York, 1958, 75 l., as JPRS/DC–L–657.)

Knoch, K., and A. Schulze. 1952. Methoden der Klimaklassifikation. Petermanns Geographischen Mitteilungen sep. no. 249. Justus Perthes Gotha. 78 p. + 10 maps.

Landsberg, H. E. 1945. Climatology. In F. A. Berry, E. Bollay, and Norman Beers [ed.] Handbook of meteorology. McGraw-Hill, New York. Sect. XII:928–997.

Landsberg, H. E., H. Lippmann, Kh. Paffen, and C. Troll. 1963. World maps of climatology. Springer-Verlag, Berlin-Göttingen-Heidelberg. 28 p. + 5 maps.

Meigs, Perveril. 1951. La repartition mondiale des zones climatiques arides et semi-arides. Annex II, Bibliogr., UNESCO/NS/AZ/37, Paris. rev. 16 p.

Meigs, Peveril. 1952a. Arid and semiarid climatic types of the world. Int. Geogr. Union, Proc., VIII Gen. Assembly, XVIIth Congr. p. 135–138.

Meigs, Peveril. 1952b. Water problems in the United States. Geogr. Rev. 42:346–366.

Papadakis, Juan. 1962. Avances recientes en el estudio hidrico de los climas. Inst. Suelos Agrotec., Buenos Aires, Publ. no. 81. 28 p. + 5 maps.

Riehl, Herbert. 1962. Jet streams of the atmosphere. Colorado State Univ. Dep. Atmos. Sci. Tech. Rep. no. 32. 117 p.

Tannehill, I. R. 1956. The hurricane. US Dep. Com.-Weather Bur., Government Printing Office, Washington. rev. 22 p.

Trewartha, Glenn T. 1961. The earth's problem climates. University of Wisconsin Press, Madison. 334 p.

Wallén, C. C. 1962. Agroclimatology in the Near East. World Meteorol. Organ. Bull. XI(3):116–123.

4 | Influence of Local Physiographic Features

MILTON L. BLANC

Environmental Data Service, Environmental Science Services Administration, Tempe, Arizona

I. INTRODUCTION

The climatological pattern of world precipitation and the general causes of aridity have been explained in chapter 3. The present chapter describes how local physiography modifies the general weather patterns and affects water supply, water need, and water use. Physiography, as used here, includes absolute and relative altitude, aspect, topography (especially as related to prevailing winds), drainage, water bodies, and other similar physical features.

Nichols (1924) showed that differences in climate which are common between areas quite distant from one another also exist between locations which are relatively adjacent. In high mountains, for example, marked differences in climate occur in short distances with changes in elevation, accompanied by corresponding differences in vegetation. Even within areas of generally uniform elevation, the climate is locally modified by small differences in topography and by relative positions of land and water bodies. The aspect of the terrain (i.e., north- or south-facing slopes) influences types of vegetation. Plants found in ravines differ from those on level or rounded uplands.

II. EFFECT OF LOCAL PHYSIOGRAPHY ON WATER SUPPLY

A. Altitude

Both absolute and relative altitude should be considered as they affect precipitation. It is difficult to generalize regarding the effect of absolute altitude because of other masking influences. In most areas there is a gradual increase in precipitation with altitude up to 1,000 m or so. The principal mechanism for producing rain involves the lifting and cooling of moisture-laden air. At higher elevations the air is normally cooler and drier, hence, many high plateaus are arid or semiarid. The effects of relative altitude on precipitation are easier to observe and will be discussed in greater detail in section B of this part.

Sternberg (1956) describes an area in Brazil which is a classical example of a regional precipitation pattern influenced by elevation. In this area, precipitation decreases rapidly from the east (the Atlantic coast) toward the west (about 950 km inland). He describes three zones: a low coastal humid forest zone; an eastward-sloping transitional zone of relatively stable agriculture; and a broader inland plateau with insufficient and highly variable precipitation. Island-like eminences rise above the general level of the inland plateau which are favored by more abundant and reliable rainfall. For example, at Meruoca (elev. 670 m) the annual precipitation is over 1,700 mm compared to 852 mm at Sobral (elev. 70 m), only 23 km distant.

The nature of precipitation also changes with altitude. At higher elevations a greater part falls as snow, much of which is stored during the winter months to become available for irrigation during the growing season.

B. Topography

Slopes facing the prevailing moisture-laden winds favor the precipitation process by lifting and cooling. A long gentle slope usually produces uniform precipitation over a large area. If the slope levels off into an upland plain the windward edge will receive amounts equal to or slightly more than the slope. Then the amounts gradually diminish with increasing distance from the slope. Where slopes face away from the prevailing winds, a very rapid decrease occurs. Valleys or plains on the leeward side of mountains or high ridges are often quite dry. Figure 4-1 shows variation of total winter precipitation with elevation across the Wasatch Mountain Range in Utah, USA.

On the western slopes of this mountain range, average annual total snowfall varies from near 125 cm in the lower valleys (1,280 m) to about 950 cm at 2,650 m (Brown and Williams, 1962). Because there is less melting at higher elevations during the winter, more snow is left for spring runoff. The average water equivalents of the spring snowpack at these same elevations range from 33 mm to near 790 mm.

The Hawaiian Islands afford many striking examples of the effect of topography on precipitation (Stidd and Leopold, 1951; Blumenstock, 1961). The primary precipitation control for the islands is their location in the belt of easterly trade winds. Mountain barriers often rise 900 to 1,500 m with some isolated peaks near 4,000 m. Figure 4-2 gives annual precipitation amounts for the island of Hawaii.

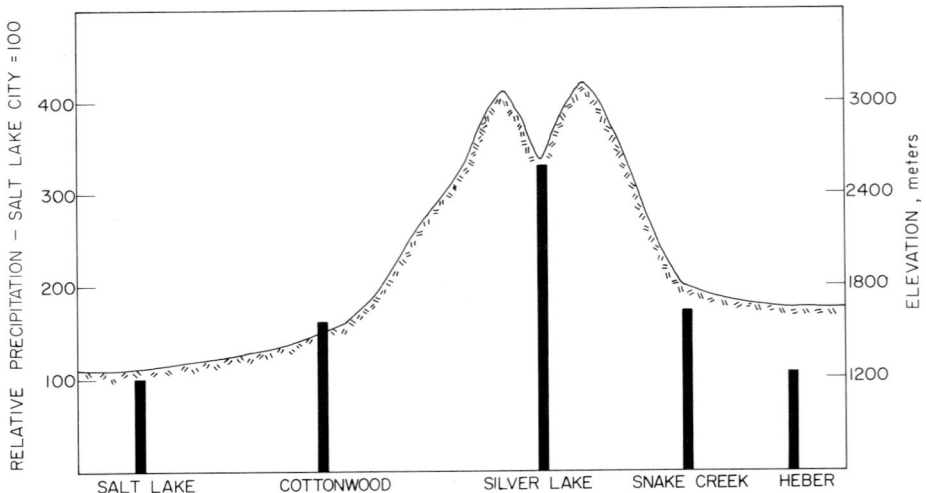

(Adapted from Williams and Peck, 1962)

Fig. 4-1. Relative winter (October-April) precipitation across a mountain range **near** Salt Lake City (adapted from Williams and Peck, 1962).

Mean Annual Precipitation, Inches

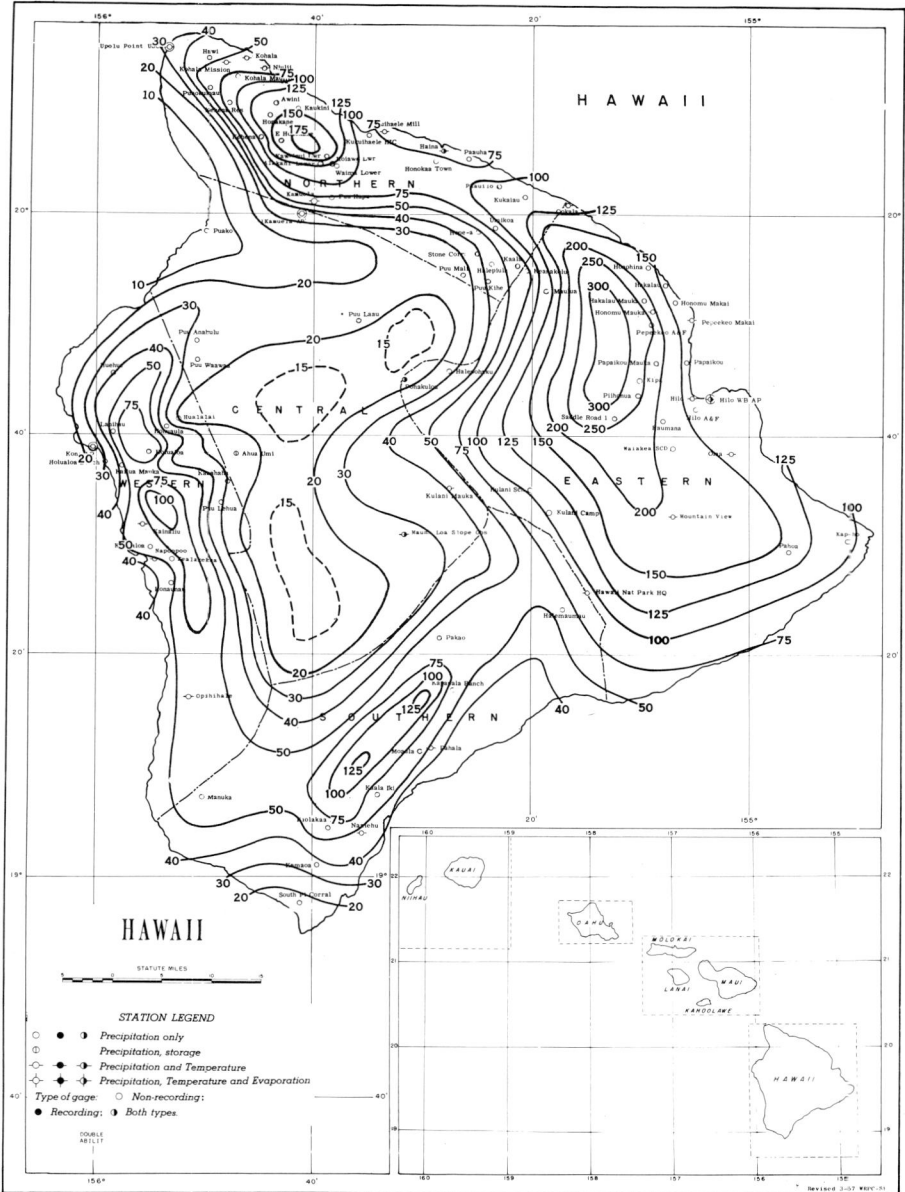

Based on period 1931–55

Isolines are drawn through points of approximately equal value. Caution should be used in interpolating on these maps, particularly in mountainous areas.

Fig. 4–2. Mean annual precipitation (inches) on island of Hawaii. Data for period 1931–1955. Smoothed isolines are drawn through points of approximately equal value. Caution should be used in interpolating, particularly in mountainous area.

On the leeward coast and also near the summit of the very high mountains (e.g., Mauna Loa, 4,200 m), the rainfall averages < 500 mm/year compared with 7,600 mm along the lower windward slopes of the high mountains and at, or near, the crest of lower slopes. On the island of Kauai (not shown), Mt. Waialeale, one of the world's wettest spots, has an annual average of 11,684 mm while the semiarid leeward coast, just 29 km away, has < 635 mm.

The situation on Mauna Loa illustrates that continued increase in height is not always accompanied by continued increase in rainfall. In this case the moisture-laden winds that rise over the lower mountains cause a rapid increase in precipitation with height to a maximum near or slightly to the leeward of the crest. However, the isolated peak of Mauna Loa permits the air to rise part way up the slopes and then flow around the flanks. Here the maximum amounts are at elevations of from 600 to 1,200 m while the summit is relatively dry.

The rate of change of annual amounts of rainfall, as influenced by topography, is exceptional in a number of places in these islands. Blumenstock (1961) computed a number of these rates in terms of increase in the annual average per unit distance along a straight line up the slope. Such gradients often exceed 400 mm/km. In an extreme case the gradient is 1,880 mm/km along the 4-km line from Hanalei Tunnel to Mt. Waialeale on Kauai.

In the Upper Rio Grande Basin there is a good general relationship between increase in altitude and in annual precipitation (Dortignac, 1956). An exception is the San Luis Valley (elev. 2,000 m). It lies on the east (rain-shadow) side of a mountain range separating it from the Rio Chama Valley of similar elevation. The San Luis Valley has less than half the annual precipitation of the Rio Chama Valley.

Topography is also related to runoff and infiltration. Slopes which receive adequate precipitation due to topography may be unsuited for agriculture because of heavy runoff and erosion. However, if much of the precipitation is in the form of snow (as at high altitudes), more moisture penetrates the soil as the snow slowly melts.

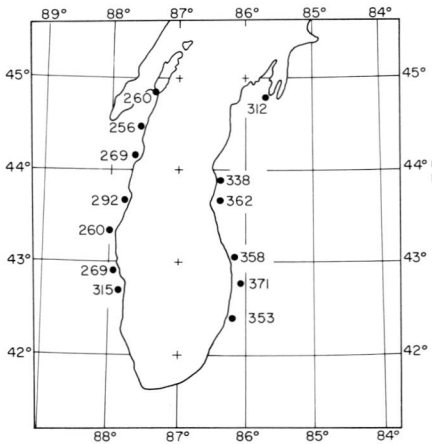

Fig. 4–3. Winter (October-March) precipitation (mm) at selected lakeshore stations in Wisconsin and Michigan, USA.

Spreen (1947) developed a method of estimating mean winter precipitation in drainage basins of central Colorado, USA, using the following topographic parameters: elevation of the station, maximum slope of the land within a 5-mile radius, exposure, and orientation of exposure. Exposure is the sum of those sectors within a 20-mile radius which do not contain a barrier 1,000 ft or more above the station. Such parameters serve to define the ease with which air masses can enter the basin and the amount of orographic lifting. The parameters also define the kind of air mass (i.e., whether wet or dry) according to the direction of the open exposure, taking into account the general climatological controls of the region.

In the Fraser Experimental Forest, Colorado (Garstka et al., 1958), investigations on the effect of aspect on snow melt showed that a south-facing slope lost its snow 30 days sooner than a north slope and that the time of most rapid snow melt was 40 days earlier. The period of snow melt was 77 days on the north slope compared to 46 days on the south slope. Steepness and aspect were much more important than elevation in determining the rate of melt. North-facing slopes had only 70% of the solar insolation measured on south slopes in late April, increasing to 95% in early June. East and west slopes were 95% of the south in late April and 104% in early June.

C. Water Bodies

Small lakes, ponds, or other similar bodies of water have relatively little effect on the moisture content of the air as far as general precipitation is concerned. However, there is sometimes a noticeable local increase in shower activity on the leeward shore. This is due in part to an increase in moisture content as the wind passes over the lake. A more important factor is the lifting of air by the lake shore itself and also by the piling up of air due to the increase in friction as it moves from the smooth lake surface to rough terrain. In winter, locally very heavy snows occur along lee shores around the Great Lakes, USA, as a result of this process. Figure 4–3 shows average winter (October to March) precipitation at selected points on the windward (Wisconsin, USA) and leeward (Michigan, USA) shores of Lake Michigan. There is a general 20% increase at the same latitude across the lake. A similar but not so large (about 14%) increase is observed in annual amounts.

Water bodies can contribute to the quantity of dew deposition. In arid regions, dew may account for an appreciable proportion of the total precipitation. Ashbel (1936) ascribes considerable importance to dew in Palestine agriculture. Slatyer and McIlroy (1961) show that in an arid region regularly overrun by moist air from nearby water bodies, the advection and radiation processes may act to give maximum possible amounts of dew.

D. Weather Modification

The possibility of increasing water supplies through some artificial process cannot be ignored. There has been great interest in this subject and many conflicting views have been presented in the literature. The American Meteorological Society (1962) issued a statement on the implications of the control of weather and climate. Certain evaluations of cloud-seeding operations have indicated that precipitation might be increased over local areas where specific conditions prevail (e.g., over some regions where forced lifting of air over mountains occurs), but present knowledge of atmospheric processes offers no real hope for large-scale changes of weather or climate by cloud seeding. Local topography is of great importance in this regard. Some locations which depend on the winter accumulation of snow for water supply during the growing season might increase this supply by increasing the snow catch on favorable slopes and high ground during the winter. Also, where the topography favors up-slope motion of moisture-laden winds a large percentage of the time, there is some possibility of increasing the frequency and/or duration of precipitation through artificial means. It must be emphasized at this stage that none of the more carefully conducted experiments in the USA designed specifically for evaluation have succeeded as yet in identifying effects on precipitation at a satisfactory level of statistical significance. This applies both to precipitation measurements at ground level and tests using radar (*see also* chapter 5).

III. EFFECT OF LOCAL PHYSIOGRAPHY ON WATER NEED AND WATER USE

In general, local physiographic features which tend to increase the amount of precipitation also tend to reduce the water requirements for plant production. Conversely, those which tend to decrease the natural precipitation also tend to increase the water requirement. Some brief examples follow but it will be apparent that these physiographic influences are less important than those discussed in the preceding part II.

A. Altitude

Altitude affects water need indirectly through the evapotranspiration process. As shown in the first paragraph of part I, section A, an increase in elevation is often accompanied by an increase in the amount of precipitation and also in the number of rainy and cloudy days. These conditions decrease evapotranspiration and reduce the water need. Even if the clouds do not produce rain, the increased humidity and decreased insolation are important factors. However, if the altitude is such that the area is above the general cloud and moist air level, then the water requirements tend to increase due to reduced humidity and increased insolation (Longacre and Blaney, 1963).

B. Topography

In a somewhat analogous manner, vegetation on windward slopes exposed to moist climatic conditions requires less moisture for evapotranspiration. Vegetation on leeward slopes or located some distance to the leeward of a ridge or mountain, and exposed to a greater amount of sunshine and lower humidity, requires more moisture for growth.

Further, the aspect of the slope is quite important in the evapotranspiration process. South-facing slopes (Lowry, 1963) receive more solar energy per unit area than level land or north-facing slopes, resulting in an increased moisture requirement.

C. Drainage

Drainage, of course, is most directly concerned in the engineering aspects of irrigation and is discussed in greater detail in later chapters (10, 43, 44, 48). However, if the local drainage greatly increases surface runoff, then less water infiltrates, the irrigation problem becomes more complicated, and an increase in the water requirement may result.

D. Water Bodies

To some extent, local water bodies increase moisture content of the air and decrease the water requirement. However, this effect is usually quite small and local and would not normally be of much importance. Obviously, local fresh water bodies afford some opportunity for obtaining easy access to irrigation water, and, in this respect they become part of the irrigation engineering problem.

IV. FROST PROTECTION

Chapter 54 is devoted to the subject of frost protection with irrigation. In many world areas this is becoming increasingly important and attention must be given to an adequate supply of water for this purpose during the frost danger season. Topography plays an important role in determining the length and severity of the frost danger season, thereby influencing the amount of water which must be provided for frost protection.

Many damaging frosts and freezes are the result of the rapid loss of heat from the soil and vegetative surfaces by radiation on clear dry nights with little or no wind. Under these circumstances the cold air flows by gravity and collects in low places. In areas where frost damage from such conditions is a serious problem, a great many vineyards and orchards are located on the slopes and hillsides but do not extend down into the valleys. Experience has shown that damaging temperatures are most frequent and severe on the lower ground due to the collection of cold air. The general climate of the area and the frequency and duration of the damaging conditions can be greatly modified by the local topography which must be taken into account when estimating the quantity of water required for frost protection purposes (World Meteorol. Organ., 1963).

LITERATURE CITED

American Meteorological Society. 1962. Statement on the implications of the control of weather and climate. Bull. Amer. Meteorol. Soc. 43(8):400–401.

Ashbel, D. 1936. On the importance of dew in Palestine. J. Palestine Orient. Soc. 16: 316–321.

Blumenstock, David I. 1961. Climate of Hawaii. US Dep. Com.-Weather Bur. Climates of the States no. 60–51. p. 7–8.

Brown, Merle J., and Philip Williams, Jr. 1962. Maximum snow loads along the western slopes of the Wasatch mountains of Utah. J. Appl. Meteorol. 1(1):123–126.

Dortignac, E. J. 1956. Watershed resources and problems of the Upper Rio Grande basin. Rocky Mountain Forest Range Exp. Sta. p. 14.

Garstka, W. U., L. D. Love, B. C. Goodell, and F. A. Bertle. 1958. Factors affecting snow melt and stream flow. US Government Printing Office, Washington. 189 p.

Longacre, Leonard L., and Harry F. Blaney. 1963. Evaporation at high elevations in California. Amer. Soc. Civil Eng. Irrig. Drainage Div. J. 89(2–1):77–78.

Lowry, William P. 1963. Observations of atmospheric structure during summer in a coastal mountain basin in northwest Oregon. J. Appl. Meteorol. 2(6):713–721.

Nichols, George Elwood. 1924. The terrestrial environment in its relation to plant life. In M. R. Thorpe [ed.] Organic adaptation to environment. Yale University Press, New Haven, Conn. p. 22–23.

Slatyer, R. O., and I. C. McIlroy. 1961. Practical microclimatology. CSIRO, (Australia) and UNESCO, Paris Ch. 6:1–3.

Spreen, William C. 1947. A determination of the effect of topography upon precipitation. Amer. Geophys. Union, Trans. 28:285–290.

Sternberg, Hilgard O'Reilly. 1956. Geography's contribution to the better use of resources. In G. F. White [ed.] The future of arid lands. Amer. Ass. Advance. Sci. Publ. 43, p. 200–220

Stidd, C. K., and Luna B. Leopold. 1951. The Geographic distribution of average monthly rainfall, Hawaii. Meteorol. Monogr. 1(3):24–33.

Williams, Philip, Jr., and Eugene L. Peck. 1962. Terrain influences on precipitation in the intermountain west as related to synoptic situations. J. Appl. Meteorol. 1(3):343–347.

World Meteorological Organization. 1963. Protection against frost damage. Tech. Note No. 51. Geneva. 62 p. 160 ref.

5 | Weather Variation and Modification

RICHARD A. SCHLEUSENER
South Dakota School of Mines & Technology
Rapid City, South Dakota

LEWIS O. GRANT
Colorado State University
Fort Collins, Colorado

I. WEATHER VARIATION

A. Introduction

Weather consists of short-term variations of the atmosphere and affects all phases of agriculture and soil use. Short-term weather variations are those that occur from hour-to-hour, day-to-day, season-to-season, and year-to-year. Climate is the synthesis of all weather during a specified time, usually measured in decades. Climate determines the type of agriculture that can be carried out in a region while weather and weather variations affect specific planning and operations within the region.

B. The Effects and Magnitude of Weather Variations

Weather variation affects nearly all aspects of agriculture in a specific region, including land preparation, planting, soil bacterial action, application of pesticides and herbicides, plant growth, and harvesting. Although these weather variations may differ from region to region, they are substantial in most regions from the tropics to the arctic. Day-by-day temperature variations in tropical and subtropical areas are usually minor, particularly near large bodies of water. They usually increase significantly, particularly in continental areas, toward the poleward storm belts located some 60° either way from the equator. Annual variations in temperature in many tropical and subtropical oceanic areas may be < 6.7 C (12 F), while daily variations > 37.8 C (100 F) have been observed in more northerly, continental areas. Most regions of the world experience annual precipitation variations of at least 100% of the mean annual precipitation. Continental areas away from the tropics may experience annual variations of 400 to 800% of the mean annual precipitation, and greater variations are not infrequent in many continental, subtropical regions.

C. Contemporary Forecasts of Weather Variations

The ability to predict weather variations would form the basis for altering almost all agricultural activities. Prediction needs to be considered on various time scales.

1. FORECAST PERIOD 24–48 HOURS

The ability to predict short-period weather variations for intervals of 24–48 hours can be of utmost value in day-to-day agricultural activities. Predictions of this nature took a large step forward in the early 20th century with the advent of air mass concepts and rapid communications to simultaneously define existing weather over large segments of the earth. These forecasts have continued to improve at a slow but steady pace since the 1930's. The advent of computer processing has accelerated improvement in forecasts of the time interval of 24–48 hours.

Shorter period forecasts of up to a few hours have improved greatly with the advent of better reporting networks, particularly the installation of weather radar which can track individual storm elements on a continuous basis. This is of special value in the prediction of severe storms, such as tornadoes and hurricanes. Most 24- to 48-hour forecasts show significant skill and are correct about 90% of the time. This accuracy is increased substantially, however, by the large number of days when weather is persistent from day-to-day and decreased when large variations are occurring and correct forecasts would be the most useful.

2. FORECAST PERIOD 2–7 DAYS

Substantial progress has been made since the early 1940's in extending the daily predictions of weather for intervals up to about 1 week. These predictions, however, are less accurate than daily forecasts and need to be presented in more general terms, such as "warming trend during the period." Nevertheless, weekly predictions have been demonstrated to be an improvement over the use of climatological averages and can be of definite value in many activities.

3. LONG-RANGE FORECAST PERIOD OF 7 DAYS TO A FEW YEARS

Outlooks for weather prediction for periods exceeding 1 week are on uncertain ground since mathematical prediction models have not proven stable for these time intervals and physical controls have not been established. Forecasts for these time intervals are usually labeled as experimental and forecasting skill for these periods has not been demonstrated.

D. The Outlook for Improved Forecasts

1. FORECASTS THROUGH 7 DAYS

Progress in predicting weather variations for periods up to 7 days cannot be expected to accelerate greatly in the near future. Slow improvements in forecasts can be expected as circulation models and their treatment by electronic computers are improved. Improved techniques for data acquisition such as meteorological rockets (Giraytys and Rippy, 1964), meteorological satellites (Tepper, 1963), and weather reconnaissance aircraft (Widger, 1963) also offer the promise of improved forecasts, particularly for remote areas where conventional data are sparse.

2. FORECASTS BEYOND 7 DAYS

a. **Statistical Techniques.** Longer range outlooks utilizing statistical techniques were explored extensively earlier in this century. They included efforts to estimate local variations from statistical relationships established with atmospheric parameters in other regions. They have included efforts to establish climatological periodicities from long-period weather information derived from tree rings (Weakly, 1965), carbon dating, etc. These efforts have improved the understanding of worldwide weather relationships but have not proven to be substantial forecast tools.

b. Mathematical Models. Since World War II, concentrated efforts have been made to improve and extend mathematical models of atmospheric circulation. The models consistently break down after several days, and under certain atmospheric conditions, break down almost immediately. Atmospheric actions and reactions are not well enough defined and probably not sufficiently stable for substantial expansion of these models in time.

c. Physical Techniques. The approach presently receiving the most attention is that involving physical controls, such as abnormal oceanic temperatures, variations in atmospheric ozone, the earth's reaction to solar variations, etc. There is a possibility that such physical controls may improve the understanding of some longer range variations—monthly, seasonal, or for a few years—more substantially than for shorter periods.

II.   WEATHER MODIFICATION

A.   Introduction

Water use in the USA has increased nearly 800% between 1900 and 1960 (Gloyna et al., 1960). Large increases in both population and water consumption per capita are combining to rapidly strain water supplies even in water-abundant areas. Water shortages are becoming chronic in many drier regions. Wollman (1960) estimates the magnitude of these shortages for 1980 and 2000 for various regions of the USA. For example, the sustained yield for the Colorado River by 1980 will be only 58% of project demand, 16% for the Rio Grande, 83% for the Great Basin area, 93% for the Upper Missouri, and only 8% for the South Pacific area of the Southwest USA.

1. HISTORICAL BACKGROUND

Proposals for and attempts to modify the weather have extended back over many centuries. Espy (1839) suggested that large fires could be built to generate updrafts which in a humid atmosphere would support cumulus clouds and would lead to rain. As had been the case after the Napoleonic Wars, many proposals again were made after the Civil War to carry out rainmaking experiments by shooting explosives into clouds. Veraart (1931) used dry ice in Holland in a manner which resembles recent cloud seeding efforts. The contemporary basis for interest in weather modification was started in the 1930's by the work of Bergeron (1933, 1935) and Findeisen (1938) (See also W. Findeisen, 1938. Meteorologisch-physikalische Begingtheiten der Vereisung in der Atmosphare. *Hauptversammlung der Lilienthal-Gesellschaft fur Luftfahrtforschung, 1938.*). Experimentation using these theories was opened by Langmuir and Schaefer in their experiments with dry ice starting in 1946 (Schaefer, 1953).

2. *THE ATMOSPHERE AS THE BASIC SOURCE OF WATER*

All available surface and underground water supplies will need to be developed to the maximum advantage in the foreseeable future. This will have to include increased attention to conservation and reuse. The increase of fresh water supplies from salt water conversion may provide substantial additional supplies in many areas. In most areas, however, the only source for actually increasing water supplies, in contrast to improved usage, is from the original source of our available water—the atmosphere.

Tremendous quantities of water are supplied to and transported in the atmosphere. McDonald (1958) concludes that even if highly optimistic estimates were made of man's ability to increase precipitation over continental areas, such activities would "... constitute a minuscule alteration of the natural process," and "... [would not] sensibly alter the prevailing water vapor content of the atmosphere." Computations by the authors indicate that the mean daily flux of water vapor over Colorado, USA, exceeds the annual precipitation by more than an order of magnitude. This is consistent with studies made in other geographic regions which have shown that the amount of water reaching the ground as precipitation is $< 10\%$ of the total water which flows into the region. Yearly periods have been considered in Illinois, USA (Huff and Stout, 1951) and daily periods in Arizona, USA (Reitan, 1957).

Although considerations of wide-spread precipitation from cyclonic storms give an indication of higher precipitation efficiency (Wexler, 1960), the foregoing considerations suggest that even a small change in the amount of water vapor precipitated would represent a fairly substantial increase in the natural precipitation.

3. *ECONOMIC CONSIDERATIONS*

There are two types of errors generally associated with evaluations of weather modification techniques. A type I error occurs if the experimenter concludes that differences existed when actually they did not. It is a standard procedure for the experimenter to design his experiment in such a fashion as to minimize the probability of this type of error. The results of the test are termed not statistically significant unless the probability of the type I error is less than, say, 5% or 1%. A type II error arises when an experimenter concludes that no effect existed when actually there was an effect. Because of the high benefit–cost ratio connected with weather modification activity, there would be an economic loss if it were concluded incorrectly that no beneficial results could be gained from weather modification.

The absence of statistically significant results from weather modification experiments normally refers to the type I error. Failure to detect statistically significant differences from such experiments does not necessarily mean that such differences may not exist. Moreover, economic considerations of the benefits that could be realized from even a small artificial change in precipitation (type II error consideration) has been a factor in the high degree of interest which continues in the subject of weather modification.

The significance of the type II error is covered in standard statistical texts (Steel and Torrie, 1960). Its application to the weather modification problem is discussed by Thom (1957), and is considered in connection with decision-making processes in weather modification activity by Berndt (1957).

B. Physical Basis for Weather Modification

1. ICE CRYSTAL PROCESS

The ice crystal process involves the rapid transfer of water molecules from supercooled liquid water droplets to adjacent ice crystals, when subcooled liquid droplets and ice crystals occur in the same cloud mass. When silver iodide or dry ice is used for cloud seeding, the assumption is made that the Bergeron, or ice crystal process, predominates in the formation of precipitation particles. This transfer of water molecules from the liquid cloud droplets to the adjacent ice crystals permits a rapid crystal growth to a size sufficient to have a substantial fall velocity from the cloud system.

2. COALESCENCE PROCESS

Under some conditions, coalescence can take place between liquid cloud particles, and between liquid and solid particles, so that substantial and rapid growth can take place, despite the fact that cloud particle sizes are commonly so small and uniform that such coalescence does not take place. Recent research by Braham (1964) has shown that ice and snow pellets are present in most convective clouds in Missouri by the time the clouds reached temperatures of −10 C. This suggests that even in areas with these cold clouds, and well removed from oceans, that coalescence processes are playing an important role. High priority needs to be given to the geographic boundaries and storm variations of the coalescence process and the ice crystal process.

3. CHANGES IN CLOUD DYNAMICS

Squires (1949) reported explosive growth following seeding of certain clouds. Evidence has been presented by Malkus and Simpson (1964) that tropical cumulus clouds experienced an explosive growth following seeding with large quantities of silver iodide. These experiments suggest the possibility of altering the dynamics of cloud growth by cloud seeding rather than alteration of only the microphysics of clouds by cloud seeding.

4. OTHER PROPOSALS FOR WEATHER MODIFICATION

Black and Tarmy (1963) have suggested the use of asphalt coating on watersheds as a mechanism for increasing rainfall by inducing increased convection. However, the basic assumptions underlying this proposal have been questioned by Howell (1964). Numerous proposals have been made for efforts to change the albedo in the polar regions by blackening large sections of snow covered areas.

5. BROAD-SCALE MODIFICATION OF WEATHER

Numerous proposals have been made for changing large-scale atmospheric circulation through systematic changes in storm energy resulting from releases of energy from weather modification efforts or changes in the energy relationships from contamination of the upper atmosphere (Nat. Acad. Sci., 1964).

6. ATMOSPHERIC WATER BY TECHNIQUES OTHER THAN FROM CLOUD CHANGES

Supplementing water supplies directly from the atmospheric water vapor source also needs attention. This includes investigations that might lead to the trapping of additional water on obstructions or on snowfields that have been manipulated to extend to seasons of the year when substantial condensation can occur directly on them.

III.   CURRENT STATUS

A.   Cloud Seeding for Increased Precipitation

The Advisory Committee on Weather Control (1957) concluded that the statistical procedures used indicated evidence for a 10% to 15% increase in precipitation from seeding in mountainous areas, but that in nonmountainous areas the same statistical procedures did not detect any increase in precipitation.

Evaluation of 41 project seasons of winter orographic cloud seeding by the National Academy of Science Panel on Weather and Climate Modification support the earlier conclusion of the Advisory Committee on Weather Control.

Analyses of five winters of randomized seeding of orographic cloud systems by the authors in the central Colorado Rockies shows a great variability in the seedability of various storm types. Increases substantially greater than the 10 to 15% suggested by the Advisory Committee appear to result with certain storm systems. No modification, or possibly even negative effects, appear to result with other cloud systems.

B.   Storm Abatement

Attempts have been made to ameliorate by cloud seeding the damage caused by a variety of storms, including hurricanes, hailstorms, and damaging winds from tropical thunderstorms. No conclusive demonstrations of success have been given from any of these programs, although some of the evidence available suggests that favorable effects may have been obtained in some cases. Results on the hurricane experiments have been reported by Simpson and Malkus (1963), and by Simpson et al. (1963). Reports on hail modification experiments include Schleusener (1962) and Schleusener et al. (1965), Grandoso and Iribarne (1963), and Hohl (1963). Reports indicating successes in reducing windstorm damage from tropical thunderstorms by cloud seeding have been presented by Lopez and Howell (1961) and Quate (1959).

C.   Supercooled Cloud Dispersal

The dispersal of supercooled fog and stratus clouds by seeding with dry ice has been demonstrated on repeated occasions. The conditions under which this technique may be used successfully are being delimited by current research (Vickers and Profio, 1964). Fog dispersal is now being used on an operational basis.

D.   Statistical Verification of Weather Modification

1.   INTRODUCTION

Large changes of total precipitation in a seeded area of 100%, 50%, or even 25% resulting from weather modification would have been easily established by the evaluation techniques that have been employed. Statistical techniques have been used extensively since the physical processes involved are extremely complicated and many phases not definable at a given time. While there have been rather consistent indications of precipitation increases from cloud seeding, they have been consistently $< 25\%$, a change which is very difficult to distinguish from natural variations which are frequently several hundred per cent.

2. TARGET VS. CONTROL PRECIPITATION RELATIONSHIP

The statistical procedure that has been most commonly used, involves establishing a regression relationship between the precipitation in the seeded area commonly referred to as "target" and an area nearby but outside the influence of the seeding operation, commonly referred to as a "control." Statistical comparisons are then made with the precipitation occurring with respect to this regression relationship during the seeded period. The President's Advisory Committee on Weather Control (1957) using a refined statistical procedure based on this type of analysis concluded that:

1) The statistical procedures employed indicated that the seeding of winter-type storm clouds in mountainous areas in the western USA produced an average increase in precipitation of 10% to 15% from seeded storms with heavy odds that this increase was not the result of natural variations in the amount of rainfall.

2) In nonmountainous areas, the same statistical procedures did not detect any increase in precipitation that could be attributed to cloud seeding. This does not mean that effects may not have been produced. The greater variability of rainfall patterns in nonmountainous areas made the techniques less sensitive for picking up a small change which might have occurred there, than when applied to a mountainous region.

3) No evidence was found in the evaluation of any project which was intended to increase precipitation that cloud seeding had produced a detectable negative effect on precipitation.

3. RANDOMIZATION

Certain statisticians (Neyman et al., 1960; Brownlee, 1960) have criticized the use of historical data to establish the regression relationships, pointing out the possibility that trends could occur between the interval when seeding took place that would differ from those in the historical period. They maintain that to eliminate uncertainties of this nature a randomization procedure should be used whereby regression relationships can be established between target and control areas during unseeded periods randomly intermixed with seeded intervals. Such randomization has and is forming the basis of most statistical verification experiments since the late 1950's. However, certain problems exist even with randomization since it has been suggested (Grant, 1963) that there may be residual effects from one period to another.

E.   Summary of State of Knowledge of Physical Processes of Cloud Seeding

1. DEMONSTRATED PHASES OF THE SEEDING PROCESS

Many phases of the physical process for modifying precipitation have been demonstrated in the laboratory and the atmosphere.

1) Numerous laboratory tests have shown that adequate numbers of ice nuclei can be prepared to provide concentrations required to nucleate atmospheric clouds at appropriate temperatures.

2) It has been demonstrated that under certain conditions clouds can be definitely modified by a change in the phase from liquid to ice.

3) Observations around the world have shown that concentrations of ice nuclei are consistently below values usually believed to be required (Ludlam, 1955) to make maximum use of cloud moisture.

4) It has been shown that these particles can be transported for long distances in the atmosphere in the desired concentrations (Boucher, 1956; Grant and Schleusener, 1961).

5) Precipitation has frequently been observed following artificial seeding.

2. UNCERTAIN PHASES OF THE SEEDING PROCESS

While certain aspects of the seeding process have been demonstrated satisfactorily, uncertainties still exist which are topics for certain current research. Satisfactory answers to the uncertainties are essential in order to exploit the full potential of weather modification.

1) The basic questions in connection with artificial seeding involves the question of what the cloud would have done without seeding.

2) The relative importance and geographical extent of the ice crystal and coalescence processes is not well defined.

3) The efficiency of precipitation from various types of clouds is not adequately known.

4) The understanding of the details of the precipitation process, including the dynamic effects of heat release, is not adequately known.

IV. OUTLOOK

The question of direction of future efforts at weather modification research and operations is the subject of expanding interest. Recent emphasis has been placed on the necessity of using available techniques and tools, not singly, but in concert (Greenfield et al., 1962); and on the desirability of large-scale studies which include not only cloud microphysics, but also the entire spectrum of atmospheric motions and processes (Nat. Acad. Sci., 1964). Earlier recommendations from the National Academy of Science (1959) concerning the desirability of including randomization as a feature of the design of weather modification field experiments appear to be incorporated into many of the weather modification research projects presently being conducted in the USA. This should continue.

The economic gains that would result from success in increasing precipitation artificially has generated a practical interest in looking only at the end product—precipitation, without paying much attention to the intermediate processes. A broad range of investigations is required to provide detailed information on cause and effect at each stage of the seeding process. It is probable that extensive programs with the primary objective of augmenting water supplies will be initiated during the next several years. These programs, if properly designed, can incorporate the capability for providing much of the needed detailed scientific information. This information, as it becomes available, can provide the basis for advancing seeding technology for better and more consistent returns.

For scientists and engineers first getting involved with weather modification, there is a strong temptation to attempt to bypass these intermediate steps and concentrate only on the input (silver iodide, in most cases) and output (precipitation). Experience has shown that faster progress will be made, in the long run, by consideration of the intermediate processes.

In addition to the specific literature references included in the text of this article a number of general references are included for those interested in additional information.

LITERATURE CITED

Advisory committee on weather control. 1957. Final report. US Government Printing Office, Washington. 422 p.

Bergeron, T. 1933. On the physics of cloud and precipitation. Proces-Verbaux des seances d l'Assoc. de Meteorol. de l'U.G.G.I. a Lisbonne. 1933. 19 p.

Bergeron, T. 1935. On the physics of cloud and precipitation. Proces-Verbaux des seances d l'Assoc. de Meterol. de l'U.G.G.I. a Paris. Part II. 19 p.

Berndt, Gerald D. 1957. An evaluation of commercial cloud seeding operations conducted during the summer months in South Dakota. Final Rep. Adv. Comm. Weather Control, Tech. Rep. 5, US Government Printing Office, Washington. p. 69–86.

Black, James F., and Barry L. Tarmy. 1963. The use of asphalt coating to increase rainfall. J. Appl. Meteorol. 2(5):557–564.

Boucher, Roland J. 1956. Operation overseed. Mt. Wash. Observ. Milton, Mass. 116 p.

Braham, Roscoe R., Jr. 1964. What is the role of ice in summer rain-showers? J. Atmos. Sci. 21(6):640–645.

Brownlee, K. A. 1960. Statistical evaluation of cloud seeding operations. J. Amer. Statist. Ass. 55:291.

Espy, J. P. 1839. Artificial rains. Nat. Gaz. Lit. Regist., Philadelphia. 19:5798.

Findeisen, W. 1938. Der Aufbau der Regenwolken. Zeitschrift fur Angewandte Meteorologie das Wetter. 55(7).

Giraytys, James, and Harold R. Rippy. 1964. The USAF meteorological rocket sounding network: Present and future. Bull. Amer. Meteorol. Soc. 45(7):382–387.

Gloyna, Ernest F., Jerome B. Wolff, John C. Geyer, and Abel Wolman. 1960. Present and prospective means for improved reuse of water. US Government Printing Office, Washington. 54 p.

Grandoso, Hector N., and Julio V. Iribarne. 1963. Experiencia de modification artificial de granizadas en mendoza. Facultade de Ciencus Exactas Y Naturales, Universidad de Buenos Aires, Argentina. 70 p.

Grant, Lewis O., and Richard A. Schleusener. 1961. Snowfall and snowfall accumulation near Climax, Colorado. Ann. West. Snow Conf., Proc. 29th 53–64.

Grant, Lewis O. 1963. Indications of residual effects from silver iodide released into the atmosphere. Ann. West. Snow Conf. Proc. 31st 109–115.

Greenfield, S. M., R. E. Huschke, Yale Mintz, R. R. Rapp, and J. D. Sartor. 1962. A rationale for weather control research. Amer. Geophys. Union, Trans. 43(4):469–489.

Hohl, P. 1963. Grossvevsuch I I I zur Bekampfung des Hagels im Tessin. Tatigkeitsbericht Nr. 15, Eldg. Volkswirtschafts—Departmentos in Bern. 1963.

Howell, Wallace E. 1964. Comment on "The use of asphalt coatings to increase rainfall." J. Appl. Meteorol. 3(5):642–643.

Huff, F. A., and G. E. Stout. 1951. A preliminary study of atmospheric moisture-precipitation relationships over Illinois. Bull. Amer. Meteorol. Soc., 32:295–297.

Lopez, Manual E., and Wallace E. Howell. 1961. The campaign against windstorms in the banana plantations near Santa Marta, Colombia. 1956–57. Bull. Amer. Meteorol. Soc. 42(4):265–276.

Ludlam, F. H. 1955. Artificial snowfall from mountain clouds. Tellus 7(3):277–290.

Malkus, J. S., and R. H. Simpson. 1964. Modification experiments on tropical cumulus clouds. Science 145:541–548.

McDonald, James E. 1958. The physics of precipitation. *In* Advances in geophysics. Academic Press, New York. 5:223–303.

National Academy of Science. 1959. The skyline conference on the design and conduct of experiments in weather modification. Nat. Res. Counc.-Nat. Acad. Sci., Washington, D. C. Publ. 742. 24 p

National Academy of Science. 1964. Scientific problems of weather modification. Nat. Res. Counc.-Nat. Acad. Sci., Washington, D. C. Publ. 1236. 56 p.

Neyman, Jerry, Elizabeth L. Scott, and Marija Vasilevskis. 1960. Statistical evaluation of the Santa Barbara randomized cloud seeding experiment. Bull. Amer. Meteorol. Soc. 41(10):531–547.

Quate, Boyd E. 1959. Final report of project wind control. Weather Engineers, Loomis, California. 110 p.

Reitan, C. H. 1957. The role of precipitable water vapor in Arizona's summer rains. University of Arizona, Tucson. Tech. Rep. Meteorol. Clim. Arid Regions no. 2. 19 p

Schaefer, V. J. 1953. Project Cirrus, Part I. Laboratory, field and flight experiments. Final Report. General Electric Res. Lab. RL–785. 170 p.

Schleusener, Richard A. 1962. The 1959 hail suppression effort in Colorado and evidence of its effectiveness. Nubila ANNO V–N 1 p. 31–59.

Schleusener, Richard A., John D. Marwitz, and William L. Cox. 1965. Hailfall data from a fixed network for the evaluation of a hail modification experiment. J. Appl. Meteorol. 4(1):61–68.

Schleusener, Richard A. 1964. Weather modification. Amer. Soc. Civil Eng., Proc., J. Hydraul. Div. 90(HY1):57–75.

Simpson, R. H., M. R. Ahrens, and R. D. Decker. 1963. A cloud seeding experiment in Hurricane Esther, 1961. Nat. Hurricane Res. Proj. Rep. No. 60. US Dep. Com., Weather Bur. Washington, D. C.

Simpson, R. H., and J. S. Malkus. 1963. An experiment in hurricane modification: Preliminary Results. Science 142:498.

Squires, P. 1949. The artificial stimulation of precipitation by means of dry ice. Australian J. Sci. Res. Series A, 2(2):232–245.

Steel, Robert G. D., and James H. Torrie. 1960. Principles and procedures of statistics. McGraw-Hill, New York. p. 70–71.

Tepper, Morris. 1963. A solution in search of a problem. Bull. Amer. Meteorol. Soc. 44(9):543–548.

Thom, H. C. S. 1957. A statistical method for evaluating augmentation of precipitation by cloud seeding. Final Rep. Advance Comm. Weather Control, Tech. Rep. 1. US Government Printing Office, Washington. p. 14–15.

Veraart, A. W. 1931. Meer Zonneschijn in Het Nevelig Noorden, Meer Regen in de Tropen (More sunshine in the cloudy north, more rain in the tropics). N. V. Seyffardt's Boek en Muziekhandel, Amsterdam.

Vickers, William W., and Richard Profio. 1964. Investigation of optimal design for supercooled dispersal equipment and techniques. Final Rep., Contract no. AF 19(628)–1692, Air Force Cambridge Lab., Tech. Oper. Res., Burlington, Mass. 56 p. 44 p. appendix.

Weakly, H. E. 1965. Recurrence of drought in the Great Plains during the last 700 years. Agr. Eng. 46(2):83.

Wexler, Raymond. 1960. Efficiency of natural rain. *In* Physics of precipitation. Amer. Geophys. Union Man. no. 5, p. 158–163.

Widger, William K., Jr. 1963. Meteorological satellites and weather reconnaissance aircraft—complementary observing systems. Bull. Amer. Meteorol. Soc. 44(9):549–563.

Wollman, Nathaniel. 1960. A preliminary report on the supply and demand for water in the United States as estimated for 1980 and 2000. 86th US Congr. Select Comm. Nat. Water Res., Comm. Print no. 32.

GENERAL REFERENCES

Ahlmann, H. 1948. The present climatic fluctuation. Geogr. J. CXII:4–6.

American Meteorological Society. 1957a. Statement on weather modification. Bull. Amer. Meteorol. Soc. 38(6):366.

American Meteorological Society. 1957b. Statement on weather forecasting. Bull. Amer. Meteorol. Soc. 38(6):406.

Battan, L. J. 1959. Cloud physics research in the Soviet Union. Bull. Amer. Meteorol. Soc. 49(9):444–464.

Battan, L. J. 1963. Relationship between cloud base and initial radar echo. J. Appl. Meteorol. 2(3):333–336.

Bergeron, T. 1955. The problems of artificial control of rainfall on the globe: I. General effect of ice nuclei in clouds. Tellus 1:15–32.

Braham, Roscoe R., Jr. 1958. Cumulus cloud precipitation as revealed by radar—Arizona 1955. J. Meteorol. 15:75–83.

Mason, B. J. 1957. Artificial stimulation of precipitation. In The physics of clouds. Oxford University Press, London. p. 268–308.

Mason, B. J. 1959. Recent developments in the physics of rain and rainmaking. Weather XIV(3)

Namias, Jerome. 1964. Problems of long-range forecasting. J. Wash. Acad. Sci. 54: 191–195.

National Academy of Sciences. 1966: Weather and climate modification problems and prospects. Vol. I: Summary and recommendations. Final report of Panel on Weather and Climate Modification to the Committee on Atmospheric Sciences, National Academy of Sciences, National Research Council, Publication No. 1350, Washington, D.C.

National Academy of Sciences. 1966: Weather and climate modification problems and Prospects. Vol. II: Research and development. Final report of panel on Weather and Climate Modification to the Committee on Atmospheric Sciences, National Academy of Sciences, National Research Council, Publication No. 1350, Washington, D.C.

National Science Foundation. 1962. Weather modification. 3rd Ann. Rep. Nat. Sci. Found. 62–27. US Government Printing Office, Washington, D.C. 78 p.

National Science Foundation. 1964. Weather modification. 5th Ann. Rep. Nat. Sci. Found. 64–19. US Government Printing Office, Washington, D.C. 40 p.

Petterssen, Sverre. 1964. Meteorological problems: Weather modification and long-range forecasting. Bull. Amer. Meteorol. Soc. 45(1):2–6.

Select Committee on National Water Resources. 1960. Weather modification, 86th US Congr., Comm. Print No. 22. US Government Printing Office, Washington, D.C. 46 p.

Smagorinsky, Joseph. 1964. Some aspects of general circulation. J. Roy. Meteorol. Soc. 90:114.

State of California. 1955. Weather modification operations in California. State Water Resources Board Bull. 16. 30 p.

University of Chicago. 1957. Cloud physics and rainmaking. Res. Programs Univ. Chicago Dep. Meteorol. 40 p.

Willette, H. C. 1950. Temperature trends of the past century. Centenary Proc. Roy. Meteorol. Soc.

section III

Water Sources for Irrigation

6 | Surface Water Supply and Development

H. S. RIESBOL

Bechtel Corporation
San Francisco, California

C. H. MILLIGAN

Utah State University
Logan, Utah

A. L. SHARP

Agricultural Research Service, USDA (retired)
West Linn, Oregon

L. L. KELLY

Agricultural Research Service, USDA
Beltsville, Maryland

I. INTRODUCTION

Adequate technical investigations and planning are prime requirements in the formulation of projects receiving their water supply from surface water, for such projects are usually large in extent and involve the commitment of substantial funds for development. Furthermore, the demand for optimum use of surface water supplies often requires that the project be designed to fill other needs, such as for flood control, power, or recreation. Techniques for assessing the adequacy of supplies, for hydrologic design of regulating works, and for balancing of water supply with water demand have been developed over the course of many years. A brief review of these techniques is the subject of this chapter.

II. MEASUREMENTS AND SOURCES OF DATA

Meteorologic, hydrologic, and physiographic measurements and data are essential to the planning, design, and operation of irrigation developments served from surface water sources. Such basic data are required for predicting the demand, the supply, and the regulation of water. Perhaps no other use of water requires basic data to as great a degree as does that for irrigation. Table 6–1 summarizes the principal Federal sources of much of the available data. Many state agencies also collect and publish a substantial amount of data.

A. Meteorologic Measurements and Data

Data on precipitation, snow accumulation, temperature, evaporation, humidity, solar radiation, and wind are used for the synthesizing or extension of streamflows and for estimates of the water required for the various demands of the project. Most of these data are collected and published by the US Weather Bureau. Their *Selective Guide to Published Climatic Data Sources* (US Dep. Com., 1963) provides a comprehensive listing of the various publications wherein the data may be found.

1. PRECIPITATION

Precipitation data, from recording, nonrecording, and storage gauges, are a necessity in planning the development of surface water supplies and operating irrigation water projects utilizing surface waters. These data are useful in correlation studies for extending or filling gaps in runoff records; forecasting runoff and floods; predicting water yields; synthesizing spillway and channel design floods; and estimating water demands of crops.

In those parts of the world where considerable amounts of streamflow result from snowmelt, measurements of the accumulated snow packs, snow surveys, provide a basis for forecasting the seasonal runoff (Monson and Codd, 1961). Additional accuracy of the forecasts may be obtained if information on soil moisture and frost conditions of the soil below the snow pack is available.

2. PAN EVAPORATION

Evaporation and transpiration are of major concern in irrigation projects. Data on evaporation from pans are very often used to estimate the evaporation loss from reservoirs, and are quite often used in the various formulas for estimating consumptive use. Though it is realized that pan data has many shortcomings, it none-the-less provides useful indexes and references for developing or transposing evaporation or consumptive use data. (*see* chapter 29)

3. OTHER METEOROLOGICAL DATA

Many other meteorological data are available that are often useful in carrying out investigations of water supply and demand, and especially in connection with the use of more recent techniques for estimating reservoir evaporation as well as the consumptive use of crops—one of the elements that determines the demand on the available water supply. Included are daily observations of air temperatures, solar radiation, humidity, and wind direction and velocity. These data are also helpful in studies of snowmelt.

B. Streamflow Measurements and Data

The single most valuable item in the planning of a surface water supply for an irrigation project is a long, continuous record of streamflow of the potential source. It is highly desirable that the period of record include a critical drouth period. Peak and minimum flows, average daily and monthly flows, and seasonal flows are required for estimating spillway and channel floods and seasonal and carryover storage. Where adequate streamflow data are not available they must be synthesized by correlation or other studies utilizing flow data from nearby streams and climatic, and physiographic data.

Few major streams, but many small streams, lack streamflow records. It is obviously impossible to gauge the flow of the many thousands of such small streams. The publications of the Agricultural Research Service (US Dep. Agr.,

Table 6–1. Principal Federal sources of data useful in planning and operation
of irrigation developments served from surface sources

Type	Agency†
Humidity	WB,* FS, TVA, IBWC, ARS
Pan evaporation	WB,* BR, IBWC, TVA, CE, ARS, FS
Precipitation	WB,* TVA,* IBWC, SCS, ARS, FS, BR, CE
Chemical quality of water	GS,* PHS,* BR, ARS, TVA, IBWC
Suspended load	GS,* BR, TVA, CE, IBWC, ARS, PHS
Reservoir sedimentation	SCS, GS, BR, TVA, CE, ARS, FS
Snow	SCS,* WB,* CE, FS, BR, TVA, ARS, MPS
Solar radiation	WB,* ARS, BR, FS
Streamflow	GS,* ARS,* IBWC,* BR, CE, FS, TVA
Temperature air	WB,* ARS, IBWC, TVA, FS
Wind	WB,* TVA, ARS, CE, FS, BR
Soils	SCS,* FS, BLM, BIA, ARS
Topography	GS*
Geology	GS*
Land use	BC,* ERS, FS, SCS, BLM, BIA, SRS

* Asterisk indicates agencies that regularly publish data. Other agencies publish data
intermittently in research or other reports.
† Key to agencies:

Department of Agriculture
  ARS - Agricultural Research Service
  ASCS - Agricultural Stabilization and
            Conservation Service
  ERS - Economic Research Service
  FS - Forest Service
  SCS - Soil Conservation Service
  SRS - Statistical Reporting Service

Department of the Interior
  BIA - Bureau of Indian Affairs
  BLM - Bureau of Land Management
  BR - Bureau of Reclamation
  GS - Geological Survey
  NPS - National Park Service

Department of Commerce
  BC - Bureau of the Census
  WB - Weather Bureau

Department of Health, Education and Welfare
  PHS - Public Health Service

Department of the Army
  CE - Corps of Engineers

Other Federal Organizations
  IBWC - International Boundary and Water
  Commission, United States and Mexico
  TVA - Tennessee Valley Authority

1963) are useful sources of data for developing estimates of seasonal and annual
yield for these ungauged areas.

C. Sediment Measurements and Data

Excessive rates of sediment deposition in storage reservoirs, in canals and
laterals, and on the land decrease the useful life of the project and increase
maintenance costs. A long record of measured sediment at the site of a proposed
reservoir provides the most useful information for estimating sediment storage
requirements of the reservoir. If such a record is not available, estimates of sediment storage requirements must be made by consideration of sediment measurements in nearby streams or of sediment deposited in existing reservoirs in the
same physiographic area.

D. Quality of Water Measurement and Data

Quality of water, as it determines the suitability of water for irrigation, is
measured by the type and composition of the dissolved and suspended constituents
other than suspended sediment. The subject of quality of water is discussed in
detail in chapter 9.

E. Physiographic Data

Data on soils, geology, topography, and land use are essential to the planning and design of the facilities on the lands to be irrigated, but are also needed for the planning and design of the major water storage facility and distribution works.

In the case of the small project where the potential water supply is from an ungauged stream, these types of data, along with meteorological data and stream-flow records from nearby streams, are needed for assessing the available water supply. Information readily available from existing maps and the Agricultural Census is often adequate for this purpose, but may have to be supplemented by more detailed inventories.

Existing topographic and geologic maps may suffice for a preliminary investigation of reservoir sites and major canal locations, but more detailed topographic and geologic information is generally necessary for more specific planning.

III. WATER DEMANDS

In the context of this chapter, water demand refers to the various items that may either deplete the surface water supply or restrict or inhibit additional withdrawals. Assuming that an adequate inventory may be made of the demands placed on the water supply by those having prior legal rights to a portion of the water supply in question and that a reasonable assessment may be made of depletions that may occur because of uncontrollable factors, the balance of the water becomes available for irrigation or such other uses as may be in the best interests of those concerned.

A. Water Right Law

Many state laws declare water to be public property. But individuals, corporations, and public and private agencies may obtain and hold a right to use water by compliance with certain principles and laws. Holders of vested water rights are protected in these rights as long as they make beneficial use of the water. Water users are expected to understand the law sufficiently to protect their rights. Hutchins (1962) has summarized the development, and Turney and Ellis (1962) have prepared an excellent bibliography of water law in the USA.

Waters allocated under law are not available for further development except through purchase, exchange, or due process of law. Hence an important phase of any project development is a thorough investigation of vested water rights and of procedures for obtaining water rights for the proposed project. Water right law may eixst in the form of rules and customs developed and accepted by the people in their everyday activities, laws or codes established by legislative bodies, and decisions and decrees established by court procedure. Water law is so complex and so variable among jurisdictions that a competent attorney should be employed to advise upon this phase of the project.

Two basic doctrines or systems underlie the water law in the USA: the riparian and the appropriation. In general, the more arid states, including Arizona, Colorado, Idaho, Montana, Nevada, New Mexico, Utah, and Wyoming, have abrogated the riparian system. The semiarid states, including California, North and South

Dakota, Nebraska, Kansas, Oklahoma, Oregon, Texas, and Washington recognize the riparian system in principle but administer most of their waters under the appropriation doctrine. The riparian system is recognized in the Eastern USA.

The essence of the riparian doctrine, which came from England, is that a land owner contiguous to a stream is entitled to have the water of the stream flow by his land undiminished in quantity and unpolluted in quality. A strict application of this rule would not permit consumptive use of water from a stream or use of water on lands not contiguous to the stream. This would not be a serious limitation on the river system in a humid region where the only use of water might be for navigation or recreation, but for an irrigation project this is a serious limitation. Many of the semiarid states which recognize the riparian doctrine have modified or limited it.

The appropriation doctrine declares that water belongs to the public, but a right to use it may be obtained by individuals or agencies provided they comply with certain procedures and principles. Beneficial use is the measure and limit of the right. Rights may be lost by failure to use the water. In times of water scarcity, prior rights of appropriators must be satisfied before subsequent appropriators are entitled to the water. The elements of an appropriation right include: quantity of flow, time of use, point of diversion, nature of use, place of use, and the priority of the right. In general, none of these elements of the right may be changed by the water user without prior approval by the state authority, because changes in these elements may infringe on prior rights. Appropriation doctrine permits storage of water, transfer of water from one basin to another, commingling of transferred water with waters of natural streams, and water right exchanges. Many state laws recognize preferential water users, with water for human consumption having first preference. Theoretically, a higher preferential use may permit condemnation of water from a lesser preferential use by eminent domain proceedings.

B. Water Demands Other than for Irrigation

Only rarely does there seem to be a source of surface water adequate to meet all possible demands. Prior rights to water are a limitation on any surface supply being considered for irrigation. Among the first demands that must be considered are those to satisfy needs for domestic water, both upstream and downstream, from the potential irrigation projects. Future rights to the use of water for this purpose, as well as prior rights, must be considered if this legal concept is included in the water laws of the area.

Some demands are supported by legal rights that cannot be abridged or have such strong social or economic justification that they cannot be ignored. These may include: maintaining flows for dilution of downstream pollution; controlling salinity in ocean bays at river mouths; maintaining adequate depths of flow for navigation; preserving habitats for fish and wildlife; and providing for water-related recreation.

The development of a surface water supply for purposes of irrigation can often be combined with other purposes, such as for hydropower, municipal and industrial supply, flood control, and the other water conservation purposes mentioned above. None of the several purposes considered may be wholly compatible, but formulation of a project under the multiple-purpose concept may permit more comprehensive development because the integrated water demands are often less than the total of the demands for each purpose considered separately and because substantial economies may result in construction of common reservoir facilities.

Changes in the land use and treatment on the watershed lands contributing to the stream system may result in modifying stream flow. To the extent that changes that can be foreseen result in decreased flows, these decreases are, in a sense, a water demand that will subtract from the potential supply available for irrigation or other use. This subject is discussed in detail in chapter 59.

C. Irrigation Water Demands

The success of an irrigation project depends in a large measure on how well the water supply meets crop requirements, both as to the total and the peak requirements. Most studies of water requirements begin with an estimate of the consumptive use of the crops grown in the project area. To obtain the total irrigation requirements, or demand, additional amounts of water are added to the estimate to allow for administrative, conveyance, and application losses and to provide extra water to maintain the salt balance.

Consumptive use is related to several climatic, soil, and plant factors and is estimated in terms of a specific crop, farm, or valley. If the unit consumptive use of each of the crops within a given area can be estimated by one of the methods described in chapter 29, the consumptive use for any other area can be synthesized.

Administrative losses result from the operation of the water storage and conveyance system. It may not always be possible or feasible to regulate the water supply in the canal exactly to irrigation demands. Consequently, some water may flow over the wasteways. Golze (1961) reported that on 25 US Bureau of Reclamation Projects in 15 western States operational losses averaged 8.3% of the water diverted.

Conveyance losses result from seepage from canal banks and bottoms. This subject is treated in detail in chapter 60. An additional volume of water must be supplied to the farm over and above consumptive use to provide deep percolation and other on-farm losses, as discussed in chapter 61.

The maintenance of a tolerable salt balance must be considered, as discussed in chapter 51, in the determination of the water requirements of an irrigation project. As water evaporates from the soil, salt and alkali are left in the soil and the soil-water solution. Overirrigation to prevent excessive salt and alkali accumulation may be necessary.

D. Losses from Reservoirs

1. EVAPORATION LOSS

Evaporation from the surface of water stored in reservoirs represents a major loss to the irrigator. The factors that control evaporation are vapor pressure differences at the water surface and in the air above, air and water temperatures, atmospheric pressure, wind, and to a minor extent, the quality of the water. These factors vary widely with altitude, latitude, topography, and other climatic and physiographic conditions. As a result, evaporation potential varies from one place to another. Maps and data published by the US Weather Bureau (US Dep. Com., 1950, 1959) delineate the variation in evaporation potential and are useful in preparing estimates of reservoir evaporation loss.

Numerous formulas have been developed from which evaporation estimates can be computed from data on the above factors. Their accuracy is presently largely dependent on assumptions and selection of coefficients based on judgment; hence, for practical purposes, evaporation is usually estimated by adjustment of actual evaporation measurements, using one of several available methods.

Direct measurements of evaporation from tanks or pans (US Dep. Com., 1955) may be adjusted to provide estimates of evaporation from the surface of water storages. Coefficients for converting pan evaporation to large water surfaces have been tentatively established. The usual coefficients for various types of pans in common use are:

Type of pan	Diameter	Setting	Coefficient
Weather Bureau Class A	4 ft	Above ground	0.70
Bureau of Plant Industry	6 ft	Set in ground	0.94
Colorado land	3 ft (square)	Set in ground	0.78
Geological Survey	3 ft (square)	Floating	0.77
Weather Bureau	4 ft (square)	Floating	0.78

When pan data are not available from places at or near the reservoir site, it is necessary to adjust the estimate. The maps of evaporation potential published by the US Weather Bureau are useful tools for making this adjustment.

The Weather Bureau has derived preliminary techniques for a more rational adjustment of pan evaporation to represent lake or reservoir evaporation, taking into account the effects of advected energy into the lake and heat transfer through the pan (US Dep. Com., 1955).

Evaporation from reservoirs may be computed in accordance with the thermal energy budget and the mass transfer theory when adequate instrumentation and basic data are available (US Dep. Int., 1952). Such data are available at but few sites and the method is presently in the testing stage. A practical field technique (Harbeck, 1962) has been developed for estimating reservoir evaporation by the mass transfer theory using records of reservoir stage, windspeed, humidity, and water surface temperature. These records, of course, are not available when the reservoir is in the planning stage.

It should be remembered that evaporation and transpiration losses have existed in the reservoir area before inundation. These estimated pre-reservoir losses should be deducted from future reservoir losses. For rough studies involving a relatively undeveloped reservoir area in arid climates, it is usually satisfactory to assume that the consumptive use loss is equal to the mean precipitation. For humid areas the pre-reservoir consumptive loss will approach potential evapotranspiration loss.

2. SEEPAGE LOSS

Another factor contributing to reservoir loss is seepage. It is customary to consider these losses in the case of existing reservoirs where significant loss is known to occur and in the case of potential reservoirs where loss is anticipated. The most common type of seepage loss is probably beneath the dam through valley fill materials too deep to be cut off economically and where it is too expensive to grout or blanket with more impermeable material. Losses through pervious formations, such as broken lava or cavernous limestone, exposed in the reservoir basin may be even more troublesome to anticipate. Estimates of the probable underflow can only be made by detailed geological investigations of the dam and reservoir sites. Seepage losses may reappear downstream, somewhat in the nature of uncontrolled reservoir releases. It may therefore often be possible to put the "losses" to beneficial use.

IV.　WATER SUPPLY

A.　Streamflow Characteristics

It is essential to understand the characteristics of streamflow in order to analyze and interpret streamflow records in the planning of irrigation projects and the design of project works. The characteristics of streamflow are largely dependent on the source of the water supply (i.e., rainfall, snowmelt, or groundwater) and the physiographic characteristics of the drainage basin.

1.　COMPONENTS OF STREAMFLOW

Streamflow may be made up of three basic components, singly or in combination. Each component is a reflection of some combination of climatic and physiographic conditions prevailing in the basin.

Base flow is the dependable low flow of the stream and provides the base upon which the immediate effects of rains, snowmelt, and nonuniform releases from storage are superimposed. The base flow of a stream is usually derived from groundwater, either through seepage or springflow. However, it may also include steady and dependable releases from storage in reservoirs or lakes, diversion from other streams, or return flows from irrigation or municipal water supply. Streams that are fed largely or entirely by base flow have a steady, uniform flow with little variation.

Interflow, sometimes referred to as subsurface storm flow, consists of water that percolates into the soil during rain, snowmelt, or irrigation, is intercepted by a plowsole or an impermeable layer at shallow depth, and returns to the surface by seepage. Interflow is particularly significant in certain mountain areas where thin layers of duff and soil overlay relatively impermeable rock strata, or in agricultural areas where shallow pervious soils overlay tight clay subsoils.

Surface runoff from rain, snowmelt, or irrigation provides the largest component of flow for most streams, and is the source of natural flood flows in most areas. In forested mountain areas, interflow may provide most of the flood volume.

2.　SIZE OF DRAINAGE BASIN

The size of the drainage basin is a major factor in determining the streamflow characteristics. A small basin may be fed largely from one source, for example, a small basin at high altitude may be fed entirely from snowmelt, or may be dominated by one physiographic feature. A large drainage basin, on the other hand, usually integrates the three basic streamflow components and a variety of physiographic variations; consequently, it will be neither completely uniform nor extremely flashy, and will usually be perennial.

3. INFLUENT AND EFFLUENT STREAMS

Streams are classified as influent or effluent. An influent stream, or stretch of a stream, is one that contributes water to the zone of saturation. The water surface of such a stream stands at a higher level than the water table or piezometric surface of the groundwater body to which it contributes water. An effluent stream, or stretch of stream, is one that receives water from groundwater from the zone of saturation. The water surface of such a stream stands at a lower level than the water table or piezometric surface of the groundwater body from which it receives water. This classification is related to the gains and losses associated with groundwater. Streams may also gain in flow simply because they receive water from tributaries producing more runoff. Conversely, as they pass through arid areas, substantial amounts of water may be lost from the beds and banks by evaporation, and by transpiration from riparian vegetation. Many streams of the Western USA are gaining streams in the upper mountain reaches but become losing streams as they cross the arid areas.

4. VARIATIONS IN STREAMFLOW

Variations in streamflow at the point of storage or diversion are a major factor in determining the dependability of the supply and the design of the irrigation works. Streamflow may vary in daily, seasonal, or annual patterns, or, usually, in a combination of all these patterns.

Daily patterns of variation are the result of natural change in evaporation and transpiration demand from day to night or of controlled changes in storage release or diversions. Changes from one day to another may result from rainstorms, periods of rapid snowmelt, or gradual changes in base flow. Daily variations are important in the design and operation of diversion works, but are not so significant for storage reservoirs. Most published streamflow data are presented in terms of mean daily discharge, which masks the important variations in streamflow that occur within a 24-hour period, particularly from small drainage basins. The hourly variations and instantaneous peak discharges are important in determining amount of flow divertible at a given heading, or in computations of flood frequencies for spillway design.

Seasonal variations in streamflow are particularly important in determining the utilization of natural flows for irrigation, and in the design of diversion and storage works. These variations are produced by seasonal fluctuations in precipitation and seasonal occurrence of snowmelt. They can also be man-created through seasonal diversions to storage or for irrigation. An excellent discussion of seasonal and annual variations is contained in the 1955 *USDA Yearbook of Agriculture* (Langbein and Wells, 1955).

The *annual* variation of streamflow is a key factor bearing on the degree of regulation required on any particular stream to meet the demands for surface water supply for irrigation. Major variations in annual streamflow occur from year to year and over long periods of years as a result of variations in climate. This is the familiar pattern of a series of flood years followed by a series of drouth years that occurs in most of the midwestern and western USA. The critical periods to be considered in planning and designing irrigation projects are those sequences of dry years that determine the maximum amount of storage water required to supplement natural flow to meet any firm demand.

The *mean annual runoff* of a stream at a selected point or gauge is the estimated or computed average for a specific period of years, usually the period of record. It is a rough measure of the dependable flow of the stream, if it could be completely regulated. Mean annual runoff varies greatly from one basin to another, depending on the climatic and physiographic characteristics of each area. The annual runoff values from which the mean is computed vary from the mean depending largely on the variations in annual precipitation over the basin. The *maximum and minimum* values of annual runoff and of monthly runoff are a measure of the variability of streamflow and of the reservoir storage that must be provided for regulation.

The *maximum and minimum mean daily discharges* or, when available, the maximum and minimum instantaneous discharges for each year of record are essential to the design of many irrigation structures. Minimum discharges help to determine the dependable divertible flows while maximum discharges are used in probability studies of flood magnitude and frequency for solution of spillway and cross drainage problems.

B. Streamflow Analysis

Development of a dependable surface water supply for an irrigation project requires interpretation, analysis, and adjustment of historical streamflow records. The proposed damsite or diversion point on a stream may be located at a considerable distance upstream or downstream from existing or historical stream gauges, in which case it is necessary to adjust the record to the diversion or storage site by accounting for unmeasured tributary inflows. Also, the record may be of short duration or contain gaps so that it is necessary to extend it through critical drought and wet periods. If a long record is available, it may have been affected by man-created depletions or accretions due to upstream developments such as diversions, storage, and importation.

Adjustment for unmeasured inflows may be accomplished on a drainage area basis by applying unit yield values, either monthly or annual, to the area that produces the unmeasured inflows. Distinction must be made between contributing areas and noncontributing areas such as deserts, highly pervious soil or rock formations, or other factors.

Streamflow records are extended by correlation to fill gaps in the record and to project it through critical dry or wet periods. These correlations are made with records for gauges on the same stream or neighboring streams having drainage areas of similar size and physical and hydrologic characteristics. A simple correlation of one gauge with another on a daily, monthly, or annual basis may provide a sufficiently high correlation coefficient. In other instances, it is necessary to prepare multiple correlations involving several stream gauges and precipitation data.

The streamflow record must also be examined to determine whether significant changes in the record have occurred with time as the result of changes in gauge location or operation, or other causes. The double-mass curve analysis (Linsley et al, 1949) is useful in determining the time at which such changes were effective and for making appropriate adjustments in the records.

Flow duration curves (Searcy, 1959) are useful for determining the percent of time during which selected minimum flows will be equalled or exceeded. Such analyses are useful in determining divertable flows.

C.   Sediment Studies

Sediment transported by streamflow affects water supply and poses a threefold problem to the irrigator:

1) Sediment deposition in reservoirs decreases the useful storage space and interferes with the operation of gauges, outlets, pumping plant intakes, and other facilities.

2) The aggradation and degradation of stream channels affect the operation of diversion dams and headings, spillway and outlet structures, and pumping plant intakes.

3) Sediment deposition in canals and laterals increases operation and maintenance costs, although it may reduce water losses from canal seepage.

Sediment yield rates are usually expressed in units of acre-feet per square mile of basin drainage area per year or tons per acre per year. Values of yield rates are obtained from sediment volumes as measured by reservoir or pond resurveys, gross erosion rates with sediment delivery ratios, or sampling of suspended sediment. Yield rates are used for small drainage areas where the sediment produced is more nearly a function of drainage basin characteristics rather than of stream hydraulics.

Suspended sampling data plus unmeasured-load estimates are the most accurate means of determining sediment inflow to a reservoir. The flow-duration, sediment-rating curve method of computing sediment yield (Miller, 1951) is commonly employed. The amount of inflowing sediment that will be trapped in a reservoir depends on the trap efficiency, which is usually close to 100% for large reservoirs. The trap efficiency decreases as the storage capacity vs. inflow ratio decreases (Brune and Allen, 1941). After the annual rate at which sediment will be trapped in a reservoir has been estimated, it is necessary to decide how many years of sediment storage will be provided. This period should at least be equal to the period of amortization plus a development period. Space is usually provided for from 50 to 100 years of sediment accumulation without encroaching on useful storage. Distribution of the total sediment in the reservoir is estimated by one of several methods (Borland and Miller, 1958) so that minimum elevations of outlets can be determined.

Estimates of stream aggradation and degradation above and below reservoirs are difficult engineering problems requiring detailed knowledge of the hydraulic and sediment transport characteristics of the stream and of the material in the stream bed.

Estimates of material that will be deposited in canals and laterals are based on the amount and character of sediment transported by the stream from which diversions are made, on the type of diversion works, and on the amount of water diverted.

V.   REGULATION OF STREAMFLOW

Some streams with very stable flows or that have low demands for water may require no regulation—the water is directly diverted for use. But under present-day conditions, full resource development requires some degree of storage.

A.   Diversion

To make water flowing in a stream available for irrigation use, it must be diverted by means of a diversion dam and headworks, a diversion headworks alone, or a pumping plant. Primitive and early irrigation developments used such works to divert the natural, unregulated streamflow directly into project canals. Today, diversions are usually coordinated with upstream storage in order to make maximum use of the available streamflow.

The approximate location of a diversion dam or heading is usually established by the overall requirements of the project plan. Although the general location of a structure is fixed by other than design considerations, the precise location can usually be shifted within certain limits to take advantage of any natural features of the site that lower construction costs.

The location of the diversion determines to some extent the amount of water that will be available to the project. In a direct diversion project, the fluctuation of natural streamflow requires more consideration than for a storage project. Seasonal high flows and peak discharge in excess of diversion capacity cannot be utilized to meet project demands. It therefore becomes an economic problem to balance the project demand and divertible supply in determining the diversion capacity to be adopted. A flow-duration curve of historic mean daily flows, adjusted for downstream rights and requirements, offers a means of determining the per cent of time, during months of diversion, that a selected divertible flow will be available. In any event, it is necessary to obtain as realistic an appraisal as basic data will allow of hourly, daily, and monthly fluctuations in streamflow and their frequency of occurrence. This determines the type and size of diversion and canal structures needed.

A diversion dam or weir is used to raise or control the water surface in the river, so that the desired flow may be diverted into the canal headworks, and to serve as a spillway for flood flow. The crest elevation is determined by the required elevation of the water surface in the canal when flowing at full capacity.

In the design of diversion dams, adequate engineering studies must take into account foundation conditions and problems, maximum flood flows, limiting backwater conditions, downstream or tailwater relations to provide for energy dissipation, downstream channel erosion and retrogression, possible shifting of upstream channel, ice pressures, and availability of materials. Special provisions may also be required for utilities such as railways, highways and pipelines, for passage of migratory fish, and for sluicing or desilting of flows heavily laden with sediment.

Sluiceways are required in most diversion dams to reduce the amount of bedload sediment entering the headworks and to assist in maintaining a channel to the headworks. Therefore, it is desirable to place the sluiceway as near the headworks as practicable. The sluiceway is of little or no value in protecting against suspended sediment load. This protection is usually provided by some form of settling basin placed in the canal just downstream from the headworks. The basin should be of such size that the velocity of water flowing through it into the regular canal section will be low enough to allow the silt to settle in the basin and only clean water to enter the canal.

The purpose of the headworks structure is to control and regulate the flow into the diversion canal. A headworks structure may be employed to divert water from a diversion dam, lake, equalizing reservoir, or natural stream. As a regulating and controlling structure, safety in design and reliability of operation are important. Proper engineering consideration must be given to location, relation to other structures such as diversion dams and canals, effects of stream curvature, foundation conditions, fluctuation in water surface elevations in the river or reservoir, and the degree of refinement required in control of diversions.

B. Storage

Reservoir storage is required to attain full utilization of most streams. The primary purpose of reservoir storage is to regulate the stream so that the natural flow can be adjusted to meet, as nearly as possible, the rate of demand for water. Reservoirs may be designed to provide regulation of daily, monthly, and seasonal flow variations within the year, i.e., annual regulation; or to provide regulation of water from years of surplus runoff to years of deficient runoff, i.e., regulation by carryover storage.

Storage reservoirs may be either onstream or offstream. The onstream reservoir is fed directly by the stream, receives its entire sediment load, and must pass or store the maximum floods. The offstream reservoir is fed by diversions from the main stream or tributaries, receives only a part of the sediment load, and must pass or store floods from a usually small drainage area.

1. RESERVOIR OPERATIONS STUDIES

The degree of regulation of streamflow that is possible or desirable with a storage reservoir can, in part, be determined by reservoir operation studies. These studies are closely related to the economic and other studies; they are but one of the useful tools for formulating the total project plan.

Before reservoir operations studies can be started, it is necessary, as previously discussed, to determine net losses of water due to evaporation and seepage at the reservoir site, and to determine areas and capacities of the potential reservoir at various elevations as adjusted for anticipated sedimentation. Likewise, it is necessary to select a historical period of stream flow, adjust the flows for changes anticipated by developments upstream, and determine the flow necessary to meet irrigation water requirements and other demands, if any.

Reservoir operations studies are required to integrate unregulated water supply with reservoir characteristics to detemine storage capacity to best serve demands for irrigation water. Studies can be made on an annual, monthly or daily basis. Annual studies will suffice for preliminary estimates. Monthly studies are generally adequate for most purposes, but daily studies may be necessary when there is need for a high degree of refinement.

The study consists essentially of inflow, demands, losses, and reservoir content, and in its simplest form is a tabular accounting of the equation: Inflow equals change in content (algebraic) plus losses, plus demands. The tabulation becomes complex as many components of these basic factors are added.

2. SELECTION OF STORAGE SITES

The selection of desirable dam and reservoir sites requires adequate basic data and competent engineering analysis. Reservoir storage characteristics that tend to increase unit costs of storage are:

1) Small storage increase in terms of incremental height of dam may require excessive height of dam.

2) Large area and shallow depth increase evaporation losses.

3) A poor dam site may require an excessively long or costly dam.

4) Poor foundation conditions cause excessive cost of dam.

5) Storage sites remote from use area require long and costly conveyance works and increased water loss.

6) The possible occurrence of floods of great magnitude requires costly investment in either spillways or flood storage space.

The engineering considerations involved in the actual design of diversion and storage dams, canal headworks, canals, siphons, and other irrigation works are exceedingly complex. It is suggested that the reader refer to appropriate engineering firms or to state and Federal agencies skilled and experienced in these problems.

VI.   ECONOMIC CONSIDERATIONS

The water users on an irrigation project usually must pay the cost of construction and of operation and maintenance of the project. If the irrigators are to be successful and prosperous, exceedingly careful analysis must be made of the economic aspects of the total project. Some of the questions that should be considered in the economic analysis are:

1) Is there a demand and market for the products of the potential project?

2) Is the chosen scale of the project such that net benefits, i.e., the amount by which benefits exceed costs, will be at a maximum?

3) Have all the alternatives been investigated to assure that each separable segment and the total project is the most economical to accomplish the objectives?

4) Has the order of development of the various parts of the project been scheduled to yield maximum benefits?

Many other less tangible considerations must also be taken into account, for example: Will the water users be in a position to pay their allocated costs? is the organization adequately structured to assure economical and continuous maintenance? and what is the impact of the irrigation project on other existing or potential developments? The methods employed in economic studies are akin to those used in agriculture and industry in allocating land, labor, capital, and management to production of goods and services in order to maximize profits. The reader is referred to chapters 10 and 11 for further treatment of the subject.

VII.   PROJECT OPERATION AND MAINTENANCE

The success of an irrigation project depends in no small measure on a good program of operation and maintenance for the entire system (Roush, 1955; US Dep. Int., 1957). The objectives of this program should be:

1) To maintain and protect all water rights.

2) To obtain and to deliver an adequate water supply to the water users when the water is needed.

3) To establish an adequate program of maintenance and improvements to keep the physical structures and properties in working order and to obtain optimum use of the water and of the water facilities.

4) To keep such records of water deliveries and project costs needed to ensure equitable water distribution, provide the evidence of beneficial use in the event of court proceedings, provide a basis for economic analysis of the operations, and satisfy any legal requirements pertaining to the particular type of organization. Records may also be required in connection with the repayment program for the original construction costs.

5) To educate water users in the means for obtaining high water use efficiency.

6) To develop sound budgets for covering the costs of operations and maintenance and to obtain the necessary funds by assessments, loans, bonds, etc., for financing.

A. Organization for Operation and Maintenance

In the Western USA, operations and maintenance are usually carried out by an organization formed by the water users. On US Bureau of Reclamation projects, the Bureau may carry on these functions until the project can be turned over to a water-user organization. Several kinds of organizations are used.

The *mutual irrigation company* is a private cooperative association of irrigators organized under state corporation law to provide water at cost to the members of the company. The mutual company has both advantages and disadvantages. Operation and maintenance costs are usually low. However, the low costs are not necessarily an indication of economical operation. Many mutual companies are too small to achieve all of the objectives of operation and maintenance.

Commercial companies, usually incorporated under state laws, are organized for the purpose of constructing and operating irrigation systems for the profit of those who provide the capital. The companies usually select their own consumers and fix the relationships with the consumer by contract. Rates and services are subject to public regulation. In general, the commercial company has not been a profitable enterprise, that is, returns on the investment have been rather low.

An *irrigation district* is a quasi-public corporation organized under state law primarily to provide water to irrigate lands within its legal boundaries. Most irrigation districts are nonprofit organizations. The organization of an irrigation district is initiated by petition and after a majority of the landowners have voted approval, it is incorporated under the state irrigation district law. The district has taxing power, may issue bonds to finance projects, and may also contract with the US Bureau of Reclamation to construct irrigation systems. Because of the greater legal powers of the irrigation district, it is in a better position to finance project developments than are the mutual or commercial companies.

The *water conservancy district* is another quasi-public corporation somewhat like the irrigation district, but with broader functions and broader powers. It may engage in multiple-purpose projects, tax all who are within the district and receive any form of benefits from water projects, and contract with government or private agencies to investigate, construct, operate, and maintain water projects. This type of organization is particularly adaptable to the large basinwide multiple-purpose projects of today.

B. Operations and Maintenance Techniques

Adequate maintenance of the physical facilities and component operation are paramount to the success of any irrigation enterprise. Many of the details of maintenance are discussed in other chapters or are readily available elsewhere. An irrigation project that receives its water from surface supplies, because of its usually greater size and scope, presents some problems of operation not present in the individual farm enterprise. Two of the major problems are discussed here.

1. WATER SUPPLY FORECASTS

Several procedures have been developed for forecasting seasonal water supply from streams fed by snowmelt. The forecast procedures express the expected streamflow as a function of antecedent hydrologic factors such as accumulated water content of the snow cover, rainfall, streamflow, and temperature. The functional relationships are developed from past data by regression analysis. Modern computers are extremely useful in developing forecast equations and preparing forecasts from the current data.

With the results of streamflow forecasts water users can be warned well in advance of pending water shortages or of expected floods, permitting plans and action to alleviate damages. If floods are expected, reservoirs can be drawn down so that more of their storage capacity can be used for flood storage. Temporary flood protection works can be constructed and areas expected to be flooded can be evacuated.

If a serious water shortage is expected, programs can be carried out to increase the water supplies. Additional sources of water, such as drilled wells, can be sought and developed. Perhaps water power companies with alternate power sources can be influenced to store water ordinarily used for power generation during the nonirrigation season. Irrigation companies can combine streams to obtain greater conveyance and irrigation efficiencies. Sections of canals with high seepage rates can be lined to prevent losses and leaky irrigation structures can be repaired. Farmers can limit the acreages planted, and cropping patterns can be changed to conserve water.

2. WATER DISTRIBUTION

The extent to which water is delivered to the users in accordance with their rights and needs, both as to time and quantity, is a measure of the success of the operation methods used. The financial success of a project depends on the financial success of the individual users. Convenience and economy of operation of the irrigation project facilities are usually secondary to the convenience and economy of water use on the farm.

There are three principal methods of water delivery: (i) continuous flow, (ii) rotation, and (iii) delivery-on-demand. Some systems may use a combination of these methods. A full discussion of these methods is given in chapter 61.

LITERATURE CITED

Borland, W. M., and C. R. Miller. 1958. Distribution of sediment in large reservoirs. Amer. Soc. Civil Eng., Proc. 84(HY2) Paper 1587. 18 p.

Brune, G. M., and R. E. Allen. 1941. A consideration of factors influencing reservoir sedimentation in the Ohio Valley Region. Amer. Geophys. Union, Trans. 22:649–655.

Golze, A. F. 1961. Reclamation in the United States. 2nd ed. Caxton Printers, Ltd., Caldwell, Idaho. 484 p.

Harbeck, G., Jr. 1962. A practical field technique for measuring reservoir evaporation utilizing mass-transfer theory. US Geol. Surv. Prof. Pap. 272–E. p. 101–105.

Hutchins, W. A. 1962. Background and modern development in water law in the United States. Nat. Resources J. 2(3):416–444.

Langbein, W. B., and J. V. B. Wells. 1955. The water in the rivers and creeks. In Water. US Dep. Agr. Yearbook of Agr. US Government Printing Office, Washington, D.C. p. 52–62.

Linsley, R. K., Jr., M. A. Kohler, and J. L. Paulhus. 1949. Applied hydrology. McGraw-Hill, New York. p. 217–219.

Miller, C. R. April 1951. Analysis of flow-duration, sediment-rating curve method of computing sediment yield. US Dep. Int., Bur. Reclam. Denver, Colorado. 55 p.

Monson, A. W., and A. R. Codd. Nov. 1961. Snow surveys and water supply forecasting in Montana. US Dep. Agr.-Soil Conserv. Serv. and Montana Agr. Exp. Sta. Bull. 562. 35 p.

Roush, F. M. 1955. Operation and maintenance of irrigation systems. Amer. Soc. Civ. Eng. Sep. no. 623. 8 p.

Searcy, J. K. 1959. Flow-duration curves. US Geol. Surv. Water-Supply Pap. 1542–A. p. 1–33.

Turney, J. R., and H. H. Ellis. Dec. 1962. State water right laws and related subjects. US Dep. Agr. Misc. Publ. 921. 199 p.

US Department of Agriculture. 1963. Hydrologic data for experimental agricultural watersheds in the United States 1956–1959. Agr. Res. Serv. Misc. Publ. 945. 650 p.

US Department of Commerce. 1950. Mean monthly and annual evaporation, from free water surface for the United States, Alaska, Hawaii, and West Indies. US Weather Bur. Tech. Pap. 13. 10 p.

US Department of Commerce. May 1955. Evaporation from pans and lakes. US Weather Bur. Res. Pap. 38. 21 p.

US Department of Commerce. 1959. Evaporation maps for the United States. US Weather Bur. Tech. Pap. 37. 13 p.

US Department of Commerce. 1963. Key to meteorological records. Documentation No. 4.11, Selective guide to published Climatic Data Sources. US Weather Bur. 84 p.

US Department of the Interior. 1952. Water-loss investigations: Vol. 1—Lake Hefner studies, tech. report. US Geol. Surv. Circ. 229. 153 p.

US Department of the Interior. 1957. Irrigation operation and maintenance. Reclam. Instr. Ser. 230, US Bur. Reclam. Denver, Colo. 35 p.

7 | Groundwater Supply and Development

HAROLD E. THOMAS
Geological Survey, US Department of the Interior
Menlo Park, California

DEAN F. PETERSON, JR.
Utah State University
Logan, Utah

I. INTRODUCTION

Groundwater is by far the most abundant freshwater resource in the habitable parts of the earth. The aggregate amount of groundwater within depths < 0.5 mile below the land surface is more than 30 times the total water in all freshwater lakes, more than 60 times the total in soil and other unsaturated rock materials, more than 300 times the water vapor in the atmosphere, and more than 3,000 times the average volume in all the rivers and rivulets in the world (Nace, 1964). If groundwater were uniformly distributed geographically, every acre of land would be underlain by 100 acre-feet, or 32.6 million gallons, of water.

But groundwater is not uniformly distributed. It may be reached at very shallow depth in some places, and at depths of several hundreds or even thousands of feet in others. Below the water table, probably less than 10% of the rocks within 0.5 mile of the land surface are of the types—especially gravel and sand, sandstone, limestone, and basalt—that yield water readily to wells. All other rock types, though less permeable, have some pore space which can be occupied by water, such as the minute voids between particles of clay or the fractures in consolidated rocks. This water may or may not drain slowly into wells in quantities sufficient for desired use. Hence, groundwater is so widely distributed that there are only a few places in the world where one could guarantee *not* to find any. At the other extreme, there are several places where the natural conditions are sufficiently uniform and well known to permit reliable predictions of the quantity and quality of the groundwater that can be obtained. Although we may have some indications of the distribution of groundwater, there are varying degrees of uncertainty as to what will be found at any specific site selected for drilling a well. Especially where the degree of uncertainty is greatest, water witching continues to flourish in our American culture, as shown by Vogt and Hyman (1959).

Groundwater is not a static resource, similar to the solid minerals, ores, and fuels. It is not excavated from a mine. It can be obtained only to the extent that it can move under the influence of gravity to wells, drains, tunnels, galleries, and other structures for obtaining groundwater. Fortunately groundwater is in motion almost everywhere. McGuinness (1963, p. 23–28), after showing some of the complexity of groundwater movement, summarizes as follows:

"Water gets into the ground wherever and whenever it is available in excess of the field capacity of the soil and can move downward by gravity through the zone of aeration to the water table, or wherever and whenever water in a surface body has a higher head than the adjacent groundwater; moves through the rocks around, over, under, and through obstacles formed by zones of lower permeability; it approaches the land surface or a body of surface water where the head is lower; and it is discharged by seepage or spring flow into streams, lakes, or the ocean or is dissipated by water-loving plants or by evaporation from the soil."

Practically all groundwater available to man moves freely through the pores of unconsolidated sediments and fractured rocks of the earth's crust although it may occur under pressure in strata which are confined by relatively impermeable layers. Sources other than from this circulating system are of little practical value; however, since they are frequently salty, may need consideration from the point of view of contamination. These noncirculating classes include "connate" waters, those trapped in sedimentary rocks at time of deposition; "magmatic" waters associated with volcanic action; and "metamorphic" water created in the recrystallizing processes forming metamorphic rocks (White, 1957). Large supplies of fresh groundwater which are part of the circulating system occur in coastal areas essentially as lower density bubbles floating on the heavier salt water.

The flow of groundwater is at rates that rarely exceed 10 ft/day except in large underground channels, and may be only a few feet per year. Because of this snail's pace, underground flow is a miserably inefficient mode of transmitting water from one place to another: All the water moving through clean, well-sorted gravel in a bed 100 ft thick and 1 mile wide with 1% gradient could be transmitted in a pipe 14 inches in diameter at the same gradient. On the other hand, groundwater basins provide efficient reservoirs. One square mile of the above aquifer could supply the discharge for the same pipe for 4 years. Groundwater reservoirs serve as nature's great delaying and storing medium for freshwater. They contain water in volumes far greater—in some instances hundreds of times greater—than the average annual rate of flow through them. Only this flow-through constitutes the renewable resource, and its discharge is visible at springs and in the dry weather flow of many streams, but invisible in evaporation from ponds, swamps, and marshes.

In the development of groundwater supplies, questions as to availability generally fall in three categories: where, how much, and how long? The following parts of this chapter discuss various aspects of these questions, and refer the reader to various documents that provide a more thorough analysis of the problems and their solutions. Part II, Exploration, is concerned primarily with the occurrence of groundwater, and thus, with its storage characteristics. Part III, Well Development, and part IV, Well Hydraulics, are concerned especially with the quantitative aspects of the flow characteristics. Part V, Reservoir Management, covers many of the broader aspects of groundwater development and use, including the utilization of groundwater reservoirs in overall water-resource management. Together, the four parts are presented in a pattern similar to that of an hourglass: Starting with overall geological considerations, narrowing progressively through groundwater basins to site selection, development, and analysis of a single well; then broadening again to embrace eventually the entire hydrologic unit that may be related to the well.

The broad field of groundwater hydrology has been covered by Todd (1959) and by Schoeller (1962) in recent books, and by Meinzer (1942) in a chapter of a general treatise on hydrology. Other discussions, ranging from highly technical to semitechnical and from short papers to entire books, are listed in the General

References at the end of this chapter. For those with a continuing or recurring interest in groundwater, the quarterly journal *Ground Water,* published for the National Water Well Association, endeavors to cover "all phases of progress in the groundwater field."

II. EXPLORATION

A desirable first step in the development of groundwater is to learn, by library research, what is known about groundwater in the region and the specific locality where the development is desired. The occurrence of groundwater in the nation has been outlined by Thomas (1952), who divides the conterminous USA into 10 groundwater regions; and it has been summarized in far more detail by McGuinness (1963), who describes the current situation in each of the 50 States and includes 135 pages of references to hydrologic studies of specific areas. Additional information concerning specific localities may be obtained from the district offices of the Water Resources Division of the US Geological Survey located in nearly every state, and from the State agencies responsible for groundwater studies, most of whom cooperate formally with the Geological Survey in its studies. Local well drillers, several Federal and State agencies, and others who have developed groundwater supplies in specific areas, are also potential sources of information on groundwater occurrence.

If the planned development is in an area where detailed hydrologic information is already available—and presumably, where some wells are already in existence —much of the exploratory work may already have been accomplished, and the principal problems remaining would then be the selection of the particular point (geographic location and depth) at which water is to be taken, and the development of the well to yield that water: These problems are discussed in subsequent sections. In most areas, however, available information and experience will not be sufficient to guarantee the outcome of a proposed groundwater development, and additional studies are desirable prior to the selection of a site for drilling a well. These studies involve use of all methods of "seeing" farther below the ground than would the unpracticed eye, and thus they involve projections based on experience in the hope of increasing the chances for successful development. The search is for (i) permeable rocks (ii) containing sufficient water (iii) of usable quality.

In the search for permeable rocks, the general rule is that the loose or unconsolidated rock materials, gravel, sand, silt, and clay, have greater porosity than do the consolidated rocks; but the pores in clay and silt are too small for permeability, and a sizable clay or silt fraction in any loose rock material makes it relatively impermeable. About 90% of all the water pumped from wells in the USA comes from gravel or sand aquifers: in the intermountain valleys of the West, the great plains of the Midwest, the coastal plains of the East and South, and river valleys widely distributed. Sandstone, the consolidated equivalent of sand, has reduced pore space because of cement between the grains, but wells obtain large yields from sandstones that are poorly cemented or well jointed, as for example in Wisconsin, Illinois, and Arizona. Many limestones are sufficiently permeable, even cavernous, to yield large volumes of water to wells, notably in New Mexico, Texas, the Ozark region, Florida, and other southeastern states (Swinnerton, 1942). Basalts similarly form excellent aquifers, notably in Hawaii, Idaho, and other states of the Pacific Northwest (Stearns, 1942). In other consolidated rocks, whether igneous, sedimentary, or metamorphic, the permeability is generally limited to that provided by fractures or joints, and is rarely sufficient to yield water in the quantities desired for irrigation. But under exceptional conditions, high permeability in these rock types may be produced by weathering, faulting, or other geologic processes; and on the other hand, many limestones and volcanic rocks are relatively impermeable (Stearns, 1942; Swinnerton, 1942). If a well is to be productive it must penetrate permeable rocks saturated with usable water.

Geologic studies are of prime importance in groundwater exploration, and should be a first step in the prediction of the distribution, depth, thickness, and other characteristics of aquifers. In some areas the geologic information available in published reports and maps may be adequate for a tentative appraisal of the groundwater conditions; in other areas, additional geologic field work may be required to evaluate the groundwater prospects properly (Fent, 1949; Leggette, 1950). In many localities aerial photographs are valuable aids for interpreting variations in permeability of soils and underlying materials (Purdue University, 1953; Ray, 1960), and they are being used increasingly for locating possible water-bearing formations (Howe, 1958).

Geophysical methods of exploration detect differences, or anomalies, of such physical properties as electrical resistivity, elasticity, density, and magnetism within the earth's crust. Even where such anomalies are clearly shown and measured, the reasons therefore may be obscure or misinterpreted unless there is other information concerning the subsurface conditions. Increasingly in recent years, experience and research are leading to rational interpretations of geophysical data in terms of geologic structure, rock type and porosity, water content, and water quality (Dobrin, 1952). Surface geophysical techniques have proved to be valuable tools for groundwater exploration in many areas where they have been conducted under the supervision of an experienced geologist (Carpenter and Bessarab, 1964), and especially where the geophysical studies have been integrated with geologic studies (McGinnis and Kempton, 1961).

Of all surface geophysical methods, the electrical resistivity method has been used most widely in groundwater investigations, partly because the equipment used is inexpensive, portable, and easy to use. Resistivities of rock formations vary over a wide range, depending upon the material, density, porosity, pore size and shape, water content and quality, and temperature. Actual resistivities are determined from an apparent resistivity, which is computed from measurements of current and potential differences between pairs of electrodes placed in the ground surface; as the spacing of electrodes is increased, the resistance between electrodes refers to rocks at increasing depths. The method is seldom effective for determining actual resistivities below a few hundred feet. It is especially well adapted for locating subsurface saltwater boundaries, because the decrease in resistance of the salt water becomes apparent on a resistivity-depth curve.

Seismic methods involve the creation of a small shock at the earth's surface and measuring the time for the resulting sound or shock wave to travel known distances. Seismic waves may be reflected or refracted at any interface where a velocity change occurs; the seismic reflections provide information on geologic boundaries thousands of feet below the surface, whereas seismic refraction methods cover only a few hundred feet in depth and are more commonly used in groundwater exploration. Seismic wave velocities are governed by the elastic properties of the rocks through which they pass. Where a seismic wave passes through rocks of contrasting properties, its velocity is changed at the boundary, which can be thus identified. Increased porosity tends to decrease wave velocity, but increased water content increases it. Because seismic methods require special equipment and trained technicians, they have been used to only a limited extent in groundwater exploration, although they are very widely used in explorations for oil and natural gas.

The gravity method involves variations in density as measured at the earth's surface. Unconsolidated materials have generally lower densities than consolidated rocks and these variations may give some indication of varying thicknesses of alluvium under an alluvial plain, or buried geologic structure. The magnetometer enables magnetic fields of the earth to be mapped, thus pointing up anomalies that may be caused by basalt or serpentine or other rocks with magnetic properties, or by other causes.

III. WELL DEVELOPMENT

Surface explorations may end with the selection of a site for a well, and commonly with some uncertainty as to where and how much groundwater can be obtained. Necessarily then, exploration continues, but in a new direction—downward. For this exploration, test holes can generally be made more cheaply than the wells that will ultimately be required for production, particularly of large capacity, and test drilling is well advised in any locality where subsurface conditions are inadequately known from previous drilling.

A. Exploration in Depth

Since test drilling serves primarily as insurance against loss of a larger investment in a production well, its cost should generally be scaled to that proposed investment. Among the least expensive tools are the small portable hydraulic rigs and jetting rigs which are in a do-it-yourself category and most suitable for shallow depths. The benefit of test drilling is lost, however, if it does not provide reliable information concerning the physical and hydrologic properties of the materials explored: Moulder and Klug have described the jetting technique and its limitations, and Stevens has discussed the examination of samples obtained in test drilling (Bentall, 1963b, p. H1–29). For the larger and deeper groundwater developments there is increasing recognition that testing of drawdown and discharge prior to installation of the permanent well is the most important phase of the development, and this requires well drillers qualified and equipped to do the job (Bennison, 1947, p. 94–122). As with standard well drilling, test drilling is commonly by (i) cable tool, in which the pounding drill bit is followed by casing to preserve the hole in unstable materials, and cuttings are removed by bailer; (ii) rotary, in which mud is circulated by pumping through the rotating drill stem to the bottom of the hole, and is of sufficient density to raise the cuttings and preserve the hole; and (iii) reverse rotary, in which the hole is kept full of water to ground level while also pumping water up through the drill stem to raise the cuttings from the bottom. By close observation during drilling, it is possible for skilled observers to pinpoint each change in rock material and its resistance to the drill, the depth (or depths) represented by sample cuttings, and the zones in which the hole is losing or making water (or mud). These detailed observations are recorded as made on a log which constitutes the field record. Samples of cuttings are also taken and preserved for further study.

The test hole provides an opportunity to obtain a great variety of geophysical data interpretable in terms of depth, thickness, continuity, structure, porosity, permeability, and degree of saturation of the rock materials, and the chemical quality of the contained water. The logging techniques, and the theory upon which they are based, are well summarized by Jones and Skibitzke (1956) and by Anderson (1951). Briefly, instruments have been developed to log continuously with depth the following data in addition to the observations mentioned previously:

1) Electric resistivity of the rock, which indicates the content, salinity, and distribution of the water in the rock; and

2) spontaneous potential, which indicates primarily the relative salinity of water in the borehole and in the adjacent rock; ordinarily recorded at the same time as resistivity to make the "electric log," which can be done only in uncased holes.

3) Natural gamma-radiation intensity, proportional to the radioactive element content of the rock; and

4) induced neutron and gamma radiation, which varies with the hydrogen and thus with the water content of the rock; these radioactivity logs are the only means of obtaining lithologic information in cased holes.

5) Temperature gradient, which varies with the thermal conductivity of the rock and its contained fluid and may also indicate geological anomalies.

6) Borehole diameter, which may vary with the degree to which the rocks resist erosion and solution.

7) Vertical flow of water, either natural or induced, which (corrected for borehole diameter variation) varies with the rate of gain or loss of water through the borehole wall.

8) Conductivity, which varies primarily with the mineral content of the water in the borehole.

Experience on the part of the driller is a vital factor in obtaining reliable information from test holes. Examination and interpretation of logs and specimens in the light of local geological information by a trained geologist may add materially to the validity of interpretation of test hole information.

The geophysical logging techniques have been developed primarily in petroleum engineering. As pointed out by Patten and Bennett (1963), the interpretive methods suitable for groundwater differ from those used in the oil industry, and an understanding of underlying theory is essential for interpreting the record data.

B. Well Construction

Large production wells are generally constructed by drilling, of which the methods most commonly in use are the cable tool or percussion, hydraulic rotary, and reverse rotary. Each method is particularly suited for drilling in some materials and not in others, and each is also adaptable to a wide variety of conditions. There are many variations in detailed techniques used, so that well drilling has become an art which may be skillfully adapted to local conditions. Construction methods differ regionally and among individual drillers. Many drillers have equipment in sufficient variety to utilize the techniques best adapted to the materials encountered in the well. The techniques of well drilling (Bennison, 1947, p. 123–232; Anderson, 1951; Moss, 1958; Stow, 1963) have progressed far beyond the do-it-yourself level of 50 years ago. Experience, sometimes bitter, has demonstrated again and again that successful, economical development of significant groundwater supplies by wells requires the services of experienced, well equipped drillers.

It is important that complete drilling logs be required for production wells even though test drilling may have been done in advance. These add to local geophysical information as well as provide a basis for design and development decisions for the particular well under construction.

Although most large wells are vertical drilled tubes ranging from tens of feet to several thousand feet in depth, large production is also obtained from a considerable variety of other types of excavations related partly to the depth from which the water is obtained, partly to the tools employed to make the excavation, and partly to hydrologic factors. For water in gravel and sand at shallow depth, a well may be dug by hand labor, generally 3 feet or more in diameter; a larger well can be dug by a clamshell, and a bulldozer can create an excavation large enough to be called a pond. Small diameter pipes with pointed screens (well points) may be driven or jetted to shallow water tables. Water may be obtained from a thin aquifer by means of a long trench or horizontal excavation to form an infiltration gallery. The collector type of well is a vertical shaft of large diameter extending into the water-bearing zone, where pipes are driven horizontally into the aquifer; collectors are especially productive where river infiltration is the chief source of the groundwater. A long horizontal tunnel at the top of the water-bearing zone, and an inclined shaft through which water is pumped to the surface, are the essential features of the Maui-type well, extensively used in Hawaii to skim fresh water from an aquifer that has salt water at greater depth.

Horizontal, unlined tunnels discharging by gravity have been extensively used since ancient times in the mideast from Turkey to West Pakistan. Nineveh, 800 B.C., was supplied by a ghanat (Bennison, 1947) and one-third of the irrigated land of Iran is presently watered by more than 20,000 ghanats, some approximating 30 miles in length (Utah State University, 1962). Mine shafts, drain tunnels, and other types of excavations yield water in quantities sufficient for irrigation in some localities. While excavations other than wells may be adapted under special circumstances, wells will doubtless continue to be, by far, the major devices used in exploiting groundwater supplies in the future.

A great many considerations go into the design and construction of a modern well. The well must have structural properties which assure that the casing, screens, and other components are of adequate strength to withstand the dynamic loads of driving and operation as well as the static loads imposed by the surrounding rocks; and hydraulic properties which assure that the water passes with minimum resistance from the aquifer into the well meanwhile excluding the aquifer sediments. Understanding of local conditions should permit much of the design to be determined in advance; however, specifications should be sufficiently flexible to take full advantage of conditions encountered during drilling and development. Extensive information has been developed on well design, specifications, and contractual considerations (Amer. Waterworks Ass., 1958; Bennison, 1947) which may be adapted to local conditions. Well diameter is usually determined by drilling practices or by the size of the equipment to be installed in the well. Theoretically, yield increases logarithmically with diameter; however, experience indicates that doubling the diameter commonly increases the specific capacity by only 10 to 20% (Johnson, 1959). Resistance to corrosion may be a factor in choosing the thickness and composition of the casing. In stable consolidated rocks, casing may be omitted. For rotary drilled wells, casings and screens may be set as desired after the hole is drilled; however, for cable-tool wells the casing must be driven as the hole progresses.

A completed well has blank casing often cemented into place wherever it penetrates unconsolidated materials that contain undesirable water or yield no water. Where a well penetrates several aquifers, it may be unwise, or even unlawful (Plumb and Welsh, 1955), to develop all of them because of transfer of water through the well between aquifers at different piezometric levels. In water-bearing zones, screens or preperforated casing may be installed or the casing may be perforated in place using a "Mills" knife to provide openings to permit ready entrance of water. The casing and especially the screen should be corrosion resistant; the openings should be of a shape to minimize clogging and of a size large enough to allow passage of grains up to the 60% to 75% size for the aquifer. The action of iron-forming or sulfate-reducing bacteria in clogging or corroding well screens has recently been recognized as a problem of major importance [Amer. Public Health Ass., 1955; Luthy, 1964; and Olin Kalmbach, Dec. 1964. (Tipton and Kalmbach, Denver, Colorado. *Personal communication.*)] Poorly constructed or abandoned wells, especially in artesian basins, can contaminate or waste water. They should be repaired or sealed to prevent wastage and transfer of water between aquifers, which may occur by water moving up outside of the casing as well as inside.

Many wells are completed with gravel packing in order to avoid entrance of sand from an aquifer containing much fine or very fine sand or to increase the permeability in the neighborhood of the well. Gravel may be fed down an auxiliary hole drilled alongside the well, or into the annular space between the hole and casing for rotary-drilled wells. The envelope of gravel should be at least 6 inches thick, and the gravel should consist of a range of grain sizes, including sand, related to the grain sizes in the aquifer and to the size of openings in the screen in order to form a filter (Taylor, 1948 p. 134; Moore and Sens, 1954 p. 55). After development, the introduced gravel, like the coarser particles in a natural aquifer, remains outside the screen as a permeable stabilizer, that permits entry of water into the well.

"Development" of a well refers to treating the completed well in such a way that the finer particles, including drilling mud, are dislodged from the surrounding materials into the well where they may be removed by pumping or bailing, thus increasing the permeability and stability of the materials adjacent to the well and reducing or eliminating continued entry of sand. For proper development the well screen or perforations are sized so that they will exclude a sufficient size fraction of the aquifer materials or artificial envelope so that instability and caving do not occur. The importance of development cannot be overemphasized. In some cases, discharge has been increased by two or three times by proper development. Development techniques depend on subjecting the surrounding aquifer to large transient fluctuating velocities sufficient to dislodge the smaller particles and include surging, pumping or overpumping at variable rates, and backwashing. Compressed air, water jets, calgon, acid, or dry ice may also be used (Bennison, 1947, p. 233–250). References applicable to the design, construction, and development of irrigation wells include Gordon (1958), Rohwer (1941), Johnston (1951), Jain (1962), and Moore and Sens (1954).

C. Testing for Yield

After development, the well should be tested for drawdown and yield in order to properly select the permanent pumping equipment and plan for the use of the well. This also provides important geophysical information and is commonly done by installing a test pump, preferably of capacity greater than that desired from the well; measuring the static water level in the well (prior to pumping); pumping at a measured steady discharge rate until there is negligible change in the pumping water level in the well; measuring that pumping level; and if desirable, repeating at new values of discharge. The test gives drawdown (static level less pumping level) and an estimate of specific capacity (yield in gallons per minute per foot of drawdown). Various types of equipment are used for the discharge and water level measurements required in the test (Johnson, 1956). In wells where test-pumping equipment is not available, some indication of the capabilities for yield may be obtained by bailing tests as described by Skibitzke (Bentall, 1963a) or slug-injection tests as described by Ferris and Knowles (Bentall, 1963a), but these tests give inconclusive results for wells of high yield. If the construction, development, and testing have been done properly, the well should produce sand-free water at a known rate and pumping lift.

Poor performance of a well relative to other wells in the vicinity may be due to differences in aquifer characteristics, or to faulty well construction. A step-drawdown test (Jacob, 1947; Rorabaugh, 1953) can be made to evaluate the entrance losses at the well, and discriminate them from formation losses.

D. Equipment

Equipment used for raising water from irrigation wells throughout the world ranges from primitive, though ingenious, devices such as the Persian wheel, to highly efficient modern pumps like the deep-well vertical turbine. Other modern types of pumps include displacement pumps, centrifugal pumps, and propeller pumps (Bennison, 1947). Under some conditions water may be raised advantageously using compressed air (Anderson, 1951). Pumps, such as centrifugal pumps which must be set above the water level, require that the inlet pressure be less than atmospheric, thus the suction lift cannot exceed about 80% of atmospheric pressure head. Vertical turbine pumps, which are more properly vertical-shaft centrifugal pumps, are installed below the minimum water level in the well and may consist of one or more stages in series. They are driven by hollow-shaft electric motors or by belt- or shaft-connected internal combustion engines of many different types (Amer. Waterworks Ass., 1955; Moore and Sens, 1954). Performance characteristics of pumping and driving equipment may be obtained from reliable manufacturers. Selection of the best pumping equipment for a large well merits careful consideration by a specialist. Provision should be made for measuring discharge regularly during operation, for which many excellent, relatively inexpensive, integrating meters are on the market. Instantaneous discharge may be measured using orifices or weirs or by other means (Anderson, 1951).

IV. WELL HYDRAULICS

As stated by Theis (1935), "When a well discharges by pumping or artesian flow, the water levels in wells in its vicinity are lowered. The lowering of the water table or other piezometric surface thus represented has been called the cone of depression or the cone of influence of the well." He then (1938) emphasized the following points:

"1) All water discharged by wells is balanced by a loss of water somewhere.

2) This loss is always to some extent and in many cases largely from storage in the aquifer. Some groundwater is always mined. . . .

3) After sufficient time has elapsed for the cone to reach the area of recharge, further discharge by wells will be made up at least in part by an increase in the recharge if previously there has been rejected recharge. . . .

4) Again, after sufficient time has elapsed for the cone to reach the area of natural discharge, further discharge will be made up in part by diminution in the natural discharge."

In this statement of the sources of water discharged by wells, two inherent characteristics of aquifers—their ability to store and to transmit water—are seen to be important factors. Groundwater hydraulics can be described as the process of combining field data on water levels, water level fluctuations, natural or artificial discharges, etc., with suitable equations or computing methods to find these hydraulic characteristics—the coefficients of storage and transmissivity—of the aquifer; it includes the logical extension of these data and computing methods to the prediction of water levels, the determination of optimum well yields, the design of well fields, and other uses, all under stated conditions.

A. Aquifer Characteristics

The cornerstone of groundwater hydraulics is the Darcy relationship, proposed by Henry Darcy in 1856, which states that volume discharge is linearly proportional to the gross cross sectional area of the conducting rocks and the piezometric gradient. The proportionality coefficient for water K, variously called permeability, coefficient of permeability, and hydraulic conductivity, depends primarily on the size of the free pore spaces and, theoretically, increases in proportion to the square of their diameters. Its basic dimensions as implied by Darcy's law are LT^{-1}; however, in practice, use of nonhomogeneous unit systems, for example, gallons, atmospheres, miles, feet, minutes, and years in a single computation may obscure its fundamental simplicity.

Flow capacity per unit width of aquifer is commonly used in practice and is expressed by the coefficient of transmissivity T, which has the basic dimension of L^2T^{-1}; but which is frequently expressed in volume rate per unit width. For a simple, homogeneous aquifer, T equals Km where m is the aquifer thickness. Well yield is almost directly related to transmissivity and thus depends both on permeability and aquifer thickness. Transmissivity values of the order of 10^{-1} ft^3/sec. per foot (63,300 gal/day per foot), or larger, will yield excellent wells but values less than 10^{-2} ft³/sec. per foot (6,330 gal/day per foot) will be poor or marginal producers of irrigation supplies.

In practice T may be thought of as a measurement of gross aquifer response in which inhomogeneities have been ironed out. Pore size of unconsolidated rocks depends not so much upon particle size as upon size distribution since the smaller particles tend to fill the pores between the larger ones. Fairly clean sand as small as 0.2 mm in diameter may produce usable supplies of water. Permeability is also affected by degree of compaction of the sediment particles and somewhat by their shape.

The groundwater flow phenomenon is invariably transient, although the degree to which this is the case varies greatly and steady-state approximations are often quite adequate for analysis. Transient analysis involves the removal or replacement of water from storage. The storage coefficient S is the volume yield per unit of drawdown in each unit of area; it is nondimensional and may include water released by pore drainage, by elastic expansion of the water, or compression of the aquifer as discussed later. Storage may not be fully reversible, especially that associated with consolidation.

Groundwater hydraulics, as based on Darcy's law, implies that flow through the pores is laminar rather than turbulent in contrast to flow in streams and conduits, and experience demonstrates that this is almost invariably the case. While fundamentally dynamic, groundwater flow becomes purely kinematic under gravity forces. By considering piezometric head as a scalar potential, groundwater flow problems may be approached using potential theory, in which the space derivatives of the potential are the hydraulic gradients, or if multiplied by K or T, the volume flows per unit cross sectional area or per unit width. This characteristic makes possible the use of an extensive body of mathematical theory equally applicable to heat and electricity and provides the basis for electrical analog models in which the similarity of the Darcy equation to Ohm's law defines analogous electric units. The nature of permeability and the use of mathematical techniques have been extensively treated by many authors, for example, Taylor (1948) and Harr (1962). Groundwater hydraulic theory in three- or two-dimensional rectangular coordinate systems is useful in studying the recharge and depletion characteristics of groundwater basins; cylindrical or polar coordinate systems are used in considering wells.

B. Nonequilibrium Formula

The nonequilibrium formula to fit the case of a well withdrawing water from storage was derived by Theis (1935) from analogy with an equivalent thermal system.

$$s = \frac{aQ}{T} \int_u^\infty \frac{e^{-u}}{u}\, \mathrm{d}u$$

where
$u = br^2 S/Tt$,
s = drawdown, in an observation well at distance r from a well discharging at a constant rate Q,
t = time since discharge started,

S = coefficient of storage,

T = coefficient of transmissivity, and

a and b = constants that depend upon the units of measurement used; equal to $1/(4\pi)$ and $1/4$, respectively, if units are homogeneous.

Where the drawdown s can be measured for several values of time t at a single observation well, or for one value of t at several observation wells (several values of r), the aquifer constants S and T can be determined graphically by developing a curve of r^2/t vs. s, and superposing it upon a type curve of the exponential integral. Alternately, approximate formulas giving direct explicit solutions for T and S may be obtained by expanding the integral in series and omitting higher terms (see Ferris, 1949).

In the mathematical model that is the basis for this formula, the aquifer is homogeneous, isotropic, of uniform transmissivity at all times and all places, and of infinite areal extent; the discharging well has infinitesimal diameter and penetrates the entire aquifer; and the water removed from storage is discharged instantaneously with decline in head. The field of groundwater hydraulics in < 30 years has progressed from the point of being overwhelmed by the unlikelihood of ever finding such an ideal aquifer in nature, to being overwhelmed by the variety of mathematical models of flow to wells and flow to natural outlets that have been devised to simulate natural transient hydraulic conditions. The most commonly used methods and equations are assembled in several general texts (Ferris, 1949; Jacob, 1950; Todd, 1959), and in special papers on aquifer tests (Peterson et al., 1952; Ferris et al., 1962; Spiegel, 1962; Walton, 1962; Bentall, 1963a, b, c; Hantush, 1966). Progress in the field is international in scope, with significant contributions from Russia (Polubarinova-Kochina, 1952). The present status of the art, and many solved cases, were summarized at a recent symposium (Maasland and Bittinger, 1963).

In the transient flow of water to infiltration galleries or tunnels, or to drains or surface streams, the mathematical model is of a line sink rather than a point sink. This aspect of groundwater hydraulics also has wide application in design of drainage systems for agricultural lands (Luthin, 1957; Van Schilfgaarde et al., 1956; *see also* chapter 50). Mathematical developments using rectangular coordinate systems were also discussed in a recent symposium (Maasland and Bittinger, 1963).

C. Degree of Confinement of Water

During construction of a well the degree of confinement of the water encountered may be inferred from its static level, or head. If the water remains at the level where first encountered, that level is deemed to be the water table, and the water, free and unconfined. If the water rises above the level at which it was encountered, it has been confined in the aquifer under artesian pressure which may or may not be sufficient to cause the completed well to flow at the land surface. If, after sealing off one water-bearing zone, another is encountered with free or unconfined water, then the lower zone is deemed to define the water table and the upper one to be perched. In the course of drilling a deep well, several water-bearing zones may be encountered, each with different head—which, however, may not be determined accurately during drilling.

When a well discharges water from a confined aquifer, the aquifer remains saturated as long as the pumping level does not drop below the top of the aquifer.

Any water removed from storage, therefore, is attributed solely to compressibility of the aquifer material and of the water (Jacob, 1940). The storage coefficient S for confined aquifers is generally small, of the order of 0.001 or less. The non-equilibrium formula is very widely used to determine the hydraulic constants of confined aquifers, with adaptations as necessary for known boundary conditions. These constants can also be determined for an aquifer from which a well is discharging with constant drawdown but varying discharge (Jacob and Lohman, 1952; Hantush, 1964).

When a well discharges water from a water table aquifer (Hansen, 1953), the water removed from storage includes that attributed to compressibility of the material and water in the saturated portion of the aquifer, and also to the gravity drainage of water in the zone through which the water table moves (the cone of depression). For most aquifers the water of gravity drainage will be orders of magnitude greater than that attributable to compressibility; thus for practical purposes the coefficient of storage will equal the specific yield, and will range from 0.05 to 0.30.

The nonequilibrium formula has been used successfully in numerous tests of water table aquifers, even though they depart significantly from the mathematical model. Although there may be appreciable delay in removal from storage by gravity drainage, the formula may be applicable if the time of pumping is sufficiently long. In the vicinity of the pumped well the observed drawdown must be adjusted because of the reduced saturated thickness during pumping, in order to obtain the transmissivity of the aquifer (Jacob, *in* Bentall, 1963a). In the early stages of pumping from a watertable aquifer, electric analog studies (Stallman, 1963) show that vertical components of flow are sufficiently large to cause errors in a test of an aquifer that is not truly isotropic.

D. Leaky Aquifers and Multiaquifer Wells

A leaky aquifer is one to which water flows from less permeable materials beyond its defined limits; the water of the leak, or "leakance," is transmitted vertically through a (generally overlying) confining bed into the aquifer at a rate assumed to be proportional to the drawdown in the aquifer. Leakance may originate from overlying groundwater or surface water, with no changes of storage within the confining bed, as in the mathematical model proposed by Hantush (1956, 1960). Depletion derived from storage accompanied by compaction is also treated as leakance by Poland (1960). In either case the water is in excess of that properly calculated as stored within the aquifer.

Multiaquifer wells are those which are screened in and produce water from two or more water-bearing zones. Static levels, pumping discharge, and drawdowns in these wells are the resultants of the contributions of all aquifers tapped. Where it is desirable to determine the hydraulic characteristics of the individual aquifers, or where "thievery" is suspected, special techniques are available (Bennett and Patten, 1960, 1962; Patten and Bennett, 1962).

E. Lateral Boundaries of Aquifer

The nonequilibrium formula describes the growth of the cone of depression, by depletion of storage, so long as nothing interferes, and if nature could fulfill the assumption of infinite areal extent of the aquifer, the growth could continue forever. Some wells tap aquifers whose nearest boundaries are tens of miles away, and they can expand their cones untrammeled for many years. Others may be within a very short distance of a significant hydrologic boundary. Definitive aquifer tests can be made in such wells, provided data are collected (perhaps in the first few seconds or minutes) before the cone of depression has expanded to the boundary (Wyrick and Floyd, 1961).

The outer boundaries of the aquifer may affect the development of the cone of depression, and thus modify the conditions predicted by the nonequilibrium formula, in either of two contrasting ways. The boundary may constitute a source of water, in which case as Theis pointed out, the water discharged by the well would be made up at least in part by an increase in the recharge to the aquifer or by a decrease in the natural discharge therefrom. Or the boundary may constitute an impermeable barrier preventing further expansion of the cone, so that, if the well discharge remains constant, parts of the cone must deepen more rapidly than would be predicted by formula.

By the theory of images (Ferris et al., 1962, p. 144–165) the effects of these boundary conditions can be simulated. Thus, for a discharging well in an aquifer hydraulically controlled by a perennial stream, the stream constitutes a line source at constant head, and its effect upon the discharging well can be simulated by an imaginary well at equal distance from this line source but on the opposite side, recharging simultaneously and at the same rate as the real well. For an impermeable boundary, across which there can be no flow of water, the image system requires an imaginary discharging well at the same distance from the barrier as the real well but on the opposite side, and with both wells on a common line perpendicular to the boundary.

The nature and location of hydrologic boundaries of aquifers in some cases can be determined from the analysis of pumping test data. During the early part of the test, the drawdown data for observation wells near the pumped well will reflect principally the pumping effects. Superposition, matching, and extrapolation of these early data (s vs. r^2/t) on the type curve, indicate the trend of drawdown if the aquifer were infinite. As the test continues, however, the measured drawdowns in each observation well will reflect the net effect of the pumped well and any boundaries that are present. The departure of the later observed data from the type-curve trace provides a basis for computing the distances from the observation well to the boundary that has caused the departure, and thus identifying that boundary.

In general, the graphic comparison of the Theis type curve with the curves developed from specific pumping test data may provide many clues as to the specific hydrologic conditions, providing the test procedures are planned so that they will conform as closely as possible with the theory. Then, the dispersion of the test data will be a measure of how the aquifer departs from the ideal.

F. Equilibrium Formula

When a well discharges water without depleting the storage within the aquifer, it has been adopted as a part of a natural hydrologic system in which there is a dynamic equilibrium between inflow and outflow; it provides part of that outflow, as would a spring of constant discharge or one in which outflow fluctuates in response to variations in the inflow. The equilibrium formula to fit this condition was developed by Dupuit and Forchheimer, but Gunther Thiem was apparently the first to use it for determining permeability of earth materials (Wenzel, 1936). The Thiem formula can be written as

$$T = \frac{aQ \log (r_2/r_1)}{s_1 - s_2}$$

in which

$T =$ coefficient of transmissivity, expressed as volume rate per unit width of aquifer,

$Q =$ rate of discharge of pumped well,

r_1 and $r_2 =$ distances from the pumped well to the first and second observation wells,

s_1 and $s_2 =$ drawdowns in the first and second observation wells, and

$a =$ a constant that depends upon the units of measurement and logarithms used, equal to $\pi/2$ for homogeneous units and natural logarithms.

The formula requires the following assumptions, which are basic also to the nonequilibrium formula: The aquifer is homogeneous, isotropic, or has uniform transmissivity at all times and places, and the discharging well penetrates the entire aquifer. However, the equilibrium formula assumes that none of the water is taken from storage, and that pumping has continued long enough to reach a steady state wherein the drawdown does not change with time. The equilibrium formula is used widely for determining transmissivity of aquifers in which well discharge is balanced by substantial recharge (Hydrol. Colloq., 1964). Although the formula is independent of the parameters of time and storage, methods have been described for determining also the coefficient of storage S under steady-state conditions (Jacob, 1950). In a water table aquifer, Ramsahoye and Lang (1961) have determined the storage coefficient by computing the volume of unwatered material in a cone of depression that is at quasi-equilibrium, and comparing it with the total volume of discharged water.

V.   RESERVOIR MANAGEMENT

The problems of reservoir management may be broadly grouped (Amer. Soc. Civil Eng., 1961) into (i) reservoir operation to maintain a firm supply, (ii) distribution, (iii) contamination and pollution, and (iv) salvage and beneficial use of water now being wasted. If wells draw first from storage and continue to deplete storage until they can rob some other source of water, where is the hope that groundwater can be a renewable resource comparable to precipitation, soil water content, and surface water, and furnish a perennial supply? This is the first problem for groundwater reservoir management. Rational groundwater reservoir management requires a good understanding of the physical characteristics of the groundwater basin, sound economic planning, and a public policy base supported by an adequate legal framework. Hydrologic analysis may then provide the key to optimum benefit.

Physical investigation includes locating and quantitatively inventorying underground storage space and then understanding the related hydrology. Several reports have described techniques for determining the hydrologic characteristics of an extensive reservoir (Guyton, 1941; Cooper and Jacob, 1946), and the hydraulics of stream-connected aquifer systems (Spiegel, 1962). Hydrologic evaluation (Amer. Soc. Civil Eng., 1961) is usually approached by applying the *hydrologic equation* in which items of supply and disposal are identified and balanced. Solution of the equation requires a hydrologic inventory designed to measure or estimate the items in the equation and leads to evaluation of quantities which may be produced, called "yield", under various conditions of development. Estimation of equation items such as precipitation or stream inflow-outflow may be based on some direct measurements; others, such as evapotranspiration, may be based on inventories of vegetative cover and evapotranspiration rates based on heat or radiation units. The concept of yield is more complicated than would appear at first glance. Yield varies with the techniques of development and management and its definition is influenced by both economic and policy objectives which must often be taken into account in the calculations. Physically, surface streams are frequently interconnected with groundwater and their discharges are thus modified by groundwater development.

Besides storage, hydrologic management of an aquifer may be limited by the rate at which water may be transferred from supply areas to use areas. Inadequate transmissivity is a common problem widely encountered. Transfer may be studied using the basic Darcy equation or by a modification of the Theis formula (Robinson and Skibitzke, 1962).

It is not practical or economic to regard all groundwater resources as permanently renewable. Several large reservoirs receive negligible replenishment, for example, those of the High Plains of Texas and some desert valleys of the Southwest USA. The result is progressive depletion of the reserve. Such reservoirs may be developed in a completely *laissez-faire* way or may receive varying degrees of public regulation designed to optimize the public benefit or achieve other objectives. In any case, once a repayment period is selected for amortization of necessary investments, scientific management of the reservoirs can insure maximum yield at minimum cost over that period; this may be accomplished in large measure by proper spacing of wells (Hantush, 1961; Lang, 1961).

If a well is to produce water perennially, it must have established an equilibrium whereby its discharge is made up by increased recharge to or decreased natural discharge from the aquifer. This could be achieved best and quickest by wells near those recharge or discharge areas. Shallow wells in unconfined aquifers would have an advantage over wells in confined aquifers in areas where there can be recharge directly from precipitation, streams, ponds, or irrigated areas. Because such groundwater operations must affect water beyond the limits of the groundwater reservoir, the authority and responsibility of management cannot be limited to the groundwater. The need for utilization of groundwater reservoirs in stream system development was pointed out years ago and discussed by numerous engineers (Conkling, 1946). Conjunctive operation of groundwater basins with surface reservoirs, including the use of groundwater reservoirs for carry-over storage, presents a very attractive arrangement in many instances and techniques have been developed for operational studies leading to optimum overall management (Chun et al., 1964). A basic element of the California Water Plan is the cyclic storage of water: recharging underground reservoirs with the surplus surface water of a wet period lasting several years, to be pumped out during subsequent dry periods. Groundwater development may often provide an attractive solution to waterlogging problems, especially where the quality of the water is adequate for water supply use. Drying up groundwater-fed seeps, ponds, and marshes by lowering the water table provides a means for converting wasted water to useful purposes (Remson and Randolph, 1958).

Quantitative predictions of aquifer response to changes in the hydrologic environment caused by man or nature provide the technical basis for groundwater management decisions. The basic physical data collected bear on this problem, particularly the fluctuations of water levels in wells, the discharge of wells and springs, the characteristics of earth materials penetrated by wells; but a true understanding of the flow system requires far more. For practically all natural systems very little is known of the overall framework, the porous media, in which the water occurs and moves. Underground reservoirs are irregular and intricate in form, and their internal hydrologic characteristics are often complex and nonuniform.

To an increasing degree, models for simulating an aquifer system are being designed utilizing both electrical analogs and analog and numerical computers. Analogs depend on the correspondence between the basic laws and continuity relationships of laminar liquid flow and those of electrical flow (Robinove, 1962). Analog computers depend on the use of various other electrical elements such as operational amplifiers to simulate the terms of appropriate algebraic or differential equations. Use of computers depends on formulation of a valid mathematical model. Use of general purpose analog and digital computers is outlined by Tyson and Weber (1964) and analog and computer methods have been applied to the Indus River groundwater basin in West Pakistan (White House-Interior Comm., 1964). Analog or computer analysis permits the inclusion of more known variables than can be processed by other mathematical techniques or by intuition, and permits the analytical result to be expressed directly in familiar and useful numerical terms. Necessarily, the validity of the model analysis is directly tied to the completeness and accuracy of the data and interpretations collected and furnished by the field hydrologist; however, models may be refined as additional aquifer data are acquired. At present, any attempted prediction of reservoir response to a proposed management practice is more likely to show up unknowns that must be solved or evaluated than to give satisfactory answers from existing data. Nevertheless, analogs and computers provide potentially powerful tools for depicting and analyzing aquifer systems, both as to the quantity and the quality of the water.

Underground pore space itself is being recognized as a valuable resource. In some confined aquifers pore space is reduced permanently by pumping water resulting in the subsidence of the overlying land surface (Green, 1964). As summarized by Poland (1960),

"Evidence to date shows that the subsidence is chiefly the result of compaction of compressible (clayey) materials in the confined deposits due to decline in artesian head. Compaction of such confined aquifer systems is a complicating factor in estimating groundwater resources. For example, in the central western part of the Central Valley of California, net withdrawal of ground water from 1943 to 1953 was about 4 million acre-feet. The volume of subsidence in the same period was about 2 million acre-feet. Thus, half the withdrawal represents water squeezed out of the aquifer system by compaction, and is available only once. This factor must be considered in estimates of available future supply."

Artificial recharge will be a prerequisite to effective utilization of underground storage space. As of now, all but a negligible proportion of groundwater replenishment is by natural processes, but quantitative data about groundwater recharge are far fewer even than those about groundwater discharge or changes in storage. While artificial recharge should ultimately benefit all wells drawing water from the aquifer, the benefit to wells remote from the recharge area may be long delayed, because of the slow rate of travel of water underground. For confined aquifers, artificial input must be through recharge wells, and this has developed many complications (Sniegocki, 1963; Sniegocki and Reed, 1963; Sniegocki et al., 1963). The achievements and problems in artificial recharge are discussed in chapter 8.

Artificial recharge is closely related to inadvertent recharge that can introduce undesirable elements into the underground storage space. Disposal of soluble wastes upon the land or underground has caused inadvertent recharge and pollution of groundwater in many places, with complications resulting from interactions among the native water, the introduced water, and the porous media (Brown, 1964). Pollution may also occur by saltwater intrusion where wells are located near the sea coast or near saline inland lakes. Depletion of the landward fresh groundwater may induce hydraulic gradients favorable to the transmission of saline water landward; special care is necessary in the development of such aquifers. Knowledge of the physical location of the freshwater-saltwater interface as well as other basin characteristics can permit optimum utilization based on spacing and depth of wells. Drawing first from deepest freshwater bodies is ordinarily most favorable for maximum yield of the resource; however, because of the smaller initial cost of shallow wells, the contrast is often unfortunately the historical case. Saltwater intrusion may be stemmed under favorable conditions by developing a piezometric mound between the freshwater and the sea by artificial recharge.

Management of groundwater resources in the best public interest almost invariably requires consideration of legal structure (Valantine, 1964). Existing water laws may be inadequate and new statutory action may be desirable. Clarification by the courts through actual judicial procedure may be necessary. Water rights doctrines are quite variable in detail from state to state, especially in their treatment of underground waters (McGuinness, 1951). Appropriate agencies, with the power and authority to regulate, practice, and construct the necessary wells and recharge works, depend on legislative authority incorporated in the various state laws relating to water supply administration and to special districts (Amer. Soc. Civil Eng., 1961, p. 123-131).

In most places the chief problem in scientific reservoir management will be to adapt the present cultural pattern to it. It would be far easier to attempt scientific management within the framework of the existing cultural pattern, but in most places this would place limitations so severe as to inhibit any significant accomplishment of the desired objectives. The existing pattern has resulted chiefly from private initiative, and in most localities can best be described as haphazard. Efficient reservoir management would require that withdrawals be from wells so spaced in location, depth, and times and rates of pumping as to take maximum advantage of the storage and flow characteristics of the reservoir under given conditions of replenishment. Thus, the spacing pattern should reflect the objectives of withdrawal in each part of the reservoir, whether to salvage water from natural discharge, withdraw the most readily replenishable water, induce additional recharge, or deplete the storage. The hydrologic problems of groundwater reservoir management go beyond the groundwater reservoir boundaries, to embrace the waters that can be used for recharge and the effects of diverting the water that would be discharged naturally from the aquifer.

LITERATURE CITED

American Public Health Association. 1955. Standard methods for the examination of water, sewage, and industrial waste. 10th ed. p. 405–410.

American Society of Civil Engineers. 1961. Groundwater basin management. Manual Eng. Pract. 40. 160 p.

American Water Works Association. 1955. American standard specifications for deep-well turbine pumps. J. Amer. Water Works Ass. 47:703–729.

American Water Works Association. 1958. American Water Works Association standard for deep wells. Amer. Water Works Ass., New York. 51 p.

Anderson, K. E. 1951. Water well handbook. Missouri Well Drillers Ass., Rolla, Missouri. 199 p.

Bennett, G. D., and E. P. Patten, Jr. 1960. Borehole geophysical methods for analyzing specific capacity of multiaquifer wells. US Geol. Surv. Water-Supply Pap. 1536–A. p. 1–25.

Bennett, G. D., and E. P. Patten, Jr. 1962. Constant-head pumping test of a multiaquifer well to determine characteristics of individual aquifers. US Geol. Surv. Water-Supply Pap. 1536–G. p. 181–203.

Bennison, E. W. 1947. Groundwater, its development, use, and conservation. Edward E. Johnson, St. Paul, Minnesota. 507 p.

Bentall, Ray. 1963a. Methods of determining permeability, transmissibility, and drawdown. US Geol. Surv. Water-Supply Pap. 1536–I. p. 243–341.

Bentall, Ray. 1963b. Methods of collecting and interpreting groundwater data. US Geol. Surv. Water-Supply Pap. 1544–H. 97 p.

Bentall, Ray. 1963c. Short cuts and special problems in aquifer tests. US Geol. Surv. Water-Supply Pap. 1545-C. 117 p.

Brown, R. H. 1964. Hydrologic factors pertinent to reservoir contamination. Ground Water 2(1):5–12.

Carpenter, G. C., and D. R. Bessarab. 1964. Case histories of resistivity and seismic ground water studies. Ground Water 2(1):21–25.

Chun, R. Y. D., L. R. Mitchell, and K. W. Mido. 1964. Groundwater management for the nation's future—optimum conjunctive operation of groundwater basins. Amer. Soc. Civ. Eng. J. Hydraulics Div. HY 4:79–95.

Conkling, Harold. 1946. Utilization of groundwater storage in stream system development. Amer. Soc. Civ. Eng., Trans. 111:273–354.

Cooper, H. H., Jr., and C. E. Jacob. 1946. A generalized graphical method for evaluating formation constants and summarizing well-field history. Amer. Geophys. Union, Trans. 27:526–534.

Dobrin, M. B. 1952. Introduction to geophysical prospecting. McGraw-Hill, New York. 435 p.

Fent, O. S. 1949. Use of geologic methods in groundwater prospecting. J. Amer. Water Works Ass. 41:590–598.

Ferris, J. G. 1949. Groundwater. *In* C. O. Wisler, and E. F. Brater [ed.] Hydrology. John Wiley, New York. p 198–272.

Ferris, J. G., D. B. Knowles, R. H. Brown, and R. W. Stallman. 1962. Theory of aquifer tests. US Geol. Surv. Water-Supply Pap. 1536–E. p. 69–171.

Gordon, R. W. 1958. Water well drilling with cable tools. Bucyrus-Erie Co., South Milwaukee, Wisconsin. 230 p.

Green, J. H. 1964. The effect of artesian-pressure decline on confined aquifer systems and its relation to land subsidence. US Geol. Surv. Water-Supply Pap. 1779–T. 11 p.

Guyton, W. F. 1941. Applications of coefficients of transmissibility and storage to regional problems in the Houston district, Texas. Amer. Geophys. Union, Trans. 21: 756–772.

Hansen, V. E. 1953. Unconfined groundwater flow to multiple wells. Amer. Soc. Civ. Eng., Trans. 118:1098–1130.

Hantush, M. S. 1956. Analysis of data from pumping tests in leaky aquifers. Amer. Geophys. Union, Trans. 37:702–714.

Hantush, M. S. 1960. Modification of the theory of leaky aquifers. J. Geophys. Res. 65(11):3713–3725.

Hantush, M. S. 1961. Economic spacing of interfering wells. Int. Ass. Sci. Hydrology Publ. 57. p. 350–364.

Hantush, M. S. 1964. Drawdown around wells of variable discharge. Amer. Geophys. Union, Trans. 69:4221–4235.

Hantush, M. S. 1966. Wells in homogeneous anisotropic aquifers. Water Resources Res. 2:273–279.

Harr, M. E. 1962. Groundwater and seepage. McGraw-Hill, New York. 315 p.

Howe, R. H. L. 1958. Procedures of applying air photo interpretation in the location of groundwater. Photogramm. Eng. 24(1):35–49.

Hydrologische Colloquium. 1964. Steady flow of groundwater to wells. Comm. Hydrol. Res., T.N.O. no. 10, The Hague, Netherlands.

Jacob, C. E. 1940. On the flow of water in an elastic artesian aquifer. Amer. Geophys. Union, Trans. 21:574–586.

Jacob, C. E. 1947. Drawdown test to determine effective radius of artesian well. Amer. Soc. Civ. Eng., Trans. 112:1049.

Jacob, C. E. 1950. Flow of groundwater. In Hunter Rouse [ed.] Engineering hydraulics. John Wiley, New York. p. 321–386.

Jacob, C. E., and S. W. Lohman. 1952. Nonsteady flow to a well of constant drawdown in an extensive aquifer. Amer. Geophys. Union, Trans. 33:559–569.

Jain, Jagat Kishore. 1962. Handbook on boring and deepening wells. Min. Food Agr., Govt. India Press, New Delhi. 115 p.

Johnson, E. E., Inc. 1956. Testing water wells for drawdown and yield. Bull. 1243, St. Paul, Minnesota. 8 p.

Johnson, E. E., Inc. 1959. Principles and practical methods of developing water wells. Bull. 1033, St. Paul, Minnesota. 29 p.

Johnston, C. N. 1951. Irrigation wells and well drilling. California Agr. Expt. Sta. Circ. 404. 32 p.

Jones, P. H., and H. E. Skibitzke. 1956. Subsurface geophysical methods in groundwater hydrology. In H. E. Landsberg [ed.] Advances in geophysics. Academic Press, New York. p. 241–300.

Lang, S. M. 1961. Methods for determining the proper spacing of wells in artesian aquifers. US Geol. Surv. Water-Supply Pap. 1545–B. p. B1–16.

Leggette, R. M. 1950. Prospecting for groundwater—geologic methods. J. Amer. Water Works Ass. 42:945–946.

Luthin, J. N. [ed.] 1957. Drainage of agricultural lands. Agronomy Vol. 7. 620 p.

Luthy, R. G. 1964. New concept for iron bacteria control in water and wells. Water Well J. 17(3):

Maasland, D. E. L., and M. W. Bittinger. 1963. Proceedings of the symposium on transient groundwater hydraulics. Civ. Eng. Sec., Colorado State Univ. Rep. CER63DEM–MWN70. 223 p.

McGinnis, L. D., and J. P. Kempton. 1961. Integrated seismic, resistivity, and geologic studies of glacial deposits. Illinois State Geol. Surv. Circ. 323. 23 p.

McGuinness, C. L. 1951. Water law with special reference to groundwater. US Geol. Surv. Circ. 117. 30 p.

McGuinness, C. L. 1963. Role of groundwater in the national water situation. US Geol. Surv. Water-Supply Pap. 1800. 1, 118 p.

Meinzer, O. E. 1942. Groundwater. In O. E. Meinzer [ed.] Hydrology. Dover, New York. (3rd printing, 1949) p. 384–477

Moore, A. W., and Howard Sens. 1954. The vertical pump. Johnston Pump Co., Pasadena, California. 392 p.

Moss, Roscoe, Jr. 1958. Water well construction in formations characteristic of the Southwest. J. Amer. Water Works Ass. 50:777–788.

Nace, R. L. 1964. Water of the world. Nat. Hist. 74(1):10–19.

Patten, E. P., Jr., and G. B. Bennett. 1962. Methods of flow measurement in well bores. US Geol. Surv. Water-Supply Pap. 1544–C. 28 p.

Patten, E P. Jr., and G. B. Bennett. 1963. Applications of electrical and radioactive well logging to groundwater hydrology. US Geol. Surv. Water-Supply Pap. 1544–D. 58 p.

Peterson, D. F., Jr., O. W. Israelson, and V. E. Hansen. 1952. Hydraulics of wells. Utah Agr. Exp. Sta. Tech. Bull. 351. 48 p.

Plumb, C. E., and J. L. Welsh. 1955. Abstract of laws and recommendations concerning water well construction and sealing in the US. California Div. Water Resources, Water Quality Invest., Rep. 9. Sacramento. 391 p.

Poland, J. F. 1960. Land subsidence in the San Joaquin Valley, California, and its effect on estimates of groundwater resources. Int. Ass. Sci. Hydrol., Publ. No. 52. p. 324–335.

Polubarinova-Kochina, P. I. 1952. Theory of groundwater movement. (in Russian) Gosteckhizdat, Moscow. 676 p. (Engl. trans. by J. M. R. de Wiest. Princeton Univ. Press. 1962.)

Purdue University. 1953. A manual on the airphoto interpretation of soils and rocks for engineering purposes. Purdue Univ. School Eng. and Eng. Mech., LaFayette, Indiana. 206 p.

Ramsahoye, L. E., and S. M. Lang. 1961. A simple method for determining specific yield from pumping tests. US Geol. Surv. Water-Supply Pap. 1536–C. p. 41–46.

Ray, R. G. 1960. Aerial photographs in geologic interpretation and mapping. US Geol. Surv. Prof. Pap. 373. 230 p.

Remson, Irwin, and J. R. Randolph. 1958. Design of irrigation ponds using pond and groundwater storage. Amer. Soc. Agr. Eng., Trans. 1:65–67, 75.

Robinove, C. J. 1962. Groundwater studies and analog models. US Geol. Surv. Circ. 468. 12 p.

Robinson, G. M., and H. E. Skibitzke. 1962. A formula for computing transmissibility causing maximum possible drawdown due to pumping. US Geol. Surv. Water-Supply Pap. 1536–F. p. 175–180.

Rohwer, Carl. 1941 (rev.) Putting down and developing wells for irrigation. US Dep. Agr. Circ. 546. 87 p.

Rorabaugh, M. I. 1953. Graphical and theoretical analysis of step-drawdown test for artesian well. Amer. Soc. Civ. Eng., Proc. 79 (Sep. 362). 23 p.

Schoeller, Henri 1962. Les eaux souterraines. Masson et Cie, Paris VIᵉ. (in French) 619 p.

Sniegocki, R. T. 1963. Geochemical aspects of artificial recharge in the Grand Prairie region, Arkansas. US Geol. Surv. Water-Supply Pap. 1615–E. 41 p.

Sniegocki, R. T., and J. E. Reed. 1963. Principles of siphons with respect to the artificial-recharge studies in the Grand Prairie region, Arkansas US Geol. Surv. Water-Supply Pap. 1615–D. 19 p.

Sniegocki, R. T., J. E. Reed, F. H. Bayley, III, and Kyle Engler. 1963. Equipment and controls used in studies of artificial recharge in the Grand Prairie region, Arkansas. US Geol. Surv. Water-Supply Pap. 1615–C. 39 p.

Spiegel, Zane. 1962. Hydraulics of certain stream-connected aquifer systems. New Mexico State Eng. Spec. Rep. 105 p.

Stallman, R. W. 1963. Electric analog of three-dimensional flow to wells and its application to unconfined aquifers. US Geol. Surv. Water-Supply Pap. 1536–H. p. 205–242.

Stearns, H. T. 1942. Hydrology of lava-rock terranes. In O. E. Meinzer [ed.] Hydrology. Dover, New York. (3rd printing, 1949) p. 678–704.

Stow, G. R. S. 1963. Modern water-well drilling techniques in use in the United Kingdom. Ground Water. 1(3):3–12.

Swinnerton, A. C. 1942. Hydrology of limestone terranes. In O. E. Meinzer [ed.] Hydrology. Dover, New York. (3rd printing, 1949) p. 656–677.

Taylor, D. W. 1948. Fundamentals of soil mechanics. John Wiley, New York. 700 p.

Theis, C. V. 1935. Relation between the lowering of the piezometric surface and the rate and duration of discharge of a well using groundwater storage. Amer. Geophys. Union, Trans. 16:519–524.

Theis, C. V. 1938. The significance and nature of the cone of depression in groundwater bodies. Econ. Geol. 33:889–902.

Thomas, H. E. 1952. Groundwater regions of the United States—their storage facilities. In Physical and economic foundation of natural resources. US Congr., House Represent., Interior Insular Aff. Comm. vol. 3, 78 p.

Todd, D. K. 1959. Groundwater hydrology. John Wiley, New York. 325 p.

Tyson, H. N., and E. M. Weber. 1964. Groundwater management for the nation's future —computer simulations of groundwater basins. Amer. Soc. Civ. Eng. J. Hydraulics Div. HY 4:59–77.

Utah State University. 1962. Iran and Utah State University, half a century of friendship and a decade of contracts. Utah State Univ., Logan. 144 p.

Valantine, V. E. 1964. Groundwater management for the nation's future—effecting optimum groundwater basin management. Amer. Soc. Civ. Eng. J. Hydraulic Div. HY 4:97–105.

Van Schilfgaarde, J., Don Kirkham, and R. K. Frevert. 1956. Physical and mathematical theories of tile and ditch drainage and their usefulness in design. Iowa State Coll. Agr. Exp. Sta. Res. Bull. 436. p. 667–706.

Vogt, E. Z., and Ray Hyman. 1959. Water witching, USA. Univ. Chicago Press, Chicago. 248 p.

Walton, W. C. 1962. Selected analytical methods for well and aquifer evaluation. Illinois State Water Surv. Bull. 49. 81 p.

Wenzel, L. K. 1936. The Thiem method for determining permeability of water-bearing materials, and its application to the determination of specific yield. US Geol. Surv. Water-Supply Pap. 679–A. 57 p.

White, D. E. 1957. Magmatic, connate, and meteoric waters. Geol. Soc. Amer. Bull. 68:1659–1682.

White House-Department of Interior Panel on waterlogging and salinity in West Pakistan. 1964. Rep. on land and water development in the Indus Plain. The White House, Washington. p. 291–363.

Wyrick, G. G., and E. O. Floyd. 1961. Microtime measurements in aquifer tests on open-hole artesian wells. US Geol. Surv. Water-Supply Pap. 1545–A. 11 p.

GENERAL REFERENCES

Ferris, J. G. 1949. Groundwater. In C. O. Wisler and E. F. Brater [ed.] Hydrology. John Wiley, New York. p. 198–272.

Ferris, J. G., and A. N. Sayre. 1955. The quantitative approach to groundwater investigations. Econ. Geol. 50(2):714–749.

Hantush, M. S. 1964a. Hydraulics of wells. In V. T. Chow [ed.] Advances in geosciences. Academic Press, New York. p. 281–429.

Hirshliefer, Jack, J. C. DeHaven, and J. W. Milliman. 1960. Water supply economics, technology, and policy. Univ. Chicago Press, Chicago. 378 p.

Jacob, C. E. 1950. Flow of groundwater. In Hunter Rouse [ed.] Engineering hydraulics. John Wiley, New York. p. 321–386.

Maas, Arthur, M. M. Hufschmidt, Robert Dorfman, H. A. Thomas, S. A. Marglin, and G. M. Fair. 1962. Design of water resource systems. Harvard Univ. Press, Cambridge, Massachusetts. 620 p.

Meinzer, O. E. 1923. The occurrence of groundwater in the United States, with a discussion of principles. US Geol. Surv. Water-Supply Pap. 489. 321 p.

Meinzer, O. E. 1923. Outline of groundwater hydrology, with definitions. US Geol. Surv. Water-Supply Pap. 494. 71 p.

Meinzer, O. E. 1932. Outline of methods for estimating groundwater supplies. US Geol. Surv. Water-Supply Pap. 638–E. p. 99–144.

National Water Well Association. Ground Water. Water Well J. Publ. Co., Urbana, Illinois. (Published quarterly beginning Jan. 1963.)

Peterson, D. F., Jr. 1957. Hydraulics of wells. Amer. Soc. Civil Eng., Trans. 122:502–517.

Thomas, H. E. 1951. The conservation of groundwater. McGraw-Hill, New York. 327 p.

Thomas, H. E. 1955. Underground sources of our water. In Water. US Dep. Agr. Yearbook of Agr. US Government Printing Office, Washington, D.C. p. 62–78.

Thomas, H. E., and L. B. Leopold. 1964. Groundwater in North America. Science 143:1001–1006.

8 Groundwater Recharge and Storage

LEONARD SCHIFF
Agricultural Research Service, USDA
Fresno, California

DEAN C. MUCKEL
Agricultural Research Service, USDA
Boise, Idaho

Groundwater reservoirs (called aquifers in some localities) are geologic formations where water is or can be stored by natural or artificial means and from which water can be artificially and economically withdrawn for beneficial use. Most of the groundwater is stored in and released from sands and gravels. Groundwater reservoirs are found in stream valleys, interior valleys, and coastal plains. They provide storage for deep percolation from precipitation and streamflow used in overlying areas, and for water artificially placed in them. Groundwater reservoirs serve as regulator conduits to convey water from areas of recharge to those of production and use (Muckel, 1961).

Recharging activities are conducted to increase the amount of groundwater available for the common benefit of all users. Artificial recharge primarily utilizes floodwaters to aid in the storage of water and in the alleviation of overdrafts or threatened overdrafts (Muckel, 1961; Todd, 1959). Artificial recharge can establish a barrier against intrusion of salt water along coastal areas (Amer. Soc. Civil Eng., 1961).

Many underground reservoirs have a capacity far greater than the largest surface reservoir. To make full use of storage capacity, groundwater reservoirs, like surface reservoirs, should be depleted during dry periods to provide storage for wet periods (Amer. Soc. Civil Eng., 1961). Refilling during periods of surplus may require recharge by artificial means to augment natural recharge. In many areas of the Western USA, groundwater reservoirs are now overdrawn (pumping draft exceeds natural recharge) and artificial recharge is used to minimize the overdrafts.

Artificial recharge has been used to some extent for disposal of sewage, industrial wastes, and irrigation return flow. Most state health codes prohibit the injection of polluted water into a groundwater reservoir or aquifer. The danger of contamination is largely eliminated if surface methods of recharge are employed, and if the groundwater mound does not build up to the surface so that a direct contact through saturated soils exists between the contaminated water and the groundwater.

Use of poor quality water for recharge, such as salt-laden water from agricultural drains or irrigation return flow, is hazardous. The salt balance of a groundwater reservoir may be adversely affected (Amer. Soc. Civil Eng., 1961).

Before using industrial wastes for recharge the possibility of contaminating the groundwaters should be carefully investigated based on the particular type of waste water under consideration (California Sanit. Eng. Res. Lab., 1954, 1953).

I.   SELECTION OF RECHARGE SITE

The following factors must be considered in selecting the proper location of sites for artificial recharge: (i) geologic structure of the groundwater reservoirs, (ii) pattern of pumping draft, (iii) movement of water within the water table, (iv) surface soils and substrata, (v) water supply—source, turbidity, and quality, and (vi) social and economic considerations.

A.   Geologic Structure of Groundwater Reservoirs

Groundwater reservoirs may be either free water or artesian reservoirs. The intake area where natural replenishment occurs may lend itself to artificial recharge (Fig. 8–1). Free water reservoirs may be supplied by application of water on the surface above. An artesian reservoir is confined by an overlying stratum of low permeability. Surface application would be beneficial if practiced where the aquifer outcropped at the upper end or extended beyond the limits of the overlying impermeable strata.

B.   Pattern of Pumping Draft

The pattern of pumping draft on the reservoir may dictate the location of spreading sites. Soil and substrata may be of low permeability near wells causing depressions in the water table and low yields. Other sites may be considered where surface soil and substrata are sandier and investigations show that water will flow to the wells in question in quantities needed and within the time required. Groundwater flows in the direction of the steepest slope that is normal to the water level contour lines. The steepness of this slope or hydraulic gradient and

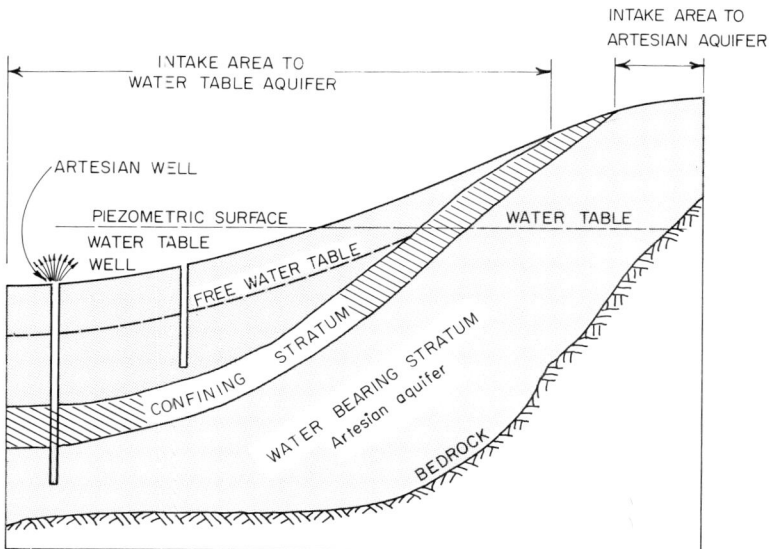

Fig. 8–1. Intake areas to artesian and free water table aquifers.

Table 8–1. Average velocities of water movement in granular materials (From *Ground Water* by C. F. Tolman. Copyright 1937 McGraw-Hill Book Co., Inc., used by permission of the publisher.)

Type of material	Grain size	Average velocity, 1% gradient
	mm	feet/day
Silt, sand, and loess	0.005 to 0.25	0.065
Sandstone and medium sand	0.25 to 0.5	1.16
Coarse sand and sandy gravel	0.50 to 2.0	6.33
Gravel	2.0 to 10.0	30.00

the permeability or hydraulic conductivity of the material through which water flows are the two key factors affecting quantity and time.

C. Movement of Water Within the Water Table

The rate of groundwater movement is usually very slow compared with that of surface water movement. Under natural conditions, hydraulic gradients of more than 10 to 20 ft/mile are seldom encountered. Through soil formations in which wells of good yield may be developed, flow velocities of 5 ft/day are usual. Table 8–1 shows average velocities of water movement in granular materials (Tolman, 1937).

D. Surface Soils and Substrata

The geology, pattern of pumping draft, and rate of movement of water underground will generally outline an area of considerably greater extent than would be required for concentrated artificial recharge purposes. A soil-stratum survey that gives the location, extent, and physical characteristics of the surface and the various underlying soil layers will be required for the purpose of selecting the best site or sites for recharge within that area. The permeability of the surface soil and of the substrata, subsurface storage space, and specific yield are important considerations.

E. Water Supply—Source, Turbidity, and Quality

Fig. 8–2. Silt and clay deposits clog soil.

The source, availability, location, quantity, and quality of water, as well as the nature of its occurrence, will influence not only the location of recharge sites but the method of recharge to be used. Water is not usually diverted into spreading areas when the sediment content exceeds about 600 ppm. Even this amount can clog bare soil (Fig. 8–2). Methods of handling sediment include the use of grasses, detention basins, flocculation and detention, filtering, scraping or filtering to remove deposited materials, and harrowing deposits of fine material to distribute with depth and subsequent removal by scraping (Schiff et al., 1959, 1961). Pumping after injecting water containing sediment into multiple purpose wells used for irrigation and recharge in Texas, USA, removed only 7% of the clay injected (Schiff et al., 1959; Schiff, 1961a).

Model wells receiving unfiltered water clogged, whereas those receiving filtered water did not clog. Filtration rates were more than twice as great when water flowed over a filter than when ponded on a filter (Schiff, 1961b). Raking the filters to a depth of about ½ inch caused deposited sediments to go into suspension and be carried away by the flowing water, increased the filtration rate and thus movement of water into the model recharge wells. This approach envisions potential use of shafts or shallow wells, with overlying filter materials, strategically spaced in waterways, and maintained by disrupting the upper portion of the filter material. Shafts may also be used in basins and maintained by scraping off clogged surface soil.

Waters that contain a high proportion of sodium salts cause infiltration problems. Waters that contain industrial wastes or that naturally carry quantities of undesirable minerals must be used with caution, so as not to contaminate the groundwater.

Laboratory and field investigations on the travel of pollution from direct recharge into underground formations (California Sanit. Eng. Res. Lab., 1954), and waste water reclamation in relation to groundwater pollution (California Sanit. Eng. Res. Lab., 1953), show a hazard exists when polluted water is injected directly into the underground aquifer by means of wells. A lesser hazard exists when surface spreading methods are employed. Sewage effluents were used in the study. Migration of chemical pollutants was found to be greater than that of bacterial pollutants. In highly fertilized agricultural areas the possibility of pollution by nitrates exists.

F. Other Considerations

Social and economic factors enter into the selection of a recharge site. The acquisition of sufficient land and rights-of-way as well as water supply is involved. Urban development of land overlying basins could well prove to be a major obstacle to recharge. In the operation of a recharge system, nuisances to urban areas may arise, such as mosquitoes, rodents, and growth of weeds, which must be eliminated or controlled.

Rights to water for recharge purposes and rights to water that has been placed underground and made available to wells within a groundwater reservoir are at present not clearly defined in most states. As a result, legal controversies may be expected.

II. MOVEMENT OF RECHARGE WATER TO THE WATER TABLE

The Darcy Law for the velocity v of saturated flow vertically into the soil surface may be written

$$v = k \ (h_s + y) / y \qquad [8\text{--}1]$$

where k is the hydraulic conductivity, h_s is the surface head, and y is the length of saturated soil column. By definition the infiltration rate f equals v for saturated flow. When y is small, h_x may constitute most of the hydraulic head and f increases almost directly proportional to increases in h_x (Schiff, 1953). When h_x is small and y is large the gradient or $(h_x + y)/y$ approaches unity and small variations in h_x have little effect on f. Frequently, soil characteristics are such that y remains small and the use of reasonably greater heads will increase infiltration rates.

Equations [8–2] and [8–4] have been based on laboratory experiments (Bodman and Colman, 1944) and analytical developments (Hall, 1955). Equation [8–1] may be written:

$$v_1 = k_1 \, (h_1 + y_1)/y_1 \qquad\qquad [8\text{–}2]$$

for saturated flow vertically into a less pervious substratum where the symbols are the same as in equation [8–1] except referenced by subscripts 1. From equation [8–2] it may be seen that v_1 increases as h_1 builds up, since generally y_1 will not lengthen proportionally.

Multiplying equations [8–1] and [8–2] by area a of the surface soil and area a_1 of the substratum, respectively, gives the quantities of water, Q and Q_1, flowing through these areas, respectively. When the length of saturated column y is sensibly constant, and all of the water infiltrating through the spreading area a moves through area a_1, Q may be equated to Q_1 and

$$a_1 = Q/[k_1 \, (h_1 + y_1)/y_1]. \qquad\qquad [8\text{–}3]$$

The average hydraulic gradients are involved in the expressions, primarily since h_1 will vary and will also affect the length y_1. As $(h_1 + y_1)/y_1$ in equation [8–3] approaches unity, a_1 approaches a maximum

$$a_1 = Q/k_1. \qquad\qquad [8\text{–}4]$$

Since v and f are computed for the total area involved and the water actually moves through pores or voids, k includes porosity. Consequently, k depends partially on characteristics of the soil particles. The particle parameter, which has been represented by definitions of texture and structure, is dependent on the size, shape, and arrangement of soil particles. These factors also determine the amount and nature of soil pores. Characteristics and stability of porosity bear an important relation to k.

If the hydraulic conductivity of substrata is lower than the hydraulic conductivity of the surface soil, water will be retarded in its movement to the water table and perched water tables will develop (Schiff, 1955, 1957). Equations [8–2], [8–3], and [8–4] apply in such situations.

III. MEASURING WATER MOVEMENT IN SOIL

Physical and chemical characteristics affecting soil permeability and measurements of hydraulic conductivity have been described by the American Society of Agricultural Engineers Drainage Research Committee (Bouwer et al., 1961) and by Bouwer (1962a), Donnan and Aronovici (1961), and Reeve and Luthin (1957). Various devices used in the field by the Groundwater Recharge Project at Fresno, California, to obtain information on water movement, include manometer-equipped infiltrometers (Schiff, 1953), the neutron probe, piezometers, moisture blocks placed at selected depths below infiltrometers to determine the transmission rate (actual velocity of water in soil), and covered infiltrometers for sampling for soil water change down to field capacity due to drainage alone (Schiff, 1964). A strain gauge tensiometer cell has been developed for transforming soil water suction to electrical resistance at considerable depths below a recharge area (Bianchi, 1962).

Soil water data obtained with a neutron probe under the Woodville Recharge Basin, California, illustrates the extreme textural variability of some spreading sites, Fig. 8–3 (Schiff, 1961a). Water penetrated to a depth of 12 feet in 2 days and to a depth of 36 feet in 5 days. Note the restrictive soil layer at the 20-foot depth. Other information obtained from measurements included field capacity, field saturation, infiltration rate, transmission rate, and specific yield.

IV.   METHODS OF RECHARGE

Modified streambed (basins and/or furrows on exposed sand bars and islands), basins (Fig. 8–4 and 8–5), ditches or furrows (Fig. 8–6), and replenishment irrigation are all methods by which water is spread on the surface of the land and movement of water to the underground supplies is accomplished by increased wetted area and/or increased time of wetting. These methods are most effective where infiltration rates are high and where water can move freely through the substrata. The basin method is by far the most common.

A basic and sometimes limiting condition in spreading water for recharge is the infiltration rate. Figure 8–7 shows the relatively high infiltration rate for a

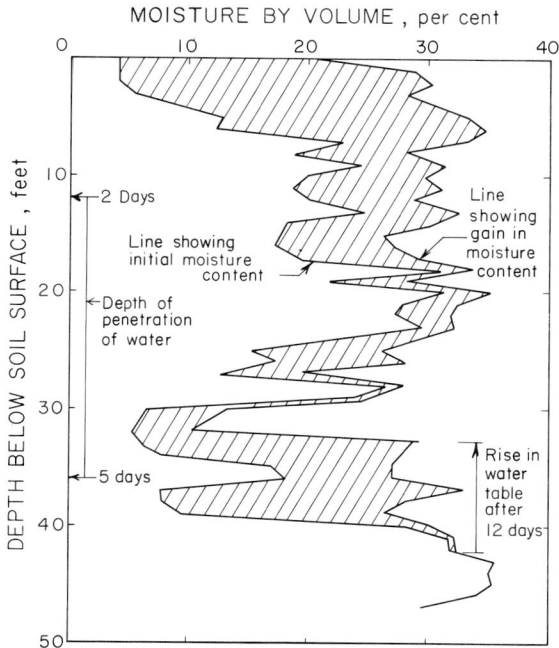

Fig. 8–3. Change in soil moisture after spreading water for 12 days in basin at Woodville, California.

fine sandy loam and the relatively low rate for a clay. Average infiltration rates (in feet per day) related to soil texture for recharge projects in California (Richter, 1955) are broadly summarized as follows: Coarse-textured soil ranged from 1.6 to 10 with an average of 4.2; medium-textured soils ranged from 1.7 to 3.6 with an average of 2.2; and fine-textured soils were < 1.0 with an average of 0.46.

Treatments of soil and/or water (whether physical, chemical, or biological) and operational procedures have been developed to increase the hydraulic conductivity and/or hydraulic gradient and to increase the infiltration rate (Bliss and

Fig. 8–4. Spreading basins as used by Kern County Land Company and North Kern Water Storage District near Bakersfield, California. Water is about 1 ft deep.

Fig. 8–5. Water spreading basins in Arroyo Seco Wash adjacent to main stream channel near Pasadena, California. (Photo courtesy of Los Angeles County Flood Control District.)

Johnson, 1952; Schiff, 1955). Current operational procedures suggest spreading water, freed of much of the silt by methods described previously, on land while infiltration rates are relatively high [rates decline with time because of microbial sealing (Allison, 1947)]; using relatively high depths of water on the surface to increase the hydraulic conductivity (Schiff, 1953); then drying this land for recovery in infiltration rate; and minimizing cultural activities that break down soil structure and compact soil. Organic residues can increase infiltration rates (Johnson, 1957). Grits and sand materials placed over aquifer material at least doubled

Fig. 8–6. Furrows in sand and gravel beds adjacent to main stream channel, Santa Ana River, Orange County, California.

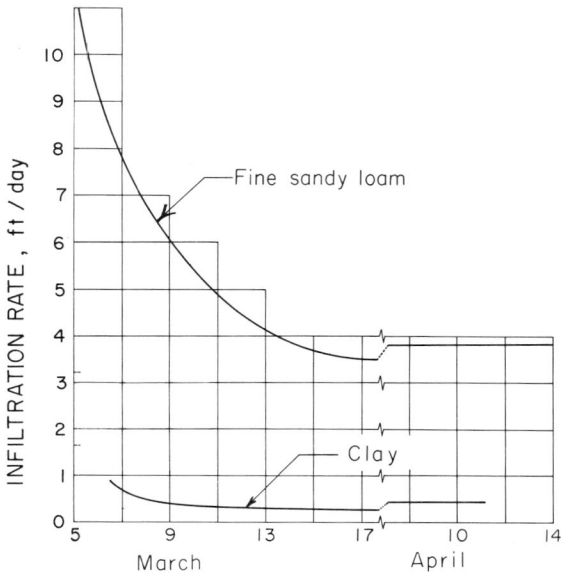

Fig. 8–7. Infiltration curves for two soils at the El Rio Spreading Ground.

infiltration rates into the aquifer material (Schiff, 1958; Schiff and Johnson, 1958). Pea gravels 1/8 or 1/4 inch in size had little effect.

In injection methods employing shafts, pits, and trenches, shallow subsurface layers of low permeability are generally removed, permitting ponded water to contact at a greater hydraulic gradient aquifer material of higher hydraulic conductivity. An injection rate of 58 ft/day was obtained in the 1/10-acre Bakersfield Pit in California (Fig. 8–8) and a rate of 65 ft/day was obtained in the first 1/7-acre Peoria Pit in Illinois, USA (Suter and Harmeson, 1960). Pits on Long Island, New York, USA, have recharged groundwater at a rate of 3 ft/day. Water meandering in canals over a part of 80-acre Rohrer Island, Dayton, Ohio, USA, has recharged groundwater at a rate of 107 acre-ft/day. At Bakersfield, California, a 1/2-acre narrow trench infiltrated as much water over a comparable period of a few months as did surface spreading on 4 acres of land nearby (Schiff, 1954).

In some installations, satisfactory injection rates using wells have been attained, but in many attempts results have been disappointing. Difficulties encountered in maintaining adequate recharge rates have been attributed to silting, bacterial and slime growths, air entrapment, and rearrangement of soil particles (Cullinan and Reeves, 1961; Steinbrugge et al., 1954). In some cases it has been found that continuous application of chlorine is necessary to maintain a satisfactory intake rate (Muckel, 1961). In others, best results have been obtained by pumping the wells for short periods each day (Cullinan and Reeves, 1961). If wells are properly constructed and operated, there is evidence that this can be a feasible method of recharge. Gravel-packed wells are generally used with a packer or seal placed around the well to prevent the upward movement of water into the confining layer above the aquifer.

The high cost of injecting clear chlorinated water in wells along the coast at Manhattan Beach in southern California (Baumann, 1952) is justified in this case, since the system of injection wells protects an inland groundwater basin from sea water intrusion. E. I. duPont de Nemours and Company is injecting treated water in a well in an experiment to determine feasibility of this approach to water conservation (Anon., 1965). Feasibility is enhanced when intrusion of contaminated water is prevented or when cool water is beneficial to industries pumping water.

A relatively new approach to recharge is replenishment or recharge irrigation. This may be defined as irrigation of crop lands with sufficient surplus water to cause percolation to the water table. The advantage of this method is that lands

Fig. 8–8. One-tenth acre Bakersfield pit in California; injection rate is 58 ft/day.

can be cropped at the same time water is being recharged. Present thinking tends to place emphasis on using this method during off-irrigation seasons. Existing irrigation distribution systems can be used. Damage to crops or soils is to be avoided or reasonable maintenance established to minimize damage, such as the use of additional nitrogen. Such an approach may be desirable under many soil or stratigraphic conditions. The method seems particularly adaptable when soils or substrata of low hydraulic conductivity are involved. Alfalfa (*Medicago mocpa*) has been used experimentally for replenishment irrigation (Hall et al., 1957). Preliminary investigations indicate prolonged irrigation produced no significant differences in crop yields of Acala 4–42 cotton (*Gossypium hirsutum*) (Haskell and Bianchi, 1963). The great areas involved in terms of irrigated lands vs. concentrated surface recharge have an intriguing appeal in recharge possibilities.

V. DESIGN OF RECHARGE FACILITIES

Based on either water management practices, including channel stabilization or the desilting of water in reservoirs, or on approaches previously mentioned, assume that a reasonably clear water is available for recharge. Facilities and areas required for 1 million acre-feet of recharge based on an average infiltration or injection rate of a number of recharge works in the USA, using water free from excessive suspended load, follow: Basins with fine-textured soil at 0.5 ft/day, medium-textured soil at 2.0 ft/day, and coarse-textured soil at 4.0 ft/day require 22,000, 5,500 and 2,800 acres, respectively, in continuous operation for 3 months; agricultural land under sufficient irrigation to cause 18 ft of percolation a year below the root zone requires 55,000 acres; channels and pits with an injection rate of 30 ft/day require 360 acres in continuous operation for 3 months; or 5,500 wells are required with an injection rate of 2 acre-ft/day or 1 sec-ft in continuous operation for 3 months (assumes no clogging).

A minimum area of land can be used for maximum recharge if the equations given and infiltration and percolation rates are considered in designing systems of recharge. In the rotational system, parts of an area are flooded when infiltration rates are high while other portions are being dried for recovery in infiltration rate. This system lends itself to soils of reasonably uniform topsoil and substrata characteristics or to soils underlain by relatively shallow less pervious substrata. In the strip system, strips or portions of an area may be used where substrata limit flow. Water will accumulate and spread out laterally on a substratum: A "subsurface lake" can be visualized. The spacing of strips may be based on the infiltration rate/percolation rate ratio (Schiff, 1954). If the ratio is 10, about the same amount of recharge could be obtained by using one-tenth of the land as by using the entire area. Strips treated to increase infiltration rates, or strips in the form of channels, or shafts in the bottom of channels may further reduce the areas required. Another system combines surface and injection spreading. Water flows from a detention basin into a spreading basin. Water flows upslope through grasses in the spreading basin for additional removal of sediments, and then into a pit. The depth, shape, and extent of substrata not only influence the design of a system but also determine the useable storage capacity of the subsurface basin, whether the water table will build up and merge with the saturated zone extending down from the soil surface and thus limit infiltration, and whether drainage problems will be created.

VI. HYDRAULICS OF GROUNDWATER MOUNDS

Mathematical equations for rising and falling groundwater mounds by artificial recharge were developed for a two-dimensional condition, where length of spreading area is much greater than the width (Baumann, 1952). The Boussinesq equation (Boussinesq, 1877) for unsteady groundwater flow reduces to the anlog of the flow of heat and for steady flow to the Dupuit equation (Dupuit, 1863), and provides a solution for the heat flow analog (Baumann, 1952).

Shapes of mounds determined for certain stratigraphic conditions using various equations (Baumann, 1952; Glover, 1961; Marmion, 1962), were compared with shapes obtained in a two-dimensional model representing a vertical section in an unconfined aquifer (Marmion, 1962). Conditions under which good agreement occurred were described. Equations given for mounds must be used with caution since they include such assumptions as a homogeneous medium, horizontal flow, and that the surface gradient applies throughout the full depth of the saturated aquifer.

A resistance network analog has been used to establish equations for the theoretical analysis of rising, falling, and stable groundwater mounds for two-dimensional or radial flow systems under recharge (Bouwer, 1962b). The basis of treatment is a numerical solution of the Laplace equation and the analogy between Darcy's law and Ohm's law. Application of the equations requires strategic assumptions.

LITERATURE CITED

Allison, L. E. 1947. Effect of microorganisms on permeability of soil under prolonged submergence. Soil Sci. 63:439–450.

American Society of Civil Engineers. 1961. Groundwater basin management. Manuals of Engineering Practice No. 40, New York. 160 p.

Anonymous. March-April 1965. Water conservation with a recharge well. Johnson Drillers J. p. 1–2.

Baumann, Paul. 1952. Groundwater movement controlled through spreading. Amer. Soc. Civ. Eng., Trans. 117:1024–1085.

Bianchi, W. C. 1962. Measuring soil moisture tension changes. Agr. Eng. 43:398–399, 404.

Bliss, E. S., and C. E. Johnson. 1952. Some factors involved in groundwater replenishment. Amer. Geophys. Union, Trans. 33:547–558.

Bodman, G. B., and E. A. Colman. 1944. Moisture and energy conditions during downward entry of water into soils. Soil Sci. Soc. Amer. Proc. (1943) 8:116–122.

Boussinesq, M J. 1877. Essai sur la théorie des eaux courantes. Memoires a l'Académie des Sciences de l'Institut de France. 23:257.

Bouwer, Herman. 1962a. Field determination of hydraulic conductivity above a water table with the double-tube method. Soil Sci. Soc. Amer. Proc. 26:330–335.

Bouwer, Herman. 1962b. Analyzing groundwater mounds by resistance network. Amer. Soc. Civ. Eng. Proc. Irrig. Drainage Div. J. 88 (IR 3):15–36.

Bouwer, H., W. W. Donnan, N. A. Evans, F. R. Hore, R. C. Reeve, G. O. Schwab, Phelps Walker, R. J. Winger, Jr., and Jan van Schilfgaarde. 1961. Measuring saturated hydraulic conductivity of soils. Amer. Soc. Agr. Eng., Spec. Publ. SP–SW–0262. 19 p.

California Sanitary Engineering Research Project. 1953. Waste water reclamation in relation to groundwater pollution. State Water Pollution Control Board. Publ. 6. 124 p.

California Sanitary Engineering Research Laboratory. 1954. Investigation of travel of pollution. State Water Pollution Control Board. Publ. 11. 218 p.

Cullinan, Thomas A., and C. C. Reeves, Jr. 1961. Aquifer clogging by silt and clay in recharge water. Water Well J. (Urbana, Illinois) XV(5):14.

Donnan, W. W., and V. S. Aronovici. 1961. Field measurement of hydraulic conductivity. Amer. Soc. Civ. Eng. Proc. Irrig. Drainage Div. J. 87 (IR 2):1–13.

Dupuit, Jules. 1863. Etudes theóriques et pratiques sur le mouvement des eaux dans les canaux decouverts et a travers les terrains pérmeables. 2nd ed.-Dunod, Paris. 304 p.

Glover, R. E. 1961. Mathematical derivations as pertain to groundwater recharge. Agr. Res. Serv. US Dep. Agr. mimeo. 81 p.

Hall, W. A. 1955. Theoretical aspects of water spreading. Agr. Eng. 36:394–399.

Hall, W. A., R. M. Hagan, and J. D. Axtell, 1957. Recharging groundwater by irrigation. Agr. Eng. 38:98–100.

Haskell, E. E., Jr., and W. C. Bianchi. 1963. Effects of prolonged irrigation on cotton. Agron. J. 55:202–203.

Johnson, C. E. 1957. Utilizing the decomposition of organic residues to increase infiltration rates in water spreading. Amer. Geophys. Union, Trans. 38:326–332.

Marmion, K. R. 1962. Hydraulics of artificial recharge in nonhomogeneous formations. Hydraulic Lab., Univ. of California, Berkeley. Water Resources Center Contrib. no. 48. 88 p.

Muckel, Dean C. 1961. Replenishment of groundwater supplies by artificial means. US Dep. Agr. Tech. Bull. 1195. 51 p.

Reeve, R. C., and J. N. Luthin. 1957. Methods of measuring soil permeability. Agronomy 7:395–445.

Richter, R. C. 1955. Geologic considerations of artificial recharge. Geol. Soc. Amer. Bull. 66:1661.

Schiff, Leonard. 1953. The effect of surface head on infiltration rates based on the performance of ring infiltrometers and ponds. Amer. Geophys. Union, Trans. 34:257–266.

Schiff, Leonard. 1954. Water spreading for storage underground. Agr. Eng. 35:794–800.

Schiff, Leonard. 1955. The status of water spreading for groundwater replenishment. Amer. Geophys. Union, Trans. 36:1009–1020.

Schiff, Leonard. 1957. The Darcy law in the selection of water-spreading systems for groundwater recharge. Ass. Int. d'Hydrologie (de l'Union Geod. et Geophys.) Publ. 41:99–110.

Schiff, Leonard. 1958. The use of filters to maintain high infiltration rates in aquifers for groundwater recharge. Int. Union Geodesy Geophys., 11th Gen. Assembly, Proc. Toronto, 1957. 2:207–211.

Schiff, Leonard. 1961a. Progress in artificial groundwater recharge. Int. Comm. Irrig. Drainage Ann. Bull. p. 69–75.

Schiff, Leonard. 1961b. Effect of filtering on model recharge wells. Amer. Soc. Civ. Eng. Proc., Irrig. Drainage Div. J. 87 (IR 4): 55–63.

Schiff, Leonard. 1964. Devices for measuring soil water movements in designing recharge facilities. Amer. Soc. Agr. Eng., Trans. 7(1):67–69.

Schiff, Leonard, and C. E. Johnson. 1958. Some methods of alleviating surface clogging in water spreading with emphasis on filters. Amer. Geophys. Union, Trans. 39:292–297.

Schiff, Leonard, et al. 1959. Proc. biennial conference on groundwater recharge. Agr. Res. Serv. US Dep. Agr. mimeo. 105 p.

Schiff, Leonard, et al. 1961. Proc. biennial conference on groundwater recharge. Agr. Res. Serv. US Dep. Agr. mimeo. 234 p.

Steinbrugge, G. W., L. R. Heiple, N. Rogers, and R. T. Sniegocki, 1954. Groundwater recharge by means of wells. Univ. Arkansas Dep. Eng., Agr. Exp. Sta. 119 p.

Suter, Max, and R. H. Harmeson. 1960. Artificial groundwater recharge at Peoria, Illinois. Illinois State Water Surv. Bull. 48:1–48.

Todd, D. K. 1959. Annotated Bibliography on artificial recharge of groundwater through 1954. US Geol. Surv. Water-Supply Pap. No. 1477. 115 p.

Tolman, C. F. 1937. Groundwater. McGraw-Hill, New York. p. 199.

9 | Quality of Irrigation Water

L. V. WILCOX

US Salinity Laboratory, ARS, USDA
Riverside, California

W. H. DURUM

Geological Survey, US Department of the Interior
Washington, D.C.

I. INTRODUCTION

Interest in the quality of irrigation water dates back a comparatively short time. In contrast, the quantity of water available for irrigation has always been of primary concern.

Hilgard (1906) was among the first to recognize the importance of irrigation water quality and proposed standards based on composition as well as on total concentration. Significant contributions have since been made by Kelley and Brown (1928), Kelley et al. (1939), Scofield (1936), Scofield and Headley (1921), Scofield and Wilcox (1931), Eaton (1935, 1936, 1950), Doneen (1949, 1954), Thorne and Thorne (1951), Wilcox (1948, 1955), and US Salinity Laboratory Staff (1954). Taylor et al. (1935) proposed an empirical equation, known as the "Salt Index," for the classification of irrigation waters. Hill (1940, 1942) and Piper (1944) in this country and Durov (1948) in the USSR developed procedures for the geochemical classification of waters. Lunin et al. (1960) investigated the use of brackish waters for supplemental irrigation in areas of relatively high rainfall. California State Water Quality Control Board (1963) presents a review of the criteria for all ordinary uses of water and a very extensive bibliography. The papers cited above are representative of the more important contributions to the subject but do not constitute a complete literature review. There are many other papers on the applied research level that are addressed primarily to water users.

Although there are differences between the several schemes for the classification of irrigation waters, there is reasonable agreement with respect to criteria and limits. This makes it possible to anticipate, with considerable confidence, the effect of a water on soils and plants. But the successful use of a water may not depend on quality alone but on other factors, including the drainage characteristics of the soil.

II. IRRIGATION WATER ANALYSIS

A. Constituents to be Determined

The total concentration and the concentration of the more important constituents must be determined to judge the quality of the water. Table 9–1 lists the determinations usually made on an irrigation water, together with the units in which they are reported.

Table 9–1. The constituents usually determined in an irrigation water analysis, their abbreviations, and the units in which they are reported

Determination	Abbreviation	Unit
Electrical conductivity	$EC \times 10^6$ at 25C	micromhos per cm
Soluble-sodium percentage	SSP	per cent
Sodium-adsorption-ratio	SAR	
Boron	B	parts per million, ppm
Dissolved solids	DS	ppm
pH		
Cations:		
Calcium	Ca	Milliequivalents/liter, meq/liter
Magnesium	Mg	meq/liter
Sodium	Na	meq/liter
Potassium	K	meq/liter
Sum of cations		meq/liter
Anions:		
Carbonate	CO_3	meq/liter
Bicarbonate	HCO_3	meq/liter
Sulfate	SO_4	meq/liter
Chloride	Cl	meq/liter
Nitrate	NO_3	meq/liter
Sum of anions		meq/liter

Table 9–2. Minor elements in large rivers of North America

Element	Micrograms/liter Median	Micrograms/liter Range	Atlantic Coast*	Gulf Coast*	Pacific Coast*
Ag	0.09	0-0.94	+	0	0
Al	238	12-2,550	0	+	0
B	10	1.4-58	0	0	0
Ba	45	9-152	−	+	−
Co	0	0-5.8			
Cr	5.8	0.72-84	+	−	−
Cu	5.3	0.83-105	0	+	0
Fe	300	31-1,670	0	+	0
Li	1.1	0.075-37	−	+	0
Mn	20	0-185	+	0	0
Mo	0.35	0-6.9	+	0	+
Ni	10	0-71	+	0	0
Pb	4.0	0-55	0	−	+
Rb	1.5	0-8.0	0	+	−
Sr	60	6.3-802	+	−	0
Ti	8.6	0-107	+	+	−
V	0	0-6.7			
Zn	0	0-215			

* + indicates significant percentage of determinations above median; 0 indicates approximately 50% of determinations above and below median; − indicates significant percentage of determinations below median

A somewhat less detailed analysis will often suffice after the general characteristics of a water are known. For many applications, conductivity is adequate as a measure of total concentration so that the determination of dissolved solids can be omitted. If it has been determined that the boron concentration is low, this constituent need not be included in subsequent analyses. Potassium and nitrate may be important from the viewpoint of fertility but can usually be omitted in routine irrigation water analyses.

Minor elements, including boron, are sometimes shown in water analyses but, in unpolluted natural water, it is unlikely that they would be present in amounts sufficient to affect the chemistry of the soil. They are important, however, from the standpoint of plant nutrition because several are essential to normal plant growth. Their occurrence in the large rivers of North America, as reported by Durum and Haffty (1961), is shown in Table 9–2.

B.   Sampling Techniques

Water samples are collected and analyzed to obtain information on which to base a quality appraisal. Representative samples and accurate analytical work are, therefore, essential.

Collecting water samples is discussed in detail by Hem (1959) and by Rainwater and Thatcher (1960) and is usually included in texts on methods of water analysis. The mechanics of the sampling program will depend on the use to which the information is to be put, the chemical characteristics of the water, and the nature of the supply to be sampled whether a well, a stream, or a reservoir.

An irrigation well, under normal operating conditions, presents no particular sampling problem. If the natural replenishment of the aquifer equals the withdrawal, there will be very little change with time in the chemical characteristics of the water. If, however, the withdrawal exceeds the replenishment, as evidenced by a falling water table, there may be a change in the composition of the water. Ordinarily, it is not possible to forecast whether there will be a change or, if there is a change, whether it will result in a deterioration or an improvement in quality. In such a situation, additional analyses are necessary.

A surface stream is more difficult to sample, especially if it is uncontrolled. If possible, the samples should be collected at a gaging station so that the analytical data can be related to the discharge or to the runoff. The details of the sampling program, including the frequency of sample collection, should be developed after a study of the stream's discharge characteristics.

A controlled stream that is fed by the discharge from a reservoir is comparatively constant in composition. Daily or weekly samples from a gaging station are usually adequate for irrigation-management needs or for salt-balance studies.

Water sampling in a large, deep reservoir may be a complex undertaking. Often, this water is not thoroughly mixed, which can be determined only by sampling and analysis. In such a situation, it would be necessary to collect samples from several depths and at a number of locations in the reservoir. Water in small reservoirs is usually homogeneous so that samples from the outlet are representative.

Samples for irrigation water analysis should be collected in 1- or 2-liter clean glass or polyethylene bottles. Samples from wells should be taken after the pump has been running for some time. Samples from surface streams should be taken from running water, and a few inches below the surface to avoid floating oil. The samples should be analyzed as soon as possible after collection because chemical changes can take place on standing. It is most important that adequate descriptions accompany the samples and become a part of the records and reports.

C.   Methods of Water Analysis

Spectrophotometry, flame photometry, gas chromatography, the Schwarzenbach reactions, and other advances have revolutionized water analysis in recent decades. Excellent and authoritative manuals employing these techniques and applicable to

Table 9–3. Selected chemical and physical properties of water and
suitable methods of analysis

Constituent or physical property	Methods of analysis
Silica, SiO_2	Molybdate blue colorimetric method.
Iron, Fe	Spectrophotometric determination with 2-2'-bipyridine.
Manganese, Mn	Spectrophotometric determination after oxidation to permanganate.
Calcium, Ca, and magnesium, Mg	Calcium titrated with ethylenediaminetetraacetic acid with murexide indicator. Magnesium calculated from difference between total Ca+ Mg and calcium titrations.
Sodium, Na, and potassium, K	Sodium and potassium determined with flame photometer using an internal standard.
Carbonate, CO_3, and bicarbonate, HCO_3	Titration with sulfuric acid to (i) phenolphthalein end point for carbonate and to methyl orange end point for bicarbonate, or (ii) electrometric titration to end point measured by pH meter as follows: carbonate to pH 8.2; bicarbonate from pH 8.2 to pH 4.5.
Sulfate, SO_4	Gravimetric determination as $BaSO_4$.
Chloride, Cl	Mohr titration.
Fluoride, F	Colorimetric determination with zirconium-alizarin.
Nitrate, NO_3	Colorimetric determination with phenoldisulfonic acid.
Boron, B	Spectrophotometric determination with carmine or potentiometric determination with mannitol.
Minor elements	Spectrograph with carbon arc.
Dissolved solids	Residue on evaporation to dryness at 105 C or 180 C. "Sum" is mathematical summation of determined constituents with HCO_3 expressed as CO_3.
Electrical conductivity, EC	Wheatstone bridge measurement.
Hydrogen ion activity, pH	Measurement with pH meter using glass electrode.
Suspended sediment	Sediment concentration is determined by filtration, drying, and weighing.

Table 9–4. Number of significant figures to be shown in the report
of an irrigation water analysis

Constituent	Number of significant figures	
	Total	To right of decimal
$EC \times 10^6$ at 25C	3	1
SAR	2	1
Boron	2	1
Dissolved solids	3	0
Cations and anions	4	2

the analysis of water for irrigation use are available (Amer. Publ. Health Ass., 1960; Ass. Offic. Agr. Chem., 1960; Rainwater and Thatcher, 1960; US Salinity Lab. Staff, 1954). Table 9–3 lists the more important constituents and suitable methods for their determination.

D. Precision and Accuracy

The two terms precision and accuracy are not synonymous. Precision is a measure of the reproducibility of a procedure, and accuracy is a measure of the error. The methods listed in Table 9–3 are of satisfactory precision in the hands of a competent analyst. The precision can be determined by repeated analyses of the same sample. Similarly, the accuracy of the methods listed is adequate and can be determined by the analysis of standard samples.

The results are reported in the units shown in Table 9–1. The accuracy should determine the number of significant figures to be reported. Discussions of this subject are given in the texts cited and need not be repeated here. Table 9–4 suggests the number of significant figures to be shown in the report of an irrigation water analysis. These are reasonably in accord with the accuracy of the methods and are proposed for purposes of uniformity.

E.   Published Analyses

The US Geological Survey publishes two series of Water-Supply Papers under the general titles: (i) *Quality of surface waters of the United States,* and (ii) *Quality of surface waters for irrigation, Western United States.* The first series reports results, in parts per million, from all major river basins in the USA. The second series reports results, in equivalents per million (numerically equal to meq/liter), from the more important streams of the Western USA from which water is diverted for irrigation.

The International Boundary and Water Commission publishes annual summaries of the chemical quality of the international streams in the Rio Grande Basin in a series of Water Bulletins under the title, *Flow of the Rio Grande, and related data.*

The US Public Health Service obtains water-quality data at key stream locations and publishes results in an annual summary, *National Water Quality Network.*

A very useful source of information is the Federal Inter-Agency Committee on Water Resources inventories of published and unpublished chemical analyses (1956).

III.   CLASSIFICATION AND INTERPRETATION OF WATER ANALYSES

A.   Purpose of Classification

The analysis of an irrigation water should provide information on the suitability of the water for irrigation use and throw light on the management practices that should be followed. The purpose of the classification and interpretation is to glean this information from the analysis.

The first step in the interpretation is selecting criteria that will yield the type of information desired. The second step is classifying criteria to evaluate interrelationships.

There are four principal hazards related to the chemical character of the water: total concentration, sodium, bicarbonate, and boron or other phytotoxic substances. Criteria that measure these hazards have been worked out and are in general use.

B.   Criteria that Determine Quality

The quality of an irrigation water is determined by the composition and concentration of the dissolved substances or solutes that are present in the water. The principal solutes are the cations calcium, magnesium, and sodium, and the anions bicarbonate, sulfate, and chloride. Boron, fluoride, and nitrate are usually present in low, but significant, concentrations. Small amounts of carbonate are found in many waters, as well as trace amounts of other less important constituents. The concentrations of the several ions show wide variations but, because of solubility limitations, sodium and chloride often predominate in the more saline waters.

1. TOTAL CONCENTRATION

Total concentration is probably the more important single criterion of irrigation water quality. The importance derives from the fact that the salinity of the soil solution is usually related to, and often determined by, the salinity of the irrigation water. Thus, plant growth may be impaired or prevented, depending on the salt content of the water.

Total concentration may be expressed in terms of parts per million (ppm) of dissolved solids, or as electrical conductivity (EC) in micromhos per centimeter (EC μmho/cm). The latter is preferred. More than half of the irrigation waters in use in Western USA have conductivity values < 750 μmho/cm (about 500 ppm dissolved solids). Saline waters with conductivity values > 2,250 μmho/cm (about 1,500 ppm dissolved solids) make up < 10% of the total number of waters and an even smaller fraction of the total quantity of water being used. There are very few waters with conductivity values > 5,000 μmho/cm (about 3,200 ppm dissolved solids) that are being used successfully, although they can be used for certain crops under very special conditions. Such waters are important, however, in that they constitute the only available supply in many arid regions.

2. SODIUM

Sodium (Na^+) is unique among the cations in its effect upon the soil. When present in the soil in exchangeable form, even at low concentrations compared with the other cations, it causes adverse chemical and physical conditions to develop. Exchangeable sodium tends to make a moist soil impermeable to air and water, and on drying, this soil is hard and difficult to till. Dense crusts form that interfere with germination and seedling emergence. The most reliable index of the sodium hazard, or the tendency of the irrigation water to form exchangeable sodium in the soil, is the sodium-adsorption-ratio, SAR (US Salinity Lab. Staff, 1954). It is a calculated value and is defined as:

$$ SAR = \frac{Na^+}{\sqrt{\dfrac{Ca^{2+} + Mg^{2+}}{2}}} \qquad [9\text{--}1] $$

where concentrations are expressed in meq/liter.

A nomogram for determining the SAR value of an irrigation water is shown in Fig. 9–1. An exchangeable-sodium-percentage (ESP) scale is included in the nomogram opposite the SAR scale. This ESP scale is empirical, but it is based on a regression equation of high statistical significance. After the SAR value of an irrigation water is determined by use of the nomogram, it is possible to estimate from the central scale the ESP value of a soil that is at equilibrium with the irrigation water. Under field conditions, the actual ESP may be slightly higher than the estimated equilibrium value. This is because the total salt concentration of the soil solution is increased by evaporation and plant transpiration which results in a higher SAR and a correspondingly higher ESP.

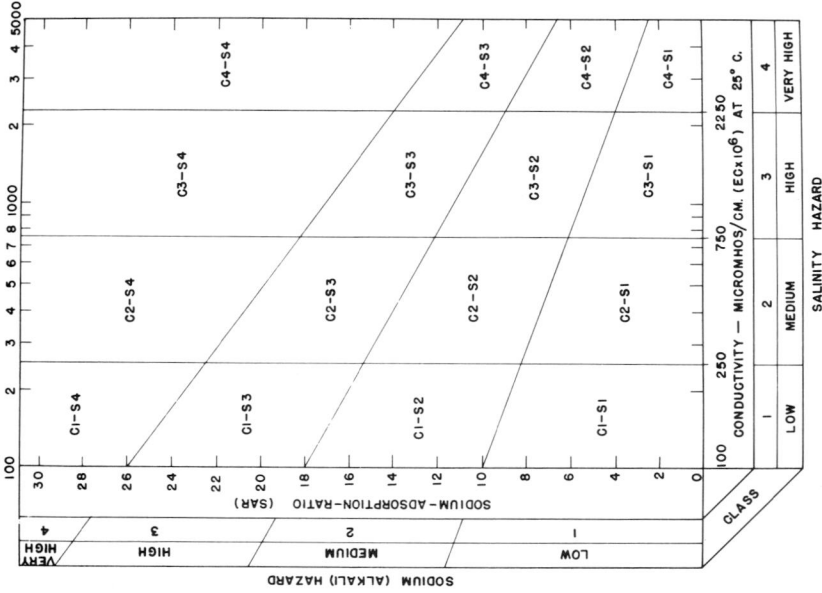

Fig. 9-2. Diagram for the classification of irrigation waters.

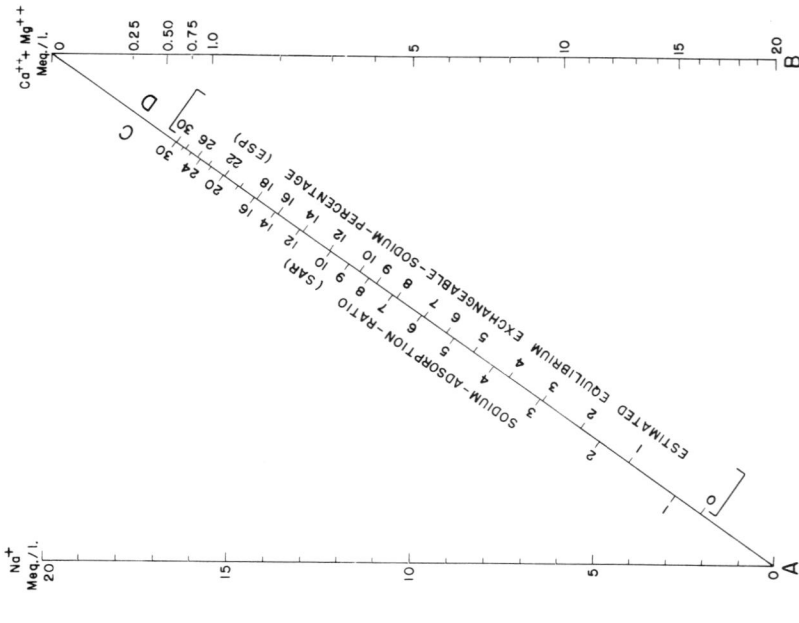

Fig. 9-1. Nomogram for determining the SAR value of irrigation water and for estimating the corresponding ESP value of a soil that is at equilibrium with the water.

3. BICARBONATE

Bicarbonate (HCO_3^-) is important primarily in its relation to calcium and magnesium. There is a tendency for calcium to react with the bicarbonate and precipitate as the normal carbonate, $CaCO_3$. The corresponding magnesium salt is more soluble so there is less tendency for it to precipitate. Magnesium may be lost from a water, however, by an indirect reaction. Magnesium enters the exchange complex of the soil, replacing calcium which reacts with bicarbonate and precipitates as $CaCO_3$. Ordinarily, magnesium will not replace calcium to any great extent but, if calcium is precipitated as it is released, the reaction proceeds toward completion.

As calcium and magnesium are lost from a water, the relative proportion of sodium is increased with an attendant increase in the sodium hazard. This hazard can be evaluated in terms of the residual sodium carbonate (RSC) as proposed by Eaton (1950) and defined as

$$RSC = (CO_3^{2-} + HCO_3^-) - (Ca^{2+} + Mg^{2+}) \qquad [9-2]$$

in which the concentrations are expressed in meq/liter. Studies by Wilcox et al. (1954) indicate that waters with > 2.5 meq/liter RSC are probably not suitable for irrigation purposes. Water containing 1.25 to 2.5 meq/liter are marginal, and those containing < 1.25 meq/liter RSC are probably safe. Good management practices and proper use of amendments, particularly gypsum, might make possible use of some of the marginal waters. A condition not provided for by the RSC concept has been encountered in recent years. If the concentrations of both Ca and HCO_3 are about equal and are high, i.e., in the order of 10 meq/liter or greater, the RSC will be low or possibly zero. Such waters will precipitate some $CaCO_3$ and should be considered at least marginal.

4. PHYTOTOXIC SUBSTANCES

a. Boron. The occurrence of boron in toxic concentrations in certain irrigation waters makes it necessary to consider this constituent when assessing the quality of water.

Plant species differ markedly in their tolerance to high concentrations of boron. In areas where boron occurs in excess in the soil or in the irrigation water, boron-tolerant crops may grow satisfactorily, whereas sensitive crops may fail. The relative boron tolerance of a number of crops as determined by Eaton (1935), with only minor modifications based on more recent field observations (Wilcox, 1960) is shown in Table 9–5. Incorporated into the table are the limits for boron in irrigation water essentially as proposed by Scofield (1936).

b. Other Substances. Very few substances other than boron occur in toxic concentrations in natural waters. However, many substances in the industrial wastes that are discharged into surface streams are probably toxic to plants. Wilcox (1959) assembled information on a number of such substances for which the phytotoxic properties are known. If the presence of pollutants is suspected, great care should be exercised in the use of the water for irrigation.

Table 9–5. Relative tolerance of plants to boron; in each group, the plants first named are considered to be more tolerant and the last named, more sensitive; the figures at the top and bottom of each column represent the limiting boron concentration in the irrigation water

Tolerant	Semitolerant	Sensitive
4. 0 ppm of boron	2. 0 ppm of boron	1. 0 ppm of boron
Athel (Tamarix aphylla)	Sunflower (native) (Helianthus annuus)	Pecan (Carya pecan)
Asparagus (Asparagus officinalis)	Potato (Solanum tuberosum)	Walnut (black and Persian or English) (Juglans spp.)
Palm (Phoenix canariensis)	Cotton (Acala and Pima) (Gossypium spp.)	Jerusalem artichoke (Helianthus tuberosus)
Date palm (Phoenix dactylifera)	Tomato (Lycopersicum esculentum)	Navy bean (Phaseolus vulgaris)
Sugar beet (Beta vulgaris)	Sweetpea (Lathyrus odoratus)	American elm (Ulmus americana)
Mangel (Beta vulgaris)	Radish (Raphanus sativus	Plum (Prunus domestica)
Garden beet (Beta vulgaris)	Field pea (Pisum sativum)	Pear (Pyrus communis)
Alfalfa (Medicago sativa)	Ragged-robin rose (Rosa)	Apple (Pyrus malus)
Gladiolus (Gladiolus spp.)	Olive (Olea europaea)	Grape (Sultanina and Malaga)(Vitis vinifera)
Broadbean (Vicia faba)	Barley (Hordeum vulgare)	Kadota fig (Ficus carica)
Onion (Allium cepa)	Wheat (Triticum vulgare)	Persimmon (Diospyros spp.)
Turnip (Brassica rapa)	Corn (Zea mays)	Cherry (Prunus avium)
Cabbage (Brassica oleracea var. capitata)	Milo (Sorghum vulgare)	Peach (Prunus persica)
Lettuce (Lactuca sativa)	Oat (Avena sativa)	Apricot (Prunus armeniaca)
Carrot (Daucus carota)	Zinnia (Zinnia elegans)	Thornless blackberry (Rubus spp.)
	Pumpkin (Cucurbita pepo)	Orange (Citrus sinensis)
	Bell pepper (Capsicum frutescens)	Avocado (Persea americana)
	Sweet potato (Ipomoea batatas)	
	Lima bean (Phaseolus lunatus)	Grapefruit (Citrus paradisi)
2. 0 ppm of boron	1. 0 ppm of boron	0. 3 ppm of boron

C. Procedures for the Classification and Interpretation of an Irrigation Water Analysis

1. THE IRRIGATION WATER TO BE USED UNDER ARID CONDITIONS

The interrelation of the two criteria electrical conductivity and sodium-adsorption-ratio (SAR) must be evaluated as the first step in the interpretation of an irrigation water analysis. This can be accomplished by means of a classification diagram such as shown in Fig. 9–2. Several such diagrams have been proposed, including those by Wilcox (1948), Thorne and Thorne (1951), and US Salinity Lab. (1954). The latter will be described briefly to illustrate the classification procedure.

a. US Salinity Laboratory Procedure for the Classification of an Irrigation Water Analysis. The diagram shown in Fig. 9–2 is used to classify the analysis. The horizontal axis represents conductivity in μmho/cm, and the vertical axis represents SAR. The curves running from left to right are given a negative slope to take into account the dependence of the sodium hazard on the total concentration. Thus, a water with a SAR of 15 and a conductivity of 200 is classed, insofar as the sodium hazard is concerned, as a S2 water. With the same SAR and a conductivity of 500, it becomes a S3 water, and with a conductivity of 3,000, the water is rated as S4. This system by which waters at a constant SAR value are given a higher sodium hazard rating with increasing total concentration is empirical, but it is supported by theory.

To classify a water analysis, the numerical values for conductivity and SAR are plotted, as coordinates, on Fig. 9–2. The position of the point determines the quality classification of the water. The significance and interpretation of these quality ratings are summarized below.

1. Salinity Classification.

C1—LOW SALINITY WATER can be used for irrigation with most crops on most soils, with little likelihood that a salinity problem will develop. Some leaching is required, but this occurs under normal irrigation practices except in soils of extremely low permeability.

C2—MEDIUM SALINITY WATER can be used if a moderate amount of leaching occurs. Plants with moderate salt tolerance can be grown in most instances without special practices for salinity control.

C3—HIGH SALINITY WATER cannot be used on soil with restricted drainage. Even with adequate drainage, special management for salinity control may be required, and plants with good salt tolerance should be selected.

C4—VERY HIGH SALINITY WATER is not suitable for irrigation under ordinary conditions but may be used occasionally under very special circumstances. The soil must be permeable, drainage must be adequate, irrigation water must be applied in excess to provide considerable leaching, and very salt-tolerant crops should be selected.

2. Sodium Classification.

S1—LOW SODIUM WATER can be used for irrigation on almost all soils, with little danger of the development of a sodium problem. However, sodium-sensitive crops, such as stone-fruit trees and avocados, may accumulate injurious amounts of sodium in the leaves.

S2—MEDIUM SODIUM WATER may present a moderate sodium problem in fine-textured (clay) soils unless there is gypsum in the soil. This water can be used on coarse-textured (sandy) or organic soils that take water well.

S3—HIGH SODIUM WATER may produce troublesome sodium problems in most soils and will require special management—good drainage, high leaching, and additions of organic matter. If there is plenty of gypsum in the soil, a serious problem may not develop for some time. If gypsum is not present, it or some similar material may have to be added.

S4—VERY HIGH SODIUM WATER is generally unsatisfactory for irrigation except at low- or medium-salinity levels where the use of gypsum or some other amendment makes it possible to use such water.

To summarize the discussion on the classification and interpretation of water analyses, first consideration should be given to the salinity and sodium hazards by referring to Fig. 9–2 and the quality-class ratings that follow the diagram. Consideration should then be given to the independent characteristics, bicarbonate and boron or other phytotoxic substances, any one of which may change the quality

Table 9–6. Tentative classification for effective salinity of irrigation waters

Soil conditions	Class 1	Class 2	Class 3
	meq/liter		
Little or no leaching of the soil can be expected	3	3 - 5	5
Some leaching but restricted; deep percolation or drainage slow	5	5 - 10	10
Open soils; deep percolation of water easily accomplished	7	7 - 15	15

rating. Finally, recommendations for use of the water must take into account such factors as infiltration rate, drainage, quantity of water used, climate, rainfall, and salt tolerance of crop. In the classification scheme described above, *average conditions* with respect to these factors are assumed. Large deviations from the average for one or more of these factors may make it unsafe to use a water that would be safe under average conditions. Similarly, under some unusual circumstances, it may be possible to use a water that would be considered unsafe under average conditions. This relationship to average conditions must be kept in mind in connection with the use of any general method for the classification of irrigation waters.

b. Procedure Proposed by Doneen (1954). Doneen defines the term "effective salinity" and bases a classification procedure on it. Effective salinity includes all of the dissolved salts of the irrigation water except calcium sulfate and calcium and magnesium bicarbonates. It is assumed that these salts precipitate and so do not contribute to soil salinity. Table 9–6 shows the limits for effective salinity for three sets of soil conditions.

Doneen emphasizes that effective salinity is only one of several factors that must be considered. He indicates that sodium, bicarbonate, and boron or other toxic constituents must be evaluated.

c. Formulas for Estimating Gypsum and Leaching Requirements as Proposed by Eaton (1954). Formulas are presented for characterizing and interpreting analyses of irrigation waters in the following terms: (i) The percentage of applied irrigation water that should pass through the root zone as drainage to insure reasonable yields (70 to 80% of yields on nonsaline land) of rotation crops of intermediate salt tolerance in a semiarid climate. (ii) The amount of calcium that should be added to irrigation waters to insure that the sodium percentage of the soil water leaving the root zone will not contain more than about 70% sodium. The use of the formulas is discussed and illustrated in a number of tables.

2. THE IRRIGATION WATER TO BE USED UNDER HUMID CONDITIONS

a. The Procedure Proposed by Lunin et al. (1960). Supplementary irrigation in the humid regions has increased rapidly during the past few decades. Quality classifications, developed for use under arid conditions, are not directly applicable to these areas. It has been found that waters of higher salt content can be used occasionally in such situations. Water quality in relation to supplementary irrigation along the Eastern seaboard (USA) has been investigated by Lunin et al. (1960).

IV. PHYSICAL QUALITY OF IRRIGATION WATER

A. Sediment

Jacobsen and Adams (1958) have described the role of sedimentation in ancient Mesopotamian agriculture and its contribution to breakup of past civilizations.

In modern times, sediments in irrigation waters clog intake screens and pumps, fill channels, reduce reservoir capacity and useful life, and deposit soil materials on crop lands. Few natural stream waters are completely free of sediments; therefore, adequate measurement and control of the suspended and bed-load sediments must be developed in irrigation planning.

Sediment particles, being more dense than water, tend to settle with respect to the water at rates that depend on the differences in density between particles and the water, the viscosity of the water, and the size, shape, and flocculation of the particles. Water temperature and the presence of other sediment particles influence the effective viscosity of the fluid and hence the fall velocities of the particles. Thus, samples of sediment in the streams involved must be taken to obtain the concentration, and the size and settling rates of particles.

The range of sediment concentrations of a river throughout the year is usually much greater than the range of dissolved solids concentrations. Maximum concentrations may be 10 to more than 1,000 times the minimum concentrations. Usually, the sediment concentration is higher during high flow than during low flow. This differs from dissolved solids concentration, which is usually lower during high flow.

Rainwater (1962), in broad national coverage, describes some facts about river composition. In 50% of the country, the prevalent concentration of dissolved solids is $<$ 230 ppm, and the discharge-weighted sediment concentration is $<$ 600 ppm. In 90% of the country, the prevalent concentrations are $<$ 900 ppm and 8,000 ppm, respectively. Mean concentrations of dissolved solids and sediment correlate very well, geographically; low dissolved solids concentrations tend to coincide with low sediment concentrations.

Annotated bibliography on hydrology and sedimentation, USA and Canada, 1955–58 (Riggs, 1962), is a compilation of the more important recent papers in the field of sedimentation.

B. Corrosion and Corrosion Control

Corrosion and encrustation of pumping equipment and distribution lines in irrigation facilities is occasionally reported as caused by the water supply. Water is corrosive to a metal pipe when it dissolves the metal as positive ions or furnishes constituents which react with the metal at the interface. As Camp (1963) points out, either process is oxidative and is accompanied by the release of an equivalent number of negative electrons which flow backward through the metal and are captured in a reduction reaction at some other point in the metal-water interface. For corrosion to take place, there must be water and moisture in contact with the metal, and the water or moisture must contain ions to form the electrolyte. There must be anodic areas at which the oxidation takes place and cathodic areas at which the reduction takes place.

In describing corrosiveness of well waters in the Western Desert of Egypt, Clarke (1963) states that steel well screens and casings deteriorate rapidly because of low pH values and oxidation-reduction potentials, and relatively high CO_2 and H_2S concentrations which prevail in the Nubian sandstone aquifers. The normal cathodic protection and use of chemical inhibitors were not considered feasible. Tests of various casing and screen materials including fiberglass, soft aluminum, and various stainless steel alloys, gave early indication of pitting in aluminum alloy but certain types of stainless steel showed no evidence of pitting. Epoxy resin-bonded fiberglass tubing was shown to be an excellent material for corrosion-free well screen.

Recently, Schaschl and Marsh (1963) observed that aerobic and anaerobic zones surrounding underground pipelines are responsible for much soil corrosion of metal. Soil water contents and types of soil were shown to be related to the rates of corrosion.

C. Sewage and Industrial Wastes

Disposal of closely-controlled sanitary sewage effluent to produce agricultural crops has a long history; however, progress in its use for industrial application has been most rapid in the food processing industry. Sanborn (1953) and others report on installations employing principles of spray or ridge and furrow irrigation for the disposal of cannery, milk processing, and other food processing waste effluents.

Among factors to be considered in the application of waste effluents are the effect of soil characteristics on its capacity for treatment of effluents, the effect of effluent composition on soil characteristics, fertilizing or soil conditioning effects of effluents, or adverse effects on groundwater quality where highly colored or undesirable constituents are applied as irrigation water.

There are two problems with sewage and detergent wastes in the USA. One involves groundwater, the other surface water. Surface waters receive wastes from sewage plants where aerobic degradation is used to clear the effluent of contaminants. Without sewage systems, septic tanks or cesspools draining to the soil are used to degrade wastes. Some investigators conclude that, should the soil be saturated, there is little chance for complete degradation since an abundance of oxygen is lacking.

For example, Wayman and Robertson (1963) conclude from laboratory studies of alkyl benzene sulfonate (ABS) that air (oxygen) is required for the degradation of both hard and soft detergents. Surface waters—rivers, lakes, and ponds—are usually aerated. Groundwaters, however, are not as well aerated, and some contain little or no oxygen. Thus the degradability of both hard and soft detergents in wastes discharged underground will be less than in a fully aerated environment.

Klein et al. (1963), in a series of controlled experiments on the fate of ABS in soils and plants, conclude the following: The major findings of these experiments were that, although ABS caused severe growth inhibition in water-culture studies (about 70% at 10 mg/liter ABS and almost 100% at 40 mg/liter), only one species (sunflower, *Helianthus annuus*) of the three grown in soil was adversely affected. Furthermore, plants irrigated with sewage containing 4.6 to 12.7 mg/liter ABS far surpassed in growth those irrigated with water, regardless of soil fertilization practices or the addition of up to 15 mg/liter ABS to sewage.

Blosser and Owens (1964), in discussing laboratory studies on soils, developed a base for establishing general parameters for field application of pulpmill effluents. These guides take into consideration cover vegetation, percolation rate, and evapo-transpiration from the leaf system. Grass yield from seed germinated and irrigated, for a period of almost 3 years, with unbleached Kraft effluent is equal to that irrigated with tap water.

Guides for field application of pulpmill effluents

BOD	< 200 lb/acre day
Color	Individual site investigation
pH	6.5–9.0
SAR	< 8 on permeable soils

The possible effects of residues from pesticides that are normally encountered in irrigation waters await facts regarding their occurrence and movement. However, a report of the President's Science Advisory Committee (1963) points out that, although pesticides may occur in minute quantities, their variety, toxicity, and persistence are affecting biological systems in nature.

Those used in greatest tonnage, and the most persistent, are the familiar chlorinated hydrocarbons DDT, dieldrin, aldrin, endrin, toxaphene, lindane, methoxychlor, chlordane, and heptachlor. Among those used extensively as herbicides are 2,4-D and 2,4,5-T for control of broadleaved weeds.

It remains to be established whether or not significant amounts of pesticides and herbicides enter waterways used for irrigation or leave by way of irrigation return flow. Certain chlorinated pesticides are known to persist for long periods in the soil, but to assume they are washed from the soil to reappear in watercourses is conjectural. Evidence to date is that herbicides such as 2,4-D and 2,4,5-T are more likely to be found persistently in irrigation return flow (in parts per trillion range) where water channels are sprayed for weed control.

V.   TREATMENT TO IMPROVE QUALITY

The quality of many irrigation waters might be improved by the addition or removal of certain constituents. Carbon dioxide (CO_2), mineral acids, gypsum, and wetting agents have been tried. Simliarly, the removal of boron from high-boron waters has been considered.

A.   Acids

A strongly alkaline water is characterized by a high pH value, relatively high concentrations of Na^+ and HCO_3^-, and relatively low concentrations of Ca^{2+} and Mg^{2+}. Free CO_3^{2-} may be present. The addition of acid to such a water would reduce the pH and, if the water were applied to a calcareous soil, it would increase the quantity of $CaCO_3$ dissolved by the water. These are desirable effects, and the usefulness for this purpose of both CO_2 and sulfuric acid (H_2SO_4) has been tested in the field.

1. CARBON DIOXIDE

It is not economcially feasible to use CO_2 from high-pressure cylinders but attempts have been made to utilize the CO_2 in the exhaust gases from diesel engines. Schoonover and Martin (1953) investigated this possibility and, in addition, conducted laboratory studies using pure CO_2. They concluded that the "use of CO_2 in irrigation water had no significant effect toward the reclamation of black alkali soil in the cases of two Tulare County irrigation systems studied." One explanation was that most of the CO_2 added to irrigation water is lost into the atmosphere.

Another source of CO_2 is plant roots. This was studied by Goertzen and Bower (1958) who concluded that "the effect of CO_2 on the replacement of adsorbed Na was measurable but small, and of less importance than the effect brought about by the hydrolysis of $CaCO_3$ on leaching."

2. SULFURIC ACID AND SULFUR

Overstreet et al. (1951) tested H_2SO_4, sulfur, and gypsum as soil amendments in the reclamation of alkali (sodic) soils. All were used successfully, but the H_2SO_4 treatments gave better and more rapid response. This suggests that it might be possible to add H_2SO_4 directly to the irrigation water to improve its quality. Consideration would have to be given, however, to corrosion of pipes and structures.

B. Gypsum

Gypsum is a widely distributed and relatively inexpensive source of Ca for use in the reclamation of sodic soils. For this purpose, it is usually applied to the soil rather than being dissolved in the irrigation water. This subject is discussed in chapter 51.

Waters of very low total salt content (in the range of 50 ppm or less) may cause the soil to disperse and the infiltration rate to decrease even though the ESP is low. Gypsum, dissolved in the water or applied to the soil, will usually control this difficulty. Gypsum is also useful in waters of low to medium total salt content that are marginal with respect to SAR or have high concentrations of HCO_3, as reported by Doneen (1949). The added Ca would tend to correct the SAR or to reduce the HCO_3 by precipitation as $CaCO_3$. In either instance, the gypsum can be dissolved in the irrigation water or applied to the soil. Gypsum dissolves slowly and is soluble only to the extent of about 30 meq/liter so that there are definite advantages to soil application.

C. Wetting Agents

If the rate of entry of water into a soil is too slow, it may limit or prevent crop production. The cause may be related to soil texture or the presence of exchangeable sodium, or both. With the development of wetting agents for use in synthetic detergent products, it was speculated that such materials, when added to irrigation water, might increase the infiltration rate. This was investigated by Lunt and Huberty (1954) who tested nine surfactants in the laboratory and a single surfactant on a field-plot scale. They concluded that the wetting agents had made no significant changes in infiltration rates.

VI.   TRENDS IN WATER QUALITY

Determining the effects of man and nature on the quality of water in any area depends on many factors, the most important of which is the availability of ade-quate historical streamflow and salinity records.

Nearly every area that is irrigated from a surface stream is served by a drainage system. The drains collect the shallow groundwater and return it to the stream. The total quantity of dissolved solids removed from the irrigated lands in the drainage water must equal or exceed the quantity brought to the lands in the irrigation water; otherwise, salinity conditions will develop. This balance between output and input of salts is referred to as "salt balance." It was proposed and defined by Scofield (1940) and is discussed in more detail in chapter 51. As a consequence of this necessity for maintaining salt balance in the irrigated lands, the total quantity of salt in the stream remains reasonably constant. The total quantity of water diminishes as a result of evapotranspiration so that the concentration of dissolved solids in the stream water increases with each irrigation diversion and drainage return. Thus, the quality of the water deteriorates, and its suitability for further beneficial use, including irrigation, is impaired. This degradation of water quality is evident in many of the river basins where irrigation is practiced and must be taken into account in connection with any future development.

In a summary of the quality of the water of the Upper Colorado River Basin, based on average yearly water and dissolved solids discharges for the water years 1914–57 and adjusted to 1957 conditions, the US Department of the Interior (1963) concludes, tentatively, that, if there had been no activities of man in the upper basin, exclusive of transmountain diversions, the long-term weighted average concentration of dissolved solids of the Colorado River at Lees Ferry, Arizona, USA, would be about 263 ppm. The indicated increase in dissolved solids concentration of 238 (501 ppm minus 263 ppm) caused by domestic, industrial, and agricultural uses of water is equivalent to 13.3 ppm for each 100,000 acre-ft of water consumed. All but about 1% of this total was presumed to be an effect of irrigation.

In describing water quality and quantity of the Missouri River, Ferguson (1956) shows that for a short-term record (about 5 years) for stations between Williston, North Dakota, USA, and Nebraska City, Nebraska, USA, the increase in average dissolved solids content of river water is quite small. Ferguson concludes that the character of the water resources of the Missouri River Basin is not yet sufficiently changed by the works of man to permit a clear view of the postdevelopment hydrology of the basin. He further concludes that changes in chemical quality of the river waters are not likely to be severe because of the fair degree of uniformity of the waters at various points along the river.

In the Rio Grande Basin, the dissolved solids content of the main stream increases progressively in a downstream direction, with the concentration at Fort Quitman, Texas, USA, being nearly 16 times the concentration at the Otowi Bridge, just below the Colorado-New Mexico state line.

According to Howard (1953), irrigation practices have effected significant changes in the quality of the San Joaquin River (USA) from the headwaters, just below Friant Dam, to the lower end of the basin. Concentrations of dissolved solids have increased downstream.

Potential problems relating to water quality and river development for irrigation are: (i) Irrigation will occur during the summer season when the stream flows are normally low; (ii) the water to be stored for irrigation will be from those areas which yield low mineral water and will not be available to dilute the waters carrying greater amounts of mineral solids; (iii) the return flow from irrigated areas will carry a higher concentration of dissolved salts and will, therefore, be less useful for dilution; and (iv) the minimum flows may not be adequate to maintain reasonable chemical quality and to carry away treated wastes. For example, Fish (1959) concludes that impoundment within Roanoke Rapids Reservoir has sharply reduced the average summer dissolved oxygen (DO) content of the river water. Under these circumstances, the available oxygen concentrations are no longer a direct function of river discharge, and this complicates the full potential of the river to absorb organic waste water.

These are but a few of the problems that develop in connection with the allocation of stream waters to the many and varied beneficial uses. It is not the purpose of this chapter to discuss these problems, although some are considered in other chapters of this monograph.

LITERATURE CITED

American Public Health Association. 1960. Standard methods for the examination of water, sewage, and industrial wastes. 11th ed. Amer. Publ. Health Ass., Inc., New York. 626 p.

Ass. Official Agricultural Chemists. 1960. Official and tentative methods of analysis of the Association of Official Agricultural Chemists. 9th ed. 832 p.

Blosser, R. O., and E. L. Owens. 1964. Irrigation and land disposal of pulpmill effluents. Water Sewage Works III (9):424–432.

California State Water Quality Control Board. 1963. Water-quality criteria. Publ. no. 3–A. 548 p.

Camp, Thomas R. 1963. Water and its impurities. Rheinhold Publ. Corp., New York. 355 p.

Clarke, Frank E. June 1963. Appraisal of corrosion characteristics of Western Desert well water, Egypt. US Geol. Surv., Open File Rep. 65 p.

Doneen, L. D. 1949. The quality of irrigation water and soil permeability. Soil Sci. Soc. Amer. Proc. (1948) 13:523–526.

Doneen, L. L. 1954. Salinization of soil by salts in the irrigation water. Amer. Geophys. Union, Trans. 35:943–950.

Durov, S. A. 1948. Classification of natural waters and graphic representation of their composition. Doklady Akad. Nauk USSR 59(1):87–90.

Durum, W. H., and Joseph Haffty. 1961. Occurrence of minor elements in water. US Geol. Survey Circ. 445. 11 p.

Eaton, Frank M. 1935. Boron in soils and irrigation waters and its effect on plants. US Dep. Agr. Tech. Bull. 448. 132 p.

Eaton, Frank M. 1936. Changes in the composition of groundwaters resultant to anaerobic sulfate decomposition and the attendant precipitation of calcium and magnesium. Amer. Geophys. Union, Trans. Part II:512–516.

Eaton, Frank M. 1950. Significance of carbonates in irrigation waters. Soil Sci. 69: 123–133.

Eaton, Frank M. 1954. Formulas for estimating leaching and gypsum requirements of irrigation waters. Texas Agr. Exp. Sta. Misc. Publ. 111. 18 p.

Federal Inter-Agency River Basin Committee. 1956. Inventory of published and unpublished chemical analyses of surface waters in western United States, 1947–55. Bu.l. no. 9. 114 p.

Ferguson, G. E. 1956. Water quality and quantity. J. Amer. Water Works Ass. 48(8): 951–962.

Fish, F. C. 1959. Effect of impoundment on downstream water quality, Roanoke River, N. C. J. Amer. Water Works Ass. 51 (1):47–50.

Goertzen, J. O., and C. A. Bower. 1958. Carbon dioxide from plant roots as a factor in the replacement of adsorbed sodium in calcareous soils. Soil Sci. Soc. Amer. Proc. 22:36–37.

Hem, J. D. 1959. Study and interpretation of the chemical characteristics of natural water. US Geol. Surv. Water Supply Pap. 1473. 269 p.

Hilgard, E. W. 1906. Soils, their formation, properties, composition, and relations to climate and plant growth. Macmillan, New York. 593 p.

Hill, Raymond. 1940. Geochemical patterns in Coachella Valley. Amer. Geophys. Union, Trans. Part I:46–49.

Hill, Raymond. 1942. Salts in irrigation waters. Amer. Soc. Civil Eng., Trans. 107: 1478–1493.

Howard, C. S. 1953. Effect of irrigation on surface water supplies. J. Amer. Water Works Ass. 45 (11):1171–1178.

Jacobsen, Thorkild, and Robert M. Adams. 1958. Salt and silt in ancient Mesopotamian agriculture. Science 128 (3334):1251–1258.

Kelley, W. P., and S. M. Brown. 1928. Boron in the soils and irrigation waters of Southern California and its relation to citrus and walnut culture. Hilgardia 3:445–458.

Kelley, W. P., S. M. Brown, and G. F. Liebig, Jr. 1939. Chemical effects of saline irrigation waters on soils. Soil Sci. 49:95–107.

Klein, Stephen A., David Jenkins, and P. H. McGauhey. 1963. The fate of ABS in soils and plants. J. Water Pollut. Contr. Fed., 35:653.

Lunin, Jesse, M. H. Gallatin, C. A. Bower, and L. V. Wilcox. 1960. Use of brackish water for irrigation in humid regions. US Dep. Agr. Inform. Bull. No. 213. 5 p.

Lunt, O. R., and M. R. Huberty. 1954. Infiltration rates, effect of wetting agents in water on infiltration rates into soils. California Agr. 8:12.

Overstreet, Roy, J. C. Martin, and H. M. King. 1951. Gypsum, sulfur, and sulfuric acid for reclaiming an alkali soil of the Fresno series. Hilgardia 21:113–127.

Piper, Arthur M. 1944. A graphic procedure in geochemical interpretation of water analyses. Amer. Geophys. Union, Trans. Part VI:914–923.

President's Science Advisory Committee. 1963. Use of pesticides. A report of The President's Science Advisory Committee, The White House, Washington, D. C. 25 p.

Rainwater, F. H. 1962. Stream composition of the conterminous United States. US Geol. Surv. Hydrol. Invest. Atlas HA–61. 3 sheets.

Rainwater, F. H. and L. L. Thatcher. 1960. Methods for collection and analysis of water samples. US Geol. Surv. Water Supply Pap. 1454. 301 p.

Riggs, H. C. 1962. Annotated bibliography on hydrology and sedimentation, United States and Canada, 1955–58. US Geol. Survey Water Supply Pap. 1546. 237 p.

Sanborn, N. H. 1953. Disposal of food processing wastes by spray irrigation. Sewage Ind. Wastes 25 (9):1034.

Schaschl, E., and G. A. Marsh. 1963. The effect of soil conditions on corrosion of steel by oxygen. Int. Congr. Metallic Corrosion, Trans. 2nd Congr. 104 p.

Schoonover, Warren R., and J. C. Martin. 1953. Alkali soil reclamation. California Agr. 7:7, 14, 15.

Scofield, Carl S. 1936. The salinity of irrigation water. Smithsonian Report for 1935 275–287.

Scofield, Carl S. 1940. Salt balance in irrigated areas. J. Agr. Research 61:17–39.

Scofield, Carl S., and Frank B. Headley. 1921. Quality of irrigation water in relation to land reclamation. J. Agr. Research 21:265–278.

Scofield, Carl S., and L. V. Wilcox. 1931. Boron in irrigation waters. US Dep. Agr. Tech. Bull. 264. 66 p.

Taylor, E. McKenzie, Amar Nath Puri, and A. G. Asghar. 1935. Soil deterioration in the canal-irrigated areas of the Punjab. Part I. Equilibrium between Ca and Na ions in base exchange reactions. Punjab Irrig. Res. Inst. Publ. IV (7): 15 p.

Thorne, J. P., and D. W. Thorne. 1951. Irrigation waters of Utah. Utah Agr. Exp. Sta. Bull. 346. 63 p.

US Department of the Interior. 1963. Quality of water. Upper Colorado River Basin, Prog. Rep. p. 41.

US Salinity Laboratory Staff. 1954. Diagnosis and improvement of saline and alkali soils. US Dep. Agr. Handbook 60. 160 p.

Wayman, C. H., and J. B. Robertson. 1963. Biodegradation of surfactants in synthetic detergents under aerobic and anaerobic conditions at 10C. Art. 120 in US Geol. Surv. Prof. Pap. 475–C. p. C–224–C–227.

Wilcox, L. V. 1948. The quality of water for irrigation use. US Dep. Agr. Tech. Bull. 1962. 40 p.

Wilcox, L. V. 1955. Classification and use of irrigation waters. US Dep. Agr. Circ. no. 969. 19 p.

Wilcox, L. V. 1959. Effect of industrial wastes on water for irrigation use. Amer. Soc. Test. Mater. Symp. Technical Development in the Handling and Utilization of Water and Industrial Waste Water. Spec. Tech. Publ. No. 273. p. 58–64.

Wilcox, L. V. 1960. Boron injury to plants. US Dep. Agr. Inform. Bull. no. 211. 7 p.

Wilcox, L. V., George Y. Blair, and C. A. Bower. 1954. Effect of bicarbonates on the suitability of water for irrigation. Soil Sci. 77:259–266.

section IV

Selection of Land for Irrigation

IV

10 | Selection and Classification of Irrigable Land

JOHN T. MALETIC[1]

*Bureau of Reclamation, US Department of the Interior
Denver, Colorado*

T. B. HUTCHINGS[1]

*Soil Conservation Service, USDA
Salt Lake City, Utah*

I. INTRODUCTION

The formulation of water resource projects is the process of selecting and evaluating the purposes to be served, the physical means of development, the size of facilities, and the area to be benefited. The process is directed toward efficient accomplishment of socially valid objectives of people. In the USA plans are usually selected to achieve greater economic efficiency, so that benefits may exceed costs by the maximum amount. At other times and places other social objectives may be more important. Different plans are needed when social goals include maximum employment opportunities, development of a self-sustaining dietary balance, economic growth, or redistribution of income. These varying objectives will be derived from the system of social organization, knowledge and beliefs, and the system of values within a given society.

The selection of lands for irrigation encompasses social, economic, and physical factors. Interdisciplinary work is involved. Cooperating groups include the basic disciplines of soil science, agronomy, engineering, hydrology, geomorphology, and sociology; the decision-making disciplines of economics and law; and the normative disciplines which test and evaluate, including logic and ethics. Value judgments arise in the choice of irrigable or nonirrigable and in further defining within the irrigable set the degree of suitability for irrigation use. The physical factors of climate, soil, topography, and drainage are, therefore, but a part of the set of all factors determining suitability of land for irrigation use.

Irrigation causes profound changes in land resources. As amount, quality, time, and place of water use change, the chemical, physical, and biological state of soil is altered and concomitant changes occur in the substrata beyond the solum. These shifts may be exploitative or conservational; the former results in loss of productivity, while the latter maintains or improves productivity of land resources. Improvement of the social well-being of people is a prime objective of irrigation. Over the long run this can be best accomplished by conservational use. In the project planning process, conservational use of land resources is equated to long-run social well-being.

[1] The authors acknowledge comments and suggestions by H. L. Parkinson, W. B. Peters, and R. J. Winger, Jr., colleagues in the Bureau of Reclamation.

II. BASIC PRINCIPLES

In developing irrigation projects land and water resources should be efficiently combined by an engineering and settlement plan that best meets defined, realistic, and attainable goals of people. Land classification, as a systematic procedure for delineating lands on the basis of suitability for irrigation use, provides a sound basis for fitting land resources into a plan of irrigation development. The landscape complex and the broad variety of economic, social, and institutional factors render impractical the specification of a rigid system. A land classification system is, therefore, required which can be adjusted to fit the environment of individual projects. The structure of such a system can be founded upon certain basic principles. These may be identified as the principles of prediction, economic correlation, arability–irrigability analysis, and permanent–changeable factors. In addition, the system should conform to the traditional principles that govern classification, i.e., some one aspect of the facts should be selected and adhered to for the entire classification, the classification should be exhaustive, and the subdivisions in the classification should be mutually exclusive.

A. Prediction

Lands selected for irrigation should be permanently productive under the change in physical regime anticipated under irrigation. The land classes must, therefore, express predictions of future soil–water–crop interactions. The introduction of irrigation shifts the natural balance established over time between water, land, vegetation, fauna, and man. Irrigation project planning, therefore, identifies and evaluates the changes and the plans are formulated to assure that a successful permanent agriculture will result.

Irrigation induces change in the physical, chemical, and biological characteristics of land. Many of the changes are interrelated and complex. Soil structure may be modified by salinization, increase in exchangeable sodium percentage (ESP), loss or gain in organic matter, and alteration of clay minerals. Minor changes in texture occur with translocation and movement of clay downward in the profile, while major changes result if the irrigation water carries a large suspended sediment load. These actions will then alter pore space configuration, inducing changes in water conductance.

Important chemical changes occur in the composition and concentration of dissolved constituents in the soil solution. Simultaneously, new cation equilibrium levels are reached between the exchange and solution phases. Dissolution or precipitation of calcium sulfate and calcium carbonate can occur, changing the concentration and distribution of these compounds in the soil. Cation-exchange capacity (CEC) may be altered by increased weathering, loss or gain of organic matter, or the addition of a suspended sediment load. With increased moisture levels, the redox potential will change, generally in the direction of limiting oxidation processes.

The irrigation water may introduce phytotoxic ions such as boron, fluoride, and lithium. Conversely, toxic ions such as selenium that are harmful to animals and man may be removed by cultivation following irrigation.

Both microrelief and the soil profile can be considerably altered by land forming. Profile modification techniques such as deep plowing, chiseling, hardpan shattering, rock removal, brush and tree removal, and the addition of amendments cause drastic changes in the characteristics and qualities of soils. Under particular substrata situations, irrigation may cause subsidence of the land surface, influencing design of the project and farm distribution systems. Irrigation also introduces new erosion hazards, and, depending on the control exercised, may modify both the microrelief and the soil.

The direction and extent of changes in land resulting from irrigation are determined largely by farm practices and the water regime. The dynamic equilibrium of groundwater and the relation of water input to output are major determinants of the permanence of irrigation agriculture. The chemical changes in salinity and ESP levels are largely controlled by these factors. It is an essential consequence of the prediction principle that land classification survey must, therefore, deal not only with the soil but the substrata conditions as well. A normal arid soil of seemingly high productive capacity can be readily transformed into a nonproductive, saline, saline-sodic, or nonsaline-sodic soil through failure to provide adequate drainage.

From consideration of the multiplicity of changes in land that can result from irrigation, land classifications organize and synthesize facts concerning such parameters as (i) drainage requirements, (ii) equilibrium salinity levels, (iii) equilibrium ESP levels, (iv) water requirements, (v) soil productivity following land forming and expected soil profile modification practices, (vi) crop production inputs and outputs, (vii) anticipated land use and management practices, (viii) chemical suitability of the water supply, (ix) quality of return flow, (x) flood hazard, and (xi) soil erosion.

B. Economic Correlation

The formulation of plans for water resource development projects is directed toward providing the best use or combination of uses of water and related land resources to meet foreseeable short-term and long-term needs by means of engineering works and nonstructural measures. The plans are designed to alter the amount, time, and place of use for water resources as a means of improving future conditions in the area affected by the project. To derive a meaningful basis for fitting land resources into a plan, provision must be made, therefore, for correlating physical land factors with economics. This can be achieved under the principle that farm income and farm cost in a given ecologic and economic environment are uniquely related to physical land factors under a given level of managerial ability and technological development. From this principle, important corollaries may be derived: (i) Land classes are defined as economic entities; (ii) relevant and mappable land characteristics are chosen at a given time and place, to comprise the set of land class determining factors; and (iii) land class differentiae and their individual range and value within a class will vary with the economic, ecological, technological, and institutional factors prevailing or expected to prevail in the area.

As will be shown in chapter 11, input–output functions provide a means for achieving correlation between the physical and economic factors. From such functions there can be derived a suitable economic parameter for definition of land class. Thus, on Federal reclamation projects land class is defined as a category of land having similar physical and economic characteristics which affect the suitability of the land for irrigation and which express a relative range in payment capacity (Bur. Reclam., 1953). The economic parameter of payment capacity is then defined as the residual available to defray the cost of water after all other costs have been met by the farm operator. Under this concept a classification model can be constructed which takes the form:

$$Y = -a + bX_1 - cX_2 - dX_3 \qquad [10\text{-}1]$$

where

Y = payment capacity (dollars),

X_1 = productivity rating (per cent),

X_2 = land development cost (dollars),

X_3 = farm drainage costs (dollars), and

a, b, c, and d = constants derived from farm budget analysis.

Through the model and farm budget analyses it is possible to isolate the physical factors for field mapping and to retain economic relevancy of results. The farm budgets integrate into the land classification the influences of institutional factors, managerial levels, farm practices, cost–price relationships, markets, social conditions, climate, and other factors that bear on the input–output relationship.

Depending on purposes to be served by the land classification, other economic parameters may be chosen to define land class. On the Canadian Prairie Provinces, net farm income is used as a measure of the producing ability of various physical categories of land (Can. Dep. Agr., 1964). For large-scale water development projects the economic bases of the land classification system should contribute toward determining whether irrigation is feasible for increasing net income, how irrigation might be planned to maximize net benefits, and the important interrelationships of investment feasibility and optimum water use as set out in chapter 11.

Working through a correlation model, relevant land characteristics can be chosen at a given time and place as the set of land class determining factors. This set will include soil, topography, and drainage. The number and range in value of the individual characteristics within the set will vary between areas to reflect the impact of the total project environment. The characteristics and their values are chosen and rules are specified for interactions among the characteristics under the assumed management practices so that class boundaries coincide with the chosen economic parameters defining land class. This involves synthesis and interpretation of facts aimed at the singular purpose of selecting lands suitable for irrigation. In this process, appraisals are made of soil productivity in relation to the land development costs and costs of production.

C. Permanent and Changeable Factor

The land changes that arise from irrigation development impose a need to identify characteristics that will remain without major change and those which will be significantly altered. This permits construction of a consistent basis for defining land class levels. Whether given characteristics can be changed will usually depend on economic considerations; farm planning and investment decisions are involved which can be analyzed by the principles of irrigation cost maximization and feasibility described in chapter 11. Typical permanent factors are soil texture; depth to sand, gravel, and cobble or bedrock; depth to lime zone, clay pans, and hardpans; and macrorelief. Changeable factors typically include salinity levels, ESP levels, microrelief, water table levels, flood hazard, brush and tree cover, and stone content.

Land classification deals with two aspects of this principle: (i) Can the change be physically accomplished? and (ii) what degree of change is economically feasible? This will largely depend on the economic setting of the project. For example, a large investment may be made to reclaim a saline-sodic soil which after improvement would yield a net farm income of \$200/acre. In another climatic and economic setting where net income after improvement would only be \$30/acre, a soil having similar saline-sodic conditions would be regarded as non-irrigable. In the latter case it would be infeasible to make the change. Moreover, in the same project area the permanent and changeable sets are not mutually exclusive. Thus, microrelief may be altered by land forming where the soil has suitable depth and uniformity; the same microrelief should not be altered if exposure of sand and gravel would reduce productivity below economically feasible production limits. During land classification survey these decisions can be guided by the physical-economic correlation model established for the project.

Under some conditions, improvements will occur in soil characteristics without a specific investment for accomplishing the change. With favorable drainage, water of good quality, and water use practices, a saline area may be reclaimed as a natural consequence of irrigation. Under such circumstances, the initial state of the soil characteristics is less pertinent to the final designation of land class.

D. Arability-Irrigability Analyses

The land classification must initially identify the land areas having adequate productivity to warrant consideration of that land for irrigation. For convenience in project analyses, such lands may be defined as arable. The condition that productivity is adequate, as measured by the economic parameter chosen to define land class, is not sufficient to determine whether the arable land should be included in the plan. In this context, irrigable land is defined as that portion of the arable land included in a specific plan of development. The selection of irrigable land is made through the plan formulation process.

Meaningful analyses of large-scale water development projects can be made when land classes represent appropriate economic entities. This facilitates application of the principles of optimum design or scale, net benefit maximization, and other criteria used in formulating single- and multiple-purpose projects. These principles are described in chapter 11. Through the formulation process, water is allocated to irrigation, hydroelectric power, municipal and industrial water supplies, and other project purposes. The irrigable area is thus selected in relation to the water allocated to irrigation and to the size and location of the distribution and drainage systems.

Fundamentally, then, the selection of lands for irrigation is a two-step process: (i) Selecting an arable area as guided by farm production economics, and (ii) selecting the irrigable area as guided by the economics of plan formulation. The steps are interrelated and often complex, requiring close interdisciplinary cooperation.

The application of plan formulation criteria generally leads to successive elimination of identifiable increments of arable land from the plan of development.

Typical adjustments include (i) The elimination of uneconomic increments such as those that are too costly to serve, drain, or provide distribution works; (ii) conformance of land area to the available water supply; (iii) elimination of tracts located above water surface delivery elevations; (iv) exclusion of isolated segments, odd-shaped tracts, and severed areas that cannot efficiently be fitted into a farm unit; (v) deletion of proposed public rights-of-way; and (vi) elimination of areas unable to meet minimal criteria for economic returns under the plan.

In establishing criteria for land classes, the land of lowest quality that may be defined as arable and consequently as irrigable needs to be carefully specified. The goals of project development will guide selection of this minimum. On Federal reclamation projects, the minimum is prescribed by law which states that irrigable lands shall be classified with respect to their power under a proper agricultural program to support a farm family and pay water charges (Bur. Reclam., 1924). Accordingly, the minimal quality lands have been defined as those capable of supporting a farm family and paying at least the annual operation, maintenance, and replacement costs expected to prevail when the project is placed into operation. With land classes defined in terms of payment capacity, the minimal land quality level is reached when the operation, maintenance, and replacement charge is equal to payment capacity. Where payment capacity exceeds the charge, the residual is identified as the amortization capacity available to pay capital costs allocated to the irrigation function.

Selection of lands on Federal irrigation projects is thus seen to be fundamentally guided by economic criteria requiring the selected irrigable lands to (i) be included in a plan having a favorable benefit–cost relationship; (ii) have sufficient amortization capacity to pay assigned construction charges; and (iii) have ability to meet anticipated annual operation, maintenance, and replacement charges. Under other goals and institutional factors, different criteria for selection would need to be applied.

III. PHYSICAL FACTORS

Climate, soil, topography, and drainage are the basic physical factors to be considered in selecting irrigable lands. These factors operate upon the choice— irrigable or not irrigable—through a set of guiding criteria aimed at achieving defined social goals at a given time and place. No single set of physical characteristics and their limiting values can therefore be specified that would universally define arable and irrigable land.

The physical characteristics may be regarded as independent variables whose interactions under an irrigation regime are to be equated to an economic parameter. In a given climatic setting:

$$E = f(S,T,D) \qquad [10–2]$$

where E = economic parameter, S = soil characteristics, D = drainage characteristics, and T = topographic characteristics. In this equation E represents a suitably chosen farm output or net residual essential in formulating irrigation project plans. The factors S, T, and D are everywhere considered but the individual characteristics of each, such as texture, structure, horizon ararngement, depth, salinity and alkalinity of S, micro- and macrorelief of T, and surface and subsurface drainage of D, are selected on the basis of relevance to prediction of E at the given time or place. For land classification purposes the quality of land for irrigation use can then be indicated by land classes that represent specified meaningful ranges in the value of E.

The functional relationship expressed by equation [10–2] may be approximately solved through use of input–output functions. For example, the variable E may be equated to Y of equation [10–1]. When this is done the influence of S, T, D, expressed in physical characteristics, may be translated into a dollar economic effect in terms of X_1, X_2, and X_3. Such an integration of economic and physical factors provides a necessary conceptual foundation for a finding of arability. The integration involves a difficult synthesis of many individual characteristics, both measured and observed, to arrive at an inferred level of quality for irrigation use. In field practice the physical factors are related to the anticipated production, cost of production, and cost of land development under defined systems of management and use.

Kellogg (1961) describes two ways for deriving the anticipated production: (i) By induction from knowledge of the interactions involved among soil characteristics, the needs of the crops, and the management practices; and (ii) by empirical observation of yields of crops produced on the soil under the specified management. For irrigation project planning the latter method is widely applied by joint studies of soil scientists and agricultural economists in fully developed irrigated areas having physical and climatic conditions similar to the area under investigation. Where irrigation experience is not available for measurement and study, recourse is made to research, operation of demonstration farms, or subjective procedures to derive the needed data. Production cost data are assembled from field studies and economic research reports applicable to the area. Land development costs are similarly collected from field studies. As a guide to field mapping, development costs can be estimated by making paper layouts of farm distribution systems including grading requirements on areas typifying the prevailing topographic features of the landscape.

Although bases for selecting irrigable land may be critically defined, the complex of interactions among the physical characteristics themselves and with management systems imposes limitations upon the attainable precision involved in fitting the economic and physical factors into land classes of defined range in the parameter E. Mathematical models such as equation [10–1] serve basically as guides in performing the synthesis indicated by equation [10–2]. The imprecision involved in application needs to be accepted; otherwise, the basic economic principles guiding choice of lands would be replaced with wholly arbitrary physical limits for land class differentiae. In this area a critical need exists for expanded interdisciplinary research aimed at improved quantification of economic land classification models. Studies involving applications of farm budgets, linear programing, and production functions appear as fruitful avenues of approach. Examples of research in this area include work by James (S. C. James, 1961. Techniques for characterizing Oregon soils for agricultural purposes in terms of physical and economic productivities. *Unpubl. Thesis. Oregon State College*, Corvallis, Oreg.) and Miller (S. F. Miller, 1965. An investigation of alternative methods of valuing irrigation water. *Unpubl. Thesis. Oregon State University*, Corvallis, Oreg.).

A. Climate

Climate exerts important influences on the selection of lands for irrigation. The character of the soil, drainage conditions, distribution of native vegetation, and crop adaptation are strongly related to climate. To a lesser extent climate also influences the relief of the land surface. These factors affect needs of the area, type of plan formulated, design of facilities, and economic impacts from irrigation. As a result, the lands selected to be irrigated will have different characteristics and qualities in different climate settings.

Interactions of climate, land, economic, and social factors operating over time express themselves in broad patterns of irrigation farm types (Dominy, 1964). While climate, and to a lesser extent land, renders stability to these farm-type patterns, economic and social factors cause dynamic pattern shifts over time. Thus, the physical environmental factors determine what will or will not grow, while economic and social factors determine what is grown. A fundamental consideration in planning an irrigation project therefore involves deriving a proper set of assumptions regarding the cropping pattern and management systems. This choice affects (i) quality of lands suitable for irrigation, (ii) water requirements and irrigation efficiency, (iii) design and capacity of the project water distribution system, (iv) selection of farm irrigation methods, (v) design and capacity of the on-farm irrigation system, (vi) surface and subsurface drainage requirements, (vii) minimum depth to which the groundwater table is to be controlled, (viii) leaching requirement and salt balance, (ix) farm inputs and outputs, and (x) project feasibility and justification. Considerable data, analyses, and sound judgment are therefore required to fit the project to the climatic setting.

Where a wide range of crop adaptability exists, the selection of a cropping pattern needs to be made in relation to possible future shifts in the pattern. Allowances for such contingencies are particularly needed in designing the distribution and drainage systems. Also, boundaries between climatic zones shift positions from year to year; hence, the boundaries represent a mean location for individual climatic years. Thus, the variability of climate from year to year is of significant importance in making judgments as to cropping patterns and system design.

The effects of climate require that the characteristics defining arable land in any given area be adjusted to fit that area. For example, the intensity of storms on the Gulf Coastal Plain, USA, limits arability to lands occurring on slopes of < 3%. In contrast, apples *(Malus sylvestris)* can be grown under irrigation on slopes up to 35% where a low intensity of rainfall occurs as in the Pacific Northwest. Similarly precipitation patterns on the Great Plains generally limit the maximum slope at about 6%. To achieve adequate control of salt levels and provide a well aerated root zone, water tables in areas of high evaporation rates devoted to citrus production, such as the Coachella Valley, USA, should be controlled at depths of about 6 ft. On the northern Great Plains, with much lower evaporation rates, field crops can be grown with dynamic equilibrium water table levels reaching a maximum height of 4 ft. In cool climates devoted to production of hay and pasture, fluctuating water tables may reach as high as 1.5 to 2.0 ft provided the groundwater has a low total soluble salt content.

The climatic influences on the farm input–output relationship have considerable effect on the quality of lands included in irrigation projects. In areas producing high income crops such as citrus, vegetables, and cotton (*Gossypium herbaceum*), expenditures as high as $550/acre can be made to develop the farm for irrigation. In areas producing general field crops similar expenditures would probably not exceed $250, while areas growing hay, pasture, and small grains would have limits of about $100/acre. (These dollar values are for Federal Reclamation Project areas and would be different for private development having different institutional constraints.)

In general, as climate favors higher farm income, greater expenditure can be made for land forming, farm distribution systems, leaching salt and exchangeable sodium, profile modification practices, and farm surface and subsurface drainage. When considered in terms of land class determining factors such as uneven microrelief; soil texture, structure, and depth; ESP and soluble salt levels; permeability of substrata; and depth to groundwater barrier, then more severe deficiencies involving such factors can be tolerated in climates favoring high farm incomes than in those favoring lower incomes.

Within the broad regional climatic patterns important localized influences occur. The average climatic conditions on individual farms or small tracts may be different from those of the region mainly because of slope and exposure. In selecting irrigable lands, this factor can be of considerable importance. In areas to be devoted to fruit production, land should be differentiated with respect to frost hazard. This is typical practice in selecting irrigable lands on projects in the Pacific Northwest where the cropping pattern is dominated by fruit. Here frost hazard lands are intricately associated with lands rarely experiencing frost. In regions where experience is lacking to guide such differentiation, detailed studies of the microclimate may be necessary. Mason (1958) describes such a study in which lands are identified as frost-free, frost-rare, intermediate, frost-liable, and frost-subject, all occurring within an area of 10 square miles.

Even though climate is recognized as having a direct influence on land class, no universal methods or standards have been developed to proportionately weigh this factor in a land classification system. Hutchings[2] and Arkley and Ulrich (1962) have used the Thornthwaite (1948) climatic classification and calculated actual, potential, and frost-free period evapotranspiration values to approximate climatic influences for land capability classes, subclasses, and units.

B. Soil

There exists in any project area a set of soil attributes that may be equated to sustained irrigation productivity. The set will be conditioned by the amount and quality of water available for irrigation and the topographic and drainage conditions expected to prevail under irrigation. The attributes include morphological, chemical, and physical characteristics and related performance qualities of the soil. The morphological characteristics include texture; depth to bedrock, hardpan,

[2] Hutchings, T. B., 1954. Arizona heat and moisture indexes for use in land capability classification. US Dep. Agr.-Soil Cons. Serv. mimeo M–521:1–41. Colorado heat and moisture indexes for use in land capability classification. US Dep. Agr.-Soil Cons. Serv. mimeo M–522:1–41. New Mexico heat and moisture indexes for use in land capability classification. US Dep. Agr.-Soil Cons. Serv. mimeo M–523:1–52. Utah heat and moisture indexes for use in land capability classification. US Dep. Agr.-Soil Cons. Serv. mimeo M–524:1–32. Hutchings, T. B., 1955. Heat and moisture indexes for use in the capability classification (Idaho). US Dep. Agr.-Soil Cons. Serv. mimeo, Boise, Idaho. 25 p.

sand, gravel, caliche, or other root-limiting influences; structure, including the shape and arrangement, the sizes, and the distinctness and durability of the visible aggregates or peds; consistence; color and mottling; kind and amount of coarse fragments; and kind, thickness, and sequence of horizons. The foregoing are measured or observed in the field.

The predictive aspects of selecting irrigable lands further requires a need for laboratory measurements of many chemical and physical characteristics. These typically include particle size distribution, bulk specific gravity, porosity, clay mineralogy, surface area, composition and concentration of the soil solution, cation-exchange capacity, exchangeable cations, surface charge density, soil reaction, gypsum, alkaline-earth carbonates, and organic matter.

The performance qualities include fertility, productivity, erodibility, and drainability of the solum. Such performance qualities may be inferred from the soil characteristics. We may also identify as performance qualities the infiltration rate, hydraulic conductivity, moisture characteristic curve, and moisture-holding capacity. The latter are typically measured and related to the soil characteristics to provide a means for extending appraisals from a test site or location to the total natural soil body typifying the test site conditions. Measurements are also made to confirm field observations of soil structure stability. For this purpose hydraulic conductivity of fragmented samples, settling volume, dispersion rates, air-water permeability ratio, and aggregate stability may be used. For special studies, factors influencing unsaturated flow such as capillary conductivity and diffusivity are measured or computed. Also, toxic ions such as boron, lithium, and selenium are measured where problems with such ions may be anticipated.

The position of the water table, piezometric pressures, barrier depths, specific yield, transmissibility of aquifer, and other physical as well as chemical attributes of the substrata are observed and measured in selecting irrigable lands. These attributes of the substrata are of equal importance to the soil attributes in making judgments regarding the response of land to irrigation. They will be discussed under the drainage factor.

The land classification is directed towards identification of soil bodies and their placement in defined classes ranked according to irrigation suitability. This involves a synthesis of soil characteristics and qualities. The synthesis is difficult and complex and cannot be avoided in any meaningful classification. Kellogg (1961) describes this process as one occurring in overlapping segments. Inferences are drawn from the soil characteristics to derive soil qualities such as productivity, fertility, tilth, and drainability which are regarded as limited interpretations. Of these the broader concept of productivity is most important. Soil productivity is defined by the Soil Survey Staff (1951) as that quality of a soil that summarizes its potential for producing specified plants or sequences of plants under defined sets of management practices. It is measured in terms of output in relation to inputs for a specific kind of soil under physically defined systems of management. In segments the productivity level is derived from a synthesis of soil fertility, tilth, and anticipated water relations. The concept of productivity and selection of a productivity level of soil are essential to the solution of equation [10–2] or of specific production functions such as equation [10–1].

We have further seen that the limiting value for characteristics and qualities used as class differentiae need to be selected and defined specifically for a given area in harmony with project development goals. Usually, however, soils suitable for sustained irrigation use in the Western USA have permeable profiles with field-measured hydraulic conductivities ranging from about 0.05 to 5.0 inches/hr, textural classes ranging from loamy sand to friable clay, cation-exchange capacities (CEC) > 3 meq/100 g, depths to root-limiting influences varying from 12 inches to 60 inches or more, water-holding capacities varying from 0.75 to 3 inches/ft of depth, salinity levels at equilibrium with the irrigation water at 8 mmho/cm or less, and equilibrium ESP not exceeding 15%. In addition, the dynamic equilibrium position of the water table is controlled by drains or naturally remains at depths that preclude severe oxygen deficits and the accumulation of soluble salts or exchangeable sodium in the root zone.

These boundary values of attributes typifying irrigable soils depend on a complex of agronomic, economic, ecologic, and technologic factors. Unusual situations arise. Soils to be used to grow salt-tolerant crops such as cotton and bermudagrass (*Cynodon dactylon*) can have higher equilibrium salt levels than the value given. Conversely, soils to be used for citrus require lower salinity equilibrium values. Soils whose clay mineral composition consists principally of illite or other non-expanding clay minerals can have ESP levels > 15% before sufficiently adverse physical conditions arise to declare them nonirrigable. Also, the electrolyte content of the irrigation water is an important factor in the relationship between ESP and permeability (Quirk and Schofield, 1955; Gardner et al., 1959; Reeve and Brown, 1960). Thus, water quality influences the choice of the limiting value for ESP. Some clay soils can have hydraulic conductivities as measured in place as low as 0.005 inches/hr and still adequately produce under irrigation. On the other hand, very sandy soils, through use of specially adapted irrigation methods, can be successfully irrigated even though the hydraulic conductivity exceeds 5.0 inches/hr. Similarly successful production has been achieved under irrigation with soils having a water holding capacity of about 0.60 inches/ft.

The CEC of some soils, particularly those developed in the tropics, will be < 3 meq/100 g and yet with properly selected and applied management systems they may produce well. Conversely, good fruit production on soils in the Pacific Northwest usually occurs on soils having CEC > 7 meq/100 g. Rice (*Oryza sativa*) production on tropical soils requires a special set of class differentiae, particularly for paddy production. The problem here is complicated because many changes occur in the soil characteristics following continuous submergence. These few illustrations show that each landscape considered for irrigation development presents a unique classification challenge. The selection and definition of class differentiae or specifications for mapping is therefore one of the most important aspects of the irrigable land selection process. In certain circumstances the development of necessary and sufficient specifications requires much more time and effort than performance of the field classification.

The Victoria clay, a grumusol developed on the Gulf Coast Prairie of Texas, USA, is a striking example of an unusual soil condition requiring careful and detailed analysis to derive a finding of arability. Various properties of this soil are given in Table 10–1 (Bureau of Reclamation, 1963. Texas Basins Project, land classification annexes. *US Dep. Int.-Bur. Reclam. mimeo*) and the morphology and a hypothesis of its genesis are given by Kunze et al. (1963). The hydraulic conductivities of the subsoil as measured in place is as low as 0.003 inches/hr while the ESP exceeds 15%. Detailed investigations (Bureau of Reclamation, 1963. Texas Basins Project, land classification annexes. *US Dep. Int.-Bur. Reclam. mimeo*)

Table 10–1. Attributes of the Victoria clay*

Depth	Description	ECs × 10³	pH 1-5	HC (d)†	CEC	ESP	Sat. per- cent.	Sand	Silt	Clay
inches		mmho/ cm		in/ hr	meq/ 100g		%	%	%	%
0-12	Dark gray-brown clay, granular, sli. moist	1.0	8.5	0.14	34	7.4	73.3	27.7	26.2	46.1
12-20	Dark gray-brown clay, angular blocky	1.4	8.5	0.08	37	8.4	76.2	26.1	26.9	47.0
20-36	Light gray-brown clay, subangular blocky, plastic	2.6	8.8	0.01	39	15.1	76.9	22.4	28.0	49.6
36-48	Pale brown clay, plastic, sticky	4.0	9.1	trace	34	26.8	86.1	18.5	27.1	54.4
48-60	Pale brown clay, plastic, sticky	4.5	8.9	0.001	31	30.3	82.8	20.2	27.3	52.5
60-84	Pale brown clay, plastic, sticky, salt crystals, gypsum	6.7	8.7	0.03	31	26.7	84.7	20.4	27.7	51.9
84-102	Pale brown clay, dense, compact, salt and gypsum crystals	6.0	8.9	0.002	31	28.6	82.1	17.8	32.2	50.0
102-120	Pale brown clay, dense compact, salt and gypsum crystals	8.0	8.2	0.02	30	24.4	75.6	19.7	33.7	46.6

* In-place hydraulic conductivity at 30- to 40-inch depth 0.006 inches/hr and at 70- to 120-inch depth 0.03 inches/hr. † Hydraulic conductivity of fragmented sample.

show that this and similar soils have been successfully irrigated on the Gulf Coast Prairie of Texas. Cotton (*Gossypium herbaceum*), sorghum (*Sorghum vulgare*), and winter vegetables are the principal crops grown. The average annual rainfall of about 30 inches/year and the use of water with a low salinity level has resulted in favorable salinity conditions in the root zone. Movement of water through the profile and substrata occurs primarily through unsaturated flow in amounts suffi- cient to maintain a favorable salt balance in the root zone. The depth to the water table is 50 ft or more. Water enters these soils mainly as a result of cracks. The cracking occurs at sufficiently high moisture levels to preclude development of highly adverse soil water stress prior to irrigation.

While favorable water relationships for crop growth can be maintained in the Victoria clay, this may not be so in other fine-textured soils. Stirk (1954) studied a group of Australian soils whose clay fraction measured as the sizes < 0.002 mm ranged from 45% to 78%. He concluded that the cracking of many clay soils of massive structure cannot be relied on to compensate for their low permeability and poor aeration, if it is intended to include such soils in areas for irrigation development. He found the influence of the cracks to be greatest when the soil water content was reduced to levels which are not always practicable with con- tinuous irrigation culture.

Extremely sandy soils occur on the Yuma Mesa of the Gila Project in Arizona, USA. Detailed investigations of the moisture relationships of the soil indicate that among other factors the major determinant of arability was the waterholding capacity of the soil. Soil attributes of developed areas of known productivity were directly compared with similar virgin lands to establish the land class differentiae (Bureau of Reclamation, 1949. Land classification report, Unit 1, Yuma Mesa Div. of the Gila Project, Arizona, Boulder City, Nevada. *US Dep. Int.-Bur. Reclam.*

Table 10–2. Estimates of available water held in Ephrata soil 4 days after irrigation as compared with estimates based on the ⅓-bar percentage

Soil depth	Estimates of available water	
	1/3 to 15-bar %	Field values, 15 bar %
inches		
0–6	0.79	1.45
6–12	0.69	1.76
12–18	0.61	1.36*
18–24	0.41	1.16*
	2.50	5.73

* Field values at 12 to 18 and 18 to 24 inches estimated from desorption curve because rocks prevented accurate sampling (data of Miller, 1964).

mimeo). The limiting value for arable land was found to be 2.5 inches of available water in the upper 4 ft—or an average of 0.625 inches/ft. To apply this limiting value, the matric suction of the soils was correlated to water-holding capacity. Field mapping was accomplished by collecting soil samples from borings made to depths of 4 or 5 ft or more within a grid of 100 holes/square mile. The moisture retention values were determined by the suction plate technique [Richards and Weaver, 1944; (K. R. Goodwin and M. N. Langley, 1946. Procedure for determining moisture retention of soils as used in land classification. *US Dep. Int.-Bur. Reclam. mimeo. Gila Project, Arizona*)]. Land class boundaries were then drawn as guided by the field and laboratory appraisals on base maps having a scale 200 feet equals 1 inch. Subsequent development of the Yuma Mesa has aptly demonstrated the efficiency of such careful and detailed laboratory and field studies in selecting the irrigable lands.

Where soils and substrata are of uniform texture with no abrupt change in physical properties of the substrata then laboratory measurements of soil water relations as made on the Yuma Mesa yield reliable results. On stratified soils, however, field measurements of the water relations are essential for reliable class appraisals. Miller and Bunger (1963) and Miller (1964) have studied the influences of coarse layers on the water retention properties of stratified soils. They found that soil water was retained at much lower suction in soil underlain by layers of sand and gravel than in similar but nonlayered soil. The water content of the soil over the layers was correspondingly greater than that in nonlayered soil. Miller compared estimates of water held in several stratified soils derived by use of the difference between the 1/3-bar percentage and the 15-bar percentage and field values and the 15-bar percentage. The results are given in Table 10–2.

Where stratified soils are involved it is thus seen that field measurements provide a more reliable means for selecting class limits in a particular area. Such field tests need to be evaluated in relation to the anticipated water table level under irrigation. Where the water table is expected to reach the bottom of the root zone, then water and matric suction levels will be different from the field tests conducted under conditions involving a deep water table.

The design of an irrigation system depends heavily on the water retention properties of the soils. The land classification and soil surveys should provide data needed to guide choice of proper design. In this area close teamwork among disciplines is involved. Soil scientists can identify the soil water characteristics and depict the areal patterns of significant groupings. To this may be matched the farm practices and associated land development and production costs through cooperative efforts with agricultural economists. Water requirements, irrigation efficiencies, and system capacities can then be developed by hydrologists and the results translated into a system design by layout engineers. On the same basis the deep percolation losses are considered by the drainage engineers and they accordingly derive a design for the drainage system to adequately control water levels. Concurrent consideration is given to the leaching requirement.

The development of salt-affected soils is an ever present threat to the maintenance of a permanently productive irrigation agriculture. The predictive aspects of land classification survey must deal squarely with this matter. In field work very close coordination of the soil and drainage aspects are thus required. A number of considerations are involved. Among these are the drainability of the substrata, quality of water, ability of the soil to transmit the needed leaching requirement, exchangeable sodium, soluble salt levels, cropping patterns, and management systems all conditioned by the climatic setting of the project.

In selecting irrigable lands the salt problem arises in two ways. First, the salt-affected soils may be initially present with investigations directed toward evaluating practicability of making the lands permanently productive. Second, in their initial state the soils are normal and the investigations are directed toward evaluating means to retain or improve productivity.

With respect to the initially salt-affected soils two questions arise in the selection process: (i) Can the soils be permanently reclaimed? and (ii) are the costs economic for the time and place? Answers to these questions are sought in different ways. The initial step is to locate and delineate the salt-affected soils and to identify their specific characteristics and qualities. Reliable diagnostic techniques (e.g. Richards, 1954) are available for this purpose and should be critically applied in the land classification and soil survey. Following diagnosis the salt-affected tracts are appraised to evaluate possibilities of reclamation and the associated costs. This may be done by conducting field experiments, studying irrigation experience on similar soils, and in many situations conducting simple field leaching tests as part of the classification process. Considerable cost is involved in such studies; therefore various techniques have been developed to screen the soils by subjecting soil samples to a progressive series of tests until the salt problem is specifically identified.

The development of a salt leaching curve through simple field tests is at present a very effective and practical procedure for evaluating reclamation possibilities. This involves developing the relationship between the percentage of initial salt remaining in the soil for various profile depts as related to depth of leaching water per unit depth of soil. An example of such a leaching curve as given by Reeve (1957) is shown in Fig. 10–1. The technique initially applied to field plots by Reeve et al. (1955) was modified by Hulsbos and Boumans (1960) who used two concentric cylinders with diameters of 50 and 70 cm, respectively, as used for infiltration rate measurements. Soil sampling is done before and after the infiltration rate determination. Measuring the volume of infiltrated water and the electrical conductivity of the saturation extract on the soil samples provides the requisite parameters for constructing the leaching curve. Replicate determinations at the test site are usually made. Water quality will influence test results; therefore, the water used should typify the water quality to be made available for irrigation use. The shape of the leaching curve provides a basis for making

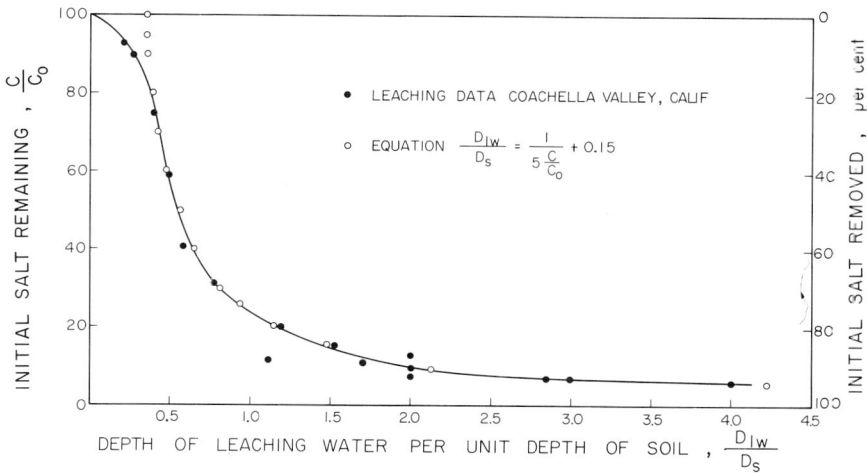

Fig. 10–1. Depth of water per unit depth of soil required to leach a highly saline soil, Coachella Valley, California. (Reeve, et al., 1955)

judgments regarding inclusion and exclusion of land from the plan of development. Concurrent with the development of the leaching curve, the soil samples can also be analyzed to determine ESP levels before and after leaching. This provides a basis for appraising possibilities of reclaiming salt-affected soils. If adequate reductions in ESP levels do not occur, it may then be necessary to conduct properly designed field experiments before deciding to include the lands in the project plan.

Soils in a given potential project area may have seemingly ideal attributes for irrigation development but yet their development may lead to disappointing or socially catastrophic failures. Kelley (1964), in a recent review, traces some of the physical factors responsible for failure. Appraisal of the irrigability of the soil thus goes beyond consideration of the soil in its nonirrigated state. The changes imposed by a new water regimen require special consideration of quality of soil in relation to quality of water and drainage conditions under anticipated project operating conditions.

The quality of water to be available for irrigation should be evaluated as part of the irrigable land selection process. Water quality sampling networks are generally established early in the investigation and conducted as needed through the operating stage of the project. Various parameters are measured, such as total dissolved solids, sodium, calcium, magnesium, potassium, carbonate, bicarbonate, sulfate, chloride, nitrate, boron, silicon, iron, pH, and suspended solids. The frequency of sampling, parameters measured at each sampling, and the location of sampling sites are matched to needs in any particular area. If necessary, data are also obtained on the quality of the groundwater. For applications to project planning, measurement of water quality needs to be accompanied by measurement of the volume of flow. This permits establishing correlation between total concentration and discharge or between individual ions and discharge. Using such relationships, reservoir operation studies can be performed to derive estimates of the mean water quality to be made available from the reservoir and the fluctuations in water quality occurring over time.

The process of estimating the available water quality is not clear cut and straight forward. When reservoir operation studies are made, assumptions regarding the size of the service area are used. However, size of the service area may depend on the quality of water available. Thus several stages of approximations may be involved as the soils of the area are fitted to the water quality.

In irrigation project planning it is not desirable to base the suitability of water supply for irrigation on general water quality rating schemes. Studies of soil and water conditions expected to prevail should be made as part of the land classification field studies. This may be illustrated by studies conducted on the Shadehill Unit, Missouri River Basin Project, located near Lemmon, South Dakota, USA. Here the irrigation water supply to be made available from the Shadehill Reservoir on the Grand River would have a sodium adsorption ratio (SAR) varying between 6 and 15, a total salinity concentration varying from 400 to 1,400 ppm, and a residual sodium carbonate (RSC) content varying from 3 to 7 meq/liter. To determine the influence of the water on the lands to be irrigated, changes in ESP levels of the soils were followed for a 9-year period under various irrigated cropping systems adapted to the area (Shadehill Development Farm Annual Report, 1962. Missouri River Basin Project, Shadehill Unit, Grand Div., Missouri-Oahe Projects. *US Dep. Int.-Bur. Reclam. and South Dakota State Coll. mimeo*). The ESP values of the 0- to 15-cm depth for individual years as measured in the experiments were correlated to the mean water quality used in that particular year.

The correlations consisted of determining appropriate constants for the ESP–SAR regression (Richards, 1954) statistical estimate of Vanselow's (1932) mass-action equation, and the double-layer equation (Eriksson, 1952; Bower, 1961). The ESP–SAR regression was determined to be:

$$\text{ESP} = \frac{100\,(-0.0579 + 0.0155\,\text{SAR})}{1 + (-0.0579 + 0.0155\,\text{SAR})} \qquad [10\text{--}3]$$

where ESP is the exchangeable sodium percentage and SAR is the sodium adsorption ratio of the irrigation water.

The statistical estimate of the mass-action equation was:

$$\text{ESP} = \left[\frac{(-7.7 \times 10^{-3}\,P) + (3.97 \times 10^{-4})}{(3.9923\,P) + (3.97 \times 10^{-4})}\right]^{\frac{1}{2}} \qquad [10\text{--}4]$$

where P is the ratio $(Ca+Mg)/(Na)^2$ as determined from concentrations of the cations in the irrigation water in mmole/liter.

The double-layer equation as applied was:

$$\text{ESP} = 100\left[\frac{r}{1.86\,(1.06 \times 10^{15})^{\frac{1}{2}}\,\Gamma} \sinh^{-1} \frac{1.86\,(1.06 \times 10^{15})^{\frac{1}{2}}\Gamma}{r + 4\,(Ca + Mg)^{\frac{1}{2}}}\right] \qquad [10\text{--}5]$$

where

$\Gamma = $ the surface charge density of the soil in meq/cm², and

$r = Na/(Ca + Mg)^{\frac{1}{2}}\,(mole/liter)^{\frac{1}{2}}$,

Na, Ca, and Mg are the concentrations of sodium, calcium, and magnesium in the irrigation water, and Ca+Mg are similarly the concentrations of calcium plus magnesium in the irrigation water in mole/liter. The value 1.86 was derived by making a least square fit of the equation to field data. Bower (1961) defines this value as a correction factor for Γ which is related to the radii of the specific cations

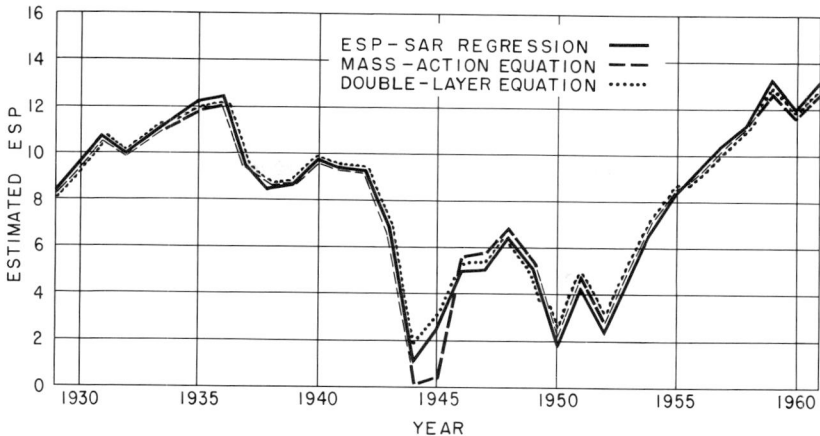

Fig. 10–2. Shadehill Unit—Predicted exchangeable sodium percentage, 0- to 15-cm depth.

in the irrigation water. Various values have been found for this factor depending upon the system studied. Bolt (1955) derived a value 1.2, Bower (1959) found a value of 1.4 to be applicable to four soils and two clays, and Pratt and Blair (1964) found the factor to range from 1.15 to 1.52.

Corollary studies showed that for the volumes of irrigation water applied, equilibrium in the 0- to 15-cm depth was attained by the end of each irrigation season. Reservoir operation studies for the historic period 1929–1961 were then conducted to identify the fluctuations of water quality resulting with time. From this study, the mean quality of the April through September releases, representing the irrigation season, was derived and used in the cation-exchange equations to obtain the predicted ESP of the soils under irrigation. The results obtained are given in Fig. 10–2. The convergence of evidence from the three equations each having different theoretical bases suggest a permanent irrigation agriculture could be maintained with the available water if used on soils comparable to or better than those on which the experiments were performed. Selection of irrigable land then proceeded by application of a land classification survey utilizing specifically adapted land class differentiae based on attributes of the soil at the experimental site. The chosen class differentiae are given in Table 10–6.

For many applications, equilibrium ESP levels can be satisfactorily estimated by using available gross correlations between ESP and SAR of irrigation waters such as given by Bower, 1961 (Fig. 2, p. 218) or by direct application of the double-layer equation using a correction factor of 1.4. Pratt and Grover (1964) found that in soils with small amounts of organic matter and dominated by montmorillonite and illite clays the general regression equations relating ESP to SAR are useful. Soils with large organic matter levels dominated by kaolinite and amphorous clay fit a different but equally useful regression of ESP to SAR, while soils containing mixtures should fall between these groups. In field practice, bicarbonate precipitation may have important influences and these may be initially appraised as described by Bower (1961). Additional study may be required in some situations, as illustrated by the Shadehill Unit example, to derive the influence of this factor.

Long-term salinity equilibrium levels can be estimated by application of the leaching requirement (LR) concept (Richards, 1954). Characteristically in the project planning process estimates of deep percolation losses are made and the drainage requirement developed. The LR is then compared to determine whether it equals or exceeds the deep percolation losses. When it appears that less than the LR may be transmitted, then additional investigations will be needed. These might involve developing engineering provisions in the plan to improve salinity levels of the water supply, adjusting cropping patterns, eliminating soils unable to meet the LR, or reducing land area to permit use of more water for leaching.

In many cases it will be quite apparent from initial appraisal of water analyses in relation to the soil, climate, drainage, and cropping conditions expected to prevail that the land of a potential project area can readily be adapted to the use of available water supply. In other cases, one or more of the particular aspects of the conditions involved will require careful investigation.

Improvements in the efficiency of the process for selecting irrigable lands are needed. These include better applications of existing theories regarding behavior of the soil-water-salt-plant system, and the development of new techniques particularly as related to the dynamics of the system. The needs have been reviewed by Moodie et al. (1964) and include (i) definition of criteria for adequate ground-water control; (ii) improved understanding of the mechanisms and tolerance of plants to salinity; (iii) critical processes and properties of the soil-water-salt system including nonequilibrium cation exchange equations, mechanisms of salt precipitation and solution in the soil, biochemical reactions, capillary pressure distributions as related to the rate of flow of water and salt, and clay mineral and surface phenomenon; (iv) dynamic behavior of the soil-water-salt system; and (v) improvement of instrumentation with emphasis on salt measurement on soils in place.

C. Topography

The principal topographic attributes determining suitability of land for irrigation are degree of slope, relief, and position. These factors influence land development needs and costs, design of on-farm water conveyance systems, erosion hazards, crop adaptability, drainage requirements, water use practices, and selection of management systems. The topographic attributes are correlated with soil and drainage conditions in selecting the irrigable lands. The quality of land for irrigation is then derived by appraising the correlated effect of topography on productivity, cost of land development, and production costs. Size and shape of the areas are usually considered as part of the topographic factor because of their common relationship to land development and production costs.

1. SLOPE

The effect of slope on the suitability of land for irrigation is primarily influenced by the crops to be grown, the duration and intensity of rainfall, erodibility of the soil, farm income potential, and the method of irrigation to be used. In selecting irrigable lands, an initial decision needs to be made as to whether or not the slope conditions should be modified by landforming. That is, whether the slope factor is to be treated as permanent or changeable. If the dominant slope is to remain, then the land is appraised for irrigation suitability primarily on the basis of productivity and labor inputs. In this case, an increase in per cent or complexity of

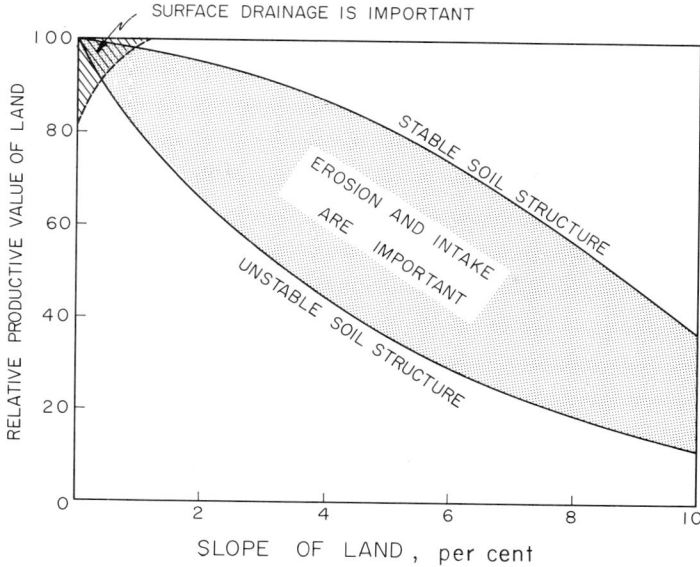

Fig. 10–3. Effect of slope on the relative productive value of land. (from Harris and Hansen, 1958)

slope generally results in a decrease of the quality of the land for irrigation. The decrease will vary with the irrigation method, being particularly pronounced with gravity methods and somewhat subdued under the sprinkler method. As slope increases, shifts in cropping patterns from row to close-growing crops are frequently made. In general, the row crops produce higher net incomes than the close-growing crops. Hence, there is a decline in net productivity. Also, the labor costs involved in water management on sloping lands are usually higher than that of nearly level lands. Surface drainage problems frequently arise on lands lacking gradient and these can usually be corrected by land forming and installing surface drainage systems. In areas producing fruit or vegetables, relief and position need to be considered with respect to frost hazard.

Harris and Hansen (1958) find that the relationship between agricultural land value and slope is governed by the relative importance of surface drainage, erosion, and water intake. The relationship is shown in Fig. 10–3. Values for a particular area will, according to Harris and Hansen, lie somewhere within the shaded area depending upon the relative importance of the factor cited. McMartin (1950) describes agricultural economic assumptions made for irrigation investigation in North Dakota, USA. He relates the expected crop yields on various slope groups

Table 10–3. Expected crop yields expressed as per cent of yield
on lands with slopes of 0–2%

Slope	Hay and pasture	Small grains	Row crops
0–2	100	100	100
2–6	100	100	90
6–10	92	90	70

* Assumes similar soil conditions within each slope group (adopted from McMartin, 1950).

as a percentage of the yield on lands having slopes of 0 to 2%. The results derived from the study are given in Table 10–3.

In land classification surveys for irrigation, no universal set of slope groups can be correlated to land class levels. The interactions of the many factors involved necessitate selection and definition of class limits for each particular project. Across the Western USA, the limiting slope value separating irrigable and non-irrigable land ranges from 35% in the Pacific Northwest to 3% in the Gulf Coastal Plain. On the Central Great Plains, this limiting value is usually set at 6%. Lands in western Colorado devoted to pasture and hay production are successfully irrigated on slopes as high as 20%. In California avocados (*Persea americana*) and grapes (*Vitis vinifera*) are also irrigated on slopes as high as 20%.

As the complexity of slope increases, there is a successive reduction in the amount of gradient that should be permitted within an irrigable land class. Complexity of slope affects the size and shape of fields, type and number of water control structures needed, size and length of water conveyance systems, and requirements for relift pumps to reach high areas. In some situations involving complex slopes the irrigation problems involved can be most efficiently handled by adopting sprinkler irrigation.

2. LAND DEVELOPMENT

In selecting lands for irrigation the more important land development factors considered are: clearing vegetation such as trees or brush; removing stones, cobble, or outcrops of bedrock; leveling land; terracing land; floating and planing; deep plowing or chiseling; subsoiling and hardpan shattering and perforation; constructing laterals, ditches, borders, dikes, basins, drops, checks, weirs, flumes, and small storage ponds; installing wells, pumps, main farm laterals, pipelines, fixed and portable sprinkler systems, nozzles and rotating devices, farm surface drains, ditch and lateral lining, grassed waterways, inlet and outlet structures, open and closed subsurface drain laterals, and inverted wells; and initial and recurring soil improvement with organic and inorganic amendments and by leaching. The requirements for land development vary from tract to tract. The method of irrigation, value of crops, climatic conditions, volume of water available for irrigation, slope of land, surface irregularities, soil properties, and drainability of the substrata are the principal variables considered in determining costs of land development. The cost will also vary in accordance with preferences of individual farmers. In the selection process, assumptions are therefore made regarding the amount of land development to be performed on a project.

Land development costs are estimated for tracts of land during the land classification survey. The magnitude of these costs are directly related to the suitability of the land for irrigation. Economic evaluations determine the total permissible land development costs that can be incurred in association with various levels of soil productivity. Estimates of land development costs are derived from engineering layouts of farm units on typical topographic situations and results extended to other areas through correlation with observable terrain features.

Land forming costs are commonly the largest single cost item in land development. This cost depends on the method of irrigation and other considerations. In field practice, an estimate of the cubic yards of earth moving or cost of land leveling for each delineation of arable land is made. Such estimates may be supported by an appropriate number of detailed grid studies designed to furnish

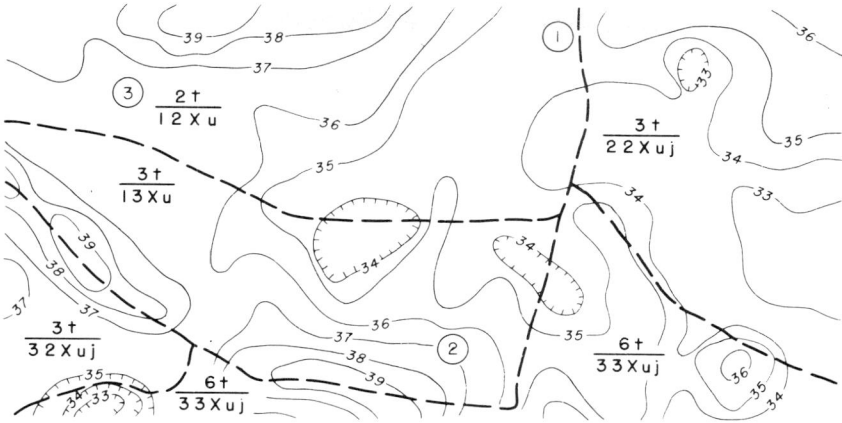

ORIGINAL TOPOGRAPHY WITH LAND CLASSES
$N\frac{1}{2}$ $NE\frac{1}{4}$ SEC. 7, T. 145 N., R. 59 W.

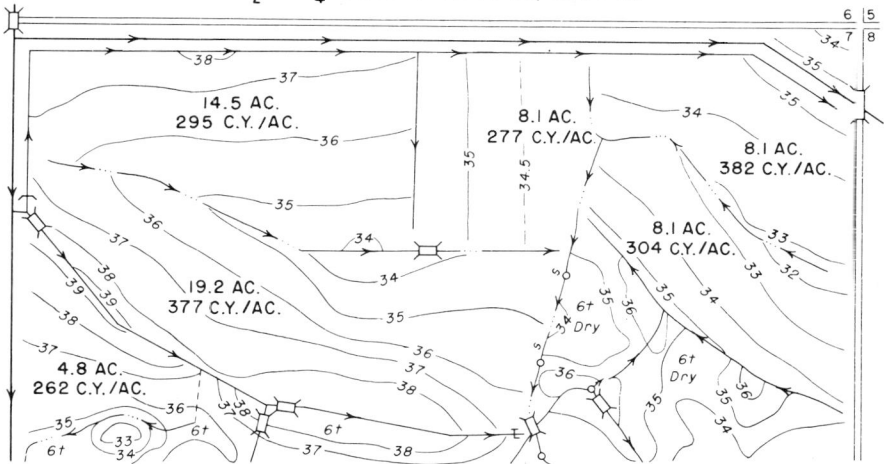

DEVELOPED TOPOGRAPHY
$N\frac{1}{2}$ $NE\frac{1}{4}$ SEC. 7, T. 145 N., R. 59 W.

EXPLANATION

PROJECT LATERAL
PROJECT DRAIN
FARM LATERAL
FARM DRAIN
FARM IRR. STRUCTURES
CULVERTS

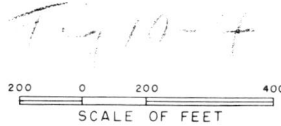

SCALE OF FEET

BUREAU OF RECLAMATION

Fig. 10–4. Comparison of original and developed topography.

land leveling requirements. Standard techniques such as the plane method, the profile method, the plan-inspection method, the contour adjustment method, or the average profile method are commonly used by agricultural engineers to calculate land leveling needs (Soil Cons. Serv., 1959, Ch. 12, Sec. 15; Butler, 1961). Unit costs per cubic yard vary considerably according to the distance of haul, ease of cutting, and the type of equipment used (Marr, 1957). These costs generally range from 12¢ to 20¢/cubic yard.

Reliable topographic maps with appropriate contour intervals are indispensable, as an aid in estimating leveling requirements. Topographic maps are particularly useful on land with low gradient where the land surface may deceive the eye and the microrelief is deceptive. They assist in determining the layout of farm fields; direction of irrigation runs; and appropriate locations for farm laterals, diversion boxes, and drains.

The original topography and land classification depicted in Fig. 10–4 provides an example of the use of topographic maps in selecting irrigable lands. The lower portion of this figure shows the field layout, final contours, cubic yards of leveling required, and the location of ditches, drains, and water control structures after land development.

Selecting the type and amount of land forming and the method of irrigation adapted to a tract has an effect on relative suitability for irrigation, irrigation efficiency, irrigation and farm labor requirements, and on the crop yields which may be anticipated. Physical limitations, such as soil depth, degree of slope, size of irrigation stream, and irregularity of topography are considered along with anticipated cropping patterns to select the methods of irrigation most likely to be used by average farm managers. The maximum costs that can be economically incurred and the selection of the type of land development which may be best adapted to project conditions will be further controlled by the productive level of the land expected to prevail under irrigation. In this regard, during land classification for irrigation, soils are evaluated on the basis of their attributes after, rather than before, land leveling.

In some areas the lands proposed for irrigation are covered with trees or brush at the time of land classification. In such circumstances, land development appraisals should include costs of removal. The primary variables that determine costs of removing small trees or brush are size and type of cover, the number of trees or intensity of brush per acre, and the hourly charges for necessary equipment and labor. Removal costs may vary from a low of $6 to $8/acre for sagebrush to over $200/acre for dense coniferous or deciduous forests (Cotner and Jameson, 1959; Bureau of Reclamation, 1957. Land clearing methods and costs in southwestern Colorado report. *US Dep. Int.-Bur. Reclam. mimeo, Durango, Colo.*). The cost of removing large trees depends on various factors such as size of trees or stumps to be removed, type and depth of root systems, density of stand, soil type, completeness of removal, size of area to be cleared, type and size of equipment, and skill of operator. Accordingly, studies of the local conditions relating to removal costs are usually required prior to classification for irrigation use.

Rock or stone removal cost estimates are somewhat analogous to tree clearing in that local conditions such as type and size of rocks, crops to be grown, and availability of special equipment for rock removal influence the cost estimates. Rock picking machines are available for hire in some areas. These machines usually can remove rocks at costs of about $3/cubic yard. Hand removal costs, depending on size and type of rock, and costs of labor are usually in excess of $5/cubic yard. Land development cost estimates for rock removal are not highly precise because of the inability to accurately estimate the volume of rocks within the plow zone. Various techniques such as excavating small sample areas and piling the rocks or estimating rocks exposed from a plow furrow have been used as a basis for estimating the volume of rock removal required.

Table 10–4. Unit costs for private land development, Cedar Rapids Division, Nebraska

Item	Unit	Unit cost (dollars)
Earthwork	yard^3	0. 20
Floating	acre	2. 25
Seeded farm drainageways, 9. 8 ft^3/sec at 0. 1 grade	foot	0. 066
Seeded waterways		
25 ft^3/sec at 0. 8 grade	foot	0. 25
90 ft^3/sec at 0. 8 grade	foot	0. 50
Farm laterals	foot	0. 0075
Diversion terraces	foot	0. 10
Siphon pipe, 12 inch	foot	2. 00
Culvert pipe, 12 inch	foot	2. 00
Siphon inlet	each	105. 00
Siphon outlet	each	105. 00
Drainage outlet pipe, 18 inch	foot	4. 00
Aluminum pipe, 20-foot length with couplers		
10-inch	foot	2. 40
9-inch	foot	2. 20
8-inch	foot	2. 00
7-inch	foot	1. 70
Aluminum pipe, 20-foot length with couplers and 40-inch gate spacings		
10-inch	foot	2. 70
9-inch	foot	2. 50
8-inch	foot	2. 30
7-inch	foot	
Concrete structures		
36-inch drop	each	95. 00
30-inch drop	each	87. 00
24-inch drop	each	80. 00
18-inch drop	each	73. 00
12-inch drop	each	65. 00
Drop turnout	each	45. 00
Turnout	each	25. 00
Stilling basin	each	60. 00
Division box	each	60. 00

Suitable land development appraisals at the time of detailed land classification impose a need to prepare tentative farm unit layouts on a topographic map. This assists in determining the approximate location and length of farm ditches and surface drains, the need for and location of water and erosion control structures, and to delineate the fields for which land leveling requirements are to be estimated. Unit cost for such development varies considerably, depending on the project setting. Unit costs are therefore typically developed for a given project area. An example of such costs for an irrigation project in Nebraska (Bureau of Reclamation, 1965. Agricultural economics appendix, feasibility report, Cedar Rapids Division, Missouri River Basin Project, Nebraska. *US Dep. Int.-Bur. Reclam. mimeo. Grand Island, Nebr.*) is given in Table 10–4.

Some topographic conditions, in addition to influencing the land development costs, cause a decrease in productivity. For example, small or odd-shaped fields usually have less net returns per acre because of productive area losses for turnrows, ditches, and drains and because the per acre costs for planting, cultivating, irrigating, and harvesting are higher than on large fields. Thus, field size and shape are important factors influencing the quality of land for irrigation.

D. Drainage

Prediction of the drainage requirement is the crucial element in selecting land for irrigation. The adequacy of this prediction and its engineering implementation are among the prime physical determinants of the success of irrigation enterprises.

Under the concept of conservational use, irrigable land must be drainable land. Systems devised for selecting irrigable lands should therefore encompass drainability evaluations. This includes investigating the substrata as well as the soil. Some lands are endowed with adequate natural drainage to sustain irrigation. This, however, needs to be derived from investigations and never simply assumed. Unfortunately, naturally drainable areas occupy but a minor portion of the landscape. Requirements for engineering works to remove excess water and salts will therefore arise in most irrigated areas.

In selecting lands for irrigation both surface and subsurface drainage are considered. Surface drainage is the removal of excess water from the surface of the land. The excess may arise from precipitation, irrigation, losses from conveyance and storage facilities, or seepage from groundwater at a higher elevation. Subsurface drainage is the removal of excess water from within the soil by downward or lateral flow through the soil and substrata . . . it involves control of groundwater levels and the corresponding salt levels in the soil. The source of the water may arise from deep percolation from precipitation or irrigation; leakage from canals, drains, or surface water bodies at higher elevations; or upward leakage from artesian aquifers. Surface and subsurface drainage problems are frequently coincident and inseparable.

Provisions for drainage in project plans are essential to assure sustained productivity and to allow efficiency in farming operations. To derive the provisions, land classification and drainage investigations are usually conducted concurrently. These are further supported by geologic, hydrologic, and economic studies of the project area. Through plan formulation, decisions are then reached regarding justifiable expenditures for drainage works as related to identified land bodies.

Development goals along with the physical and economic setting of the potential project determine the amount of drainage works which can be provided. These factors thus control the quality of land selected for development. For conditions on Federal Reclamation Projects in the Western USA, maximum justifiable expenditures for drainage may vary from about $50 to $700/acre, depending on the project setting. The economic limitation is applied by (i) selecting and defining physical requirements for crop growth in the area, (ii) conducting land classification and drainage investigations to identify areas within which the requirements can be met, and (iii) applying plan formulation criteria involving benefit–cost studies, ability to repay construction allocations, and capacity to meet operation, maintenance, and replacement charges.

1. RECOGNITION OF DRAINAGE PROBLEMS

Existing drainage problems may be identified by careful field observations. The following generally indicate adverse drainage conditions: (i) water standing in topographic depressions for prolonged periods; (ii) occurrence of salt-affected soils with barren surfaces; (iii) soils containing high concentrations of soluble salts in surface layers or having a distinct surface crust; (iv) shallow water table; (v) mottling or presence of gleyed horizons; (vi) crop symptoms of unfavorable drainage such as stunted growth or late maturity, disease, and shallow root development; and (vii) presence of phreatophytic vegetation.

Recognizing a potential drainage problem requires field observations and measurements coupled with careful engineering analyses. Drainage investigations are designed to identify (i) possible sources of excess water, (ii) adequacy of outlets for conducting excess water out of the project area, (iii) capacity of the soil and substrata to conduct water, (iv) volume of excess water to be removed, and (v) design required to achieve most efficient drainage. The investigations are initiated by collecting, reviewing, and analyzing relevant data regarding geology and landforms, soils, topography, well logs, water levels and fluctuations, precipitation and surface flow, and similar factors.

In general a potential drainage problem may exist when (i) soil and substrata have hydraulic conductivities of < 1 inch/hr; (ii) soil or unconsolidated substrata consist of textures of fine sandy loam or finer and contain exchangeable sodium usually > 15%; (iii) soils contain excess soluble salts for the climate under which they developed; (iv) impervious strata of shale or sandstone occur within depths of 10 ft or less and have an unevenly weathered or eroded surface obstructing both lateral and vertical water movement; (v) vertical or nearly vertical barriers occur such as faults, dikes, and intrusions accompanied by topographic and substrata conditions that intercept groundwater movement, constrict flow, and favor high water table conditions; (vi) obstructions to surface drainage occur such as road and railroad embankments; (vii) lands lie adjacent to a large proposed unlined canal, particularly the lower lying lands enclosed by a U-shaped canal alignment; (viii) lands border natural drainage channels such as stream bottoms and low terraces; (ix) lands lie adjacent to lakes or reservoirs whose water elevation may rise sufficiently to unfavorably influence the groundwater levels; (x) tracts are subjected to sidehill runoff or seepage from higher lands; and (xi) irrigation water, groundwater, or both contain an ionic composition that may induce imperviousness in the soil or substrata through chemical reactions. Recognizing such factors assists in determining the type and detail of drainage investigations needed in a particular area to select irrigable lands and design drainage facilities.

2. SURFACE DRAINAGE

Provision for achieving adequate and timely removal of excess surface water is an essential requisite of successful irrigation. To a certain extent land preparation, involving land forming or grading, and proper water management will satisfy a large part of the surface drainage requirement. With the furrow, corrugation, and border methods of irrigation some excess surface runoff is usually unavoidable. Therefore, a surface drainage outlet is needed for most irrigated farms. The outlet may be a natural drainage course or a constructed drain. The farm waste water is conducted to the outlet by shallow or minor channels and waterways that collect the excess water from the lower ends of fields, terraces, and terminals of irrigation ditches. The farm surface drains are often a few feet or less in depth and comparatively inexpensive to construct. Provision for reuse of the waste water may be made on some farms.

In selecting land for irrigation, surface drainage influences land quality by its effect on the investment cost required to install remedial measures on the farm. Since costs for surface drains are relatively inexpensive, this factor is seldom limiting. When recognized as a land class determining factor, surface drainage requirements can be described in terms of cost and related to design and length of the ditch. For example, on a particular project the class level with respect to surface drainage may be described as:

"On farm surface drainage not to exceed cost of $14/acre; 225 linear feet, maximum 1 foot deep, 6-foot bottom, 3:1 side slope; or 85 linear feet, maximum 2 feet deep, 6-foot bottom, 3:1 side slope."

Lands of lower quality would have correspondingly larger investment costs as determined by the design and length of channel needed to properly remove excess water from the farm or tract of land being classified.

Planning irrigation projects will also include designing a surface drainage system. Generally such systems will assure a drainage outlet for each farm unit. Also protective works for canal and lateral distribution systems are designed to prevent damage from storm runoff. Such protective works serve a dual function by also preventing flood damage to the lands. In some cases runoff from sidehills may be so severe that special works such as detention dams will need to be provided before lands can be declared irrigable. Close integration of the planning and design of project surface drainage systems and canal protective works is necessary for critical selection of irrigable lands.

3. SUBSURFACE DRAINAGE

In selecting lands for irrigation, subsurface drainage investigations are conducted as a separate but closely coordinated function with the land classification survey. The procedures and techniques applied in conducting the investigations are matched to the soil, topographic, geologic, and hydrologic setting of the project area. To establish investigative needs, data are initially attained regarding such factors as location and condition of outlets and natural drainage ways; general geology, particularly the stratigraphy; areal distribution of landforms; general water table levels; anticipated cropping patterns and practices; degree of land preparation to be accomplished, especially as related to irrigation efficiency; topographic features; history and influences of flooding; quality of irrigation and groundwater; soil conditions; and drainage practices and experience in the area under study or a similar area.

With such a background of data available, a program of additional investigations can be laid out. This may involve identifying the thickness, position, continuity, and characteristics of the strata; specific soil conditions including profile descriptions and measurements of infiltration rate, permeability, and salt status; depth of barrier layers; sources of excess water from precipitation, deep percolation from irrigation, seepage, hydrostatic pressure from artesian aquifers, or a combination of these sources; and groundwater conditions including position, extent, fluctuation, direction of movement, and areas of discharge. Details regarding techniques for such investigations as applied to irrigated lands are given by the Soil Conservation Service (Soil Cons. Serv., 1959, Ch. 2, Sec. 16 and Ch. 3, Sec. 16) and the Bureau of Reclamation (Bureau of Reclamation, 1964. Land drainage techniques and standards. *Reclam. Instr. US Dep. Int.-Bur. Reclam. Ser. 520;* and Bureau of Reclamation, 1964. Groundwater engineering techniques and standards. *Reclam. Instr. US Dep. Int.-Bur. Reclam. Ser. 530).*

The principal objective of the investigation is to delineate the land areas that should receive irrigation water and to further develop within such areas the design and cost of essential drainage works needed to support an irrigated agriculture.

In planning large projects, the design and costs are usually developed for identifiable project segments thereby allowing application of plan formulation criteria. This permits an economic choice to be made regarding the maximum permissible drainage costs that may be used in a given project situation. Such studies provide the basis for final selection of irrigable lands.

A major factor influencing the choice of irrigable lands is the spacing of drains needed to sustain irrigated agriculture. The spacing may be derived by applying the transient flow theory [Dumm, 1964, 1954; Dumm and Winger, 1964; and (L. D. Dumm, 1962. Drain-spacing method used by the Bureau of Reclamation. *Paper presented at US Dep. Agr.-Soil Cons. Serv. drainage workshop, Riverside, California*)] or the steady-state theory (Donnan, 1947; Donnan et al. 1954). A general application of the steady-state theory to land classification is proposed by Fly (1961). His technique provides a general basis for initial recognition of the extent of a subsurface drainage problem. Its application would be limited to simple landscapes, where parallel relief drains could be used. Because of the assumptions made in defining the drainage classes, the proposed technique could perhaps be best applied by deriving different relations between barrier depths and permeability based on assumptions applicable to a specific area.

Applications of drainage spacing formulae require data on the depth of soil above a barrier to vertical water movement, permeability of the soil in the profile, thickness of the root zone above the maximum allowable water table, specific yield of the soil and substrata between the bottom of the drain and the maximum allowable water table, and the quantity of water that will percolate below the root zone. Barrier location, permeability, and specific yield are measured during field investigations. For the drain spacing calculations, lateral hydraulic conductivities are measured. Many barriers can be located by direct field observations. On the other hand, in stratified, unconsolidated materials vertical hydraulic conductivities may need to be measured to select an appropriate barrier depth. In-place tests are preferred to laboratory measurements. Techniques for making such measurements are described in publications by Winger (1960) and the Bureau of Reclamation (Bureau of Reclamation, 1964. Land drainage techniques and standards. Reclamation Instructions, US Dep. Int. Bur. Reclam. Ser. 520). The specific yield may be estimated from a curve relating specific yield and hydraulic conductivity (Winger, 1960).

The maximum allowable water table depth is selected on the basis of crops to be grown, evaporative conditions, salt levels of the groundwater, capillary conductivity of soil, length of growing season, amount and distribution of rainfall, and related agronomic considerations. Generally the maximum dynamic equilibrium depth may range from 1.5 to 6 ft as discussed under the section on climate in part III on physical factors. Based on current, incomplete knowledge, it appears that for most field crops and conditions a 4-ft depth is satisfactory. Only meager research data are available regarding the effects of fluctuating water tables on crop yields. A concerted research effort on this problem is needed to identify not only the effects of the varying water levels on yield but also the effects generated by various levels of irrigation and groundwater quality.

Selection of the amount of deep percolation is an equally complex problem. This value, in addition to being influenced by the skill of the irrigator, depends upon the types of crops to be grown, soil, climate, irrigation efficiency, and the frequency and amount of irrigation. Also special consideration needs to be given to evaluating the LR to assure that the amount of deep percolation is at least adequate to meet the LR. In practice this amount is estimated through consideration of cropping patterns, soil attributes, consumptive use, water extraction pattern, effective rainfall, and frequency and depth of irrigations. Additional research is needed to provide more critical bases for deriving the deep percolation losses.

Predicting the dynamic equilibrium position of the water table is an important aspect of the irrigable land selection process. It provides a basis for not only evaluating drain spacing but also indicates and helps select areas where drainage may not be a problem. Procedures for estimating such groundwater levels are given by Dumm (1964) and Maasland (1959, 1961). Where irregular patterns of recharge from irrigation and rainfall are involved, Dumm's incremental step procedure can be applied. The method deals with variations in recharge that arise from both precipitation and irrigation. In Maasland's model a regular distribution of recharge is assumed and recharge during the nonirrigation season must be small compared to that during the irrigation season. In general, satisfactory predictions can be made with these models when working with simple topographic landscapes having well defined boundary conditions. The models are not adaptable to strongly rolling landscapes having complex natural drainage systems, or on broad, gently sloping alluvial fans where boundary conditions are difficult to identify. Under conditions of complex topography, it is sometimes helpful to assume that the water table will rise to the root zone and then apply Dumm's method to estimate the drain spacing needed to prevent the water table from rising above the selected maximum allowable water table depth.

Selecting irrigable lands is thus seen to involve many observations, measurements, and analyses. A portion of these activities can be integrated into land classification surveys. Observation and logging of the characteristics of the soil and substrata can be made during the field survey. Usual practice in detailed surveys involves describing the soil and substrata to depths of 10 ft or more. By making careful descriptions at test sites where measurements are made, a basis becomes available for extending results to areas having the same or similar characteristics. The critical nature of hydraulic conductivity values in drainage studies makes it desirable that descriptions of the soil and unconsolidated substrata include factors that influence this property such as texture, structure, consistence, color, layering, salinity, ESP, cleavage planes, visible pores, and bulk density. In addition to developing requisite field data, the land classification survey should provide a basis for excluding lands that are not capable of transmitting the essential LR beyond the root zone. This may involve development of a minimum standard to be met by all land designated as irrigable. See for example the land characteristic standard for hydraulic conductivity given in Table 10–5.

The concept of a landform provides a very useful and practical basis for organizing and conducting drainage investigations and land classification surveys on irrigation projects. Several different definitions and classifications of landforms have been devised for engineering use (Belcher, 1948; McLerran, 1957; Lueder, 1959). For selecting irrigable lands and deriving drainage requirements, the definition and classification system of Lueder is perhaps most applicable. He defines a unit landform as a terrain feature or terrain habit, usually of third order, created by natural processes in such a way that it may be described and recognized in terms of typical features wherever it may occur, and which, when identified, provides dependable information concerning its own structure and either composition or texture. Some examples of unit landforms are (i) glacial origin—till plain, marginal moraine, and drumlin; (ii) glacio-fluvial origin—outwash plain, terrace, delta, esker, and kame; (iii) fluvial origin—floodplain, terrace, delta, fan, and bajada; (iv) moraine origin—coastal plain and beach; (v) lacustrine—lakebed; and (vi) aeolian—loessial plain and dune. A broad but useful collection of facts are obtained by identifying the unit landform. Thus till plains, particularly when derived from shale, can typically be expected to present serious drainage problems under irrigation. Loessial plains will have highly uniform soils and drainage explorations may be concerned with outlets, buried profiles of fine texture that may be barrier layers and depth to bedrock. Alluvial fans developed over horizontal beds of marine shales with strongly undulating surfaces can be expected to cause serious drainage problems. Very detailed exploration involving development of a depth-to-shale map may be necessary on such a landform. Under dryland conditions excellent soils often develop on such fans but their subsequent irrigation may give rise to expensive drainage problems. Lakebeds underlain by till can be expected to require an intricate system of outlet, collector, and parallel relief drains. On the other hand, river terraces, underlain with coarse sands and gravels frequently give rise to very favorable drainage conditions.

Figures 10–5, 10–6, 10–7, 10–8, 10–9, 10–10 depict, in an idealized way, the relationships of landform and drainage requirements. Conditions of ideal natural drainage are shown on Fig. 10–5. Here a loessial mantle 10 or more feet in thickness overlies highly fractured basalt. A subsurface drainage problem would not develop on such land. These lands have been observed to occur in southeastern Idaho, USA, and in other basalt regions of the Northwest USA.

In eroded and refilled valleys there is always the possibility of shale or clay dikes impeding drainage into the natural drains. Figure 10–6 shows a typical example of adequate natural drainage on one side of a river and impeded drainage because of a shale dike on the other side. The high water table area upslope from the dike could be drained by locating subsurface drains in the more permeable aquifer just upslope of the dike. If the aquifer is more than about 12 ft below the ground surface and the aquifer is permeable, a pumped well, located in the upslope aquifer would control the water table. If there is slowly permeable clay material between the dike and the natural outlet, subsurface drains should be located in the more permeable surface soils above the clay as shown.

Moderately permeable lake silts as shown in Fig. 10–7 seldom have adequate natural subsurface drainage. The narrow river terraces would be drained by the river, but spaced subsurface drains would be required for the flat areas above the terraces.

On wide river terraces of moderately permeable material and underlain by a slowly permeable barrier, the groundwater usually occurs at the toe of each break in slope. Deep open or closed interceptor drains located at the toe of each break, as shown in Fig. 10–8, provide the necessary drawdown to keep the water from surfacing at the break. If the terrace is wide, additional equally spaced drains would be required to control the water table between the interceptors.

Flat valleys between long wide ridges present potential drainage problems. If the valleys are not drained, they eventually develop high water tables and sometimes cannot be used, even for pasture. If the valleys are narrow, a deep drain down the middle usually provides adequate drainage; but for the wider valleys, interceptor drains at the breaks in slope with spaced relief drains between as

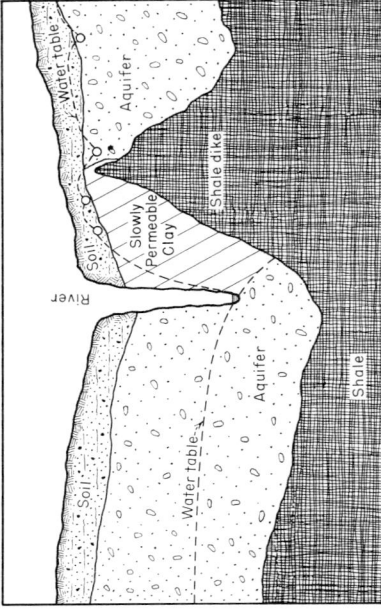

Fig. 10–6. Subsurface drainage when shale dike barrier impedes drainage into deep natural drain.

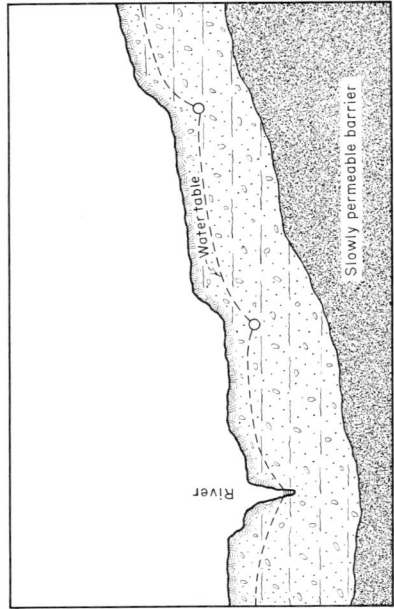

Fig. 10–8. Interceptor drains at breaks in slope for river terraces.

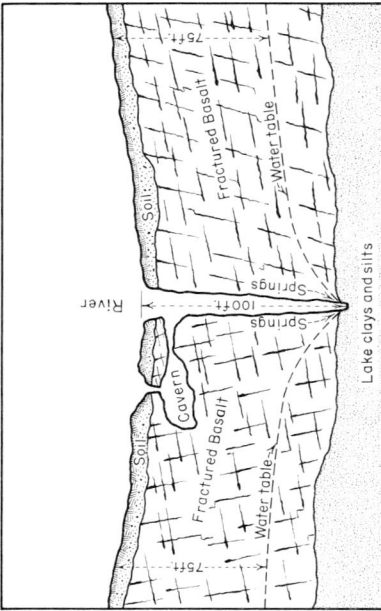

Fig. 10–5. Ideal drainage conditions found in some basalt regions of the Northwest USA.

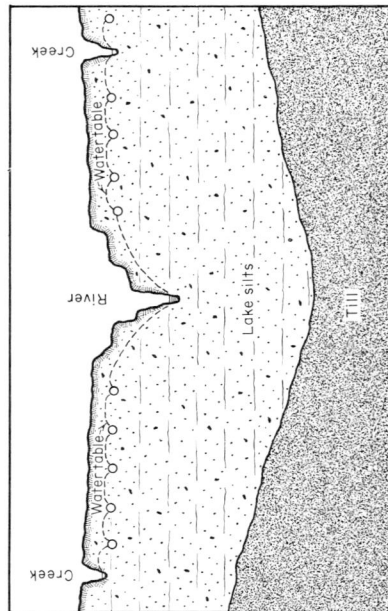

BUREAU OF RECLAMATION

Fig. 10–7. Subsurface drainage in moderately permeable lake silts underlain by a till barrier.

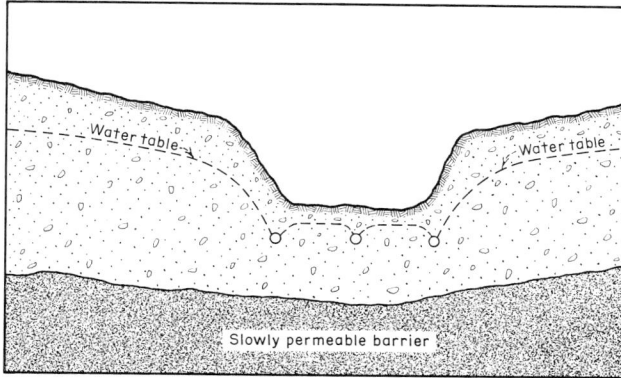

Fig. 10–9. Subsurface drainage for flat valleys between long wide ridges over shallow barrier.

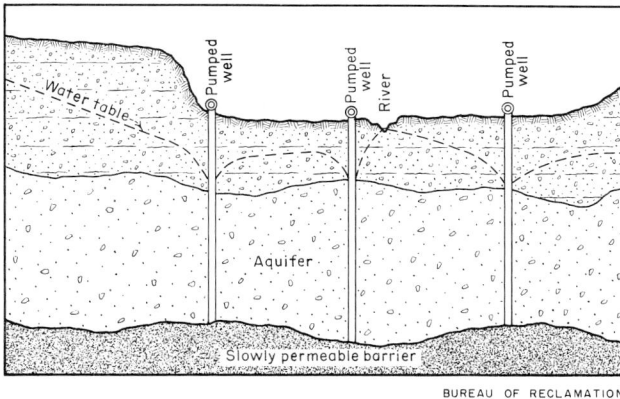

Fig. 10–10. Subsurface drainage by means of pumped wells.

shown in Fig. 10–9 are required. Experience has shown that pipe drains installed before the high water table develops provide the needed drainage even in un-stable materials. If open drains are used, the soil must be stable when saturated.

When an area requiring drainage is underlain by a permeable aquifer, pumping from fully penetrating wells provides adequate and economical drainage as shown by Fig. 10–10. If the aquifer is under pressure, the piezometric water table can be lowered by pumping from the aquifer. However, to lower the perched water table above the piezometric water table, the material over the aquifer must have sufficient vertical permeability to move the required deep percolation below the root zone within a period of 2 or 3 days.

Derivation of drainage requirements involves complex processes that are not fully understood. While much progress has been made in soil physics, hydraulics, and drainage engineering, the prediction techniques are just barely emerging into a quantitative stage. The complexity of the processes involved and the expense and difficulties of measurements below the soil requires, and will continue to require, application of considerable sound judgment and experience.

IV. FIELD METHODS

A. Review of Methods

A variety of concepts and procedures have evolved for selecting lands for irrigation use. All concepts recognize, to a varying degree, the importance of the physical factors. Thus appraisal of soil, topography, and drainage conditions are common to all rational methods of selection. Integration of these factors with the equally important factors of climate, economics, sociology, and engineering is neglected in some, assumed or implied in others, while several explicitly recognize and accomplish such integration. Within the conceptual framework of any scheme wide differences in procedure are also encountered. These differences relate primarily to selection of relevant characteristics and the detail of observation, measurement, and delineation.

A brief survey of land classification for irrigation purposes is given by Jacks (1946). He stresses that one of the natural factors of the environment, moisture, is artificially transformed and the transformation may modify another natural factor, soil. Geographical, economic, and social factors such as land, extent and distribution of land types, accessibility of water supply, markets and transport are recognized as important factors in determining the feasibility of supplying irrigation water to potentially productive land and selecting successful systems of farming.

A soil rating system developed specifically for selecting irrigable lands on the Canadian Prairie is described by Bowser and Moss (1950). Soils are rated on the basis of soil profile, type of geologic deposition, soil texture, salinity, stoniness, wind and water erosion, and topography. Each factor is given a numerical index value. Multiplication of indices provides an overall rating. Based on the rating the soils are placed in five groups ranging from very good to unsuitable for irrigation use. A similar multiple factor index rating system is proposed by Storie (1964).

The principles of planning multiple-purpose water resource developments are described by Sain (1951). He presents a discussion of the type and extent of data required for evaluating arable and irrigable land. Within a series of Reclamation Instructions describing principles and techniques of multiple-purpose project planning (Bur. Reclam., 1953) the method used to select irrigable lands is presented. The conceptual basis for the method involves selecting land capable of adequately supporting a farm family and paying water charges on a sustained basis. The land classification survey is thus directed toward selecting, within a framework of economic, social, and institutional factors, lands suitable for irrigation and organizing the facts required to fit land resources into an integrated plan of development.

Application of a reconnaissance survey in Angola, Africa, employing a numerical rating system based on factors affecting productivity, is described by the Hydrotechnic Corporation (1955). In this scheme an empirical scale of values relates soil texture, depth of soil, slope of land, surface configuration of land, and drainage hazard to the agricultural value of the land under irrigation. The product of the factors provides an overall land value which is then used to place land in four classes ranging from most favorable for irrigated agriculture to unsuitable for irrigated use. Additional developments and improvements of this scheme are presented by Harris and Hansen (1958).

The use and application of natural soil bodies as delineated in basic soil surveys to select lands for irrigation as a part of land classification is described by Maletic (1961). In this approach interpretations of soil survey results are made specifically with respect to irrigation suitability. The interpretations are expressed as land classes as defined by the Bureau of Reclamation (1953) following appropriate integration of results with economics, drainage, hydrologic, and plan formulation factors. Molenaar (1956) broadly outlines the use of soil surveys and land classification surveys for selecting irrigable lands and resolving associated problems of land use. Interpretations of soil and other survey results are recommended as the basis for preparing land use classification surveys. The land classification is intended to show land areas best adapted to various uses from the point of view of sustained production and greatest benefit to the users of the land and water and to the nation.

A reconnaissance survey directed toward selecting irrigable lands in the Lower Mekong Basin, Laos, is described by Marinet et al. (1961). They applied the pedogenic classification of Aubert and Duchaufour (1956) based on the various modes and intensities of evolution of the soils. Using this classification of soils as a base, categories of suitability of land for irrigation were defined. Principal criteria were soil fertility, soil depth, fitness of soil for crop diversification, micro- and macrorelief, variation in groundwater level, and flood hazard. They recognized six land classes. Classes 1, 2, and 3 were considered as areas to which it would be reasonable to limit development; Class 4 lands required further study to determine feasibility of development. Class 5 lands were not recommended for development at present because of cost and limited choice of crops, and Class 6 was land unsuited for development.

Brough et al. (1956) developed a land type concept which involved an integration of soil series into soil groups and land classes as mapped by the Bureau of Reclamation. The land types were reported to be relatively homogenous with respect to agricultural capabilities and management requirements under irrigation farming. They concluded that land types gave more information on specific land conditions than do land classes or soil groups alone.

On the Bow River Project (Can. Dep. Agr., 1960) lands were classified with respect to their irrigation suitability according to probable future net income per acre. Various soil areas were first grouped according to their physical and chemical suitability for sustained production under irrigation based on a soil survey. Subsequent groupings were then made on the basis of topographic factors. Through farm budget studies the net return per acre of the various categories was determined. Finally the physical land categories were grouped according to their relative earning capacities. Four land classes were delineated with Class 4 having the highest and Class 1 the lowest earning capacities.

Wydler (1960) also applied a soil survey interpretative technique to select lands for irrigation in the Rio Negro Lower Valley, Argentina. Soils were first classified as Brown, Sierozem, Sub-Humic Gley, saline and alkali, and sediments. These were then placed into four major categories reflecting the physical suitability of the lands to grow crops under irrigation. The position and movement of groundwater, drainage conditions, and salinity were major determinants in making the irrigation suitability grouping.

Desaunettes (1960) proposes a two-factor classification to select irrigable lands. In this concept the value and use capabilities of a soil for irrigation are assumed to be completely defined by the useful soil depth and water-holding capacity. Five cartographic classes are used in this scheme.

The foregoing review, while not exhaustive, provides a cross section of land classification methods in use. The preferred procedures for water resource development projects explicitly recognize development goals within a defined framework of economic, social, and institutional factors. Such procedures regard land classification survey as one of the principal instruments of plan formulation leading to the allocation of water among selected parcels of land and among project purposes. When properly accomplished, a land classification survey provides a critical basis for evaluating the quality of the resource development and where appropriate for selecting reasonable water charges. It also provides basic data necessary to plan and operate irrigation projects such as the determination of benefits, water requirements, land use, distribution system layout, land development programs, and drainage requirements. Procedures based on poorly defined or arbitrary groupings merely reflecting intuitive judgments of soil productivity do not provide a sound basis for fitting land resources into either single-purpose irrigation or multiple-purpose development plans.

B.   The Standard Soil Survey

The standard soil survey is a system for defining, classifying, and delineating each kind of soil and making predictions of soil behavior under defined management. A standard soil survey is made on a designated area such as a county, a project, or soil conservation district. All the soils within the area are mapped and classified without regard for existing or expected farm boundaries, but each soil is delineated and defined so that the information will be available to design farm field layouts, irrigation, and drainage systems.

Soil is a natural body that supports plants and has properties due to the integrated effect of climate and living matter acting on parent material and conditioned by relief over periods of time. Each kind of soil has its peculiar set of characteristics and qualities.

The morphology expressed in the profile of each soil reflects the combined effects of the relative intensities of the soil forming factors responsible for its development. A soil cannot be defined by any one characteristic, but by combinations of characteristics considered collectively. When accurately defined, a specific soil can be distinguished from all other kinds of soil.

Within the survey area, these unique soil units are recognized and defined and delineated on maps to show location, size, and shape. On high intensity, detailed soil surveys, phases of a soil type are the common mapping unit. They are named, classified, and correlated within the general system of soil nomenclature, classification, and correlation (Soil Surv. Staff, 1951, 1960).

The response of each kind of soil under defined systems of management is appraised. Reasonable alternatives of use and management and the expected returns are determined. Reliable predictions are attained by integrating basic data from field and laboratory research, from field experiments and the experience of farmers. The decisions among alternative combinations of management practices are based on inputs and outputs with expected costs and prices used to calculate returns.

The basic data from the standard soil survey can be interpreted in many different ways to provide either primary or supplemental information for specific needs. Each interpretation should be designed for its unique purpose with the greatest possible simplicity without loss of needed exactness. When this is done, reliable predictions of soil behavior become useful tools in land classification, land appraisal, tax assessment, hydrologic estimates, planning and zoning, recreation developments, wildlife developments, and soil engineering appraisals in addition to the many agricultural groupings to show relative productivity, intake rates, permeabilities, etc. Interpretations of standard detailed soil surveys are made to select irrigable lands.

When available prior to the initiation of a land classification survey, the soil survey provides a clear definition of the natural soil bodies potentially available for irrigation service. Careful study of the areal pattern of soil bodies, landforms, and geology provides a basis for determining detail of additional studies to be performed, types of problems to be encountered, potential cropping patterns, selection of sampling sites, and intensity of laboratory analyses required. Organization of field work is thus enhanced and cost of performing land classification surveys are reduced. In addition, the soil survey provides useful data regarding dryland areas adjoining the proposed irrigation development. Such areas are frequently integrated into and profoundly influenced by the irrigation project.

During land classification field work the soil survey provides a basis for remembering and correlating many characteristics and qualities, thereby facilitating appraisal of land class level however defined. Following irrigation development many soil changes occur. Productive soils may become unproductive and conversely unproductive soils may be made productive. The soil survey assists in assessing these changes. Research on identified natural soil bodies provides a land classifier with an important tool for making the multiple and frequently complex predictions essential to sound irrigation development. Research scientists should remain ever mindful of these practical needs by providing explicit descriptions of the particular natural soil body on which their research is performed.

Critical applications of soil survey to the selection of irrigable lands requires careful synthesis and interpretation. The boundaries of natural soil bodies will rarely coincide with class boundaries ranking land for irrigation suitability because many factors external to soil are relevant to such interpretation. As Kellogg (1961) has stated, the location, size of tract, and other economic characteristics of land are highly significant in land classification. Since areas of kinds of soil have natural boundaries and are commonly found in contrasting economic environments, then a grouping of like soils may place together areas with unlike economic characteristics. Where soil surveys are made prior to the planning of an irrigation scheme, then the mapping legend can provide for recognition of phases to facilitate interpretation and provide a higher degree of correlation between the soil survey and the land classification survey. The fundamental requirement in selecting irrigable lands is to define, for the time and place, what is to constitute a finding of irrigability and then to establish principles and procedures for consistent application of that definition. In this process, the natural soil bodies, because of their high information content, provide a critical basis for deriving essential predictions.

C. The Irrigation Suitability Classification

The Irrigation Suitability Classification of the Bureau of Reclamation is designed to select lands for irrigation (Bur. Reclam., 1953) under laws and policies governing water resources developments in the Western USA. In this system, land class is defined as a category of lands having similar physical and economic attributes which affect the suitability of land for irrigation; it is an expression of the relative level of payment capacity. The amount of money remaining to the farm operator after all costs, except water charges, have been met and after an allowance has been made for family living, is identified as the payment capacity. The economic and physical factors are correlated through the relationship of soil, topographic, and drainage factors to productive capacity, cost of production, and land development costs for a given project setting.

Class 1 has the highest level of irrigation suitability, hence the highest payment capacity. Class 2 has intermediate suitability and payment capacity. Class 3 has the lowest suitability and payment capacity. Class 4 designates special use classes such as 4F fruit, or it is used to designate land with excessive deficiencies which special engineering and economic studies have shown to be irrigable. Class 5 is used as a temporary designation for lands requiring special studies before a final land class designation can be made, and Class 6 is land not suitable for irrigation development.

Subclasses result from appending the letters "s" for soils, "t" for topography, and "d" for drainage to indicate the reason areas are placed in a class lower than 1. The subclasses of the land classes are s, t, d, st, sd, td, and std.

Present land use is designated for application in subsequent economic analyses. Broad groupings are used such as "L" for nonirrigated cultivated, "P" nonirrigated permanent grassland, "G" brush or timber and "C" irrigated cultivated.

Informative appraisals are made to provide additional useful information regarding the land resource for use by economists, hydrologists, and engineers concerned with development of the plan. This involves an appraisal of the interaction of productivity and cost of production, and land development costs. A numeral 1, 2, 3, 4, and 6 designation is used for these appraisals which relate to the appropriate land class level. The interaction of these factors determines the land class. A farm water requirement appraisal is sometimes made using the letters A, B, and C to indicate low, medium, and high requirements, respectively. Also drainage appraisals are made usually relating to the 5- to 10-foot depth of the soil and substrata. The letters X, Y, and Z are used to indicate good, restricted, and poor drainability conditions in the 5- to 10-foot zone. Additional informative appraisals are made such as "g" for slope, "u" for undulations, "f" for flooding,

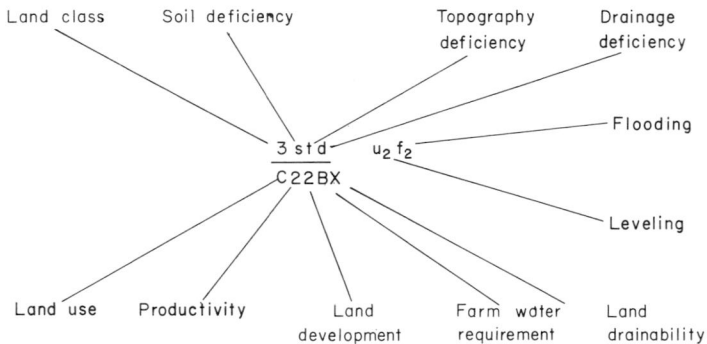

Fig. 10–11. Example of the mapping symbol used in the Irrigation Suitability Classification (Bur. Reclam., 1953).

"k" for shallow depth to sand, gravel, cobble, and so forth. All the symbols are combined in the mapping unit designation as shown in Fig. 10–11. With this method of symbolization it is essential to hold the extent of appraisals and delineations to the minimum in the interests of efficiency and economy. The informative appraisals are made in full consideration of purposes to be served; irrelevant factors are not symbolized or shown.

The land classes in the system are not everywhere the same. For each project, land class determining factors are selected and identified consonant with the prevailing climatic and economic setting of the project. Tables 10–5, 10–6, 10–7, and 10–8 provide examples of such definitions as used in the USA. The specifications for the Oahe Unit given in Table 10–5 were developed for conditions on an ancient glacial lake plain in South Dakota where subsurface drainage is critical. The Shadehill Unit specifications (Table 10–6) were prepared for an area to be irrigated with a water supply whose SAR will vary between 6 and 15, salinity will vary from 400 to 1,400 ppm, and residual sodium carbonate (RSC) will vary between 3 to 7 meq/liter. The Kokee Project specifications (Table 10–7) were developed for a project in Hawaii where sugarcane (*Saccharum officinarum*) is expected to be the dominant crop. The Chief Joseph Dam Project specifications (Table 10–8) indicate adjustments made for an area in the Pacific Northwest to be devoted to apple production.

The sequence and dependency of actions followed to select irrigable lands are portrayed in the flow network shown in Fig. 10–12.

The Bureau of Reclamation system of classification cannot and should not be directly applied by following a set of general land class determining factors. Each potential project setting presents its particular land classification challenge. Land classification surveys should therefore be designed, and land classes defined to meet development goals and economic requirements existing within the physical setting of the project.

D. The Land Capability Classification

Efficient irrigation requires the right combinations of soil–water–crops and management. The integration of these factors must take place on the fields of farms and ranches. It is there that one must know soil-by-soil and field-by-field the effective usable soil depth, the water retention and supplying capacity of each soil, the basic intake rates under specified management, the permeability of the soil, the amount and kinds of salt and the sodium status of the soil, the relief and gradient of the soil surface, the crops to be grown, and the quality of water. Such data are needed to determine the amount of water to be applied to replenish that used by crops and to maintain a proper salt balance, and to determine rate of water application to prevent erosion and water loss by runoff and deep percolation.

The Soil Conservation Service of the US Department of Agriculture, in its work with individual landowners and operators, uses the standard soil survey to supply these data and the land capability classification to interpret them as basic tools in planning the right combinations of soil-water-crops and management.

The descriptions of soil taxonomic units and associated management and response data gathered in the standard soil survey provide the information needed to classify each kind of soil into its appropriate land capability unit, a category in the Land Capability Classification.

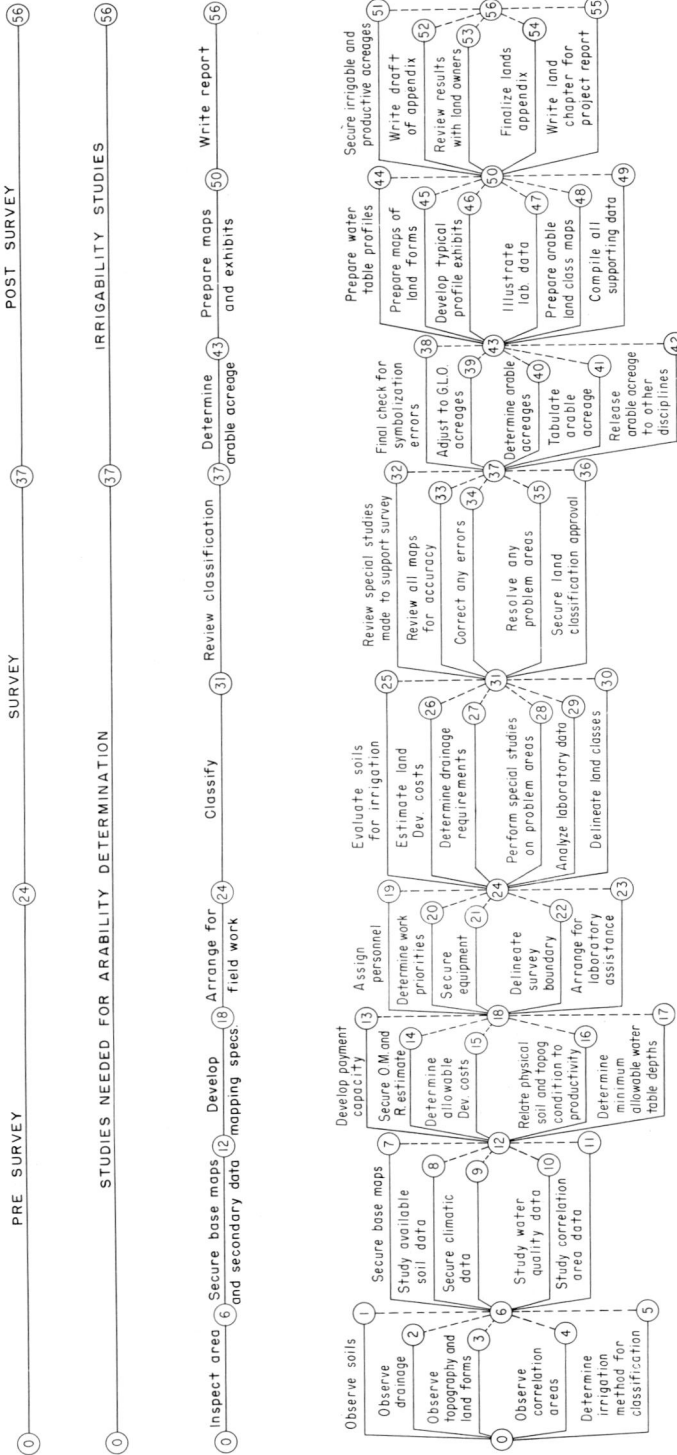

Fig. 10–12. US Bureau of Reclamation Land Classification Work Performance Network.

Table 10–5. Detailed land classification specifications, Oahe Unit, South Dakota, Region 6

Land characteristics	Class 1--Arable	Class 2--Arable	Class 3--Arable
Soils			
Texture (2 ⁓μ clay)	Sandy loam, loam, and silt loam	Loamy sand to clay loam, and silty clay loam inclusive; clay loam and silty clay loam occurring as a single horizon below the surface associated with medium textures in other portions of profile	Loamy sand to silty clay inclusive
Depth to incoherent sand	36 inches or more of free working soil of fine sandy loam to silt loam, or 42 inches of sandy loam	24 inches or more of free working soil of fine sandy loam through clay loam, or 30 inches of loamy sand	18 inches or more of free working soil of sandy loam through clay loam, or 24 inches of loamy sand
Hydraulic conductivity, undisturbed	For all classes. Not <0.20 inch/hour in 0- to 2-ft zone and not <0.06 inch/hour in the 2- to 4-ft zone. In a layered profile having varying rates, the minimum will be controlling. These values will be applied in the field by excluding from the arable area all soils with a solonetz horizon and all soils with "clay pan" in the upper 2 ft of soil which display all or most of the following characteristics: textural class silty clay or silty clay loam; apparent density >1.4 as indicated by sheen appearance of cleavage surfaces, compaction, or porosity; consistence hard when dry and sticky when wet; few or no visible pores. Questionable areas will be related to known soils on which field permeability (Winger tests) have been determined.		
Salinity	Salinity is not considered to be a deficiency in lands having suitable permeability and adequate drainage.		
Alkalinity	For all classes. Exchangeable sodium may not exceed native gypsum in excess of 10%. If <10% clay, exchangeable sodium may be in excess of gypsum by 20%. The distinguishing characteristics of this factor in laboratory analyses are: very slow disturbed permeability rate, usually low total salt with wide pH spread. In the field, this condition is recognized in association with the solodized horizon (claypan) of most nonarable soil types.		
Topography			
Slope	<2% in general gradient	<5% in general gradient.	<8% in general gradient
Irrigation pattern	400-ft minimum run, 8-acre minimum size	300-ft minimum run, 5-acre minimum size	150-ft minimum run, 2-acre minimum size
Surface leveling	0 to 200 cubic yards excavation/acre 0.0- to 0.24-ft average cut and fill	200 to 450 cubic yards excavation/acre, 0.24- to 0.48-ft average cut and fill	450 to 750 cubic yards excavation/acre, 0.49- to 0.85-ft average cut and fill
Cover, trees, 6- to 15-inch diameter	0 to 12 trees/acre	12 to 30 trees/acre	30 to 60 trees/acre
Drainage			
Surface, outlets	0 to 200 cubic yards excavation/acre	200 to 450 cubic yards excavation/acre	450 to 750 cubic yards excavation/acre
Internal	Project drainage is assumed for all arable lands.	Presence of water table should be mentioned in profile notes, when encountered.	
Class 6	This includes all lands that do not meet minimum specifications for Class 3		

Table 10-6. Detailed land classification specifications, Shadehill Unit, MRBP, South Dakota

Land characteristics	Class 1--Arable	Class 2--Arable	Class 3--Arable
Soils			
Clay mineral	Not > 80% Beidellite, with minimum of 20% non-swelling type clay minerals	Same as Class 1	Same as Class 1
Surface area	Total surface area not >120 m²/g and internal surface area not over 80 m²/g	Same as Class 1	Same as Class 1
Texture; limiting mechanical composition	Clay, 24%; silt, 50%; sand, 40%	Same as Class 1	Same as Class 1
Texture class	VFSL, FSL, SL, and L with <20% clay	LFS, VFSL, FSL, SL, and L with > 20% and >20% and <24% clay	LS, LFS, VFSL, SL, L, > 20% clay, SCL <24% clay
Depth to sand, gravel or sand and gravel matrix	36 inches or more friable FSL or finer, or 42 inches of SL	24 inches or more friable FSL or finer, or 30 inches of LFS	18 inches or more of LFS or finer, or 24 inches of LS
Depth to shale or sand-stone bedrock	6 ft or more with L and FSL or overburden; 5 ft or more with coarser overburden	Same as Class 1	6 ft or more with L, VFSL, and SCL over-burden; 5 ft or more with coarser overburden
Exchangeable sodium level	ESP <20%; observable solonetizic development absent	Same as Class 1	Same as Class 1
Total soluble salts	Equilibrium concentrate of saturation extract under irrigation not > 2 mmho/cm	Same as Class 1	Same as Class 1
Water retention	Total available water not <3, 5 inches of water in 4-ft depth and not <2 inches of water in upper 2-ft depth (1/10 atm not <13 in 0- to 2-ft zone and 10 in 2- to 4-ft	Same as Class 1	Same as Class 1
Topography			
Slope	No slopes > 2%	Same as Class 1	Same as Class 1
Irrigation pattern	400-ft minimum length of run, 8-acre minimum size of fields	300-ft minimum length of run, 5-acre minimum size of fields	150-ft minimum length of run, 2-acre minimum size of fields
Irrigation method	Land suitable for border irrigation; readily adaptable to furrow or currugation methods	Same as Class 1	Same as Class 1
Land forming	Land forming 0-75 cubic yards/acre 0.0-ft to 0.1-ft average cut and fill	Land forming 75-175 cubic yards/acre 0.1-ft to 0.25-ft average cut and fill	Land forming 175-350 cubic yards/acre 0.25-ft to 0.45-ft average cut and fill
Land development costs	Land forming $0 to $14; irrigation ditches and structures $0 to $10	Land forming $14 to $32; irrigation ditches and structures $0 to $10	Land forming $22 to $63; irrigation ditches and structures $0 to $10
Drainage			
Subsurface drainage	Permanent water table controlled at a depth of 5 ft or more in L and VFSL and at 4 ft or more in other textures at cost <$14/acre	Permanent water table controlled at a depth of 5 ft or more in L and VFSL and at 4 ft or more in other textures at cost not >$32/acre	Permanent water table controlled at a depth of 5 ft or more in L, SCL and VFSL and at 4 ft or more in other textures at cost not >$63/acre
Surface drainage	On-farm surface drainage cost not <$14/acre; 225 linear feet maximum 1-ft deep, 6-ft bottom, 3:1 side slope; 85 linear feet maximum 2-ft deep, 6-ft bottom, 3:1 side slope	Surface drain cost not >$32/acre; 530 linear feet maximum 1-ft deep, 6-ft bottom, 3:1 side slope; 200 linear feet maximum 2 ft-deep bottom, 3:1 side slope	Surface drain cost not >$63/acre; 1,000 linear feet maximum 1-ft deep, 6-ft bottom, 3:1 side slope; 400 linear feet maximum 2-ft deep, 6-ft deep bottom, 3:1 side slope

Table 10–7. Land classification specifications, Kokee Project, Hawaii

Class differentia Subclass--land characteristic	Class 1--Arable	Class 2--Arable	Class 3--Arable	Class 6--Nonarable
Soils				All lands that do not meet the minimum requirements for Class 3 lands
Texture	Sandy loam to silty clay	Sandy loam to permeable clay	Loamy fine sand to moderately permeable clay	
Depth to basalt	3 ft or more	2 ft or more	1.5 ft or more	
Depth to coral formation	3 ft or more	2 ft or more	1.5 ft or more	
Depth to strongly gleyed layer or marl	No gleyed layers present	2 ft or more	1 ft or more	
Depth to red, yellow, gray highly weathered rock	3 ft or more	2 ft or more	2 ft or more	
Exchangeable sodium percentage	15 %	15%	15% low humic latosol, alluvial, man-made; 10% dark mg. clays and gray hydromorphs	
Salinity, $EC_s \times 10^3$	4 mmho/cm at equilibrium	8 mmho/cm at equilibrium	8 mmho/cm at equilibrium	
Topography				
Surface slope	0–2%	2–12% reasonably large bodies sloping in the same plane; well to moderately well suited to use of mechanical equipment	12–20% moderately suited to use of mechanical equipment; where complex rolling surfaces are not present, slope may be slightly greater where necessary to obtain regular field pattern	
Undulations	Only minor planing and smoothing operations required	Planing and smoothing may be required but no cuts >3 inches	Planing and smoothing required but no cuts >6 inches	
Width of field	100 ft or more	100 ft or more	May be <100 ft depending on irrigation pattern	
Minimum length of run	300 ft	150–300 ft in small tracts; upper part of range required for large tracts	75–150 ft in small tracts; same range as Class 2 for large tracts	
Elevation (m.s.l)	1,000 ft	1,000–1,750 ft	1,000–1,750 ft	
Rock	0–5% by volume within root depth	5–15% by volume within root depth	15–20% by volume within root depth	
Clearing	Low brush, clearing $15 or less	Brush and scattered trees, clearing $15–$30	Dense tree cover, clearing $30–$45	
Drainage				
Internal drainage (natural)	Well aerated; no limit to water movement or root development. Cultural practices can be conducted over a wide range of soil water conditions.	Well to moderately well aerated; water movement and root development somewhat impeded. Cultural practices can be conducted over a moderately wide range of water.	Moderately well aerated; water movement and root development moderately restricted. Cultural practices can be conducted over a narrow soils water range.	
Surface	No flooding anticipated	Infrequent overflows of short duration requiring moderate costs for improvement	Frequent overflows, improvement costs not >$100/acre	
Subsurface	Some drains required at spacing >600 ft	Drains 4 ft deep;200–600 ft spacing	4 ft depth; 80–200 ft spacing	
Controlled depth of water table	4 ft or more	2–4 ft	2–4 ft	

* Applicable chiefly to sugar cane or crops of similar adaptabilities and culture.

Table 10–8. Detailed land classification specifications, Chief Joseph Dam Project, Washington

Land characteristics	Class 4F (1) Arable (For apple production)	Class 4F (2) Arable (For apple production)
Soils		
Texture	Good sandy loam to silt loam	Loamy fine sand to friable clay loam
Alkalinity of soil paste	<pH 8, 0 in surface foot, and <pH 9, 0 to 4 ft, unless soil is very permeable and underlain with open coarse material at <5 ft	<pH 8, 0 in surface foot and <pH 9, 2 to 4 ft unless soil is very permeable and underlain with open, coarse materials at 5 ft or less
Salinity	Total salts not > 0, 1% at equilibrium with irrigation	Total salts not > 0, 2 % at equilibrium with irrigation
Topography		
Slope	Up to 15% in general gradient; may be up to 25% where sprinkler pressure is provided in the project system	Up to 30% in general gradient; may be up to 35% where sprinkler pressure is provided in the project system
Surface; does not apply if project pressure provided	Up to 10-inch average cut and fill where soil depth is adequate and permanent injury not sustained	May be very irregular, so that sprinkler irrigation required, but increased production costs and decreased productivity not to exceed $15/acre annually
Cover, loose cobble and vegetation	Up to 65 cubic yards of rock and cobble per acre; cost of removing light brush cover not significant	Up to 100 cubic yards of rock and cobble per acre, or comparable cost in tree or brush removal
Size and shape of fields	200-ft minimum length, 4-acres minimum size	100-ft minimum run, 1-acre minimum size
Drainage		
Water drainage	No drainage problem is anticipated	Drainage conditions such that good soils aeration may be maintained to depth of 4 ft or more; farm surface drainage construction not to exceed 500 ft of shallow open drain per acre
Air drainage	Slope and position of land such that air drainage or air movement is not impeded	Slope and position of land such that adequate air drainage seems probable

No Class 4F (3)

Class 4H--Includes all land in suburban development.

Class 6--Includes all land not meeting minimum specifications for above classes.

The Land Capability Classification is designed to emphasize the hazards and limitations in the different kinds of soil. It provides three major categories of soil groupings—the capability class, subclass, and unit. Through the process of soil correlation and classification in the standard soil survey, it is possible to coordinate the land capability classification between districts and projects and to a considerable extent between states.

The capability class is the most inclusive category. The risks of soil damage or limitations in use become progressively greater from Class 1 to Class 8. The soils within a capability class are similar only with respect to degree of limitation in use for agricultural purposes or hazard to the soil when it is so used. Specific statements about suitable kinds of crops or other management needs cannot be made at the class level.

Capability classes are distinguished from each other by a summation of the degree of permanent or continuing limitations or risks of soil damage that affect their management requirements for long sustained use for the common cultivated crops, pastures, range, woodland, and wildlife purposes. Hazards such as excess water, stones, soluble salts, or exchangeable sodium are not considered permanent limitations to use where the removal of these limitations is feasible. That is, where the characteristics and qualities of the soil are such that removal is physically possible and is within current or foreseeable economic possibilities.

Soils considered feasible for improvement by draining, leaching, diking, or reshaping are classified according to their continuing limitations in use, or the risks of soil damage or both, after the improvements have been installed. However, a favorable ratio of input to output is one of several criteria used for placing any soil in a class. If an unfavorable input-output ratio would result from the installation of practices, the improvements would not be planned and the soil would be classed on the basis of the continuing limitations and hazards without such improvements. A comprehensive treatise on the land capability classification and the pertinent assumptions therein are discussed by Klingebiel and Montgomery (1961).

The capability subclass is based on kinds of dominant limitations. The *erosion* subclass (e) is made up of soils whose dominant hazard or problem in use is susceptibility to erosion. Erosion susceptibility and past erosion damage are the major soil factors for placing soils in this subclass. The *excess water* subclass (w) is composed of soils where excess water is the dominant hazard or limitation in use. The *soil limitation* subclass (s) includes soils with such limitations within the rooting zone as low moisture supplying capacity, shallow rooting zones, salinity, excess exchangeable sodium, and low fertility. The *climatic limitation* subclass (c) is made up of those soils where temperature or lack of moisture are the only major hazards or limitations in use.

Where two or more kinds of limitations have comparable degrees of restriction the subclasses have the priority e, w, s. The c subclass includes soils with no limitation other than climate. Where important, two or more kinds of limitations may be shown with the dominant first.

National specifications guide placement of soil taxonomic units into land capability classes and subclasses, but the land capability unit is defined locally within the limits circumscribed by class and subclass criteria.

The capability unit is a grouping of soil taxonomic units that are alike or nearly alike in suitability for plant growth and about equal in response to the same management.

Table 10-9. Land capability classification, Central Utah County Area, Utah

Class	Subclass and unit	Soils description	Slopes, nature and %	Water supplying capacity, estimated inches	Intake family & estimated permeability, inches	Climate
I Soils with few or no continuing limitations or hazards	I-1	Well drained, neutral to moderately calcareous, moderately to highly fertile soils with loam surface textures, moderately slow to moderately permeable loam or clay loam subsoils underlain by carbonate horizons; 60 inches effective rooting depth	Level and nearly level 0 to 1	9 to 11 available for 5-ft profile	$\dfrac{1.5}{0.2\text{-}2.0}$	150-170 days frost-free period 3 out of 4 years; evapotranspiration >20 inches; annual precipitation 16-18 inches
	I-2	Well drained, neutral to moderately calcareous moderately fertile soils with very fine sandy loam or light sandy clay loam surface textures and rapidly permeable very fine sandy loam subsoils underlain by carbonate horizons; rooting depth of 60 inches	Level and nearly level 0 to 1	7.5 to 9.0 available for 5-ft profile	$\dfrac{3.0}{2\text{-}6}$	150-170 days frost-free period 3 out of 4 years; evapotranspiration >20 inches; annual precipitation 16-18 inches
II Soils requiring moderately intensive treatments or having a slightly reduced choice of plants	IIe1	Well and moderately well drained, neutral to moderately calcareous, moderately to highly fertile soils with loam surface textures, moderately slow to moderately permeable, loam or clay loam subsoils underlain by strong carbonate horizons; effective rooting depth of 60 inches; slight to moderate erosion hazard	Gently sloping 1 to 3	9 to 11 available for 5-ft profile	$\dfrac{1.5}{0.2\text{-}2.0}$	150-170 days frost-free period 3 out of 4 years; evapotranspiration >20 inches; annual precipitation 16-18 inches
	IIe2	Well drained, moderately calcareous moderately fertile soils with silty clay loam surface textures and slowly permeable clay loam or silty clay loam subsoils containing thin strata of coarser textures; effective rooting depth is 60 inches; slight erosion hazard	Gently sloping 1 to 3 with slightly uneven surface	11 available for 5-ft profile	$\dfrac{0.5}{0.2\text{-}0.63}$	150-170 days frost-free period 3 out of 4 years; evapotranspiration >20 inches; annual precipitation 16-18 inches
	IIw1	Somewhat poorly drained, moderately & strongly alkaline, watertable >40 inches, moderately and strongly calcareous, moderately to highly fertile soils with loam or silt loam surface and moderately to rapidly permeable silt loam to fine sandy loam subsoils; moderately saline and moderately saline-alkali soils contain <15% exchangeable sodium above 10 inches and <25% above 40 inches; $EC_e \times 10^3 < 8$ & containing gypsum	Gently sloping 1 to 3, minor surface irregularities	10 available for 5-ft profile	$\dfrac{2.0}{2.0\text{-}6.0}$	130-150 frost-free days, 3 out of 4 years; evapotranspiration 15 to 20 inches
	IIc1	Well and moderately well drained mildly alkaline, moderately fertile soils with loam surface textures and loam or very fine or fine sandy loam, moderate to rapidly permeable subsoils underlain by thinly stratified loams to fine sand with gravel in places below 3 ft; slight wind erosion hazard	Level and nearly level 0 to 1 uniform	8 to 10 available for 5-ft profile	$\dfrac{1.5}{2.0\text{-}6.0}$	130-150 frost-free days, 3 out of 4 years; evapotranspiration 15 to 20 inches

Class	Subclass	Description	Slope / Surface	Rooting zone		Climate
III Soils with severe limitations that reduce the choice of plants or require special practices or both	IIIe1	Well and somewhat excessively drained neutral and moderately alkaline, moderately fertile soils with gravelly loam or gravelly fine sandy loam surface and subsoils that are underlain between 20 and 30 inches with gravelly or cobbly sand or light sandy loam; moderate water erosion hazard; root zone inhibited by gravel and cobble	Moderately sloping 3 to 6; slightly uneven surface	6 available for rooting zone	$\dfrac{4.0}{4.0-6.0}$	150-170 days frost-free period 3 years out of 4; evapotranspiration > 20 inches; precipitation 14 to 18 inches
	IIIw1	Deep moderately calcareous, moderately fertile soils with silty clay loam surface soils and light silty clay slowly permeable subsoils; water tables at 30 to 60 inches unless drained; soluble salts $EC_e \times 10^3 = <8$ and exchangeable sodium percentage <15% above 10 inches and <40% above 40 inches; containing some gypsum; flood occasionally (1 year in 10)	Level or nearly, <1	11 for 5-ft profile	$\dfrac{0.5}{0.05-0.62}$	130-150 days frost-free period 3 years out of 4; evapotranspiration 15 to 20 inches; precipitation 14 to 16 inches
	IIIs1	Moderately well drained, moderately alkaline, moderately fertile soils with fine sandy loam surface textures and very rapidly permeable loamy fine sand subsoils underlain by a lime horizon usually below 36 inches; rooting depth 5-ft; exchangeable sodium percentage <15 above 10 inches and <40 above 40 inches; $EC_e \times 10^3 <8$	Gently sloping 1 to 3 alluvial fans and lake terraces	4 in a 5-ft profile	$\dfrac{4.0}{6.0}$	150-170 days frost-free period 3 years of 4 years; evapotranspiration > 20 inches; precipitation 14 to 16 inches
IV Soils with severe limitations that restrict choice of plants, require very careful management or both	IVe1	Well and somewhat excessively drained neutral and moderately alkaline, moderately fertile soils with gravelly loam or gravelly fine sandy loam surface and subsoils that are underlain between 20 and 30 inches with gravelly or cobbly sand or light sandy loam; moderate to severe water erosion hazard; root zone somewhat limited by gravel and cobble	Strongly sloping 6 to 10, alluvial fans colluvial slopes	6 within 5-ft rooting zone	$\dfrac{4.0}{6.0}$	150-170 days frost-free period 3 out of 4 years; evapotranspiration > 20 inches; precipitation 14 to 16 inches
	IVw1	Moderately to strongly calcareous, poorly drained, moderately fertile soils with silt loam or silty clay loam surface textures and light silty clay subsoils; distinct carbonate horizons usually within 16 inches; sodium percentage <15 above 6 inches and <40 above 20 inches; $EC_e \times 10^3 <16$ above 20 inches; difficult to drain and reclaim; occasional flooding	Nearly level to gently sloping 0 to 3, lake terraces	9 within 5-ft rooting zone	$\dfrac{0.5}{0.05-0.62}$	130-150 days frost-free period 3 out of 4 years; evapotranspiration 15 to 20 inches; precipitation 14 to 16 inches
	IVs1	Gravelly well to excessively drained, neutral to moderately alkaline soils with gravelly or cobbly sandy loam surface soils and gravelly or cobbly loam or sandy loam subsoils underlain at 20 to 30 inches with gravelly or cobbly sand; low to moderate fertility; difficult to till; erosion hazard is high	Strongly sloping to steep dominantly 6 to 15 up to 25 on Pleasant Grove soils	4 in 5-ft profile; 1.5 to 2 in main Pleasant rooting zone	$\dfrac{4.0}{4.0-6.0}$	150-170 days frost-free period 3 out of 4 years; evapotranspiration > 20 inches; annual precipitation 16 to 18 inches

Table 10-9. (Continued from previous page)

Class	Subclass and unit	Soils description	Slopes, nature and %	Water supplying capacity, inches	Intake family & estimated permeability inches	Climate
VI Soils with severe limitations making them generally unsuited to cultivation and limiting their use largely to pasture	VIw1	Somewhat poorly drained, strongly to very strongly alkaline soils with silt loam or silty clay loam surface textures and heavy silty to light silty clay subsoils and substrata to depths of 6-ft with high shrink-swell ratios; salinity $EC_e \times 10^3$ up to 16; sodium percentage <15 above 6 inches, <60 above 40 inches	Gently sloping 1 to 3 with slightly uneven surface	1 to 3 limited by salt content	$\dfrac{<0.5}{0.05\ \text{or less}}$	130–150 days frost-free period 3 years out of 4 evapotranspiration 15 to 20 inches; precipitation 14 to 16 inches
	VIs1	Well drained neutral to moderately alkaline soils with very cobbly or very stony loam or very stony sandy loam surface textures; subsoils range from cobbly or stony sand; irrigation not feasible because of coarse fragments and/or steep slopes; erosion hazard moderate to high	Sloping to steep 3 to 35	4 to 5 in 5-ft profile	$\dfrac{4.0}{4.0\text{–}6.0}$	150–170 days frost-free period 3 out of 4 years; evapotranspiration >20 inches; annual precipitation 16 to 18 inches
VIII Soils and land types that have limitations that preclude use for commercial plant production	VIIIw1	Very strongly saline-alkali soils with very slow permeability and mixed strongly saline-alkali alluvial land	Level to gently sloping 0 to 3 uneven surface	1 to 3 limited by high salt content	$\dfrac{<0.5}{<0.05\ \text{(dominantly)}}$	130–150 days frost-free period 3 years out of 4; evapotranspiration 15 to 20 inches; precipitation 14 to 16 inches

The soils in a capability unit (i) produce similar kinds of cultivated crops, pasture plants, or trees with similar management practices, (ii) require similar management under the same kind and condition of vegetative cover, and (iii) have comparable potential productivity. The range in production among the soils within the capability unit does not vary more than 25%. Land capability units take into account any characteristics or qualities of soils and environment that are locally important. Soils in Class 1 have no continuing hazards or limitations and no subclasses are required.

Monetary values for costs and returns can be assigned to management practices and associated yields and economic comparison made between units. The units may be arrayed in sequence to show net output.

For any given project or district the number of land capability units within one class or subclass depends upon the diversity of soil and climatic conditions.

Any favorable fixed or recurring soil or landscape features may limit the safe and productive use of a soil, but most frequently combinations of soil characteristics determine land capability.

Table 10–9 shows the criteria used to place mapped and defined soil taxonomic units in the Central Utah County Area, Utah, USA, into land capability units. The land capability units shown were chosen to illustrate the placement of soils with different kinds and intensities of limitations. They are representative but not inclusive of all soils in the area.

The Land Capability Classification is only one of the interpretations of the soil survey included in a published soil survey report. Detailed descriptions of each taxonomic unit are in the report. The description sets forth for each kind of soil its observed and measured characteristics, properties, and inferred qualities along with their permissible ranges. By specifying the named soils within each capability unit, the limitations for each unit can be obtained in detail by referring to the soil descriptions. The brief soil descriptions in the table indicate the principal profile characteristics and qualities of the soils in the capability unit.

The soil survey report (Ulrich, 1962) is representative of a complete report. It gives suggested alternative cropping systems, tillage practices, irrigation specifications and supplemental management for each capability unit suited to irrigation.

LITERATURE CITED

Arkley, R. J., and R. Ulrich. 1962. The use of calculated actual and potential evapotranspiration for estimating potential plant growth. Hilgardia 32:443–462.

Aubert, G., and P. Duchaufour. 1956. Projet de classification des sols. Int. Congr. Soil Sci., 6th (Paris, France) Rep. V–97. Vol. E:597–604.

Belcher, D. J. 1948. The engineering significance of landforms. Nat. Res. Counc., Highway Res. Board Bull. 13. p. 9–29.

Bolt, G. H. 1955. Ion adsorption by clays. Soil Sci. 79:267–276.

Bower, C. A. 1959. Cation-exchange equilibrium in soils affected by sodium salts. Soil Sci. 88:32–35.

Bower, C. A. 1961. Prediction of the effects of irrigation waters on soils. Teheran Symp. Salinity problems in the arid zones. UNESCO Arid Zone Research 14:215–221.

Bowser, E. W., and H. C. Moss. 1950. A soil rating and classification for irrigation lands in western Canada. Sci. Agr. 30:165–171.

Brough, O. L., A. L. Walker, E. R. Franklin, P. M. McMains, and V. Divers. 1956. Columbia Basin Project—relative land productivity and income. Washington Agr. Exp. Sta. Bull. 570. p. 2–7.

Bureau of Reclamation. 1924. Reclamation law: Classification of lands. Subsec. D, 2nd deficiency act (43 Stat. 702, 43 USC 462).

Bureau of Reclamation. 1953. Land classification handbook. US Dep. Int.-Bur. Reclam. Publ. V, Part 2. 53 p.

Butler, E. D. 1961. Land leveling in the Arkansas Delta. Agr. Eng. 42(3):128–130.

Canada Department of Agriculture. 1960. Land classification of the Bow River Project. Prairie Farm Rehabilitation Admin. p. 1–59.

Canada Department of Agriculture. 1964. Handbook for the classification of irrigated land in the prairie provinces. Prairie Farm Rehabilitation Admin. p. 1–92.

Cotner, M. L., and D. A. Jameson. 1959. Costs of juniper control: Bulldozing vs. burning individual trees. US Dep. Agr. Rocky Mountain Forest Range Exp. Sta. Pap. No. 43. 14 p.

Desaunettes, J. R. 1960. Binary classification of soils as a function of their value. Int. Congr. Soil Sci., Proc. 7th (Madison, Wis., USA). IV:379–387.

Dominy, F. E. 1964. Irrigation development in the western United States as related to the great soil groups. Int. Congr. Soil Sci., Proc. 8th, Comm. 6 (Bucharest, Rumania) (In press)

Donnan, W. W. 1947. Model tests of tile-spacing formula. Soil Sci. Soc. Amer. Proc. (1946) 11:131–136.

Donnan, W. W., G. B. Bradshaw, and H. F. Blaney. 1954. Drainage investigations in Imperial Valley, California, 1941–51 (a 10-year summary). US Dep. Agr.-Soil Cons. Serv. Publ. SCS–TP–120.

Dumm, L. D. 1954. Drain-spacing formula. Agr. Eng. 35(10):726–730.

Dumm, L. D. 1964. The transient-flow concept in subsurface drainage: Its validity and use. Amer. Soc. Agr. Eng., Trans. 7(2):142–146.

Dumm, L. D., and R. J. Winger, Jr. 1964. Designing a subsurface drainage system in an irrigated area through use of the transient-flow concept. Amer. Soc. Agr. Eng., Trans. 7(2):147–151.

Eriksson, E. 1952. Cation-exchange equilibrium in clay minerals. Soil Sci. 74:103–113.

Fly, C. L. 1961. The soil drainability factor in land classification. Amer. Soc. Civil Eng. J. Irrig. Drainage 87(IR3), Part 2, 47–62.

Gardner, W. R., M. S. Mayhugh, J. O. Goertzen, and C. A. Bower. 1959. Effect of electrolyte concentration and exchangeable-sodium percentage on the diffusivity of water in soils. Soil Sci. 88:270–274.

Harris, K., and V. W. Hansen. 1958. Relative productive value of land. Utah State Univ. Agr. Appl. Sci. Eng. Exp. Sta. 29 p.

Hulsbos, W. C., and J. H. Boumans. 1960. Leaching of saline soils in Iraq. Netherlands J. Agr. Sci. 8(1):1–10.

Hydrotechnic Corporation. 1955. Hydro agricultural and hydro electric program, Cuanza, Lucala, and Bengo River Basins, Province of Angola, Africa. Hydrotechnic Corp., New York. V:1–65.

Jacks, G. V. 1946. Land classification for land-use planning. Imp. Bur. Soil Sci. Tech. Comm. 43:64–69.

Kelley, W. P. 1964. Maintenance of a permanent irrigation agriculture. Soil Sci. 98(2): 113–117.

Kellogg, C. E. 1961. Soil interpretation in the soil survey. US Dep. Agr.-Soil Cons. Serv. p. 1–27.

Klingebiel, A. A., and P. H. Montgomery. 1961. Land capability classification. US Dep. Agr.-Soil Cons. Serv. Agr. Handbook no. 210. 21 p.

Kunze, G. W., H. Oakes, and M. E. Bloodworth. 1963. Grumusols of the coast prairie of Texas. Soil Sci. Soc. Amer. Proc. 27:412–421.

Lueder, D. R. 1959. Aerial photographic interpretation. McGraw Hill, New York. 462 p.

Maasland, M. 1959. Water table fluctuations induced by intermittent recharge. J. Geophys. Res. 64:549–559.

Maasland, M. 1961. Water table fluctuations induced by irrigation. Amer. Soc. Civil Eng. J. Irrig. Drainage. 87(IR2):39–58.

McLerran, J. H. 1957. Glossary of pedologic (soils) and landform terminology for soil engineers. Nat. Res. Counc., Highway Res. Board Spec. Rep. 25, Publ. 481. p. 1–32.

McMartin, W. 1950. The economics of land classification for irrigation. J. Farm Econ. Ass. XXXII:553–700.

Maletic, J. T. 1961. Using soil survey information in land classification for irrigation. Central Treaty Organ. Land Classification Sem., Proc., Ankara, Turkey. Annex J, p. 1–11.

Marinet, J., J. Seguy, P. Andre. 1961. Development of the plain of Vientiane. Agr. Soil Study, Soc. Grenobloise d'etudes et d'Appl. Hydrol., Grenoble, France (in French). 232 p.

Marr, J. C. 1957. Grading land for surface irrigation. State Coll. Washington Ext. Bull. 526. 55 p.

Mason, B. 1958. An example of climatic control of land utilization. Climatology and microclimatology. Canberra Symp. Proc., UNESCO 11:188–194.

Miller, D. E. 1964. Estimating moisture retained by layered soil. J. Soil Water Cons. 6(19):235–237.

Miller, D. E., and W. C. Bunger. 1963. Moisture retention by soil with coarse layers in the profile. Soil Sci. Soc. Amer. Proc. 27:586–589.

Molenaar, A. 1956. Projects involving development for irrigation. UN Publ. no. 11.F.3: 415–417.

Moodie, C. D., R. C. Reeve, L. S. Willardson, N. A. Evans, C. E. Houston, C. R. Maierhofer, J. T. Maletic, D. R. Nielson, and D. W. Thorne. 1964. Salinity and alkali problems. Panel No. 2. J. Amer. Soc. Civil Eng. 90(IR4), Part I, 41–49.

Pratt, P. F., and F. L. Blair. 1964. Application of the double-layer equation to sodium-calcium exchange equilibria in bentonites. Soil Sci. Soc. Amer. Proc. 28:32–35.

Pratt, P. F., and B. L. Grover. 1964. Monovalent-divalent cation-exchange equilibria in soils in relation to organic matter and type of clay. Soil Sci. Soc. Amer. Proc. 28:32–35.

Quirk, V. P., and R. K. Schofield. 1955. The effect of electrolyte concentration on soil permeability. J. Soil Sci. 6:163–178.

Reeve, R. C. 1957. The relation of salinity to irrigation and drainage requirements. Int. Comm. Irrig. Drainage, 3rd Congr. 10:175–187.

Reeve, R. C., and C. A. Brown. 1960. Use of high salt waters as a flocculant and source of divalent cations for reclaiming sodic soils. Soil Sci. 90:139–143.

Reeve, R. C., A. F. Pillsbury, and L. V. Wilcox. 1955. Reclamation of a saline and high boron soil in the Coachella Valley of California. Hilgardia 24, 4:0–91.

Richards, L. A., ed. 1954. Diagnosis and improvement of saline and alkali soils. US Dep. Agr. Handbook 60. 160 p.

Richards, L. A., and L. R. Weaver. 1944. Moisture retention by some irrigated soils as related to soil moisture tension. J. Agr. Res. 69:215–235.

Sain, S. K. 1951. Multiple-purpose river projects. Formulation and economic appraisal of development projects. UN Publ. 11.B.4.11:701–741.

Soil Conservation Service Staff. 1959. National engineering handbook. US Dep. Agr.-Soil Cons. Serv. Handbook 18. p. 12.1–12.59, 2.1–2.40, 3.1–3.60.

Soil Survey Staff. 1951. Soil survey manual. US Dep. Agr. Handbook 18. 501 p.

Soil Survey Staff. 1960. Soil classification, a comprehensive system, 7th approximation. US Dep. Agr.-Soil Cons. Serv. 265 p.

Stirk, G. B. 1954. Some aspects of soil shrinkage and the effect of cracking upon water entry in to the soil. Australian J. Agr. Res. 5:2.

Storie, R. E. 1964. Soil and land classification for irrigation development. Int. Congr. Soil Sci., Proc. 8th, Comm. 5 (Bucharest, Rumania). (In press)

Thornthwaite, C. W. 1948. An approach towards a rational classification of climate. Geogr. Rev. 28:55–94.

Ulrich, R. 1962. Soil survey of Madera Area, California. US Dep. Agr.-Soil Cons. Serv. and California Agr. Exp. Sta. Ser. 1951, No. 11. 155 p. + 91 maps.

Vanselow, A. P. 1932. Equilibria of the base-exchange reaction of bentonites, permutites, soil colloins, and zeolites. Soil Sci. 33:95–113.

Winger, R. J., Jr. 1960. In-place permeability tests and their use in subsurface drainage. Int. Comm. Irrig. Drainage, 4th Congr. (Madrid, Spain). R.23:11.418–11.469.

Wydler, R. 1960. Soil survey on Rio Negro Lower Valley (Argentine) for irrigation purposes. Int. Congr. Soil Sci., Trans. 7th (Madison, Wis., USA). IV:380–384.

11

Economics of Irrigation Policy and Planning

HARRY A. STEELE

Water Resources Council
Washington, D. C.

GEORGE A. PAVELIS[1]

Economic Research Service, USDA
Washington, D.C.

In this chapter, economics is regarded as the science of guiding managerial decisions oriented to prespecified managerial objectives. A socio-economic discussion of irrigation as a means for improving or stabilizing agricultural production requires such a broad definition because the decision-making unit can be a nation considering whether to commit its limited resources to irrigation projects as an instrument for promoting its national production, land settlement, or interregional development policies. Or, it can be an individual farmer considering whether to commit his limited capital and other resources to irrigation as a production practice. Complex technical, legal, and social problems are associated with irrigation in either case, and it is important to recognize that economic science does not require that they all be resolved within a strict monetary context.

Specific economic matters discussed in this chapter include: (i) The feasibility and profitability of irrigation as a production practice on individual farms; (ii) some general economic problems in irrigation project planning; and (iii) some special problems in multipurpose project planning. A concluding section deals very briefly with some current policy issues in irrigation as they relate to farm output requirements, farm production efficiency, and economic development in general.

I. ECONOMICS OF FARM IRRIGATION

Farm income is determined by many physical and economic variables, some of which relate directly to farm resources and others of which do not originate on the farm. Major on-farm variables are inherent soil productivity, the farm labor supply, general capital requirements for crop or livestock enterprises feasible in a locality, and the operator's managerial abilities. Common external factors are weather, crop and livestock prices, and market costs of capital goods or other items used in production.

The effect on production of factors that might be controlled can be expressed in terms of an input-output function which relates physical output Y to inputs of variable resources or resource classes. The latter are stated in general as X_1 (land),

[1] The authors appreciate the suggestions of Karl Gertel, Norman E. Landgren, Robert C. Otte, and Mark M. Regan, colleagues in the Economic Research Service; and those of Prof. Emery N. Castle, Head, Department of Agricultural Economics, Oregon State University.

X_2 (labor), X_3 (capital), and X_4 (management). The general land variable means land from its areal, soil, and soil moisture standpoints, each of which can be set out as subvariables. Equipment, buildings, and other specialized capital inputs likewise may be incorporated in total capital requirements. Since managerial performance can be measured only in relation to success in achieving management objectives, the management variable is often treated as a residual factor to which is imputed the output effects not fully explained by the others. Time and technology as additional determinants of production possibilities are recognized by assuming that each defined input-output function represents a unique set of possibilities corresponding to a given level of technology, be it primitive, modern, or futuristic.

But even with technology given, time enters farm planning decisions in other ways. Production decisions may still be more or less completely long-run. Theoretically, this permits variation or adjustments in any input to control production. In this case output Y may be stated explicitly in various ways, ranging from the linear function $Y = a + b X_1 + c X_2 \ldots z X_n$, to the more complicated exponentials $Y = a X_1{}^b X_2{}^c \ldots X_n{}^z$. Successively shorter run situations are represented by more of the X_i being fixed or invariable, with their effects on output then subsumed in the constant a. Essentials of irrigation economics are illustrated here by first reducing a general input-output relation to a definite three-variable case per unit land area, taking Y as annual yield, I as average annual capital investment, and W as applied irrigation water: [2]

$$Y = f(I, W) = I^{1/4} W^{5/12}. \qquad [11\text{--}1]$$

The mathematical and economic properties of this and many alternative functions have been investigated by Beringer (1961), Carlson (1939a), Cobb and Douglas (1938), Heady and Dillon (1961), and Moore (1961). We have hypothesized equation [11–1] merely as a point of departure for illustrating the usefulness of economic principles in irrigation planning, investment, and water application decisions.

Such decisions are mainly concerned with whether irrigation is a feasible means for increasing net income; how irrigation might be planned to maximize net benefits; and important interrelations of investment feasibility and optimum water use.

A. Long-Run Feasibility and Minimum Costs [3]

Since the feasibility of irrigation requires that total benefits be greater than total costs, cost minimization for each given yield in equation [11–1] will maxi-

---

[2] We assume W to be the only water available merely to simplify discussion. In relation to W_c as seasonal consumptive use per unit area, $W = W_c/E_i$, with E_i as the efficiency of water application. The focus is on W as total applied water because it is most directly related to irrigation costs. Each variable in equation [11–1] can be expressed in any convenient physical or monetary units.

[3] For brevity in illustrating principles only, we are assuming an infinite planning period and permanence of the initial capital investment. Under these conditions, interest is the only cost of capital, and future returns as well as water and other costs (including interest) would be discounted to the present at the long-run interest rate. The annual values herein would thus remain proportional if discounted to present values as the fundamental basis of evaluation. See Lutz and Lutz (1951) or other standard texts for details on handling finite-period planning, capital replacement or maintenance, and other time complications.

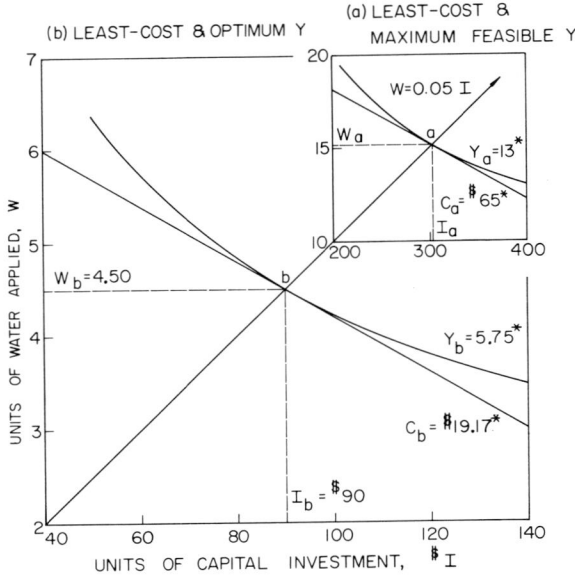

(b) LEAST-COST & OPTIMUM Y

(a) LEAST-COST &
MAXIMUM FEASIBLE Y

* YIELD AND COST CONTOURS RISE IN SEPARATE VERTICAL
DIMENSIONS. THE DIAGRAM IS DRAWN IN TWO-DIMENSIONAL
FORM TO SIMPLIFY AND EMPHASIZE POINTS OF TANGENCY.

Fig. 11–1. Least-cost capital investment and water application.

mize the likelihood of feasibility. Minimizing annual cost, regardless of yield, requires that capital investment and average annual water application be in proportions such that the ratios of their partial derivatives or incremental contributions to increased yield are equal to the ratios of their cost per unit. With P_i as the long-run borrowing interest rate on (or average annual cost of) capital and P_w as the unit "price" or cost of applying water, the condition for an optimum combination of capital and water in equation [11–1] is expressed as

$$\frac{a\,Y \,/\, a\,I}{a\,Y \,/\, a\,W} = dW \,/\, dI = 0.6\,W \,/\, I = P_i \,/\, P_w; \text{ and} \qquad [11\text{–}2]$$

$$W = I\,P_i \,/\, 0.6\,P_w, \text{ or } W = 0.05\,I. \qquad [11\text{–}3]$$

In equation [11–3] we assume a borrowing interest rate P_i of 8% and a unit water application cost P_w of $2.67.[4] Equations [11–2] and [11–3] show that the most basic facts for planning the optimum or least-cost use of capital investment and water (or any other pair of inputs) in farm irrigation are the input-output relation and per unit capital and water costs. For example, equation [11–3] indicates that one-twentieth unit of water will be applied per unit of investment to minimize total costs of obtaining any specified Y. This is illustrated in Fig. 11–1. The arrow is the locus of a continuum of points where equation [11–2] is satisfied, but only two points (a, b) corresponding to yields and total costs of interest here are

[4] This and all subsequent capital quantities, prices, costs and benefits are in USA dollars. These are the only terms necessarily stated in the same units.

plotted. The yield and cost curves are assumed to rise in the third dimension, and yield contours can be plotted from equation [11–1] by fixing Y and then either I or W. Though Fig. 11–1 is truncated, the I and W intercepts of the linear cost contours are determined by allocating completely any budgeted or otherwise given total cost quantity to interest on capital investment and then completely to water application, on the basis of the given rate of interest and average water application costs.

The practical values of first working out the minimum cost method of obtaining given yield responses from irrigation are that it can save a lot of work later, permits early analysis and comparison of farm irrigation potentials, and identifies soil and yield-response situations where irrigation is most likely to be feasible. Determining feasibility in fact means determining whether total benefits exceed total costs. This requires a net benefit relation and additional information on the probable unit price or value of output P_y. Using the previous symbols and equations, annual net benefits N are the difference between total annual benefits B and total annual costs C:

$$N = B - C = Y P_y - (IP_i + WP_w), \qquad [11–4]$$

which reduces[5] to

$$N = YP_y - (0.520\ Y^{3/2} + 0.868\ Y^{3/2}). \qquad [11–5]$$

Since minimum total costs for any response Y are inherent in equation [11–5] or equivalent relations, break-even prices, yields, or input costs as thresholds of economic feasibility are readily determined. For example, price P_y as the average benefit of irrigation would need to exceed average total costs per unit of increased yield Y for irrigation to be feasible. Average total costs per unit Y in equation [11–5] are C/Y or $1.388\ Y^{1/2}$, and rise from zero yield. Irrigation would be feasible at any yield provided $P_y > (C/Y) = 1.388\ Y^{1/2}$.

Alternatively, feasibility limits can be considered from an output standpoint with prices given. For $P_y = 5$, $(C/Y) = 5$ at Y of about 13, and irrigation would be infeasible, as it would result in net losses for $Y > 13$, even with costs minimized. The maximum feasible yield as estimated from $P_y = 5$ and the foregoing benefit-cost relations is denoted as $Y_a = 13$ in Fig. 11–1 and 11–2.

While grossly simplified, this treatment of irrigation cost minimization and feasibility illustrates the dependence of economic feasibility on three basic factors. First is the response of yield, not only to water application but also to capital requirements of different types of farm irrigation works or systems, associated land preparation, or any other initial investments required before the actual application of water can begin. Analyzing capital investment and water application through input-output relations (as in equation [11–1]) allows relative substitution of one input variable for another, according to their productivity and unit costs. For example, more careful land preparation, requiring increased investment, can increase average response to water, meaning that a given yield benefit can be obtained for less water.

Second, costs have both relative and absolute importance in feasibility, because ratios between unit input costs, when related to incremental contributions of inputs

---

[5] By substitution of equation [11–3] into equation [11–1], then transposition of equation [11–1] for I in terms of Y, and substitution of both this result and equation [11–3] into equation [11–4].

to yield, determine proportions in which inputs are combined most efficiently (*see* equation [11–2]). A lower interest rate as the unit cost of capital investment, other costs being unchanged, will tend to encourage investment in more elaborate works and systems, as well as increase the likelihood of feasibility. Similarly, lower water application costs, as reflected in lower wage rates or energy costs, will tend to favor intensive water application over capital investment and will also make for feasibility.

The third set of feasibility factors, crop and livestock prices, are prime indicators of irrigation benefits for individual farms and, when compared with costs (as in equation [11–4]), are a major determinant of the economic justification of irrigation. Consumer demand, existence of suitable markets, and quality standards for agricultural products must be considered in planning irrigation. These are reflected in expected prices at the farm level. It is possible to describe these prices as favorable or unfavorable only with respect to expected yield benefits and irrigation costs, since "high" prices do not insure feasibility. An excess of irrigation benefits over total costs is both necessary and sufficient for feasibility, regardless of the yield benefits, prices, and per unit costs involved.

Maximizing net benefits is a separate matter. In this case the minimization of costs for various yield levels represents the necessary condition for net benefit maximization. Favorable prices as prices over minimum average output costs then can be regarded as a condition sufficient for permitting net benefit maximization. The latter has several facets, depending on whether all inputs can be varied (the long-run problem), whether at least one input is a fixed quantity (an intermediate-run problem), and whether benefit maximization may be a qualified objective in view of risk and similar considerations. In its extreme form the short-run would be a situation in which all inputs were fixed quantities, in which case net benefits would be determined only by per unit prices and costs. This essentially is what economists call a market problem and is not covered here.

B. Long-Run Maximum Net Benefits

Guidelines for maximizing net benefits in the long-run by optimizing input combination: follow from the foregoing cost minimization and feasibility considerations. With average benefits P_y per unit yield response again assumed at $5, we rewrite equation [11–5] as

$$N = 5Y - (0.520\ Y^{3/2} + 0.868\ Y^{3/2}) = 5Y - 1.388\ Y^{3/2}. \qquad [11\text{–}6]$$

Incremental net benefits are then differentiated and set to zero as

$$dN\ /\ d\,Y = 5 - 2.983\ Y^{1/2} = 0. \qquad [11\text{–}7]$$

Solving equation [11–7] for Y indicates that net benefits are at a maximum per unit of land irrigated when Y is about 5.75 units. Denote this as Y_b. If maximization of annual net benefits is the objective of irrigating, and if all inputs (capital investment and water) can be employed as needed, a yield Y_b of 5.75 units is the optimum yield to attain. From equation [11–6], total benefits B_b would be $28.75, total costs C_b would be $19.17, and net benefits N_b a maximum of $9.58. Total C_b would be divided (*see* equation [11–5]) as interest of $7.20 on $90 of capital and as $11.97 for 4.5 units of applied water. This is the optimum combination of the two inputs for $Y_b = 5.75$.

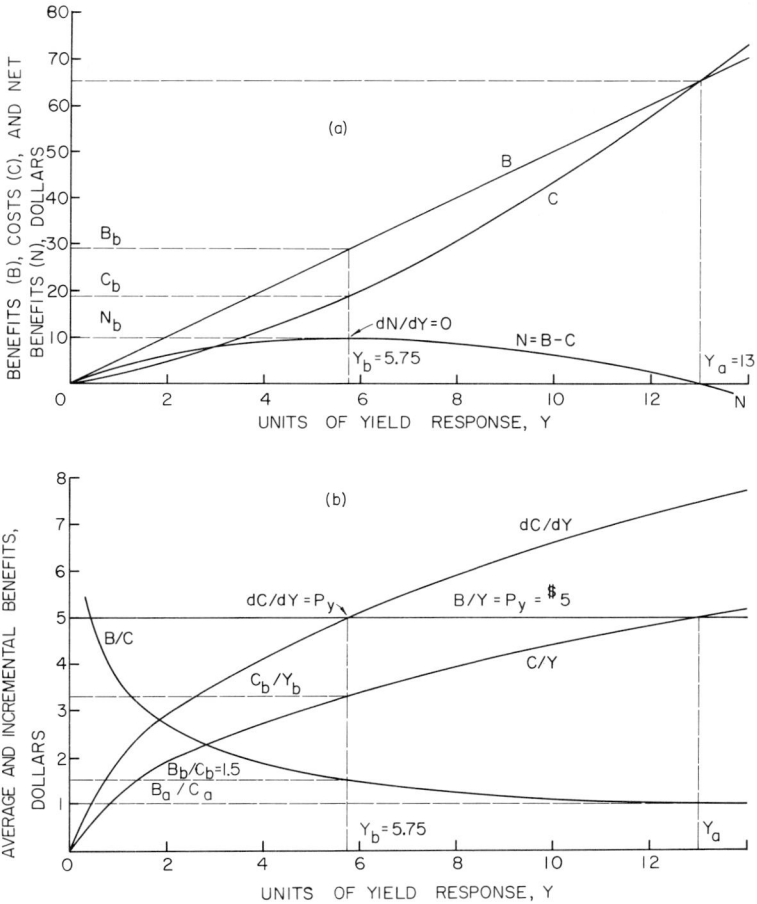

Fig. 11–2. Least-cost irrigation response and maximum long-run benefits.

Figure 11–2 shows the graphic solution of equation [11–6] for maximum net benefits, with dN/dY in the upper section zero at $Y_b = 5.75$, where incremental costs of dC/dY in the lower section are made equal to incremental benefits $P_y = 5$. Also plotted in the bottom section is the relation between minimum cost yield and the benefit-cost ratio B/C emphasized in project planning. The importance of the ratio as a long-run feasibility indicator is verified, since irrigation is feasible ($B/C > 1$) through the yield range where $B > C$, or $N > 0$. It also shows that both the ratio B/C and net benefits N are a maximum for any given Y produced at minimum cost. It is important to remember that the ratio B/C is most useful for long-run irrigation planning when maximized with reference to specified yield responses and costs. That is, the theoretical upper limit of B/C (as ∞ in Fig. 11–2), may deviate considerably from its optimal value (1.5 at $Y_b = 5.75$) for maximum net benefits.

C.   Investment Analysis and Optimum Water Use

Determining the feasibility of investment in irrigation is basically an inter-mediate-run problem, since capital as one input is predetermined with the re-maining inputs then allocated on the basis of feasibility or net benefit criteria analogous to those already outlined. There is the theoretically equivalent case, for example, where water application cannot exceed a water-rights-based level of maximum availability, with economic planning then keyed to determining the optimum investment for obtaining different yield responses, or maximizing net benefits—both subject to restricted water use.

The pervasiveness of limited capital or investment capability is due in large measure to capital being the unconsumed portion of past income, capital goods, or gifts in nature which may have numerous alternative uses for producing future income. The economic analysis of alternative income-producing enterprises has thus emphasized future income-yielding potential in relation to capital investment as current consumption foregone. Only major elements of such an analysis applied to irrigation will be outlined here. Irrigation planning with capital or other re-sources fixed is of great practical importance, yet it constitutes a special case within the decision framework already described. While we deal only with restricted capital here, the methods are pertinent to budgeting or enterprise analyses where many resources may be limiting to the farmer.

Consider an instance where a farmer needs or has $4,500 in capital to provide facilities for irrigating 50 units of land. The average fixed investment I_f of $90 per unit of irrigated area [6] is simply inserted as a constant in relations such as equation [11–1], which would be modified to

$$Y = 90^{1/4}\ W^{5/12} = 3.080\ W^{5/12}, \text{ and} \qquad [11\text{–}8]$$

$$W = 0.067\ Y^{12/5}. \qquad [11\text{–}9]$$

These two equations show the relation between yield response and water appli-cation, as well as the inverse water requirements of given yield responses. Whether the given capital investment of $90 is justified hinges on the net returns from water application. The latter are computed as the surplus S of total benefits B over the still variable water application costs C_v $(= W\ P_w)$, since the surplus is the source of interest payment for capital use and capital repayment. Again as-suming average benefits at $P_y = 5$ and unit water application costs P_w of $2.67, both irrigation feasibility and capital repayment are enhanced by maximizing S or net returns to water:

$$S = B - C_v = YP_y - WP_w = 5Y - 0.180Y^{12/5}. \qquad [11\text{–}10]$$

Also, $dS/dY = 0$ at $Y_c' = 5.75$ units (Fig. 11–3a). Total benefits B_c would again be $28.75, variable water costs V_c again $11.97, and the surplus S_c available for interest payment a maximum of $16.78. The associated optimum water require-ment W_c is computed directly from optimum Y_c in 11.9 as about 4.5 units.

---

[6] Taken as the previously derived optimum long-run investment only to show that, though incremental costs of obtaining a greater yield response through additional water application alone depend on the assumed levels of capital or other variables, optimum intermediate-run water use can coincide with optimum long-run use.

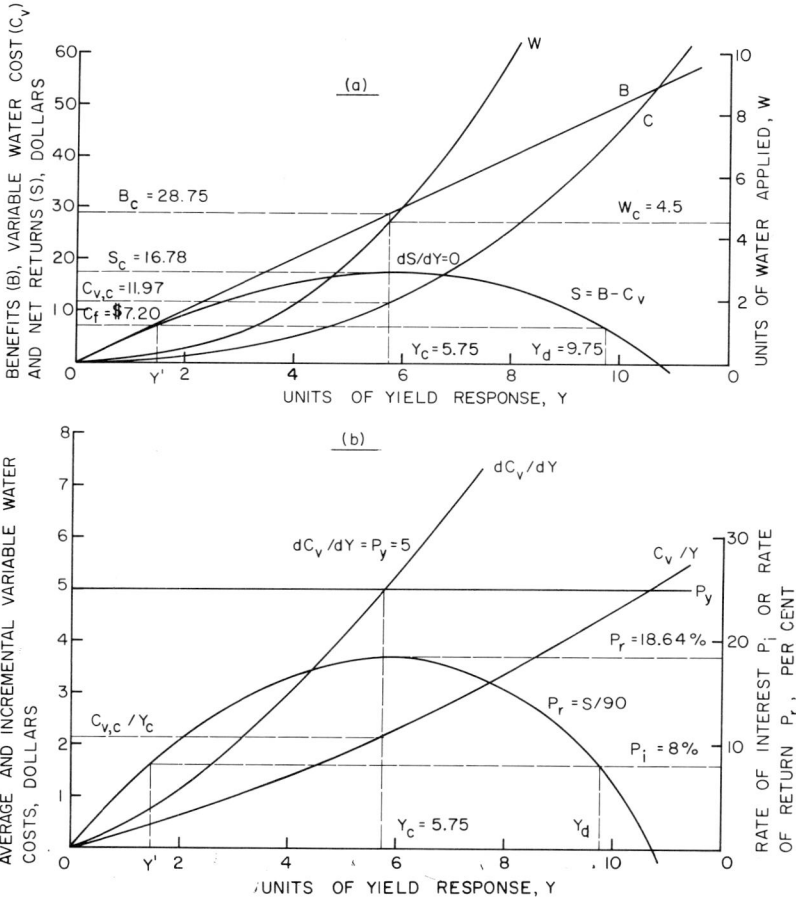

Fig. 11–3. Maximization of net returns to water application (capital investment fixed at $90 per unit area).

Maximizing S by optimizing Y and W maximizes the percentage rate of return P_r ($= 100S / I_f$) on a fixed investment, as expounded in the work of Fisher (1930), Lutz and Lutz (1951), and many others. The rate P_r is internal or indigenous to the irrigation enterprise. And the feasibility of a given investment depends on P_r being greater than P_i, with P_i the external or borrowing rate of interest on capital referred to earlier. Maximum P_r would be $16.78/90.00 or 18.64% in our example, indicating that a fixed capital investment of $90 per unit irrigated area would be justified if the market rate of interest P_i were $< 18.64\%$. The dependence of a maximum internal rate of return on optimum yield and water use is shown in Fig. 11–3, with strict reference to an investment I_f or $90 per unit area irrigated.

The equivalence of this approach to investment feasibility with methods that relate benefits B and variable costs C_r to fixed costs of investment C_f is shown by again referring to the net benefit relation equation [11–4]. We simply write.

$$N = -I_f P_i + (YP_y - WP_w) = -C_f + (B - C_v) = -C_f + S. \quad [11\text{–}11]$$

This emphasizes fixed costs as expense incurred regardless of whether any water is actually applied. These costs must be covered before any net benefits can accrue. Inspection shows that net benefits will be positive where $S > C_f$ or $S > I_f P_i$ and, where I is fixed as in the above case, positive if $P_i < P_r$. In other words, irrigation will be feasible if the market rate of interest on the fixed investment is less than the internal rate of return on the same investment.[7] This condition holds between Y' and Y_d in Fig. 11–3b if P_i is taken at 8%. Even if feasible, however, investment in irrigation may be the best use of limited capital only if its internal rate of return P_r is maximized (at Y_c in Fig. 11–3), and if this maximum rate is not exceeded in alternative lines of investment on or off the farm. Application of the rate-of-return principle to comparisons of alternative private and public resource development activities, including irrigation, is discussed extensively in current works on project economics (*see* Eckstein, 1957, 1958; McKean, 1958).

D. Benefit Maximization and Risk Factors

So far we have treated climatic risks only as they are inherent in the input-output equation [11–1], and have abstracted completely from situations where the object of irrigation may be minimization of drouth hazard as much as maximization of S as surplus returns to water application. The two objectives blend partly in that, over a limited range of yield response, surplus returns S normally will increase with reduction of drouth hazards. But beyond the yield response where S is maximized, further reduction of drouth concomitantly reduces S, because irrigation costs mount more rapidly than additional yield response and gross benefits. Parallel plotting of yield response and probabilities of remaining drouth against S and corresponding rates of return P_r will indicate relations in particular cases as in Fig. 11–4. This diagram is based on the preceding problem, but supposes 100% remaining drouth[8] at zero yield Y_0, 30% remaining drouth for the maximum-surplus yield Y_1, and 13% remaining drouth for yield Y_3—beyond which irrigation will be of no net monetary benefit. At Y_3 we assume again a borrowing interest rate P_i of 8%. Fixed interest costs of $7.20 on the $90 investment then exhaust S.

The complementarity of drouth reduction and maximization of S (and P_r) is obvious in Fig. 11–4 through the range Y_0 to Y_1, but a competitive relation is evident between Y_1 and Y_3. This says simply that any irrigator will strive for a response of at least Y_1, but that some rationally might seek responses of between Y_1 and Y_3, depending on their aversion to risks of remaining drouth. Risk aversion can be defined as the extent to which sacrifices of net returns will be subjectively balanced against more intensive irrigation for drouth reduction *per se*—in achiev-

---

[7] In present value terms, this says that irrigation will be feasible if the surplus S capitalized at the market rate of interest exceeds the fixed investment.

[8] In line with the assumption of zero yield without irrigation in equation [11–1] and subsequent functions. Where average dryland yields permitted some profit, irrigation benefit and cost relations would have a common origin at average yields and corresponding drouth probabilities under 100%. All magnitudes in Fig. 11–3 and 11–4 would then be interpreted as increases over preirrigation averages, but the same maximizing principles would apply.

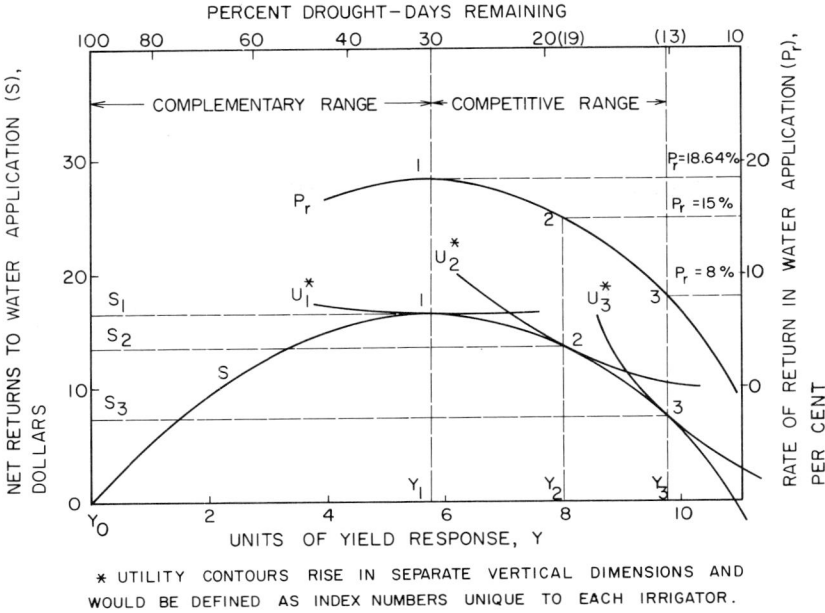

Fig. 11–4. Balancing net returns and drought reduction for maximum utility.

ing the greatest degree of personal utility or satisfaction from irrigating. In Fig. 11–4 maximum utility contours of U_1, U_2, and U_3 must be imagined as rising in the third dimension. They refer, respectively, to an irrigator completely indifferent to remaining drouth risk beyond a yield response of Y_1, one moderately averse to remaining risk, and one strongly averse to risk. All of the resulting equilibria are economically justified and optimum in terms of the given personal objectives. Irrigator 1 will ignore remaining drouth in maximizing returns to water application at Y_1, while irrigator 2 will strike some kind of balance between maximizing net monetary returns and further drouth reduction, as at yield Y_2. Irrigator 3 as the extreme case will only break even monetarily at Y_3 but at the advantage of more substantial drouth reduction.

The question of balancing maximum average returns against minimized drouth is largely academic where irrigation is essential for sustained agricultural production. But where the practice is regarded as supplementing rainfall, as in the relatively humid eastern USA, risk avoidance assumes considerable importance in irrigation planning. This is because systems must be designed and economically justified more for particular farm situations considering periodic rather than yearly use, and perhaps independently of large-scale water development projects. Sharp increases in irrigation noted in recent years in the eastern USA have stimulated considerable economic research, much of which is oriented to statistical determination of drouth recurrence and its relation to irrigation feasibility. Particularly suggestive studies are those of Ehlers (1960) and Knetsch (1959), wherein annual and intraseasonal drouth frequencies, optimum water application, and optimum associated fertilizer use have been analyzed simultaneously.

II. ECONOMICS OF PROJECT PLANNING

The cost-minimization and benefit-maximization principles outlined above can, with necessary adaptations, be carried over to decisions involving irrigation project formulation and evaluation. Empirical difficulties in this transfer are formidable, however, and further complicated by major conceptual differences in planning farm irrigation systems, single-purpose irrigation projects of varying scope, and multipurpose water development programs involving irrigation. Such differences relate mainly to the need to consider monetary, intangible, and other social benefits or costs for many beneficiaries over a project's entire life and, as well, to the practical impossibility of analyzing in detail every engineeringly feasible set of alternatives. Concepts and procedural problems will first be discussed with reference to irrigation projects—leading into generalized principles appropriate to the analysis of multipurpose water resource development.

A. Conceptual Problems in Project Planning

Conceptual problems in project planning are essentially of three types: qualitative problems of identification or definition, problems of quantitative measurement, and problems of prediction. Major qualitative questions are whether a project is to be evaluated from a local, regional, or national standpoint; what project effects are to be included as benefits or costs; and what effects can be expressed quantitatively or perhaps only descriptively. Logical and uniform approaches to these problems have evolved through research, interdisciplinary communication, and the experience of various public agencies engaged in irrigation or other resource development activities. Increasingly projects are being evaluated simultaneously from national, regional, state, and local viewpoints, and there is wider agreement on what project effects are to be analyzed. Also, operational interpretations of both primary and secondary benefits and costs have been set forth, principally by Holje et al., (1956), and Kimball and Castle (1963).

The long life of irrigation projects and the fact that planning takes place in a dynamic and uncertain setting give rise to numerous problems of prediction. Included here are basic questions on how far into the future one should reasonably look, what future prices and costs will be, and how future benefits should be weighted (through discounting) when compared to investment costs that must be incurred immediately. These questions are crucial to long-run feasibility. Accordingly, much current discussion of irrigation economics centers on such matters as interest and discount rates to be used in comparing irrigation projects with other public investments; demand, price, and cost projections; and realistic assumptions on the general state of a nation's or a region's economy over the period of a project's existence. But despite the importance of these issues, their significance is overshadowed by the complexities of multipurpose water resource planning, which provides an operational basis for economically balancing irrigation with other water uses.

B. Multipurpose River Basin Planning

Many countries have reached a stage of economic development where multipurpose land and water resource planning is the vehicle for obtaining optimum benefits from water resource development or management. In the USA, for example, the Senate Select Committee on National Water Resources has shown definitively that, in some river basins, water supplies are insufficient to meet even current demands for all uses and that in the near future other regions face the same prospects, especially when pollution abatement requirements are introduced into regional supply-demand appraisals. In these situations the efficiency of water use, reallocations or transfers among uses, and the development of new supply sources through multipurpose project planning are important and interrelated economic problems.

From an economic standpoint, the essence of multipurpose planning is that irrigation, power production, flood control, navigation, recreation, pollution abatement, or other water resource development objectives can be realized most efficiently if considered jointly, even if separate projects were engineeringly practical and economically feasible. Multipurpose planning is a complex field of engineering and economic analysis involving joint design or production principles, alternative allocations of joint costs, and various methods of project financing, all of which are further complicated by considerations of equitableness and related principles of welfare economics. The latter have been covered in some detail by Krutilla (1961) and various other writers.

1. FORMULATING MULTIPURPOSE PROJECTS

A major objective of economic analysis as applied to multipurpose projects is to formulate proposals that are of optimum design or scale, or that show promise of producing maximum net tangible benefits. Such a maximum is not an all-important choice indicator, however, as there may be cases where intangible benefits are considered important enough to suggest additional investments not balanced by additional monetary benefits. Or, intangible adverse effects may suggest reduced investment at the expense of monetary benefits. This general principle and related formulation criteria have been expressed in much of the literature of water resource economics, and have been made a matter of current agency policy in the USA by the President's Water Resources Council (1962).

Two important necessary conditions for maximum net benefits set forth by the Council are: (i) Tangible benefits from each separable purpose or all purposes combined should exceed separable or total costs; and (ii) there is no more economical means, evaluated on a comparable basis, of accomplishing the same purpose or purposes which would be precluded from development if the plan were undertaken.

The sufficient condition for net benefit maximization is that the scope of the project as a whole be carried to the point where aggregate benefits from the last increment of scale are equal to the costs of including that increment. Within this scope, incremental benefits from each purpose are equated.

This rule may be implemented in various ways, depending partly on the nature and complexity of given projects, but mainly on the availability of data required to construct total benefit and cost schedules and then to derive average and incremental benefits and costs. Figure 11–5 illustrates a simple case involving optimum total capacity of a reservoir serving irrigation i and municipal-industrial water needs m. The curves B_i and B_m in the upper section denote gross benefits for each purpose in relation to respective allocations of the total capacity that might be planned. Incremental benefits for each purpose are the derivatives of their respective gross benefits and are denoted by B'_i and B'_m in the lower section. Total incremental benefits B'_t stemming from either or both purposes are synonymous with B'_m at capacities under S_x; thereafter, B'_t is composited from B'_i and B'_m by horizontal adding. The total benefit curve B_t in the upper section is the integral of B'_t. Total cost as related to total capacity is given by C_t and its derivative or incremental costs by C'_t. Corresponding aggregated net benefits as total benefits less costs are denoted as N_t.

Aggregate net benefits N_t would be maximized in Fig. 11–5 if S_t units of total storage were planned, with S_i units allocated to irrigation and S_m units to municipal-industrial purposes. At these optimal capacities, incremental costs of storage are shown (in the lower section) to be equal to incremental total benefits as well as to incremental benefits for each purpose. Optimal positions on total benefit and cost functions are indicated by the small circles in the upper section, which denote points at which slopes of tangents to the functions would be equal, according to the condition $C'_t = B'_t = B'_i = B'_m$.

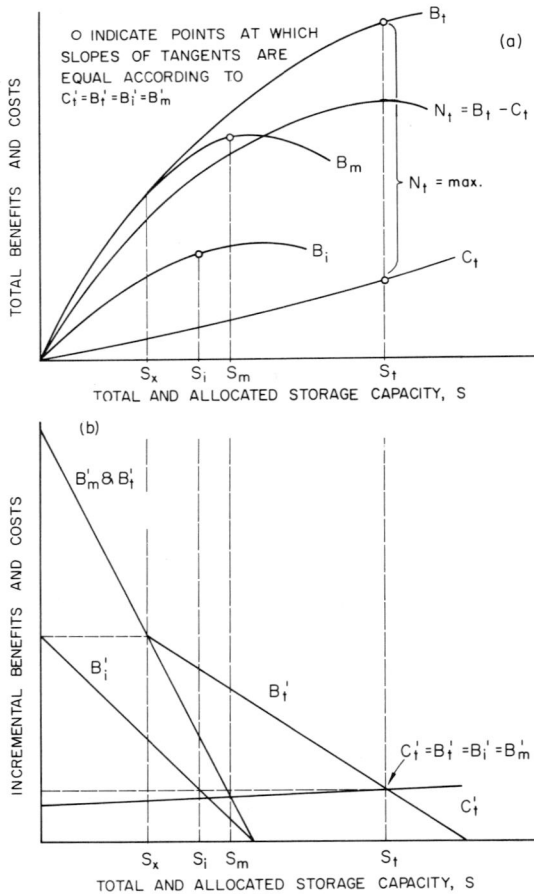

Fig. 11–5. Net benefit maximization and optimum storage for two purposes.

2. COST ALLOCATION AND REIMBURSEMENT CRITERIA

Because of its effects on the charges made for project benefits to recover capital investments, and thus on the incidence of net benefits, cost allocation is another important phase of multipurpose project planning. Even if costs attributed to some purposes are considered nonreimbursable as a matter of public policy, there remains the problem of setting these out accurately, so as to avoid understating or overstating reimbursable costs. Where various purposes are served jointly by given increments of scale (or capacity), an unequivocally accurate division of costs among purposes is impossible under most conditions. This is a result not only of the complexity of the typical multipurpose project but is due also to theoretical limitations. The latter are especially applicable where joint purposes are served in fixed proportions, as noted by Carlson (1939b).

Such difficulties do not mean that cost allocation is totally arbitrary, however, and a number of systematic and rational methods for allocating joint costs have been developed, although the "remaining benefits" method is now used most widely by resource development agencies in the USA. These methods have been reviewed extensively by the US Interagency Committee on Water Resources (1958), Gertel (1951), Golzé (1961), Huffman (1953), and Linsley and Franzini (1964). From these and other works we can list some generally accepted criteria for rational and equitable allocation and reimbursement:

1) Any allocation method should be consistent with proper project formulation to permit the inclusion of all justified purposes, including any purpose that produces benefits sufficient to cover its separable cost.

2) All purposes served by the project should be treated comparably and required to share in the allocation of project costs, regardless of applicable cost sharing and reimbursement requirements.

3) The minimum allocation to each purpose should correspond to its separable or incremental cost. This is the difference in total project cost with and without the purpose in question. The maximum allocation to any purpose should not exceed the benefits or the alternative costs, whichever is smaller. Within this range various purposes should share proportionately in both the advantages and the costs associated with the use of joint facilities.

4) Standards for cost sharing and reimbursement should promote optimum development and use of resources in proper sequence from the public viewpoint. Charges should be high enough to prevent waste, low enough to encourage full use, and sufficiently flexible to achieve an effective program.

5) Charges and assessments should be designed to bring about a reasonable association between benefit and cost distribution patterns.

6) Requirements for cost sharing and reimbursement for comparable purposes should be uniform for the various projects and agency programs.

7) Charges and assessments should be consistent with welfare and policy objectives. If a matter of defined policy, they may take account of such objectives as widespread benefit distribution, the conservation of resources for future generations, and economic growth or stability.

III.   SOME GENERAL POLICY ISSUES IN IRRIGATON

Questions of national and international irrigation policy center on creating the conditions and means whereby the rate of further irrigation development is in phase with sustained maximum rates of national economic development and trade policies. This assumes prior consideration of two basic factors: (i) The present capacity of a nation's agricultural economy to meet future domestic and export demands for agricultural products; and (ii) the availability of water supplies in relation to prospective water needs in the nonagricultural as well as the agricultural sectors of basin economies. Public irrigation projects are the direct instruments of irrigation policy. Some indirect but no less important influences on irrigation are private reallocations of water and other resources intended to increase or stabilize income—all in the face of continuing shifts in consumer demand, changes in market prices or costs, and variations in actual water supplies. These private influences as well as project planning, interact with various price support and other agricultural programs. Consequently, irrigation policy is interrelated with agricultural policy in general, as shown definitively by Ruttan (1965).

In countries or regions where current agricultural output is insufficient to meet impending needs, water resource planning can be oriented closely to irrigation and such other measures as agricultural flood control and drainage, though not to the exclusion of power generation, navigation, or other means for economic development. The difference is mainly one of degree in more developed areas, areas that already may be experiencing intense competition among irrigation and other water uses. In all situations, enlightened irrigation development must proceed with clear recognition of benefits possibly foregone elsewhere in a basin or in other sectors of basin economies, as studied by Gertel and Wollman (1960), Wollman (1962), and others. Such circumstances are increasingly common and have made multipurpose planning not only mandatory politically but an essential of any policy for economically balancing future irrigation and other water uses.

New irrigation developments may be opposed because they increase farm production and thereby possibly add, for certain crops, to agricultural surplus problems. It is generally recognized that the need for farm products is only one factor to be considered in irrigation development. Other objectives that must be considered are reduction in flood and drouth hazards, encouragement of family farms, and promotion of regional economic growth.

Another question is whether projects will increase the overall efficiency of crop and livestock production. This argues for simultaneous analyses of the productivity of land, water, and other farm resources either with or without irrigation in all major producing areas, and with respect to prospective rather than current output requirements. There is no justification for the arbitrary exclusion of irrigation or any other production technique from these analyses.

Finally, the practicalities of multipurpose planning indicate that river basins are best developed as entities, with all resources developed to serve a variety of purposes, including irrigation, flood prevention, recreation, domestic and industrial water supply, and so on. If a development plan fails to include justifiable irrigation, future irrigation development may be precluded entirely or at least made more expensive. This underscores the need for a long-range policy of balancing comprehensive development of land and water resources with resource adjustment programs to achieve continuing maximum social benefits from resource use.

LITERATURE CITED

Beringer, C. 1961. An economic model for determining the production function for water in agricuture. California Agr. Exp. Sta., Giannini Found. Res. Rep. no. 240. 20 p.

Carlson, S. 1939a. Input and output. *In his* A study of the pure theory of production. p. 10–28. Kelley and Milliman, Inc., New York. (As republished in 1956).

Carlson, S. 1939b. Joint production and joint costs. *In his* A study of the pure theory of production. p. 74–102. Kelley and Milliman, Inc., New York.

Cobb, C. W., and P. H. Douglas. 1938. A theory of production. Amer. Econ. Rev. 18(Supp.):139–156.

Eckstein, O. 1957. Evaluation of Federal expenditures for water resource projects. *In* Federal expenditure policy for economic growth and stability, p. 657–667. US Congr. 85(1). Joint Econ. Comm. print. US Government Printing Office, Washington.

Eckstein, O. 1958. The benefit-cost criterion. *In his* Water resource development; the economics of project evaluation. p. 47–109. Harvard Univ. Press, Cambridge, Massachusetts.

Ehlers, W. F. 1960. Economic implications of drought probabilities for humid area irrigation. Amer. Farm Econ. Ass., Procedures, Ames, Iowa. J. Farm Econ. 42(5): 1518–1519.

Fisher, I. 1930. The investment opportunity principles. *In his* Theory of interest. p. 150–177. Kelley and Milliman, Inc., New York. (As republished in 1954).

Gertel, K. 1951. Recent suggestions for cost allocation of multiple-purpose projects in the light of the public interest. J. Farm Econ. 33(1):130–134.

Gertel, K., and N. Wollman. 1960. Price and assessment guides to western water alloca tion. Amer. Farm Econ. Ass., Proc., Ames, Iowa. J. Farm Econ. 42(5):1332–1344.

Golzé, A. R. 1961. Reclamation in the United States. 2nd ed. The Caxton Printers, Ltd., Caldwell, Idaho. 486 p.

Heady, E. O., and John L. Dillon. 1961. Agricultural production functions. Iowa State Univ. Press, Ames. 667 p.

Holje, H. C., R. E. Huffman, and C. F. Kraenzel. 1956. Indirect benefits of irrigation development; methodology and measurement. Montana Agr. Exp. Sta. Bull. 517. 70 p.

Huffman, R. E. 1953. Irrigation development and public water policy. Ronald, New York. 336 p.

Kimball, N. D., and E. N. Castle. 1963. Secondary benefits and irrigation project planning. Oregon Agr. Exp. Sta. Tech. Bull. 69. 35 p.

Knetsch, J. L. 1959. Moisture uncertainties and fertility response studies. J. Farm Econ. 41(1):70–76.

Krutilla, J. V. 1961. Welfare aspects of benefit-cost analysis. Resources for the Future, Inc. Washington, D. C. Reprint no. 29. 10 p.

Linsley, R. K., and J. B. Franzini. 1964. Planning for water-resource development. *In their* Water resources engineering. p. 605–628. McGraw-Hill, New York.

Lutz, F., and Vera Lutz. 1951. Criteria of profit-maximization. *In their* The theory of investment of the firm. p. 16–48. Princeton Univ. Press, Princeton, New Jersey.

McKean, R. N. 1958. Special problems in the analysis of water resource projects. *In his* Efficiency in government through systems analysis; with emphasis on water resource development. p. 103–182. John Wiley and Sons, Inc., New York.

Moore, C. V. 1961. A general analytical framework for estimating the production function for crops using irrigation water. J. Farm Econ. 43(4–1):876–888.

President's Water Resources Council. 1962. Policies, standards, and procedures in the formulation, evaluation, and review of plans for use and development of water and related land resources. US Congr. 87(2). Senate Doc. 97. US Government Printing Office, Washington. p. 8.

Ruttan, V. W. 1965. The economic demand for irrigated acreage. Johns Hopkins Press (for Resources for the Future, Inc.), Baltimore, Md. 139 p.

US Interagency Committee on Water Resources, Subcommittee on Evaluation Standards. 1958. Proposed practices for economic analysis of river basin projects. US Government Printing Office, Washington.

Wollman, Nathaniel [ed.] 1962. The value of water in alternative uses. Univ. of New Mexico Press, Albuquerque. 426 p.

section V

Soil-Water Relations

V

12 Nature of Soil Water

PAUL R. DAY
University of California
Berkeley, California

G. H. BOLT
State Agricultural University
Wageningen, Netherlands

D. M. ANDERSON
US Army Cold Regions Research &
Engineering Laboratory
Hanover, New Hampshire

I. INTRODUCTION

Within the soil, water is involved in many different processes. Some seem to be almost purely physical, others seem to be predominantly chemical in nature, and still others appear to be both simultaneously. This circumstance has made difficult the formulation of a comprehensive theory which encompasses such diverse phenomena as soil water movement, water uptake by plants, and water interaction with plant nutrients, to name only three examples.

Experimental and theoretical work directed toward achieving a satisfactory understanding of soil water in its various forms and roles has been carried on actively for more than 60 years. Empirical measurements and qualitative interpretations have given way gradually to studies of fundamental mechanisms and to methods of expressing soil water phenomena mathematically. In modern perspective, quantitative measurement and mathematical expression are essential to the understanding of soil water and to the application of the knowledge to practical agriculture. This chapter deals with the properties of soil water which should be considered in any comprehensive theory of its behavior.

II. THE FORMS AND OCCURRENCE OF SOIL WATER

Soil water may be encountered as a liquid, a solid (ice), or a vapor. The condensed forms of water (liquid or solid) often exceed 30% of the soil weight, while the weight of water vapor rarely exceeds 5 ppm of the soil weight, and comprises < 5% of the weight of the soil atmosphere. At temperatures above freezing, the condensed phase has the properties of a liquid, although in the layers immediately adjacent to the soil particles these properties may differ considerably from those of a bulk liquid (*see* part III in this chapter).

Liquid water plays a vital role in many soil processes. It functions, for example, as a solvent, as a leaching agent, as a reactant, as a medium for chemical reactions, and as a plasticizing agent. Because of its power as a solvent, soil water is never pure, but contains many dissolved substances. Even in soils considered to be nonsaline, the total electrolyte concentration of soil water varies between 1 and 20 meq/liter. Ions most commonly found in solution are Na, Ca, Mg, K, NH_4, HCO_3, Cl, CO_4, and NO_3. In addition some dissolved organic substances and the gases N_2, O_2, and CO_2 in combinations up to 25 ppm are commonly found. Evaporation tends to concentrate soluble substances at the soil surface and leaching tends to carry them downward into the soil profile. The result is that the concentration of the soil solution is rarely, if ever, uniform.

The manner in which liquid water is distributed within the soil mass depends to a considerable extent on the individual soil properties. Soil is a porous, particulate, and (to a certain extent) colloidal medium. The solid phase consists of inorganic and organic particles having a wide range of particle sizes. The composition, size distribution, and arrangement of particles determine the other soil properties, to a large extent. The particles which form the framework (or matrix) have a low degree of solubility, with the exception of certain minerals such as calcite and gypsum. On the other hand, soil particles are generally hydrophyllic; that is, water adheres to them. Thus, although they are only sparingly soluble, they nevertheless have an affinity for water. The extent and nature of the affinity will be considered later (*see* part III in this chapter).

The area of solid surface accessible to water ranges from < 1000 cm^2/g of soil, for coarse sands, to $> 1,000,000$ cm^2/g for clay soils. The latter figure may include large amounts of "internal" surface, i.e., interlayer surface which may accommodate only thin layers of water. When only limited amounts of water are present (e.g., when the soil is air-dry), the interlayer spaces are already full of water and the external surfaces of the particles are covered with thin layers of water. The interstitial space not occupied by liquid water contains atmospheric gases, including water vapor. When greater amounts of water are introduced the layers present on external surfaces grow in thickness, and at the same time rings and wedges of water form at the points of contact of adjacent soil particles. The air-water interfaces have shapes which are determined by the configuration of the matrix and by the laws of capillarity. In the fully saturated soil, all of the interstices are occupied by water.

The water present in the soil in its various stages of wetting has been described functionally by the terms hygroscopic, capillary, and gravitational water (Briggs, 1897), and geometrically by the terms pendular and funicular water (Versluys, 1917). More recently, it has been found that there are functional relationships between the water content and physical variables such as vapor pressure, matric suction (*see* part IV of this chapter) and capillary or hydraulic conductivity (chapter 13). These functions are essentially continuous, showing that there are no sharp distinctions between the different stages of wetting. Thus, the older terms have been found to be too arbitrary for quantitative usage.

Inasmuch as the properties of liquid soil water are influenced by the close proximity of the solid phase, they will receive further attention in part III of this chapter.

The behavior of water in frozen soils is less well understood than in nonfrozen ones, but a number of studies have been made. The following are some of the relevant facts: Freezing can be initiated only after the soil has reached a temperature considerably lower than the freezing point (the "supercooling" effect). Furthermore, the freezing point is always lower than that of pure water. The ice which is formed becomes a part of the soil matrix, but not all of the soil water will freeze at the same temperature. In fact, as the temperature is lowered, the amount of ice increases.

Alternate freezing and thawing are responsible for interesting structural changes in soil; the segregation of soil particles of various sizes is one particularly interesting example (Corte, 1963). Frost heaving in soils, the formation of massive ground ice, and many other interesting phenomena need to be considered in a thorough discussion of frozen soils. However, freezing phenomena are seldom critically involved in irrigation, and will not be discussed further.

III. PROPERTIES OF SOIL WATER

The liquid phase of the soil shares a large amount of interfacial area with the solid phase on the one hand and the gas phase (when present) on the other. *The interactions between phases exert a profound influence on the movement and retention of water by the soil, and the behavior of water in the soil appears anomalous when compared with that of water in bulk.* The nature and causes of the unusual behavior will be touched on in this part of the chapter and treated in greater detail in later ones.

Much is known about the structure and properties of pure, free, water (Bernal and Fowler, 1933; Dorsey, 1940; Buswell and Rodebush, 1956; Low, 1961). Only a brief statement can be made here. Liquid water is a partially hydrogen-bonded substance made up of two parts of the isotopes of hydrogen to one part of the isotopes of oxygen. It is unique in having the highest dielectric constant of any known liquid and therefore it is the best general solvent for electrolytes. Moreover, its specific heat, melting point, and boiling point are unusually high and on freezing it expands instead of contracting as is the case with practically all other substances.

When substances are dissolved in water, the properties of the solution and those of the component *water* in the solution depart from those of pure water. For example, when electrolytes are added to water, the water molecules are oriented and compressed by the electrostatic fields of the individual ions, forming hydration hulls around them, and decreasing the dielectric constant of the solution. The dielectric constant decreases further when the solution is made more concentrated. Other properties undergo similar changes. In the soil, the influence of dissolved solutes on the properties of the water (their osmotic effect) is generally regarded as little different, in principle, from that in ordinary solutions. Expositions summarizing the rather considerable knowledge of the properties of aqueous solutions have been given by Gurney (1936) and more recently by Harned and Owen (1958).

It is generally acknowledged that the state of the water in the solid-liquid interfacial regions of the soil probably differs from that in the interior of the liquid phase because of interactions between the two phases. There are at least four possible kinds of interactions, which might extend some distance outwards from the interface: (i) hydration of adsorbed cations; (ii) formation of an electrical double layer at the solid surface, producing a high osmotic pressure near the interface; (iii) formation of hydrogen bonds, linking the water molecules to the mineral or organic surfaces of the particles; (iv) existence of London-Van der

Waals forces. It should be noted that the above forces, and more specifically those listed under (iii) and (iv) constitute the so-called *adhesion forces* between liquids and solids. It is clear from these considerations that any theory regarding the "nature of soil water" must involve the nature of the *soil*, including its content of adsorbed ions and dissolved solutes, as well as the nature of water.

In discussing the influence of the above forces on the behavior of water in soil, *the most outstanding fact is the rapid decrease of the partial molar free energy (or water potential) which occurs with decreasing water content (see* parts V and VI of this chapter). It should be noted that the partial molar free energy of water in soil is a property of the system, and not simply of the water. It is a macroscopic property, not rigorously assignable to any specific point of the liquid phase, but characteristic of a macroscopic region of the soil. (Its frequent use as a point function is justified only on statistical grounds, and is not valid for random volume elements of microscopic size.) The partial molar entropy and the latent heat of vaporization are also system properties, and they, like the partial molar free energy, are dependent on the water content.

In addition to the thermodynamic properties of the system, special consideration should be given to the properties of water in the vicinity of the solid surface, where conditions may be different than in a bulk water phase. Clay particles, in particular, are characterized by an electrostatic field which extends outwards from the particle surface. The extent of the field, when the particle is surrounded by water, depends strongly on the nature of the adsorbed ions (particularly on their valence), and on the concentration of salts dissolved in the water. The distribution of the ions is governed, in turn, by the electrostatic field. The concentration of adsorbed ions varies roughly from 2M to 5M at the solid surface, to 1M or less at 5A distance, and to less than a few tenths molar at 15A from the surface. The properties of water in the vicinity of the solid are undoubtedly influenced by the high ionic concentration in the region. R. K. Schofield (1935) was the first to interpret the behavior of water in soil in terms of the electrical double-layer theory. The application of this theory to the calculation of the water potential was tested experimentally on certain clay suspensions by Bolt and Miller (1955) and by Warkentin and Schofield (1962).

Hydrogen bonding and London-Van der Waals forces have also been cited as special features of the clay-water interface. Particular attention has been given to the hydrogen bonding mechanism by Low (1961) and others. Anderson and Low (1958) presented data indicating that the water adjacent to the solid phase of Wyoming bentonite has a lower density than that of water in bulk, and that the zone of decreased density extends outwards as far as 60A from the solid surface. Bloksma (1957), Kemper (1960), and Low (1958) have shown that ions are less mobile in such layers. Furthermore, water becomes more viscous near clay surfaces than in the bulk water phase, as demonstrated by the activation energy measurements of Low (1960), by experiments on the diffusion of deuterium hydroxide in clay pastes (Kemper et al., 1964), and by the nuclear magnetic resonance measurements of Pickett and Lemcoe (1959).

Low (1961) proposed the following hypothesis to account for the unusual properties of water in the vicinity of clay surfaces: Hydrogen bonding occurs between the water and the oxygen atoms of the clay surface, causing the adjacent water molecules to assume a tetrahedral coordination, and inducing additional layers of water to do likewise by an extension of the hydrogen bonding mechanism. According to this hypothesis, the decreased density of the water, the greater viscosity, and the reduced rates of ionic diffusion are caused by the presumed quasicrystalline structure of the water, which may pervade all or most of the liquid phase. Low suggests that the osmotic activity of the exchangeable cations (as postulated in the electrical double-layer theory) and the intensified hydrogen bonding mechanism (as postulated in his own theory) are both capable of decreasing the partial molar free energy of the water, and that the "relative importance of the two factors is still unknown."

There is considerable disagreement in the literature and among the present authors concerning the distance to which the abnormal properties extend into the liquid from the solid phase, and concerning the validity of the Low hypothesis. Additional research will be necessary before these questions can be settled.

IV. FORCES ACTING ON SOIL WATER

The forces acting on the soil water can be classified, for convenience, into (i) matric forces (those which result from the presence of the solid phase), (ii) osmotic forces (those caused by dissolved solutes), and (iii) body forces (inertial forces and gravitational force).

A. Matric Forces

When water is added to the soil from below, the soil becomes wet for a considerable distance above the water table. The movement of water upwards is similar to the rise of liquids in capillary tubes, and the two phenomena are assumed to have a common origin.

Capillarity involves molecular forces located in the boundary layers between phases. The mutual interaction of molecules in an air-water interface produces a stress distribution in the surface like that of a stretched elastic membrane (Bakker, 1928). If a narrow slit could be made in the surface layer, the exposed edges would tend to retract from each other. The amount of force that would be required per unit length of slit to prevent separation is called the surface tension. Its magnitude for water against air is 72 dynes/cm at 25C. Many authors prefer to replace the surface tension by the surface free energy, which has numerically the same magnitude, but is expressed in ergs/cm^2.

Other forces which must be considered in an analysis of capillarity are those which reside in the liquid-solid interface. The adhesion of liquids to solids can be described by the amount of mechanical work required to separate them when they are pulled apart at right angles to one another (Adam, 1941). This quantity (the work of adhesion) is related to the liquid-solid contact angle (θ) by the equation $W = \sigma (1 + \cos \theta)$, where σ is the surface tension. Clean glass and quartz, which have an angle of contact with water of $0°$, have a work of adhesion of 2σ ergs/cm^2. Since the work of cohesion of water is also 2σ it has been concluded that water adheres to such solids as strongly as water does to itself. On the other hand, those solids which do not have a strong affinity for water give large contact angles, and in some cases negative heights of rise.

When water rises to a height h in a cylindrical capillary tube of radius r, with zero contact angle, the upward force due to surface tension is $2\pi r\sigma$ and the downward force caused by the weight of the water column above the free water surface is $\pi r^2 h\, \rho g$. Equating these, we obtain

$$\rho g h = 2\sigma/r.$$

Note that $\rho g h = P_2 - P_1$, where P_1 is the hydrostatic pressure immediately below the interface, and P_2 is the pressure in the tube on a level with the free water surface. Hence,

$$P_2 - P_1 = 2\sigma/r.$$

Since the hydrostatic pressure immediately above the interface does not differ appreciably from that in the free water surface, the preceding equation describes a lowering of the hydrostatic pressure across the curved meniscus, an effect known as the Laplace pressure jump. The amount of the pressure decrease is commonly referred to as the suction.

In the case of a cylindrical tube, the liquid surface has a hemispherical shape. In general, however, the configuration of a liquid-air interface caused by contact between liquid and solid depends on the geometry of the solid, the angle of contact, the surface tension, and often the amount of liquid present. Laplace's equation,

$$P_2 - P_1 = \sigma[(1/R_1) + (1/R_2)],$$

relates the suction to the configuration of the interface, when its principal radii of curvature, R_1 and R_2, are known.

In the soil, the irregular shape and size of the pore spaces and the complicated way in which they interconnect has prevented the derivation of a comprehensive theory of soil water based on capillarity. However, many interesting developments have had their origin in the general knowledge of capillary theory (Miller and Miller, 1956). Rings and wedges of water are thought to adjust their curvatures in accordance with the prevailing pore water pressure, and *vice versa*. Furthermore, the removal of some of the water causes a readjustment in interfacial configuration leading to a decrease in pore water pressure so that soil water suction increases with decreasing water content.

A second mechanism of water binding by the soil matrix arises from the electrostatic fields of the solid surfaces (diffuse double-layer theory). According to this theory the adsorbed ions are distributed outwards from the surface in accordance with the Boltzmann distribution. The distribution is affected by the amount of water in the system and by the boundary conditions. Three cases may be distinguished: (i) An excess of water is present, placing the outer boundary of the liquid layer beyond the range of the electric field; (ii) a limited amount of water is present and the liquid layer is terminated at a distance from the particle surface less than the distance to which the field would extend if more water were present; and (iii) a limited amount of water is present, and the liquid layer extends to the surface of another particle.

The first case (i) serves to define the "equilibrium solution," i.e., the solution outside the range of influence of the solid surface. The electric field serves as a constraint, preventing the adsorbed ions from escaping, and therefore playing a role like that of the membrane in an osmometer. Hence, at equilibrium inside the double layer, the pressure at any point is greater than that in the equilibrium solution by an amount at least equal to the difference in osmotic pressure between this point and the equilibrium solution. The term "at least" is used here since there are other forces acting on the water in the direction of the solid surface (e.g., the electric forces acting on the water dipoles) which, at equilibrium, also cause an increase of the pressure.

In case (ii) the ionic concentration is everywhere greater than in the equilibrium solution of case (i). Consequently, the osmotic pressure will be everywhere greater than in (i). Equilibrium requires again that the pressure increases when moving from the outer plane of the liquid layer towards the solid surface, by an amount at least equal to the change in osmotic pressure experienced. This increase in pressure inside the liquid layer may thus be regarded as an internal compensation for the effect of the changing ionic concentration.

As in all other parts of the liquid in case (ii), the outer layer has a concentration greater than that of the equilibrium solution defined in (i). Since this layer has the additional feature that the electric field strength there is zero (the surface charge of the solid is neutralized completely within this outer layer, which is therefore called the "neutral plane") one concludes that the water present in this layer *cannot* be in equilibrium with the equilibrium solution defined in (i) *unless* the pressure of the water in the neutral plane exceeds that of the equilibrium solution, by an amount *equal* to the difference in osmotic pressure between "neutral plane" and equilibrium solution. Thus "free" equilibrium solution (or pure water), when brought into the neutral plane (the latter being subjected to atmospheric gas pressure) will be taken up spontaneously, expanding the liquid layer (Kruyt, 1952).

Thus, a soil system with water layers of finite thickness on the solid surfaces behaves as though the water were present at a negative pressure. This pressure is then equal in magnitude (but opposite in sign), to the osmotic pressure of the solution present in the neutral plane, i.e., at the air-water interface or, in the case of completely filled pores, in the plane of symmetry between opposing solid surfaces, where the field strength is also zero (case iii).

Space does not permit a more extensive discussion of the double-layer theory (see Bolt and Miller, 1958, for further details).

Much less quantitative information is available on the other adsorptive forces acting on soil water, which have already been referred to briefly in part III of this chapter.

B. Osmotic Forces

Dissolved (neutral) salts that are present in addition to the adsorbed ions are, in first approximation, not constrained by the electric fields of the solid surfaces. They thus will not affect directly the translocation of water in the soil (neglecting diffusion transport), unless other constraining barriers (e.g., semipermeable membranes) are present. Since the roots of higher plants and living organisms of various kinds possess such membranes, the exchange of water between them and soil is affected by osmotic forces. Furthermore, the exchange of water between soil and atmosphere is affected by the osmotic pressure of dissolved solutes. Since the solutes do not penetrate the gas phase, the equilibrium vapor pressure of soil water is determined by the combined matric and osmotic forces (*see* part VI, section A of this chapter).

Examining the matter more closely, the electric fields of the particles can cause a decrease in the concentration of the dissolved salts in the close proximity of solid surfaces, without notably affecting the concentration of water (the so-called negative adsorption). In effect this means that a porous body like soil can act as a semipermeable barrier for salt solutions, the available cross section for water transport exceeding that for transport of the dissolved salts. As a result, the osmotic forces due to salt concentration gradients may induce water transport in soil. The amount transported in this manner will depend on both the "degree of semipermeability" and on the gradient of the osmotic pressure. For systems with rather wide pores filled completely with water (saturated flow in well-aggregated or coarse-grained soil) the thickness of the layer with a decreased salt concentration is small in comparison to the total-transporting cross section. As a result this "degree of semipermeability," which could perhaps be called the osmotic efficiency of the system, is very small. The ensuing effect of the osmotic forces on the transport of water is then negligible.

C. Body Forces

Soil water is always subject to the force of gravity. In problems dealing with the movement of soil water, the inertial forces can generally be neglected, since the accelerations are generally small. However, centrifugal force, which is one type of inertial force, is sometimes used in experiments on water retention. In such cases, the centrifugal force is used as an equilibrating force against the matric forces, as in the case of moisture-equivalent determinations.

V. SOIL WATER SUCTION AND ITS RELATIONSHIP TO WATER CONTENT

It is a matter of common observation that when free water is brought in contact with a partially saturated soil, the water enters the soil spontaneously. This can be demonstrated through the use of a tensiometer, a simple form of which is shown in Fig. 12–1. The instrument consists of a porous filter cylinder and a narrow manometer connected by a suitable tubing and filled with water. When the water in the manometer comes in equilibrium with that in the beaker, the beaker is removed and replaced by a moist, but not saturated, sample of soil which is then covered to prevent evaporation. The level of water in the manometer arm will fall until equilibrium is obtained, provided that the soil is not too dry, in which case the water will reach the bottom of the U-bend. The difference between the equilibrium pressure observed in the presence of soil, and that observed when the instrument is immersed in free water, is called the soil water suction (matric suction). It may be expressed in centimeters of water head, bars or millibars, atmospheres, etc.

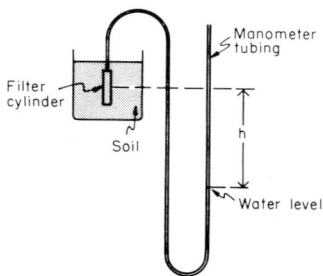

Fig. 12–1. Example of soil water suction measurement by means of a simple tensiometer (Suction $= S_m = \rho g h$).

Conversely, some of the water in a saturated sample may be extracted by means of a suction filter operating at a known suction (e.g., 100 cm of water) and additional amounts may be removed by applying greater suctions. Also, water will re-enter the soil from the filter cavity when the amount of suction is decreased. The ready displacement of water in this manner demonstrates that water is mobile in the soil, that it can be translocated by regulating the suction, and that the soil water content is functionally related to the suction.

A simple method of demonstrating the relationship between water content and suction was developed by Haines (1930), and is shown in Fig. 12–2. The results of such an experiment are shown in Fig. 12–3, with sand as the test material.

The hysteresis effect shown in Fig. 12–3 is probably a real relationship and not merely a delayed equilibrium, as sometimes claimed. Hysteresis has always been a troublesome aspect of soil water theory, and serious theoretical interpretations have been attempted only recently (Poullovassilis, 1962; Philip, 1964). Hysteresis has a strong bearing on the amount of water that can be stored in the soil, and in periods of intermittent rainfall it is a major factor in determining the shape of the moisture profile (Youngs, 1958).

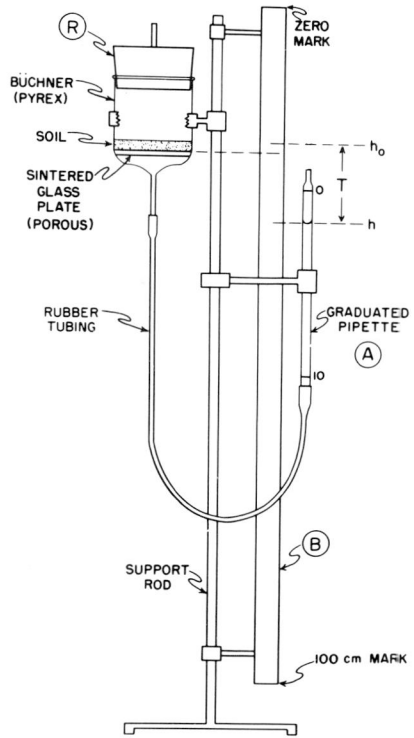

Fig. 12–2. Modified Haines method for determining curves of water content vs. suction ($S_m = \rho g h$).

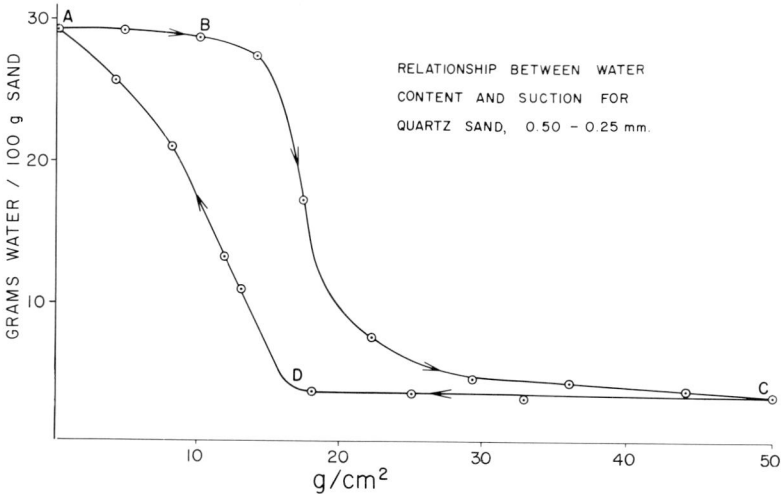

RELATIONSHIP BETWEEN WATER CONTENT AND SUCTION FOR QUARTZ SAND, 0.50 – 0.25 mm.

Fig. 12–3. Curves of water content vs. suction, showing hysteresis: ABC = drying; CDA = wetting (Suction is indicated in grams force per cm^2 = S_m/g.).

VI. THE POTENTIAL CONCEPT

A. Definitions of Potentials

The tendency of the water in the manometer (part V of this chapter) to enter the soil until the pressure deficit in the manometer equals the suction of the soil water demonstrates the binding of water by the soil matrix. Obviously the water ceases to enter the soil at the moment the net force acting on all water molecules equals zero. Although the absence of a net force acting on the water is a valid criterion for equilibrium, in practice this criterion cannot be handled conveniently, as it would require proof that the *vector* sum of all forces acting on the water equals zero. (This would be difficult in view of the fact that the different forces act in different directions.)

It has been found much more convenient to describe the effect of a force on an infinitesimal body of water by means of the potential energy of this body of water in the existing force field. This potential energy is defined as the negative integral of the force over the path taken when moving the body of water from a predefined standard location to the point under consideration. When expressed per unit amount[1] of water the potential energy thus defined is referred to as the potential of water (ψ_k) in the particular force field considered, or

$$\psi_k = - \int \mathbf{F}_k \cdot d\mathbf{l}$$

in which $\mathbf{F}_k$ is the force per unit amount due to the field "k." When $\mathbf{F}_k$ is expressed in dynes/g mass, one finds the potential in ergs/g. This potential is a scalar, which implies that the sum of the effects of different force fields can now be found by taking the *algebraic* sum of the "component" potentials corresponding to the different fields acting, or

$$\Psi = \sum_k \psi_k = \sum_k - \int \mathbf{F}_k \cdot d\mathbf{l},$$

in which Ψ designates the "total" potential (i.e., potential due to the combined effects of all force fields). Thus in a system *at equilibrium*, where the net force on an infinitesimal body of water equals zero at all locations, the total potential must be invariant with location, since

$$\Psi = \sum_k - \int \mathbf{F}_k \cdot d\mathbf{l} = - \int \left(\sum_k \mathbf{F}_k \right) \cdot d\mathbf{l} = 0.$$

The reverse of the preceding statement, i.e., "equilibrium exists if the total potential Ψ is invariant with location" is strictly true only if the additional conditions of constant temperature and constant electrolyte (or solute) concentrations are fulfilled. Accepting these conditions, the usefulness of the above analysis becomes evident if one realizes that although at equilibrium the total potential remains constant, the component potentials may vary, thus compensating one another. Especially since the value of certain component potentials may be de-

---

[1] Water potentials in this chapter are expressed as *energy per unit mass* while in chapter 16 they are expressed as *energy per unit volume*.

rived directly from experiment the equilibrium condition may allow the calculation of other component potentials.

Leaving the temperature condition out of consideration, the necessity (in principle) of constant electrolyte concentration for equilibrium merits further attention. Three situations should be distinguished here.

1) All parts of the system are in "open" connection with each other (no semipermeable barriers present). Then constant electrolyte (and other solute) concentrations must be established in order to attain true equilibrium. If concentration differences exist, however, these will disappear through diffusion, usually without materially affecting the transport of water. Thus for this case the condition for "hydraulic" equilibrium could be given in close approximation as *constancy* of the total potential *without* taking the *electrolyte concentration* into account.

2) Certain parts of the system are connected only via perfect semipermeable barriers (e.g., vapor phase). Since solute transfer is now impossible, differences in concentration may be compensated by complementary differences in other component potentials e.g., pressure difference, as in an osmometer at equilibrium. The condition for equilibrium is then solely *constancy* of the *total* potential.

3) Certain parts are connected through imperfect semipermeable barriers (cf part IV, section B, of this chapter). Now differences in salt concentration will induce in principle both diffusion and mass transfer of the solution, the combined effect of which cannot be compensated permanently by complementary differences in other component potentials (since diffusion will persist), whereas neglect of the concentration gradient on the transport of water is often not warranted. The equilibrium condition for these systems must be: *constancy* of the *total* potential and *constancy* of the *solute* concentrations.

It is convenient to distinguish the following component potentials:

1) Gravitational potential (ψ_g), the potential of water in the gravity field. This potential is found immediately from the application of the potential definition. Thus $F_g = -g$, and

$$\psi_g = \int g\,\mathrm{d}z = gz$$

in which z is the height of the water above a standard level chosen at a convenient location.

2) Pressure potential (ψ_p), the potential due to the pressure in the soil water. Using the definition of potential one may derive this potential by introducing the force component per unit mass in the direction l corresponding to the direction of the pressure gradient, $\mathrm{d}P/\mathrm{d}l$. Thus, $F_p = -(1/\rho)\ (\mathrm{d}P/\mathrm{d}l)$, and

$$\psi_p = \int \frac{1}{\rho} \frac{\mathrm{d}P}{\mathrm{d}l} \times \mathrm{d}l = \frac{1}{\rho} \Delta P.$$

In soil, the pressure potential is usually incorporated into the matric potential defined below.

3) Matric potential (ψ_m), the potential originating in the solid phase. As was pointed out in part IV of this chapter, the water in soil is subject to several groups of forces caused by the presence of the solid phase (the matrix). Of these, the "capillary forces" manifest themselves as a lowering of the water pressure under curved menisci (part VI, section B, of this chapter). In addition, the solid surface exhibits adsorption forces, among others, those related to the presence of adsorbed ions. Since usually it is not possible to distinguish these solid phase effects from each other it is necessary to combine them into a single component potential, named matric potential. It should be remembered that the above pressure potential is usually an integral part of the matric potential.

4) *Osmotic potential* (ψ_s). In addition to the above potentials, the effects of the dissolved salts in the soil water must be taken into account. Designating this effect as osmotic potential, one finds, in analogy with the pressure potential, that

$$\psi_s = -\frac{1}{\rho} \Pi,$$

in which Π = osmotic pressure due to the dissolved salts or other solutes. It can be shown that the osmotic pressure should be that of the equilibrium solution (i.e., dialyzate) of the soil, since overlapping of the osmotic potential and the matric potential must be avoided.

Recombining the component potentials, one thus finds for soil water:

$$\Psi = \psi_g + \psi_m + \psi_s.$$

In those cases in which the matric influence is expressed entirely by a pressure deficiency, ψ_m may be replaced by ψ_p.

B.   Experimental Evaluation of the Potentials

Applying the above concept to the system shown in Fig. 12–1, one may state immediately that in case of complete equilibrium the total potential of water in the soil Ψ must be equal to that of the water in the manometer Ψ'. Thus,

$$\psi'_g + \psi'_p + \psi'_s = \psi_g + \psi_m + \psi_s.$$

The gravity and osmotic terms being equal, one finds that

$$\psi'_p = \psi_m.$$

Using the equation given for the pressure potential in part VI, section A, of this chapter, one finds that

$$\frac{1}{\rho} \times (P - P_o) = \psi_m,$$

where P and P_o are the manometer readings when the tensiometer is placed in the soil and in free water, respectively. As $(P_o - P)$ was designated the matric suction S_m, this implies that

$$\psi_m = \frac{-1}{\rho} S_m.$$

The tensiometer thus gives a direct measure of the matric potential, provided that diffusional equilibrium has been established. If not, small deviations may be expected, depending on the permeability of the tensiometer cup and of the surrounding soil for the dissolved solute. For coarse-grained soils and a coarse filter cup this effect is usually negligible. If, on the other hand, the material of the filter cup contains such fine pores that the solute is effectively retained (i.e., a semipermeable membrane) the pressure difference registered by the manometer will include the contribution of the osmotic potential. Assuming the tensiometer to be filled with pure water ($\psi'_s = 0$) one then finds at equilibrium:

$$\Psi' = \Psi, \text{ or}$$
$$\psi'_g + \psi'_p + \psi'_s = \psi_g + \psi_m + \psi_s, \text{ or}$$
$$\psi'_p = \psi_m + \psi_s, \text{ and}$$
$$\frac{1}{\rho}(P - P_o) = \psi_m + \psi_s.$$

The pressure difference $(P_o - P)$, observed with the "semipermeable" tensiometer cup, has been termed the total moisture stress, or total suction S_t.

The best example of a semipermeable membrane is the gas phase, which does not transmit the dissolved salts. The water is then transmitted as vapor and one arrives at the determination of the vapor pressure as a measure of the total suction. In that case,

$$\Psi'_{(vapor)} = \Psi,$$

or, neglecting again the gravity term,

$$\Psi'_{(vapor)} = \psi_m + \psi_s.$$

The pressure potential of water in the vapor form is found as

$$\psi'_{p\,(vapor)} = \int \frac{dp}{\rho_v},$$

in which p = vapor pressure ρ_v = density of water vapor. Using the vapor pressure of free pure water p_o as standard state, and assuming the gas law to be applicable, this becomes:

$$\psi'_{p(vapor)} = \int_{p_o}^{p} \frac{p\,RT}{M} \times \frac{dp}{p},$$

in which R/M = gas constant expressed per gram of gas. Thus,

$$\psi'_p = \frac{RT}{M} \times \ln p/p_o$$

$$\simeq \frac{25 \times 10^9}{18 \times 0.4343} \log p/p_o \text{ (ergs/g)}$$

$$= 3.2 \times 10^6 \log p/p_o \text{ (cm water head)}.$$

This equation is useful in estimating the relation between relative vapor pressure and total suction. Thus a relative vapor pressure of 0.90 (90%) corresponds to a total suction of 146,000 cm water head, whereas 99% relative vapor pressure still amounts to a suction of about 14,000 cm water head. These figures indicate the limitations of the vapor pressure method for determining suction values. For agricultural practice the range of greatest interest is between 100 and 15,000 cm of water head, corresponding to relative vapor pressures between 99% and 100%. Rather elaborate instrumentation is necessary to acquire sufficient accuracy in this range (Richards and Ogata, 1958). The application of the tensiometer, on the other hand, is limited to the suction range between 0 and 800 cm of water head, dissolved gas in the water of the tensiometer giving rise to the formation of gas bubbles around this value of the suction, thus interrupting the continuity of the liquid in the tensiometer.

Determination of the suction values in the range between 1,000 and 20,000 cm of water head is usually accomplished by means of the pressure membrane apparatus. In that case the water in the soil is subjected to an excess gas pressure, and connected to water under atmospheric pressure via a finely porous membrane. At equilibrium the total potential is again equal on both sides of the membrane. One finds again that

$$\Psi' = \Psi, \text{ or}$$

$$\psi'_g + \psi'_p + \psi'_s = \psi_g + \psi_m + \psi_s + \psi_{p_e} \cdot$$

Now the potential of the water in soil under the influence of an external gas pressure in excess of 1 atm contains an extra term ψ_{p_e}, i.e., the pressure potential due to this external gas pressure. Assuming gravity and osmotic components to be identical on both sides of the membrane this implies

$$\psi'_p = \psi_m + \psi_{p_e} \cdot$$

The water outside the separating membrane being under atmospheric pressure, $\psi'_p = 0$, and thus

$$\psi_m = -\psi_{p_e} \cdot$$

Applying the equation for the pressure potential due to external gas pressure this gives:

$$\psi_m = -(1/\rho) \times P_e.$$

Thus the matric suction equals the applied gas pressure in the apparatus, expressed in appropriate units.

C.   The Isothermal Equilibrium of Water in Soil

The condition for equilibrium of water in a soil profile is that the total potential be equal throughout, since the net force per gram of water must be zero ($-d\Psi/dl = 0$ in every direction). Assuming diffusional equilibrium to be established in the profile and using the tensiometer as the test instrument one may derive the distribution of the matric potential. The gravity potential is found by the measurement of the height.

In the special case of equilibrium above a water table, where the latter is taken as the datum level for ψ_g and ψ_m, the equilibrium is determined by

$$0 = \psi_g + \psi_m = gz + \frac{P - P_o}{\rho} = gz - \frac{S_m}{\rho}, \text{ or}$$

$$z - S_m/\rho g = 0,$$

where $S_m/\rho g$ represents the value of the suction expressed in centimeters of water head. Figure 12–4 shows this relationship between suction and height above the water table at equilibrium.

Fig. 12–4. Distribution of water content W, suction S_m, and gravitational potential above water table at equilibrium, after drainage from the saturated state ($z =$ cm; $\psi_g/g =$ cm; $W =$ percentage of water, by volume).

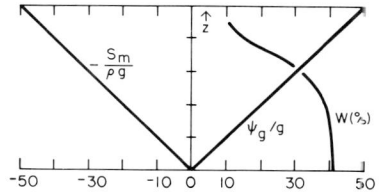

Because of the relationship between suction and water content (part V of this chapter), the curve of z vs. w in Fig. 12–4 is essentially the same as that of suction vs. water content curve in Fig. 12–3.

The equilibrium of water near a solid surface is likewise governed by the principle of uniformity of the total potential. Neglecting gravity because of the small dimensions of the system considered, one finds that the matric potential is constant. In this case it is interesting to separate the pressure component ψ_p from the adsorption component ψ_a. The latter potential decreases to large negative values towards the solid surface, its decrease being compensated by a local increase of the pressure potential. Thus,

$$\psi_m = \psi_a + \psi_p = \text{constant.}$$

Figure 12–5 shows a schematic representation of the potentials at various distances out from the surface.

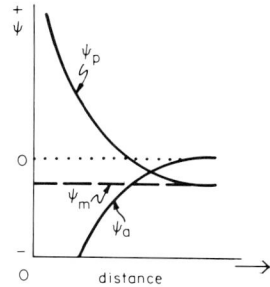

Fig. 12–5. Distribution of component potentials close to the solid surface

VII. CONCLUSIONS: IMPLICATIONS OF SOIL WATER THEORY WITH RESPECT TO IRRIGATION SCIENCE

The nature of soil water has been reviewed briefly with special emphasis on the potential concept. Although it is not practical to attempt to measure directly the forces which act on the water of the soil, it is relatively easy to measure its potential, by the various techniques mentioned. The variations of the potential from point to point in the soil furnish sufficient information to calculate the total driving force, which is zero at equilibrium and finite under nonequilibrium conditions.

Because of the relationship between water content and suction, which can be determined experimentally, the distribution of water in the soil in equilibrium with the gravitational field can be predicted. Further uses of the suction/water content relationship will be found in chapters 13 and 15.

Calculations of energy relationships and potential distributions in the soil furnish the best means yet devised for explaining and predicting by a single theory most of the phenomena relating to soil water. Further extensions of the theory, especially for dynamic conditions, will be given in subsequent chapters.

LITERATURE CITED

Adam, N. K. 1941. The physics and chemistry of surfaces. 3rd ed. Oxford Univ. Press, London. 436 p.

Anderson, D. W., and P. F. Low. 1958. The density of water adsorbed by lithium-, sodium-, and potassium-bentonite. Soil Sci. Soc. Amer. Proc. 22:99–103.

Bakker, G. 1928. Kapillarität und Oberflächenspannung. Wien-Harms Handbuch der Experimentalphysik. Akademische Verlagsgesellschaft, Leipzig. 458 p.

Bernal, J. D., and R. H. Fowler. 1933. A theory of water and ionic solution, with particular reference to hydrogen and hydroxyl ions. J. Chem. Phys. 1:515–548.

Bloksma, A. H. 1957. The diffusion of sodium and iodide ions and of urea in clay pastes. J. Coll. Sci. 12:40–52.

Bolt, G. H., and R. D. Miller. 1955. Compression studies of illite suspensions. Soil Sci. Soc. Amer. Proc. 19:285–288.

Bolt, G. H., and R. D. Miller. 1958. Calculation of total and component potentials of water in soil. Amer. Geophys. Union, Trans. 39:917–928.

Briggs, L. J. 1897. The mechanics of soil moisture. US Dep. Agr. Bur. Soils, Bull. 10. 24 p.

Buswell, A. M., and W. H. Rodebush. April 1956. Water. Sci. Amer. p. 2–10.

Corte, A. E. 1963. Particle sorting by repeated freezing and thawing. Science 142: 499–501.

Dorsey, N. E. 1940. Properties of ordinary water substance. Reinhold Publ. Co., New York. 673 p.

Gurney, R. W. 1936. Ions in solution. Cambridge Univ. Press, Cambridge (England). 206 p.

Haines, W. B. 1930. Studies in the physical properties of soil: V. The hysteresis effect in capillary properties and the modes of moisture associated therewith. J. Agr. Sci. 20: 97–116.

Harned, H. S., and B. B. Owen. 1958. The physical chemistry of electrolytic solutions. 3rd Ed. Reinhold Publ. Corp., New York. 803 p.

Kemper, W. D. 1960. Water and ion movement in thin films as influenced by the electrostatic charge and diffuse layer of cations associated with clay mineral surfaces. Soil Sci. Soc. Amer. Proc. 24:10–16.

Kemper, W. D., D. E. L. Maasland, and L. K. Porter. 1964. Mobility of water adjacent to mineral surfaces. Soil Sci. Soc. Amer. Proc. 28:164–172.

Kruyt, H. R. 1952. Colloid science. Elsevier Publ. Co., Amsterdam. I:255. 389 p.

Low, P. F. 1958. The apparent mobilities of exchangeable alkali metal cations in bentonite-water systems. Soil Sci. Soc. Amer. Proc. 22:395–398.

Low, P. F. 1960. Viscosity of water in clay systems. Clays and Clay Mineral. 8:170–182.

Low, P. F. 1961. Physical chemistry of clay-water interaction. Advance. Agron. 13: 269–327.

Miller, E. E., and R. D. Miller. 1956. Physical theory for capillary flow phenomena. J. Appl. Phys. 27:324–332.

Philip, J. R. 1964. Similarity hypothesis for capillary hysteresis in porous materials. J. Geophys. Res. 69:1553–1562.

Pickett, A. G., and M. M. Lemcoe. 1959. An investigation of shear strength of the clay-water system by radio-frequency spectroscopy. J. Geophys. Res. 64:1579–1586.

Poulovassilis, A. 1962. Hysteresis of pore water, an application of the concept of independent domains. Soil Sci. 93:405–412.

Richards, L. A., and G. Ogata. 1958. Thermocouple for vapor pressure measurement in biological and soil systems at high humidity. Science 128:1089–1090.

Schofield, R. K. 1935. The interpenetration of the diffuse double layers surrounding soil particles. Int. Congr. Soil Sci., Trans. 3rd (Oxford, Great Brit.) 1:30–33.

Versluys, J. 1917. Die Kapillarität der Böden. Int. Mitt. f. Bodenk. 7:117–140.

Warkentin, B. P., and R. K. Schofield. 1962. Swelling pressure of Na-montmorillonite in NaCl solution. J. Soil Sci. 13:98–105.

Youngs, E. G. 1958. Redistribution of moisture in porous materials after infiltration 1. Soil Sci. 86:117–125; 2. Soil Sci. 86:202–207.

13

The Dynamics of Soil Water

Part I—Mechanical Forces

E. E. MILLER

University of Wisconsin
Madison, Wisconsin

A. KLUTE

University of Illinois
Urbana, Illinois

In standard irrigation practice, water transport within the soil may be classified in three phases: (i) *infiltration*, while water is being applied; (ii) *redistribution*, after application ceases, usually culminating in a quasi-equilibrium or slow-moving distribution; and (iii) *withdrawal*, which is mainly absorption of water by plant roots to supply transpiration requirements, although evaporation at the soil surface or drainage to lower levels may be significant in certain situations. These are all dynamic processes of such importance to irrigation practice that a detailed exposition of present knowledge of soil water movements—complicated and incomplete though it may be—is an essential part of this monograph.

The great bulk of past work on this subject, covering more than 50 years, has assumed isothermal conditions and the Darcy proportionality of flow to mechanical driving force. Although this idealization is rarely an accurate representation of conditions in the field, it has been the most practical point of departure in the long effort to understand, and hence better to control, the tremendously complicated water flow behavior of soils. Even these oversimplified systems are sufficiently complicated that only with the very recent development of sophisticated computer methods and of nondestructive monitoring techniques have there been signs that some of the long standing questions and arguments may soon be resolved. The emphasis in the first section of this chapter is therefore basically concerned with isothermal Darcy law systems. Part II of chapter 13 by Cary and Taylor discusses the investigation of water transfer under temperature and solute gradients which has become increasingly active, following the theoretical investigations of Philip and De Vries (1957), published about a decade ago.

The first part is divided into four sections:

I—*Characterization of flow through moist soils.* This includes the pertinent physical properties of the soil and its two fluids, water and air, and a discussion of the nature of isothermal water movement within soils.

II—*Analysis of flow systems.* The discussion of section I is translated into mathematical language in the form of a macroscopic partial differential equation with associated boundary conditions. Methods of solution are surveyed in broad terms.

III—Applications to irrigation. Darcy flow systems which have been investigated experimentally and/or theoretically and are pertinent to irrigation are discussed in considerable detail.

IV—Unsolved problems. Inadequacies of the Darcy law approach are summarized as a guide to more effective and realistic employment of this technique.

I. CHARACTERIZATION OF FLOW THROUGH MOIST SOILS

A. The Two Fluids: Water and Air

Growth of an irrigated crop ordinarily requires that the soil be unsaturated or "moist"; i.e. it contains both water and air. In chapter 12, many of the physical and chemical properties of the soil water were discussed at some length, but for present purposes, let us review the mechanical properties of water and air in bulk.

1. **Water.** The density ρ is about 1 g cm^{-3}; the surface tension σ ranges from 30 to 70 dyne cm^{-1} (the higher values being associated with higher degrees of purity); the contact angle γ is commonly near zero for soil materials; and the viscosity η is about 1 centipoise (0.01 dyne sec cm^{-2}).

2. **Air.** The density is about $1/1000$ that of water, and the viscosity is not quite $1/50$ that of water. (In practice, therefore, it is usually reasonable to assume that the pressure of the air throughout moist soil is the same as that of the atmosphere bathing the soil surface.)

B. Activation of Flow Processes

As we have seen in chapter 12, the water in a moist soil is in a lower energy state than is free water at the same elevation. Stated operationally, when the free water inside a tensiometer pressure cup is in equilibrium with a moist soil surrounding the cup, the pressure of this water is less than that of the air surrounding the soil. The functional relationship between the water content of the soil and the tensiometer pressure, variously called "water retention curve" or the "moisture characteristic" of the soil, is nonlinear (*see* Fig. 13I–2A) and is strongly hysteretic.

With this background of soil moisture statics, we shall now investigate the movement of water which occurs when a moist soil is not in equilibrium. Broadly speaking, we may consider such movements to be (i) *driven* by energy gradients, (ii) *supplied* by changes in the water content and/or flow across boundaries, and (iii) *impeded* by the minuteness and tortuosity of the intermeshing paths of flow through the partly filled pores.

C. Macroscopic vs. Microscopic Viewpoints

At the very heart of soil water analysis is the successful development of a *macroscopic* approach. The basic principles which govern the mechanical behavior of water and air within individual pores (or at least within the larger ones) are reasonably well understood from classical physics. However, the actual pore geometry of any given specimen of soil is too complicated to be described in microscopic detail. Another specimen of the "same" soil will, of course, be utterly different in microscopic detail. The "sameness' of two specimens can only be expressed in statistical or macroscopic terms. We are already quite familiar with such macroscopic concepts as fractional pore volume or "total porosity" θ_s; fractional water volume θ; and saturation $S = \theta/\theta_s$. For dynamic systems, a macroscopic flow-velocity vector will be needed, because the microscopic velocity vector varies drastically from point to point within any given pore. A logical choice for this vector is the simple vector average of the microscopic velocities over the total volume of all three phases (water, air, and solid). This vector average $\vec{v}$ can easily be shown to have the property that the volume rate of transfer of water across a macroscopic plane, per unit area of the plane, is just the component of $\vec{v}$ in the direction perpendicular to that plane. This provides an equivalent operational definition, so that $\vec{v}$ can be measured in the laboratory.

A tensiometer inserted into a soil spans hundreds or thousands of pore diameters; thus, the tensiometer pressure p is patently a macroscopic quantity, whatever may be its relation to the microscopically-varying pressure within the individual pores. In discussing flow it is necessary to apply calculus to this macroscopic pressure, to consider its gradient, for example. Employing calculus is permissible only if it is possible to choose macroscopic differential elements which are big enough to include a representative assortment of pores, but which at the same time are small enough that, within a single element, the values of macroscopic variables such as p and $\vec{v}$ are approximately constant. Basically, this is the same problem that is encountered in treating molecular distributions as continua (in the development of the classical concepts of dielectric constant, specific heat, thermal conductivity, etc.) except that molecules are several orders of magnitude smaller than representative pores, so that considerably more leeway is available for "smearing out" molecular distributions into continua.

D. Driving Forces: Mechanical Energy Terms

The forces which activate the motion of water in soils may be classified as mechanical (pressure, gravitational, centrifugal), molecular (thermal, osmotic, absorptive), electrical, etc. As mentioned above, the first section of this chapter largely concentrates on the mechanical forces for which an elaborate technology has been developed.

It is generally helpful in mechanics to deal with forces indirectly, by interposing the energy concept. An energy distribution, being a scalar field, is usually much easier to handle than is a vector field of forces. Since the flow velocities in soils are ordinarily so low that kinetic energy is completely negligible, we shall be concerned only with *potential* energy and, for the most part, only with its macroscopic, mechanical components.

E. Differences of Total Head, H, Activate the Flow: Pressure Head, h, Governs Wetness

The total potential energy (mechanical) per unit "quantity" of water, described in the previous chapter as a constant for static systems, also governs the movement of water in dynamic systems. Water may be expected to flow from regions of higher total potential to adjacent regions of lower total potential. All three mechanical choices for expressing the unit quantity of water—volume, mass, and weight—have been used widely in the literature. By electing, for this section, to use *weight* as the unit of quantity, we obtain for the total potential the well-known hydraulic head, which we shall call total head H.[1]

The reference level or datum for H, as for any potential, is purely arbitrary. However, to discuss the state of wetting of a particular point in the soil, we will also define a localized head which is referenced *to the level of that point*, and we will call this the *pressure* head h because it represents the *pressure* of the free water within a tensiometer cup in equilibrium with the soil at that point. We will then describe the height of the point in question, taken positive where the point is above the datum, as a *gravitational* head z so that

$$H = h + z. \qquad [13I-2]$$

For an unsaturated, wettable soil the pressure head h is inherently negative. Since we are free to place the datum level where we like, we have chosen in Fig. 13I–1, which illustrates equation [13I–2], to place the datum above the soil, so that in this drawing the signs of h, H, and z are all alike, i.e., all are negative. Since we may wish on occasion to discuss a soil contained in a pressure cell, Fig. 13I–1 covers such a general case. The manometer that measures h opens into the reservoir of gas which permeates the soil. The pressure of this gas, relative to the atmosphere outside the cell, is shown as h_c, in waterhead units. Except when a pressure cell is used, h_c will be zero and need not be mentioned.

Within large water-filled pores of any soil which is not too dry, the pressure head represents (in head units and relative to the pressure in the gas-filled spaces) the actual microscopic pressure inside the water-filled spaces. In such pores, this gas-water pressure difference $(p_w - p_g)$ must be balanced mechanically by curvature of the interface, thus,

$$\rho g h = (p_w - p_g) = 2\sigma R^{-1} \qquad [13I-3]$$

where σ is the gas-water interfacial tension, and R^{-1} is the interface curvature. Technically, R^{-1} is the Gauss mean curvature. In the special case where the interface is a spherical segment, R^{-1} is the reciprocal of the sphere radius. Since h and $(p_w - p_g)$ are inherently negative for unsaturated soils and σ is considered positive, equation [13I–3] requires that R^{-1} be taken negative for the normal, gas-side-concave condition of the interface. In petroleum technology, the term $-(p_w - p_g)$ is called capillary pressure and is inherently positive (although the subscripts would be changed; w being translated into wetting and g being changed into nw for nonwetting, e.g., oil).

This relation can be regarded as a general microscopic differential equation

---

[1] Symbolizing the total potential energy of an element of water by E_T and its volume by V, we can represent its mass by ρV, and its weight by $\rho g V$. If we describe the total energy per unit volume as a pressure, p_T, and assign to the energy per unit mass the symbol ψ_m (which was used without the subscript as a general symbol for energy per unit "quantity" in chapter 12) we see that

$$\left\{ \frac{E_T}{V} \equiv p_T \right\} = \rho \left\{ \frac{E_T}{\rho V} \equiv \psi_m \right\} = \rho g \left\{ \frac{E_T}{\rho g V} \equiv H \right\}$$

$$\text{or } p_T = \rho \psi_m = \rho g H. \qquad [13I-1]$$

Note also that H is ψ_m/g, as one would expect in comparing the weight and mass bases. The interrelation between alternative sets of units will be carried farther in the Darcy law discussion below.

ARBITRARY DATUM LEVEL

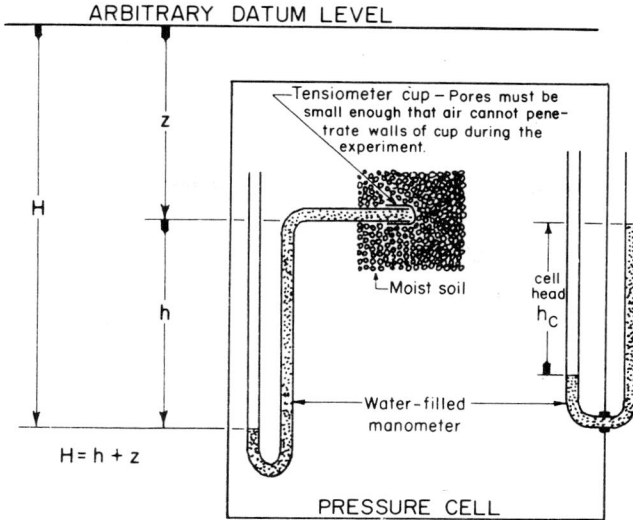

Fig. 13I–1. Relation of total (mechanical) head H to its components: pressure head h and gravitational head z. Heads are taken positive upward, hence all three are negative in this drawing. (Only a portion of the soil flow system is shown; it is not necessarily at equilibrium.)

governing the shape of the interface (Miller and Miller, 1955a, 1955b, 1956). Since the interface shape sets the microscopic boundaries of the water-filled spaces within the pores, it obviously determines (for large-pored media) the macroscopic water content θ or the fractional saturation S. Multiplicities in the allowable solutions of this equation for the interface shape give rise to irreversible "Haines jumps" from one solution to another solution at the same pressure. These irreversible microscopic jumps show up macroscopically as hysteresis in the soil water characteristic, $\theta_H(h)$. (When H appears *as a subscript* it specifies that the relation is hysteretic; when not used as a subscript, H means total head.)

For progressively smaller pores or crevices, the actual microscopic pressure inside water-filled spaces within the pores begins to deviate from the pressure within an associated tensiometer cup, as various adsorptive forces begin to encroach upon a significant proportion of the water-filled space. Other things being equal, however, the macroscopic water content is still governed by the macroscopic pressure head h measured by the tensiometer. However, the basically hysteretic (i.e., time-scale invariant) relation of θ to h may become noticeably disturbed by time-dependent effects caused by such mechanisms as the swelling and shrinking in clay components of the soil structure.

F. Mechanisms of Water Transport in Soil

The primary modes of transport of water in soils are (i) viscous flow through liquid-filled pores and crevices, including quasiviscous flow reaching into the layers of water adsorbed on the solid surfaces, and (ii) diffusion of vapor through air-filled pores. In principle both of these modes should contribute to the flow in combination. Liquid flow is the dominant mode in fairly moist, nearly isothermal soils. Vapor flow does not achieve dominance until soils become quite dry, although the presence of large temperature gradients favors the contribution of this mechanism. For typical soil water situations, either of these modes of transport would be expected to produce rates of flow proportional to potential energy gradients—for the first mode because kinetic energy is negligible (no turbulence); for the second mode because the rates of molecular diffusion and heat conduction (the latent heat of evaporation and condensation must be supplied by conduction) are proportional to corresponding gradients. Any combination of the two modes should also exhibit such proportionality.

Anomalies observed in colloidal materials have sometimes been interpreted as evidence for an abnormally high viscosity of water in the layers closest to solid surfaces. An abnormal local viscosity can only produce a failure of macroscopic flow proportionality if it is *also* a nonproportional (non-Newtonian) type of "viscosity". The reader will observe that in the second section of this chapter an assumption of proportionality between flux and gradient is made for a number of other types of transport in soils.

G. Darcy's Law (Flow Rate Proportional to Energy Gradient)

The first published relation between flux of water and energy gradient was obtained empirically in 1856 by Henry Darcy after a study of the saturated sand filters used in the fountains at Dijon, France (Darcy, 1856; Hubbert, 1956). While Darcy's law was one-dimensional and was not written in differential form, it clearly required a proportionality between flow rate and the gradient of total energy.

Slichter (1899) generalized Darcy's law for saturated media into a three-dimensional macroscopic differential equation of the form,

$$\vec{v} = - K_s \vec{\nabla} H \qquad [13I\text{--}4]$$

where K is a constant, frequently called the hydraulic conductivity, and the subscript, s, is used to denote saturation. The unit for K is best written as a velocity per unit hydraulic gradient.[2]

---

[2] A considerable source of confusion associated with the otherwise very convenient head units, is that $\vec{\nabla} H$ comes out as a length over a length, which is dimensionless. It is helpful to remember that H represents a *height of water*, i.e., a mechanical potential energy per unit weight of water. Hence $\vec{\nabla} H$ is the energy gradient per unit weight, the unit of which is tied to the standardized earth value of g, and is called the unit hydraulic gradient. (For example, if water is kept for a long time lightly ponded on the upper surface of a vertical tube of soil which drips all the while from its lower surface, the flow proceeds within the tube under unit hydraulic gradient.) Another unit for the Darcy constant which is also in wide use, especially in the petroleum literature but also in soils, is called "permeability" and is usually written with a small letter k. It is based upon the energy per unit volume (pressure) option of equation [13I–1], except that it has conventionally been multiplied by the bulk viscosity of the liquid in an attempt to obtain a universal flow constant for a given medium, independent of the liquid with which it is saturated. (For complicated media such as soils this attempt is usually not very successful, but it is quite useful in some petroleum applications.) Following the notation of the previous footnote on units, we write,

$$\vec{v} = -K \vec{\nabla} H = -(k/\eta)\vec{\nabla} p_T = -(k/\eta)\rho \vec{\nabla} \psi_m = -(k/\eta)\rho g \vec{\nabla} H \quad [13I\text{--}5]$$

hence $K = k(\rho g/\eta)$. Notice that k/η will be in units of velocity divided by pressure gradient, pressure being energy/volume. Therefore, in absolute units, the permeability k becomes a length squared.

Richards (1931) postulated that Slichter's (1899) concept could usefully be extended to unsaturated states by assuming that the conductivity K, as well as the water content, could be treated as (nonhysteretic) functions of the *pressure* head h. Richards' generalization has been used for soils with considerable success in the last three decades, although it must be admitted that even today the number of careful tests to which it has been subjected is disconcertingly small. Some of these will be discussed later in this chapter. The analogous permeability equation[2] has been used quantitatively by petroleum technologists much more extensively and is widely accepted as a practical success.

An obvious deficiency in the simple-function postulate is that it fails to take into account the hysteresis of soil water characteristics, which is known to be quite large. Experimentally, this defect has usually been side-stepped by limiting the choice of experiments to those in which the pressure head changes monotonically, either always drying, or always wetting, so that the use of single-valued functions is permissible. Since hysteresis is inherent in the wetting-drying processes of irrigation and since it produces important and complicated effects during the redistribution phase after water application ceases, we have previously used a subscript H to denote hysteresis of the moisture characteristic $\theta_H(h)$ and will now extend this usage to the conductivity relationship (Miller and Miller, 1956). With this modification, Richards' modification of Darcy's law may be written

$$\vec{v} = -K_H(h)\vec{\nabla} H. \qquad [13I-6]^3$$

The relation $K_H(h)$ for soils is highly hysteretic, though the relation $K_H(\theta)$ or its equivalent $K_H(S)$ is only slightly hysteretic for those few media which have thus far been examined (G. C. Topp, 1964. Hysteretic moisture characteristics and hydraulic conductivity for glass-bead media. *Ph.D. thesis, Univ. Wisconsin, Madison*). (*See* Topp and Miller, 1966). This observation is helpful, but the problem of dealing with hysteresis is still serious because the highly hysteretic water characteristic still enters basically into any analysis of hysteretic flow systems.

With decreasing water content, $K_H(\theta)$ decreases drastically in size because the large pores are emptied first, forcing the small pores and crevices to carry the flow through much smaller and considerably more tortuous channels. That the range of K which is of practical concern is indeed large is apparent from the fact that experimental K values are commonly plotted on logarithmic paper containing many decades; e.g., in Fig. 13I–2B, eight decades are employed.

II.   ANALYSIS OF FLOW SYSTEMS

A.   Heat Flow Analogs

For *saturated* media, Slichter's (1899) generalization of Darcy's law was formally identical with Fourier's classical equation for the conduction of heat in solids, so that the many solutions which had been worked out for steady-state

---

[3] If one wishes to take account of mechanical body forces other than gravity which may act on the fluid, e.g., centrifugal forces, the gradient of total head, $\vec{\nabla} H$, in equation [13I–6] may be replaced by $(\vec{f} - \vec{\nabla} h)$ where $\vec{f}$ is the body force per unit weight which in equation [13I–6] would be represented by $-\vec{\nabla} z$.

heat flow became ready-made solutions for the analogous saturated steady-state flow systems.

For *unsaturated* porous media, the conductivity becomes a nonlinear function of either h or S so that unsaturated steady-state solutions are considerably more difficult to analyze. The water storage characteristic, $\theta(h)$ of an unsaturated medium is analogous to the heat storage capacity of a solid, except, of course, that it is a nonlinear and hysteretic relationship. The non-steady-state heat flow problems which have been solved by the classical methods for treating boundary-value problems are qualitatively analogous to corresponding non-steady-state systems of unsaturated porous media; in quantitative terms, the latter are made vastly more difficult by their hysteretic and nonlinear characteristics. Furthermore, there is no heat flow analog for the gravitational force field.

Fortunately, recent improvements in computer techniques make possible a brute-force attack on nonlinear systems so that even such formidable obstacles are not insuperable.

B. Derivation of Differential Equation by Elimination of $\vec{v}$

The generalized Darcy equation relates flow velocity to driving forces. A second constraint on flow velocity is provided by the condition of conservation of matter, expressed in terms of the water storage characteristic $\theta_H(h)$. Combining these constraints into a single condition. we may eliminate flow velocity as a separate variable, thereby obtaining a differential equation which contains only the space and time derivatives of a single variable. The condition of conservation of matter is simply that water, entering the boundaries of a small region, must accumulate within it, i.e.

$$- \operatorname{div} \vec{v} = \partial\theta/\partial t \,. \qquad\qquad [13\text{I--}7]$$

The inward flow at the boundaries of a macroscopic differential element is represented by $- \operatorname{div} \vec{v}$, the volume fraction of water by θ, and time by t.

Eliminating $\vec{v}$ between the conservatation equation and the Darcy equation, we obtain a relation between various derivatives of h and θ. However, h and θ are already interrelated by the water characteristic of the medium; hence the differential equation can be written either in terms of h or in terms of θ. Each variable has unique advantages. Instead of choosing between them, it is convenient to use both. Therefore we have:

$$\operatorname{div} [K_H(h) \; \vec{\nabla} (h + z)] = \left(\frac{d\theta}{dh}\right) \frac{\partial h}{\partial t} \qquad\qquad [13\text{I--}8]$$

$$\operatorname{div} \left[\; K_H(\theta) \; \left\{\vec{\nabla} z + \left(\frac{dh}{d\theta}\right) \vec{\nabla}\theta\right\} \right] = \frac{\partial\theta}{\partial t} \,. \qquad\qquad [13\text{I--}9]$$

In the petroleum literature the saturation S is normally used instead of θ in equations analogous to [13I–8] and [13I–9], and this alternative has its advantages.

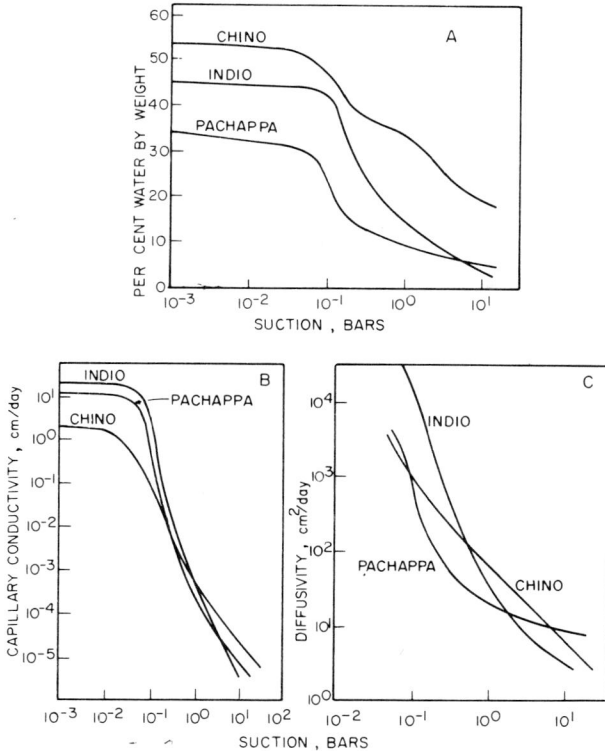

Fig. 13I–2. (A) Moisture characteristics $\theta(h)$, (B) conductivity $K(h)$, and (C) diffusivity $D(h)$, for three soils: Pachappa sandy loam, Indio loam, and Chino clay. From W. R. Gardner, 1960a.

C. Boundary Conditions

In addition to differential equations describing the behavior of the interior of the soil, it is necessary also to consider the conditions which govern the interfaces between different types of soil, as well as the conditions which can be applied at exterior boundaries. It will be noticed below that the physical reasoning employed is consistent with that used previously in developing the differential equation.

1) *The pressure head must be continuous* across permeable interfaces between regions containing different media. If a discontinuity were impressed on an interface at $t = 0$, the resulting infinite flow rate would promptly convert the discontinuity into a finite gradient. Notice that it does *not* follow that water content or saturation are continuous across an interface. This is one of the clear advantages of retaining equation [13I–8] in which *pressure head* is the dependent variable.

2) *The normal component of the macroscopic flow velocity vector is continuous* across any interface. This is obviously required by conservation of matter and remains true for an impermeable boundary or impermeable interface, in which case, of course, the normal component is zero.

D.   Some Unusual Problems Relating to Boundary Conditions

At the edges of a flow system, one may specify such conditions—pressure head as a function of time, normal component of flow rate, etc.—as may be appropriate. In some cases, however, the governing condition may change in character at some stage of the process. For example, one might specify a constant rainfall rate and also require that any excess water on the surface be kept drained away. At first, infiltration might be faster than the specified rainfall rate, so that the rainfall rate would be the governing boundary condition. Later, when the infiltration rate dropped below the rainfall rate, the condition that pressure head is zero would become the governing condition. Such mixed boundary conditions are obviously a little more difficult to treat analytically.

With nonlinear systems, it is quite easy to assign an arbitrary boundary condition which turns out to be impossible. For example, one may specify a rate of withdrawal at a surface which is too large to be sustained by the pressure head gradients, even if the medium were kept bone-dry at the surface. Nonsensical results of a calculation may sometimes be explained by such an accidental oversight.

E.   Simpler Versions of the Differential Equation

Solutions of the general differential equations [13I–8] or [13I–9] for specified boundary conditions can only be obtained by numerical, brute-force methods because of the empirical functions involved. Simplifications or approximations have been widely used to obtain analytical solutions for special cases, especially in the days before computers were generally available. Examples of such simplifications and approximations are the following:

1) Whenever the effect of gravity and other body forces can be neglected, the $\overrightarrow{\nabla} z$ term drops out of equation [13I–9]. Childs and Collis-George (1950) pointed out that the product $K_H(\theta) \times dh/d\theta$ is a function of θ. It is given the special name "moisture diffusivity" $D_H(\theta)$. Equation [13I–9] then becomes

$$\text{div } [D_H(\theta) \overrightarrow{\nabla} \theta] = \partial\theta/\partial t. \qquad [13I\text{–}10]$$

The term diffusivity and the symbol D are used because the form of equation [13I–10] is the same as that of Fick's law of diffusion; *there is no implication that molecular diffusion is or is not involved as a mechanism.* Because of its convenience, D is commonly used instead of K whenever equation [13I–10] will suffice. Given the water storage characteristic, K can be calculated from D, and conversely. However, it is not convenient simply to abandon K altogether in favor of D. Where gravity is a factor, D has no advantage. Furthermore, D is somewhat less convenient than K whenever hysteresis enters because D suffers discontinuities at each reversal of the direction of pressure change (Klute et al., 1964) while K is continuous. In fact, as mentioned previously $K(\theta)$ is not only continuous, it is virtually hysteresis-free. As shown in Fig. 13I–2C, $D(h)$ blows up in the region of saturation where $dh/d\theta$ becomes infinite (for suctions < 0.1 bar in this figure), whereas $K(h)$ is well-behaved. However, the variation of D from 0.1 to 10 bars is about two orders of magnitude less than the variation of K over this range.

2) The difficult problem of dealing with hysteresis drops out of monotonic systems—one stays with either the wetting or the drying characteristic for D, K, or θ, and hence the subscript H can be omitted.

3) For a "steady-state" system the water content at all points is independent of time so the right-hand sides of equations [13I–8, –9, and –10] are zero.

4) Empirical expressions for such characteristics as $\theta(h)$, $D(\theta)$, $K(\theta)$, etc., sometimes allow analytical solutions of particular problems. Three representations of $D(\theta)$ which have been used are the constant (W. R. Gardner, 1956), and the linear and exponential approximations (Scott and Hanks, 1962).

F.   Methods of Using the Flow Equation

In the third section of this part a number of pertinent applications of the flow equation are discussed in some detail, but it is useful first to outline the types of application which have been used, since space does not permit examples of all types.

1) The most general and modern technique is that of numerical solution employing a high speed digital computer. Without this technique the potential impact of soil water technology would be severely restricted.

2) Approximate solutions (or exact solutions of approximate equations) have been found to be very useful either for special cases of unusual simplicity (e.g., one-dimensional infiltration), or as a tool for checking theory against experiment, especially in earlier times.

3) Qualitative behavior of porous media can often be interpreted usefully from the flow equation and its boundary conditions without recourse to quantitative calculations. An example of this is a discussion by W. H. Gardner (1962) of the interesting behavior of a wetting front at a boundary between two soils of different texture.

4) Scale modelling is a useful technique when properly applied. This technique has been much more widely applied to petroleum reservoirs than to soils though some work has been done (W. R. Gardner, 1958, 1962).

5) With judicious selection, and use of physical insight, it is often useful to ride roughshod over questions of mathematical propriety to obtain order-of-magnitude estimates of flow system behavior, to single out dominant features, or to locate the real obstructions of a flow system for which a more accurate analysis is not possible, or at least is not convenient. Van den Honert (1942) has applied such methods to a discussion of the entire soil-plant-air flow system. Such methods are also quite appropriate for discussing such poorly defined practical concepts as "field capacity."

III.   APPLICATION TO IRRIGATION

A.   Plan of Discussion

In general, it is convenient to proceed from the simple to the more difficult types of flow systems. Accordingly, most of this section will be concerned with flow systems containing soil but no plants; a brief portion at the end will deal with the more difficult case of systems which contain growing plants. The discussion of the flow systems without plants will also begin with the simplest type, steady-state systems, and end with non-steady-state systems. The treatment of non-steady-state systems will begin with the "monotonic" (always wetting or always drying) systems and will close with the hysteretic redistribution of water after infiltration.

B. Soil Water Flow Systems Without Plants

1. STEADY STATE FLOW SYSTEMS

a. **Upward Flow Through a Uniform Soil.** Steady-state flow in the upward direction is related to evaporation of water from the top of a soil profile where it is removed by evaporation. This type of flow through a uniform soil column has been studied by Philip (1957a), W. R. Gardner (1958), Gardner & Fireman (1958), Moore (1939), Schleusener & Corey (1959), King & Schleusener (1961), and Wind (1955). An analysis based on Darcy's law can be developed by rearranging equation [13I–6] and integrating from the water table ($z = 0$, $h = 0$), giving

$$z = - \int_o^h \frac{K(h)}{K(h) + v_z} \ dh. \qquad [13I\text{--}11]$$

For steady-state, one-dimensional flow the flux v_z is a constant, independent of the depth z, so the integration may be completed if $K(h)$ is known. W. R. Gardner (1958) has suggested that much of the known conductivity data can be fitted by the function:

$$K(h) = a/[(-h)^n + b] \qquad h \leq 0 \qquad [13I\text{--}12]$$

where a, b, and n are constants. Small values of n (approximately 2) seem to be associated with fine-textured materials and larger values of n (3 or 4) with coarse materials. Schleusener and Corey (1959) observed that a function of the form

$$K(h) = C \ (h_d/h)^n \qquad [13I\text{--}13]$$

with n in the range 10 to 25 would fit the available conductivity data. In equation [13I–13], C, h_d, and n are constants.

An analysis based on equation [13I–11] and using either [13I–12] or [13I–13] leads to the conclusion that the evaporation rate should approach a finite limit as the pressure head at the upper boundary approaches an infinitely large negative value (*see* Fig. 13I–3). The limiting flux is nearly proportional to L^{-n}, L being the depth to the water table (W. R. Gardner, 1958) (*see* Fig. 13I–4).

At the limiting flux condition, the nonlinearity of K produces a very steep pressure head gradient at the top of the column. Below this region, as depth increases, the pressure head rapidly approaches the static equilibrium distribution. The prediction (i) of a limiting flux, and (ii) of the effect on it of water table depth are not entirely supported by experimental results. Experimental results shown by Gardner appear to confirm the concept of a limiting flux. However, data reported by Schleusener and Corey (1959) do not confirm the limiting flux. Instead, as the potential rate of evaporation was made greater the rate of evaporation from various soil columns passed through a maximum and decreased. The more shallow the water table, the greater the maximum rate of evaporation. The explanation of this phenomenon is not entirely clear. Schleusener and Corey advanced a theory which involved concepts of hysteresis. These same authors concluded from experimental observations that the decrease of evaporation rate as the potential evaporation was increased could not be explained by temperature

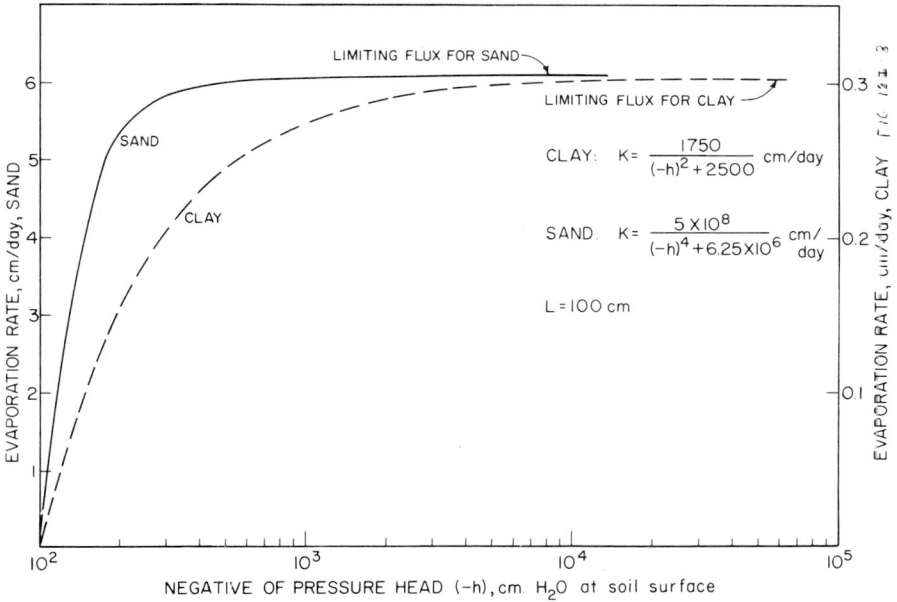

Fig. 13I–3. Theoretical evaporation rate in steady-state upward flow from a water table as a function of the pressure head at the soil surface for two conductivity functions representative of a sand and a clay.

gradients in the column. King and Schleusener (1961) reported additional experiments using cyclic evaporative conditions in which the same decrease of evaporation rate was observed.

Philip (1957b) and W. R. Gardner (1958) have considered the contribution of vapor movement to the evaporation rate. The soil column can be divided into a dry upper zone and a moist lower zone. In the upper zone, water movement is largely in the vapor phase. This region rarely extends into the soil more than a few centimeters. In the lower zone the movement is predominantly in the liquid phase. In the steady state, the flux is the same in each zone. For movement of water in either the liquid or the vapor phase, K falls rapidly and nonlinearly with increasing soil dryness, going to zero for both types of transport when no moisture is present. This nonlinearity produces a limiting flow rate, *with* or *without* the vapor-transfer surface zone.

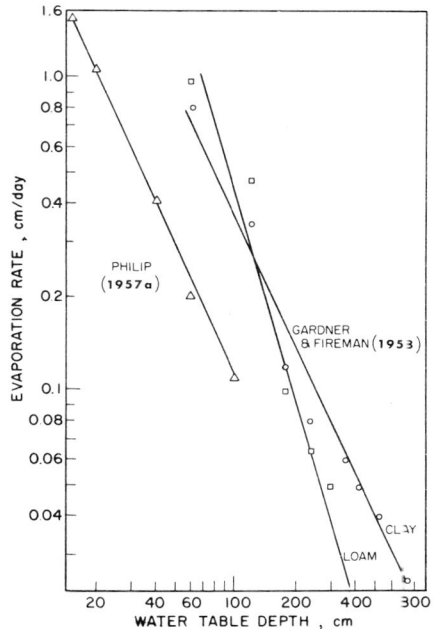

Fig. 13I–4. The effect of water table depth on evaporation rate. Adapted from Wiegand and Taylor, 1961.

Although both nonlinearities go to zero, the liquid flow rate falls far below the vapor flow rate in the drier ranges. Gardner's analysis shows that because of this, the presence of the vapor flow zone increases the limiting flux somewhat but never more than 20%.

b. Upward Flow Through a Stratified Soil. Willis (1960) extended Gardner's analysis of the steady-state flow from a water table to the analysis of soil columns composed of two textural layers. Darcy's law, [13I–5], was integrated, using the appropriate conductivity function for each layer and matching pressure heads at the boundaries between layers. It was concluded from the analysis that the presence of inhomogenities in the soil column would have the most effect on the evaporation rate when the water table was relatively shallow (in Willis' system, < 3 m).

2. NONSTEADY STATE FLOW SYSTEMS

a. Monotonic (Always Wetting or Always Drying) Systems.

1. Wetting Phase (Infiltration).

Water may enter a soil from a ponded layer on the soil surface, as in basin or furrow irrigation; it may fall on the surface at various rates, as in sprinkler irrigation or rain; or it may move up into the soil from below, as in tiled subsurface irrigation or in certain cases of soils underlain by highly permeable aquifers.

The outcome of an infiltration experiment may be specified in terms of the water content in the soil as a function of depth and time. Less complete but simpler descriptions may be given in terms of the time dependence of rate of entry or of cumulative total entry.

Gravity-free infiltration. The simplest infiltration problem is that in which gravity is negligible, approximated experimentally by horizontal infiltration along a narrow soil column. This system has been the subject of intensive mathematical and experimental analysis (Ferguson & W. H. Gardner, 1963; W. R. Gardner & Mayhugh, 1958; Hanks & Bowers, 1962; Nielsen et al., 1962; Philip, 1955; Rawlins & W. H. Gardner, 1963; Scott and Hanks, 1962; Bruce and Klute, 1956; Kirkham and Feng, 1949). An analysis of this flow system can be developed by assuming the validity of the diffusivity equation. In terms of one dimension x along the column, equation [13I–9] reduces to:

$$\frac{\partial \theta}{\partial t} = \frac{\partial}{\partial x} \left(D\left(\theta\right) \frac{\partial \theta}{\partial x} \right). \qquad [13I-14]$$

If we assume in addition that the column is infinitely long, that the initial water content throughout the column is constant at some value, θ_o, and that for all times greater than zero the water content at the entry face ($x = 0$) is maintained constant at some higher[4] water content θ_x, then for any monotonic and physically sensible function, $D(\theta)$, equation [13I–14] with these boundary conditions has a solution with the general property that the rate of advance of the position of any given water content is proportional to $t^{-1/2}$. It follows that the infiltration rate at the surface must also be proportional to $t^{-1/2}$. The time integrals of these two quantities are, respectively, the total distance traveled by a given water-content

---

[4] If θ_s were *lower* than θ_o, the equation would be describing the analogous drying problem, and we shall use it for just this purpose in the subsequent discussion of drying.

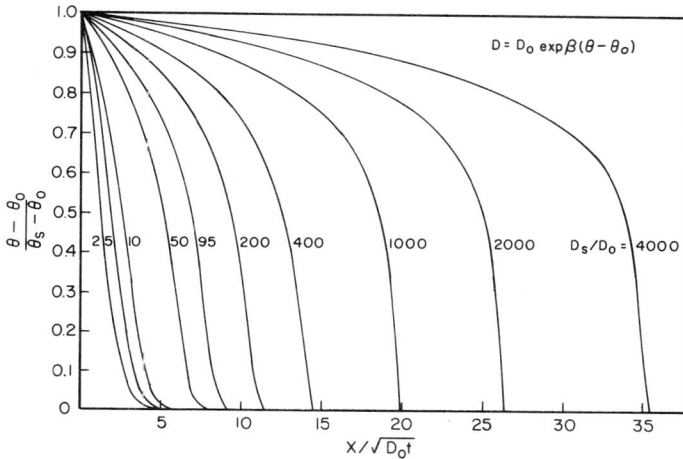

Fig. 13I–5. Solutions of the nonlinear diffusion equation for the gravity-free-infiltration case described in the text. From Gardner and Mayhugh, 1958.

contour and the cumulative infiltration; hence, both of these are proportional to $t^{+1/2}$. Accordingly, all of the curves $\theta(x, t_i)$ showing the distribution of water content with depth at various selected times t_i will coalesce into a single curve $\theta(xt^{-1/2})$ whose shape is determined by the diffusivity function. A family of plots of this form for various diffusivities is shown in Fig. 13I–5 in which θ $[x/(D_o t)^{1/2}]$ has been calculated, assuming D functions of the form

$$D = D_o \exp \beta \, (\theta - \theta_o) = D_o \exp \left[(\ln D_s/D_o)(\theta - \theta_o)/(\theta_s - \theta_o) \right]. \qquad [13I\text{–}15]$$

The diffusivities corresponding to the initial and surface water contents are denoted as D_o and D_s, respectively. Note that the several curves in Fig. 13I–5 do *not* represent different times, but rather different values of the ratio D_s/D_o. For small values of D_s/D_o, the curves resemble the "error curves" of ordinary molecular diffusion while for large values of D_s/D_o, a steep and very well-defined wetting front is displayed.

Coarse-textured soils generally seem to exhibit a more rapid dependence of D on θ than fine-textured soils; it is also observed that for a fixed pair of values, θ_s and θ_o, the wetting fronts for coarse soils are usually sharper than those for fine soils. In the absence of a careful survey, it can usefully be remarked that, taken together, these two observations are at least consistent with Fig. 13I–5. (Note that for two "similar media," e.g., a small size and a large size of monodisperse glass beads, D_s/D_o would be identical (Miller and Miller, 1956), so the wetting fronts should be equally sharp for such special pairs of media.)

The "wetting front" phenomenon is commonly observed when water infiltrates into fairly dry soil. This front is, in effect, a moving boundary for the dynamic zone of the soil water. If the soil is initially very dry and high in colloidal content, there may be rather large temperature gradients in the vicinity of the wetting front due to the heat of wetting of the soil and the latent heat of evaporation of the water which tends to move primarily by vapor transfer at the extreme front (Anderson and Linville, 1961; Anderson et al., 1963). However, it appears that the soil must at least be drier than the 15 atm water content before this temperature disturbance becomes significant (Anderson et al., 1963).

When the water content gradient at the front becomes extremely steep, there will be large changes of pressure head over distances equivalent to a small multiple of the pore diameter. The macroscopic features of the flow, which should represent averages over many pores, will in this case be averages over a few pores. However, there is no clear evidence that discrepancies between theory and experiment arise from this source.

A number of tests of the validity of the diffusivity equation [13I–14] and boundary conditions for horizontal infiltration have appeared in the literature. In these tests the water content distribution in space and time and cumulative infiltration as a function of time were measured on experimental flow systems which were assumed to conform to these conditions. Such tests may consist of plotting cumulative infiltration Q vs. $t^{+1/2}$, infiltration rate R vs. $t^{-1/2}$, or the position x of a given water content vs. $t^{+1/2}$ (Fig. 13I–6). Alternatively, $\theta(xt^{-1/2})$ may be plotted for various time epochs t_i. For water sources at or near zero pressure head, data were obtained that were in reasonable agreement with the theory (Gardner & Mayhugh, 1958; Ferguson & Gardner, 1963; Miller & Gardner, 1962; Jackson, 1963a, 1963b; Nielsen et al., 1962). Deviations from theoretical predictions were reported by Rawlins and Gardner (1963) and Stewart (G. L. Stewart, 1962. Water content measurement by neutron attenuation and applications to unsaturated flow of water in soil. *Unpubl. Ph.D. thesis. Washington State Univ., Pullman*).

However for water sources at pressure heads of −50 and −100 cm of water, Nielsen et al., (1962) have observed deviations from theory as shown in Fig. 13I–6. In view of the considerable preponderance of favorable results and of the ease with which experimental artifacts may arise, the authors are inclined to accept for the present the favorable results for infiltration at or near zero pressure head until further clarification is available. In those cases where deviations have been observed, it is difficult to assign a cause to the deviation. There are many possible reasons for departure from the square root-of-time behavior, some of which are associated with the assumptions in the flow equation, some with the boundary

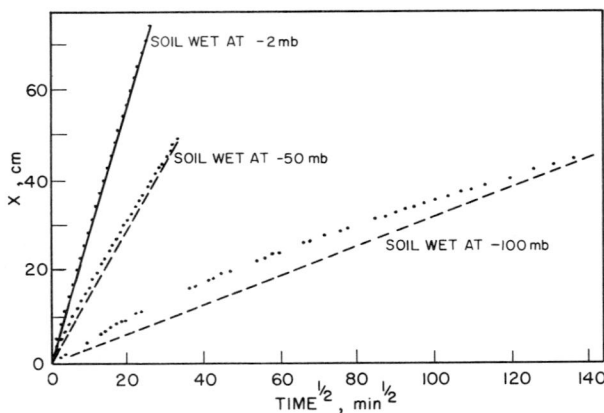

Fig. 13I–6. Distance to the wetting front vs. $t^{1/2}$ for horizontal columns of Columbia silt loam wet at pressure heads of −2, −50, and −100 cm of water. From Nielsen et al., 1962. [*Note:* One millibar (mb) is approximately equal to 1 cm of water.]

conditions, and some with systematic measurement errors. It appears that additional study of this problem is needed. In particular, measurements of the space-time distributions of both pressure head and water content are needed.

Gravity-aided infiltration. The downward entry of water into soil has also been extensively studied, both mathematically and experimentally (Hanks and Bowers, 1962; Marshall and Stirk, 1949; Miller and Gardner, 1962; Miller and Richard, 1952; Nielsen et al., 1961; Parr and Bertrand, 1960; Philip, 1957d).

The influence of such factors as the initial water content, conductivity, stratification, and structure on the infiltration rate and distribution of water in the profile have been investigated in detail.

In a series of papers, Philip (1957c, 1957d, 1957e, 1957f, 1957g, 1957h, 1958) has used a series method for solving the diffusivity form of the flow equation [13I–8], i.e.:

$$\frac{\partial \theta}{\partial t} = \frac{\partial}{\partial z}\left[D(\theta)\,\frac{\partial \theta}{\partial z}\right] - \frac{\partial K(\theta)}{\partial z}. \qquad [13I\text{--}16]$$

In Philip's analysis, the spatial coordinate z was chosen positive in the downward direction. As in the gravity-free case, the medium was assumed to be infinite in extent with initial water content θ_o throughout and with the surface maintained at θ_s after time zero. The solution obtained by Philip was in the form:

$$z = \sum_{n=1}^{\infty} f_n(\theta)\, t^{n/2} \qquad [13I\text{--}17]$$

z being the depth to a particular water content θ; t the time; and the coefficients $f_n(\theta)$ being calculated from a knowledge of the diffusivity and conductivity functions.

Figure 13I–7 shows the soil water profiles calculated by Philip for infiltration into Yolo light clay. At small times the position of any fixed θ increases as $t^{1/2}$, just as it does for absorption into a horizontal column. This is to be expected, since at small times the hydraulic gradient due to capillarity is much larger than that due to gravity. At intermediate times the position of a fixed θ gradually departs from proportionality to $t^{1/2}$. The profile (except at θ_o) moves down continuously at an ever decreasing velocity. At large times the profile moves down the column at a velocity which approaches the constant value, $(K_s - K_o)/(\theta_s - \theta_o)$, where K_s and K_o are the conductivities at θ_s and θ_o, respectively.

The predictions of the theory for vertical infiltration have not been tested as carefully or as frequently as for the horizontal infil-

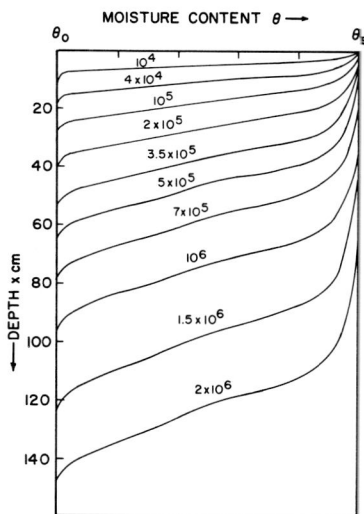

Fig. 13I–7. Computed soil water profiles during infiltration into Yolo light clay. The number of each profile is the time in seconds at which the profile is realized. From Philip, 1957b.

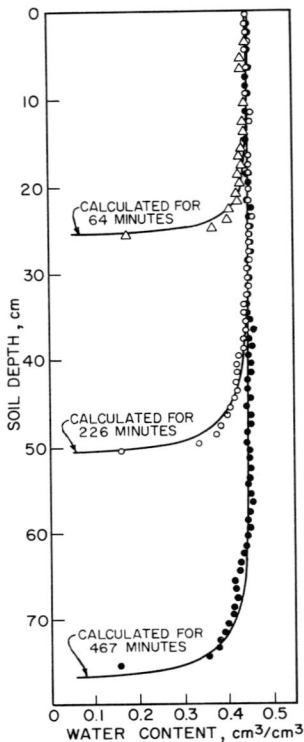

Fig. 13I–8. Water content profiles during vertical infiltration into air-dry Columbia soil. The points are experimental data and the solid curves were calculated by Philip's method. Adapted from Davidson et al., 1963.

tration case. Youngs (1957) and Davidson et al., (1963) (see Fig. 13I–8) used Philip's method of solution of the flow equation to compare theoretically predicted and experimentally observed water content profiles. The agreement was quite good. Qualitatively at least, many other observations (Marshall and Stirk, 1949; Miller and Richard, 1952; Nielsen et al., 1961; Staple and Lehane, 1954) agree with the predictions of the theory. Bodman and Coleman (1944) measured the water content distribution in a vertical soil column during infiltration. The general features exhibited by their measurements, described as "zones of saturation, transition, transmission, and wetting," are shown in Fig. 13I–9. The mathematical analysis predicts all of these zones except the transition zone. This omission is presumably due to an aberration of the diffusivity rather than to a deficiency of Philip's analysis. Such a departure could easily be caused by alterations of soil structure or of packing near the surface. Alternatively, Philip (1957f) suggests that in this region the relation between water content and pressure is dependent on the depth because of the smaller probability that air bubbles will be entrapped in the layer near the top surface (these layers having been exposed to a sharper wetting front).

Philip's solution, equation [13I–17], also provides a series describing the cumulative infiltration volume per unit area of soil surface:

$$Q = \sum_{n=1}^{\infty} a_n(\theta) \ t^{n/2}. \qquad [13I\text{--}18]$$

The coefficients $a_n(\theta)$ are obtained from $K(\theta)$ and $D(\theta)$.

The first two terms of equation [13I–18] are sufficient for an approximate description for the cumulative infiltration:

$$Q \cong St^{1/2} + At. \qquad [13I\text{--}19]$$

The corresponding approximation for the surface infiltration rate R is then

$$R \cong (1/2)St^{-1/2} + A. \qquad [13I\text{--}20]$$

Philip's general expressions, equations [13I–17], and [13I–18], fail at large t, but the two-term approximations, equations [13I–19] and [13I–20], do not. Philip (1957e) has taken another approach to consider the behavior at large time, finding that the infiltration rate approaches the limiting value K_s. In the truncated equation [13I–20], A must clearly be K_s. An increase in initial water content decreases the gradients in the early stages and hence the early infiltration rates; however, the ultimate infiltration rate K_s is unchanged, as indicated in Fig. 13I–10. Philip

WATER CONTENT

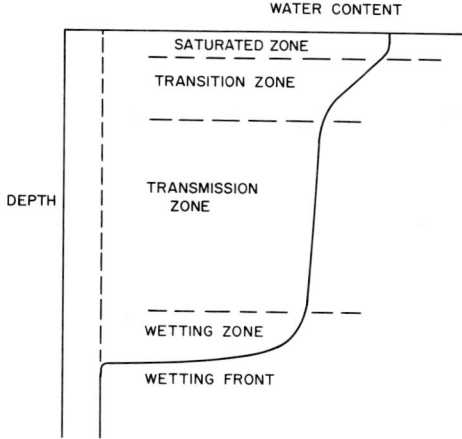

Fig. 13I–9. Infiltration zones delineated by Bodman and Coleman, 1944.

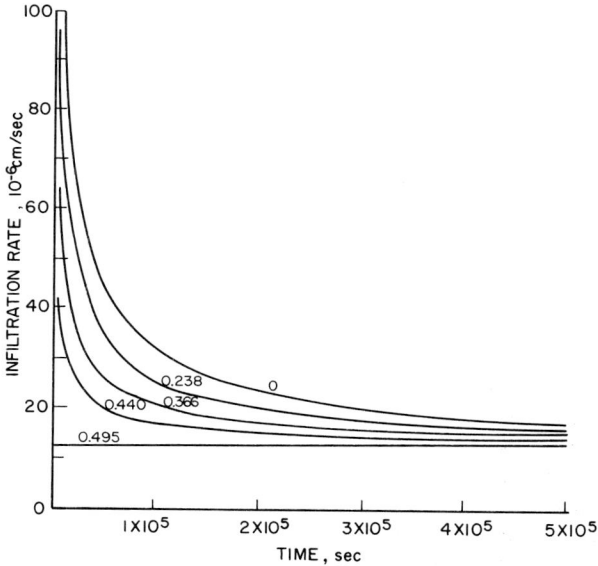

Fig. 13I–10. Computed infiltration rate curves for Yolo light clay showing effect of initial moisture content, θ_o. The numbers on the curves are values of θ_o. From Philip, 1957b.

showed that the ultimate velocity of a given water content down the profile is $(K_s - K_o)/(\theta_s - \theta_o)$. Since K increases with θ at a faster-than-linear rate, it is clear that an increase in θ_o produces a bigger percentage effect on $(\theta_s - \theta_o)$ than on $(K_s - K_o)$. Hence. an increase of initial water content produces an increase in the rate of advance down the profile.

The reader is referred to the papers by Philip for a more detailed discussion of these matters as well as for the effect of the depth of ponded water and the shapes of the D and K functions on the infiltration process.

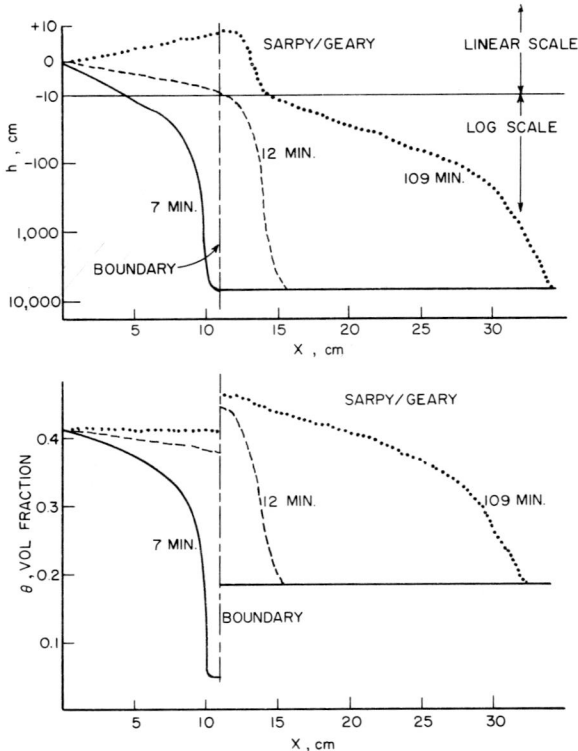

Fig. 13I–11. Calculated pressure head and water content profiles for Sarpy loam over Geary silt loam. From Hanks and Bowers, 1962.

Infiltration into stratified profiles. The effect of two-layer soil stratification on infiltration has been analyzed by Hanks and Bowers (1962), who have solved equation [13I–16] numerically. Some of their results are shown in Fig. 13I–11 and 13I–12. At the layer boundaries, their curves show discontinuities in water content, but not in pressure, in agreement with the discussion of "boundary conditions" earlier in this chapter. At large times, positive pressure heads may develop in a coarse soil overlying a fine soil (Fig. 13I–11), because the conductivity of a wet, coarse soil is higher than that of a wet, fine soil. Looking at the infiltration rates for these cases shown in Fig. 13I–12, one observes that the coarse-over-fine sequence has at first the same infiltration rate as the coarse soil alone, since the fine soil has not yet been reached. When the wetting front arrives at the boundary between layers, the infiltration rate begins to fall off and eventually approaches that of the fine soil alone. The infiltration rate for the fine-over-coarse sequence was not noticeably different from that of the uniform fine material, since the flow was dominated by the lower conductivity of the layer of fine soil.

D. E. Miller and W. H. Gardner (1962) have reported experiments showing the effect on infiltration rate of a thin foreign layer sandwiched into an otherwise uniform profile. The presence of layers, either coarser or finer than the rest of the soil, was shown to cause a decrease of infiltration rate which began when the wet front reached the layer. For a coarse layer, sandwiched into a column of fine

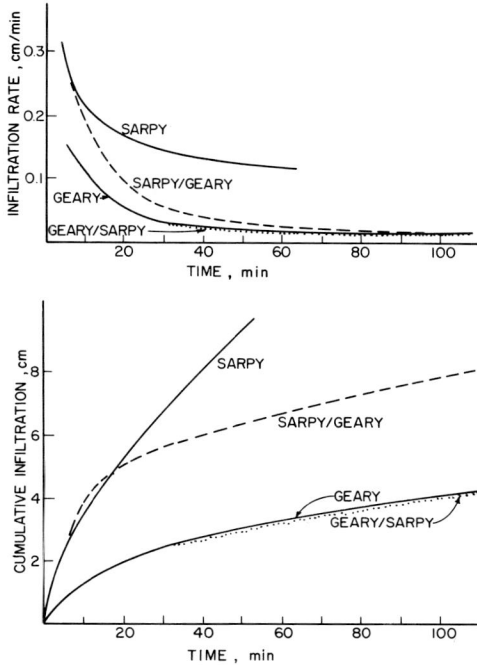

Fig. 13I–12. Influence of layer sequence on infiltration rate and cumulative infiltration. From Hanks and Bowers, 1962.

material, the infiltrating water cannot enter the coarse material until the pressure head in the water at the wet front is high enough to permit substantial wetting of the coarse layer. In the case of a fine material sandwiched into a coarse material, water enters the fine material as soon as the wet front reaches the boundary, but since the resistance to flow through the fine material is usually larger than through the coarse material, a reduction in infiltration rate results. In general, all these results seem to be in qualitative accord with field and laboratory observations of the effect on infiltration rate of the structure and texture of the soil surface layers and of practices which modify them, but detailed comparisons have not been made between theory and experiment.

Sprinkler-controlled infiltration. Sprinkler irrigation or rainfall can supply water at rates too low for ponded water to be maintained on the soil surface. Infiltration under these conditions has been analyzed by Youngs (1960a) and by Rubin and Steinhardt (1963, 1964) and Rubin et al. (1964). The latter authors obtained numerical solutions of equation [13I–16] subject to the condition of a constant flow rate, R, at the soil surface, where R is the rainfall rate. Water content-depth-time profiles were calculated for various rainfall rates and several porous materials. Comparisons between theoretical and experimental water content profiles showed fair agreement, but there were discrepancies for reasons that could not be ascertained. More experimental work needs to be done on this flow system.

The available theory predicts that if R is less than the ultimate rate of infiltration for the ponded case, then infiltration can proceed indefinitely without the development of ponding on the soil surface. If R is greater than this limit, ponding will develop after a finite time. For rainfall rates too slow for ponding to occur, the water content at the surface should approach a limit θ_L such that $K(\theta_L) = R$. Rubin et al. (1964) showed that at long times the wetting front should move downward with a constant velocity given by $[R - K(\theta_i)]/[\theta_L - \theta_i]$ where θ_i is the initial water content in the profile.

Field conditions. Actual infiltration behavior in the field will, of course, be more complex than the simplified, idealized model on which the mathematical theory is based. In the field, one can expect to encounter nonuniform initial water content, hysteresis, nonuniform soil profiles, nonisothermal conditions, changes in soil properties with time, divergent rather than one-dimensional flow, and back-pressure effects (Wilson and Luthin, 1963) from air flow ahead of the wetting front. For these and other reasons, field results can be expected to differ more or less from the predictions of the model. Nielsen et al. (1961) compared calculated moisture profiles with experimental profiles for two silt loam soils in the field. Fair agreement between theoretical and experimental curves was obtained in one soil but rather poor agreement in the other. Green et al. (1964) used the numerical procedure of Hanks and Bowers (1962) to solve the flow equation and obtain estimates of the infiltration rate in a two-horizon soil profile with three nonuniform initial water content profiles. The calculated infiltration rates were in fair agreement with the experimentally observed rates. The lower the initial water content of the soil profile, the greater were the deviations between theory and experiment.

Much remains to be done in the field application of the theory of infiltration, to sort out the significant parameters and features of the flow system that influence infiltration. Nonetheless, the analyses based on the relatively simple models described above have verified field observations in many respects and have led to a deeper understanding of the significant features of the infiltration process.

2. Drying Phase (Evaporation and Drainage).

The evaporation of water from a soil is a rather complex process in which water may be moving in either the vapor or liquid phase and which, in the general case, will involve the combined transport of heat and water. The heat required to vaporize the water must be transported to the evaporation site which requires a nonzero temperature gradient. It is the purpose of this section briefly to describe the application of the isothermal flow theory as embodied in equation [13I–14] to the phenomenon of evaporation from soil. The authors do not mean to imply that the results of such an analysis always constitute an adequate description of the drying process, but it is felt that they can form a point of departure for a more complete analysis which embodies the simultaneous transport of heat and water.

The evaporation of water from a soil has been discussed above as a steady-state flow phenomenon. For non-steady-state evaporation, it is convenient to distinguish two rather well-marked stages, (i) a constant-rate period and (ii) a falling-rate period. When the soil is relatively wet, the rate of evaporation is controlled by external conditions, i.e. by the energy available for evaporation. For a constant environment, the evaporation rate is constant, even though the water content at the soil surface is falling steadily. When the water content at the soil surface begins to approach equilibrium with the atmosphere, the flux at the surface begins to decrease and the water content at the soil surface becomes essentially constant. These two stages are well known experimentally (Fisher, 1923; W. R. Gardner, 1958; Pearse et al., 1949; Wiegand and Taylor, 1961). The transition from the initial constant-rate period to the subsequent falling-rate period can sometimes be rather abrupt (Philip, 1957a) although in many cases the transition is more gradual.

To apply isothermal flow theory to the non-steady-state drying of soil, it has been customary to assume a uniform soil column of constant initial water content with negligible influence of gravity. The latter seems a reasonable assumption for soils and relatively short times. The diffusivity form of the flow equation [13I–14] is assumed, using diffusivity values associated with desorption.

The soil column may be chosen to be of finite length L or of semi-infinite length, i.e. $0 < x < \infty$. For the finite column, one end is assumed blocked off, i.e. $\partial \theta/\partial x = 0$ at $x = L$ while for the semi-infinite column, the water content is assumed to remain θ_i at infinity, θ_i being the initial water content of the soil column.

To represent the constant-rate period of evaporation, the surface flux, i.e. $[D(\theta) \cdot \partial\theta/\partial x]$ at $x = 0$, is held constant after $t = 0$. The constant-rate period is considered to last until some time t, at which the water content at the boundary $x = 0$ has decreased to a certain critical value, θ_b.

Covey (1963) has published a theoretical analysis of the finite and semi-infinite problem during the constant-rate period. As one would expect, the finite and semi-infinite columns behave in the same manner in the early stages of evaporation. According to Covey's analysis the significant parameter for the process is $\beta q_o L/D_i$, where β is a constant in the exponential diffusivity function, $D = \gamma \exp (\beta\theta)$, q_o is the evaporation rate, and D_i is the diffusivity at the initial water content θ_i. When $\beta q_o L/D_i > 5$, the column behaves as a semi-infinite column throughout the first stage of drying. Small values of this parameter correspond to slow rates of drying, short column lengths, and/or high initial water contents. In such a case, the column dries nearly uniformly with a flat profile of water content. If the parameter is large, the exposed surface dries much more rapidly than the under-lying material. The slower the rate of evaporation q_o the longer the constant initial rate may be maintained.

The falling-rate period may be represented by a constant water content θ_b, at the boundary of the flow system $(x = 0)$. The initial condition for this period should be the water content distribution $\theta(x,t_1)$ attained at the end of the constant-rate period, but in fact those who have analyzed this flow system (Gardner, 1959; Klute et al., 1965) have used a constant initial water content. It is probable that the effect of the nonuniform initial water content distribution will not persist very long and the behavior will approach that for a constant initial water content. For a semi-infinite column, the cumulative evaporation is proportional to $t^{1/2}$. Fig. 13I–13 is a reduced plot of the water content distribution during desorption.

The corresponding falling-rate problem for a column of finite length has been analyzed by W. R. Gardner (1959) and Klute et al. (1965). In this case, the flux during the early stages is proportional to $t^{-1/2}$, but after the water content at the closed end $(x = L)$ of the system begins to decrease, the flux decreases sharply and is no longer proportional to $t^{-1/2}$. Fig. 13I–14 shows the effect of column length on cumulative evaporation from Pachappa sandy loam as a function of time. W. R. Gardner (1959) and W. R. Gardner and Hillel (1962) found quite good agreement between experimental and theoretical cumulative evaporation curves.

Vapor movement will occur primarily in the vicinity of the soil surface, and if it can be assumed to be isothermal vapor movement, the square root of time

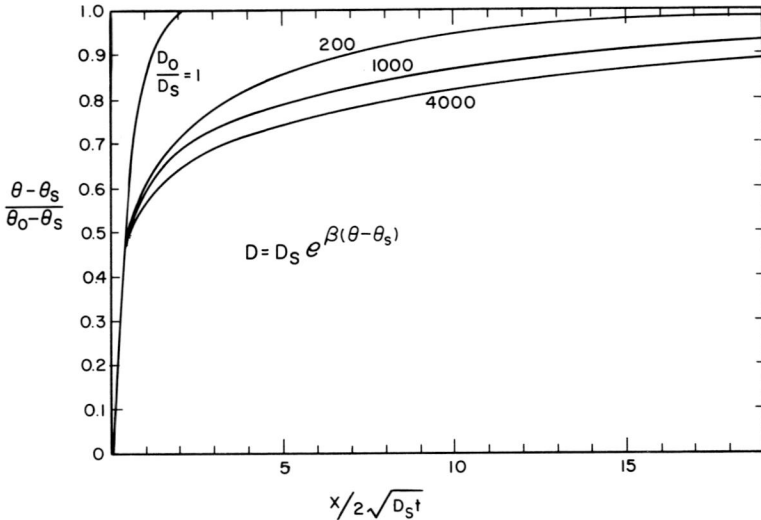

Fig. 13I–13. Computed relative water content distribution during falling-rate period of drying of a semi-infinite soil column plotted as a function of the reduced variable $x/2(D_s t)^{1/2}$. The numbers on the curves are values of the diffusivity ratio D_o/D_s corresponding to the initial water contents θ_o and the water content at $x = 0$, θ_s. Adapted from Gardner, 1959.

dependence of the evaporation rate will still be obtained. Jackson (1964a,b,c) has shown that isothermal vapor movement appears to be described by a non-linear diffusion equation of the same form as equation [13I–14].

The higher the initial rate of evaporation the shorter the duration of the con-stant-rate period. It has been speculated that water might be conserved under such conditions. However, W. R. Gardner and Hillel (1962) present analyses and experimental data which show that the higher drying rate will in fact result in a higher cumulative loss at any given time.

On the basis of the available evidence it appears reasonable to conclude that the isothermal flow equation predicts many of the essential features of the con-stant-rate and falling-rate periods of evaporation. Philip (1957a) concluded from a consideration of steady-state heat and water flow that the isothermal model may be applied to the evaporation phenomenon as long as the surface soil is not extremely dry and departures from the isothermal model will not occur until the late stages of the drying process. This conclusion was supported by the analysis and data of W. R. Gardner and Hillel (1962).

Under field conditions, the boundary conditions described above may not be applicable. Nonisothermal conditions will occur, external evaporative conditions will fluctuate, the soil profile may not be uniform, and there may be plant roots in the profile. The extent to which these factors are important and methods of incorporating them in the theory have not yet been fully explored.

The literature on evaporation is voluminous. Excellent bibliographies may be found in W. R. Gardner (1960b) and Wiegand and Taylor (1962) along with further discussion of the phenomenon.

In addition to evaporation and transpiration, flow to lower horizons or drains

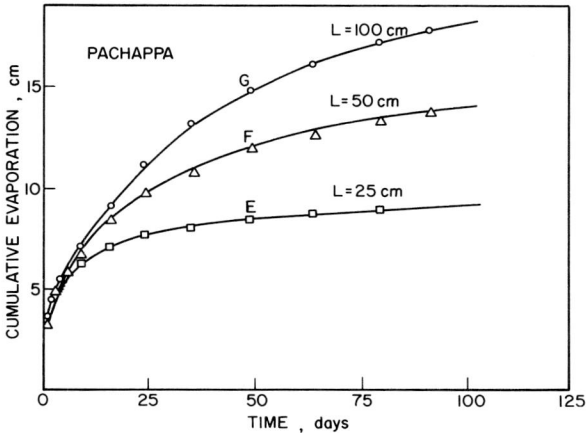

Fig. 13I–14. Cumulative evaporation as a function of time for columns of Pachappa sandy loam 25, 50, and 100 cm in length. Initial water contents were 27.8, 24.8, and 23.7% by weight, respectively. The smooth curves are theoretical curves fitted to the experimental data by choosing values of D_o and D_i/D_o.

may remove some of the water from the soil profile. For irrigated soils, the necessity of adequate drainage for salinity control is well known. The theory of drainage has been treated in a monograph (Luthin, 1957). Much of the work dealing with drainage has involved saturated flow systems in which the Laplace equation, with the hydraulic head as a dependent variable, was assumed valid. We shall not discuss such flow systems here.

The direct application of the theory of unsaturated flow to drainage situations where unsaturated flow may play a significant role, such as the falling water table case, has been prevented by the mathematical difficulties of obtaining solutions of the equation. Various approximate treatments of the falling water table case have been reported (Childs, 1947; Kirkham and Gaskell, 1951; Isherwood, 1959; Maasland, 1959; Collis-George and Youngs, 1958). These have assumed that the non-steady-state falling water table may be treated as a succession of steady-state flows, and that there is a constant specific yield or volume of water released per unit surface area per unit fall of the water table. Childs (1960) presents arguments to show that the latter assumption is, in some instances, not valid, and Luthin and Worstell (1957) present data on a laboratory system that supports this conclusion. Childs and Poulovassilis (1962) present theory and experimental data to show that the shape of the soil water profile above a falling water table depends on the rate of fall, being somewhat more "stretched out" over a falling water table than in the static case.

The one-dimensional drainage of an initially saturated soil column with a water table at its lower end has been studied by a number of workers. Day and Luthin (1956) applied a one-dimensional form of equation [13I–7] and, using separately measured conductivity and water capacity data, attempted to predict the space-time distribution of pressure head in the system. Discrepancies between theory and experiment were found which could have been due to inadequate knowledge of the conductivity and water capacity or to inaccuracies in the numerical solution of the flow equation.

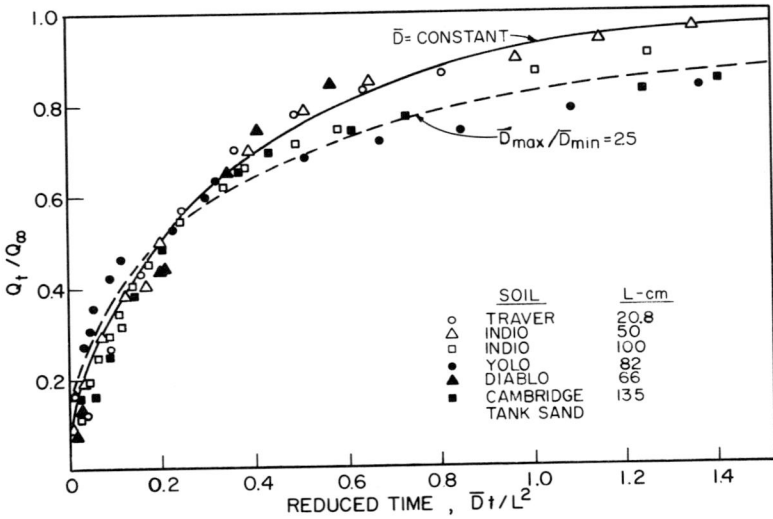

Fig. 13I–15. Fractional cumulative outflow as a function of time for several soils compared with theoretical outflow curves. From W. R. Gardner, 1962.

Approximate treatments of the above one-dimensional drainage case have been given by Youngs (1960b) and W. R. Gardner (1962). Youngs used the capillary tube model to arrive at an equation for the time dependence of the outflow volume. W. R. Gardner used a simplified version of the equation of flow (in which it was assumed that the hydraulic head was a linear function of water content) to derive an equation for the outflow volume as a function of time. Figure 13I–15 shows some of Gardner's data for cumulative outflow from several soils and column lengths. Reasonable agreement between theoretical and experimental outflow was obtained by both Youngs and Gardner. Neither approximate treatment permits the detailed description of the water content and pressure head distribution within the soil column. However, Gardner suggests that the total water content of the soil profile is a better parameter of drainage status than the water table position.

b. Hysteretic Flow Systems.

The theoretical analysis of water flow in soil during the redistribution which follows the cessation of infiltration is of interest because of its application to the interpretation of field capacity. Staple (1962) and Youngs (1958a,b; 1960a) have studied this flow system for vertical flow.

1. Redistribution, Gravity-aided. The flow equation for this system may be written

$$\left(\frac{d\theta}{dh}\right)_H \frac{\partial h}{\partial t} = \frac{\partial}{\partial z}\left(K_H(h)\frac{\partial h}{\partial z}\right) + \frac{\partial K_H(h)}{\partial z} \qquad [13\text{I}-21]$$

where $(d\theta/dh)$ is a hysteretic function of h. After the cessation of infiltration at time t_1 the hydraulic gradient at the soil surface is zero if water loss at the soil surface is prevented. For the conservation of matter, the integral

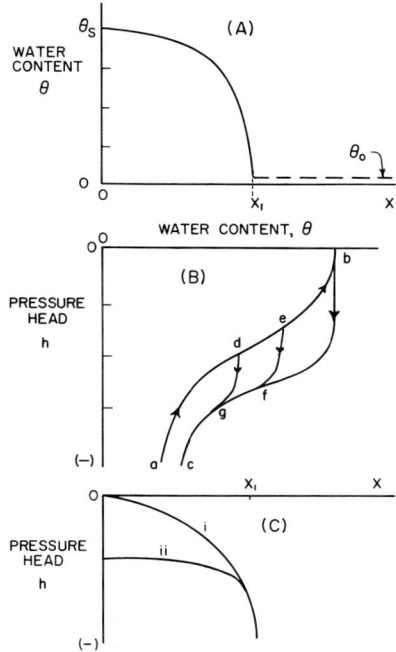

Fig. 13I–16. (A) A hypothetical water content profile at the end of horizontal infiltration, at time t_1. (B) Hypothetical water content pressure head curve for the same porous material. (C) The pressure head distribution along the column just before the end of infiltration (curve i) and immediately after the start of redistribution (curve ii). Adapted from Youngs, 1958a.

$$\int_{x=0}^{x=\infty} \theta \, dx$$

must be constant for all values of $t > t_1$. During redistribution, the soil beyond the position which was reached by the wetting front at time t_1 continues to wet, so for this portion of the soil the water capacity function $(d\theta/dh)_{II}$ and the conductivity $K_H(h)$ are always those appropriate to a wetting process. In most of the region of the flow system which was behind the wetted front at time t_1, the soil abruptly stops wetting at t_1 and begins to drain. At time t_1 the soil water profile would be similar to that shown in Fig. 13I–16a and the corresponding profile of pressure head would be curve i of Fig. 13I–16c. A typical water content-pressure head relation is shown Fig. 13I–16b. During infiltration before time t_1 all points of the flow system wet up along curve adeb of Fig. 13I–16b. When drainage begins, each point follows a different draining scanning curve, e.g., curves ef and dg of Fig. 13I–16b. If these curves are nearly parallel to the p axis, the water capacity $d\theta/dh$ will be small so that a considerable decrease in pressure head can occur with very little change in water content. Thus, when redistribution begins, the pressure head distribution quickly drops to curve ii in Fig. 13I–16c. Because each point of the flow system follows a different scanning curve, the conductivity function and the water capacity function are no longer uniquely determined by the pressure head or water content, but vary with position. A complete analysis of this flow system, i.e. a numerical or analytical solution of equation [13I–21] including comparison with experimental observations, has not been published. Youngs has, however, developed an approximate analysis of the transition water content and pressure at which the material changes from wetting to draining. Some experimentally observed water content distributions for two materials and

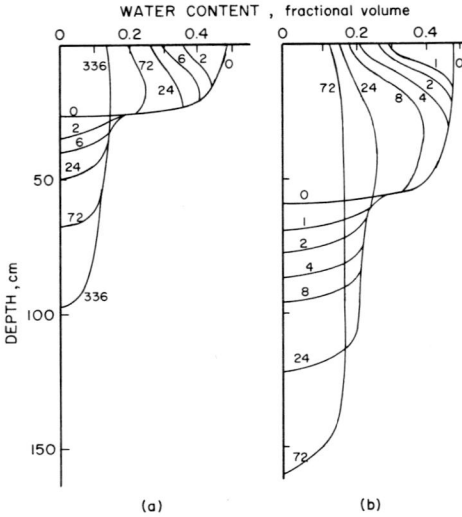

Fig. 13I–17. Experimental water content profiles during the redistribution of water after infiltration in slate dust. The numbers near the curves are the time in hours after cessation of infiltration. Depth of initial wetting, (a) 27 cm, (b) 59 cm. From Youngs, 1958b.

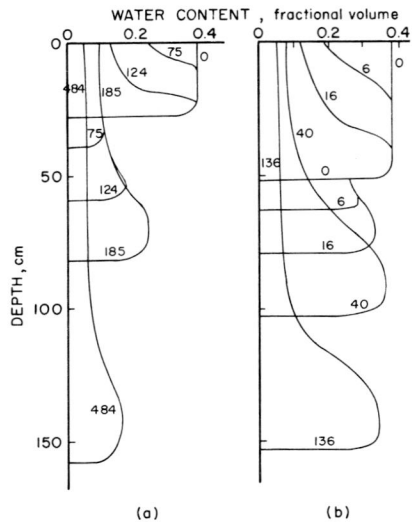

Fig. 13I–18. Experimental water content profiles during the redistribution of water after infiltration in "Ballotini" grade 15. Depth of initial wetting, (a) 27.5 cm, (b) 51.5 cm. The numbers near the curves are the time in hours after cessation of infiltration.

for two depths of infiltration are shown in Fig. 13I–17 and 18. During the early period of redistribution, the greater the initial depth of infiltration, the greater the transition water content. This effect was predicted by Youngs (1958a,b) in his analysis. For shallow depths of infiltration in materials with a limited range of pore size (and hence "abrupt" $\theta(h)$ curves) the redistribution takes place very slowly. As the depth of infiltration increases, the water content at the transition point increases and redistribution proceeds more rapidly. The formation of the negative water content gradient in the wetting zone during redistribution (Fig. 13I–18) has not received a satisfactory explanation. For a material with a more gradual $\theta(h)$ curve (e.g. slate dust, Fig. 13I–17) the initial transition water content will be higher and the negative gradient in the wetting zone does not appear to develop.

The experimental water content redistribution data obtained by Staple (1962) for one depth of wetting (about 4 inches) on a silt loam soil are in qualitative accord with the above concepts.

2. *Redistribution Phase in Relation to "Field Capacity"*. The concept of "field capacity", representing a practical upper limit for "available" water in soil, has long been used by soils workers. However, confusion has arisen as to the precise meaning and hydraulic interpretation of the term. Great effort has been directed toward the development of a laboratory method for determining "field capacity".

The field capacity was defined by Viehmeyer and Hendrickson (1949) as ". . . the amount of water held in soil after excess water has drained away and the rate of downward movement of water has materially decreased, which usually takes place within 2 or 3 days after a rain or irrigation in pervious soils of uniform structure and texture." They appear to have envisaged (i) a downward infiltration of water, either from irrigation or rainfall, into a deep uniform profile without a water table followed by (ii) a redistribution as water drains from the upper part of the profile, at first at a relatively rapid rate and then more slowly. If such a "field capacity" could be defined, it would be useful in establishing the upper limit of available water, a concept which has considerable practical interest.

Application of the flow principles outlined in this chapter leads to the conclusion that the time rate of change of water content at any particular point in the soil, e.g., the upper part of a profile, is a continuous function of (i) the properties of the soil as described by the capacity and conductivity functions, and (ii) the boundary conditions that are applicable to the particular flow system. The dynamic nature of "field capacity" has long been recognized (Coleman, 1944; Marshall and Stirk, 1949; Miller and McMurdie, 1953; Richards, 1955; Veihmeyer and Hendrickson, 1949), as well as the fact that it is not an equilibrium value. In some soils it may be difficult to recognize a point at which redistribution may be considered negligible, i.e. there may not be any sharp breaks in the curve of water content vs. time which could be identified with the field capacity.

There are two reasons for the reduction in the rate of movement that occurs during redistribution: (i) The hydraulic gradient becomes smaller and (ii) the conductivity becomes smaller. Either one of these factors can cause the flux to become negligibly small. The hydraulic gradient may become small because of boundary conditions and/or because of internal redistribution of the water in the soil. Because the boundary conditions play a part in determining when the flow becomes negligible, it is *impossible* to devise a laboratory method that is appropriate to all soils and field conditions. Equilibrium measurements, such as the water content of a sample drained on a pressure plate to a pressure difference of 1/10 bar, 1/3 bar, etc., will be correlated with "field capacity" only to the extent that these tensions are related to the conductivity function of the soil. The concept of an upper limit of available water involves plant and environmental factors which, being external to the soil, cannot properly be appraised by measurements made in the laboratory on samples removed from the profile.

C.   Soil Water Systems Containing Plants

The flow of water to plant roots is of importance in the study of water availability to plants. This problem has been studied by W. R. Gardner (1960a), Philip (1957b), and Covey (Winton Covey, May 1958. A numerical solution of the equation of diffusion of water toward an inner cylinder. *Unpubl. mimeo-Spec. Prob. Agron. 685. Texas A & M Univ.*). There are at least two approaches that may be used in applying the flow theory to water uptake by roots.

1.   SINGLE-ROOT MODEL

In this approach the details of the flow about a single root are examined. It is customary to assume a cylindrical root in a soil mass of finite or perhaps infinite radial extent and to assume the flow equation [13I–8] or [13I–9] in radial form with gravity neglected.

W. R. Gardner (1960a) has used the single-root model in an infinite medium to discuss the dynamic aspects of availability of water to plants. He assumed that soil water would not be taken up unless the pressure head of the soil water was higher than that of the water in the root. A constant flux condition was imposed at the root surface and the diffusivity and conductivity were assumed constant. The flow equation then becomes:

$$\frac{\partial \theta}{\partial t} = \frac{D}{r} \frac{\partial}{\partial r} \left(r \frac{\partial \theta}{\partial r} \right) \qquad\qquad [13I-22]$$

with initial and boundary conditions:

$$\theta = \theta_o \qquad @ \ t = 0$$

$$-\left[2\pi a D \frac{\partial \theta}{\partial r} \right] = q \qquad @ \begin{array}{l} r = a \\ t > 0 \end{array} \qquad\qquad [13I-23]$$

where a is the root radius and q is the constant flux of water at the root expressed as volume of water per unit length of root per unit time. The solution of equation [13I-22], subject to these conditions is

$$h - h_o = \Delta h = \frac{q}{4\pi K} \left(\ln \frac{4Dt}{r^2} - \gamma \right) \qquad\qquad [13I-24]$$

where K is the capillary conductivity and h_o is the pressure head corresponding to θ_o. The proportionality between D and K has been used to obtain the solution in terms of pressure head. The root has been approximated by a line source or in this case a sink.

Inspection of equation [13I-24] with $r = a$, shows that the decrease of pressure head Δh near the root surface is proportional to the rate of uptake of water and inversely proportional to the capillary conductivity of the soil. Together with the average pressure head of the soil water, these constitute the most important factors affecting the pressure head at the root surface. The diffusivity, time, and root radius appear in the logarithmic terms and hence Δh is much less sensitive to these factors.

According to equation [13I-24] the distance at which any given pressure head occurs increases as the square root of the time. Thus the distance from which water may eventually move to the root may be estimated from the time allowed for movement. With the times, root radii, and diffusivities to be expected it is found that water can move distances of the order of a few centimeters.

The dependence of pressure head and water content on distance from the root were calculated by W. R. Gardner (1960a) for several soils and were found to be quite flat until an initial pressure head of −15 bars was approached. On the basis of this he concluded that the assumption of a constant diffusivity and conductivity were justified.

Figure 13I–19 shows the pressure head at the root surface as a function of average water content in the soil for two soils, Pachappa sandy loam and Chino clay. The uptake rate of 0.1 ml/cm per day is consistent with the data of Ogata et al. (1959). In a soil such as Pachappa it is difficult to investigate the effect of the average water content of the soil on the plant processes, since the pressure head at the root decreases very rapidly with a small decrease in water content. In the Chino soil the −1- to −15-bar pressure head range corresponds to approximately a 12% water content range and thus it is easier to evaluate the effect of soil water content on plant processes that might be influenced by the pressure head of the soil water.

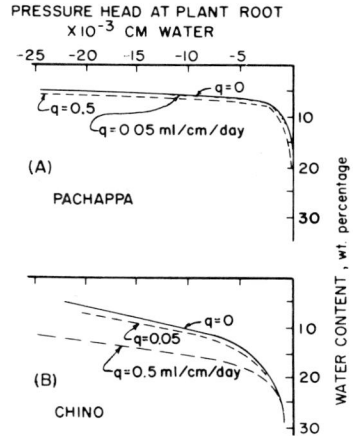

Fig. 13I–19. (A) Calculated pressure head at the plant root for several rates of water uptake as a function of average soil water content for Pachappa sandy loam. (B) The same for Chino clay. From Gardner, 1960a.

W. R. Gardner used equation [13I–24] in a theoretical analysis of the effect of the rate of transpiration (as measured by q) on the pressure head of the soil water and on the water content of the soil at wilting. It was assumed that wilting would occur when the pressure head in the soil water adjacent to the root reached −20 bars. For low rates of uptake the pressure head in the soil water was very close to −20 bars, but as the flux increased to the range 0.2 to 0.5 ml/cm per day, the pressure head in the soil water increased to −8 or −10 bars (for the soils investigated). The corresponding effect on water content was small in Pachappa sandy loam, but the wilting point of Chino clay increased from 16 to 23% as the transpiration rate approached 0.5 ml/cm per day. Thus, the pressure head of the soil water and the water content at wilting are influenced by dynamic considerations.

The reader is referred to the papers by W. R. Gardner and J. R. Philip for additional details of these analyses. One might question the application of macroscopic flow theory to regions of flow in the vicinity of the root that are rather microscopic in nature. However, these analyses lead to results that seem at least qualitatively reasonable.

2. ROOT SYSTEM MODEL

In this approach no attempt is made to treat a single root. Instead the plant roots are assumed to be continuously distributed through the soil, the uptake of water being thereby treated as a macroscopic source term G in the flow equation:

$$C(h) \frac{\partial h}{\partial t} = \text{div} \left[K(h) \vec{\nabla} H \right] + G. \qquad [13I–25]$$

The function G represents the volume of water produced per unit time per unit soil volume. For root uptake, G must be restricted to negative values. These values will depend on position and/or time because of nonuniform and changing root distribution in the profile and possibly for other reasons. In this kind of analysis one must be able to specify the source term G in addition to the conductivity and water capacity functions in order to obtain solutions of the equation of flow. Ogata et al. (1959) have used this concept in an analysis of some data on transpiration by alfalfa (*Medicago sativa*), but the technique has not been widely used.

IV. UNSOLVED PROBLEMS

Despite rapid progress now being made in untangling the complexities of purely mechanical flow systems, the practical application of present Darcy technology to field soil systems is not in a satisfactory state. There are basically two reasons for this inadequacy: (i) the complexity of the technique, and (ii) competing effects whose nature and relative importance are not yet clear.

Only a limited number of idealized Darcy systems have been studied quantitatively. Actual field soil profiles are so varied in flow properties that a comprehensive survey and the development of simplified parameters for classification will be required before Darcy technology can be used quantitatively in management. Falling in the same class are (i) the variability of soils from point-to-point in a field in the absence of practical methods for averaging out such macroscopic variations; (ii) the variability of the soil character with the changing seasons and with managment operations, both mechanical and chemical; and (iii) complex environmental conditions—erratic rainfall, the variable water demand of crops, etc. All of these difficulties could presumably be handled in principle by computer methods and more adequate measurements of soil properties. In practice something much simpler is needed. Hopefully, computer methods may be used for sorting out the effects of the greatest importance and finding simple quantitative ways of approximating these effects for practical application.

Before such a large undertaking can be justified at full scale, it will be necessary to investigate in much more quantitative detail the conditions under which other factors are more important than those describable by Darcy technology. Outstanding among such other factors is the effect of the severe temperature gradients which occur in the surface layers of field soils. The present status of this subject is discussed in part II of this chapter. Unfortunately our quantitative knowledge concerning the practical significance of this effect under field conditions is quite inadequate at the moment. Another strong competitor among the other factors is the quantitative water use behavior of growing plants in the field. Although some work on this subject has been done, as outlined in section III of this part and further work is being very actively pursued at this moment, the subject is still far from a quantitative and predictive stage of development. Whenever water conduction occurs mostly through clay filaments within larger pores, salt sieving may be of such importance that gradients of solute concentration become more important than gradients of total mechanical head. From this point, one could easily add a long and familiar list of other disturbing and poorly quantified effects, ranging from the gradual solution of entrapped bubbles to population explosions among microorganisms.

The key question in all of these effects is quantitative—under what conditions and to what extent are they of practical importance? There are two ways of proceeding: The direct method of isolating or concentrating on a given effect by itself; and the indirect method of trying to find perturbations of quantitative Darcy predictions which can be safely attributed to one particular disturbing effect. The last few years have brought this indirect method almost within reach. It is to be hoped that great progress will be made in the coming decade in roughing out the importance of various non-Darcy effects by whichever of the two methods can be brought to bear on particular cases. Concurrently, work must also proceed, seeking out unknown but potentially practical ways of combining basically different types of effects to produce useful predictions for purposes of practical management. It should be remembered that although our present knowledge leaves much to be desired, it has contributed very considerably to a general understanding of the principles controlling soil water movement in the field.

LITERATURE CITED

Anderson, D. M., and A. Linville. 1961. Temperature fluctuations at a wetting from: I. Characteristic temperature time curves. Soil Sci. Soc. Amer. Proc. 26:14–18.

Anderson, D. M., G. Sposito and A. Linville. 1963. Temperature fluctuations at a wetting front: II. The effect of initial water content of the medium on the magnitude of the temperature fluctuations. Soil Sci. Soc. Amer. Proc. 27:367–369.

Bodman, G. B., and E. A. Coleman. 1944. Moisture and energy conditions during downward entry of water into soils. Soil Sci. Soc. Amer. Proc. (1943) 8:116–122.

Bruce, R. R., and A. Klute. 1956. The measurement of soil moisture diffusivity. Soil Sci. Soc. Amer. Proc. 20:458–462.

Childs, E. C. 1947. The water table, equipotentials, and streamlines in drained land: V. The moving water table. Soil Sci. 63:361–376.

Childs, E. C. 1960. The nonsteady state of the water table in drained land. J. Geophys. Res. 65:780–782.

Childs, E. C., and W. Collis-George. 1950. Permeability of porous materials. Proc. Roy. Soc. (London) Ser. A 201:392–405.

Childs, E. C., and A. Poulovassilis. 1962. The moisture profile above a moving water table. J. Soil Sci. 13:271–285.

Coleman, E. A. 1944. The dependence of field capacity upon the depth of wetting of field soils. Soil Sci. 58:43–50.

Collis-George, N., and E. G. Youngs. 1958. Some factors determining water table height in drained homogeneous soil. J. Soil Sci. 9:332–338.

Covey, Winton. 1963. Mathematical study of the first stage of drying of a moist soil. Soil Sci. Soc. Amer. Proc. 27:130–134.

Darcy, Henry. 1856. Les Fontaines Publique de la Ville de Dijon. Victor Dalmont, Paris. 570 p.

Davidson, J. M., D. R. Nielson, and J. W. Biggar. 1963. The measurement and description of water flow through Columbia silt loam and Hesperia sandy loam. Hilgarcia 34:601–616.

Day, P. R., and J. N. Luthin. 1956. A numerical solution of the differential equation of flow for a vertical drainage problem. Soil Sci. Soc. Amer. Proc. 20:443–447.

Ferguson, Hayden, and W. H. Gardner. 1963. Diffusion theory applied to water flow data using gamma ray absorption. Soil Sci. Soc. Amer. Proc. 27:243–245.

Fisher, E. A. 1923. Some factors affecting the evaporation of water from soil. J. Agr. Sci. 13:121–143.

Gardner, W. H. 1962. How water moves in the field. Crops and soils 15:7–11.

Gardner, W. R. 1956. Calculation of capillary conductivity from pressure plate outflow data. Soil Sci. Soc. Amer. Proc. 20:317–321.

Gardner, W. R. 1958. Some steady solutions of the unsaturated moisture flow equation with application to evaporation from a water table. Soil Sci. 85:228–232.

Gardner, W. R. 1959. Solutions of the flow equation for the drying of soils and other porous media. Soil Sci. Soc. Amer. Proc. 23:183–187.

Gardner, W. R. 1960a. Dynamic aspects of water availability to plants. Soil Sci. 89: 63–73.

Gardner, W. R. 1960b. Soil water relations in arid and semiarid conditions. Plant water relationships in arid and semiarid conditions, reviews of research. UNESCO 15:37–61.

Gardner, W. R. 1962. Approximate solution of a nonsteady state drainage problem. Soil Sci. Soc. Amer. Proc. 26:129–132.

Gardner, W. R., and M. Fireman. 1958. Laboratory studies of evaporation from soil columns in the presence of a water table. Soil Sci. 85:244–249.

Gardner, W. R., and D. I. Hillel. 1962. The relation of external evaporative conditions to the drying of soils. J. Geophys. Res. 67:4319–4325.

Gardner, W. R., and M. S. Mayhugh. 1958. Solutions and tests on the diffusion equation for the movement of water in soil. Soil Sci. Soc. Amer. Proc. 22:197–201.

Green, R. E., R. J. Hanks, and W. E. Larson. 1964. Estimates of field infiltration by numerical solution of the moisture flow equation. Soil Sci. Soc. Amer. Proc. 28:15–19.

Hanks, R. J., and S. A. Bowers. 1962. Numerical solution of the moisture flow equation for infiltration into layered soils. Soil Sci. Soc. Amer. Proc. 26:530–535.

Honert, T. H. van den. 1942. Water transport in plants as a catenary process. Faraday Soc. Discuss. 3:146–153.

Hubbert, M. King. 1956. Darcy's law and the field equations of the flow of underground fluids. Amer. Inst. Mining, Met., Petrol. Eng. Trans. 207:222–239.

Isherwood, J. D. 1959. Water table recession in tile-drained land. J. Geophys. Res. 64:795–804.

Jackson, R. D. 1963a. Porosity and soil water diffusivity. Soil Sci. Soc. Amer. Proc. 27:123–126.

Jackson, R. D. 1963b. Temperature and soil water diffusivity. Soil Sci. Soc. Amer. Proc. 27:363–366.

Jackson, R. D. 1964a. Water vapor diffusion in relatively dry soil: I. Theoretical considerations and sorption experiments. Soil Sci. Soc. Amer. Proc. 28:172–176.

Jackson, R. D. 1964b. Water vapor diffusion in relatively dry soil: II. Desorption experiment. Soil Sci. Soc. Amer. Proc. 28:464–466.

Jackson, R. D. 1964c. Water vapor diffusion in relatively dry soil: III. Steady state experiments. Soil Sci. Soc. Amer. Proc. 28: 466–470.

King, L. G., and R. A. Schleusener. 1961. Further evidence of hysteresis as a factor in the evaporation from soils. J. Geophys. Res. 66:4187–4191.

Kirkham, Don, and C. L. Feng. 1949. Some tests of the diffusion theory and laws of capillary flow in soils. Soil Sci. 67:29–40.

Kirkham, Don, and R. E. Gaskell. 1951. Falling water table in tile and ditch drainage. Soil Sci. Soc. Amer. Proc. 15:37–42.

Klute, A., F. D. Whisler, and E. J. Scott. 1964. Soil water diffusivity and hysteresis data from radial flow pressure cells. Soil Sci. Soc. Amer. Proc. 28:160–163.

Klute, A., F. D. Whisler, and E. J. Scott. 1965. Numerical solution of the flow equation for water in a horizontal finite soil column. Soil Sci. Soc. Amer. Proc. 29:353–358.

Luthin, J. N. [ed.] 1957. Drainage of agricultural lands. Agronomy Vol. 7. 620 p.

Luthin, J. N., and R. V. Worstell. 1957. The falling water table in tile drainage: I. A laboratory study. Soil Sci. Soc. Amer. Proc. 21:580–584.

Maasland, M. 1959. Water table fluctuations induced by intermittent recharge. J. Geophys. Res. 549–559.

Marshall, T. J., and G. B. Stirk. 1949. Pressure potential of water moving downward into soil. Soil Sci. 68:359–370.

Miller, D. E., and W. H. Gardner. 1962. Water infiltration into stratified soil. Soil Sci. Soc. Amer. Proc. 26:115–118.

Miller, E. E., and R. D. Miller. 1955a. Theory of capillary flow: I. Practical implications. Soil Sci. Soc. Amer. Proc. 19:267–271.

Miller, E. E., and R. D. Miller. 1955b. Theory of capillary flow: II. Experimental information. Soil Sci. Soc. Amer. Proc. 19:271–275.

Miller, E. E., and R. D. Miller. 1956. Physical theory for capillary flow phenomena. J. Appl. Phys. 27:324–332.

Miller, R. D., and J. L. McMurdie. 1953. Field capacity in laboratory columns. Soil Sci. Soc. Amer. Proc. 17:191–195.

Miller, R. D., and Felix Richard. 1952. Hydraulic gradients during infiltration in soils. Soil Sci. Soc. Amer. Proc. 16:33–38.

Moore, R. E. 1939. Water conduction from shallow water tables. Hilgardia 12:383–426.

Nielson, D. R., J. W. Biggar, and J. M. Davidson. 1962. Experimental consideration of diffusion analysis in unsaturated flow problems. Soil Sci. Soc. Amer. Proc. 26:107–112.

Nielson, D. R., Don Kirkham, and W. R. van Wijk. 1961. Diffusion equation calculations of field soil water infiltration profiles. Soil Sci. Soc. Amer. Proc. 25:165–168.

Ogata, G., L. A. Richards, and W. R. Gardner. 1959. Transpiration of alfalfa determined from soil water content changes. Soil Sci. 89:179–182.

Parr, J. F., and A. R. Bertrand. 1960. Water infiltration into soils. Advance. Agron. 12:311–363.

Pearse, J. F., T. R. Oliver, and D. M. Newitt. 1949. The mechanisms of the drying of solids: Part I. The forces giving rise to movement of water in granular beds during drying. Inst. Chem. Eng., Trans. (London) 27:1–8.

Philip, J. R. 1955. Numerical solution of equations of the diffusion type with diffusivity concentration dependent. Trans. Faraday Soc. 51:885–892.

Philip, J. R. 1957a. Evaporation, and moisture, and heat fields in the soil. J. Meteorol. 14:354–366.

Philip, J. R. 1957b. The physical principles of soil water movement during the irrigation cycle. Int. Congr. Irrig. Drainage., Proc. 3rd 8.125–8.154.

Philip, J. R. 1957c. Numerical solution of equations of the diffusion type with diffusivity concentration dependent: II Australian J. Phys. 10:29–42.

Philip, J. R. 1957d. The theory of infiltration: 1. The infiltration equation and its solution. Soil Sci. 83:345–357.

Philip, J. R. 1957e. The theory of infiltration: 2. The profile at infinity. Soil Sci. 83:435–448.

Philip, J. R. 1957f. The theory of infiltration: 3. Moisture profiles and relation to experiment. Soil Sci. 84:163–178.

Philip, J. R. 1957g. The theory of infiltration: 4. Sorptivity and algebraic infiltration equations. Soil Sci. 84:257–264.

Philip, J. R. 1957h. The theory of infiltration: 5. The influence of the initial moisture content. Soil Sci. 84:329–339.

Philip, J. R. 1958. The theory of infiltration: 6. Effect of water depth over soil. Soil Sci. 85:278–286.

Philip, J. R., and D. A. de Vries. 1957. Moisture movement in porous materials under temperature gradients. Amer. Geophys. Union., Trans. 38:222–232.

Rawlins, S. L., and W. H. Gardner. 1963. A test of the validity of the diffusion equation for unsaturated flow of soil water. Soil Sci. Soc. Amer. Proc. 27:507–510.

Richards, L. A. 1931. Capillary conduction of liquids through porous mediums. Physics 1:318–333.

Richards, L. A. 1955. Water content changes following the wetting of a bare soil in the field. Soil Sci. Soc. Florida 15:142–148.

Rubin, J., and R. Steinhardt. 1963. Soil water relations during rain infiltration: I. Theory. Soil Sci. Soc. Amer. Proc. 27:246–250.

Rubin, J., and R. Steinhardt. 1964. Soil water relations during rain infiltration: III. Water uptake at incipient ponding. Soil Sci. Soc. Amer. Proc. 28:614–620.

Rubin, J., R. Steinhardt, and P. Reiniger. 1964. Soil water relations during rain infiltration: II. Moisture content profiles during rains of low intensities. Soil Sci. Soc. Amer. Proc. 28:1–5.

Schleusener, R. A., and A. T. Corey. 1959. The role of hysteresis in reducing evaporation from soils in contact with a water table. J. Geophys. Res. 64:469–475.

Scott, E. J., and R. J. Hanks. 1962. Solution of the one-dimensional diffusion equation for exponential and linear diffusivity functions by power series applied to moisture flow in soils. Soil Sci. 94:314–322.

Schlichter, Charles S. 1899. US Geol. Surv., Ann. Rep. 19–II:295–384.

Staple, W. J. 1962. Hysteresis effects in soil moisture movement. Can. J. Soil Sci. 42:247–253.

Staple, W. J., and J. J. Lehane. 1954. Movement of moisture in unsaturated soils. Can. J. Agr. Sci. 34:329–342.

Veihmeyer, F. J., and A. H. Hendrickson. 1949. Methods of measuring field capacity and wilting percentages of soils. Soil Sci. 68:75–94.

Wiegand, Craig L., and S. A. Taylor. 1961. Evaporative drying of porous media. Agr. Exp. Sta. Utah State Univ. Logan. Spec. Rep. 15.

Wiegand, Craig L., and S. A. Taylor. 1962. Temperature depression and temperature distribution in drying soil columns Soil Sci. 94:75–80.

Willis, W. O. 1960. Evaporation from layered soils in the presence of a water table. Soil Sci. Soc. Amer. Proc. 24:239–242.

Wilson, L. G., and J. N. Luthin. 1963. Effect of air flow ahead of the wetting front on infiltration. Soil Sci. 96:136–143.

Wind, G. P. 1955. A field experiment concerning capillary rise of moisture in a heavy clay soil. Neth. J. Agr. Sci. 3:60–69.

Youngs, E. G. 1957. Moisture profiles during vertical infiltration. Soil Sci. 84:283–290.

Youngs, E. G. 1958a. Redistribution of moisture in porous materials after infiltration: 1. Soil Sci. 86:117–125.

Youngs, E. G. 1958b. Redistribution of moisture in porous materials after infiltration: 2. Soil Sci. 86:202–207.

Youngs, E. G. 1960a. The hysteresis effect in soil moisture studies. Int. Congr. Soil Sci., Trans. 7th (Madison, Wis., USA) 1:107–113.

Youngs, E. G. 1960b. The drainage of liquids from porous materials. J. Geophys. Res. 65:4025–4030.

13

The Dynamics of Soil Water

Part II—Temperature and Solute Effects

JOHN W. CARY

Agricultural Research Service, USDA
Kimberly, Idaho

S. A. TAYLOR *(deceased, June 1967)*

Utah State University
Logan, Utah

I. INTRODUCTION

Changes in the soil temperature and solute concentration occur continually under natural field conditions. Examples of the variations in solute concentration are given in chapter 14, particularly Fig. 14–13 and 14–14. In addition to this type of change induced by moisture flow, the concentration of the soil's solution oscillates between saturation under dry conditions to very dilute values after a rain or irrigation. Soil temperature, like solution concentration, continually changes. Its variation is conveniently classified as diurnal (Fig. 13II–1) and seasonal (Fig. 13II–2). The diurnal thermal changes are generally significant to a depth of 20 or 30 cm; the soil zone which contains the greatest proportion of plant roots. The seasonal temperature wave extends well below the zone of most crop roots.

These dynamic thermal and osmotic changes create gradients of physical properties in the soil that may influence the movement of soil water. Thus the questions arise: in what manner may these changes affect the flow of soil water, and under what conditions are these affects significant? As in the description of other physical systems, the primary driving force for water is pictured as a water potential gradient. Its effectiveness is defined as the hydraulic conductivity. Changes in the soil's solutes and temperature may affect both the water potential gradient and the hydraulic conductivity. In addition, the flow of heat along a temperature gradient and the flow of solutes along a concentration gradient may affect the net water flow.

II. CHANGES IN CONDUCTIVITY

Osmotic and thermal changes have been shown to alter the conductivity of moisture in soil. Gardner et al. (1959) have reported the effects of exchangeable sodium percentage and soil solution concentration on the weighted-mean diffusivity for two soils. For normal soils with low sodium, the diffusivity doubled as the concentration was increased from 2 meq/liter to 100 meq/liter. However, when the exchangeable sodium was high, diffusivities dropped by orders of magnitude as the salt concentration was reduced. Quirk and Schofield (1955) have studied the permeability of several soils as affected by the concentration of salt in the water entering the soil. Over a period of 5 hours, the permeability of a calcium soil decreased by 25% when pure water was entering the surface, but by only 5 percent over the same time period when 1×10^{-3} M $CaCl_2$ solution was allowed to enter the soil. Again the effect on sodium soils was many times greater. Changes in water conductivities caused by the dissolved salt result principally from the effect of concentration on the expansion of the electric double layer around the soil's colloidal particles. The expansion of the double layer increases as the concentration of salts decreases; thus there is a tendency for the particles to swell or to disperse and clog pores. The viscosity of liquid film in the neighborhood of the double layer is higher than that of water (Low, 1960); thus, a dispersion of soil colloids may have the additional secondary effect of directly reducing the mobility of the soil water.

Jackson (1963) has measured the effect of temperature on the weighted-mean diffusivity for three soils. As the temperature rose from 5 to 45C, the diffusivities doubled in an approximately linear fashion. This temperature dependence was described by the temperature dependence of the ratio of surface tension to the viscosity of water and seems to be adequate to describe thermally induced changes in the diffusivity at the low matric suctions associated with infiltration. At greater suctions, however, soil water conductivity appears to be more temperature dependent than is free water viscosity. Meeuwig (1964) found that the temperature dependence of the soil water viscosity was two or three times that of free water viscosity in the three soils studied. Although the effect was most in the dry soils, a difference in temperature dependence of the viscosity was clearly evident at soil water suctions as low as 100 mbars.

III. CHANGES IN POTENTIAL

The energy status of the soil water is influenced by temperature and salt concentration. The question is: Do gradients of such quantities as kinetic energy, osmotic pressure, vapor pressure, and surface tension give rise to a significant transfer of soil water? Experiments have shown that under certain conditions they do.

The effects of thermal and osmotic gradients on water vapor diffusion through the soil's gas phase are easily recognized since the transfer of such water is directly proportional to the vapor concentration (vapor pressure) gradient. This relation (Fick's first and second laws) may be written as

$$q = -D(dc/dx) = -(D/RT)(dp/dx) \qquad [13II-1]$$

and

$$\partial c/\partial t = D(\partial^2 c/\partial x^2) \qquad [13II-2]$$

where q is the vapor flux, D a constant diffusion coefficient for the vapor, c the concentration of vapor, x the direction of flow, R the universal gas constant, T the absolute temperature, p the vapor pressure of the moisture, and t the time. Jackson (1964) has rewritten equation [13II-1] in a form similar to that of the diffusion equation for soil water as discussed in part I of chapter 13. For two soils the diffusion coefficient varied between 10^{-5} and 10^{-4} cm^2/sec. in the dry range with

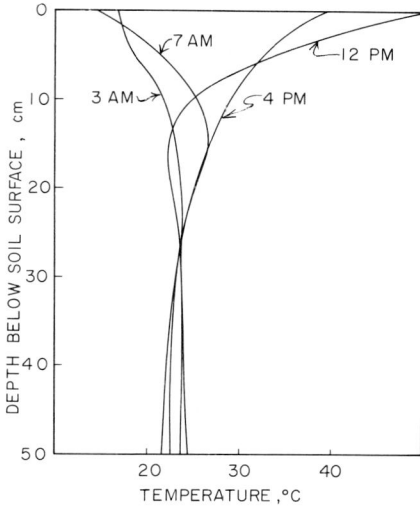

Fig. 13II–1. Daily summer temperature changes in a fallow field soil. Data from Onchukov (1957) or see Carson and Moses (1963).

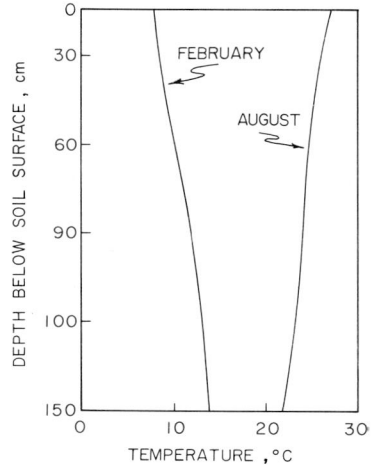

Fig. 13II–2. Average soil temperatures under a fallow soil in the winter and summer. Data from Qashu and Zinke (1964) or see Carson and Moses (1963).

volumetric water contents from 0.01 to 0.08. Roughly, this means that if the soil water content changed from 1 to 6% over a 10-cm distance, the induced water flux would be 0.2 mm/day. Such flow would be primarily in the form of vapor diffusion under nearly isothermal conditions. The vapor pressure gradient causing the transport would be caused partly by the difference in matric suction on the soil water and partly by the concentration changes in the soil solution. Under field conditions, temperature gradients would have further influenced the vapor pressure differences, since a continuous thermal wave passes through the soil. Matthes and Bowen (1963) have written equation [13II–2] with D as a variable and p as a function of temperature. In this form they were qualitatively able to predict changes in soil water contents responding to both changes in T and changes in D as it varied with soil density.

The vapor pressure of water is very sensitive to changes in temperature. The change in vapor pressure as the temperature drops from 26C to 25C would be approximately equivalent to that which occurs between pure water and a solution with an osmotic pressure of 80 bars. Moreover, the apparent diffusion coefficients for water vapor under a thermal gradient are surprisingly large. Experimentally, dT/dx is generally measured, and dp/dx is inferred from the known relation between vapor pressure of pure water and temperature. Using this approach, values of D calculated from equations [13II–1] and [13II–2] are always greater than one would expect from known values of the diffusion coefficient of water vapor into air, corrected for porosity and diffusion path length. Philip and De Vries (1957) in summarizing the literature prior to 1957 pointed out that the thermally induced transfer of water through soil was from 4 to 18 times greater than that pedicted by this simple theory. Part of the trouble seems to arise from the fact that the macroscopic thermal gradient is less than the microscopic thermal gradients across the internal air spaces due to the heterogeneity of thermal conductivities of the soil's constituents. Thus, the over-all thermal gradient does not correctly describe the internal vapor pressures. This has been discussed by Woodside and Kuzmak (1958). Also, a part of the discrepancy is caused by a thermally driven liquid phase component of transfer from the warm to cool which accompanies the vapor diffusion. Philip and De Vries (1957) developed equations to account for this liquid phase flow based on liquid-air surface tension differences induced by the thermal gradient. This was a notable advance in the theory, though it is now recognized to have one serious shortcoming; thermally induced water flow does not become zero in the absence of internal liquid-air interfaces at saturation (Corey and Kemper, 1961; Taylor and Cary, 1960). This problem has been explored by Deryaguin and Melnikova (1958) and more recently by Cary (1965) using current theories of hydrogen bond distributions in water.

From time to time, some thought has been given to describing thermal water transport, coupled with osmotic and suction-induced flow, with a free energy expression analogous to Darcy's law; for instance,

$$q_w = - K' (d\mu/dx) \qquad [13II-3]$$

where q_w is the flux of water, K' is a conductivity for water, μ is the chemical potential of the soil water, and x is distance. However, one soon finds this equation must be limited to isothermal conditions, both for theoretical reasons and intuitively because the free energy of pure water decreases as the temperature rises. Moreover, μ is a function of both soil matric potential (suction) as indicated by a tensiometer and the solute potential (or osmotic suction). This has led to further difficulties because in some cases the solute potential is effective in causing water flow through soil and in some cases it is not. Still other problems arise from the opposite temperature dependences of the soil's matrix suction and its relative humidity (Kijne and Taylor, 1964).

Under normal field conditions, the net water flow is the sum of liquid and vapor transfer as affected by simultaneous changes in matric potential, temperature, and solute potential. The failure of simple expressions to completely describe the flow has caused recourse to a more basic consideration for analytical treatment of the problem. In particular, a system which is not in equilibrium, will, under natural conditions, spontaneously readjust such that entropy becomes a maximum. Any natural spontaneous change is an irreversible process which creates entropy. Since the creation of entropy is a time and rate dependent quantity, it seems reasonable to suppose that the natural transport of mass and energy in the soil system will occur simultaneously in a way such that the entropy proceeds toward a maximum. Based on this philosophy and certain assumptions (Taylor and Cary, 1964), it can be argued that the flux of water in the soil may be described by

$$q_w = \sum_{j=1}^{n} L_{wj} \left[\psi_g - \frac{d(\mu_j)_T}{dz} \right] - L_{wh} \frac{1}{T} \frac{dT}{dz} \qquad [13II-4]$$

where q_w is the net water flux, j represents any one of the n various mass species in the soil which can move, ψ_g the gravitational potential, $d(\mu_j)_T/dz$ the gradient of chemical potential of the "j" mass species at temperature T, L phenomenological coefficients, the subscripts w and h signify water and heat, and dT/dz is the thermal gradient. For the particular case of $dT/dz = 0$ in a solute-free soil, equation [13II–4] reduces to any of the well known forms of Darcy's law or the water diffusion equation which have received much study.

In the special case where $dT/dz = 0$, the solute potential of the soil solution is created by a single solute and gravity effects are negligible, equation [13II–4] becomes

$$q_w = - L_{ww} RT \frac{\mathrm{dln}p}{\mathrm{d}x} - L_{wk} \frac{\mathrm{d}\mu_k}{\mathrm{d}x} \qquad [13II–5]$$

using the identity $(\mathrm{d}\mu_w)_T/\mathrm{d}x \equiv RT \ (\mathrm{dln}p)/\mathrm{d}x$ where p is the aqueous vapor pressure. The flux of solute is described by an analogous equation

$$q_k = - L_{kw} RT \frac{\mathrm{dln}p}{\mathrm{d}x} - L_{kk} \frac{\mathrm{d}\mu_k}{\mathrm{d}x} \qquad [13II–6]$$

and the tendency for water to move with solute and solute to move with water is expressed by the interaction coefficients such that

$$L_{wk} = L_{kw}. \qquad [13II–7]$$

The water and solute transmission coefficients are expressed by L_{ww} and L_{kk}. To illustrate the meaning of the two terms in equation [13II–5], consider the following situations. First, in a soil where no osmotic gradient exists, $\mathrm{d}\mu_k/\mathrm{d}x = 0$ and so the moisture flow would be described by the first term. In this case, it might be convenient to rewrite equation [13II–5] as

$$q_w = - L_{ww} \bar{V} \frac{\mathrm{d}P}{\mathrm{d}x} = - K \frac{\mathrm{d}P}{\mathrm{d}x} \qquad [13II–8]$$

where the vapor pressure gradient has been replaced by its equivalent in a pressure gradient as given by tensiometers, e.g. Taylor and Kijne (1964). If there were a solute concentration gradient in the solution, but the water-filled pores were so large that no osmotic water movement occurred, then the flow would still be described by equation [13II–8] with dP/dx taken to reflect only the matric suction. In this case $\mathrm{d}\mu_k/\mathrm{d}x \neq 0$ and $\mathrm{dln}p/\mathrm{d}x$ would reflect the sum of both matric suction and osmotic components of the energy gradients. For equation [13II–5] to describe the flow, L_{wk} would have a value such that the sum of the two terms in [13II–5] would equal the single term in equation [13II–8]; e.g., the solute force would be substracted out. On the other hand, if the water-filled pores in the soil were so small that they behaved as a perfectly semipermeable membrane then $L_{wk} = L_{kw} = 0$, and the flow of water would be described by the first term of equation [13II–5] or by equation [13II–8] with P representing the sum of both matric and solute potentials (e.g. the total suction). In intermediate cases where some solute can "leak" through the soil water films, the solute potential gradient becomes only partly effective in causing moisture flow and is characterized by the value of L_{wk}. It is these intermediate cases in which equation [13II–8] fails. This type of analysis is receiving increased attention in plant-water relations (Dainty, 1963), as well as in soil water and solute relations (Abdel Aziz and Taylor, 1965).

The effect of osmotic potential gradients on liquid phase soil water flow is related to the salt sieving phenomenon. Equation [13II–6] describes the isothermal transfer of salt in soil when both solute and matric potential gradients exist. The first term describes the amount of solute which would be carried along with the flow of water in the absence of any solute potential gradient. The second term accounts for the diffusion of solute along its own potential (concentration) gradient. Under conditions where the solute is, to some degree, excluded from the water flow path (e.g. salt sieving), the value of L_{kw} decreases accordingly. This is the condition which can give rise to osmotically induced flow of water. That is, the more the solute is excluded from some portion of the water flow path, the more effective a solute concentration gradient will be in developing a liquid phase water flow. In fact, where the Onsager reciprocity relation holds, a functional relation (equation [13II–7]) is fixed between solute sieving and osmotically induced water flow. Detailed studies of the mechanics and conditions which lead to this type of transfer have been made by Kemper and Maasland (1964), Kemper and Evans (1963), and Kemper (1961). Their data indicate that solute sieving and osmotic water flow in clays may be significant in water films up to at least 100A thick. Taylor and Cary (1960) have experimentally demonstrated small solute-induced flow of water flows through a tightly packed sample of saturated loam soil. The simultaneous movement of water and KCl has also been demonstrated in a silt loam soil and in kaolinite at an average water potential of −31.2 joules/kg (312 millibars suction) in response to a salt concentration difference between 0.2 M and 0.3 M across a soil plug 2.86 cm long (Abdel Aziz and Taylor, 1965). Concurrently, a hydraulic head difference of 10 cm of water induced a flow of both salt and water and the results were in accordance with equations [13II–5], [13II–6], and [13II–7]. Burns and Dean (1964) have demonstrated osmotically induced water flow in soil. Their experiment showed a flux of water into $NaNO_3$ bands and a subsequent redistribution of nitrate. However, Kemper (1961) has pointed out that for normal soils, osmotic gradients are probably not very effective in moving water until the matric suction rises above 1 bar. Exceptions could occur near the surface when fertilizer granules are present or where salt is accumulated by the evaporative process (Cary, 1964; Doering et al., 1964).

When thermal and suction gradients are creating water flow in the absence of significant solute effects, equation [13II–4] may be written as

$$q_w = L_{ww} \left[\phi - \frac{d(\mu_w)_T}{dx} \right] - L_{wh} \frac{1}{T} \frac{dT}{dx} \qquad [13II–9]$$

where the first term describes the flow of moisture due to gravitational and water potential (matric suction) gradients and the second term accounts for the flow induced by a thermal gradient. Though equation [13II–9] is a natural consequence of the time rate of entropy change in the soil system, it may be derived independently on the basis of flow mechanisms as shown by Philip and De Vries (1957). There is also a simultaneous heat flow equation which goes with [13II–9] and a reciprocal relation between heat and water flow coefficients similar to that shown in equation [13II–7]. In this case, the reciprocal relation provides a theoretical tool for exploring molecular energy transport (Cary, 1965; Taylor and Cary, 1964). Methods for handling the heat transfer equation in soil have been treated in a monograph edited by Van Wijk (1963). Laboratory methods are now available for separating the vapor phase flow and the liquid phase flow components from the net water transfer (Jackson, 1965; and Cary, 1965), thus the liquid and vapor phase components of each of the two terms in equation [13II–9] may be studied separately. This allows the transfer coefficients L_{ww} and

L_{wh} to be evaluated in terms of more commonly used numbers such as the soil's water conductivity, vapor diffusion coefficients, heat of vaporization, etc. (Cary, 1965). However, these coefficients depend on soil properties, and there is no *apriori* way of writing down their exact values. Some experimental measurements are required for each soil.

A thermal gradient in soil may cause large net quantities of water transfer. Even at saturation a small water flux may be observed (Taylor and Cary, 1960). Cary (1965) has shown that in one loam soil with a soil water suction of only −6.6 joules/kg (66 mbars suction), a thermal gradient of 0.5C per cm would cause as much water transfer as a hydraulic head gradient of 2 cm of H_2O/cm. At this water content, 80% of the thermally induced water transfer was in the liquid phase. When this soil's water potential was lowered to −45 joules/kg (450 mbars suction), a thermal gradient of 0.5C/cm moved as much water as a hydraulic head gradient of 250 cm of water per cm. At a water potential of −220 joules/kg (2.2 bars suction), Taylor (1962) reported a temperature difference of 1°/cm would cause as much water flow as a water potential difference of 140 joules/kg/cm (1.4 bars/cm); and if the soil was at a potential $< -12.6 \times 10^3$ joules/kg (126 bars), a temperature difference of one degree caused about the same water flow as a water potential difference of 8×10^3 joules/kg (80 bars). Presumably the flow at the lower potentials was largely in the vapor phase. As may be noted in Fig. 13II–1, gradients of 0.5C/cm or more may be expected under field conditions.

Perhaps the most striking example of thermal water transfer in the field is that which results from a frost zone at the soil surface. For example, Willis et al. (1964) have presented field data showing a water table drop of more than 120 cm as the moisture moved upward under the thermal gradient into a frost zone. Other striking examples of this type of flow have been reported by Ferguson et al. (1964) and by Meyer (1960). Evidently the freezing zone sharply reduces the downward gravitational and matric suction flow which leaves the net result determined by the thermal gradient which operates in the same direction over a period of several months. The fact that diurnal thermal waves passing through the soil surface layers in the summer do not cause such an obvious transfer of water does not mean they are less important, but only that steep water potential gradients must develop to account for any net water flow against the thermal gradients. When soil water moves up through the profile and into the atmosphere, an energy requirement of 570 to 800 cal/g of water must be eventually supplied for the phase change. Such large energy requirements produce temperature gradients in the soil, and thus, soil water flow must ultimately be closely related to heat transfer (Wiegand and Taylor, 1962; Anderson and Linville, 1962).

IV.   SUMMARY

From the experimental evidence available in the literature, it is possible to make some statements about water flow in response to solute and temperature gradients under field conditions.

1) Changes in temperature and solute concentration of the soil solution may affect water flow in any, or all, of three ways—by changing the conductivity of the soil (Gardner et al., 1959); by changing the water potential (Kijne and Taylor,

1964); or by inducing water to flow along solute or thermal gradients (Taylor and Cary, 1964).

2) Normal variations in either the temperature or the solute concentration of field soils may change the hydraulic conductivity by a factor of at least two. In the event of a high sodium adsorption ratio, changes in the concentration of salts may cause changes in the conductivity to exceed two or three orders of magnitude.

3) Thermal gradients directly cause water to flow from warm areas to cool areas in both the liquid and vapor phase. This transfer will occur at any water content, but it becomes progressively more important as the soil becomes drier. In general one may expect normal thermal gradients in the root zone to be about as effective as the gravitational potential in moving water when the soil is near saturation. At field capacity these thermal gradients will be about 10 times more important than gravity and will become 1,000 times more effective than gravity before the permanent wilting percentage is reached. Although the largest thermal gradients occur in daily cycles through the principle rooting area of a crop, thermally induced flow is quite striking during winter as water moves up into the frozen zone.

4) Salt concentration gradients may cause water to flow through soil in both the liquid and vapor phase. Transport in the vapor phase results from gradients in the solution's vapor pressure. While liquid phase, solute induced flow has been experimentally demonstrated in saturated and unsaturated loam soils, normal solute gradients in productive field soils probably do not move significant amounts of water until the water potential is reduced to the neighborhood of −100 joules/kg (1 bar suction) except when salts are accumulated at the surface due to evaporation or when fertilizer granules are present.

LITERATURE CITED

Abdel Aziz, M. H., and S. A. Taylor. 1965. Simultaneous flow of water and salt through unsaturated porous media: I. Rate equations. Soil Sci. Soc. Amer. Proc. 29:141–143.

Anderson, D. M., and A. Linville. 1962. Temperature fluctuations at a wetting front: I. Characteristic temperature-time curves. Soil Sci. Soc. Amer. Proc. 26:14–18.

Burns, G. R., and L. A. Dean. 1964. Movement of water and nitrate around bands of sodium nitrate in soils and glass beads. Soil Sci. Soc. Amer. Proc. 28:470–474.

Carson, J. E., and H. Moses. 1963. The annual and diurnal heat exchange cycles in the upper layers of soil. J. Appl. Meteorol. 2:397–406.

Cary, J. W. 1964. An evaporation experiment and its irreversible thermodynamics. Int. J. Heat Mass Transfer 7:531–538.

Cary, J. W. 1965. Water flux in moist soil: Thermal versus suction gradients. Soil Sci. 100:168–175.

Corey, A. T., and W. D. Kemper. 1961. Concept of total potential in water and its limitations. Soil Sci. 91:299–302.

Dainty, J. 1963. Water relations of plant cells. p. 279–326. In Advances in Botanical Research. Vol. 1. Academic Press, London.

Deryaguin, B. V., and M. K. Melnikova. 1958. Mechanism of moisture equilibrium and migration in soils. Int. Symp. on Water and its conduction in soils. Highway Res. Board. Spec. Rep. 40. p. 43–54.

Doering, E. J., R. C. Reeve, and K. R. Stockinger. 1964. Salt accumulation and salt distribution as an indicator of evaporation from fallow soils. Soil Sci. 97:312–319.

Ferguson, H., P. L. Brown, and D. D. Dickey. 1964. Water movement and loss under frozen soil conditions. Soil Sci. Soc. Amer. Proc. 28:700–703.

Gardner, W. R., M. S. Mayhugh, J. O. Goertzen, and C. A. Bower. 1959. Effect of electrolyte concentration and exchangeable sodium percentage on diffusivity of water in soils. Soil Sci. 88:270–274.

Jackson, R. D. 1963. Temperature and soil water diffusivity relations. Soil Sci. Soc. Amer. Proc. 27:363–366.

Jackson, R. D. 1964. Water vapor diffusion in relatively dry soil: III. Steady state experiments. Soil Sci. Soc. Amer. Proc. 28:467–470.

Jackson, R. D. 1965. Water vapor diffusion in relatively dry soil: IV. Temperature and pressure effects on sorption diffusion coefficients. Soil Sci. Soc. Amer. Proc. 29:144–148.

Kemper, W. D. 1961. Movement of water as affected by free energy and pressure gradients: II. Experimental analysis of porous systems in which free energy and pressure gradients act in opposite directions. Soil Sci. Soc. Amer. Proc. 25:260–265.

Kemper, W. D., and N. A. Evans. 1963. Movement of water as affected by free energy and pressure gradients: III. Restriction of solutes by membranes. Soil Sci. Soc. Amer. Proc. 27:485–490.

Kemper, W. D., and D. E. L. Maasland. 1964. Reduction in salt content of solution on passing through thin films adjacent to charged surfaces. Soil Sci. Soc. Amer. Proc. 28:318–323.

Kijne, J. W., and S. A. Taylor. 1964. The temperature dependence of soil water vapor pressure. Soil Sci. Soc. Amer. Proc. 28:595–599.

Low, P. F. 1960. Viscosity of water in clay systems. Nat. Conf. Clays Clay Minerals, 8th Pergamon Press, New York. p. 170–182.

Matthes, R. K., and H. D. Bowen. 1963. Water vapor transfer in soil by thermal gradients and its control. Amer. Soc. Agr. Eng., Trans. 6:244–248.

Meeuwig, R. O. 1964. Effects of temperature on moisture conductivity in unsaturated soil. Ph.D. Thesis. Utah State Univ. 100 p. Univ. Microfilms. Ann Arbor, Mich. (Diss Abstr. 25:3180)

Meyer, A. F. 1960. Effect of temperature on groundwater levels. J. Geophys. Res. 65: 1747–1752.

Onchukov, D. N. 1957. The phenomenon of heat and moisture transmission in soils and subsoils. (In Russian). Moskov. Tekhnol. Inst. Pishck. Promysk. Trudy. 1957:55–63.

Philip, J. R., and D. A. de Vries. 1957. Moisture movement in porous materials under temperature gradientes. Amer. Geophys. Union, Trans. 38:222–232.

Qashu, H. K., and P. J. Zinke. 1964. The influence of vegetation on soil thermal regime at the San Dimas Lysimeters. Soil Sci. Soc. Amer. Proc. 28:703–706.

Quirk, J. P., and R. K. Schofield. 1955. The effect of electrolyte concentration on soil permeability. J. Soil Sci. 6:163–178.

Taylor, S. A. 1962. Influence of temperature upon the transfer of water in soil systems. Mededel. Landbouwhogeschool, Ghent: 27:535–551.

Taylor, S. A., and J. W. Cary. 1960. Analysis of the simultaneous flow of water and heat or electricity with the thermodynamics of irreversible processes. Int. Congr. Soil Sci., Trans. 7th. (Madison, Wis., USA) 1:80–90.

Taylor, S. A., and J. W. Cary. 1964. Linear equations for the simultaneous flow of matter and energy in a continuous soil system. Soil Sci. Soc. Amer. Proc. 28:167–172.

Taylor, S. A., and J. W. Kijne. 1964. Evaluating thermodynamic properties of soil water. 1st Int. Symp. Humidity Moisture Contr. Sci. Ind. Pap. 14. Reinhold, New York. 3:325–342.

Wiegand, C. L., and S. A. Taylor. 1962. Temperature depression and temperature distribution in drying soil columns. Soil Sci. 94:75–79.

Wijk, W. R. van [ed.] 1963. Physics of plant environment. Interscience Publ., New York. p. 102–143.

Willis, W. O., H. L. Parkinson, C. W. Carlson, and H. J. Haas. 1964. Water table changes and soil moisture loss under frozen conditions. Soil Sci. 98:244–248.

Woodside, W.. and J. M. Kuzmak. 1958. Effect of temperature distribution on moisture flow in porous materials. Amer. Geophys. Union., Trans. 39:676–680.

14

Miscible Displacement and Leaching Phenomenon

J. W. BIGGAR AND D. R. NIELSEN[1]

University of California
Davis, California

Miscible displacement of one fluid by another is a universal process which occurs in time and space often without recognition. When one fluid displaces another, each miscible with the other, a mixing takes place between the two fluids which depends on the properties of the fluids and the material in which the fluids are contained. In soils the fluids are aqueous solutions of various salts. The presence of salts in the solvent modifies its properties and its relation to the medium. Different salts interact with each other and the porous material in the solvent environment. And finally the transport processes vary in relation to the surfaces and pore structure of the medium and the degree of pore saturation. In these ways, the mixing which takes place during miscible displacement becomes an important aspect of the movement of water and its dissolved constituents through soils.

In the case of water movement in soils, the hydraulic conductivity is generally considered a most important soil parameter. For many soils the hydraulic conductivity is sensitive to the presence of constituents dissolved in the water. Even when it is not, the mixing which takes place between two solutions during displacement is not described by a knowledge of the hydraulic conductivity. This is a consequence of the pore structure of the medium and its control over the flow paths in a manner not described by the hydraulic conductivity. As a result, the transport of the salt or dissolved constituent must be considered as well as the movement of the solvent in any displacement process. Such considerations become particularly important in irrigated agriculture when it is desirable to know the concentration and location of a dissolved constituent in the soil profile, the removal of undesirable constituents, the reactions of constituents with each other and the soil matrix during the displacement, and the transport of water and solutes to plant roots.

A general illustration of the mixing which takes place during displacement is provided by the elution curve obtained when a solution containing chloride ion (Cl^-) is displaced by a solution not containing chloride from a foot-long column of saturated sand. The displacement occurred at an average flow velocity of 0.473 cm/hour and is numerically equal to the flux divided by the water-filled porosity. If no mixing occurred between the displacing and displaced solutions, an elution curve illustrated by the broken vertical line would be obtained. Such a curve is obtained for piston displacement or complete displacement of one solution by another. The sigmoid shape of the elution or breakthrough curve indicates that mixing has occurred; otherwise the dilution of the Cl solution would not take place until an amount of displacing solution equivalent to that held originally in

[1] This chapter was partially supported by funds from Research Grant No. DA–AMC–36–039–63–G5 for the Army Electronics R & D Activity, Fort Huachuca, Arizona, USA; and the Water Resources Center, University of California, USA.

the column had passed from the effluent end. The volume of effluent which corresponds to the piston flow curve is referred to as the pore volume of effluent or V/V_o where V_o is the volumetric water capacity of the column determined gravimetrically at 105C and V the volume of effluent. When V equals V_o, the pore volume of effluent is unity.

The processes which contribute to mixing in a nonreactive porous material are basically two. The first is the variation in the pore velocity of the solution as it moves through the column. Since there is a distribution of pore sizes and shapes, there is also a distribution of velocities, the velocities having both magnitude and direction. For a homogeneous random pore distribution it is expected that the range of velocities can be represented by a mean velocity. This mean velocity would be the velocity of the moving boundary between the displacing and displaced solution. Such a representation of the various velocities is inadequate to describe the mixing at the boundary in question except for very simple porous materials and velocities beyond the range which occurs in soils.

The second process which contributes to mixing is that of diffusion. This process, a result of random motion of the ions or molecules, occurs preferentially in one direction or another depending on the activity gradients which provide the driving force. The existence then of a sharp boundary between two miscible solutions is of a transitory nature whether mass movement is occurring or not. The progress of these two processes occurring simultaneously during displacement suggests the superposition of one upon the other to obtain a total impression of the mixing which occurs. Such a procedure ignores the effect the geometry of the porous material has on variations of the pore velocity and ultimately the area at the boundary between the displacing and displaced solutions through which diffusion takes place. The activity gradients depend in part then on the variations in the pore velocity which is governed by the nature of the porous material. At the present time an adequate description of these processes and their interactions is not available. However, much can be learned and concepts developed despite the imperfect nature of a theoretical description.

The area enclosed by the vertical line, the upper part of the Cl breakthrough curve (BTC), and the line C/C_o equal to 1 in Fig. 14-1 is related to the amount of original water in the column that is not easily displaced. It has been referred to as holdback by Danckwerts (1953) and generally increases in magnitude as the porous material increases in complexity. The shape and position of the breakthrough curves (BTC) obtained during displacement illustrate the processes involved in mixing. Consider in Fig. 14-2 the displacement of 0.10N $CaCl_2$ water by 0.01N $CaSO_4$ from water-saturated, stable Aiken clay loam aggregates of three size ranges, at an average flow velocity of 3 cm/hour. The size ranges of 0.25 to 0.5, 0.5 to 1.0, and 1.0 to 2.0 mm create three ranges of macropores between the aggregates. The pores within the aggregates themselves are probably of comparable size and shape regardless of the aggregate size since the aggregates were all obtained from the same soil mass. A bimodal pore size distribution with large pores between the aggregates and small pores in the aggregates exists for any one aggregate range. If the shapes of the curves are determined mainly by the velocity distribution, and the total number of contacts between aggregates decreases with increasing aggregate size, the proportion of total flow through the aggregates will decrease with increasing aggregate size. Thus, mixing in the column becomes less complete and is dominated more by the larger pores between aggregates. This means that the Cl⁻ ion concentration will decrease initially in the effluent with less volume put through as aggregate size increases, the BTC will be flatter, and the volume required to reach C/C_o equal to 0.0 increases as illustrated in **Fig. 14–2**.

That inherent properties of the porous material are manifested in the mixing by velocity which occurs during miscible displacement is illustrated in **Fig. 14–3** and **14–4** by comparing leaching of Cl⁻ ion from Yolo loam and Columbia silt loam at two velocities under saturated conditions. At almost identical average flow velocities of 4.40 and 4.78 cm/hour for the Yolo and Columbia, respectively, the shapes and positions of the leaching curves are significantly different. Both soils exhibit considerable mixing by velocity as shown by their positions relative to the curve for piston flow (vertical broken line). However, the Yolo soil exhibits greater holdback which is associated with a larger number of slowly conducting pores. When comparison is made between BTC's obtained at 2 velocities in only the Columbia soil, the curve obtained at 0.104 cm/hour is displaced to the left and rotated clockwise but does not cross the BTC obtained at 4.78 cm/hour. The displacement to the left and rotation are manifestations of the contribution ionic diffusion makes to mixing in addition to mixing by velocity which occurs in the column. In contrast to Columbia soil, the BTC in Yolo soil obtained at a velocity of 0.098 cm/hour does cross the BTC obtained at 4.40 cm/hour. Since the times of displacement and, thus, the diffusion time would be comparable in both soils at the slow velocity, the crossover might not be expected in the Yolo soil. However, because of the important difference in pore structure and consequently flow velocity distribution, the larger number of slowly conducting pores in Yolo soil act as sources for Cl⁻ ion and a greater volume of water must pass through the soil to reduce the Cl⁻ ion to a low concentration.

An extension of these considerations may be useful at this time with respect to these two soils. The distribution of Cl⁻ ion in the soil pores will depend on the electric field surrounding the individual soil particles. For certain well-defined systems the ion distribution may be described by a diffuse double-layer model. Such ion distributions in a heterogeneous mixture of various sized particles possessing electric fields of varying intensity will undoubtedly occur and will be of varying extent with respect to their extension into the solution phase. In addition, the extension of the anion layer based on repulsion of the ion in the surface force fields depends on the velocity of the solution phase. Consequently, the anion distribution will vary from pore to pore, depending on the size and shape as well as other factors. For steady-state conditions of constant average ion concentrations and flow, an average distribution may be assumed. But where the ion concentration is changing, and the average flow velocity is modified, the interactions of velocity and ion distribution need not change in a uniform manner. In fact, there is no assurance that Darcy flow takes place in all pores and this is probably one of the causes of variations in the microscopic pore water velocity.

The position and shapes of the Cl BTC for Yolo and Columbia soil reflect the nonuniformity of ion distribution and pore velocity when the average flow velocity is changed. In this way the more general description of the contributions to mixing by variations in pore velocity, ionic diffusion, and the coupling between these processes accounts for processes occurring during miscible displacement in soils. Comparisons of BTC's for several porous and nonporous granular materials in **Fig. 14–5** appear to substantiate these considerations. With increasing complexity

Fig. 14–2. Chloride breakthrough curves in 3 size ranges of Aiken clay loam aggregates saturated with water.

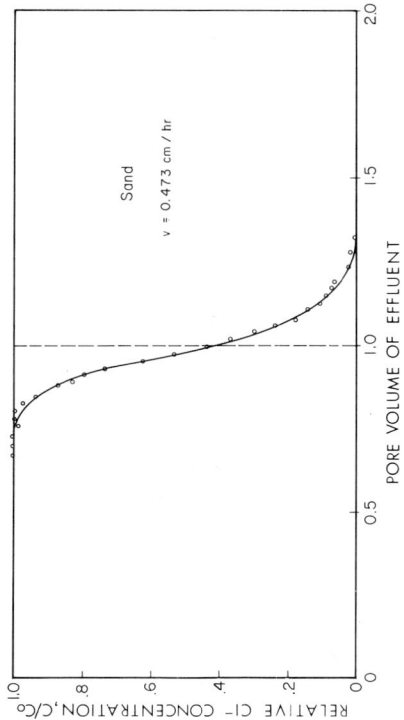

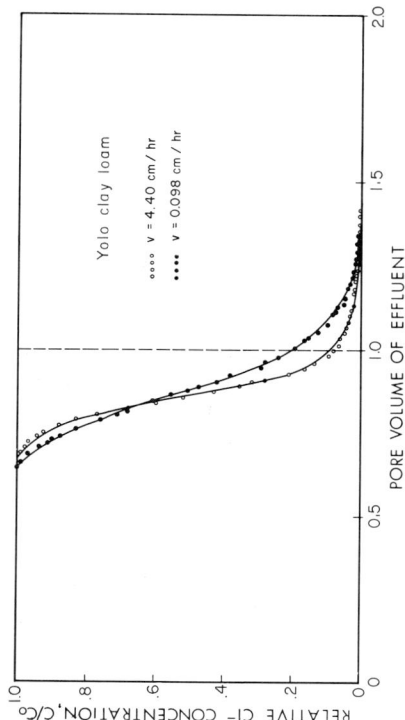

Fig. 14–1. An elution or breakthrough curve of concentration of chloride vs. pore volume of effluent in sand leached with chloride-free water. C_o represents the original chloride concentration of the sand solution.

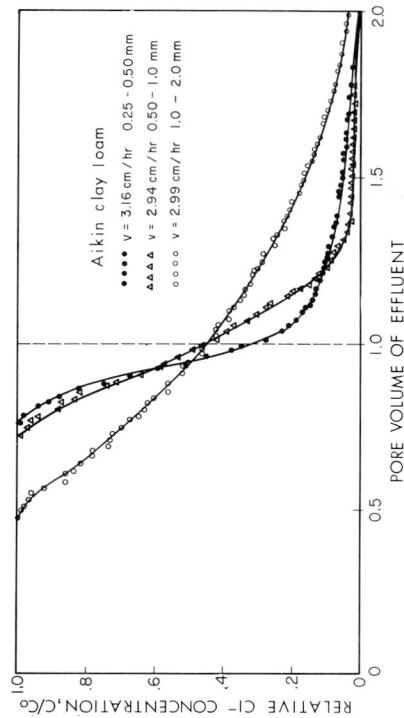

Fig. 14–4. Chloride breakthrough curves in water-saturated Columbia silt loam for 2 average flow velocities.

Fig. 14–3. Chloride breakthrough curves in water-saturated Yolo clay loam for 2 average flow velocities.

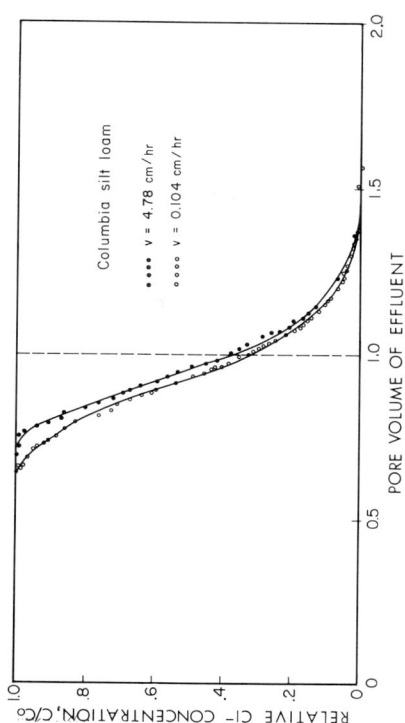

of porous material progressing from the simple glass bead to Yolo loam, the effluent curves are displaced further to the left and the shapes become more skewed.

The contribution of diffusion to the mixing is more easily demonstrated in a 390μ glass bead medium to minimize the interaction of ions with solid surfaces. A series of experiments conducted on the same column and illustrated in Fig. 14–6 show the displacement at two velocities of $CaCl_2$ solution by Na_2SO_4 solution matched as to density. The assumption is made that Darcy's law is capable of describing the flow. For the smaller flow velocity, Cl^- ion disappears in the effluent with a smaller effluent volume and the curve is flatter. On the basis of purely velocity dispersion, the curves are expected to be identical. Thus, a dispersion coefficient calculated for the greater velocity is incapable of describing dispersion at the slower velocity. The transport of the ions behind the average displacement front (longitudinal diffusion) and the additional mixing in the column at the slower velocity are the result of ionic diffusion.

I. FLUID PROPERTIES

The properties of the fluid must also be considered in any displacement process. Several investigators have suggested that the viscosity and density of water in soil may be modified by the proximity of mineral surfaces. On the basis of gross fluid properties it is not uncommon to experience the displacement of soil solutions by solutions of different density and viscosity. Examples of this phenomenon include the infiltration of rainwater into soil, the displacement of soil solution by irrigation water of different concentration and the addition of fertilizers to the native soil solution.

The existence of concentration or activity gradients of inorganic salts in aqueous solutions responsible for solute transfer by diffusion guarantees that the displacing and displaced solutions do not generally have identical densities nor viscosities no matter how close their values. When two superposed solutions of unequal density are accelerated in a direction perpendicular to their interface, the surface may be stable or unstable. Differences in densities provide unbalanced acceleration forces while the viscosities account for unequal drag forces. For example, unstable flow occurs for particular velocities vertically downward when a denser, more viscous fluid displaces a less dense, less viscous fluid. Here the unbalanced forces tend to accelerate the denser solution into the less dense solution below, with the viscous drag of the lower fluid unable to counterbalance this acceleration. With this action "fingers" of the more viscous fluid invade certain pore sequences occupied by less viscous fluid. The stability of the displacement depends on the viscosities and densities of the fluids, the permeability of the porous material and the direction and magnitude (velocity) of the displacement. If Darcy's law is obeyed for steady movement of velocity v vertically upwards through a medium of permeability k, the interface between the two solutions will be unstable for

$$(\mu_2 - \mu_1)\, v + k(\rho_2 - \rho_1)\, g < 0 \qquad\qquad [14–1]$$

where ρ is the density, μ the viscosity, and g the gravitational acceleration. The subscripts 1 and 2 refer to the displaced and displacing solutions, respectively. The same inequality, except for signs owing to the directional sense of the velocity,

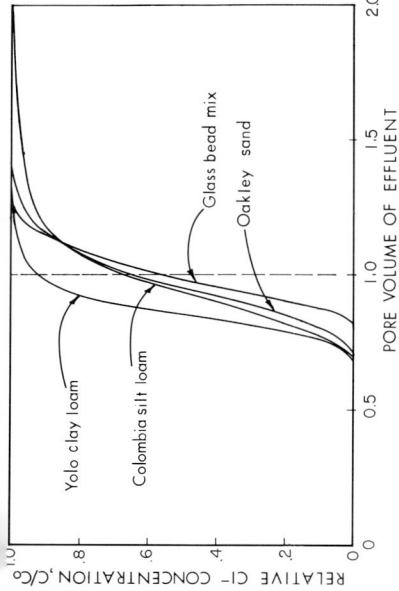

Fig. 14-6. Chloride breakthrough curves in water-saturated glass beads at 2 flow velocities.

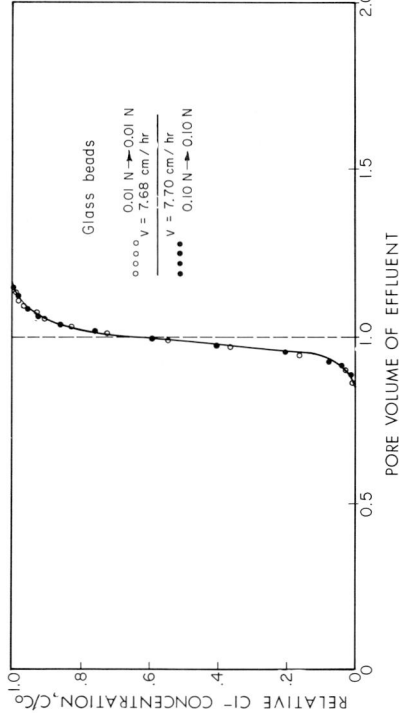

Fig. 14-8. Chloride-36 breakthrough curves in saturated 390μ glass beads for unequal NaCl concentrations. Displacements were made vertically upward. Data are taken from Biggar and Nielsen (1964).

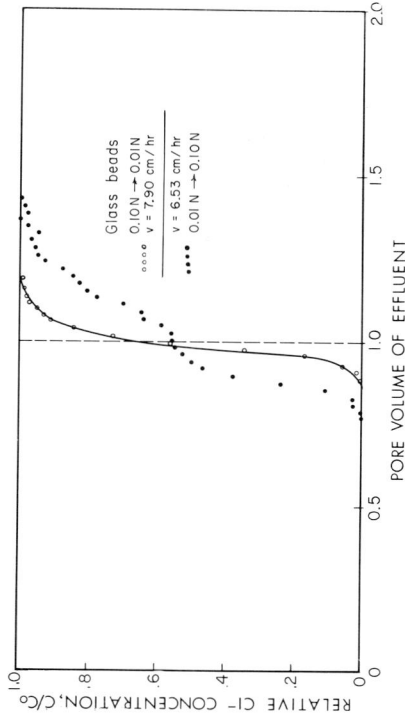

Fig. 14-5. Chloride breakthrough curves for 4 water-saturated porous materials obtained at approximately equal average flow velocities. Data are taken from Nielsen and Biggar (1961).

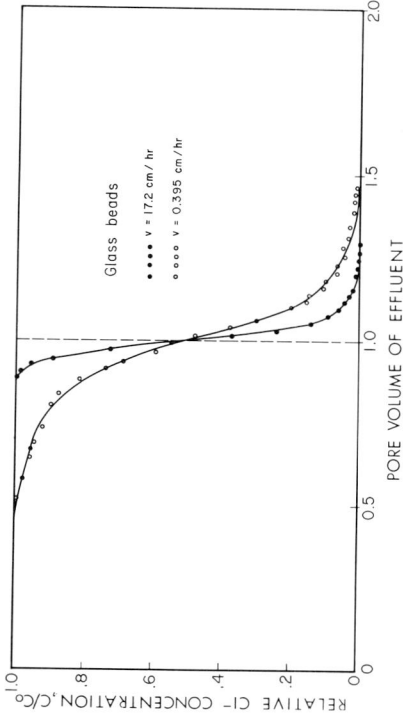

Fig. 14-7. Chloride-36 breakthrough curves in saturated 390μ glass beads for equal NaCl concentrations. Displacements were made vertically upward. Data are taken from Biggar and Nielsen (1964).

applies to movement vertically downward. Inequality [14–1] has been used extensively for immiscible fluids and to a lesser extent for miscible fluids. For inequality [14–1] to apply to miscible fluids it is assumed that the thickness of the interface region within an individual pore is small. For miscible fluids, the thickness of the interface region is not constant but tends to increase owing to molecular diffusion. Under such circumstances, the viscosity and density of the interfacial fluid are neither those of the displaced fluid nor displacing fluid. Furthermore, the pore geometry is not included in equality [14–1] except by an average value of k.

At the present time solutions of mathematical theories of miscible displacement do not explicitly consider the effect of nonuniform density and viscosity differences on mixing during miscible displacement. Only as these properties enter into the diffusion coefficient and velocity of fluid flow in an average manner are they included.

To examine experimentally the effect of the properties of solution on the mixing during displacement 390μ glass beads were selected to provide a porous medium of minimum interaction with the solution. The density and viscosity of the solutions were modified by using different concentrations of salt. Sodium chloride solutions of 0.01N and 0.1N concentration identified by Cl^{36} were displaced from a 30-cm long column by NaCl solutions of the same concentrations. The densities of 0.01N and 0.1N NaCl solutions reported elsewhere are 0.997511 and 1.00094 g/ml with corresponding viscosities of 0.8945 and 0.8928 centipoises, respectively. Displacements were made vertically upward.

In Fig. 14–7, Cl^{36} BTC are indistinguishable for matched concentrations of 0.01N and 0.1N at 7.7 cm/hour. Self-diffusion coefficients for Cl^{36} in free solution are 1.98×10^{-5} and 1.94×10^{-5} cm^2/second, respectively. At these concentrations the self-diffusion coefficients are not sufficiently dependent on concentration to manifest themselves in these experiments.

For stable flow vertically upward, inequality [14–1] must apply. With the values of density and viscosity of the two solutions used in these experiments, it can be deduced from inequality [14–1] that, whenever a dilute NaCl solution is displaced by an equal or more concentrated NaCl solution, flow will be stable. This

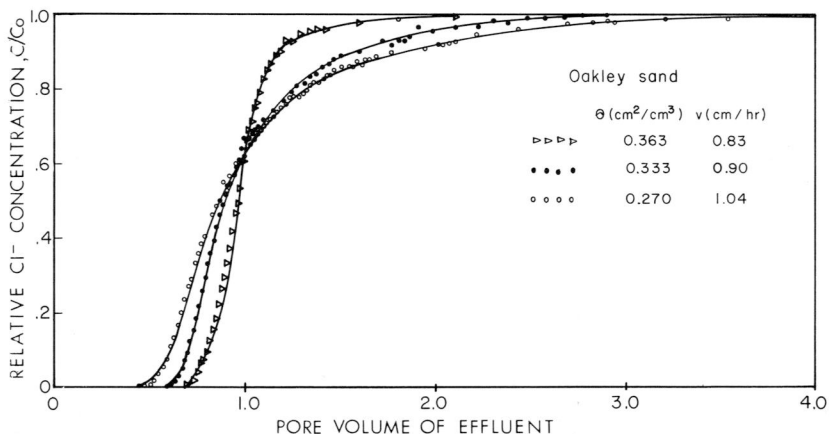

Fig. 14–9. Chloride breakthrough curves in Oakley sand at 3 water contents for approximately equal flow velocities. Data are taken from Nielsen and Biggar (1961).

is evident from Fig. 14–8. Conversely when a dilute solution displaces a more concentrated one, unstable flow exists as shown in the same figure. For the latter case, the curves are no longer smooth but are irregular as evidenced by the nature of the data points. The instability is caused when fingers of displacing fluid run ahead of the average displacement front.

It is recognized that more complex soil materials introduce other complications associated with the presence of force fields responsible for adsorption and exchange. These force fields modify both the density and viscosity of solutions when compared to measured values in the absence of external fields. For the case considered here, where the surface activity of the beads is small, minute differences in the viscosity and density of the solutions produce measurable changes in the mixing. Neither the velocity distribution as discussed previously nor the apparent diffusion coefficients will yield a correct description of this special case of miscible displacement. The work of Burns and Dean (1964) is an example of unstable flow although they refer to it as "drop out."

II.   UNSATURATED FLOW

At this point it is useful to examine miscible displacement in unsaturated porous materials. Not only is the unsaturated case more prevalent in field soils but studies of mixing at water contents less than saturation allow a more complete description of flow in saturated soils.

Breakthrough curves obtained for equal flow velocities from Oakley sand at three water contents are presented in Fig. 14–9. In this case 0.1N $CaCl_2$ replaces 0.01N $CaSO_4$. The drier the soil, the greater is the volume of effluent required to reach a maximum of C/C_o equal to 1. Desaturating the soil has progressively shifted the BTC's to the left at the initial breakthrough which may be explained by an incomplete displacement of one solution by another. Desaturation eliminates the larger flow paths and increases the proportion of water which does not readily move within the soil. The proportion of film water is significantly increased and the ion distribution within the various sized pores is modified according to changes in film thickness and pore saturation. To the extent that solid-liquid interaction determines the structure of the water in the region of mineral surfaces, and the degree of desaturation, the flow and related ion distribution will be reflected in the shapes of the BTC even though the average flux is the same for all three water contents.

For this particular soil which has an exchange capacity of only 4 meq/100 g, the existence of diffuse double layers and the influence of these on pore ion distribution is probably small. It must be assumed that the increased skewness and tailing effect is caused by the presence of these films and packets of water through which flow is extremely slow. That such a condition can exist is illustrated by BTC obtained from a column of glass beads. The interaction with the bead due to charge of the ion and the existence of varying water structure is expected to be minimal. Glass beads having diameters of 200μ and 490μ were mixed in proportions of 1:6 on a weight basis, respectively. The curves shown in Fig. 14–10 were obtained on the same column but at different water contents at flow velocities of 4.94 and 9.06 cm/hour. For the drier sample, the BTC is more skewed. Comparison of the BTC of the beads and Oakley sand indicates identical general shapes

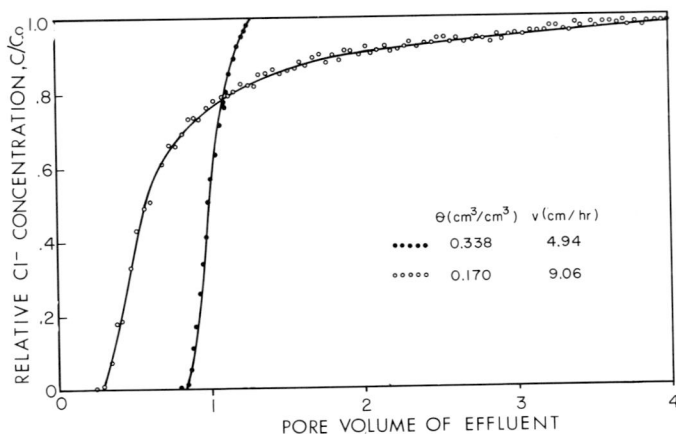

Fig. 14–10. Chloride breakthrough curves in saturated and unsaturated glass beads. Data are taken from Nielsen and Biggar (1961).

when displacement occurs in the unsaturated columns. Since the ion particle interaction is probably even less for the beads than the Oakley soil, the curve shape and tailing provide additional evidence of the incomplete displacement of one fluid by another in the intermediate and smaller pores.

III.   EXCHANGE PROCESSES

In the previous discussion attention has been confined to the mixing of miscible solutions in which the identifiable ion exhibits either a minimum interaction or repulsive effect in the presence of the medium. Of greater importance in many agricultural problems is the condition where one ion replaces another in an exchange process. In these problems the chemical processes of exchange are usually assumed to be of major consequence and the flow processes to be of minor importance.

The theory of ion exchange in porous materials (resins) has been intensively studied by scientists interested in industrial application. A critical review of the science of ion exchange column performance by Helfferich (1962) illustrates the state of development and the confusion which faces the soil scientist if he is to adopt a theory suitable for ion exchange column performance in soil.

The problems that must be considered by soil scientists have in general been considered by the resin chemist. These problems include the inherent differences in the ions, processes of diffusion, rates of reaction, hydrodynamic effects, and instability of flow to name a few. A modest attempt has been made to adapt some of the theories to ion exchange during flow of solutions in soils. Some of these will be considered here with the realization that the investigations are inadequate.

Greacen investigated the rate equations of Boyd et al. (1947) in relation to symmetrical and nonsymmetrical exchange in Yolo and Aiken soil aggregates (E. L. Greacen, 1949. Cation exchange rates and permeability in columns of soil. *Unpubl. Ph.D. thesis. University of California, Berkeley.*). Rate-determining processes based on diffusion of the ions, leaching rate, and exchange rate were considered. Agreement between exprimental data and theory was satisfactory for some of the aggregate sizes and flow velocities particularly in the mid-region of the breakthrough curves. Rible and Davis (1955) adapted the chromatographic theory of De Vault to ion exchange processes in soil columns. Their basic equation is

$$X = V / [\alpha + M f' (C_1)]$$

where V is volume of solution containing $C_1{}^\circ$ equivalents per liter of ion B_1, X is the depth, C_1 is the concentration of ion B_1 in the adjacent pore space replacing ion B_2 on the soil particles, α is the pore volume per unit length of column, M is the amount of exchange per unit length, and $f'(C)$ is the derivative of $f(C_1)$, an experimentally observed function of the C_1 concentration. Rible and Davis (1955) employed a simplified form of the statistical exchange equation of Davis (1950) to express the exchange isotherm. The equation has the form

$$k_{1\text{-}2} = \frac{(B_2)^{r_1} [B_1]^{r_2} \{ [B_1] + [B_2] \}^{r_1 - r_2}}{(B_1)^{r_2} [B_2]^{r_1}}$$

where quantities in parenthesis are ion activities in the solution phase and those in brackets are moles of the ionic species in the exchanger phase, r_1 and r_2 are the valences of B_1 and B_2, respectively, and $k_{1\text{-}2}$ is the exchange constant for replacement of B_2 by B_1.

The agreement between experiment and theory for Mg $\rightarrow$ Ca exchange and Na $\rightarrow$ Ca exchange in Yolo sandy loam soil was good. It was noted by Rible and Davis (1955) that both Na and Mg penetrated deeper into the soil column than predicted by theory. Also Ca was able to effect almost complete replacement of Na and Mg in the upper layers. With the combined effect of Na penetrating more deeply into subsurface layers and the fact that Ca almost completely replaces Na in the surface layers, the problem of developing Na-affected soils to greater depths combined with the difficulty of getting Ca deep into the profile to reclaim such soils remains as a serious problem in reclamation practice. Speculation at this time might suggest that the deep penetration of Na may result partly from the nonuniformity of displacement by velocity dispersion. Likewise by controlling the water and amendment used in the reclamation process according to principles revealed in miscible displacement experiments, it should be possible to bring about deeper penetration of Ca into the subsoil.

Bower et al. (1957) investigated equations of column ion exchange developed by Thomas (1944), Hiester and Vermeulen (1952) for predicting Na, Ca, and Mg exchange in soil columns. The mass conservation equation was combined with a rate equation for exchange and solved for particular experimental conditions to predict the exchange in Chilcott and Traver soils. For uni-divalent exchange, certain approximations were made in order to arrive at a satisfactory solution.

The equations contain a series of parameters which cannot easily be evaluated independently. Consequently a value for one parameter is obtained for one point on the experimental elution curve from which theoretical curves can be constructed by means of graphs and tables, Opler and Hiester (1954). The agreement between the predicted and measured elution curves was exceptionally good for both soils. Additional testing of the equations in soils will establish the generality of the theory for a number of other experimental conditions.

In an attempt to describe the leaching of fertilizer salts in soils, Thomas and Coleman (1959) modified a chromatographic equation proposed by Walter (1945).

For unsymmetrical exchange of a strongly held cation by a more weakly held cation ($k' < 1$ as in K – Ca or K – Al exchange) the equation has the form

$$q = \frac{L}{L - k'} \left[\frac{(k' \ P \ C_o \ V)^{1/2}}{L} \right] - k'P$$

where q is the concentration of applied cation adsorbed in meq/cm, C_o the initial concentration of the applied cation in solution in meq/ml, V the volume of solution containing the applied cation at concentration C_o, P the cation-exchange capacity per unit length of column in meq/cm, k' the modified Gapon equilibrium constant, and L the distance down the column. For exchange of a weakly held cation by a more strongly held cation both of the same valence, the above equation reduces to a simpler form.

The equations were unable to predict with precision the distribution of K in the Ca-, Al-, or Na-saturated soils. There were indications that the predicted K distribution was better in the unsymmetrical exchange experiments than those of K replacing Na. This is so even though it might be expected that the equation for symmetrical exchange should be superior. Apart from the difficulties that are encountered in the exchange process during flow, it would appear that the modified Walter equation is totally inadequate to describe ion exchange in soil columns.

The formulation of the column exchange theories of De Vault and Walter as treated by Rible and Davis (1955) and Thomas and Coleman (1959) assumes the ion distribution between exchange and solution phases during flow to be governed by the chemical reactions of exchange. The equations treated by Greacen (1949 Ph.D. thesis) and Bower et al. (1957) consider the rate dependent processes of ion transfer from mobile solution to exchange sites. In all the above cases the mixing which takes place as a result of flow velocity distribution within the column was not considered. Furthermore, the important case of exchange in partially water-saturated soils was not examined. Recent work has yielded additional information relative to the distribution of ions between the exchange and solution phases in saturated and unsaturated soil columns at two flow velocities, Biggar and Nielsen (1963). Magnesium chloride solutions of 0.1N concentration displaced 0.1N calcium acetate solutions from Ca-saturated Oakley sand. In Fig. 14–11 the BTC were obtained at a velocity of 5.83 cm/hour.

The solid line is calculated from the equation of Rible (J. M. Rible, 1952. Ion exchange in soil columns. *Unpubl. Ph.D. thesis. University of California, Berkeley.*) as

$$C/C_o = \{ C_o - [(\phi \ K \ C_o x)/(V - \epsilon x)]^{1/2} \} /C_o (1 - K)$$

where ϕ is the exchange capacity per unit length, V the volume of effluent, ϵ the pore volume per unit length, x the length of column, and

$$K = \frac{(Ca) \ [Mg]}{(Mg) \ [Ca]} = 0.51,$$

the equilibrium exchange constant.

The Cl BTC is a typical curve for a nonreactive material and is different than the Mg curve which reflects the process of exchange. Magnesium appears in the effluent well in advance of what might be expected since only 65 meq have been

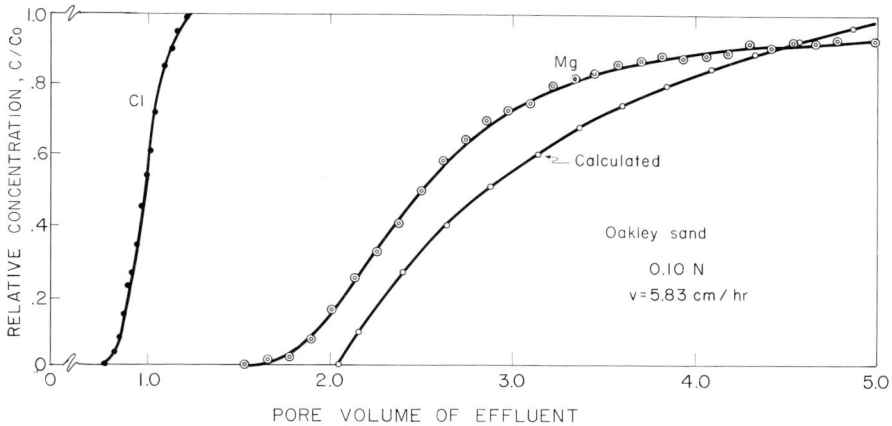

Fig. 14–11. Chloride and magnesium breakthrough curves in water-saturated Oakley sand. The volumetric water capacity and cationic exchange capacity of the sample are 416 ml and 85.7 meq, respectively. Data are taken from Biggar and Nielsen (1963).

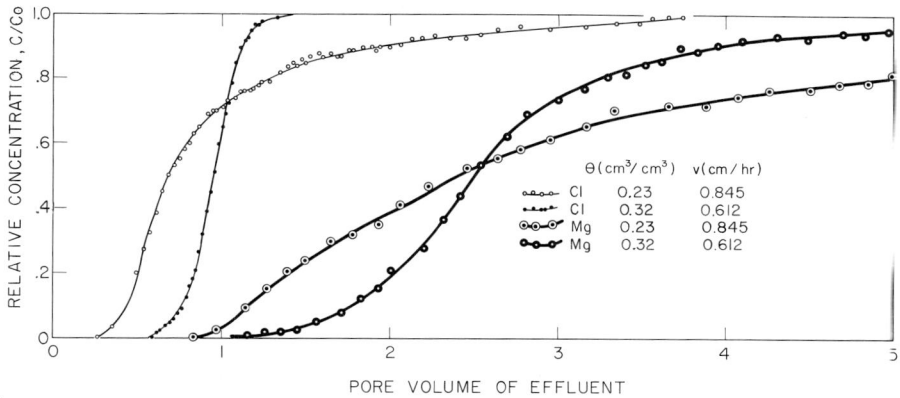

Fig. 14–12. Chloride and magnesium breakthrough curves in Oakley sand for 2 water contents. Data are taken from Biggar and Nielsen (1963).

applied to a column having a total capacity of 127.3 meq. If the calculated curve in Fig. 14–11 is sufficiently accurate to account for the differences in exchange properties of Mg and Ca, then the separation of the experimental and calculated curve may be attributed to mixing caused by the flow velocity distribution within the sample.

Comparison of the Mg BTC of Fig. 14–11 and 14–12 for saturated columns and velocities of 5.83 and 0.612 cm/hour shows the earlier breakthrough at the lower velocity. Had nonequilibrium conditions prevailed at the greater velocity, decreasing the velocity should have delayed the appearance of Mg. The early arrival of Mg in the effluent has been attributed to diffusion and the differences in dispersion-exchange interactions that occur at the different velocities.

In Fig. 14–12 comparison is made between BTC for Mg^{2+} and Cl^- ions at approximately the same velocity but two water contents, 0.32 cm^3/cm^3 (saturated) and 0.23 cm³/cm³. Displacement under unsaturated conditions drastically modifies both the Cl and Mg BTC. Not only do both ions appear in the effluent after less volume has been applied but the shapes of the curves change significantly. The change in shape of the cation curve cannot be attributed to variable exchange properties of the ions or soil but must occur as a result of changing flow paths which involves the mixing by velocity and diffusion in relation to exchange.

At the present time theoretical models are unable to describe the change in shapes and positions of either the Cl or cation BTC in unsaturated flow systems. However, the unsaturated curves suggest that mechanisms and management practices may be developed to improve reclamation procedures involving replacement of Na by Ca.

Soil systems often involve several cation and anion salts, precipitation reactions, and variable soil materials and waters of various qualities. Dutt and Tanji (1962) developed a computer program for calculating the cation status of the soil solution in which a water of some chemical composition is brought to equilibrium by calculation with a section or sample of soil. The solution is then displaced hypothetically to a new section and once again equilibrated. By successive displacements a BTC is obtained as well as the corresponding exchangeable cation analysis if desired. Gypsum salts in the column have been accounted for by this procedure.

Hashimoto et al. (1964) have developed equations solvable on a computer which have been used to describe mixing of both anions and cations in clinoptilolite-sand and Hanford soil columns where cation exchange is possible. At the flow velocities of 36 to 100 cm/hour considered in these experiments, calculations were capable of describing the mixing over a wide range of velocities. Coefficients of dispersion for cations exhibited the same dependence on velocity as the anions, despite the exchange that occurred.

IV.   THEORETICAL DEVELOPMENT

Various models have been proposed to account for the mixing which takes place during miscible displacement. Taylor (1953) considered the displacement of two miscible fluids of equal density and viscosity in a capillary tube of radius a at a constant average velocity v_o. Using the parabolic formula for laminar flow and realizing that the fluid at the wall does not move whereas the velocity at the center is a maximum, the velocity v at a point distance r from the axis of the tube is

$$v = 2v_o(1 - r^2/a^2).$$

Mixing is by convection only and for a concentration of C at $t = 0$, $x = 0$ where x is the distance along the axis, then at time t the concentration is

$$C = C(x - vt, r).$$

Combining dispersion owing to the velocity distribution with mixing by molecular diffusion where D the diffusion coefficient is constant, Taylor (1953) obtained a solution to the equation

$$D\left[\frac{1}{r} \frac{\partial C}{\partial r} + \frac{\partial^2 C}{\partial r^2} \right] = \frac{\partial C}{\partial t} + 2v_o \left(1 - \frac{r^2}{a^2} \right) \frac{\partial C}{\partial x}$$

assuming longitudinal diffusion negligible compared to radial diffusion. Depending on the velocity and radius of the capillary it is possible to approach piston displacement since the effect of radial diffusion counteracts the mixing by velocity. Studies of dispersion in a capillary tube serve to illustrate some of the mechanisms involved but cannot describe additional variations caused by the geometry and surfaces of porous material. Comparisons have been made by Blackwell et al. (1959).

Beran (M. J. Beran, 1955. Dispersion of soluble matter in slowly moving fluids. *Unpubl. Ph.D. thesis. Harvard University, Cambridge, Mass.*) and Scheidegger (1954) considered dispersion in a homogeneous isotropic porous medium as a random walk process which assumes a fluid particle undergoes a succession of statistically independent straight steps in equal small intervals of time. Statistically a Gaussian probability distribution function with a variance proportional to time is applicable, and the dispersion is described by a diffusion type equation. The solute concentration distribution in the effluent (BTC) resulting from continual displacement of the original fluid ($C = 0$) by a solution $C = C_o$ is then described by the equation

$$C/C_o = \tfrac{1}{2} \text{ erfc } [(x - vt)/(4Dt)^{1/2}]$$

where erfc is the complimentary error function and D the coefficient of dispersion. Molecular diffusion is not included in this theory. Since the time interval is broken up into equal intervals, the theory precludes the physical possibility of unequal rest and motion times of a particle even though the velocity is different at different points. Therefore the dispersion is isotropic. A consequence of the equation is that when $x = vt$ or one pore volume has been displaced the ratio C/C_o must be equal to 0.5. Such a requirement is contrary to a large body of both laboratory and field data.

Saffman (1959, 1960) in developing a dispersion theory investigated a model in which the porous material is considered as an assembly of randomly oriented and distributed interconnected straight pores. The path of a fluid particle is regarded as a random walk in which the length, direction, and duration of each step are random variables. As in the treatment of Scheidegger (1954) the contribution of diffusion is not included.

Qualitative agreement was obtained between the model and experimental results of Von Rosenberg (1956) and De Josselin de Jong (1958). The contribution of diffusion was recognized to be highly important at lower velocities and the effects on dispersion of non-Darcy flow at higher velocities considered. It is significant that non-Darcy flow was ignored at lower velocities. All theories of dispersion consider the macroscopic flow process to be described by Darcy's law in the range of velocities applicable to most porous media.

Saffman (1959, 1960) also considered the problem of lateral dispersion as related to longitudinal dispersion, the former occurring normal to the average direction of flow. This problem has been investigated by Harleman and Rumer (1963) in packed plastic sphere columns at velocities somewhat greater than normally found in soils. At Reynolds numbers of 1.0 and 0.1 the ratios of longitudinal to lateral dispersion were 18.3 and 5.8, the ratios depending on the velocity. At Reynolds numbers $< 10^{-2}$ the contribution of diffusion appears to be of major significance.

De Josselin de Jong (1958), recognizing the difference in transverse and longitudinal dispersion, proposed a statistical model based on a schematic pore canal sequence that accounts for liquid movement in detail. The probability distribution of the particles is continuous with respect to the direction of the path rather than a discontinuous distribution suggested by previous workers. The theory which combines both transverse and longitudinal dispersion into one equation assumes the flow is described by Darcy's law and molecular diffusion is sufficient to cause piston flow in all pores. De Jong's analysis is a particular case of the more general formulation by Saffman.

Rifai et al. (1956) used a stochastic model which assumes the fluid is a continuous medium of which each point has a particular flow path. The fluid, divided into a rest phase and motion phase, moves through the pores as piston displacement. The rest phase, assumed to be long compared to the motion phase, has a probability of occurrence independent of time and position. From the standpoint of displacement of a solute in a fluid flowing in porous media the model seems untenable (Nielsen and Biggar, 1962).

Bear (1961) considered mixing in porous media by velocity only but assumed a model that describes displacement as a second-rank tensor. The concentration distribution resulting from this displacement is normal and the coefficient of dispersion was described as a fourth-rank tensor. Such an analysis may be satisfactory for mixing at some intermediate velocities but since it ignores diffusion as a contributing process to mixing, it has limited use in soils problems.

Nielsen and Biggar (1962) examined a differential equation proposed by Lapidus and Amundson (1952) for the description of mixing by both velocity and diffusion. For an average flow velocity v and constant molecular diffusivity D in a unidirectional model the equation is written as

$$\frac{\partial C}{\partial t} = D \frac{\partial^2 C}{\partial x^2} - v \frac{\partial C}{\partial x} \ .$$ [14–2]

The solution of equation [14–2] subject to the conditions

$$
\begin{array}{llll}
C = C_o & x = 0 & t > 0 \\
C = 0 & x > 0 & t = 0 \\
C = 0 & x \to \infty & t > 0
\end{array}
$$

may be found to be

$$\frac{C}{C_o} = \frac{1}{2} \left[\text{erfc} \left(\frac{x - vt}{\sqrt{4Dt}} \right) + \exp \frac{vx}{D} \text{erfc} \left(\frac{x + vt}{\sqrt{4Dt}} \right) \right] .$$ [14–3]

Equation [14–3] has been considered by other authors as well (e.g., Aronofsky and Heller, 1957, and Ogata and Banks, 1961). It was possible to fit the equation to experimental BTC for soil and glass beads (Nielsen and Biggar, 1962) providing D was allowed to vary. The calculated diffusivities indicate the interaction or coupling which occurs between the mixing by diffusion and velocity variations. They also illustrate the inadequacy of assuming a constant average v and the difficulties of separating these two processes. Additional considerations of this point were discussed under unstable flow in this and previous articles (Biggar and Nielsen, 1964).

In summary it may be concluded that most progress in the analytical description of mixing in porous media has been made in the range of velocities greater than those found in most field soils and for porous materials as nonreactive as sands. However, even for these media the relation of the dispersion coefficient to grain size (pore distribution) is not well defined. At the present time the question may be asked what one wishes to obtain from the equations—prediction or under-standing. If it is the latter, then the equations leave much to be desired. If it is the former, then for some range of velocities, simple porous materials, and a reasonable magnitude of error some satisfaction may be obtained.

V. LEACHING FIELD SOILS

Miscible displacement concepts have thus far been applied most successfully in agriculture to the leaching of excess salts from saline soils. This is so because very few other applications such as movement of fertilizer nutrients, pesticides, and organic matter have been investigated.

Van der Molen (1956) applied Gluechauf's theory of exchange to the desalini-zation of Dutch soils and found some general agreement between the theory and observed values. Gardner and Brooks (1957) distinguished between immobile and mobile salt moving with the same velocity as the leaching front. The theory of Hiester and Vermeulen (1952) was adopted and tested in laboratory columns and a field plot of Pachappa sandy loam. The theory when fitted to the experi-mental data yields parameters which may be related to the dispersion coefficient given by Scheidegger (1954). The observation was made on several soils that $1 + 2 B$ pore value replacements entering the soil reduce the salinity of the soil to 20% of its original value. B was defined as the equilibrium ratio of concentra-tion of immobile salt to concentration of mobile salt. For four soils (texture range —sandy loam to clay) and sand at various initial water contents, various column lengths, bulk densities, and initial salt levels, B had an average value of 0.208. This suggests that 1.42 pore values of water must enter the soil to reduce the salinity by 80%. It was concluded the flow process rather than diffusion was mainly responsible for mixing.

Brooks et al. (1958) tested the applicability of the Thomas-Hiester-Vermeulen equations for predicting changes in the composition of dissolved and exchangeable cations in soils under field conditions. Agreement between theory and experiment was quite good for a high sodium, high salt (27.5 meq/Na, 2.5 meq/liter Ca + Mg) water applied to a disturbed soil plot initially low in exchangeable sodium.

Biggar and Nielsen (1962b) suggested that leaching soils at a water content below saturation could produce more efficient leaching and thereby reduce the amount of water required as well as reduce drainage problems in areas of high water tables. They also suggested that leaching by rainfall is often more efficient because of the leaching under unsaturated conditions. Such considerations arise from the manner in which pore water velocity and diffusion contribute to the mixing process in a porous material. Wilson et al. (1961) found that rainfall leached salt from a diatomaceous earth soil more efficiently than larger quantities of water applied by ponding. The subsoil structure containing large vertical cracks conducted the ponded water more rapidly than the adjacent soil blocks. During winter rainfall much more of the water was transmitted through the soil blocks rather than the cracks, producing more efficient salt removal.

In a second field experiment established on Levis silty clay soil classified as moderately alkali and high in salts, Nielsen et al. (1965) compared leaching by ponding and intermittent sprinkling. The latter treatment applied the water in such a manner that the intake rate exceeded the application rate, thus avoiding ponding on the surface. The average infiltration rate under continuous ponding was 0.02 cm/hour—a slowly permeable soil. The plots were covered to minimize evaporation. Under these conditions it was found that 10.3 inches of water applied by sprinkling reduced the salt content of the upper 2 ft to the same degree as 29.5 inches applied by continuous ponding.

In a more extensive field experiment Miller et al. (1965) made a detailed study of how solutes move through a Panoche clay loam profile free of vegetation. A comparison was made of the movement of a surface-applied chloride salt under two methods of water application, (i) continuous ponding and (ii) intermittent 2-inch applications. Under the latter treatment leaching occurred at lower soil water contents since 2 inches was applied whenever the soil water content reached 0.30 cm^3/cm^3 at the 1-ft depth (about 1 week of drainage). Figures 14–13 and 14–14 illustrate the Cl distribution measured in the soil profiles under the two methods of irrigation for various amounts of water that have entered the soil surface. We are interested in the redistribution of salt in the profile as well as complete removal from the profile.

The redistribution pattern is entirely different under the two leaching methods. Under continuous ponding the salt is spread more uniformly through the profile with the flattening of peaks and increase in base width of the salt profile. In contrast, the 2-inch intermittently ponded method moves the salt out more as a slug with sharper peaks and narrower base width. Such a behavior is predicted by the BTC obtained under unsaturated conditions (*see* Fig. 14–9 on Oakley sand at three water contents). This behavior was also evident in sandstone columns and slug flow studied by Corey et al. (1963). On the average the soil water content under intermittent ponding was 0.06 cm^3/cm^3 less than the continuous ponding

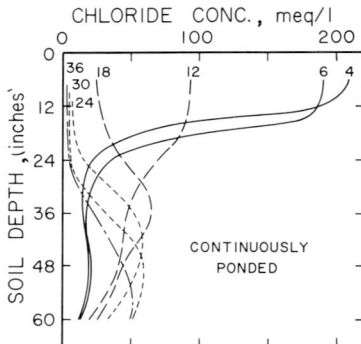

Fig. 14–13. Chloride concentration of the soil solution vs. profile depth of Panoche clay loam continuously ponded with water. The number near each curve signifies the total inches of water that entered the soil surface at the time of measurement. Data are taken from Miller, Biggar, and Nielsen (1965).

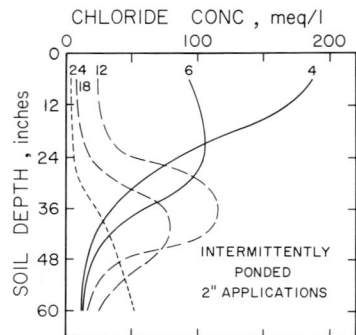

Fig. 14–14. Chloride concentration of the soil solution vs. profile depth of Panoche clay loam intermittently ponded with 2 inches of water. The number by each curve signifies the total inches of water that entered the soil surface at the time of measurement. Data are taken from Miller, Biggar, and Nielsen (1965).

treatment which approached saturation. In addition the average flow rate in the intermittent method (0.2 to 0.01 inches/hour) is less than under continuous ponding (2.16 to 0.2 inches/hour). The slower rate which allows more time for diffusion combined with the unsaturated flow conditions accounts for the difference in leaching characteristics under the two methods of application.

Interpretation of these results into practical leaching conditions under field conditions based on the salt distribution after 24 inches have entered the soil under 2-inch intermittent ponding and 36 inches under continuous ponding reveals that one-third less water was required by the former method. Therefore, if a high water table is a problem or irrigation water limiting, the intermittent method could be most useful in soils similar to Panoche clay loam. That rainfall may and does provide a more efficient leaching than continuous ponding is readily explained by these results.

On the other hand, the distribution of some chemical through the profile may be the desired result rather than the movement of a narrow band. For these requirements it may be preferable to pond the water.

VI. FUTURE INVESTIGATIONS

Both experimental and mathematical developments in miscible displacement theory have proved fruitful in the solution of applied problems. These developments have suggested explanations of certain heretofore unexplainable observations as well as suggesting improved management practices. Application extends into the areas of petroleum, chemical, civil, and sanitary engineering as well as irrigation science. The implications of fluid and solute mixing in porous media need not be restricted to water and inorganic ions. For instance, the movement of emulsified DDT at 50 ppm concentration in an alcohol-water mixture through Oakley sand (Fig. 14–15) is an example of other problems of current concern. The transport of bodies in the colloidal particle size range and below (microorganisms and viruses) through soils has not been extensively investigated. Gaseous exchange in the soil undoubtedly involves dispersion mechanisms.

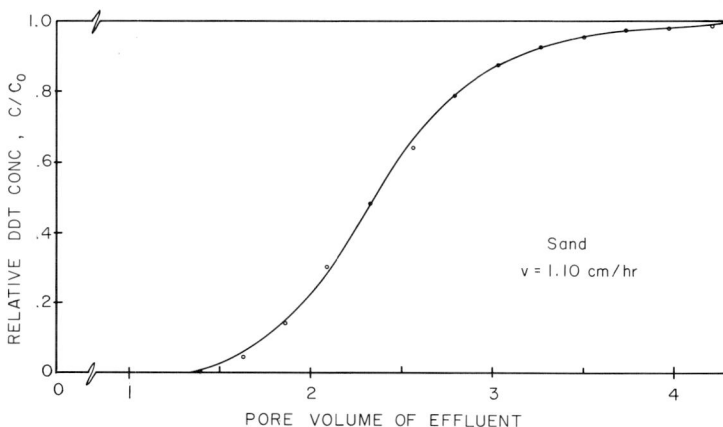

Fig. 14–15. A DDT breakthrough curve in a saturated sand.

The spreading of a tracer in a porous material during miscible displacement was observed by Slichter (1905) during groundwater flow studies in aquifers. Mixing by velocity and diffusion of the tracer was correctly hypothesized by Slichter and precluded interest both in miscible displacement and tracer work decades later. Presently, tracers are used for particularly difficult flow problems when other methods for ascertaining the flow characteristics fail. Generally, their use yields only qualitative information owing to the complexity of the problem. Selection of the tracer material is usually based upon semi-empirical criteria. Kaufman and Orlob (1956) investigated several materials for use in tracing water. Although chloride appeared to be superior to all others, tritium has been widely used for this purpose. Biggar and Nielsen (1962a) concluded that tritiated water was not a good tracer of the net movement of water through many porous media because of the exchange that tritium undergoes with the adsorbed water. Regardless of the tracer material selected, the reliability of tracer methods will depend on a knowledge of the mixing behavior of the tracer during the displacement process.

Additional study should be given to the anion distribution, extraction, and movement in soils as they relate to miscible displacement processes. In this regard, Bolt (1961) has assumed certain anion distributions in selected porous media. Lagerwerff has touched on the problem (1964). Kemper (1960), in treating salt sieving, assumes certain average distributions within and between pores of various sizes. Taylor (1962) considers the simultaneous transport of water, heat, and salts using models that depend on average distributions. However, in this latter development, the macroscopic model is all that is required. It will be interesting and useful to examine these problems within the framework of miscible displacement.

LITERATURE CITED

Aronofsky, J. S., and J. P. Heller. 1957. A diffusion model to explain mixing of flowing miscible fluids in porous media. Amer. Inst. Mining, Met., Petrol. Eng., Trans. 210: 345–349.

Bear, J. 1961. On the tensor form of dispersion in porous media. J. Geophys. Res. 66: 1185–1197.

Biggar, J. W., and D. R. Nielsen. 1962a. Miscible displacement: 2. Behavior of tracers. Soil Sci. Soc. Amer. Proc. 26:125–128.

Biggar, J. W., and D. R. Nielsen. 1962b. Improved leaching practices—save water, reduce drainage problems. California Agr. 16:5.

Biggar, J. W., and D. R. Nielsen. 1963. Miscible displacement: 5. Exchange processes. Soil Sci. Soc. Amer. Proc. 27:623–627.

Biggar, J. W., and D. R. Nielsen. 1964. Chloride–36 diffusion during stable and unstable flow through glass beads. Soil Sci. Soc. Amer. Proc. 28:591–595.

Blackwell, R. J., J. R. Rayne, and W. M. Terry. 1959. Factors influencing the efficiency of miscible displacement. Amer. Inst. Mining, Met., Petrol. Eng., Trans. 217:1–8.

Bolt, G. H. 1961. The pressure filtrate of colloidal suspensions: I. Kolloid. Z. 175:33–39.

Bower, C. A., W. R. Gardner, and J. O. Goertzen. 1957. Dynamics of cation exchange in soil columns. Soil Sci. Soc. Amer. Proc. 21:20–24.

Boyd, G. E., A. W. Adamson, L. S. Myers. 1947. The exchange adsorption of ions from aqueous solutions by organic zeolites: II. J. Amer. Chem. Soc. 69:2836–2848.

Brooks, R. H., J. O. Goertzen, and C. A. Bower. 1958. Prediction of changes in the cationic composition of the soil solution upon irrigation with high-sodium waters. Soil Sci. Soc. Amer. Proc. 22:122–124.

Burns, G. R., and L. A. Dean. 1964. The movement of water and nitrate around bands of sodium nitrate in soils and glass beads. Soil Sci. Soc. Amer. Proc. 28:470–474.

Corey, J. C., D. R. Nielsen, and J. W. Biggar. 1963. Miscible displacement in saturated and unsaturated sandstone. Soil Sci. Soc. Amer. Proc. 27:258–262.

Danckwerts, P. V. 1953. Continuous flow systems: distribution of residence times. Chem. Eng. Sci. 2:1–13.

Davis, L. E. 1950. Ionic exchange and statistical thermodynamics: I, II. J. Colloid Sci. 5:71–79.

De Josselin de Jong, G. 1958. Longitudinal and transverse diffusion in granular deposits. Amer. Geophys. Union, Trans. 39:67–74.

Dutt, G. R., and K. K. Tanji. 1962. Predicting concentrations of solutes in water percolating through a column of soil. J. Geophys. Res. 67:3437–3439.

Gardner, W. R., and R. H. Brooks. 1957. A descriptive theory of leaching. Soil Sci. 83: 295–304.

Harleman, D. R. F., and R. R. Rumer. 1963. Longitudinal and lateral dispersion in an isotropic porous medium. J. Fluid Mech. 16:385–394.

Hashimoto, I., K. B. Deshpande, and H. C. Thomas. 1964. Peclet numbers and retardation factors for ion exchange columns. Ind. Eng. Chem. Fundamentals 3:213–218.

Helfferich, F. 1962. Theories of ion-exchange column performance: A critical study. Angew. Chem. Int. Ed. 1:440–453.

Hiester, N. K., and T. Vermeulen. 1952. Saturation performance of ion-exchange and adsorption columns. Chem. Eng. Progr. 48:505–516.

Kaufman, W. J., and G. T. Orlob. 1956. An evaluation of groundwater tracers. Amer. Geophys. Union, Trans. 37:297–306.

Kemper, W. D. 1960. Water and ion movement in thin films as influenced by the electrostatic charge and diffuse layer of cations associated with clay mineral surfaces. Soil Sci. Soc. Amer. Proc. 24:10–16.

Lagerwerff, J. V. 1964. Extraction of clay-water systems. Soil Sci. Soc. Amer. Proc. 28: 502–506.

Lapidus, L., and N. R. Amundson. 1952. Mathematics of adsorption in beds. J. Phys. Chem. 56:984–988.

Miller, R. J., J. W. Biggar, and D. R. Nielsen. 1965. Chloride displacement in Panoche clay loam in relation to water movement and distribution. Water Resources Res. 1: 63–73.

Nielsen, D. R., and J. W. Biggar. 1961. Miscible displacement in soils: 1. Experimental information. Soil Sci. Soc. Amer. Proc. 25:1–5.

Nielsen, D. R., and J. W. Biggar. 1962. Miscible displacement: 3. Theoretical considerations. Soil Sci. Soc. Amer. Proc. 26:216–221.

Nielsen, D. R., J. W. Biggar, and J. N. Luthin. 1965. Desalinization of soils under controlled unsaturated flow conditions. Int. Comm. Irrig. Drainage, 6th Congr., New Delhi, India. Question 19, p. 15–24.

Ogata, A., and R. B. Banks. 1961. A solution of the differential equation of longitudinal dispersion in porous media. Geol. Surv. Pap. 411–A. US Government Printing Office, Washington, D. C. 7 p.

Opler, A., and N. K. Hiester. 1954. Tables for predicting the performance of fixed bed ion exchange and similar mass transfer processes. Stanford Res. Inst., Stanford, California. 111 p.

Rible, J. M., and L. E. Davis. 1955. Ion exchange in soil columns. Soil Sci. 79:41–47.

Rifai, M. N. E., W. J. Kaufman, and D. K. Todd. 1956. Dispersion phenomena in laminar flow through porous media. Sanit. Eng. Res. Lab., Div. Civ. Eng. Rep. 3, Univ. of California, Berkeley. 157 p.

Saffman, P. G. 1959. A theory of dispersion in a porous medium. J. Fluid Mech. 6: 321–349.

Saffman, P. G. 1960. Dispersion due to molecular diffusion and microscopic mixing in flow through a network of capillaries. J. Fluid Mech. 7:194–208.

Scheidegger, A. E. 1954. Statistical hydrodynamics in porous media. J. Appl. Phys. 25: 997–1001.

Slichter, C. S. 1905. Field measurements of the rate of movement of underground waters. US Geol. Surv. Ser. 0, US Government Printing Office, Washington, D. C. 122 p.

Taylor, G. I. 1953. Dispersion of soluble matter in solvent flowing slowly through a tube. Proc. Roy. Soc. (London) 219A:186–203.

Taylor, S. A. 1962. The influence of temperature upon the transfer of water in soil systems. Mededel., Landbouwhogeschool, Ghent 27:535–551.

Thomas, H. C. 1944. Heterogeneous ion exchange in a flowing system. J. Amer. Chem. Soc. 66:1664–1666.

Thomas, G. W., and N. T. Coleman. 1959. A chromatographic approach to the leaching of fertilizer salts in soils. Soil Sci. Soc. Amer. Proc. 23:113–119.

Van der Molen, W. H. 1956. Desalinization of saline soils as a column process. Soil Sci 81:19–27.

Van Rosenberg, D. V. 1956. On the mechanics of steady-state single phase fluid displacement from porous media. Amer. Inst. Chem. Eng. J. 2:55–58.

Walter, J. E. 1945. Multiple adsorption from solutions. J. Chem. Phys. 13:229–234.

Wilson, L. G., J. N. Luthin, and J. W. Biggar. 1961. Drainage-salinity investigation of the Tulelake lease lands. Calif. Agr. Exp. Station Bull. 779, Berkeley, California. 56 p.

15 Measurement of Soil Water

J. W. HOLMES
*Commonwealth Scientific & Industrial
Research Organization
Adelaide, S. A., Australia*

S. A. TAYLOR *(deceased, June 1967)*
*Utah State University
Logan, Utah*

S. J. RICHARDS
*University of California
Riverside, California*

I. WATER CONTENT OF SOILS

Soil water has been the subject of numerous and extensive investigations. Shaw and Arble (1959) published a bibliography of over 600 references to literature describing both direct and indirect methods for evaluating soil water relations. Haise (1955) has prepared a useful brief summary. (*See also* Table 15–1.)

A. Gravimetric

Drying samples in the oven at 105 C until constant weight is attained is the basic method. The water content so determined is usually not the total water content. More water is given off, from clay soils particularly, at higher temperatures. For purposes of definition, a soil possesses zero water content when it has been dried in a vacuum, over P_2O_5 as a desiccant, and therefore has come to equilibrium with an atmosphere of very small water vapor pressure, at normal temperature.

Soil samples may be dried quickly by impregnating with absolute alcohol which is then burned away (Bouyoucos, 1937). Dielectric heating is also used and commercial apparatus is available.

In other quick tests, the water in the sample may be used to dilute another fluid whose properties depend on the dilution with water. The conductivity of an acetone-ethyl alcohol mixture was used for this purpose (Hancock and Burdick, 1957). The dielectric constant of a dioxane extract of the soil water has also been used (Van der Marel, 1959; Spauszus, 1955). There are many methods for measuring water content and humidity of samples in an industrial context, and the interested reader should refer to a recent symposium on the subject (Wexler, 1965).

There are several designs of augers suitable for taking soil samples from depth. Those widely used include the Jarrett Auger (Australia) and the Iwan Auger (USA), both of which include space within the cutting head for the sample and provide an uncontaminated sample. The Veihmeyer (1929) tube and the Stace and Palm (1962) tube are useful for obtaining undisturbed core samples. Powered augers may be used. There are so many models now for sale commercially that descriptions of such apparatus no longer appear in the scientific literature.

B. Gamma-Ray Attenuation

Rather than taking a sample of soil, it is often desired to measure the moisture status *in situ*, without disturbance to the system. A method for laboratory soil columns uses the absorption of gamma rays in the column (Ferguson and Gardner, 1962; Gurr, 1962). Gurr's apparatus comprised a source of 25mc of cesium–137, whose emitted gamma rays were collimated and passed through the sample contained in a square section box. The detector was a scintillation counter with a collimated entry window. The water content was given by

$$\theta = a - b \ln q \qquad\qquad [15\text{--}1]$$

where a and b are parameters which have to be determined experimentally, and q is the observed counting rate (counts/min) at the water content θ. With good collimation, the diameter of the sample so measured is about 1.5 cm for a column 16 cm long. The standard error of measurement appears to be about 2%.

Table 15–1. Measurement of soil water content

Recommended for measuring	Operating conditions		Principal advantages or disadvantages
	Favoring	Limiting	
Gravimetric method			
Water content in samples	1) Infrequent sampling 2) Uniform, medium textured soils	1) Rocky or gravelly soil 2) Frequent sampling	1) This is the basic method for measuring water content 2) It is time consuming and requires perforating the soil to secure samples
Gamma-ray attenuation method			
Water content in the laboratory	1) Columns of soil in which nondestructive water determination is required	1) Columns of soil are required 2) Measurements must be done in the laboratory 3) Variations in bulk density strongly influence results	1) A nondestructive method 2) Limited to soil columns 3) Requires precise radiation equipment 4) Radiation hazard
Neutron scatter method			
Water content in the field	1) Frequent sampling at same location 2) Nondestructive measurement needed 3) Uniform, medium textured soils 4) Water content of the soil profile is desired	1) Shallow soils 2) Rocky or gravelly soils 3) Organic soils 4) Laminated soil 5) Water content at a given depth is desired	1) Can repeatedly determine water in soil profile at same location 2) May need field calibration 3) Requires special equipment to determine water at surface 4) Cannot be used to determine water content at a given depth

C.   Neutron Scatter

The neutron moisture meter is much used in the field for measuring water content (Belcher et al., 1950; Gardner and Kirkham, 1952; Spinks et al., 1951). It is a bore-hole logging technique in which the counting rate of slow neutrons, detected in the BF_3 tube in the probe (*see* Fig. 15–1), is approximately a linear function of the water content of the soil expressed as a volume fraction, cc/cc. The calibration is not unique, however; it depends on the abundance in the soil of nuclei which absorb slow neutrons strongly and on the total hydrogen content of which the water content is a component (Holmes and Jenkinson, 1959). For use in research, it is advisable to calibrate each soil separately, rather than to rely on a "universal calibration."

The calibration curve is of the form

$$r_\theta = r_1 \times \theta \times F \qquad [15\text{–}2]$$

where r_θ is the counting rate at the water content θ, r_1 is the counting rate in an access tube in pure water, and F is an empirical factor ($F = 1$ when $\theta = 1$) which depends on the soil solids composition and weakly on θ. The use of r_1 as a measured parameter normalizes subsequent measurements at a water content θ to a standard response with the probe in water. The usual corrections should be made for background and dead time. Commercial equipment is available.

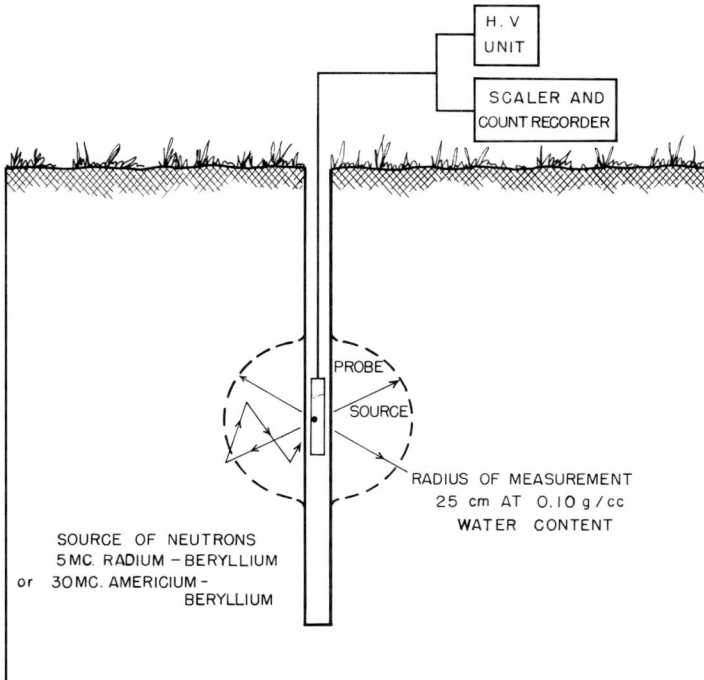

Fig. 15–1. The neutron moisture meter in use for measuring soil water content.

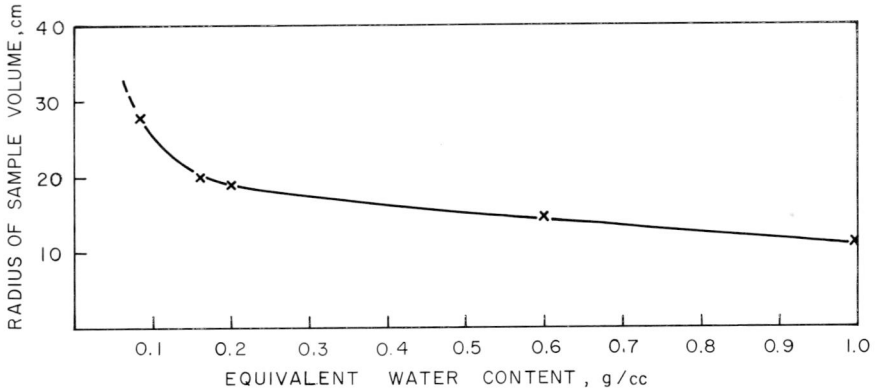

Fig. 15–2. Sample size vs. water content measured by the neutron moisture meter.

The size of the sample volume depends strongly on the water content, as shown in Fig. 15–2 (Holmes and Turner, 1958; Mortier et al., 1960). In making probe measurements in an access hole, the probe must be far enough below the surface of the soil for the neutrons to be slowed and scattered in an effectively infinite medium. If not, calibration would not be reproducible. Surface measurement with semi-infinite medium geometry is desirable. It is fraught with difficulties of inter-pretation, however. Van Bavel (1961) concluded that the method was unsuitable for determination of water content at the surface of soils in which there was a water gradient from the surface. For materials with uniform water content, the method may be useful. Pawliw and Spinks (1957) investigated the influence of various "reflectors" placed over the BF_3 tube and probe on the counting rate and slope of the calibration curve. Their recommendations are the best available.

D.   Number of Samples Required

The water content of soil varies from one location to another. Plants remove water from the soil in an uneven manner. Water is applied unevenly because of local undulations, cracks, discontinuities, changes in soil structure, and pore size distribution. In addition, soils vary in texture while many are laminated. A random sampling of the soil picks up these variations. This requires that a large number of samples must be taken to determine accurately the water content of any plot or field. Thus, Taylor (1955) estimated a coefficient of variability of about 10% for field sampling. Staple and Lehane (1962) found standard deviations varying from 0.61 to 1.27 inches of water (6.9 to 13.3%) in a profile 4 ft deep. Hewlett and Douglass (1961) pointed out that the variations in measuring water content on a volume basis result from errors in both water content and bulk density. They found coefficients of variation ranging from 7 to 23% of the mean value, depend-ing on the methods used to secure the samples. The least variation (7%) was obtained when both water content and bulk density were measured on the same undisturbed sample and the greatest variation was obtained when water content and bulk density were determined on different samples that were taken throughout the plot at random.

The number of samples that need to be taken to give any required degree of precision depends on the local variability. There are a number of ways of estimating this number (Staple and Lehane, 1962; Hewlett and Douglass, 1961), depending on how much is known about the variations of the particular plot. A typical example would require about 10 samples to determine the average water content within 1% and between 25 and 30 samples would be needed to measure within 0.5%.

II. SOIL WATER POTENTIAL

It is well known that different types of soil retain different amounts of water in field conditions. Fine-textured soils have a greater attraction for water and usually have a larger total porosity than coarse-textured soils. Within the species of clay minerals, montmorillonite absorbs more water than illite, which in turn absorbs more than kaolinite. Such differences in behavior may be described thermo-dynamically. At a given point in space and time, the total potential of soil water, μ_θ, depends on the interfacial properties and composition of the soil solids

which absorb the water, the kind and amount of solutes in the water, the ambient temperature T, and the total pressure P (Taylor et al., 1961). Expressed as a sum of differential terms, we have

$$d\mu_\theta = \left(\frac{\partial \mu_\theta}{\partial \theta}\right)_{T,P,n_1} d\theta + \sum_{j=1}^{k} \left(\frac{\partial \mu_\theta}{\partial n_j}\right)_{T,P,\theta} dn_j - \overline{S}_\theta \, dT + \overline{V}_\theta \, dP \qquad [15\text{--}3]$$

where n_i is the content of chemical species i, θ is the water content, $\overline{S}_\theta$ and $\overline{V}_\theta$ are the partial specific entropy and volume of water in the system, and T is measured in Kelvin degrees.

The first term of equation [15–3] describes the interaction between the water and the soil solids and for this reason has been called the matric potential. The partial differential $(\partial \mu_\theta / \partial \theta)_{T,P,ni}$ is the slope of the soil water retention curve and may be called the specific water potential. It is the reciprocal of the specific (or differential) water capacity.

The second term of equation [15–3] expresses the influence of composition on the water potential. For a particular soil of constant mineralogical and organic matter content, under a given state of packing and in the absence of swelling or shrinking, this term includes only solute effects. It is the solute potential sometimes called osmotic potential.

To measure matric and solute potentials, temperature and pressure should be held constant. If measurements are always made at the same temperature (e.g., 25C) and atmospheric pressure (e.g., 1.013 × 10^6 dynes/cm²), we may dismiss the last two terms of equation [15–3]. By changing the sign of the matric and solute potentials, expressing them in units of energy per unit volume (pressure), and renaming them suctions (see chapter 22) we consider first the measurement of matric or soil water suction. (See also Table 15–2.)

Table 15–2. Measurement of soil water potential (or suction)

Recommended for measuring	Operating conditions		Principal advantages or disadvantages
	Favoring	Limiting	
	Suction table		
Water content at a given matric potential in laboratory	1) Undisturbed cores are available 2) Few measurements needed 3) Measurements need not be repeated 4) Moist soil	1) Soils difficult to sample 2) Large number of measurements needed 3) Measurements can not be made in dry soil	1) Simple inexpensive equipment 2) Requires samples to be brought to laboratory 3) Covers only wet portion of soil water range 4) Requires long time for equilibrium
	Pressure equipment		
Relation between water content and matric potential in laboratory	1) Undisturbed cores 2) Few measurements needed 3) Measurements need not be repeated	1) Soils difficult to sample 2) Large numbers of measurements needed	1) Covers field range of soil water 2) Requires samples be brought to laboratory 3) Requires long time for equilibrium 4) Requires special equipment
	Tensiometers		
Matric potential in the field	1) Moist soil 2) Repeated measurements at same location	1) Dry soil	1) Quick, accurate determination of water potential in the field 2) Requires field servicing 3) Temperature sensitive 4) Crops may be trampled
	Psychrometers		
Total water potential in the laboratory	1) Precise laboratory conditions 2) Samples are easily obtained	1) Samples difficult to get 2) Nonprecise laboratory conditions	1) A quick method to get total water potential in soil or plant material 2) Requires special precise laboratory equipment and procedures 3) Requires scrupulous cleanliness
	Freezing point		
Total water potential	Not recommended at this time		

A. Suction Table

By the use of a suction table (Jamison and Reed, 1949; Leamer and Shaw, 1941; Russell, 1942; Tanner et al., 1954) for which a variety of porous materials, including bisque-fired ceramics, are suitable, soil samples placed thereon may be brought to water content equilibrium with the suction applied through the water in the porous surface of the plate. The suction plate itself is always backed by a reservoir of water at a reduced pressure (see Fig. 15–3) which may be easily established by using a hanging water or mercury column of the desired length. Another way is to use a vacuum manifold and a capillary or needle reducing valve to increase the pressure to that established by a bleed or bubbler mechanism (Holmes, 1955; Rawlins, 1962). To ensure that samples approach a true equilibrium water content, they should be weighed regularly and replaced until the

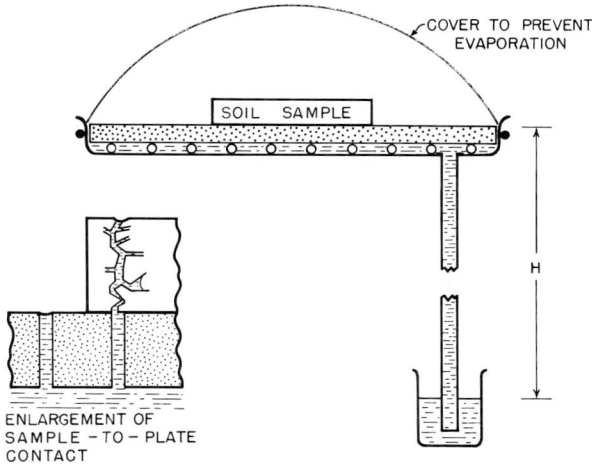

Fig. 15–3. Suction plate apparatus. The suction is H cm of water column.

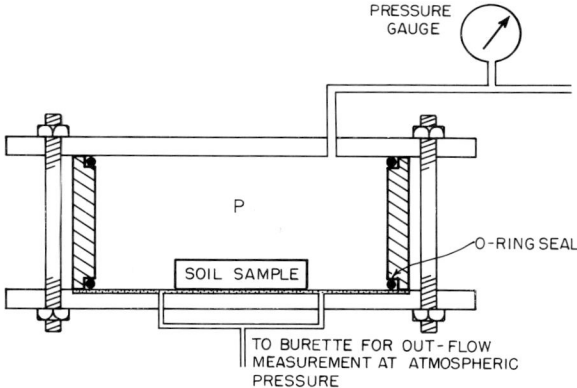

Fig. 15–4. Pressure membrane apparatus. The soil water suction at equilibrium is P, the applied pressure, expressed as cm of water column.

weight does not change, or the volume of water released (or imbibed) should be measured until its rate of change approaches zero. The time to approach equilibrium commonly takes about 10 days, though it may be longer with a clay sample.

B.   Pressure Equipment

If suctions > 1 atm (in practice, about 0.8 atm) must be established, the soil samples are brought to equilibrium inside a pressure membrane or pressure plate cell (Richards, 1947, 1949). The construction of such an apparatus, which should be designed to operate safely to 15 atm pressure (or more for special purposes), is shown in Fig. 15–4. The water content of the sample attains an equilibrium, just as it does on the suction plate, by drainage or imbibition of water through the porous membrane whose pores must be sufficiently small to remain water filled at all working pressures. The influence of pressure on the specific volume of water is negligible; insofar as structural relations are alike, results from suction plate and pressure membrane are entirely concordant. A number of pressure plate and pressure membrane cells are described by Tanner and Elrick (1958) and Tanner and Hanks (1952). Commercial apparatus is available.

C. Tensiometers

The tensiometer is used to measure the matric suction of soils in the field. Its principle is similar to that of the suction table in that the soil water comes to suction equilibrium through the porous cup with a reservoir of water inside the instrument of sufficient size for its pressure to be measured (*see* Fig. 15–5). A mercury-filled manometer may be used for this or a Bourdon-tube or diaphragm vacuum gauge.

Generally, water must flow through the porous cup of the instrument to attain equilibrium. Water is therefore either added to or withdrawn from the soil in the immediate neighborhood of the cup, causing a change in the suction of the soil itself. This problem can be minimized by suitable design. Denoting by S the gauge sensitivity of the tensiometer, we want S to be as large as possible. It is given by

$$S = dP/dQ$$

where P is the pressure inside the tensiometer and Q is the total volume of water inside the tensiometer. If the tensiometer is partially filled with water, and the remainder of its volume contains air and water vapor, S may become quite small.

The rate of approach to an equilibrium gauge reading when the soil water suction has changed suddenly at time $t = 0$, from P_0 to P_1 is given by

$$P = P_1 - (P_1 - P_0) \exp(- KSt) \qquad [15–4]$$

where P is the pressure inside the tensiometer at time t, and it is understood that $P = P_0$ at $t = 0$; i.e., the tensiometer is in equilibrium with the soil water suction before it is changed. There is one important assumption in deriving equation [15–4]: The cup conductance K is small compared with the equivalent con-

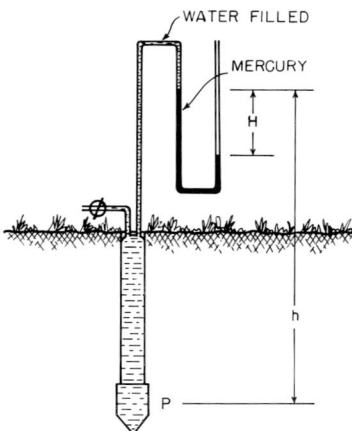

Fig. 15–5. Tensiometer. The suction of the soil at equilibrium with the porous cup at P is $- P = (13.6 \ H - h)$ cm of water column.

ductance of the soil in which the cup is embedded. If this is not so, evidently we must introduce terms to describe the effect of flow of water in the soil itself on the pressure inside the tensiometer (Gardner, 1960). Tensiometers with an extremely small time constant $1/KS$ have been designed by Miller (1951), Leonard and Low (1962), and Klute and Peters (1962). They used the concept of balancing the internal pressure against an externally controlled pressure, accurately measured, using null-point indication.

When installing a tensiometer, it is important that good contact be made between the cup and the soil so that approach to equilibrium is not hindered by contact impedance. It is best to install the tensiometer in an auger hole several times larger than the diameter of the tensiometer itself. The cup should be pushed into the soil at the bottom of the hole, if it is soft enough. If the soil is too hard, either it may be softened by water or a hole is made to receive the cup, using a metal tool. The soil surrounding the tensiometer standpipe should be refilled and compacted well. All above-ground parts of the tensiometer should be kept shielded from the sun to avoid a "thermometer" effect. The readings should be examined carefully to detect the effect of a leak, and the tensiometer should be kept filled with previously de-aired (boiled) water. A leaky tensiometer can give very misleading results.

One principle of design that applies to all the suction measuring devices, concerns the porous membrane. The cellophane sheet, porous ceramic, sintered glass, sintered metal, PORVIC (a porous P.V.C. sheeting), or any other, all must be leak proof up to the highest pressure desired for the apparatus. The maximum-sized pore of the material should therefore not exceed $0.15/\tau$ cm in radius, where τ is the desired matric suction in centimeters of water, so that the air entry value be not exceeded.

D. Psychrometers

The soil water potential may be expressed in terms of the water vapor pressure p with which it is in equilibrium, at constant temperature, by the relation

$$\Psi = \Delta\mu_\theta = \overline{R}T \ln (p/p_o) \qquad [15\text{--}5]$$

where p_o is the vapor pressure of pure, free water at the same temperature and total pressure, and $\overline{R}$ is the specific gas constant for water vapor.

A technique to measure the vapor pressure over soil samples placed in an equilibrium chamber so sensitive that the usual range of field water may be investigated has been developed by Richards and Ogata (1958, 1961), Monteith and Owen (1958), and Korven and Taylor (1959).

A precisely controlled constant temperature bath is essential with temperature fluctuations not exceeding 0.001C. The sample of soil needs 0.5 to 6 hours to reach temperature and vapor equilibrium with the chamber in which it is exposed. Stability of control of the water bath is most difficult to achieve. There are electronically controlled temperature baths available on the commercial market capable of this precision. However, they are subject to failures and sometimes respond to variations in the supply voltage. The bath should be monitored with a temperature recorder to ensure that both short and long-time fluctuations are within acceptable limits.

The principle of the method is to measure the difference between the temperature of the water bath, i.e. of the soil sample properly equilibrated, and the temperature of a minute drop of pure water which presents a surface freely evaporating into the chamber. This temperature difference is measured with a thermocouple pair, of which one junction is in contact with the evaporating surface of pure water, and the other is held at the reference temperature of the bath.

Two different methods have developed for providing the wet, evaporating surface in the sample chamber. One method (Richards and Ogata, 1961) inserts a droplet of pure water held in a fine, silver wire loop, that is itself the thermojunction with its supporting filaments. When thermal equilibrium and a steady evaporation rate have developed, the appropriate temperature difference is measured with a sensitive potentiometer circuit. The other method (Monteith and Owen, 1958) inserts a bare, dry thermojunction into the sample chamber, allows thermal and vapor equilibrium to be established, then cools the dry junction below the dewpoint by passing an electric current through it (the Peltier effect). When sufficient vapor has condensed to provide enough condensate upon the junction, the current is stopped, and the temperature of the cooled junction is measured as the water film evaporates.

Both methods are useful. However, they are not completely equivalent. The wet-loop method has the disadvantage that it introduces an extraneous drop of water into the sample chamber, which may measurably change the soil water potential. The method may thus be noticeably influenced by the rate of water absorption by the sample (Rawlins, 1964). The Peltier cooling method minimizes this difficulty if the amount of condensate is kept small. In practice the film usually evaporates again in 1 or 2 min. The Peltier cooling method has the added advantage that the thermojunction can be sealed into the soil sample chamber and left indefinitely without disturbing the system.

The Peltier cooling method has been unsatisfactory in relatively dry soils, with relative vapor pressure < 0.96 because sufficient vapor cannot be made to condense on the cooling junction. The wet-loop method can be used in relatively dry soils, but the hazard of affecting the soil water potential of the sample increases with the dryness.

Scrupulous cleanliness is needed to avoid nuclei for condensation of water vapor. It is desirable that all parts of the sample tube be built from material of the same heat capacity and conductivity to avoid differential temperatures which may produce a site for nucleation and condensation of vapor.

Both methods require the highest possible standard of technique in measuring the electrical potential difference between the wet and dry thermojunctions. It is

Table 15–3. Values of the free energy of KCl calibration solutions
for four different temperatures (Kijne, 1964)

Molarity KCl solution	a_θ at 25C*	$\Delta\mu_\theta$, – joules/kg at			
		12.00C	16.00C	21.05C	28.80C
0.1	0.99667	439.854	446.024	453.814	465.768
0.2	0.99343	868.484	880.667	887.597	919.650
0.3	0.99025	1291.048	1309.158	1332.022	1367.108
0.5	0.98389	2138.905	2168.909	2206.787	2264.915
1.0	0.96814	4265.372	4325.206	4400.741	4516.660

* a_θ is the relative activity of the water.

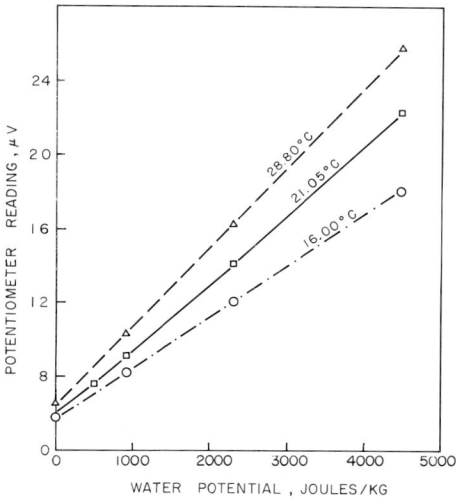

Fig. 15–6. Calibration curves for a wet-loop psychrometer.

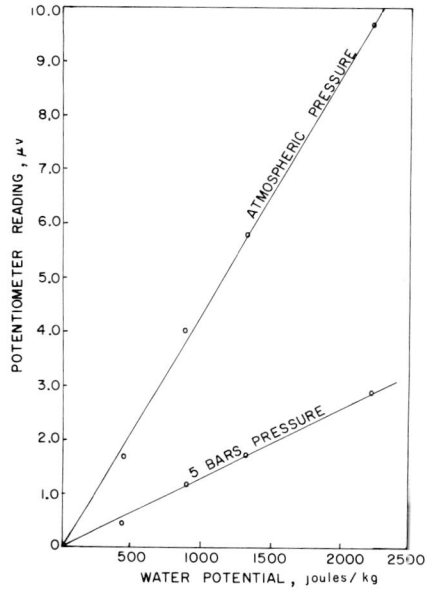

Fig. 15–7. Calibration curves for a Peltier thermocouple psychrometer.

sometimes convenient to record the potential differences, using a stable d-c amplifier, whose output is fed to a recorder.

It is necessary to calibrate every thermojunction individually, since small variations in construction, particularly in the heat conduction properties of the supporting wires, spoil reproducibility between individual devices. Potassium chloride solutions are useful for calibration. Recent measurements of the free energy of KCl solutions were obtained specially for this work (Kijne, 1964) and are given in Table 15–3. It is important that the geometry of the sample chamber and sample be the same during calibration as during actual measurements of the soil water potential.

Examples of calibration curves relating the relative vapor pressure with the thermoelectric output are shown in Fig. 15–6 for the wet-loop method and in Fig. 15–7 for the Peltier cooling method.

E. Freezing Point

It is possible to measure the temperature at which water in moist soil freezes (Taylor et al., 1961; Richards and Campbell, 1949). The freezing point thus measured should give an estimate of the combined influence of matric and osmotic suctions. In practice, it is found that soil must be cooled several degrees below the freezing temperature to get freezing to take place. The amount of this undercooling is variable from sample-to-sample and time-to-time. Likewise, the apparent freezing point seems to be associated with the degree of undercooling. This anomaly is not well understood, and no methods have been successful in correcting or avoiding it.

It seems likely that the heat content of the soil water varies with the thickness of the film of adsorbed water. There are four problems, and possibly others, that need to be solved before freezing point depression can be used as a routine and reliable method for measuring the potential of water in the soil. First, as water "freezes out" of the soil, both the heat content and the potential of the remaining water changes. Thus, exact and reproducible freezing temperatures cannot be obtained. Second, the adsorbed and exchangeable ions in the soil exert an influence on the observed freezing point. This influence is of unknown magnitude but is probably greater in dry than in moist soils. Third, the soil water potential exhibits a significant temperature dependence. The available data are insufficient to provide a quantitative correction for this. Fourth, as water freezes out, the potential of the remaining water increases along the water retention curve. This in turn causes the freezing point to become progressively lower. It is not possible to observe the freezing point of water in soil without freezing out a finite quantity of it. As soil gets drier, it becomes increasingly difficult to concentrate sufficient water at the nucleation point to get crystallization.

III. CALIBRATED METHODS FOR MEASUREMENT OF EITHER WATER CONTENT OR MATRIC POTENTIAL (OR SUCTION)

Gardner (1898) proposed the use of electrical resistance between buried electrodes as an index to soil water. He later questioned the technique because of unreliable contact between electrodes and soil, but modifications of this proposal are still in use. Since soil suction (matric potential) can be observed directly in the field with tensiometers over a limited range only, there is need for a secondary standard type of instrument which can be calibrated and used to measure suction over wide ranges. Pressure plate and pressure membrane equipment may be used to calibrate instruments over the range of suction commonly occurring in agricultural soils, provided the sensing elements can be enclosed within the pressure chamber in such a way that resistance readings may be made when equilibrium has been achieved.

Edlefson and Anderson (1941) used a four-electrode technique to eliminate the uncertain contact resistance between soil and electrodes. The procedure appears to be too complicated for wide-scale usage, although more recent workers are using a similar technique for evaluating salinity levels in soils that remain at or near saturation. (See Table 15–4 for a summary of methods.)

A. Electrical Resistance Blocks

Bouyoucos and Mick (1940) proposed that electrodes be placed in a block of porous material which in turn is placed in a soil. The water conditions in the block change with corresponding changes in water conditions in the soil, and changes within the block are reflected by changes in resistance between the electrodes. Plaster-of-paris is the porous medium most often used. Its pore structure is such that a favorable range in resistance values is found as water conditions vary, although resistance values change rather slowly with water content in the wetter range of values. The fact that the calcium sulfate maintains essentially a saturated concentration for the water retained by such a block makes this type

Table 15–4. Calibrated methods for measurement of water content
or matric potential (or suction)

Recom-mended for measuring	Operating conditions		Principal advantages or disadvantages
	Favoring	Limiting	
	Electrical resistance blocks		
Matric potential in the field; not recommended for water content measurements	1) Nonsaline soils 2) Dry portion of field water range 3) Repeated measurements at same point	1) Saline soils 2) Moist and wet soils	1) Repeated nondestructive measurements at same point in soil are possible 2) Moderately good relation between potential and block reading 3) Inexpensive units 4) Calibration changes 5) Very sensitive to electrolytes 6) May cause trampling of crops 7) Poor relation between soil water content and block readings
	Penetrometer		
Not recommended at this time			
	Electric capacitance		
Not recommended at this time			
	Electrothermal methods		
Not recommended at this time			
	Porous absorbers		
Not recommended at this time			

of block less sensitive to changes in the soil solution concentrations. Glass fibers, ceramic, and nylon cloth have been tried as the porous body of electrical resistance units. Fiberglass units, as developed by Coleman and Hendrix (1949), along with several models built from plaster-of-paris, have been commercially available. The various models using plaster-of-paris differ primarily in external shape and in material used in the electrode. By using concentric cylindrical electrodes made from screen wire, advantage is claimed that all of the electric current flow remains within the block (Croney et al., 1951). Flat blocks with large screen electrodes parallel to the sides of the block also retain most of the electric current flow within the block and exhibit somewhat less lag than cylindrical blocks (Taylor et al., 1961).

Calibration curves may be determined for resistance blocks, using either water content or soil suction as the independent variable (Bourget et al., 1958; Aitchison and Butler, 1951). Calibration curves for eight types of resistance units and their variance with repeated cycling are shown by Cannell (1958). Ideally, such blocks should be calibrated to read suction and used to complement the tensiometer for obtaining suction readings in the range above 0.8 bar. For reproducible results, the selection and handling of the material for casting the blocks must be carefully standardized. Some consideration should be given to the material forming the porous block in relation to permanence. Some materials placed in the soil are shorter lived than others. Impregnation of the plaster blocks with resin has been found to prolong their useful life (Bouyoucos, 1961).

Even though resistance units may be imperfect in some of their characteristics, they appear to be best-suited for general use in studying soil water relations. As mentioned, they can be calibrated to indicate either water content or suction, they are relatively inexpensive, hence more stations can be established on a given budget, and, even though calibration data may not be as accurate as wished, the fact that changes are or are not occurring at various locations in the root zone of crops is helpful information. Such studies as depth of rooting or depth of water penetration are important, even though precise values may not be possible.

B. Electric Capacitance

Relative to other substances, water has a high dielectric constant. It is to be expected that units for measuring the water content of a soil might be based on the change in capacitance of a condensor arranged so that a typical soil sample would be located between or near the plates of such a capacitor. Any change in water content of the soil would then be reflected by a corresponding change in capacity. Fletcher (1940) was one of the early investigators, but as yet the difficulties in assessing the dielectric properties of water under the influence of the force fields found in soil have not been adequately overcome (Childs, 1943).

C. Thermal Conductance

Thermal properties of soils change with soil water content (van Duin and de Vries, 1954; de Vries and Peck, 1958). An electrothermal method for measuring soil water was first proposed by Shaw and Baver (1939). Later Bloodworth and Page (1957) proposed the use of thermistors imbedded in porous materials for adapting the electrothermal technique for measuring soil suction. Although the technique appears to be insensitive to salinity levels, it has not as yet found wide usage.

D. Penetrometer

Numerous instruments have been devised for measuring the force required to push a probe into a soil (Bodman, 1949). Often a simple design is used to indicate only the maximum force needed to penetrate a given depth with a given probe. Other more complicated models are designed to record the force applied at each depth of penetration. In many instances, penetrometers were designed to evaluate soil compaction, soil profile development, or to correlate with root penetration. One such study is reported in the paper by Shaw, Haise, and Farnsworth (1943). In every attempt of such evaluations, soil water conditions were found to influence markedly the response of the instruments. While qualitative interactions of soil properties and water relations on penetrometer records are generally recognized, quantitative functions for separating these variables are not yet available. In spite of restrictions imposed by such interactions, the "feel" of a rod or shovel as it is forced into a soil is a rough estimate of soil wetness or, more specifically, a means of estimating the depth of wetting after a water application. Based on the penetrometer principle, at least one instrument has been proposed as a guide for irrigation management. Allyn and Work (1941) showed that the force needed (as measured by a small portable instrument) to penetrate a soil core taken from a deep, fine-textured soil was an indication of the need for irrigation of deep-rooted trees.

E.   Porous Absorbers

Several attempts to establish a soil water index have been made using porous ceramic absorbers. The soil point proposed by Livingston and Riichrio (1920) had certain favorable features. The use of this device involved the placing of a pre-dried and weighed cone-shaped ceramic "point" into a soil. After an arbitrary interval, the cone was removed, brushed free of soil, and weighed. The quantity or rate of water taken up was related to the "water-supplying power" of the soil. In theory, this type of index should evaluate the same soil factors involved in the root action of plants, but the method has not been explored for its practical use-fulness. Richards and Weaver (1943) proposed absorbers of ceramic which were left in contact with specific profile locations at the bottom of access tubes. Peri-odically, the absorbers were separated from the soil and weighed. The change in weight at any time was considered to be related to the change in water content occurring in the soil. The observation of a hysteresis loop did not disqualify the use of such a block since it corresponded to similar cycles in the soil water system. Davis and Slater (1942) had earlier proposed a weighed sorption block using gypsum.

IV.   INTERCONVERSION OF WATER CONTENT AND MATRIC SUCTION MEASUREMENTS

Many curves have been published to show a functional relationship between water content and matric suction for specific soils. When measured with other variables, such as bulk density and soil structure controlled as nearly as possible to field conditions, such curves give useful information about given soils. Such curves have been called water retention or moisture characteristic curves. The derivative or slope of a retention curve is, by definition, the specific (or differ-ential) water capacity variable which appears universally in differential equations used to predict water transfer in soils.

Unfortunately, no single curve can be taken as characteristic for a given soil sample. A well-established hysteresis occurs which means measured values of water retention at given suction are different, depending on whether the soil is on a wetting or a drying cycle. In fact, any point within the hysteresis loop can be a hydraulically stable point if the wetting or drying cycle is reversed at less than extreme values (Richards, 1939). Theoretical studies on soil water transfer usually resolve the problem of hysteresis by working either on a single drying or wetting curve. In field studies, this is not possible and attempts to correlate tensi-ometer readings with water content samples have not always proven satisfactory (Stolzy et al., 1959).

The relation between the soil matric potential and water content as it depends on the wetting and drying history of the sample is shown in Fig. 15–8. The limit-ing curves represent the values that are obtained with a sample that is first satu-rated, then dried to air dryness and rewetted gradually. The probable scanning curve A–C represents a curve that might result if the sample is allowed to wet to a value represented at A (near field capacity) before drying out. If the originally wet soil were dried only to point B and water applied, a curve similar to B–A would result. It is thus apparent that many possible scanning curves could result,

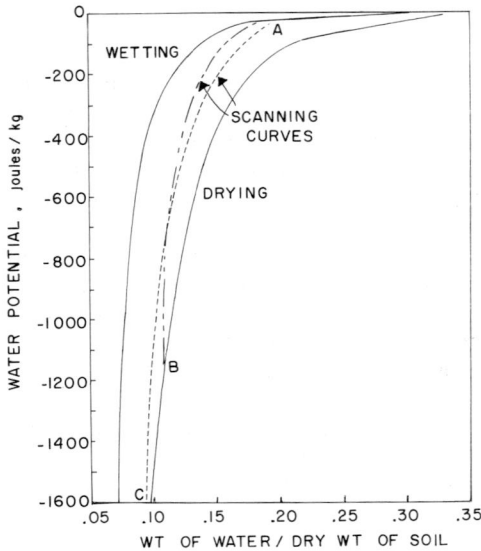

Fig. 15–8. Hysteresis curve for Millville silt loam at 25C. A–C is a
scanning curve for drying and B–A
is a scanning curve for wetting.
The limiting curves were obtained
by drying from saturation and wetting from air dryness.

and the true relation could fall anywhere between the limiting curves. When
measured in the field with a tensiometer, the true water potential is indicated.

Hysteresis is not the only factor that may cause variations between measured
water content and potential. In so-called field calibrations of instruments, where
soil samples are taken for relating water content to instrument readings, local
variability in soil texture and structure will also cause variability in the calibration
curves. In general, rather than resorting to calibration curves for converting water
content to water potential, or vice versa, it is recommended that when one or
the other variable is under study, instruments should be used which respond
to the particular variable involved.

V. HYDRAULIC CONDUCTIVITY OF SATURATED AND UNSATURATED
SOILS: SOIL WATER DIFFUSIVITY

Core samples of soil, taken from the field, may be used in laboratory tests for
measurement of the hydraulic conductivity (see Table 15–5 for a summary).
One-dimensional flow is easily impressed upon a sample by confining it, except
for the ends, in an impermeable casing. The conductivity of a saturated sample
is then given by

$$K = (\Delta Q/\Delta t)(X/AH) \tag{15–6}$$

where $\Delta Q/\Delta t$ is the steady rate of flow of water through the sample, A is the cross
sectional area of the sample, X is the sample length, and H is the difference in
hydraulic head between the two ends.

A great deal of care is needed to avoid spurious results. Factors which influence
the observed flow are blocking of pores by entrapped air (Christiansen, 1944),
leakage of water through cracks or faults in the specimen (Smith and Browning,
1947), or along the walls of the permeameter apparatus, and the blocking of pores

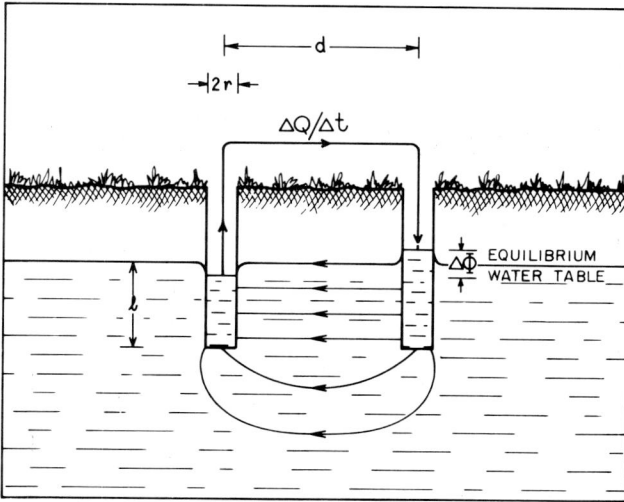

Fig. 15–9. The two-well method (Childs) for measuring hydraulic conductivity.

by bacterial growth and its byproducts (Allison, 1947). It is desirable to be able to impose a small suction of perhaps 4 cm water to drain cracks and the fissure at the boundary between sample and its container, thus minimizing the contribution to flow of spurious channels. The technique is reviewed in detail in *Drainage of Agricultural Lands* (Luthin, 1957). An oscillating permeameter which minimizes the problem of entrapped and occluded air is described by Childs and Poulovassilis (1960).

The water used in conductivity experiments must have a suitable electrolyte content if the sample contains appreciable amounts of clay. Bodman and Fireman (1950) showed that the hydraulic conductivity of Yolo clay loam and Aiken clay loam became very small as the electrolyte concentration was reduced. This result is true generally, with electrolyte species of the cation also important. A sufficiently strong concentration of electrolyte can keep a sodium soil flocculated and prevent the decrease in conductivity which occurs when a weak solution is used as the permeating fluid (Fireman and Bodman, 1940).

The usefulness of small samples tested in the laboratory is restricted to those soils which are homogeneous on the size scale of the sample. Frequently, it is desirable to measure conductivity in the field *in situ*.

By the method of Childs (1952) and Childs et al. (1953), two wells, each with diameter 2r, are drilled in the soil with a spacing d between centers (*see* Fig. 15–9) to a depth l below the equilibrium water table. The hydraulic conductivity is given by

$$K = \cosh^{-1}(d/2r) \times \frac{\Delta Q}{\Delta t} \times \frac{1}{(1 + l_f)\,\Delta\phi}. \qquad [15\text{–}7]$$

The first term is constant, depending on the geometry of the installation. The addition of l_f to the measured length of the wells allows some compensation for the curved streamlines below the bottoms of the wells. The quantity $\Delta Q/\Delta t$ is the steady-state rate of flow between the wells achieved by pumping from one well

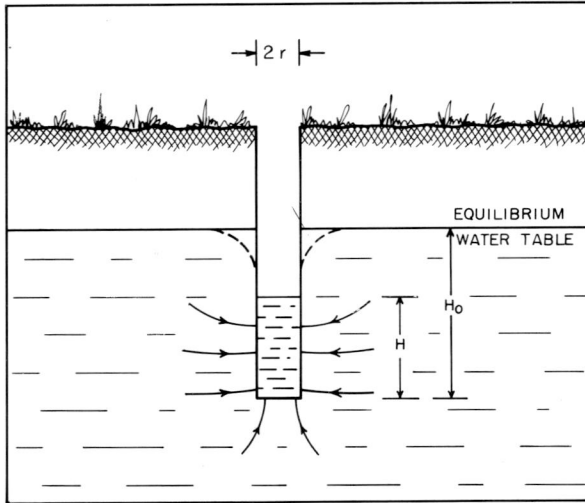

Fig. 15–10. The auger-hole method for measuring hydraulic conductivity.

and delivering the pumped water into the other, and $\Delta\phi$ is the difference in hydraulic head so caused.

Single-well methods are, in essence, non-steady-state methods. However, there is a widely known treatment of the well steadily discharging at the rate $\Delta Q/\Delta t$, which assumes pseudo-equilibrium. At a sufficient distance from the well, the water table changes very slowly. The Dupuit formula for conductivity is

$$K = \frac{\Delta Q \ln (d_2/d_1)}{\Delta t \, \pi \, (H_2{}^2 - H_1{}^2)}. \qquad [15\text{–}8]$$

The distances d_1 and d_2 from the discharging well refer to the location of test-wells in which the hydraulic heads H_1 and H_2 may be measured, referred to the bottom of the discharging well as datum. Peterson et al. (1952) describe the approximations necessary to this treatment of the well problem.

The auger-hole method is often used as a non-steady-state method (*see* Fig. 15–10), and it needs only simple equipment (Maasland and Haskew, 1957). If the water level in the hole is allowed to come to equilibrium level, which is measured, and then some of the water is quickly baled or pumped out, the rate of flow back into the auger hole is given by

$$(dQ/dt) = KA \, (H_o - H) \qquad [15\text{–}9]$$

where A is a geometrical factor of the same kind as the first term in equation [15–7], H is the instantaneous height of water in the auger hole, and H_o is the equilibrium level.

Using the known dimensions of the auger hole, equation [15–9] becomes, when solved for K,

$$K = \frac{\pi \, R^2}{A \, (H_o - H)} \times \frac{dH}{dt}. \qquad [15\text{–}10]$$

Exponents of the auger-hole technique appear to prefer to use this differential form of the equation, and the development of the method requires a suitable numerical value to be assigned to the first term which Maasland and Haskew (1957) refer to as the C-factor, in the equation

$$K = C(dH/dt).$$ [15–11]

The same authors give useful graphs, computed from formulas derived by Ernst (1950) for C as a function of (H_o/R) and $(H_o - H)/R$. Beers (1958) has also prepared similar graphs and a nomograph for ease with the calculation. Comparison of worked examples reveals that there may be a discrepancy of about 10% between the two authors' graphs. It is clear that C, assumed to be constant, actually depends on the changing head of water H in the auger hole. For this reason, it is recommended that the maximum lowering of water in the auger hole should not exceed $0.2\ H_o$. The auger-hole method is claimed to be very useful in routine investigations.

There are objections to this approximate treatment of the auger-hole problem when it is considered as an exercise in potential theory. Kirkham worked out the exact theory of flow into an auger hole, both for penetration to the impermeable floor (Kirkham and Van Bavel, 1949) and for a partially penetrating hole (Kirkham, 1958, 1959). The form of the result is a formidable combination of Bessel functions, and no convenient graphs have yet been prepared by calculation of these expressions for the appropriate parameters.

The integrated form of equation [15–10] may be written as

$$K = \frac{R^2}{A} \times \frac{\ln(z_1/z_2)}{(t_2 - t_1)}$$ [15–12]

where z is the depth from the equilibrium water level to the measured water level at time t; i.e., $z = (H_o - H)$. Kirkham (1946) proposed that a piezometer tube should be used to measure hydraulic conductivity. It has important advantages over the auger-hole method. Layered soil may be investigated in stages by placing the piezometer opening in the layer of interest. In addition, by changing the length of the cavity, anisotropy in conductivity may be observed (Reeve and Kirkham, 1951). Luthin and Kirkham (1949) measured the A-factor in equation [15–12] for a variety of geometries by electrical analogue. Donnan and Aronovici (1961) proposed a modification of the Kirkham-tube method in which a screened cavity is used to measure in soils and earth materials which slump below the water table.

By combining the two-well method and the Kirkham-tube method, Childs et al. (1957) were able to assess quantitatively the anisotropy of soils which generally appear to have a K larger in the horizontal than in the vertical direction. Maasland devotes a chapter in *Drainage of Agricultural Lands* (Luthin, 1957) to a discussion of anisotropy in hydraulic conductivity.

Thus far, a requirement of all the methods has been that a water table exist in the soil so that appropriate potential gradients can be arranged for measurement. To predict what the hydraulic conductivity of a soil would be under future water table conditions, Bouwer (1961, 1962) developed the double-tube method. Two concentric tubes are placed in the soil (*see* Fig. 15–11) with careful attention to the correct technique for doing this. Water in the tubes infiltrates into the relatively dry soil, and measurements begin when the soil below the tube bottom

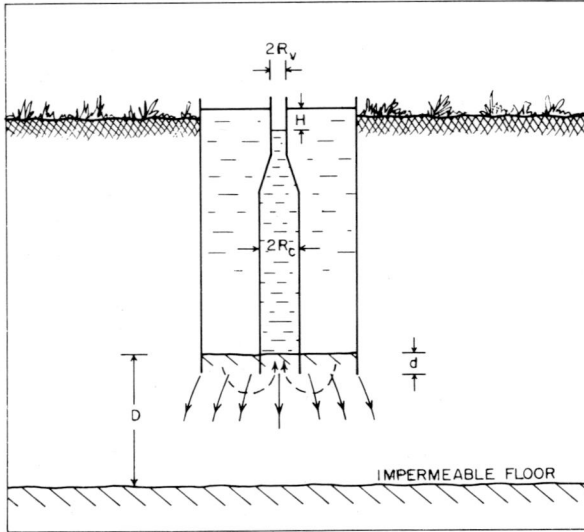

Fig. 15–11. Geometry of the double-tube method for measuring hydraulic conductivity.

has been saturated to a sufficient depth, in practice $> 2 R_c$. Then, a small difference H between the water level in the inner tube and that in the outer annulus will cause water to flow through the soil between outer and inner tubes, in addition to downward infiltration. Two sets of observations are made to distinguish between these two flows. First, the fall in water level in the inner tube is measured as a function of time, with the outer water level maintained constant. Then both water levels are arranged to fall at an equal rate. The hydraulic conductivity is given by

$$K = \frac{R_v^2 \, \Delta H_t}{F_F \, R_c} \Big/ \int_o^t H \, dt. \qquad [15\text{–}13]$$

where F_F is a geometry factor which must be obtained from Bouwer's (1961) graphs and depends on d / R_c and D / R_c. The sets of curves giving F_F were obtained from an electrical analogue experiment. The quantity ΔH_t is the difference between the two measurements of H at time t (specified to be the same t in each case), the one when both water levels fall together, the other when the outer water level is maintained at a constant height. Other symbols are given in Fig. 15–11.

Table 15–6 shows a comparison between the hydraulic conductivity determined by the double-tube method and by laboratory permeameter tests on samples. The agreement is good, making allowance for the natural inhomogeneity of soils and suggests that this new method may prove useful.

Whereas the hydraulic conductivity of saturated soil is constant, when the soil becomes unsaturated, the conductivity depends strongly on its water content. It is convenient, again, to consider separately steady-state and non-steady-state methods. Marshall (1959) described a variety of steady-state methods. Apparatus is similar, in principle, to that used for determinating the conductivity of saturated soil cores in the laboratory. The ends of the specimen are enclosed by porous discs, through which the appropriate suction is applied, and water flow achieved.

Table 15–5. Measurement of hydraulic conductivity

Recommended for measuring	Operating conditions		Principal advantages or disadvantages
	Favoring	Limiting	
One-dimensional flow of water through a core			
Hydraulic conductivity	1) Undisturbed cores 2) Unidimensional flow 3) Laboratory measurement 4) Uniform soil	1) Nonuniform soil 2) Inaccessibility of laboratory	1) Method is simple 2) It requires soil cores 3) It is markedly affected by small heterogeneities, cracks, and crevices in the soil
Two well permeameter			
Field measurement of hydraulic conductivity below the water table	1) Large sample 2) Non-uniform soil 3) Anisotropic soil	1) Water table required	1) It requires considerable field equipment, & is best used when combined with piezometer method
Auger hole			
Field measure of hydraulic conductivity below the water table	1) Large sample	1) Water table required 2) Non-uniform soil 3) Anisotropic soil	1) It requires only simple equipment
Piezometer method			
Field measurement of hydraulic conductivity below the water table	1) Large sample 2) Non-uniform soil 3) Anisotropic soil 4) Investigation of layers	1) Water table required	1) It requires considerable field equipment, & is best combined with two-well permeameter

Table 15–6. Comparison between K from double-tube method and K from soil samples (Bouwer, 1962)

	K, double-tube	K, soil samples
	cm/min	
Coarse sand, mean particle size approximately 0.5 mm, no clay	0.64	1.00* 0.91* 0.53†
Loamy sand, 65% fine sand, 31% silt, 4% clay	0.037	0.057* 0.049* 0.028†
Adelanto loam	0.036	0.032‡

* Duplicate test of mixed sample.   † Permeameter test of water-deposited sample.   ‡ Average of 6 horizontal and 5 vertical undisturbed core samples.

To eliminate unknown contact impedance at the boundaries between sample and porous plates, the potential gradient may be measured in the sample itself with small tensiometers. In all experiments, it is essential to provide for ready access of air to ensure appropriate draining of the pore space.

Steady-state experiments require a relatively long time for equilibrium to be established. Gurr (1962) developed a more rapid method which employs quasi-steady-state. The water flux is computed from changes in water content of the soil column, as it is allowed to evaporate from one end only, into the atmosphere.

By assuming the soil water content to be a single-valued linear function of soil matric potential, Gardner (1956) was able to solve the flow equation for the movement of water out of a single short core of soil of length L in the pressure membrane or pressure plate apparatus. He found that

$$\ln (Q_o - Q) = \ln \frac{8Q_o}{\pi^2} - \alpha^2 Dt \qquad [15\text{--}14]$$

where Q_o is the total outflow that occurs from any increment of pressure increase in the system, Q is the outflow that has occurred in time t, and α is $n\pi/2L$ where n is an integer 1, 3, 5 . . . which can be taken as unity after a few minutes of outflow time.

If the experimentally determined values of $(Q_o - Q)$ are plotted against t on semilog paper, a straight line should result with slope B except for small values of t. Then

$$D = 4BL^2/\pi^2 \qquad [15\text{--}15]$$

and

$$K = \frac{4BL^2}{\pi^2} \frac{d\theta}{dP} \cong \frac{4BL^2}{\pi^2} \frac{\theta_o}{P_o} \qquad [15\text{--}16]$$

where θ_o is the change in water content of the sample that corresponds to the increment increase in pressure P_o. Gardner (1956) gives examples of the use of equation [15–16] to calculate the unsaturated conductivity from pressure membrane outflow observations.

The chief difficulties with this method are keeping the soil in uniform contact with the plate or membrane and avoiding the influence of impedance of the membrane out of the sample. A method to correct for the latter problem has been proposed by Miller and Elrick (1958). Their method requires that curves be plotted and matched with theoretical curves for similar media. For details of the application, consult the original paper of Miller and Elrick or other recent works (Butijn and Wesseling, 1959; Elrick, 1963; Kunze and Kirkham, 1962; Rijtema, 1959). There are still a number of difficulties with the pressure outflow method which can only be solved by further research. Sometimes the theoretical curves do not match the experimental data, and the reasons are not yet completely understood (Bruce and Klute, 1963). Some of the more obvious difficulties may be associated with probably unreliable outflow data: (i) because of the long time required to reach equilibrium, the total water outflow Q_o is difficult to measure; (ii) at high pressures in the cell, air bubbles may diffuse through the membrane and cause errors in measuring Q; and (iii) high pressures may have a direct influence on the water potential.

A frequently used method for measuring diffusivity was proposed by Bruce and Klute (1956). Water is allowed to infiltrate a horizontal column of soil for a given time period; then at a particular time, the water content distribution within the column is determined. The diffusivity is calculated from the equation

$$D = -\frac{1}{2t_{(part.)}}\frac{dX}{d\theta}\int_{\theta_o}^{\theta} X\,d\theta \qquad\qquad [15\text{--}17]$$

where the subscript (part.) means at a particular time and distance X. From a plot of water content θ vs. distance X, the slope $dX/d\theta$ $and \int_{\theta_o}^{\theta} X d\theta$ are evaluated at a series of values of θ.

Gardner (1960) suggested that a tensiometer might be used to measure unsaturated conductivity or diffusivity of soil *in situ*. The theory would be analogous to that for the pressure outflow method except the equations would be in cylindrical form, as considered by Richards and Richards (1962).

None of the currently used methods for measuring the hydraulic conductivity in unsaturated soil are completely satisfactory. All of them are influenced more or less by hysteresis and inability to transfer the experimentally determined values from the particular sample to the field soil or other sample. Considerable research effort is being spent to develop better methods and to overcome some of these problems.

VI.   INFILTRATION AND INFILTRATION RATE

The rate of infiltration of water through soil vertically downwards is numerically equal to the hydraulic conductivity when the soil is saturated to a sufficient depth, because the gradient for flow is just gravity (unit gradient). But if the soil is relatively dry at shallow depth, the rate of infiltration may greatly exceed the hydraulic conductivity because the capillary and adsorption phenomena at the wetting front make the gradient for flow behind the wetting front larger than the gravity gradient. The total infiltration i may be approximated by the equation

$$i = St^{1/2} + At \qquad\qquad [15\text{--}18]$$

where S is described as the sorptivity by Philip (1957), and A is another function of the soil related to the unsaturated conductivity.

Parr and Bertrand (1960) reviewed the phenomena of water infiltration into soils, to 1959 literature, paying some attention to methods of measurement. The infiltration rate depends on the same factors of the soil physical condition as does the hydraulic conductivity. In addition, the washing of unstable soil particles over and into the soil surface exerts an important influence on the measurement. For practical purposes, soils intended to be sprinkler irrigated should be measured with a sprinkler infiltrometer; those that will be flood irrigated should be measured with a flooding infiltrometer.

Handbook 60 (Richards, 1954) describes a technique for measurement using a 12-inch diameter infiltrometer ring which is pressed or hammered into the soil. Loss of water from the ring by flow paths in disturbed and fractured soils is minimized by deeper insertion of the ring. Nevertheless, this and lateral spread of water cause observations to be erratic. An improvement in technique (Marshall and Stirk, 1950) is achieved by using two concentric rings. The outer annular space serves as a guard through which the infiltration of water is more rapid than through the central ring in which the infiltration is measured. For best results, the water levels in the two compartments should be kept accurately the same.

Bouwer (1963) showed that there may be appreciable leakage from one reservoir to the other through the soil if there is a head difference between them.

The nozzle through which the water issues into the infiltrometer ring should be so constructed that minimum damage is done to the natural structure of the soil. Sometimes it is desirable to stabilize the structure of the soil with a soil conditioner, or to put a protecting layer of fine gravel on the soil surface. The quality of the water used should be chosen to keep the clay colloids flocculated (Quirk and Schofield, 1955). If dispersion of the clay should be initiated, the infiltration rate will fall rapidly. Natural rain is low in electrolyte content and may cause dispersion of clay in salt-affected soils. If this is the effect to be studied, the experiment should be appropriately managed.

For sprinkler application, the modified North Fork apparatus (Rowe, 1940) is widely used in the USA. The nozzles eject water upward and the spray describes an arched trajectory to fall upon an area of soil 2.5 ft². Although the sprinkler infiltrometer is an attempt to simulate rainfall conditions (Duley and Kelly, 1941), the drop size distribution is not the same as it is in rain. A successful rainfall simulator has not yet been designed which gives a good reproduction of raindrop sizes. Yet, as Rose (1960) has shown, the rate of detachment of soil particles from aggregates depends on the momentum per unit area of the raindrop flux.

VII.   RELATION OF *IN SITU* TO DISTURBED SAMPLE MEASUREMENT

When soil is disturbed, the bulk density of the sample is changed from the natural field density. The pore size distribution, therefore, is changed and the relation between the water content and matric suction is different (Taylor and Box, 1961).

Sampling equipment designed to take undisturbed cores may reduce the effect, but a sample in place at some depth within the soil profile experiences an earth pressure which confines it. A swelling clay is particularly vulnerable to changes in its water characteristics when removed from the profile location. It may be desirable to measure the water retention properties in a triaxial apparatus (Bishop and Henkel, 1962) to reproduce the confining pressures of the profile, in place, for depths > 2 m.

Croney and Coleman (1954) showed that remolded clay soils possess virgin consolidation curves with a much larger retention of water than in subsequent second-cycle measurements. The amount of consolidation depends on the degree of remolding and the work history of the specimen which is difficult to reproduce and is quite arbitrary. This effect should be distinguished from hysteresis which is reproducible and irreversible.

VIII.   CONDUCTIVITY AND DIFFUSIVITY AS DERIVED PROPERTIES

In concluding this chapter on measurement of soil water, we direct attention to methods of calculating conductivity and diffusivity. Childs and Collis-George (1950) showed that the unsaturated conductivity of sands and slate dust could be calculated from the measured water-release curves when translated to pore-size distribution. The result agreed acceptably with the measured conductivity. The basis of the calculation is to assume that each pore may be characterized by an effective radius for flow which may be substituted in Poiseuille's equation for rate of flow through a uniform capillary tube. The whole conductivity is given by the sum of all contributions, which, for convenience, are grouped into a finite number of classes.

Marshall (1958) improved the technique of computation and recently reviewed his own and other attempts to refine the method to obtain a wider application to unconsolidated media and soils with loam to heavier textures (Marshall, 1962). His own equation for conductivity K, with the units centimeters per second is

$$K = (2.7 \times 10^2) \, \epsilon^2 \, n^{-2} \, [h_1^{-2} + 3h_2^{-2} + 5h_3^{-2} + \ldots + 2(n-1) \, h_n^{-2}].$$
[15–19]

In this expression, ϵ is the porosity, h_1 to h_n are the characteristic suctions (in centimeters of water) at which pores of the 1st class, 2nd class, etc., just empty of water, and n is the total number of such classes. The classes are chosen to be each of equal total porosity and of mean pore size decreasing from the first as the largest, with corresponding air entry suction h_1.

The soil water diffusivity D is given by the expression

$$D = K(dP/d\theta) = -K(dh/d\theta)$$
[15–20]

and it is possible to calculate D from the two quantities on the right hand side, the conductivity, which itself may be computed from the measured pore-size distribution or retention curve determination, and the specific water potential, $dP/d\theta$ also obtained from the retention curve. However, there has been little work yet done to check the accuracy of such computations by comparison with experiment, for a range of soil materials.

LITERATURE CITED

Aitchison, G. D., and P. F. Butler. 1951. Gypsum block moisture meters as instruments for the measurement of tension in soil water. Australian J. Appl. Sci. 2:257–266.

Allison, L. E. 1947. Effect of microorganisms on permeability of soil under prolonged submergence. Soil Sci. 63:439–450.

Allyn, R. B., and R. A. Work. 1941. The availometer and its use in soil moisture control: II. Soil Sci. 51:391–406.

Beers, W. F. J. van. 1958. The auger hole method: A field measurement of the hydraulic conductivity of soil below the water table. Int. Inst. Land Reclam. Impr. Bull. 1. Wageningen, The Netherlands. 32 p.

Belcher, D. J., T. R. Cuykendall, and H. S. Sack. 1950. The measurement of soil moisture and density by neutron and gamma ray scattering. US Civil Aeronautics Admin., Tech. Develop. Rep. 127. 20 p.

Bishop, A. W., and D. J. Henkel. 1962. The measurement of soil properties in the triaxial test. 2nd ed. Edward Arnold Ltd. 228 p.

Bloodworth, M. E., and J. B. Page. 1957. Use of thermisters for the measurement of soil moisture and temperature. Soil Sci. Soc. Amer. Proc. 21:11–15.

Bodman, G. B. 1949. Methods of measuring soil consistency. Soil Sci. 68:37–56.

Bodman, G. B., and M. Fireman. 1950. Changes in soil permeability and exchangeable cation status during flow of different irrigation waters. Int. Congr. Soil Sci., Trans. 4th. (Amsterdam, Neth.) 1:397–400.

Bourget, S. J., D. E. Elrick, and C. B. Tanner. 1958. Electrical resistance units for moisture measurements: Their moisture hysteresis, uniformity, and sensitivity. Soil Sci. 86:298–304.

Bouwer, H. 1961. A double tube method for measuring hydraulic conductivity of soil *in situ* above a water table. Soil Sci. Soc. Amer. Proc. 25:334–339.

Bouwer, H. 1962. Field determination of hydraulic conductivity above a water table with the double-tube method. Soil Sci. Soc. Amer. Proc. 26:330–335.

Bouwer, H. 1963. Theoretical effect of unequal water levels of the infiltration rate determined with buffered cylinder infiltrometers. J. Hydrol. 1:29–34.

Bouyoucos, G. J. 1937. Evaporating the water with burning alcohol as a rapid means of determining moisture content of soils. Soil Sci. 44:377–381.

Bouyoucos, G. J. 1961. Soil moisture measurement improved. Agr. Eng. 42:136–138.

Bouyoucos, G. J., and A. H. Mick. 1940. An electrical resistance method for the continuous measurement of soil moisture under field conditions. Michigan State Coll. Agr. Exp. Sta. Tech. Bull. 172. p. 1–38.

Bruce, R. R., and A. Klute. 1956. The measurement of soil moisture diffusivity. Soil Sci. Soc. Amer. Proc. 20:458–462.

Bruce, R. R., and A. Klute. 1963. Measurements of soil moisture diffusivity from tension plate outflow data. Soil Sci. Soc. Amer. Proc. 27:18–21.

Butijn, J., and J. Wessling. 1959. Determination of the capillary conductivity of soils at low moisture tensions. Neth. J. Agr. Sci. 7:155–163.

Cannell, G. H. 1958. Effect of drying cycles on changes in resistance of soil moisture units. Soil Sci. Soc. Amer. Proc. 22:379–382.

Childs, E. C. 1943. A note on electrical methods of determining soil moisture. Soil Sci. 55:219–223.

Childs, E. C. 1952. The measurement of the hydraulic permeability of saturated soil *in situ:* I. Principles of a proposed method. Roy. Soc. (London), Proc. A215:525–535.

Childs, E. C., A. H. Cole, and D. H. Edwards. 1953. The measurement of the hydraulic permeability of saturated soil *in situ:* II. Roy. Soc. (London), Proc. A216:72–89.

Childs, E. C., and N. Collis-George. 1950. The permeability of porous materials. Roy. Soc. (London), Proc. A201:392–405.

Childs, E. C., N. Collis-George, and J. W. Holmes. 1957. Permeability measurements in the field as an assessment of anisotropy and structure development. J. Soil Sci. 8:27–41.

Childs, E. C., and A. Poulovassilis. 1960. An oscillating permeameter. Soil Sci. 90:326–328.

Christiansen, J. E. 1944. Effect of entrapped air upon the permeability of soils. Soil Sci. 58:355–365.

Coleman, E. A., and T. M. Hendrix. 1949. The fiberglass electrical soil moisture instrument. Soil Sci. 67:425–438.

Croney, D., and J. D. Coleman. 1954. Soil structure in relation to soil suction. J. Soil Sci. 5:75–84.

Croney, D., J. D. Coleman, and E. W. H. Currer. 1951. The electrical resistance method of measuring soil moisture. Brit. J. Appl. Phys. 2:85–91.

Davis, W. E., and C. S. Slater. 1942. A direct weighing methods for sequent measurements of soil moisture under field conditions. J. Amer. Soc. Agron. 34:285–287.

Donnan, W. W., and V. S. Aronovici. June 1961. Field measurement of hydraulic conductivity. J. Irrig. Drainage Div. Amer. Soc. Civil Eng. Proc. 87(IR2):1–13.

Duin, R. H. A. van, and D. A. de Vries. 1954. A recording apparatus for measuring thermal conductivity and some results obtained with it in soil. Neth. J. Agr. Sci. 2:168–175.

Duley, F. L., and L. L. Kelly. 1941. Surface condition of soil and time of application as related to intake of water. US Dep. Agr. Circ. 608. 30 p.

Edlefsen, N. E., and A. B. C. Anderson. 1941. The four electrode resistance method for measuring soil moisture content under field conditions. Soil Sci. 51:367–376.

Elrick, D. E. 1963. Unsaturated flow properties of soils. Australian J. Soil Res. 1:1–8.

Ernst, L. F. 1950. A new formula for the calculation of the infiltration factor in the auger-hole methods. Mimeo Rep. Landbouwproef-station, Bodenkundig Inst. T.N.O., The Netherlands.

Ferguson, H., and W. H. Gardner. 1962. Water content measurement in soil columns by gamma ray absorption. Soil Sci. Soc. Amer. Proc. 26:11–14.

Fireman, M., and G. B. Bodman. 1940. Effect of saline irrigation water upon permeability. Soil Sci. Soc. Amer. Proc. (1939) 4:71–77.

Fletcher, J. E. 1940. Dielectric methods for determining soil moisture. Soil Sci. Soc. Amer. Proc. (1939) 4:84–88.

Gardner, F. D. 1898. The electrical method of moisture determination in soils: results and modifications in 1897. US Dep. Agr. Bur. Soils. Bull. 12. 24 p.

Gardner, W. R. 1956. Calculation of capillary conductivity from pressure plate outflow data. Soil Sci. Soc. Amer. Proc. 20:317–320.

Gardner, W. R. 1960. Measurement of capillary conductivity and diffusivity with a tensiometer. Int. Congr. Soil Sci., Trans. 7th (Madison, Wis., USA) 1:300–304.

Gardner, W., and D. Kirkham. 1952. Determination of soil moisture by neutron scattering. Soil Sci. 73:391–401.

Gurr, C. G. 1962. Use of gamma rays in measuring water content and permeability in unsaturated columns of soil. Soil Sci. 94:224–229.

Haise, H. R. 1955. How to measure the moisture in the soil. p. 362–371. In Water. US Dep. Agr. Yearbook.

Hancock, C. K., and R. L. Burdick. 1957. Rapid determination of water in wet soils. Soil Sci. 83:197–205.

Hewlett, J. D., and J. E. Douglass. 1961. A method for calculating error of soil moisture volumes in gravimetric sampling. Forest. Sci. 7:265–272.

Holmes, J. W. 1955. Water sorption and swelling of clay blocks. J. Soil Sci. 6:200–208.

Holmes, J. W., and A. F. Jenkinson. 1959. Techniques for using the neutron moisture meter. J. Agr. Eng. Res. 4:100–109.

Holmes, J. W., and K. G. Turner. 1958. The measurement of water content of soils by neutron scattering: A portable apparatus for field use. J. Agr. Eng. Res. 3:199–204.

Jamison, V. C., and I. F. Reed. 1949. Durable asbestos tension tables. Soil Sci. 67:311–318.

Kijne, J. W. 1964. Temperature dependence of soil moisture potential. Ph.D. dissertation, Utah State Univ., Logan. (University Microfilm, Ann Arbor, Michigan.) No. 64–13, 742.

Kirkham, D. 1946. Proposed method for field measurement of permeability of soil below the water table. Soil Sci. Soc. Amer. Proc. (1945) 10:58–68.

Kirkham, D. 1958. Theory of seepage into an auger hole above an impermeable layer. Soil Sci. Soc. Amer. Proc. 22:204–208.

Kirkham, D. 1959. Exact theory of flow in a partially penetrating well. J. Geophys. Res. 64:1317–1327.

Kirkham, D., and C. H. M. van Bavel. 1949. Theory of seepage into auger holes. Soil Sci. Soc. Amer. Proc. (1948) 13:75–82.

Klute, A., and D. B. Peters. 1962. A recording tensiometer with a short response. Soil Sci. Soc. Amer. Proc. 26:87.

Korven, H. C., and S. A. Taylor. 1959. The Peltier effect and its use for determining the relative activity of soil water. Can. J. Soil Sci. 39:76–85.

Kunze, R. J., and D. Kirkham. 1962. Simplified accounting for membrane impedance in capillary conductivity determinations. Soil Sci. Soc. Amer. Proc. 26:421–426.

Leamer, R. W., and B. Shaw. 1941. A simple apparatus for measuring noncapillary porosity on an extensive scale. J. Amer. Soc. Agron. 33:1003–1008.

Leonard, R. A., and P. F. Low. 1962. A self-adjusting, null-point tensiometer. Soil Sci. Soc. Amer. Proc. 26:123–125.

Livingston, B. E., and K. Riichrio. 1920. The water supplying power of the soil as related to the wilting of plants. Soil Sci. 9:469–485.

Luthin, J. N. [ed.] 1957. Drainage of agricultural lands. Agronomy, Vol. 7. 620 p.

Luthin, J. N., and D. Kirkham. 1949. A piezometer method for measuring permeability of soil in situ below a water table. Soil Sci. 68:349–358.

Maasland, M., and H. C. Haskew. 1957. The auger-hole method of measuring the hydraulic conductivity of soil and its application to tile drainage problems. Water Conserv. Irrig. Comm. Bull. 2. New South Wales, Australia.

Marshall, T. J. 1958. A relation between permeability and size distribution of pores. J. Soil Sci. 9:1–8.

Marshall, T. J. 1959. Relations between water and soil. C.A.B. Commonwealth Bur. Soils. Tech. Comm. No. 50.

Marshall, T. J. 1962. Permeability equations and their models. Inst. Chem. Eng. Conf. Interaction Fluids Particles (London) p. 299–303.

Marshall, T. J., and G. B. Stirk. 1950. The effect of lateral movement of water in soil on infiltration measurements. Australian J. Agr. Res. 1:253–265.

Miller, R. D. 1951. A technique for measuring soil moisture tensions in rapidly changing systems. Soil Sci. 72:291–301.

Miller, E. E., and D. E. Elrick. 1958. Dynamic determination of capillary conductivity extended for non-negligible membrane impedance. Soil Sci. Soc. Amer. Proc. 22: 483–486.

Monteith, J. L., and P. C. Owen. 1958. A thermocouple method for measuring relative humidity in the range 95 to 100. J. Sci. Instr. 34:443–446.

Mortier, P., M. de Boodt, W. Dansercoer, and L. de Leenheer. 1960. The resolution of the neutron scattering method for soil moisture determination. Int. Congr. Soil Sci., Trans. 7th (Madison, Wis., USA) 1:321–328.

Parr, J. F., and A. R. Bertrand. 1960. Water infiltration into soils. Advance. Agron. 12: 311–363.

Pawliw, J., and J. W. T. Spinks. 1957. Neutron moisture meter for concrete. Can. J. Tech. 34:503–513.

Peterson, D. F., O. W. Israelsen, and V. E. Hansen. 1952. Hydraulics of wells. Utah State Agr. Coll. Bull. 351. 48 p.

Philip, J. R. 1957. The theory of infiltration: 4. Sorptivity and algebraic infiltration equations. Soil Sci. 84:257–264.

Quirk, J. P., and R. K. Schofield. 1955. The effect of electrolyte concentration on soil permeability. J. Soil Sci. 6:163–178.

Rawlins, S. L. 1962. A commercially available, highly sensitive control for porous-plate apparatus. Soil Sci. Soc. Amer. Proc. 26:207.

Rawlins, S. L. 1964. Systematic error in leaf water potential measurements with a thermocouple psychrometer. Science 146:644–646.

Reeve, R. C., and D. Kirkham. 1951. Soil anisotropy and some field methods for measuring permeability. Amer. Geophys. Union, Trans. 32:582–590.

Richards, L. A. 1947. Pressure membrane apparatus—construction and use. Agr. Eng. 28:451–454.

Richards, L. A. 1949. Methods of measuring soil moisture tension. Soil Sci. 68:95–112.

Richards, L. A. [ed.] 1954. Diagnosis and improvement of saline and alkali soils. US Dep. Agr. Handbook 60. 160 p.

Richards, L. A., and R. B. Campbell. 1949. The freezing point of moisture in soil cores. Soil Sci. Soc. Amer. Proc. (1948) 13:70–94.

Richards, L. A., and G. Ogata. 1958. Thermocouple for vapor pressure measurement in biological and soil systems at high humidity. Science 128:1089–1090.

Richards, L. A., and G. Ogata. 1961. Psychrometric measurements of soil samples equilibrated on pressure membranes. Soil Sci. Soc. Amer. Proc. 25:456–459.

Richards, L. A., and P. L. Richards. 1962. Radial-flow cell for soil water measurements. Soil Sci. Soc. Amer. Proc. 26:515–518.

Richards, L. A., and L. R. Weaver. 1943. The sorption-block soil moisture meter and hysteresis effects related to its operation. J. Amer. Soc. Agron. 35:1002–1011.

Richards, S. J. 1939. Soil moisture content calculations from capillary tension records. Soil Sci. Soc. Amer. Proc. (1938) 3:57–64.

Rijtema, P. E. 1959. Calculation of capillary conductivity from pressure plate outflow data with non-negligible membrane impedance. Neth. J. Agr. Sci. 7:209–215.

Rose, C. W. 1960. Soil detachment caused by rainfall. Soil Sci. 89:28–35.

Rowe, P. B. 1940. Influence of woodland chaparral on water and soil in central California. US Dep. Agr. Flood Control Coordinating Comm. Misc. Publ. No. 1. 70 p.

Russell, M. B. 1942. Pore-size distribution as a measure of soil structure. Soil Sci. Soc. Amer. Proc. (1941) 6:108–116.

Shaw, M. D., and W. C. Arble. 1959. Bibliography on methods for determining soil moisture. Pennsylvania State Univ. Eng. Res. Bull. B–68. 152 p.

Shaw, B., and L. D. Baver. 1940. An electrothermal method for following moisture changes of the soil in situ. Soil Sci. Soc. Amer. Proc. (1939) 4:78–83.

Shaw, B. T., H. R. Haise, and R. B. Farnsworth. 1943. Four years' experience with a soil penetrometer. Soil Sci. Soc. Amer. Proc. (1942) 7:48–55.

Smith, R. M., and D. R. Browning. 1947. Some suggested laboratory standards of subsoil permeability. Soil Sci. Soc. Amer. Proc. (1946) 11:21–26.

Spauszus, S. 1955. Rapid determination of the moisture content of soil by determination of the dielectric constant. Zeits. f. Pflanzenern. Dung. Bodenk. 70:23–26.

Spinks, J. W. T., C. A. Lane, and B. B. Torchinsky. 1951. A new method for moisture determination in soil. Can. J. Tech. 29:371–374.

Stace, H. C. T., and A. W. Palm. 1962. A thin-walled tube for core sampling of soils. Australian J. Exp. Agr. Anim. Husb. 2:238–241.

Staple, W. J., and J. J. Lehane. 1962. Variability in soil moisture sampling. Can. J. Soil Sci. 42:157–164.

Stolzy, L. H., L. W. Weeks, T. E. Szuszkiewicz, and G. A. Cahoon. 1959. Use of neutron equipment for estimating soil suction. Soil Sci. 88:313–316.

Tanner, C. B., S. J. Bourget, and W. E. Holmes. 1954. Moisture tension plates constructed from alundum filter discs. Soil Sci. Soc. Amer. Proc. 18:222–223.

Tanner, C. B., and D. E. Elrick. 1958. Volumetric porous (pressure) plate apparatus for moisture hysteresis measurements. Soil Sci. Soc. Amer. Proc. 22:575–576.

Tanner, C. B., and R. J. Hanks. 1952. Moisture hysteresis in gypsum moisture blocks. Soil Sci. Soc. Amer. Proc. 16:48–51.

Taylor, S. A. 1955. Field determinations of soil moisture. Agr. Eng. 36:654–659.

Taylor, S. A., and J. E. Box. 1961. Influence of confining pressure and bulk density on soil water matric potential. Soil Sci. 91:6–10.

Taylor, S. A., D. D. Evans, and W. D. Kemper. 1961. Evaluating soil water. Utah Agr. Exp. Sta. Bull. 426. 67 p.

Van Bavel, C. H. M. 1961. Neutron measurement of surface soil moisture. J. Geophys. Res. 66:4193–4198.

Van der Marel, H. W. 1959. Rapid determination of soil water by dielectric measurement of dioxane extract. Soil Sci. 87:105–119.

Veihmeyer, F. J. 1929. An improved soil sampling tube. Soil Sci. 27:147–152.

Vries, D. A. de, and A. J. Peck. 1958. On the cylindrical probe method of measuring thermal conductivity with special reference to soils: II. Analysis of moisture effects. Australian J. Phys. 11:409–423.

Wexler, A. 1965. Humidity and moisture measurement and control in science and industry. 4 vol. Reinhold, New York.

section VI

Plant-Water Relations

VI

16 | Nature of Plant Water

H. B. CURRIER

University of California
Davis, California

I. INTRODUCTION

It is a truism, usually stated in monographs on plant-water relations, that the importance of water for the plant tends to be overlooked because there is so much of it present. Still, it has to be remembered that the content varies widely in higher plants from over 90% of the fresh weight for young actively growing plants to as low as 5% for some air-dry seeds.

Since it is usually assumed that life ceases if metabolism is brought to a standstill, this would mean that life, as we know it, would be impossible in the complete absence of water. It has, in fact, been suggested that even air-dry plants or plant parts (e.g. spores, seeds, etc.) cannot survive if their water content drops below a certain minimum value—5 to 6% according to some investigators (*see* Levitt, 1956). In at least some cases, this has proved to be quantitatively erroneous, for some seeds have been dried *in vacuo* at temperatures up to 100C until essentially no water could be detected, yet they germinated normally on reimbibing water (Harrington and Crocker 1918). The complete removal of water from living plants has, of course, not yet been achieved. Even if this were accomplished without injuring the plant, it would leave unanswered the philosophical question: Does life disappear with the last traces of water and reappear with imbibition, or can the cells be considered dormant though still alive?

Water is distributed throughout the plant to all of its organs and tissues. It is present normally both in the liquid state (protoplasts, cell walls) and in the gaseous state (intercellular spaces). The liquid system extends from cell-to-cell as a branched but continuous network from the root epidermis throughout the breadth and length of the root, stem, and leaf, and into the surface (epidermal) cells of the leaf. This continuity, important for xylem translocation of water and other aspects of water physiology is due (i) to the association of water molecules by hydrogen bonding; (ii) to open connecting strands (plasmodesmata) between protoplasts, not blocked by lipoid membranes; and (iii) to the continuous system of hydrophilic cell walls with microcapillary channels.

At temperatures below the freezing point of the plant, water may also be present in the solid state, but such ice normally forms a discontinuous system confined to the intercellular spaces as long as the plant is alive (Levitt, 1956)

II. ROLES OF PLANT WATER

In general, and considering the whole organism, water in plants performs the following roles:

1) Water is a structural constituent held in cells by osmotic and imbibitional

forces, its movement constrained by membranes, less by cell walls, except for walls rendered hydrophobic by cutin, wax, and suberin. It adds bulk, affects form through turgor development, and is responsible for many of the physical-structural characteristics of the plant.

2) Water particularly hydrates enzymes and thus affects metabolism.

3) Water solubilizes substances, permitting metabolic reactions to occur, which can proceed only in an aqueous environment. It is true that specific reactions may actually occur at interfaces between solid surfaces and water, but even then the water must be considered the medium, since the reactions do not occur or occur at an extremely slow rate at solid-air interfaces in the absence of water, e.g. in dry seeds.

4) Water is the medium in which all substances, except long-distance transport of gases, must be moved from the region of supply (the soil or other cells) to the region of use, the protoplasm. The movement is not only in xylem and phloem but cell-to-cell through both symplast (interconnected system of protoplasts) and apoplast (interconnected system of cell walls). Though the cell water is in the liquid state, its freedom of movement is somewhat restricted due to the gel nature of the cell wall, the protoplasm, and in some cases even the vacuole. This gel structure interferes with bulk flow (i.e. with mass movement) of the water but it has essentially no effect on its diffusibility. In the case of protoplasm, the gel structure may also fail to stop flow, as shown by cytoplasmic streaming, due presumably to the complex sol-gel transformations that may take place or to the existence of protoplasmic components with different properties.

5) With a high specific heat and high heat of vaporization, water acts as a buffer against high- or low-temperature injury. In transpiration it cools leaves exposed to the sun's heat.

6) Water is both a metabolic reactant (e.g., photosynthesis) and product (e.g., polymerization).

III. WATER IN THE CELL COMPONENTS AND SPECIFIC ROLES

In mature plant cells, water is distributed between cell wall sap, protoplasmic sap, and vacuolar sap. Wall water, held mainly by imbibitional forces, is relatively pure. Protoplasmic water is held both by solutes and imbibants (colloidal substances). The potential of the vacuolar water is reduced by solutes such as sugars and organic acids, but sometimes also by colloids, e.g., pectins.

In laboratory practice it is difficult to determine precisely the relative amounts of water in the three cell components identified above (Crafts et al., 1949). But

Table 16–1. Amount of air space in various organs

Organ	Species	% by vol.	Reference
Leaf, turgid	Ligustrum japonica	22–26	Katayama, 1962
Leaf, flaccid	Ligustrum japonica	27–37	Katayama, 1962
Leaf	Vinca rosea	33.2	Turrell, 1965
Root, 0.5–1.0 cm from tip	Oryza sativa	3.6	Katayama, 1962
Root, 2.0–2.5 cm from tip	Oryza sativa	10.9	Katayama, 1962
Tuber	Solanum tuberosum	1.3	James, 1953, p. 145
Bulb	Allium cepa	10.0	Currier, unpubl.

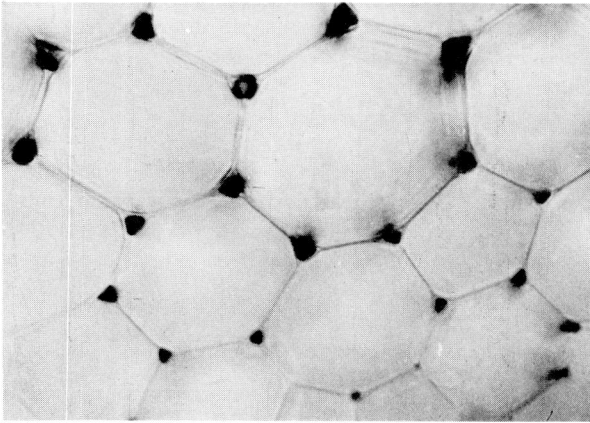

Fig. 16–1. Cross section of young living pea stem showing black-appearing intercellular air. ×750.

in general, the greater part of water in mature cells is stored in vacuoles, since the volume of this component is usually greater than the combined volume of the other two. In meristematic cells, protoplasmic water is the largest fraction. In exceptional instances, e.g. certain terrestrial mosses and leaves such as *Eucalyptus globulus* (noted in a later section), the cell wall constitutes the largest relative volume of the cell and wall water would predominate. Consideration will now be given to water in the individual cell components.

A. Intercellular Spaces

The fact that gas-filled intercellular channels exist as structural and functional components of plants is sometimes not sufficiently recognized. One of the gases is water vapor. It is the confined gas that produces opaqueness in plant tissues, provides buoyancy in hydrophytes, and accounts for a certain amount of structural flexibility in stems and other organs.

The student in elementary botany soon finds, to his disadvantage, that the intercellular spaces, with gas cell wall interfaces, appear black in the microscope (Fig. 16–1), making observation of fresh tissues difficult. Often it is helpful to vacuum infiltrate tissues prior to observation.

The proportionate volume of this gas storage and two-way transport system varies from tissue to tissue and from organ to organ (*see* Table 16–1). Unger (1854) found in leaves of 44 species a range from 7.7% in *Camphora officinalis* to 71.3% in *Pistia texensis*. The volume of air space in phloem, xylem, and meristematic tissue is extremely small, compared to its volume in pith, cortex, and mesophyll. Spaces are lacking in epidermal and cork tissue except for stomata, hydathodes, and lenticels.

In addition to the volume of air spaces, their surface area is of importance (Turrell, 1965). Ratios of internal surface to external surface in leaves was found by Turrell (1936) to be as high as 31.3 in *E. globulus;* however, values generally were lower in other plants.

Higher plants vary in respect to the continuity of their intercellular space system. Redies (1962) distinguished between those species having continuous (homobar) and those with unconnected (heterobar) systems.

The air-space system opens to the upper atmosphere via stomata of leaves, and, to a lesser extent, via lenticels in the bark of older stems. The terminal portion of the transpiration stream consists of the passage of water vapor through the spaces in the mesophyll, ending in passage through the stomatal openings. In the root, openings to the soil environment occur to some extent through lenticels of older roots. Normally the spaces do not open from young roots; only if there were fresh wounds would this happen and even then only for a short time because of wound healing.

1. WETTABILITY OF THE SPACE SURFACES

Polar liquids normally do not wet the intercellular wall surfaces (Häusermann, 1944). This explains why liquid water or aqueous solutions rarely exist in the spaces. Exceptions include hydathodes, glandular tissue, and wound tissue. In addition, water may enter stomata and move into the spaces during a hard rain or forceful watering; as a consequence of disease, treatment with toxic chemicals, or cold injury, cell sap may leak into the spaces giving the tissue a water-soaked appearance.

The feature of unwettability is important to the plant, in that flooding of the spaces at times of water abundance is avoided. Such flooding would interfere markedly with gaseous interchanges of cells. If the relative humidity is near 100% it is possible that condensation would occur with the lowering of temperature. This water doubtless is deposited as isolated droplets and would be evaporated subsequently in transpiration or absorbed by the cells, depending on the direction of the energy gradient.

The interfacial tension between the lipoid surface and water can be reduced experimentally by adding a small amount of a surface-active agent. Aqueous leaf sprays penetrate open stomata and move in the intercellular spaces of the leaf if an efficient surfactant is present in the solution (Dybing and Currier, 1961).

2. TRANSPORT IN THE INTERCELLULAR SPACES

For the present discussion the problem is in understanding the behavior and functional importance of the water vapor in the system. The vapor has a density of about $1/1,330$ that of liquid water at 20C, so that it comprises an insignificant proportion of the total water content on a weight basis. Considering a plant with 80% water, 10% by volume of air space, and water-saturated air in the spaces at 20C, only about $1/50,000$ of the plant water would be present as vapor (Handbook Chem. Phys., 1960, table p. 2486).

The important consideration is that water vapor escaping into the spaces from the protoplasts is eventually transpired, tending to increase water stress. Diffusion is doubtless the mechanism involved, but flow effects due to stems bending, etc. may exert an influence. Only under conditions of extreme water deficit would atmospheric water vapor move into leaves. The amounts entering are generally very small with little being transported to the remainder of the plant (Slatyer, 1956; Vaadia and Waisel, 1963). Even here there is no evidence for movement predominantly via the intercellular spaces.

B. The Cell Wall

1. THE WALL AS HYDROPHILIC IMBIBANT

Plant cell walls (excluding cutinized and suberized portions) are hydrophilic gels composed of a submicroscopic framework embedded in an amorphous matrix. The framework of microfibrils approximately 25 mμ in diameter, is mainly of cellulose (β-1,4-D-glucan), but other polysaccharides such as xylans and mannans usually are present (Frey-Wyssling, 1959; Roelofsen, 1959). The amorphous sub-stances, not forming microfibrils, include pectins (polygalacturonic acid and de-rivatives), lignin, certain hemicelluloses (various polysaccharides, uronic acids), gums, mucilages, and other materials. The relative proportions of cellulose, hemi-celluloses, and pectins are quite variable from species to species (Setterfield and Bayley, 1961). In tissues exposed to the atmosphere, mainly epidermis and cork, cutin and suberin provide resistance to water movement.

The microfibrils are arranged in various patterns. In primary walls the arrange-ment resembles an interwoven network. Secondary walls, with a greater proportion of cellulose and lignin, possess layers of microfibrils arranged in dense parallel packing. The framework is so arranged that irregular pores, on the average about 10 mμ in diameter, are formed. These microcapillaries are large enough to permit the ready movement of solute particles at least as large as sucrose molecules. Even within microfibrils, pores sufficiently large (about 1 mμ) to permit the entrance of water are believed to exist and to cause swelling of the fibrils (North-cote, 1958). In most instances the cell wall is to be considered a permeable semi-rigid membrane surrounding the protoplast. Cell walls constitute the "free space" of plant tissue (*see* Levitt, 1957) in which ions and molecules can move freely by diffusion or by flow.

2. COMPOSITION OF WALL SAP

The nature of the fluid present in the walls of living cells can only be surmised, since there is no satisfactory method for extracting the sap. Attempts involve such cellular destruction that no confidence can be placed in the results. Tests using dead tissue may be meaningless. While there has been considerable study of the water-absorbing and water-retaining properties of wood, cotton, and other non-living cell wall systems, in this chapter attention is focused on living plants exclusively. In general, the osmotic concentration of the wall sap must be low, especially in actively growing plants, because of the accumulative capacity of the protoplasts. Plants growing in saline soils may be exceptions. In the absorbing root it is likely that some of the mineral ions being absorbed are present in the wall. Here the concentration of solutes usually must be no greater than that in the root medium; however, leakage of ions from cells at times might increase the concentration.

3. FORCES RESPONSIBLE FOR RETENTION OF WATER IN WALLS

Hydrogen bonding constitutes the principal mechanism by which water hydrates wall substances. It accounts for adsorption due to polar groups, mainly OH but also COOH and C=O to a lesser extent; and, in a more gross manner, it accounts for capillary attraction and filling of the microcapillary spaces in the wall. There are no clear-cut lines between these; when the energy content is plotted against these effects that reduce water potential, a smooth curve is evident (*see* Crafts et al., 1949).

The capillary water, which constitutes most of the cell wall water, is relatively free, mechanically confined. Some of it, in the larger pores, is mobile. Water adsorbed on surfaces of micelles of dry wood was found to be only about 8% of the total (Stamm, 1944). The more tightly surface-held water is considered "bound" (*see later* in this chapter). Walls dried at 105C are extremely hygroscopic; Kollmann (1951) reported an imbibition pressure of 2,000 kg/cm^2 under these conditions. Hydrophily of wall constituents increases: lignin < cellulose < hemicelluloses (Frey-Wyssling, 1959); that of cutin and suberin would be placed before lignin.

4. ROLES PLAYED BY CELL WALL WATER

It has been suggested that water in the apoplast replaces the aqueous environment of the earliest organisms. This water acts as a translocation medium, as a "buffer" opposing rapid hydration change of the protoplast, and as an initial protection against protoplasmic freezing and desiccation.

Probably the role of the wall as a translocation pathway has been underestimated. Strugger (1939) earlier stressed the apoplast as the path of the "extrafasicular transpiration stream;" an important recent reference is Weatherley (1963). A rapid movement in cell walls following dipping of leaves into water containing a fluorescent dye tracer was demonstrated by Dybing (C. D. Dybing, 1958. Foliar penetration by chemicals. *Unpubl. Ph.D. Thesis, Univ. of California, Davis*). (*See* Dybing and Currier, 1961.)

The concept that cell wall water acts as a buffer to counteract rapid change in hydration has been presented recently by Gaff and Carr (1961). They show that such water constitutes as much as 40% of the total leaf water in *E. globulus* and suggest that this water may be a factor in drouth resistance.

Freezing of plant tissues usually begins in cell walls. The heat of fusion released by ice formation decreases the likelihood of freezing of the protoplast. Further, the growth of ice crystals in the intercellular spaces, slowly drawing water from the protoplast, prevents protoplasmic freezing which is usually fatal.

C. The Protoplast

Water in the protoplasm, in addition to hydrating proteins, inorganic salts, and other diverse substances, is compartmentalized by systems of living membranes. The two principal ones are the plasmalemma at the wall-protoplast interface, and the tonoplast at the protoplast-vacuolar boundary. These membranes are permeable to water but not so permeable as the cell wall. Plastids and mitochondria are also limited by semipermeable membranes and constitute in themselves osmotic systems. These particulates must have a much smaller water content than the hyaloplasm, since lipids account for such a large fraction of their dry matter.

Water in the protoplasm changes constantly with metabolic rates and environmental conditions. Permeability of the membranes fluctuates. Particulates change volume when supplied ATP (Packer, 1963; Lehninger, 1964), probably due to some energized effect on the protein components (e.g., folding, unfolding); however, the possibility of membrane changes cannot be excluded. Salts and other molecules, including sugars and amino acids, are transported actively across the membranes, causing water to follow. There is no conclusive evidence that water itself is actively transported in plant cells (Levitt, 1947, 1953, 1954; Ordin et al., 1956). The condition of water in the transpiration stream (apoplast) will change in response to evaporative demand and absorption rates. Water in the protoplasm tends toward a steady-state equilibrium with extraprotoplasmic water in the plant and will fluctuate with similar responses to the environment.

Vacuolarcontraction is an example of an intracellular shift in water content. In this phenomenon (Fig. 16–2), the cytoplasm swells, and the vacuole shrinks correspondingly in volume. The fact that water can come from the vacuole is demonstrated by the reaction in cells immersed in paraffin oil. The mechanism, not well understood, may involve a redistribution of salt, an increase in cytoplasmic matric potential, or both (Crafts et al., 1949, p. 125).

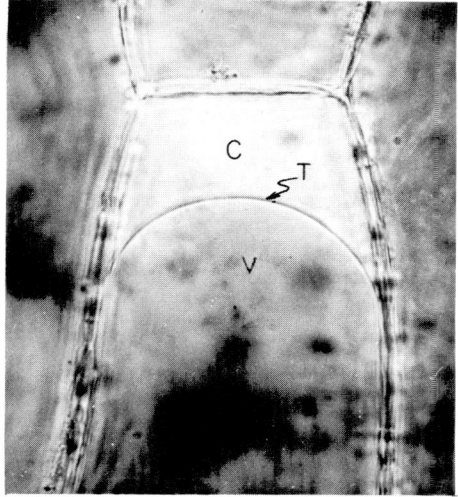

Fig. 16–2. Vacuolarcontraction in one end of a living epidermal cell of an onion bulb scale. (*C*) swollen cytoplasm; (*T*) toroplast; (*V*) vacuole containing anthocyanin. ×1300.

IV. ENERGY STATUS OF PLANT WATER

A. Terminology

The energy state of water is expressed by water potential ψ,[1] defined as the difference between the partial specific Gibbs free energy (chemical potential) of the water in the system and that of pure free water at the same temperature (Slatyer and Taylor, 1960; Taylor and Slatyer, 1961; Slatyer, 1962; Aslyng, 1963). This is expressed by the relation $\psi = \mu_w - \mu_w{}^\circ / V_w$, where V_w is the partial molal volume of water. The term water potential replaces the expression diffusion pressure deficit (DPD) that has been used commonly by physiologists in the USA. Osmotic pressure (OP) is designated as Π and turgor pressure (TP) as P. It is clear that ψ values will be negative, and the expression DPD = OP – TP is replaced by $\psi = P - \Pi$.

B. Factors Affecting Water Potential ψ

Water moves into, within, and from plant to atmosphere along free energy gradients, from higher to lower ψ. In cells the ψ value of water is determined by solutes (sugars, acids, inorganic ions, etc.); imbibants (matric substances such as proteins, polysaccharides, etc.); turgor (either positive or negative); and, to a

---

[1] Water potentials in this chapter are expressed as *energy per unit volume* while in chapter 12 they are expressed as *energy per unit mass*.

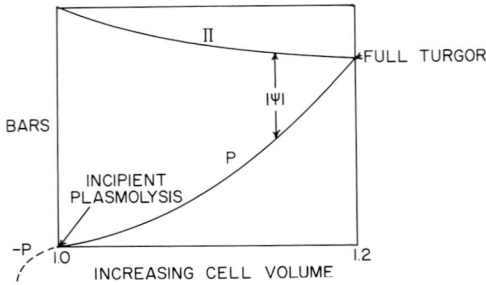

Fig. 16–3. Integration of cell water quantities with changing cell volume. Π = osmotic pressure, P = turgor pressure, $-P$ = tension, Ψ = water potential. Here Ψ is used in the absolute sense $|\Psi|$ because of change of sign, also because it represents a difference between the chemical potential of water in two systems, a difference which becomes smaller as turgor pressure increases.

lesser extent, by temperature. Thus there are several components which contribute to the resultant water potential, as expressed by $\Psi = \Psi_s + \Psi_m + \Psi_p + \Psi_t$, where Ψ_s is osmotic potential, or the contribution of solutes to the water potential; Ψ_m is matric potential, the influence of imbibants; Ψ_p is pressure potential, the influence of turgor pressure; and Ψ_t the influence of temperature.

The so-called cell water quantities, Ψ, P, and Π, vary in magnitude with change in cell volume, as shown in Fig. 16–3. Secondarily influencing these primary factors which determine Ψ, are metabolism, active solute transport, growth, and hysteresis. Contributing factors such as electroosmosis and membrane pore pressure gradients (Ray, 1960), simply affect the rate at which the system comes into equilibrium. Hydrostatic pressure and intercellular pressure can both be included in the turgor pressure term. Wall pressure, as distinct from turgor pressure, is a useless concept and should be discarded.

Under steady-state conditions $\Psi_{\text{vacuole}} = \Psi_{\text{cytoplasm}} = \Psi_{\text{cell wall}}$. Gradients arise from one or more of the following factors or events: (i) *solutes*—loss or gain by passive or active transport, synthesis or hydrolysis of compounds, ionic dissociation and association; (ii) *imbibants*—change in matric potential due to pH; synthesis, hydrolysis, hysteresis; and (iii) *turgor pressure*—transpiration, changes in wall elasticity, plasticity, and cell enlargement. In each case a loss of water will decrease the Ψ of the water remaining in the cell; a gain will have the opposite effect. The proportionate influence of solutes, imbibants, and turgor, differs from cell wall to protoplasm to vacuole, from cell to cell, and in tissues and organs.

The turgor pressure component makes the water relations of plant cells uniquely different from those of animal cells. Possession of walls having considerable tensile strength permits the development of as much as 50 bars of positive pressure. Thick compact walls in vessels and tracheids, reinforced by various forms of wall sculpturing, allow many bars of tension ($-P$) to be maintained. Even some parenchyma cells, e.g., in conifer leaves, are thought to be able to exist under such tension levels. Scholander et al. (1965) measured xylem tensions of 4 to 5 atm for species growing in a damp forest to 80 atm in the desert. There were diurnal fluctuations of 10 to 20 atm.

In vessel sap of transpiring plants Ψ is lowered by tension and a small amount of solute. In phloem sap it is a solute effect, mainly due to sucrose. In meristematic cells, lacking a large central vacuole, turgor is relatively low, and Ψ is influenced largely by solutes and imbibants, with the latter proportionately more important. This has been shown for wilting squash plants (*Curcurbitia pepo* L.) where younger leaves and buds, with a lower Π and higher colloid content, are still turgid when older leaves, with higher Π and less colloid, wilt (Stocking, 1945).

C.   Free vs. Bound Water

Due to the presence of hydrophilic colloids, the energy of some of the water molecules is thought to be reduced to such a low level that other properties are also changed. This fraction has been called bound water. One major difference suggested by Gortner (1938) is an inability to act as a solvent. Some of it has even been thought of as existing in the crystalline form (Gortner, 1938; Klotz, 1958). Nevertheless, there is no sharp line between free and bound water and all gradations between the two exist. As water is removed from cells, the remaining water is held more and more tightly. A separation point on the desorption curve at which the activity is sufficiently low (e.g. 0.8) may be selected, below which the water could be considered bound, but the curve is smooth and selection of the point is arbitrary. Bound water is more abundant in hardened plants and is considered a factor in frost and drouth resistance (Levitt, 1956, 1959, 1962; Scarth and Levitt, 1937).

D.   Energy of Intercellular Space Water

The degree to which the vapor pressure (VP) of the water vapor in the intercellular spaces approaches a steady-state equilibrium with wall water depends both on the part of the plant in question and on the rate of transpiration. In organs other than transpiring leaves and stems, such equilibrium must be approached closely. Walter (1963), generalizing and with mesophytes in mind, believes that usually the hydrature (relative humidity) of the spaces is 99% or above, corresponding to $\Psi = -13$ bars or more. In transpiring leaves of water-stressed plants, on the other hand, and especially near the stomatal aperture, the VP deficit may be appreciable (Klemm, 1956). One may assume that xerophytic leaves under high water stress could have a Ψ of -100 bars; this corresponds to a hydrature of 93% at 20C in the spaces (see Crafts et al., 1949, p. 56), assuming equilibrium between the cells and spaces. If no such equilibrium exists, the hydrature would be lower. This is Klemm's view, that with increasing stress, the Ψ would drop to a value much lower than the Ψ of the protoplast water, hence the real barrier in transpiration under this condition is the mesophyll cell wall. Walter considers that Klemm's report of a hydrature as low as 60%, equivalent to 700 atm, needs confirmation. Better understanding of the matter must await the development of improved measuring methods.

The energy content of the space water vapor is increased markedly by a rise in temperature and decreased by a temperature drop (Table 16–2). If the intercellular air could be assumed to be saturated with water, a rise in temperature

Table 16–2. Effect of increase in temperature (T) on vapor pressure (VP) gradients

Water in intercellular air*			Water in external air			VP gradient
T	RH	VP	T	RH	VP	
° C	%	mm Hg	° C	%	mm Hg	mm Hg
20	100	17. 55	20	60	10. 53	7. 02
30	100	31. 85	30	34. 2	10. 89	20. 96

* T = temperature;   RH = relative humidity;   VP = vapor pressure.

from 20 to 30C could result in a three-fold increase in VP gradient, as the data show, with resultant effects on transpiration rate.

E. Energy of Cell Wall Water

Because of a steady-state equilibrium between the various cellular components, the Ψ of the water in the wall closely approaches that of the water in the protoplast. The energy drop, however, is in the direction of the wall because of evaporation into the spaces. At full turgor Ψ will closely approximate that of pure water. At zero turgor it will decrease to approximate the Π of the protoplast numerically, but with a negative sign.

Walter (1963) discusses the changes in cell wall water due to turgor fluctuation. Changes in water content are small in most instances. The stretched wall of a fully turgid cell would be expected to have less water than the wall of a flaccid cell. He reasons that stretching reduces the micropore space, which in turn reduces the total water content somewhat. Conversely, when the P decreases to flaccidity, the wall becomes thicker, the matric potential greater, and the total water content correspondingly increases. But now the wall water is in equilibrium not with cell sap with $\Psi = 0$, but with the osmotic value of the cell sap which counteracts any increase as turgor decreases.

F. Examples of Changing Ψ Gradients

Two examples of changing internal water gradients as influenced by environmental factors will be discussed. The first refers to a steady-state equilibrium in a plant at predawn when the transpiration rate is very low. After sunrise, with the opening of stomata and increasing transpiration, loss of water lowers Ψ first in the wall, then in the protoplasm and then in the vacuole. The turgor is reduced but this is counteracted by the lower Ψ (greater water absorbing power) which sets up a gradient that moves water from the adjacent cells into the transpiring cells. The gradient extends to the vessels, and the tension produced extends via the continuous xylem sap to the root; this increases the gradient from soil to xylem, thereby enhancing absorption. Parenchyma cells associated with vessels throughout the root-stem-leaf tend now to lose water to the vessels, because the tension in the vessel lumina has decreased the Ψ of the water in the enclosed sap. In late afternoon the transpiration rate has decreased, tension in the vessels is lower, and the events described occur in reverse order. Diurnal shrinkage and swelling of stems, as measured by a dendrograph, are reflections of the processes referred to.

The second example is similar to the first, but it involves the steady-state equilibrium between water in sieve-tube sap and vessel sap in a stem, involving again the effect of transpiration. The distance between mature functioning sieve elements and vessels across the cambium can be as little as 10 μ; in monocots, with no cambium, the distance could be less. We can assign a relatively high Π value to the sieve tube, 15 bars; and a low value, 1 bar, to the vessel. Thus at steady state there is a steep Π gradient but no Ψ gradient. It is clear, as here, that sometimes Π gradients are useless in predicting water movement. Referring to Fig. 16–4, when the transpiration rate is minimal, before dawn, cell water

Plant not transpiring
$$\begin{cases} \Pi = & 1 & 6 & 12 \\ P = & 0 & 5 & 11 \\ \psi = & -1 & -1 & -1 \end{cases}$$

Plant transpiring
$$\begin{cases} \Pi = & 1 & 6 & 12 \\ P = & -3 & 2 & 8 \\ \psi = & -4 & -4 & -4 \end{cases}$$

DIAGRAMMATIC 3-DIMENSIONAL VIEW OF A DICOT HERBACEOUS STEM

A. EPIDERMIS
B. COLLENCHYMA
C. PARENCHYMA CELLS
D. ENDODERMIS
E. PARENCHYMA CELLS
F. SCLERENCHYMA FIBERS
G. SIEVE TUBES AND COMPANION CELLS

H. CAMBIUM
I. PITTED VESSEL
J. XYLEM FIBERS
K. SPIRAL VESSEL
L. ANNULAR VESSEL
M. PITH
N. RAY

M   L K J   I   H   G   F   E   D C B A   S
XYLEM   PHLOEM   CORTEX
              PERICYCLE

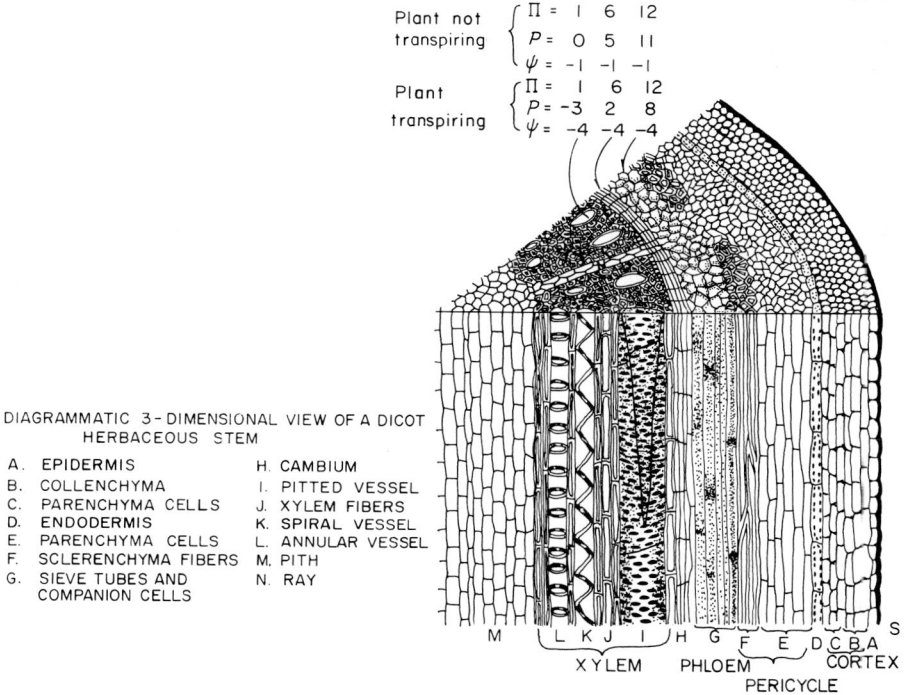

Fig. 16–4. Theoretical steady-state values of II, P, and Ψ in vessels, cambium, and sieve tubes in the absence of transpiration (top set of figures) and during transpiration (lower set of figures). Non-transpiration values of P and of Ψ may be slightly greater or less depending on height above ground, root pressure activity, and diurnal time course. A radial-cross section of stem is diagrammatically shown. Drawing by Wilson Stewart. Reprinted courtesy Holt, Reinhart and Winston, Inc., New York, from Fuller, The Plant World, 1941.

quantity values indicated by the top figures may be in effect. A steady-state water potential of –1 bar is shown for all tissues. When transpiration increases, with increasing light, tension in the vessels becomes greater. This sets up a gradient causing water to move from cambium to xylem, phloem to cambium, cortical parenchyma to phloem. When tension levels at –3 bars, values of II, P, and Ψ for vessels, sieve tubes, and cambium are those shown in the lower set of figures. There is now a new steady-state equilibrium value of –4 bars for Ψ.

The examples show that changes in transpiration rate have a pronounced effect on movement of water between vascular tissues and parenchyma, affecting all tissues of leaf, stem, and root.

In addition to transpiration effects in lowering Ψ, any influence that increases root resistance or permeability to water will also increase tension and lower Ψ. Such influences would include lowering of temperature and anoxia in the root environment. Gradients can also result from solute changes in sieve tubes due to "source" and "sink" activities. Active transport of assimilate from mesophyll (source) into the sieve element increases II, and water is absorbed, increasing P. In the root, when sucrose is moved out into parenchyma (sink) and converted to starch, the water freed from solute influence increases the Ψ of sieve tube sap. Such extra water will move along Ψ gradients in any direction; there is evidence that it can be absorbed by the transpiration stream, an aspect of the plant's circulatory system.

LITERATURE CITED

Aslyng, H. C. 1963. Soil physics terminology. Int. Soc. Soil Sci. Bull. 23:1–4.

Crafts, A. S., H. B. Currier, and C. R. Stocking. 1949. Water in the physiology of plants. Ronald Press, New York. p. 64, 70–71, 125.

Dybing, C. D., and H. B. Currier. 1961. Foliar penetration by chemicals. Plant Physiol. 36:169–174.

Frey-Wyssling, A. 1959. Die pflanzliche Zellwand. Springer Verlag, Berlin. p. 125–127, 294.

Gaff, D. F., and D. J. Carr. 1961. The quantity of water in the cell wall and its significance. Australian J. Biol. Sci. 14:299–311.

Gortner, R. A. 1938. Outlines of biochemistry. John Wiley, New York. p. 301.

Handbook of Chemistry and Physics. 1960. Chemical Rubber Publ. Co., Cleveland, Ohio. p. 2486.

Harrington, G. T., and W. Crocker. 1918. Resistance of seeds to desiccation. J. Agr. Res. 14:525–532.

Häusermann, E. 1944. Über die Benetzungsgrösse der Mesophyllinterzellularen. Schweiz. bot. Ges. Ber. 54:541–578.

James, W. O. 1953. Plant respiration. Clarendon Press, Oxford, England. p. 145.

Katayama, T. 1962. Investigations on measuring methods of intercellular spaces of several plants. Crop Sci. Soc. Japan, Proc. 30:150–154.

Klemm, G. 1956. Untersuchungen über den Transpirationswiderstand der Mesophyll-membranen und seine Bedeutung als Regulator für die stomatäre Transpiration. Planta 47:547–587.

Klotz, I. M. 1958. Protein hydration and behavior. Science 128:815–822.

Kollmann, F. 1951. Technologie des Holzes. Vol. 1. Springer Verlag, Berlin. 1050 p.

Lehninger, A. L. 1964. The mitochondrion. W. A. Benjamin, New York. p. 180–204.

Levitt, J. 1947. Thermodynamics of active water absorption. Plant Physiol. 22:514–525.

Levitt, J. 1953. Further remarks on the thermodynamics of active (nonosmotic) water absorption. Physiol. Plant. 6:240–252.

Levitt, J. 1954. Steady state versus equilibrium thermodynamics in the concept of active water absorption. Physiol. Plant. 7:592–594.

Levitt, J. 1956. The hardiness of plants. Academic Press, New York. p. 22–26, 71, 143.

Levitt, J. 1957. The significance of apparent free space (AFS) in ion absorption. Physiol. Plant. 10:882–888.

Levitt, J. 1959. Bound water and frost hardiness. Plant Physiol. 34:674–677.

Levitt, J. 1962. A sulfhydryl-disulfide hypothesis of frost injury and resistance in plants. J. Theo. Biol. 3:335–391.

Northcote, D. H. 1958. The cell walls of higher plants: Their composition, structure, and growth. Biol. Rev. 33:53–102.

Ordin, L., T. H. Applewhite, and J. Bonner. 1956. Auxin-induced water uptake by *Avena* coleoptile sections. Plant Physiol. 31:44–53.

Packer, L. 1963. Structural changes correlated with photochemical phosphorylation in chloroplast membranes. Biochem. et Biophys. Acta 75:12–22.

Ray, P. M. 1960. On the theory of osmotic water movement. Plant Physiol. 35:783–795.

Redies, H. 1962. Über "homobare" und "heterobare" Interzellularsysteme in höhere Pflanzen. Beitr. Biol. Pflanzen 37:411–445.

Roelofsen, P. A. 1959. The plant cell wall. Handbuch der Pflanzenanatomie. Borntraeger, Berlin. p. 54–57.

Scarth, G. W., and J. Levitt. 1937. The frost-hardening mechanism of plant cells. Plant Physiol. 12:51–78.

Scholander, P. F., H. T. Hammel, E. D. Bradstreet, and E. A. Hemmingsen. 1965. Sap pressure in vascular plants. Science 148:339–346.

Setterfield, G., and S. T. Bayley. 1961. Structure and physiology of cell walls. Ann. Rev. Plant Physiol. 12:35–62.

Slatyer, R. O. 1956. Absorption of water from atmospheres of different humidity and its transport through plants. Australian J. Biol. Sci. 9:552–558.

Slatyer, R. O. 1962. Internal water relations of higher plants. Ann. Rev. Plant Physiol. 13:351–378.

Slatyer, R. O., and S. A. Taylor. 1960. Terminology in plant and soil water relations. Nature 187:922–924.

Stamm, A. J. 1944. Surface properties of cellulosic materials. p. 449–550. In L. E. Wise [ed.] Wood chemistry. Reinhold Book Co., New York.

Stocking, C. R. 1945. The calculation of tensions in *Cucurbita pepo*. Amer. J. Bot. 32: 126–134.

Strugger, S. 1939. Die lumineszenzmikroskopische Analyse des Transpirationsstromes in Parenchymen. Biol. Zbl. 59:409–442.

Taylor, S. A., and R. O. Slatyer. 1961. Proposals for a unified terminology in studies of plant-soil-water relationships. UNESCO Arid Zone Res. 16:339–349.

Turrell, F. M. 1936. The area of the internal exposed surface of dicotyledon leaves. Amer. J. Bot. 23:255–264.

Turrell, F. M. 1965. Internal surface-intercellular space relationships and the dynamics of humidity maintenance in leaves. In Arnold Wexler [ed.] Humidity and moisture: II. Applications. Reinhold Book Co., New York. p. 39–53.

Unger, F. 1854. Beiträge zur Physiologie der Pflanzen: I. Bestimmung der in den Interzellulargängen der Pflanzen enthaltenen Luftmenger. Kon. Akad. Wiss. Wien Math.-Natw. 12:367–378.

Vaadia, Y., and Y. Waisel. 1963. Water absorption by the aerial organs of plants. Physiol. Plantarum 16:44–51.

Walter, H. 1963. Zur Klärung des spezifischen Wasserzustandes in Plasma und in der Zellwand bei der höheren Pflanze und seine Bestimmung. Ber. deut. bot. Ges. 76:40–71.

Weatherley, P. E. 1963. The pathway of water movement across the root cortex and leaf mesophyll of transpiring plants. p. 85–100. In A. J. Rutter and F. H. Whitehead [ed.] The water relations of plants. John Wiley. New York.

17

Water Absorption, Conduction, and Transpiration

PAUL J. KRAMER

Duke University
Durham, North Carolina

ORLIN BIDDULPH

Washington State University
Pullman, Washington

FRANCIS S. NAKAYAMA

US Water Conservation Laboratory, ARS, USDA
Phoenix, Arizona

Plant water relations include three interrelated processes: water absorption, ascent of sap, and water loss by transpiration. Another physiologically important area of plant water relations is the plant water balance, which is controlled by the relative rates of water absorption and water loss. This chapter deals briefly with all four aspects of plant water relations.

The water in plants forms a continuous system through the water-saturated cell walls, termed the hydrodynamic system by Meyer (1956), and any disturbance of the free energy or water potential status in one part of this system is transmitted throughout the plant. The continuity of water in plants not only provides a pathway for movement of solutes, but also plays an important role in linking water absorption to water loss.

I. THE ABSORPTION OF WATER [1]

The absorption of water will be dealt with under three major headings: (A) the absorbing system, (B) the mechanisms involved in absorption, and (C) factors affecting absorption.

A. The Water Absorbing System

The role of roots in water absorption was dealt with at length in articles by Kramer (1956b) and Slatyer (1960) and therefore can be treated rather briefly here. Water also is absorbed through aerial structures (Gessner, 1956; Stone, 1957), but such absorption constitutes such a small proportion of the total that it is of little importance for most cultivated crops.

1. THE ABSORBING ZONE

Studies on absorption of water by young roots suggest that most rapid absorption

---

[1] Prepared by Paul J. Kramer.

occurs a short distance behind the tip where the xylem is well differentiated, but suberization has not progressed far enough to seriously reduce permeability. This also is the region where most extensive development of root hairs usually occurs. The importance of root hairs may have been overemphasized because they would be of little use in soils near field capacity and they usually collapse in dry soils.

Although young roots constitute the primary absorbing surface of rapidly growing annuals they constitute a relatively small fraction of the total surface in most perennial plants. During dry weather and in the winter few or no unsuberized root tips are found on the root systems of most plants and water absorption must occur through the suberized roots. Although suberized roots are much less permeable than unsuberized roots their large surface probably makes them a relatively important absorbing surface.

2. RADIAL MOVEMENT OF WATER

The radial pathway of water movement from root surface to xylem is of considerable interest. It might move from vacuole to vacuole, through the cytoplasm of adjacent cells, through the cell walls, or through the intercellular spaces. Russell and Woolley (1961) estimated from the relative permeabilities and cross sectional areas that most of the water movement in roots could occur through the cell walls, at least as far as the endodermis. The endodermis is of particular interest because suberization of the radial walls of endodermal cells tends to make them impermeable to water, hence at this point water and solutes are believed to pass through the protoplasts rather than around them.

The endodermis is pierced by branch roots and eventually is split off during secondary growth. It usually is assumed that the principal barrier to the entrance of water and solutes into roots which have undergone secondary thickening is the layer of suberized tissue which occurs in the outer surface. The cambial layer also may be a barrier to salt movement. As mentioned earlier this region is not as impermeable as once was supposed and considerable water and salt must be absorbed through suberized regions of root systems.

3. ROOT GROWTH

The most important feature of the root systems of annual crops is their rapid extension into previously unoccupied soil. It is this continuous invasion of new soil masses which enables plants to continue growing for days or weeks without rain or irrigation. Movement of water through the soil toward roots is too slow to maintain a satisfactory water balance in transpiring plants (Gardner, 1960; Gardner and Ehlig, 1962; Philip, 1957). Root extension and continual branching and rebranching bring the roots of crop plants into contact with the water in hitherto untouched soil and keep plants growing long after a static root system would have exhausted all the available water.

B. Mechanisms Involved in Water Absorption

It is assumed in this discussion that water moves into and through plants along gradients of decreasing water potential Ψ. Attempts have been made to involve nonosmotic and electro-osmotic phenomena as causes of water movement. However, it is doubtful that these forces, if they operate, play a significant role in water absorption by either cells or roots (Dainty, 1963).

It will be assumed that the difference in water potential between the surround-ing medium and the roots are produced either by the osmotic effects of solutes accumulated in the root xylem of slowly transpiring plants or by the tension developed in the hydrodynamic system of more rapidly transpiring plants. Renner (1915) termed these two types of absorption "active" and "passive" because in the former the roots play an active part while in the latter they act simply as passive absorbing surfaces.

1. ACTIVE ABSORPTION AND ROOT PRESSURE

In slowly transpiring plants growing in warm, well-aerated soil with a water content near field capacity, positive pressure, usually called root pressure, often develops in the xylem. This results in guttation or exudation of liquid from the tips and margins of leaves and occasionally from lenticels of twigs, and in exuda-tion or so-called bleeding from wounds and stumps of plants, although not all instances of bleeding should be attributed to root pressure (Kramer, 1949, 1956a). It is believed that root pressure develops because roots are capable of acting as osmometers. Salt is accumulated by an active transport mechanism in the xylem sap of roots of many species to a concentration considerably above that of the external medium. Water then moves along the resulting potential gradient from the soil solution across the differentially permeable membrane formed by the root tissues, often developing pressures of 1 or 2 atm in the xylem sap.

Root pressure is detectable only during periods of low transpiration because the active absorption mechanism has a very low absorbing capacity, often $< 5\%$ of the water required by a rapidly transpiring plant. Furthermore, it cannot operate when Ψ_{soil} is < -1 or -2 bars.

2. PASSIVE ABSORPTION BY TRANSPIRING PLANTS

As the rate of transpiration increases, water tends to be removed from the xylem more rapidly than it is absorbed, Ψ_{plant} is reduced, and the pressure on the xylem sap falls to zero. If transpiration continues, tension is developed in the xylem sap and is transmitted through the hydrodynamic system to the surfaces of the roots, reducing water potentials in the root tissues and producing a gradient along which water moves from soil to root xylem. Under conditions of rapid transpiration and high xylem tension, water probably is literally pulled into the roots by mass flow (Levitt, 1956). The cortical parenchyma and other tissues of roots offer considerable resistance to the mass movement of water, resulting in a tendency for water absorption to lag behind water loss. This absorption lag causes development of appreciable water deficits and tensions in the hydrodynamic system of rapidly transpiring plants, even when they are growing in moist soil.

3. RELATIVE IMPORTANCE OF ACTIVE AND PASSIVE ABSORPTION

Although active absorption and the associated root pressure have received much attention because of the uncertainty concerning the mechanisms involved, they probably are of little importance in the water economy of plants. Active absorption apparently can supply only a small fraction of the total water required by freely transpiring plants and probably does not even operate in rapidly transpiring plants in which high tensions exist. Probably over 90% of the total water absorption is brought about by the passive mechanism.

C. Factors Affecting Water Absorption

The absorption of water is affected by a variety of plant and environmental factors. Some of these will be discussed in more detail in other chapters, and therefore will only be mentioned briefly in this chapter.

1. PLANT FACTORS

Under normal growing conditions the rate of water absorption is controlled primarily by the rate of transpiration because loss of water reduces Ψ_{leaf} and establishes gradients along which water moves into the roots. However, because of the resistance to the entrance of water in the roots, absorption tends to lag behind transpiration. This absorption lag is great enough to result in a measurable water deficit during periods of high transpiration, even in plants growing in soil near field capacity.

The distribution and efficiency of root systems also are important because they determine the volume of soil moisture potentially available to a plant. One of the best insurances against drouth injury is a deep, wide spreading, and many branched root system which can remove all of the readily available water from the soil to a considerable depth. Thus varieties of plants with deep and abundantly branched root systems are more drouth resistant than those with shallow and sparsely branched root systems.

The efficiency of a root system of given size depends on its permeability to water, which tends to decrease as suberized layers develop during differentiation and maturation. Thus all degrees of permeability occur in a given root system which consists of roots of widely varying ages and degrees of suberization. It seems probable that much water absorption by perennial plants and mature annuals occurs through the less permeable, older, suberized roots simply because they comprise most of the absorbing surface of such plants.

2. SOIL FACTORS

Aeration, temperature, available water, and concentration of the soil solution are important soil factors because of their effects on both water absorption and root growth. The growth of root systems is discussed in chapter 21, hence we can confine this discussion to the direct effects of environmental factors on water absorption.

a. Available Water. The availability of soil water to plants is a complex function of soil water potential and the hydraulic conductivity of the soil. The nature and dynamics of soil water are covered in chapters 12 and 13. As long as Ψ_{plant} can be reduced below Ψ_{soil} by desiccation or increase in osmotic pressure, net water movement will be in the direction of the plant. The water potential, conductivity, and retention properties are specific for a given soil material so that generalizations regarding water availability are difficult if not dangerous to make (Gardner, 1960; Gardner and Ehlig, 1962; Peters, 1957; Philip, 1957).

b. Concentration of Soil Solution. In arid regions where evapotranspiration exceeds rainfall, salt accumulation often increases until the osmotic pressure of the soil solution II_{soil} becomes a limiting factor. It usually is assumed that the injurious effects are largely osmotic, caused by decreased availability of water leading to a reduction in Ψ_{soil} solution $- \Psi_{roots}$. However, it has been shown that as Π_{soil} or $\Pi_{substrate}$ increases, the Π_{plant} also increases up to at least 5 or 6 bars (Bernstein, 1961; Eaton, 1942; Slatyer, 1961; Walter, 1955). Thus, within limits, $\Psi_{soil} - \Psi_{plant}$ can be approximately maintained over a considerable concentration range.

As Bernstein (1961) points out, this situation creates a dilemma because most of the experimental data support the view that plant growth in saline soils and concentrated solutions is controlled by the water potential or osmotic pressure of the soil or solution. Bernstein suggests that growth is slowed down by the accumulation of salt necessary to build up the Π_{plant} and maintain turgor. The writer believes the reduction in Ψ_{plant} which results from salt accumulation probably has important effects on permeability and metabolism. Evidently more research will be required to explain the effects of high concentrations of soil solution on water absorption and plant growth.

c. Soil Aeration. It is well known that inadequate aeration interferes with root growth and water absorption of most crop plants except rice. However, the effects of poor soil aeration are complex and not fully understood. One effect is decreased root permeability, which interferes with passive absorption of water. Another effect is reduced salt accumulation in roots, which in turn reduces or stops active absorption.

The chronic aeration deficiencies found in many fine-textured soils probably are more important factors in reducing crop yields than the less common but more severe effects encountered in saturated soils. However, the effects of chronic aeration deficiency on water absorption probably are less important than their effects on root growth, salt absorption, and synthetic activities. Russell (1952) has a good discussion of the numerous effects of soil aeration on plant growth.

d. Soil Temperature. Water absorption is significantly reduced at soil temperatures below approximately 20C. The reduction is much greater in warm season crops such as bermuda-grass (*Cynodon dactylon*), cotton (*Gossypium hirsutum*), and water melons (*Citrullus vulgaris*), than in cool season crops such as bluegrass (*Poa pratensis*) and collards (*Brassica oleracea*). The principal cause of reduced water absorption is the reduction in root cell permeability and increase in viscosity of water itself which increases the resistance to passive movement of water through roots (Kramer, 1938; Kuiper, 1964). A decrease in rate of movement of water in the soil also is of some importance. The decreased metabolic activity of the roots probably is a minor factor in absorption, but decreased root growth at low temperatures may significantly reduce the absorbing surface.

The use of cold water for irrigation may sometimes significantly affect plant growth. Schroeder (1939) reported that watering greenhouse cucumbers (*Cucumis sativus*) with cold water resulted in injury and it seems probable that some of the water used for irrigation in the Western USA is cold enough to reduce water absorption and retard growth (*see* chapter 53). Readers are referred to the article by Richards et al. (1952) for further discussion of the effects of temperature.

II.   WATER CONDUCTION[2]

A.   General

The most satisfactory theory of water conduction through the body of the plant was developed by Dixon (1914) and is generally known under the name of the cohesion theory. Kramer (1959) has recently summarized the work on which the theory is based. The essential feature is that the evaporation of water, primarily from the leaves, sets up imbibitional forces in the cell walls which are transmitted through the hydrodynamic system and cause the ascent of sap.

Physically, the theory is based both on observations that water molecules have attractive forces for each other and that water columns confined in small capillaries cohere with a tensile strength sufficient to lift them to the evaporating surfaces. Dixon (1914) showed that the water columns were so firmly anchored in the walls of the leaf mesophyll cells that they could support the pull of a water column at least 30 m in height. He visualized the water columns as hanging suspended from the evaporating surfaces of the leaf cells.

Water takes the following path in moving through plants from soil to air: (i) the living root cells, (ii) the conductive system of the xylem, (iii) the living leaf cells, and (iv) the intercellular spaces of the leaf, the stomata and an air layer around the leaf. In steps (i), (ii), and (iii) water moves as a continuous cohering liquid system; in step (iv) it moves as a vapor.

Through each step of the water pathway, from the intake to the exit, the transport is governed by a difference in water potential, with the tissues in between offering a resistance to flow according to their structural characteristics (Gradmann, 1928; Van den Honert, 1948). Differences in water potential in living tissues have been called by various authors suction force, water-absorbing power, suction pressure, suction tension, effective (net) osmotic pressure, turgor deficit, water potential, or most frequently diffusion pressure deficit (DPD). Values usually are expressed in atmospheres or bars. The resistance met is essentially a resistance to streaming through capillary pores of various dimensions. At present it cannot be decided how much of the transport is through cell walls and how much through the protoplasts although Russell and Woolley (1961) and Strugger (1949) believe considerable movement occurs through cell walls. In (ii), the xylem system, the path of least resistance is the lumen of vessels or tracheids.

The resistance to water movement resides largely in the plant when Ψ_{soil} is high, i.e. > -0.6 bar. Resistance in the soil becomes limiting when Ψ_{soil} is < -1 or -2 bars (Gardner and Ehlig, 1962).

B.   Roots

Intact roots of transpiring plants show zones of preferential entrance of water. Normally most water enters through the root hair zone, or that zone where xylem is newly differentiated but where suberization is at a minimum. As transpiration

---

[2] Prepared by Orlin Biddulph.

stresses become more severe, progressively more water passes through the more mature regions of the root (Brouwer, 1953). It is as though the larger potential differences result in movement through pores which are inoperative at smaller potential differences.

The radial pathway across roots from epidermis to xylem appears to be somewhat different for water movement by diffusion and by mass flow. Studies on the diffusion of water into detached roots, coleoptiles, and other tissues show that the resistance to diffusion is uniformly distributed in the tissue. Neither the outer boundary nor the inner passage across cell walls, membranes, or vacuoles provides unusual resistance or conductivity. Inward diffusion occurs as a solid sheet, fulfilling the expectations of diffusion according to Fick's second law of diffusion (Bonner, 1959).

Water moving by mass flow appears to find certain pathways of low resistance which lead more or less directly through the plant to the atmosphere. By the use of tracer water (THO and H_2O^{18}) it has been shown that as much as 2 days of active transpiration are required to reduce the original tissue water in roots to one-fourth of its initial value (Biddulph et al., 1961; Cline, 1953; Vartapetyan and Kursanov, 1959). Since the volume of water transpired under these circumstances would be many times the volume of the roots, there cannot be a successive replacement of all water in the root tissue with nutrient solution water as the transpiration stream ascends. Rather there must exist in roots channels of relatively low resistance through which the transpiration stream moves, bypassing a large volume of stationary water. The resistance to water flow encountered in a root system is greater than in its corresponding stem, or in its leafy shoot (Jensen et al., 1961; Kramer, 1938).

C. The Xylem System

The cross sectional area of the lumina of vessels and tracheids is large enough that cavitation, or rupturing, of the water columns within them occurs under high tensions, on wounding, or on freezing and thawing (Scholander et al., 1955, 1957). Cavitation in the lumina renders them inoperative as pathways for water transport because the cohesive continuity of the capillary column is interrupted. If air enters vessels, the air-water menisci move only to the first transverse pit boundaries where they are arrested by their inability to pass the micropores. The stream of water is then forced to move around the air-filled cavities and pursue a path through the microcapillaries of the cell wall. The resistance to flow is increased materially by the blocked lumina, resulting in an increased pressure difference across the area of higher resistance (Scholander et al., 1955, 1957). It appears that the xylem system of the stem has adequate conductive capacity to sustain the loss of considerable conduction caused by blocking of the vessel lumina, without disaster. There is no satisfactory explanation of how continuity of the water columns is re-established in vessels blocked by gas.

The flow of a labeled transpiration stream through the stem has been shown to result in a rapid replacement of all the original water in the xylem cylinder, and most of that in the surrounding cortex. (Biddulph et al., 1961). The replacement pattern is similar to that which can be expected if flow of the tracer is through tubes with porous walls so that a reversible exchange of water molecules can occur between the moving stream and the medium surrounding the tubes.

In a general way it may be concluded that the stem as a whole functions in water conduction, but the major part of the flow is through the tissues with the lowest resistance, i.e., the vessels.

The dynamics of sap flow in plants was discussed recently by Huber (1956) who compared relative conducting surfaces, expressed as the cross-section of conducting tissue in square millimeters per gram of leaf fresh weight. He estimated the relative conducting surfaces to be as follows: aquatics, 0.02; desert succulents, 0.10; trees, 0.5; and nonsucculent desert plants, 3.4 mm²/g of leaf fresh weight.

Huber (1956) also estimated the specific conductivity, expressed as flow cf water in milliliters per square centimeter of cross-section of conducting tissue per meter of length per atmosphere of pressure per hour. Some specific conductivities are: trunks of deciduous trees, 65 to 128; roots of deciduous trees, 292 to 5,388; conifers, 20; lianas, 236 to 1,273. Roots of trees and stems of lianas characteristically have wide vessels and high conductivities.

The velocity of sap flow has been measured by use of dyes, radioactive tracers, and thermoelectric methods (see references in Kramer and Kozlowski, 1960). Rates up to 40 m/hour were reported for ring-porous trees such as oak and 60 m/hour for herbaceous species, but only 1 to 6 m/hour for conifers and diffuse-porous deciduous trees. Lianas are said to have even higher velocities of sap flow than ring-porous trees. Flow generally is confined to the latest increment of growth in ring-porous trees, whereas the rate is lower in the diffuse-porous trees, but flow occurs in several annual rings. The older wood is considered to be largely nonfunctional in rapid sap rise.

The validity of the cohesion theory has been questioned by a number of workers, most of whom are cited by Greenidge (1957). These questions are based chiefly on uncertainty concerning the actual cohesive forces existing in xylem sap, the instability of water columns under tension, and plugging of large numbers of xylem elements by gas bubbles. Greenidge (1957) and Loomis et al. (1960) agree that xylem sap can sustain tensions of 20 to 30 atm, which are sufficient to pull water to the top of tall trees. It also seems probable that water columns confined within the hydrophilic walls of xylem elements are much more resistant to shock than water columns in glass tubes.

The xylem system of most plants appears to have such a large excess of conducting capacity that a considerable part of it can be blocked by air bubbles or overlapping horizontal cuts without disaster to the plant (Postlethwait and Rogers, 1958; Preston, 1952, 1954; Scholander et al., 1957). In spite of its weaknesses the cohesion theory is the only acceptable explanation of the rise of sap in tall trees.

D. Leaves

Water takes the path of least resistance through leaves, moving directly from vascular tissues, through intervening mesophyll tissue to stomata and then to the atmosphere. Incoming tracer water fails to mix completely with the indigenous tissue water, so that even after three photoperiods, i.e., 72 hours, about one-third of the indigenous water appears to remain in the leaf (Biddulph et al., 1961). In such tracer studies Vartapetyan (1960) and Vartapetyan and Kursanov (1961) were unable to show that the apparent retention of indigenous water was due to an exchange of tracer water in plants for unlabeled atmospheric water.

Unpublished data of F. S. Nakayama showed that, when leaves are exposed to labeled atmospheric vapor, exchange of the tracer with unlabeled leaf water can occur. Furthermore, the exchange rate was a function of the moisture status of the root media, the rate being related directly to the water content. Until the magnitudes of the exchange rates can be resolved, it is difficult to denote the preferential paths of water movement through the various plant organs. However, it seems reasonable to assume from other data (Jensen et al., 1961; Klemm, 1956; Wilson and Livingston, 1937) that the resistance to water flow as a liquid through the vascular tissue and parenchyma of leaves is greater than in the stems, but less than in the roots.

E. Leaf-Air Interface

Water moves as vapor from the surface of the chlorenchyma cells bounding the substomatal cavities to the air circulating around the leaves. The resistance to diffusion in the adhering air layer is responsible for the slow movement of water at this step in the path. There is a gradual transition from essentially still air in the substomatal cavities and the vicinity of the stomata to the freely circulating air away from the leaf surface. The thickness of the stationary air layer surrounding a leaf has been calculated to be as follows: in still air, 10 mm; in slow wind, 2.8 mm; in strong wind, 0.4 mm (Van den Honert, 1948). Therefore, the rate of water transmission in isothermal evaporation is governed by the diffusion resistance of an air layer 0.4 to 10 mm in thickness depending on the rate of air movement. This resistance also is a function of stomatal pore size and quantitatively may amount to 14.2%, 32.3%, 52.3%, and 65.5% of the total resistance to gaseous diffusion from the leaf in calm air for pore radii of 2, 5, 10 and 15 μ, respectively (Lee and Gates, 1964). Beyond the boundary of the stationary air layer (macrovapor cup) over the leaf surface, it can be assumed that the water has truly escaped from the plant, since, the plant ceases to have an influence on water movement beyond that boundary.

The differences in DPD or water potential between the liquid water in the cell wall colloids and the vapor phase in the free air surrounding the leaf are frequently large, as shown in Table 17–1. The resistance to water flow through the water-air

Table 17–1. Approximate water potential relations in the soil-plant-atmosphere continuum (after Philip, 1957)

Soil-plant-atmosphere continum	"Normal" transpiration		"Temporary" wilting		"Permanent" wilting	
	bars					
Free atmosphere	-1250		-1250		-1250	
Turbulent sublayer	-1000 to -1250		-1000 to -1250		-1000 to -1250	
Laminar sublayer	-250 to -1000		-250 to -1000		-250 to -1000	
Stomatal pore	-200 to -250		-200 to -250		-200 to -250	
Intercellular space	-12 to -200		-27 to -200		-35 to -200	
Mesophyll cell	-12		-27		-35	
Leaf vein	-9		-24		-32	
Xylem vessel	-5 to -9		-20 to -24		-28 to -32	
Endodermis	-5		-20		-28	
Cortex	-4		-19		-27	
Root surface	-3		-17		-25	
Soil	0		-1		-3	

Table 17–2. The relative resistance to water movement in
various parts of its pathway

Water displacement through	Displacement rate mg/cm²/hr/bar	Relative resistance/cm²
1 m nonconiferous wood*	100,000	1
1 m conifer wood*	20,000	5
Protoplast, Salvinia*	3.3	30,000
Surface water-air†		
in still air†	0.007	14,000,000
in slow wind†	0.025	4,000,000
in strong wind†	0.191	520,000

* Huber and Höfler, 1930.   † Van den Honert, 1948.

interface and the stationary air layer around the leaf is, however, also very large so that this resistance is most generally the rate-determining step in the whole pathway of water movement within and from the plant.

An approximation of the resistance to water movement in several parts of the water path is shown in Table 17–2. The gaseous diffusion part of the path offers much the greatest resistance. It is in this part of the path that the stomata are found and it is here that water loss can be most effectively controlled. This is done by variation in the stomatal aperture, which in turn controls the resistance to diffusion.

III.   TRANSPIRATION[3]

It is evident from the results of early investigations in transpiration that the water lost from a plant by transpiration is far in excess of the water used in the plant for normal plant growth. With the unpleasant realization that water for agronomic crops is becoming restricted even in areas where water supply was believed to be plentiful, an understanding of the transpiration process has become not only of academic but also of economic importance. The reader is referred to several reviews and research reports of the past two decades to appreciate the problems encountered and the progress made in this field (Kramer, 1959; Meyer and Anderson, 1952; Milthorpe, 1959; Preston, 1954; Slatyer, 1960).

A. Cells and Tissues

Transpiration from plants can be considered primarily as a physical evaporation process in which the rate of water loss is conditioned to a degree by the anatomy and physiological behavior of the plant. This discussion does not include lenticular or cuticular transpiration because water loss by these processes is very small in relation to stomatal transpiration. The transformation of liquid water to the vapor phase occurs at the moist surface of the mesophyll cells, with consequent diffusion of the water vapor into intercellular leaf spaces and diffusion through the stomates into the atmosphere.

1. MESOPHYLL

The mesophyll is made up of loosely connected parenchyma cells with a relatively large volume of continuous air space in relation to the volume of the cells.

---

[3] Prepared by Francis S. Nakayama.

The area of the internally exposed cell surface also is large, in the range of 5 to 30 times the external leaf surface area, and Turrel (1936) suggested using the internally exposed surface instead of the external surface area to obtain a better indication of the potential transpiration loss of leaves, other external factors being the same.

The mesophyll cell walls are hydrophobic and the surfaces are covered with a liquid layer (Hausermann, 1944; Lewis, 1948; Scott, 1964). This means that the property of a thin layer of liquid water on the mesophyll surface may not be the same as that of a free water surface and consequently the rates of evaporation from the internal surfaces may not be similar. The magnitude of this difference has not been determined and its consequence cannot be evaluated at present.

Some investigators claim that partial dehydration of cells results in a decrease in permeability of the protoplasm to water movement and also a decrease in water conductivity of cell walls (Klemm, 1956; Meidner, 1955; Shimshi, 1963a). This would result in less water being available to evaporate into the internal leaf space. Hygen (1951, 1953) presents data which show that the rate of transpiration changes over a wide range in the water content of the leaf. Other investigators (Gregory et al., 1950; Milthorpe and Spencer, 1957; Williams and Amer, 1957) have presented data indicating that the transpiration rate is not influenced by the leaf water content in the range from fully turgid to wilted leaves. This difference in opinion has not been satisfactorily resolved.

2. GUARD CELLS

It is unlikely that the evaporation process *per se* can be directly affected by a biological process, but any biological control of transpiration must come about by the control of the path of liquid and vapor movement. In this respect the guard cells function as a regulator of the stomatal opening through which water vapor diffuses from the intercellular leaf spaces into the atmosphere. It is unfortunate that, under conditions favorable to rapid transpiration, plants usually must develop relatively high water stress before stomates close tightly enough to reduce water loss materially. Heath (1959), Miller (1938), and Stålfelt (1956, 1957) have presented comprehensive reports on structure and behavior of the guard cells. Miller's summary indicates that in still air the effectiveness of the guard cells in reducing vapor loss begins only when the stomates are nearly closed to completely closed. Ting and Loomis (1963) also reported that diffusion is not significantly reduced until the stomates are nearly closed. Stålfelt (1932) noted a rapid increase in transpiration just at the start of stomatal opening. The influence of the stomatal aperture is more apparent at high transpiration rates than at low rates.

Heath (1959) reviewed a number of postulates and experiments which have attempted to elucidate the mechanism of guard cell movement. This movement is undoubtedly controlled by turgor, but with a number of complex interactions. When cells are under a water stress, movement is particularly sensitive to other factors such as light, CO_2 concentration, shock, and wind.

3. WATER POTENTIAL GRADIENTS WITHIN LEAVES

There has been very little study made on the movement of liquid and vapor water in the leaf proper. The work with tracer water mentioned in the section on water conduction in leaves indicates that a significant part of the water in the leaf does not participate directly in the transpiration process. That is, the water loss through the stomates in transpiration is replaced primarily from the leaf vein through the xylem instead of from the mesophyll cells further removed from the evaporating surface. Williams (1950) stated that the epidermal water is supplied by the lateral movement of water from the main leaf veins and not from the mesophyll layer. The conductivity of the veins would probably be similar to that of the stem xylem. In the mesophyll layer, water conduction may occur as a thin film layer from cell wall to cell wall in addition to movement through the cell proper. Thus investigations to explain the pathway of water movement in leaves pose a very interesting challenge for the researcher.

There has been little work done in determining the water potential distribution in the leaf. Thut (1939) showed that the vapor pressure in the intercellular spaces of transpiring leaves can be less than that at saturation and Whiteman and Koller (1964) reported saturation deficits as high as 320 bars in leaves of a desert halophyte. In regard to the liquid phase, Slavik (1959) demonstrated the presence of an osmotic pressure gradient in the leaf cells in which the osmotic pressure increases from the midveins to the leaf margins. These results show a non-equilibrium condition in regard to water in a transpiring leaf.

B. Mechanism of Water Loss

Most of the loss of water from plants occurs in two stages, the evaporation from moist cell walls into the intercellular spaces and diffusion from the intercellular spaces into the outside air.

1. EVAPORATION FROM WATER SURFACES

In leaves a vapor pressure gradient exists from the water surface at the mesophyll cells through the intercellular spaces and to the atmosphere through the stomatal openings. Empirical equations developed by Leighly (1937) and Martin (1943) describe with some success the evaporative loss of water vapor from leaves and leaflike structures. For a detailed discussion of evapotranspiration the reader is referred to chapter 27.

2. DIFFUSION THROUGH STOMATAL PORES

Most of the water lost by transpiration escapes through the stomates. The diffusive capacity of the stomates is very high. Transpiration from a leaf with a stomatal area of only 1% can be 50% of evaporation from an equivalent free water surface. Early concepts concerning the diffusion of gases through stomatal pores are summarized by Meyer and Anderson (1952).

Bange's (1953) treatment of the diffusion path for water in the leaf in terms of a series of diffusion resistances, internal and external to the leaf, gives a good picture of the diffusion process occurring in the leaf. These include the resistance between the evaporating surface and the pore, resistance through the pore, resistance in the microshell over the pore, and the resistance in the macroshell over the entire leaf. This type of analysis assists in explaining the relations between transpiration rate and stomatal aperture. At small stomatal apertures and in still air the resistance of the stomates determines transpiration, but, as the aperture increases, the resistance in the microshell becomes the limiting factor. Wind will break the diffusion macrovapor shell and thus the transpiration rate becomes more a function of the stomatal aperture again. Further discussion of this topic can be found in papers by Kuiper (1961), Lee and Gates (1964), and Slatyer and Bierhuizen (1964).

C. Factors Affecting Transpiration

Review articles by Stocker (1956) and Kramer (1959) discuss the factors affecting transpiration (*also see* chapters 26, 27, and 28). For convenience, a brief summary is given here.

1. ENVIRONMENTAL FACTORS

Solar radiation is the most important factor affecting transpiration because this is the source of energy necessary for the transformation of liquid water to vapor (*see* chapter 26). Visible light affects transpiration through its effect on the guard cells. Other factors that are of immediate importance to the plants are temperature, humidity, and wind. Any one or a combination of these factors which affect the water vapor concentration gradient will affect transpiration. Thus maximum transpiration will occur at high wind and leaf temperatures, and low humidity. Soil water potential and the hydraulic conductivity also affect transpiration.

2. PLANT FACTORS

Anatomical features such as thick layers of cutin and location of stomates in deep furrows reduce transpiration significantly, but such characteristics of a plant are not reliable indicators of its rate of transpiration (Meyer and Anderson, 1952; Oppenheimer, 1960). Physiological responses, however, such as stomatal closure, leaf rolling, curling, and orientation aid in regulating water loss. Root distribution has an effect on water absorption and consequent transpiration losses. The root–shoot ratio also is an important factor to be considered in the water relations of transpiring plants (Kramer, 1949). Further discussion of environmental factors will be found in chapter 27.

3. CHEMICAL TREATMENT

Several different types of commonly used agricultural chemicals such as herbicides and fungicides have been observed to affect transpiration (Blandy, 1957; Horsfall and Harrison, 1939). More recently, a systematic study of the effect of metabolic inhibitors on transpiration has been conducted (Shimshi 1963a, 1963b; Smith and Buchholtz, 1962; Stoddard and Miller, 1962; Waggoner et al., 1964; Zelitch, 1963; Zelitch and Waggoner, 1962a, 1962b). In these studies a significant decrease in transpiration rate occurred following treatment of the leaves with dilute concentrations of chemical. In some experiments the water loss data were accompanied by measurements of the stomatal aperture.

An alternative method, which is based on the creation of a physical barrier at the leaf-air interface, has been tried. This involved the treatment of the leaf with an emulsion-type spray or the soil with the monolayer producing chemical (Gale, 1961; Neales and Kriedman, 1962; Oertli, 1963; Olsen et al., 1962; Roberts, 1961). The results of the studies are conflicting but evidence indicates that this approach is not as promising as the chemical treatment. Slatyer and Bierhuizen (1964) compared the two general types and concluded that the metabolic inhibitor type antitranspirant worked better than the physical barrier type. It is apparent, in either case, that an effective antitranspirant must not only reduce water loss, but must permit the exchange of other gases essential for plant growth.

LITERATURE CITED

Bange, G. G. J. 1953. On the quantitative explanation of stomatal transpiration. Acta Bot. Neerl. 2:255–297.

Bernstein, L. 1961. Osmotic adjustment of plants to saline media: I. Steady state. Amer. J. Bot. 48:909–918.

Biddulph, O., F. S. Nakayama and R. Cory. 1961. Transpiration and ascension of calcium. Plant Physiol. 36:429–436.

Blandy, R. V. 1957. The effect of certain fungicides on transpiration rates and crop yield. Int. Congr. Crop Protect., Proc. 4th. p. 1513–1516.

Bonner, J. 1959. Water transport. Science 129:447–450.

Brouwer, R. 1953. Water absorption by the roots of *Vicia faba* at various transpiration strengths: I. Analysis of the uptake and the factors determining it. Koninkl. Ned. Akad. Wetenschap., Proc. Ser. C56:106–115.

Cline, J. F. 1953. Absorption and metabolism of tritium oxide and tritium gas by bean plants. Plant Physiol. 28: 717–723.

Dainty, J. 1963. Water relations of plant cells. 1:279–326. *In* R. D. Preston [ed.] Advances in botanical research. Academic Press, New York.

Dixon, H. H. 1914. Transpiration and the ascent of sap in plants. Macmillan and Co., Ltd., London. vii + 216 p.

Eaton, F. M. 1942. Toxicity and accumulation of chloride and sulfate salts in plants. J. Agr. Res. 64:357–399.

Gale, J. 1961. Studies on plant antitranspirants. Physiol. Plant. 14:777–786.

Gardner, W. R. 1960. Dynamic aspects of water availability to plants. Soil Sci. 89:63–73.

Gardner, W. R., and C. F. Ehlig. 1962. Some observations on the movement of water to plant roots. Agron. J. 54:453–456.

Gessner, F. 1956. Die Wasseraufnahme durch Blätter und Samen. 3:215–246. *In* W. Ruhland [ed.] Encyclopedia of plant physiology. Springer, Berlin.

Gradmann, H. 1928. Untersuchungen über die Wasserverhaltnisse des Bodens als Grundlage des Pflanzenwachstums. Jahrb. Wiss. Bot. 69:1–100.

Greenidge, K. N. H. 1957. Ascent of sap. Ann. Rev. Plant Physiol. 8:237–256.

Gregory, F. G., F. L. Milthorpe, H. L. Pearse and H. J. Spencer. 1950. Experimental studies of the factors controlling transpiration: 1. Apparatus and experimental technique. 2. The relation between transpiration rate and leaf water content. J. Exp. Bot. 1:1–28.

Hausermann, E. 1944. The amount of wetting of mesophyllic intercellular spaces. Ber. schweiz. bot. Ges. 54:541–578.

Heath, O. V. S. 1959. The water relations of stomatal cells and the mechanisms of stomatal movement. 2:193–250, 727–730. *In* F. C. Steward [ed.] Plant physiology. (Academic Press, New York.)

Horsfall, J. G., and A. L. Harrison. 1939. Effect of Bordeaux mixture and its various elements on transpiration. J. Agr. Res. 58:423–443.

Huber, B. 1956. Die Gefässleitung. 3:541–582. *In* W. Ruhland [ed.] Encyclopedia of plant physiology. Springer, Berlin.

Huber, B., and K. Höfler. 1930. Die Wasserpermeabilität des Protoplasmas. Jahrb. Wiss. Bot. 73:351–511.

Hygen, G. 1951. Studies in plant transpiration: I. Physiol. Plant. 4:57–183.

Hygen, G. 1953. Studies in plant transpiration: II. Physiol. Plant. 6:106–133.

Jensen, R. D., S. A. Taylor, and H. H. Wiebe. 1961. Negative transport and resistance to water flow through plants. Plant Physiol. 36:633–638.

Klemm, G. 1956. Untersuchungen über den Transpirationswiderstand der Mesophyllmembranen und seine Bedeutung als Regulator für die stomatäre Transpiration. Planta 47:547.

Kramer, P. J. 1938. Root resistance as a cause of the absorption lag. Amer. J. Bot. 25:110–113.

Kramer, P. J. 1949. Plant and soil water relationships. McGraw-Hill, New York. viii + 340 p.

Kramer, P. J. 1956a. Physical and physiological aspects of water absorption. 3:124–159. *In* W. Ruhland [ed.] Encyclopedia of plant physiology. Springer, Berlin.

Kramer, P. J. 1956b. Roots as absorbing organs. 3:188–214. *In* W. Ruhland [ed.] Encyclopedia of plant physiology. Springer, Berlin.

Kramer, P. J. 1959. Transpiration and the water economy of plants. 2:607–726. *In* F. C. Steward [ed.] Plant physiology. Academic Press, New York.

Kramer, P. J., and T. T. Kozlowski. 1960. Physiology of trees. McGraw-Hill, New York. ix + 642 p.

Kuiper, P. J. C. 1961. The effects of environmental factors on the transpiration of leaves, with special reference to stomatal light response. Meded. Landbouwhogeschool Wageningen 61:1–49.

Kuiper, P. J. C. 1964. Water uptake of higher plants as affected by root temperature. Meded. Landbouwhogeschool Wageningen 64:1–11.

Lee, R., and D. M. Gates. 1964. Diffusion resistance in leaves as related to their stomatal anatomy and microstructure. Amer. J. Bot. 51:963–975.

Leighly, J. 1937. A note on evaporation. J. Ecol. 18:180–198.

Levitt, J. 1956. The physical nature of transpirational pull. Plant Physiol. 31:248–251.

Lewis, F. J. 1948. Water movement in leaves. Discuss. Faraday Soc. 3:159–162.

Loomis, W. E., R. Santamaria-P., and R. S. Cage. 1960. Cohesion of water in plants. Plant Physiol. 35:300–306.

Martin, E. 1943. Studies of evaporation and transpiration under controlled conditions. Carnegie Inst., Washington. 550 p.

Meidner, H. 1955. Changes in the resistance of the mesophyll tissue with changes in the leaf water content. J. Exp. Bot. 6:94–99.

Meyer, B. S. 1956. The hydrodynamic system. 3:596–614. *In* W. Ruhland [ed.] Encyclopedia of plant physiology. Springer, Berlin.

Meyer, B. S., and D. B. Anderson. 1952. Plant physiology. 2nd ed. D. van Nostrand, New York. v + 784 p.

Miller, E. C. 1938. Plant physiology. 2nd ed. McGraw-Hill, New York. xxi + 1,201 p.

Milthorpe, F. L. 1959. Transpiration from crop plants. Field Crops Abstr. 12:1–9.

Milthorpe, F. L., and E. J. Spencer. 1957. Experimental studies of the factors controlling transpiration: III. The inter-relations between the transpiration rate, stomatal movement, and leaf water content. J. Exp. Bot. 8:413–437.

Neales, T. F., and P. E. Kriedman. 1962. Reduction of plant transpiration by cetyl alcohol. Nature 195:1221–1222.

Oertli, J. J. 1963. Effects of fatty alcohols and acids on transpiration of plants. Agron. J. 55:137–138.

Olsen, S. R., F. S. Watanabe, W. D. Kemper, and F. E. Clark. 1962. Effect of hexadecanol and octadecanol on efficiency of water use and growth of corn. Agron. J. 54:544–545.

Oppenheimer, H. R. 1960. Adaptation to drouth: Xerophytism. *In* Plant water relationships in arid and semiarid conditions. UNESCO Arid Zone Res. 15:105–138.

Peters, D. B. 1957. Water uptake of corn roots as influenced by soil moisture content and soil moisture tension. Soil Sci. Soc. Amer. Proc. 21:481–484.

Philip, J. R. 1957. The physical principles of soil water movement during the irrigation cycle. Int. Congr. Irrig. Drainage, Proc. 3rd. 8:125–154.

Postlethwait, S. N., and B. Rogers. 1958. Tracing the path of the transpiration stream in trees by the use of radioactive isotopes. Amer. J. Bot. 45:753–757.

Preston, R. D. 1952. Movement of water in higher plants. Ch. 4. *In* A. Frey-Wyssling [ed.] Deformation and flow in biological systems. North Holland Publ. Co., Amsterdam. xii + 552 p.

Preston, R. D. 1954. The transpiration of plants. Proc. Leeds Phil. Lit. Soc. 6:154–167.

Renner, O. 1915. Die Wasserversorgung der Pflanzen. Handwörterbuch Naturwissenschaften 10:538–557.

Richards, S. J., R. M. Hagan, T. M. McCalla. 1952. Soil temperature and plant growth. Ch. 5. *In* B. T. Shaw [ed.] Soil physical conditions and plant growth. Academic Press, Inc., New York. xv + 491 p.

Roberts, W. J. 1961. Reduction of transpiration. J. Geophys. Res. 66:3309–3312.

Russell, M. B. 1952. Soil aeration and plant growth. Ch. 4. In B. T. Shaw [ed.] Soil physical conditions and plant growth. Academic Press Inc., New York. xv + 491 p.

Russell, M. B., and J. T. Woolley. 1961. Transport processes in the soil-plant system. p. 695–721. In M. X. Zarrow, H. Beevers, I. Tessman, L. E. Trachman, and J. L. White [ed.] Growth in living systems. Basic Books Inc., New York.

Scholander, P. F., W. E. Love, and J. W. Kanwisher. 1955. The rise of sap in tall grapevines. Plant Physiol. 30:93–104.

Scholander, P. F., B. Ruud, and H. Leivestad. 1957. The rise of sap in a tropical liana. Plant Physiol. 32:1–6.

Schroeder, R. A. 1939. The effect of root temperature upon the absorption of water by the cucumber. Missouri Agr. Res. Sta. Res. Bull. 309. 27 p.

Scott, F. M. 1964. Lipid deposition in intercellular space. Nature 203:164–165.

Shimshi, D. 1963a. Effect of chemical closure of stomata on transpiration in varied soil and atmospheric environments. Plant Physiol. 38:709–712.

Shimshi, D. 1963b. Effect of soil moisture and phenylmercuric acetate upon stomatal aperture, transpiration, and photosynthesis. Plant Physiol. 38:713–721.

Slatyer, R. O. 1960. Absorption of water by plants. Bot. Rev. 26:331–392.

Slatyer, R. O. 1961. Effects of several osmotic substrates on the water relationships of tomato. Australian J. Biol. Sci. 14:519–540.

Slatyer, R. O., and J. F. Bierhuizen. 1964. The effect of several foliar sprays on transpiration and water use efficiency of cotton plants. Agr. Meteorol. 1:42–53.

Slavik, B. 1959. Gradients of osmotic pressure of cell sap in the area of one leaf blade. Biol. Plant. 1:39–47.

Smith, D., and K. P. Buchholtz. 1962. Transpiration rate reduction in plants with atrazine. Science 136:263–264.

Stålfelt, M. G. 1932. Der stomatare Regulator in der pflanzlichen Transpiration. Planta 17:22–85.

Stålfelt, M. G. 1956. Die stomatäre Transpiration und die Physiologie der Spaltoffnungen. 3:351–426. In W. Ruhland [ed.] Encyclopedia of plant physiology. Springer, Berlin.

Stålfelt, M. G. 1957. The water output of the guard cells of the stomata. Physiol. Plant. 10:752–773.

Stocker, O. 1956. Die Abhangigkeit der Transpiration von den Umweltfaktoren. 3:436–488. In W. Ruhland [ed.] Encyclopedia of plant physiology. Springer, Berlin.

Stoddard, E. M., and P. M. Miller. 1962. Chemical control of water loss in growing plants. Science 137:224–225.

Stone, E. C. 1957. Dew as an ecological factor. Ecology 38:407–422.

Strugger, S. 1949. Praktikum der Zell- und Gewebephysiologie der Pflanzen. 2nd ed. Springer, Berlin. vii + 225 p.

Thut, H. F. 1939. The relative humidity gradient of stomatal transpiration. Amer. J. Bot. 26:315–319.

Ting, I. P., and W. E. Loomis. 1963. Diffusion through stomates. Amer. J. Bot. 50:866–872.

Turrell, F. M. 1936. The area of the internal exposed surface of dicotyledon leaves. Amer. J. Bot. 23:255–264.

Van den Honert, T. H. 1948. Water transport in plants as a catenary process. Discuss. Faraday Soc. 3:146–153.

Vartapetyan, B. B. 1960. Further investigation of the water metabolism of plants with the help of heavy water H_2O^{18}. Fiziol. Rastenii. 7:395–397.

Vartapetyan, B. B., and A. L. Kursanov. 1959. A study of the water metabolism of plants using water containing heavy oxygen, H_2O^{18}. Fiziol. Rastenii. 6:144–149.

Vartapetyan, B. B., and A. L. Kursanov. 1961. Water exchange between plant tissues and liquid water and vapor in the environment. Fiziol. Rastenii. 8:569–575.

Waggoner, P. E., J. L. Monteith, and G. Szeicz. 1964. Decreasing transpiration of field plants by chemical closure of stomata. Nature 201:97–98.

Walter, H. 1955. The water economy and the hydrature of plants. Ann. Rev. Plant Physiol. 6:239–252.

Whiteman, P. C., and D. Koller. 1964. Saturation deficit of the mesophyll evaporating surfaces in a desert halophyte. Science 146:1320–1321.

Williams, W. T. 1950. Studies in stomatal behavior: IV. The water relations of the epidermis. J. Exp. Bot. 1:114–131.

Williams, W. T., and F. A. Amer. 1957. Transpiration from wilting leaves. J. Exp. Bot. 8:1–19.

Wilson, J. D., and B. E. Livingston. 1937. Lag in water absorption by plants in water culture with respect to changes in wind. Plant Physiol. 12:135–150.

Zelitch, I. [ed.] 1963. Stomata and water relations in plants. Connecticut Agr. Exp. Sta. Bull. 664.

Zelitch, I., and P. E. Waggoner. 1962a. Effect of chemical control of stomata on transpiration and photosynthesis. Nat. Acad. Sci., Proc. 48:1101–1108.

Zelitch, I., and P. E. Waggoner. 1962b. Effect of chemical control on transpiration of intact plants. Nat. Acad. Sci., Proc. 48:1297–1299.

18

Measurements of Internal Water Status and Transpiration

R. O. SLATYER

Division of Land Research, CSIRO
Canberra, A. C. T., Australia

E. SHMUELI

National & University Institute of Agriculture
Rehovot, Israel

I. INTRODUCTION

This chapter provides an interpretative account of the most promising techniques presently available for transpiration and internal water status measurement. Space limitations restrict this treatment to four main types of measurements of most interest. These are tissue water content, tissue water potential Ψ, vacuolar osmotic pressure Π, and stomatal aperture and transpiration. Additional detail can be found in the references cited.

II. MEASUREMENTS OF TISSUE WATER CONTENT

Tissue water content measurements are generally made on leaves but have also been reported for most tissues and organs, including particularly stem sections and fruit. For nonphotosynthetic tissue, expression of the water content on a dry weight basis, analogous to that used for soils, is generally quite satisfactory. The tissue is generally sampled and weighed to give a fresh weight W_f, oven dried to constant weight at a temperature 85 to 90C, and reweighed to give a dry weight W_d, and the water content $(W_f - W_d)$ expressed as a percentage of the dry weight, thus: $100 (W_f - W_d)/W_d$. Expression of the results on a fresh weight basis [i.e., $100 (W_f - W_d)/W_f$] is unsatisfactory for almost all purposes because the denominator is not a constant. For leaf tissue, expression on a dry weight basis may also be unsatisfactory for the same reason since the dry weight changes diurnally due to photosynthesis and, over longer periods, due to growth.

For leaf tissue a much better denominator appears to be turgid water content, obtained by floating the tissue in water until the water deficit existing at the time of sampling is eliminated. The turgid weight W_t may be then introduced to the "relative turgidity" expression developed by Weatherley (1950, 1951), and now often called "relative water content." Thus:

$$\text{relative water content (or relative turgidity)} = 100 \, \frac{(W_f - W_d)}{(W_t - W_d)}.$$

It can also be introduced to the "wasser defizit" or water deficit term developed by Stocker (1929):

$$\text{water deficit} = 100 \frac{(W_t - W_f)}{(W_t - W_d)}.$$

Both of these terms are usually expressed as percentages (as shown), and each complements the other (i.e., relative water content = 100 – water deficit).

The validity of both methods depends on obtaining reproducible and reliable estimates of turgid water content W_t. In this regard Stocker (1929) suggested placing intact leaves, after sampling and weighing, in small humid chambers with their petioles in water until they became turgid. They were then removed, weighed (W_t), oven dried, and reweighed (W_d). Although this procedure has been used with success by some investigators, the elimination of the water deficit was observed to take 2 to 3 days in some cases. During this time important metabolic changes may occur in the tissue.

To reduce the time required, Weatherley (1950, 1951) proposed using leaf tissue segments and discs, instead of entire leaves, and floating them on water in closed petri dishes to hasten water uptake. Barrs and Weatherley (1962) subsequently showed that water uptake could be divided into two fairly distinct phases. The first is associated with elimination of the passive water deficit, and the second with renewed uptake associated with continued tissue growth (*see* Fig. 18–1). They considered these two phases to be independent (i.e., phase two did not commence until phase one was completed). Phase one was generally completed in 4 hours, and phase two could be minimized by metabolic inhibitors. Although the first point is open to argument, the rapid elimination of the water deficit and the relatively slow uptake in phase two means that, for most purposes, a reproducible estimate of W_t can be obtained by removing the tissue after phase one is completed. The time can be determined by prior experimentation. For more precise work, metabolic inhibitors can be used to suppress phase two uptake (Barrs and Weatherley, 1962). Turgid water content determinations are also subject to error from injection of water into the intercellular spaces, changes in dry weight during the floating period, and surface drying after floating. Most of these effects can be minimized (Slatyer and Barrs, 1965).

Much additional information on these methods can be found in papers by Weatherley (1950, 1951), Werner (1954), Farbrother (1957), Slatyer (1955,

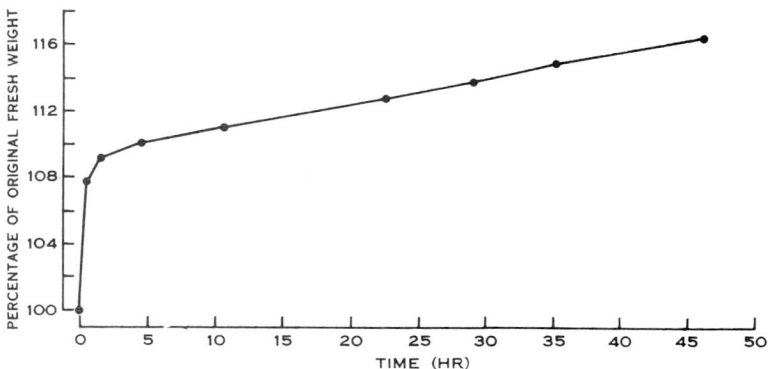

Fig. 18–1. Water uptake by floating leaf discs of *Ricinus communis* (castor oil) showing two fairly distinct phases, after Barrs and Weatherley (1962).

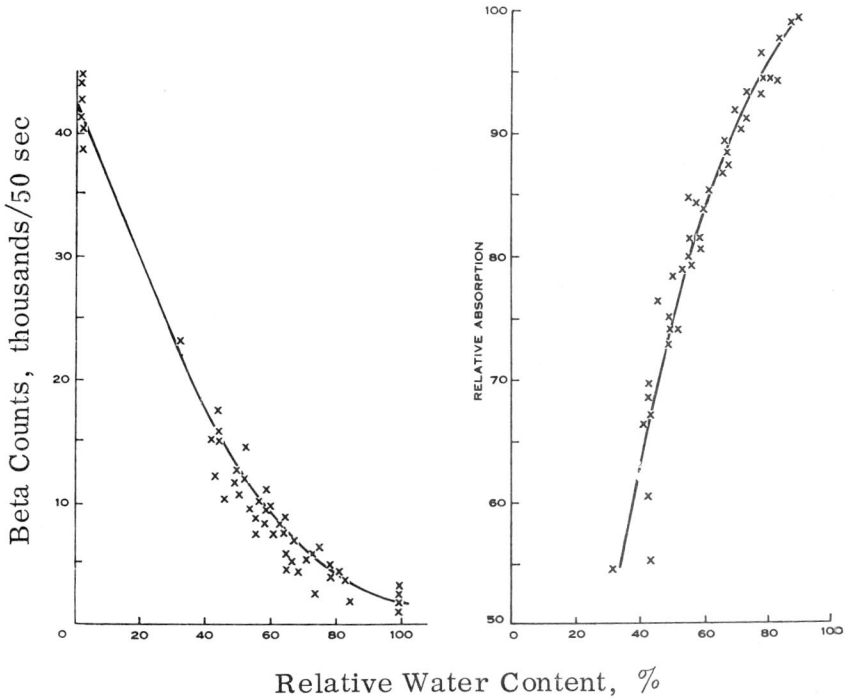

Relative Water Content, %

Fig. 18–2. Relationship between beta gauge readings and relative water content for leaves of *Sorghum vulgare* (grain sorghum) after Whiteman & Wilson (1963).

1961), Catsky (1965), Barrs and Weatherley (1962), Hewlett and Kramer (1963), and Slatyer and Barrs (1965).

The only effective nondestructive method so far developed for water content measurements is a beta gauge, similar to the type used industrially, which effectively measures leaf thickness by placing a beta source (such as C^{14}, Pm^{147}, Tc^{99}, or Tl^{204}) on one side of the leaf and a detector on the other side. This technique was first used for leaves by Yamada et al. (1954) and Mederski (1961). It has been developed further by Whiteman and Wilson (1963) and Nakayama and Ehrler (1964) and employed by a number of workers. Although leaf water content is not always proportional to leaf thickness, because of differential contraction during water removal and leaf density changes caused by air entering the system, the gauge readings give close agreement with relative turgidity observations. This technique is an extremely important tool in water relations investigations. Changes in leaf thickness (and density) from point-to-point and with time as leaf age increases limit its usefulness. Thus measurements need to be made at the one point, or on a sampling basis. In the first case the instrument must, in many applications, be removed from the proximity of the leaf between readings, so that care must be taken to set it up in the same position each time. Also, because of ontogenetic changes in the leaf and accumulation of photosynthate, periodic recalibrations against relative turgidity are required. An example of the relationship between beta gauge readings and relative water content is given in Fig. 18–2, from Whiteman and Wilson (1963).

III.   MEASUREMENTS OF WATER POTENTIAL ψ

At equilibrium, in a cell or tissue segment, the water potential ψ is assumed to be constant throughout the tissue and also in the vapor or liquid with which the tissue is surrounded. The most convenient procedure for determination of ψ therefore is to determine the value of the liquid or vapor with which the tissue is in equilibrium. This usually is achieved by one of two main techniques, involving either the direct determination of the equivalent vapor pressure with a thermocouple psychrometer or similar device, or determination of the concentration of the vapor or solution in which the tissue neither gains nor loses weight. The former technique generally uses a thermocouple psychrometer of the type developed by Spanner (1951) or Richards and Ogata (1958), and since modified and used by Monteith and Owen (1958), Korven and Taylor (1959) and Ehlig (1962), among others.

The Spanner instrument is designed to measure a function of vapor pressure of the air in a small chamber containing the leaf tissue after equilibrium has been established. It utilizes the Peltier effect to establish a minute wet bulb element in the chamber, the output from which is measured. The Richards and Ogata instrument introduces an actual wet bulb element to the chamber. If this is done after leaf-air equilibrium is reached (about 4 hours in many applications), the system is somewhat similar to that of the Spanner instrument. However, if it is introduced to the chamber at the same time as the leaf tissue, the thermocouple output reflects the change in temperature of the wet bulb element as water evaporates from it and flows to the plant surface. A relatively constant wet-bulb depression associated with a relatively steady-state flow develops and is used for calibration purposes. No further details are required here except to mention that problems exist in both techniques because leaf tissue sometimes apparently behaves differently than the wet filter paper generally employed to calibrate the instrument using solution of known vapor pressure. It is thought that these differences arise in part because there is much less water in the psychrometer when leaf tissue is being used. Consequently the amount of water which appears to be adsorbed on the psychrometer walls during equilibration may significantly reduce the water content and water potential ψ of the tissue at the time when the measurement is made. Also the tissue is continually generating small quantities of heat through respiration (Barrs, 1964). Further, the Richards technique, when used for steady-state measurements rather than equilibrium measurements, has the additional problem that flow to the plant surface follows a longer diffusion pathway than wet filter paper. The diffusion pathway involves vapor flow from the outer leaf surface to the effective evaporation sinks. As the leaf dries, the resistance to diffusive flow across this segment increases and may dominate the total resistance, so that calibrations based on wet filter paper will no longer apply (Rawlins, 1964). These problems can be reduced by using pre-equilibrated leaf tissue, rather than wet filter paper, for calibration purposes. But the existence of such sources of error indicates that these techniques should be used with caution.

The liquid or vapor exchange techniques have been used for many years (*see* Crafts et al., 1949). In practice several tissue samples of segments, discs, strips, or individual cells are immersed in a series of graded osmotic solutions or in vapor over a range of solutions for which the osmotic pressure, vapor pressure, and,

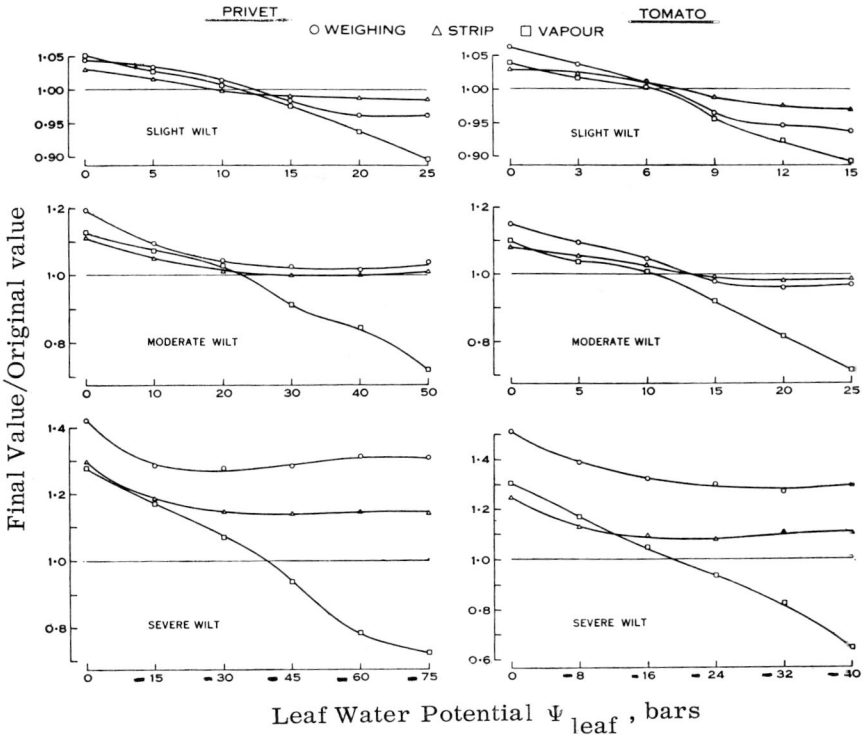

Fig. 18–3. Comparison of standard vapor equilibration technique (Slatyer, 1958) for determination of Ψ_{leaf} with liquid exchange techniques involving relative change in tissue weight or length ("strip") (Crafts et al., 1949), after Slatyer (1958).

hence, water potential characteristics are known. Water exchange occurs between the tissue and the surrounding liquid or vapor. After sufficient time has elapsed to allow measurable changes in weight, length or volume (or in the density or refractive index of the solutions), the samples are removed and reweighed or remeasured. The results are expressed as shown in Fig. 18–3 after Slatyer (1958). The tissue water potential ψ is read off from the intercept on the diagram.

Of the above methods, those using liquid exchange may involve errors whenever ψ is lower than the plasmolytic value of some of the cells of the tissue because further shrinkage of the protoplast, and plasmolysis, of such cells leads to entry of external solution without any further change of volume or weight. When all the cells are plasmolyzed (at ψ values lower than the plasmolytic value when the measurements are commenced), the tissue may actually increase in volume and weight regardless of the external solution concentration, so that no determination can be made. Thus, as soon as plasmolysis commences in some of the cells, there is a tendency to underestimate ψ (the apparent intercept in Fig. 18–3 moving further along the abscissa), and when all cells are plasmolyzed no measurements can be made (Slatyer, 1958).

Minor sources of error, which can arise in the same way as in relative water content determinations, are associated with removal of surface water and injection of external solution into intercellular spaces. However, these can usually be corrected by appropriate experimental technique. By comparison, those liquid exchange methods involving determination of a solution characteristic, such as refractive index or density (Ashby and Wolf, 1945; Shardakov, 1957), are not invalidated by the above factors since there is still a net water influx or efflux depending on the initial values of Ψ_{tissue} compared with $\Psi_{solution}$. Thus injection and surface wetting are of no consequence.

The other main source of error, common to all liquid exchange methods, occurs when the cell membranes are sufficiently permeable to the solute used in the external solution that, instead of a simple water exchange phenomenon operating. there is a net volume exchange. For example, a volume efflux, consisting of water alone, is associated with a volume influx consisting of solute and water. At the (quasi-) equilibrium situation of zero volume flow, tissue volume (and weight) is therefore greater than would occur if there had been no solute uptake. In a way similar to the previous example, this shifts the apparent point of intercept in Fig. 18–3 along the abscissa, leading to a low estimate of Ψ_{tissue}. The effect can be reduced to some extent by plotting change of water content, rather than change of fresh weight, against external solution concentration, particularly if the solute has a high molecular weight as with sucrose (Weatherley, 1955). However, this approach does not eliminate the problem, and it is ineffective when the tissue is plasmolyzed. This phenomenon is accentuated if the solute and water interact in their passage through the cell membranes so that the solute influx tends to drag some water in the same direction (Kedem and Katchalsky, 1958; Dainty, 1963). A term, the reflection (or selectivity) coefficient, can be introduced to account for the effects of a permeable solute when they are apparent. These effects are detectable only when solute permeability is rapid. They appear to be relatively insignificant in some storage tissues but may be quite important in some leaf tissue.

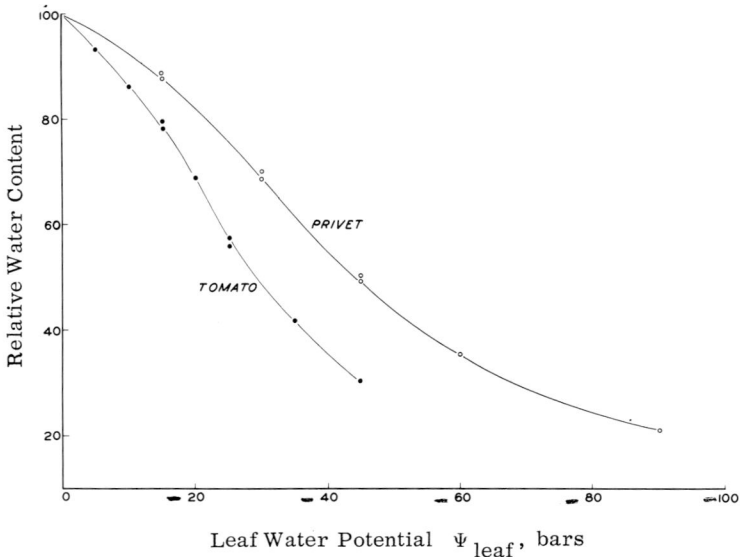

Fig. 18–4. Relationship between relative water content and Ψ_{leaf} for tomato (*Lycopersicon esculentum*) and privet (*Ligustrum* sp.) tissue, after Weatherley and Slatyer (1957).

The vapor exchange technique is free of the problems associated with solute uptake. In some respects it provides the only absolute method of determining Ψ. However, very accurate temperature control is required, and precautions are necessary to avoid errors caused by (i) loss of weight during tissue manipulation and (ii) respiration (Kreeb, 1960; Kreeb and Onal, 1961). These authors have proposed the use of humid transfer chambers to minimize the former source of error and a correction to account for the latter source.

As an alternative to the direct methods of water potential measurement, indirect measurements can be of considerable value in some situations. The most commonly used indirect measurement involves establishing a relative water content/Ψ curve for the tissue under study (*see* Fig. 18–4) and using this to calibrate Ψ in terms of relative water content. This procedure, first suggested by Weatherley and Slatyer (1957), has been used by Slatyer (1960, 1961) and is implicit in the general use by most researchers of relative water content measurements. Its accuracy depends on the stability of the calibration and the sensitivity of Ψ to changes in water content. The calibration can be expected to vary with changes of internal osmotic pressure as well as of matrix characteristics. All these factors change with age. However, where a plant is grown in a fairly uniform environment and leaves of the same physiological age can be sampled during the experiment, quite reliable results can be obtained. The solution lies in frequent rechecking of the calibration curve. The sensitivity of Ψ to changes in water content is generally low at Ψ values from zero to about -5 bars (*see* Fig. 18–4), but sensitivity increases at higher Ψ values. For this reason indirect methods may be unsatisfactory under low stress condition.

IV.   MEASUREMENTS OF VACUOLAR OSMOTIC PRESSURE II

Measurements have been made far more frequently on vacuoles than on any other tissue region. Probably this occurs because the vacuole occupies a large proportion of many cells, because its water potential is considered to be influenced only by pressure and solute concentration so it is a relatively simple system, and partly because of analogies with classical cell osmometers. Also, such measurements are key ones in understanding the dynamics of water and solute exchanges, for the vacuole is both a sink and source for water and solute movement between cells and their surroundings.

Two main procedures have been used to measure vacuolar osmotic pressure: one involves extraction of the vacuolar sap and the other measurements on intact cells in more or less undisturbed assemblages or tissue segments.

The first approach is subject to errors of variable magnitude in sap extraction. The normal procedure is to rupture the cell structure by freezing (or some other procedure) and then to express the sap under pressure. But the material so obtained contains cytoplasmic and extracellular sap, too, which is generally of lower osmotic pressure. In consequence the first aliquot may be dilute compared with the final one (Gortner et al., 1916; Mason and Phillis, 1939), but this is not always observed. In addition rapid chemical changes, such as inversion of sugars, may occur after the sap has been expressed, and it is important to reduce these as much as possible by low temperature storage. The osmotic pressure of the sap is then determined by measuring one of the colligative properties of the solution, generally its vapor pressure or freezing point depression. A number of techniques for this purpose have been described in recent years (Weatherley, 1961; Vaadia and Marr, 1961; Van Andel, 1953), and the thermocouple psychrometers previously described for water potential measurements can also be used when sufficient sap is available. Commercial equipment is also available. A useful general account of these techniques is given by Crafts et al. (1949). In summary, it may be stated that the sap extraction method has been of considerable value for many studies and appears to be particularly useful when tissue of high vacuolar content is under study. For tissue of low vacuolar content however, significant errors may be introduced.

The alternative procedure involves measuring osmotic pressure at incipient plasmolysis, and adjusting it to the volume of the normal cell. The plasmolytic method involves microscopic observations on the cells in thin strips of tissue immersed in a series of graded osmotic solutions. An external solution which reduces the turgor of the cell to zero is considered to be isotonic with the vacuolar osmotic pressure and the cell is assumed to be in a condition of *limiting plasmolysis*. In practice a slightly more advanced stage is detected, known as *incipient plasmolysis*, in which the protoplast visibly commences to shrink away from the cell wall. This too is taken as isotonic, even though a discrepancy exists of a magnitude equal to the adhesion between the protoplast and the wall. Measurements of cell, or protoplast, volume are made in the original and plasmolyzed states and a proportional adjustment is made to give the osmotic pressure at the original condition. The plasmometric method involves similar general procedures except that severe plasmolysis is induced by strongly hypertonic solutions, followed by measurement of the respective volumes and proportional correction.

Although these techniques permit observations on single cells and can be conducted in conjunction with studies of cell permeability, they are subject to the errors reported earlier for the liquid exchange measurements of Ψ. These are often insignificant when solute penetration is very slow, but should always be checked. Also, errors are introduced by inaccurate volume estimation, adhesion between the cell wall and the protoplast, and other sources (Crafts et al., 1949; Mercer, 1955). Volume measurement can be a particularly difficult proposition since vacuolar volume is to be measured and its shape cannot necessarily be assumed to be regular. Protoplast volume is frequently measured instead. This is satisfactory when the cytoplasm is minimal in volume but unsatisfactory in cells with significant amounts of cytoplasm or when cap plasmolysis occurs since the cytoplasm swells to occupy the space between the vacuole and the wall.

Some of these effects can be eliminated by the procedure of Bernstein and Nieman (1960), who proposed adapting free space measurements to indicate the point of incipient plasmolysis. Since the apparent free space increases rapidly as soon as the tissue is plasmolyzed, this increase is revealed as a sharp discontinuity in the curve of exodiffusion against external osmotic concentration.

V. MEASUREMENTS OF STOMATAL APERTURE

Stomatal aperture is of particular significance as an indicator of water deficits in plants because it influences both photosynthesis and transpiration by its effect on CO_2 and water vapor transport, and under adequate light conditions is generally closely related to leaf water deficit. Many methods are available for determining stomatal aperture, the most quantitative being those which estimate stomatal diffusive resistance (*see* for example, Kuiper, 1961; Gaastra, 1963; Monteith, 1963; Slatyer and Bierhuizen, 1964). However, for the purposes of this chapter most interest is associated with those methods which can be used on plants growing in the field. Because of the number of techniques, the only compact method of presentation appears to be as given in Table 18–1, where the main advantages and disadvantages are briefly indicated.

It is apparent that the purpose for which measurements are being made determines, to a considerable degree, the technique to be selected. If laboratory measurements are to be made in conjunction with transpiration and energy balance studies, for example, the best techniques are almost certainly in group 5. By comparison, if a qualitative, rapid, field technique is required group 4 is probably most appropriate. For many purposes viscous air flow porometers (group 3) are suitable, but it must be remembered that the data are not readily converted into diffusive resistances.

The main disadvantages of direct observation are that possible changes in aperture may occur during manipulation and that very few stomata can be examined at one time. However, in other respects, this method, which is the calibration standard, is absolute and reliable. By comparison, the surface film impression procedure, while most attractive conceptually, presents difficulties if the impression does not include the actual pore but merely imprints a guard cell outline or an outline at a different level to the minimum pore diameter.

Anybody wishing to measure stomatal aperture should become familiar with the direct microscopic technique as well as with other procedures, so that an absolute reference is available. A thorough review of the whole subject is given by Heath (1959).

VI. MEASUREMENTS OF TRANSPIRATION

There are two main methods for measuring transpiration from individual leaves or plants; the first involves determining the change in weight of the system when other sources of weight loss are eliminated or accounted for, and the second involves determining the rate of vapor loss. An energy balance method can also be used but is seldom necessary. Other methods, generally more indirect and less quantitative, have not achieved wide use; accounts of them can be found in texts by Miller (1938), Crafts et al. (1949), and Meyer and Anderson (1952).

The first of the main methods is analogous to the water balance method used for the whole plant community. Applied to an entire plant it reaches its most accurate and meaningful form when change of weight of plant, substrate, and container is determined and evaporation from the soil (or other substrate) is minimized or otherwise accounted for. Alternatively, change in the volume of water in the substrate can be measured by soil water determinations if the plant is rooted in soil (*see* for example, Ashton, 1956), or by determining the volume of water added to bring the culture solution level up to a given point, if the plant is in water culture.

Over short periods of several days changes in water content within the plant can be neglected if the measurements are made at the same time each day, say at sunrise, or can be estimated from plant water content determinations. Over longer periods changing fresh weight of the plant has to be accounted for, generally by periodic harvests of replicated groups of plants. These procedures are

Table 18-1. Comparison of methods of studying stomatal aperture

Method	Main advantages	Main limitations	Applications	References
1. Direct microscopic observation				
1.1 Direct observation of attached or detached leaves using medium magnification without cover glass	Direct observations of living material	Possible effect of light, temperature, etc., on stomatal aperture; quick manipulation necessary; only a few stomata can be examined simultaneously	Thin leaves, medium size to large stomata; attached leaves may be repeatedly observed on the same spot	Stålfelt (1956)
1.2 Leaf segments in liquid paraffin; observation using high magnification without cover glass	As in 1.1	As in 1.1; observation cannot be repeated on the same segment	Moderately thick leaves; stomata of any size	Stålfelt (1956)
1.3 Opaque illumination	As in 1.1	As in 1.1; pronounced effects of environmental factors; primarily a laboratory method	Moderately thick to thick leaves; stomata of any size; attached leaves may be repeatedly observed on the spot	Nadel (1935, 1940)
1.4 Camera adaptation for still and time-lapse photography	As in 1.1	As in 1.3: expensive method	Method in early stages of development; feasibility of timelapse photography questionable	Elkins and Williams (1961)
2. Microscopic examination of impressions				
2.1 Rubber impressions	Observations of living material; numerous stomata may be examined simultaneously; examination is not tied to sampling time; positive impression observed; may be obviously repeated on the same spot of the leaf	Possible effect of impression material on stomatal aperture	Leaves of any thickness; stomata of any size; stomata in the plane of epidermal surface; cuticle should not dissolve in impression material	Shmueli (1953) Zelitch (1961)
2.2 Collodion impressions	As in 2.1	Cannot be repeated on the same spot; negative impression observed	As in 2.1	Long and Clements (1934) Shmueli (1953)

	Method				References
2.3	Examination of epidermis following fixation	In suitable material and with proper skill, observed material is in state close to that of living material; also other advantages specified for 2.1	Possibility of change in stomatal aperture; only for leaves with easily separated epidermis; possible shrinkage of the epidermis in fixation agent	Leaves with easily separated epidermis; stomata of any size	Lloyd (1908) Ashby (1931)
3.	Viscous air flow porometer				
3.1	Gas flow under small pressure gradients	A very large number of stomata examined simultaneously; rapid test; resistance to flow can be evaluated quantitatively; does not damage living material	Only suitable for leaves with a continuous system of intercellular spaces; results of the tests on one side of the leaf influenced by stomatal aperture on the other side and by nature of intercellular system; stomatal aperture may change during the procedure	Can be applied to leaves regardless of stomatal size and structure of stomata, epidermis and cuticle	Ashby (1931) Heath (1959)
3.2	Resistance porometer for laboratory measurement	As in 3, 1; tests may be repeated on the same spot of the leaf; gas composition can be changed	As in 3, 1; ratio of stomatal density on two sides of leaf should preferably be 1:1 to 1:3	As in 3, 1; tests should be made under constant environment	Wilson (1942) Heath (1959) Heath and Meidner (1961)
3.3	Resistance porometer for field measurements	As in 3, 1, but gas composition cannot be changed	As in 3, 2; tests must be made in short time	As in 3, 1 for field conditions; unsuitable in leaves with damaged cuticle or leaves damaged by insects (especially sucking ones)	Shimshi (1963) Alvim (1965) Moreshet (1964) Bierhuizen, Slatyer, and Rose (1965)
4.	Liquid infiltration				
4.1	Flow of low-surface tension liquids, infiltration	As in 3, 1; tests very rapid	As in 3, 1; results may be affected by wettability of epidermis and intercellular spaces; liquid may injure living material; tests cannot be repeated on the same spot; results may be affected by subjective evaluation	Useful field technique	Molisch (1922) Schorn (1929) Shmueli (1953) Alvim and Havis (1954) Haley (1960)

(Continued on next page)

Table 18-1. Continued

Method	Main advantages	Main limitations	Applications	References
4.2 Flow of water under vacuum or pressure		Immersion in water may cause change of stomatal aperture; pressure may cause deformation and tearing of delicate tissue; air passage may be clogged with resins; laboratory procedure		Dengler (1912)
5. Diffusion methods				
5.1 From vapour flow transpiration measurements	A very large number of stomata examined simultaneously; simulation of natural processes of gas exchange	Does not distinguish between resistance in stomatal pore and elsewhere	Laboratory method	Meidner and Spanner (1959) Gaastra (1959) Kuiper (1961) Slatyer and Bierhuizen (1964)
5.2 Hydrogen diffusion porometer	As in 3	Possibility of changes in stomatal aperture	As in 5.1	Gregory and Armstrong (1936) Heath (1959)
5.3 Cobalt paper measurements	A very large number of stomata examined simultaneously	Stomatal aperture may change during the procedure; calibration against a color chart necessary; subjective evaluation of change of color	As in 3.3	Milthorpe (1955)

straightforward and do not need separate descriptions. Details can be found in many papers dealing with transpiration such as Slatyer (1957).

Extrapolation of transpiration so measured to transpiration from assemblages of plants under natural conditions should only be attempted when the plant is rooted in soil in what is essentially a microlysimeter located in the plant community itself (Pasquill, 1949). When the measurements are made in the laboratory or on a greenhouse bench the environment may be so different that transpiration so measured bears little relationship to that out-of-doors. Under these circumstances, far better estimates of plant community transpiration can be obtained by using empirical formulae, if direct field measurements cannot be made.

The problem of extrapolation is more pronounced when transpiration of a single leaf is measured by detaching a leaf and measuring its loss of weight, often called the "cut-shoot" method (Huber, 1927; Maximov, 1929; Eckardt, 1960), or by placing it in a small potometer and measuring the change of weight or volume of the system (see Meyer and Anderson, 1952). While these techniques do provide a measure of transpiration from the leaf employed, under the environmental conditions imposed and subject to physiological changes in the tissue, even the extrapolation to considerations of transpiration from the whole plant is fraught with difficulties because the leaf cannot readily be oriented with respect to incident energy and air movement in a way that is typical of an array of leaves on a whole plant, and its water supply is also influenced. It can be appreciated that any change in the energy load and wind structure immediately influences the energy balance of the leaf and, hence, transpiration.

It is not intended here to condemn the use of single leaf measurements for studies on single leaves, but instead to point out the serious sources of error which can be introduced if extrapolation to whole plants, or even groups of plants, is attempted. This criticism applies particularly to the cut-leaf method which has over the years been used extensively to indicate relative transpiration rates of different species or of the same species at different times of day or seasons. Every effort should be made to discourage this procedure, which is not only subject to errors due to the altered environment in which the leaf is placed for weighing (so that the individual leaf is not in an environment typical of a leaf identically oriented on the plant, before cutting), but also to extrapolation errors as applied to the whole plant and to errors due to physiological changes in the tissue caused by cutting and associated with changes in stomatal aperture. The last-mentioned sources of error were investigated for range of species and found to cause marked differences in most cases (Ivanoff, 1928; Weinmann and LeRoux, 1948; Franco and Magalhaes, 1963). Subsequently several workers (Anderson et al., 1954; Decker and Wien, 1958; Brun, 1961) have shown that there is a brief (≈ 2 to 5 minutes) surge in transpiration directly following cutting, presumably associated with an increase in stomatal aperture, during which transpiration may increase by 10% to 20% under normal conditions, followed by a progressive decline as water content and stomatal aperture decrease. Because of these factors the cut-shoot method, if used at all, must be employed with great caution.

The second main method for transpiration measurement, involves determining the rate of vapor flow away from the plant. It involves enclosing a whole plant, or portion of a plant, and determining the rate of change in water content of the air in through the chamber so formed, if the system is closed (Decker and Wetzel, 1957) or of the difference in water content of the air entering and leaving the chamber, if the system is open (Glover, 1941; Heath and Meidner, 1961; Decker et al., 1962; Bierhuizen and Slatyer, 1964).

This approach enables a more sensitive monitoring of transpiration than the change-of-weight methods since, with suitable sensing instruments, it has a response time of < 1 min. When used with controlled environmental conditions in the laboratory the open system approach appears to provide the best means presently available for detailed studies of the transpiration process; details of suitable equipment for this purpose, particularly for use with single leaves, have been given by Gaastra (1959), Heath and Meidner (1961), Kuiper (1961), and Bierhuizen and Slatyer (1964), among others. However, if extrapolation is attempted, problems similar to those just considered arise due to the different environment inside and outside the chamber.

These problems still exist when the measurements are made out-of-doors unless precautions are taken to simulate external conditions, particularly leaf temperature, bulk humidity, and wind speed (that is, the factors affecting the leaf-air vapor pressure difference Δe and the external diffusive resistance r_a associated with the boundary layer sheathing the leaf). The possible magnitude of these effects can be appreciated when it is recalled that, for example, if leaf temperature increases by 5C, the value of Δe may increase by 25%; if air movement through the chamber is inadequate to control bulk humidity at the external level, Δe will be reduced; if the ventilating air stream is increased until bulk humidity and leaf temperature reach external levels, r_a may be reduced significantly and affect the total diffusive resistance in the water vapor pathway (*see* Slatyer and Bierhuizen, 1964). In addition, depending on the spectral transmission characteristics of the chamber walls, the radiation fluxes themselves may change. While all of these factors may be controllable and the measured transpiration may closely approximate natural transpiration, it is apparent that considerable thought must be given to the experimental layout before any measurements are made.

In the case of the closed system, bulk humidity increases continuously as soon as the chamber is closed, as does leaf and air temperature. While a circulating fan may keep r_a at a level similar to that outside, it is apparent that Δe changes continuously in a direction which depends on whether the increase in bulk humidity e_a compensates for the increase in the vapor pressure at the leaf e_l caused by increasing leaf temperature. Extrapolation of the rate of increase in bulk humidity to zero time therefore may not indicate original rate of transpiration. This type of system is clearly quite unsuited for use out-of-doors and measured transpiration may bear little relation to natural transpiration.

LITERATURE CITED

Alvim, P. de T. 1965. A new type of porometer for measuring stomatal opening and its use in irrigation studies. UNESCO Arid Zone Res. 25:325–329.

Alvim, P. de T., and E. R. Havis. 1954. An improved series for studying stomatal opening as illustrated with coffee. Plant Physiol. 29:97–98.

Andel, O. M. van. 1953. The influence of salts on the exudation of tomato plants. Acta. Bot. Neerl. 2:445–521.

Anderson, N. E., C. H. Hertz, and H. Rufelt. 1954. A new fast recording hygrometer for plant transpiration measurements. Physiol. Plant. 7:753–767.

Ashby, E. 1931. Comparison of two methods of measuring stomatal aperture. Plant Physiol. 6:715–719.

Ashby, E., and R. A. Wolf. 1945. A critical examination of the gravimetric method of determining suction force. Ann. Bot. 11:261–268.

Ashton, F. M. 1956. Effects of a series of alternating high or low soil water contents on the rate of apparent photosynthesis in sugar cane. Plant Physiol. 31:266–274.

Barrs, H. D. 1964. Heat of respiration as a possible source of error in the estimation, by psychrometric methods, of water potential in plant tissue. Nature 203:1126–1137.

Barrs, H. D., and P. E. Weatherley. 1962. A re-examination of the relative turgidity technique for estimating water deficits in leaves. Australian J. Biol. Sci. 15:413–428.

Bernstein, L., and R. H. Nieman. 1960. Apparent free space of plant roots. Plant Physiol. 9:25–46.

Bierhuizen, J. F., and R. O. Slatyer. 1964. An apparatus for the continuous and simultaneous measurement of photosynthesis and transpiration under controlled environmental conditions. CSIRO Tech. Pap. 24. 16 p.

Bierhuizen, J. F., R. O. Slatyer, and C. W. Rose. 1965. A porometer for laboratory and field operation. J. Exp. Bot. 16:182–191.

Brun, W. A. 1961. Photosynthesis and transpiration from upper and lower surfaces of intact banana leaves. Plant Physiol. 36:399–405.

Catsky, J. 1965. Leaf disk method for determining water saturation deficit. UNESCO Arid Zone Res. 25:353–360.

Crafts, A. S., H. B. Currier, and C. R. Stocking. 1949. Water in the physiology of plants. Chronica Botanica, Waltham, Mass. 239 p.

Dainty, J. 1963. Water relations of plant cells. Advance Bot. Res. 1:279–326.

Decker, J. P., W. G. Gaylor, and F. D. Cole. 1962. Measuring transpiration of undisturbed tamarisk shrubs. Plant Physiol. 37:393–397.

Decker, J. P., and B. F. Wetzel. 1957. A method for measuring transpiration of intact plants under controlled light, humidity, and temperature. Forest Sci. 3:350–354.

Decker, J. P., and J. D. Wien. 1958. Carbon dioxide surges in green leaves. J. Solar Energy Sci. Eng. 2:39–41.

Dengler, F. 1912. Eine neue Methode zum Nachweis der Spaltoffnungsbewegungen bei Coniferen. Ber. Deut. Bot. Ges. 30:452–462.

Eckardt, F. 1960. Ecophysiological measuring techniques applied to research on water relations of plants in arid and semiarid regions. UNESCO Arid Zone Res. 15:139–171.

Ehlig, C. F. 1962. Measurement of energy status of water in plants with a thermocouple psychrometer. Plant Physiol. 37:288–290.

Elkins, C. B., and G. C. Williams. 1961. Camera adaptations for still and time-lapse photography of stomatal behaviour. Ass. South Agr. Workers, Proc. 58th Annu. Congr. p. 231.

Farbrother, H. G. 1957. On an electrical resistance technique for the study of soil moisture problems in the field. Empire Cotton Growing Rev. 34:71–92.

Franco, C. M., and A. C. Magalhaes. 1963. Disadvantages of the rapid weighing method for measuring transpiration. Phyton 20:87–96.

Gaastra, P. 1959. Photosynthesis of crop plants as influenced by light, carbon dioxide, temperature, and stomatal diffusion resistance. Mededel. Landbouwhogeschool 59:1–68.

Gaastra, P. 1963. Climatic control of photosynthesis and respiration. p. 113–140. In L. T. Evans [ed.] Environmental control of plant growth. Academic Press, New York.

Glover, J. 1941. A method for the continuous measurement of transpiration of single leaves under natural conditions. Ann. Bot. 5:25–34.

Gortner, R. A., J. V. Lawrence, and J. A. Harris. 1916. The extraction of sap from plant tissues by pressure. Biochem. Bull. 5:139–142.

Gregory, P. G., and J. I. Armstrong. 1936. The diffusion porometer. Roy. Soc. London, Proc., Ser. B. 121:27–42.

Halevy, A. 1960. The influence of progressive increase in soil moisture tension on growth and water balance of leaves and the development of physiological indicators for irrigation. Amer. Soc. Hort. Sci., Proc. 76:620–630.

Heath, O. V. S. 1959. The water relations of stomatal cells and the mechanism of stomatal movement. 2:193–250. In F. C. Steward [ed.] Plant physiology—a treatise. Academic Press, New York.

Heath, O. V. S., and H. Meidner. 1961. The influence of water strain on the minimum intercellular space carbon dioxide concentration, Γ, and stomatal movement in wheat leaves. J. Exp. Bot. 12:226–242.

Hewlett, J. D., and P. J. Kramer. 1963. The measurement of water deficits in broadleaf plants. Protoplasma 62:381–391.

Huber, B. 1927. Zur methodik der Transpirations bestimmung am Standort. Ber. Deut. Bot. Ges. 45:611–618.

Ivanoff, L. 1928. Zur methodik der Transpirations bestimmung am Standort. Ber. Deut. Bot. Ges. 46:306–310.

Kedem, O., and A. Katchalsky. 1958. Thermodynamic analysis of the permeability of biological membranes to nonelectrolytes. Biochim. Biophys. Acta 27:229–246.

Korven, H. C., and S. A. Taylor. 1959. The Peltier effect and its use for determining relative activity of soil water. Can. J. Soil Sci. 39:76–85.

Kreeb, K. 1960. Uber die Gravimetrische Methode zur Bestimmung der Saugspannung unde das Problem des negativen turgors: I. Planta 55:274–282.

Kreeb, K., and M. Onal. 1961. Uber die Gravimetrische Methode zur Bestimmung der Saugspannung und das Problem des negativen turgors: II. Planta 56:406–415.

Kuiper, P. J. C. 1961. The effects of environmental factors on the transpiration of leaves with special reference to stomatal light response. Mededel. Landbouwhogeschool. 61:1–49.

Lloyd, F. E. 1908. The physiology of stomata. Carnegie Inst., Washington, D. C. 82:1–142.

Long, F. L., and F. E. Clements. 1934. The method of collodion films for stomata. Amer. J. Bot. 21:7–17.

Mason, T. G., and E. Phillis. 1939. Experiments bearing on the extraction of sap from the vacuole of the leaf of the cotton plant and their bearing on the osmotic theory of water absorption by the cell. Ann. Bot. 3:531–544.

Maximov, N. E. 1929. The plant in relation to water. Allen and Unwin, London. 451 p.

Mederski, H. J. 1961. Determination of internal water status of plants by beta ray gauging. Soil Sci. 92:143–146

Meidner, H., and D. C. Spanner. 1959. The differential transpiration porometer. J. Exp. Bot. 10:190.

Mercer, F. M. 1955. The water relations of plant cells. Linnean Soc., Proc., New S. Wales. 80:6–29.

Meyer, B. S., and D. B. Anderson. 1952. Plant physiology. D. van Nostrand, Princeton, New Jersey. 784 p.

Miller, E. C. 1938. Plant physiology. McGraw-Hill, New York. 1,201 p.

Milthorpe, F. L. 1955. The significance of the measurement made by the cobalt chloride paper method. J. Exp. Bot. 6:17–19.

Molisch, H. 1922. Das Offen-und Geschlossensein der Spaltoeffnungen, veranschaulicht durch eine neue Methode. Z. Bot. 4:106–122.

Monteith, J. L. 1963. Gas exchange in plant communities. p. 95–112. In L. T. Evans [ed.] Environmental control of plant growth. Academic Press, New York.

Monteith, J. L., and P. E. Owen. 1958. A thermocouple method for measuring relative humidity in the range 95–100 per cent. J. Sci. Inst. 35:443–446.

Moreshet, S. 1964. A portable wheatstone bridge porometer for field measurements of stomatal resistance. Israel J. Agr. Res. 14:27–30.

Nadel, M. 1935. On the influence of various liquid fixatives on stomatal behavior. Palestine J. Bot. Hort. Sci. 1:22–47.

Nadel, M. 1940. Sur la mesure de l'ouverture des stomates. Palestine J. Bot. Rehovot Ser. 3:2–64.

Nakayama, F. S., and W. L. Ehrler. 1964. Beta ray gauging technique for measuring leaf water content changes and moisture status of plants. Plant Physiol. 39:95–98.

Pasquill, F. 1949. Some estimates of the amount and diurnal variation of evaporation from a clay land pasture in fair spring weather. Quart. J. Roy. Meteorol. Soc. 75:249–256.

Rawlins, S. 1964. A systematic error in leaf water potential measurements with a thermocouple psychrometer. Science 146:644–646.

Richards, L. A., and G. Ogata. 1958. A thermocouple for vapor pressure measurement in biological and soil systems at high humidity. Science 128:1089–1090.

Schorn, M. 1929. Untersuchungen uber die Vervendbarkeit der Alkoholfixierungsund Infiltrations Methode zur Messung von Spltoffnungsweiten. J. Wiss. Bot. 71:783–840.

Shardakov, V. S. 1957. Principles for determining the watering periods of the cotton plant in relation to the magnitude of the suction force of leaves (from Russian). 1:5–32. *In* Physiological questions in cotton and grasses. Akad. Nauk. Uzbek. SSR (Tashkent).

Shimshi, D. 1963. Effect of soil moisture and phenylmercuric acetate upon stomatal aperture, transpiration, and photosynthesis. Plant Physiol. 38:713–721.

Shmueli, E. 1953. Irrigation studies in the Jordan Valley: I. Physiological activity of the banana in relation to soil moisture. Bull. Res. Counc. Israel. 3:228–247.

Slatyer, R. O. 1955. Studies of the water relations of crop plants grown under natural rainfall in Northern Australia. Australian J. Agr. Res. 6:365–377.

Slatyer, R. O. 1957. The influence of progressive increases in total soil moisture stress on transpiration, growth and internal water relationships of plants. Australian J. Biol. Sci. 10:320–336.

Slatyer, R. O. 1958. The measurement of diffusion pressure deficit in plants by a method of vapor equilibration. Australian J. Biol. Sci. 11:349–365.

Slatyer, R. O. 1960. Aspects of the tissue water relationships of an important arid zone species (*Acacia aneura* F. Muell) in comparison with two mesophytes. Bull. Res. Counc. Israel. 8D:159–168.

Slatyer, R. O. 1961. Internal water balance of *Acacia aneura* F. Muell in relation to environmental conditions. UNESCO Arid Zone Res. 16:137–146.

Slatyer, R. O., and H. D. Barrs. 1965. Modifications to the relative turgidity technique with notes on its significance as an index of the internal water status of leaves. UNESCO Arid Zone Res. 25:331–342.

Slatyer, R. O., and J. F. Bierhuizen. 1964. Transpiration from cotton leaves under a range of environmental conditions in relation to internal and external diffusive resistances. Australian J. Biol. Sci. 17:115–130.

Spanner, D. C. 1951. The Peltier effect and its use in the measurement of suction pressure. J. Exp. Bot. 11:145–168.

Stalfelt, M. G. 1956. Die Stomatare Transpiration und die Physiologie der Spaltcffnungen. Handel. Pflanzenenphys. 3:351–426.

Stocker, O. 1929. Wasserdeficit von Gefassplanzen in verschiedenen Klimazonen. Planta 7:382–387.

Vaadia, Y., and A. G. Marr. 1961. Rapid cryoscopic technique for measuring osmotic properties of drop size samples. Plant Physiol. 36:677–680.

Weatherley, P. E. 1950. Studies in the water relations of the cotton plant: I. The field measurement of water deficits in leaves. New Phytol. 49:81–87.

Weatherley, P. E. 1951. Studies in the water relations of the cotton plant: II. Diurnal and seasonal variations in relative turgidity and environmental factors. New Phytol. 50:36–51.

Weatherley, P. E. 1955. On the uptake of sucrose and water by floating leaf disks under aerobic and anaerobic conditions. New Phytol. 54:13–28.

Weatherley, P. E. 1961. A new micro-osmometer. J. Exp. Bot. 11:258–268.

Weatherley, P. E., and R. O. Slatyer. 1957. The relationship between relative turgidity and diffusion pressure deficit in leaves. Nature 179: 1085–1086.

Weinmann, H., and M. LeRoux. 1948. A critical study of the torsion balance method of measuring transpiration. S. Africa J. Sci. 42:147–153.

Werner, H. O. 1954. Influence of atmospheric and soil moisture conditions on diurnal variations in relative turgidity of potato leaves. Nebraska Agr. Exp. Sta. Bull. 176. 39 p.

Whiteman, P. C., and G. L. Wilson. 1963. Estimation of diffusion pressure deficit by correlation with relative turgidity and Beta radiation absorption. Australian J. Biol. Sci. 16:140–146.

Wilson, C. C. 1942. The porometer method for the continuous estimation of stomata. Plant Physiol. 22:582–589.

Yamada, Y., S. Tamai, and T. Miyaguchi. 1954. The measurement of the thickness of leaves using S–35. 2nd Japanese Conf. Radioisotopes, Proc.

Zelitch, I. 1961. Biochemical control of stomatal opening in leaves. Nat. Acad. Washington, Proc. 47:1423–1433.

19

Physiological Processes as Affected by Water Balance

YOASH VAADIA

Negev Institute of Arid Zone Research
Beersheva, Israel

YOAV WAISEL

Tel Aviv University
Tel Aviv, Israel

I. INTRODUCTION

This chapter is not intended as a comprehensive review of the relevant scientific literature, as several are already available (Kramer, 1959; Stocker, 1960; Oppenheimer, 1960; Vaadia et al., 1961). Instead, it will attempt to analyze the objective difficulties facing researchers, and to provide a brief account of current knowledge in this field.

Provision of a clear summary of the effects of water balance on physiological processes would be desirable, but it is well to state at the outset that this subject cannot be treated as fully as it deserves in this chapter. The considerable literature on the subject is fraught with many ambiguities, contradictions, and lack of sufficient basic information.

II. PROBLEMS OF DEFINITION AND MEASUREMENT

Studies on the influence of plant water deficits on physiological processes are hampered by several experimental difficulties; a major obstacle is the definition of water balance itself. It is known that in the daily course of transpiration, water deficits develop within the plant even when soil water is ample (Kramer, 1949). Such deficits are usually replenished at night. Thus, even under controlled conditions, plants are subjected to a diurnal periodicity of water deficits which may affect metabolism in a manner different from that of a constant deficit. Studies on the effect of water deficits on specific metabolic pathways are usually of short duration, and this limits their usefulness.

Furthermore, water deficits are not easily measured. The concept of "relative turgidity" provides one criterion (Weatherly, 1950). This, essentially, is the relative water content of water deficient tissues as compared to nondeficient tissues. Since most of the water in the tissues of growing higher plants is in the vacuolated and nonliving (cell wall) regions (Carr and Gaff, 1961), a change in relative turgidity is not a sensitive measure of change in the living protoplasm itself.

As discussed in chapter 18, the accurate measurement of tissue and leaf water potentials is not simple. Since relative turgidity can be determined easily, attempts have been made to establish its relationship to leaf water potentials (Weatherly and Slatyer, 1957). If water potentials could be shown to vary uniquely with relative turgidity, the latter could serve to estimate leaf water potentials. The relationship, however, is not exclusive; it varies with plant species, leaf age, previous history, and other factors (Whiteman and Wilson, 1963). Thus, estimation of leaf water potentials from relative turgidity data is somewhat unreliable.

Available methods for estimating leaf water potentials share the common drawback of using detached tissue samples, and therefore result in tissue equilibrium values in which the tensions developed during transpiration are absent. Excision of a transpiring leaf may result in a considerable increase of the water loss from that leaf (Rufelt, 1963; Aston and Vaadia, 1963; Willis et al., 1963). Thus, cutting modifies the water relations of the leaf, and in all likelihood, its water potentials as well. A further difficulty in interpreting results of water potential measurements in plant tissues is that they are made under the assumption that all the parts of the cell or tissue are at the same water potential. This assumption can be seriously questioned, particularly under conditions of rapid transpiration (Carr and Gaff, 1961). There are indications that considerable tensions can develop in some parts of corn plant (*Zea mays* L.) leaves under conditions of normal transpiration, without any symptoms of leaf wilt or curl (Shimshi, 1963). Direct measurement of leaf water potentials in such tissues are not likely to show high values.

Apart from the problems of measurement and definition of water deficits, the expression of results requires caution and presents some difficulties. Weatherly (1951) and others point out that the water content of tissues expressed on dry weight basis decreases during the growing season because dry weight increases with age. These changes occur in the absence of severe water deficits (Shah and Loomis, 1965). Thus, since dry weight is not a constant base for comparison, it should be used with caution. A demonstration of this type of difficulty is given by the data of Gates and Bonner (1959). They showed no differences in the nucleic acid contents of leaves of water stressed and control tomato plants (*Lycopersicon esculentum*) when expressed on dry weight basis, but consistent differences when expressed on per lamina basis.

The above discussion illustrates some of the objective difficulties that we face in establishing quantitative functional relations between plant water deficits and physiological processes. Despite difficulties in measuring and expressing results, there has recently been an overdue revival of interest in the way that water deficits influence physiological processes and reduce measurable plant growth.

III. CELL WATER RELATIONS AND GROWTH

It is generally agreed that one way in which water deficits decrease growth is to decrease turgor (Kramer, 1949; Crafts et al., 1949; Ordin et al., 1956; Slatyer, 1961). Several workers showed that cell elongation is correlated with cell turgor pressure. Cell enlargement theories consider turgor pressure as a primary cause in the process, either through actual stretching of the cells or through increased water uptake by the cells as a result of increased extensibility of the cell wall and decreased turgor pressure (Broyer, 1950; Burström, 1954, 1956; Clark and Levitt, 1956; Cleland and Bonner, 1956).

Turgor pressure of the guard cells regulates stomatal opening (Kettelaper, 1963) and, thereby, gas exchange between leaf and atmosphere (Kuiper, 1961).

Turgor pressure has been invoked to explain the downward translocation of metabolites in plants (Zimmerman, 1960) and is thus presumed to exercise considerable influence on the distribution of photosynthetic and other products in the plant (Pallas, 1960; Wilson and McKell, 1961). It is, therefore, unfortunate that it is not yet possible to measure turgor pressure reliably and directly and that all the above information was obtained without quantitative turgor pressure measurements.

Changes in turgor pressure are generally associated with changes of water potential. Therefore, it is very important to establish whether the observed effects of reduced growth under water deficits is due to increasing osmotic pressure and decreasing water potential, or to reduced turgor pressure.

Ordin (1960) tried to establish whether the decreased growth—particularly as expressed in the incorporation of C^{14} in cell wall fractions—was affected by the higher osmotic pressures of cells maintained under water stress or by the decreased turgor pressure of the cells. He studied the incorporation of C^{14} into various cell wall fractions of *Avena* coleoptiles exposed to different water potentials obtained by pretreatment in mannitol and salt solutions. The data showed that increased internal osmotic pressure does not disturb the metabolism of noncellulosic polysaccharides but does interfere with the formation of cellulose. Turgor pressure affected both cell wall metabolism and cell elongation. On the basis of these and other observations, Ordin suggested that some aspects of cellulose synthesis may be involved in the elongation response to turgor. Moreover, the work showed that high osmotic pressures as well as low turgor pressure affected synthesis, so that an explanation is required for the manner in which positive cytoplasmic pressure on the wall or high internal concentrations can affect synthesis. Ordin proposes some possibilities.

The relative importance of reduced turgor pressure or increased osmotic concentrations on growth has been a topic in several recent papers (Slatyer, 1961; Bernstein, 1961, 1963) on investigations of plant adjustment to high concentration salinity and nonelectrolytic osmotic substrates. Decreased turgor pressure has long been considered part of the explanation for decreased growth of plants under high salinity (Wadleigh and Gaugh, 1948). It has been argued that under high salinity the gradient for water uptake is smaller and water absorption is slower

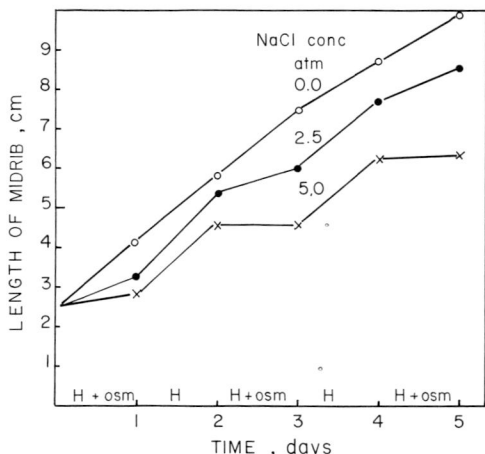

Fig. 19–1. The effect of NaCl at various concentrations on leaf growth of bean plants. H = Hoagland solution; $H + Osm$ = Hoagland plus NaCl solution at the concentrations indicated on the curves (adapted from Brouwer, 1963).

Table 19–1. Growth and osmotic potentials (OP) of cotton plants at 4 different levels of salinity (NaCl) in the nutrient solution (adapted from Bernstein, 1961)

Added NaCl	Top fresh weight	Root fresh weight	OP of expressed saps and OP_a*					
			Leaves		Stems		Unrinsed roots	
			OP	OP_a*	OP	OP_a*	OP	OP_a*
atm	g	g			atm			
0	268	94	10.2	9.8	10.3	9.9	4.9	4.5
3	185	75	12.5	9.1	11.5	8.1	8.8	5.4
6	165	69	17.9	11.5	13.9	7.5	11.2	4.8
12	107	46	21.3	8.9	22.4	10.0	16.7	4.3

* OP_a = differences between sap and culture solution.

Table 19–2. Osmotic potentials (OP) observed in tomato shoots 28 hr after the roots were immersed in various osmotic substrates (adapted from Slatyer, 1961)

Substrate	Low concentration treatments (5 atm)			High concentration treatments (10 atm)		
	OP of substrate	OP of expressed sap	Difference	OP of substrate	OP of expressed sap	Difference
	atmospheres					
Potassium nitrate	5.7	14.9	9.2	10.7	20.8	10.1
Sodium chloride	5.7	16.7	11.0	10.7	21.6	10.9
Sucrose	5.7	15.2	9.5	10.7	19.7	9.0
Control	0.7	10.9	10.2	0.7	10.9	10.2

(Hayward and Spurr, 1944). Results such as those shown in Fig. 19–1 are usually explained on this basis. In this case, because of relatively short exposure to the salt solution, the plants had not yet achieved osmotic adjustment.

Both Eaton (1942) and Slatyer (1961) showed, however, that there is some adjustment of plant parts, such as leaves and stems, to increased salt concentration and hence to the osmotic pressure of the external solution. Bernstein (1961) has added the important information that such an adjustment also occurs in roots. Such adjustments occur rapidly, within a day (Slatyer, 1961; Bernstein, 1963), and result in considerable and sometimes complete restoration of the osmotic water uptake gradient (Boyer, 1965). Tables 19–1 and 19–2 and Fig. 19–2 show typical results of osmotic adjustment of plants. These results indicate that the

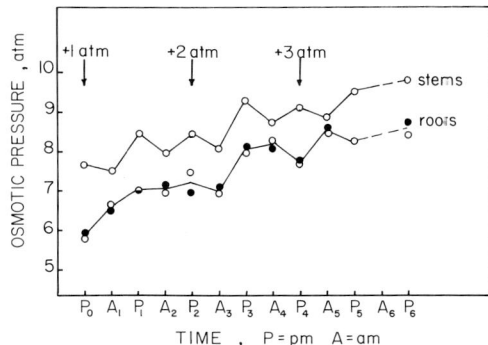

Fig. 19–2. Time course of osmotic adjustment of roots and stems of bean plants. Sodium chloride added to base Hoagland solution to give osmotic pressures indicated at arrows. Harvest at 6 PM and 6 AM (adapted from Bernstein, 1963).

estimated absolute turgor pressures should be similar for both normal plants growing in solutions of low osmotic pressure and for adjusted plants growing in solutions of higher osmotic pressure. Since the water potential of adjusted plants is lower than that of plants not subjected to salinity stresses, their turgor could not be their full turgor pressure at any time.

If observations of osmotic adjustment in osmotic solutions are correct, one should expect the transpiration rates of adjusted plants to be similar to those of normal plants. Slatyer (1961) reports that the transpiration rates of plants transferred to solutions of 5 atm KNO_3 or sucrose are equal to those of the control after 24 hours. This is shown in Fig. 19–3 and 19–4. More detailed demonstration of similar transpiration rates in salt-adjusted and control cotton plants (*Gossypium herbaceum*) is provided by Boyer (1965).

On the other hand, results showing incomplete recovery of transpiration rates— in fact, indicating continued decrease of transpiration—are also available (Brouwer, 1963). It is possible that in these cases the absorbing roots were either not fully adjusted or damaged. Damage will usually occur if, in adjustment studies, the concentration of the medium is abruptly, and not gradually, increased (Bernstein, 1961).

The water relations of adjusted plants must be further investigated since factors other than deficient water supply appear to control growth. This would provide better understanding of the influence of water potential and its components, osmotic potential and turgor pressure, as well as others (Slatyer and Taylor, 1960) on plant growth. The complexity of this relationship is indicated in a brief and incomplete paper by Meyer and Gingrich (1964), among others. They showed that a slight osmotic stress (1 bar with Carbowax 6000) applied to only some of

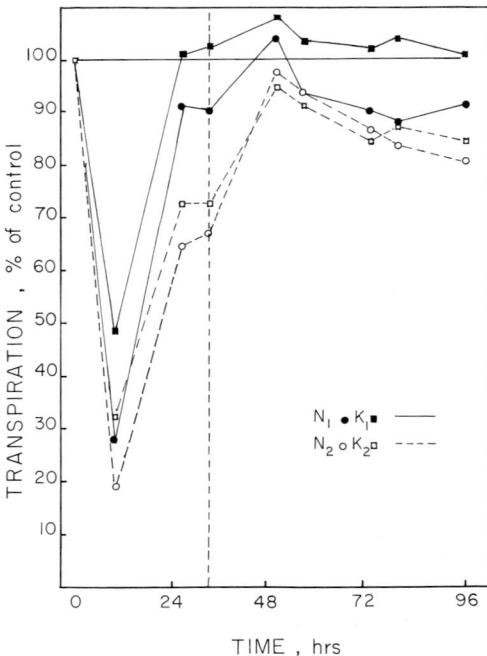

Fig 19–3. Changes in transpiration, expressed as percentage of the control of plants transferred to 5– (N_1; K_1) and 10– (N_2; K_2) atm solutions of NaCl (N) and KNO_3 (K). Plants transferred at time 0 and returned to Hoagland solution after 28 hr (adapted from Slatyer, 1961).

a wheat plant's roots (*Triticum vulgare*) significantly affected certain metabolic fractions throughout the plant. In this case the results are not attributable to changes of the plants' turgidity or water status. They do indicate that some specific effect in the root subsequently disturbs metabolism throughout the plant. However, toxicity of Carbowax cannot be ruled out.

When adjusted plants are transferred to solutions of lower concentration they may exhibit greater growth than normal (Slatyer, 1961). Similar observations have been made after relief of soil water stress (Owen, 1958; Kemper et al., 1961). This accelerated growth may be the result of increased tissue turgor pressure due to enhanced water uptake. It is possible that both increased soil suction and increased concentration of the soil solution cause an increase in tissue osmotic pressures during soil water depletion cycles. Bauman (1957) showed that the osmotic pressure of leaves of several crops increased by 10 to 20 atm (without visible symptoms of wilting) when irrigation was withheld, while the increase was insignificant in the irrigated control.

In soils, however, limited water supply must play an important role in limiting growth. The hydraulic conductivity of the soil decreases during a depletion cycle. Thus, if water supply to the roots is to remain constant, the decrease in plant water potential must exceed the decrease in soil water potential. Since changes of hydraulic conductivity associated with small changes of soil water potential are large, this is not possible. Nevertheless, plant adjustment to decreasing soil water potentials must be considered in future experiments because it may involve an influence on growth that is not directly related to the plant's water supply at a given moment.

Quantitative experiments intended to distinguish between the effects of plant adjustment and water supply are difficult but not impossible. They are essential

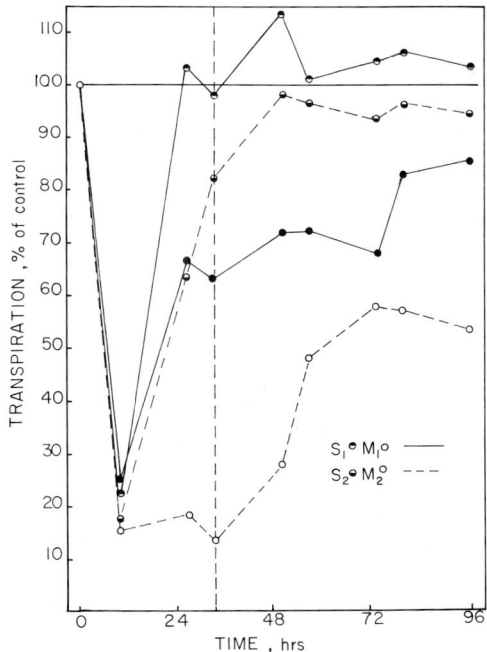

Fig. 19–4. Changes in transpiration, expressed as percentage of the control of plants transferred to 5– (M_1; S_1) and 10– (M_2; S_2) atm solutions of mannitol (M) and sucrose (S). Other details as in Fig. 19–3 (adapted from Slatyer, 1961).

for assessing the effects of water supply limitation on growth. The problem is to establish quantitatively the extent to which the cited increases in tissue osmotic pressure (Bauman, 1957) represent an osmotic adjustment similar to that reported by Slatyer (1961) and Bernstein (1961, 1963), and to what extent it is due to reduced water contents associated with deficits developed and to metabolic shifts essentially different from the osmotic adjustment process.

If, as appears likely, the observed reduction in growth of osmotically-adjusted plants is not due to water supply limitation and reduced turgor, what other factors could inhibit growth processes?

IV. PROTOPLASMIC HYDRATION

Protoplasmic hydration is a frequently cited reason for changes in physiological activity under water deficits. (Iljin, 1953; Kramer, 1959; Levitt, 1956, 1962). The importance of protoplasmic hydration is best exemplified in seeds where physiological activity increases very rapidly with increased water content (Owen, 1958). However, many physiological processes occur at very low water contents, e.g., all the metabolic and morphogenetic processes occurring during vernalization (Waisel et al., 1962). Water deficits in growing tips, both in division and elongation regions, reduce growth rates. Nevertheless, although these cells suffer high water stresses, their tolerance to dehydration is very high, exceeding any other type of active cell (Wilson, 1948; Waisel, 1962a,b). Plant hardiness, specifically drouth and frost hardiness, seem, among other factors, to be related to the relative amount of bound water in the tissue, i.e. the more water held at low potentials in the tissue, the more tolerant the plant. Bound water is considered important since it prevents the protein molecules of active sites from combining and coagulating. Exclusion of bound water from cytoplasm under freezing facilitates the oxidation of protein —SH groups to —S—S— bonding, which leads to irreversible coagulation (Levitt, 1962).

There is much experimental evidence indicating the importance of protoplasmic hydration, but the detailed evaluation of mode of action is rare. Proteins are the major nonaqueous component of protoplasm and their hydration will largely determine their spatial configuration. Klotz (1958) concluded that hydrated protein molecules have water envelopes and that in them the molecules have a configuration similar to that of ice. Configuration of the hydration water can be markedly influenced by various factors, including water potential. A change in this potential may result in reorientation and changes in the configuration of various protein molecules. These changes and the possible alteration of distances between enzyme groups may modify enzyme reaction rates and offset the normal balance of specific reactions and metabolites. The pattern of tissue physiological activity would therefore change. As already pointed out by Skoog (1955) the nature of a tissue's metabolism is correlated with the ratio between various metabolites. Changes in these ratios are, as will be shown, usually observed under water stresses.

This cursory and hypothetical picture of the influence of protoplasmic hydration on growth processes was used by Van Overbeek (1959) to hypothesize the mode of auxin action. He postulated that association between the auxin molecule and protein causes a physiochemical change in protein hydration, thus altering reaction rates and the tissue's physiological pattern.

V.   PHOTOSYNTHESIS

The relation between water stresses and photosynthesis is of prime importance for agriculture and has been studied extensively (*see* Kramer, 1959; Vaadia et al., 1961). In most investigations, water deficits were regulated by limiting the soil water supply to the plants; only a few provided rigorous environmental control, and even less describe actual water deficits in plant tissues (Gaastra, 1959; Brix, 1962; Moss et al., 1961). The results therefore do not always lend themselves to valid comparison.

Decreased CO_2 supply and diffusive capacity are usually cited as the most common factors limiting photosynthesis under a given set of conditions. Several recent papers have considered these problems (Gaastra, 1959; Kuiper, 1961). Gaastra (1959) showed that CO_2 saturation concentration for photosynthesis was approximately 0.1%. He further showed that with low light intensity, considerable stomatal closure may occur without affecting photosynthetic rate, and that with light saturation and normal atmospheric CO_2 content, the rate of CO_2 diffusion into the leaf determines photosynthetic rate. Since plant water balance largely determines the degree of stomatal opening, it controls photosynthesis by regulating the rate of CO_2 diffusion into the plant. Gaastra also showed that apart from diffusion resistance through air and stomates there is measureable resistance to CO_2 transfer in the mesophyll. Mesophyll resistance can be as great, or greater, than the stomatal resistance in normal air and saturating light intensities.

The nature of mesophyll resistance is yet to be determined and is not necessarily only diffusional. It could well differ in leaves of different water stress or age, and may well include a direct metabolic inhibitory effect of water stress on photosynthesis.

Gaastra's findings neatly explain the reason for transpiration being more dependent than photosynthesis on stomatal opening. Transpiration depends on the algebraic sum of resistances in the stomates and surrounding air, whereas photosynthesis depends on both these and mesophyll resistance. Gaastra's work shows that investigations on the effect of water stress on photosynthesis may yield conflicting results if they are conducted at different light intensities. Experiments showing no decrease in photosynthesis (until plants started to wilt) with increasing water deficits (Upchurch et al., 1955) were probably run at limiting light intensities. Because of stomatal closure, CO_2 diffusion was not reduced sufficiently to affect photosynthetic rates. At saturating light intensities, photosynthesis generally decreases as tissue water deficits increase.

It is not clear to what extent the physical considerations such as outlined above are responsible for decreased photosynthesis under water stress. It is possible that a direct effect of water stress on the biochemical processes themselves would also contribute to the observed reduction. As yet, however, this has not been clearly demonstrated. One possible demonstration is provided in Fig. 19–5. Here, photosynthetic rates may be four times as great in leaves with small water deficits as compared with water-deficient leaves. This may indicate either that mesophyll resistance in water-deficient leaves is considerable or a direct effect on the biochemical processes themselves.

Fig. 19–5. Relationship between net photosynthesis, stomatal width, and water
deficit for leaves of *Asarum euro-
paeum*. Water deficit expressed as the
difference between weight of leaf saturated with water and actual leaf
weight in percentage of the former.
Light source: Warm white fluorescent
lamps, 10,000 lux (data from Pisek
and Winkler, 1956, as presented by
Gaastra, 1963).

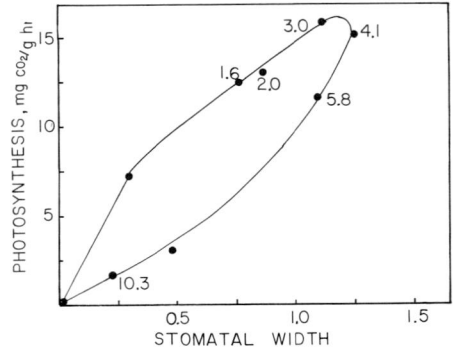

Shimshi (1963) provided another example. He showed that at a given soil water content the stomates had greater control of transpiration than of photosynthesis; this fully agrees with Gaastra's theory. With low soil water treatments, however, photosynthesis was much more reduced than transpiration. This could be interpreted as a direct effect of leaf water deficits on photosynthesis, although the possibility of increased mesophyll resistance with increasing water deficits cannot be ruled out.

Several observations indicate that water deficits affect photosynthetic metabolism less than elongation. Owen (1958) showed that short periods of water stress (up to 10 atm) caused no reduction in dry matter production or net assimilation rates of CO_2. However, leaf area of the stressed sugar beet (*Beta vulgaris* L.), broad bean (*Vicia faba*), and lettuce (*Lactuca sativa*) was significantly smaller than the control. Hagan et al. (1957) and Gingrich and Russell (1956, 1957) have shown that tissue growth as measured by dry weight is less affected by water deficits than is elongation.

Several investigators (Ashton, 1956; Upchurch et al., 1955) showed that photosynthetic rates decrease rapidly or even approach zero when wilting begins (Fig. 19–6), and, furthermore, that recovery of original rates after re-irrigation takes several days (Ashton, 1956). Similar results have been obtained for transpiration (Kramer, 1950); it appears that this phenomenon largely depends on damaged root system or other tissues, which in turn have a degree of control on the stomatal reaction, and hence transpiration and photosynthesis. Maize (*Zea mays* L.) stomates remain closed for a long time after short periods of water stress, but sorghum (*Sorghum vulgare*) stomates recover quickly from much longer stress periods (Glover, 1959). Fast recovery, high water use efficiency, and other factors (Wit, 1958) account for sorghum's superiority as a dry region crop.

In conclusion, it may be said that observed rates of photosynthesis under water deficits are generally reduced. Such reductions are related to the water balance of the tissue, both through the effects that such deficits may have on CO_2 diffusion rates, and through the possible effects of changes in protoplasmic hydration under these conditions. At present, there is no way of predicting the extent to which photosynthesis may be reduced under given water deficits and environmental conditions.

VI. RESPIRATION

Experimental results on the effect of water deficits on respiratory rates of plant tissues are confusing. Figure 19–6 illustrates this well; it shows that with increasing water deficits, respiratory rates decrease, later increase, and at still higher deficits, decrease again (*see also* Stocker, 1960). It appears that little can be gained by making simple measurements of respiration in water-deficient tissues. The confusion may be resolved, but only after detailed analysis of the various factors involved in the experimental values.

Plant protoplasm is continuously changing, due to the synthesis of compounds and structures and their degradation. Synthetic processes require energy; degradation releases energy in the form of heat or high energy bonds that generally energize formation processes. Heat evolved during biochemical reactions represents an irreversible loss of energy to the plant. Heat loss may result from respiratory inefficiency and other causes. The measurement of respiratory rates tells little about the plant's energy status. If, during periods of water deficits, degradation reactions are accelerated, the heat evolved by the tissue will increase. If, under such conditions, respiratory rates remain constant, the energy level of the tissue will decrease and anabolic processes will begin. More accurate assessment of energy changes occurring in a tissue must include direct measurement of heat evolved by the tissue as well as respiratory rates. Such measurements would give an indication of the plant's efficiency of energy utilization under various conditions. Unfortunately, measurement of heat losses from plants is not simple. Zholkevitch (1961) attempted to do so by using microcalorimetry.

Both *Ranunculus* and *Rumex* were subjected to two levels of water deficits,

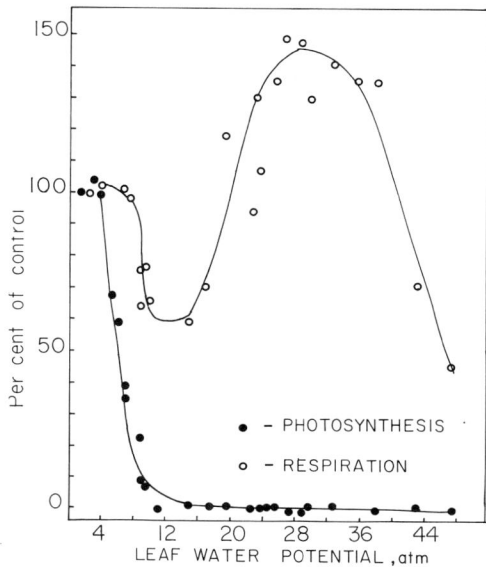

Fig. 19–6. The effect of water stress on the rates of photosynthesis and respiration in loblolly pine seedlings. Water stress is expressed as the water potential (in atm) of the leaves. Rates of photosynthesis and respiration are given as the percentage of the rates with soil water at field capacity (data of Brix, 1962).

Table 19–3. Energy balance of respiration (adapted from Zholkevitch, 1961)

Plant	Treatment	Water content	E_{resp}	Q	Q/E
		%wet wt	cal/100g/hr		
Ranunculus repens	1 Irrigated	81.5	160	119	0.74
	nonirrigated	81.2	152	83	0.54
	2 Irrigated	81.7	188	182	0.97
	Nonirrigated	76.1	263	352	1.34
Rumex confertus	3 Irrigated	83.0	140	137	0.98
	Nonirrigated	72.4	322	245	0.76
	4 Irrigated	83.6	178	173	0.97
	Nonirrigated	66.1	330	435	1.32

one being greater, as indicated by the tissues' water contents (Table 19–3). The ratio of Q (heat evolved) to E_{resp} (respiratory energy produced) decreases with slight water deficits (treatments 1 and 3), but as water deficits increase (treatments 2 and 4), the heat evolved from the tissue exceeds the maximum respiratory energy available to the plant. This represents rapid degradation of the tissue. Zholkevitch's data are insufficient for a complete interpretation of respiratory efficiency, but they are the first that attempt to include a measure of the plant's energy balance as a function of water deficits. This approach, coupled with other measurements such as oxidative phosphorylation in deficient and nondeficient tissues may yield important data on the basic nature of the metabolic shifts induced by water deficits. One must add to the above considerations, which all occur in the dark, the importance of light to the tissue energy balance. Light will modify the energy balance both because of irradiated heat and because photophosphorylation, as well as respiratory oxidative phosphorylation, probably occurs in intact plants (Arnon, 1961). Wilson and Huffaker (1964) have analyzed the gross effects of water stress on phosphorus metabolism and changes in levels of various phosphorylated compounds in subterranean clover (*Trifolium subterraneum*). Little change could be observed in phosphorylated intermediates until near wilting but substantial reductions were noted at greater water deficits. The main contribution of this work is in pointing to a new direction which should be pursued more actively.

Respiration rates in different plant species respond differently to water deficits (Brix, 1962). In Fig. 19–6 respiratory rates of loblolly pine (*Pinus taeda* L.) vary with increasing deficits. However, in tomato plants (*Lycopersicon esculentum* Mill.) grown under similar conditions, respiratory rates decline gradually with increasing deficits. According to Takaoki (1957), halophytic and glycophytic species' respiration behaves differently when exposed to a series of solutions of increasing salt concentrations. This only shows that respiration rates in themselves are not meaningful measures of the metabolic status of the plant.

VII. PLANT METABOLISM

Many papers report specific changes in composition and metabolism in water deficient plants, but again, many contradictions can be found. However, since it is clear that some reactions are more sensitive to water stresses than others (Ordin, 1958 & 1960), some shifts in metabolism can be expected under water deficits.

Table 19–4. Some experimental observations on changes in carboyhdrate compostion of water deficient plants*

Species	Tissue	Stated nature of water stress	Carbohydrate fraction	Change observed	Reference
Common bean	leaves stems	high soil water suction	sucrose	decrease	Wadleigh & Ayers (1945)
Common bean	leaves stems	one and four cycles of drouth	starch sugar	decrease	Woodhams & Kozlowski (1954)
Tobacco	leaves	high soil water suction	sugar	increase	Petrie & Arthur (1943), Van Bavel (1953)
Cotton	leaves	different irrigation regimes	hexose sugars, starch	increase decrease	Eaton & Ergle (1948)
	stems		hexose sugars, starch	decrease	
Cotton	leaves	nonirrigated	sugar	increase	Yarosh (1959)
Pumpkin	leaves roots	low soil water content	sugar	increase	Zholkevitch & Koretskaya (1959)
Sugar beet	storage	different water regimes	sucrose	decrease	Haddock (1959)
Sugar beet	storage tissue	one drouth prior to harvest	sucrose	increase	Hunter & Yungen (1952), Wiersma †
Sugar beet	storage tissue	high salinity	sucrose	increase	Waisel & Bernstein (1959)

* Adapted from C. B. Shah, 1962. Physiological studies on the growth responses of the sugar beet to moisture stress. Ph. D. diss., Univ. Calif., Davis.
† D. Wiersma, 1956. Soil moisture conditions and sugar accumulation in the sugar beet. Ph. D. thesis, Univ. Calif., Davis.

A. Carbohydrate Metabolism

Some of the experimental observations on changes in carbohydrate composition of water-deficient plants are summarized in Table 19–4.

In well watered plants, photosynthates are primarily used in metabolism and growth processes, or are stored in different organs as starch or sugars. Since photosynthesis is usually reduced by water deficits while respiration may frequently increase, the observation of less assimilates can be expected. Water deficits are also reported to decrease the total content of organic acids and phosphorylated compounds in the leaves, and to inhibit transamination reactions and growth processes (Zholkevitch and Koretskaya, 1959), thereby slowing down carbohydrate utilization. The balance of stored carbohydrate in water-deficient leaves will depend on relative magnitude of the depression of growth and photosynthesis, changes of respiration, and variations in translocation. For these reasons, the results appearing in the above table need not necessarily be conflicting. Metabolic pathways may be expected to be modified under water and salinity stresses.

B. Nitrogen Metabolism

Relatively more progress has been made in studies on nitrogen than on carbohydrate metabolism in water-deficient tissues. Advances in nitrogen research and increased interest in nucleic acid and protein metabolism have stimulated their study under water deficits.

Protein content in various plant tissues declines with age, due to increased proteolysis and/or decreased synthesis (Robinson, 1956 & 1960; Palmcrantz, 1957; Pirie, 1959). Similar observations are known for water-deficient tissues (Mothes, 1933; Petrie and Wood 1938a). Petrie and Wood (1938b) and Yarosh (1959) observed that protein synthesis decreased during water stresses. Yarosh reported a ratio of 0.14:0.24 of nonprotein to protein nitrogen in stressed cotton plants and a ratio of 0.11:0.19 in irrigated ones. He attributed this to both decreased synthesis and increased hydrolysis under water deficits. Ivanov (1959) showed lower amino acid contents in stressed corn plants. Zholkevitch and Koretskaya (1959) observed a decrease in organic acids and an inhibition of amination reactions during drouth in pumpkin (*Cucurbita pepo*). Prusakova (1960) found larger amounts of tryptophan in control plant wheat leaves than in water-stressed leaves.

Since ribonucleic acid (RNA) is considered to be intimately involved in protein synthesis (Bonner, 1959; Watson, 1963), the effects of age and water stress on RNA metabolism are relevant to the above-mentioned results. Gates and Bonner (1959) showed that the amounts of both deoxyribonucleic acid (DNA) and RNA per tomato leaf were lower under water deficits. Their data suggested that the observed decrease was not attributable to decreased synthesis, since the rate of P^{32} incorporation into nucleic acids remained unchanged for irrigated and stressed plants. They therefore concluded that enhanced degradation occurs under water deficits. Kessler (1961) reached similiar conclusions; he showed that RNA-ase activity is enhanced under water stress, resulting in a lower RNA content. Shah and Loomis (1965) amplified on this point. Instead of the P^{32} of Gates and Bonner, they studied the incorporation of labelled uracil into RNA of stressed and control sugar beet leaves (*Beta vulgaris* L.). Their results suggest both decreased synthesis and increased destruction of RNA in water-deficient tissues. A parallelism between the reduction with age and water deficits in protein content and RNA was also demonstrated. Benzyladenine application helped to maintain RNA and protein levels during soil water depletion, as shown in Fig. 19–7.

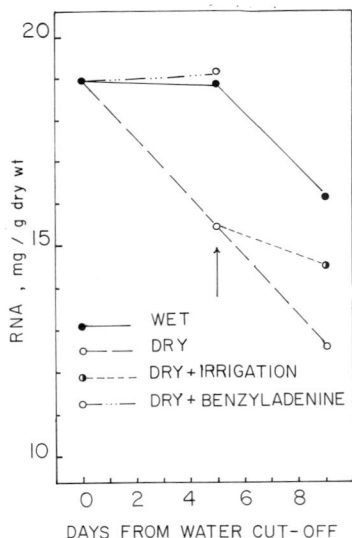

Fig. 19–7. Ribonucleic acid (RNA) in blades of recently matured sugar beet blades in the course of water depletion after irrigation (data from Shah and Loomis, 1965).

Benzyladenine or kinetin is known to retard nucleic acid breakdown in excised tissues (Richmond and Lang, 1957; Mothes et al., 1958). As a class of growth factors, the kinins are very important regulators of metabolism (Miller, 1961; Osborne, 1962). It has been recently shown that sunflower root (*Helianthus annus*) exudate contains compounds having kinetin-like properties (Kende, 1964). Itai and Vaadia (1965) have compared the kinin activity of sunflower plants after brief water stress with that of nonstressed plants.

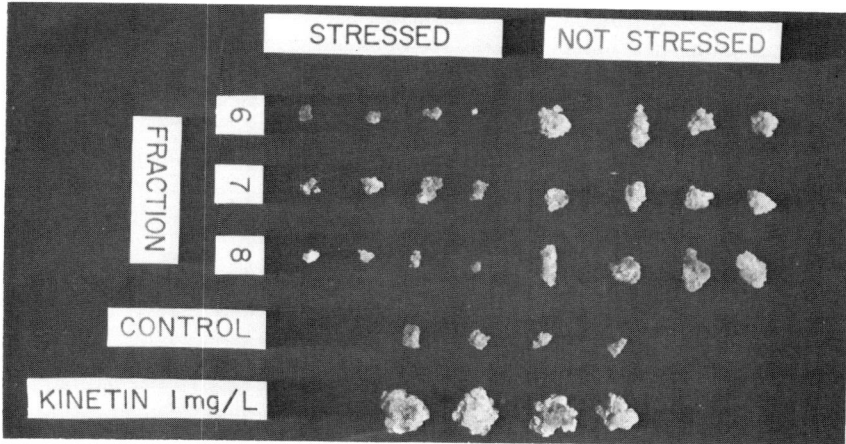

Fig. 19–8. The effect of three chromatographic fractions obtained from root exudate of water stressed and nonstressed sunflower plants on callus growth. The activity of kinin-free (control) substrate and 1 mg/liter kinetin are presented for comparison (Itai and Vaadia, 1965).

Materials eluted from chromatograms of exudate of nonstressed plants stimulate callus growth (a standard bioassay for kinins), while that obtained from previously stressed plants do not (Fig. 19–8). Similar results were obtained with the ability of root exudate to retard chlorophyll degradation. Weiss and Vaadia (1965) have shown that endogenous kinins can be found in high concentrations in root apices and suggested that they are synthesized in root tips. Livnè and Vaadia (1965) showed that kinins promote stomatal opening of detached barley leaves (*Hordeum vulgare*) and tobacco leaf (*Nicotiana* L.) discs. All these observations are in line with the hypothesis that kinins may serve as chemical messengers controlling shoot response to varying conditions in the rhizosphere. The quantitative relations between stresses and biosynthesis and/or translocation of root factors to shoots is currently under investigation (Itai and Vaadia, 1965). Alvim (1960) has shown that flowering in coffee (*Coffea arabica*) has a water stress requirement. His data further suggest that the hormonal balance of the plant is influenced by water deficits.

VIII. PLANT COMPOSITION

This topic has been discussed in some detail (Stocker, 1960; Evenari, 1962); since an exhaustive review would be pointless, only the less contradictory results of some apparent agronomic significance will be cited.

The enhanced conversion of starch to sugars in water stressed tissues is well known. How this is related to observations in sugar crops is not very clear, but it is known, for example, that sugar levels in grapes (*Vitis vinifera* L.) reach higher levels earlier in water stressed vineyards (Vaadia and Kasimatis, 1961). This observation is further confirmed by the fact that materials applied to foliage for retarding transpiration, delay grape maturity and decrease the rate of sugar accumulation in the fruit (Gale and Poljakoff-Mayber, 1965). Higher sugar percentage, although not higher yields, were obtained from sugar beets grown under salinity stress (Waisel and Bernstein, 1959).

The percentage of nitrogen in grains is generally increased under stress (Petinov and Ivanov, 1957). Konovalov (1958) found that the grain of stressed wheat plants contained 2% more nitrogen than the controls. This increase has been shown to be due to translocation of nitrogen from the vegetative parts to the seeds, and does not conflict with observations cited earlier that protein levels in leaves decrease under water deficits.

It has been generally observed that levels of alkaloids, rubber, anthocyanins, essential oils and fats increase with water deficits (*see* Evenari, 1962). There are no good explanations for these phenomena, but it is usually assumed that these materials are synthesized from decomposition products arising from enhanced hydrolysis of proteins and carbohydrates. These observations are important to the farmer for they may dictate irrigation practice for better quality crops. For example, withholding irrigation prior to harvest in certain crops may become a desirable practice under certain conditions. Crops where this may be useful include grapes, sugar beets, certain fruits, oil crops such as olives (*Olea europaea* L.), as well as others. Establishment of such practices must first be evaluated by field trials and from the economic standpoint.

IX. CONCLUSIONS

Water deficits generally reduce plant growth rates. The reasons for such reductions are still not completely known. It is quite clear that much of the reduction in growth is associated with reduction of turgor and cell wall development. However, it appears unlikely that turgor can be considered the key to the general response. Data are cited where reductions in growth were accompanied by no changes or only slight changes in turgor. Attempts to find a consistent metabolic pattern under stress have yielded much information, but what actually happens is still unclear.

At least two physiological aspects of the problem have apparently not received the attention they deserve. The first of these involves energy transformations in plants subjected to stress. Simultaneous measurements of respiration and photosynthesis are now commonly made, but the thermodynamic efficiency of the plant is neither known nor measured. Knowledge of the internal energy balance of the plant under stress in conjunction with certain key metabolic reactions should clarify some of the current contradictions. The second, which appears to have been neglected in water stress research, is that of hormonal balance. If the availability to the shoots of essential root factors is drastically reduced, it should be expected that the relationship between shoot development and water balance cannot be a simple one.

It is not our intention to state that the work so far done has not clarified many questions. It has. Nor is it intended to imply that the two above-mentioned aspects are the only ones. Many questions remain unposed. However, the two aspects posed, coupled with the considerable knowledge we have already, should provide very fruitful lines of research and progress.

LITERATURE CITED

Alvim, P. de T. 1960. Moisture stress as a requirement for flowering of coffee. Science 132:354.

Arnon, D. I. 1961. Cell free photosynthesis and the energy conversion process. p. 489–569. *In* W. D. McElroy and Bentley Glass [ed.] Life and light. John Hopkins Univ. Press, Baltimore, Md.

Ashton, F. M. 1956. Effects of a series of cycles of alternating low and high soil water contents on the rate of apparent photosynthesis in sugar cane. Plant Physiol. 31:266–271.

Aston, M. J., and Y. Vaadia. 1963. Resistance to water flow in roots. Plant Physiol. 38: supp. p. IX.

Bauman, L. 1957. Uber die Beziehungen Zwischen Hydratur and Ertrag. Ber. deut. Bot. Ges. 70:67–78.

Bernstein, L. 1961. Osmotic adjustment of plants to saline media: I. Steady phase. Amer. J. Bot. 48:909–918.

Bernstein, L. 1963. Osmotic adjustment of plants to saline media: II. Dynamic phase. Amer. J. Bot. 50:360–370.

Bonner, J. 1959. Protein synthesis and the control of plant processes. Amer. J. Bot. 46:58–62.

Boyer, J. S. 1965. Effects of osmotic stress on metabolic rates of cotton plants with open stomata. Plant Physiol. 40:229–234.

Brix, H. 1962. The effect of water stress on the rates of photosynthesis and respiration in tomato plants and loblolly pine seedlings. Physiol. Plant. 15:10–20.

Brouwer, R. 1963. The influence of the suction tension of the nutrient solutions on growth, transpiration, and diffusion pressure deficits of bean leaves. Acta Bot. Neerl. 12:248–260.

Broyer, T. C. 1950. Some gross correlations between growth enlargement and the solute and water relations of plants, with special emphasis on the relation of turgor pressure to distention of cells. Plant Physiol. 25:420–432.

Burström, H. 1954. Studies on growth and metabolism of roots: XI. The influence of auxin and coumarin derivatives on the cell wall. Physiol. Plant 7:548.

Burström, H. 1956. Die Bedeutung des Wasserzustandes für das Wachstum. 3:665–668. *In* W. Ruhland [ed.] Encyclopedia of plant physiology. Springer-Verlag, Berlin.

Carr, D. J., and D. F. Gaff. 1961. The role of cell water in the water relations of leaves. Proc. UNESCO Arid Zone Research 16:117–125.

Clark, Jasper A., and J. Levitt. 1956. The cell walls of drouthed soybean plants are less elastically extensible than those of nondrouthed plants. Physiol. Plant. 9:598–606.

Cleland, R., and Y. Bonner. 1956. The residual effect of auxin on the cell wall. Plant Physiol. 31:350–54.

Crafts, A. S., H. B. Currier, and C. R. Stocking. 1949. Water in the physiology of plants. Chronica Botanica Co., Waltham, Massachusetts. 240 p.

Eaton, F. M. 1942. Toxicity and accumulation of chloride and sulfate salts in plants. J. Agr. Res. 64:357–99.

Eaton, F. M., and D. R. Ergle. 1948. Carbohydrate accumulation in the cotton plant at low moisture levels. Plant Physiol. 23:169–187.

Evenari, M. 1962. Plant Physiology and arid zone research. UNESCO Arid Zone Research 18:175–195.

Gaastra, P. 1959. Photosynthesis of crop plants as influenced by light, carbon dioxide, temperature, and stomatal diffusion resistance. Mededel. Landbouwhogeschool Wageningen 59:1–68.

Gaastra, P. 1963. Climatic control of photosynthesis and respiration. p. 113–138. *In:* L. T. Evans [ed.] Environmental control of plant growth. Academic Press, New York.

Gale, J., and A. Poljakoff-Mayber. 1965. Antitranspirants as a research tool for the study of the effects of water stress on plant behavior. UNESCO Arid Zone Research 25:269–274.

Gates, C. T., and J. Bonner. 1959. The response of young tomato plant to a brief period of water shortage: IV. Effects of water stress on the ribonucleic acid metabolism of tomato leaves. Plant Physiol. 34:49–55.

Gingrich, J. R., and M. B. Russell. 1956. Effect of soil moisture tension and oxygen concentration on the growth of corn roots. Agron. J. 48:517–520.

Gingrich, J. R., and M. B. Russell. 1957. A comparison of effects of soil moisture tension and osmotic stress on root growth (corn). Soil Sci. 84:185–194.

Glover, J. 1959. The apparent behavior of maize and sorghum stomata during and after drouth. J. Agr. Sci. 53:412–416.

Haddock, J. L. 1959. Yield, quality, and nutrient content of sugar beets as affected by irrigation regimes and fertilizers. Proc. Amer. Soc. Sugar Beet Tech., 10:344–355.

Hagan, R. M., M. L. Peterson, R. P. Upchurch, and L. G. Jones. 1957. Relationship of soil moisture stress to different aspects of growth in Ladino clover. Soil Sci. Soc. Amer. Proc. 21:360–365.

Hayward, H. E., and Winifred B. Spurr. 1944. Effects of isoosmotic concentrations of inorganic and organic substrates on the entry of water into corn roots. Bot. Gaz. 106:131–139.

Hunter, A. S., and J. A. Yungen. 1952. The response of sugar beets to fertilizers, spacing, and irrigation on eastern Oregon soils. J. Amer. Soc. Sugar Beet Tech. 7:180–188.

Iljin, W. S. 1953. Causes of death of plants as a consequence of loss of water: Conservation of life in dessicated tissues. Bull. Torrey Bot. Club 80:166–177.

Itai, Ch., and Y. Vaadia. 1965. Kinetin-like activity in root exudate of water stressed sunflower plants. Physiol. Plant. 18:941–944.

Ivanov, V. P. 1959. Effect of foliar nutrition and soil moisture on the growth and development of maize. Fiziol. Rast. (Amer. Inst. Biol. Sci. Engl. trans.) 6:368–373.

Kemper, W. D., C. W. Robinson, and H. M. Golus. 1961. Growth rates of barley and corn as affected by changes in soil moisture stress. Soil Sci. 91:332–338.

Kende, H. 1964. Preservation of chlorophyll in leaf sections by substances obtained from root exudate. Science 145:1066–1067.

Kessler, B. 1961. Nucleic acids as factors in drouth resistance in higher plants. Recent Advance. Bot. 2:1153–1159.

Kettelapper, H. J. 1963. Stomatal physiology. Ann. Rev. Plant Physiol. 14:249–270.

Klotz, I. M. 1958. Protein hydration and behavior. Science 128:815–822.

Konovalov, Yu. B. 1958. Effect of soil moisture deficiency on grain ripening in spring wheat. Fiziol. Rast. 6:189–196.

Kramer, P. J. 1949. Plant and soil water relationships. McGraw-Hill, New York.

Kramer, P. J. 1950. Effects of wilting on the subsequent intake of water by plants. Amer. J. Bot. 37:280–284.

Kramer, P. J. 1959. The role of water in the physiology of plants. Advance. Agron. 11:51–57.

Kuiper, P. J. C. 1961. The effects of environmental factors on the transpiration of leaves with special reference to stomatal light response. Mededel. Landbouwhogeschool Wageningen. 61:1–49.

Levitt, J. 1956. The hardiness of plants. Academic Press, New York. 278 p.

Levitt, J. 1962. A sulfhydryl-disulfide hypothesis of frost injury and resistance in plants. J. Theor. Biol. 3:355–391.

Livnè, A., and Y. Vaadia. 1965. Stimulation of transpiration rate in barley leaves by kinetin and gibberellic acid. Physiol. Plant. 18:658–664.

Meyer, R. E., and J. R. Gingrich. 1964. Osmotic stress: Effects of its application to a portion of wheat root systems. Science 144:1463–1464.

Miller, C. D. 1961. Kinetin and related compounds in plant growth. Ann. Rev. Plant Physiol. 12:395–408.

Moss, D. W., R. B. Musgrave, and E. R. Lemon. 1961. Photosynthesis under field conditions: III. Some effects of light, carbon dioxide, temperature, and soil moisture on photosynthesis, respiration, and transpiration of corn. Crop Sci. 1:83–87.

Mothes, K. 1933. Die Vaccuminfiltration im Ernährungsversuch. (Dargestellt an Untersuchungen über die Assimilation de Ammoniaks). Planta 19:117–138.

Mothes, K., I. Böttger, and R. Wellgiehn. 1958. Untersuchungen über den Zusammenhang Zwischen Nukleinsäure und Eiweisstoffwechsel in grünen Blättern Naturwissens. 45:316.

Oppenheimer, H. R. 1960. Adaptation to drouth: Xerophytism: Plant water relationships in arid and semiarid conditions. UNESCO Arid Zone Research 15:105–138.

Ordin, Lawrence. 1958. The effect of water stress on the cell wall metabolism of plant tissues. Vol. IV:553–564. In Radioisotopes in Scientific Research. Pergamon Press, London.

Ordin, Lawrence. 1960. Effect of water stress on cell wall metabolism of Avena coleoptile tissue. Plant Physiol. 35:443–451.

Ordin, L., T. H. Applewhite, and J. Bonner. 1956. Auxin-induced water uptake by Avena coleoptile sections. Plant Physiol. 31:44.

Osborne, D. 1962. Effect of kinetin on protein and nucleic acid metabolism in Xanthiun leaves during senescence. Plant Physiol. 37:595–602.

Owen, P. C. 1958. Growth of sugar beets under different water regimes. J. Agr. Sci. (London) 51:133–136.

Pallas, J. E. 1960. Effects of temperature and humidity on foliar absorption and translocation of 2,4-dichlorophenoxyacetic acid and benzoic acid. Plant Physiol. 35:575–580.

Palmcrantz, P. J. 1957. L and D-leucy glycine splitting dipeptidases in Elodea densa. Arkiv. Kemi. 11:455–461.

Petinov, N. S., and V. P. Ivanov. 1957. The effect of shortened day length in the germination period on water relations and on acceleration of development in maize. Fiziol. Rast. (Amer. Inst. Biol. Sci. Engl. trans.) 4:171–183.

Petrie, A. M. K., and J. I. Arthur. 1943. Physiological ontogeny in the tobacco plant: Effect of varying water supply on the drifts in dry weight and leaf area and on various components of leaves. Australian J. Exp. Biol. Med. Sci. 21:191–200.

Petrie, A. M. K., and J. G. Wood. 1938a. Studies on nitrogen metabolism of plants: I. The relation between the contents of proteins, amino acids, and water in the leaves. Ann. Bot. N.S. 2:33–60.

Petrie, A. M. K., and J. G. Wood. 1938b. Studies on nitrogen metabolism of plants: III. On the effect of water content on the relationship between proteins and amino acids. Ann. Bot. N.S. 2:887–898.

Pirie, N. W. 1959. Leaf proteins. Ann. Rev. Plant Physiol. 10:33–52.

Prusakova, L. D. 1960. Influence of water relations on tryptophan synthesis and leaf growth of wheat. Fiziol. Rast. (Amer. Inst. Biol. Sci. Engl. trans.) 7:139–148.

Richmond, A. E., and A. Lang. 1957. Effects of kinetin on protein content on survival of detached Xanthium leaves. Science 125:650–651.

Robinson, E. 1956. Proteolytic enzymes in growing root cells. J. Exp. Bot. 7:296–300.

Robinson, J. R. 1960. Metabolism of intracellular water. Physiol. Rev. 40:112–149.

Rufelt, H. 1963. Rapid changes in transpiration of plants. Nature 197:985.

Shah, C. B., and R. S. Loomis. 1965. Ribonucleic acid and protein metabolism in sugar beets during drouth. Physiol. Plant. 18:240–254.

Shimshi, D. 1963. The effect of soil moisture and phenylmercuric acetate upon stomatal aperture, transpiration, and photosynthesis. Plant Physiol. 38:713–721

Skoog, F. 1955. Growth factors, polarity, and morphogenesis. Anée Biol. 59:1–13.

Slatyer, R. O. 1961. Effects of several osmotic substrates on the water relationships of tomato. Australian J. Biol. Sci. 14:519–540.

Slatyer, R. O., and S. A. Taylor. 1960. Terminology in plant soil water relations. Nature 187:922–924.

Stocker, O. 1960. Physiological and morphological changes in plants due to water deficiency. UNESCO Arid Zone Research 15:63–104.

Takaoki, T. 1957. Relationships between plant hydrature and respiration: II. Respiration in relation to the concentration and the nature of external solutions. J. Sci. Hiroshima Univ. Sec. B. 8:73–80.

Upchurch, R. P., M. L. Peterson, and R. M. Hagan. 1955. Effect of soil moisture content on the rate of photosynthesis and respiration in Ladino clover. Plant Physiol. 30:297–303.

Vaadia, Y., and A. Kasimatis. 1961. Vineyard irrigation trials. Amer. J. Enol. Viticult. 12:88–98.

Vaadia, Y., F. C. Raney, and R. M. Hagan. 1961. Plant water deficits and physiological processes. Ann. Rev. Plant Physiol. 12:265–292.

Van Bavel, C. H. M. 1953. Chemical composition of tobacco leaves as affected by soil moisture conditions. Agron. J. 45:611–614.

Van Overbeek, J. 1959. Auxins. Bot. Rev. 25:271–350.

Wadleigh, C. H., and A. D. Ayers. 1945. Growth and biochemical composition of bean plants as conditioned by soil moisture tension and salt concentration. Plant Physiol. 20:106–132.

Wadleigh, C. H., and H. G. Gaugh. 1948. Rate of leaf elongation as affected by the intensity of the total soil moisture stress. Plant Physiol. 23:485–495.

Waisel, Y. 1962a. Ecotypic variation in the flora of Israel: IV. Seedling behavior of some ecotype pairs. Phyton. 18:151–156.

Waisel, Y. 1962b. Presowing treatments and their relation to growth and to drought, frost, and heat resistance. Physiol. Plant. 15:43–46.

Waisel, Y., and R. Bernstein. 1959. The effect of irrigation with saline water on the yield and sugar content of forage and sugar beet. Bull. Res. Counc. Israel. 7D:90–92.

Waisel, Y., H. Kohn, and J. Levitt. 1962. Sulfhydryls—A new factor in frost resistance: IV. The relation of GSH oxidizing activity to flower induction and hardiness. Plant Physiol. 74:272–276.

Watson, J. D. 1963. Involvement of RNA in the synthesis of proteins—The ordered interaction of three classes of RNA controls the assembly of amino acids into proteins. Science 140:17–26.

Weatherly, P. E. 1950. Studies in the water relations of the cotton plant: I. The field measurement of water deficits in the leaves. New Phytol. 49:81–97.

Weatherly, P. E. 1951. Studies on the water relations of the cotton plant: II. Diurnal and seasonal fluctuations and environmental factors. New Phytol. 50:36–51.

Weatherly, P. E., and R. O. Slatyer. 1957. Relationship between relative turgidity and diffusion pressure deficit in leaves. Nature 179:1085–1086.

Weiss, H., and Y. Vaadia. 1965. Kinetin-like activity in root apices of sunflower plants. Life Sci. 4:1323–1326.

Whiteman, P. C., and G. L. Wilson. 1963. Estimation of diffusion pressure deficit by correlation with relative turgidity and beta-radiation absorption. Australia J. Biol. Sci. 16:140–146.

Willis, A. J., E. W. Yemm, and S. Balasubramaniam. 1963. Transpiration phenomena in detached leaves. Nature 199:265–266.

Wilson, A. M., and R. C. Huffaker. 1964. Effects of moisture stress on acid soluble phosphorous compounds in *Trifolium subterraneum*. Plant Physiol. 39:555–560.

Wilson, A. M., and C. M. McKell. 1961. Effect of soil moisture stress on absorption and translocation of phosphorous applied to leaves of sunflower. Plant Physiol. 36:762–765.

Wilson, C. C. 1948. Diurnal fluctuations in growth in length of tomato stem. Plant Physiol. 23:156–157.

Wit, C. T., de. 1958. Transpiration and crop yields. Inst. Biol. Scheik. Wageningen Mededel. 59:1–88.

Woodhams, D. H., and T. T. Kozlowski. 1954. Effects of soil moisture stress in the carbohydrate development and growth in plants. Amer. J. Bot. 41:316–330.

Yarosh, N. P. 1959. Effects of water supply on biochemical changes in cotton leaves and seeds. Fiziol. Rast. (Amer. Inst. Biol. Sci. Engl. trans.) 6:211–214.

Zholkevitch, V. N. 1961. Energy balance of respiring plant tissues under various conditions of water supply. Fiziol. Rast. (Amer. Inst. Biol. Sci. Eng. trans.) 8:407–416.

Zholkevitch, V. N., and T. F. Koretskaya. 1959. Metabolism of pumpkin roots during drouth. Fiziol. Rast. (Amer. Inst. Biol. Sci. Engl. trans.) 6:690–700.

Zimmerman, M. H. 1960. Transport in the phloem. Ann. Rev. Plant Physiol. 11:167–190.

20

Shoot and Root Growth as Affected by Water Availability[1]

D. B. PETERS

Agricultural Research Service, USDA
and University of Illinois
Urbana, Illinois

J. R. RUNKLES

Texas A&M University
College Station, Texas

I. INTRODUCTION

Water is a major constituent of living tissues, often amounting to as much as 90 or 95% of the fresh weight. Water provides plants with their mechanical strength through the mechanism of cell turgor, is raw material in metabolism and in the synthesis reactions that occur in living plants, and serves as the transportation agent within the plant, moving raw materials and metabolized products between the various plant parts.

Plants are strongly affected by the water supply available in the root zone. More specifically they are affected by the amount of water within the plant tissues. The supply in the plant tissues is determined by the energy status and conduction properties of soils and by the rate of water loss from the plant surfaces. Therefore, the role of water in regulating growth of plants must be analyzed in terms of the water rate processes occurring throughout the entire soil-plant-atmosphere system. In addition, certain phases of the growth cycle are conditioned strongly by the plant itself.

II. GERMINATION

The germination process is embryo growth with the generation of sufficient force to rupture whatever embryo covers are present. The first change that occurs when seeds germinate is the imbibition of water. In the dry seeds of maize (*Zea mays*) and bean (*Phaseolus vulgaris*) all of the tissues are shrunken, cell vacuoles are small, the nucleus is irregular, and the cell contents are plasmolyzed; but with the absorption of water the cells become turgid. During the first 10 to 12 hours no elongation of cell walls can be noted, nor can chemical changes be detected. Elongation of the cells of maize can be observed about 20 hours after the beginning of imbibition. The coleorhiza breaks the pericarp and extends about 2 mm beyond the surface. The radicle elongates to fill the extending coleorhiza and soon

---

[1] Contribution from the Soil and Water Conserv. Res. Div., Agr. Res. Serv., US Dep. Agr.; the Illinois Agr. Exp. Sta.; and the Texas Agr. Exp. Sta.

breaks through the sheath which does not develop further. In the early stages of germination the pressure developed by imbibition may be as high as 1,000 atm (Shull, 1914).

During the water-filling period the O_2 requirement is low, but at the onset of cell elongation and cell division there is a rapid increase in O_2 requirement. For our purposes here, germination will be confined to the early stages of imbibition and onset of cell elongation and cell division. For some excellent reviews of the physiology of the germination process the reader is referred elsewhere (Toole et al., 1956; Koller et al., 1962).

The start of germination is the coupling of imbibition and respiration (intake of water and O_2). Seeds vary in their morphological construction in an infinite variety of ways to control the entrance of water and O_2. The scarification of some seeds is a practice to enhance the entrance of water and O_2 into seeds. Direct evidence of coat function in the case of water relations between the embryo and its environment is available (Toole et al., 1956), but the case for O_2 control is not so clear. However, there is ample evidence that in O_2-limiting situations germination is inhibited.

The process of germination is herein strictly limited to the initial absorption of water (and other substances) which in itself triggers biochemical changes and produces cell expansion of sufficient force to rupture the embryo cover. Also included are the initial cell division phases. The visual evidence of germination in agriculture practice is emergence of the seedling from the ground, but it appears that the emergence phase is more sensitive to environmental influences than is the defined germination phase.

Most evidence seems to indicate that the critical level of water suction to inhibit germination completely is of a much higher order than usually expected. For example, Owen (1952) found a germination percentage > 20 for seeds germinated over a 0.70M solution of NaCl representing a suction of 320m of water. The permanent wilting point is generally accepted to be at a suction of 160m of water. While stress *per se* is usually not a factor in initiation of embryo growth, it does manifest itself in the rate of radicle emergence and, as a consequence, affects the per cent germination. The lowered germination percentage was attributed to the increased chance for microbial infection (Owen, 1952).

Associated with stress (water suction) in soils and seed coats is the reduced permeability factor. The reduction in permeability (of both soils and seed coats) with increasing water suction contributes to the delay in germination. In an experiment which greatly contrasted the liquid-seed contact, Sedgley (1963) found that germination of seeds was not influenced by matric suction. He did find, however, that the rate of germination was somewhat pressure dependent.

Several experimenters have tried to separate the effect of matric pressure from osmotic pressure on seed germination, but in general the results have been inconclusive. Fertilizer salts placed with seeds at planting time in soil that was at or near field capacity had little effect on germination (Chapin and Smith, 1960). Germination at or near the wilting point was greatly reduced or even prevented. Other studies showed that an increase in osmotic pressure reduced the germination rate, but with some materials, large pressures were necessary before marked changes occurred (Collis-George and Sands, 1962). They found that soil water suction did not appreciably change total germination but did influence the rate of germination.

It therefore appears that initial phases of germination (imbibition, cell elongation, and cell division) are not too strongly affected by soil water suction *per se* but that the time required for complete germination is considerably extended by increased suction. The occasional osmotic effect appears to be more related to toxicity than to osmotic pressure. The time delay of germination under increased soil water suction is probably due to the limited rate of conduction of water to the embryo.

The dominant environmental factor in seed germination in agricultural practice is water; however, the role of temperature cannot be minimized. Temperature is concerned not only with the respiratory mechanism but is also interrelated with the light reactions which control germination in some species. The reader is referred to the two previously cited reviews (Toole et al., 1956; Koller et al., 1962) for a more complete analysis of the role of temperature in germination. Went (1953) has presented an excellent review of many of the general responses to temperature conditions.

The requirement for oxygen in the early stages of germination is meager, but at the onset of elongation and cell division the respiration rate increases. Thus it is not surprising that poor soil aeration often limits germination and more specifically seedling emergence. Poor soil aeration is usually the result of high bulk density coincident with high water contents (Millington, 1959).

III. EMERGENCE

Establishing plant seedlings is a 2-stage process, germination and seedling emergence. Whereas germination is relatively insensitive to normal soil water conditions, seedling emergence is highly sensitive to soil water and other soil conditions prevailing at the time of emergence.

Water pressure, either suction or osmotic, strongly influences both the per cent and rate of emergence. The data of Ayers (1952) point this out conclusively as shown in Fig. 20–1.

Other investigators have reported similar results (Doneen and MacGillivray, 1943; Helmerick and Pfeifer, 1954; Hunter and Erickson, 1952). Triplett and Tesar (1960) showed that the detrimental effects of lowered water contents on seedling emergence could be somewhat alleviated by compaction of the seedbed. This would indicate that in the early stages of seedling growth the transport of water to the small developing roots is a contributing factor in regulating water absorption and resultant growth. The data of Danielson and Russell (1957) tend to substantiate this belief.

Seedling emergence is strongly conditioned by the nature of the forces holding the soil particles together during the germination period. From the strict mechanical standpoint, the way in which seedlings emerge from the soil controls, to a degree, the success of emergence under compacted conditions. Seeds may be classified into two principal groups: (i) those in which the cotyledons emerge from the seed and (ii) those in which the cotyledons remain permanently within the seed. Most seeds of dicots and seeds of some monocots, such as onion (*Allium cepa*), belong to the first group, while the seeds of grasses and of some dicots, such as peas (*Pisum sativum*) and oak (*Quercus* sp.), belong in the second. In

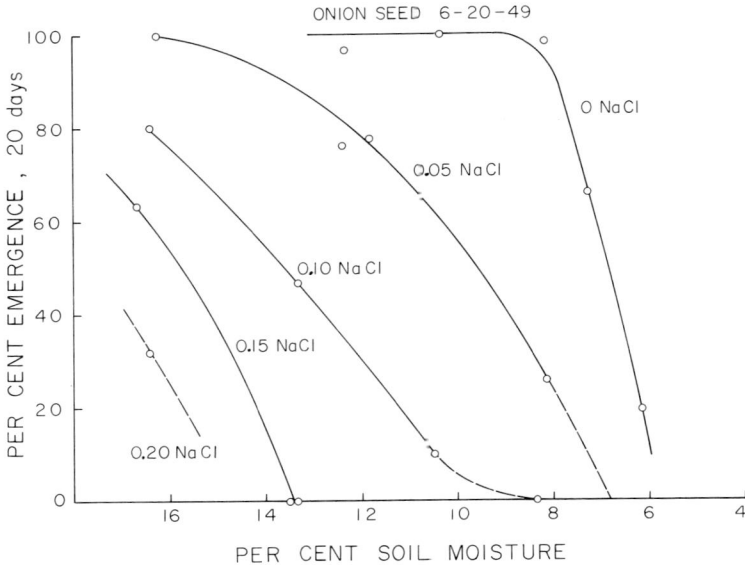

Fig. 20–1. Germination of onion seed as a function of soil moisture and added salt (after Ayers, 1952).

general, seeds of the second group are less influenced by mechanical soil stress conditions than are members of the first group.

In view of the nature of some seedlings and the relative susceptibility of the young seedlings to environmental influences, it is not surprising to find that seedling emergence is strongly influenced by soil conditions. The mechanical strength of the soil can be large under certain soil conditions such as compaction and drying. Hanks and Thorp (1957) have reported variations of modulus of rupture as much as 0 to 1400 mbars. They also found that soil crust strength, limiting emergence, was a function of water. It would be very difficult to separate the contribution due to water and that due to soil strength, inasmuch as the two are closely interrelated. Hanks and Thorp (1957), however, did find that at constant water content, seedling emergence decreased with increasing crust strength for the plants and soils studied.

The period of seedling emergence is characterized by high respiration rates and biochemical activity, thus it follows that the rate of seedling emergence is conditioned by the temperature regime and the aeration status. The effect of temperature on seedling emergence is especially striking, as has been shown by several investigators (Glendening, 1942; Willis et al., 1957).

IV. ROOT GROWTH

Weaver (1926) states that the general characters of the root species are often as marked and distinctive as those of the aerial vegetative parts, and although they can be greatly modified when subjected to different environmental conditions, they still retain the characteristic impress of the species in its usual habitat. The root system of a given plant, therefore, may vary in its structure, extent, weight, number, and direction of roots, according to the conditions under which it is grown. This discussion is concerned principally with the root system in relation to soil water and soil structural conditions.

Water influences root systems in three general ways: (i) direction of root growth, (ii) lateral extent and depth of penetration, and (iii) relative weight of tops and roots.

Little needs to be said here about the direction of root growth beyond the well-established fact that roots will turn and follow water in the soil when they are in direct contact or in very close proximity to it (Hunter and Kelley, 1946).

The lateral extent and depth of penetration is controlled principally by their genetic character but is subject to modification by environmental conditions. A compilation of the extent of root systems of cultivated plants may be found in Miller's textbook on plant physiology (Miller, 1931).

The relative weight of roots and tops received considerable attention in the early literature but of late has been little studied. In general, investigators (Tucker and Van Seelhorst, 1898) reported that the weight of both grain and straw increases with increasing water content, but the greatest weight of roots is produced in soil with the smallest amount of water; this is relative to the weight of tops.

Under water stress conditions the aerial portion of plants is usually affected more than is the root system. In other words, the aerial portions dehydrate more readily than do roots. Consequently, in a consideration of the plant system as a whole when stress conditions occur, growth usually is reduced at a faster rate in the aerial portions than in roots.

In those studies in which only the root systems are considered, most investigators have found that increasing water suction has reduced the rate of growth. Moreover, the growth is related not only to plant stress but to the water content of the soil, indicating that the rate of water movement to the root is a factor in water absorption and, consequently, in growth.

The data of Peters (1957) clearly illustrate the effect of soil water suction and soil water content on water absorption and root growth. Corn (*Zea mays* L.) seedlings in controlled-growth chambers were covered with soils (soil and sand

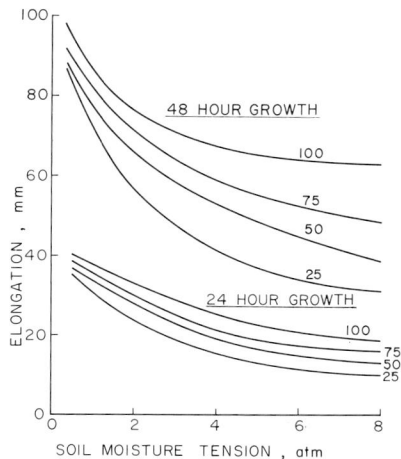

Fig. 20–2. Elongation of corn roots as function of soil-moisture tension (after Peters, 1957).

mixed in various proportions) that had been brought to definite soil water suction values. The results are presented in Fig. 20–2. The number on each curve in Fig. 20–2 indicates the per cent of soil in the mixture. At the same suction the roots grew better in the soil having the highest water content. Thus, soil water suction was not the only factor affecting growth; water content was also important, probably because of the greater unsaturated hydraulic conductivity in the fine soil (higher water content). Gingrich and Russell (1957) compared a series of soil samples and mannitol solutions of corresponding stresses. Their results show that the roots grew better in osmotic solutions than in soil samples having the same water stress. There was, however, a definite osmotic effect. Several investigators (Ayers et al., 1943; Eaton, 1941) have reported similar effects of excess salts in reducing root growth in soils. Moreover, the carefully controlled experiments of Hayward and Spurr (1944) in which rate of water absorption was measured by potometers when corn roots were subjected to different salines, sucrose and mannitol at isometric pressures, lend considerable support to the view that the major effect of the salts usually present in saline soil solutions is osmotic. Wadleigh (1946) has proposed a scheme for assessing the total effective stress (suction + osmotic) into an integrated soil water stress. However, recent work by Bernstein (1961) casts doubt on the validity of adding the stress components since he found that some plants rapidly adjust the internal osmotic stress to the external osmotic stress.

While all the evidence indicates that root growth is diminished as water suction increases, it is not to be implied that such reduced growth prohibits the root from performing its essential function, i.e. the absorption of water and plant nutrients. In many plant species it is frequently found beneficial for the plants to experience some water stress in the early stages of growth to produce a maximum harvestable yield. It is commonly thought that an early stress period enhances the development of an extended and diffuse root system which is capable of handling the increased demands at later stages of growth. Conversely, if plants develop in a wet regime early in the season it is possible they will suffer at later stages of growth when demand for water and nutrients is high. At this time, no reliable guidelines are available which would serve to distinguish those water conditions which best serve to develop a root system capable of supplying the aerial portion with its needs during all stages of growth. Such as are available are the result of irrigation practice and experiments in a type location and usually specific for a given plant species. For such information, literature on irrigation practice is usually available in a type location (see chapters 21 and 30 to 39).

The information we now possess shows conclusively that the rate and extent of root development is controlled by water stress (suction and/or osmotic). Overall growth diminishes with increasing water stress, but root growth is less influenced than is shoot growth.

While we are concerned herein principally with the effects of water on growth, brief mention should be made of the other factors influencing growth processes—temperature, nutrients, aeration, and physical impedance. Each of these factors is controlled by, or interacts with, the water regime.

Briefly, temperature affects the rate of the growth process *per se* through its control over the rate of respiration, photosynthesis, and translocation. In general, high temperatures favor more rapid translocation and accelerated respiratory activity. When soil temperatures are low, a retarded rate of translocation may restrict growth. When soil temperatures are high, the ratio of respiration to protein synthesis may be so increased that the carbohydrate balance is depleted, and the growth of roots and underground storage organs is decreased. Consequently, growth may be restricted at either end of the temperature-growth curve and be maximized at some median temperature. In this regard there has developed the concept of cardinal temperatures. In general, the growth of a given species in relation to temperature shows three cardinal points: the minimum, the lowest temperature at which growth may occur; the optimum, the temperature at which growth is maximum; and the maximum, the highest temperature at which growth will occur. An excellent review of cardinal temperatures and plant growth has been presented by Richards et al. (1952).

Because of the high heat capacity of water, the temperature of and conduction of heat in a soil is controlled principally by the water regime. Therefore, in irrigation practice some degree of control over soil temperatures can be obtained.

The effect of nutrient status on root growth is correlated principally with the interaction between the shoots and roots. It is generally observed that increasing the N content of the soil will decrease the root-top ratio. With high N more of the carbohydrates are used for the synthesis of tissue and less translocated to the roots. On the other hand, P appears to favor a more profuse root system (Miller, 1931).

The placement and composition of fertilizer bands has been shown to affect the rooting habit of corn (Duncan and Ohlrogge, 1958). Data reported by Jordan appear to indicate that fertilizer applications tend to concentrate the roots in the upper soil zones to the extent that in severe drouth conditions the plants suffer from water (R. W. Jordan, 1961. Studies and measurements of soil respiration. *Unpubl. M.S. Thesis. University of Illinois, Urbana*). The data of fertilizer effects on root development and water absorption are too scanty to draw accurate conclusions. However, it does appear that fertilizer practice must be correlated with expected water regime, especially in irrigation practice.

The absorption of ions is related to water suction. Olsen et al. (1961) reported reduced absorption of phosphate as water suction increased. Others (Peters and Russell, 1960) have found similar results with rubidium. The reduction in ion uptake is usually attributed to the decreased rate of ion diffusion as the water content decreases. Since root and top growth is intimately related to nutrient status, it is possible that growth may be inhibited by the ion-soil water interaction.

An inverse relationship exists between soil water content and aeration. As a consequence, most situations, where aeration limits growth, can be traced to excess water, either alone or coupled with high soil density. A marked interaction between soil water suction and aeration has been observed on the growth of corn seedlings (Gingrich and Russell, 1956). Increases in soil water suction from 1 through 12 atm brought progressively smaller increases in root growth, being most sensitive in the range between 1 and 3 atm. At low water suction values, O_2 concentration was most critical, needing to be above 10.5% for maximum growth.

In any intensive farming practice, such as irrigation, the danger always exists that the growth of roots may be restricted by an external factor such as mechanical impedance. Considerable evidence indicates that roots are unable to penetrate certain bodies of soil and that the root system is drastically changed by mechanical soil conditions. Inasmuch as all the internal environmental factors (water, aeration, nutrients, temperature) are conditioned by the impedance factor, it is exceedingly difficult to determine the cause of reduced growth as mechanical conditions change.

Barley (1962) has recently reported experiments in which mechanical impedance was the cause of reduced growth. In one experiment the apex of the root was compressed in a way that permitted the cells to elongate and differentiate while subject to stress. Growth in length proceeded without interruption, although at a reduced rate. In another experiment the entire length of the root was compressed. Growth in length was stopped for a number of hours and then resumed at a low rate. In the first experiment the translocation of growth materials was not impeded and apparently the growing points could exert sufficient force to continue growing. In the second experiment the translocation tissues were compressed and growth stopped until the plant could readjust.

Numerous experiments have been reported which show that increases in soil bulk density decrease root growth. Phillips and Kirkham (1962) found that corn seedling root growth decreased linearly as bulk density increased from 0.94 g/cc to 1.30 g/cc.

Trause and Humbert (1961) have established critical soil bulk densities for the principal soils under Hawaiian sugarcane (*Saccharum officinarum*) cultivation. Their studies indicated a decrease in rooting efficiency with increasing soil density as measured by uptake of Rb^{86}. Zimmerman and Kardos (1961) studied a number of soils and found, by pooling all soils for each crop, a highly negative correlation between bulk density and penetrating root weight for both soybeans (*Glycine max*) and sudan-grass (*Sorghum sudanese*). There seems to be little doubt that increasing soil bulk densities decrease root growth, but it is probable that in many instances the reduced growth is due to the effect of increasing bulk density on other factors. Taylor and Gardner (1963) found that the penetration of taproots by cotton plants (*Gossypium hirsutum*) was more highly correlated with soil strength measurements than with water contents and bulk density. They concluded that soil strength, not bulk density, was the critical impedance factor controlling root penetration in the sandy soils of the Southern Great Plains, USA.

Barley (1963) also showed that soil strength plays an important role in root growth. Corn radicles were grown in a fine-grained cohesionless material. The shearing strength of the material was controlled by subjecting cylindrical packs to known uniform pressures in a modified triaxial test cell, and the influence of the shearing strength on the rate of root elongation was examined. Radicles were prevented from elongating by an ambient effective pressure of 0.6 kg/cm² which corresponded to an initial shearing strength of 0 3 kg/cm². Barley's (1963) results would indicate that small negative pressures (water suction) in the pore water may increase the shearing of some soils sufficiently to reduce the growth of roots.

It would appear, then, that in addition to the deleterious effects of increasing bulk density on the other internal environmental factors there is an actual mechanical stress factor which reduces root growth. The evidence thus far appears to indicate that the principal effect of mechanical stress is in upsetting the translocation of growth substances.

V. SHOOT GROWTH

The quality and quantity of shoot growth in plants is determined by genetics and environment acting through the internal biochemical and physiological processes. An environmental factor, such as water, influences shoot growth indirectly through its effect on these internal processes.

Water is a component of all plants and it serves four general functions in them:

1) Water is a constituent of active protoplasm comprising 75 to 95% of the fresh weight of stems and leaves, 85 to 95% of the fresh weight of fruits, and 5 to 11% of the fresh weight of dry seeds (Kemper et al., 1961). Decreasing the water content of most plants reduces the rate of many physiological processes. Plants have a certain water content beyond which increased dehydration results in death.

2) Water is a reactant or reagent in many physiological processes. In photosynthesis water is a reagent and is equally as important as CO_2.

3) Water is a solvent for gases and organic and inorganic solutes which enter plants and are translocated from cell to cell and organ to organ within the plant.

4) Water maintains the turgidity of the plant which is responsible for movements of leaves and floral parts and general maintenance of the form of the plant. Turgidity is important in the opening of stomata and is essential for growth.

The amount of water required for these functions is small, compared to the total water passing through the plant in a given period of time. It is estimated that a plant uses < 5% of the total water which passes through it (Kramer, 1959).

Since plant cell walls are saturated, water in plants is continuous in the hydrostatic sense. Water movement or adjustment within plants is along the line of water potential gradients. When the rate of absorption of water by plants equals, or exceeds, the rate of transpiration, no serious internal deficit of water exists, and the plant water potential between plant organs reaches a minimum. However, when the rate of transpiration exceeds the rate of absorption for any appreciable period of time, an overall internal water deficit develops and the plant water potential gradient between plant organs increases so internal competition for water exists between these organs. The internal competition for water between leaves and fruit of the cotton plant was investigated by Anderson and Kerr (1943). If the soil water was limiting, they found a full size cotton boll to shrink in size during the day when the parent plant was visibly wilted and to regain its size during the night. As the soil water suction increased (water content decreases), they found the cotton boll to respond in size to diurnal fluctuations but exhibit an overall gradual shrinking. The degree in shrinking of the boll was found to be proportional to the severity of wilting of the parent plant. Similar competition for internal water between leaves and fruits of other plants was observed by Chandler (1914); Rokach (1953); Lloyd (1920); and Bartholomew (1926). Wilson (1948) found that the shoot growing tip and the section below the first node exhibited diurnal reversals in dehydration, but priority for water remained in the growing tip. In periods of drouth, it is frequently observed that the older leaves lose their water to the younger leaves as the former usually die first. Shaded leaves also suffer from lack of water and die during periods of drouth because they cannot develop as low a water potential as unshaded leaves.

Water in plants is dynamic in nature, largely as a result of opposing processes of transpiration and absorption. The loss of water from plant cells causes a dehydration of the protoplasm and cell wall and results in a reduction or cessation of cell division or enlargement, or both. The enlargement phase of growth seems to be influenced more by an internal water shortage than the cell division phase. Maximum cell enlargement can apparently only be attained if water is not limiting for any appreciable period of time. The relationship of shoot growth to soil water suction is illustrated by experiments of Slatyer (1957) who determined stem elongation, growth, relative turgidity, and transpiration of cotton, tomato (*Lycopersicon esculentum*) and privet (*Ligustrum* sp.) in relation to total soil water stress during a single drying cycle. His results are shown in Fig. 20–3. The response pattern of each species showed a close relationship to water stress and in each species growth (as total dry weight) did not continue beyond a stress value such that there was zero turgor pressure in the tissue of adult leaves. In privet and cotton, stem elongation also ceased at this value but in tomato it continued until higher stress levels developed. Gates (1955) investigated the response of tomatoes during a single drying cycle of short duration and found that growth was reduced during the drying cycle but on rewatering the growth rate increased over the control (unwilted plant). However, there was no indication that this recovery effect was complete at the final harvest. This increased rate of growth following a temporary water stress has been observed in other plants (Kemper et al., 1961). The relationship of shoot growth to soil water regime has been investigated in a number of experiments by allowing the soil water content to cycle through predetermined levels. The growth response was measured after the plant was subjected to a number of these cycles. Wadleigh and Ayers (1945) determined shoot growth of bean plants in relation to soil water regime and found that growth was decreased as the water suction at time of rewatering was increased even though in some regimes the soil water suction was always above the wilting percentage. Similar observations were made with corn, tomatoes, sunflower (*Helanthus annuus*), and cotton, respectively by Haynes (1948), Salter (1954), Frei (1954), and Wadleigh and Gauch (1948). For a comprehensive review of the relationship of shoot growth to soil water regime the reader is referred to the excellent review by Hagan et al. (1959).

Since the internal water relations of plants are determined by both absorption and transpiration, it is important to consider the interrelations of soil water suction and the evaporation demands of the aerial environment in shoot growth relations. Weatherly (1950) measured the diurnal internal water deficit of cotton in relation to soil water suction and the evaporation power of the atmosphere and found that when the soil water suction was above a certain value, the water balance of the plant was largely controlled by the evaporative demands of the atmosphere, but if it were below this value it was controlled by both soil water suction and aerial evaporative demands. Recently Van Bavel et al. (1963) have reported that transpiration of sudangrass growing in a well watered soil was controlled by the evaporative demands of the atmosphere. Peters (1957) measured the shoot growth of corn in relation to soil water suction and the evaporative demands of the atmosphere. He found that shoot growth was profoundly influenced by soil water suction at low relative humidity. However, at low evaporative demands of the atmosphere (high relative humidity) growth was influenced only to a minor degree by soil water suction. Bierhuizen and De Vos (1959) have observed that the

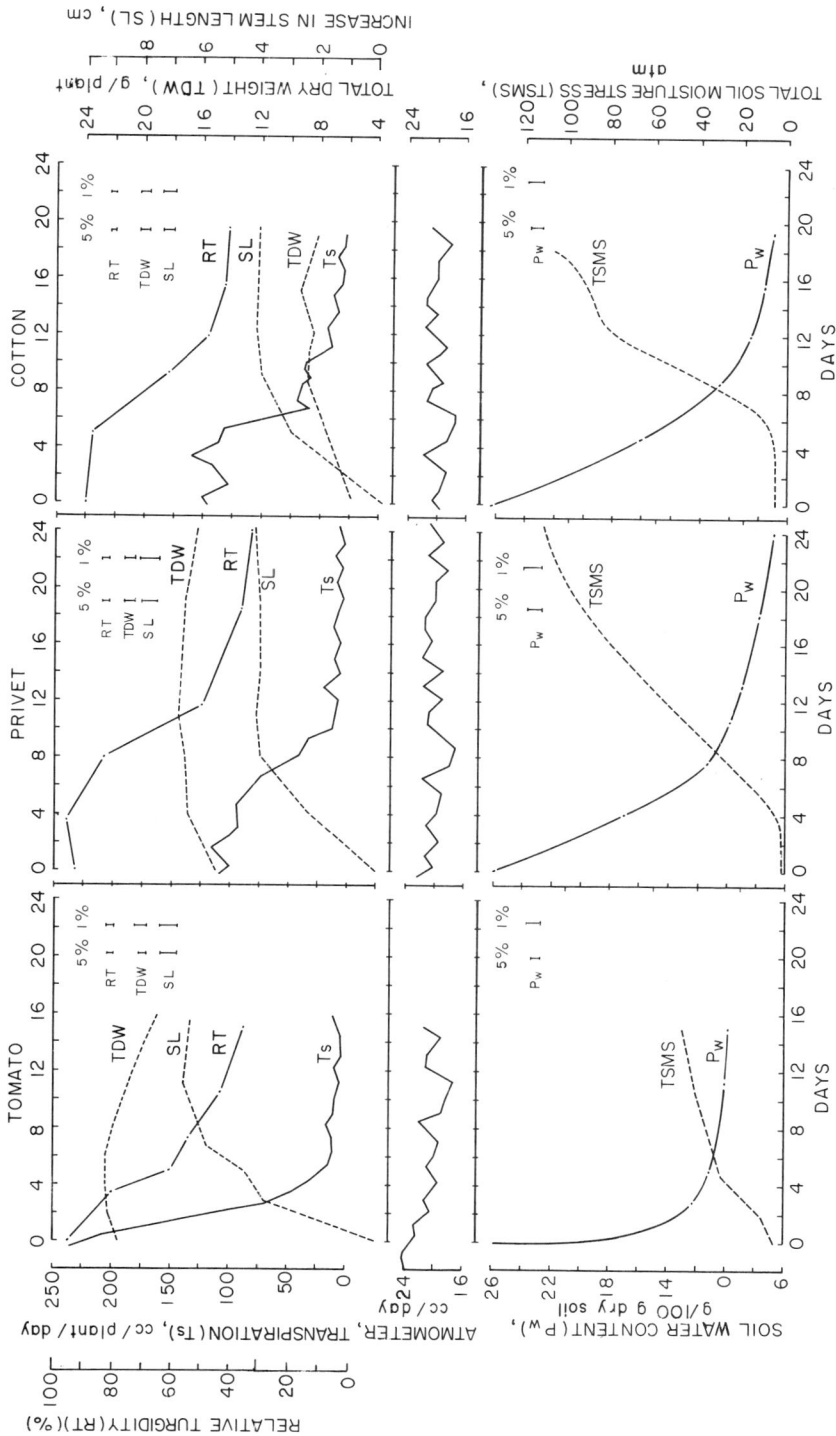

Fig. 20-3. Growth (TDW), increase in stem length (SL), relative turgidity (RT), and transpiration (T_s) of tomato, privet, and cotton, in relation to soil water content (P_w) and total soil moisture stress ($TSMS$). Daily atmometer evaporation is shown in the body of the diagram. When applicable, minimum significant differences at 5% and 1% levels are indicated (after Slatyer, 1957).

production of vegetables was greatest under a high soil water suction when the evaporative demand of the atmosphere was low.

VI. SHOOT-ROOT RATIO

The shoot-root ratio is defined as the ratio of dry weight of shoots formed to the dry weight of roots formed during the growth period under consideration. In the early part of this century, the shoot-root ratio was used rather extensively to characterize plant response to the imposed environment changes but it has been used relatively little in recent years.

Each plant species has a distinctive root system with regard to nature and extent of branching just as it has a distinctive vegetative shoot system. However, the type and extent of development can be altered considerably by environment. The shoot-root ratio varies among crop plant species from relatively large values in some field crops in the humid regions to very low values in some grasses which occupy the shortgrass prairie (Weaver, 1920, 1926, 1961).

Water is one of the more important environmental factors that influence shoot-root ratio. Harris (1914) studied the shoot-root ratio of corn, wheat (*Triticum vulgare*), and peas (*Pisum sativum*) in relation to water content of the rooting medium. His results with corn based on green weight are shown in Table 20–1. As seen in the Table, the shoot-root ratio decreases with decreasing water content. The weight of tops also exhibits a linear decrease with water content. However, root weight is at a maximum at about 20% water. Root weight decreases with increasing water content above this value. The reduction in root growth is a result of poor aeration in the rooting medium. Decreased root growth as a result of poor aeration has been observed by others (Erickson, 1946).

Miller and Duley (1925) studied the influence of varying water supply during the growth period on the shoot-root ratio. Their results show that reducing the soil water content during the last one-third of the growth period of corn resulted in a narrowing of the ratio as a result of both reduced top growth and increased root growth. A low water content during the first one-third of the growth period also resulted in a lowering of the ratio. Harris (1914) has made a similar observation with wheat.

The shoot-root ratio of plants is also influenced by the nutrient status of the rooting medium. Harris (1914) determined the shoot-root ratio of wheat grown to maturity as affected by soil fertility and found the ratio to increase with increas-

Table 20–1. Green weight of corn tops and roots grown with different amounts of water for a period of 17 days (after Harris, 1914).

Water	No. of plants	Green weight		
		Tops	Roots	Ratio tops/roots
%		grams		
38	5	3.63	4.05	.90:1
30	5	3.54	4.21	.84:1
20	5	3.36	5.18	.65:1
15	5	2.35	4.90	.48:1
11	5	1.56	4.30	.36:1

ing fertility, especially N. Similar results have been reported by Turner (1922). With a low concentration of N in the rooting medium, the absorbed N is utilized in root growth with little translocation to the shoots. This results in a low shoot-root ratio. As the nitrate concentration of the rooting medium is increased, some of the absorbed N is translocated to the shoots, increasing growth in the latter and the shoot-root ratio.

VII. REPRODUCTION AS AFFECTED BY SOIL WATER AVAILABILITY

Some vegetative shoots of plants continue to grow indefinitely, but sooner or later some of them become transformed into reproductive shoots. The time required for the vegetative shoot to be transformed into a reproductive one varies considerably between plants, between shoots on a given plant, and to some extent between environments. Soil water availability plays a minor role in determining the time when a vegetative shoot becomes reproductive. However, it does function as a major parameter in the overall reproductive stage of development.

Water deficits at the flowering and pollination stage of development in many plants can irreversibly damage yields. Flowering and pollination is a relatively short development period in comparison with vegetative or fruit maturity. The influence of water deficits for a short duration during flowering and pollination on yield of corn is illustrated by experiments of Robins and Domingo (1953). They found a 2-day deficit during this period reduced the yield 22%, and a 6-day deficit reduced the yield 50%. Similar results have been reported by Denmead and Shaw (1960). These researchers showed that water deficits during flowering and pollination reduced corn yields 50%, whereas water deficits during the vegetative and ear development reduced yields only 25% and 21%, respectively. The flowering and pollination stage of development is associated with an increase in rate of water use (Grimes et al., 1962; Musick and Grimes, 1961); therefore, internal water deficits develop rapidly in plants unless water is available at low suction in the soil (*also see* chapter 32).

Once pollination has taken place and the fruit set in the reproductive cycle the remaining important phase is fruit development. This part of the cycle lasts considerably longer than the flowering and pollination phase. Water deficits during fruit development have a very pronounced effect on size of individual fruit as well as total yield. Hedrick (1909) found that the size of apples (*Malus sylvestris*) was markedly reduced by decreased soil water content. Hendrickson and Veihmeyer (1929) have reported that the rate of enlargement and total size of peaches (*Prunus persica*) was increased by adequate soil water during the stage of fruit development. Extended water deficits during fruit development result in large percentage of small size fruit (Feldstein and Childers, 1957). It was reported by Grimes et al. (1962) that adequate water is needed in winter wheat during the milk stage of the grain to prevent shriveling.

The quality of yield of several crops is also influenced to a considerable degree by soil water availability. Kimball (1933) and Lewis et al. (1912) have shown that the color, size, and shape of apples are improved by increasing the soil water content. Jones and Colver (1912) have made a similar observation on apricots (*Prunus armeniaca*), cherries (*Prunus* sp.), peaches and plums (*Prunus domestica*). Haddock (1961) has observed that the mealiness of Russet Burbank potatoes

(*Solanum tuberosum*) was improved by increasing the soil water content. Eaton and Ergle (1952) have found that decreasing the soil water resulted in reduced fiber length and reduced seed and lint index of cotton. Sturkie (1947) demonstrated that the oil percentage of cotton seed was also reduced by a shortage of soil water (*also see* chapters 31 to 39).

VIII. SUMMARY

Water is essential to the growth and well-being of plants from germination through maturity. The initial phases of germination (imbibition, cell elongation, and cell division) are not too strongly affected by soil water suction *per se*, but the time required for complete germination is considerably extended by increased soil water suction. The time delay of germination under increased soil water suction is probably due to the limited rate of conduction of water to the embryo. Water pressure, either suction or osmotic, strongly influences both the per cent and rate of seedling emergence. Seedling emergence is also strongly conditioned by the nature of the forces holding the soil particles together during the germination period.

Water influences root systems in three general ways: (i) direction of root growth; (ii) lateral extension, extent of branching, and depth of penetration; and (iii) relative weight of top and roots. It has been adequately demonstrated that roots turn and follow water in the soil. The information available shows conclusively that the rate and extent of root development is controlled by water stress (suction and/or osmotic). Growth diminishes with increasing water stress, but root growth is less influenced than is shoot growth.

The condition of water in plant shoots is dynamic and is largely a result of opposing processes of transpiration and absorption. When an internal water deficit occurs in plants as a result of increased transpiration or decreased absorption, competition for water exists from cell-to-cell and organ-to-organ within the plant. Young, actively growing tissue is able to obtain water at the expense of the older, more mature tissue. Stem elongation and growth (total dry weight) are profoundly influenced by soil water suction. Increasing soil water stress (suction and/or osmotic) decreases both rate of stem elongation and total growth. On rewatering a plant that has been stressed, the rate of growth increases over that of an unstressed plant, but the recovery is usually not complete. Shoot growth is also conditioned by the interrelations of the evaporative demands of the atmosphere and soil water stress. At low relative humidity, shoot growth is influenced by soil water suction, but at high relative humidity it is influenced only to a minor degree.

Water is a major parameter in the overall reproductive development of most agronomic and horticultural crops. The pollination and flowering stage of reproduction seems to be most critical, with water stress at this stage causing considerably lower yields than those of a nonstressed crop.

LITERATURE CITED

Anderson, D. B., and T. Kerr. 1943. A note on the growth behavior of cotton bolls. Plant Physiol. 18:261–269.

Ayers, A. D. 1952. Seed germination as affected by soil moisture and salinity. Agron. J. 44:82–84.

Ayers, A. D., C. H. Wadleigh, and O. C. Magistad. 1943. The interrelationships of salt content and soil moisture with the growth of beans. J. Amer. Soc. Agron. 35:798–810.

Barley, K. P. 1962. The effects of mechanical stress on the growth of roots. J. Exp. Bot. 13:95.

Barley, K. P. 1963. Influence of soil strength on growth of roots. Soil Sci. 96:175.

Bartholomew, E. T. 1926. Internal decline of lemons: III. Water deficit in lemon fruit caused by excessive leaf evaporation. Amer. J. Bot. 13:102–117.

Bernstein, Leon. 1961. Osmotic adjustment of plants to saline media. Plant Physiol. 36:22.

Bierhuizen, J. F., and N. M. de Vos. 1959. The effect of soil moisture on the growth and yield of vegetable crops. Rep. Suppl. Irrig., Comm. VI. Int. Soil Sci. Soc. p. 83–93.

Chandler, W. H. 1914. Sap studies with horticultural plants. Missouri Agr. Exp. Sta. Res. Bull. 14. p. 485–552.

Chapin, J. S., and F. W. Smith. 1960. Germination of wheat at various levels of soil moisture as affected by application of ammonium nitrate and muriate of potash. Soil Sci. 89:322–327.

Collis-George, N., and Jocelyn E. Sands. 1962. Comparison of the effects of the physical and chemical components of soil water energy on seed germination. Australian J. Agr. Res. 13:575–584.

Danielson, R. E., and M. B. Russell. 1957. Ion absorption by corn roots as influenced by moisture and aeration. Soil Sci. Soc. Amer. Proc. 21:3–6.

Denmead, O. T., and R. H. Shaw. 1960. The effect of soil moisture stress at different stages of growth on the development and yield of corn. Agron. J. 52:272–274.

Doneen, L. D., and John H. MacGillivray. 1943. Germination (emergence) of vegetable seed as affected by differential soil moisture conditions. Plant Physiol. 18:524–529.

Duncan, W. G., and A. J. Ohlrogge. 1958. Nutrient uptake from fertilizer bands: II. Root development. Agron. J. 50:605.

Eaton, F. M. 1941. Water uptake and root growth as influenced by inequalities in the concentration of substrate. Plant Physiol. 16:545–564.

Eaton, F. M., and D. R. Ergle. 1952. Fiber properties and carbohydrate and nitrogen levels of cotton plants as influenced by moisture supply and fruitfulness. Plant Physiol. 27:541–562.

Erickson, L. E. 1946. Growth of tomato roots as influenced by oxygen in nutrient solution. Amer. J. Bot. 33:551–561.

Feldstein, J., and N. F. Childers. 1957. Effect of irrigation on fruit size and yield of peaches in Pennsylvania. Amer. Soc. Hort. Sci., Proc. 69:126–130.

Frei, E. 1954. Transpiration and growth of sunflower plants as a function of the soil moisture tension. Int. Congr. Soil Sci., Trans. 5th (Leopoldville, Bel. Congo) 2:74–81.

Gates, C. T. 1955. The response of the young tomato plant to a brief period of water shortage: I. The whole plants and its principal parts. Australian J. Biol. Sci. 8:196–214.

Gingrich, J. R., and M. B. Russell. 1956. Effect of soil moisture tension and oxygen concentration on the growth of corn roots. Agron. J. 48:517.

Gingrich, J. R., and M. B. Russell. 1957. A comparison of effects of soil moisture tension and osmotic stress on root growth. Soil Sci. 84:185.

Glendening, George E. 1942. Germination and emergence of some native grasses in relation to litter cover and soil moisture. J. Amer. Soc. Agron. 34:797.

Grimes, D. W., G. M. Herron, and J. T. Musick. 1962. Irrigating and fertilizing winter wheat in southwestern Kansas. Kansas Agr. Exp. Sta. Bull. 442. 8 p.

Haddock, J. L. 1961. The influence of irrigation regime on yield and quality of potato tubers and nutritional status of plants. Amer. Potato J. 38:423–434.

Hagan, R. M., Y. Vaadia, and M. B. Russell. 1959. Interpretation of plant responses to soil moisture regimes. Advance. Agron. 11:77–98.

Hanks, R. J., and F. C. Thorp. 1957. Seedling emergence of wheat, grain sorghum, and soybeans as influenced by soil crust strength and moisture content. Soil Sci. Soc. Amer. Proc. 21:357.

Harris, F. S. 1914. The effect of soil moisture, plant food, and age on the ratio of tops to roots in plants. J. Amer. Soc. Agron. 6:65–75.

Haynes, J. L. 1948. The effect of availability of soil moisture upon vegetative growth and water use in corn. J. Amer. Soc. Agron. 40:385–395.

Hayward, H. E., and W. B. Spurr. 1944. Effects of isosomotic concentration of inorganic and organic substrates on entry of water into corn roots. Bot. Gaz. 108:131–139.

Hedrick, V. P. 1909. A comparison of tillage and sod mulches in an apple orchard. New York Agri. Exp. Sta. Bull. 314. p. 77–132.

Helmerick, R. H., and R. P. Pfeifer. 1954. Differential varietal responses of winter wheat germination and early growth to controlled limited moisture conditions. Agron. J. 46:560–562.

Hendrickson, A. H., and F. J. Veihmeyer. 1929. 1. Irrigation experiments with peaches in California. Calif. Agr. Exp. Sta. Bull. 479. 56 p.

Hunter, Albert S., and Omer J. Kelley. 1946. Extension of plant roots into dry soil. Plant Physiol. 21:445–451.

Hunter, J. R., and A. E. Erickson. 1952. Relation of seed germination to soil moisture tension. Agron. J. 44:107.

Jones, J. S., and C. W. Colver. 1912. The composition of irrigated and non-irrigated fruits. Idaho Agr. Exp. Sta. Bull. 75. 53 p.

Kemper, W. D., C. W. Robinson, and H. M. Gclus. 1961. Growth rates of barley and corn as affected by changes in soil moisture stress. Soil Sci. 91:332–338.

Kimball, D. A. 1933. The influence of soil moisture differences on apple fruit color and conditions of the tree. Sci. Agri. 13:566–575.

Koller, D., A. M. Mayer, A. Poljakoff-Mayber, and S. Klein. 1962. Seed germination. Ann. Rev. Plant Physiol. 13:437–464.

Kramer, P. J. 1959. Transpiration and the water economy of plants. p. 607–726. *In* F. C. Steward [ed.] Plant Physiology. Vol. II. Academic Press, New York.

Lewis, C. I., E. J. Kraus, and R. W. Rees. 1912. Orchard irrigation studies in the Rogue River Valley. Oregon Agr. Exp. Sta. Bull. 113. 47 p.

Lloyd, F. E. 1920. Environmental changes and their effect upon boll-shedding in cotton. Ann. New York Acad. Sci. 29:1–131.

Miller, Edwin C. 1931. Plant Physiology. McGraw-Hill, New York. 1,201 p.

Miller, M. F., and F. L. Duley. 1925. The effect of varying moisture supply upon the development and composition of the maize plant at different periods of growth. Missouri Agr. Exp. Sta. Res. Bull. 76. 36 p.

Millington. R. J. 1959. Establishment of wheat in relation to apparent density of the surface soil. Australian J. Agr. Res. 10:487–494.

Musick, J. T., and D. W. Grimes. 1961. Water management and consumptive use by irrigated grain sorghum in western Kansas. Kansas Agr. Exp. Sta. Tech. Bull. 113. 20 p.

Olsen, S. R., F. S. Watanabe, and R. E. Danielson. 1961. Phosphate absorption by corn roots as affected by moisture and phosphorus concentration. Soil Sci. Soc. Amer. Proc. 25:289–294.

Owen, P. C. 1952. The relation of germination of wheat to water potential. J. Exp. Bot. 3:188–193.

Peters, D. B. 1957. Water uptake of corn roots as influenced by soil moisture content and soil moisture tension. Soil Sci. Soc. Amer. Proc. 21:481–484.

Peters, D. B., and M. B. Russell. 1960. Ion uptake by corn seedlings as affected by temperature, ion concentration, moisture tension, and moisture content. Int. Congr. Soil Sci., Trans. 7th (Madison, Wis., USA) 3:457–466.

Phillips, R. E., and D. Kirkham. 1962. Mechanical impedance and corn seedling root growth. Soil Sci. Soc. Amer. Proc. 26:319–323.

Richards, S. J., R. M. Hagan, and T. M. McCalla. 1952. Soil temperature and plant growth. *In* Soil Physical conditions and plant growth. Agronomy 2:303–481.

Robins, J. S., and C. E. Domingo. 1953. Some effects of severe soil moisture deficits at specific growth stages in corn. Agron. J. 45:618–621.

Rokach, A. B. 1953. Water transfer from fruits to leaves in the Shamouti orange tree and related topics. Palestine J. Bot. Ser. 8. p. 143–151.

Salter, P. J. 1954. The effects of different water regime on the growth of plants under glass. J. Hort. Sci. 29:258–268.

Sedgley, R. H. 1963. The importance of liquid-seed contact during the germination of Medicago Tribuloides Desr. Australian J. Agr. Res. 14:646–653.

Shull, C. A. 1914. Measurement of the internal forces of seeds. Kansas Acad. Sci., Trans. 27:65–70.

Slatyer, R. O. 1957. The influence of progressive increases in total soil moisture stress on transpiration, growth, and internal water relationship of plants. Australian J. Biol. Sci. 10:320–336.

Sturkie, D. G. 1947. Effects of some environmental factors on the seed and lint of cotton. Alabama Agr. Exp. Sta. Bull. 263. 87 p.

Taylor, Howard M., and Herbert R. Gardner. 1963. Penetration of cotton seedling tap roots as influenced by bulk density, moisture content, and strength of soil. Soil Sci. 96:153–156.

Toole, E. H., S. B. Hendricks, H. A. Borthwick, and V. K. Toole. 1956. Physiology of seed germination. Ann. Rev. Plant Physiol. 7:299.

Trause, A. C., and R. P. Humbert. 1961. Some effects of soil compaction on the development of sugar cane roots. Soil Sci. 91:208.

Triplett, G. B., and M. B. Tesar. 1960. Effects of compaction, depth of planting, and soil moisture tension on seedling emergence of alfalfa. Agron. J. 52:681–684.

Tucker, M., and C. van Seelhorst. 1898. Der Einfluss, welchen der Wassergehold und der Reichtum des Bodens auf die Ausbilding dur Worzein und der Oberindischer Organe der Hoferpflanze Ausuber. J. F. Landw. 46:52–63.

Turner, T. U. 1922. Studies of the mechanism of the physiological effects of certain mineral salts in altering the ratio of top growth to root growth in seed plants. Amer. J. Bot. 9:427–445.

Van Bavel, C. H. M., L. J. Fritschen, and W. E. Reeves. 1963. Transpiration by sudangrass as an externally controlled process. Science 141:269–270.

Wadleigh, C. H. 1946. Integrated soil moisture stress upon a root system in a large container of saline soil. Soil Sci. 61:225.

Wadleigh, C. H., and A. D. Ayers. 1945. Growth and biochemical composition of bean plants as conditioned by soil moisture tension and salt concentration. Plant Physiol. 20:106–132.

Wadleigh, C. H., and H. G. Gauch. 1948. Rate of leaf elongation as affected by the intensity of the total moisture stress. Plant Physiol. 23:485–495.

Weatherly, P. E. 1950. Studies in the water relations of the cotton plant: II. Diurnal and seasonal variations in relative turgidity and environmental factors. New Phytologist 50:36–51.

Weaver, J. E. 1920. Root development in the grassland formation. Carnegie Inst. Washington, D.C. No. 292. 151 p.

Weaver, J. E. 1926. Root development of field crops. McGraw-Hill, New York.

Weaver, J. E. 1961. The living network in prairie soils. Bot. Gaz. 123: 16–28.

Went, F. W. 1953. The effect of temperature on plant growth. Ann. Rev. Plant Physiol. 4:347.

Willis, W. C., W. E. Larson, and D. Kirkham. 1957. Corn growth as affected by soil temperature and mulch. Agron. J. 49:323–327.

Wilson, C. C. 1948. Diurnal fluctuations of growth in length of tomato stem. Plant Physiol. 23:156–157.

Zimmerman, R. P., and L. T. Kardos. 1961. Effect of bulk density on root growth. Soil Sci. 91:280.

21 | Root Systems in Relation to Irrigation

R. E. DANIELSON
Colorado State University
Fort Collins, Colorado

I. INTRODUCTION

Most cultural operations associated with the production of higher plants are conducted to provide a favorable environment for the development and activity of the root system. Present knowledge allows us to intelligently alter the root environment, but we still lack the ability to make practical changes in the foliar environment except in a minor way. Crop species are selected for adaptation to specific climatic regions, particularly air temperature and light. The soil environment is then adjusted by various physical and chemical treatments including irrigation.

Paramount to modern crop production, then, is a knowledge of root system characteristics and their response to alteration of environment. The many problems associated with soil properties such as water, aeration, fertility and fertilizer placement, acidity, salinity, and all the others, as well as with land preparation and tillage, are of importance in root development and activity essential to producing a plant consistent with its desired use.

It is the purpose of this chapter to consider plant root systems and their relation to crop production under irrigated agriculture. Certainly, irrigation is a practice carried out almost exclusively with an aim to provide a more desirable root environment.

II. ROOT MORPHOLOGY AND ANATOMY

The root systems of plants vary greatly as to type, extent, degree of branching, function, etc. These variations are the result of both inherited genetic properties of the species or variety and of the environmental and cultural conditions influencing the entire plant. It is not the purpose of this chapter to discuss in detail the general botany, anatomy, and physiology of plant root systems, but for some readers a brief review might be useful to consider the relationships between plant roots and irrigation.

A general classification of plant root systems distinguishes the primary roots, which are derived directly from the seed, from the adventitious roots, which arise adventitiously from another organ, usually the stem. The initial primary root develops from the rudimentary root or radicle of the seed embryo. If it continues to develop and forms a central structure from which many lateral roots diverge, it is considered to be a taproot. When laterals are relatively numerous and develop at a rate approximately equal to that of the initial root, the resulting primary root system is referred to as fibrous. Adventitious roots, which include those developing other than from the primary root, may arise from such tissue as stems, leaves, bulbs, tubers, rhizomes, or stem cuttings. The origin is usually in the vicinity of meristematic regions of the organ from which they arise. In some plants, e.g. *Gramineae*, some of the adventitious roots are present in the embryo and develop from the stem soon after germination. Such roots are frequently identified as seminal roots. A typical example is corn (*Zea mays* L.) where seminal roots appear soon after the coleoptile has penetrated the seed coat. Further adventitious roots originate after emergence from above-ground nodes of the stalk and develop into the familiar "brace" roots. In perennial plants, particularly the dicotyledons, the tap root and older laterals may undergo secondary growth to conduct food and water in addition to anchoring the plant. Some roots may develop for a specialized function such as the fleshy storage roots.

Root laterals originate deep in the main root tissue at some distance from the apical meristem. The taproot is frequently referred to as the primary root and its branches as secondary roots. Branches of secondary roots are called tertiary roots. Root systems may contain branches up to the fifth order which is reported by Cannon (1954) to be the highest order known to occur. The lateral roots are considered to branch, under optimum environmental conditions, to the "ultimate" order which is a specific feature of the species. If environmental conditions are suboptimal, root branching occurs only to the "highest" order which may be a lower order than the ultimate for the species.

The anatomical and histological aspects of roots and their development are reviewed by Esau (1953), Hayward (1938), and Stocking (1956). A selected list of references on the morphology and anatomy of roots has also been prepared by Miller (1960). Only a brief review of the general root anatomy, characteristic of roots in general, will be presented here.

A typical young root (Fig. 21–1) has three well defined regions consisting of the epidermis, cortex, and stele or vascular cylinder. The epidermis consists of closely packed, thin-walled, elongated cells. It is through this outermost cellular layer, having no intercellular air spaces, that the water and nutrients absorbed by the roots must pass. Maturation of the root results in the epidermis becoming impermeable due to suberization or cutinization of the cell walls. One of the outstanding characteristics of the young epidermal cells of most roots is the development of root hairs. These elongated, thin-walled, protuberances normally develop in greatest density in the region immediately back of the zone of cell elongation, usually from one to several millimeters from the root tip. Measurements made by Dittmer (1938) on three species of field-grown grasses, however, showed living root hairs scattered over the entire surface of all roots of each of the species.

The cortex includes in its outermost region a layer of cells known as the exodermis. As the root matures, these cells may become suberized in the region basal to that of maximum absorption and thus become relatively impermeable. The cortical tissue beneath the exodermis is composed of large parenchyma cells which typically separate during cell enlargement resulting in the establishment of intercellular spaces. These spaces constitute an extensive intercellular air system within the root which may be involved in the transport of oxygen to the cells for respirational activity. The breakdown of cortical parenchyma cells subsequent to their development has been observed in some plants resulting in further development of intercellular voids. These developments are thought to be related to root aeration (Beal, 1918; Bryant, 1934; McPherson, 1939). Root contraction, a phenomenon which occurs quite commonly in the early growth of roots of perennial plants is frequently connected with structural changes in the cortex (Thoday and

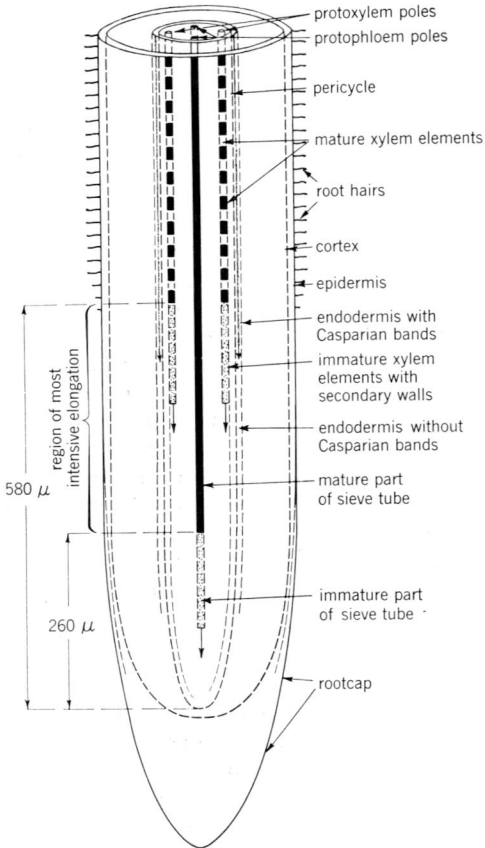

protoxylem poles
protophloem poles

pericycle

mature xylem elements

root hairs

cortex

epidermis

endodermis with
Casparian bands

immature xylem
elements with
secondary walls

endodermis without
Casparian bands

mature part
of sieve tube

immature part
of sieve tube

rootcap

region of most
intensive elongation

580 μ

260 μ

Fig. 21–1. Diagram of longitudinal section of a young tobacco root tip. (Reprinted with permission from K. Esau, *Plant Anatomy*, 1953, John Wiley and Sons, Inc., New York.

Davey, 1932). The cortex includes a layer of cells in its innermost region referred to as the endodermis. This tissue has received considerable attention because the thin-walled cells almost invariably develop a strip running lengthwise through their radial and transverse walls which is thickened and cutinized or suberized or contains other substances of a hydrophobic nature. This casparian strip (Fig. 21–2) resembles a narrow rubber band embedded in the primary wall and extended around each cell. It is considered to be quite impermeable to the passage of water and solutes and thus has been the subject of considerable controversy concerning its function in influencing the passage of these materials from the soil solution to the vascular cylinder.

Internal to the cortical endodermis is the stele or vascular cylinder. The outermost tissue of this region, the pericycle, has meristematic properties and thus is commonly the source of initial development of branch roots. The vascular system consists of phloem strands, distributed closely within the pericycle, and the xylem, either alternating with the phloem or occupying the central core of the root. The phloem, consisting of sieve tubes and companion cells, is considered to be the tissue through which organic substances synthesized in the leaves are translocated to the root meristematic regions. The xylem with its complex network of tracheids and vessels serves principally in conducting water and mineral solutes upward from the absorbing regions of the root.

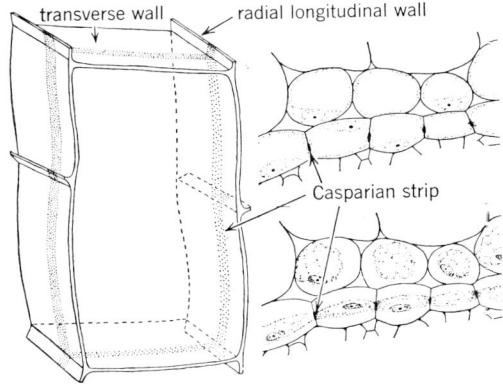

Fig. 21–2. Drawing representing an endodermal cell of a root and showing the location of the casparian strip. (Reprinted with permission from K. Esau. *Plant Anatomy*, 1953, John Wiley and Sons, Inc., New York.

In longitudinal section the young root tip consists of a root cap, commonly regarded as a protective structure aiding the root in its penetration through the soil; a zone of maximum meristematic activity; a region of rapid cell elongation; and a region where cell differentiation and maturation is most pronounced. Absorption of water from the soil takes place most rapidly in the younger portions of the root immediately basal to the meristematic region. It is in this region that the root hairs develop greatest density. The xylem elements have sufficiently matured to serve in conducting the water up the root to the stem, whereas permeability of epidermal and endodermal cells has not yet decreased.

Considerable study has been made and contradictory conclusions have evolved concerning the so-called "apparent free space" of root tissues. The apparent free space is considered (Briggs and Robertson, 1957; Bernstein and Nieman, 1960) to represent that portion of the root volume into which there is free penetration of solute or solvent from the external solution. Hope and Stevens (1952) and Briggs and Robertson (1957) have concluded that the tonoplast, the cytoplasmic membrane adjacent to the vacuole, is the limiting boundary of the free space and thus transport through the cytoplasm is to be expected. Levitt (1957) implies that this concept is erroneous and that the fraction of the cell wall available for free diffusion is probably identical to the apparent free space. The work of Smith (1960) indicating that free diffusion can take place directly into the xylem without meeting a barrier at the casparian strip is refuted by Bernstein and Gardner (1961) where Smith's data are used to show the opposite conclusion, namely, that a barrier does exist.

III. THE ROOTING HABIT OF CROP PLANTS

There is considerable variation in the normal root habit among the various species and varieties of plants. Thus, irrigation methodology has been and must be peculiar to the specific crop under production. It must be recognized, however, that other than genetic factors may influence root development. Root extent and configuration may be altered by the physical, chemical, and biological properties of the soil, and by cultural practices. Ten Eyck (1899, 1900, 1905) concluded that cultural methods, evolved from the experience of successful farmers, are a direct result of the different gross root development characteristics of the various species. Actually, modern production practices are probably more related to the type of top growth than to the root system of crops. Plant root systems are in part responsible for the success of those techniques commonly practiced because they produce the most desired results. Unfortunately, the development of irrigation methods for a given crop have largely evolved to fit the other agronomic practices used.

A. Types of Root Systems

Cannon (1949) has described, diagrammed, discussed, and classified 10 distinct types of root systems based on development and extension of the primary and adventitious components (Fig. 21–3). Types I through VI are primary roots and types VII through X are of adventitious origin. A given plant may develop a root system comparable to one or more of these types.

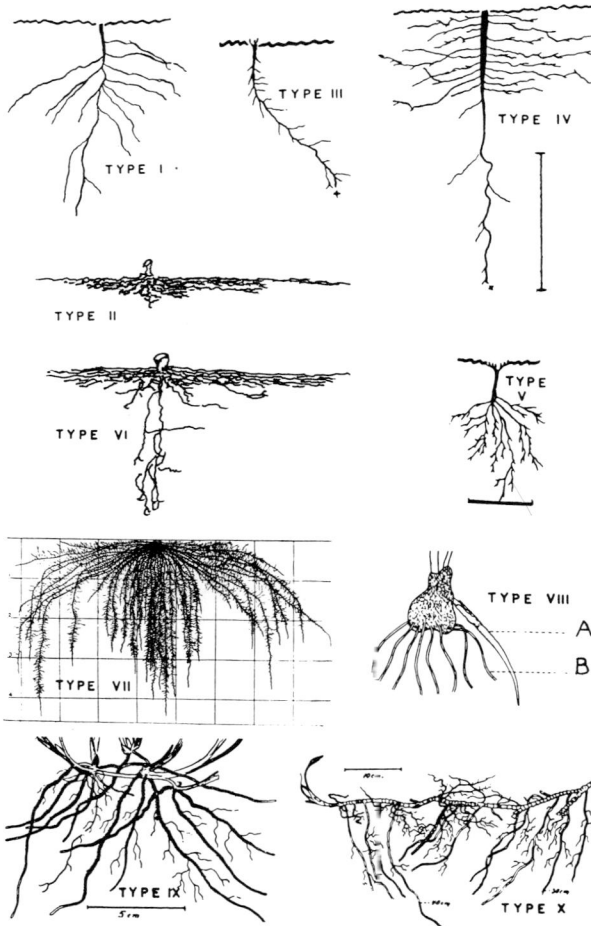

Fig. 21–3. Drawings of characteristic plant root system types. Types I through VI represent primary root systems and types VII through X represent adventitious root systems (after Cannon, 1949).

Cannon (1949) points out that the primary root is usually positively geotropic while the adventitious counterparts are diatropic or only weakly geotropic. Thus, the two systems are affected differently by gravity and their vertical and horizontal orientation in the soil is more a result of other environmental factors in one case than the other. This pattern of growth influences irrigation practices. In some cases, the taproot and its branches make up the entire root system; in others, the existence of the primary root is relatively short and the mass of surface laterals of adventitious origin comprise the majority of the subterranean organs. The former case is typical of the gymnosperms and dicotyledons while the latter is common to the monocotyledons. Intergrades exist, however, and it is not uncommon to find a mass of fibrous surface roots with a deeply penetrating taproot.

Much information concerning the normal root type and extent characteristic of a particular species has been provided by early field studies involving laborious excavation and mapping. Among the earliest works are those of King (1892) in the USA and those published by Fruwirth (1895) in Europe. The investigations of Cannon (1911) were limited to desert plants but were perhaps the first to provide the detail necessary to explain the findings. Troughton (1957) recently pointed out that to fully compare the root characteristics of plant species, all stages of plant growth should be studied under all environments favorable to growth. As this is impractical, most investigations have involved either comparisons of plants grown under identical conditions or over a wide range of environments where the species are naturally adapted. The latter, a survey method, limits root studies of a given plant to those conditions that approach the optimum for its growth and thus the range of conditions may not be comparable. The technique is ideal, however, in studying the ecological significance of root development. A great majority of root studies have been of the survey type, and undoubtedly, the most extensive investigations of this kind have been conducted by Weaver and his associates of Nebraska (Weaver, 1919, 1920, 1925, 1926, 1954; Weaver et al., 1922; Weaver and Bruner, 1927; Jean and Weaver, 1924) and by Pavlychenko (1937a).

Much of the recent work on root development has involved comparisons of specie variation under uniform environment or the influence of modifications in environment or culture on the roots of a given species. Laird (1930) found that the density of the fibrous root system developed by sod-forming grasses varied among species when compared under similar soil and management conditions. Dittmer's classical work shows distinct genetic variations in root and root hair development among grasses as well as many other plant families (Dittmer, 1938, 1948, 1949).

Varieties and strains within a given species also have genetically controlled variations in rooting type and development. Typical examples are those reported within species of oats (*Avena*) (Derick and Hamilton, 1942), wheat (*Triticum*) (Janssen, 1929; Webb and Stephens, 1936; Worzella, 1932), barley (*Hordeum*) (Hess, 1949), corn (*Zea*) (Kiesselbach and Weihing, 1935) and alfalfa (*Medicago*) (Garver, 1922, 1946; Burton, 1937; Smith, 1951). Many variations of root systems have been correlated with certain characteristics of the plant such as drouth resistance, winter hardiness, or lodging tendencies. Furthermore, such characteristics are inherited and tend to be expressed in spite of environmental factors. Kiesselbach and Weihing (1935) demonstrated that corn hybridization resulted in an exhibition of heterosis by the root system as well as the stalks. The hybrid

Table 21–1. Relative stalk and root development of selfed lines of corn and their first and second generation hybrids (Kiesselbach and Weihing, 1935)

	Stalk height	Leaf area	Maximum root depth	Maximum root spread	Combined length of main roots
Avg. of inbred parents	100	100	100	100	100
F_1 generation	128	143	155	116	145
F_2 generation	114	116	129	100	121

vigor of the F_1 generation, compared to inbred parents, as well as the decline in the F_2 generation, is indicated by relative values of stalk and root development for mature plants presented in Table 21–1.

B.   Techniques of Root System Investigations

Observation and measurement of plant root systems requires laborious and time-consuming effort. Numerous techniques and modifications have been devised to speed the process and reduce labor requirements. Complete separation from the soil of an entire root system of a well-established plant is almost an insurmountable task. Thus, root studies on mature plants have involved modified sampling techniques. Methodology has been developed with specific objectives in mind and no method is without certain limitations.

The most precise and successful method for studying the morphological character and extent of root systems is the so-called "trench" method. The cut trench is oriented to expose a vertical soil wall containing a desired cross section of the plant roots to be investigated. The soil is then carefully and tediously removed from the trench face so that the exposed roots may be counted, measured, mapped, or photographed. This method has been used extensively for ecological studies (Weaver and Bruner, 1927). The procedure requires much labor and is not well adapted to quantitative investigations because the finer roots are frequently lost. The use of water spray to wash away the soil has advantages in some profiles (Stoeckeler and Kluender, 1938; Tharp and Muller, 1940). Pavlychenko (1937b) developed a "soil block" or "monolith" method whereby an entire soil block is removed from the site of the growing plant, and then soaked and washed. It has the advantage of reduced breakage of the finer roots and is better suited to quantitative measurements. Various modifications and improvement techniques have been reported by Weaver and Darland (1949), Weaver and Voigt (1950), and Upchurch (1951).

Soil cores are sometimes used to study plant root systems but are inadequate for complete mapping of root continuity. The core technique is useful in evaluating root density as influenced by soil horizons. Laird (1930) forced steel cylinders into the soil to secure root samples. Fehrenbacher and Alexander (1955) and Bloodworth et al. (1958) used a powered core sampler and subsequently washed the roots free of soil with special techniques. Boehle et al. (1961, 1963) successfully soaked and washed roots free from soil cores taken with a tractor drawbar attachment. Other developments for sampling and washing roots have been published by Jacques (1937), Fribourg (1953), and Williams and Baker (1957).

Direct observation of plant root systems, either by excavation or by core samplings, has provided very important information on the extent, distribution, and arrangement of the roots. The relative activity of the roots with respect to nutrient uptake is, however, not directly obtained in such investigations. Thus, techniques have been devised for measuring the absorption of chemical elements by roots to indirectly measure root system extent, rate of growth, and absorptive capacity. The methodology essentially consists of locating an absorbable chemical substance at desired locations in the soil and measuring the concentration of the substance in the foliage as a function of time. The absorbed substance must remain in the immediate location where placed and be readily translocated to the aerial portions of the plant where measured. Early attempts at measuring the extent of corn root systems by such a method were made by Sayre and Morris (1940) using lithium chloride.

The advent of readily available radioactive isotopes has resulted in numerous attempts to use this type of tracer in the absorption method. Radioactive phosphorus, P^{32}, has been found to meet the requirements in evaluating root systems of a number of different crops including grapes (*Vitis*) (Lott et al., 1950); corn (Hall et al., 1953; Murdock and Engelbert, 1958; Nye and Foster, 1961); grasses (Burton et al., 1954; Oswalt et al., 1959); alfalfa (Lipps et al., 1957); sorghums (*Sorghum*) (McClure and Harvey, 1962; Nakayama and Van Bavel, 1963); and vegetable crops (Hammes and Bartz, 1963). The quantitative measurement of the isotope in the foliage allows relative evaluation of the root activity in the soil zone where the material was placed. A rather complete discussion of the tracer technique in measuring root activity and growth has been provided by Hall et al. (1953).

IV. ABSORPTION OF WATER BY PLANT ROOTS

An important function of plant root systems is that of water absorption from the soil. Variability among plant species is frequently a major factor in crop production in arid regions. The capacity of root systems in their entirety to meet transpirational demands is in part determined by their gross morphological nature and extent, and in part by the specific impedance exhibited at a given absorption point. Rather complete coverage of the literature pertaining to mechanisms of water uptake by roots has been included in reviews by Kramer (1945) and Slatyer (1960) and in chapter 17 of this monograph. It is only necessary at this point to indicate that cultural practices, environmental conditions, and physiological stresses in the plant itself exert an influence on the absorptive potential of the root system as a whole. Soil temperature, for example, influences not only the rate of development of highly absorbing root branches, but also the resistance of established roots to the passage of water (Kramer, 1938). The influence of various environmental factors on the permeability of root cells has been demonstrated by Ariz et al. (1951) and by Mees and Weatherley (1957a, 1957b). Ordin and Gairon (1961) have reported increases in the self-diffusion rate of tritiated water through root tissue due to increases in the internal water stress of the tissue itself, and Skidmore and Stone (1964) have demonstrated marked diurnal fluctuations in root impedance to water absorption and transmission even in solution culture under controlled environment conditions.

A.   Zones of Maximum Absorption

The effectiveness of the plant root system in absorbing water from the soil will, in part, be related to the total root surface area through which water may be taken up. There is much evidence to show that the absorption rate varies along the longitudinal axis of roots; thus total root surface alone does not completely govern the gross absorption rate. The evidence has been obtained under somewhat artificial conditions since actual uptake rates for a given root segment are nearly impossible to obtain under normal conditions of growth in soil culture.

Rosene (1937, 1941a) carefully measured rates of water absorption by young onion roots (*Allium cepa* L.) using very small potometers to measure intake rate for a small longitudinal segment of the root at any desired distance from the apex. A general increase in absorption rate per unit of surface area from the apex to the base was noted for roots < 5 cm in length. Maximum absorption for older roots occurred closer to the root tip. Roots longer than 8 cm absorbed water at the greatest rate within 40 mm from the apex. The development of lateral roots caused a definite shift of the maximum absorption region toward the apex. A similar longitudinal gradient of intake velocity was measured with onion roots excised from the bulb. Although considerable differences in rate of water absorption exist between the distal and more proximal regions of the root, no such difference could be observed in the osmotic pressure of the external solution required to prevent water uptake (Rosene, 1941b).

Potential absorption capacity of a root system depends to a large degree on development of root hairs (Dittmer, 1937, 1938, 1949). The hairs contribute the major portion of the total surface area of the root systems measured. Rosene (1943, 1954), Rosene and Walthall (1954), and Rosene et al. (1949) have measured absorption velocities into surface sections of individual root hairs of various species and found values comparable to those for hairless epidermal cells from equivalent regions of the root.

An important amount of the transpirational requirements of plants is probably absorbed through the older root regions (Kramer, 1956; Slatyer, 1960). However, the greatest potential uptake occurs where the conducting tissues of the xylem have developed and where the epidermal and endodermal cells have not become suberized. Thus, the histological development of the root controls the relative uptake rates along the length of the root. When roots are limited in their elongation rate by any of various environmental factors, maturation and suberization extends closer to the tip, leaving a shorter region for maximum absorption.

B.   Water Transport from Soil to Root

Water absorption by plant roots depends on a continuous and adequate rate of transfer from the soil matrix to the root surface. Since the conductivity of soil to water flow decreases rapidly as the water content decreases, considerable potential gradients would be expected to develop over short distances from the root surfaces of transpiring plants. Philip (1957, 1958) suggests that vapor phase transfer is an important factor in supplying the transpirational demands of the plant under moderate absorptive conditions where sufficiently large soil water gradients develop near the root surfaces. Such a vapor gap could possibly be increased in effectiveness by shrinkage of the root as desiccation of the cells accompanies

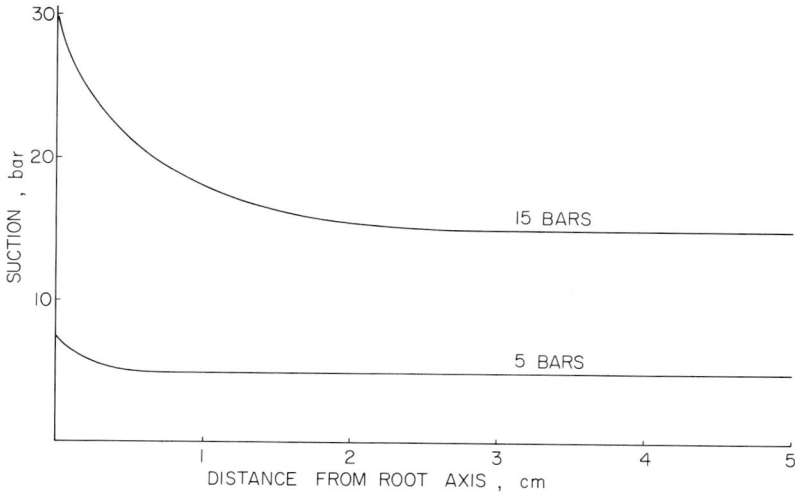

Fig. 21–4. Soil water suction curves calculated, using capillary conductivity values for Pachappa sandy loam soil, to provide a constant uptake of 0.1 cm³/day for each 1 cm length of root when initial suction was 5 and 15 bars (after Gardner, 1960).

drying of the soil. Philip concludes that such a discontinuity of water films would serve as an ideal semipermeable membrane allowing water to cross in the vapor phase but preventing solute transport. Bonner (1959) has supported the vapor gap concept, but Bernstein et al. (1959) present contrary arguments that soil water diffusivities for vapor movement are far too small to account for more than possibly 10% of the water transfer even at soil water conditions in the wilting range for plants.

Actual measurements of the hydraulic conductivity of soil at low soil water levels are, as yet, limited; however, some of the available data has been used by W. R. Gardner (1960) to develop quantitative predictions of the role of water movement in relation to its availability to plant roots. Solutions to the equation of water flow in soil were obtained, using certain arbitrary but reasonable assumptions and boundary conditions, to evaluate the soil suction necessary at the root-soil boundary to maintain a constant flux of water to the root. Figure 21–4 indicates the expected changes in suction in relation to distance normal to a root surface. The curves were calculated from known capillary conductivities and water desorption curves for Pachappa sandy loam assuming a reasonably sized root with constant absorption rate of 0.1 cm³/day for each 1 cm of length. The two curves represent conditions where the soil water suction is uniform at 15 and 5 bars outside of the region of root influence. The calculations indicate that the suction values at the root surface are probably not very great until the average soil suction reaches several bars. Soils with higher clay content and capillary conductivities can tolerate higher average suctions before root surface suction values become significantly high. As Gardner points out, the rate of water absorption per unit length of root is proportional to the total transpiration rate of the plant and inversely proportional to the effective length of the root system. Assuming a hypothetical value of 20 bars suction at the root surface as the wilting condition for a plant, Gardner's calculations indicate this "wilting point" would be reached when the average soil water suction is approximately 12 and 7 bars, respectively, for absorption rates of 0.1 and C.5 cc/cm of root per day.

W. R. Gardner and Ehlig (1962a) measured the transpiration rate of Birdsfoot trefoil (*Lotus corniculatus* L.) growing in soil under greenhouse conditions. Transpiration remained essentially constant, and almost equal, for plants growing in sand, loam, and clay until initial wilting caused a sharp drop. Initial wilting occurred at average soil suctions of approximately 2, 3, and 9 bars for the sand, loam, and clay soils, respectively. The investigators concluded that capillary conductivity of the soil is the major influence on water availability to the plant in the range where wilting occurs and impedance within the plant is of minor consequence. In further investigations, Gardner and Ehlig (1962b) measured the net transpiration rate of pepper (*Capsicum frutescens* L.) plants growing in greenhouse containers and calculated water potential gradients from soil-to-plant using measurements of soil water suction and leaf tissue vapor pressure. Calculations indicated that the impedance to water movement remained relatively constant until approximately 0.6 bars of average soil suction was reached. Since the capillary conductivity values for many soils of different textures are similar in this suction range, it was assumed that the 0.6-bar value would be the suction at which soil conductivity becomes limiting for many soils. At lower suctions, impedance occurs almost completely within the plant. As soil water suction increases, the impedance to movement through the soil to the root surface is of greater and greater significance.

V.   INFLUENCE OF SOIL AND CULTURAL FACTORS ON ROOT DEVELOPMENT

The root system characteristic for a given species of plant will develop only when the environment is in proper balance for the genetically controlled characters to be expressed. Chief environmental factors are those connected with the soil profile in which the root system is contained. However, the foliar environment and the cultural methods used in production influence indirectly the extent, distribution, and vigor of the subterranean plant parts. Many of these factors affecting root development have been previously reviewed by Wiersma (1959).

Since so many physical, chemical, and biological factors affect plant root activity, exacting quantitative relationships have not been established to predict the influence of any one factor on the ultimate root character obtained. There have, however, been sufficient investigations to allow reasonable conclusions. Two approaches have been used: (i) studies of root development of specific species under natural field conditions and correlation of the findings with soil and climatic data; (ii) studies where soil and climate is controlled and modified as desired to evaluate the effect of a particular factor while others are maintained as uniform as possible.

A.   Soil Factors

1.   TEXTURE AND STRUCTURE

The soil matrix exerts an influence directly through the mechanical impedance it offers to plant root development. However, since the physical condition of the soil has an important influence on other factors involved in root development, it becomes difficult to measure the direct effects of physical impedance *per se*. Many investigations have shown a significant relation of soil texture and structure to development and activity of roots, but evaluation of the mechanical resistance exerted by the soil is limited and frequently only obtained by inference.

Extensive studies identifying the nature of root systems of various species as related to textural and structural features of natural soil profiles have provided important contributions to present knowledge of soil effects on rooting habit (Weaver and Crist, 1922; Carlson, 1925; Fitzpatrick and Rose, 1936; Weaver and Darland, 1949; Lamba et al., 1949). Unfortunately, the specific cause-and-effect relationships are not available from such studies. Lutz (1952) provided an excellent review of the literature prior to 1952, and made a special effort to separate the direct effects of mechanical factors in his discussion. A review of the status of research on soil compaction, and a discussion of needed research, was published by Raney et al. (1955). An extremely useful annotated bibliograph on soil compaction has been provided by Gill et al. (1959) and contains references to soil compaction-root system relationships.

Soil texture appears to exert its influence on root development chiefly in relation to water and nutrient availability. By far, the most striking examples of textural influence are those where abrupt changes occur vertically through the soil profile. A change in particle size distribution greatly alters water movement in unsaturated soil so that sand layers tend to be drier and clay layers wetter than would be the case in uniform profiles. A typical example of this effect is shown in Fig. 21–5 (Muller, 1946). Root branching of the guayule (*Parthenium argentatum* A. Gray) plants decreased rapidly as the loam soil in the surface foot (30 cm) graded to sand below. Proliferation of branching occurred along thin, horizontal veins of clay with greater water-holding capacity at 2 and 4.2 feet. Muller reported great difficulty in distinguishing between chemical and physical factors responsible for root responses to soil variation. Fertility, as well as soil water, may be involved. In the Netherlands, restricted root growth in sandy layers underlying clay or highly organic surface layers occurs with a great variety of crops on widely differing particle size ranges of sand (Goedewaagen, 1955; Hidding and Van Den Berg, 1960). Subsoil tillage resulted in better root penetration, but the benefits were only temporary.

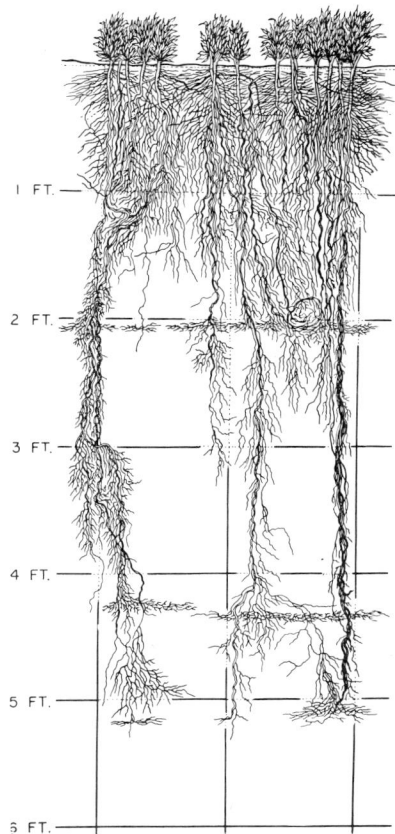

Fig. 21–5. Root development of guayule in sandy soil. Note horizontal development along thin veins of clay at approximately 2 and 4.2 feet and vertical aggregation in an old root channel on the right (after Muller, 1946).

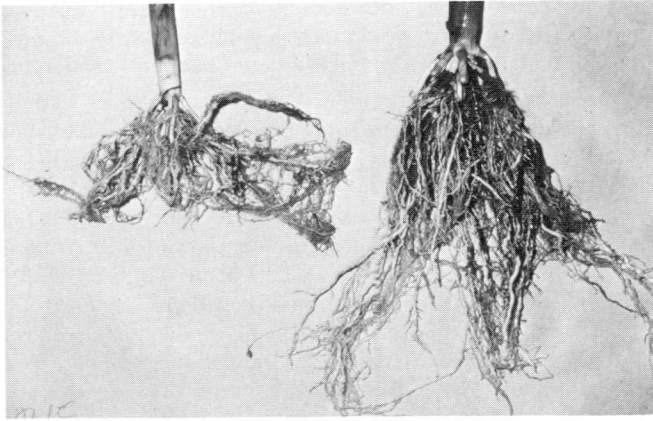

Fig. 21–6. Effect of soil compaction on root development of corn. The restricted root system on the left resulted from intentional compaction of the surface soil prior to planting.

Soil tilth can vary greatly for a given soil. Root growth, restricted in zones of "poor" physical condition, may be improved by fertilizer placement or other methods of altering the root environment (Fehrenbacker and Snider, 1954; Fehrenbacker and Rust, 1956). Phillips and Kirkham (1962) found that mechanical impedance, as measured by bulk density and needle penetrometers, was most highly correlated with reduction of root growth of corn; and Brown et al. (1932) reported that the turning of the taproot of cotton (*Gossypium hirsutum* L.) by a compact subsoil was due directly to the hard soil layer and not to soil water, acidity, fertility, or aeration. Still other investigations have indicated no particularly detrimental influence of soil compaction (Van Diest, 1962; Nutman, 1933) or even slight beneficial effects (Hubbell and Staten, 1951). Moderate compaction of very loose soils in Australia has been found to greatly increase yields and to provide significant benefits in relation to plant use of residual manganese from earlier fertilizer applications (Passiouri and Leeper, 1963). Explanation of these results, and of others (Miller and Mazurak, 1958), are based on improved root–soil–solution contact.

Soil compaction affects the development of plants and their root systems differently on different soils (Rosenberg and Willits, 1962) and plant species differ in their ability to penetrate restricting layers (Zimmerman and Kardos, 1961; Flocker et al., 1960). The author has found that compaction of field plots by excessive tillage operations and traffic on clay loam soil seriously inhibits root development and yield of corn, especially during the early stages of growth. A typical example of the effect on root morphology is shown in Fig. 21–6. Sugar beet (*Beta Vulgaris* L.) yields were essentially unaffected by the same soil treatments. The shape of the beet roots, however, was markedly altered by soil compaction (Fig. 21–7). Physical manipulation of the soil resulting in breaking up of compacted zones, reduction in traffic over the soil, and the use of fibrous-rooted green manure crops have in general provided beneficial effects to root development on problem soils (Pendleton, 1950; De Roo, 1957, 1961).

Various attempts have been made to evaluate the threshold values of soil bulk density at which root penetration is prevented (Veihmeyer and Hendrickson,

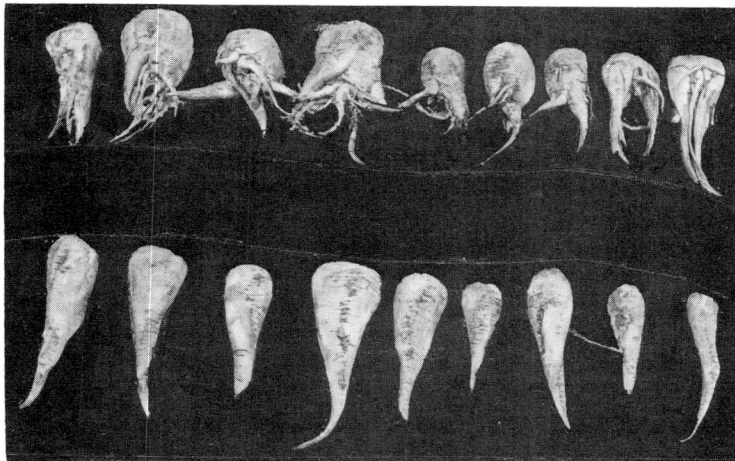

Fig. 21–7. Effect of soil compaction on root development of sugar beets. Preplanting compaction of surface soil resulted in excessive root branching of the beets shown in the top row.

1948; Bertrand and Kohnke, 1957; Zimmerman and Kardos, 1961; De Roo, 1961). Reported values vary for different soils and plant species and are difficult to interpret because environmental conditions are not comparable and the actual restraint imposed on the developing roots is not known.

Direct effects of soil mechanical impedance on root growth have been studied under controlled laboratory and environmental conditions (Gill and Miller, 1956). Actual values of mechanical restraint in a granular material at the root surface were not known but relatively greater compression resulted in relatively slower root development. The partial pressure of O_2 in the root atmosphere also influenced corn root elongation through the media (Fig. 21–8). Similar interactions between O_2 and confining pressure were obtained by Barley (1962). Since soil compaction necessarily affects soil aeration (Bertrand and Kohnke, 1957; Flocker et al., 1959), it is of considerable significance to have actual data showing the influence of O_2 content on the penetrating ability of the root. Wiersum (1957) varied the rigidity of sand used as a growth medium and found that the depth of penetration by roots of oats (*Avena sativa* L.) and turnip (*Brassica rapa* L.) was decreased as the rigidity of the sand increased. Wiersum also grew a wide variety of species in cintered glass filter disks of different pore diameter ranges. He concluded that a root is only able to penetrate a rigid pore having a diameter exceeding that of the young root. Wax materials have been used to simulate plastic soil pans in attempts to evaluate the penetrating ability of roots (Taylor and Gardner, 1960; H. R. Gardner and Danielson, 1964). The method allows the plants to be grown in soil, where environmental conditions are controlled, and at the same time evaluate the degree of root penetration into a standard test layer underlying the soil. Soil factors such as water, aeration, and the presence of soluble salt were shown to influence root penetrating ability.

It must be concluded that, in general, root growth as such is negatively correlated with soil density and rigidity as normally found in field soils. The direct influence of mechanical impedance, however, has not been satisfactorily evaluated.

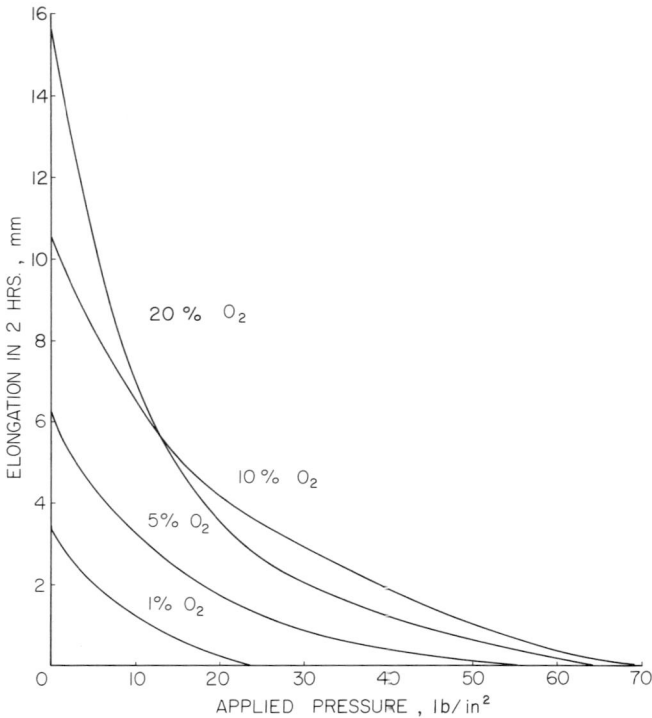

Fig. 21–8. Corn root elongation at various O_2 concentrations as a function of pressure applied to a confined layer of glass beads in which the seedlings were growing (after Gill and Miller, 1956).

Early measurements by Pfeffer (1893; *see* review by Gill and Bolt, 1955) have shown that a plant root is capable of exerting considerable forces against confining material. However, if the plant must exert a sizeable force in order to elongate and expand radially, it must necessarily do work in the process. Perhaps the energy thus expended could be otherwise used in the production of a more desirable product for the grower. A realistic approach to the problem must involve considerations of how best to maintain or improve soil tilth and limit even temporary deterioration of physical condition. Irrigation itself is often directly influential in soil compaction and in reorientation of the soil particles resulting in crust formation. On some soils, the irrigation method adopted should take these factors into consideration.

2. TEMPERATURE

Soil temperature exerts an influence on the development and activity of plant roots through its effect on the many reactions involved in growth. It also affects the many physical, chemical, and biological activities in the soil external to the root. A discussion of these temperature-influenced growth activities, together with an exceptionally thorough coverage of the earlier literature pertaining to soil temperature, root development studies, has been provided by Hagan (1952).

For many species the optimum temperature for root development is lower than that for shoot development, and cool climate species of plants have optimum temperatures for root development considerably lower than do warm climate species. In all probability the soil temperature is of greater importance than air temperature in influencing the adaptability of a given plant to a specific climate.

It is impossible to cite precise temperature values which will provide the most desirable root growth for a particular species since interactions with other environmental factors are too pronounced. However, in a very general sense, it would appear from the results reviewed by Hagan (1952) that small grains make the greatest root growth at approximately 20C while corn and soybeans (*Glycine max* Piper) prefer about 25C as their root temperature. Cotton would appear to make the best root growth at near 30C. Optimal growth of excised roots apparently occurs at slightly lower temperatures (Went, 1953). Practically all of the reported attempts to obtain values of optimum root temperature for root development of temperate region crops indicate values between 18 and 25C. Thus, they are considerably below the air temperature values conducive to maximum foliage production.

Soil temperature effects on nutrient uptake and plant nutrition in general, as well as growth and development of plant tops, have received considerable attention in recent years. The effects on root development, however, have been less thoroughly investigated. Barney (1951) found that root production of pine seedlings grown under greenhouse conditions increased from 5 to 25C but decreased again at 30 and 35C. The root growth of strawberry (*Fragaria vesca* L.) plants under greenhouse sand culture was reported by Roberts and Kenworthy (1956) to decrease slightly as temperature increased from 7 to 24C. Foliage production, however, increased progressively over the same temperature range. The strawberry roots tended to be more slender and profusely branched when grown at the higher temperatures. Root development of oats where soil temperature was controlled by placing pots in water baths in the greenhouse, was studied at various stages of plant development by Nielsen et al. (1960). At all stages of maturity root yield tended to decrease with increases in temperature from 5 to 26.5C. Again the difference in root temperature effect on root and foliage growth was indicated by increased top-root ratios with increasing temperature. Later work, Nielson et al. (1961), provided opposite results for corn where root growth was shown to increase from 5 to 26.5C. Bromegrass (*Bromus inermis* Leyss) root yields increased from 5 to 19.5C and then decreased at 26.5C, and potato (*Solanum tuberosum* L.) root production varied considerably as a function of temperature depending upon fertilizer treatment.

Various studies conducted in controlled environment chambers, where root and foliage temperatures were nearly identical, have also provided data to indicate the beneficial effect of relatively low temperature on gross root production. Alfalfa (*Medicago sativa* L.) root growth decreased from 15.5 to 32C (Gist and Mott, 1957), barley (*Hordeum vulgare* L.) developed maximum root yields at approximately 15C but growth was lower at 26.5 than at 7C (Power et al., 1963), and red clover (*Trifolium pratense* L.) had an apparent maximum root development at 15.5C but also was less at 26.5 than at 10C (Robinson et al., 1959).

Soil temperature has been shown to exert an important influence on the rate of water absorption by plant roots. In general, the studies indicate a decrease in water uptake as the water and root temperature decreases. Kramer (1942) measured the effect of gradual cooling of the soil on water absorption by many species of plants. Rate of absorption declined for all species, but the effect was greater for those types adapted to warm climates. Kramer assumed that the differences observed between species was related to variations in the protoplasm viscosity and permeability changes due to lower temperature. Gradual cooling of solution culture media causes a gradual reduction in water absorption rates of the tomato (*Lycopersicon esculentum* Mill.), sunflower (*Helianthus annuus* L.), and bean (*Phaseolus vulgaris* L.) (Böhning and Lusanandana, 1952). In each case relatively slow decreases in temperature from 25 to 5C produced no visible sign of wilting of the foliage. When the solution temperature was suddenly decreased from 25 to 5C, each of the three species wilted severely and only gradually recovered during subsequent days. The bean plants were most seriously affected and some permanent damage resulted. Alfalfa has also been shown by Ehrler (1963) to develop definite water deficits resulting in wilting when the nutrient solution bathing the root system is cooled. Schroeder (1939) reported leaf and fruit injury to cucumbers (*Cucumis sativus* L.) due to water deficiencies in the plant caused by lowered soil temperature. Again the injury was greater the more rapid the temperature change. The resistance to water movement through intact plants, as well as isolated root, stem, and leaf portions, has been demonstrated to be inversely related to water temperature by Jensen et al. (1961) and Jensen and Taylor (1961). Activation energies calculated at various temperatures were found to be greater for water flow through roots than through stems and leaves, and were sufficiently high to imply a more complex transport mechanism than simple viscous or diffusional flow.

Certain investigations have provided information pertaining to the direct effects of adverse soil temperature on root anatomy and physiology. Nightingale (1935) observed that the cortex of fruit tree roots at £4C turned brown and sloughed off. The central cylinder was woody, mechanically strong, and lacked succulence. At 18C and lower the roots were typically white and succulent. Root temperatures of 30 and 32C apparently caused a seriously limited digestion of starch in the old roots. The chemical analyses of Roberts and Kenworthy (1956) indicate that root respiration is increased by high root temperatures and maturation processes are more rapid. Earlier maturation and suberization of cells due to higher temperatures was also reported by Barney (1951). Alfalfa "scald" was shown by Erwin and Kennedy (1957) to be a result of xylem necrosis and root collapse which was reproducible in the greenhouse at root temperature of 39C. Davis and Lingle (1961) have suggested from studies with tomato that excessively low root temperatures retarded phloem transport, resulting in a congestion of photosynthesized substances in the shoot which may depress metabolic activity and salt accumulation. Vinokur (1957) has reported quantitative variation between individual amino acids in lemon (*Citrus medica* L.) seedlings due to differences in nutrient solution temperature and assumes that temperature acts differently on individual metabolic processes in the root.

Soil temperature may be controlled to some extent by the cultural practice used in crop production. The greatest effect occurs in early spring when seedling roots are shallow and soil temperature in general is low. Burrows (1963) has measured the effect of tillage-related microrelief on soil temperature and the germination and seedling development of corn. In general, the highest root zone temperature resulted from seed placement on ridges and from row orientation from north to south. Tillage methods resulting in surface mulches influence soil temperature (Willis et al., 1957; Army and Hudspeth, 1960; Burrows and Larson, 1962). The result is chiefly that of delaying soil warming in the spring and the effect on plant development depends on latitude and the climatic conditions involved (Van Wijk et al., 1959).

3. SOIL WATER AND DRAINAGE

The influence of soil water on the development and physiological functions of plant root systems is undoubtedly exerted more frequently and consistently than any other soil factor. The nature of the soil solution, as determined by its composition and amount, is an effective governor of most complex processes involving the plant root. The specific relationships of soil water availability on root growth are covered in chapter 20 where references to numerous other reviews on this subject are made. The literature dealing with root development under excessively high water levels has been previously reviewed by Van't Woudt and Hagan (1957).

The relation of drainage, and the establishment and maintenance of water tables in the soil profile, to root development involves both factors of water availability in regions above the water table and factors of anaerobic saturation below. Many investigations have provided results which are difficult to evaluate because adequate measurements of soil water stress throughout the entire profile have not been obtained. In most instances the presence of a static water table provides a beneficial effect in supplying water to roots. Optimum crop production can be obtained even when normally deep-rooted plants are restricted to shallow regions by a water table if the water is of good quality and adequate fertility is supplied (Tovey, 1963). Species vary in regard to root growth under water table conditions. Some will extend roots into the saturated soil, others exhibit abrupt stoppage of growth at the water level, and still others develop root systems to any appreciable extent only in the moist but well-aerated region above the water table. These responses are undoubtedly related to the aeration requirements of the roots and the ability of the plant to translocate oxygen from well-aerated regions.

A most important factor associated with root development in soil containing a water table involves the fluctuation of the saturated zone. A rapid fall of the water level induces drouth conditions because the previously restricted root system is inadequate to absorb sufficient water from the resulting drier soil. When the water table rises a portion of the established root system is inundated, resulting in death if the condition is maintained for a critical period of time. Results of studies in this regard are discussed in the review of Van't Woudt and Hagan (1957). Gilbert and Chamblee (1959) demonstrated that a water table fluctuating between two levels may result in less total root production than a constant water table at either level. Reconstruction of a portion of their data, obtained from greenhouse studies, is presented in Table 21–2. Variation between species is apparent, particularly in the surface 15-cm depth of soil. Apparently, root development in the fluctuating zones during periods of drainage was at the expense of those in the surface, and as new roots were periodically killed, the result was to decrease total production. The extent of root damage caused by a fluctuating water table depends on the duration of inundation, water quality, and temperature. Recent work of Tovey (ca. 1964) indicates that well-established 5-year-old alfalfa root systems are not impaired by submergence periods of 4 days or less. The studies were conducted in field lysimeters where the water table was maintained at 60 cm except when raised to the soil surface each 14 days to impose

Table 21–2. Root yield of three forage species as affected by depth of water table; values represent dry weight in grams per pot after 8 months of growth
(after Gilbert and Chamblee, 1959)

Species	Depth of water table		
	15 cm	50 cm	15-50 cm*
	Roots in surface 15 cm		
Ladino clover	5. 27 (96. 6%)	7. 04 (89. 8%)	4. 75 (96. 6%)
Orchardgrass	11. 70 (81. 4%)	11. 08 (64. 4%)	9. 23 (88. 9%)
Tall fescue	10. 38 (91. 7%)	6. 36 (63. 0%)	6. 48 (76. 5%)
	Total roots at all depths		
Ladino clover	5. 29	7. 84	4. 92
Orchardgrass	12. 80	17. 20	10. 38
Tall fescue	11. 32	10. 10	8. 47

* Alternating at 15 cm for 1 week, then at 50 cm for 3 weeks.

the treatments. Duration of submergence in each cycle varied from 1 to 11 days. The effect after one growth season is photographically presented in Fig. 21–9.

The detrimental effect of root inundation by water is undoubtedly related largely to factors connected with aeration and disease. Extended periods of surface flooding by irrigation of soils with layers of low permeability may cause near saturation of surface zones while roots deeper in the profile may be in relatively dry soil. Oxygen diffusion from the atmosphere is prevented under such conditions and anaerobic conditions could be expected to occur throughout the root zone. Kemper and Amemiya (1957) reported an average O_2 concentration of 6% in the root zone of alfalfa following 8 days of flooding for leaching purposes. Root investigations were not conducted but foliage growth appeared to be benefited rather than impaired by the flooding conditions.

4. AERATION

The maintenance of conditions for adequate soil aeration is an important requisite to root development (*see* chapter 47). The discussion here concerns aeration and its interrelation to the various other soil factors influencing plant roots.

Aeration is frequently a contributing factor to root responses influenced by soil structure, temperature, and moisture. Deficient O_2 and excessive CO_2 are directly involved in root development under conditions of soil compaction or excessive water. Studies involving controlled aeration have frequently demonstrated that the effect of such conditions may be altered considerably if adequate root aeration is provided. The relation of O_2 concentration to root elongation under a confining pressure as indicated in Fig. 21–8 is a typical example. Reduced root production at high soil temperature is no doubt related to increased cell respiration under such conditions. Although precise studies have been inadequate for proof, it appears possible that "excessive" aeration may occur in many normal soils. Anderson and Kemper (1964) have suggested root respiration in excess of that required for normal root function may decrease yield of corn by unnecessary oxidation of the products produced in photosynthesis. Further study is needed in this regard.

5. FERTILITY

Proper mineral nutrition of roots, like all plant parts, is essential to their development and growth. Soil fertility is consequently an additional factor in their

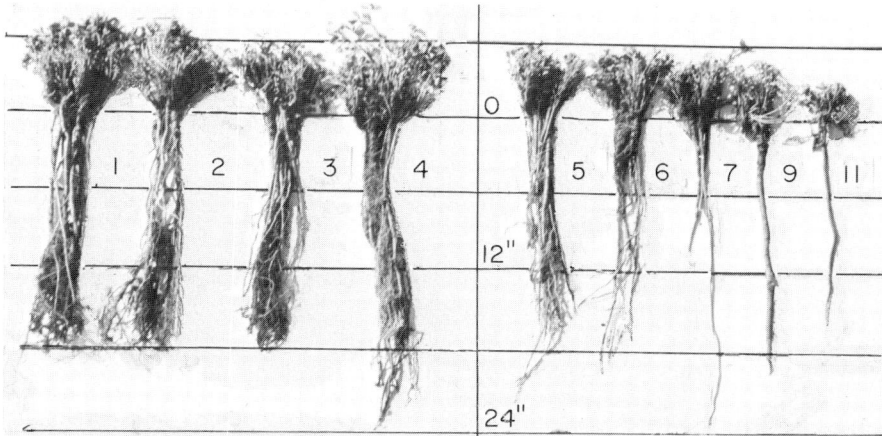

Fig. 21–9. Alfalfa roots showing the effect of raising the water table from 60 cm to the soil surface each 14 days during a growing season and maintaining it at the surface for periods ranging from 1 to 11 days (after Tovey, ca. 1964).

response to environment. Details of the specific functions of various elements to the processes of root elongation and branching, and thus to the general morphology of the root system, have not been clearly ascertained. It is clear, however, that root distribution in the soil profile is influenced by variation in the available nutrient supply.

In general, it appears that deficiencies in soil nitrogen are of greater significance in limiting the development of foliage than of roots; whereas, at high concentrations of N in the soil solution, roots are restricted to a greater degree than shoots. This general increase in root production with N supply up to a limit, followed by a decline as N continues to increase, has been discussed by Troughton (1957). It has recently been demonstrated again by Lorenz and Rogler (1964) who reported Russian wildrye (*Elymus junceus* Fisch.) receiving heavy N applications (400 lb/acre) usually developed lower root weight than those fertilized at lesser rates, but always more than those receiving no N fertilizer. Results obtained from root investigations in fertilizer experiments have provided contradictory conclusions in regard to N effects. This is perhaps due to the great variation in the natural N supplying potential of soils throughout the growing season.

Oswalt et al. (1959) found that supplemental N decreased depth of penetration of roots of orchardgrass (*Dactyllis glomerata*) and bromegrass. It increased root diameter, decreased rate of elongation, and caused feeding near the soil surface for a longer time period. Almost identical results were obtained by Bosemark (1954) with wheat. He investigated the cause and related growth inhibition at higher N levels to the combined action of reduced cell multiplication and elongation. Thus, he proposed a relation of N level to natural auxin production in the root. Holt and Fisher (1960), on the other hand, reported heavy N fertilization resulted in slightly deeper root penetration of bermudagrass (*Cynodon dactylon*). The activity of corn roots, as measured by Ca^{45} and water absorption at a depth of 61 to 66 cm, was depressed by N irrespective of the depth of placement (Younts and York, 1956). Linscott et al. (1962) found considerably greater root weights and deeper penetration in N fertilized corn plots early in the season, but essentially no difference at maturity. Additional soil N has also been shown to increase root production of various grasses (Wright, 1962; Haas, 1958; McKell et al., 1962) and of wheat (Kmock et al., 1957). Soybean roots will develop higher order branching of lateral roots in response to localized placement of N and P according to Wilkinson and Ohlrogge (1962). They searched for mechanisms responsible and found that extracts obtained from roots produced under N and P fertilization tended to promote lateral branching of other roots. Additional research is needed to determine the relation of soil nutrients to the development of growth-promoting substances in the plant root.

Phosphorus has received less attention than has N in its effect on root responses. Again, the available results are conflicting. Troughton (1957) cites experimental evidence for negative, positive, and neutral effects of added P on root development. He discredits the old opinion that P has some special benefit to root growth.

Early work by Brenchley and Jackson (1921) indicated that the number of "white roots" developed on barley plants was distinctly increased by P, while similar results were not found with wheat. More recently, Duncan and Ohlrogge (1958) have reported a more rapid development and greater branching of the smaller corn roots in those soil bands where both N and P fertilizer had been placed. The stimulation was slight when P alone was provided and absent when only N was used. Restricted root branching of grasses in a claypan region has also been shown by Fox et al. (1953) to be related to low P availability. Sommer (1936) studied the relation of P concentration in solution cultures to root system development of various plants, and concluded that root-top ratios were in general decreased by increasing P. Root development was not stimulated by high phosphate in solution.

Evidence of potassium influence on root development and function appears to be quite meager. In most studies it has been included in conjunction with other major nutrients. Haas (1958), however, evaluated root production by grass species and reported no response to K fertilizer.

Comparisons of root systems developed under differing levels of mixed fertilizer or soil amendment application have, in general, indicated favorable responses to improved fertility or soil reaction when such treatment is also beneficial to shoot growth. Frequently, root penetration has been increased by deep placement of fertilizer or lime if native subsoil fertility or reaction is inadequate. Bushnell (1941) demonstrated greater depth of rooting with potatoes by deep fertilizer placement, and corresponding results were reported by Kohnke and Bertrand (1956) with corn. Pohlman (1946) markedly increased alfalfa root development and nodulation in the subsoil regions where soil acidity was decreased by application of lime. Similar results were obtained with alfalfa by Fox and Lipps (1955) and Englebert and Truog (1956); and Woodruff and Smith (1946) concluded the lack of lime in subsoil claypan conditions to be of greater significance than physical condition in restricting corn root penetration.

The importance of micronutrients in root physiology and growth needs additional attention. Important contributions could be expected from such research. The critical function of zinc in meristem activity of tomato roots has been demonstrated by Carlton (1954) using solution culture techniques. Similar studies, with tomato and other species, by Haynes and Robbins (1948) showed that boron was essential in root environment, and that it could be satisfactorily provided through only a portion of the root system.

B. Cultural Factors

Those practices commonly employed in the modern culture of crop plants have, to an appreciable extent, evolved through a long succession of failures and successes of farmers. In recent times the results of research and technology have had a marked effect in changing these practices. However, the changes have chiefly involved the means of obtaining the desired results, and not with objectives themselves.

Frequently, the objectives desired from a cultural practice are connected with development and function of the plant root system. Specifically, such terms as seedbed preparation, crop rotation, soil fertilization, weed control, irrigation, drainage, plant spacing, and summer fallowing merely connote an attempt to improve or maintain the various soil factors affecting root development previously discussed. The influence on the plant is largely directed through the influence on soil structure, depth, water, aeration, fertility, and temperature. As changes in cultural methodology take place, through advances in agricultural machinery and chemicals, the resultant influence on soil environment, in regard to its effect on plant roots, must constantly be evaluated.

Certain cultural practices, however, influence plant root systems directly, regardless of the environmental changes they produce. A portion of the root branches may be severed by cultivation or fertilizer placement. The result is usually undesirable, and the damage appears to be most significant when it occurs in the early periods of plant development. Lateral root pruning decreases the effective surface area and reduces the capacity for water and nutrient absorption. In addition, it greatly increases the potential for disease and insect infestation. Simmonds and Sallans (1933) studied the effect of root amputation on wheat plants. They reported that loss of seminal roots during the seedling stage greatly reduced yield and had a tendency to delay maturity. The severity of the injury decreased as root removal was delayed. Crown root cutting had quite the opposite effect in that removal became more important the later they were cut. Hunt and Miller (1961) evaluated the effect of seminal root removal on very young grass seedlings prior to transplanting. Reduced root and top development was apparent as much as 60 days later. Lateral root development by alfalfa was shown by Klebesadel (1964) to be considerably increased when the taproot was severed at 14 cm below the soil surface. He proposes a possible significance of such effects to studies involving frost heaving damage. Undoubtedly, the stress imposed on the plant by root pruning largely depends on subsequent soil water and nutrient availability. Elazari-Volcani (1936) found, for example, that more than 50% of the root system of young citrus trees could be removed before wilting occurred when the trees were well supplied with water.

The removal of plant foliage, or a portion of the foliage, has been shown in many studies to exert immediate and sometimes lasting effects on the root system. Investigations have concerned mostly those perennial plants harvested periodically, such as legumes or grasses used for pasture or cut for hay. Defoliation of the plant influences initiation of new roots, elongation of existing ones, and the translocation of nitrogen and carbohydrate reserves resulting in weight losses. The degree of the effect depends on plant development at time of foliage removal, extent and frequency of removal, and environmental conditions prior to and subsequent to the removal. The results of most investigations indicate a partial or complete stoppage of root development following substantial shoot removal. In certain cases, however, the removal of specific plant parts, particularly seed-forming organs, have resulted in increased root development.

Frequent and early cutting of alfalfa was reported by Nelson (1925) and Graber et al. (1927) to deplete root reserves resulting in retarded growth and consequent low yields of tops. Frequent cutting at succulent stages of growth prevented root weight increases throughout the growing season. When tops were removed at full bloom or later, root systems showed continued weight increases. Hildebrand and Harrison (1939) studied cutting frequency effects on transplanted alfalfa cultures in the greenhouse, and found the period and height of cutting were directly related to root weights obtained at the conclusion of the experiment. Dotzenko and Ahlgren (1950) also found that dry weight and per cent total polysaccharides in alfalfa roots increased as stage of maturity at cutting increased. The yield of hay in the subsequent year was positively correlated with amount of total polysaccharide stored in the roots the previous winter.

The effect of defoliation on root growth of perennial grasses has been discussed in considerable detail by Troughton (1957) and Baker (1957a, 1957b). Numerous studies were cited showing restricted root development resulting from top removal, especially when conducted at early stages of development. A rather extensive study of root growth following foliage removal was conducted on eight grass species by Crider (1955). His investigations included field as well as greenhouse cultures and involved a technique whereby roots were marked with carbon black so that apical growth during a particular time interval could be measured. Under all conditions, and with each species, repeated defoliation completely stopped root growth. For most species root growth was stopped following the initial clipping. Crider concluded that root growth stoppage occurred when 40 to 50% by volume of the foliage was cut away. Results similar to those of Crider's have been obtained by Oswalt et al. (1959), Thurman and Grissom (1954), and Wright (1962).

Some investigators have reported no particularly detrimental effect of defoliation on specific root responses, and under certain circumstances it may be beneficial. Salmon et al. (1925) observed no particular differences in rooting habit of alfalfa following 9 years of repeated harvest at different stages of maturity. Hagan and Peterson (1953) concluded from water extraction records that, if the botanical composition remains constant, there is no material difference in root distribution due to harvest frequencies of 2-, 3-, 4-, or 5-week intervals. Mowing of grasses to prevent seed formation may improve root development (Laird, 1930); and floral bud removal from cotton was shown by Eaton (1931) to increase root growth. Overbearing of coffee (*Coffea arabica*) causes dieback of roots (Nutman, 1933) and corn roots make very little growth while ears are developing (Loomis, 1935).

Under most conditions, it must be concluded that defoliation at early growth stages results in a depletion of root reserves. Subsequent root development is largely restricted to initiation of new roots near the crown. Thus, the need for adequate fertility and irrigation of the surface soil following hay harvest or close grazing, to minimize the effects of disease and insect infestation (Colville and Torrie, 1962) and loss of stand (Nelson, 1944) is apparent.

VI. ROOT DEVELOPMENT AND IRRIGATION PRACTICES

The irrigation of agricultural land, except when conducted specifically for salt removal or subsurface water storage, is to provide adequate soil water for development of absorbing root surfaces necessary to meet the water and nutrient requirements of the crop. The depth, extent, and configuration of the root system is unimportant providing the absorptive potential is attained and sufficient anchorage of the plant is provided. Adequate root activity for high yield is possible with relatively shallow root systems if proper environmental conditions are provided. Under careful management, good production is frequently attained on soils restricted in effective depth by a gravel layer or water table, or under the confining conditions of greenhouse or hydroponic culture.

Maintenance of proper soil environment is extremely difficult, however, when the volume available for root exploitation is restricted. Frequent and rather precise attention is required to provide adequate water, aeration, and fertility. The influence of root system development—its extent, distribution, and character— becomes important when the surface soil condition is not maintained for optimum root function. Such is the more common situation in commercial crop production. As the surface soil is reduced in water content and fertility, the plant with an extensive root system may continue to meet nutrient and transpirational demands from greater depths. Thus, the interval between irrigations and fertilizer applications can be extended, efficiency is increased, and the natural fertility of the subsoil may be utilized.

A. Extent and Activity of Root Systems

High efficiency in the use of irrigation water requires a minimum of penetration below the root zone. Knowledge of rooting depth is, therefore, important to the design of irrigation systems and to the determination of proper time and amount of application. The potential depth of root extension varies among plant types. Certain species are shallow rooted; others can extend to great depths. A listing of the vertical and lateral extent of rooting observed for various crop plants, trees, and shrubs has been provided by Miller (1938). Such data, however, are of limited use in irrigation planning because of the wide range in values reported. The influence of soil factors too often prevents the potential root depth from being realized. Doneen and MacGillivray (Table 21–3) have separated vegetable crops into three rooting depth categories as a guide for irrigation.

Root distribution with depth is perhaps a more important factor in proper irrigation than is maximum penetration. The works of Jean and Weaver (1924) and Weaver (1926) reveal that the general "working depth" of roots of many annual crops is limited to 150 cm or less. Even when the soil profile is uniform and adequately supplied with water and nutrients, root development decreases rapidly with depth. A great proportion of the total root weight is frequently found to occur very near the surface (Pumphrey and Koehler, 1958; Bloodworth et al., 1958; Foth, 1962; Lorenz and Rogler, 1964). Root weight is a questionable criterion for root activity, however. Measurements of water and nutrient absorption are probably more meaningful. Hall et al. (1953) used depth placements of

Table 21–3. Depth of rooting of truck crops*

Shallow rooted (down to 60 cm)	Moderately deep rooted (down to 120 cm)	Deep rooted (down to 180 cm)
Broccoli (Brassica oleracea var. Italica)	Beans (Phaseolus vulgaris)	Artichoke (Aynara scalymus)
Brussel sprouts (Brassica oleracea var. gemmifera)	Beets (Beta vulgaris)	Asparagus (Asparagus officinalis)
Cabbage (Brassica oleracea var. capitata)	Carrots (Daucus carota)	Cantaloupe (Cucumis melo var. cantalupensis)
Cauliflower (Brassica oleracea var. botrutis)	Chard (Beta vulgaris var. cicla)	Lima bean (Phaseolus limesis
Celery (Apium graveolens)	Cucumber (Cucumis sativus)	Parsnip (Pastinaca sativa)
Lettuce (Lactuca sativa)	Eggplant (Solanum melongena)	Pumpkin (Cucurbita pepo)
Onion (Allium cepa)	Pea (Pisum sativum)	Squash, winter (Cucurbita maxima)
Potato (Solanum tuberosum)	Pepper (Capsicum frutescens)	Sweet potato (Ipomoea batatas)
Radish (Raphanus sativus)	Squash, summer (Cucurbita maxima)	Tomato (Lycopersicon esculentum)
Spinach (Spinacia oleracea)	Turnip (Brassica rapa)	Watermelon (Citrullus vulgaris)

* Doneen, L. D. and J. H. MacGillivray, 1946. Suggestions on irrigating commercial truck crops. Unnumbered leaflet. California Agr. Exp. Sta.

radioactive P to indicate root activity under field conditions. Uptake decreased rapidly with depth for corn, cotton, and peanuts (*Arachis hypogeae* L.) throughout the growing season. The maximum zone of absorption by tobacco (*Nicotiana tobacum* L.) roots tended to shift downward to 30 to 45 cm as the crop matured. Similar studies by Hammes and Bartz (1963) disclosed major root activity of vegetable crops in the surface soil. Nakayama and Van Bavel (1963) reported approximately 90% of the water and P^{32} uptake by sorghum (*Sorghum vulgare* Pers.) occurred from the upper 60% of the root zone in all of various irrigation treatments. Utmost attention, then, must be given to the normally high concentration and activity of roots in the surface soil when irrigation scheduling is considered. A general "rule of thumb" frequently used is that 40, 30, 20, and 10% of the total transpirational requirement is supplied respectively from each successively deeper one-quarter of the root zone. Hunter and Kelley (1946) found that roots of alfalfa and guayule (*Parthenium argentatum*) extracted water from the surface soil at fairly high suctions while water was available in the deeper root regions at much lower suctions. Apparently distance from the plant, rather than depth, is the important factor since Davis (1940) reported similar suction gradients developed in a horizontal direction away from the crown of corn plants. Roots near the plant removed water much more rapidly and dried the soil to "below the wilting point" while roots of apparently equal concentration 120 cm away were in moist soil.

Deep rooting and lateral extension into the subsoil should be encouraged, however, so that maximum utilization of the soil and water resources may be obtained. Sufficient irrigation water should be applied at each application to bring the soil to field capacity throughout the desired rooting depth. Frequent, light irrigations, when water is inadequate below, will restrict root penetration (Bierhuizen, 1961) and increase the degree of drouth damage if dry periods subsequently occur. The possibility of inducing deeper rooting by intentional drying of the surface soil is a topic of frequent concern. Root elongation and branching is restricted by increasing soil water suction. Is it, then, to be expected that if root development is restricted at the surface by drying soil, comparable increases will result at deeper depths where water is available? Experimental data to substantiate such reasoning appears to be lacking; however, Weaver (1925) observed corn plants were shallower rooted and exhibited less branching when the soil was irrigated and manured than when it was not. Nakayama and Van Bavel (1963), on the other hand, reported water removal by sorghum from 120 to 180 cm was not appreciably different for zero and five irrigations. Russell and Danielson (1956) found soil water at the end of the season between 60 and 150 cm was similar when plots received 0, 18.5, and 35 cm of precipitation and irrigation. Taylor and Slater (1955) state that plants should not be allowed to suffer for water in the mistaken belief that root systems will be forced downward into deep soil because growth of both roots and shoots is retarded during any period when water is deficient in a portion of the soil occupied by roots. It was demonstrated by Hunter and Kelley (1946) that water and nutrients could be absorbed at lower levels and conducted through 80 cm of dry soil to the foliage. Growth, however, was considerably less than when available water was present throughout the profile. Since root growth is related to the general vigor of the plant, maximum development in any region of the soil should take place when optimum conditions are maintained in all regions. Undoubtedly, under certain circumstances, restriction of surface soil water may allow a crop to survive later drouth periods because of decreased evapotranspiration resulting from restricted foliage development.

Continued development of root branches, and their elongation through the soil is of primary importance in water relations of plants. Capillary conductivity of the soil becomes increasingly lower as soil water decreases, and impedance to water absorption increases with age of root tissue. Root extension continuously brings young root tips into contact with soil water. Dittmer (1937) estimated that over 114 thousand new roots per day, over a 4-month growing period, were produced by a single rye (*Secale cereale* L.) plant, and that daily new root length averaged 3.1 miles (49.9 km). Using Dittmer's measurement, Kramer and Coile (1940) calculated the amount of water made available by root extension and concluded that under some conditions it might amount to all the water required by the plant.

B. Genetic Variability of Plants

Irrigation planning for maximum effect must take into consideration certain peculiarities of the species or varieties grown. The primary root may develop to considerable depth before seminal and adventitious roots get started (Lea, 1961). Irrigation planning must provide for adequate surface soil water for these late roots. Crown roots developing from the above-ground nodes on stems of cereals are important in the water relations of the plant during later stages of growth. Pavlychenko (1937a) has pointed out the need for moist soil surfaces during the establishment of these roots. Perennial crops and winter grains continue to develop roots through the autumn and early winter months, (Weaver et al., 1924). Late season irrigation is necessary if soil water is inadequate.

Those irrigation practices established and used in the production of a given crop species may need to be altered if new varieties or strains are developed having different rooting habits or water requirements. Modern trends in plant breeding are to develop plant types for specific purposes. A typical example is the dwarf tomato adapted to machine harvesting. Weihing (1935) found a high correlation between the extent of root and shoot systems of corn; similar results could be expected for the tomato. Changes in root vigor of corn results from hybridization (Spencer, 1940) and irrigation practices may need to change as other hybrids are introduced.

Development of varieties and strains by plant breeders has largely involved selections based on foliar characteristics of the plant. Root characteristics undoubtedly are responsible in part for improved yield and quality. Perhaps additional attention to root characteristics in the breeding program could be used to develop plants more efficient in utilizing soil water and nutrients. Selection of plant types for their ability to germinate under high osmotic pressures is being used in variety improvement (Dotzenko and Dean, 1959; Dotzenko and Haus, 1960). Similar attention to subsequent root development should allow selections for other desirable features. Techniques for rapid root characterization need to be developed so that evaluation of large numbers of genotypes may be readily made.

Consideration must be given to the conditions under which irrigated varieties are selected. Nurseries maintained at optimum levels of soil water by the ultimate in irrigation design and facilities may result in selections not necessarily the most suitable for commercial production under limited irrigation water supplies.

LITERATURE CITED

Anderson, W. B., and W. D. Kemper. 1964. Corn growth as affected by aggregate stability, soil temperature, and soil moisture. Agron. J. 56:453–456.

Ariz, W. H., R. J. Helder, and R. van Nie. 1951. Analysis of the exudation process in tomato plants. J. Exp. Bot. 2:257–297.

Army, T. J., and E. B. Hudspeth, Jr. 1960. Alteration of the microclimate of the seed zone. Agron. J. 52:17–22.

Baker, H. K. 1957a. Studies on the root development of herbage plants: II. The effect of cutting on the root and stubble development, and herbage production, of spaced perennial ryegrass plants. J. Brit. Grassland Soc. 12:116–126.

Baker, H. K. 1957b. Studies on the root development of herbage plants: III. The influence of cutting treatments on the root, stubble, and herbage production of a perennial ryegrass sward. J. Brit. Grassland Soc. 12:197–208.

Barley, K. P. 1962. The effects of mechanical stress on the growth of roots. J. Exp. Bot. 13:95–110.

Barney, C. W. 1951. Effects of soil temperature and light intensity on root growth of loblolly pine seedlings. Plant Physiol. 26:146–163.

Beal, C. C. 1918. The effect of aeration on the roots of Zea mays. Indiana Acad. Sci., Proc. 1917:177–180.

Bernstein, L., and W. R. Gardner. 1961. Perspective on function of free space in ion uptake by roots. Science 133:1482–1483.

Bernstein, L., W. R. Gardner, and L. A. Richards. 1959. Is there a vapor gap around plant roots? Science 129:1750, 1753.

Bernstein, L., and R. H. Nieman. 1960. Apparent free space of plant roots. Plant Physiol 35:589–598.

Bertrand, A. R., and H. Kohnke. 1957. Subsoil conditions and their effects on oxygen supply and the growth of corn roots. Soil Sci. Soc. Amer. Proc. 21:135–140.

Bierhuizen, J. F. 1961. Plant growth and soil moisture relationships. UNESCO Arid Zone Research 16:309–315.

Bloodworth, M. E., C. A. Barleson, and W. R. Cowley. 1958. Root distribution of some irrigated crops using undisrupted soil cores. Agron. J. 50:317–320.

Boehle, J., Jr., L. T. Kardos, and J. B. Washko. 1961. Effect of irrigation and deep fertilization on yields and root distribution of selected forage crops. Agron. J. 53:153–158.

Boehle, J., Jr., W. H. Mitchell, C. B. Kresge, and L. T. Kardos. 1963. Apparatus for taking soil-root cores. Agron. J. 55:208–209.

Böhning, R. H., and B. Lusanandana. 1952. A comparative study of gradual and abrupt changes in root temperature on water absorption. Plant Physiol. 27:475–488.

Bonner, J. 1959. Water transport. Science 129:447–450.

Bosemark, N. O. 1954. The influence of nitrogen on root development. Physiol. Plant. 7:497–502.

Brenchley, W. E., and V. G. Jackson. 1921. Root development in barley and wheat under different conditions of growth. Ann. Bot. (London) 35:533–556.

Briggs, G. E., and R. N. Robertson. 1957. Apparent free space. Ann. Rev. Plant Physiol. 8:11–30.

Brown, H. B., E. C. Simon, and A. K. Smith. 1932. Cotton root development in certain South Louisiana soils. Louisiana Agr. Exp. Sta. Bull. 232. 24 p.

Bryant, A. E. 1934. Comparison of anatomical and histological differences between roots of barley grown in aerated and in nonaerated culture solutions. Plant Physiol. 9:389–391.

Burrows, W. C. 1963. Characterization of soil temperature distribution from various tillage-induced microreliefs. Soil Sci. Soc. Amer. Proc. 27:350–353.

Burrows, W. C., and W. E. Larson. 1962. Effect of amount of mulch on soil temperature and early growth of corn. Agron. J. 54:19–23.

Burton, G. W. 1937. The inheritance of various morphological characters in alfalfa and their relation to plant yields in New Jersey. New Jersey Agr. Exp. Sta. Bull. 628. 35 p.

Burton, G. W., E. H. De Vane, and R. L. Carter. 1954. Root penetration, distribution and activity in southern grasses measured by yields, drought symptoms, and P[32] uptake. Agron. J. 46:229–233.

Bushnell, J. 1941. Exploratory tests of subsoil treatments inducing deeper rooting of potatoes on Wooster silt loam. J. Amer. Soc. Agron. 33:823–828.

Cannon, W. A. 1911. Root habits of desert plants. Carnegie Inst. Washington, D.C. Publ. 131. 96 p.

Cannon, W. A. 1949. A tentative classification of root systems. Ecology 30:542–548.

Cannon, W. A. 1954. A note on the grouping of lateral roots. Ecology 35(2):293–295.

Carlson, F. A. 1925. The effect of soil structure on the character of alfalfa root systems. J. Amer. Soc. Agron. 17:336–345.

Carlton, W. M. 1954. Some effects of zinc deficiency on the anatomy of the tomato. Bot. Gaz. 116:52–64.

Colville, W. L., and J. H. Torrie. 1962. Effects of management on food reserves, root rot incidence, and forage yields of medium red clover, *Trifolium pratense* L. Agron. J. 54:332–335.

Crider, F. J. 1955. Root growth stoppage following clipping. US Dep. Agr. Tech. Bull. 1102. 23 p.

Davis, C. A. 1940. Absorption of soil moisture by maize roots. Bot. Gaz. 101:791–805.

Davis, R. M., and J. C. Lingle. 1961. Basis of shoot response to root temperature in tomato. Plant Physiol. 36:153–162.

Derick, R. A., and D. G. Hamilton. 1942. Root development in oat varieties. Sci. Agr. 22:503–508.

De Roo, H. C. 1957. Root growth in Connecticut tobacco soils. Connecticut Agr. Exp. Sta. Bull. 608. 36 p.

De Roo, H. C. 1961. Deep tillage and root growth. A study of tobacco growing in sandy loam soil. Connecticut Agr. Exp. Sta. Bull. 644. 48 p.

Dittmer, H. J. 1937. A quantitative study of the roots and root hairs of a winter rye plant (*Secale cereale*). Amer. J. Bot. 24:417–420.

Dittmer, H. J. 1938. A quantitative study of the subterranean members of three field grasses. Amer. J. Bot. 25:654–657.

Dittmer, H. J. 1948. A comparative study of the number and length of roots produced in nineteen angrosperm species. Bot. Gaz. 109:354–358.

Dittmer, H. J. 1949. Root hair variations in plant species. Amer. J. Bot. 36:152–155.

Dotzenko, A. D., and G. H. Ahlgren. 1950. Response of alfalfa in an alfalfa-bromegrass mixture to various cutting treatments. Agron. J. 42:246–247.

Dotzenko, A. D., and J. G. Dean. 1959. Germination of six alfalfa varieties at three levels of osmotic pressure. Agron. J. 51:308–309.

Dotzenko, A. D., and T. E. Haus. 1960. Selection of alfalfa lines for their ability to germinate under high osmotic pressure. Agron J. 52:200–201.

Duncan, W. G., and A. J. Ohlrogge. 1958. Principles of nutrient uptake from fertilizer bands: II. Root development in the band. Agron. J. 50:605–608.

Eaton, F. M. 1931. Root development as related to character of growth and fruitfulness of the cotton plant. J. Agr. Res. 43:875–883.

Ehrler, W. L. 1963. Water absorption of alfalfa as affected by low root temperature and other factors of a controlled environment. Agron J. 55:363–366.

Elazari-Volcani, T. 1936. The influence of a partial interruption of the transpiration stream by root pruning and stem incisions on the turgor of citrus trees. Palestine J. Bot. Hort. Sci. 1:94–96.

Engelbert, L. E., and E. Truog. 1956. Crop response to deep tillage with lime and fertilizer. Soil Sci. Soc. Amer. Proc. 20:50–54.

Erwin, D. C., and B. W. Kennedy. 1957. A root collapse of alfalfa associated with the interaction of high soil temperatures and water-saturated soil. Phytopathology 47:10.

Esau, K. 1953. Plant anatomy. John Wiley, New York. p. 470–529.

Fehrenbacher, J. B., and J. D. Alexander. 1955. A method for studying corn root distribution using a soil-core sampling machine and shaker-type washer. Agron. J. 47:468–472.

Fehrenbacher, J. B., and R. H. Rust. 1956. Corn root penetration in soils derived from various textures of Wisconsin-Age glacial till. Soil Sci. 82:369–378.

Fehrenbacher, J. B., and H. J. Snider. 1954. Corn root penetration in Muscatine, Elliott, and Cisne Soils. Soil Sci. 77:281–291.

Fitzpatrick, E. G., and L. E. Rose. 1936. A study of root distribution in prairie claypan and associated friable soils. Amer. Soil Surv. Ass. Bull. 17:136–145.

Flocker, W. J., H. Timm, and J. A. Vomocil. 1960 Effect of soil compaction on tomato and potato yields. Agron. J. 52:345–348.

Flocker, W. J. J. A. Vomocil, and F. D. Howard. 1959. Some growth responses of tomatoes to soil compaction. Soil Sci. Soc. Amer. Proc. 23:188–191.

Foth, H. D. 1962. Root and top growth of corn. Agron. J. 54:49–52.

Fox, R. L., and R. C. Lipps. 1955. Influence of soil profile characteristics upon the distribution of roots and nodules of alfalfa and sweetclover. Agron. J. 47:361–367.

Fox, R. L., J. E. Weaver, and R. C. Lipps. 1953. Influence of certain soil-profile characteristics upon the distribution of roots of grasses. Agron. J. 45:583–589.

Fribourg, H. A. 1953. A rapid method for washing roots. Agron. J. 45:334–335.

Fruwirth, C. 1895. Ueber die ausbildung des wurzelsystems der hülsenfrüchte. Forsch. Geb. Agr. Phys. 18:461–479.

Gardner, H. R., and R. E. Danielson. 1964. Penetration of wax layers by cotton roots as affected by some soil physical conditions. Soil Sci. Soc. Amer. Proc. 28:457–460.

Gardner, W. R. 1960. Dynamic aspects of water availability to plants. Soil Sci. 89:63–73.

Gardner, W. R., and C. F. Ehlig. 1962a. Some observations on the movement of water to plant roots. Agron. J. 54:453–456.

Gardner, W. R., and C. F. Ehlig. 1962b. Impedance to water movement in soil and plant. Science 138:522–523.

Garver, S. 1922. Alfalfa root studies. US Dep. Agr. Bull. 1087. 28 p.

Garver, S. 1946. Alfalfa in South Dakota—Twenty-one years research at the Redfield Station. South Dakota Agr. Exp. Sta. Bull. 383. 79 p.

Gilbert, W. B., and D. S. Chamblee. 1959. Effect of depth of water table on yield of Ladino clover, orchardgrass, and tall fescue. Agron. J. 51:547–550.

Gill, W. R., and G. H. Bolt. 1955. Pfeffer's studies of the root growth pressures exerted by plants. Agron. J. 47:166–168.

Gill, W. R., H. R. Haise, and R. M. Hagan. 1959. Annotated bibliography on soil compaction. Amer. Soc. Agr. Eng., St. Joseph, Michigan. 32 p.

Gill, W. R., and R. D. Miller. 1956. A method for study of the influence of mechanical impedance and aeration on the growth of seedling roots. Soil Sci. Soc. Amer. Proc. 20:154–157.

Gist, G. R., and G. O. Mott. 1957. Some effects of light intensity, temperature, and soil moisture on the growth of alfalfa, red clover, and birdsfoot trefoil seedlings. Agron. J 49:33–36.

Goedewaagen, M. A. J. 1955. Root development on grassland profiles near Klundert and Velsen consisting of a clay cover of different thickness and a sandy or peaty subsoil. Verslag. Landbouwk. Onderzoek. 61.7:19–34.

Graber, L. F., N. T. Nelson, W. A. Luekel, and W. B. Albert. 1927. Organic food reserves in relation to the growth of alfalfa and other perennial herbaceous plants. Wisconsin Agr. Exp. Sta. Res. Bull. 80. 128 p.

Haas, H. J. 1958. Effect of fertilizers, age of stand, and decomposition on weight of grass roots and of grass and alfalfa on soil nitrogen and carbon. Agron. J. 50:5–9.

Hagan, R. M. 1952. Soil temperature and plant growth. p. 336–447. In B. T. Shaw [ed.] Soil physical conditions and plant growth. Academic Press, New York.

Hagan, R. M., and M. L. Peterson. 1953. Soil moisture extraction by irrigated pasture mixtures as influenced by clipping frequency. Agron. J. 45:288–292.

Hall, N. S., W. F. Chandler, C. H. M. van Bavel, P. H. Reid, and J. H. Anderson. 1953. A tracer technique to measure growth and activity of plant root systems. North Carolina Agr. Exp. Sta. Tech. Bull. 101. 40 p.

Hammes, J. K., and J. F. Bartz. 1963. Root distribution and development of vegetable crops as measured by radioactive phosphorus injection technique. Agron. J. 55:329–333.

Haynes, J. L., and W. R. Robbins. 1948. Calcium and boron as essential factors in the root environment. J. Amer. Soc. Agron. 40:795–803.

Hayward, H. E. 1938. The structure of economic plants. Macmillan, New York. p. 39–56.

Hess, N. 1949. Beobachtungen über das Wurzelsystem der Wintergerste. Bodenkulter 3:211–214.

Hidding, A. P., and C. van den Berg. 1960. The relation between pore volume and the formation of root systems in soils with sandy layers. Int. Congr. Soil Sci., Trans. 7th (Madison, Wisc., USA). I:369–373.

Hildebrand, S. C., and C. M. Harrison. 1939. The effect of height and frequency of cutting alfalfa upon consequent top growth and root development. J. Amer. Soc. Agron. 31:790–799.

Holt, E. C., and F. L. Fisher. 1960. Root development of coastal bermudagrass with high nitrogen fertilization. Agron. J. 52:593–596.

Hope, A .B., and P. G. Stevens. 1952. Electric potential differences in bean roots and their relation to salt uptake. Australian J. Sci. Res. B5:335–343.

Hubbell, D. S., and G. Staten. 1951. Studies on soil structure. New Mexico Agr. Exp. Sta. Tech. Bull. 363. 51 p.

Hunt, O. J., and D. G. Miller. 1961. Effects of amputation of primary seminal roots on seedling grasses. Agron. J. 53:276–277.

Hunter, A. S., and O. J. Kelley. 1946. A new technique for studying the absorption of moisture and nutrients from soil by plant roots. Soil Sci. 62:441–450.

Jacques, W. A. 1937. A new type of root sampler. New Zealand J. Sci. Tech. 19–A:267–270.

Janssen, G. 1929. Effect of date of seeding of winter wheat on plant development and its relationship to winter-hardiness. J. Amer. Soc. Agron. 21:444–466.

Jean, F. C., and J. E. Weaver. 1924. Root behavior and crop yield under irrigation. Carnegie Inst. Washington, D.C. Publ. 357. 66 p.

Jensen, R. D., and S. A. Taylor. 1961. Effect of temperature on water transport through plants. Plant Physiol. 36:639–642.

Jensen, R. D., S. A. Taylor, and H. H. Wiebe. 1961. Negative transport and resistance of water flow through plants. Plant Physiol. 36:633–638.

Kemper, W. D., and M. Amemiya. 1957. Alfalfa growth as affected by aeration and soil moisture stress under flood irrigation. Soil Sci. Soc. Amer. Proc. 21:657–660.

Kiesselbach, T. H., and R. M. Weihing. 1935. The comparative root development of selfed lines of corn and their F_1 and F_2 hybrids. J. Amer. Soc. Agron. 27:538–541.

King, F. H. 1892. Natural distribution of roots in field soils. Wisconsin Agr. Exp. Sta. Rep. p. 112–120.

Klebesadel, L. J. 1964. Modification of alfalfa root system by severing the tap root. Agron. J. 359–361.

Kmock, H. G., R. E. Ramig, R. L. Fox, and F. E. Koehler. 1957. Root development of winter wheat as influenced by soil moisture and nitrogen fertilization. Agron. J. 49:20–25.

Kohnke, H., and A. R. Bertrand. 1956. Fertilizing the subsoil for better water utilization. Soil Sci. Soc. Amer. Proc. 20:581–586.

Kramer, P. J. 1938. Root resistance as a cause of the absorption lag. Amer. J. Bot. 25:110–113.

Kramer, P. J. 1942. Species differences with respect to water absorption at low soil temperatures. Amer. J. Bot. 29:828–832.

Kramer, P. J. 1945. Absorption of water by plants. Bot. Rev. 11:310–355.

Kramer, P. J. 1956. Roots as absorbing organs. III:188–214. In W. Ruhland [ed.] Encyclopedia of plant physiology. Springer, Berlin.

Kramer, P. J., and T. S. Coile. 1940. An estimation of the volume of water made available by root extension. Plant Physiol. 15:743–747.

Laird, A. S. 1930. A study of the root systems of some important sod-forming grasses. Florida Agr. Exp. Sta. Bull. 211. 27 p.

Lamba, P. S., H. L. Ahlgren, and R. J. Muckenhirn. 1949. Root growth of alfalfa, medium red clover, bromegrass, and timothy under various soil conditions. Agron. J. 41:451–458.

Lea, J. D. 1961. Studies on the depth and rate of root penetration of some annual tropical crops. Tropical Agr. 38:93–105.

Levitt, J. 1957. The significance of "apparent free space" (A.F.S.) in ion absorption. Physiol. Plant. 10:882–888.

Linscott, D. L., R. L. Fox, and R. C. Lipps. 1962. Corn root distribution and moisture extraction in relation to nitrogen fertilization and soil properties. Agron. J. 54:185–189.

Lipps, R. C., R. L. Fox, and F. E. Koehler. 1957. Characterizing root activity of alfalfa by radioactive tracer techniques. Soil Sci. 84:195–204.

Loomis, W. E. 1935. The translocation of carbohydrates in maize. Iowa State Coll. J. Sci. 9:509–520.

Lorenz, R. J., and G. A. Rogler. 1964. Effect of row spacing and nitrogen fertilizer on production of irrigated Russian wildrye (*Elymus junceus* Fisch.): II. Relative crown and root development. Agron. J. 56:7–10.

Lott, W. L., D. P. Satchell, and N. S. Hall. 1950. A tracer element technique in the study of root extension. Amer. Soc. Hort. Sci., Proc. 55:27–34.

Lutz, J. F. 1952. Mechanical impedance and plant growth. Agronomy 2:43–71.

McClure, J. W., and C. Harvey. 1962. Use of radiophosphorus in measuring root growth of sorghum. Agron. J. 54:457–459.

McKell, C. M., M. B. Jones, and E. R. Perrier. 1962. Root production and accumulation of root material on fertilized annual range. Agron. J. 54:459–462.

McPherson, D. C. 1939. Cortical air spaces in the roots of *Zea Mays* L. New Phytologist 38:190–202.

Mees, G. C., and P. E. Weatherley. 1957a. The mechanism of water absorption by roots: I. Preliminary studies on the effects of hydrostatic pressure gradients. Proc. Roy. Soc. (London) Ser. B. 147:367–380.

Mees, G. C., and P. E. Weatherley. 1957b. The mechanism of water absorption by roots: II. The role of hydrostatic pressure gradients across the cortex. Proc. Roy. Soc. (London) Ser. B. 147:381–391.

Miller, E. C. 1938. Plant physiology. 2nd ed. McGraw-Hill, New York. p. 121–187.

Miller, R. H. 1960. A selected reference list on the morphology and anatomy of roots. Scholar's Bibliogr. Ser., Scholar's Libr., New York. 21 p.

Miller, S. A., and A. P. Mazurak. 1958. Relationships of particle and pore size to the growth of sunflowers. Soil Sci. Soc. Amer. Proc. 22:275–278.

Muller, C. H. 1946. Root development and ecological relations of guayule. US Dep. Agr. Tech. Bull. 923. 114 p.

Murdock, J. T., and L. E. Engelbert. 1958. The importance of subsoil phosphorus to corn. Soil Sci. Soc. Amer. Proc. 22:53–57.

Nakayama, F. S., and C. H. M. van Bavel. 1963. Root activity distribution patterns of sorghum and soil moisture conditions. Agron. J. 55:271–274.

Nelson, M. 1944. Maintenance of alfalfa stands. Arkansas Agr. Exp. Sta. Bull. 447. 30 p.

Nelson, N. T. 1925. The effects of frequent cuttings on the production, root reserves, and behavior of alfalfa. J. Amer. Soc. Agron. 17:100–113.

Nielsen, K. F., R. L. Halstead, A. J. MacLean, S. J. Bourget, and R. M. Holmes. 1961. The influence of soil temperature on the growth and mineral composition of corn, bromegrass, and potatoes. Soil Sci. Soc. Amer. Proc. 25:369–372.

Nielsen, K. F., R. L. Halstead, A. J. MacLean, R. M. Holmes, and S. J. Bourget. 1960. The influence of soil temperature on the growth and mineral composition of oats. Can. J. Soil Sci. 40:255–263.

Nightingale, G. T. 1935. Effects of temperature on growth, anatomy, and metabolism of apple and peach roots. Bot. Gaz. 96:581–639.

Nutman, F. J. 1933. The root-system of *coffea arabica*: Part II. The effect of some soil conditions in modifying the normal root system. Empire J. Exp. Agr. 1:285–296.

Nye, P. H., and W. N. M. Foster. 1961. The relative uptake of phosphorus by crops and natural fallow from different parts of their root zone. J. Agr. Sci. 56:299–306.

Ordin, L., and S. Gairon. 1961. Diffusion of tritiated water into roots as influenced by water status of tissue. Plant Physiol. 36:331–335.

Oswalt, D. L., A. R. Bertrand, and M. R. Teel. 1959. Influence of nitrogen fertilization and clipping on grass roots. Soil Sci. Soc. Amer. Proc. 23:228–230.

Passiouri, J. B., and G. W. Leeper. 1963. Soil compaction and manganese deficiency. Nature 200:29–30.

Pavlychenko, T. K. 1937a. Quantitative study of the entire root systems of weed and crop plants under field conditions. Ecology 18:62–79.

Pavlychenko, T. K. 1937b. The soil-block washing method in quantitative root study. Can. J. Res. C15:33–57.

Pendleton, R. A. 1950. Soil compaction and tilling operation effects on sugar beet root distribution and seed yields. Amer. Soc. Sugar Beet Tech., Proc. 6:278–285.

Pfeffer, W. 1893. Druck und arbeitsleistung durch wachsende planzen. Abhandl. Königlich Sächsischen Ges. Wiss. 33:235–474.

Philip, J. R. 1957. The physical principles of soil water movement during the irrigation cycle. Intn. Comm. Irrig. Drainage, Proc. 3rd Congr., p. 8.125–8.154.

Philip, J. R. 1958. The osmotic cell, solute diffusibility, and the plant water economy. Plant Physiol. 33:264–271.

Phillips, R. E., and D. Kirkham. 1962. Soil compaction in the field and corn growth. Agron. J. 54:29–34.

Pohlman, G. G. 1946. Effect of liming different soil layers on yield of alfalfa and on root development and nodulation. Soil Sci 62:255–266.

Power, J. F., D. L. Grunes, W. O. Willis, and G. A. Reichman. 1963. Soil temperature and phosphorus effects upon barley growth. Agron. J. 55:389–392.

Pumphrey, F. V., and F. E. Koehler. 1958. Forage and root growth of five sweetclover varieties and their influence on two following corn crops. Agron. J. 50:323–326.

Raney, W. A., T. W. Edminster, and W. H. Allaway. 1955. Current status of research in soil compaction. Soil Sci. Soc. Amer. Proc. 19:423–428.

Roberts, A. N., and A. L. Kenworthy. 1956. Growth and composition of the strawberry plant in relation to root temperature and intensity of nutrition. Amer. Soc. Hort. Sci., Proc. 68:157–168.

Robinson, R. R., V. G. Sprague, and C. F. Gross. 1959. The relation of temperature and phosphate placement to growth of clover. Soil Sci. Soc. Amer. Proc. 23:225–228

Rosenberg, N. J., and N. A. Willits. 1962. Yield and physiological response of barley and beans grown in artificially compacted soils. Soil Sci. Soc. Amer. Proc. 26:78–82.

Rosene, H. F. 1937. Distribution of the velocities of absorption of water in the onion root. Plant Physiol. 12:1–19.

Rosene, H. F. 1941a. Comparison of rates of water intake in contiguous regions of intact and isolated roots. Plant Physiol. 16:19–38.

Rosene, H. F. 1941b. Control of water transport in local root regions of attached and isolated roots by means of the osmotic pressure of the external solution. Amer. J. Bot. 28:402–410.

Rosene, H. F. 1943. Quantitative measurement of the velocity of water absorption in individual root hairs by a microtechnique. Plant Physiol. 18:588–607.

Rosene, H. F. 1954. The water absorptive capacity of root hairs. Int. Bot., 8th Congr., (Paris) 11:217–218.

Rosene, H. F., and A. M. J. Walthall. 1954. Comparison of the velocities of water influx into young and old root hairs of wheat seedlings. Physiol. Plant. 7:190–194.

Rosene, H. F., A. M. J. Walthall, and A. M. Jockel. 1949. Velocities of water absorption by individual root hairs of different species. Bot. Gaz. 111:11–21.

Russell, M. B., and R. E. Danielson. 1956. Time and depth patterns of water use by corn. Agron. J. 48:163–165.

Salmon, S. C., C. O. Swanson, and C. W. McCampbell. 1925. Experiments relating to time of cutting alfalfa. Kansas Agr. Exp. Sta. Tech. Bull. 15:1–50.

Sayre, J. D., and V. H. Morris. 1940. The lithium method of measuring the extent of corn root systems. Plant Physiol. 15:761–764.

Schroeder, R. A. 1939. The effect of root temperature upon the absorption of water by the cucumber. Missouri Agr. Exp. Sta. Bull. 309. 27 p.

Simmonds, P. M., and B. J. Sallans. 1933. Further studies on amputations of wheat roots in relation to disease of the root system. Sci. Agri. 13:439–448.

Skidmore, E. L., and J. F. Stone. 1964. Physiological role in transpiration rate of the cotton plant. Agron. J. 56:405–410.

Slatyer. R. O. 1960. Absorption of water by plants. Bot. Rev. 26:331–392.

Smith, D. 1951. Root branching of alfalfa varieties and strains. Agron. J. 43:573–575.

Smith, R. C. 1960. Influence of upward water translocation on uptake of ions in corn plants. Amer. J. Bot. 47:724–729.

Sommer, A. L. 1936. The relationship of the phosphate concentration of solution cultures to the type and size of root systems and the time of maturity of certain plants. J. Agr. Res. 52:133–148.

Spencer, J. T. 1940. A comparative study of the seasonal root development of some inbred lines and hybrids of maize. J. Agr. Res. 61:521–538.

Stocking, C. R. 1956. Histology and development of the root. III:173–187. In W. Ruhland [ed.] Encyclopedia of plant physiology. Springer, Berlin.

Stoeckeler, J. H., and W. A. Kluender. 1938. The hydraulic method of excavating the root systems of plants. Ecology 19:355–369.

Taylor, H. M., and H. R. Gardner. 1960. Use of wax substrates in root penetration studies. Soil Sci. Soc. Amer. Proc. 24:79–81.

Taylor, S. A., and C. S. Slater. 1955. When to irrigate and how much water to apply. In Water. US Dep. Agr. Yearbook. p. 372–376.

Ten Eyck, A. M. 1899. Preliminary work in a study of the root systems of plants grown under natural field conditions. North Dakota Agr. Exp. Sta. Bull. 36. p. 333–362.

Ten Eyck, A. M. 1900. A study of the root systems of cultivated plants grown as farm crops. North Dakota Agr. Exp. Sta. Bull. 43:535–560.

Ten Eyck, A. M. 1905. The roots of plants. Kansas Agr. Exp. Sta. Bull. 127. p. 199–252.

Tharp, B. C. and C. H. Muller. 1940 A rapid method for excavating root systems of native plants Ecology 21:347–350.

Thoday, D., and A. J. Davey. 1932. Contractile Roots: II. On the mechanism of root-contraction in *Oxalis incarnata*. Ann. Bot. (London) 46:993–1006.

Thurman, R. L., and P. Grissom. 1954. Relationship of top and root growth of oats. Agron. J. 46:474–475.

Tovey, R. 1963. Consumptive use and yield of alfalfa grown in the presence of static water tables. Nevada Agr. Exp. Sta. Tech. Bull. 232. 65 p.

Tovey, R. ca. 1964. Water table fluctuation effect on alfalfa production. Univ. Nevada, Coll. Agr. Tech. Bull. 1. 11 p.

Troughton, A. 1957. The underground organs of herbage grasses. Commonwealth Bur. Pastures Field Crops Bull. 44. 163 p.

Upchurch, R. P. 1951. The use of the trench-wash and soil-elution methods for studying alfalfa roots. Agron. J. 43:552–555.

van Diest, A. 1962. Effects of soil aeration and compaction upon yield, nutrient takeup and variability in a greenhouse fertility experiment. Agron. J. 54:515–518.

van Wijk, W. R., W. E. Larson, and W. C. Burrows. 1959. Soil temperature and the early growth of corn from mulched and unmulched soil. Soil Sci. Soc. Amer. Proc. 23:428–434.

van't Woudt, B. D., and R. M. Hagan. 1957. Crop responses at excessively high soil moisture levels. In Drainage of agricultural lands. Agronomy. 7:514–578.

Veihmeyer, F. J., and A. H. Hendrickson. 1948. Soil density and root penetration. Soil Sci. 65:487–493.

Vinokur, R. L. 1957. Influence of temperature of the root environment on root activity, transpiration, and photosynthesis of leaves of lemon. (In Russian) Plant Physiol. (Engl. trans.) 4:268–273.

Weaver, J. E. 1919. The ecological relations of roots. Carnegie Inst. Washington, D.C. Publ. 286. 128 p.

Weaver, J. E. 1920. Root development in the grassland formation. A correlation of the root systems of native vegetation and crop plants. Carnegie Inst. Washington, D.C. Publ. 292. 151 p.

Weaver, J. E. 1925. Investigations on the root habits of plants. Amer. J. Bot. 12:502–509.

Weaver, J. E. 1926. Root development of field crops. McGraw-Hill, New York. 291 p.

Weaver, J. E. 1954. North American prairie. Johnsen Publ. Co., Lincoln, Nebraska. 348 p.

Weaver, J. E., and W. E. Bruner. 1927. Root development of vegetable crops. McGraw-Hill, New York. 351 p.

Weaver, J. E., and J. W. Crist. 1922. Relation of hardpan to root penetration in the Great Plains. Ecology 3:237–249.

Weaver, J. E., and R. W. Darland. 1949. Soil-root relationships of certain native grasses in various soil types. Ecol. Monogr. 19:303–338.

Weaver, J. E., F. C. Jean, and J. W. Crist. 1922. Development and activities of roots of crop plants. Carnegie Inst. Washington, D.C. Publ. 316. 117 p.

Weaver, J. E., J. Kramer, and M. Reed. 1924. Development of root and shoot of winter wheat under field environment. Ecology 5:26–50.

Weaver, J. E., and J. W. Voigt. 1950. Monolith method of root-sampling in studies on succession and degeneration. Bot. Gaz. 111:286–299.

Webb, R. B., and D. E. Stephens. 1936. Crown and root development in wheat varieties. J. Agr. Res. 52:569–583.

Weihing, R. M. 1935. The comparative root development of regional types of corn. J. Amer. Soc. Agron. 27:526–537.

Went, F. W. 1953. The effect of temperature on plant growth. Ann. Rev. Plant Physiol. 4:347–362.

Wiersma, D. 1959. The soil environment and root development. Advance Agron. 11: 43–51.

Wiersum, L. K. 1957. The relationship of the size and structural rigidity of pores to their penetration by roots. Plant Soil 9:75–85.

Wilkinson, S. R., and A. J. Ohlrogge. 1962. Principles of nutrient uptake from fertilizer bands: V. Mechanisms responsible for intensive root development in fertilized zones. Agron. J. 54:288–291.

Williams, T. E., and H. K. Baker. 1957. Studies on the root development of herbage plants: I. Techniques of herbage root investigations. J. Brit. Grasslands Soc. 12:49–55.

Willis, W. O., W. E. Larson, and D. Kirkham. 1957. Corn growth as affected by soil temperature and mulch. Agron. J. 49:323–328.

Woodruff, C. M., and D. D. Smith. 1946. Subsoil shattering and subsoil liming for crop production on claypan soils. Soil Sci. Soc. Amer. Proc. 11:539–542.

Worzella, W. W. 1932. Root development in hardy and nonhardy winter wheat varieties. J. Amer. Soc. Agron. 24:626–637.

Wright, N. 1962. Root weight and distribution of blue panicgrass, *Panicum antidotale* Retz., as affected by fertilizers, cutting height, and soil-moisture stress. Agron. J. 54:200–202.

Younts, S. E., and E. T. York, Jr. 1956. Effect of deep placement of fertilizer and lime on yield and root activity of corn and crimson clover. Soil Sci. 82:147–155.

Zimmerman, R. P., and L. T. Kardos. 1961. Effect of bulk density on root growth. Soil Sci. 91:280–288.

section VII

Water-Soil-Plant Relations

VII

22

Comparative Terminologies for Water in the Soil-Plant-Atmosphere System

A. T. COREY

Colorado State University
Fort Collins, Colorado

R. O. SLATYER

Division of Land Research, CSIRO
Canberra, A. C. T., Australia

W. D. KEMPER

Agricultural Research Service, USDA
and Colorado State University
Fort Collins, Colorado

I. INTRODUCTION

Portions of the soil-plant-water system have been studied by investigators having training in a variety of disciplines. Because of their training and also because they have different applications in mind, they have developed different analyses to describe water systems.

The terminology used in reference to a water system, whether in the soil, the plant, or in both, depends to a large extent on the analysis used to describe the system. It is not always possible to discuss the terms employed without reference to the associated analyses. In some cases, a given term used in one analysis may have an exact counterpart in another, so that the two terms refer to the same concept. Unfortunately, this is not always the case. Just as it is often misleading to translate one language into another by making a word-for-word substitution, it may be misleading to substitute a term used in fluid mechanics, for example, for one used in soil science or plant physiology.

There are several reasons why one should be explicit in selecting terms to be used in describing water in the soil-plant system. Engineers, for example, may think of fluid systems as continua, being concerned primarily with the flow of fluids in bulk (Lamb, 1948). In contrast, plant physiologists may be interested in the movement of certain constituents that make up the solutions in the plants and soil. Many authors fail to state clearly the dimensions and units of the quantities to which they refer. Some terms are used in reference to related quantities that have different dimensions, or the same quantity may be referred to by a variety of terms. Terms or units may also be used in a technical sense which differs from the historical or usual meaning of such terms or units.

The discussion of terminologies presented here is confined to commonly used terms of fundamental significance which are used to describe the state, content, and motion of fluids in the soil-plant system.

To facilitate understanding the literature dealing with water in soil-plant systems, the following classification and comparison of terms is presented.

II.   TERMS REFERRING TO WATER CONTENT

A.   Water Content of Soils

The term water content has been used in reference to the content of solution as well as the content of pure water, and moisture content has been used in similar ways. It is not always clear in which sense particular authors use these terms. One standard method of determining water (or moisture) content is by measuring the water lost from the soil on crying at 105 to 110C. In this case, water content usually refers to the content of nearly pure water since most of the salts in the solution remain in the dry soil. Water content is usually expressed as a ratio or percentage in one of three ways, the quantity measured being dimensionless in each case:

1) *Volume water content*—the amount of water lost from the soil on drying at 105 to 110C expressed as the volume of water per bulk volume of soil (Int. Soc. Soil Sci., 1963).

2) *Water content on dry weight basis*—the amount of water lost from the soil on drying at 105 to 110C expressed as the weight (or mass) of water per weight (or mass) of dry soil (Int. Soc. Soil Sci., 1963).

3) *Saturation*—the amount of solution contained in the soil expressed as the volume of solution per volume of soil pore space (Collins, 1961). This should not be confused with the term *saturated* which means all of the pore space is occupied with liquid.

B.   Water Content of Plants

Plant water content measurements are seldom made of entire plants and are usually confined to specific plant organs, particularly leaves. The tissue is generally sampled and weighed to obtain the fresh weight, W_f, oven dried at a temperature of 85 to 90C to constant weight, and reweighed to obtain the dry weight, W_d. These determinations enable water content to be expressed on either a fresh or dry weight basis, usually as a percentage, as follows:

1) *water content on a dry weight basis:* $100 \ (W_f - W_d)/W_d$.

2) *water content on a fresh weight basis:* $100 \ (W_f - W_d)/W_f$.

Relative water content is related more directly to the physiological adequacy of water in plants. It is determined by floating the freshly sampled (and weighed) tissue on water until it becomes turgid when it is reweighed, W_t, prior to oven drying (*see* chapter 18). Relative water content is then expressed as a percentage, i.e. $100 \ (W_f - W_d)/(W_t - W_d)$. It has also been termed "relative turgidity."

An alternative expression for relative water content is *water deficit* which is $100 \ (W_t - W_f)/(W_t - W_d)$. The two terms are complementary, i.e. relative water content equals 100 minus the water deficit.

III.   TERMS REFERRING TO THE STATE OF FLUIDS

A.   Solution in Bulk

The condition of the water solution at points within a soil or plant system may be described as a potential in units of energy. Potential is usually defined with respect to the magnitude of forces acting on and within the considered system as compared to the magnitude of these forces in some standard condition.

Potentials, as used in connection with irrigation by engineers, soil scientists, and agronomists, involve a profusion of names and dimensions resulting in considerable confusion. These terms will be defined and explained in the following sections and are summarized in Table 22–1.

Potentials may have the dimensions of energy per volume, per mass, or per weight (Childs, 1957). They may be defined with respect to forces acting at "points" within a system. When a potential is defined at a point within a fluid, a special meaning must be assigned to the term "point" (Streeter, 1948). A point in this sense is a volume element having dimensions very small compared to the smallest significant dimension of the space which the fluid occupies and large dimensions compared to the mean free path of the molecules or ions comprising the fluid. If any significant dimension of the space occupied by the fluid is too small, volume elements meeting both of these requirements do not exist in which case the terms defined below may not be applicable.

The following paragraphs summarize potentials as affected by pressures, adsorptive forces, and gravity.

1. PRESSURE POTENTIALS

These are a measure of the compressive force of the fluid which is available for reversible work of expansion. They may be expressed in the dimensions of energy per volume, energy per mass, or energy per weight.

The dimensions energy per volume is equivalent to force per area, or simply "pressure." Pressure represents the slope of a curve obtained by plotting the reversible work of expansion as a function of volume of an element of fluid at a point (Sears, 1950). Because pressure depends on the volume, it should not be assumed that pressure represents the potential energy of expansion per volume of fluid, although it does have these dimensions (Rouse, 1945). Pressure potentials often appear in the literature with adjectives so that the following may have the same meaning: *pressure intensity, static pressure, neutral pressure, pore pressure, stress, solution pressure,* and *turgor pressure. Hydrostatic pressure* often has a similar meaning, although it sometimes implies a static distribution of pressure, i.e. the pressure varies linearly with depth in the fluid system.

In the most general case, deformation of fluid elements may produce tangential components of surface force (shear) and also normal components of surface force that are nonisotropic and not available for reversible expansion of fluid elements. In this case, the physical meaning of the scalar pressure is not easily visualized. For a Newtonian viscous fluid undergoing only steady laminar flow, the value of pressure p is given by

$$p = \tfrac{1}{3}(\sigma_x + \sigma_y + \sigma_z)$$ [22–1]

where σ_x, σ_y, σ_z represent the normal components of surface force intensity in three mutually perpendicular coordinate directions (Streeter, 1948). A normal force acting toward a point (compression) is considered here as positive, although this is not always the convention.

In the particular case of a static fluid, pressure has a simple physical interpretation, i.e. the average intensity (force/area) of the surface force acting on the fluid element at a point. Although intensity of surface force varies from point-to-point in a static fluid, it is equal in all directions and does not have a tangential component. In this particular case, therefore, surface force intensity and pressure are synonymous.

a. **Differences in Usage of "Pressure" in Fluid Dynamics and in Thermodynamic Analyses.** The concept of pressure (classical pressure) as described above, whether it is used with or without adjectives, is the common usage in the literature dealing with fluid dynamics or with flow in porous media if the latter is written employing a fluid dynamic rather than a thermodynamic analysis.

The concept of pressure used in thermodynamic and fluid dynamic developments differs as a result of the types of fluid elements considered.

Fluid dynamic analyses consider forces acting on differential volume elements in the interior of the fluid system. It is not necessary to assume that the external pressure is everywhere the same. The internal pressure may be different from the external pressure as a result of position, the force of gravity, adsorptive forces, and interfacial or membrane forces.

Classical thermodynamic analysis considers surface forces acting externally on the boundary of a fluid system. It is assumed that the external pressure is everywhere the same and equal in magnitude to the pressure at all interior points. These assumptions are valid only when gravity, adsorptive forces, and interfacial forces have a negligible effect (Planck, 1945). It has been convenient to make such assumptions to mathematically formulate the laws of thermodynamics in terms of equations of state, because this formulation presupposes that the volume of the fluid system is uniquely related to the pressure and temperature.

In irreversible thermodynamics, fluid systems in which pressure varies from point-to-point are analyzed by dividing the systems into subsystems for each of which the pressure is assumed to be uniform. In the limit, as a system must be divided into differential elements(to apply the second law), the concept of pressure employed in irreversible thermodynamics becomes synonymous with that employed in fluid dynamics (Prigogine, 1955).

In some sciences the *boundary pressure* is designated by an adjective, depending on the kind of boundary which is external to the fluid system under consideration, e.g., *atmospheric pressure* in engineering and soil science and *wall pressure* in plant science.

b. **Potentials Resulting from Force Fields Acting Toward Solid Surfaces (Adsorption Potential).** A special use of pressure terms is frequently made in engineering and soil physics literature dealing with flow in porous media to describe the influence of force fields acting toward solid surfaces (Edlefsen and Anderson, 1943). This usage is often misunderstood and thus needs clarification.

An adsorption potential may be defined by a work integral analogous to that used to define the gravitational potential (*see* part IV of this chapter), i.e.:

$$\psi_a \equiv \int_{\bar{s}_o}^{\bar{s}_i} - \bar{a} \cdot d\bar{s} \qquad [22\text{--}2]$$

where $\bar{a}$ is the adsorptive force vector. Such an integral cannot generally be evaluated, however, because the functional relationship between the adsorptive force and distance from a solid surface is unknown. Present methods of measurement do not distinguish between pressure and adsorption potentials. Reported pressure potentials commonly include the adsorption potential (*see* Table 22–1). In plant physiology literature the adsorption potential has been called an *imbibition potential* or *imbibition pressure*.

For the portion of the fluid system close enough to solid surfaces to be affected by their adsorptive force fields, "pressure" is thus affected both by pressure in the fluid dynamic sense and by whatever adsorptive forces are acting at that point. Thus, pressure within fluids affected by adsorptive forces could be called *apparent pressure*. Tensiometers (*see* chapter 15) really measure the apparent pressure (Childs, 1957). However, it may be argued that because the adsorptive forces are substantial only in a region very close to solid surfaces where the concept of a scalar pressure may not be strictly applicable, the difference between pressure as used in fluid dynamics and apparent pressure is entirely academic.

For systems in which the pressure of air in contact with a solution is everywhere atmospheric, this apparent pressure is synonymous (except for sign) with the terms *tension, suction, matric suction*, etc., frequently appearing in the soils literature.

c. Dimensions Used for Pressure Potentials

1. Energy Per Volume or Force Per Area. The dimensions force per area for pressure potential are commonly used. Use of the dimensions force per area leads to expressing pressure potentials as pressures in the usual sense.

2. Energy Per Mass. These dimensions are sometimes employed for the pressure potential (*see* Int. Soc. Soil Sci., 1963). It is the pressure, or apparent pressure, divided by the fluid density, providing the density is a constant throughout the system. It may also be defined by the work integral

$$\psi_p \equiv \int_{\bar{s}_o}^{\bar{s}_i} \frac{\bar{\nabla \mathbf{p}}}{\rho} \cdot \bar{d\mathbf{s}} \qquad [22\text{–}3]$$

where **p** is the pressure, ρ is the fluid density, $\bar{s}$ is a displacement vector, $\bar{s}_o$ is the displacement vector representing the datum position, and $\bar{s}_i$ is the position for which the potential is to be evaluated. When ρ is everywhere a constant, the value of ψ_p is simply p/ρ.

Unless ρ is everywhere a constant, or at least varies in space only in the direction of the resultant of all body forces (i.e. vertically when gravity is the only appreciable body force), the integral defining the pressure potential is path dependent. When the integral is path dependent, the pressure component of the field of force is not conservative, and the definition of the pressure potential is ambiguous.

3. Energy Per Weight (Length). When these dimensions are employed, the quantity referred to is often called "pressure head." It can be defined by a work integral analogous to that given in equation [22–3], but with fluid density being replaced by the specific weight ρg, g being the magnitude of the force per mass due to gravity.

Such a potential has the same limitation as ψ_p in that the integral is not always independent of path. If the density is uniform, the pressure head is simply $p/\rho g$.

2. GRAVITATIONAL POTENTIAL

Gravitational potential is the potential resulting from the position of the solution in the gravitational force field.

a. Force Per Area. The gravitational potential is often written with these dimensions so it can be added to pressure to obtain the "piezometric pressure" (*see* below). In this form the gravitational potential can be defined by the work integral

$$\psi_g \equiv \int_{\bar{s}_o}^{\bar{s}_i} -\rho\bar{g} \cdot d\bar{s} \qquad [22\text{--}4]$$

and for uniform ρ and $\bar{g}$, the value of ψ_g is $\rho g h$, where h is the elevation difference between the datum at $\bar{s}_o$ and the position $\bar{s}_i$ where ψ_g is evaluated, and $\bar{g}$ is the gravitational force vector. The reason for the negative sign in front of $\rho\bar{g}$ in equation [22–4] is that potentials are customarily defined as work done *against* a force field. The definition of ψ_g is ambiguous when ρ varies (other than in a barotropic manner), because the integral given in equation [22–4] is path dependent.

b. Energy Per Mass. In this form, the gravitational potential can be added to the pressure potential to obtain a quantity sometimes referred to as "total potential," Ψ,

$$\Psi = \psi_p + \psi_g = (p/\rho) + gh. \qquad [22\text{--}5]$$

When the gravitational potential is expressed in dimensions of energy per mass, it can be defined by a work integral analogous to equation [22–4] but with the fluid density omitted. In this case, the gravitational potential has the value gh, provided g is uniform throughout the system. A gravitational potential having these dimensions can always be defined, because the gravitational force field (expressed as force per mass) is conservative, i.e. the work integral

$$\int_{\bar{s}_o}^{\bar{s}_i} -\bar{g} \cdot d\bar{s}$$

is independent of path, regardless of density variations.

c. Elevation Head. The elevation of the element in question (above a reference elevation) is called the elevation head.

3. PIEZOMETRIC POTENTIAL

This is the sum of the apparent pressure and gravitational potentials (Int. Soc. Soil Sci., 1963). It is also called the *hydraulic potential*.

It should be emphasized that the pressure and gravitational potentials can be added only when both are expressed in the same dimensions. There is no way of defining an exact piezometric potential unless the fluid density is uniform throughout the system (or varies only in a barotropic manner). Whereas the pressure potential having the dimensions of force per area and the gravitational potential having the dimensions of energy per mass can be defined, regardless of density variations, there are no dimensions with which both can be defined in the general case. If the changes in density are negligible compared to changes in the piezometric potential, use of the average density in equations [22–6], [22–7], and [22–8] may be justified.

If the fluid density is uniform, the quantity measured by a tensiometer and the piezometer are synonymous. This assumes that the apparent pressure, rather than the classical pressure, is added to the gravitational potential. When the density of the fluid is uniform throughout the system, three variations of the piezometric potential are possible with dimensions as follows:

a. Energy Per Volume. In this case, potential is called *piezometric pressure* which is given by

$$p^* \equiv p + \rho gh. \qquad [22\text{–}6]$$

b. Energy Per Mass. These dimensions are used by some soil physicists who call it the "total potential." This use of the term "total potential" should not be confused with the more common concept of *"total water potential"* which refers to the *water constituent only* and includes "osmotic potential." The potential referring to the *solution in bulk* is given by

$$\frac{p^*}{\rho} \equiv (p/\rho) + gh. \qquad [22\text{–}7]$$

a. Energy Per Weight (Length). This potential is called *piezometric head* and is given by

$$H \equiv (p/\rho g) + h. \qquad [22\text{–}8]$$

Piezometric head is similar to *hydraulic head* except that the latter term includes a velocity component.

4. POTENTIAL DUE TO EXTERNAL GAS PRESSURE

The potential due to external gas pressure is that portion of the pressure potential which results from the pressure of the atmosphere in contact with the solution (Int. Soc. Soil Sci., 1963). It can be expressed with the same dimensions as the pressure potential.

5. SUCTION POTENTIAL

Suction potential is the difference between the external gas pressure potential and the (apparent) pressure potential of the liquid in a 2-phase fluid system. When the external gas pressure is atmospheric, it is synonymous except for sign with the (apparent) pressure potential. The special significance of this potential is that the water content of soil can be related to it, regardless of the external gas pressure.

Since this potential is related to the degree of liquid saturation in soils, an analogous potential is sometimes defined thermodynamically by the relation

$$\psi_s \equiv \int_{\theta_o}^{\theta_i} \frac{\partial G_m}{\partial \theta} \, d\theta \qquad [22\text{–}9]$$

where ψ_s is the suction potential, θ is the liquid content per volume of soil, θ_o is the liquid content when the soil is completely saturated, θ_i is the liquid content of the soil for which ψ_s is defined, and G_m is the Gibbs potential (specific free energy) of the *bulk solution* (Taylor and Slatyer, 1960).

The suction potential has appeared in the literature under many names depending on the background of the author and the dimensions in which it is expressed:

a. Force Per Area. When expressed in these dimensions, the suction potential has been called *capillary pressure* or *suction pressure*. Frequently, to avoid a negative sign (*see below*), it is also expressed as *tension, soil water tension, suction, soil water suction, matric suction,* and *soil water stress*. Other terms occasionally used include *suction force, suction pressure,* and *suction tension*. Some of these names have also been used when suction potential is expressed in other dimensions, thus adding to confusion.

The sign convention applying to suction potential (energy per mass) is usually opposite of that applying to suction (force per area). Thus, the International Society of Soil Science (1963) has defined matric potential so that it is negative when the pressure of the soil solution is less than atmospheric and matric suction so that it is positive under an equivalent situation.

b. Energy Per Mass. The common names for the suction potential expressed in these dimensions are *suction potential, matric potential,* and *capillary potential*.

c. Energy Per Weight. When expressed in these dimensions, the word "head" is usually added, e.g., *capillary pressure head* or *suction head*. Sometimes *tension* is used without the word "head" to express this quantity. Sometimes the logarithm of suction head is employed (especially in relating suction head to saturation) in which case the quantity has been called pF in the older literature. The sign convention with respect to suction head is usually the same as that applying to suction and opposite of that applying to suction potential.

B. Water as a Constituent in Solutions

Although water may move in both soils and plants as a consequence of the mass transport of the bulk solution, it can also move in response to molecular transport or to a combination of mass and molecular transport. Furthermore, water can undergo changes from liquid to vapor and the reverse. Scalar quantities, which are a function of the condition of the water constituent in a water solution and which are indicative of the tendency of water to undergo molecular transport or a change in phase, are described below.

The terms referring to the water constituent (like those referring to the bulk solution) are often called potentials. In some cases, however, the use of the term "potential" in reference to the water constituent may not conform to the classical concept of a potential as defined by a work integral.

1. OSMOTIC POTENTIAL

This is usually the maximum pressure potential which can be developed in a solution when separated from pure water by a rigid membrane permeable only to water (Meyer and Anderson, 1939) or the negative of this depending on the sign convention used (Int. Soc. Soil Sci., 1963).

In order to determine the osmotic pressure of a solution, it would be necessary to enclose it in a rigid membrane permeable only to water, to immerse this membrane in pure water, and to exert just enough pressure on the solution to prevent any increase in its volume due to the entrance of water. In other words, the osmotic pressure potential is the maximum pressure potential which would develop in a solution if it were placed under the necessary conditions.

The word potential is ordinarily added only when the quantity referred to has the dimensions of energy per mass, although this is not universally true. A more commonly used term is simply *osmotic pressure*. In the latter case, the quantity referred to has the same dimensions as pressure, i.e. force per area. Osmotic pressure is also called *osmotic suction, osmotic power, osmotic value, osmotic concentration*, and *solute suction*. The latter two names are a consequence of the fact that the primary factor controlling osmotic pressure is the concentration of solutes in the water solution.

Osmotic pressure is a function of all the constituents making up the solution, but the condition that is described by the quantity defined is the *condition of the water constituent only, not the bulk solution*.

Versions of the osmotic potential having dimensions of energy per weight (length) are not in common usage. The term osmotic head is used if expressed as energy per weight.

The sign convention with respect to osmotic potential has not been universally standardized. The convention adopted by the International Society of Soil Science (1963) is such that water in the soil solution has a lower osmotic potential but a higher osmotic suction than pure water.

2. TOTAL WATER POTENTIAL

This is the sum of the "apparent" pressure potential and the osmotic potential. This definition assumes the sign convention for osmotic potential as given by the International Society of Soil Science (1963). Another way of stating the same thing is the sum of *negative* suction, external gas, and *negative* osmotic potentials. The same name is sometimes given also to a scalar which includes a gravitational component (Int. Soc. Soil Sci., 1963).

The special significance of the total water potential (without the gravitational term) is that it can be related to the vapor pressure of water in the soil or plant solution. It can also be related to the tendency of water in a solution to pass through a membrane permeable only to water.

As defined above, a higher potential is associated with a higher water vapor pressure.

a. Force Per Area. When these dimensions are used, the word "potential" is sometimes replaced by "suction" and the sign convention is usually reversed. Thus, *total suction* or *total water suction* is the sum of the suction or matric suction and the osmotic suction.

Other names often having the same significance are *diffusion pressure deficit, osmotic equivalent, net osmotic pressure, suction force, suction pressure, suction tension, water absorbing power, turgor deficit, water suction, soil water stress*, and *total stress*.

b. Energy Per Mass. Whenever the word "potential" appears explicitly in the name of the quantity defined, it usually has these dimensions. Furthermore, quantities specifically designated as potentials, usually follow a sign convention such that a higher potential is associated with a higher water vapor pressure as previously stated.

Other names sometimes given for the total water potential having dimensions of energy per mass are: *water potential, specific free energy,* and *total water potential.*

a. Energy Per Weight (Length). These dimensions have rarely been used in connection with the total water potential.

d. Energy Per Mole. Total water potential is often expressed in dimensions of energy per mole by physical chemists and those using a chemical analysis. It is, therefore, sometimes called the *partial molar free energy of water* or the *chemical potential of water.*

Both the chemical potential of water and the specific free energy may be related to the Gibbs free energy by the equation

$$\frac{\partial G}{\partial m_w} = \frac{1}{M_w} \frac{\partial G}{\partial n_w} = \frac{1}{M_w} \mu_w \qquad [22\text{--}10]$$

where G is the Gibbs free energy, m_w is the mass, n_w is the number of moles of water, M_w is the molecular weight of water, and μ_w is the chemical potential. The specific free energy is represented in equation [22–10] by the partial derivative $\partial G / \partial m_w$ and is the same quantity designated by G_m in equation [22–9].

C. Summary of Factors Primarily Affecting State of Fluids in Engineering, Soil, and Plant Systems

The state of fluids in engineering works (for storage, conveyance, distribution, etc.), in soils, and in plants is affected to different degrees by pressures, adsorptive forces, gravity, and solutes. The following paragraphs summarize factors affecting water in these systems with the most common usage of terms indicated.

1. ENGINEERING SYSTEMS

While water is being stored, it is generally not moving, and the piezometric potential is constant. The component potentials which vary and describe the system are the *pressure* potential and *gravitational* potential, the sum of which is the piezometric potential.

Conveyance and distribution of water involve the relatively rapid movement of water. Under these conditions there is often a conversion of piezometric potential energy to kinetic energy. Kinetic energy may be expressed per unit volume ($\rho V^2 / 2$), per unit mass ($V^2 / 2$) or per unit weight ($V^2 / 2g$), where ρ is density, V is velocity, and g is the force of gravity. In these forms, kinetic energy is readily combined with the respective piezometric, pressure, and gravitational energy units to describe water in conveyance and distribution systems. The term *hydraulic head* is commonly used to denote the sum of the pressure, gravitational, and kinetic heads ($V^2 / 2g$).

Excessive kinetic energy is capable of destroying earth channels. To prevent such destruction, structures which cause dissipation of kinetic energy (by converting it to heat) are often a necessary part of conveyance and distribution systems.

2. SOIL SYSTEMS

The velocity of movement in porous media, such as soil, is relatively slow, and the kinetic energy of the water is negligible, compared to the variations in *pressure and gravitational potentials* occurring in the system. Movement of water in soil systems involves a practically quantitative conversion of piezometric (gravitational plus pressure) potential energy to heat.

When the area of solid water interface becomes very large (i.e. clays), adsorption potentials may influence an appreciable portion of the water. However, the adsorption potential is generally included in the *"apparent" pressure potential* because measuring instruments, such as tensiometers, do not distinguish between pressure and adsorption potentials.

Osmotic potential becomes an important factor in soil water characterization only when solutes are restricted appreciably more than the water. This may occur at soil-root interfaces in regions of the soil such as those with low water contents near the evaporating surface and as water goes from solution of one concentration to solution of another via the gaseous phase.

Temperature gradients also cause movement of water from warmer to colder regions through the liquid and vapor phases of the soil water.

3. PLANT SYSTEMS

In plant cells the energy status of water is determined by three major factors: turgor pressure, imbibants, and solutes. Contributing factors such as membrane pore pressure gradients and electro-osmosis simply add to the pressure and solute components.

The *diffusion pressure deficit* (DPD) is still most commonly used. It has been understood that DPD expresses the energy (or free energy) of water in the system, although it is not defined or used according to standard thermodynamic practice. DPD is commonly related to osmotic pressure (OP) and to turgor pressure (TP) as follows:

$$DPD = OP - TP. \qquad [22\text{--}11]$$

Seeking to introduce a more physico-chemical approach, Broyer (1947) proposed *influx specific free energy* (IF), *efflux specific free energy* (EF), and *net influx specific free energy* (NIF) related as follows:

$$NIF = \Sigma IF - \Sigma EF \qquad [22\text{--}12]$$

giving an expression roughly comparable to equation [22–11] above.

Water potential (Taylor and Slatyer, 1960), commonly designated Ψ, is increasingly used to replace DPD. It is defined as the difference between the partial specific Gibbs free energy of water in the system and that of free pure water at the same temperature which is arbitrarily taken as zero.

Confusion will continue to arise because as the water potential decreases, it will generally require expression in increasingly negative values, whereas values of DPD increase positively. The water potential (absolute value) will increase with added turgor pressure, but decrease with more imbibants or solutes. For example, a water potential of –20 bars is equivalent to a DPD of +20 bars.

D. Tabular Summary and Conversion Factors

1. TABULATED LIST OF POTENTIAL TERMINOLOGY

Table 22–1 presents most of the terms which have been used to describe potential arranged according to whether energy is considered per unit volume, per unit mass, or per unit weight.

Table 22–1. Potentials describing the state of the bulk solution (or water) in soil-plant systems*

	Referring to the bulk solution		
Potentials	Terms having dimensions as indicated		
	energy/volume (force/area)	energy/mass	energy/weight (length)
Pressure	p pressure intensity (e) static pressure (g) solution pressure (p) neutral pressure (e) pore pressure (e) stress (s) hydrostatic pressure (g)	p/ρ pressure potential (s)(e)	p/pg pressure head (e)(s)
Gravitational	ρgh gravitational potential (g)	gh gravitational potential (s) (e	h elevation head (e)(s)
Piezometric	$p + \rho gh$ piezometric pressure (e)	$p/\rho + gh$ piezometric potential (g) hydraulic potential (s)(e)	$p/\rho g + h$ piezometric head (e)(s) hydraulic head (s)(e)
External gas pressure	p_a external gas pressure (s)	p_a/ρ potential due to external gas pressure (s)	$p_a/\rho g$ head due to external gas pressure (s)
Suction	$p_a - p$ capillary pressure (s)(e) suction (s) matric suction (s) soil-water suction (s) soil-water stress (s)(p) suction force (p) suction pressure (p) suction tension (p) tension (s) water absorptive power (p)	$(p_a - p)/\rho$ capillary potential (s)(e) suction potential (s) matric potential (s)	$(p_a - p)/\rho g$ capillary pressure head (s) (e) suction head (s) matric suction head (s) tension (s)

	Referring to the water constituent only		
Potentials	Terms having dimensions as indicated		
	energy/volume (force/area)	energy/mass	energy/mole
Osmotic	osmotic pressure (g) solute suction (s) osmotic suction (s) osmotic power (p) osmotic value (p) osmotic concentration (p)	osmotic potential (s)(p) solute potential (s)	
Total water	water potential (p, s) total suction (s) total water suction (s) water suction (s) suction force (s) suction pressure (s) suction tension (s) diffusion pressure deficit(p) osmotic equivalent (p) water absorbing force (p) net osmotic pressure (p) turgor deficit (p) soil water stress (s) total stress (s)	water potential (s)(p) total water potential (s) specific free energy of water (s) (p)	partial molar free energy of water (s) (p) chemical potential of water (g)

* Usage by disciplines is indicated by (g), (p), (s), and (e), which stand for general, plant physiology, soil science, and engineering, respectively.

Table 22–2. Conversion factors for units used to express potentials, pressures, and tensions (or suctions)*

	Energy/volume (pressure)				Energy/mass (potential)				Energy/weight (head)		
one dyne/cm²	one cbar	one bar	one atm	one erg/g	one joule/kg	one cal/g	one cal/mole H₂O	one cm	one foot	
1.000	10^4	10^6	1.013×10^6	1.00	1.00×10^4	4.18×10^7	2.32×10^6	9.8×10^2	2.99×10^4	dynes/cm²
10^{-4}	1.000	100	101.3	1.00×10^{-4}	1.00	4.18×10^3	2.32	9.8×10^{-2}	2.99	cbars
10^{-6}	0.01	1.000	1.013	1.00×10^{-6}	1.00×10^{-2}	4.18	2.32	9.8×10^{-4}	2.99×10^{-2}	bars
0.987×10^{-6}	0.987×10^{-2}	0.987	1.000	0.99×10^{-6}	0.99×10^{-2}	41.4	2.30	9.7×10^{-4}	2.96×10^{-2}	atm
1.00	1.00×10^4	1.00×10^6	1.01	1.000	1.000×10^4	4.18×10^7	2.32×10^6	9.8×10^2	2.99×10^4	ergs/g
1.00×10^4	1.00	100	101	1.000×10^{-4}	1.00	4.18×10^3	232	9.8×10^{-2}	2.99	joules/kg
2.39×10^{-8}	2.39×10^{-4}	2.39×10^{-2}	2.42×10^{-2}	2.39×10^{-8}	2.39×10^{-4}	1.00	0.0556	2.34×10^{-5}	7.1×10^{-4}	cal/g
4.31×10^{-7}	4.31×10^{-3}	0.431	0.435	4.31×10^{-7}	4.31×10^{-3}	18	1.00	4.22×10^{-4}	1.28×10^{-2}	cal/mole H₂O
1.02×10^{-3}	10.2	1.02×10^3	1.03×10^3	1.02×10^{-3}	10.2	42.6×10^4	2.37×10^3	1.00	30.48	cm
3.34×10^{-5}	0.334	33.4	33.8	3.34×10^{-5}	0.334	1.41×10^3	78	0.0328	1.00	feet

* For those cases in which density of the water is a necessary factor in the conversion, the accuracy is limited to the number of significant figures shown to allow for density variations at temperatures from 0 to 35 C.

2. CONVERSION FACTORS FOR POTENTIALS

The factors presented in Table 22–2 may be used for converting potentials described in one set of units and dimensions to potentials described in other units and dimensions.

IV. TERMS REFERRING TO MOVEMENT OF FLUIDS

A. Movement of Solution in Bulk

By *motion in bulk* is meant that all constituents of the solution have the same *net* velocity in the same direction at particular points within the solution. In other words, a particular volume element of solution does not change in composition as it moves in space.

When solutions flow in bulk they sometimes may be considered to constitute continua. In such cases it is often possible to define a scalar quantity, the gradient of which represents the resultant driving force tending to produce motion. When such a scalar can be uniquely defined by a work integral, the driving forces are said to constitute a conservative force field. This condition may be approximately satisfied for soil-plant solutions, provided their densities do not vary too greatly in space, and the portions of solution considered are not separated by interfaces or by membranes that selectively restrict certain constituents of the solution.

The gradients of the scalar potentials may have dimensions of force per volume, mass, or weight, depending on the dimensions of the potentials (which have been previously defined). An equation relating the rate of bulk flow in soils to the potential gradient is known as *Darcy's law*. This equation is of the form

$$q = -K \frac{\Delta\phi}{\Delta L} \qquad [22\text{–}13]$$

where q is the component of *volume flux* in the direction of L, ϕ is a potential, and K is a coefficient depending on the dimensions and units of ϕ and the units used to measure the length L and the flux q.

1. TERMS FOR RATE OF BULK FLOW THROUGH SOIL

In most studies of flow through soils, it is the flow through a macroscopic element of the bulk soil which is measured. The rate is expressed as volume of discharge per time per area of bulk medium. The flux thus measured has the dimensions of velocity but should not be confused with the actual velocity of the fluid.

In equation [22–13], the symbol q is used to designate the rate of flux having dimensions of velocity, but sometimes the flux is designated by q/t where q (in this case) is a volume per area and t is time. Other authors designate the flux by Q/at, where Q is a volume, a is the cross-sectional area, and t is time. Names in common usage for the flux are: *volume flux, mass flux, flow velocity, superficial velocity, seepage velocity, filter velocity,* or *velocity.*

2. TERMS FOR THE DRIVING FORCE PRODUCING FLOW

The following are alternative ways of expressing the component of the driving force in the direction of the measured component of flux, the direction being designated by x (Childs, 1957):

1.) force per volume

$$- \Delta p^* / \Delta x$$

where p^* is the piezometric pressure, $p + \rho g h$.

2.) force per mass

$$- \Delta \phi^* / \Delta x$$

where ϕ^* is the piezometric potential, $(p/\rho) + gh$.

3.) force per weight (dimensionless)

$$- \Delta H / \Delta x$$

where H is the piezometric head, $(p/\rho g) + h$.

For systems in which the air pressure is nearly constant throughout the media, p can be replaced by the negative of the suction. The advantage of doing this is that suction can be related to the water content.

3. THE COEFFICIENT OF PROPORTIONALITY IN DARCY'S EQUATION

a. **Length Per Time (Velocity).** Darcy wrote his original equation as

$$q = - K \, (\Delta H / \Delta x) \qquad [22\text{--}14]$$

where $\Delta H / \Delta x$ is dimensionless. Consequently, the coefficient K in his equation had the same dimensions as the volume flux q, i.e., velocity. This coefficient is a function of fluid as well as soil properties, and today it is usually known as *hydraulic conductivity* (Int. Soc. Soil Sci., 1963). It is sometimes called *permeability*, however, especially by groundwater engineers. A great variety of units are in common usage for expressing hydraulic conductivity.

b. **Length Squared.** Another coefficient in common usage, this is designed to eliminate (as far as possible) the effect of fluid properties on the quantity defined. This coefficient is usually called *permeability* and is a function primarily of properties of the solid matrix only (Richardson, 1961, sect. 16–3, p. 16–6, 16–7). Permeability is related to hydraulic conductivity by the relationship

$$K = k\rho g / \eta \qquad [22\text{--}15]$$

where K is the hydraulic conductivity, η is the fluid viscosity, and k is the permeability. The coefficient k is sometimes called *intrinsic permeability* or *specific permeability*.

The following units are in common usage for the permeability k:

1) *centimeters squared*—obtained when all quantities in Darcy's equation are expressed in centimeter-gram-second (cgs) units.

2) *microns squared*—10^{-8} cm^2, the order of magnitude of the permeability in many soils.

3) *darcy*—obtained when the flux is expressed in cm/sec, the driving force in atm/cm, and the viscosity in centipoises. One darcy is equivalent to 0.987 microns squared.

4) *millidarcies*—0.001 darcy, a unit which is convenient for many oil-producing rocks and fine-textured soils.

c. **Coefficients Having Other Dimensions.** The coefficients described above are the most commonly used, but a few authors employ coefficients that are related to permeability k by:

$$K' = k/\eta \qquad\qquad [22\text{–}16]$$

or

$$K'' = k\rho/\eta. \qquad\qquad [22\text{–}17]$$

Such coefficients are sometimes called *permeability, intrinsic permeability,* or *conductivity* so that the reader must be alert to distinguish between these and the more commonly used coefficients of the same name.

4. TERMS USED IN REGARD TO PARTIALLY SATURATED SOILS

a. **Wetting Phase.** That fluid phase which in a two-fluid system (e.g., air-water) preferentially wets the solid surface of the medium. In an air-water system, water is normally the wetting phase.

b. **Nonwetting Phase.** That fluid phase which in a two-fluid system is displaced from direct contact with the solid surfaces by the more strongly adsorbed fluid.

c. **Effective Permeability.** A coefficient in Darcy's equation analogous to permeability and distinguished from the latter only in that it refers to a particular fluid phase in a fluid system involving more than one fluid phase, e.g. the effective permeability to water in an air-water system or the effective permeability to air in a similar system. It is expressed in the same units as intrinsic permeability.

d. **Capillary Conductivity.** A coefficient in Darcy's equation analogous to hydraulic conductivity but referring to water only in a fluid system consisting of both air and water. It has the same dimensions and may be expressed in the same units as hydraulic conductivity.

e. **Relative Permeability.** A dimensionless quantity, k_r given by

$$k_r = k_e/k, \qquad\qquad [22\text{–}18]$$

where k_e is the effective permeability at the particular moisture content for which k_r is evaluated, and k is the permeability of the medium when the medium is completely saturated with the fluid to which k_r refers. The relative permeability varies from unity when the medium is fully saturated to zero when the fluid referred to is absent (Richardson, 1961, sect. 16–7, p. 16–47, 16–52).

5. TERMS USED IN THE DIFFUSIVITY EQUATION

The terms "diffusivity equation" or "diffusion equation" are used to describe phenomena in which the flow takes place in response to a concentration gradient of flowing material. In partially saturated, homogeneous, isothermal, and isotropic soil, in which air pressure is constant, movement in horizontal planes may be described by a diffusion type equation derived from Darcy's law as follows. The hydraulic conductivity $K(\theta)$ is a function of the water content θ. Thus

$$q = -K(\theta)\,\frac{\partial\phi}{\partial x} \qquad\qquad [22\text{–}19]$$

for one-dimensional flow which may be written as

$$q = -K(\theta)\,\frac{\partial\phi}{\partial\theta}\,\frac{\partial\theta}{\partial x} \qquad\qquad [22\text{–}20]$$

when ϕ is a unique function of θ.

Under these conditions $K(\theta)(\partial\phi/\partial\theta)$ may be expressed as a single function $D(\theta)$ of the water content. Darcy's law then becomes the diffusion equation,

$$q = -D(\theta)\frac{\partial\theta}{\partial x}.$$ [22–21]

Assuming that the fluid is incompressible, and considering the requirements for continuity, one obtains,

$$\frac{\partial\theta}{\partial t} = \frac{\partial}{\partial x}[D(\theta)\partial\theta/\partial x],$$ [22–22]

which is commonly known as the "*diffusivity equation*" even though the flow is bulk flow. The water content may be expressed as a percentage of the bulk soil volume, pore volume, or dry weight of soil without changing D dimensionally. Terms used in regard to this equation (Int. Soc. Soil Sci., 1963) are:

a. **Soil Water Diffusivity.** The coefficient of $D(\theta)$ in equations [22–19] and [22–20] (or the flux of water per unit gradient of moisture content, provided that the water gradient corresponds in direction and is directly related to the potential gradient).

b. **Differential Water Capacity.** The absolute value of the rate of change of the water content with soil-water suction, $\partial\theta/\partial\phi$ (Int. Soc. Soil Sci., 1963). It is possible to obtain the soil water diffusivity D by dividing the hydraulic conductivity by the differential water capacity (care being taken to be consistent with units).

B. Molecular Diffusion of Water

Molecular diffusion in its purest sense refers to the random thermal motion of particles which results in net movement of components from regions of higher to regions of lower concentration.

In general, the tendency to diffuse depends on the concentration, temperature, pressure, gravity, and viscosity of the fluid, as well as on any other body force that may be acting in particular cases, such as the attraction of neighboring particles or masses.

1. ISOTHERMAL DIFFUSION

Isothermal diffusion, in general, takes place in response to a chemical potential gradient, du/dx, in accord with the equation

$$q = \frac{D_o c}{RT}\frac{du}{dx}$$ [22–23]

where D_o is *the diffusion coefficient* at infinite dilution in water, c is the concentration of the water (cc/cc solution), R is the gas constant, and T is the temperature in degrees Kelvin. The chemical potential of water in solutions is affected by pressure, solute content, and adsorptive forces, and may be determined from the water vapor pressures (p) in equilibrium with those solutions according to the equation

$$u - u_o = RT\ln(p/p_o)$$ [22–24]

where u_o and p_o are values taken in a standard state (usually pure water). Diffusion of water takes place in the liquid and gaseous phases in soil and plants. In both phases the *diffusion coefficients* will be in units of *length squared per time*, and will be reduced from the bulk phase coefficients by the tortuosity factors, fractions of the total volume made up by the particular phase, etc.

If the bulk solution is moving, diffusion within the lattice in response to chemical potential gradients will still occur. However, the frame of reference to be used for the distance coordinate is the moving semicrystalline lattice of the water itself, rather than the porous media.

2. NON-ISOTHERMAL DIFFUSION

When temperature gradients exist within partially saturated porous media, they cause vapor pressure gradients in the vapor phase. These vapor pressure gradients result in diffusion from regions of high to regions of low temperature. Movement of water from regions of high to regions of low temperature also occurs in the liquid phase, apparently in response to thermal energy differences (*see* chapter 13, part II). Thermally induced movement of water is accompanied by the transfer of energy and solutes, and the movements are often appreciably "coupled" or interdependent. Thermodynamics of irreversible processes have been used to handle such complex movement. Since the terminology is complex and is still in the process of development, it is not discussed here. Some of the most recent developments, including terminology, are outlined in chapter 13, part II.

C. Osmotic Flow

Osmotic flow differs from both bulk flow and molecular diffusion but it is related to both. It differs from bulk flow in that the composition of fluid elements may change as they move in space, and it differs from molecular diffusion in that the net transport is not always due to the random motion of individual particles of the solution. Osmotic flow results when the solid matrix restricts a certain constituent or constituents of a solution more than others. The movement of the other constituent particles, however, is not random. As in bulk flow, adjacent fluid particles and fluid elements exert forces, such as viscous and pressure forces, on each other so that the motion takes on some of the characteristics of bulk flow.

Osmotic flow is usually a much more complex process than either bulk flow or simple diffusion. As of the present time, no equation describing osmotic flow for the general case has gained universal acceptance.

V. SUMMARY

The multiplicity of terms used to describe potentials and permeability units in plant, soils, and engineering literature has led to considerable confusion. However, in most cases, the terms have been used correctly. Careful analysis and conversion to desired units allow comparisons to be made between the results of many outstanding and informative studies.

It does not appear likely that individual investigators in the wide range of disciplines involved will, in the near future, adopt uniform terminology. Consequently, it will continue to be necessary to understand basic physical concepts to correlate and integrate research contributions in this subject area.

LITERATURE CITED

Broyer, T. C. 1947. The movement of materials into plants: I. Osmosis and the movement of water into plants. Bot. Rev. 13:1–58.

Childs, E. C. 1957. The physics of land drainage. Agronomy 7:41–45.

Collins, R. E. 1961. Flow of fluids through porous materials. Reinhold, New York. Ch. 2, p. 22.

Edlefsen, N. E., and A. B. C. Anderson. 1943. Thermodynamics of soil moisture. Hilgardia 15:101–110, 270–271.

International Society of Soil Science. 1963. Soil physics terminology. Bull. 23. p. 2–5.

Lamb, Sir Horace. 1948. Hydrodynamics. 6th ed. Dover, New York. Ch. 1, p. 1.

Meyer, B. S., and D. B. Anderson. 1939. Plant physiology. 4th ed. D. van Nostrand, New York. Ch. 8, p. 94.

Planck, Max. 1945. Treatise on thermodynamics. 3rd ed. (Translated from the 7th German ed.) Dover, New York. Ch. 1, Sec. 6, p. 3–4.

Prigogine, I. 1955. Introduction to thermodynamics of irreversible processes. Charles C. Thomas, Springfield, Illinois. Ch. 3, p. 32.

Richardson, J. G. 1961. Flow through porous media. In V. L. Streeter [ed.] Handbook of fluid dynamics. McGraw-Hill, New York.

Rouse, H. 1945. Elementary mechanics of fluids. 5th ed. John Wiley, New York. Ch. 5. p. 114.

Sears, F. W. 1950. Thermodynamics. Addison-Wesley Press, Cambridge, Massachusetts. Ch. 3, p. 26.

Streeter, V. L. 1948. Fluid dynamics. McGraw-Hill, New York. Sec. 2–3, p. 3.

Taylor, S. A., and R. O. Slatyer. 1960. Water-soil-plant relations terminology. Int. Congr. Soil Sci., Trans. 7th (Madison, Wis., USA) 1:394.

23

Factors Affecting Plant Responses to Soil Water

G. STANHILL

Volcani Institute of Agricultural Research
Rehovot, Israel

YOASH VAADIA

Negev Institute of Arid Zone Research
Beersheva, Israel

I. INTRODUCTION

Research into plant responses to soil water appears to be of only recent origin despite the antiquity, importance, and almost world-wide occurrence of irrigation. Even now this research is on a comparatively small scale.

The first experimental demonstration of the essential part that soil water plays in plant growth is generally attributed to Van Helmont (1577–1644) in his classical experiment with a potted willow tree (*Salix* sp.). Almost 100 years later John Woodward (1699) published the results of his investigation into the quantitative relationship between the increase in plant weight of spearmint (*Mentha spicata* var. *viridis*) and the amount of water transpired. He clearly showed that the composition of the water applied affected the plant response and this might well be considered the first demonstration of nutritional effects on the efficiency of water use. At the same time Stephen Hales (1677–1761) was publishing the first quantitative demonstration of the effect of both climatic and plant factors on the intensity of water loss by transpiration.

Field research into crop water relationships started early in this century after the introduction of irrigation farming in the arid western and southwestern states of the USA (Widstoe and Merrill, 1912). The foundations of modern research into irrigation water requirements followed Veihmeyer's work (1927) which showed the impossibility of maintaining a constant soil water content around the roots of transpiring plants. Veihmeyer and Hendrickson's view (1929, 1950) that soil water is equally available for crop growth and transpiration over the entire range of available soil water between field capacity and permanent wilting percentage was widely accepted until the early 1940's. However, a number of later experiments showed that plant growth and transpiration decreased before the wilting point was reached and, in some cases, a clear dependence between these plant responses and soil water potential was established (Wadleigh and Ayers, 1945). An analysis (Stanhill, 1957) of some 80 experiments on the effect of the so-called "available soil water" (that held between field capacity and wilting point) on growth showed that the majority of results did not favor Veihmeyer's views that water was equally available for growth.

These findings, however, were not able to provide a basis for drawing up irrigation programs that would maximize yields and minimize water application, and it became increasingly clear that the relation between plant growth, crop yield, and soil water varied considerably, depending on the crop, the soil, and the climatic conditions of cultivation (Hagan and Vaadia, 1960).

II. PLANT WATER STATUS AND CROP RESPONSES

It is obvious that the responses of crops to soil water are ultimately mediated via the plant's water status. For this reason, much recent research deals with the effect of soil water and other environmental factors on plant water balance and on tissue water potentials. However, the question discussed in chapter 19 on the nature of the relationship between plant water balance and growth remains unanswered. Nevertheless, the dynamic approach which considers the plant as an aqueduct for water flow from soil to atmosphere has done much to resolve the old question of soil water availability. It now appears that soil water availability cannot be theoretically or practically defined in simple terms of ranges of soil water contents or soil water potentials. Correlations between plant growth and such quantities do not have any general significance but rather only apply to the specific situation investigated. It is commonly assumed, and there are data to indicate that maximum crop yields in terms of green matter or dry weight are obtained whenever plant water potentials ψ_p are kept high—corresponding to low water stress (Kramer, 1963). Low plant water potentials resulting in reduced tissue hydration may affect plant growth in several ways: retard enzymic reactions, close or partially close stomata, decrease cell expansion, and reduce leaf area. The latter is normally the major determinant of crop growth (Watson, 1952). Partial or complete stomatal closure restricts gaseous diffusion between plant and atmosphere which, under rapid growth conditions, may limit photosynthesis by reducing the CO_2 supply. This and other factors involved are treated in detail in chapter 19. The above suffices to demonstrate that an understanding of the factors affecting plant water balance are basic to the question of plant response to soil water.

III. WATER POTENTIALS AND FLUXES IN THE SOIL-PLANT-ATMOSPHERE CONTINUUM

In 1928 Gradmann suggested the use of an electrical analog to describe water flow in the soil-plant-atmosphere system, and this idea was developed by Honert (1948). Since then, many workers have used the method to analyze the flux of water and/or CO_2. With this approach, the flux of water and/or CO_2 through the various parts of the system is seen as the current flowing in response to potential gradients with the permeability or conductivity of the various parts of the system treated as the inverse form of resistances. Dainty (1960) has pointed out that there are two advantages in using this electrical analog. First, the three quantities—current (flux), potential (driving force), and resistance or conductivity—are kept very distinct, whereas in other formulations they tend to lead to mental confusion. The analog is thus of value in helping to clearly formulate a physical concept of the system. However, the prime use of the approach is its practical value in calculating the various fluxes from known potentials and resistances or in analyzing the importance of the various resistances from measured fluxes and potentials.

The following three equations may be used to describe water fluxes in the various segments of the soil-plant-atmosphere system. Note that flows from soil to roots and from shoot to atmosphere are assigned positive values:

$$Q_{s\text{-}r} = -\frac{\Psi_r - \Psi_s}{R_s + R_r} \qquad [23\text{--}1]$$

$$Q_{r\text{-}l} = -\frac{\Psi_l - \Psi_r}{R_r + R_l} \qquad [23\text{--}2]$$

$$Q_{l\text{-}a} = \frac{\Psi_a - \Psi_l}{R_l - R_a} \qquad [23\text{--}3]$$

where Q denotes flux, Ψ denotes water potential, and R denotes resistances in the soil (s), roots (r), leaf (l), and surrounding free air (a).[1] It should be noted that these three equations cannot be equated unless the system is in a steady state with respect to a water flow. However, this condition seldom exists under typical irrigated agriculture. Instead the water content of plants undergoes substantial diurnal change. Although these equations are simplified, they offer a means of at least qualitatively assessing the factors that may affect plant water potentials and identifying the parameters to be evaluated.

Equation [23–1] shows that if Q is to be maintained constant during a soil water depletion cycle $\Psi_r - \Psi_s$ must remain constant as Ψ_s decreases, providing the resistances do not change. However, it is known that at least R_s (reciprocal of hydraulic conductivity) increases considerably as the soil dries out (see chapter 13). This requires that Ψ_r decreases more than Ψ_s, if water flow is to remain constant in the system. There is evidence that as the soil dries out R_s increases and exceeds the value of R_r (Gardner and Ehlig, 1962; Lemon, 1963).

The conclusion which can be drawn from equation [23–1] for irrigation practice is that in soils where R_s is lower for a given Ψ_s, more water can be utilized by the plant for a given increment of $\Psi_r - \Psi_s$. This means that plants with a deep and extensive root system in a soil with a relatively high unsaturated conductivity will tolerate lower values of Ψ_s than plants lacking such a root system under otherwise similar conditions. It has been suggested by several workers that R_s at the soil-root interface is very high during periods of rapid water absorption (Gardner and Ehlig, 1962). This is because it is expected that under conditions of rapid flow rate the soil water potentials in the immediate vicinity of the absorbing root surface increase much more rapidly than the average soil water potential. This would suggest that even in wet soils the supply of water to the roots may be limiting, providing the flow rate is high due to high evaporating conditions. This may well explain the reason for the mid-day wilting sometimes observed during particularly hot spells, even in recently irrigated crops.

Equation [23–2] describes water movement within the plant from the roots

---

[1] R_a in equation [23–3] is a function of the potential differences. It should not be confused with the conventional diffusion resistance r_a (sec/cm). R_a and r_a are related to each other as follows:

$$R_a = r_a [\Delta \Psi / e_s - e_a]$$

where e_s and e_a are the vapor pressures in the leaf and above the laminar layer in the surrounding free air and $\Delta \Psi$ is the potential difference. This relation assumes a linear vapor pressure gradient between the leaf surface and the free air justified by the molecular nature of vapor diffusion through the laminar layer bounding the leaf surface.

to the leaves. It should be noted there are few quantitative data for either water potentials or resistances in the plant system. Some of the better data, which show rather low resistances in the xylem, the main water conducting system, have been that of Dixon (1914). Jensen et al. (1961) have shown that the resistance to water flow between root and leaf is considerably smaller than for water entry into the roots (*see also* Rawlins, 1963).

Plant water potentials vary between 0 and −20 to −30 bars for irrigated crops. Generally the osmotic and water potentials decrease with height. They are highest in the roots and lowest near the growing points of the shoots. Figure 23–1 shows the variation in osmotic potentials of wheat (*Triticum vulgare*) roots and leaves at various values of soil water potentials. The measurements were taken daily during a drying cycle from plants growing under constant environmental conditions (Closs, 1958). It should be noted that the roots appear to be much more sensitive than the leaves to increases in soil water potential. Although the data do not show plant water potentials, they do suggest a low resistance to water flow between roots and leaves.

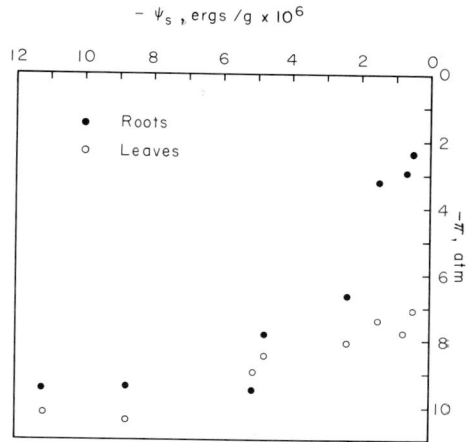

Fig. 23–1. Osmotic potentials of wheat roots and leaves as a function of soil water potential (data from Closs, 1958).

Shinn and Lemon (1962) provide some data concerning equations [23–1] and [23–2]. They show (Fig. 23–2) the distribution of Ψ_s and Ψ_p (plant water potential) in the soil plant system at the time of maximum values of flux occurring during the midday period of four summer days of similar evaporation potential. As the soil dried out (left to right in Fig. 23–2) Ψ_p and Ψ_s increase. The gradient of Ψ_p in the plant is not steep despite the fact that the transpiration from the upper leaves was probably much greater than that from the lower leaves (the radiation balance was only 17% of that at the upper leaves). These data support the conclusion that the resistances to water flow within the plant are low.

Probably the most important link in the chain between soil and atmosphere is that between the leaf and atmosphere described in equation [23–3]. It is here that water is lost from the soil-plant system to the atmosphere, the main pathway being through the stomates and through the same diffusion path CO_2 enters the leaves and is fixed in photosynthesis.

Honert (1948) assumed steady-state conditions and used equations [23–1, −2 and −3] to demonstrate that under normal conditions the atmospheric resistance R_A is the greatest of all the resistances in the path of water movement. The potential difference between the leaf and the atmosphere is by far the greatest (Ψ_a at 25C and 50% relative humidity is equivalent to nearly −1,000 bars). Therefore, the resistance at this interface must also be the greatest if the fluxes in the various segments of the system are identical. van den Honert concluded that the rate-controlling step in the water pathway lies in this leaf atmosphere segment. Such a

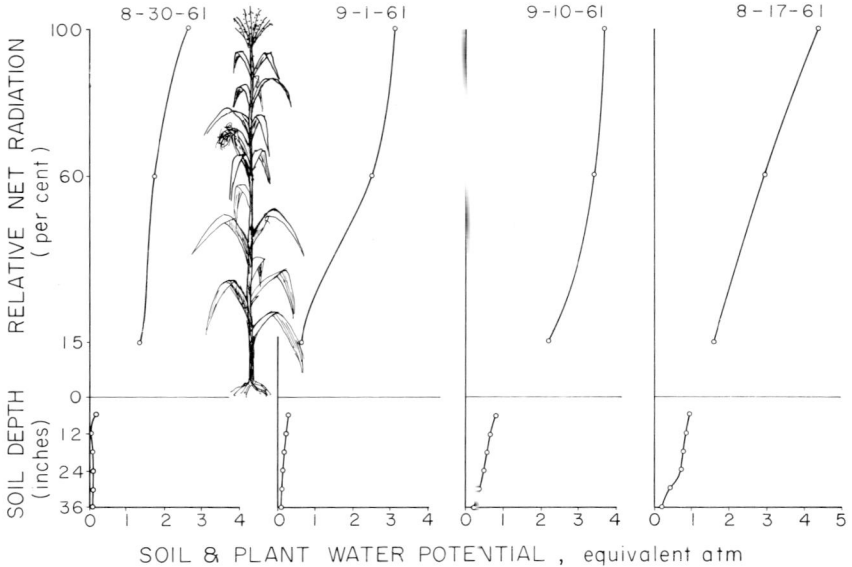

Fig. 23–2. Distribution of soil and plant water potential in corn plants during time of maximum evaporation on four days of an irrigation drying cycle (after Shinn and Lemon, 1962).

conclusion from this simplified model is dangerous because evaporation is not linearly related to potential difference. Rawlins (1963) has modified the model for the logarithmic relationship between evaporation and water potential difference, concluding that control may possibly still be greatest in the vapor phase. Under normal non-steady-state field conditions, it is most likely that all the other resistances in the pathway, particularly R_s, usually influence transpiration indirectly through their effect on stomatal resistance to water vapor diffusion via their effect on leaf water potentials (Rawlins, 1963).

For example, if the water supply to the roots is limiting, Ψ_p will increase, and this will lead to stomatal closing or reduced aperture. Although the mechanism of this control is not certain, it is clear that the internal water status of the plant is one of the factors of greatest significance (Ketallapper, 1959). The exact relationship between the size of stomatal opening and its resistance to the diffusion of water vapor (and CO_2) is uncertain, mainly because of the difficulty of obtaining direct measurements under natural growing conditions (Heath, 1959), and this makes the development of any empirical relationship difficult. It is also partly due to the complex geometry of the stomatal opening which makes any theoretical derivation of the relationship either unrealistically simple or, in practice, impossibly complex.

Rawlins (1963) has recently reviewed the evidence for a significant nonstomatal resistance to water flux in the leaf. Three mechanisms have been suggested to explain the observations of a number of investigators that the cell walls in the substomatal cavities, at which the change from the liquid to the vapor state occurs, are not, in fact, moist and therefore provide a significant resistance and a major drop in potential. The classical explanation is that of incipient drying. A second possibility is that chemico-physical changes in the cell wall can lead to increased resistance to water flow, similar to those known to occur in root cells under certain conditions. The third explanation (Boon-Long, 1954) is that the process of evaporation leads to a concentration of solutes on or at the evaporating surface which then causes an osmotic potential on the cell membranes much higher than indicated by the mean values of the leaf tissues. The existence of such leaf resistances under irrigated field conditions is yet to be demonstrated, and it still seems likely that Honert's (1948) conclusion that the most important plant resistance to water loss is provided by the stomata is correct under most conditions of interest to irrigated agriculture. Teleological considerations support this view for only a variable and controlling stomatal resistance, dependent on water potential within the plant, can allow the plant to maintain a relatively stable and favorable internal moisture environment, which to a certain degree, is independent of and insulated from the large external variations in atmospheric and soil water content.

A comparison of these two major resistances to water flux, both of which affect the movement in the vapor phase, suggest that under irrigated field conditions the stomatal and atmospheric resistances are of comparable orders of magnitude (Gaastra, 1963). The almost complete lack of quantitative field data does not allow any more detailed statement.

The proportion between these two resistances is of practical importance where meteorological indices, such as potential evapotranspiration, evaporation pans, or atmometers, are used as a guide to crop water requirements. So long as the aerodynamic resistance is larger than the stomatal, the relation between measured cr calculated open water loss and crop evapotranspiration will remain constant for a given crop as the same limiting factor operates for both surfaces. However, under conditions of either soil or atmospheric water stress the stomatal resistance will tend to increase and may become the limiting factor to crop water loss, reducing its rate relative to that of the open water surface. It is clear that here, too, quantitative data on the size of these resistances applicable to field conditions is needed.

It is now relevant to briefly consider the flux of CO_2 from the free atmosphere to the site of photosynthesis within the chloroplasts of the leaves. Equation [23–4] shows this flux in an analogous fashion to that used for water flux

$$Q_{CO_2} = \frac{C_a - C_c}{r_c + r_m + r_l + r_a} \qquad [23\text{–}4]$$

where Q_{CO_2} is the flux of carbon dioxide and C_a and C_c are the concentrations of carbon dioxide in the free air and chloroplasts, respectively, and the resistances r are in the conventional units of such diffusion equations. The subscripts, c, m, l and a refer to the chloroplasts, mesophyll, leaf (stomata), and atmosphere (laminar layer), respectively. Variations in the CO_2 content of the free air above a growing crop are considerable, both within and between days and show long-term seasonal differences (Huber and Pommer, 1954). It also appears probable that there are significant yearly differences. The CO_2 content at the site of photosynthesis within the chloroplast cannot be directly measured, but is believed to be close to zero under the saturating light conditions at which the greater part of dry matter production occurs (Gaastra, 1963).

The resistances to the CO_2 flux between the free air and the cell walls of the substomatal cavity ($r_a + r_l$) are proportional to those for water vapor, where the proportionality is the ratio of the diffusion coefficients of the two gases. Another similarity is that the stomatal resistance is not independent of the CO_2 content as the stomata open increasingly as CO_2 concentrations fall below normal levels (Heath, 1959). However, there is a further important resistance in the case of the CO_2 flux provided by the mesophyll cells (r_m). The length of this diffusion path is the same order of magnitude as that in the air, but in resistance terms it is of much greater significance, as diffusion is some 10,000 times slower through the liquid cell contents than through the air. There is some evidence that this resistance increases with plant water potential (Heath and Meidner, 1961).

The resistances of chloroplasts to CO_2 flux (r_c) can be considered as made up of the resistance to the dark process of photosynthesis whereby CO_2 is fixed by an acceptor molecule. Here the resistance can be defined as the reciprocal of the rate constant of this process. The second resistance is in the light process whereby the CO_2 is reduced to a carbohydrate and is inversely proportional to the light intensity.

It is now clear that both water and CO_2 flux and, therefore, to a great extent, transpiration and dry matter production, can be treated as essentially similar diffusion phenomena, although of opposite sign, with common diffusion resistances in the vapor phase, one of which, the stomatal, is largely controlled by the plant water potential.

Although it is clear that both transpiration and dry mater production will be affected by the soil water status (via its effect on plant water potential), the relative response of these two processes, i.e. the efficiency and often profitability of water use, will depend on the relative importance of stomatal resistance in the total resistance chain of the two fluxes. This relative importance will obviously vary from crop to crop and also no doubt varies for a single crop, according to its growth stage. Crops possessing a high mesophyll resistance (i.e. a long mean diffusion path between the substomatal cavity and chloroplasts), a high chloroplast resistance (i.e. a low rate constant for the dark process), or growing under light-limiting factors for the light process of photosynthesis, could be expected to show much greater transpiration than growth reduction with increasing soil water potential. The optimum irrigation treatment for such species might well allow considerable soil water potentials to build up before irrigation, economizing in water use without reducing yields at the same time. Irrigation experiments with crops such as guayule (*Parthenium argentatum*) (Wadleigh et al., 1946) confirm this expectation.

Whatever the value of the mesophyll and chloroplast resistance, it is obvious that stomatal resistance will always be of greater relative significance to transpiration than to the growth process. This fact explains why the experimentally determined optimum irrigation treatment generally allows some depletion of the available soil water before irrigation. Such treatments allow economies in water application and yet do not significantly reduce yields. It is for this reason, no doubt, that the actual amounts of water applied to irrigated crops in areas where water is limited or expensive are generally much below the calculated potential evapotranspiration values which assume that stomatal resistance does not limit water loss.

The above considerations suggest, at least theoretically, that it should be possible to calculate the optimum irrigation treatment which would minimize water loss and maximize dry matter production for each crop. Such a calculation would require absolute and relative values of the various resistances to water and CO_2 flux for each crop, their response to soil water potentials, and data on the atmospheric water and CO_2 potentials.

IV. THE CONCEPT OF SOIL WATER AVAILABILITY

The dynamic approach to water in the soil-plant-atmosphere system outlined in the preceding section can be said to have settled the controversy concerning the availability of soil water for growth and transpiration. Equations [23–1, –2, –3, and –4] can be used to explain the difficulty in providing a general definition of soil water availability. It is clear from equation [23–1] that soil water cannot be defined in terms of a range of soil water contents or potentials which are available or readily available. With low rates of evaporative demand and hence low flow values, plant water potentials Ψ_p need not be much different from soil water potentials Ψ_s, even if the resistance to water flow in the soil r_s is high. Conversely, under conditions of high evaporative demand, Ψ_p may be much greater than Ψ_s, even if resistance to water flow in the soil is low.

These conclusions are illustrated well in Fig. 23–3 for transpiration and Fig 23–4 for growth. Figure 23–3 shows that under low evaporative demand the transpiration rate was not affected by increasing soil water potentials up to values of 10 bars. Under conditions of high evaporative demand the transpiration rate was reduced at soil water potentials of < 1 bar (Denmead and Shaw, 1962).

Similar results are shown in Fig. 23–4 for three vegetable crops growing on two soil types under two levels of evaporative demand (Bierhuizen and Vos, 1958). Under conditions of high evaporative demand (line A), the relative yield decreased from 100% at a soil water potential of 0.2 bars to 30 to 40% for a soil water potential of 2.5 bars. Under low evaporative conditions (line B), the decrease was less, and for comparable soil water potentials the yield was only reduced to 60 to 70%.

In discussing the importance of factors that modify the response of the plant water balance to soil water it is necessary to consider the relationship between plant water potentials and plant response in the light of the responses which are desirable for a crop yield. This subject has been reviewed elsewhere (Vaadia et al., 1961) and is also discussed in chapter 19. Here we need only point out that the agronomist will favor different manifestations of growth response in different crops. Thus, elongation of plant organs, increase in fresh or dry weight, and vegetative or reproductive growth may each be the desired growth response. Such responses are the result of the complex combination of many physiological processes, each of which may be affected differently by plant water deficits. The reduction in fresh weight, caused by such deficits, may be proportionately much greater than the loss of dry weight (Bierhuizen, 1961). With certain crops, water deficits during maturation are not harmful and in some cases may be beneficial. An example is a crop grown for seed (Hagan et al., 1957). During the seed or flower maturation period the vegetative growth in these crops may be drastically reduced by water deficits without a measurable decrease in the yield of the harvested organs. The extensive literature on the effect of the stage of maturity on plant response to soil water has recently been reviewed (Salter and Goode, 1967).

The fact that the various measurable aspects of growth do not respond in the same manner to plant water deficits and plant water potentials further complicates the problem of formulating a simple relationship between soil water and plant growth. Another important complication, in some cases, is that growth may be

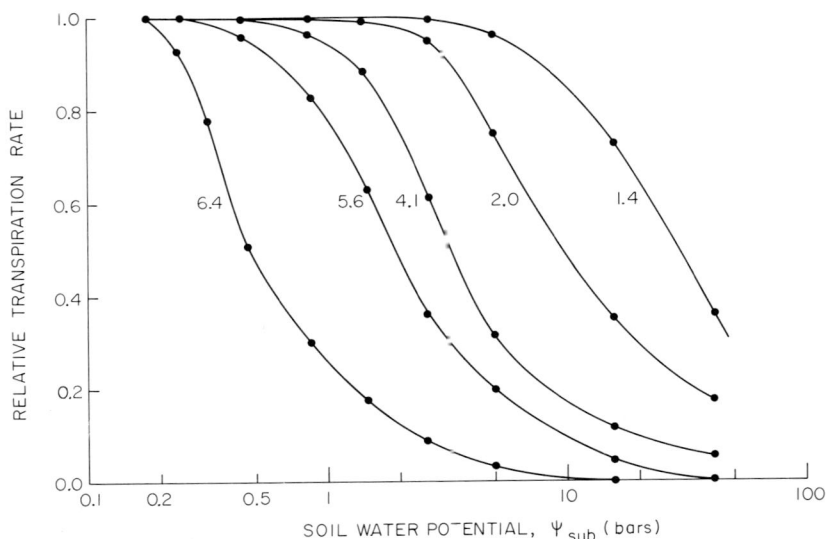

Fig. 23–3. Relative transpiration rate as a function of soil water potential for different evaporative conditions on 5 days on which the transpiration rate differed as given in the body of the figure in millimeters per day (after Denmead and Shaw, 1962).

limited by nondiffusive resistances. Thus, if light is a limiting factor (r_c in equation [23–4]) rather than CO_2 diffusion (r_a, r_s, and r_m in the same equation), only very small growth responses to soil water can be expected and, indeed, this has been found experimentally (Abd El Rahman et al., 1959).

Under agricultural conditions, light may be limited by climatic factors leading to low levels of incident radiation or there may be insufficient leaf area to utilize the incident light. Insufficient leaf area may arise either because of an incomplete ground coverage by the crop or a low leaf area index. Any factor that will lead to a low leaf area index (soil nutrient deficiency, especially nitrogen, suboptimal plant population, insect or disease damage, etc.) may be expected to reduce plant sensitivity to soil water status. The many experimental results showing a large positive interaction in yield response between fertilizer (especially nitrogen) or plant spacing and irrigation treatment may be partially explained in this way. The relationship is, however, not simply related to leaf area. Reduced sensitivity to the soil water status may result if leaf area is reduced relative to root area. If root growth is also reduced by nutrient deficiency, disease and insect damage, defoliation, etc., the sensitivity of the crop may not be affected as the ratio of evaporating to absorbing surface is essentially maintained. Plant sensitivity to the soil water status will be increased if leaf area is greatly increased relative to root area. The reason being that the plant has the capacity to lose more water than may be absorbed from the soil.

In the same manner as variations in leaf area affect a plant's sensitivity to the soil water status, variations in the extent of rooting will have similar and usually opposite effects. Reduction of the root system due to nutrient deficiency, disease and insect damage, waterlogging, and any factor which affects the supply of photosynthate from the leaves may increase the sensitivity of the plant. This is

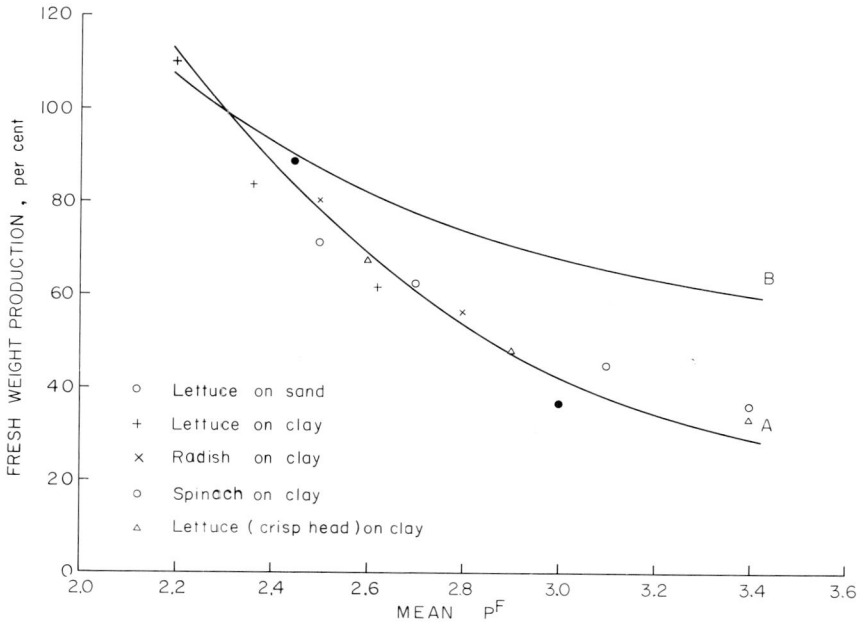

Fig. 23–4. Relative fresh weight production of various crops on various soils for different evaporative conditions. Curve A from high evaporative conditions and curve B low— the individual points of curve B were omitted as their variation was similar to that of A (after Bierhuizen and Vos, 1958).

the result of lowered root/leaf area ratios. Good soil conditions, fertilization, etc. which increase the extent of the root system may decrease the plant's sensitivity to soil water because of the ability to fully utilize all available water. The effects of different fertilizer and irrigation treatments are partially explainable on the basis of root/leaf area ratios. Nitrogenous fertilizers and frequent shallow irrigation decrease the root area while increasing leaf area and thus will increase the sensitivity of the plant to soil water status due to low root/leaf ratios. Phosphatic fertilizers and infrequent deep irrigations which tend to increase the root/leaf area ratio may decrease a plant's sensitivity.

One last factor that also must be considered as a possible influence on plant response to soil water is the size of the irrigated area. In arid and semiarid zones there is usually a very significant contrast between the microclimate of irrigated fields and surrounding dry areas (see discussion of advection in chapter 26). The effect of the intensity of the evaporating conditions on plant transpiration and growth response to soil water has already been discussed. Thus we should expect that in arid zones plant responses to soil water will be less marked as the size of the irrigated area increases, and hence, the intensity of the evaporating conditions decreases. However, the situation is more complex if we consider the results from a small[2] plot irrigation experiment with its patchwork of different treatments

_____

[2] Small in this connection means a plot size which is insufficiently large for the microclimate above the point of measurement to come into equilibrium with the surface conditions of that plot. In arid areas this distance will be of the order of magnitude of 100m to the upwind border.

Table 23-1. Plant, soil, and climatic factors favoring a marked and slight plant response to soil water

Marked plant response to soil water	Slight plant response to soil water
Plant factors	
Slow growing, sparse, and shallow root system; thin, smooth, and flat leaves completely covering soil surface; fresh weight of vegetative tissue or primary photosythates (starch, sugar) harvested as yield	Fast growing dense and deep root system; thick, hairy and curled leaves with small total area; dry matter of reproductive tissue or complex photosynthates (oils, etc.) harvested as yield
Soil factors	
Shallow soil with low hydraulic conductivity; low absolute water content at high water potentials (coarse textured soils with low organic matter content); low osmotic potential (salt or fertilizer content high); unfavorable for root development (too high or low soil temperatures, poor aeration, root pests and diseases); high nutrient status	Deep soil with high hydraulic conductivity; high water content at high water potentials (fine textured with high organic matter content); high osmotic potential; favorable for root development; low nutrient status
Climate factors	
High evaporation potential; advective energy supply; aerodynamically rough crop surface	Low evaporation potential; no advective energy supply; low, closed stiff crop canopy, small turbulence

and microclimates. Here the microclimate of an individual plot will depend on the irrigation treatment received by the plot itself, that received by the upwind plots, and lastly, the evaporating conditions in the nonirrigated surrounding area. The general effect will be that the effect of individual treatments on the microclimate will be reduced so that the plant response to soil water may well be quite different in small plot experiments from large-scale field practice. For this reason caution is needed before extrapolating plant responses to soil water from small plot, irrigation experiments to large-scale commercial practice.

This caution applies not only to the absolute values of transpiration and yield thus obtained but also to the yield response vs. water application curve.

In an attempt to summarize this chapter, the plant, soil, and climatic factors favoring a marked and slight plant response to soil water will be listed (Table 23-1), without any attempt at any order of importance. Similar tables have been outlined previously (Hagan and Vaadia, 1960).

LITERATURE CITED

Abd El Rahman, A. A., P. J. C. Kuiper, and J. F. Bierhuizen. 1959. Preliminary observations on the effect of light intensity and photoperiod on transpiration and growth of young tomato plants under controlled conditions. Mededel. Landbouwhogesch. Wageningen 59. 11:1–12.

Bierhuizen, J. F. 1961. Plant growth and soil moisture relationships. UNESCO Arid Zone Res. 16:309–315.

Bierhuizen, J. F., and N. M. de Vos. 1958. The effect of soil moisture on the growth and yield of vegetable crops. Rep. Conf. Suppl. Irrigation, Comm. VI. Int. Soc. Soil Sci., Copenhagen. p. 83–92.

Boon-Long, T. S. 1954. Transpiration as influenced by osmotic concentration and cell permeability. Amer. J. Bot. 28:53–62.

Closs, R. L. 1958. Transpiration from plants with a limited water supply. In Climatology and microclimatology. UNESCO Arid Zone Res. 11, Proc. Canberra Symp. p. 168–171.

Dainty, J. 1960. Electrical analogues in biology. Models and analogues in biology. Symp. Soc. Exp. Biol. No. 14. Cambridge Univ. Press, London. p. 140–151.

Denmead, O. T., and R. H. Shaw. 1962. Availability of soil water to plants as affected by soil moisture content and meterological conditions. Agron. J. 45:385–390.

Dixon, H. H. 1914. Transpiration and the ascent of sap in plants. MacMillan, London. 216 p.

Gaastra, P. 1963. Climatic control of photosynthesis and respiration. p. 113–140. In L. T. Evans [ed.] Environmental control of plant growth. Academic Press, New York.

Gardner, W. R., and C. F. Ehlig. 1962. Some observations on the movement of water to plant roots. Agron. J. 54:453–456.

Hagan, R. M., M. L. Peterson, R. P. Upchurch, and L. G. Jones. 1957. Relationships of moisture stress to different aspects of growth in Ladino clover. Soil Sci. Soc. Amer Proc. Vol. 21:360–365.

Hagan, R. M., and Y. Vaadia. 1960. Principles of irrigated cropping. UNESCO Arid Zone Research 15:215–225.

Heath, O. V. S. 1959. The water relations of stomatal cells and the mechanism of stomatal movement. Ch. 3, p. 193–250. In F. C. Steward [ed.] Plant physiology: A treatise. Vol. 2. Academic Press, New York.

Heath, O. V. S., and H. Meidner. 1961. The influence of water-strain on the minimum intercellular space carbon dioxide concentration and stomatal movement in wheat leaves. J. Exp. Bot. 12:226–242.

Hendrickson, A. H., and F. J. Veihmeyer. 1929. Irrigation experiments with peaches in California. California Agr. Exp. Sta. Bull. 479–1. p. 3–56.

Honert, T. H. van den. 1948. Water transport in plants as a catenary process. Discuss. Faraday Soc. 3:146–153.

Huber, B., and J. Pommer. 1954. Zur frage eines jahresgeitlichen ganges im CO_2— Gehalt der atmosphare. Angew. Bot. 28:53–62.

Jensen, R. D., S. A. Taylor, and H. H. Wiebe. 1961. Negative transport and resistance to water flow through plants. Plant Physiol. 36:633–638.

Ketallapper, H. J. 1959. The mechanism of stomatal movement. Amer. J. Bot. 46:225–231.

Kramer, P. J. 1963. Water stress and plant growth. Agron. J. 55:31–35.

Lemon, E. R. 1963. Energy and water balance of plant communities. p. 55–78. In L. T. Evans [ed.] Environmental control of plant growth. Academic Press, New York.

Rawlins, S. L. 1963. Resistance to water flow in the transpiration stream. In Stomata and water relations in plants. Connecticut Agr. Exp. Sta. Bull. 664. p. 69–85.

Salter, P. J., and J. E. Goode. 1967. Crop responses to water at different stages of growth. Commonwealth Bur. Hort. Plantation Crops, Res. Rev. 2. 246 p.

Shinn, J. H., and E. R. Lemon. Sept. 1962. Studies of water relations in a corn field in Ellis Hollow, Ithaca, N.Y. Interim Rep. 62–68. US Dep. Agr. Res. Rep. 356. 72 p.

Stanhill, G. 1957. The effect of differences in soil moisture on plant growth: A review and analysis of soil moisture regime experiments. Soil Sci. 84:205–214.

Vaadia, Y., F. C. Raney, and R. M. Hagan. 1961. Plant water deficits and physiological processes. Ann. Rev. Plant Physiol. 12:265–292.

Veihmeyer, F. J. 1927. Some factors affecting the irrigation requirements of deciduous orchards. Hilgardia 2:125–191.

Veihmeyer, F. J., and A. H. Hendrickson. 1950. Soil moisture in relation to plant growth. Ann. Rev. Plant Physiol. 1:285–305.

Wadleigh, C. H., and A. D. Ayers. 1945. Growth and biochemical composition of bean plants as conditioned by soil moisture tension and salt concentration. Plant Physiol. 20:106–132.

Wadleigh, C. H., H. G. Gauch, and O. C. Magistad. 1946. Growth and rubber accumulation in guayule as conditioned by soil salinity and irrigation regime. US Dep. Agr. Tech. Bull. 925. 34 p.

Watson, D. J. 1952. The physiological basis of variation in yields. Advance. Agron. 4:101–145.

Widstoe, J. A., and L. A. Merrill. 1912. The yields of crops with different quantities of irrigation water. Utah. Agr. Coll. Exp. Sta. Bull. 117:69–118.

Woodward, J. 1699. Some thoughts and experiments concerning vegetation. Phil. Trans. Roy. Soc. 21. No. 253, p. 193–227.

24

Nutrient Availability in Relation to Soil Water[1]

FRANK G. VIETS, JR.

Agricultural Research Service, USDA
Fort Collins, Colorado

I. INTRODUCTION

Water and the 13 mineral nutrients generally regarded as essential for higher plants are complexly intertwined in their effects on growth and reproduction. All are essential and yet so interdependent that one cannot be considered without the others during their transport from the soil to the roots, absorption by the roots, and translocation in the plant. A deficiency of one or more nutrients can arrest root development so that water extraction from the soil is impaired, particularly from deeper zones. Nutrient deficiency can restrict the development of stems and leaves so that transpiration will be less than the potential evapotranspiration set by the heat available in the environment.

Conversely, availability of water is of great significance to the plant's need for and ability to absorb nutrients, and the soil's ability to supply them. An extreme deficiency of soil water can cause wilting and ultimate death of the plant; but before such obvious effects set in, the status of nutrients in the soil and the plant's ability to get them may be impaired. Some effects of soil water on nutrient availability are readily apparent, such as the occurrence of boron-deficiency symptoms of sensitive crops on some dry soils and the failure of soluble sidedressed fertilizers to produce expected response until rainfall occurs or the soil is irrigated. In the main, however, the effects of water on nutrient uptake are not easily discerned except by careful growth measurements and chemical analysis of plants.

This chapter discusses the effect of soil water content and suction on the pools of nutrient elements in the soil, their transport to and absorption by the root and, finally, the relation of available water to total nutrient requirements of crops. Strictly speaking, the amount of nutrient the plant can absorb from the soil is termed nutrient availability. It depends on the kind of plant, a multitude of conditions affecting plant growth and root extension, the position of the nutrient in relation to the root system, and the chemical and microbial reactions of the nutrient in the soil. Availability of a nutrient is often confused with the amount of nutrient that can be extracted from the soil with some solvent like water or a salt solution, or with the amount added to a soil in a fertilizer. Extractable nutrients and additions of soluble fertilizers are more properly called indexes of availability since an inference is required that a plant can absorb them predictably.

---

[1] Contribution of the Soil and Water Conservation Research Division, Agricultural Research Service, U. S. Department of Agriculture.

II. EFFECTS OF WATER ON THE POOLS OF ESSENTIAL NUTRIENTS

A nutrient must be in an "available" form before a root can absorb it. That is a simple statement, but to give it precise meaning is impossible now. All ions in the soil solution like NO_3^-, Cl^-, and SO_4^{2-} are generally regarded as available for absorption by the root. Ions that can be extracted with weak ion exchangers, such as NH_4OAc, are also regarded as available, but the availability depends on the degree of saturation of the exchange complex. However, roots can also absorb elements such as Fe, Zn, and Cu from soils and minerals that have extremely low solubilities in water and weak extracting agents. Viets (1962a) discussed these arbitrary pools of nutrients and their significance in nutrient availability. Chemists measure these pools and combinations of them and correlate the data with response of crops to fertilizers and nutrient uptake in developing soil tests.

Water, containing CO_2, organic acids, and natural chelating agents produced by microbes and roots, is the solvent for the minerals in the soil. Water content of the soil has both short- and long-term effects on the distribution of elements among these pools in the soil. The short-term effects are simply matters of distribution of ions between the solution and the adsorbed phase or solid phase. For example, all of the NO_3^- is in the solution phase except in very dry soils. Little of the P is in the solution phase at any soil water content, but the P from the adsorbed or solid phase can maintain the concentration of the solution phase as P is removed by roots. The long-term effects of water on distribution of nutrients among the pools of soluble, adsorbed, and solid forms are related to microbial activity and the effect of water and oxygen on this microbial activity.

One of the problems of assessing either the short- or the long-term effects of soil water content or suction on the pools of nutrients is the extraction of the true soil solution existing in the water suction range from a fraction of a bar to about 15 bars, the approximate wilting point. Reitemeier and Richards (1944) showed that the concentrations of Mg, Na, K, and Cl in pressure membrane extracts of a soil extracted at water contents of 2.85, 1.78, and 1.24 times the 15-atm percentage increased with a decrease in the initial water content. Successive aliquots of extract obtained from a soil extracted at a given moisture content did not change materially in concentration. This procedure was not suited to measurement of P in solution because of adsorption on the membrane. Measurement of Cu, Zn, Mn, Co, and Fe in the soil solutions would also pose the same and other problems.

Much more is known about extractable nutrients under soil conditions of extreme wetness or extreme dryness. Under wet conditions, the concentration of some elements increase and others decrease. Both chemical and microbial reactions are involved and are affected by the concentration of dissolved oxygen. If oxygen demand by organisms and roots is high, the redox potential drops and the solution concentrations of NH_4^+, P, Fe, Mn, and perhaps Mo, increase (Kee and Bloomfield, 1962; Robinson, 1930). Concurrently, NO_3^- is denitrified and if the reducing conditions continue, SO_4^{2-} is reduced to toxic H_2S.

Under dry conditions, as met with when soils are air-dried, some soils will fix NH_4^+ against extraction with cation exchangers like NaOAc. Some soils will also fix K^+, but many soils will release K^+ to the exchangeable form (Hanway and Scott, 1957). On many soils, it is not safe to estimate available nutrients on air-

or oven-dried samples when predictions are to be made of available nutrients on soils that never dry (Hanway et al., 1961), or on soils that are kept waterlogged as in lowland rice (*Oryza sativa* L.) production.

III. WATER AND THE TRANSPORT OF NUTRIENTS

Water and the transport of nutrients dissolved in it has two important aspects with respect to plant nutrition. The first aspect is the mass transport of soluble ions in moving water. All of the NO_3^- and Cl^-, most of the SO_4^{2-}, and part of the boron (B) are dissolved in the soil water and are free to move with it. There must be an equivalence of soluble cation to move with them. These anions and associated cations are readily leachable in percolating waters from excessive irrigation or rain. In furrow irrigation, salts can accumulate in ridges between rows as water moves into them and is subsequently evaporated. Advantage of this is taken in escaping salinity in soils by planting crops near the irrigation furrow so that salts move or have moved past the plants into the ridge. Accumulation of NO_3^- in ridges where there are no roots frequently accounts for poor response to N fertilizers and low N recovery by the crop. Fall rains leaching this accumulated NO_3^- into the root zone can account for low sugar content in sugar beets (*Beta vulgaris* L.) (Stout, 1961). Another manifestation of the role of water is that topdressed or sidedressed applications of soluble fertilizers are usually ineffective until rain or irrigation water washes them into the root zone.

The other elements are affected much less or not at all by the movement of percolating water. These nutrients are termed nonmobile, a term that has relative meaning only. Most of the ammonium nitrogen, K^+, Ca^{2+}, and Mg^{2+} is held on the exchange complex of clay or organic matter in reversible equilibrium with the ions in the soil solution. These elements, relatively speaking, are unleachable in most irrigated soils. Copper, zinc, and cobalt are so tightly held by mineral soils that only traces are found in the soil solution. In fact, these elements can be removed from water by passing the water through soil.

The second aspect of the significance of water in nutrition is related to water content, soil water suction, and the thickness of water films as they affect ion transport to the root surface. A wet soil simply has more cross sectional area filled with water through which ions can diffuse than a dry soil of the same texture and bulk density. One of the problems in trying to simplify this discussion of the role of water in transport, as well as size of the nutrient pools previously noted, is whether an ion must pass through the solution (aqueous) phase or can be absorbed directly from the solid phase by exchange of an ion on the root surface for one on the soil particle as proposed by Jenny and Overstreet (1939). The question is not settled for all ions, and perhaps for none, that roots absorb ions only from the solution phase. Dean and Rubins (1945) could find no evidence for a contact effect involving phosphate. Since plants can get nutrients from compounds like MnO_2, $Fe(OH)_3$, CuS, and Zn compounds soluble only in 0.1N HCl (Boawn et al., 1957) there can be little question but that microregion phenomena (Glauser and Jenny, 1960) implicating slow equilibrium and diffusion in the solution phase are involved.

Barber (1962) gelled the ideas of many others into a concept of ion diffusion and mass flow as quantitatively additive mechanisms to account for transport of

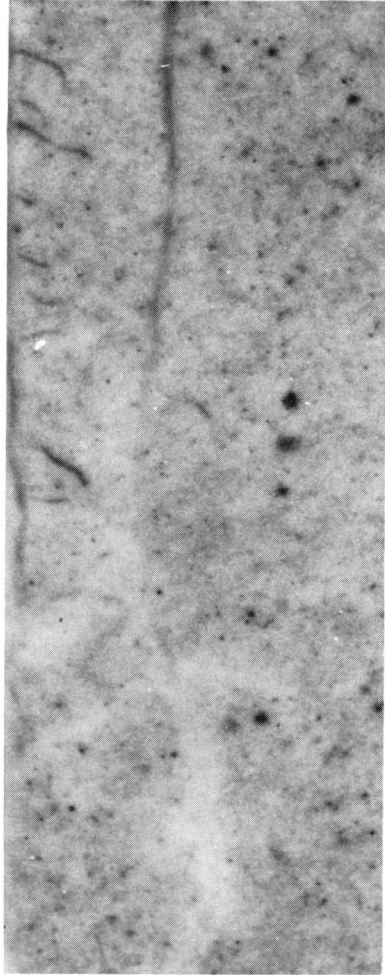

Fig. 24–1. Autoradiograph of Mo distribution about a corn root when the soil was uniformly labelled with Mo. Lavy and Barber (1964) interpreted the clear areas near the root tip as areas of removal by diffusion to the root tip. The dark areas above the root tip were interpreted as areas of accumulation in which mass flow transported Mo to the root surface faster than it was absorbed.

ions from adsorption surfaces or the soil solution to the root surface. Diffusion is the movement of ions in response to a concentration gradient of the ion species toward the root which continually absorbs ions and lowers the concentration at the root surface. Diffusion can be away from the root if mass flow moves ions to the root surface faster than they are absorbed. See Fig. 24–1. Mass flow is the transport of ions by bulk or viscous flow of soil water to the root surface along potential gradients for water. Barber calculated that for some ions mass flow could account for all transport, whereas for others, diffusion would have to be involved to account for the amount absorbed. Rate of transport by mass flow will depend on soil water suction in direct proportion to the extent that soil water suction limits water absorption by the root.

Diffusion is directly related to soil water content through its effects on tortuosity and the cross sectional area containing water. Brown (1953) studied adsorption of Ca, Mg, K, and Na by H-saturated resin membranes embedded in seven soils at different moisture contents. He found the amount of ion exchange in 96 hours to be markedly dependent on soil water content. For example, in a Crowley silt loam, uptake of the cations by the resin at different water contents relative to that at the moisture equivalent was: saturation, 340%; 75% available water, 64%; 50% available water, 50%; 25% available water, 36%; and 0 available water (15 bars tension), 28%. Porter et al. (1960) found that the diffusivity of Cl⁻ in each of three Ca-saturated and three sodic soils decreased as volumetric water content decreased in the suction range of 1/3 to 15 bars. Olsen et al. (1965) showed that the diffusion coefficient for phosphate decreased about eightfold for a twofold drop in volumetric moisture content for two soils. The apparent diffusion coefficients of Rb (Place and Barber, 1964) and molybdate (Lavy and Barber, 1964) depended on the water content of the soil.

Root extension decreases enormously the distances over which ions must be transported by diffusion or mass flow. Extension is confined exclusively to the direction of root growth as the absorbing portion of the root increases little in diameter while it is capable of absorption. Distance covered by root extension is great compared to distances through which ions diffuse or water moves in the soil, particularly at suctions > 1 bar. The autoradiographs of Walker and Barber (1961) of roots of 12-day-old corn plants (*Zea mays* L.) show zones of depletion of Rb[86] attributed to diffusion extending about 5 mm from root surfaces. The diffusion curves of Olsen et al. (1962) predict little depletion of soluble P by diffusion at distances > 1 mm from the root surface. In contrast, some examples of rates of root elongation are: 1.9 to 5.1 cm/day for grain sorghum (*Sorghum vulgare* Pers.) (Nakayama and Van Bavel, 1963), 1.27 cm/day for 70 days for wheat (*Triticum aestivum* L.) (Weaver et al., 1924), 5.1 to 6.5 cm/day for 3 or 4 weeks for corn (Weaver, 1925), and 2.84 cm/day for a 50-day period for peanuts (*Arachis hypogaea* L.) (Lea, 1961). Rate of root elongation is highly dependent on available water because root growth is in part a hydration process. For example, Peters (1957) showed that elongation was favored by low soil water suction and high water content at a given suction. Increase in soil water suction also increases the shearing strength of soils and may increase the mechanical impedance to root extension, particularly in dense soils (Barley, 1963). Other factors affecting rate of root extension are nutrient deficiencies, salinity, and soil temperature.

Even though roots appear to thoroughly permeate the soil and can extract all of the water uniformly in most soils down to the wilting range, the effective volume of soil that contributes relatively immobile nutrients such as P or Zn to the plant is low. Wiersum (1962) states that this may be 1% to 5% of the soil volume. Barber et al. (1963) claim that roots grow to < 3% of the available nutrients in the soil.

In summary, all evidence indicates that low soil water suction should be most favorable for rapid uptake of nutrients by plants. The specific factors favorably affected are the size of the nutrient pool, the diffusion rate of ions, the extension of root systems, and the mass flow of water. However, as water saturation of the soil is approached, O_2 diffusion may be so limited that physiological processes in roots of most species are inhibited.

IV. NUTRIENT UPTAKE AT NEARLY CONSTANT SOIL WATER SUCTION

One of the experimental difficulties in studying nutrient availability in relation to soil water over the whole range of available water is that transpiration and water absorption cause the suction to continually increase. When water is applied, another drying cycle starts. The result is that nutrient uptake can be measured only for water regimes, for example, 25% vs. 50% depletion of the available soil water, and little is known about nutrient uptake at a specific suction.

In attempts to hold water potential constant, some investigators have studied nutrient uptake of nontranspiring seedlings transplanted into boxes containing soil wetted to predetermined water contents or suction. Some growth is invariably involved that requires water uptake and the availability of water affects elongation, so that at best it is a system in which soil water suction changes are minimized rather than eliminated. Danielson and Russell (1957) studied the 24-hour uptake of Rb[86] by corn seedlings in closed boxes over a soil water suction range of 1/3 to 12 atm. Uptake of Rb[86] and water content of the seedlings decreased almost logarithmically with an increase in suction, uptake decreasing sharply at about 1 atm. Uptake of Rb was proportional to water content of the soil over the whole suction range. Osmotic suction induced with mannitol in solution did not produce similar depressing effects. They concluded, "It is probable that the rate of ion diffusion as influenced by the thickness of the moisture films rather than the moisture stress *per se* is responsible for controlling the ion absorption rate." Olsen et al. (1961), using identical technique, showed that P[32] uptake by corn was proportional to water content in each of four soils. Wiersum (1958) also found that uptake of Rb[86] by excised roots of *Vicia faba* depended on the water content of sand. Dean and Gledhill (1956) devised a technique of pressing excised rye roots (*Secale cereale*), grown in quartz sand and then washed free of sand, against the smooth face of soil at different water contents to study the uptake of P[32] for periods up to 8 hours. Roots continually lost water to the soil regardless of its water content and gave somewhat anomalous results for P[32] uptake.

Other techniques to study uptake of nutrients at a near-constant water content involve a system in which the plant gets all or most of its water through some part of the root system other than the part studied. Mederski and Wilson (1960) grew corn for 25 days in a split-medium system in which the surface roots in quartz sand were separated by a water-impermeable barrier from the lower part of the root system in soil of preestablished water content. Most of the transpired water came from the frequently watered sand and water absorption from the soil was greatly minimized. They found that dry weight of tops and roots increased linearly with increasing soil water and that both total uptake and percentage content of P, K, and Mg increased with increasing soil water. Hobbs and Bertramson (1950) divided the root systems (split-root technique) of tomatoes (*Lycopersicon esculentum*) longitudinally and put half in surface soil and half in quartz sand or subsoil. Little uptake of native or fertilizer B occurred from the surface soil if it was kept dry and the plant got its water from the subsoil or sand. Boron deficiency symptoms were induced by keeping the surface soil dry, but no deficiency symptoms were present if the surface soil was kept wet.

Some studies with split-root or split-medium techniques in which water content was held almost constant dealt with nutrient uptake from a part of the soil at or drier than the wilting point. In such experiments, there is always the probability of water transfer from wet soil through the plant to dry soil surrounding the roots as pointed out by Breazeale (1930), and affirmed by Volk (1947) and Hunter and Kelley (1946a). Under such experimental conditions, uptakes of K (Breazeale, 1930) and of K and N (Volk, 1947) were shown. Hunter and Kelley (1946a), Volk (1947), and Boatwright et al. (1964) obtained no significant uptake of P from dry soils. Roots have been observed to grow into dry soils (Hunter and Kelley, 1946a) but growth rates were extremely slow. Hunter and Kelley (1946b), using specially constructed apparatus, showed that guayule (*Parthenium argentatum*) could absorb P^{32} from moist soil more than 122 cm below the surface and transport it to the plant top through 122 cm of soil at or below the wilting percentage.

When nutrient uptake has been observed from soils where the soil mass was at or below the wilting point, the amount of uptake has been small or only a fraction of that of soils near field capacity. In studies on nontranspiring systems that involved equal soil water throughout the whole system mentioned above (*see also* Gingrich and Russell, 1956, 1957), the effects of soil water suction were usually studied up to 10 or 12 bars. At this maximum suction, root elongation was sharply reduced. At the wilting point, root elongation would probably have been zero. A reasonable conclusion is that root extension into and ion uptake from dry soils is largely due to water transported in the root system to soil particles around the root. The soil around the root is not, therefore, as dry as the wilting point.

Under field conditions, plants often have part of their roots in dry soil and part in wet soil. This situation is magnified by skip-row irrigation or failure to wet the soil between furrows in every-row irrigation. This dry soil is not active in nutrition of the plant and usable soil volume is thereby reduced. The importance of this reduction depends on whether the whole volume of soil is needed to supply nutrients.

V.   NUTRIENT AVAILABILITY AND WATER REGIMES

Continual removal of water from the soil by plants and evaporation produces a linear or almost linear decrease in water content and a logarithmic increase in water suction until rain or irrigation restores the water balance, usually quickly. Since constant water content and suction can be maintained by subirrigation or some system of autoirrigation only at low suction where the hydraulic conductivity is high, the best that can be done in studying plant reaction to variations in water suction characteristic of natural growth conditions of most crops is to pit one suction or water content regime against another; e.g., extraction to 1 bar vs. 10 bars or 25% vs. 75% available water depletion. Such experiments are difficult to interpret, but are useful. One of the complexities is that rapid nutrient absorption immediately after irrigation or rain may completely obscure slower nutrient uptake at higher soil water suctions. If the plant can fulfill its capacity for nutrient absorption when the suction is low, it may have little capacity to absorb them when the stress is high.

No hour-by-hour or day-by-day studies of nutrient uptake in relation to simultaneous measurement of suction have been conducted. If the plant can supply its nutrient needs during periods of low suction, then periods of reduced nutrient availability during water stress may be of little importance to it. Short-term studies of absorption of some nutrients by roots indicate that they have a capacity for nutrient absorption far in excess of the plant's total seasonal needs. Another complexity of water regime experiments is that ones with lower average suction generally produce greater growth, although there are some important exceptions where seed or fruit production rather than total growth are the criterion of judgment (Stanhill, 1957). Increased growth is generally associated with increased root production which, in turn, increases the accessibility of the native immobile nutrients. Availability of nutrients in fertilizers is affected less by soil water regime because fertilizers are incorporated or banded in the surface soil where root density is greatest. Increased growth also dilutes the concentration in the plant which may increase the roots' capacity to absorb nutrients and the diffusion gradient in the soil solution to the root surfaces (Grunes, 1959).

If low water suction increases growth, then almost invariably the total nutrient uptake is increased and so, by strict definition, nutrient availability has been increased. If nutrient concentration (percentage of the dry weight or ppm) drops as plant growth is increased, then certainly nutrient availability has not kept pace with growth. Conversely, an increase in nutrient concentration means that nutrient availability has increased more than growth. Both concentration and total uptake data are needed in making interpretations. Richards and Wadleigh (1952) summarized the available data on soil water-nutrient availability relations and concluded that decreasing water supply produced a definite increase in N concentration, a definite decrease in the K concentration, and variable effects on the Ca, Mg, and P concentrations in the plant.

Some of the best information on soil water in relation to nutrient availability has been obtained where fertilizers were labeled with radioisotopes such as P^{32}. Such experiments, however, are subject to the effects of differential growth induced by the differences in water regime. This differential growth may affect the relative proportions of nutrients coming from the fertilizer, compared to the soil (Grunes, 1959). Haddock (1952) showed that high soil water (< 750 cm water suction at 15 cm depth) produced higher soluble P in beet petioles than drier moisture regimes. With broadcast application of superphosphate to a calcareous soil, there was little effect of soil water regime on the ratio of fertilizer to soil P absorbed. With P sidedressed 10 cm deep and 15 cm to the side of the row at thinning, $< 24\%$ of the leaf P came from the fertilizer on a wet regime, but $> 40\%$ came from the fertilizer on a dry regime. Presumably, the availability of the soil P was reduced in the drier soil causing the plants to extract more of their P from the fertilized zone of higher soluble P concentration. Power et al. (1961) found that the soil water regime had a large effect on the ratio of fertilizer P (P^{32}-tagged) to soil P taken up by spring wheat. The four water regimes were two amounts of soil water at seeding in factorial combination with two amounts of irrigation water to imitate seasonal precipitation. Water regime did not affect the amount of P taken up from the fertilizer, but there was more soil P taken up from the wetter soils so that the relative contribution of the soil was greater the wetter the soil. Lipps et al. (1957) investigated the uptake of P^{32} placed at different depths in the soil by subirrigated 3-year-old alfalfa (*Medicago sativa* L.).

The water table was stable at about 240 cm. Up to late June when the whole profile contained available water, P^{32} uptake from the first 46 cm was very active and there was virtually none from the lower depths. In July and August when soil water in the upper layers was depleted, uptake of P^{32} from the wetter, lower depths became very active. All of these studies are consistent with the concept that roots cannot absorb much nutrient from dry soils because of lack of root activity and slow rates of ion diffusion and water movement.

Results contrary to those cited were obtained by Fawcett and Quirk (1960, 1962) who found that uptake of P^{32}, applied at three different rates to a soil very low in total P, by wheat in small beakers was independent of soil water stress unless roots were damaged by plant wilting. They took due notice of the effect of high water suction on reduced diffusion rate of P and hypothesized that the lack of effect of soil water stress was due to soluble and adsorbed P being higher in the small pores.

VI.   NUTRIENT AVAILABILITY IN RELATION TO EXCESS WATER

If infiltrated rainfall or irrigation water exceeds the storage capacity of the root zone, this water can either percolate through the soil, if internal drainage of the profile is good, or else accumulate and waterlog the soil. Excessive percolation affects nutrient availability through loss of soluble nutrients. Quantitatively, nitrate is the largest anionic loss but losses of sulfate can be of nutritional significance. The cations lost are K, Na, Ca, and Mg. Phosphate and the micronutrient cations are almost completely retained. Some leaching of soluble components is desirable in arid and semiarid soils to avoid accumulation of salts. The amounts of excess water needed to maintain a desired salt level can be calculated and is known as the leaching requirement.

If the soil cannot drain, the air porosity of the soil is reduced and all or most of the oxygen may be used depending on root and microbial activity. The effect of waterlogging and poor aeration on the solubility of the nutrient elements has been discussed. The significance of the effect on soluble-nutrient pools to the plant and nutrient uptake or availability depends on the plant's growth tolerance of this poorly aerated root environment. For mesophytes, poor aeration means less growth, root development, and nutrient uptake of most elements and, hence, reduced nutrient availability.

Rice affords an interesting opportunity to study nutrient availability under submerged vs. well-aerated conditions since it can be grown under the two types of culture. Shapiro (1958) working with 11 different soils in containers with the same varieties of rice found that flooding increased yield, total P uptake and concentration of P in the plant, and total N uptake. Fertilizer P and N (as urea) were utilized better under flooded conditions. Average recovery of fertilizer N applied at three rates on seven soils was 48.4% on flooded soils and 36.0% on unflooded soils. On seven unfertilized soils, average yields under flooded vs. unflooded conditions were 4.34 vs. 3.33 g/pot, total N in the tops was 40.8 vs. 27.3 mg, and the N percentage was 0.94 vs. 0.81. Response to P fertilizer is frequently obtained on nonflooded rice fields that will not respond to P if flooded. Nearpass and Clark (1960) found that rice on five soils took up less S when flooded than when unflooded. The depressing effect of flooding on S uptake was increased by adding dried, ground rice hulls to increase SO_4^{2-} reduction; SO_4^{2-} can be reduced in flooded soils and S deficiency may result.

This depressing effect of organic matter on yields under flooding can be offset by addition of SO_4^{2-}. In similar studies of rice growth in upland vs. submerged culture, Clark et al. (1957) concluded that the better growth of flooded rice on some soils is due to higher availability of Mn.

Excessive water, directly or indirectly, aggravates lime-induced chlorosis, an Fe deficiency and nutritional disturbance of sensitive crops such as fruit trees, beans, corn, and sorghum grown on calcareous soils. Chlorosis is generally worse on wet soils, and plants often become greener if the soil is allowed to dry. Chlorosis can be corrected by soluble Fe compounds applied to the foliage or Fe chelates applied to the soil. The role of water in iron chlorosis has not been clarified but the HCO_3^- associated with high CO_2 in wet soils has been implicated (Brown, 1961).

VII. MOISTURE REGIME AND NUTRIENT CONTENT IN PLANT PARTS

Chemical composition of petioles and leaves (foliar analysis) is often used as a guide to plant nutritional status and to future fertilizer practices. Mineral content of tissues has been studied in relation to water regimes in the soil. From such data, inferences frequently cannot be drawn about nutrient availability *per se* because plant size and total nutrient uptake may be affected by irrigation regime, but such data are nevertheless useful as an indication of the effect of water on nutrition. Haddock (1952, 1953) and Kelley and Haddock (1954) showed that the soluble P concentration of sugar beet petioles was highest, and soluble N content lowest, on high water regimes. For example, petioles contained 13,000, 900, and 500 ppm of soluble P if beets were irrigated when gypsum blocks, placed at the 15-cm depth in the row, read about 1,000, 10,000, and 100,000 ohms, respectively (Haddock, 1952). Soluble P in petioles should be kept above 1,000 ppm on a dry weight basis for good beet yields. High sugar content of the root and purity of the juice are related to low soluble N or NO_3^- in beet petioles about 1 month before harvest. This example of sugar beets illustrates that soil water availability does affect the nutrition of the plant and can affect nutrient contents in tissues commonly used for foliar analysis.

VIII. SIGNIFICANCE OF NUTRIENT AVAILABILITY TO THE EXPRESSION OF SOIL WATER SUCTION EFFECTS ON PLANT GROWTH

The evidence presented thus far indicates that nutrient availability is highest for most crops when soil water suction is low. The size of the soluble and exchangeable nutrient pools, transport to the root by diffusion and mass flow, root extension, and the nutritional demands of the plant, are greatest at low water suction. The plant growing in a fertile soil can therefore meet its needs for more nutrients when water conditions are made more favorable by the ability of its roots to get them and the soil to supply them. This is especially true of the nonmobile nutrients like P, K, and the micronutrients. Cases where significant positive interactions of nonmobile nutrient availability with soil water supply have been demonstrated are rare except on sands where the nutrient reserves of all elements

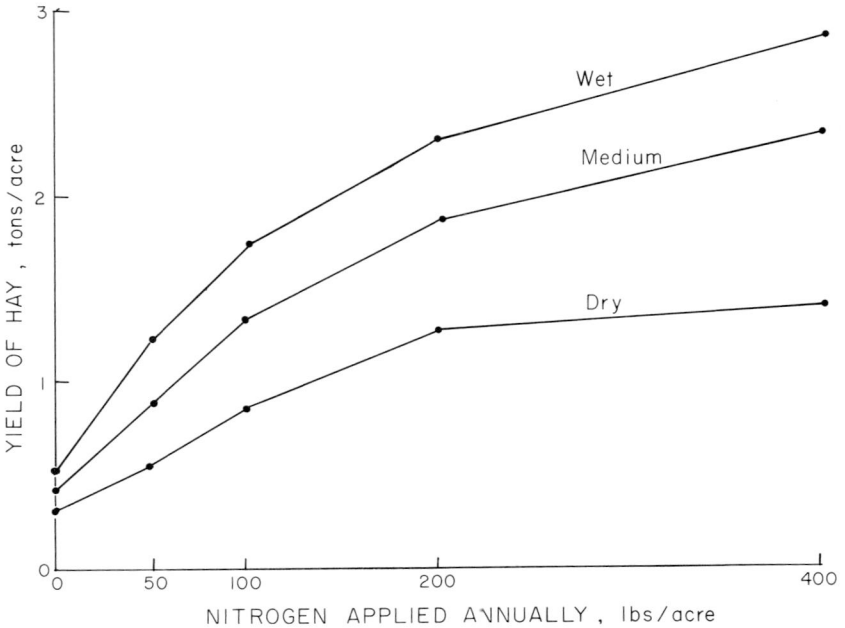

Fig. 24–2. Annual yield of western wheatgrass hay was increased by N fertilization, the increase being greatest at low soil water suction (1 lb/acre = 1.12 kg/hectare).

may be extremely low. For example, rarely has it been shown that the exchangeable K, soluble P, or the rate of application of fertilizers containing these elements, need be higher for ample water than for limited water.

However, N presents a different story. The reasons for this are at least three: N deficiency is more common, most N is absorbed from the highly mobile NO_3^- form, and the N requirements of the plant are high. Simply stated, a corn crop that yields 2,800 lb of grain removes about 65 lb of N per acre, whereas a crop with 11,200 lb of grain contains more than four times as much N. Positive interactions of N with soil water regime are more common than lack of interaction. Figures 24–2 and 24–3 show a typical example, western wheatgrass (*Agropyron smithii*) for hay in Montana, USA on a clay soil (Schumaker and Davis, 1961). The moisture regimes were established by irrigating when the available water in the top 1 foot of soil dropped to: (i) dry, 25%; (ii) medium, 50%; and (iii) wet, 75% of field capacity. Without fertilizer N, there was only a small benefit from keeping water suction low, but as more fertilizer N was used the beneficial effect of low water suction became greater (Fig. 24–2). The same effects are less apparent in the total N uptake for the 4-year period (Fig. 24–3).

The effects of soil water supply on nutrient availability and requirements discussed in the preceding paragraphs pertained to the short-term effects of the water regime: What happens in 1 or 2 years? The longer-term effects of soil water supply may be quite different. If a larger crop is grown, more nutrients will be removed from the field and some deficiency may develop. The order in which deficiency appears depends on the soil and the cropping system. When nutrient availability does not keep pace with increased crop yield brought about by irriga-

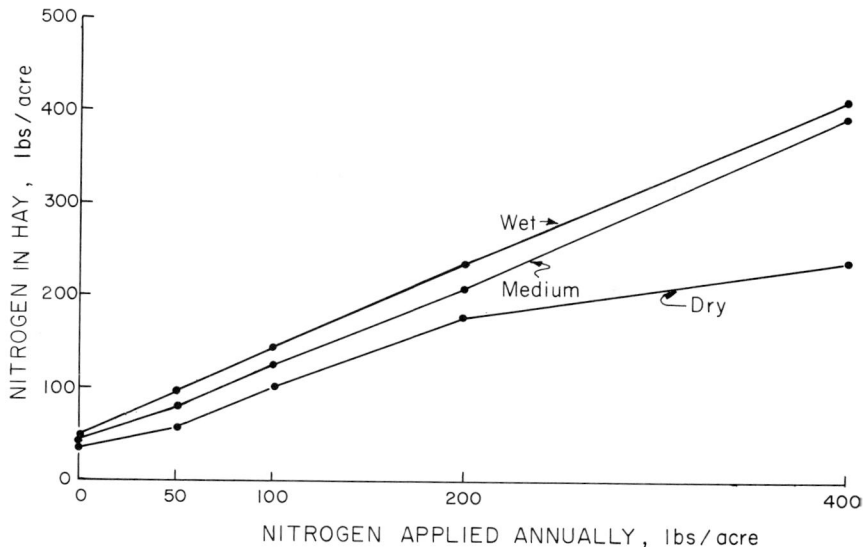

Fig. 24–3. Total N in western wheatgrass hay for a 4-year period was increased by annual applications of N fertilizer and by maintenance of low soil water suction (1 lb/acre = 1.12 kg/hectare).

tion, as in the transition from nonirrigated to irrigated farming, the potential yield of the new environment will not be realized. Furthermore, after several years, the quality of the crop with respect to minerals essential to livestock may decline. This is especially apparent in N or crude protein content of grain or forage. Without N fertilization, corn grain produced under irrigation may contain only 1% N, whereas the normal content is about 2%. The only answer is to adjust nutrient availability by fertilization so that the yield and quality of crop is almost equal to the potential of the climate with its new man-made soil water regimes, due regard being given to the economics of such attainment. Correction of nutrient deficiencies with appropriate fertilizers is one way to economically increase the efficiency of water used in evapotranspiration (Viets, 1962b).

LITERATURE CITED

Barber, Stanley A. 1962. A diffusion and mass-flow concept of soil nutrient availability. Soil Sci. 93:39–49.

Barber, S. A., J. M. Walker, and E. H. Vasey. 1963. Mechanisms for the movement of plant nutrients from the soil and fertilizer to the plant root. J. Agr. Food Chem. 11:204–207.

Barley, K. P. 1963. Influence of soil strength on growth of roots. Soil Sci. 96:175–180.

Boatwright, G. O., H. Ferguson, and P. L. Brown. 1964. Availability of P from superphosphate to spring wheat as affected by growth stage and surface soil moisture. Soil Sci. Soc. Amer. Proc. 28:403–405.

Boawn, L. C., F. G. Viets, Jr., and C. L. Crawford. 1957. Plant utilization of zinc from various types of zinc compounds and fertilizer materials. Soil Sci. 83:219–227.

Breazeale, J. F. 1930. Maintenance of moisture equilibrium and nutrition of plants at and below the wilting percentage. Arizona Agr. Exp. Sta. Tech. Bull. 29. p. 137–177.

Brown, D. A. 1953. Cation exchange in soils through the moisture range, saturation to the wilting percentage. Soil Sci. Soc. Amer. Proc. 17:92–96.

Brown, John C. 1961. Iron chlorosis in plants. Advance. Agron. 13:329–369.

Clark, Francis, D. C. Nearpass, and A. W. Specht. 1957. Influence of organic additions and flooding on iron and manganese uptake by rice. Agron. J. 49:586–589.

Danielson, R. E., and M. B. Russell. 1957. Ion absorption by corn roots as influenced by moisture and aeration. Soil Sci. Soc. Amer. Proc. 21:3–6.

Dean, L. A., and V. H. Gledhill. 1956. Influence of soil moisture on phosphate absorption as measured by an excised root technique. Soil Sci. 82:71–79.

Dean, L. A., and E. J. Rubins. 1945. Absorption by plants of phosphorus from a clay-water system: Methods of ensuing observations. Soil Sci. 59:437–448.

Fawcett, R. G., and J. P. Quirk. 1960. Effect of water-stress on the absorption of soil phosphorus by wheat plants. Nature (London) 188:687–688.

Fawcett, R. G., and J. P. Quirk. 1962. The effect of soil-water stress on the absorption of soil phosphorus by wheat plants. Australian J. Agr. Res. 13:193–205.

Gingrich, J. R., and M. B. Russell. 1956. Effect of soil moisture tension and oxygen concentration on the growth of corn roots. Agron. J. 48:517–520.

Gingrich, J. R., and M. B. Russell. 1957. A comparison of effects of soil moisture tension and osmotic stress on root growth. Soil Sci. 84:185–194.

Glauser, R., and H. Jenny. 1960. Two-phase studies on availability of iron in calcareous soils: I. Experiments with alfalfa plants. Agrochimica 4:263–277.

Grunes, D. L. 1959. Effect of nitrogen on the availability of soil and fertilizer phosphorus to plants. Advance. Agron. 11:369–396.

Haddock, Jay L. 1952. The influence of soil moisture condition on the uptake of phosphorus from calcareous soils by sugar beets. Soil Sci. Soc. Amer. Proc. 16:235–238.

Haddock, Jay L. 1953. Sugar beet yield and quality as affected by plant population, soil moisture condition, and fertilization. Utah Agr. Exp. Sta. Bull. 362. p. 1–72.

Hanway, J. J., S. A. Barber, R. H. Bray, A. C. Caldwell, L. E. Englebert, R. L. Fox, M. Fried. D. Hovland, J. W. Ketcheson, W. M. Laughlin, K. Lawton, R. C. Lipps, R. A. Olson, J. T. Pesek, K. Pretty, F. W. Smith, and E. M. Stickney. 1961. North Central Region potassium studies: I. Field studies with alfalfa. N. Cent. Reg. Publ. No. 124, Iowa Agr. Exp. Sta. Res. Bull. 494. p. 163–188.

Hanway, J. J., and A. D. Scott. 1957. Soil potassium-moisture relations: II. Profile distribution of exchangeable K in Iowa soils as influenced by drying and rewetting. Soil Sci. Soc. Amer. Proc. 21:501–504.

Hobbs, J. A. and B. R. Bertramson. 1950. Boron uptake by plants as influenced by soil moisture. Soil Sci. Soc. Amer. Proc. (1949) 14:257–261.

Hunter, A. S., and Omer J. Kelley. 1946a. A new technique for studying the absorption of moisture and nutrients from soil by plant roots Soil Sci. 62:441–450.

Hunter, Albert S., and O. J. Kelley. 1946b. The extension of plant roots into dry soil. Plant Physiol. 21:445–451.

Jenny, H., and R. Overstreet. 1939. Cation interchange between plant roots and soil colloids. Soil Sci. 47:257–272.

Kee, Ng Siew, and C. Bloomfield. 1962. The effect of flooding and aeration on the mobility of certain trace elements in soils. Plant Soil XVI: 108–135.

Kelley, Omer J., and J. L. Haddock. 1954. The relation of soil water levels to mineral nutrition of sugar beets. Amer. Soc. Sugar Beet Tech., Proc. 8:344–356.

Lavy, T. L., and S. A. Barber. 1964. Movement of molybdenum in the soil and its effect on availability to the plant. Soil Sci. Soc. Amer. Proc. 28:93–97.

Lea, J. D. 1961. Studies on depth and rate of root penetration of some annual tropical crops. Tropical Agri. 38:93–105.

Lipps, R. C., R. L. Fox, and F. E. Koehler. 1957. Characterizing root activity of alfalfa by radioactive tracer techniques. Soil Sci. 84:195–204.

Mederski, H. J., and J. H. Wilson. 1960. Relation of soil moisture to ion absorption by corn plants. Soil Sci. Soc. Amer. Proc. 24:149–152.

Nakayama, F. S., and C. H. M. van Bavel. 1963. Root activity distribution patterns of sorghum and soil moisture conditions. Agron. J. 55:271–274.

Nearpass, D. C., and Francis E. Clark. 1960. Availability of sulfur to rice plants in submerged and upland soil. Soil Sci. Soc. Amer. Proc. 24:385–387.

Olsen, S. R., W. D. Kemper, and R. D. Jackson. 1962. Phosphate diffusion to plant roots. Soil Sci. Soc. Amer. Proc. 26:222–227.

Olsen, S. R., W. D. Kemper, and J. C. van Schaik. 1965. Self-diffusion coefficients of phosphorus in soil measured by transient and steady-state methods. Soil Sci. Soc. Amer. Proc. 29:154–158.

Olsen, S. R., F. S. Watanabe, and R. E. Danielson. 1961. Phosphorus absorption by corn roots as affected by moisture and phosphorus concentration. Soil Sci. Amer. Proc. 25:289–294.

Peters, D. B. 1957. Water uptake of corn roots as influenced by soil moisture content and soil moisture tension. Soil Sci. Soc. Amer. Proc. 21:481–484.

Place, G. A., and S. A. Barber. 1964. The effect of soil moisture and Rb concentration on diffusion and uptake of Rb86. Soil Sci. Soc. Amer. Proc. 28:239–243.

Porter, L. K., W. D. Kemper, R. D. Jackson, and B. A. Stewart. 1960. Chloride diffusion in soils as influenced by moisture content. Soil Sci. Soc. Amer. Proc. 24:460–463.

Power, J. F., G. A. Reichman, and D. L. Grunes. 1961. The influence of phosphorus fertilization and moisture on growth and nutrient absorption by spring wheat: II. Soil and fertilizer P uptake in plants. Soil Sci. Soc. Amer. Proc. 25:210–213.

Reitemeier, R. F., and L. A. Richards. 1944. Reliability of the pressure-membrane method for extraction of soil solution. Soil Sci. Soc. Amer. Proc. 25:119–135.

Richards, L. A., and C. H. Wadleigh. 1952. Soil water and plant growth. p. 73–251. In B. T. Shaw [ed.] Soil physical conditions and plant growth. Academic Press, New York.

Robinson, W. O. 1930. Some chemical phases of submerged soil conditions. Soil Sci. 30:197–217.

Schumaker, G., and S. Davis. 1961. Nitrogen application and irrigation frequencies for western wheatgrass production on clay soil. Agron. J. 53:168–170.

Shapiro, Raymond E. 1958. Effect of flooding on availability of phosphorus and nitrogen. Soil Sci. 85:190–197.

Stanhill, G. 1957. The effect of differences in soil moisture status on plant growth: A review and analysis of soil moisture regime experiments. Soil Sci. 84:205–214.

Stout, Myron. 1961. A new look at some nitrogen relationships affecting the quality of sugar beets. J. Amer. Soc. Sugar Beet Tech. 11:388–398.

Viets, Frank G., Jr. 1962a. Chemistry and availability of micronutrients in soils. J. Agr. Food Chem. 10:174–178.

Viets, Frank G., Jr. 1962b. Fertilizers and the efficient use of water. Advance. Agron. 14:223–264.

Volk, Gaylord, M. 1947. Significance of moisture translocation from soil zones of low moisture tension to zones of high moisture tension by plant roots. J. Amer. Soc. Agron. 39:93–106.

Walker, John M., and Stanley A. Barber. 1961. Ion uptake by living plant roots. Science 133:881–882.

Weaver, J. E. 1925. Investigations on the root habits of plants. Amer. J. Bot. 12:502–509.

Weaver, J. E., J. Kramer, and M. Reed. 1924. Development of root and shoot of winter wheat under field environment. Ecology 5:26–50.

Wiersum, L. K. 1958. Influence of water content of sand on rates of uptake of Rb86. Nature (London) 181:106–107.

Wiersum, L. K. 1962. Soil structure, rooting and plant nutrition. Landbouwk, Tijdschr. 74:961–972.

25

Microbial Activity in Relation to Soil Water and Soil Aeration

FRANCIS E. CLARK

Agricultural Research Service, USDA
Fort Collins, Colorado

W. D. KEMPER

Agricultural Research Service, USDA
and Colorado State University
Fort Collins, Colorado

I. INTRODUCTION

That either extremely wet or extremely dry soils greatly restrict microbial activity is well known. Between such extremes, the water contents that provide optimal or near optimal conditions for microorganisms in soil are not precisely defined. What is optimum for one group of organisms or for one transformation is not necessarily optimum for another; likewise, what is optimum in one soil and at one temperature or reaction or other variable does not necessarily hold for other soils or for other soil conditions.

To speak of an optimum soil water content for microorganisms that falls somewhere between extreme wetness and extreme dryness implies that some restrictions must be operative at the two extremes that in some fashion make the soil habitat unfavorable for microbial activity. Under conditions of drouth, the restrictive factor for microorganisms, as for higher plants, is simply the unavailability of water, or the high level of energy with which the water is held by the soil. Microbes, as well as all other forms of life, require water to grow.

Excess water in soil restricts microbial activity almost wholly because it prevents the movement of oxygen into and through soil in sufficient quantity to meet the oxygen demands of soil organisms. Under conditions of oxygen stress, microorganisms may fail to carry out their normal pathways of metabolism, or they may turn to alternative pathways, and in so doing make certain plant nutrients either more or less available than they are in well-aerated soil. Also, under conditions of oxygen stress, soil organisms compete very effectively with plant roots for any available oxygen. Questions for initial consideration, therefore, are those of minimal water requirements for microorganisms and the magnitude of their oxygen demand in soil.

II. MICROBIAL ACTIVITY IN DROUTHY SOIL

Different microorganisms vary in the extent to which they are active in drouthy soil and, consequently, all microbial transformations are not uniformly curtailed during drying out of soil. Ammonification, for example, can proceed under more stringent drouth than can nitrification.

There is abundant evidence that microorganisms can carry out at least some decomposition of organic residues in soils at the permanent wilting percentage, and to some extent at even lesser water content. According to Dommergues (1959) a material such as ordinary sugar can be decomposed in soil at a water content roughly one-half that of the permanent wilting percentage, whereas for decomposition of cellulose, the critical water content is just slightly less than that of the wilting percentage. At 25C the threshold water content for the decomposition of many plant residues is approximately that of 80% relative humidity (Bartholomew and Norman, 1947). Decomposition seldom, if ever, occurs in fibers and foodstuffs at 75% relative humidity (Miller and Clark, 1955). From the practical standpoint, it should be emphasized that as soil water content drops below the permanent wilting percentage, decomposition processes drop almost to a standstill. At 81% relative humidity, the rate of decomposition of oat (*Avena sativa*) straw is < 1% of that measured at the optimum water content (Bartholomew and Norman, 1947).

Ammonification may occur in soils at water contents below their permanent wilting percentages. In studies using a soil holding 42% water at field capacity and 21% at the permanent wilting percentage Robinson (1957) observed ammonification to occur slowly at 10.5% water, or at one-half the wilting percentage. No ammonification was observed in this soil when air-dry. Nitrification, although it can occur slowly in soil at 15 bars suction, (Justice and Smith, 1962) does not occur in soil of appreciably lower water content. In the field, therefore, drouthy conditions can lead to some accumulation of ammonia. The amount so accumulated is of the order of only a few pounds per acre, and with replenishment of the soil water, rapid conversion of the accumulated ammonia to nitrate can be expected.

The observation that field soils sometimes have a high nitrate content during drouth has led some workers to contend that nitrification does occur in extremely dry soil. Wetselaar (1961) investigated this possibility in some detail. He concluded that the high nitrate accumulation found in or near the surface of field soil after a prolonged dry period is due to capillary movement upward, during soil drying, of nitrate formed in the underlying soil during periods of favorable water content.

The number of bacteria drops markedly as soil dries, but seldom are individual species of bacteria entirely eliminated from soils kept dry for long periods. When such soils are rewetted, such varied phenomena as nonsymbiotic nitrogen fixation, nitrification, or sulfur oxidation almost invariably proceed without any soil reinoculation. Calder (1957) noted that the nitrate productivity of soil stored dry for 3 years was almost unimpaired. The soybean (*Glycine max*) nodule organism has been observed to survive 19 years of storage in air-dry soil (Sen and Sen, 1956). Such an observation does not mean that legume nodule bacteria invariably withstand soil drying. In some poorly buffered field soils cropped to annual legumes and subjected to drouth in the off season, there may be insufficient carryover of viable rhizobia to effect nodulation of the newly planted crop. In such instances, inoculation with rhizobia must be undertaken annually at time of planting.

III. AVAILABLE OXYGEN AND MICROBIAL OXYGEN DEMAND IN WET SOILS

Wet soils are unfavorable for most bacteria simply because filling pore spaces with water diminishes soil aeration. The restrictive factor is lack of oxygen and not the excess water in itself. The amount of oxygen present in a soil microsite at any given time is primarily a function of the biochemical oxygen demands of soil organisms and plant roots and of the rate of oxygen movement into the soil atmosphere and through the soil water barriers surrounding the respiring organisms.

A. Rate of Oxygen Use in Soil

The microbial oxygen demand in fallow soils to which fresh organic residues have not recently been applied is of the order of 1 to 2 lb/acre per hour. Following addition of residues, such as stable or green manures, oxygen consumption can be expected to increase to a rate of 2 to 4 lb/acre per hour. Rate of oxygen use in soil containing growing plants is of the order of 3 to 6 lb/acre per hour. The difference in oxygen demand in fallow and cropped soil is not wholly due to plant root respiration. Several studies (Barker and Broyer, 1942; Lundegårdh, 1927), have shown that roughly half the oxygen used by plant roots is actually used by microorganisms closely associated with the root surfaces. These associated microbes subsist on organic materials sloughed or exuded from the roots.

B. Oxygen Movement into the Soil Atmosphere

Oxygen moves into the soil largely by diffusion from the atmosphere through and into the air-filled soil pores. Ordinarily, the oxygen pressure difference necessary to cause adequate movement of oxygen to the root zone need equal only 1 to 4% oxygen. Accordingly, oxygen contents in the soil atmosphere are generally of the order of 17 to 20%. However, if the soil surface is covered by water, or if air in the pores of the surface soil is displaced by water, diffusion of oxygen into the soil is reduced to a negligibly slow rate. This occurs during heavy rainfall and during flooding such as when basin irrigation is practiced, and to a lesser degree in furrow and sprinkler irrigation. When soils being irrigated have low infiltration rates, the question often arises as to how long water may be ponded on soils before oxygen levels in the soil atmosphere are reduced to levels detrimental to plant growth. In providing an estimate of this time, it should be remembered that only the surface (or top few inches) of the soil is saturated under most basin irrigation conditions. Consequently, pathways are open for oxygen to diffuse to soil pores in the rhizosphere from the reservoir of soil air contained at deeper depths in the soil.

The amount of oxygen in this reservoir is determined by the depth of the water table, the fraction of the soil volume which is occupied by air, and the initial oxygen content of this soil air. How long this reserve of oxygen will last may be estimated by dividing the amount of oxygen in the reservoir by the rate of use. If rate of oxygen use is expressed in pounds per acre per hour, air-filled pores are considered as the percentage of the soil volume occupied by air when the irrigation or water infiltration is completed and are expressed as a percentage of the soil volume, depth to the water table in feet, and oxygen in the soil air in per cent the time in hours that this reservoir of oxygen would last is given by:

$$\text{Time} = \frac{0.38 \text{ (depth to water table) (air-filled pores) } (O_2 \text{ in soil air)}}{\text{Rate of oxygen use by plants and microorganisms}}.$$

This equation gives the time until exhaustion of the reserve oxygen supply. Data of Harris and Van Bavel (1957), Woodford and Gregory (1948), and others indicate that levels of oxygen in the soil atmosphere of $< 10\%$ may inhibit plant growth. Therefore, an equation which gives a safer value of time for which the soil surface may be sealed is:

$$\text{Time} = \frac{0.38 \text{ (depth to water table) (air-filled pores) } (\% \ O_2 \text{ in soil air} -10)}{\text{Rate of oxygen use by plants and microorganisms}}.$$

Thus, if the water table were 7 feet below the surface, air-filled pores constituted 20% of the soil volume, the initial oxygen content of the soil air were 19%, and oxygen use were at the rate of 5 lb/acre per hour, the surface could probably be sealed off by irrigation for about 96 hours before the soil atmospheric oxygen content was lowered to a damaging level for plant roots.

Any such calculation should not be accepted blindly. It should be recognized that the surface may be sealed for some time after all the surface water has apparently entered the soil. The length of this time will be a function of the soil structure and the drying conditions prevalent. It will often exceed 24 hours. Other variables such as soil temperature should also be considered. A warm soil increases and a cold soil decreases the rate of oxygen use. In a warm soil containing an abundant supply of easily decomposed organic matter, the rate of oxygen use may be so rapid that the reserve oxygen supply is exhausted within the course of 1 or 2 days. Finally, the equation given assumes that there is no restriction to oxygen movement within the soil. If oxygen consumption is mostly in the root zone just below the wetted surface, then oxygen content there may be below 10% when the average content of the entire volume above the water table is above 10%. Stated otherwise, an oxygen pressure difference is required to move oxygen from the lower depths to the root zone.

C. Oxygen Movement from the Soil Atmosphere to Respiring Microorganisms and Root Cells

Oxygen diffuses through air-filled pores about 10,000 times as fast as through water. Loci for respiration reactions in microorganisms and root cells are surrounded by water, and consequently the diffusion rate of oxygen through water is often the limiting factor in respiration.

An approximation of the root-soil-water-air geometry is presented in Fig. 25–1. The shearing and expanding action of the root has shifted soil particles to destroy most of the large air-filled pores immediately adjacent to the root, resulting in a more complete encirclement of the root by soil particles and the included small pores. Under low soil water suctions, water can be expected to fill all pore space adjacent to the root.

Not shown in Fig. 25–1 is the mantle of microorganisms that exists on the root surface. The distribution of bacteria with respect to the root and its surrounding

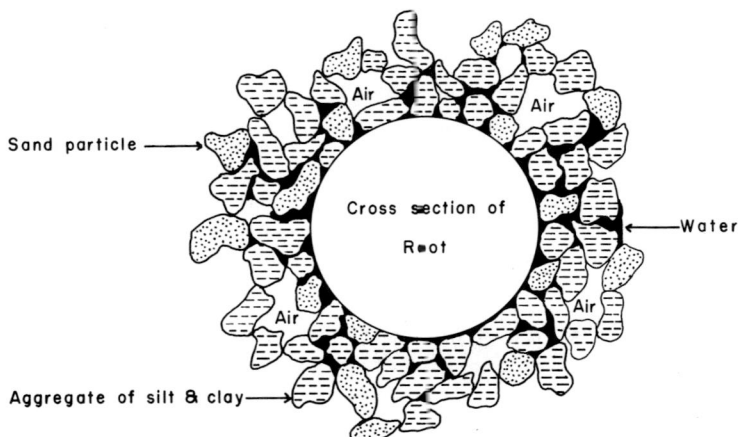

Fig. 25–1. Diagrammatic cross-section of a root surrounded by water film and soil particles.

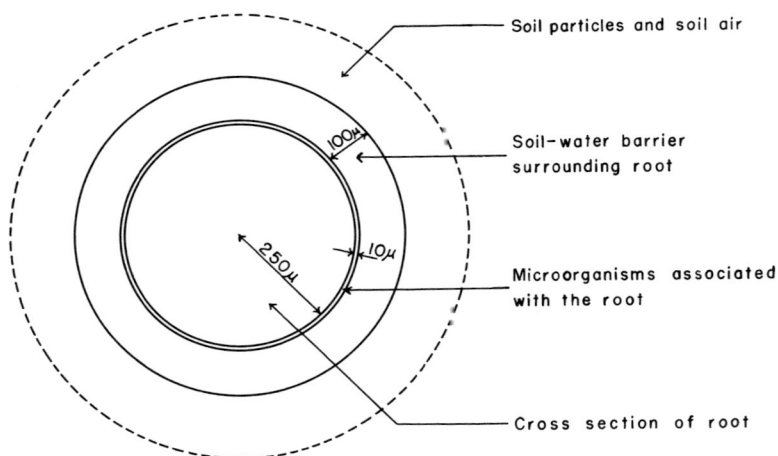

Fig. 25–2. Idealized geometry of root and surrounding water film as used to make calculations of oxygen diffusion.

film of water and soil particles is shown diagrammatically in Fig. 25–2. This idealized geometry is used in making calculations of oxygen diffusion to root cells.

The resistance to diffusion of oxygen from the soil air to the root surface is often expressed in terms of a film of water of uniform thickness surrounding the root. Commonly, when the diffusion coefficient of oxygen in water is used, equivalent water film thicknesses of the order of 1 cm are calculated. These values are somewhat misleading since they are about 50 times larger than the actual soil water barrier thicknesses. This apparent anomaly results from using the diffusion coefficient of oxygen in water rather than the smaller effective diffusion coefficient of oxygen in water in the porous media, and the logarithmic relation between diffusion coefficient and film thickness. It is probable that tortuosity of the diffusion path and reduction in cross section due to the solid matrix each reduces the

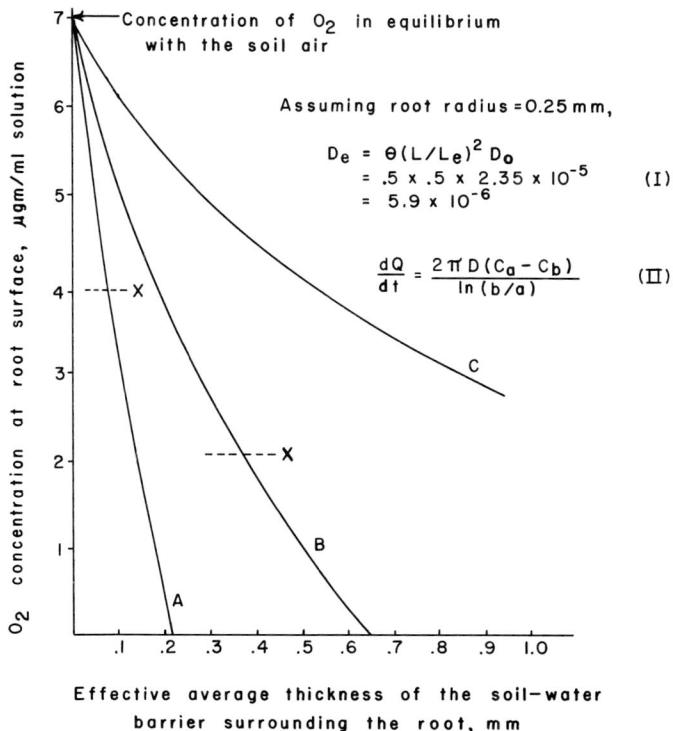

Fig. 25–3. Calculated critical thicknesses (marked as X) of moisture film at root surface which lead to oxygen deficiency of root cells. Curve A: calculated from data of Lemon and Wiegand (1962). Curve B: calculated from data of Anderson and Kemper (1964). Curve C: calculated from curve B as the probable curve for roots without bacteria. The X point (not shown) would be reached when moisture film reached a thickness of 2 mm.

effective diffusivity coefficient of oxygen by a factor of about one-half. Thus, diffusion coefficients of oxygen through the soil water barrier enveloping the root will be about 0.6×10^{-5} at 23C. When this value of the diffusion coefficient is used, the concentrations of oxygen in soil solution at the root surface have the values indicated in Fig. 25–3.

Lemon and Wiegand (1962), calculated the critical oxygen contents, in solution at the root surface, below which oxygen is in short supply to the internal root cells. The calculations used in constructing Fig. 25–3 indicate that their critical levels would be reached when soil water barrier thicknesses of 0.1 to 0.5 mm are present.

During and immediately following irrigation, it is likely that the soil water barrier thicknesses are at least this thick and therefore large enough to cause oxygen contents at the root surfaces to drop below the critical levels. Indications that such temporary reductions in oxygen consumption reduce plant growth or yield have been obtained by Letey et al. (1962) and Stolzy et al. (1964).

Because of their location inside the soil water barrier, as shown in Fig. 25–2, the soil microorganisms at the root surface have first call on the supply of incoming oxygen. As noted a few paragraphs earlier, their use of oxygen is about equal to the oxygen use by the roots themselves. If, then, one halves the highest rate of oxygen use (2×10^{-10} g O_2/cm root per second, and plotted as curve B in Fig. 25–3) observed by Anderson and Kemper (1964) for roots plus associated bacteria and plots the resulting calculations, curve C in Fig. 25–3 results. At 1×10^{-10} g O_2/cm root per second, critical oxygen levels at the root surface would not have been reached until the soil water barrier was about 2 mm thick.

Obviously, microbial activity does greatly affect oxygen availability to the root cells. Unfortunately, about all that can be done is to call attention to the magnitude of the bacterial effect, as there appears no practical way to eliminate this activity. Consequently, for all practical purposes, the use of oxygen by roots and their associated microorganisms may equally well be considered as the oxygen requirement of the plant roots themselves.

IV. INFLUENCES OF MICROORGANISMS ON PLANT NUTRIENTS IN WET SOILS

Rates of oxygen use by microorganisms and the problems of maintaining an adequate concentration of the gas in soil and at the root surface are of interest not only because microorganisms compete very effectively with plant roots for any short oxygen supply in wet soils, but also because under conditions of oxygen stress microorganisms may turn to pathways of metabolism that greatly affect the availability of certain plant nutrients. Minerals particularly susceptible to microbial transformations in wet soils are nitrogen in the nitrate form (Bremner and Shaw, 1958; Wijler and Delwiche, 1954) and several elements of variable valence, notably iron, manganese, and sulfur (Clark and Resnicky, 1956; Clark et al., 1957; Nearpass and Clark, 1960; Takai et al., 1956).

Most arable soils, in the absence of a growing crop, accumulate nitrate nitrogen, sometimes to as much as 100 or more pounds per acre. Ideally, this nitrate nitrogen, as well as nitrate added directly to soil as fertilizer or formed from added ammonium nitrogen, should remain in the soil until removed by growing plants. It may, however, be removed from soil by leaching. It may also be lost from the soil by the process of bacterial denitrification. This process occurs when the supply of soil atmospheric oxygen becomes inadequate for microbial metabolism. In such an instance, the nitrate nitrogen is converted to gaseous nitrogen, which in turn escapes to the atmosphere. To avoid such loss, soils containing nitrate nitrogen should not be kept flooded for periods sufficiently long that the soil oxygen supply becomes exhausted. The rapidity with which oxygen is consumed in soils sealed off at the surface has been discussed above.

Soil organisms variously influence the availability of iron, manganese, and sulfur. With flooding and reduced oxygen suction, iron and manganese are changed from less soluble to more soluble forms. This change may be either beneficial or harmful to plants. Thus, in a soil high in available manganese, flooding may increase manganese to a toxic level, whereas in a soil deficient in available manganese, flooding may be beneficial. The better growth of rice (*Oryza sativa*) in paddy than in upland culture is believed to be caused in part by the higher iron and manganese solubility in the paddy soil. Sulfate is also subject to reduction in waterlogged soil. Sulfur that is not in the sulfate form is unavailable to plants, and if reduced to hydrogen sulfide, it is toxic.

There are many anomalous observations concerning the availability of plant nutrients in soils undergoing change in moisture status. Thus, air-drying of some soils markedly increases their supply of available manganese. With other soils, iron deficiency develops in plants coincidentally with the application of irrigation water. Adequate explanations of these and similar observations are still lacking. Our purpose here is simply to emphasize that microbial oxidations and reductions of several elements of variable valence are closely linked to the water and aeration status of soil.

V. MICROBIALLY INDUCED PHYSICAL CONDITIONS AFFECTING THE SOIL WATER

It is now well known that microbial activity in soil contributes to the formation and stabilization of soil aggregates (McCalla and Haskins, 1960; Swaby, 1949). The binding action exerted by microorganisms on soil particles is due partly to filamentous structures such as fungal hyphae, and partly to microbially produced gums and polysaccharides. A prerequisite for the formation of these substances is a food supply of decomposable organic matter, such as crop residues. Once microbial cells and gums have been produced, soil drying is necessary to cause the formation of the stable individual aggregates.

If stable aggregates are present, numerous large-size pores will be established in a soil coincidentally with its cultivation (Anderson and Kemper, 1964). The presence of many such pores increases the rate at which water will percolate into and through soil. The maximum number of large-size pores is produced when cultivation and rewetting of the soil occur during the peak level of aggregate stability. This peak level usually occurs about 1 month after the incorporation of organic matter into the soil. The exact time will be a function of the soil temperature, the soil water content, and the suitability of the incorporated organic matter as food material for microorganisms.

To the extent the microorganisms produce stable aggregates and thereby help to maintain large-size pores, or stated conversely, help to prevent slaking down and sealing off at the soil surface, they function to increase the rate of water intake into soil. Under some special conditions, however, microorganisms may also serve to seal off a soil surface and to prevent water intake.

In continuous ponding of water on a soil surface, as is sometimes done in order to recharge a groundwater supply, growth of microorganisms within the soil pores may seal off the soil surface. Once this occurs, it becomes necessary to drain, dry, and then cultivate the soil, to break the seal. In some sandy soils (Jamison, 1946; Prescott and Piper, 1932) certain fungi that grow in or near the soil surface produce nonwetting or water-repelling substances. These materials effectively coat a great many sand grains and permit the entry of water into the soil through only a few channels, leaving a large volume of the soil quite dry. Total infiltration is markedly reduced, sometimes by as much as 90%.

LITERATURE CITED

Anderson, W. B., and W. D. Kemper. 1964. Effects of aggregate stability, moisture and soil temperature on growth of corn. Agron. J. 56:453–456.

Barker, H. A., and T. C. Broyer. 1942. Notes on the influence of microorganisms on growth of squash plants in water culture with particular reference to manganese nutrition. Soil Sci. 53:467–477.

Bartholomew, W. V., and A. G. Norman. 1947. The threshold moisture content for active decomposition of some mature plant materials. Soil Sci. Soc. Amer. Proc. (1946) 11:270–279.

Bremner, J. M., and K. Shaw. 1958. Denitrification in soil: II. Factors affecting denitrification. J. Agr. Sci. 51:40–52.

Calder, E. A. 1957. Features of nitrate accumulation in Uganda soil. J. Soil Sci. 8:60–72.

Clark, F. E., and J. W. Resnicky. 1956. Some mineral element levels in the soil solution of a submerged soil in relation to rate of organic matter addition and length of flooding. Int. Congr. Soil Sci., Proc. 6th (Paris) C:545–548.

Clark, F. E., D. C. Nearpass, and A. W. Specht. 1957. Influence of organic additions and flooding on iron and manganese uptake by rice. Agron. J. 49:586–589.

Dommergues, Y. 1959. L'activite de la microflore tellurique aux faibles humidites. Compt. Rend. Acad. Sci., Paris. 248:487–490.

Harris, D. G., and C. M. van Bavel. 1957. Growth, yield, and water absorption of tobacco plants as affected by composition of the root atmosphere. Agron. J. 49:11–14.

Jamison, V. C. 1946. Penetration of irrigation and rain water into sandy soils of central Florida. Soil Sci. Soc. Amer. Proc. (1945) 10:25–28.

Justice, K. J., and R. L. Smith. 1962. Nitrification of ammonium sulfate in calcareous soil as influenced by combinations of moisture, temperature, and levels of added nitrogen. Soil Sci. Soc. Amer. Proc. 26:246–250.

Lemon, E. R., and C. L. Wiegand. 1962. Soil aeration and plant root relations: II. Root respiration. Agron. J. 54:171–175.

Letey, J., L. H. Stolzy, and G. B. Blank. 1962. Effect of duration and timing of low soil oxygen content on shoot and root growth. Agron. J. 54:316–319.

Lundegårdh, H. 1927. Carbon dioxide evolution of soil and crop growth. Soil Sci. 23: 417–453.

McCalla, T. M., and F. A. Haskins. 1960. Microorganisms and soil structure. Missouri Agr. Exp. Sta. Res. Bull. 765. p. 32–45.

Miller, P. R., and F. E. Clark. 1955. Water and the microorganisms. p. 25–35. In Water. Yearbook of Agriculture. U. S. Government Printing Office, Washington, D.C.

Nearpass, D. C., and F. E. Clark. 1960. Availability of sulfur to rice plants in submerged and upland soil. Soil Sci. Soc. Amer. Proc. 24:335–387.

Prescott, J. A., and C. S. Piper. 1932. The soils of the South Australian Mallee. Roy. Soc. S. Australia, Trans. 56:118–122.

Robinson, J. B. D. 1957. The critical relationship between soil moisture content in the region of the wilting point and the mineralization of natural soil nitrogen. J. Agr. Sci. 49:100–105.

Sen, A., and A. N. Sen. 1956. Survival of *Rhizobium japonicum* in stored air-dry soils. J. Indian Soc. Soil Sci. 4:215–220.

Stolzy, L. H., O. C. Taylor, W. M. Dugger, and J. D. Mercereau. 1964. Physiological changes and ozones susceptibility of tomato plants after short periods of inadequate oxygen diffusion to roots. Soil Sci. Soc. Amer. Proc. 28:305–308.

Swaby, R. J. 1949. Relationship between micro-organisms and soil aggregation. J. Gen. Microbiol., 3:236–254.

Takai, Y., T. Koyama, and T. Kamura. 1956. Microbial metabolism in reduction process of paddy soils. Soil Plant Food 2:63–66.

Wetselaar, R. 1961. Nitrate distribution in tropical soils. Plant Soil 15:110–133.

Wijler, J., and C. C. Delwiche. 1954. Investigations on the denitrifying process in soil. Plant Soil 5:155–159.

Woodford, E. K., and F. G. Gregory. 1948. Preliminary results obtained from an apparatus for the study of salt uptake and root respiration of whole plants. Ann. Bot. (New Series) 12:325–370.

section VIII

Evapotranspiration

VIII

26

Microclimatic Factors Affecting Evaporation and Transpiration

H. L. PENMAN

Rothamsted Experimental Station
Harpenden, Herts., England

D. E. ANGUS

Division of Meterological Physics, CSIRO
Victoria, Australia

C. H. M. VAN BAVEL

US Water Conservation Laboratory, ARS, USDA
Phoenix, Arizona

I. LIST OF SYMBOLS

Many symbols listed for this chapter also apply to chapter 29 (see also chapter 29 for variations).

c	Specific heat of air at constant pressure	cal g^{-1} °C^{-1}
d	Zero-plane displacement	cm
e	Vapor pressure	mbars
g	Gravitational constant	cm sec^{-2}
h	Relative humidity	dimensionless
k	von Karman constant	dimensionless
p	Total pressure	mbars
q	Specific humidity	g g^{-1} (of moist air)
r	Reflection coefficient	dimensionless
u	Wind velocity	cm sec^{-1}
u_*	Friction velocity	cm sec^{-1}
v	Horizontal component	cm sec^{-1}
w	Vertical component	cm sec^{-1}
z	Elevation above surface	cm
z_0	Roughness factor	cm
B	Transport number	cal cm^{-2} mbar^{-1} sec^{-1}
E_*	Evaporation rate	g cm^{-2} sec^{-1}
E	Evaporation—energy flux equivalent	cal cm^{-2} sec^{-1}
H	Net radiative flux	cal cm^{-2} sec^{-1}
K	Turbulent transport coefficient	cm^2 sec^{-1}
L	Diffusion length	cm
Q	Sensible heat flux	cal cm^{-2} sec^{-1}
Ri	Richardson number	dimensionless
R	Radiative flux	cal cm^{-2} sec^{-1}
S	Sensible heat flux in soil	cal cm^{-2} sec^{-1}
T	Temperature	°C or °K
α	Constant	dimensionless
β	Bowen ratio	dimensionless
γ	Psychrometric constant	mbars °C^{-1}
δ	Difference in	
ϵ	Constant	dimensionless

θ	Temperature amplitude	°C
κ	Thermal diffusivity	cm^2 sec^{-1}
λ	Latent heat of vaporization	cal g^{-1}
ρ	Density of air	g cm^{-3}
σ	Constant	dimensionless
τ	Shear stress	dyne cm^{-2}
τ	Period	seconds
χ	Absolute humidity	g cm^{-3}
Γ	Adiabatic lapse rate	°C cm^{-1}
Δ	de/dT (saturated water vapor)	mbars °C^{-1}

Subscripts

a	Pertaining to the ambient air
a	Associated with sensible heat transfer (exclusively in E_a)
d	Downward longwave
e	Pertaining to external resistance
s	Pertaining to soil
s	Pertaining to leaf resistance
u	Upward longwave
w	Of the wet bulb
B	Net longwave
H	Sensible heat
I	Incoming shortwave
M	Momentum
O	Pertaining to open water
T	Pertaining to transpiring crop
V	Water vapor

II. INTRODUCTION

A growing plant takes up soil water at its roots and transmits the water to the leaves as liquid. Then, if the stomata are open, further movement is as vapor through the stomata, and final dispersal depends on turbulence and mixing processes in the atmosphere around and above the plant. Other water vapor may reach the atmosphere by direct evaporation from the soil, so bypassing the plant, and for some time after rain, overhead irrigation, or dew deposition there is a third source as water evaporates from wet leaves. In general, while this intercepted water evaporates, the drain on soil water is reduced accordingly. Whatever the source, the unavoidable energy requirement for evaporation is the same, at about 590 cal/g of water evaporated. All of this energy must come from outside the plant because the energy content of a good crop as fuel is equivalent to no more than the amount of water transpired in a few days in summer. Once it is out of the leaves the water vapor is part of the atmosphere, and is there subject to the mixing processes that tend to produce uniformity of composition and of temperature.

Broadly, then, the evaporation process has two aspects. First, it is part of an energy balance, out of which can come quantitative estimates of water use by growing crops if there can be adequate assessment of the strengths of the sources of energy and of the other sinks for it. Second, it is part of a transport process, in which the net upward flux can be estimated when relevant physical measurements over the crop are substituted in the appropriate aerodynamic equations. The aerodynamic aspect is self-contained, and is conveniently considered first, whereas the energy balance aspect needs some aerodynamic ideas to make it complete.

III. AERODYNAMICS OF EVAPORATION

A. General

With rare exceptions the atmosphere is turbulent. A suitable system of anemometers will reveal that in an air stream having a steady mean direction and a mean velocity $\bar{u}$, there are fluctuations in forward velocity, $\pm u'$, there are fluctuating sideways components, $\pm v'$ (corresponding to changes in direction), and there are up-and-down motions, $\pm w'$. By definition, the average values of u', v', and w', over a fraction of an hour are zero, $\bar{v}$ is zero, and, if the surface is horizontal, then w is zero, too. Sensitive thermometers and hygrometers will also reveal fluctuations in air temperature T' and in air humidity q', χ', or e'. When the observations are made at different levels (z_1, z_2, z_3) above the ground, two kinds of information can be obtained. First, based on mean values over periods of 10 to 100 min, vertical profiles of wind speed, temperature, and humidity can be derived. Second, working at any particular level the fluctuations can be recorded at intervals of the order of, for example, 1 sec.

B. Fluctuation Theory

If at any point in the atmosphere the air has density ρ (g cm^{-3}), specific humidity q (g water/g air), and vertical velocity w (cm sec^{-1}), the instantaneous rate at which water is being carried upward is $\rho w q$ g cm^{-2} sec^{-1} and hence the average rate is $\overline{\rho w q}$, where the bar represents a time average. This can be split by writing

$$\overline{\rho w q} = \overline{\rho w} \times \overline{q} + \overline{(\rho w)\,'q'}, \qquad [26\text{--}1]$$

where over a site of good horizontal uniformity the value of $\overline{\rho w}$ is zero so the evaporation rate reduces to

$$E_{\circ} = \overline{(\rho w)'q'} \text{ g cm}^{-2} \text{ sec}^{-1}. \qquad [26\text{--}2]$$

Australian experience (McIlroy, 1961) is that equation [26–2] can be used if the observations are made at about 1.5 m above crop level and the time constants of the instruments are 0.1 sec or shorter. The closer to the surface, the faster the sensors must be. The basic requirement is for a linear response anemometer to measure $(\rho w)'$, a linear response hygrometer to measure q', and devices that will first multiply the two outputs, then integrate the products and display the result on a dial or chart. The difficulties in instrument design were great, but they have been overcome and the instrument has passed initial field trials (Dyer, 1961). The important property of the technique is that its performance is independent of the nature of the surface. The period of observation can vary from about 5 min (a minimum imposed by the turbulent structure of the atmosphere at 1.5 m height) up to the limit of the instrument's ability to run without attention, at present about 1 hour. Figure 26–1 shows traces obtained with an earlier instrument during a 5-min run over a level pasture. Table 26–1 gives some comparisons of instrument estimates and direct estimates of evaporation.

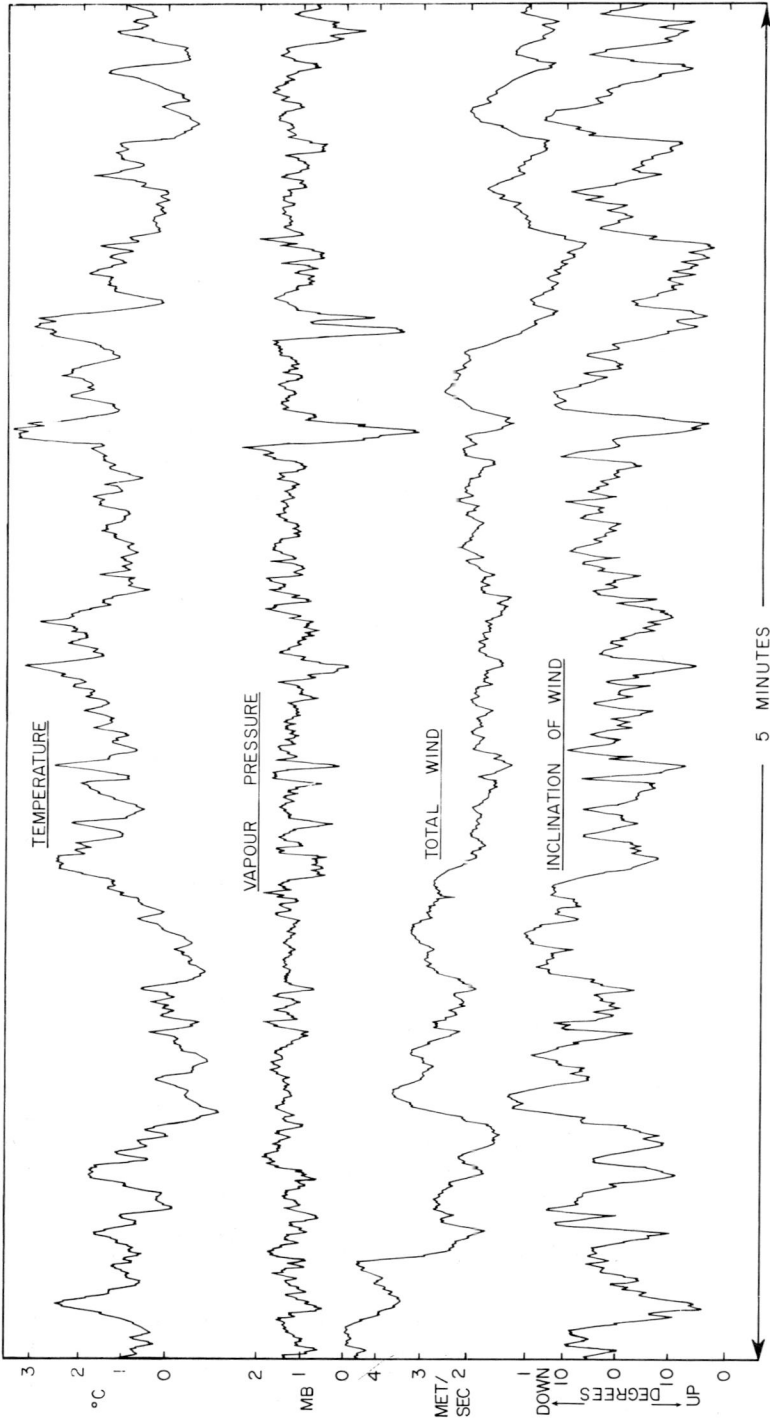

Fig. 26–1. Actual traces of air temperature, vapor pressure, wind speed, and wind inclination, measured at Aspendale, Australia.

Table 26–1. Mean daily comparisons of measured $(H - S)$ and $(E + Q)$ measured separately by the eddy correlation technique (Dyer, 1961); units are cal cm^{-2} min^{-1}

Date (1960)		no. of 5-min runs	(H - S)	(E + Q)	Wind speed, m sec^{-1}
January	4	11	0.558	0.539	5.0
	5	20	0.718	0.643	7.3
	6	9	0.815	0.872	4.3
	25	8	0.340	0.343	3.6
	27	14	0.579	0.548	3.2
	29	22	0.639	0.618	4.5
February	2	8	0.629	0.601	5.0
	3	20	0.611	0.525	5.6
	12	17	0.483	0.486	5.0
	19	7	0.464	0.480	3.9
	22	12	0.387	0.373	6.3
	26	32	0.600	0.483	6.9
	29	7	0.396	0.333	4.0
March	1	20	0.537	0.512	4.0
	2	33	0.572	0.563	4.3
	4	22	0.337	0.382	3.8
	8	4	0.378	0.430	2.0
	9	16	0.417	0.386	5.3
	10	13	0.431	0.436	5.0
	15	12	0.475	0.432	5.0
	17	9	0.503	0.515	4.5
	18	23	0.422	0.422	3.9
	22	14	0.601	0.496	5.5
	23	23	0.201	0.204	3.6
	24	14	0.429	0.432	4.3
	28	7	0.314	0.332	4.1
April	1	10	0.203	0.212	4.5
	11	17	0.189	0.193	-

For the method to be valid it must be applied to extended homogeneous areas. As the height of measurement increases, so does the stringency of the demand for a sufficiently large site. Dyer and Pruitt (1962) found at Davis, California, USA that at 3 m height a 100- by 800-m irrigated pasture surrounded by dry land was not large enough for the flux at 3 m to equal that at the surface.

C.   Profile Theory

1.   GENERAL

The increase of wind speed with height over an extended uniform surface is represented by

$$u = \frac{u_*}{k} \ln \frac{z - d}{z_0} \qquad [26\text{–}3]$$

when the atmosphere is near a state of neutral stability. In conditions of greater or lesser stability more complex expressions are needed (Deacon, 1949) and there is a continuing quest for a generalized wind profile (Priestley, 1959; Sheppard, 1958). In evaporation studies equation [26–3] is used even where it is known to be inexact, with corrections for stability applied at the end of the computations.

In equation [26–3], u is the velocity at height z above ground, z_0 is a roughness constant, d is needed to allow for the effect of the crop in effectively displacing ground level upward, k is a universal constant (von Karman, $k = 0.41$), and u_* is a constant of the particular profile, given by

$$u^2_* = \tau/\rho, \qquad\qquad [26\text{-}4]$$

where τ is the shearing stress in the moving air, and ρ is, as before, its density. It is assumed—and with good reason—that over a period of 10 min or longer the average value of τ is invariant with height up to about 3 m. For greater heights, longer averaging times are necessary. This implies that the downward flux of momentum is constant in the same layer, and with this condition satisfied, formal analysis is possible. By definition

$$\tau/\rho = K_M \partial u/\partial z \qquad\qquad [26\text{-}5]$$

where K_M is the transport constant for momentum and is frequently known as the coefficient of eddy diffusivity. Combining equations [26-3], [26-4], and [26-5] yields

$$K_M = k\, u_*\, (z - d). \qquad\qquad [26\text{-}6]$$

The next assumption is that the mechanism that transports momentum also transports heat and water vapor, and that the eddy diffusivities for heat K_H and water vapor K_V are equal to each other and to K_M. Measurements (Pasquill, 1949; Rider, 1954; Swinbank, 1955) show that the assumed identity is no more than a good working approximation. One of the formal consequences of the assumption is that the profiles of temperature and vapor pressure should have the same shape as the profile of wind speed, which can be tested by plotting u against T and e at a number of values of z. The result should be a straight line, whatever the shape of the profile (Penman and Long, 1960). The formal transport equations are

$$Q = - \rho c K_H \partial T / \partial z \text{ cal cm}^{-2} \text{ sec}^{-1} \text{ and} \qquad\qquad [26\text{-}7]$$

$$E_* = - K_V \partial \chi / \partial z \text{ g cm}^{-2} \text{ sec}^{-1} \qquad\qquad [26\text{-}8]$$

where Q is the sensible heat transfer, c is the specific heat of air, E_* is the evaporation, and χ is the absolute humidity in g cm^{-3} of air. In practice it is the vapor pressure e that is measured, related to χ by

$$\chi = e\rho\epsilon/p \qquad\qquad [26\text{-}9]$$

where p is the total pressure (e and p to be expressed in identical units) and ϵ is the ratio of the densities of water vapor and dry air at the same temperature and pressure ($= 0.622$). So the evaporation equation becomes

$$E_* = - K_V \frac{\rho\epsilon}{p} \frac{\partial e}{\partial z} \text{ g cm}^{-2} \text{ sec}^{-1}. \qquad\qquad [26\text{-}10]$$

When $K_H = K_V = K_M$, and equation [26-3] is satisfied, the aerodynamic transport equations reduce to

$$Q = \rho c\, k^2 \frac{(T_1 - T_2)(u_2 - u_1)}{\left(\ln \dfrac{z_2 - d}{z_1 - d}\right)^2} \text{ cal cm}^{-2} \text{ sec}^{-1} \text{ and} \qquad\qquad [26\text{-}11]$$

$$E_* = \frac{\rho\epsilon}{p} k^2 \frac{(e_1 - e_2)(u_2 - u_1)}{\left(\ln \dfrac{z_2 - d}{z_1 - d}\right)^2} \text{ g cm}^{-2} \text{ sec}^{-1} \qquad\qquad [26\text{-}12]$$

where T_1, u_1, e_1 are measured at z_1, and T_2, u_2, e_2 are measured at z_2. If E_* and Q are to be expressed in the same units—a requirement for any discussion of energy balance—one of the possible transformations is to express the evaporation as an energy equivalent E by multiplying by the latent heat of evaporation λ at temperature T. Then

$$E = \lambda\, E_* = \lambda\, \frac{\rho\epsilon}{p}\, k^2\, \frac{(e_1 - e_2)(u_2 - u_1)}{\left(\ln \dfrac{z_2 - d}{z_1 - d}\right)^2} \quad \text{cal cm}^{-2}\ \text{sec}^{-1}. \qquad [26\text{–}13]$$

2. BOWEN RATIO AND PSYCHROMETER CONSTANT

In advance of later needs it is convenient now to note that when Q and E are in the same unit their ratio takes a simple form:

$$\frac{Q}{E} = \frac{cp}{\epsilon\lambda}\, \frac{T_1 - T_2}{e_1 - e_2} \qquad\qquad [26\text{–}14]$$

$$= \gamma\, \frac{T_1 - T_2}{e_1 - e_2} = \beta$$

in which β is the Bowen ratio, expressing the ratio of sensible heat transfer to that of latent heat transfer (Bowen, 1926) and γ is a constant which will be recognized as that of the August-Apjohn derivation of the wet and dry bulb psychrometer equation.

3. CHOICE OF UNITS

In practice, temperatures may be in °C or °F; wind speeds may be in cm sec⁻, km hr⁻¹, miles hr⁻¹, miles day⁻¹, or even in knots; heights will usually be in cm, but as long as d and z are in the same unit the denominators of equations [26–11], [26–12], and [26–13] are independent of unit; vapor pressures may be in mm Hg, inches Hg, or mbars. For the present analysis the units will be °C, cm sec⁻¹, cm, and mbars, and the important factors for calculations become

$$\chi/e\ = 2.17 \times 10^{-4}/(T + 273) \text{ g cm}^{-3}\text{ mbar}^{-1},$$

$$\rho c k^2\ = 4.95 \times 10^{-5} \text{ cal cm}^{-3}\ °\text{C}^{-1} \text{ at 20C},$$

$$\frac{\rho\epsilon}{p}\, k^2\ = 1.24 \times 10^{-7} \text{ g cm}^{-3}\text{ mb}^{-1} \text{ at 20C, 1,000 mbar},$$

$$\lambda\, \frac{\rho\epsilon}{p}\, k^2\ = 7.3 \times 10^{-5} \text{ cal cm}^{-3}\text{ mbar}^{-1},$$

$$\lambda\ = 586 \text{ cal g}^{-1} \text{ at 20C, and}$$

$$\gamma\ = 0.66 \text{ mbar } °\text{C}^{-1} \text{ at 20C and 1,000 mbar}.$$

As an example, for $z_2 \doteq 200$ cm, $z_1 = 100$ cm, assuming $d = 20$ cm, $u_2 - u_1 = 100$ cm sec⁻¹, $T_1 - T_2 = 0.2$C, $e_1 - e_2 = 1$ mbar, then

$$(u_2 - u_1)/[\ln(z_2 - d)/(z_1 - d)]^2 = 152, \text{ and}$$

$$Q = 1.50 \times 10^{-3} \text{ cal cm}^{-2}\text{ sec}^{-1}$$

$$= 5.4 \text{ cal cm}^{-2}\text{ hr}^{-1}$$

$$= 0.09 \text{ cal cm}^{-2}\text{ min}^{-1}$$

$$E_* = 1.88 \times 10^{-5} \text{ g cm}^{-2} \text{ sec}^{-1}$$

$$= 0.675 \text{ mm hr}^{-1} \text{ (1 g cm}^{-2} \text{ is equal to a depth of 10 mm water)}$$

$$= 40.0 \text{ cal cm}^{-2} \text{ hr}^{-1}$$

$$= 0.666 \text{ cal cm}^{-2} \text{ min}^{-1}$$

Note that the Bowen ratio β is $5.4/40$ or 0.13, i.e., the energy transferred as latent heat is about 7.5 times that transferred as sensible heat.

4. DETERMINATION OF THE DISPLACEMENT AND ROUGHNESS CONSTANTS

It is necessary to have values of u, found during neutral conditions, at several heights (preferably four or more), and find by trial which value of d gives a straight line when u is plotted against $\ln(z - d)$. The intercept of this straight line on the axis at $u = 0$ gives the value of $\ln z_0$, and the slope is $(u_2 - u_1)/\ln[(z_2 - d)/(z_1 - d)]$. This can be written explicitly in modified forms of equations [26–11], [26–12], and [26–13] so that equation [26–13], for example, becomes

$$E = \lambda \frac{\rho\epsilon}{p} k^2 \frac{e_1 - e_2}{u_2 - u_1} \left[\frac{u_2 - u_1}{\ln(z_2 - d)/(z_1 - d)} \right]^2 \quad \text{cal cm}^{-2} \text{ sec}^{-1} \quad [26–15]$$

in which $(e_1 - e_2)/(u_2 - u_1)$ is the slope of the line obtained in the test of equality of profile shape, described in the paragraph before equation [26–7]. When there are data from more than two observation levels these graphical techniques have much to commend them, because they are the best way of using all the data, and they also check the validity of the assumptions made in deriving the equations.

For short crops, such as cut grass, d will usually be negligible, and z_0 will be about 1 cm. For tall crops, with a stiff woody stem, the values of d and z_0 are almost independent of wind speed and their sum is even more constant, but tall pliable crops such as cereals show clear diurnal variation in d and z_0 and in their sum (Inoue, 1955; Penman and Long, 1960). Typical values of z_0 are given by Deacon (1953) and Sutton (1953).

5. CORRECTION FOR STABILITY

During periods when stability is far from neutral, notably at night, a correction is needed. The extent of the daytime correction is often relatively small, particularly over irrigated areas. The commonest correction involves the Richardson number

$$\text{Ri} = \frac{g}{T} \frac{\partial T/\partial z + \Gamma}{(\partial u/\partial z)^2} ,$$

where Γ is the adiabatic lapse rate of $1C/100$ m. In the present context Γ can be disregarded and the differentials replaced by finite differences to give

$$\text{Ri} \approx \frac{g}{T} \frac{(T_2 - T_1)(z_2 - z_1)}{(u_2 - u_1)^2} \quad [26–16]$$

or putting $g = 981 \text{ cm sec}^{-2}$, $T = 290K$,

$$\text{Ri} = 3.4 \, (T_2 - T_1)(z_2 - z_1)/(u_2 - u_1)^2$$

Hence, for the example given in part III, section C, subheading 3,

$$Ri \approx -6.8 \times 10^{-3}.$$

The correction factor to be applied to E_a or Q calculated from equations [26–10], [26–11], and [26–12] approximates to $(1 + \sigma \, Ri)^{-1}$ or $(1 - \sigma' Ri)$ with $\sigma \approx \sigma' \approx 10$. It has been found most suitable to divide by $(1 + 10 \, Ri)$ when Ri is positive (inversion conditions, usual at night), and to multiply by $(1 - 10 \, Ri)$ when Ri is negative (lapse conditions by day), as in the example, where the multiplying factor would be 1.07.

An alternative system for correcting stability has been put forward by Deacon and Swinbank (1958). It obviates the need to know the exact form of the stability factor mentioned above. Under strong lapse conditions (Ri more negative than –0.05), the heat flux Q can be calculated from the temperature profile alone (Penman and Long, 1960), leading to

$$Q = 0.17 \mid \Delta T \mid^{3/2}$$

where Q is in cal cm^{-2} sec^{-1}, and ΔT (°C) is the difference in temperature between 2 m and 0.5 m.

6. USE OF AERODYNAMIC ESTIMATES

Equations [26–11] and [26–12] have been widely applied, not always with simultaneous direct estimates of evaporation, to check the validity of equation [26–12]. Equation [26–11] may be verified with an estimate of Q based on fluctuation theory, no direct measurement being possible. The main contributions came from Thornthwaite and Holzman (1939) and Pasquill (1950). There is a review of theory in the First Lake Mead Report (US Navy Electron. Lab., 1950).

Within the limits stated, equation [26–12] is applicable over any kind of surface, and does not involve any surface property that has to be measured, the only surface property appearing explicitly (d) being estimated from the profile measurements. It is implicitly assumed that the value of d is the same for all three quantities (momentum, heat, and vapor), or rather that $d + z_o$ is the same for all because at $z = d + z_o$, $u = 0$ (equation [26–3]), i.e., this level is the virtual sink for momentum. It is not entirely reasonable to assume that this coincides with the virtual source of vapor, but experimental demonstration of any difference is very difficult.

Profiles will seldom be found to conform to the neutral, ideal case of forced convection and stability corrections are needed as a rule.

IV.   ENERGETICS OF EVAPORATION

A.   General

At times when irrigation is needed and applied, the energy used in evaporation is a major component in the energy exchanges between the ground, the air, and sky above it. When all quantities are expressed in the same units (e.g., in calories per unit area per unit time so that E is represented by equation [26–13]) then the complete energy balance equation at the surface is

$$R_I(1 - r) + R_d - R_u - E - Q - S = 0, \qquad\qquad [26\text{–}17]$$

and this is conveniently separated into radiant energy terms and nonradiant energy terms to give

$$R_I(1 - r) + R_d - R_u = E + Q + S. \qquad [26\text{--}18]$$

B.   Radiant Energy

The quantity R_I is the short-wave radiation income on a horizontal surface and r is the reflection coefficient of the surface, so that the total short-wave radiation retained by the surface is $R_I(1 - r)$. The quantities R_d and R_u are the downward and upward fluxes of long-wave radiation, both large, with R_u almost invariably the larger. It is convenient to combine the two into a single quantity R_B, called the net back radiation (from surface to sky). The balance of short- and long-wave radiation gives the heat budget as income (H), usually known now as the net radiation

$$H = R_I(1 - r) - R_B. \qquad [26\text{--}19]$$

C.   Sensible and Latent Heat Energy

The right side of equation [26–18] contains a latent heat term (E the energy for evaporation) and two sensible heat terms (Q the sensible heat transfer *to* the atmosphere, and S the sensible heat transfer *to* the soil or other subsurface material). At the surface, the energy budget as expenditure is

$$H = E + Q + S. \qquad [26\text{--}20]$$

In advance of a more detailed discussion later (see part VI) it may be noted here that where there is advection of energy in the form of warmer air coming from a dry zone upwind of an irrigation area, equation [26–20] is still true, but it is more appropriate to regard the "income" as $H - Q$ (remembering that Q will now be negative, heat being transferred *from* the air to the ground) and to regard the "expenditure" as $E + S$. A negative value of Q will be revealed in an *increase* of air temperature with height above the crop. The parameter S may also be of either sign.

There are modern instruments for measuring H directly (Fritschen, 1963; Funk, 1959) and there is less need now to estimate it indirectly from other weather parameters using equation [26–19]. How far above the surface equation [26–20] may be used, depends on the uniformity of the site (*see again* part VI). Assuming H is known in one way or another and supposing it to be for a continuous cover of transpiring crops, then the important part of equation [26–20] may be written

$$H = E + Q, \qquad [26\text{--}21]$$

disregarding the change in heat content of the soil as negligible (*see* part IV, section F).

D.   Bowen Ratio

Taking equation [26–14] from part III, section C, subheading 2, and combining it with equation [26–21] gives, with somewhat deceptive simplicity

$$E = H/(1 + \beta), \tag{26-22}$$

where the Bowen ratio β, requiring no measurement of wind speed or surface properties, involves only the ratio of the temperature gradient to the vapor pressure gradient between any two levels of measurement and will fluctuate as these change. Details and illustrative data may be found in Suomi and Tanner (1958).

Over a homogeneous irrigated area, daytime values of β range from about zero to 0.2 or so, but if there is strong advection β will be negative. For large negative values approaching -1, use of equation [26-22] demands great accuracy in measuring $T_1 - T_2$ and $e_1 - e_2$. At $\beta = -1$ equation [26-22] becomes meaningless, but equation [27-14] becomes very meaningful, because this is the condition of the wet bulb in the wet and dry bulb psychrometer, and it is possible to use the psychrometer equation to demonstrate another important aspect of the energy balance.

E. The Significance of Wet Bulb Profile

The psychrometer equation is

$$e = e_w - \gamma \ (T_a - T_w), \tag{26-23}$$

where T_a is the temperature of the air (dry bulb), T_w is the temperature of the wet bulb, e_w is the saturation vapor pressure at T_w, and e is the actual vapor pressure being sought.

Rewriting equations [26-11] and [26-13] in the shortest possible form that keeps the units of Q and E the same, then

$$Q = \gamma B (T_1 - T_2) \tag{26-24}$$

$$E = B(e_1 - e_2), \tag{26-25}$$

where B involves wind speeds, heights, and roughness (see equation [26-13]). Writing δe for $e_1 - e_2$ and δT for $T_1 - T_2$, then

$$Q = \gamma \ B \ \delta T_a$$

and from equation [26-23] rewritten for δe,

$$E = B \ [\delta e_w - \gamma(\delta T_a - \delta T_w)].$$

Now set the slope of the saturation vapor pressure curve, de/dT, at the mean wet bulb temperature as Δ, so that $\delta e_w = \Delta \delta T_w$. Then

$$E = B[(\Delta + \gamma)\delta T_w - \gamma \delta T_a]$$

$$= B \ (\Delta + \gamma)\delta T_w - Q$$

or

$$E + Q = B \ (\Delta + \gamma)\delta T_w. \tag{26-26}$$

Hence, the total energy transfer to the atmosphere is dependent on the gradient of wet bulb temperature in the same way as the sensible heat transfer is dependent on the gradient of dry bulb temperature (equation [26-24]) but with a different coefficient which does involve the mean wet bulb temperature. Values of $\Delta + \gamma$ may be obtained from Appendix V, Table 6 in Slatyer and McIlroy (1961).

F. Soil Heat Flow

The quantity S in equation [26–17] and [26–20] was omitted from equation [26–21] and the derived equation [26–22] If it must be taken in, then equation [26–22] can be written

$$E = (H - S)/(1 + \beta), \qquad [26–27]$$

but the need to consider S will depend entirely on the time scale of importance, and of these only two really matter, namely hour-to-hour changes within a day, and month-to-month changes within a year.

During a period of fine settled weather and adequately maintained soil water supply, the daily cycle of heat content of the soil is such that what goes in by day comes out by night, and the net change over 24 hours is negligible compared with the daily value of H or E, hence the neglect in equation [26–22]. Though it is advisable to make direct measurements in experiments, it is often quite sufficient to make an estimate from idealized conditions. When a sinusoidal temperature wave of amplitude θ is applied at the surface of the soil, with a period τ (which may be 1 day or 1 year), it penetrates the soil, the amplitude decreasing with depth, and the time of the maximum temperature getting later with depth. If the soil is uniform, it is easy to show from standard theory that the maximum heat content of the soil occurs at $\tau/8$ after the surface experiences its maximum temperature, i.e., at about 3 PM during the day, and about mid-August in the northern hemisphere. The excess heat content at this time, above its average value, is

$$\Sigma S = (\rho c)_s\, \theta\, [(\kappa \tau/2\pi)]^{1/2} \qquad [26–28]$$

where $(\rho c)_s$ is heat capacity of the soil in cal °C^{-1} cm^{-3}, κ is thermal diffusivity of the soil in cm^2 sec^{-1}, and τ is the period of the wave in sec. Using reasonably representative values of the soil constants, and taking $\theta = 5C$ (representing a range of 10C), the values of ΣS are

$$\text{Daily } \Sigma S = 21 \text{ cal cm}^{-2}$$

$$\text{Annual } \Sigma S = 395 \text{ cal cm}^{-2}.$$

So, for a temperature range of 10C, the heat content of the soil will increase from dawn to about 3 PM by about 42 cal cm^{-2}—the equivalent of about 2/3 mm of evaporation—and between mid-February and mid-August it will increase by 790 cal cm^{-2}—the equivalent of about 13 mm evaporation. The effect of the cycle of heat storage is to make the maximum value of the evaporation rate occur a little later than the time of the maximum value of net radiation, to decrease the value of the evaporation maximum very slightly and increase the minimum by the same amount with no significant effect on total.

The main departures from the idealized state occur in daily changes, partly because the daily temperature wave at the surface is far from sinusoidal, and partly because periods of uniform weather have to begin and end sometime. On the first fine day after a wet period there will be a net gain in stored heat, and the evaporation calculated from equation [26–22] by ignoring S will be an overestimate of what would be measured or calculated from equation [26–2] or equation [26–12]. Conversely, on the first cold day after a warm period, neglect of S will lead to underestimates of true evaporation.

Table 26–2. Mean daily energy balance (Pine plantation, Munich, June 28–July 7, 1952)

Item	Quantity	
	cal cm^{-2}	equiv. mm
Net radiation	+586	10.0
Heat flow into ground	+ 3	0.05
Heat flow to air	+197	3.35
Evaporation	+386	6.6

Table 26–3. Mean 4-hourly energy balance (Grassland, Harpenden, England, August 1960) (a) August 3–6, soil dry after drouth. (b) August 16–19, soil wet after rain; all components in cal cm^{-2} for the 4 hours

Period (hours)	(a)					(b)				
	H	S	Q	E	β	H	S	Q	E	β
0–4	− 9.6	−3.3	− 7.4	1.1	− 6.7	−10.4	−5.1	− 5.9	0.6	−9.3
4–8	13.0	−0.8	2.2	11.6	0.19	8.1	−5.0	− 8.4	21.5	−0.39
8–12	120.7	8.2	67.6	44.9	1.50	117.6	5.6	20.9	91.1	0.23
12–16	107.6	7.4	53.4	46.8	1.14	101.4	4.6	2.8	94.0	0.03
16–20	13.1	0.9	− 4.5	16.7	− 0.27	8.1	−0.9	−15.9	24.9	−0.63
20–24	−11.2	−2.6	− 9.2	0.6	−15.3	−14.6	−6.4	− 6.2	− 2.0	−3.1
Total cal cm^{-2}	233.6	9.8	102.1	121.7	--	210.2	−7.2	−12.7	230.1	--
Equivalent mm	3.95	0.17	1.73	2.06		3.55	−0.12	− 0.21	3.90	
R_I mm	6.75					6.75				

G. Examples of Energy Balances of Cropped Surfaces

1) Baumgartner (1956). Measurements of the energy balance components were made near Munich, Germany over an area of pine trees during a 9-day period of exceptionally fine weather. The average daily values are in Table 26–2, first in energy units, and then in equivalent depths of water evaporated.

2) J. L. Monteith (*unpublished data*). At Rothamsted, in southeast England, hourly components E, H, and S were measured, and values of Q were calculated by difference.

Table 26–3 gives the means for 3 out of 4 days in each of the periods 3–6 and 16–19 August 1960. The crop was grass, and the first period was toward the end of a drouth when the estimated soil water deficit was approximately 100 mm. The second period was a day or two after heavy rain, and it is a fair presumption that transpiration was not restricted by soil water deficit. The notable contrast between sections a and b of Table 26–3 are: (i) for the dry soil, E is only about one-half of H, and Q is of the same order as E, i.e., β is near unity; (ii) for the wet state, equivalent to a recent irrigation, the value of E slightly exceeds that of H, and Q is negligible (and negative), i.e., β is effectively zero.

Other results for other crops from the same source show that over periods of several weeks of complete cover and adequate water supply the average ratio of E/H rarely departs from unity by more than a few per cent.

3) L. J. Fritschen and C. H. M. van Bavel (1964). A field of sudangrass (*Sorghum sudanese*) at Tempe, Arizona, USA was irrigated on July 9, 1962 when the grass was about 75 cm high and growing vigorously. The hourly balance

Table 26–4. Hourly energy balance, sudangrass, Tempe, Arizona, July 12, 1962;
all energy components are in cal cm^{-2} for the hour

Hour	H	S	Q	E	u	β
					cm sec^{-1}	
1	− 2.4	− 3.0	− 3.0	3.6	133	−0.83
2	− 2.4	− 3.6	− 2.4	3.6	85	−0.66
3	− 2.4	− 3.0	− 3.0	3.6	121	−0.83
4	− 2.4	− 3.0	− 3.0	3.6	86	−0.83
5	− 2.4	− 3.0	− 1.2	1.8	89	−0.66
6	− 1.8	− 3.6	− 2.4	4.2	81	−0.57
7	0.0	1.2	− 9.0	7.8	98	−1.15
8	7.8	2.4	− 7.3	13.2	106	−0.59
9	23.4	2.4	− 3.0	24.0	85	−0.12
10	36.6	2.4	− 5.4	39.6	92	−0.14
11	47.4	2.4	− 6.0	51.0	275	−0.12
12	55.2	3.0	− 7.2	59.4	357	−0.12
13	58.8	3.0	− 7.2	63.0	343	−0.11
14	58.2	0.6	− 5.4	63.0	384	−0.08
15	51.6	− 1.2	− 6.6	59.4	375	−0.11
16	40.2	− 1.8	− 10.8	52.8	321	−0.20
17	25.2	− 3.0	− 21.0	49.2	392	−0.43
18	9.0	− 2.4	− 22.2	33.6	531	−0.66
19	− 2.4	− 4.2	− 17.4	19.2	505	−0.91
20	− 4.8	− 3.6	− 16.2	15.0	527	−1.08
21	− 4.2	− 3.6	− 9.6	9.0	381	−1.06
22	− 3.0	− 3.6	− 6.0	6.6	181	−0.91
23	− 3.6	− 3.0	− 6.6	6.0	195	−1.10
24	− 4.2	− 3.0	− 7.2	6.0	190	−1.20
Total cal cm^{-2}	+377	−31	−190	+598	--	
Equivalent mm	6.44	− 0.53	− 3.25	10.2	$R_I = 12.4$	

is for 12 July (Table 26–4), a cloudless day with a total solar radiation R_I of 725 cal cm^{-2}. Net radiation and wind speed were measured at 175 cm aboveground. Net radiation was measured with a small, unventilated net radiometer (Fritschen, 1963), soil heat flux with heat flux plates, and evaporation with a weighable lysimeter (Van Bavel and Myers, 1962).

Two facts are noteworthy. First, there was sensible heat extracted from the air (i.e., Q was negative) throughout the period and the transfer was demonstrably affected by wind speed from 1600 to 1900 hours. Second, the contribution of soil heat was not negligible, amounting to the equivalent of 0.5 mm evaporation for the day.

4) Hourly values of the components over grass at Aspendale, Australia, are presented in Table 26–5a. Net radiation was measured by a Funk type radiometer, the ground heat flux by heat flux plates, and evaporation as the average value from two weighed lysimeters of the installation described by McIlroy and Angus (1963). Sensible heat flow was found as a difference. The occasion was early summer, and a hot dry wind commenced during the night. The maximum temperature that day was 40C with intermittent clouds between 1400 and 1600 hours, followed by a "cool change" (common in that region in summer) at 1630 hours, which dropped the temperature 17C in < 15 min. The effect of the dry wind in producing negative values of β (especially from 1400 to 1600 hours), and then of the cool change in almost reversing its sign, is marked.

Table 26–5b gives values about 4 months later in the fall. In this case Q was

Table 26–5 *a*. Grassland, Aspendale, Australia, December 1, 1962; all values in cal cm^{-2} for the hour

Hour	H	S	Q	E	β
1	− 4.3	−0.3	− 4.8	0.8	−6.2
2	− 4.0	−0.8	− 4.0	0.8	−5.2
3	− 3.6	−0.9	--	--	
4	− 4.5	−0.9	− 6.5	3.0	−2.2
5	− 3.3	−1.0	− 6.7	4.5	−1.5
6	0.7	−0.9	− 4.3	5.9	−0.72
7	7.9	0.0	− 9.3	17.2	−0.54
8	18.8	0.9	− 2.1	20.1	−0.11
9	32.9	2.0	+ 1.9	29.1	0.06
10	40.4	2.8	−12.3	49.9	−0.25
11	50.9	3.1	−14.1	61.9	−0.23
12	60.5	3.9	− 9.8	66.4	−0.15
13	68.6	3.9	−10.6	75.2	−0.14
14	65.7	3.6	− 8.7	70.8	−0.12
15	43.9	3.3	−22.7	63.4	−0.36
16	33.3	2.4	−25.7	56.6	−0.45
17	34.6	2.1	− 0.3	32.8	−0.01
18	16.8	0.9	− 5.0	20.9	−0.24
19	0.7	0.2	− 6.2	6.7	−0.92
20	− 5.4	−0.5	− 7.9	3.0	−2.6
21	− 5.4	−0.5	− 6.3	1.4	−4.4
22	− 4.9	−0.6	− 5.8	1.4	−4.0
23	− 4.5	−0.7	− 5.3	1.4	−3.6
24	− 2.9	−0.6	− 3.8	1.4	−2.6
Total cal cm^{-2}	432.9	21.4	−180.3	594.6	
Equivalent mm	7.40	0.36	− 3.08	10.16	

Table 26–5 *b*. Grassland, Aspendale, Australia, March 28, 1963; values in cal cm^{-2} for the hour

Hour	H	S	Q	E	(S + E + Q)	β
12	36.8	4.8	--	16.0	--	--
3	39.6	5.3	14.6	18.6	38.5	0.78
14	37.7	4.9	9.3	24.2	38.4	0.38
15	32.0	3.9	10.8	18.3	33.0	0.59
16	23.2	3.2	3.8	15.3	22.3	0.25
17	12.1	1.3	− 5.8	16.4	11.9	−0.36
18	1.1	0.1	--	11.1	--	--

measured directly for a number of 10-min periods during each hour by means of the eddy correlation technique described in part III, section B. This is believed to be one of the few occasions where all four components of the energy balances have been measured by completely independent methods; the agreement between source and sink strengths is very good (*see* columns 2 and 6).

V.   A WORKING APPROXIMATION

A.   General

To use equation [26–12] there must be humidity and wind observations at two levels, and unless an arbitrary value of *d* is to be used, wind observations are needed at four or more levels to derive a value of *d*. Similarly, to exploit equations

[26–22] and [26–14], there is need of temperature and humidity measurements at two levels. The precision needed in all these measurements is great, and it is no use attempting to make them unless the equipment is good enough to measure accurately to better than 0.1C, better than 0.1 mbar (preferably to $\pm$ 0.02C and $\pm$ 0.02 mbar), and to about 1 cm sec^{-1} in wind speed. For research, with the resources of a good physics laboratory behind it, these objectives are attainable, but for routine use on irrigation areas, or for climatological surveys there is need for something simpler. There is a requirement for use of single-level observations, such as are made in a standard meteorological station, with an inaccuracy in the final result that is no worse than that in estimating rainfall, or quantity of irrigation water applied. A formal analysis of such a prediction model will be given in terms of single-level observations. The final expressions reveal several important qualitative features of the microclimatology of evaporation in irrigation agriculture.

B. Combination of Aerodynamic and Energy Balance Approaches

1. DERIVATION OF EXPRESSION FOR EVAPORATION

If in equations [26–11] and [26–13] the lowest level is taken as the virtual sink for momentum, i.e., $z_1 = z_o + d$, then the equations reduce to

$$Q = \rho c k^2 \frac{(T_o - T_2)\, u_2}{[\ln(z_2 - d)/z_o]^2} \text{ cal cm}^{-2} \text{ sec}^{-1} \qquad [26–29]$$

$$E = \lambda \frac{\rho \epsilon}{p} k^2 \frac{(e_o - e_2)\, u_2}{[\ln(z_2 - d)/z_c]^2} \text{ cal cm}^{-2} \text{ sec}^{-1}, \qquad [26–30]$$

where T_o and e_o are the virtual values of temperature and vapor pressure at the level where the virtual value of u is zero, and their combination represents a virtual relative humidity h_o, such that the saturation vapor pressure at T_o is e_o/h_o. Following the precedent of equations [26–24] and [26–25] in part IV, section E, [26–29] and [26–30] may be written as

$$Q = \gamma B_o (T_o - T_2) \qquad [26–31]$$

$$E = B_o (e_o - e_2). \qquad [26–32]$$

Using the slope of the saturation vapor pressure curve Δ as before (but now Δ is taken at the mean dry bulb temperature, or, in practice, at T_2) we have approximately

$$Q = \frac{\gamma B_o}{\Delta} \left(\frac{e_o}{h_o} - \frac{e_2}{h_2} \right) ,$$

where h_2 is the relative humidity at height z_2. Expanding,

$$Q = \frac{\gamma B_o}{\Delta h_o} \left[e_o - e_2 + e_2 \left(1 - \frac{h_o}{h_2} \right) \right]$$

$$= \frac{\gamma E}{\Delta h_o} - \frac{\gamma}{\Delta h_o} \times B_o e_2 \left(\frac{h_o}{h_2} - 1 \right) . \qquad [26–33]$$

For simplicity write

$$E_a = B_o e_2 (h_o/h_2 - 1), \qquad [26–34]$$

Table 26–6. Values of Δ/γ (dimensionless) at 1,000 mbar, temperature in °C

Temperature	Δ/γ
0	0. 67
5	0. 92
10	1. 23
15	1. 64
20	2. 14
25	2. 78
30	3. 56
35	4. 53
40	5. 70
45	7. 10
50	8. 77

because it has the structure of an evaporation term and, except for h_o, it is made up of quantities measured in the air at height z_2. Then

$$Q = \frac{\gamma}{\Delta h_o} (E - E_a)$$

$$= H - E, \text{ from equation } [26–21]$$

or

$$E = (\Delta H + \gamma E_a/h_o)/(\Delta + \gamma/h_o)$$

$$= \frac{[(\Delta/\gamma)H] + [(1/h_o)E_a]}{(\Delta/\gamma) + (1/h_o)} . \qquad [26–35]$$

Equation [26–35] (Penman, 1961) is a slightly modified form of one given earlier by Penman (1948) and independently by Ferguson (1952). The form of [26–35] has the advantage that Δ/γ is a dimensionless, temperature dependent constant, values of which are given in Table 26–6.

Equation [26–35] represents a weighting of radiation supply and aerodynamics in evaporation, since E_a involves u_2 in addition to a number of constants (see equations [26–30] and [26–34]). In irrigation agriculture, at the time when irrigation is most used, the weighting factor Δ/γ will be greater than the factor $1/h_o$, and hence it is permissible to have a little more uncertainty in E_a than in H. If the surface is saturated (open water, or a crop after rain or overhead irrigation) so that $h_o = 1.00$, and if there is no advection, experience shows that H is approximately equal to E_a, and hence, whatever the weighting factors. to a good first approximation the evaporation rate per day is close to the net radiation per day (see Table 26–3 and comments). When there is advection of sensible heat E_a exceeds H and the evaporation rate may be dominated by upwind aricity.

At the other extreme, when the surface is dry, h_o may be less than h_2 (equation [26–34]), E_a may be negative and E decreases toward zero. Over this range the site becomes an exporter of sensible heat, i.e., a source of advected energy for other sites downwind of it (see part V).

2. OPEN WATER EVAPORATION

The simplest form of equation [26–35] is when $h_o = 1.00$, a condition that may safely be presumed for an open water surface. Then $E_a = B_o (e_a - e_2)$ and with the appropriate values of B_o and H, equation [26–35] predicts the evaporation rate from an extended open water surface.

3. PLANT TRANSPIRATION

When h_o is not very much less than unity, as it might well be in an irrigated crop (this is a problem for experiment) it may be set equal to unity in the weighting factor to give

$$E_T = \frac{[(\Delta/\gamma)H_T] + E_a}{(\Delta/\gamma) + 1} \qquad [26\text{-}36]$$

where H_T is the net radiation over a green crop completely covering the ground, and E_a may be written either as

$$E_a = B_T e_2 \left(\frac{h_o}{h_2} - 1 \right) = B_T(h_o e_a - e_2), \qquad [26\text{-}37]$$

e_a being the saturation vapor pressure of the air at height z_a or as

$$E_a = B_T e_a(h_o - h_2) = B_T \left(\frac{h_o - h_2}{1 - h_2} \right) e_a(1 - h_2) = B_T \alpha(e_a - e_2) \qquad [26\text{-}38]$$

where B_T is written as a reminder that its value will depend on the aerodynamic roughness of the surface of the transpiring crop. The term $(e_a - e_2)$ is the saturation deficit at level z_2. The new quantity $\alpha = (h_o - h_2)/(1 - h_2)$ is expected to depend on plant characteristics, but will approach unity as h_o approaches unity. Near the limit, with $\alpha \approx 1.0$, equation [26-36] defines a *potential transpiration* rate expected to be applicable to actively growing vegetation, completely shading the ground, and sufficiently extensive for edge effects to be negligible. Equation [26-36] accounts for the influence of sensible heat from other areas (advection) through the E_a term. Extensive tests by Van Bavel (*unpublished data*) at Tempe, Arizona with alfalfa (*Medicago sativa* 'Moapa') demonstrated this experimentally. An exhaustive discussion and error analysis of equations [26-35] and [26-36] was given by Tanner and Pelton (1960).

In general, the reflection coefficient is much the same for all *complete* crop covers (Monteith, 1959) and hence, without important error, it may be assumed that H_T is nearly the same for all at a given time and place, so that the main discriminant between crops will lie in the product αB_T in equation [26-38]. The rougher the crop, the greater is B_T, but if this increases the evaporation rate then $(e_a - e_2)$ may decrease, so that $B_T(e_a - e_2)$ may have a conservative character, provided e_a and e_2 are not influenced by air moving in from adjacent sites with a differing situation. But α is not independent of roughness, and deserves separate consideration.

C. Leaf Morphology as a Factor in Actual Evaporation

Because the physical conditions have been idealized to represent a virtual sink for momentum and a virtual source of vapor at height $z_o + d$, it is not unreasonable to idealize this virtual source as one huge horizontal leaf at that level. The water vapor starts from inside the epidermis at a relative humidity of unity (or within 1% of it) and passes through the stomata to air at relative humidity h_o. Assuming uniform temperature, the internal, leaf-associated vapor pressure drop is $(e_o/h_o - e_o)$. From here the vapor is carried away by turbulent diffusion, passing level z_2 where the vapor pressure is $e_2 = e_o h_2$. The external vapor pressure drop is $(e_o - e_2)$. Assuming the flux is the same at all levels, the ratio of the pressure drops must be the ratio of the "resistances" to evaporation, conveniently expressed as effective lengths of still air that would have the same resistance to transport by molecular diffusion. Giving the symbol L_s for the leaf component. and L_e for the external component, then

$$\frac{L_s}{L_e} = \frac{e_o/h_o - e_o}{e_o - e_2}$$

or

$$\frac{L_e}{L_e + L_s} = \frac{h_o(e_o - e_2)}{e_o - h_o e_2} .$$ [26–39]

It is possible to show that this ratio is not very different from α when sensible heat flow Q is relatively unimportant. The quantity L_e is inversely proportional to B_T and can be calculated from it; the quantity L_s can be measured, (see Van Bavel et al., 1965), or for some leaves, calculated from known stomatal geometry and populations. For both, the order of magnitude is a few millimeters (Penman and Schofield, 1951) provided stomates are open. If the roughness of a crop can be imagined as suddenly increased, then any increased rate of evaporation that results can only be produced by a decrease in L_e, because L_s is a constant of leaf structure, and hence, if B_T is increased then L_e/L_s, α, and, consequently, h_o must decrease too. So the product αB_T has a conservative quality. In comparisons of different crops it might be expected that, if L_s is about the same for all, then there will be similar internal compensation for differences in roughness, leading to a general conclusion that water use at a given time and place might well be the same for all crops, provided they are short, leafy, completely shading the ground and have access to an adequate supply of water. This is a hypothesis to be tested against facts, but already there is a lot of field evidence to support the hypothesis—much of it collected by Penman (1963). There is evidence in conflict, and of the many possible causes of discrepancy, three are worth comment (see part V, section E).

D. Soil Water Availability as a Factor in Actual Evaporation

In general, the normal behavior of plants is to have their stomata open in daylight, and closed in the dark, but there are plants that keep the stomata open in the dark if the soil water content is great enough, and such plants may transpire a little more than normal plants because of extended transpiration opportunity. Conversely, others are reported subject to midday wilting even on wet soils, and these may transpire less than normal plants. There is much yet to be learned about the response of leaf stomata to soil water suction and about actual transpiration before ideas involving leaf structure can be made usefully quantitative. Some of the problems are considered in the report of the Madrid conference organized by UNESCO (1961). Dutch work (Makkink and Heemst, 1956) suggests that the range of suction over which the actual rate of water is equal to the potential rate is smaller, the greater the potential rate, with the magnitude of the effect varying with soil type.

E. Modifying Factors

1. INTERCEPTION

Since the same energy cannot be used twice, while intercepted water is evaporating, the transpiration stream must be reduced accordingly. This argument suggests that the evaporation rate from wet leaves will not be greater than from dry leaves, but evidence is not conclusive. Bernard (1954) measured a small increase from grass in the Congo basin; Burgy and Pomeroy (1958) could detect no difference in laboratory experiments, nor was there any in the field (McMillan and Burgy, 1960). It seems that the special situation produced by sprinkler irrigation has no very great effect on the rate of water use, but the problem deserves further field study.

2. CANOPY FLOW

The meteorological concept so far applied has been of air blowing over the top of the crop, but few crops have a canopy close enough to prevent air movement through the crop. In effect, evaporation is "forced," and E_a in equation [26–38] must be an underestimate of the drying power of the air. The magnitude of the effect will depend on the shape of the plant, and on plant spacing in the field. For row crops, ranging from cereals to orchard trees, the effect may depend on the orientation of the rows with respect to the direction of the prevailing wind, and if the row spacing is wide (of the same order as the height of the crop, or bigger) then the nature of the space could be important. In agricultural crops the space will be bare, kept so by management; in orchards there may be grass or some other short green crop. The microclimatology of many of these varied situations has yet to be started.

3. DEGREE OF COVER

The evaporation rate from wet bare soil is usually about the same as that from a short crop cover, but the drying is very superficial so that in weather that would make irrigation desirable the rate of evaporation decreases very rapidly to near zero within a few days; bare soil in summer behaves as though it were self-mulching (Buckingham, 1907). The shading and sheltering effect of row crops on the bare areas between them will decrease the initial rate of soil evaporation after wetting, but will also diminish the self-mulching action so that over a dry period of 7 to 14 days (depending on climate) the shaded and sheltered bare soil might lose a little more water than the same soil fully exposed to sun and wind (Mather, 1954).

For an annual crop the water use for some time after emergence will be dominated by the evaporation from bare soil. As the transpiration becomes more important there is a phase in which the rate of water loss is proportional to leaf area, i.e., to plant population per unit area. As the leaves on single plants begin to shade each other, and plants begin to overlap, there is a gradual transition to a state in which *area of ground shaded* becomes dominant. Limited British experience on irrigated sugar beet (*Beta vulgaris* L.) and potatoes (*Solanum tuberosum*) suggests that "full cover" may be presumed to have been achieved when the plants have met in the rows, even though bare soil can still be seen between the rows. It is possible that at this stage the crop has its maximum aerodynamic roughness, and thereafter the increased area of ground shaded has no important effect on transpiration because the increased interception of radiant energy is compensated by a decreased effect of turbulent transfer.

VI. A NOTE ON ADVECTION

The energy balance approach (equation [26–18]) is exact when applied at the surface of the ground, as a method of calculating evaporation rates. However, it is necessary to make the meteorological observations at some height above the surface. A complete formal analysis (D. E. Angus, 1963. The influence of meteorological and soil factors on the rate of evapotranspiration of a crop. *Ph.D. thesis, Univ. of California, 128 p.*) shows that application of the Bowen ratio may lead to errors if there are *horizontal* gradients of temperature or humidity. It is easy to overemphasize the effect of advection, but when the air temperature is partly determined by the import of hot air from a dry area upwind or of cool air from a wet area, the vertical temperature and humidity gradients cannot be used without error in computing a Bowen ratio. The difference in behavior occurs because the transport constants, K_H and K_V in equations [26–7] and [26–8] do not account for any mass flow between the surface and height of measurement but only for turbulent exchange at that height. The complete argument shows that use of the uncorrected Bowen ratio leads to an overestimate of the sensible heat flux, and an underestimate of the latent heat flux at the surface. When correction is complete, two points are important. First, the estimate of evaporation is true only for the place where it is made; the value will be bigger upwind and smaller downwind, and the choice of position of the meteorological station in relation to the boundary of the irrigated area may have more important effects than neglect of possible errors in the Bowen ratio. As already shown (equation [26–26]), the parameter that governs the total energy transfer (sensible plus latent heat) is the wet bulb temperature. If, as is often approximately the case, $(H - S)$ is the same both in an irrigated area and upwind of it, the additional sensible heat brought in from upwind is the same as the additional latent heat carried out; under these conditions the horizontal gradient of wet bulb temperature is zero. In this sense, the presence of hot dry air over such a surface supplies no *additional* energy, but merely causes a redistribution of the surface available energy which goes into E and Q. This effect is very marked near the upwind boundary, where Q may in fact be negative, but diminishes with increasing distance downwind, as the air becomes cooler and moister.

From this the second point emerges. There is some evidence to suggest that the range over which advective effects vary with downward distance (called border or edge effects) is of the same order as the desirable fetch for reliable measurements of wind velocity profiles, which is about 100 times the height of the highest observation level above the crop. Additionally, however, for a given horizontal gradient, the effect is known to be roughly proportional to the wind speed, so apparent conflicts in experience may arise from this source. Argus (D. E. Angus, 1963. The influence of meteorological and soil factors on the rate of evapotranspiration of a crop. *Ph.D. thesis, University of California. 128 p.*) in California, USA, found that border effects were still important at nearly 200 m downwind when measurements were made at 100 cm height over irrigated pastures at a wind speed of 15 m sec^{-1} (almost a full gale) whereas Monteith (*private communication*) collected evidence in Israel (inferences from the appearance of the crops) that border effects were unimportant beyond about 100 m from the boundary, an impression that is supported by a generalized diagram of Stanhill (1961, Fig. 2). Data collected by Van Bavel and Fritschen (*private communication*) indicate that in Arizona during periods of low wind speed (1 to 2 m sec^{-1}) edge effects over wet, bare soil were minor within 1 m from the upwind boundary. Working at Aspendale, Australia, McIlroy and Angus (1964) found evaporation rates in excess of the latent heat equivalent of net radiation even with the surroundings thoroughly moist for many kilometers all around.

The scarcity of reliable energy balance data prevents any firm assessment of the importance of advection, but even when more are available it may not be possible to extract from them a general method for the prediction of advection effects; local rules based on local research may be the only solution.

LITERATURE CITED

Baumgartner, A. 1956. Warme- und Wasserhaushalt eines jungen Waldes. Ber. Deut. Wetterdienstes Nr. 28, 5, p. 53.

Bernard, E. A. 1954. Sur diverses consequences de la methode du bilan d'energie pour 1 'evapotranspiration. C. R. Ass. Int. Hydrologie Sci. Rome 3:161–167.

Bowen, I. S. 1926. The ratio of heat losses by conduction and by evaporation from any water surface. Phys. Rev. 27:779–789.

Buckingham, E. 1907. Studies on the movement of soil moisture. US Dep. Agr. Bur. Soils Bull. 38. 61 p.

Burgy, R. H., and C. R. Pomeroy. 1958. Interception losses in grassy vegetation. Amer. Geophys. Union, Trans. 39:1095–1100.

Deacon, E. L. 1949. Vertical diffusion in the lowest layers of the atmosphere. Quart. J. Roy. Meteorol. Soc. 75:89–103.

Deacon, E. L. 1953. Vertical profiles of mean wind in the surface layer of the atmosphere. Great Brit. Meteorol. Office. Geophys. Mem. No. 91. 68 p.

Deacon, E. L., and W. C. Swinbank. 1958. Comparison between momentum and water vapor transfer. UNESCO Arid Zone Res., Climatol. Microclimatol. Proc. Canberra Symp. 11:38–41.

Dyer, A. J. 1961. Measurements of evaporation and heat transfer in the lower atmosphere by an automatic correlation technique. Quart. J. Roy. Meteorol. Soc. 87:401–412.

Dyer, A. J., and W. O. Pruitt, 1962. Eddy-flux measurements over a small irrigated area. J. Appl. Meteorol. 1:471–473.

Ferguson, J. 1952. The rate of evaporation from shallow ponds. Australian J. Sci. Res. 5:315–330.

Fritschen, L. J. 1963. Construction and evaluation of a miniature net radiometer. J. Appl. Meteorol. 2:165–172.

Fritschen, L. J., and C. H. M. van Bavel. 1964. The energy balance as affected by height and maturity of sudangrass. Agron. J. 56:201–204.

Funk, J. P. 1959. Improved polythene-shielded net radiometers. J. Sci. Instr. 36:267–270.

Inoue, E. 1955. Studies of the phenomena of waving plants ("Honami") caused by wind, Part 1. J. Agr. Meteorol. (Japan) 11:87–9C.

Makkink, G. F., and H. D. J. van Heemst. 1956. The actual evapotranspiration as a function of the potential evapotranspiration and the soil moisture tension. Neth. J. Agr. Sci. 4:67–72.

Mather, J. R. 1954. The measurement of potential evapotranspiration. Publ. Climatol., Seabrook, New Jersey. Vol. 7, No. 1. 225 p.

McIlroy, I. C. 1961. Effect of instrumental response on atmospheric flux measurements. Div. Meteorol. Phys. CSIRO (Australia). Tech. Pap. No. 11. 30 p.

McIlroy, I. C., and D. E. Angus. 1963. The Aspendale multiple weighed lysimeter installation. Div. Meteorol. Phys. 1. CSIRO (Australia). Tech. Pap. No. 14. 27 p.

McIlroy, I. C., and D. E. Angus. 1964. Grass, water, and soil evaporation at Aspendale. Agr. Meteorol. 1:201–224.

McMillan, W. D., and R. H. Burgy. 1960. Interception loss from grass. J. Geophys. Res. 65:2389–2394.

Monteith, J. L. 1959. The reflection of short-wave radiation by vegetation. Quart. J. Roy. Meteorol. Soc. 85:386–392.

Pasquill, F. 1949. Eddy diffusion of water and heat near the ground. Roy. Soc., Proc. (London) Ser. A 198:116–140.

Pasquill, F. 1950. Some further considerations of the measurement and indirect evaluation of natural evaporation. Quart. J. Roy Meteorol. Soc. 76:287–301.

Penman, H. L. 1948. Natural evaporation from open water, bare soil, and grass. Roy. Soc., Proc. (London) Ser. A 193:120–145.

Penman, H. L. 1961. Weather, plant, and soil factors in hydrology. Weather 16:207–219.

Penman, H. L. 1963. Vegetation and hydrology. Commonwealth Bur. Soils. Harpenden. Tech. Comm. No. 53. p. 124.

Penman, H. L., and I. F. Long. 1960. Weather in wheat. Quart. J. Roy. Meteorol. Soc. 86:16–50.

Penman, H. L., and R. K. Schofield. 1951. Some physical aspects of assimilation and transpiration. Symp. Soc. Exp. Biol. 5:115–129.

Priestley, C. H. B. 1959. Turbulent transfer in the lower atmosphere. Univ. Chicago Press, Chicago. 297 p.

Rider, N. E. 1954. Eddy diffusion of momentum, water vapor, and heat near the ground. Phil. Trans. Roy. Soc. (London) Ser. A 246:481–501.

Sheppard, P. A. 1958. Transfer across the earth's surface and through the air above. Quart. J. Roy. Meteorol. Soc. 84:205–224.

Slatyer, R. O., and I. C. McIlroy. 1961. Practical microclimatology. UNESCO (Paris) 300 p.

Stanhill, G. 1961. A comparison of methods of calculating potential evapotranspiration from climatic data. Israel J. Agr. Res. 11:159–171.

Suomi, V. E., and C. B. Tanner. 1958. Evapotranspiration estimates from heat-budget measurements over a field crop. Amer. Geophys. Union Trans. 39:298–304.

Sutton, O. G. 1953. Micrometeorology. McGraw-Hill, New York. 333 p.

Swinbank, W. C. 1955. An experimental study of eddy transports in the lower atmosphere. Div. Meteorol. Phys. CSIRO (Australia), Tech. Pap. No. 2. 30 p.

Tanner, C. B., and W. L. Pelton. 1960. Potential evapotranspiration estimates by the approximate energy balance method of Penman. J. Geophys. Res. 65:3391–3413.

Thornthwaite, C. W., and B. Holzman. 1939. The determination of evaporation from land and water surfaces. Monthly Weather Rev. 67:4–11.

UNESCO. 1961. Plant-water relations in arid and semiarid conditions. UNESCO Arid Zone Res. 16. Proc. Madrid Symp. 352 p.

US Navy Electronics Laboratory. 1950. Water loss investigations: A review of evaporation theory and development of instrumentation. US Navy Electron. Lab. Rep. 159. 71 p.

Van Bavel, C. H. M., and L. E. Myers. 1962. An automatic weighing lysimeter. Agr. Eng. 43:580–583, 586–588.

Van Bavel, C. H. M., F. S. Nakayama, and W. L. Ehrler. 1965. Measuring transpiration resistance of leaves. Plant Physiol. 40:535–540.

27

Plant Factors Affecting Evapotranspiration[1]

DAVID M. GATES

*Missouri Botanical Garden
and Washington University
St. Louis, Missouri*

R. J. HANKS

*Agricultural Research Service, USDA
Fort Collins, Colorado*

Plant factors that influence evapotranspiration are many and complex. Because of the complexity of the subject and space limitations, we have chosen to emphasize basic principles rather than detailed reporting of all of the literature relating to this subject. The subject will be discussed first in terms of single plants and second in terms of plant communities. See chapters 17 and 23 for discussions of internal plant factors affecting water flow through plants and transpiration and chapters 18 and 29 for summaries of measurements of internal water status, transpiration, and evapotranspiration.

I. SINGLE PLANT FACTORS THAT INFLUENCE EVAPOTRANSPIRATION

A plant leaf loses water at a rate that depends on the concentration gradient of water vapor between the saturated cell walls of the mesophyll and that in the free air beyond the plant. The rate is regulated by the diffusion resistance in the substomatal, stomatal, and boundary layer regions of the diffusion path. The water vapor density at the saturated mesophyll cell walls depends on the temperature of the leaf which in turn depends on the energy budget of the leaf. These two mechanisms, the energy budget and the transpiration rate, operate interdependently to reach an equilibrium for given environmental conditions.

The transpiration rate E per unit area of leaf surface can be expressed in the following manner:

$$E = \frac{\rho_l(T_l) - \rho_a(T_a)}{R} \qquad [27\text{-}1]$$

where

$\rho_l(T_l)$ = water vapor concentration at the mesophyll cell walls at the leaf temperature T_l,

$\rho_a(T_a)$ = water vapor concentration in the free air beyond the leaf at the air temperature T_a, and

---

[1] Joint contribution from the Univ. of Mich. Biol. Sta. and the Northern Plains Br. Soil and Water Conserv. Res. Div. Agr. Res. Serv. US Dep. Agr.

R = resistance of the diffusion path from cell walls to free air beyond the leaf.

If the water supply to the plant is not adequate, a deficit will occur which will result in the water vapor density at the mesophyll cell walls dropping below the saturation value. This will result in a transpiration rate less than the maximum possible rate which occurs when the water supply is adequate. To estimate the maximum rates of transpiration, the following formulation may be made:

$$E = \frac{{}_s\rho_l\,(T_l) - \text{rh}\; {}_s\rho_a(T_a)}{R} \qquad [27\text{--}2]$$

where

$\qquad {}_s\rho_l(T_l)$ = water vapor concentration at saturation at the leaf temperature T_l,
$\qquad {}_s\rho_a(T_a)$ = water vapor concentration at saturation at the air temperature T_a,
$\qquad$ and
$\qquad$ rh = relative humidity of the air at temperature T_a.

The diffusion resistance within the leaf depends on leaf morphology and light intensity. Under sufficient illumination when the stomata are fully open, the diffusion resistance will depend on the species of the plant cover. The external resistance which occurs in the boundary layer adjacent to the leaf surface will depend on the windspeed but will be rapidly increased at very low windspeeds. Recently, Lee and Gates (1964) calculated values of diffusion resistance for three plant species. The resistance of the substomatal cavity amounts to only a few per cent of the total resistance. The resistance of the boundary layer around single leaves in still air may amount to 20% of the total resistance for small stomata openings but may represent > 50% of the total for large stomatal openings. Under moderately windy conditions the boundary layer is nearly nonexistent, and the diffusion resistance to transpiration primarily occurs in the stomatal aperture.

Typical values of diffusion resistance under windy conditions and full sunlight for single leaves are 3.0 to 10.0 sec cm^{-1} and will run from 4.0 to 12.0 sec cm^{-1} under still air conditions. Higher values of diffusion resistance will occur when the light intensity is low or when a dense canopy of leaves is formed.

A leaf will assume a given temperature depending on the energy transferred to or from it by radiation, convection, and transpiration. The energy budget of a single leaf may be written in the following form:

$$a_S\, S + a_s\, s + r_s\, a_s\,(S + s) + a_t\,(R_a + R_g) - \epsilon\sigma\, T_l^4$$
$$\pm\, C \pm E \pm P = 0 \qquad [27\text{--}3]$$

where

$\qquad S$ = direct solar radiation,
$\qquad s$ = diffuse sky light,
$\qquad R_a$ = long-wave thermal radiation from the atmosphere,
$\qquad R_g$ = long-wave thermal radiation from the ground or underlying plants,
$\qquad a_S$ = absorptance of leaf to direct sunlight,
$\qquad a_s$ = absorptance of leaf to diffuse sky light,
$\qquad r_s$ = reflectance of underlying surface to direct sunlight and diffuse sky light,
$\qquad a_t$ = absorptance of leaf to long-wave thermal radiation,
$\qquad \epsilon$ = emittance of leaf surface to long-wave thermal radiation,
$\qquad \sigma$ = Stefan-Boltzmann constant for blackbody radiation,

T_l = temperature of the leaf,
C = gain or loss of energy by convection,
E = gain or loss of energy through condensation or transpiration, and
P = energy gained through photosynthesis or lost by respiration.

The convection gain or loss of energy per unit area from a broad leaf is proportional to the difference in temperature, ΔT, between the leaf and the air as follows:

$$C = h_c\,\Delta T = 6.0 \times 10^{-3}\,(\Delta T/L)^{1/4}\,\Delta T \qquad [27\text{–}4]$$

where C is in cal cm^{-2} min^{-1}, ΔT is the difference between leaf and air temperature in °C, h_c is the convection coefficient in cal cm^{-2} min^{-1} °C^{-1}, and L is the width of the leaf. For conifers the heat transfer by convection and radiation has been discussed by Tibbals et al. (1964).

Equation [27–3] demonstrates that the leaf temperature will depend on the incident radiation as well as on convection and transpiration. The temperatures of sunlit leaves will often be 10C higher than air temperature (Gates, 1963). The temperatures of shaded leaves will often be several degrees lower than air temperature. The warmer leaves will lose the greatest amount of water; however, the shaded leaves will usually outnumber the sunlit leaves by a considerable factor. If a crop forms a dense canopy then all of the incident sunlight will interact with leaves and warm them. The problem of estimating evapotranspiration from a crop or stand of plants has been particularly difficult because of the complex nature of the canopy. However, one can make certain estimates based on general considerations which are of value.

II.   COMMUNITY PLANT FACTORS THAT INFLUENCE EVAPOTRANSPIRATION

A.   General

If a canopy is at an average temperature approximating air temperature, the maximum transpiration E can be estimated if one assumes a mean diffusion resistance R for the leaves of the canopy according to the following formula:

$$E = \frac{{}_s\rho_a\,(T_l) - \text{rh}\ {}_s\rho_a\,(T_a)}{R} = \frac{(1 - \text{rh})}{R}\ {}_s\rho_a\,(T_a). \qquad [27\text{–}5]$$

Expressed in this manner the diffusion resistance will be an apparent resistance only approximating the actual resistance.

Actually, the canopy temperature will change with the net radiation and the diffusion resistance will be light dependent at low light levels. In addition, the total diffusion resistance will depend on the windspeed and the amount of eddy diffusion in the canopy. The maximum evapotranspiration will occur when there is a small amount of wind, just enough to wipe away the boundary layer on the leaves.

A plot of transpiration E vs. resistance R gives a series of rectangular hyperbolae for various air temperatures and relative humidities as shown in Fig. 27–1. The temperature of sunlit leaves is usually several degrees above air temperature, and the temperatures of shade leaves are below air temperature. A reasonable

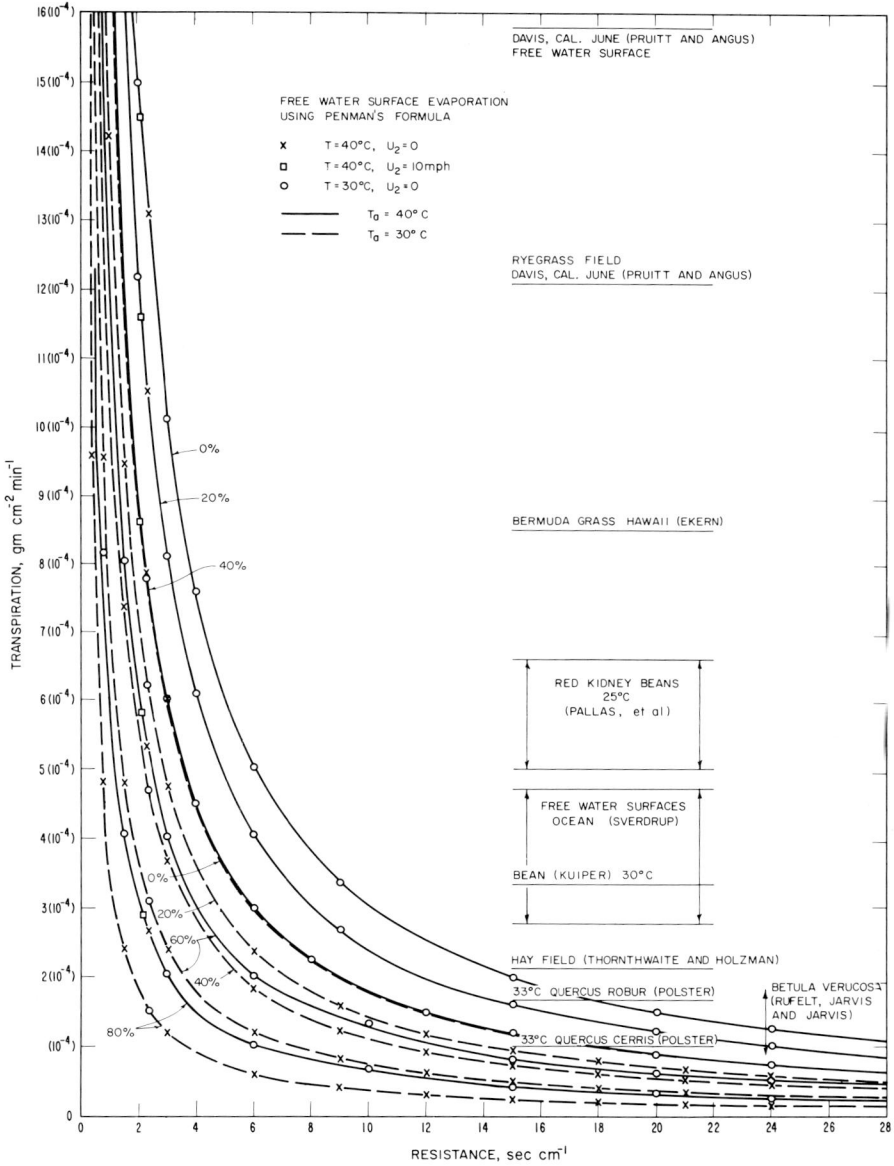

Fig. 27–1. Transpiration as a function of "apparent" diffusion resistance if the transpiring surface is at the same temperature as the air. Values for measured transpiration rates are shown at the right side of the figure and when projected to the appropriate air temperature curve to the left will give the approximate "apparent" resistance for the particular surface. Air temperatures are given only when available from the original reference.

approximation is to consider the canopy temperature to be near the air temperature. It is on this basis that Fig. 27–1 is derived. Shown on the same graph is the evapotranspiration from each of several plants, stands, and free water surfaces. The temperature of each of these surfaces is given in Fig. 27–1 only when it was available in the original reference.

The measurements for English oak (*Quercus robur*) and European turkey oak (*Quercus cerris*) were done by Polster (1963) *in situ* in the canopy of the tree. The birch (*Betula verucosa*) was measured by Rufelt et al. (1963) in the laboratory, as was red kidney beans (*Phaseolus vulgaris* L.) by Pallas et al. (1961) and Kuiper (1961). The observations of the hayfield (Thornthwaite and Holzman, 1942), bermudagrass (*Cynodon dactylon*) (Ekern, 1965), and ryegrass (*Lolium perenne*) (Pruitt and Angus, 1961) were done *in situ* with lysimeters or similar instruments. The values for free water surfaces of the ocean would represent cool temperatures, probably about 15C and high relative humidities. The free water surface observed at Davis, California, USA, represents nearly the upper limit achieveable under hot, very dry conditions. Using the formula for the evaporation from a free water surface E_o as given by Penman (1948), one can compute the evaporation as a function of the temperature and relative humidity as follows:

$$E_o = 0.35(1 + 9.8 \times 10^{-3} u_2)(p_s - p_a) \qquad [27–6]$$

where p_s = vapor pressure at the surface temperature in millimeters of Hg, p_a = vapor pressure of the air, and u_2 = windspeed in miles per hour at a height of 2 m above the surface. The computed points for zero wind conditions are shown in Fig. 27–1 and indicate an "apparent" diffusion resistance of 2.2 sec cm^{-1}. For a windspeed $u_2 = 10$ miles/hr, the resistance falls to 2.05 sec cm^{-1}, and the points are also shown in Fig. 27–1. The drop in diffusion resistance with increased temperature is very slight. By projecting each line representing the evaporation rate from the plant or water to the left until it intersects the appropriate air temperature and relative humidity curve, the approximate "apparent" resistance can be obtained. The air temperature was known for several of the examples shown in Fig. 27–1, but the relative humidities were not. This set of curves will give the maximum transpiration for a given "apparent" resistance, air temperature, and relative humidity. Lower temperature curves could also have been shown but would add confusion to the present diagram. If wind blows across a water surface or through a canopy, it will increase the evaporation and diminish the "apparent" diffusion resistance. Plants will generally possess resistances between 6 and 12 sec cm^{-1} and hence will produce evapotranspiration rates $< 3.0 \times 10^{-4}$ g cm^{-2} min^{-1} (0.17 inch/day). Notable exceptions would appear to be ryegrass and Bermudagrass. For conditions wherein the temperature of the evaporating surface is substantially above or below air temperature, the comparison using Fig. 27–1 is not valid, and a more intricate procedure is required (Gates, 1964).

For an opaque surface, the ground surface composed of soil, water, or a plant canopy, the energy budget given in equation (27–3) may be written in simpler form, since there is no transmitted energy and the only loss of incoming radiation is by reflection. If R_n is the net radiation, short-wave and long-wave included, and if energy consumed in photosynthesis is neglected, then for steady-state conditions not involving storage of heat, one can write:

$$R_n = (1 - r)(S + s) + \epsilon R_a - \epsilon \sigma T_s^4 = C + ET \qquad [27–7]$$

where r = reflectance of the whole surface, R_a = thermal long wave downward radiation from the atmosphere, T_s = surface temperature, ET = evapotranspiration and the other symbols are the same as before. For most surfaces, including water and plants, $\epsilon \cong 0.97$.

The transport of water vapor, ET, and the transport of energy by convection or eddy diffusivity C are controlled by the same process; namely by a vapor gradient or by a temperature gradient. Bowen (1926) suggested expressing the ratio of the two quantities as a characteristic of a particular surface. Hence,

$$\frac{C}{ET} = \beta = \gamma \frac{(T_s - T_a)}{(p_s - p_a)} \qquad [27\text{–}8]$$

where β is Bowen's ratio, γ is a proportionality constant, p_s = vapor density at the surface, and p_a = vapor density in the air. The same expression can be written in terms of the vapor pressure rather than the vapor density:

$$R_n = (1 + \beta)\ ET. \qquad [27\text{–}9]$$

If β is a constant, then the evaporation rate from a surface will be directly proportional to the net radiation at the surface. Combining equations [27–5] and [27–9], one can write:

$$\frac{R_n}{1 + \beta} = \frac{(1 - \text{rh})\ {}_s\rho_a\ (T_a)}{R} \qquad [27\text{–}10]$$

or

$$\beta = \frac{R\ R_n}{(1 - \text{rh})\ {}_s\rho_a\ (T_a)} - 1. \qquad [27\text{–}11]$$

When $\beta = 0$, all of the net radiation at the surface is dissipated by evapotranspiration. When $\beta = \infty$, the evaporation rate is zero, the vapor pressure difference is zero, and all the net radiation at the surface is dissipated by convection or eddy diffusion. Negative values of β may occur under certain conditions whereby either C or ET is negative and contributing energy to the surface. A negative value for ET implies condensation or dew formation. For negative values, equation (27–10) no longer holds and would need to be reformulated. Penman (1948) has reported the Bowen ratio for evaporation from a pan of open water surrounded by short vegetation to have values from 0 to 0.5 with a few negative values. Small values of β require that the net radiation not be too large for a given vapor pressure difference. During midday, when the net radiation may become large, approximately 1.0 cal cm^{-2} min^{-1}, the value of β may be as large as 6.0 and eddy diffusion will take away more energy than evaporation. Gates (1964) discussed the values of Bowen's ratio for single leaves and for crops and showed values of 6.0 near midday for single horizontal leaves. For well-watered crops, the value is frequently about 0.1.

A horizontal leaf may receive a large net radiation and hence possess a high value of β, whereas a vertical leaf may receive a very low net radiation and possess a small value of β. A plant canopy is made up of leaves of many different orientations, but for the canopy as a whole the net radiation received may be fairly large. However, due to the density of the canopy, the movement of air is inhibited and many shaded leaves may have temperatures lower than air temperature. Hence the low value of β for crops.

III. SPECIFIC PROPERTIES OF THE PLANT COMMUNITY
THAT INFLUENCE EVAPOTRANSPIRATION

Many properties of the plant canopy influence evapotranspiration. Some of the more important ones will be discussed in this part. Emphasis will be on plants that are of agronomic importance where grown under adequate soil water conditions.

A. Plant Species

Plant species differ as to time of the year when growth is made, rooting depth, density, spacing, height, orientation, etc. and, because of these differences, influence evapotranspiration. However, it is the primary object of this section to discuss differences in evapotranspiration due to species when other factors are constant.

Different plant species that are short, dense, and uniformly vegetated, actively growing, infinite in extent, and transpiring under unlimited soil water have virtually identical evapotranspiration. Penman (1956) has made two broad generalizations supporting this contention. They are:

"1) For complete crop covers of different plants having about the same color, i.e., the same reflexion coefficient, the potential transpiration rate is the same irrespective of plant or soil type.

2) This potential transpiration rate is determined by the prevailing weather."

(Note potential transpiration above refers to the condition with nonlimiting soil water.) For many crops this generalization appears to hold fairly well. The results of many investigations show equal evapotranspiration for well-watered, large alfalfa (*Medicago sativa*) fields compared with well-watered large grass meadows. Many crops do not have complete cover, at least for part of the season, so one would expect evapotranspiration to be less. Moreover, some crops such as pineapple (*Ananas sativus*) do not fit this generalization at all. In North Carolina, USA, Van Bavel and Wilson (1952) concluded that "a closed vegetation cover under equal meteorological conditions disposes of soil water with equal rapidity, regardless of the botanical composition."

A dense pineapple crop, with good soil water conditions, has a much smaller evapotranspiration than does a dense grass grown under identical conditions (Ekern, 1965). This difference is apparently due to the unique stomatal characteristics of the pineapple plant. Pineapple stomata are generally closed during the day and open at night, which is the reverse of most plants. Evapotranspiration from pineapple is greater after planting when the extent of plant cover is less than at full growth. This is exactly the opposite found for most other agronomic crops.

Angus (1959) has discussed the influence of crop characteristics on evapotranspiration. He considers the length of time the crop is in leaf to be the main effect of crop type on evapotranspiration. The percentage of ground cover also has an important effect on water use. Although he states: ". . . it has not been demonstrated by reliable experiments that there is any marked difference in the potential evapotranspiration rates of various plants," he gives reasons (but not data) why different crops cannot be expected to have exactly the same evapo-transpiration. The reasons given include: (i) different reflection; (ii) different insulating properties; (iii) different influences on turbulence; and (iv) transpiration usually ceases during night because of stomatal closure (implies that different crops may have different stomatal closure patterns.)

Burton (1959) has indicated several instances of improved water use efficiency through plant breeding. His evaluation is based on the premise of equal evapo-transpiration for different plants with all other conditions equal. The instances cited include: (i) development of early or late maturing varieties; (ii) development of varieties adapted to take advantage of the rainy season (effect on evapo-transpiration); (iii) development of varieties that remain green during periods of high moisture stress; and (iv) development of varieties and cultural practices for influencing rooting depth.

B. Influence of Light Reflected from Plants

Reflection is a factor that has been mentioned by Penman (1956) and others as a factor that influences evapotranspiration. Since reflection from different crops varies with color, degree of crop cover, crop structure, etc., one would expect evapotranspiration to be influenced to the degree that reflection influences net radiation (*see* equation 27–7).

Montieth (1959) and Haise et al. (1963) have recently made measurements of reflection. Table 27–1 shows some of the data obtained. These studies indicate that reflection from most dense crops varies from 20 to 30%. The reflection from bare soil varies from 11 to 23%. Changes in plant color appear to have very little influence on reflection.

Differences in evapotranspiration caused by differences in reflection are quite modest—certainly < 25%. For the humid region where evapotranspiration is primarily dependent on net radiation, an extreme change in reflection of 15 to 30% would cause a change in evapotranspiration of < 25% (assuming back radiation equal to 25% of incoming radiation).

Table 27–1. Solar reflectance by several crops and surfaces

Crop or surface	Reflectance (0. 3 - 2. 5 mμ)	
Soil, silt loam, dry (before cultivation)	0. 23	Haise et al. *
Soil, silt loam, dry (after cultivation)	0. 15	Haise et al. *
Soil, clay loam, wet	0. 11	Montieth (1959)
Soil, clay loam, dry	0. 18	Montieth (1959)
Grass	0. 24-0. 26	Montieth (1959)
Meadow, 18 inches high	0. 25	Haise et al. *
Alfalfa	0. 16-0. 22	Haise et al. *
Barley, full cover, boot stage to maturity	0. 21-0. 22	Haise et al. *
Wheat (spring)	0. 14-0. 20	Montieth (1959)
Wheat (winter)	0. 14-0. 27	Montieth (1959)
Corn, 2 to 7 feet tall	0. 16-0. 17	Haise et al. *
Sugar beets	0. 14-0. 24	Montieth (1959)
Sugar beets, 1/4 cover to full cover	0. 13-0. 29	Haise et al. *
Potatoes	0. 17-0. 27	Montieth (1959)

* H. R. Haise, R. J. Hanks, M. E. Jensen. Solar reflectance from soil and crop surfaces. Unpubl. report.

C.   The Influence of Plant Cover on Evapotranspiration

The influence of plant cover on evapotranspiration has been recognized by most investigators. The factor is particularly important for row crops that start out from a bare soil and approach 100% cover at maturity.

Two principal factors associated with degree of crop cover can be expected to influence evapotranspiration. The first is related to reflection. Reflection of light from bare soil, especially wet soil, is usually lower than that from a dense crop. Based on reflection alone, evapotranspiration would be expected to increase as the per cent of cover decreased.

The second factor is related to the relative ease with which water is evaporated from a bare soil compared with the transpiration from a crop. Evaporation from most bare soils decreases rapidly 1 or 2 days after an irrigation or a rain. The flow of water in the soil to the soil surface rapidly becomes less than potential evaporation. Under the same conditions, transpiration may not be limited as much as 2 weeks later. Thus, soil water content may be sufficient to maintain transpiration but not evaporation from the soil and, consequently, transpiration would increase as the per cent of cover increased.

Of the two factors affecting evapotranspiration, the second factor is by far the most important. Almost without exception, studies have shown that evapotranspiration increases as the per cent of cover increases up to 50% cover. Tanner and Lemon (1962) have further shown that evapotranspiration was higher for a crop with a wet soil surface than one with a dry soil surface.

Table 27–2 (computed by Tanner and Lemon, 1962, from data of Shaw and Fritschen, 1960) shows this influence for corn (*Zea mays* L.) The data show that evapotranspiration for the short crop (with the soil bare or covered with plastic) was less than one-half of that where the plants were taller and covered much more area. Denmead et al. (1962) indicated that the energy absorbed by the soil was about equal to that absorbed by a corn crop for the season.

Complete cover is not always necessary for maximum evapotranspiration. Marlatt (1961) found that if the per cent cover was > 50%, evapotranspiration from mowed orchardgrass (*Dactylis glomerata*) planted in rows of varying width (to vary per cent cover) either was equal to or slightly greater than evapotranspiration at full cover. Tanner (1963) found similar results for irrigated snapbeans (*Phaseolus vulgaris* L.) and potatoes (*Solanum tuberosum*) in Wisconsin, USA. Swan et al. (1963) found a good relation between evapotranspiration and net radiation for both crops provided the per cent cover was > 50%.

D.   Influence of Plant Population, Row Spacing, and Orientation on Evapotranspiration

Plant population has an influence on evapotranspiration similar to that of crop cover. Tanner, et al. (1960) gave an excellent summary of this influence as follows:

"On a given type of soil, the total ET depends . . . on the water available to the plants, as well as that available at the soil surface, and upon the total R_n above the corn and at the soil surface. When water is readily available both to the plant and at the soil surface, maximum ET obtains. Under these conditions, population and other practices

Table 27–2. Comparison of ET/R_n for crops with different cover
(Shaw and Fritschen, 1960)

Crop	Surface	ET/R_n
Corn, 6/10 to 6/26/59	3-18 inches high-soil covered w/plastic	0.17
Corn, 7/10 to 8/26/59	42-84 inches high-soil covered w/plastic	0.52
Corn, 6/10 to 6/26/59	3-18 inches high-bare soil	0.28
Corn, 7/10 to 8/26/59	42-84 inches high-bare soil	0.65

which affect the R_n at the soil surface have little effect on total ET. For example, the corn on high population planting will intercept more energy than that on a low population resulting in higher transpiration. However, the evaporation from the soil will be in proportion to the energy transmitted to the soil. Thus, the evaporation from the low population field is higher than that from the high population, making up for the difference in the transpiration . . . .

"If the soil surface is dry . . . so that the evaporation is limited by the moisture supply at the soil, a different picture develops. In this case, the transpiration from the high population corn will again exceed that from the low population. The difference in evaporation on the two populations will be small compared to the difference in transpiration because it is limited by the moisture supply and not by the radiation. Thus, there is less ET and greater sensible heat loss with low population."

The same authors also found no difference between east-west and north-south drilled rows. Measurements of the incident solar energy transmitted to the soil through the crop canopy were about the same as, and often less than, the net radiation exchanged at the soil because of long-wave radiation from the plant to the soil during the day. Thus, because of thermal radiation and the near infrared transmission of the crop, visual estimates of plant cover, based on the amount of cover, appear to be of limited value for estimating radiation interception.

Bowers et al. (1963) showed that although net radiation over the crop was only slightly more on 40-inch sorghum (*Sorghum vulgare* Pers.) rows compared with 20-inch rows, there was no difference between populations of 105 kiloplants and 13 kiloplants/acre. However, the soil with 13 kiloplants/acre absorbed 189 cal/cm^2 per day compared with 94 cal/cm^2 per day for the higher population (net radiation about 350 cal/cm^2 per day). These data, taken when plant cover was maximum, indicate that evapotranspiration would have to be considerably greater for the 105 kiloplants/acre than for 13 kiloplants/acre, provided soil water conditions were such that soil evaporation was limited but transpiration was not. Unfortunately, no soil water measurements were made.

Pineapple is very different from most other crops. Ekern (1965) reports that the maximum use by pineapple occurs when the plants are small and direct evaporation from the soil is great. The summer evapotranspiration rate following planting in October is about one-half to one-third of the evapotranspiration at planting, in spite of the fact that free water evaporation from a pan doubles from winter to summer. Black paper and black polyethylene have been used since the early 1920's in Hawaii, USA to decrease evaporation from the bare soil between pineapple plants.

Peters and Russell (1959) have shown that evapotranspiration can be decreased by one-half by using plastic mulch on the soil surface (thus minimizing evaporation directly from the soil). Evapotranspiration was influenced more in wet years than in dry years. Harrold et al. (1959) measured the relative amounts of evaporation and transpiration by the same technique with weighing lysimeters.

Table 27–3. Evapotranspiration ET, transpiration T, and evaporation from the soil E_s data from lysimeter study in Coshocton, Ohio, 1957 (Harrold et al., 1959)

Period	ET	T	E_s	T/E_s
May 1 - June 5	3.69	0	3.69	0
June 6 - June 30	4.24	1.77	2.47	0.72
July 1 - July 31	6.67	4.14	2.53	1.64
August 1 - August 31	3.99	2.40	1.59	1.51
September 1 - September 9	0.54	0.16	0.38	0.42
Totals, May 1 - September 9	19.13	8.47	10.66	0.80

They found that the seasonal evaporation from the soil in Ohio, USA in 1957 accounted for about 56% of the evapotranspiration. Table 27–3 gives the data for various periods during the year. The data show evaporation to be greater than transpiration early in the season when the plants are small and intercept little of the net radiation. Later, as the extent of plant cover increases, transpiration is greater than evaporation.

Willis et al. (1963) used plastic mulch (90% cover) in a similar experiment in a semiarid area at Mandan, North Dakota. They found very little difference in evapotranspiration, but yields were much greater on the mulched plots, indicating that most of the water was lost by transpiration on the mulched plots. In semiarid areas, it appears difficult to decrease total evapotranspiration, but methods that vary the relative proportion of evaporation to transpiration are important.

E.   Influence of Plant Height on Evapotranspiration

There appears to be very little conclusive field evidence that plant height, as such, influences evapotranspiration if other factors are equal. This is contrary to the results of many greenhouse and isolated lysimeter studies where plant height is directly related to evapotranspiration. Plant height (and size) appears to influence evapotranspiration by greater interception of advected heat (sensible heat exchange with the air). Thus, plant height would appear to affect evapotranspiration if the taller plants were isolated in a field or if the plants were in isolated containers in the greenhouse. Viets (1962) has made an extensive review of the influence of fertilizers on evapotranspiration and states that:

"Since agriculture in desert and semiarid areas is subject to advection and even crops in humid areas for short periods of time are similarly subject to turbulent heat transfer, no one can categorically say that fertilization and a larger crop may not use more water. On the other hand, it is safe to conclude that in a field a crop twice as large does not require twice as much water, and its consumptive use is about the same or is only slightly increased."

F.   Influence of Rooting Depth and Extent on Evapotranspiration

Dreibelbis and Amerman (1964) compared the evapotranspiration from a shallow-rooted grass with that of a deep-rooted grass for 5 years at Coshocton, Ohio, USA. The evapotranspiration from the deep-rooted grass averaged 3.5 inches more/year than evapotranspiration from the shallow-rooted grass. The importance of rooting depth may be greater in arid or semiarid climates. The

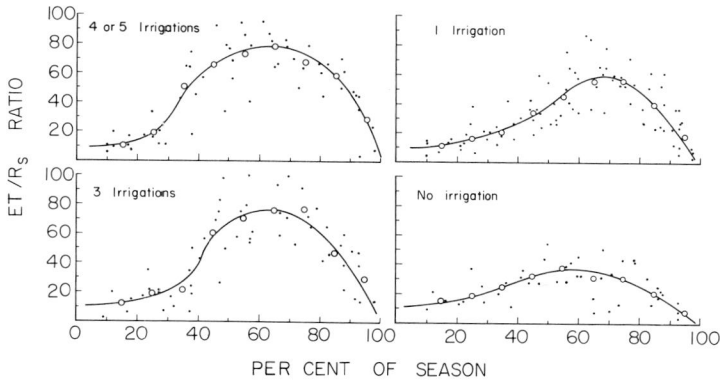

Fig. 27–2. Measured ET/R_s curves for cotton grown on Hidalgo sandy clay loam under five soil moisture regimes. Averages of 10% increments indicated by O. (data of L. N. Namken).

extent and depth of rooting determine the volume of soil from which plants can extract water. Evapotranspiration is greater the more extensive and deep the rooting characteristics of the plant, provided soil water is limiting in the upper part of the profile. Where fertilizers are needed and water is available in the lower profile depths, fertilization, in general, increases extent and depth of rooting and, thereby, evapotranspiration under semiarid conditions (Viets, 1962). Rooting depth and extent would be expected to have must less influence on evapotranspiration under humid conditions.

G. Influence of Stage of Growth on Evapotranspiration

Many investigators, particularly those who developed empirical relations for estimating evapotranspiration, have indicated that stage of growth is a very important factor on evapotranspiration. Most of these studies have shown a gradual increase in evapotranspiration from planting time to maturity, at which time evapotranspiration is equal to potential evapotranspiration. After maturation, evapotranspiration generally decreases.

The gradual increase in evapotranspiration from planting to maturity can be explained on the basis of per cent cover as discussed earlier. The decrease in evapotranspiration after maturation is probably a plant-dependent factor (Lemon et al., 1957). However, in many studies, soil water is not maintained at a high level after maturation. The data from L. N. Namken (unpublished report), shown in Fig. 27–2, support a plant-dependent factor that causes the decrease in evapotranspiration after maturation. The evapotranspiration falls off sharply after maturation for all levels of irrigation.

Fritschen and Van Bavel (1964) found that when sudangrass (Sorghum sudanese) approached maturity and developed seedheads, the evapotranspiration was much less than at an earlier stage of growth. Apparently, the seedheads absorbed the radiant energy, converted it into sensible heat, and also provided a very effective aerodynamic barrier against the transfer of sensible heat to the lower transpiring surfaces. Before the seedheads were formed, most of the radiant energy was converted into latent heat via evapotranspiration.

IV. POSSIBILITIES OF CHANGING PLANT FACTORS TO
ALTER EVAPOTRANSPIRATION

One practical method for decreasing evapotranspiration is to plant crops so that the per cent of cover is minimized. Use of varieties or plants that are short season or take advantage of natural periods of high rainfall are other possibilities. In orchards, the elimination of cover crops would also decrease evapotranspiration.

A possibility for the future would be the development of useful crops of suc- culent plants (such as pineapple) that close their stomata during the day. The possibilities of decreasing evapotranspiration by as much as one-half seem practical.

Another possibly for the future would be the development of chemicals to induce stomata closure and thereby decrease transpiration. Zelitch (1964) and Gale (1961) have tested several chemicals that show promise for this purpose. Attempts by other investigators have been discouraging, however.

V. INFLUENCE OF THE PLANT FACTOR IN EQUATIONS USED
TO PREDICT EVAPOTRANSPIRATION

Many equations have been used to predict evapotranspiration. All of them either include a plant factor or apply only to stages of growth where plant cover is above a certain value.

Penman (1956) developed an equation that included a plant factor to correct evapotranspiration estimated for a freely evaporating surface. He attributed the plant factor to be related to stomatal opening and day length.

Criddle (1958) has described a procedure, developed with Blaney (1959), based on irrigation data from the Western USA that includes a plant factor. The factor varies with crop as well as season of the year. Since the equation is empirical, the factor also depends on the location where it was determined.

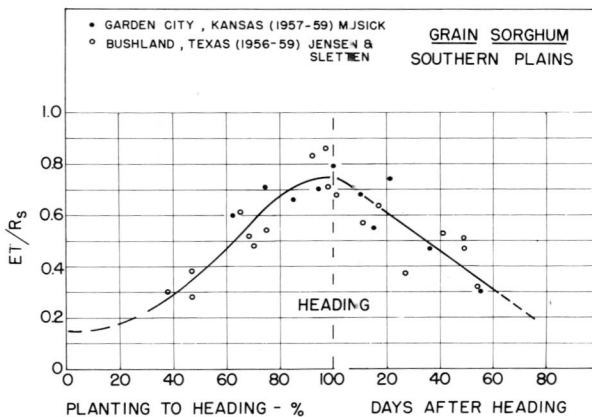

Fig. 27–3. Variation in the ET solar radiation ratio (ET/R_s) for grain sorghum in rela- tion to stage of plant growth expressed as a percentage of the period from planting to heading and days after heading (data of Jensen and Haise).

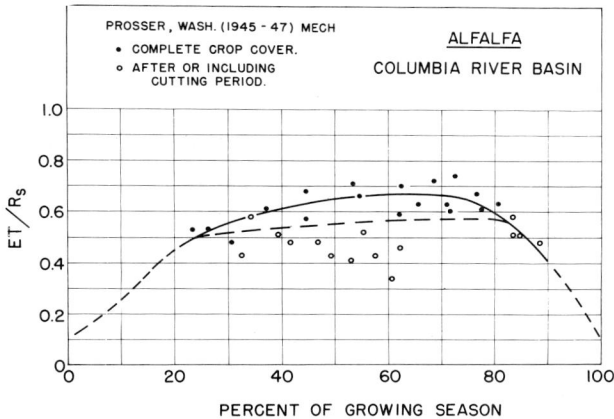

Fig. 27–4. Variation in the ET/solar radiation ratio (ET/R_s) for alfalfa in relation to percentage of growing season at Prosser, Washington (data of Jensen and Haise).

Jensen and Haise (1963) have developed an equation to estimate evapo-transpiration from meteorological data which includes a plant factor. This equation applies to irrigated conditions where soil water is not limiting (at least for transpiration). Figures 27–3 and 27–4 show the ratio of ET/R_s for various stages of growth throughout the season for typical irrigated crops (R_s = total solar radiation and is equal to S + s of equation [27–3]). The ratio of ET/R_s depends on the crop as well as the location from which the data were taken. However, the curves for the same crop at different locations are not greatly different.

Van Bavel and Harris (1962) found that ET = 0.8 R_n for irrigated corn and bermudagrass in North Carolina. This equation includes no plant factor. However, the data were taken only after the bermudagrass cover was complete and after the corn reached 1 ft height. Another equation of a similar type has been developed by Swan et al. (1963) for irrigated snapbeans and potatoes in Wisconsin. They found ET = 1.1 R_n for snapbeans and ET = 1.2 R_n for potatoes, provided plant cover was > 50%.

VI. SUMMARY

For single plants or single leaves, the plant and leaf characteristics have a large influence on evapotranspiration. Evapotranspiration from a plant-soil surface is proportional to the vapor pressure gradient between the surface and the free air above the surface and inversely proportional to the resistance of the diffusion pathway. The vapor pressure at the plant or soil surface depends on the net radiation at the surface which in turn is a function of the surface energy budget. The ratio of evapotranspiration to net radiation is a constant, related to Bowen's ratio, for a given crop and crop maturity. The more dense the crop, the smaller is Bowen's ratio, and the greater the fraction of net radiation consumed by evapotranspiration and the less by convection.

Plant factors undoubtedly influence evapotranspiration from a crop. The greatest difference among crops occurs during the growth period when the crop cover is < 50% complete. During this time, evapotranspiration of most irrigated crops is less than where cover is greater, because evaporation from bare soil decreases faster than does transpiration by crops. There appears to be little difference in evapotranspiration among many crops after cover is > 50% until maturity. Pineapple, however, is a notable exception. Because pineapple stomata are open at night and closed during the day, evapotranspiration is less for complete cover than for bare soil.

LITERATURE CITED

Angus, D. E. 1959. Water and its relation to soils and crops: C. Agricultural water use. Advance. Agron. 11:19–35.

Blaney, H. F. 1959. Monthly consumptive use requirements for irrigated crops. Amer. Soc. Civ. Eng. Proc., Irrig. Drainage Div. J. 84(IR 1):1–12.

Bowen, I. S. 1926. The ratio of heat losses by conduction and by evaporation from any water surface. Phys. Rev. 27:779–787.

Bowers, S. A., R. J. Hanks, and F. C. Stickler. 1963. Distribution of net radiation within sorghum plots. Agron. J. 55:204–205.

Burton, G. W. 1959. Soil-plant relationships: C. Crop management for improved water use efficiency. Advance. Agron. 11:104–110.

Criddle, W. D. Jan. 1958. Methods of computing consumptive use of water. Amer. Soc. Civ. Eng. Proc., Paper 1507. 84(IRI):1–27.

Denmead, O. T., L. J. Fritschen, and R. H. Shaw. 1962. Spatial distribution of net radiation in a corn field. Agron. J. 54:505–510.

Dreibelbis, F. R., and C. R. Amerman. 1964. Land use, soil type, and practice effects on the water budget. J. Geophys. Res. 69:3387–3393.

Ekern, P. C. 1965. Evapotranspiration of pineapple in Hawaii. Plant Physiol. 40:736–739.

Fritschen, L. J., and C. H. M. van Bavel. 1964 Energy balance as affected by height and maturity of sudangrass. Agron. J. 56:201–204.

Gale, J. 1961. Studies on plant antitranspirants. Plant Physiol. 14:777–786.

Gates, D. M. 1963. Leaf temperature and energy exchange. Arch. Meteorol., Geophys., u. Bioklim. 12:321–336.

Gates, D. M. 1964. Leaf temperature and transpiration. Agron. J. 56:273–277.

Haise, H. R., R. J. Hanks, and M. E. Jensen. 1963. Solar reflection from various crops and conditions. Agron. Abstr. p. 7.

Harrold, L. L., D. B. Peters, F. R. Dreibelbis, and J. L. McGuinness. 1959. Transpiration evaluation of corn grown on a plastic-covered lysimeter. Soil Sci. Soc. Amer. Proc. 23:174–178.

Jensen, M. E. and H. R. Haise. 1963. Estimating evapotranspiration from solar radiation. Amer. Soc. Civ. Eng. Proc., J. Irrig. Drainage Div. 89:15–41.

Kuiper, P. J. C. 1961. The effects of environmental factors on the transpiration of leaves, with special reference to stomatal light response. Meded. Landbouwhogeschool, Wageningen. 61:1–49.

Lee, R. and D. M. Gates. 1964. Diffusion resistance in leaves as related to their stomatal anatomy and microstructure. Amer. J. Bot. 51:963–975.

Lemon, E. R., A. H. Glaser, and L. E. Satterwhite. 1957. Some aspects of the relationship of soil, plant, and meteorological factors to evapotranspiration. Soil Sci. Soc. Amer. Proc. 21:464–468.

Marlatt, W. E. 1961. The interactions of microclimate, plant cover, and soil moisture content affecting evapotranspiration rates. Colo. St. Univ., Dep. Atmos. Sci. Tech. Pap. 23.

Monteith, J. L. 1959. The reflection of short-wave radiation by vegetation. Quart. J. Roy. Meteorol. Soc. 85:386–392.

Pallas, J. E., D. G. Harris, C. B. Elkins, and A. R. Bertrand. 1961. Research in plant transpiration. US Dep. Agr. Prod. Res. Rep. 70. 37 p.

Penman, H. L. 1948. Natural evaporation from open water, bare soil and grass. Roy. Soc., Proc. 193:120–145.

Penman, H. L. 1956. Evaporation: An introductory survey. Neth. J. Agr. Sci. 4:9–29.

Peters, D. B., and M. B. Russell. 1959. Relative water losses by evaporation and transpiration in field corn. Soil Sci. Soc. Amer. Proc. 23:170–172.

Polster, H. 1963. Photosynthesis and growth of a five-year-old stand of poplar trees in relation to water economy of the site. In A. J. Rutter and F. H. Whitehead [ed.] The water relations of plants. John Wiley, New York. p. 257–271.

Pruitt, W. O., and D. E. Angus. 1961. Comparisons of evapotranspiration with solar and net radiation and evaporation from water surfaces. Chapter VI. First Annu. Rep. Investigation of energy and mass transfers near the ground including influences of the soil-plant-atmosphere system. US Army Electronics Proving Grounds Tech. Program. Univ. California, Davis. p. 74–107.

Rufelt, H., P. G. Jarvis, and M. S. Jarvis. 1963. Some effects of temperature on transpiration. Plant Physiol. 16:177–185.

Shaw, R. H., and L. J. Fritschen. 1960. Transpiration and evapotranspiration of corn as related to meteorological factors. Iowa St. Univ. Final Rep. (Pt. 1). US Weather Bureau Contract CWB–9560.

Swan, J. B., M. D. Groskopp, and C. B. Tanner. 1963. Net radiation and evapotranspiration from irrigated snapbeans. Agron. Abstr. p. 67.

Tanner, C. B. 1963. Basic instrumentation and measurements for plant environment and micrometeorology. Soils Bull. no. 6, Univ. of Wisconsin, Madison. 156 p.

Tanner, C. B., and E. R. Lemon. 1962. Radiant energy utilized in evapotranspiration. Agron. J. 54:207–212.

Tanner, C. B., A. E. Peterson, and J. R. Love. 1960. Radiant energy exchange in a corn field. Agron. J. 52:373–379.

Thornthwaite, C. W., and B. Holzman. 1942. Measurement of evaporation from land and water surfaces. US Dep. Agr. Tech. Bull. 817.

Tibbals, E. C., E. K. Carr, D. M. Gates, and F. Kreith. 1964. Radiation and convection in conifers. Amer. J. Bot. 51:529–538.

van Bavel, C. H. M., and D. G. Harris. 1962. Evapotranspiration rates from Bermudagrass and corn at Raleigh, North Carolina. Agron. J. 54:319–322.

van Bavel, C. H. M., and T. V. Wilson. 1952. Evapotranspiration estimates as criteria for determining time of irrigation. Agr. Eng. 33:417–420.

Viets, F. G., Jr. 1962. Fertilizers and the efficient use of water. Advance. Agron. 14:223–264.

Willis, W. O., H. J. Haas, and J. S. Robins. 1963. Moisture conservation by surface and subsurface barriers and soil configuration under semiarid conditions. Soil Sci. Soc. Amer. Proc. 27:577–580.

Zelitch, Israel. 1964. The effect of control of leaf stomata on transpiration. In Research on water. Amer. Soc. Agron. Spec. Publ. Ser. no. 4, p. 104–113.

28

Soil and Cultural Factors Affecting Evapotranspiration

P. C. EKERN, JR.

Water Resources Research Center,
University of Hawaii
Honolulu, Hawaii

J. S. ROBINS

Washington State University
Prosser, Washington

W. J. STAPLE

Soil Research Institute,
Canada Department of Agriculture
Ottawa, Ontario, Canada

I. INTRODUCTION

Chapters 26 and 27 of this section elaborate on the effects of microclimatic factors and plant factors, respectively, on evapotranspiration. Certain soil and cultural factors, by modifying the microclimate or the transpirational behavior of crops, alter evapotranspiration. Our discussion will expand on these soil and cultural factors, which have not been considered in the two preceding chapters.

II. MECHANISMS FOR LIMITING EVAPOTRANSPIRATION

The macroclimate of an area imposes a general evaporational potential for the region (chapter 26). The actual rate of evapotranspiration which occurs at a specific site within this region can be raised or lowered from this potential rate by alteration of the microclimate—namely, the radiant energy balance and the interchange with the turbulent lower layers of air which transport sensible heat and water vapor to and/or from the site. The intent of much of agronomic practice is to minimize the amount of evapotranspiration required for production of harvestable yield.

In the absence of plants, evaporation from the land constitutes the sole vapor loss mechanism. In this case, soil controls the rate of evaporation by (i) transmitting water to the site of evaporation too slowly to meet the climate induced potential; (ii) reducing the absorption of energy; or (iii) restricting vapor or heat transmission from or to the evaporative site (Hide, 1954; Lemon, 1956; Philip, 1957). The upper surface of the soil often is the critical zone which determines the characteristics of each of these three mechanisms (Landsberg and Blanc, 1958).

When growing plants are present on the land, both transpiration by plants and evaporation from soils may operate as vapor sources (chapter 27). In this case, the two sources supplement each other in attempting to supply water rapidly enough to satisfy the climatic potential. The source of water reaching the evaporative surface is extended throughout all of the plant's root zone. Absorption of solar radiation, production (or consumption) of sensible heat, and the nature of the vapor gradient above the soil surface are greatly altered (Philip, 1964). Even a small increment of the incident radiation is utilized in photosynthesis (Lemon, 1960).

But despite these added complications the vegetated land surface controls evapotranspiration by the same general mechanisms as does the soil; namely, by (i) limiting the water supply to the site of evaporation, (ii) reducing the absorption of energy, and (iii) restricting vapor or heat transmission. We shall therefore look at several soil and cultural factors that have been observed or have potential to alter these phenomena.

III. SOIL FACTORS

A. Limiting Water Supply to Evaporative Sites

The drying rate of a bare soil is proportional to the water content and inversely proportional to time; and a drying front advances into the soil linearly with time (Wiegand, 1962). When the capacity of the soil to conduct water to the surface does not equal the evaporative demand, the surface dries and a parabolic water distribution develops within the soil. The rate at which the bare soil supplies water at the evaporative site is controlled by (i) water content, and (ii) hydraulic conductivity of the soil. These parameters interact to characterize the water diffusivity function which controls liquid flux rates (chapter 13).

Soil texture, surface configuration, and profile discontinuities dominate control of soil water storage (Parr and Bertrand, 1960). In general, coarse-textured or well-aggregated soils retain less water at any given water suction than do fine-textured ones (Franzmeier et al., 1960; Haise et al., 1955; Richards and Weaver, 1944; Tamboli, 1961), thus the water content contribution to the diffusivity function is texturally controlled. The lower water content in coarse-textured soils results in a smaller supply of water near the surface readily available for delivery to the evaporative site than that which occurs in fine-textured ones. Under sustained evaporative demands, therefore, the surface layers of coarse-textured materials exhibit a more rapid reduction in soil water diffusivity than do those of fine-textured soils.

Coarse-textured soils generally have higher hydraulic conductivities at low water suction (near saturation) than do fine-textured ones (Gardner and Mayhugh, 1958; Jackson, 1963; Willis, 1960). The reverse is generally true at higher soil water suctions and many soils have nearly identical capillary conductivities at about 1-bar soil suction. Thus, in coarse, as compared to fine-textured, soils added water is transmitted downward more rapidly when soil is wet and upward more slowly as the surface layers dry. Under conditions of intermittent wetting and drying of the surface, coarse-textured soils, as a result of hysteresis, generally lose less water by surface evaporation than do fine-textured ones (King and Schleusener, 1961; Wiegand, 1962; Gardner and Hillel, 1962; Staple, 1962). This holds true whether the soil is under crop or fallow.

Profile stratification, especially with coarse-textured layers but also with dense or very fine-textured layers of low conductivity, usually results in retention of more water after wetting than if the profile were uniform in conductivity with depth (Miller and Bunger, 1963; Nielsen, et al., 1959). If such layers are within about 100 cm (3 ft.) of the surface the added water supply and higher water diffusivity from the perched water table result in considerably larger evaporative losses than where the profile is uniform.

Structural condition of the surface soil layers (upper 10 to 15 cm) may alter water transmission to the surface and promote self mulching (Fox, 1964). This soil layer is critical since the site of evaporation is seldom more than 5 cm below the surface and is generally at or very near the surface. If any or all of this layer contains a large percentage of sizeable secondary aggregates (>1.0 mm in diameter) upward flow will be reduced because of pore discontinuities, the surface will dry quickly and evaporation rate will be reduced. This phenomenon is readily observed where crops such as potatoes (*Solanum tuberosum*) and field corn (*Zea mays* L.) are furrow irrigated with high ridges. Seldom are the tops of such ridges wetted during irrigation, and evaporation from the ridge is considerably less per unit area than from the wetted furrow (Van't Woudt, 1957). Layers of partially decomposed crop residue (Jamison, 1960) and soils which are coarse textured at the immediate surface but finer textured deeper in the profile display similar behavior. Mulches of gravel or other coarse material act similarly (Van Wijk and Derkson, 1961; Benoit and Kirkham, 1963). Evaporation increases if cracks develop as the soil dries (Adams and Hanks, 1964).

When a growing crop is present on the land, absorption of water from the soil throughout the entire root zone and subsequent translocation within the plant to evaporative sites at the leaf greatly increases the ability of the soil-plant system to maintain an adequate supply of water for the combined evapotranspiration. The fraction of water loss which can occur as direct soil evaporation is to a large extent determined by the planting pattern, leaf area index, and extinction coefficient of the crop canopy.

B. Reducing Energy Absorption

Color, slope aspect, surface roughness, and crop residue on the surface may alter energy absorption by soil surfaces. Chapter 27 presents a summary of recent data on this subject.

Soils of light color reflect a higher percentage of incident radiation than do darker colored ones (Graham and King, 1931; Hapke and Van Horn, 1963; Monteith and Szeicz, 1962; Ekern, 1965). As soil organic matter and water content increase, colors darken and reflectance decreases.

Slope aspect and degree greatly alter energy absorption per unit area due to the angle of impingement of the incident radiation. North-facing (poleward) slopes receive less incident radiation per unit of surface than south-facing ones, but little difference exists between east- and west-facing slopes. The effect of bed orientation on the soil temperature may well depend not on the radiation received but rather on preferential evaporational chill on the upwind side of the ridge (Shadbolt et al., 1961). The relative influence of a given slope varies from one season to another as angle of incidence of direct radiation changes (chapter 4).

When the immediate surface soil layer is cloddy, reflectance is reduced and thus absorption is increased (chapter 27). Roughness due to ridging probably has little net influence on either short-wave or sensible heat absorption by the soil, though local hot spots are produced.

Mulches of crop residue are more reflective than most soils, thus they reduce the energy absorption (Hanks, et al., 1961). This effect is most pronounced during fallow periods or early in the crop growth cycle. In spring the reduced energy absorption beneath the mulches results in cooler soil temperatures and lower evaporation than those which occur beneath a bare soil kept continously wet (Moody et al., 1963). However, the net effect of such mulches may be of little consequence in reducing evaporation when the surface is intermittently wetted since the soil surface under the mulch remains wet longer and the mulch is not an effective vapor barrier for the reduction of evaporation (McCalla and Army, 1961). The lowered soil temperature under such a mulch reduces early plant growth (Allmaras et al., 1964). This would not only tend to limit production but may delay plant development, extend the growing season, and thus prolong the period of evapotranspiration.

When the land is covered with a growing crop, the effects of soil factors on energy absorption are largely overshadowed by plant effects. Much of the incident radiation is intercepted, reflected, absorbed, or altered by the plant canopy. Reflectance of the plant becomes a dominant factor in absorption of solar radiation. Land slope and slope aspect effects on energy absorption are not altered materially by presence of plants.

The presence of a crop canopy extending into the lower atmospheric boundary profoundly influences the turbulent transfer processes. Principles of these exchange and transport processes are covered in chapters 26 and 27.

C.   Restriction of Heat or Water Vapor Exchange

The net radiation absorbed by a soil is used for three ends: (i) to raise the temperature of the soil mass, (ii) to vaporize water, or (iii) to heat the atmosphere directly. The square root of the thermal diffusivity indicates the depth of penetration of the diurnal temperature wave, and the volumetric heat capacity of the soil indicates the amount of heat required to change the temperature of the layer (Philip, 1957). The product of the two indicates the rate of heat flux into the soil under an imposed temperature regime at the soil surface, hence the name thermal contact coefficient for this quantity (Businger and Buettner, 1961). The relative thermal contact coefficients of the air and the underlying ground surface determine the division of the net radiation into sensible heat flow to the soil or into convective transfer to the air. The sensible heat transferred into the soil may then be returned in part to the air as latent heat of vaporization (Van Wijk and Gardner, 1960). Much of the thermal response of soil to radiation occurs in the surface 25 mm (Federer and Tanner, 1963).

Density, mineral composition, and water content are factors that largely control the thermal conductivity of soils. Increased density and/or water content permit more rapid heat transmission, largely due to the increased cross-sectional contact area which promotes rapid true conductive transfers (Jackson and Kirkham, 1958).

However, the effective thermal conductivity may be increased considerably by latent transfer of water vapor within the soil (Cary and Taylor, 1962).

When the soil surface is dry, the thermal contact coefficient of the soil is relatively low and much of the net radiation is used to heat the air. Since evaporation cannot occur rapidly from the dry soil, in the absence of evaporational chill, the immediate soil surface becomes hot. Despite the high temperature of the surface, evaporation is still restricted since the dry air is an effective thermal insulator.

When the soil surface is moist, the thermal contact coefficient of the soil is relatively large and more of the net radiation enters the soil. Since water is available, a large part of this heat is used in evaporation rather than in heating the soil.

Surface plant residues generally insulate the surface in addition to reducing energy absorption (Kohnke and Werkhoven, 1963). This reduces heat exchange between soil and atmosphere and results in a much cooler soil in late winter, spring, and early summer, and a slightly warmer soil in the fall and early winter.

A crop cover intercepts much of the radiant energy before it reaches the soil and the crop canopy assumes a major role in turbulent transfer processes as well. Hence as the crop canopy gradually increases, soil effects on the energy available for evapotranspiration become progressively smaller and finally are very small at best (Denmead et al., 1962).

Direct evaporation from the soil depends on external conditions as long as the rate of liquid transfer of water to the surface can keep pace with demand (Gardner and Hillel, 1962). In the falling-rate phase of soil drying, the evaporation rate depends on the balance struck between the flow of energy to the vaporization site and the rate of escape of the vapor from the soil (McCormick, 1962).

As soon as the immediate surface dries, the dry layer becomes a strong vapor flow retardant (Hanks, 1958). Flow through this layer is by vapor diffusion and the rate of such depends on the porosity of the layer (Benoit and Kirkham, 1963; Hanks and Woodruff, 1958).

Although the principle action of a crop residue on the soil surface is a reduction in energy absorption and heat exchange, some minor restriction to vapor flow is created by the presence of the mulch (Hanks and Woodruff, 1958; Benoit and Kirkham, 1963). Thick surface mulches decrease water evaporation rates by the combined action of reduction of energy transmission to the soil and reduction in the rate of vapor escape from the soil (McCalla and Army, 1961; Carter and Fanning, 1964).

D. Summary of Soil Factors

In the authors' opinion, the greatest hope for practical control of direct evaporation from the soil is through physical or chemical alteration of the zone of water storage in the soil profile, the site of vaporization within the soil, and the thermal contact coefficient of the immediate surface layers. For example, evaporation might be greatly reduced by the maintenance of a mulch of waterproofed clods which would permit rapid water infiltration, yet dry rapidly and serve as a thermal and vapor barrier. Effectiveness of such mulch might be considerably enhanced if it were light-colored, thus highly reflective. On a regional basis, it must be remembered that unless treatments imposed to reduce evaporation also reduce the net radiation, the hot air often will result in increased evaporation as it moves downwind. More detailed discussion of possibilities for evapotranspiration control is given in chapter 62.

IV. CULTURAL FACTORS

A. Irrigation Practices

The fraction of the available soil water extracted from the root zone between irrigations may significantly alter evapotranspiration (chapters 17 and 27). Unless such withdrawals deplete the root-zone water content sufficiently to cause loss of plant turgor, the reduction in transpiration is generally slight (Gardner and Ehlig, 1963). When the interval between irrigations is extended, the soil surface dries and restricts evaporation. Other water savings from such less frequent irrigations stem from the reduced losses during conveyance and application (*see* sections XI and XIII of this monograph). Under dryland conditions, there is a characteristic parabolic relationship between the rate of soil-water depletion and the stage of plant growth (Amemiya et al., 1963). Thus the principal water savings from lengthening the interval between irrigations occur early in the season when the crop cover is least. Evaporation from bare soil is only about half as great as transpiration from a full plant canopy (Doss et al., 1962). However, high soil-water content, which extends the period of peak water depletion by transpiration, increases the rate and amount of seasonal water use. The seasonal reduction obtained by the maximum tolerable extension of the irrigation interval is perhaps 10% of the total evapotranspiration (Musick et al., 1963). Unfortunately, highest yield, though generally obtained from frequently irrigated high fertility fields, often does not coincide with the greatest water-use efficiency nor with the most economical return when irrigation costs are assessed against such yield increases (Musick et al., 1963; Stanberry et al., 1963).

Minor reductions in evaporation from the soil can be gained by limiting the percentage of the soil surface that is wetted. Again the effect is limited largely to periods of incomplete vegetative cover. Such reductions seldom exceed 20% even when only one-fourth to one-third of the surface is wetted. Net seasonal water savings are likely to be no more than 5%. Alternate furrow irrigation is an effective method to limit the wetted surface yet supply water to much of the crop root zone. A shallow root zone can be wetted without wetting the entire surface by applying irrigations in small furrows placed very near the row. Significant reduction in soil evaporation can be obtained for vineyards and tree crops by irrigating in furrows or basins immediately adjacent to the individual plants.

Irrigation of fallow land is often practiced to recharge the soil reservoir during the noncrop season. This practice may be to utilize off-season labor, to assure a fully charged soil profile at planting, to compensate for inadequate rate of supply from wells or surface supplies during the growing season, or to utilize off-season flows which would otherwise not be used. As with the storage of off-season water under dryland fallow, the efficiency of such storage is very low, particularly if there is a long time for evaporation of the stored water before the crop is planted (Haas and Willis, 1962).

B. Tillage and Cultivation

Stirring of the soil by tillage or cultivating implements may be for any of several purposes: (i) to prepare a suitable crop seedbed including incorporation of crop residues, (ii) to eliminate undesired vegetation, (iii) to fracture or bury soil crusts to permit seedling emergence or improved water penetration, (iv) to construct furrows for irrigation water control, or (v) to apply fertilizer (Larson, 1964). Tillage practices play their part in evapotranspiration control largely through their effects on the water storage or on the rooting zone of the crop. Surface depression storage, transient storage within the profile, and runoff reduction all help to increase the volume of water available for evaporation. Aggregates 1 mm in diameter are critically related to the transmission of water to the seed or root zone. The relative size of secondary aggregates can be controlled by the type of tillage tool and water content of the soil at the time of tillage (Lyles and Woodruff, 1962; Siddoway, 1963). It is generally agreed that some soil stirring is necessary for successful irrigated culture of most short-season agricultural crops, although recent advances in chemical weed control methods may soon eliminate all need for tillage in certain perennials (Rao et al., 1960).

It is now generally agreed also that stirring of the soil, whether during fallow periods or under crop, effects little if any reduction in evaporation unless significan weed growth is eliminated. Severe root pruning by cultivation may temporarily reduce transpiration but such effect is limited and short-lived. Deep stirring ($>$ 8 or 10 cm) may, in fact, increase water loss when land is in fallow or when crop cover is sparse.

Deliberate establishment of preferred microrelief is a tillage practice of increasing importance and frequently has a direct or indirect effect on evapotranspiration. This is true whether such tillage be for improved efficiency in operation (Edminster, 1959), for drainage (Phillips, 1963), for water spreading to induce additional infiltration (Hauser and Cox, 1962; Hall et al., 1957), for irrigation (Hansen, 1960), or for protection from erosion (Phelan, 1960). Deliberate impounding of water, using the Zingg system with a shed area on the terrace ridges twice the size of the level bench, helps increase the soil-stored water under dryland farming (Hauser and Cox, 1962). The introduction of vegetation as vertical mulches will not be effective in increasing the entry of water unless the permeability of the deep layers within the soil profile is adequate (Hauser and Taylor, 1964; Kingsley and Shubeck, 1964). The infiltration of water impounded in the depressions in the microrelief can leach salts from these zones (Benz et al., 1964), particularly if evaporation is reduced at the same time by a surface mulch (Carter and Fanning, 1964). The microrelief in furrow irrigation produces specific zones of salt accumulation in the ridge as the wetting front advances with its burden of dissolved salts (Bernstein and Fireman, 1957). The location of the zones can be directed by control of the slope of the ridge so that the seeding site does not lie in the salted area. Though the initial intent of these practices may not be for evapotranspiration control, many of them do alter the heat and water budgets simultaneously with the increase they bring to production efficiency.

C. Water Table Control

Chapters 45 and 50 of this monograph outline the subjects of water-table control and drainage as related to irrigation management, respectively. The comments to follow relate only to water-table control as it affects evapotranspiration.

Height of the groundwater table will appreciably affect evapotranspiration only if the capillary fringe extends to or very near the soil surface (Marshall, 1959; Willis, 1960; Williamson, 1963). Evaporation from the capillary fringe will be higher than where the surface dries intermittently. After full crop canopy is present, the capillary fringe may provide much of the water required to sustain transpiration and crop growth, thus eliminating need for irrigation. In fact, crops grown on some peat soils are irrigated solely by supplying water to maintain the water table at a depth of 30 to 60 cm (Boelter, 1964; Harris et al., 1962).

Generally, presence of a water table, nearer than 100 cm to the soil surface in the case of coarse-textured mineral soils, or nearer than 200 cm in fine-textured soils, will result in a capillary fringe which extends to the soil surface. In such cases, as opposed to those where the water table is at a great depth, evapotranspiration may be greater by 10% or 15% during the crop growth season and by 50% or more during the noncrop interval. The more serious aspect of the shallow water table, however, is the unfavorable salt balance and root environment that generally results (chapters 50 and 51).

D. Surface Mulching and Shading

Effects of plant residue and soil mulches have been discussed. Mulches such as polyethylene film or asphalt sprays which form a relatively impermeable vapor barrier on the soil surface change the pattern of heat flow into the soil and evaporation from it (Spice, 1963; Collis-George et al., 1963; Waggoner et al., 1960). When lateral flow of vapor is precluded, a complete barrier of this type sharply reduces evaporation from the soil (Army and Hudspeth, 1960). Partial barriers have limited net effect unless 90% or more of the surface is covered, and reductions are more pronounced when covering is in wide rather than narrow strips (Willis, 1962; Hanks and Bowers, 1963).

Soil temperature under such mulches is generally increased considerably (Black, 1963). This promotes (i) more rapid transpiration if water supply to the crop is even slightly limiting, (ii) increased vapor flow from perforations or gaps in the barrier, and (iii) enhanced plant growth, especially early in the season (Willis et al., 1963). It is more likely that improved plant growth is due to increased temperature rather than to favorable water status resulting from reduced evaporation.

Shading as a technique for reducing evapotranspiration is discussed in chapter 62. Artificial shading will materially reduce the energy available for evapotranspiration. However, commercial application of this technique is of consequence only in greenhouse culture of highly specialized crops where the aim is control of fruiting or other growth characteristics rather than conservation of water. Effects of natural shading of the soil or lower leaves on evapotranspiration are discussed in chapters 17 and 27.

E. Summary of Cultural Factors

In general, crop quality is commensurate only with maximum consumptive use of water; hence, little possibility exists for reduction in water use without an accompanying reduction in crop quality. Perhaps the best opportunity for water savings lies in the removal of all other factors which inhibit plant growth, thus ensuring the maximum possible yield for the water expended. The elimination of unwanted vegetative growth (weeds) remains a real part of the measures to increase the efficiency of production.

In the early stages of plant growth, when evaporation from the soil contributes a substantial part of the water use, drainage of a shallow water table reduces the rate of water loss and may also promote more rapid vegetative growth. In these early stages, an insulative mulch which chills the soil reduces evaporation and can also reduce transpiration, whereas a partial cover of a vapor barrier mulch increases soil temperature and can increase evaporation. However, once the plant canopy has closed, direct evaporation from the soil makes up only a small portion of the total evapotranspiration from most crops, and cultural practices which affect only direct evaporation from the soil have but a small effect on the total evapotranspiration.

LITERATURE CITED

Adams, J. E., and R. J. Hanks, 1964. Evaporation from soil shrinkage cracks. Soil Sci. Soc. Amer. Proc. 28:281–284.

Allmaras, R. R., W. C. Burrows, and W. E. Larson. 1964. Early corn growth as affected by soil temperature. Soil Sci. Soc. Amer. Proc 28:271–285.

Amemiya, M., L. N. Namken, and C. J. Gerard. 1963. Soil water depletion by irrigated cotton as influenced by water regime and stage of plant development. Agron. J. 55:376–378.

Army, T. J., and E. B. Hudspeth, Jr. 1960. Alternation of the microclimate of the seed zone. Agron. J. 52:17–22.

Benoit, G. R., and D. Kirkham. 1963. The effect of soil surface conditions on the evaporation of soil water. Soil Sci. Soc. Amer. Proc. 27:495–498.

Benz, L. C., F. M. Sandoval, R. H. Mickelson, and E. J. George. 1964. Microrelief influences in a saline area of ancient glacial Lake Agassiz: II. On shallow groundwater. Soil Sci. Soc. Amer. Proc. 28:567–570.

Bernstein, L., and M. Fireman. 1957. Laboratory studies on salt distribution in furrow irrigated soil with special reference to the preemergence period. Soil Sci. 83:249–263.

Black, J. F. 1963. Weather control: Use of asphalt coatings to tap solar energy. Science 139:226–227.

Boelter, D. H. 1964. Water storage characteristics of several peats in situ. Soil Sci. Soc. Amer. Proc. 28:433–435.

Businger, J. A., and K. J. K. Buettner. 1961. Thermal contact coefficient. J. Meteorol. 18:422.

Carter, D. L., and C. D. Fanning. 1964. Combining surface mulches and periodic water applications for reclaiming saline soils. Soil Sci. Soc. Amer. Proc. 28:564–567.

Cary, J. W., and S. A. Taylor. 1962. Thermally driven liquid and vapor phase transfer of water and energy in soil. Soil Sci. Soc. Amer. Proc. 26:417–420.

Collis-George, N., B. G. Davey, D. R. Scotter, and D. R. Williamson. 1963. Some consequences of bituminous mulches. Australian J. Agr. Res. 14:1–11.

Denmead, O. T., L. J. Fritschen, and R. H. Shaw. 1962. Spatial distribution of net radiation in a corn field. Agron. J. 54:505–510.

Doss, R. D., O. L. Bennett, and D. A. Ashley. 1962. Evapotranspiration by irrigated corn. Agron. J. 54:497–498.

Edminster, T. W. 1959. Land forming and smoothing for efficient production. Agr. Eng. 40:84–86.

Ekern, P. C. 1965. The fraction of sunlight retained as net radiation in Hawaii. J. Geophys. Res. 70:785–793.

Federer, C. A., and C. B. Tanner. 1963. Thermal response of a dry surface to shading cycles of short period. Soil Sci. Soc. Amer. Proc. 27:266–269.

Fox, W. E. 1964. A study of bulk density and water in a swelling soil. Soil Sci. 98:307–316.

Franzmeier, D. P., E. P. Whiteside, and A. E. Erickson, 1960. Relationship of texture classes of fine earth to readily available water. Int. Congr. Soil Sci., Trans. 7th (Madison, Wis., USA) I:354–360.

Gardner, W. R., and C. F. Ehlig. 1963. Influence of soil water on transpiration by plants. J. Geophys. Res. 68:5719–5723.

Gardner, W. R., and D. I. Hillel. 1962. The relation of external conditions to the drying of soils. J. Geophys. Res. 67:4319–4325.

Gardner, W. R., and M. S. Mayhugh. 1958. Solution and tests of the diffusion equation for the movement of water in soil. Soil Sci. Soc. Amer. Proc. 22:197–201.

Graham, W. G., and K. M. King. 1961. Short wave reflection coefficient for a field of maize. Quart. J. Roy. Meteorol. Soc. 87:425–428.

Haas, H. J., and W. O. Willis. 1962. Moisture storage and use by dryland spring wheat cropping systems. Soil Sci. Soc. Amer. Proc. 26:506–509.

Haise, H. R., H. J. Haas, and L. R. Jensen. 1955. Soil moisture studies of some Great Plains soils: II. Field capacity as related to 1/3-atmosphere percentages and "minimum point" as related to 15- and 26-atmosphere percentages. Soil Sci. Soc. Amer. Proc. 19:20–25.

Hall, W. A., R. M. Hagan, and J. K. Axtell. 1957. Recharging ground water by irrigation. Agr. Eng. 38:98–100.

Hanks, R. J. 1958. Water vapor transfer in dry soil. Soil Sci. Soc. Amer. Proc. 22:372–374.

Hanks, R. J., and S. A. Bowers. 1963. Two dimensional electric simulator for nonsteady state soil moisture flow. Soil Sci. Soc. Amer. Proc. 27:240–241.

Hanks, R. J., S. A. Bowers, and L. D. Bark. 1961. Influence of soil surface conditions on net radiation, soil temperature, and evaporation. Soil Sci. 91:233–238.

Hanks, R. J., and N. P. Woodruff. 1958. Influence of wind on water vapor transfer through soil, gravel, and straw mulches. Soil Sci. 86:160–164.

Hansen, V. E. 1960. New concepts in irrigation efficiency. Amer. Soc. Agr. Eng., Trans. 3:55–57, 61, 64.

Hapke, B., and H. Van Horn. 1963. Photometric studies of complex surfaces with applications to the moon. J. Geophys. Res. 68:4545–4570.

Harris, C. I., H. T. Erickson, N. K. Ellis, and J. E. Larson. 1962. Water table level control in organic soil as related to subsidence rate, crop yield, and response to nitrogen. Soil Sci. 94:158–161.

Hauser, V. L., and M. B. Cox. 1962. Evaluation of the Zingg conservation bench terrace. Agr. Eng. 43:462–464, 467.

Hauser, V. L., and H. M. Taylor. 1964. Evaluation of deep tillage treatments on a slowly permeable soil. Amer. Soc. Agr. Eng., Trans. 7:134–136.

Hide, J. C. 1954. Observations on factors affecting evaporation of soil moisture. Soil Sci. Soc. Amer. Proc. 18:234–239.

Jackson, R. D. 1963. Porosity and soil-water diffusivity relations. Soil Sci. Soc. Amer. Proc. 27:123–126.

Jackson, R. D., and D. Kirkham. 1958. Method of measurement of the real thermal diffusivity of moist soil. Soil Sci. Soc. Amer. Proc. 22:479–482.

Jamison, V. C. 1960. Water movement restriction by plant residue in a silt loam soil. Int. Congr. Soil Sci., Trans. 7th (Madison, Wis., USA). I:464–472.

King, L. G., and R. A. Schleusener. 1961. Further evidence of hysteresis as a factor in the evaporation from soils. J. Geophys. Res. 66:4187–4191.

Kingsley, Q. S., and F. E. Shubeck. 1964. The effects of organic trenching on runoff. J. Soil Water Cons. 19:19–22.

Kohnke, H., and C. H. Werkhoven. 1963. Soil temperature and soil freezing as affected by an organic mulch. Soil Sci. Soc. Amer. Proc. 27:13–17.

Landsberg, H. E., and M. L. Blanc. 1958. Interaction of soil and weather. Soil Sci. Soc. Amer. Proc. 22:491–495.

Larson, W. E. 1964. Soil parameters for evaluating tillage needs and operations. Soil Sci. Soc. Amer. Proc. 28:118–122.

Lemon, E. R. 1956. The potentialities for decreasing soil moisture evaporation loss. Soil Sci. Soc. Amer. Proc. 20:120–125.

Lemon, E. R. 1960. Photosynthesis under field conditions. II. An aerodynamic method for determining the turbulent carbon dioxide exchange between the atmosphere and a corn field. Agron. J. 52:697–703.

Lyles, L., and N. P. Woodruff. 1962. How moisture and tillage affect soil cloddiness for wind erosion control. Agr. Eng. 43:150–153, 159.

McCalla, T. M., and T. J. Army. 1961. Stubble mulch farming. In A. G. Norman [ed.] Advances in Agronomy. Academic Press, New York. 13:125–196.

McCormick, P. U. 1962. Mathematics penetrates the drying industry. Ind. Eng. Chem. Fund. 54:51–52.

Marshall, T. F. 1959. Relations between water and soil. Tech. Comm. 50, Commonwealth Bur. Soils, Harpenden. 91 p.

Miller, D. E., and W. C. Bunger. 1963. Moisture retention by soil with coarse layers in the profile. Soil Sci. Soc. Amer. Proc. 27:586–589.

Monteith, J. L., and G. Szeicz. 1962. Radiative temperature in the heat budget of natural surfaces. Quart. J. Roy. Meteorol. Soc. 188:496–507.

Moody, J. E., J. N. Jones, Jr., and J. H. Lillard. 1963. Influence of straw mulch on soil moisture, soil temperature, and the growth of corn. Soil Sci. Soc. Amer. Proc. 27:700–703.

Musick, J. T., D. W. Grimes, and O. M. Herron. 1963. Irrigation water management and nitrogen fertilization of grain sorghums. Agron. J. 55:295–298.

Nielsen, D. R., D. Kirkham, and W. R. van Wijk 1959. Measuring water stored temporarily above field moisture capacity. Soil Sci. Soc. Amer. Proc. 23:408–412.

Parr, J. F., and A. R. Bertrand. 1960. Water infiltration into soils. In A. G. Norman [ed.] Advances in Agronomy. Acad. Press, New York. 12:311–363.

Phelan, J. T. 1960. Bench leveling for surface irrigation and erosion control. Amer. Soc. Agr. Eng., Trans. 3:14–17.

Philip, J. R. 1957. Evaporation, and moisture and heat fields in the soil. J. Meteorol. 14:354–366.

Philip, J. R. 1964. Sources and transfer processes in air layers occupied by vegetation. J. Appl. Meteorol. 3:390–395.

Phillips, R. L. 1963. Surface drainage systems for farmlands. (Eastern U.S. and Canada). Amer. Soc. Agr. Eng., Trans. 6:313–317, 319.

Rao, A. A. S., R. C. Hay, and H. P. Bateman. 1960. Effect of minimum tillage on physical properties of soils and crop response. Amer. Soc. Agr. Eng., Trans. 3:8–10.

Richards, L. A., and L. R. Weaver. 1944. Moisture retention by some irrigated soils in relation to soil moisture tension. J. Agr. Res. 69:215–235.

Shadbolt, C. A., O. D. McCoy, and T. M. Little. 1961. Soil temperatures as influenced by bed direction. Amer. Soc. Hort. Sci., Proc. 78 488–495.

Siddoway, F. H. 1963. Effects of cropping and tillage methods on dry aggregate soil structure. Soil Sci. Soc. Amer. Proc. 27:452–454.

Spice, H. R. 1963. Polythene films in agriculture. World Crops 15:239–249.

Stanberry, C. O., H. A. Schreiber, M. Lowry, C. L. Jensen. 1963. Sweet corn production as affected by moisture and nitrogen variables. Agron. J. 55:159–161.

Staple, W. J. 1962. Hysteresis effects in soil moisture movement. Can. J. Soil Sci. 42:247–252.

Tamboli, P. M. 1961. The influence of bulk density and aggregate size on soil moisture retention. Iowa state Univ. Diss. Abstr. 22:952.

Van Wijk, W. R., and W. J. Derksen. 1961. Surface temperatures in a soil consisting of two layers. Agron. J. 53:245–246.

Van Wijk, W. R., and W. R. Gardner. 1960. Soil temperature and the distribution of heat between soil and air. Int. Congr. Soil Sci., Trans. 7th (Madison, Wis., USA). I:195–202.

Van't Woudt, B. D. 1957. Infiltration behavior under furrow and sprinkler irrigation. Agr. Eng. 38:310–311, 319–320.

Waggoner, P. E., P. M. Miller, and H. C. De Roo. 1960. Plastic mulching. Connecticut Agr. Exp. Sta. Bull. 634. 44 p.

Wiegand, C. L. 1962. Drying patterns of a sandy clay loam in relation to optimal depth of seeding. Agron. J. 54:473–476.

Williamson, R. E. 1963. The management of soil salinity in lysimeters. Soil Sci. Soc. Amer. Proc. 27:580–583.

Willis, W. O. 1960. Evaporation from layered soils in the presence of a water table. Soil Sci. Soc. Amer. Proc. 24:239–242.

Willis, W. O. 1962. Effect of partial covers on evaporation from soil. Soil Sci. Soc. Amer. Proc. 26:598–601.

Willis, W. O., H. J. Haas, and J. S. Robins. 1963. Moisture conservation by surface or subsurface barriers and soil configuration under semiarid conditions. Soil Sci. Soc. Amer. Proc. 27:577–580.

29

Measurement of Evapotranspiration

C. B. TANNER

University of Wisconsin
Madison, Wisconsin

I. LIST OF SYMBOLS

Symbols listed in chapter 26 also apply to this chapter. Listed below are symbols different from those in chapter 26.

A	Area (catchment)	cm^2
C	Constant (in empirical formulas)	(varies with formula)
C_s	Surface drag coefficient	dimensionless
C_v	Constant, ρ/p	cm^{-2}sec^2
D	Diffusivity (molecular) of water vapor in air	cm^{2}sec^{-1}
M	Miscellaneous heat flux density terms	cal cm^{-2}sec^{-1}
ET	Evapotranspiration, surface depth or rate	cm, cm/time
ET$_a$	Actual evapotranspiration, surface depth or rate	cm, cm/time
ET$_p$	Potential evapotranspiration, surface depth or rate	cm, cm/time
P	Precipitation (irrigation), surface depth or rate	cm, cm/time
V_i	Volume of intercepted water	cm^3
V_L	Volume of "leakage" from catchment (not measured in V_r)	cm^3
V_r	Volume of surface and subsurface runoff from catchment	cm^3
ΔV_s	Volume change in water stored above watertable	cm^3
ΔV_w	Volume change in ground water storage	cm^3
δ	Path length of molecular diffusion at surface	cm
ν	Kinematic viscosity	cm^{2}sec^{-1}
φ	Diabatic influence function	dimensionless
Φ	Function of φ, $\Phi = \int_o^z [(\varphi - 1)\mathrm{d}z/(z + z_o)]$	dimensionless
d_L	Daylength fraction	dimensionless
t_c	Time constant	sec

II. INTRODUCTION

Evapotranspiration consists of the conversion to vapor and mixing with the atmosphere of the liquid water at the earth-atmosphere boundary; this may be soil moisture, ponded water, water intercepted on surfaces, and water in plants. The flux of water vapor to the atmosphere constitutes a major portion of the water in the hydrologic cycle of most regions, and its measurement is the concern of this chapter.

It is well to recognize early that none of the various methods for measuring evapotranspiration have complete preference. They differ in short- and long-term accuracy and in convenience and cost; consequently, the choice of method depends on application. For example, evapotranspiration estimates used to plan regional irrigation and water resource developments usually require only annual or, at most, monthly evapotranspiration. An error of 15 to 25% frequently is allowable and empirical formulae may be used provided they are calibrated from actual measurements. Evapotranspiration over shorter periods (3 to 10 days) and with less error (10 to 15%) is necessary to obtain engineering data on the time pattern of water use by crops and for irrigation criteria, particularly on sandy soils with low soil water storage and in humid regions where rainfall and weather is variable. On fine-textured soils with large water storage in semiarid regions with conservative climate (little day-to-day change), evaporation estimates over 2-week periods or longer will often suffice. At the other extreme, detailed research on water transfer in the soil-plant-atmosphere continuum and on reaction of plants to water stress may require measurements of evapotranspiration over periods as short as a few minutes or as long as 1 or 2 days and errors as low as 5%.

Measurements of ET can be divided into three classes for convenience: first, the water balance or hydrologic methods; second, micrometeorological methods; and third, empirical methods. The methods of the first two classes have a rational basis whereas empirical methods, if reasonable confidence is obtained, must be "calibrated" by relating the empirical ET indexes to actual ET measurements.

III. WATER BALANCE METHODS

The water balance methods include natural catchment (also irrigation and drainage districts) hydrology, soil water depletion sampling, and lysimeters. All of these methods can be discussed from the water balance equation

$$\text{ET} = P - (V_r + V_L + V_i + \Delta V_w + \Delta V_s)/A \qquad [29\text{--}1]$$

A. Catchments

Entire books are devoted to the science and art of catchment hydrology and the reader should refer to these for detailed discussions; we will consider only a few precautions to be observed. When equation [29–1] is applied to large natural catchments and to drainage and irrigation districts it is hoped that $V_L = 0$, and is necessary that P, V_r and A are measured with the required precision. The ΔV_w, ΔV_s, and V_i terms must either be measured or the average ET must be computed over a sufficiently long period that they can be neglected in comparison to other terms. The largest uncertainty usually is in the $V_L = 0$ assumption and in measuring A. Leakage is frequently encountered in drainage and irrigation districts and in associated storage and distribution facilities. Also, considerable judgment is required in selecting natural catchments without leakage so that V_L is negligible. The area is defined by the groundwater contours which must be measured with ground wells or piezometers and cannot be assumed equal to that defined by topography. Because of the problems involved in obtaining exact measurements of ET, independent meteorological methods should be used as useful supporting information to the hydrological measurement.

B. Soil Water Depletion

Soil water depletion measurements also are based on the water balance equation [29–1] where ΔV_s is measured and frequently an attempt is made to either

Table 29–1. Drainage from a 3-ft. profile (Robins et al., 1954)

Period, by days	0-8th	1st-8th	2nd-8th	3rd-8th	4th-8th	6th-8th
Drainage, mm	33	22	15	10	7	2

measure or control V_r. The ΔV_w and V_L terms express continued depletion of water from the sampling zone by drainage, and the error in ET is as large as this unmeasured drainage. The size of the drainage error can be substantial as was recognized by King (1910) and illustrated more recently by others such as Robins et al. (1954) and Nixon and Lawless (1960). Robins et al. found considerable drainage from the 0- to 3-foot zone of a fine sandy loam for 8 days following irrigation as summarized in Table 29–1. Different periods following irrigation and the total drainage from the 0- to 3-foot zone for the respective periods are listed. In Wisconsin, USA, and probably in most humid areas, measurement of soil water depletion rarely provides reliable ET measurements because of frequent rains and resulting drainage.

Provided that drainage errors are acceptable, measurements of soil water depletion can be used. Soil water depletion may be measured by any of the techniques discussed in chapter 15. The neutron moisture meter usually is preferred over other measurements. Sampling known soil volumes with tubes is the second choice and is preferred to water weight determination and separate bulk density determinations (Taylor et al., 1961). The neutron meter has several advantages: first, water volume fractions are determined directly; second, a larger volume of soil usually is sampled; and last, the same soil volume is measured repeatedly.

The accuracy of sampling can be illustrated by example. Taylor et al. indicate that a probable relative error of 10% or more may be expected in measuring volume fraction differences by soil volume sampling. Thus the water changes in a 60-cm depth of soil at a water volume fraction of 0.25 can be determined with ± 1.5 cm error. To measure the evapotranspiration with 10% error, the total evapotranspiration must be about 15 cm this indicates soil volume sampling will not give reliable values over periods much shorter than 2 or 3 weeks. Neutron meter sampling of water change can reduce errors to ±0.01 in the volume fraction or an error of 6 mm in the above example. Error estimates made for water depletion sampling in the 0- to 150-cm depth of a uniform Plainfield sand planted to alfalfa-brome (*Medicago sativa*) (*Bromus inermis*) usually are in 5- to 7-mm range with only six access tubes. These examples emphasize (i) that soil water depletion sampling cannot be used over periods much shorter than about 1 week and usually is useful only over longer periods; and (ii) serious error due to drainage may result and there is no way to insure that drainage will be negligible, particularly where frequent and/or heavy precipitation may occur.

C. Lysimetry

Lysimetry is the only hydrological method in which the experimenter has complete knowledge of all the terms in equation [29–1]. Thus lysimetry has great importance, not only for gathering ET information, but also as an independent check on the suitability of micrometeorological methods and for calibrating empirical formulas used for estimating ET. Slatyer and McIlroy (1961), Pelton

(1961), and McIlroy and Angus (1963) give additional data on lysimeter construction and use not included here.

A lysimeter is a device in which a volume of soil, which may be planted to vegetation, is located in a container to isolate it hydrologically from the surrounding soil. Lysimeters are constructed to make $V_L = 0$ and either to permit measurement of V_r or to make $V_r = 0$. Though lysimeters are well defined hydrologically, they must be representative samples of the surrounds if they are to provide useful ET measurements. Representativeness of soil (thermal, moisture, and mechanical properties) and of the vegetation (height, density, physiological well-being) is necessary. The intended use of the lysimeter will dictate the design and operation necessary to obtain suitable representativeness. Major factors affecting design include: First, whether measurements of ET_p (maximum possible under given micrometeorological conditions) or measurements of ET_a (including periods of drouth) are needed; second, the structure of the vegetation (grass, hay, row crops, density, etc.) and of the roots (deep, shallow, etc.); and last, the period over which the ET is to be measured (hours, days, or months). These factors influence the design of the lysimeter with respect to depth and water control, area, and the method of measuring water loss from the lysimeter. Design and operation is also affected by other factors that will be discussed.

1. LYSIMETER DEPTH AND WATER CONTROL

If the lysimeter is to measure ET_a, several precautions are necessary to insure that the root environment in the lysimeter is representative of the surrounding soil. Water distribution is the most important factor since it affects the water availability to the plants, the soil aeration and the thermal regime (thermal effects discussed later). We will illustrate the effect of the lysimeter on the water regime with Van Bavel's (1961) discussion, referring to Fig. 29-1 which represents the initial water condition following rainfall or irrigation.

"At that time a zero-pressure plane is present at the bottom and thereby the moisture tension as well as moisture content are different from those in the surrounding soil [a rare exception would be an impervious layer or coarse layer at the same depth as the lysimeter bottom—C. B. Tanner]. This may have two effects: First, more water may be available for evapotranspiration during a prolonged dry spell; and, second, the development of the root system of crops grown in the lysimeter may differ from that in the surrounding area.

"In so-called constant water table lysimeters, the same applies except that then the water table is the zero-pressure plane. Only when information is sought about the evapotranspiration rate under conditions of a permanent and high water table will a constant water table lysimeter duplicate outside conditions.

"The difficulty may be reduced by making the lysimeters deep, so that they extend well below the root zone as was done, for example, in the Coshocton lysimeters. Obviously, it is more difficult to construct a deep lysimeter. A more basic solution is to maintain tension at the bottom of a lysimeter deep enough to allow normal root development. This requires a rigid porous support for which the bubbling pressure is higher than the highest tension one wishes to establish. Tension can then be maintained to equal or approximate the tension in the surrounding profile at the same depth. This principle has been recognized for some time but it is not yet widely understood or applied. An example of its application is the Davis, California weighing lysimeter."

"Monolith" lysimeters, constructed by casing a block of soil *in situ*, have been proposed to insure that the water distribution in lysimeters is representative. If the lysimeter is installed in sandy soils, a "filled" lysimeter can be used in which

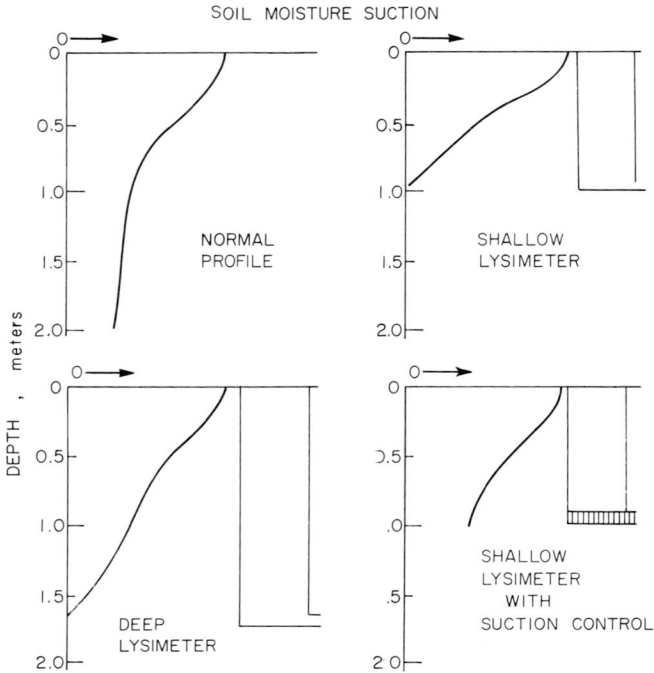

Fig. 29–1. Idealized relation of soil water suction vs. depth in lysimeters (Van Bavel, 1961).

the soil is returned in the same profile order and packed to the same density as the natural soil. Certainly monolith lysimeters appear desirable, particularly for well-aggregated, fine-textured soils; however, the author knows of no data comparing the differences with carefully filled lysimeters. The possibility remains that filled lysimeters may be adequate on many fine-textured soils. There is no question regarding the need either for deep lysimeters or lysimeters constructed with suction control devices at the bottom. Lysimeters fitted with suction devices should be as deep as the maximum rooting depth. Where suction control is not used, the depth must allow for the "capillary fringe" below the maximum rooting depth—thus, greater depths are required for fine-textured than for coarse-textured soils.

When only ET_p is required, the lysimeter requirements are much less critical. Filled lysimeters are suitable provided plant growth is representative. Because the surrounding soil and that inside the lysimeter must be watered in excess of ET, lysimeters must be deep enough (or have water suction control) that a good root system with adequate aeration develops. Root development in the lysimeter need not be identical to that outside, but must not restrict representative vegetative growth. The depth must be great enough (or have suction control) that the water content in the surface 30 to 40 cm is representative of the surrounds to avoid thermal mismatch.

2. LYSIMETER AREA

The surface dimensions of the lysimeter are dictated largely by the structure of the vegetation and also by the construction at the walls. The measured volume of water loss is divided by the "effective" area of the lysimeter to obtain the ET (surface depth). The measured area of the lysimeter container equals the effective area only when the ET from the lysimeter is representative of the average ET from an equal area of the surrounds. The possible magnitude of error is illustrated by a hill of corn (*Zea mays* L.) planted in tanks with 0.1 m^2 area, 0.5 m^2 area, and 1.0 m^2 area. The volume of water loss may be nearly the same but the ET based on the lysimeter area would vary as 10:2:1. The ET rates calculated may be proportional to each other, but we must make a decision as to the effective area in order to find the representative rate; *replication does not help solve this problem*. The representativeness of the vegetation sample per unit area of the lysimeter as compared to that of the surrounds is related to the scale of inhomogeneity of vegetation and the size of the lysimeter. The lysimeter area must be large compared to the scale of inhomogeneity in the vegetation. If the plant cover is relatively homogeneous spatially, such as with uniform grass or forages, the area of the lysimeter can be smaller than when the scale of inhomogeneity of the plant cover is large (e.g., that found with shrubbery, corn, shorter row crops, and with "patchy" surfaces such as nonuniform soils and poorly grazed pastures). The size of the lysimeter also must be large compared to the scale of inhomogeneities of soil surface and subsurface water properties (e.g., roughness, albedo, soil water retention).

This requirement of lysimeter size has been neglected frequently, particularly with regard to row crops. When small, round lysimeters are used with row crops that provide little cover (e.g., the celery (*Apium graveolens*) experiment of Drinkwater and Janes, 1957), proportionality between ET from the lysimeter and the field is doubtful because the amount of soil exposed per unit of plant row is unrepresentative and because the sampling area is inadequate. There is some advantage to rectangular lysimeters for row crops in order to achieve uniform length row samples and representative ratios of exposed soil area to vegetation.

The lysimeter area also should be large compared to the uncropped area at the border (walls and air-gap between double walls). This is necessary, not only because this area contains no plants, but also because the walls and the air gap have different thermal and water properties than the soil and will affect the heat exchange. Thin-wall containers made either from steel or plastic-fiberglass are preferable to concrete to keep the wall-gap thickness minimal.

3. LYSIMETER THERMAL PROPERTIES

The lysimeter container and soil may have different thermal properties than the surrounding soil. If the water distribution in the lysimeter differs from that outside, the heat transfer and storage will be affected. The surface layer (30 to 40 cm) is of greatest importance where hourly measurements are made, though the seasonal soil heat flux is affected by much deeper layers. If the lysimeter is shallow (even though water suction control is used), discontinuities in thermal properties at the tank bottom can cause error in weekly or monthly measurements, and temperature regulation at the bottom to match the surrounds may be necessary for high accuracy.

Thermal mismatch error decreases when the lysimeter is covered with vegetation because the soil heat balance is decreased. Relative error in daily measurements is less than in hourly measurements. King et al. (1956) found that with sparse alfalfa-brome cover (following cutting), the ratio of ET from a floating lysimeter to that given by energy balance measurements was much less for daylight hours than when abundant foliage was present. Tanner and Pelton (*unpublished data*) later found for the same lysimeter that after the alfalfa-brome was cut, hourly ET from the lysimeter was less (as much as 30 to 40%) during the daytime, but at night was much greater than the energy balance measurements; however, the total 24-hour ET measured by both methods agreed well for full-cover and following cutting. The disparity between daytime and nighttime measurements with sparse cover was due largely to thermal unrepresentativeness in which greater heat storage and conduction occurred in the lysimeter than in the surrounds. This thermal error was not observable on an hourly basis when full cover existed and the heat exchange at the soil was small. With poor cover, the errors balanced over a 24-hour period. King et al. (1956) measured ET only during the daytime and thus missed the effect of thermal mismatch of the lysimeter. Pruitt and Angus noted a small disagreement between floating lysimeters and a larger weighing lysimeter that may have been caused by similar thermal errors (W. O. Pruitt, and D. E. Angus, 1961).

The thermal representativeness of a lysimeter can be tested by measuring the soil heat flux inside and outside the lysimeters and also by comparing the ratio of lysimeter ET to that given by micrometeorological methods where applicable. The measurements should extend over the full diurnal heat cycle. Short-period matching can be tested during periods of strong radiation variation (e.g., a 30-min to 1-hour period of heavy cloud cover between periods of bright sunshine, or vice versa). Tests should be made under conditions of full cover, sparse cover, and no cover. Long-term mismatch can be detected by dissimilarity of mean soil temperature profiles. It is important to recognize that dissimilarity in mean soil temperature can introduce errors other than soil heat storage since the soil temperature affects plant temperature, vapor pressure, and temperature gradients (thus heat exchange) in the soil-plant-atmosphere layers.

4. LYSIMETER MANAGEMENT

The lysimeter must be sited in identical surrounds and with representative fetch (*see* part IV, micrometeorological methods). Nearby obstructions or nonevaporating surfaces, including balance access structures and recording instruments, paths leading to the lysimeter, roads, and exposed roofs of underground shelters should be avoided. The lysimeter and the surrounds should be planted, fertilized, watered, and otherwise managed in the same manner.

Water management should be planned to avoid unrepresentative salt accumulation. This can occur if the lysimeter drainage is either unrepresentative or if the drainage is recirculated with the lysimeter irrigation water.

Condensation and evaporation on walls of weighing lysimeters can cause errors. McIlroy and Sumner (1961) found that the error due to variable condensation was intolerable when the gap between the lysimeter container and retaining tank was sealed but was acceptable when the gap was left open for vapor exchange to the atmosphere. Dehumidifying the air surrounding the tank is inconvenient but may prove necessary to eliminate condensation errors.

When possible, replication of lysimeters is desirable; however, replication does not eliminate the systematic errors discussed in connection with lysimeter area or due to an unrepresentative lysimeter.

5. NONWEIGHING LYSIMETERS

The difference between the water supplied to the lysimeter and that which drains is measured to give ET. When the lysimeter is used to measure ET_a, drainage may be infrequent and the $P - (V_r)/A$ term of equation [29–1] is averaged over time periods between drainage occurrence. The ET error is $\Delta V_x/A$ (soil water profile may not be the same at two drainage periods). If the soil water changes in the lysimeter $(\Delta V_x/A)$ are measured with a neutron meter, weekly ET_a determinations can be made with reasonable accuracy. When ET_p is measured, irrigation is used so that drainage is more frequent. If daily irrigations are followed, the moisture profile is relatively constant so that estimates over periods of 1 week are reasonable. Even so, neutron meter measurements are desirable.

A great many designs of nonweighing lysimeters have been proposed (e.g., see reviews by Kohnke et al., 1940; Harrold and Dreibelbis, 1958), though many do not meet the requirements outlined. Nonweighing lysimeters currently providing valuable data range in size from the large-area, deep, monolith lysimeters at Coshocton, Ohio, USA (Harrold and Dreibelbis, 1958) to the small-area, shallow lysimeters constructed from oil drums (Gilbert and Van Bavel, 1954). The Coshocton lysimeters are used for ET_a and the oil-drum type for ET_p.

The constant water table lysimeter (e.g., Mather, 1950) has been widely employed (less so, currently) for ET_p measurements. The water required to maintain the water table level at a given depth is metered to give ET_p. Water table lysimeters are not reliable unless they are used to represent surrounds with a water table at the same level. If the water table is controlled at depths which allow reasonably normal root distribution, water movement from the water table into the root zone frequently is inadequate to meet the demand.

6. WEIGHING LYSIMETERS

Because nonweighing lysimeters cannot provide short-period estimates (e.g., hourly or daily) that are needed for many studies, several types of weighing lysimeters have been developed. A lysimeter container is placed inside a second tank that retains the surrounding soil so that the inside container is free for weighing. Weighing lysimeters differ not only in the mode of weighing but also in features of construction that affect the accuracy.

The most common type of weighing lysimeter employs mechanical balances to measure the weight loss. The balances usually are of special design and the reader must consult the literature for details—a few of the proven units are listed here. In the simplest arrangement, the soil container is attached to a portable overhead balance and lifted free of its supports when weighed. Among recent, commonly cited arrangements of this type is the monolith lysimeter described by Makkink (1953). The exposure of such installations as Makkink's leaves much to be desired for representative measurements. The large Coshocton weighing lysimeters (Harrold and Dreibelbis, 1958) were one of the first installations with large containers supported on a mechanical balance for continuous weighing. A monolith block of soil, 1.90 m by 4.26 m in area and 2.44 m deep, is encased in a concrete container. The retaining tank also is concrete. The balance is recording and sensitive to 0.25 mm ET. Recently Harrold and Dreibelbis (1963) have found that diurnal temperature changes produced variations in the physical properties of the grease used as a seal between the walls that in turn caused diurnal errors in the ET measurements. This resulted in overestimating dewfall and other errors in the hourly measurements. Three other sources of errors could be present: (i) wall-gap-wall area is about 65% of the soil area; (ii) the seal of the air gap between the walls may cause condensation errors; and (iii) the large underground installation adjacent to the lysimeter may influence the thermal regime. The Coshocton lysimeters recently have been modified to avoid the grease seal and to minimize wall-gap error.

The Davis, California lysimeter (Pruitt and Angus, 1960) is an excellent example of a large weighing lysimeter with minimal wall-gap area, water suction control, temperature control at the container bottom, and dehumidified air surrounding the container. The container is 6.1 m in diameter and 0.9 m deep, the wall-gap-wall area is 3% of the enclosed soil, and the balance has a continuous recording accuracy of 0.03 mm ET. Pruitt and Angus (1961) found that soil water in the lysimeter was unrepresentative at the wilting percentage with a perennial ryegrass (*Lolium perenne*) cover. Thus the lysimeter is limited either to crops with very shallow root systems or to medium (or high) soil water conditions.

Two other examples of smaller, continuous weighing lysimeters are the CSIRO unit (McIlroy and Angus, 1963) and the Tempe, Arizona unit (Van Bavel and Meyers, 1962). The CSIRO unit consists of containers and retaining tanks formed from reinforced concrete pipe. The container is 1.6 m inside diameter and 1.1 m deep, the wall-gap-wall area is 30% of the soil area, and the balance has a sensitivity of 0.025 mm ET. The Tempe lysimeter is square (1 m by 1 m area by 1.5 m deep), is formed from 3.1-mm thick steel, and the wall-gap-wall area is 5% of the soil area. Water suction control is used. A strain cell is used to readout weight changes with the major load counterbalanced by weights. The sensitivity is 0.01 to 0.02 mm ET.

Though suitably designed lysimeters with mechanical-electrical balances are preferred, the initial cost is high and several designs of hydraulic weighing systems have been proposed as alternatives. These are of two types: First, the lysimeter container is floated and weight changes are measured from changes in buoyancy; second, the lysimeters are placed on some form of hydraulic load cell and the changes in weight are measured from changes in the load cell pressure. King et al. (1956) describe a floating lysimeter formed from 1.6-mm thick steel container 1.52 m in diameter and 1.8 m deep that includes either steel or styrofoam buoyancy chambers. The container was placed in a retaining tank and floated in water introduced in the intervening space. The wall-gap-wall area is 8% of the soil area. The weight changes were measured by recording the water level in the gap between the walls with a sensitive recorder (0.025 mm). This design has been used by others without serious modification. A modification proposed recently by McMillan and Paul (1961) uses $ZnCl_2$ solution (1.9 sp gr) so that buoyancy chambers are not needed. Recently King et al. have found that the greater thermal expansion of $ZnCl_2$ solutions, compared to water, results in unduly large errors unless compensation is employed (K. M. King, E. I. Mukammal, and V. Turner, Oct. 1964. Errors involved in using zinc-chloride solutions in floating lysimeters. *Presented 6th Agr. Meteorol. Conf., Lincoln, Neb.*). Two floating lysimeters have been constructed by Russians for weighing large monoliths: One was described by Federov in 1954 (*see* Pelton, 1961 for drawings and references) and the other by Popov in 1952 (*see* Molga, 1958 and Krimgold, 1957 for drawings and references). The Russian lysimeters are complex and in addition, thermal errors could be larger than those associated with the simpler systems. Hybrid systems using both buoyancy and mechanical balances have also been used (e.g., Aslyng and Kristensen, 1961).

Several lysimeters have been proposed in which hydraulic load cells have been used to measure weight variations. Though commercial load cells are available for this purpose, designs proposed to date incorporate simple load cells formed from flexible bags with larger bearing areas. The simple flexible bags with large bearing areas yield low hydraulic pressures that are read manometrically. Friction errors are decreased and the load cell and readout arrangements are inexpensive. Though load cell lysimeters appear very promising, they have not been used or tested widely.

A very simple hydraulic load cell lysimeter originated in Hawaii with the separate work of R. D. Miller and P. C. Ekern. R. D. Miller (*personal communication, 1958*) placed water filled, inflatable air mattresses under soil and read the pressure with a water manometer. The utmost in simplicity was tested with no provision to mechanically isolate the soil over the mattress from the surrounding soil and with this arrangement, the shear strength of the soil caused large errors. Ekern (1958) constructed the first workable hydraulic load cell lysimeter by supporting a 1.5-m by 1.5-m square container 0.45 m deep on two automobile inner-tubes, partially inflated with water. The inner-tube stems were connected either to a manometer or a water level recorder. The large loading area ratio (area of tank per unit of contact area between inner tubes and container) gave a mechanical (hydraulic) amplification and this, with mechanical amplification in the recorder, resulted in a sensitivity of 0.025 mm.

The lysimeter constructed by Ekern illustrates the principle of other hydraulic load cell lysimeters which differ mainly in engineering design. If the loading area ratio is unity, the pressures are small and pressure changes (water height equivalent) are equal to ET or P. If the loading area ratio is large, the hydraulic amplification is large but the gauge pressures also are large. Manometric readout usually has been used; however, temperature errors caused by density changes in the manometer are obviously large. Manometers have advantage over other pressure transducers for measuring pressure changes in a large gauge pressure because sensitivity in readout of the changes is retained with manometers regardless of the gauge pressure range while the sensitivity of other transducers decreases with increasing gauge pressure range.

In 1960, Tanner and Swan (*unpublished results*) constructed a lysimeter with a container 3 m by 6 m in area and 0.6 m deep. The container was placed in a retaining tank with a 7-cm wall-gap thickness (8% of container area). Several load cells were tested including vinyl and rubber bags and industrial rubber pipe —the last has proven best. Laboratory and field tests indicate that a 12-inch diameter, nylon-butyl flexible pipe partially inflated with water (or antifreeze) can readily support a load of soil with a 1.3-m wide and 2-m deep section over the flexible pipe. Since gauge pressures developed are as large as 1.5 bars, a differential mercury manometer with large cup area is used to "buck" the major portion of the absolute pressure, and height changes in a low density fluid above the mercury at the low pressure side of the differential manometer is measured. To achieve mechanical stability of the tank, a nylon-rubber pipe under each quadrant of the lysimeter is used. Each pipe is connected to a differential mercury manometer and these, in turn, are connected to a common manometer tube or pressure transducer. The manometer is installed underground in a well and the temperature is either measured or controlled for temperature correction. This system has promise for large-area lysimeters up to 2 m deep, and with suitable temperature compensation, can provide hourly as well as daily measurements.

Glover and Forsgate (1962) proposed a similar design in which a rectangular lysimeter container 1.2 m by 2.4 m in area and 1.2 m deep was supported on three interconnected, water-filled "rubber bolsters." They thermostated the above ground portion of the manometer column; however, it is clear that temperature control of the below ground column is also necessary. Similar lysimeters with sheet metal cells (termite resistant) have been used by Glover with partial success (*Personal communication from G. D. Thompson, Natal, S. Africa, 1964.*). Other workers have described similar weighing systems for very small lysimeters or pots which are not large enough to provide representative sampling.

Commercial hydraulic load cells can be used with small lysimeters but cost is prohibitive for large lysimeters. Also strain-gauge load cells can be used to advantage on large and small lysimeters (Frost, 1962); however, the accuracy is suited mainly to daily (not hourly) measurements.

7. LYSIMETER SUMMARY

Weighing lysimeters are necessary for daily or hourly measurements. Mechanical balance weighing systems with automatic readout have proven to be more reliable over a wider range of energy balance and wind conditions than either the floating lysimeters or the hydraulic, load-cell lysimeters in their present range of development. Floating lysimeters are more economical than mechanical weighing lysimeters; however, unless buoyancy is achieved by floating them in dense liquids so that rooting volume is comparable, they can be used only for potential ET measurements. Hourly measurements with floating lysimeters may be in error except when the soil heat flux is very small compared to other energy balance terms, though they are well suited to daily measurements. Hydraulic, load-cell lysimeters have greater possibilities than the floating lysimeter. Thermal errors in the lysimeter will be as low as in the mechanically weighed lysimeters and, with manometer thermostating, working precision should approach mechanical balances. The load cell lysimeter's greatest advantage lies in relatively inexpensive construction of large-area lysimeters.

8. STEM FLOW MEASUREMENTS AND TRANSPARENT ENCLOSURES

Stem flow velocity measurements (e.g., Bloodworth et al., 1956; Skau and Swanson, 1963) may be regarded as a form of lysimetry with the possibility of measuring only transpiration. To be useful, the measured stem flow must be related uniquely to transpiration from the plant and, as with lysimeters, the stem flow velocity must be scaled to a representative transpiration rate for the surrounds. These problems have not been explored adequately, though some data regarding stem flow calibration is given in Decker and Skau (1964) and other important work on stem flow is referenced in the above quoted articles.

A chamber which encloses an area of vegetation and/or soil above ground with material which is transparent to radiation but which prevents water exchange with the atmosphere also is analogous to a lysimeter. Example chambers are those of Musgrave and Moss (1961), Decker et al. (1962), and Sellers and Hodges

(1962). Though useful for many studies, the space inside the chamber is not representative of the unenclosed surrounds since the radiation exchange and turbulent transfer within the enclosure is altered. Businger (1963) discusses many of the climate alterations that occur.

9. LYSIMETER TESTS OF METEOROLOGICAL METHODS

Micrometeorological methods are based on assumptions that may not be met in application. Consequently, neither micrometeorological methods nor empirical formulas can be used with confidence unless they have been tested for a given range of atmospheric and surface conditions. Adequately designed weighing lysimeters provide an independent measurement of ET over short periods that can be used for tests. Provided the lysimeter is representative so that the weight losses and gains per unit area are *proportional* to the losses and gains per unit area of the surrounds, the ratio of lysimeter ET to that given either by the micrometeorological methods or empirical formula should be constant. If the ratio is unity, we are assured that they are measuring the same quantity. If the ratio is constant but is not unity, the choice of which method provides the best absolute measure must then be made. It is desirable for completeness that the tests be run over a wide range of meteorological conditions, with bare soil and with vegetation cover.

Provided only relative ET rates are needed, some measurements can be made with small, representative lysimeters of unknown effective area. The measurements of Denmead and Shaw (1962) and some pot experiments (e.g., Gardner and Ehlig, 1963) are of this type. Though the water lost per unit lysimeter area may not represent the ET from an extended area, the relative loss rates may be used to derive important principles. However, we wish to emphasize that the data gathered from pot experiments as varied as those of Gardner and Ehlig or of Briggs and Shantz (1916a, b) cannot be used to define either relative ET rates that may exist in field plantings or the relative amounts of radiation and sensible heat exchange that occurs in the field because the heat exchange is not representative of field conditions.

IV. MICROMETEOROLOGICAL MEASUREMENTS

Micrometeorological methods provide a measurement of the flux density of water vapor in the boundary layer of the atmosphere and are based on the principles discussed in chapter 26. Though these methods have limitations as to where and how they can be used as well as instrumental difficulties, they have important advantages. When applicable, micrometeorological methods can measure the ET over very short time periods (e.g., a few minutes), can provide flux measurements of interest other than ET (e.g., CO_2 and heat), and can provide allied environmental information important to plant studies (e.g., temperature, humidity, etc.).

The micrometeorological methods in greatest use are the profile methods (also called aerodynamic or mass transport methods), the energy balance method, and combinations of each. Eddy fluctuation methods have been used and are preferred, in principle, though they have not been employed widely because of instrument difficulties.

A.   General Assumptions of the Profile and Energy Balance Methods

For successful application of the profile and Bowen ratio equations [26–11] through [26–14], certain assumptions must be fulfilled. These equations are derived assuming (i) steady-state, (ii) adiabatic conditions, (iii) one-dimensional transport (i.e., no horizontal gradients), and (iv) a homogeneous surface (i.e., the heat, water, and momentum sources and sinks are indistinguishable).

The assumption of steady-state would appear to exclude transient conditions at the surface (e.g., vapor pressure or temperature changes brought about by radiation variation). However, no serious error occurs in agronomic work, though it may appear in research with more sensitive requirements. Dyer (1963) discusses the rate of profile adjustment with time.

Adiabatic conditions must obtain for equation [26–3] to hold. Surface heating or cooling produces wind profile curvature and changes in profile slope. To describe the diabatic wind profile, corrections of the adiabatic equation are used. The corrections usually employ the Richardson number given by equation [26–16] which indicates the importance of buoyancy as compared to frictional forces in producing turbulence. When frictional effects are large, $(u_2 - u_1)^2$ is large relative to $(T_2 - T_1)$ and the Richardson number is small. Equation [26–3] modified to express the diabatic profile is of the form

$$u = [(\tau/\rho)^{1/2}/k]\{[\ln(z - d)/z_o] + \Phi\} \qquad [29\text{–}2]$$

where Φ is a function of the Richardson number (Ri).

By using equation [29–2] in place of equation [26–3], flux equations are found to be the same as in equations [26–11], [26–12], and [26–13], except that the right-hand term is multiplied by $(1/\varphi)^2$.

$$Q = -\frac{\rho c k^2 (K_H/K_M)(T_2 - T_1)(u_2 - u_1)}{[\ln(z_2 - d)/(z_1 - d)]^2} \times \frac{1}{\varphi^2} \qquad [29\text{–}3]$$

$$E_* = E/\lambda = -\frac{(\rho\epsilon/p)k^2(K_V/K_M)(e_2 - e_1)(u_2 - u_1)}{[\ln(z_2 - d)/(z_1 - d)]^2} \times \frac{1}{\varphi^2} \qquad [29\text{–}4]$$

where φ is the diabatic influence function related to Φ by $d\Phi/d\ln(z + z_o) = \varphi - 1$ [e.g. Panofsky, 1963 and Lettau (H. H. Lettau, 1962. Notes on theoretical models of profile structure in the diabatic surface layer. *Dep. Meteorology, Univ. of Wisconsin, Final Rep., Contract DA–36–039–SC–80282.*)].

Usually φ is written as $\varphi = (1 + b\text{Ri})^a$ where a and b are constants. Lettau has summarized these constants for several proposed models and found that $a = -1/4$ and $b = -18$ is the best choice. It is seen that for small Ri, $\varphi^2 = (1 + 9\text{Ri})$ which is essentially the correction suggested in chapter 26; for large Ri, the full function should be used. The diabatic correction accounts for *wind profile changes caused by thermal stratification*, and enters in the estimation of $u_* = (\tau/\rho)^{1/2}$ but does not enter into equation [26–14].

Application of equations [29–3], and [26–14] (K_H/K_V assumed to be unity in [26–14]) requires information on K_H/K_M and K_H/K_V. Some of the earlier work applied diabatic corrections only to the transport coefficients, neglecting the wind profile corrections. Present consensus among most workers is that $K_V/K_M \approx 1$. Also, most data indicate that the error in assuming $K_H = K_M = K_V$ is negligible if $-0.05 <$ Ri < 0.05. It is clear that K_H/K_M and K_H/K_V are most likely to approach unity when winds are strong and the Richardson number is small, and that error in assuming $K_H = K_M = K_V$ is minimized by working close to the surface where frictional effects are most pronounced.

The descriptions of the experimental sites where the transport coefficients have been determined indicate that $K_H \approx K_M \approx K_V$ has been found over mown uniform grass and other uniform surfaces whereas $K_H \neq K_M \approx K_V$ has been found over more patchy surfaces. If surfaces are inhomogeneous, some spots are primary sources of heat, others may be vapor sources and momentum sinks. If the scale of the wind eddy is of the same size or smaller than the inhomogeneities, heat and water (or momentum) can be added at different times to the wind stream. If heat and water (or momentum) are separated initially in the eddies, then buoyancy would act selectively resulting in greater upward acceleration of hotter and drier eddies so that the correlation between water and heat fluctuation or momentum and heat fluctuation would be relatively lower than that between water and momentum. This results in measurement of $K_H \neq K_M \approx K_V$ by either gradient or fluctuation methods.

Surface inhomogeneity not only can affect the ratio of (K_H/K_M) or (K_H/K_V) but may also create sampling problems (e.g., a sensor may rest in position systematically hotter or drier than the average). Many of the agricultural surfaces of interest are far from homogeneous (patchy pasture, row crops, etc.) and spacial sampling may be required. However, even with spacial sampling, energy balance and aerodynamic measurements may be invalid over patchy surfaces where heat, vapor, and momentum sinks and sources are separated. Data are needed on these surfaces to find the extent of application of the micrometeorological measurements of heat and mass transfer.

If experimental conditions do not fulfill reasonably the assumption of zero horizontal gradients, substantial errors can result. As air moves from a surface of given roughness, wetness, and temperature to a different surface, the velocity, air temperature, and vapor profiles change from those representing properties of the first surface to those resulting from the properties of the second. The depth of the representative boundary layer grows with the downwind distance traversed over the new surface. Profile measurements representing the new surface must be made within the boundary layer to approximate adequately one-dimensional transport from the surface. This means there must be sufficient "fetch" to permit the profile to develop until at the highest measurement, the profile adjustment is sufficiently complete that error will be small. Brooks (1961) reviewed wind tunnel work to estimate the distance downwind that the influence of obstacles such as tree rows, etc. extends. He indicated that this is at least eight times the height of the obstacle and that this distance must be added to that fetch required if no obstacles were present. A commonly used rule-of-thumb is that the fetch to measurement height should be at least 50:1 and preferably 100:1 with no upwind obstacles. Brook's review supports the 100:1 ratio.

Table 29–2. Fetch/height requirements for 90% profile adjustment (Dyer, 1963)

Height, m	0.5	1	2	5	10
Fetch, m	70	170	420	1,350	3,300
Fetch/height	140	170	210	270	330

Dyer (1963) analyzed the rate of adjustment of heat, water, and CO_2 profiles with distance from a discontinuity assuming no change in roughness length at the discontinuity. He calculated fetch requirements for a 90% profile adjustment as given in Table 29–2. Calculations for other adjustment levels and heights also are plotted by Dyer. The adjustment of temperature, water, and CO_2 profiles clearly cannot be tested by equilibrium of wind profiles unless roughness changes along with other surface properties.

A divergence resulting in a net vertical mass flow also can occur because of wind gradients in the canopy (Tanner, 1957, called this the "clothesline effect"). If the wind enters the crop structure at the edge of a field and the velocity decreases with the distance inward, a mass flow upward through the crop and horizontally occurs. This mass flow carries heat and moisture that cannot be indicated by gradients. Thus orchards and other tall and widely spaced crops are not suited to vertical measurements—the main criteria is not height but whether the structure permits horizontal airflow through the canopy. However, ET measurements can be made on these surfaces if horizontal flux as well as the vertical flux is measured. This is much more difficult.

To summarize, profile measurements made well above the surface where air is better mixed may not be in equilibrium with the surface unless the fetch is very large. Also gradients are smaller and more difficult to measure and effects of thermal stratification become more serious with increasing height. If measurements are made near the surface, gradients are steeper and more easily measured, the profiles are at equilibrium with the surface with a shorter fetch, and thermal stratification is less important. However, gradient sampling is affected by the inhomogeneity of the surface and spacial integration may be necessary. Thus the experimenter is faced with problems of spacial sampling near the surface where the gradients are measured most reliably or with increased instrument, fetch-size, and thermal stratification problems at higher levels. Lastly, if the surfaces are patchy, we should neither assume $K_H \approx K_M \approx K_V$ to apply nor assume the aerodynamic methods and Bowen ratio method to have reasonable accuracy unless tested against independent measurements.

B.   Profile Methods

In order to apply the profile equations [29–3] and [29–4], it is necessary to find d, $(u_2 - u_1)$, $(e_2 - e_1)$, and $(T_2 - T_1)$. Though measurements at two heights appear to suffice, more measurements are needed for two reasons: First, the wind must be measured at a minimum of three levels to establish d, and second, since the wind is measured at a different place and by different sampling methods than either e or T, several levels of measurement are needed to test for similarity as described in Chapter 26.

The zero-plane displacement is found from the adiabatic wind profile as discussed graphically in connection with equation [26–15]; however, graphical procedures usually are not sufficiently objective or accurate. A statistical procedure for hand calculation of d is given by Lettau (1957, p. 334) and is illustrated more completely by Tanner (1963). An efficient procedure for electronic computers is given by Robinson (1962). To find d, several heights of wind measurements are needed and very good accuracy is required in measuring the second-order properties of the profile. A small error in d may cause a large error in $[\ln(z_2 - d)/(z_1 - d)]^2$ (and $E_\circ$, Q, or τ) unless z_2 and z_1 are large. Often data from given levels is poor due to instrument malfunction, and small drifts may be expected over a day's run. Accordingly, many levels must be used and graphed so that suspicious data can be discarded. Because data for a given level may be discarded, there is little advantage in spacing levels geometrically.

In addition to methods applying equations [29–3] and [29–4], three other modifications of the aerodynamic method have been proposed that offer advantages for special uses. The first of these is the method proposed by Deacon and Swinbank (1958), and a second is the Dalton method. A third method (Tanner, 1960b) is useful in connection with the energy balance and is discussed as a combination method.

1. DEACON AND SWINBANK METHOD

This method has proven to be useful over fairly homogeneous surfaces (ryegrass, rough pasture, etc.) provided there is reasonably long fetch. The method employs two wind measurements high above the crop surface and a third measurement very near the surface where wind is not influenced greatly by stability. If z_2 and z_1 in equation [29–2] are large compared to d, we can find τ for a neutral period from two wind measurements. Moreover, we can find the drag coefficient from τ and the wind measurement u_x, near the surface since

$$\tau = C_s \rho u_x^2. \qquad [29–5]$$

We then find $E_\circ$ from Δe using the incremental form of equation [26–8], assuming $K_H = K_M$.

$$E_\circ = -C_s(\rho \epsilon / p) u_x^2 (e_2 - e_1)/(u_2 - u_1). \qquad [29–6]$$

2. DALTON METHOD

This method, often called the bulk aerodynamic method, is the oldest of the aerodynamic methods and is based on an equation similar to [26–25]

$$E_\circ = f(u_z)(e_o - e_z). \qquad [29–7]$$

Frequently the empirical form $f(u) = (a + bu_z)$ is used. The Dalton equation has been applied frequently to lakes and pans where e_o is assumed to be the saturation vapor pressure corresponding to the surface temperature. The constant "a" in the $f(u)$ wind expression is often zero as in equations [29–8] and [29–9]. Because of pan lips increase turbulence, "a" is larger with pans than with water bodies (Kohler, 1954).

Sverdrup (1937) presents an equation to account for the thickness δ of the air layer at a smooth water surface in which transfer is by molecular diffusion

$$E_\circ = \frac{(\rho \epsilon k/p) u_* (e_o - e_z)}{\left[\ln \dfrac{z + z_o}{z_o + \delta} + \dfrac{k u_* \delta}{D} \right]}. \qquad [29–8]$$

Work at Lake Hefner and at Lake Eucumbene indicates $z_o = 0.6$ cm and $u_* \delta / \nu = 28$ to 30 (Webb, 1960). These data can be used with equations [29–2] and with [29–8] to convert the wind function and the vapor pressure from one height to another. The Hefner data (Kohler, 1954) reduce to $E_\circ = (2.57/p) u_2 (e_o - e_2)$ and Slatyer and McIlroy (1961) have converted Webb's data to $E_\circ = (2.3/p) u_2$

$(e_o - e_2)$ where u_2 and e_2 are the wind (miles/day) and vapor pressure at 2 m when e and p are measured in the same units and E_a is mm/day.

Sheppard (1958) developed equation [29–9] for water bodies, which also considers molecular diffusion at the interface.

$$E = (\rho \epsilon k / p) u_* (e_o - e_z) / \ln (k u_* z / D) \qquad [29–9]$$

where $z \gg D / k u_*$. Fritschen and Van Bavel (1963) during one test have found equation [29–9] more accurate than [29–8].

C.   Energy Balance Method

Evapotranspiration can be found from the disposition of the energy at a surface. The complete energy balance of a volume is complicated (Tanner, 1960a) though in principle it can be measured. Ordinarily the simple vertical energy balance as expressed by equation [26–20] is employed. A group of miscellaneous terms such as photosynthesis, and heat storage in the plant-air layers are not given in equation [26–20] but are usually small and can be either estimated or neglected without introducing significant error in the total daily energy budget. We find the evapotranspiration from equations [26–14] and [26–20] as

$$E = \frac{(H - S - M)}{1 + \gamma (K_H / K_V)(T_2 - T_1)/(e_2 - e_1)} = (H - S - M)/(1 + \beta) \qquad [29–10]$$

where M represents the miscellaneous terms mentioned and (K_H / K_V) is shown. All terms except M and (K_H / K_V) are measured, M is estimated (or neglected) and (K_H / K_V) is usually taken as unity.

We use only similarity in [26–14] so that only $(T_2 - T_1)$ and $(e_2 - e_1)$ are needed rather than a complete profile; moreover, $(T_2 - T_1)$ and $(e_2 - e_1)$ can be determined from the same air sample.

D.   Combination Methods

Several combinations of the energy balance and aerodynamic methods have been used. Some workers have regarded the wind profile and temperature profile measurements required for Q to be more easily made than the measurements required for Bowen's ratio. As a result, many measurements of E have been made by combining the aerodynamic measurement of Q with the energy balance,

$$E = H - S - M - Q. \qquad [29–11]$$

The Q is found from equation [29–3] and the other terms are measured (or neglected if small). This same approach can be used when Q is measured by simplified aerodynamic and by eddy fluctuation methods.

The wind at a single height can be used in a simple aerodynamic method to complement energy balance measurements (Tanner, 1960b). If Φ is small and if surface properties are constant, equation [29–2] shows that u is proportional to $(u_2 - u_1)$. Accordingly E_a and Q is found from

$$E_a = C_v u (e_2 - e_1) \qquad [29–12]$$

$$Q = C_r \lambda \gamma u (T_2 - T_1) \qquad [29–13]$$

where C_v is gotten either from the wind profile or from [29–14]. During either a neutral period or when the Richardson number is small, [29–11], [29–12] and [29–13] are used to establish

$$C_v = (H - S - M)/\lambda u[(e_2 - e_1) + \gamma(T_2 - T_1)]. \qquad [29\text{–}14]$$

During adiabatic periods equation [29–14] becomes $C_v = (H - S - M)\lambda u(e_2 - e_1)$. Once C_v is known we can measure $(T_2 - T_1)$ and find Q. Then E is found from [29–11] and measurements of H and S. Equation [29–12] is particularly useful in conjunction with the energy balance method to estimated E_* for those times (low energy) when the Bowen's ratio fails (e.g., $\beta \rightarrow -1$).

Another combination method is that discussed in detail in part V of chapter 26; Slatyer and McIlroy (1961) and Budyko (1956) have developed equations that are essentially the same (see McIlroy and Angus, 1964 and Sellers, 1964, for a discussion of the McIlroy and the Budyko formulation respectively).

E. Comparison of Profile and Energy Balance Methods

The relative merits of these two procedures are not evaluated simply. We must recognize that profile measurements give information on the wind structure and turbulence whereas the energy balance does not; also the "surface properties" (d and z_o) are determined. However, our primary concern is the determination of ET rather than studying wind structure so that different criteria must be used.

1. FULFILLING ASSUMPTIONS OF EQUATIONS

Both the profile and Bowen ratio (energy balance) methods require a fetch at least 50 to 100 times the measurement height. However, the profile method requires several levels of measurement when d and z_o are established and when profile similarity is tested, demanding greater instrumentation heights and greater fetch. Both methods assume similarity; but, in addition, the profile method requires that d and z_o remain unaltered during the period of measurement. Chapter 26 mentions changes in d and z_o with wind and Lemon (1963) gives other examples. The fact that the energy balance involves only similarity is a distinct advantage. Lastly, both the profile method and the Bowen ratio assume surface homogeneity for suitable sampling and for equality of transfer coefficients. For this reason both can be regarded as quantitative for many agricultural surfaces only after testing. The profile method can be tested by comparing $\lambda E_* + Q$ from equations [29–3] and [29–4] to $H - S - M$. Both the profile and energy balance method can be tested by comparing $E_* = E/\lambda$ to good lysimetric measurements of E_*.

2. MEASUREMENT DIFFICULTIES

The most easily measured quantities (for about 5% to 10% error in the final estimates of E_*) are H which is measured directly, S, $(T_2 - T_1)$, and $(u_2 - u_1)$. The measurement of $(e_2 - e_1)$ is substantially more difficult. The greatest experimental difficulty is in the wind profile measurement of d and z_o both because of instrumentation (at least 6 to 10 levels are needed for reliability), and because high accuracy is required if the second-order characteristics of the profile are to be established with necessary precision. The problem of height reference for u vs. T or e profiles is avoided in the Bowen ratio method.

Errors introduced in the estimates simply from errors in measurements also must be evaluated. The error introduced into E and Q through errors in d may be large if z_1 and z_2 are near the surface. The relative error in the profile method introduced through errors in $(u_2 - u_1)$, $(T_2 - T_1)$, and $(e_2 - e_1)$ is in proportion to the relative error of any term; also an error in assuming $(K_H/K_M) = (K_V/K_M) = 1$ produces equal error in Q or E. The error in E from equation [29–10] is in proportion to the error in H. Because S and M are usually small relative to H, fairly large relative errors in S and M can be tolerated. The relative error in β is in proportion to the relative error in (K_H/K_V) or in $(T_2 - T_1)$ and $(e_2 - e_1)$. However, because E is found from $(1 + \beta)$, the relative error in E is proportional to a relative error in β only when $\beta = -0.5$ (half the heat in E is derived from sensible heat transferred to the surface). If β is < -0.5 the error in E is larger than in β and equation [29–10] becomes indeterminant when the E is equal to Q supplied to the surface ($\beta = -1.0$). If β is > -0.5, the error in E is less than the error in β. This is a substantial argument in favor of the energy balance–Bowen ratio method since β is rarely if ever < -0.5 on agricultural surfaces *provided the total energy balance* is large (e.g., H > 0.15 mm/hour equivalent ET). It is true that in the evening, the morning, and often during the night, the Bowen ratio cannot be determined with suitable accuracy. During these periods ET is low and either can be guessed or can be estimated with the simple combination method.

Neither the profile nor Bowen ratio methods may be reliable during periods of very light winds, regardless of the magnitude of the energy balance. In general, the energy balance–Bowen ratio method is easier to use over most agronomic surfaces and will be of higher accuracy, particularly when E is large compared to Q. The profile method or the combination method in equation [29–12] is better suited to periods of low energy balance than the Bowen ratio method. When the profile method is used, it is recommended that both Q and E be determined and that H and S be measured, permitting objective tests of $Q + E$ vs. $H - S$. This total information is much preferable to measuring either the Bowen ratio Q, or E alone. Measuring $H - S$ helps set limits on amounts and establishes a degree of reasonableness.

3. TESTS OF METHODS

Both the aerodynamic and the energy balance–Bowen ratio methods have been tested by several workers (vs. $R_n - S$ or lysimeter measurements of E). Agreement generally has been good over grass, pasture and hay crops, soil, cut hay with soil exposed, and other similar surfaces with small scale inhomogeneities (e.g., House et al. 1960; Tanner, 1960a).

Recent studies of W. O. Pruitt made over uniform perennial ryegrass are of particular interest because several of the methods discussed were tested against measured ET losses from a large lysimeter under a wide range of atmospheric stability conditions $(-2.0 < \text{Ri} < 0.2)$ (*Personal communication, June 18, 1964, Davis, California*). Atmospheric measurements were made at heights of 25, 50, and 100 cm. Assuming $(K_V/K_M) = 1$, E_o from equation [26–12] was about 1/8 that measured by the lysimeter when $-2.0 < \text{Ri} < -1.6$. Correction for the diabatic profile equation [29–4] resulted in an E_o estimate 7/10 that of the lysimeter. When the profile was neutral, E_o from equation [26–12] was the same as from the lysimeter. Pruitt also tested equation [29–6]. Even when Ri < -1.6, equation

[29–6] gave E_a estimates that were within 15% of the lysimeter measurement. Measurement of E_a by equation [29–10], assuming $(K_H/K_V) = 1$ and $M = 0$, matched the lysimeter closely over wide ranges of stability. Good agreement obtained between measurements for the 25- to 50-cm layer and the 50- to 100-cm layer indicating adequate similarity for the Bowen ratio method. The other methods gave differences of 20% or more between the two layers. Pruitt also found that E_a from equations [29–11] and [29–13] was close to the lysimeter except when Ri was large. With $-2.0 < $ Ri < -1.6 an over estimate of 30% obtained.

Van Bavel (*personal communication, March 20, 1964*) has tested the combination approach (equation [26–35]) over a 100-m by 100-m, 2-cm deep water surface with B_o found from equation [26–3] for a measured $z_o = 0.001$ cm. Hourly values agreed within 15% and daily values within 2% of a lysimeter. Fritschen and Van Bavel (1963) found the energy balance equation [29–10] and the Sheppard method [29–9] gave good results over shallow water but the Sverdrup method [29–8] was in error over 34%.

Fewer data are available regarding tests over row crops. The energy balance has been tested against a lysimeter over corn by Graham and King (1961). Their daytime comparison was good though fairly large scatter existed with hourly measurements (no statistical analysis given); however, they mention that the lysimeter measurements could have been low during the morning hours and high in the afternoon which would increase the spread of data points. Mukammal et al. compared some aerodynamic methods (Deacon, 1949; Monin and Obukhov, 1954; and equation [26–12]) and the energy balance equation [29–10] with lysimeter measurements on corn about 2 m tall (E. I. Mukammal, K. M. King, and H. F. Cork. Oct., 1964. Use of aerodynamic techniques in estimating evapotranspiration from a cornfield and the difficulties encountered. Presented 6th Agr. Meteorol. Conf., Lincoln, Neb.). The fetch was only about 25 times the greatest measurement height and $-0.07 < $ Ri < 0. All aerodynamic measurements were $1/3$ to $1/2$ the lysimeter. Though the short fetch may have contributed large error, it is of interest that equation [29–10] gave estimates with maximum error of 10%. Monteith and Szeicz (1960) found reasonably good agreement between lysimeter and energy balance measurements (10- to 20-min sampling period) over sugar beets (*Beta vulgaris*); fairly large scatter was found but the comparison of 20 half-hour periods gave mean values of the two methods that agreed within about 6%. Monteith (1963, p. 100) found good agreement between lysimeter and aerodynamic measurements (4-hour sampling periods) over beans (*Phaseolus vulgaris*); a correction for the diabatic profile was needed in this instance.

F.   Instrumentation

Instrumentation for the profile and energy balance methods cannot be discussed in the necessary detail for the beginning experimenter. The reader is referred to Lettau and Davidson (1957), Long (1957), House et al. (1960), Tanner (1960a), Dyer (1961), Slatyer and McIlroy (1961), Lemon (1963), and Tanner (1963) for examples of instrumentation. Two items are of general importance to instrumentation that should be mentioned briefly: The first is sensor radiation error and the second is sampling times.

1. RADIATION ERROR

Temperature sensors either may be very small so that there is little radiation error under natural ventilation or they may be shielded from radiation and then force ventilated (in some instances shielding with natural ventilation is used). Regardless of the design, the instruments should be tested periodically for radiation error to see if it is within acceptable limits. The simplest and most common test is to make temperature measurements during steady atmospheric conditions, periodically shading the sensors from *solar* radiation. To test gradient sensors, they are placed side by side at the same height and then the sensors are alternately and simultaneously shaded with the shade held 2 or 3 m from the sensors. Tests with heat lamps and sources cannot be substituted for a solar radiation test because of the spectral selection of shields.

2. SAMPLING TIMES

In selecting sampling times we must recognize that the gradients used in the previous equations are not instantaneous values but are time means over periods not shorter than about 10 min. Usually sampling periods are not longer than 1 hour though Webb (1964) indicates how longer time means can be corrected in applying equation [29–10]. A similar analysis to Webb's shows that long-period means used with equations [29–3] and [29–4] result in larger error than with the Bowen ratio. If sensors are small with very fast response, the data must be integrated and require either frequent or continuous sampling. If the sensors have exponential time response, as most do with a time constant of t_c, Shannon's and Hartley's theorems (Klein et al., 1958, ch. 10) indicate that the sensor signal must be sampled at periods of $2t_c$ or less if complete information of the sensor output is to be recorded. This does not mean that the driving signal (e.g., temperature) is sampled in complete detail since a long time-constant sensor has narrow band-width and attenuates higher driving frequencies. The highest frequency passed by the sensor can be approximated practically as $(4\ t_c)^{-1}$ cycle/sec. Because only long-period means are required, there is considerable advantage in employing sensors with longer time constants. The time response requirements of the data logging system also is decreased.

G. Eddy Correlation Methods

Ultimately eddy correlation methods should prove to be the most accurate of the micrometeorological methods and least dependent on the surface conditions. They require only a fetch sufficiently long to permit measurement in the boundary layer. Since the frequency of the eddy fluctuations is proportional to windspeed and inversely proportional to height, the response speed of the sensors must be increased for measurements near the surface. Very fast sensors must be used for measurements in the boundary layer with short fetch. The main drawback to eddy correlation measurements is the design and construction of fast-response directional sensors, expense of the entire system, and the technical competence required to operate and maintain the equipment.

The eddy correlation equation for evaporation is given by equation [26–2]; the analogous heat transfer equation is

$$Q = \rho c\ \overline{w'T'}. \qquad\qquad [29\text{--}15]$$

Because eddy correlation measurements are sufficiently complicated in instrumentation to preclude their general use in agriculture, we will give here only a brief qualitative description and references to the two examples of workable systems.

The first of these systems, called the "Evapotron" has been developed by the CSIRO (Australia) Division of Meteorological Physics. Though many publications have been issued on developments of this system, the main ones can be found in Dyer (1961). The system employs a heat-transport anemometer with directional characteristics (not the conventional hot wire) and wet- and dry-element thermometers made from very fine wire to give fast response and low radiation error. Dyer found that when the instruments were at the 1.5-m height (wind was about 2.5 m/sec) the sum of $(E + Q)$ from the eddy correlation measurements was only about 80% of the required total $(H - S)$. At 3 to 4 m height, agreement was satisfactory. The instrument time constants (0.1 to 0.3 sec) limited measurements close to the surface. Chapter 26 includes other comment on the Evapotron.

The sonic anemometer is a second type of eddy fluctuation instrument. Because the velocity of sound over a fixed distance depends on both the temperature and the wind speed, the sonic anemometer can be used to measure Q through equation [29–15]. Sonic anemometers are instrumented to measure either the transit time of pulses or the phase shift in continuous waves. The pulsed type anemometer is described by Suomi (1957), who also gives other references. The continuous wave anemometer is described by Kaimal and Businger (1963).

V. EMPIRICAL METHODS

Many empirical formulas relating climatological measurements and evapotranspiration have been developed. These relations usually apply to a given locale, vegetation, stage of crop development, and season. "Constants" in the formulas frequently must be changed to meet changes in these variables. Empirical methods are based on radiation, mean temperature, simple Dalton approaches (saturation deficit-wind product), water tanks and pans, and atmometers. Since this chapter is concerned with measurement rather than semiquantitative guessing, empirical methods are not reviewed completely. The following reviews are recommended in appraising extensive and frequently misleading literature: Rider et al. (1958), Deacon et al. (1958), Rijtema (1959), Stanhill (1961a), and Slatyer and McIlroy (1961). Empirical formulas have greatest usefulness when correlated with "potential" evapotranspiration. Empirical methods can be grouped into four classes (i) those depending primarily on the relation of ET to radiation, (ii) temperature methods, (iii) humidity methods, and (iv) evaporimeters. It is emphasized that these formulas are best suited to potential (not actual) ET estimates and are discussed in this connection. Many workers have applied corrections to ET_p estimates to estimate ET_a.

A. Potential Evapotranspiration

Potential evapotranspiration is defined here as that occurring at a "wet" surface ($h_o = 0$) and is limited by the heat supplied under given micrometeorological conditions and not by the water supply to the surface. The ET_p is influenced by the fetch and surrounds and by the surface structure and consequently will change across a given field. If the fetch and surrounds do not change markedly from year to year, the ET_p is quite conservative—i.e., the ET_p will be about the same over the same period on successive years.

1. FETCH AND SURROUNDS

In arid regions considerable ET results from sensible heat transported from dry surroundings into irrigated fields as well as from the heat supplied by solar radiation whereas in humid regions, radiation is the primary source of heat (e.g., see Tanner and Lemon, 1962). Stanhill (1961a) illustrated the ratio of ET_p from irrigated alfalfa (*Medicago sativa*) to that from a class "A" pan sited in an upwind fallow field as the distance (fetch) into the field varies. Sensible heat exchange decreased with increasing fetch so that (ET_p/E_{pan}) was 1.0, 0.7, and 0.6 at fetches of 20, 45, and 250 m, respectively. Halstead and Covey (1957) also illustrate the influence of area size and fetch on ET_p.

Empirical estimates of ET_p are applied best to large areas. This is necessary both to minimize pronounced edge effects (small fetch) and to make ET_p estimates that are usefully applied to field-size vegetation areas. Because both fetch (area) and the surrounds influence ET_p, empirical formulas should be calibrated locally when possible. Empirical estimates are not suited to small plots where the "clothesline" heat exchange (Tanner, 1957), is large and is influenced by vegetation height and density (e.g., Van Bavel et al., 1963). Moreover, it is neither necessary nor desirable to construct ET_p formulas for uniform moist areas of "infinite" extent. As indicated by equation [26–36], the ET_p from such a surface is $(H - S) [\Delta/(\Delta + \gamma)]$ and represents the lower limit of ET_p from any wet surface (refer to McIlroy and Angus, 1964).

2. SURFACE STRUCTURE

The surface structure may influence ET_p by affecting albedo, roughness, degree of cover, etc. (chapter 26, part V). Literature on the effects of vegetation type is voluminous and concepts are not resolved. When fetch is sufficient to avoid marked edge effects, and when water supply is adequate and ET_p obtains, the ET_p from many types of short and tall vegetation completely shading the ground is commonly found to be about the same (e.g., within 20%). However, large differences between vegetation types have been observed, particularly over short periods. The ET_p from wet bare soil is about the same as that from vegetation. Other experiments, such as those cited in chapter 26, indicate that complete vegetation cover is not required. Marlatt et al. (1961) found ET from mowed orchard-grass (*Dactylis glomerata* L.) planted in rows of varying width did not decrease below that at 100% cover until the cover was < 50%. Stanhill (1958a) found the ratio of ET from irrigated carrots to pan evaporation (ET/E_{pan}) was approximately 1.1 until 40% cover was achieved. Thereafter (ET/E_{pan}) was about 1.7 and was nearly constant. The author and J. B. Swan took measurements in Wisconsin on irrigated snap beans which show that the ratio ET/H changed from 0.6 to 1.0 as the plant leaves extended to cover 55% of the row spacing and averaged 1.07 for the remainder of the growing period. Data for irrigated Russett Burbank potatoes (*Solanum tuberosum*) gave similar results. The ET from extended open water surfaces may differ from vegetation because of a different heat storage (equivalent to S) and from increased net radiation. If the water is so shallow that heat storage is small compared to H, the evaporation will exceed the ET_p of vegetation because H is larger.

The concept of a ET_p which depends primarily on the existing meteorological conditions, thus permitting estimates of ET_p from meteorological parameters, is useful. This concept is justified because the ET from many agricultural surfaces under similar meteorological conditions is comparable provided that water supply is adequate and that fetch is sufficient to avoid marked edge effects (plot size). However, we should recognize the possibility that the ET_p from different vegetation types may not be the same and, as much as possible, calibrate empirical methods to particular surfaces.

3. CONSERVATIVENESS

Day-to-day variations of ET_p in humid regions are large; however, ET_p averaged over a longer period (e.g., 4 weeks) is about the same on successive years. In irrigated arid regions with more conservative climate, ET_p is more conservative than in humid regions as seen by Table 29–3. The conservative nature of the average ET_p over longer periods permits the empirical correlation of local climate indices with ET_p and even with ET_a. *Any meteorological index that indicates a variation of the evaporative conditions will help estimate the relatively small variations of the ET_p from the expected average.* Empirical estimates of ET_p improve as the period over which the ET_p is averaged increases. The standard deviation of climatological parameters is inversely proportional to the square root of the number of days in the averaging period (Brooks and Carruthers, 1953). For example, Stanhill (1961a) found the error of estimate of several empirical methods increased two to three times as the period of estimate was decreased from 1 month to 1 week. Further increases would be found if daily estimates were made. In general as the basis for empirical methods becomes more rational, the estimation period is decreased and they can be used over a wider range of locations and climates. It is noted that a mass plot of cumulative estimates vs. ET_p is the least sensitive test of a method.

The conservativeness of monthly ET is illustrated for a humid region by the Coshocton, Ohio, USA and for an irrigated region by the Davis, California, USA data, in Table 29–3. The California data are ET_p from irrigated perennial ryegrass and the Coshocton data are ET_a from meadow (varied grasses and legumes). Note that precipitation at Coshocton is more variable than ET_a.

Table 29–3. Lysimeter measurements of monthly ET_p (cm) at Davis, California[*] and of ET_a (cm) and P(cm) at Coshocton, Ohio (Harrold and Dreibelbis, 1958). The monthly averages and coefficient of variation (cv) are listed for the growing season.

Measured		Month					
		April	May	June	July	Aug.	Sept.
Perennial ryegrass, Davis	ET_p	12.1	16.1	21.2	21.3	17.6	12.7
California (4 years)	cv	0.13	0.09	0.03	0.03	0.04	0.14
Meadow, Coshocton, Ohio	ET_a	6.7	11.8	14.0	14.1	9.3	7.7
(8 years)	cv	0.30	0.26	0.17	0.14	0.24	0.26
Precipitation, Coshocton,	P	10.4	10.4	12.2	8.9	6.0	8.6
Ohio (8 years)	cv	0.18	0.40	0.29	0.27	0.59	0.77

[*] Personal communication, W. O. Pruitt, November 1964.

The coefficient of variation of ET_a from alfalfa-brome at Coshocton (9 years) was less than for meadow. Variability decreases as the root system increases and available water reservoir increases. The annual evapotranspiration from the different crops (meadow, grass, corn, alfalfa-brome) at Coshocton usually differed $< 10\%$. McIlroy and Angus (1964) also present data showing the conservativeness of potential evapotranspiration.

4. CALIBRATION OF EMPIRICAL FORMULAS

Because ET_p depends on the local meteorological conditions, field size, and surrounds, and to a lesser extent upon vegetation, the empirical methods are most reliable when "calibrated" for a given vegetation in a given locale and tested for the period over which estimated ET_p averages are most reliable. Empirical formulas cannot be applied with confidence either to other vegetation types or to other locations with appreciable differences in local climate and surrounds without calibration. Lysimeters are particularly useful in calibrating empirical methods.

5. RATIO OF ACTUAL TO POTENTIAL EVAPOTRANSPIRATION

Transpiration of plants (especially near wilting) and evaporation from soils depends on the amount of water in a soil. Several methods have been proposed for relating available soil water to ET_a/ET_p so that ET_a can be estimated from empirical estimates of ET_p. Figure 29–2 illustrates the major proposals for empirically relating the ET_a/ET_p ratio to available soil water. Line A represents the Veihmeyer and Hendrickson (1955) approximation in which ET_a/ET_p is unity until the soil wilting percentage is reached; the remaining water is unavailable for evapotranspiration. Line B represents the Thornthwaite and Mather (1955a) approximation in which the ratio ET_a/ET_p decreases linearly from field water capacity to the wilting percentage. Havens (1956) followed the linear concept of Thornthwaite but used oven-dryness as the lower limit as shown in line C. Pierce (1958) developed the approximation shown in curve D. Butler and Pres-

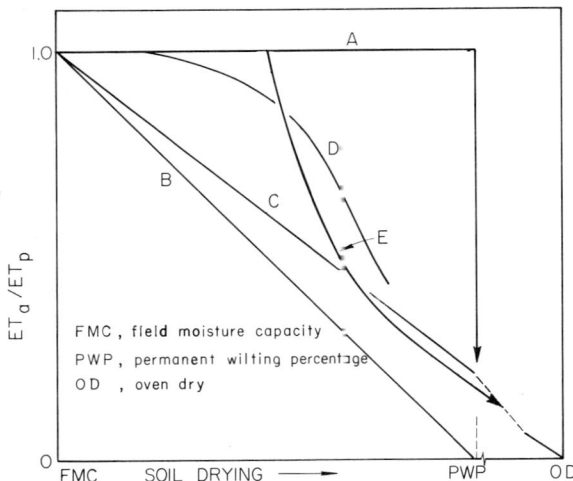

Fig. 29–2. Proposed relations of actual evapotranspiration ET_a to potential ET_p as affected by soil water content.

cott (1955) established for monthly measurements $d(ET_a/ET_p)/dw = C[2.4 - (ET_a/ET_p)]$ where w is available water (rainfall and storage in centimeters) and C is a crop constant; this equation results in a curve similar to D. And lastly, Penman (1949), Marlatt et al. (1961), Holmes and Robertson (e.g., see Holmes, 1961) have indicated that curves similar to E are the best approximation. Variations occur in establishing, the point where the ratio decreases from unity and in establishing the lowest limit of the ratio and the precise curve shape (Marlatt assumed that the falling curve joined B); however, these workers are in agreement on the qualitative aspects of curve E. Several experiments and theory indicate the general shape of curve E for both bare soils (e.g., Gardner and Hillel, 1962) and for plants (Gardner and Ehlig, 1963). Gardner and Ehlig demonstrated that the transpiration rate is governed mainly by meteorological factors until the plants begin wilting; thereafter the transpiration decreases in proportion to the remaining available water more linearly than drawn for curve E. They found the lower limit of available water to be less than the 15-bar percentage, corresponding more nearly to the 30- to 50-bar range.

Since the ratio ET_a/ET_p for the soil and for the plant will differ in time because of root proliferation and increasing cover, "crop factors" accounting for the type of crop and stage of development also have been used. Crop factors include gross effects of crop type and stage of development (height, roughness, degree of cover) and root system development. If "soil water factors" and "crop factors" are to be used, they are best found by calibration against lysimeters, and applied to the conditions under which they were developed.

B. Radiation Methods

Radiation methods for estimating ET_p are of two classes: Those based on rational energy balance arguments but which make empirical approximations to utilize existing weather data, and those which are completely empirical. The more rational methods, when "calibrated" in a given area for a given crop, are among the best available, particularly for short-period estimates. Because radiation methods are tied more closely to energy supply, they show greatest promise for short-term as well as long-term estimates.

1. ENERGY BALANCE METHODS

Methods based on the energy balance include the earlier formulations of Penman (1948), Ferguson (1952), Budyko (1956; see Sellers, 1964), and McIlroy (Slatyer and McIlroy, 1961). In principle, these formulas are not empirical and become so only when either B, $(e_a - e_2)$, or E_a are estimated from empirical formula. B has been estimated (Penman, 1948) by empirical formula of the form $B = C_1(C_2 + u)$. McIlroy has found $B = C_1(C_2 + u)$ is more satisfactory where C_1 depends on the type of vegetation (Personal communication, I. C. McIlroy, November 1964). Tanner and Pelton (1960) illustrate a third approximation $B = Cu$ where C is gotten from an estimate of z_o. Empirical estimates of E_a can be made through correlations with atmometers (Stanhill, 1962a, 1962b). Empirical estimates of $(e_a - e_2)$ may be made by assuming the daily dewpoint is equal to the minimum temperature, a fairly good approximation in humid regions. The assumption of $h_o = 1$, while empirical, does not result in large error for most well-watered vegetation. Also the assumption of $S = 0$ does not cause large error

Table 29–4. Relation of E_o (H_o from empirical formula) and E_T (H_T measured) from Penman method to measured ET_p (mm/day) from irrigated fields

Location and surface	Period		Regression equation¶	r	cv*	n	Reference
	Measurement	Estimate					
Hancock, Wis.	June-Sept.	Daily	$ET_P = 1.19\ E_T + 0.66$	0.93	0.14	40	Tanner &
Alfalfa-brome	June-Sept.	Daily	$ET_P = 1.07\ E_o + 0.89$	0.92	0.16	40	Pelton, 1960
Davis, Calif.	Jan.-May	Daily	$ET_P = 0.97\ E_T - 0.08$	0.95		76	Pruitt †
Perennial	July-Dec.	Daily	$ET_P = 1.04\ E_T - 0.03$	0.97		142	Pruitt †
ryegrass	Jan.-May	Daily	$ET_P = 0.75\ E_o + 0.01$	0.95		76	Pruitt &
	July-Dec.	Daily	$ET_P = 0.88\ E_o + 0.38$	0.96		142	Angus, 1961
	Jan.-May	Mthly	$ET_P = 1.06\ E_T - 0.25$	0.99		20	Pruitt‡
	July-Dec.	Mthly	$ET_P = 1.09\ E_T - 0.08$	0.99		22	Pruitt‡
	Jan.-May	Mthly	$ET_P = 0.83\ E_o - 0.33$	0.99		20	Pruitt‡
	July-Dec.	Mthly	$ET_P = 0.88\ E_o + 0.02$	0.99		22	Pruitt‡
Gilat, Israel	Year	Wkly	$ET_P = 0.96\ E_o + 1.12$	0.76	0.36	48	Stanhill,
Alfalfa	Year	Mthly	$ET_P = 0.97\ E_o + 0.96$	0.96	0.12	12	1961a

* Coefficient of variation (cv) = standard error of estimate divided by mean ET_P
† Pruitt, W.O. August 1962. Water Resources Center Contract Research Report. University of California, Davis.
‡ Pruitt, W.O. February 1964. Water Resources Center Annual Progress Report. University of California, Davis.
¶ E_O Evaporation estimate for open water by Penman method using empirical estimate of H_O for water and empirical B of Penman (1948).
E_T Evaporation estimate for vegetation using measured H and empirical B of Penman (1948).

and can be estimated (chapter 26). Example results from several regions for the Penman formula are listed in Table 29–4 to illustrate the usefulness of these methods. The empirical wind function, proposed by Penman (1948), was used and h_o was assumed to be unity. Though no figures were available, the coefficient of variation for daily estimates at Davis would be much greater than for the monthly estimates. The relative error in monthly estimates is indicated by Pruitt's data shown in Fig. 29–3; the small annual variation in the ratio would decrease if $(H - S)$ were used rather than H. When calibrated for a particular surface (mainly to account for B) this type of estimate should have application over a much wider range of climatic conditions than other methods.

2. DIRECT REGRESSION EQUATIONS

The second type of radiation-based estimate utilizes linear regressions with either R_I or H measurements (radiation estimates from sunshine percentage or cloud cover frequently must be used). The correlation with H is reasonable because much of the energy required for ET_P, even in arid areas, is derived from radiation. The basis for the solar radiation correlation is similar. The daily solar and net radiation are highly correlated over a given surface though the relation will change with the seasonal march of climatic and surface conditions. Since sensible heat flux is not estimated (refer to equation [26–36]) the ET_p estimates from radiation will not be as general or as satisfactory as the Penman type of estimate.

The ratio of ET_p to either R_I or H changes with temperature. For example, ET_p/R_I is about 0.60 to 0.70 in Israel (Stanhill, 1961b) and USA (summer months); about 0.45 to 0.50 in the Avon Valley (Stanhill, 1961b) and about 0.30

Table 29–5. Relation between ET_p and radiation (mm/day) for sites listed in Table 29–4

Period		Regression equation	r	cv*	n	Reference
Measure-ment	Esti-mate					
July-Sept.	Daily	$ET_P = 1.12\ H - 0.11$	0.94	0.13	40	Tanner & Pelton, 1960
Jan.-May	Daily	$ET_P = 0.51\ R_I - 0.86$	0.95		92	Pruitt &
July-Dec.	Daily	$ET_P = 0.71\ R_I - 1.35$	0.95		136	Angus, 1961
Jan.-May	Daily	$ET_P = 0.77\ H + 0.15$	0.92		64	Pruitt†
July-Dec.	Daily	$ET_P = 0.98\ H + 0.05$	0.96		87	Pruitt†
Jan.-May	Monthly	$ET_P = 0.54\ R_I - 1.25$	0.98		20	Pruitt‡
July-Dec.	Monthly	$ET_P = 0.65\ R_I - 1.27$	0.98		24	Pruitt‡
Jan.-May	Monthly	$ET_P = 0.88\ H - 0.30$	0.99		20	Pruitt‡
July-Dec.	Monthly	$ET_P = 0.99\ H + 0.10$	1.0-		18	Pruitt‡
Year	Weekly	$ET_P = 0.70\ R_I - 0.87$	0.77	0.29	48	Stanhill,
Year	Monthly	$ET_P = 0.72\ R_I - 1.04$	0.91	0.20	12	1961a

* Coefficient of variation (cv) = standard error of estimate divided by mean ET_P.
† Pruitt, W. O. August 1962. Water Resources Center Contract Research Report. University of California, Davis.
‡ Pruitt, W. O. February 1964. Water Resources Center Annual Progress Report. University of California, Davis.

during July and August at Point Barrow (Mather and Thornthwaite, 1958). The ratio ET_p/H is 0.8 or larger during the summer months in the USA while it is only 0.4 at Point Barrow. The effect of temperature can be ascribed in part to the fact that the vapor pressure at the surface increases more rapidly than the temperature thus tending to increase vapor pressure gradients relative to temperature gradients as the surface temperature increases as indicated by $[\Delta/(\Delta + \gamma)]$ in equation [26–36]. Sensible heat transfer to a wet surface from dryer surrounds also may increase with increasing temperature. Consequently, many empirical formulas using radiation include a temperature term. This is done indirectly by including a temperature-sensitive term such as the slope of the saturation vapor pressure curve or directly.

Briggs and Shantz (1916a, b) early showed the high correlation between radiation and evapotranspiration, though in their experiment, "clothesline" advection was large. Several experiments relating both actual evapotranspiration and ET_p to radiation are summarized by Tanner and Lemon (1962). Table 29–5 includes some linear regression equations between ET_p and radiation to illustrate some tests. Improvement is noted as the period of estimate is increased. Pruitt's monthly values of ET_p/R_I and ET_p/H are shown also in Fig. 29–3. Corrections for the seasonal march of temperature are much more important to the ET_p/R_I relation than to the ET_p/H relation (Fig. 29–3).

Several empirical formulas involving radiation are reviewed in the publications cited earlier. We will consider here the less complex ones that do not involve crop and soil "factors". Makkink (1957) proposed the formula

$$ET_p = 0.61\ R_I[\Delta/\Delta + \gamma)] + 0.12. \qquad [29–16]$$

Equation [29–16] has proven reasonably successful for estimating potential evapotranspiration in the Netherlands (Rijtema, 1959). In Israel the correlation was

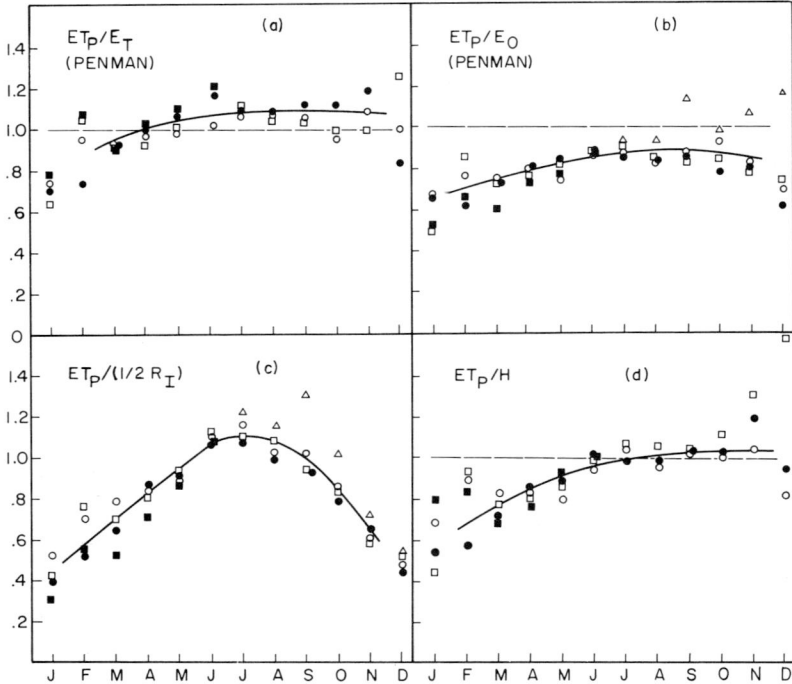

Fig. 29–3. Ratio of monthly ET_p to different estimates of ET_p from methods as given. (a, b) Penman method; (c, d) solar and net radiation; (e, f) temperature methods; (g) Class A pan. Scale is reduced on solar radiation and Thornthwaite method. Symbols are △ 1959, O 1960, □ 1961, ● 1962, ■ 1963. ET_p was from irrigated perennial ryegrass and ranged from 0.5 mm in December to 7.0 mm in June. Data from Pruitt (see page 568). Fig. 29–3 continued on next page.

good though the absolute estimates needed to be increased about 50% (Stanhill, 1961a). Jensen and Haise (1963) summarized data from 20 sites in Western USA and found

$$ET_p = R_I \,(0.025 \, T + 0.08).\qquad\qquad [29\text{–}17]$$

As mentioned earlier, McIlroy (e.g., Slatyer and McIlroy, 1961) has pointed out that the lower limit of ET_p from any wetted surface is

$$ET_p \,(\text{minimum}) = (H - S)\,[\Delta/(\Delta + \gamma)].\qquad\qquad [29\text{–}18]$$

Bryson and Kuhn (1962) found from air mass calculations for northern Canada that the Bowen ratio could be approximated empirically by $\beta = 0.48/\Delta$ where Δ is in millibars per degree C. This gives the approximate relation

$$ET_p = (H - S)\,[\Delta/(\Delta + 0.48)].\qquad\qquad [29\text{–}19]$$

We can compare these temperature functions since $\Delta/(\Delta + \gamma)$ and $\Delta/(\Delta + 0.48)$ are reasonably linear over 0 to 35C. For equation [29–16] the relation is 0.61 $\Delta(\Delta + \gamma) \approx (0.006T + 0.2)$; for equation [29–17] it is $(0.025T + 0.08)$; for equation [29–18] it is $\Delta/(\Delta + \gamma) \approx (0.01T + 0.3)$; and for equation [29–19]

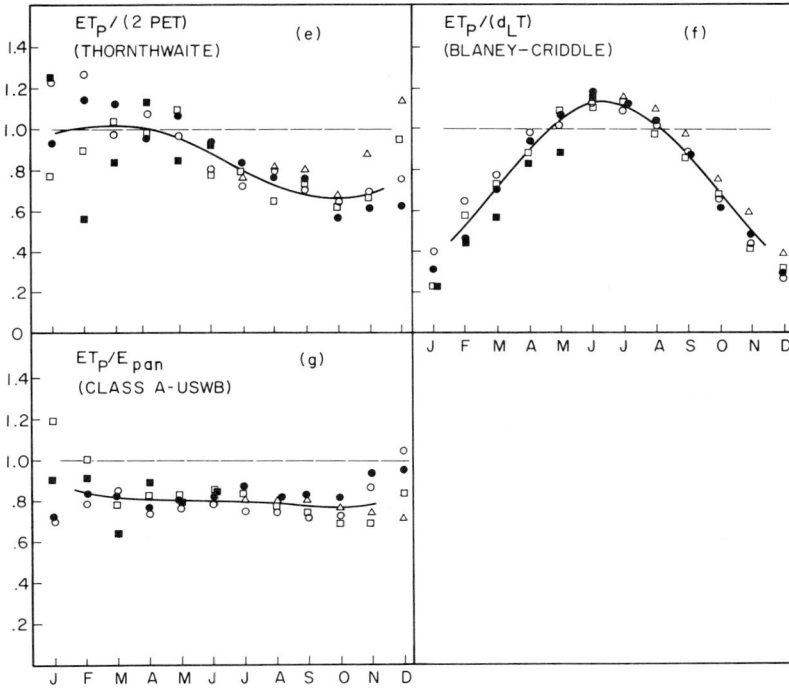

Fig. 29–3. Continued

it is $\Delta/(0.48 + \Delta) \approx (0.01T + 0.5)$. There is a substantial difference between the Makkink and the Jensen and Haise corrections. Though the Bryson and Kuhn term is somewhat larger than $\Delta/(\Delta + \gamma)$, the correspondence indicates that the ET_p for the large area analyzed was near minimal. It is clear from the Table 29–5 and the above comparison of temperature corrections that empirical radiation formulae require local calibration. They cannot be applied without calibration over as wide a range of climate as the energy balance methods.

C. Temperature Methods

Several methods of estimating ET_p from mean temperature have been proposed. They were developed for monthly and growing season estimates but have been used for daily estimates. Temperature methods most widely used employ temperature either alone or in combination with relative humidity. Availability of temperature records and simplicity of temperature methods is the main reason for their wide-spread use. The basis of temperature methods and their limitations have been discussed by Wijk and Vries (1954) and Pelton et al. (1960). It is sufficient to list here four points about temperature methods. First, they rely heavily on the mutual correlation of ET_p and temperature to radiation. Second, radiation and ET_p may be quite variable on a daily basis even though mean air temperature changes little from day-to-day (excluding marked frontal passages). Consequently temperature methods are best suited to monthly estimates and are not reliable for short-period estimates except for those rare regions with conservative short-period ET_p. Third, in regions with a large annual temperature range, the air temperature lags radiation much more than ET_p, causing the methods to underestimate during warming months and overestimate during the cooling months. They are most suited to estimates during the summer months and for annual estimates. Fourth, the method requires calibration for a given region to derive local seasonal and monthly coefficients which include temperature lag, vegetation, and local climate corrections.

1. LINEAR FORMULAS

Though temperature methods were employed following the work of Briggs and Shantz (1916a, b), the more important developments began with Lowry and Johnson (1942) who used accumulative degree-days based on maximum temperature above 0C, and with Blaney and Morin (1942) who used mean temperature and relative humidity. Later the Blaney and Morin approach was modified by Blaney and Criddle (1950) to omit relative humidity and add a daylength term. Hargreaves (1956) developed an equation which included mean temperature, daylength, and relative humidity (at noon). These formulas were developed in Western USA for irrigated crops and evaporation pans and for monthly and growing season estimates. They included coefficients for each crop (and stage of growth) or for pans or water surfaces. The formulas, except for that of Lowry and Johnson are of the form

$$ET_p = C_1 d_L T (C_2 - C_3 h) \qquad [29\text{--}20]$$

where C_1 is a crop, pan, or pond constant; d_L is a daylength measure; T is mean temperature; h, when used, is the relative humidity; and C_2 and C_3 are constants. The Blaney-Criddle equation is used more widely than other equations represented by [29–20]; it is

$$ET_p = C \, d_L T \qquad [29\text{--}21]$$

where C is a monthly "consumptive use" coefficient (varies monthly with crop, and changes with location and dimension of ET_p) and d_L is the fraction of the annual daylight hours occurring in the month. Values of C are reported in engineering literature for several locations, but are best determined locally.

2. NONLINEAR FORMULAS

Thornthwaite (1948) developed a more complicated expression relating mean temperature to ET_p that includes a daylength term and a "station" constant based on long-term mean monthly temperatures. The empirical equation developed by Thornthwaite is

$$ET_p' = 1.6(10T/I)^a \qquad [29\text{--}22]$$

where T is the monthly mean air temperature (°C); I is a heat index for the station which depends on long-term, mean monthly air temperatures; a is a function of I (both a and I can be found from tables, e.g., Thornthwaite and Mather, 1955b); and ET_p' is in cm. ET_p' is an "unadjusted" value based on a 12-hour day and 30-day month and is corrected by d_L (actual day length in hours/12) and days in a month N to give the adjusted $ET_p = ET_p' (d_L)(N/30)$.

The Thornthwaite and the Blaney-Criddle method have been used most widely and with varying success. Monthly estimates are reasonable provided formulas have been calibrated to a given locale. Temperature lag corrections rarely have been applied explicitly to temperature methods though the need was recognized very early by Prescott (1943) and was emphasized again by Wijk and Vries (1954). Temperature lag corrections are included, often unwittingly, in the monthly coefficients of equation [29–21] and in "crop constants" applied to equation [29–22] (e.g., Pierce, 1958). One can expect that inclusion of other factors, such as humidity in Hargreaves formula, will improve the correlation— inclusion of more factors, provided they are at all pertinent, usually improves empirical correlations.

Some temperature methods developed for evaporation from water surfaces have been applied to vegetation surfaces since a high correlation exists between ET_p and evaporation from shallow water bodies. An example is that of Cochrane (1956) which also utilizes relative humidity and wind.

3. TEMPERATURE-VAPOR PRESSURE FORMULAS

Halstead (1951) estimates ET_p by the formula

$$ET_p = C \, d_L \, (\chi_{max} - \chi_{min}) \qquad [29\text{–}23]$$

where d_L is relative daylength based on a 12-hour day, χ_{max} is the saturation absolute humidity (g m^{-3}) corresponding to the maximum air temperature, χ_{min} similarly corresponds to the minimum air temperature, and C is a constant ($C = 1$ when mean ET_p is mm/month). Halstead approximates the water vapor gradient above the surface with $(\chi_{max} - \chi_{min})$. The dewpoint (absolute humidity) is approximated by T_{min}. The surface temperature and absolute humidity is related to the diurnal air temperature march and thus to T_{max}. Hamon (1961) proposed the formula

$$ET_p = 0.14 \, d_L^2 \, \chi_{mean} \qquad [29\text{–}24]$$

where ET_p is mm/day.

The diurnal temperature range such as in equation [29–23] usually is correlated better than mean temperature with daily radiation over short periods. Also Deacon et al. (1958) mention that maximum temperature is preferable to mean temperature because of the nonlinear dependence of ET_p on air temperature. The vapor pressure expression in equations [29–23] and [29–24] also includes a mean temperature weighting factor that is absent in simple temperature formulas.

4. FORMULA TESTS

Lysimeter tests of equations [29–21] and [29–22] are available though similar tests of equations [29–23] and [29–24] have not been made. Tests of equation [29–22] are given in Table 29–6. The method improves as the estimation period increases. Though the correlation of the daily estimates found by Pruitt is high, the coefficient of variation would be large. The difference in the regression equations in Table 29–6 between the locations and departure of ET_p/PET from unity Fig. 29–3 indicates the need for local seasonal calibration constants. Tests in Israel (Stanhill, 1961a, b) and California indicate that equation [29–21] is as good, or slightly better than equation [29–22] unless monthly coefficients are used with equation [29–22]. The deviation from a unity ratio found with the

Table 29–6. Comparison of Thornthwaite method (PET) with measured ET_p (mm/day) over surfaces in Table 29–4

Period	Estimate	Regression equation	n	r	cv*	Reference
July-Sept.	Daily	$ET_p = 0.52\ PET + 1.64$	48	0.39	0.30	Pelton, et al. (1960)
June-Dec.	Daily	$ET_p = 1.38\ PET + 0.74$	175	0.89	-	Pruitt†
Jan.-May	Monthly	$ET_p = 1.94\ PET + 0.05$	20	0.97	-	Pruitt‡
July-Dec.	Monthly	$ET_p = 1.49\ PET + 0.08$	24	0.99	-	Pruitt‡
Yearly	Weekly	$ET_p = 1.35\ PET + 1.76$	48	0.73	0.38	Stanhill,
Yearly	Monthly	$ET_p = 1.48\ PET + 1.85$	12	0.94	0.16	(1961a)

* Coefficient of variation   (cv) = standard error of estimate divided by mean ET_p.
† Pruitt, W. O. 1960. Correlation of Climatological Data With Water Requirements of Crops. Ann. Rpt. 1959-60. University of California, Davis.
‡ Pruitt, W. O. February 1964. Water Resources Center Annual Progress Report. University of California, Davis.

Blaney-Criddle graph indicates the seasonal crop factor C needed for equation [29–21] at Davis.

D.   Humidity Methods

Most of the humidity methods are modifications of the Dalton equation where the vapor pressure gradient in equation [29–7] is replaced by the saturation deficit to give

$$ET_p = f(u_z)\ (e_c - e_z) \qquad [29\text{–}25]$$

which is similar to the E_a term in Penman's (1948) equation. The main difference between many of the methods is in the $f(u_z)$, which may be changed with the month and with location, and inclusion of "topographic" and "surface type" constants.

In addition to formulas represented by [29–25], Prescott (1949) has used the form $ET_p = C\ (e_a - e_z)^{0.75}$. Also there have been relative humidity formulas similar to the humidity term in equation [29–20]. Van der Bijl outlines a method proposed by Skvortsov in 1950, which uses the frequency of wet-bulb temperature reversals per minute and the absolute humidity (W. Van der Bijl, 1957. Evapotranspiration problem: First contribution. *Kansas State Coll., Manhattan, Dep. Physics Contr. Cwb-8806, Rep. 1956-1957.*).

Van der Bijl (*see* above) and Wang (1963) list at least a dozen references to humidity methods and other references are given in the reviews listed earlier. A few tests available show that humidity methods have no advantage over temperature methods in accuracy (e.g., Stearns and Carlson, 1960; Rijtema, 1959) and a further disadvantage that humidity data are less available than temperature data.

E.   Pans and Atmometers

One of the earliest approaches to estimating evaporation from water bodies and ET_p from soils and vegetation was through correlations to the measured evaporation from pans (or tanks) and from atmometers (small, wetted, porous surfaces). The general belief has been that the evaporation from pans and atmometers and from water bodies, wetted soils, and vegetation (amply supplied with water) should be well correlated since evaporation from both is governed by micrometeorological conditions. Over appropriate time periods there is indeed a high correlation to ET_p even though the pans and atmometers are poor ana-logues of a vegetation surface. This high correlation reinforces the arguments that ET_p is relatively conservative and that any method for estimating departures from the average can be used successfully when calibrated under specific conditions.

1. EVAPORATION PANS

Many types of evaporation pans and tanks have been employed though the US Weather Bureau Class "A" pan is most widely used in the USA. The different pans vary in diameter, depth of water, depth of water surface below the tank lip, and insulation at the sides and bottom. Pans also differ in placement; some are placed in the soil (sunken pans), others on the soil surface, and others at some height above the surface. All of these differences affect the heat storage, radiation exchange, and the transfer coefficients for heat and vapor. The micrometeorology of pan evaporation and pan design is discussed by Kohler (1954), Kohler et al. (1955), Deacon et al. (1958), Rider et al. (1958), and Mukhammal (1961). Data illustrating the effect of pan size and installation are provided in McIlrcy and Angus (1964). The energy exchange of a pan is different from vegetation in several ways. The heat storage of pans (particularly deep pans) is much greater than vegetated surfaces so that the temperature of the water is generally lower than vegetation during the day when most ET occurs, but is higher at night. This can effect the hourly and daily mean vapor pressure difference between surface and air, the evaporation, and the sensible heat exchange. If the tank is elevated above the surface, the additional radiation exchange at the sides and the sensible heat exchange at the surface and sides comprise a much greater portion of the total heat used for evaporation than is the case with vegetation.

In spite of these objections, the correlation of ET_p to pan evaporation over weekly to monthly periods is relatively high. The correlation to ET_a when water is limiting may be quite low (e.g., see Fritschen and Shaw, 1961; Stanhill, 1962b). Penman (1948) developed a seasonal relation to ET_p from grass to pan evapora-tion. He found ET_p/E_{pan} was 0.6 for November through February; 0.7 for March, April, September, and October; and 0.8 for May through August where E_{pan} was from small, sunken pans. Others (e.g., Rijtema, 1959) have found an excellent correlation between ET_p from short grass and E_{pan} over 10-day to monthly periods. Stanhill (1961a) found the monthly ET_p (average mm/day) from alfalfa to be related to that from a US Weather Bureau Class "A" pan by $ET_p = 0.70 E_{pan} + 0.47$ ($r = 0.95$; cv $= 0.15$) and for weekly periods $ET_p = 0.75 E_{pan} + 0.36$ ($r = 0.77$; cv $= 0.36$). Data for a British sunken tank were similar. Stanhill (1962b) also gives regression equations with similar correlations for five other irrigated crops. Pruitt and Angus (1961) found the daily ET_p (mm/day) from irrigated perennial rye grass to be related to Class "A" pan evaporation by $ET_p = 0.67 E_{pan} + 0.45$ ($r = 0.94$) for January through May and to be $ET_p = 0.77 E_{pan} + 0.03$ ($r = 0.90$) for July through December. Pruitt found for monthly averages that $ET_p = 0.79 E_{pan} + 0.08$ ($r = 0.99$) for January through May and

$ET_p = 0.76\,E_{pan} - 0.02$ $(r = 0.98)$ for July through December (W. O. Pruitt, Feb. 1964. Water Resources Center annual progress report. *University of California, Davis*). (The monthly ratios are given in Fig. 29–3.) The correlation and the coefficients of variation show the advantages of longer period averages. No estimate of error was given for the daily values measured by Pruitt and Angus though scatter was large. McIlroy and Angus (1964) found that ET_p was better correlated with water evaporation than radiation. Though a linear regression between ET_p and E_{pan} is most commonly used, Prescott in several publications (e.g., Butler and Prescott, 1955) has used the form $ET_p = CE_{pan}{}^{0.75}$.

The major problem in using the pan is that the time period for which reasonable estimates can be made is relatively long, particularly in humid regions where climate is variable. Further, the "calibration" of a pan to a given locale is mandatory if for no other reason than that the siting of the pan affects the pan evaporation greatly. For example, Pruitt and Angus found the cumulative evaporation from a US Weather Bureau Class "A" pan from June through December to be about 135 cm when the pan was sited in a large grass field, 150 cm with the pan surrounded by a 14.6-m fetch of grass in a dryland area and 175 cm with the pan in a dryland area with no fetch. The ratio is then 1:1.1:1.3 depending on these surrounds. This illustrates the problem of uniform siting which occurs in addition to other problems in extending pan "calibrations" from one area to another. It is desirable that pan surroundings be kept moist for 20 to 50 m in order to avoid serious siting problems (see McIlroy and Angus, 1964).

2. ATMOMETERS

Several styles of atmometers have been employed, though the more common ones are the Piche atmometer made from a flat, horizontal disk of wetted blotting paper with both sides exposed to the air (*see* Vries and Venema, 1954 for further description and references), the black and white spherical porous porcelain atmometers of Livingston (1935) and the Bellani black-plate atmometer (Livingston, 1915; flat, black, porous porcelain as the upper face of a nonporous hemisphere). Other atmometers are described in the discussion section of Mukammal (1961) and by Stanhill (1958b). Evaporation from the porous surface is due to radiative and convective transfer of heat to the porous surface and to conduction through the water from the supply system and from nonporous surfaces. Vries and Venema (1954) discuss the energy balance of the Piche atmometer and Mukammal discusses (with further references) the Bellani atmometer. The net radiation flux density per unit evaporating surface of an exposed Bellani is of the same magnitude as a crop surface whereas that of the Piche (or spherical atmometers) is about one-half because evaporation occurs at both the upper and lower surfaces. The transfer of sensible heat from the air is much larger with atmometers than with vegetation. This arises because the atmometer is placed at some height above the vegetation surface and the wind and convective transfer increase with height above the surface. As Vries and Venema point out, heat exchange of atmometers is more analogous to a single leaf exposed at the atmometer height than to the energy balance of the vegetation at the surface. The disparity at night is particularly great.

Even though the energy balance of atmometers differs greatly from vegetation, there is a reasonably good correlation between the evaporation from atmometers and from irrigated vegetation [Robertson and Holmes, 1958; Stanhill, 1958b; Pruitt (W. O. Pruitt, Sept. 1960. Correlation of climatological data with water requirements of crops. *Annual Rep. Univ. of California, Davis.*)]. Pruitt found that the correlation of the black, spherical atmometer of Livingston with daily ET_p from irrigated perennial ryegrass was substantially less than that between Class "A" pan evaporation and daily ET_p. Pruitt also found that the difference in evaporation between black and white spherical atmometers was highly correlated with ET_p. Halkais et al. (1955) also found high correlation between monthly ET from irrigated crops and the difference in evaporation from black and white atmometers and Janes (1960) found high correlations for periods of about 1 week. Halkais et al. and Janes also found a high correlation of black-white bulb differences with radiation that Livingston (1935) had anticipated. The difference between similarly exposed black and white units depends mainly on radiation since the convective transfer to each is about the same. The high correlation between radiation and the black-white differences probably accounts for the good correlation with ET_p found by Pruitt, Halkais et al., and Janes.

If atmometers are to be used, they must be sited at standard heights over large area standard surfaces since the height, the type of surface (dry or wet, rough or smooth), the surrounds and fetch seriously affect the response. Pruitt found that the evaporation difference between white and black atmometers was less affected by placement than single atmometers as is expected because convective heat exchange is similar. Provided atmometers are "calibrated" at a given locale they can be used for ET_p estimates over suitably long periods. Because the problems of siting and maintenance of cup condition are severe and because convective heat exchange to atmometers is such a large proportion of the total energy balance relative to that of plants, pans appear preferable to single atmometers. The use of the evaporation from both black and white atmometers offers improved possibility provided cup condition can be maintained, since more information is available than with black atmometers used singly. If black and white units are used, the geometry of flat-plate units with one evaporating surface would seem preferable to spherical bulbs or to two-surface units like the Piche atmometer.

F. Summary

The reliability of the empirical methods for ET_p estimation is greatly improved when the methods are "calibrated" against measurements from specific crops in a given region. The estimate also improves as the period of estimate is increased. Methods such as those of Penman, McIlroy, Ferguson, and Budyko, which are based on the energy balance, appear most valuable, and have widest applicability of all methods, particularly when sensible heat transfer from the surrounds is large. Shallow and sunken pans and methods utilizing radiation (with temperature corrections) are the next best choice. Radiation methods may be especially useful for short-period estimates in humid regions where climate is variable. Properly installed pans and radiation methods, when calibrated, are much preferred over calibrated and uncalibrated mean temperature methods. Atmometers, because of maintenance and siting problems, are not as useful for general practice as either pans or temperature methods. Humidity methods appear least suited to general use of any methods.

LITERATURE CITED

Aslyng, H. C., and K. J. Kristensen. 1961. Water balance recorder. Amer. Soc. Civ. Eng., Proc. 87 (IR1):15–21.

Blaney, H. F., and W. D. Criddle. 1950. Determining water requirements in irrigated areas from climatological and irrigation data. US Dep. Agr. Soil Conserv. Serv., SCS–TP–96. 48 p.

Blaney, H. F., and K. V. Morin. 1942. Evaporation and consumptive use of water empirical formulas. Amer. Geophys. Union, Trans. 23:76–83.

Bloodworth, M. E., J. B. Page, and W. R. Cowley. 1956. Some applications of the thermoelectric method for measuring water flow rates in plants. Agron. J. 48:222–228.

Briggs, L. J., and H. L. Shantz. 1916a. Hourly transpiration rate on clear days as determined by cyclic environmental factors. J. Agr. Res. 5:583–648.

Briggs, L. J., and H. L. Shantz. 1916b. Daily transpiration during the normal growth period and its correlation with the weather. J. Agr. Res. 7:155–212.

Brooks, F. A. 1961. Need for measuring horizontal gradients in determining vertical eddy transfers of heat and moisture. J. Meteorol. 18:589–596.

Brooks, C. E. P., and N. Carruthers. 1953. Handbook of statistical methods in meteorology. H. M. Stationery Office, London. 412 p.

Bryson, R. A., and P. M. Kuhn. 1962. Some regional heat budget values for northern Canada. Geogr. Bull. 17. p. 57–66.

Budyko, M. I. 1956. The heat balance of the Earth's surface. (Translated by N. A. Stepanova, 1958). 259 p. Office Tech. Serv., US Dep. Com., Washington 25, D. C.

Businger, J. A. 1963. The glasshouse (greenhouse) climate. In W. R. van Wijk [ed.] Physics of plant environment. North-Holland Publ. Co. Amsterdam. p. 277–318.

Butler, P. F., and J. A. Prescott. 1955. Evapotranspiration from wheat and pasture in relation to available moisture. Australian J. Agr. Res. 6:52–61.

Cochrane, N. J. 1956. Discussion of "Evaporation from and stabilization of salt on sea water surface." Amer. Geophys. Union, Trans. 37:787.

Deacon, E. L. 1949. Vertical diffusion in the lowest layers of the atmosphere. Quart. J. Roy. Meteorol. Soc. 75:89–103.

Deacon, E. L., C. H. B. Priestley, and W. C. Swinbank. 1958. Evaporation and the water balance: Climatology reviews of research. UNESCO Arid Zone Res. 10:9–34.

Deacon, E. L., and W. C. Swinbank. 1958. Comparison between momentum and water vapor transfer: Climatology and microclimatology. UNESCO Arid Zone Res. 11:38–41.

Decker, J. P. and C. M. Skau. 1964. Simultaneous studies of transpiration rate and sap velocity in trees. Plant Physiol. 39:213–215.

Decker, J. P., W. G. Gaylor, and F. D. Cole. 1962. Measuring transpiration of undisturbed tamarisk shrubs. Plant Physiol. 37:393–397.

Denmead, O. T., and R. H. Shaw. 1962. Availability of soil water to plants as affected by soil moisture content and meterological conditions. Agron. J. 45:385–390.

Drinkwater, W. O., and B. E. Janes. 1957. Relation of potential evapotranspiration to environment and kind of plant. Amer. Geophys. Union, Trans. 38:524–528.

Dyer, A. J. 1961. Measurements of evaporation and heat transfer in the lower atmosphere by an automatic eddy-correlation technique. Quart. J. Roy. Meteorol. Soc. 87:401–412.

Dyer, A. J. 1963. The adjustment of profiles and eddy fluxes. Quart J. Roy. Meterol. Soc. 89:276–280.

Ekern, P. E. 1958. Dew measurements at PRI Field Station (Wahiawa). Pineapple Res. Inst. News 6:134–135.

Ferguson, J. 1952. The rate of evaporation from shallow ponds. Australian J. Sci. Res. 5:315–330.

Fritschen, L. J., and R. H. Shaw. 1961. Evapotranspiration for corn as related to pan evaporation. Agron. J. 53:149–150.

Fritschen, L. J., and C. H. M. van Bavel. 1963. Experimental evaluation of models of latent and sensible heat transport over irrigated surfaces. Int. Ass. Sci. Hydrol., Gen. Assembly of Berkeley. Publ. 62:159–171.

Frost, K. R. 1962. A weighing evapotranspirometer. Agr. Eng. 43:160.

Gardner, W. R., and C. F. Ehlig. 1963. The influence of soil water on transpiration of plants. J. Geophys. Res. 68:5719–5724.

Gardner, W. R., and D. I. Hillel. 1962. The relation of external evaporative conditions to the drying of soils. J. Geophys. Res. 67:4319–4325.

Gilbert, M. J., and C. H. M. van Bavel. 1954. A simple field installation for measuring maximum evapotranspiration. Amer. Geophys. Union, Trans. 35:937–942.

Glover, J., and J. A. Forsgate. 1962. Measurement of evapotranspiration from large tanks of soil. Nature 195:1330.

Graham, W. G., and K. M. King. 1961. Fraction of net radiation utilized in evapotranspiration from a corn crop. Soil Sci. Soc. Amer. Proc. 25:158–160.

Halkais, N. A., F. J. Veihmeyer, and A. H. Hendrickson. 1955. Determining water needs for crops from climatic data. Hilgardia 24:207–233.

Halstead, M. H. 1951. Theoretical derivation of an equation for potential evapotranspiration. Publ. in Climatol. Interim Rep. No. 16. 4:10–12.

Halstead, M. H., and W. Covey. 1957. Some meterological aspects of evapotranspiration. Soil Sci. Soc. Amer. Proc. 21:461–464.

Hamon, W. R. 1961. Estimating potential evapotranspiration. Amer. Soc. Civ. Eng. Proc. 87(HY3):107–120.

Hargreaves, G. H. 1956. Irrigation requirements based on climatic data. Amer. Soc. Civ. Eng., J. Irrig. Drainage Div. Pap. 1105–IR3. 10 p.

Harrold, L. L., and F. R. Dreibelbis. 1958. Evaluation of agricultural hydrology by monolith lysimeters. US Dep. Agr. Tech. Bull. 1179. 166 p.

Harrold, L. L., and F. R. Dreibelbis. 1963. Evidence of errors in evaluation of dew amounts by the Coshocton lysimeters. Int. Ass. Sci. Hydrol., Gen. Assembly of Berkeley. Publ. 65:425–531.

Havens, A. V. 1956. Using climatic data to estimate water in soil. New Jersey Agri. 38:6–10.

Holmes, R. M. 1961. Discussion of "A comparison of computed and measured soil moisture under snap beans." J. Geophys. Res. 66:3620–3622.

House, G. J., N. E. Rider, and C. P. Tugwell. 1960. A surface energy-balance computer. Quart. J. Roy. Meteorol. Soc. 86:215–231.

Janes, B. 1960. Estimation of potential evapotranspiration from vegetable crops from net solar radiation. Amer. Soc. Hort. Sci., Proc. 76:582–589.

Jensen, M. E., and H. R. Haise. 1963. Estimating evapotranspiration from solar radiation. Amer. Soc. Civ. Eng., Proc. 89 (IR4):15–41.

Kaimal, J. C., and J. A. Businger. 1963. A continuous wave sonic anemometer-thermometer. J. Appl. Meteorol. 2:156–164.

King, F. H. 1910. Physics of agriculture. Publ. by author, Madison, Wisconsin. 604 pp.

King, K. M., C. B. Tanner, and V. E. Suomi. 1956. A floating lysimeter and its evaporation recorder. Amer. Geophys. Union, Trans. 37:738–742.

Klein, M. L., H. C. Morgan, and M. H. Aronson. 1958. Digital techniques for computation and control. Instruments Publ. Co., Pittsburgh, Pennsylvania. 392 p.

Kohnke, H., F. R. Dreilbelbis, and J. M. Davidson. 1940. A survey and discussion of lysimeters and a bibliography on their construction and performance. US Dep. Agr. Misc. Publ. 372. 68 p.

Kohler, M. 1954. Lake and pan evaporation. In Water-loss investigations: Lake Hefner studies technical report. Geol. Surv. Prof. Pap. 269:127–148.

Kohler, M. A., T. J. Nordenson, and W. E. Fox. 1955. Evaporation from pans and lakes. US Weather Bur. Res. Pap. No. 38. 21 p.

Krimgold, D. B. 1957. Discussion of "A floating lysimeter and evaporation recorder." Amer. Geophys. Union, Trans. 38:766.

Lemon, E. R. 1963. The energy budget at the Earth's surface: Part I. US Dep. Agr., Agr. Res. Serv. Prod. Res. Rep. No. 71, Publ. USAEPGSIG–6–41–62. 33 p.

Lettau, H. H. 1957. Computation of heat budget constituents of the earth/air interface. 1:305–327. In Exploring the atmosphere's first mile. Pergamon Press, New York.

Lettau, H. L., and B. Davidson. 1957. Exploring the atmosphere's first mile. Pergamon Press, New York. 376 p.

Livingston, B. E. 1915. A modification of the Bellani porous-plate atmometer. Science 41:872–874.

Livingston, B. E. 1935. Atmometers of porous porcelain and paper, their use in physiological ecology. Ecology 16:438–472.

Long, I. F. 1957. Instruments for micrometeorology. Quart J. Roy. Meteorol. Soc. 83: 202–214.

Lowry, R. L., and A. F. Johnson. 1942. Consumptive use of water for agriculture. Amer. Soc. Civ. Eng., Trans. 107:1243–1302.

Makkink, G. F. 1953. Een nieuw lysimeter-station. Water (Amsterdam) 37:159–163.

Makkink, G. F. 1957. Ekzameno de la formula de Penman. Neth. J. Agr. Sci. 5:290–305.

Marlatt, W. E., A. V. Havens, N. A. Willits, and G. D. Brill. 1961. A comparison of computed and measured soil moisture under snap beans. J. Geophys. Res. 66:535–541.

Mather, J. R. 1950. Manual of evapotranspiration. Publ. in Climatol., Interim Rep. 10.3:1–29.

Mather, J. R., and C. W. Thornthwaite. 1958. Microclimatic investigations at Point Barrow, Alaska, 1957–1958. Publ. in Climatol. 11:63–239.

McIlroy, I. C., and D. E. Angus. 1963. The Aspendale multiple weighed lysimeter installation. CSIRO, Div. Meteorol. Physics Tech. Pap. No. 14. Melbourne, Australia. 29 p.

McIlroy, I. C., and D. E. Angus. 1964. Grass, water, and soil evaporation at Aspendale. Agr. Meteorol. 1:201–224.

McIlroy, I. C., and J. Sumner. 1961. A sensitive high capacity balance for continuous automatic weighing in the field. J. Agr. Eng. Res. 6:252–258.

McMillan, W. B., and H. A. Paul. 1961. Floating lysimeter uses heavy liquid for buoyancy. Agr. Eng. 42:498–499.

Molga, M. 1958. Agricultural Meteorology, Part II. (Trans. from Polish, 1962). Office Tech. Serv. US Dep. Com. 351 p.

Monin, A. S., and A. M. Obukhov. 1954. Principal laws of turbulent mixing in the air layer near the ground. Trudy Geofis. Inst. 24 163–187.

Monteith, J. L. 1963. Gas exchange in plant communities. p. 95–112. In L. T. Evans [ed.]. Environmental control of plant growth Academic Press, New York. 449 p.

Monteith, J. L., and G. Szeicz. 1960. The carbon dioxide flux over a field of sugar beets. Quart. J. Roy. Meteorol. Soc. 86:205–214.

Mukammal, E. I. 1961. Evaporation pans and atmometers. Proc. Hydrol. Symp. 2:84–105. (Queen's Printer, Ottawa, Canada Cat. No. R32–361/2).

Musgrave, R. B., and D. N. Moss. 1961. Photosynthesis under field conditions: I. A portable, closed system for determining net assimilation and respiration of corn. Agron. J. 55:37–41.

Nixon, P. R., and G. P. Lawless. 1960. Translocation of moisture with time in unsaturated soil proviles. J. Geophys. Res. 65:655–661.

Panofsky, H. A. 1963. Determination of stress from wind and temperature measurements. Quart. J. Roy. Meteorol. Soc. 89:85–94.

Pelton, W. L. 1961. The use of lysimetric methods to measure evapotranspiration. Proc. Hydrol. Symp. 2:106–134. (Queen's Printer, Ottawa, Canada Cat. No. R32–361/2).

Pelton, W. L., K. M. King, and C. B. Tanner. 1960. An evaluation of the Thornthwaite and mean temperature methods for determining potential evapotranspiration. Agron. J. 52:387–395.

Penman, H. L. 1948. Natural evaporation from open water, bare soil, and grass. Roy. Soc. London, Proc. Ser. A. 193:120–146.

Penman, H. L. 1949. The dependence of transpiration on weather and soil conditions. J. Soil Sci. 1:74–89.

Pierce, L. T. 1958. Estimating seasonal and short-term fluctuations in evapotranspiration from meadow crops. Bull. Amer. Meteorol. Soc. 39:73–78.

Prescott, J. A. 1943. A relationship between evaporation and temperature. Roy. Soc., Trans. South Australia 67:1–6.

Prescott, J. A. 1949. A climatic index for the leaching factor in soil formation. J. Soil Sci. 1:9–19.

Pruitt, W. O., and D. E. Angus. 1960. Large weighing lysimeter for measuring evapotranspiration. Amer. Soc. Agr. Eng., Trans. 3:13–15, 18.

Pruitt, W. O., and D. E. Angus. 1961. Comparisons of evapotranspiration with solar and net radiation and evaporation from water surfaces. First Ann. Rep. USAEPG Contract DA–36–039–SC–80334, 1961. University of California, Davis. p. 74–107.

Rider, N. E., Chairman. 1958. Measurement of evaporation and humidity in the biosphere and soil moisture. World Meteorol. Organ. Tech. Note 21. 49 p.

Rijtema, P. E. 1959. Calculation methods of potential evapotranspiration. Inst. Land Water Manage. Res. Tech. Bull. No. 7 (Wageningen, Netherlands). 10 p.

Robertson, G. W., and R. M. Holmes. 1958. A new concept of the measurement of evaporation for climatic purposes. Int. Union Geod. Geophys. Int. Ass. Sci. Hydrol. 3:399–406.

Robins, J. S., W. O. Pruitt, and W. H. Gardner. 1954. Unsaturated flow of water in field soils and its effect on soil moisture investigations. Soil Sci. Soc. Amer. Proc. 18:344–347.

Robinson, S. A. 1962. Computing wind profile parameters. J. Atm. Sci. 19:189–190.

Sellers, W. D. 1964. Potential evapotranspiration in arid regions. J. Appl. Meteorol. 3:98–104.

Sellers, W. D., and C. N. Hodges. 1962. The energy balance of nonuniform soil surfaces. J. Atm. Sci. 19:482–491.

Sheppard, P. A. 1958. Transfer across the Earth's surface and through the air above. Quart. J. Roy. Meteorol. Soc. 84:205–224.

Skau, C. M., and R. H. Swanson. 1963. An improved heat pulse velocity meter as an indicator of sap speed and transpiration. J. Geophys. Res. 68:4743–4749.

Slatyer, R. O., and I. C. McIlroy. 1961. Practical microclimatology. CSIRO, Plant Ind. Div. Canberra (UNESCO sponsored). 328 p.

Stanhill, G. 1958a. Evapotranspiration from different crops exposed to the same weather. Nature 182:125.

Stanhill, G. 1958b. An irrigation gauge for commercial use in field and glasshouse practice. J. Agr. Eng. Res. 3:292–298.

Stanhill, G. 1961a. A comparison of methods of calculating potential evapotranspiration from climatic data. Israel J. Agr. Res. 11:159–171.

Stanhill, G. 1961b. The accuracy of meteorological estimates of evapotranspiration in arid climates. J. Inst. Water Eng. 15:477–482.

Stanhill, G. 1962a. The use of the Piche evaporimeter in the calculation of evaporation. Quart. J. Roy. Meteorol. Soc. 88:80–82.

Stanhill, G. 1962b. The control of field irrigation practice from measurements of evaporation. Israel J. Agr. Res. 12:51–62.

Stearns, F. W., and C. A. Carlson. 1960. Correlations between soil-moisture depletion, solar radiation, and other environmental factors. J. Geophys. Res. 65:3727–3732.

Suomi, V. E. 1957. Sonic anemometer-University of Wisconsin. In Exploring the atmosphere's first mile. 1:256–266.

Sverdrup, H. U. 1937. On the evaporation from the ocean. J. Mar. Res. 1:3–14.

Tanner, C. B. 1957. Factors affecting evaporation from plants and soils. J. Soil Water Conserv. 12:221–227.

Tanner, C. B. 1960a. Energy balance approach to evapotranspiration from crops. Soil Sci. Soc. Amer. Proc. 24:1–9.

Tanner, C. B. 1960b. A simple aero-heat budget method for determining daily evapotranspiration. Int. Congr. Soil Sci. Proc. 7th (Madison, Wis., USA) 1:203–209.

Tanner, C. B. 1963. Basic instrumentation and measurement for plant environment and micrometeorology. University of Wisconsin Press, Madison. 300 p.

Tanner, C. B., and E. R. Lemon. 1962. Radiant energy utilized in evapotranspiration. Agron. J. 54:207–212.

Tanner, C. B., and W. L. Pelton. 1960. Potential evapotranspiration estimates by the approximate energy balance method of Penman. J. Geophys. Res. 65:3391–3413.

Taylor, S. A., D. D. Evans, and W. D. Kemper. 1961. Evaluating soil water. Utah Agr. Exp. Sta., Bull. 426. 67 p.

Thornthwaite, C. W. 1948. An approach toward a rational classification of climate. Geog. Rev. 38:55–94.

Thornthwaite, C. W., and J. R. Mather. 1955a. The water budget and its use in irrigation. In Water. U.S. Dep. Agr. Yearbook. p. 346–357.

Thornthwaite, C. W. and J. R. Mather. 1955b. The water balance. Publ. in Climatol. 8:1–104.

Van Bavel, C. H. M. 1961. Lysimetric measurements of evapotranspiration rates in the eastern United States. Soil Sci. Soc. Amer. Proc. 25:138–141.

Van Bavel, C. H. M., L. J. Fritschen, and W. E. Reeves. 1963. Transpiration by sudan-grass as an externally controlled process. Science 141:269–270.

Van Bavel, C. H. M., and L. E. Myers. 1962. An automatic weighing lysimeter. Agr. Eng. 43:580–583.

Veihmeyer, F. J., and A. H. Hendrickson. 1955. Does transpiration decrease as the soil moisture decreases? Amer. Geophys. Union, Trans. 36:425–448.

Vries, D. A. de, and H. J. Venema. 1954. Some considerations on the behaviour of the Piche evaporimeter. Vegetatio 5–6:225–234.

Wang, J. Y. 1963. Agricultural meteorology, Pacemaker Press, Milwaukee, Wisconsin. 693 p.

Webb, E. K. 1960. An investigation of the evaporation from Lake Eucumbene. CSIRO (Australia) Div. Meteorol. Phys. Tech. Pap. No. 10. 75 p.

Webb, E. K. 1964. Further note on evaporation with fluctuating Bowen ratio. J. Geophys. Res. 69:2649–2650.

Wijk, W. R. van, and D. A. de Vries. 1954 Evapotranspiration. Neth. J. Agr. Sci. 2:105–119.

section IX

Predicting Irrigation Needs

IX

30

Soil, Plant, and Evaporative Measurements as Criteria for Scheduling Irrigation[1]

HOWARD R. HAISE

Agricultural Research Service, USDA
Fort Collins, Colorado

ROBERT M. HAGAN

University of California
Davis, California

I. INTRODUCTION

The basic purpose of irrigation is to supply plants with water as needed to obtain optimum yield and quality of a desired plant constituent. As Taylor (1965) puts it, "Irrigation should take place while the soil water potential is still high enough that the soil can and does supply water fast enough to meet the local atmospheric demands without placing the plants under a stress that would reduce yield or quality of the harvested crop." Since the 1950's, a number of new approaches and commercially available devices have been developed for scheduling irrigations which permit the irrigation farmer to evaluate the supply of water to crops and, thus, to greatly improve irrigation practices.

Criteria most suitable for scheduling irrigations vary from one situation to another. Where water is scarce or expensive, irrigations should be scheduled to maximize crop production per unit of applied water; where good land is scarcer than water, irrigations should be scheduled to maximize crop production per unit of planted area. Criteria discussed in this chapter can guide in establishing irrigation schedules which will favor optimum crop yields and efficient water use. It should be recognized, however, that other considerations may dominate in some situations. For example, irrigation schedules may be modified to minimize irrigation costs, facilitate other farm operations, overcome problems of slow penetration of irrigation water, control groundwater level, accomplish leaching of salts, or accommodate schedule of water delivery to the farm. In all cases, the criteria selected for irrigation should permit favorable crop yields, optimum use of water, and proper attention to other factors involved.

The purpose of this chapter is to consider various soil, plant, and evaporative techniques as criteria for establishing irrigation schedules for a given crop and climatic area. The following discussion applies principally to farmers who have irrigation wells or operate in projects where water is available on demand.

[1] Contribution from the Northern Plains Branch, Soil and Water Conservation Research Division, Agricultural Research Service, USDA, and from the Department of Water Science and Engineering, University of California, Davis, Calif.

II. SOIL WATER INDICATORS AND MEASUREMENTS

A. Soil Appearance and Feel

In the absence of more sophisticated methods to assess irrigation need, the use of US Dep. Agr.–Soil Conservation Service Irrigation Guides to establish approximate irrigation schedules or noting the appearance of the soil in drouth areas of the field by using a probe or shovel (*see* Fig. 30–1) to examine rooting depth and wetness of the soil could improve practices where under- or over-irrigation practices prevail. Many irrigators are reluctant to expend even the minimal effort required to assess irrigation need by such examinations.

B. Soil Water Content

Numerous irrigation experiments on agricultural crops to determine the permissible water deficit in the rooting depth without adversely affecting yield and quality of harvestable product have been summarized (Stanhill, 1957; Taylor, 1965). A common criterion allows depletion of various fractions of available water in the crop rooting depth, i.e. 25% 50%, and 75% prior to irrigating a crop. In selecting the water deficit to be permitted prior to irrigation, consideration must be given to the effects of other soil factors, plant root and shoot characteristics, climatic conditions, and management decisions involving cultural and harvest operations (Hagan and Vaadia; 1960). The expanding root system of annual crops complicates the problem.

Soil water content may be determined by gravimetric sampling or by use of the neutron soil water meter.

1. GRAVIMETRIC SAMPLING

Soil sampling and oven drying to determine gravimetric water content is laborious and time consuming. To use percentage soil water determinations on a dry weight basis, bulk density measurements are needed to convert soil water content to a volume basis or to the surface depth of water needed to replenish the soil profile at each irrigation. Use of soil water content alone to schedule irrigations requires data for each individual soil in relation to permissible soil water deficits and is limited to the soil type studied and to areas having similar climatic conditions. Some researchers utilize gravimetric procedures to study plant response to irrigation, but the method is seldom accepted by farmers for routine use.

2. NEUTRON SOIL WATER METER

The neutron scattering soil water meter directly measures the soil water content on a volume basis, but radiation hazards, high cost, and maintenance problems generally restrict its use to a research tool (*see* Fig. 30–1 and chapter 15). Lightweight commercial rate meters and depth probes may be used to advantage on large farms, such as those of corporations, if skilled technicians are available. The neutron probe measures repeatedly the volume content of water in nearly the same soil volume through the wall of a metal access tube. Knowing the relation of volume water content to the permissible water deficit for a particular crop (with necessary corrections for growth stage, climatic conditions, etc.), irrigations can be scheduled when a predetermined meter reading at a given depth or depths is reached. About 1 min is required to obtain a reading at each depth.

Fig. 30–1. Devices and instruments used to schedule irrigations including: (A) shovel, soil auger, and Oakfield probe. (B) resistance unit and meter, (C) tensiometers, (D) dendrometer, and (E) neutron rate meter and depth probe.

These approaches fail to consider that water content and its potential vary with soil texture (Fig. 30–2). Note that soil suction is higher at 25% available water depletion for the loam soil than for 50% depletion of the sandy loam. Soil water suction values for all soils at 50% depletion range from approximately 0.3 to 2.5 bars, depending on texture. Irrigation schedules based on fractional water use for a given soil, climatic and experimental site have limited application. Use of soil water suction to schedule irrigations described in section C overcomes much of the difficulty in applying results from one area to another when the relationship of crop response to soil water availability based on energy concepts is known.

C. Soil Water Suction

Plant response to irrigation is better correlated with soil water potential or suction than with soil water content. Thus measurements of soil water potential or suction provide a useful approach for scheduling irrigations.

Table 30–1. Soil water suction at which water should be applied for maximum yields of various crops grown in deep, well drained soil fertilized and managed for maximum production (adapted from Taylor, 1965)[*]

Crop	Soil suction, bars	Literature reference
	Vegetative crops	
Alfalfa	1.50	S. A. Taylor & associates, Logan, Utah
Beans (snap, lima)	0.75 - 2.00	Vittum et al. (1963)
Cabbage	0.60 - 0.70	Vittum et al. (1963) Pew (1958)
Canning peas	0.30 - 0.50	S. A. Taylor & associates, Logan, Utah
Celery	0.20 - 0.30	A.W. Marsh, personal communication Marsh (1961)
Grass	0.30 - 1.00	Vissar (1959)
Lettuce	0.40 - 0.60	A.W. Marsh, personal communication Vissar (1959) Pew (1958)
Tobacco	0.30 - 0.80[†]	Jones et al. (1960)
Sugar cane	0.25 - 0.30	Anon. (1954)
Sweet corn	0.50 - 1.00	S. A. Taylor & associates, Logan, Utah Vittum et al. (1963)
	Root crops	
Onions, early	0.45 - 0.55	Pew (1958)
Onions, bulbing	0.55 - 0.65	Pew (1958)
Sugar beets	0.40 - 0.60	S. A. Taylor & associates, Logan, Utah
Potatoes	0.30 - 0.50	S. A. Taylor & associates, Logan, Utah Vittum et al. (1963) Pew (1958)
Carrots	0.55 - 0.65	Pew (1958)
Broccoli, early	0.45 - 0.55	Pew (1958)
Broccoli, postbud	0.60 - 0.70	Pew (1958)
Cauliflower	0.60 - 0.70	Pew (1958)
	Fruit crops	
Lemons	0.40	A.W. Marsh, personal communication
Oranges	0.20 - 1.00	Stolzy et al. (1963)
Deciduous fruit	0.50 - 0.80	A.W. Marsh, personal communication Vissar (1959)
Avocadoes	0.50	Richards et al. (1962)
Grapes, early	0.40 - 0.50	A.W. Marsh, personal communication
Grapes, mature	1.00	A.W. Marsh, personal communication
Strawberries	0.20 - 0.30	A.W. Marsh, personal communication Marsh (1961)
Cantaloupe	0.35 - 0.40	Marsh (1961) Pew (1958)
Tomatoes	0.80 - 1.50[†]	Vittum et al. (1958) Vittum et al. (1963)
Bananas	0.30 - 1.50[‡]	Schmueli (1953)

(Continued on next page)

The arithmetic integration of soil moisture tension (soil water suction) measured at various depths in the root zone to obtain a single integrated value has been proposed (Taylor, 1952). The number of instruments required and the difficulties in interpretation make this approach of questionable practical value. The use of instruments placed at the depth of maximum root activity seems to be satisfactory. Taylor (1965) has compiled ranges in values of water potential or suction required for optimum growth of many common crops (Table 30–1) based on instruments placed at the depth of maximum root activity for crops growing on soils low in salt content and well fertilized. The range of values given for each crop recognizes the effects of high and low evaporative demand and the need to consider weather conditions when selecting the suction value at which irrigation is to be applied.

Table 30-1. Continued.

Crop	Soil suction, bars	Literature reference
	Grain crops	
Corn, vegetative	0.50	S.A. Taylor & associates, Logan, Utah
Corn, ripening	8.00 - 12.00	S.A. Taylor & associates, Logan, Utah
Small grains, vegetative	0.40 - 0.50	S.A. Taylor & associates, Logan, Utah
Small grains, ripening	8.00 - 12.00	S.A. Taylor & associates, Logan, Utah
	Seed crops	
Alfalfa, prebloom	2.00	Taylor et al. (1959)
Alfalfa, bloom	4.00 - 8.00	
Alfalfa, ripening	8.00 - 15.00	
Seed carrots, 60-cm depth	4.00 - 6.00 §	Hawthorne (1951)
Onions, 7-cm depth	4.00 - 6.00 §	Hawthorne (1951)
Seed onions, 15-cm depth	1.50	
Lettuce, productive	3.00§	Hawthorne & Pollard (1956)
Coffee requires short periods of low potential to break bud dormancy, followed by high water potential		Alvim (1960)

* Where two values for soil water suction are given, the lower suction value is used when the evaporative demand is high and the higher value when it is low; intermediate values are used when the atmospheric demand for evapotranspiration is intermediate. (these values are subject to revision as additional experimental data become available.)
† Based on converting 50% available water to water potential (soil suction) equivalents using curves for appropriate soil textures of Fig. 30-2.
‡ Based on converting 70% available water to water potential (soil suction) equivalents using curves for clay soils of Fig. 30-2.
§ Resistance values were converted to water potential from calibration of similar plaster resistance units.

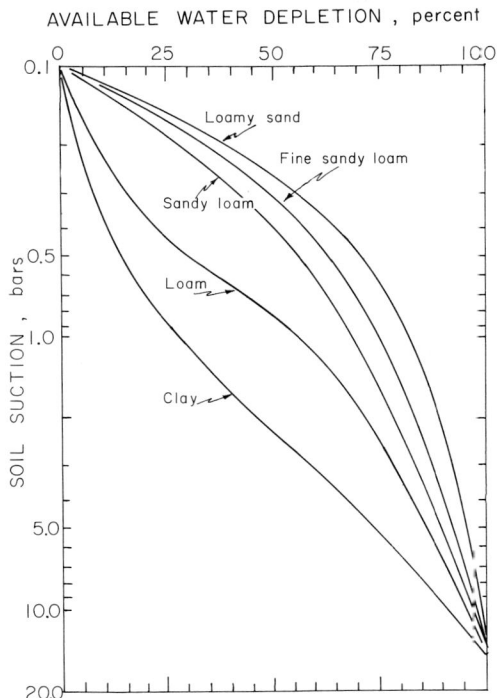

Fig. 30-2. Water retention curves for several soils plotted in terms of percent available water removed, redrawn with change in scale from Richards and Marsh (1961) and Taylor (1965).

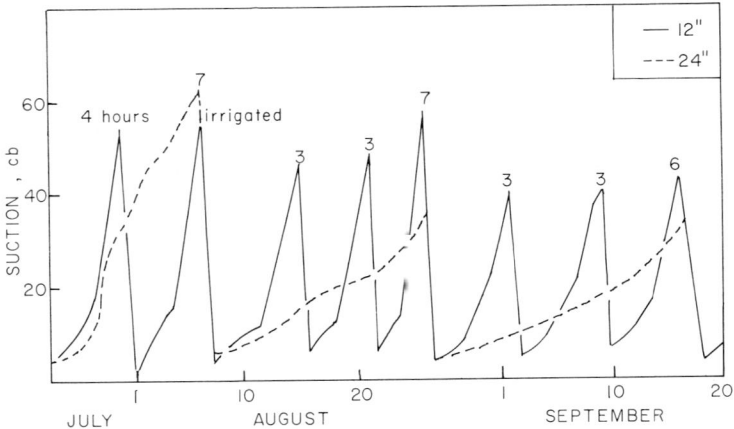

Fig. 30–3. Soil suction data copied from an avocado grower's record for a soil having limited drainage below the second foot (from Richards and Marsh, 1961).

1. TENSIOMETERS

The tensiometer described in chapter 15 and illustrated in Fig. 30–1 measures soil water suction and can be used to schedule irrigations without reference to soil water content. The porous cup and vacuum gauge, when filled with water and placed in the soil, registers soil water suction up to 0.8 bar. The vacuum gauge can be read at a glance. One can irrigate when the vacuum gauge registers the prescribed limits provided the permissible soil water suction for a given crop and the prevailing conditions are known.

Richards and Marsh (1961) suggest the use of tensiometers not only to schedule irrigations but to indicate the amount of water to apply. Data presented in Fig. 30–3 show that placement of tensiometers in the active root zone and near the bottom of the root zone (corresponding to depths of 30 and 60 cm in the case illustrated) provided information which permitted control of deep percolation losses. Irrigations applied when the soil water suction at the 30-cm depth approached 0.5 bar did not penetrate to the 60 cm depth until the application time was nearly doubled on July 5, August 26, and September 16. Between irrigations, soil water suction at the lower depth gradually increased indicating lack of deep percolation.

Tensiometers adequately cover the range of suctions required for most shallow rooted and quick growing vegetable crops such as potatoes (*Solanum tuberosum* L.), tomatoes (*Lycopersicon esculentum*), and lettuce (*Lactuca sativa* L.); root crops such as potatoes; and some tree crops. Where small grains and forages such as alfalfa (*Medicago sativa* L.) are grown, resistance blocks may be required to extend the soil water suction range for most efficient scheduling of irrigations (*see* following section). Typical requirements for installation of tensiometers and resistance blocks are given in Table 30–2.

Since 1956 approximately 38,000 commercial tensiometers have been sold in the USA, largely in California and Arizona, and about 7,000 in foreign countries. In California, county farm advisors estimate that orchard growers currently use more than 10,000 instruments. In Tulare County alone, about 2,000 tensiometers are now used to schedule irrigation of citrus, walnut (*Juglans* spp.), avocado

Table 30–2. Typical requirements for installation of tensiometers and/or resistance blocks in areas with relatively uniform crops and/or soil conditions

Number of stations	3 to 4
Number of depths	2 to 3
Depth placement	Top - zone of maximum root activity
	Bottom - near bottom of active root zone
	Intermediate - midway between top and bottom positions.
	(a shallow depth may be temporarily needed where seed-lings are being established)
Number of readings between irrigations	4 to 5 - values should be plotted so readings can be ex-trapolated to anticipate irrigation need
Site conditions	Select representative soil and vigorous crop area
Precautions	Avoid trampling area near installation; service tension-meters regularly

(*Persea spp.*), and pear (*Pyrus communis*) groves. Some are used in vineyards. An increasing number of tensiometers are being installed in cemeteries and golf courses to indicate the need for irrigation of turf. These figures serve to emphasize that tensiometers are practical and useful in scheduling irrigations, particularly where high value crops are grown.

It must be recognized that the precise selection of the suction value to be used in a given situation as the criterion for irrigation need should take into account the possible effects of soil, plant, and climatic factors known to affect the relation between soil water suction and plant water potential. The magnitudes of these effects and their importance in irrigation scheduling deserves further study.

2. ELECTRICAL RESISTANCE UNITS

Properly calibrated resistance units are capable of measuring the soil water suction (Taylor, 1952; Haise and Kelley, 1946). Sensitivity of the instrument in the lower suction ranges is generally inadequate for precise measurements. Resistance units used in conjunction with tensiometers can measure the entire range of available water. Increased sensitivity in the lower suction range was achieved by imbedding a nylon unit in plaster (Bouyoucos, 1954). Block durability is enhanced by impregnating with a special nylon solution to retard disintegration (Bouyoucos, 1953). In irrigation practice, the tensiometer is most frequently used in coarse- to medium-textured soils, and the resistance units in medium- to fine-textured soils.

Resistance readings for given soil water contents will be lower in salt-affected soils than in those of low salt content. Cylindrical and screen electrodes parallel to the flat axis reduce salt effect but not entirely (Taylor et al., 1961). Units are often unsatisfactory in coarse-textured soils due to contact problems and the resultant lag in response to changes in soil suction, particularly where evaporative demand is high.

Practical experience in utilizing the resistance units for scheduling irrigations is quite limited. The authors are aware of several instances where sugar beet (*Beta vulgaris* L.) companies have attempted to control irrigation on growers' fields using plaster blocks. The fact that the practice was not continued either reflects the farmers' unwillingness to be "bothered," the common general belief that the irrigator can do as well from experience, or the restrictions of the project water delivery system.

In the Willamette Valley, Oregon, dairy farmers who irrigate pastures by the sprinkler method were at first enthusiastic about a tapered probe equipped with resistance units (consisting of cylindrical wire electrodes imbedded in annular plaster of paris rings) at four points in a 60-cm depth. Field installations had demonstrated that most pastures were being irrigated more frequently than necessary. Although the use of resistance probes resulted in improved water management practices with greater efficiency of water use and lower pumping costs, a recent communication from Marvin Shearer (Irrigation Specialist, Oregon State Univ., Corvallis, March 16, 1965) reports that dairy men thought too much time was involved in taking readings and keeping records. Greater acceptance, however, was found among farmers who grew crops for commercial freezing and canning.

A similar situation developed in the pump-irrigated area on the high plains of Texas. Results of water management research on sorghum (*Sorghum vulgare*) and wheat (*Triticum vulgare*) (Jensen and Musick, 1960, 1962) demonstrated conclusively that plaster resistance units could be used effectively to schedule irrigations. Encouraged by the research findings, a vocational agriculture instructor initiated a card system whereby periodic readings taken by a farmer could be mailed to and tabulated in a central location. The farmer was then notified when to irrigate by phone or return mail. Failure to use this service in an area where limited groundwater is available on demand but where efficient water application is essential is further evidence of the farmers' reluctance to take the necessary time required for resistance readings or the belief that he can do as well by experience.

In contrast, the California Agricultural Extension Service estimates that approximately 5,000 resistance units are in use, mostly in orchards. Units are generally installed and operated in accordance with criteria set forth in Table 30–2, previously discussed. Here is evidence that high crop values and scarcity of water may result in a greater effort by the farmer to achieve maximum efficiency of water used.

Ewart (1951) reported on the use of plaster blocks as a guide for irrigating sugarcane (*Saccharum officinarum*) in Hawaii. Three units installed at root zone depth several furrows apart in each of 5 acres were used to control irrigations. Irrigation was applied when resistance approached 5,000 ohms (equivalent to about one-third depletion of available water). Two men taking readings could cover about 100 acres in 1/2 to 1 hour. Currently more attention is being given to meteorological approaches (discussed later) presumably because the collecting and processing of these data are considered to be easier.

III. PLANT WATER INDICATORS AND MEASUREMENTS

It is now widely recognized that plant growth is directly related to the water balance in plant tissues. As water deficits develop, physiological processes are disturbed, and growth and yield subsequently are reduced. The relative rates of water absorption and loss by plants determine their internal water balance which represents "the integrated interaction of plants with environment" (Mederski, 1961). If the internal water balance could be simply and easily determined, plant water stress might well be used as a criterion in determining irrigation need.

There are three general approaches using the plant as an indicator of water

Table 30–3. A summary of various plant measurements and techniques investigated to schedule irrigations for various crops

Plant measurement and technique	Plant organ or constituent measured	Plants investigated	Literature reference
Visual indicators			
Color	Leaf	Cotton	Bilbro et al. (1960)
		Beans	Robins & Domingo (1956)
		Beans	Howe & Rhoades (1961)
		Cotton	Hoover & Booher (1952)
		Cotton	Bilbro et al. (1960)
		Cotton	Petinov (1961)
		Beans	Burman & Painter (1964)
Plant movements	Leaf angle	Sorghum	Henderson*
		Beans	Henderson*
Growth indices	Fruit, leaf, stem, trunk	Pears	Aldrich & Work (1934)
		Dates	Aldrich et al. (1946)
		Sugarcane	Mallick & Venkantaraman(1957)
		Sugarcane	Oppenheim & Elze (1937)
		Orchard	Verner (1962)
		Orchard	Verner et al. (1962)
		Apples	Magness et al. (1935)
		Oranges	Oppenheim & Elze (1937)
		Apples	Ladin (1959)
		Peaches	Uriu et al. (1964)
		Cotton	Stockton et al. (1955)
		Cotton	Stockton & Doneen (1957)
		Cotton	Stockton et al. (1961)
Water content			
Selected leaves or tissues		Sugarcane	Clements & Kubota (1942)
Absolute water content		Sugarcane	Clements et al. (1952)
		Sugarcane	Tanimoto (1961)
		Sugarcane	Chang et al. (1963)
Relative water content		Cotton	Namken (1965)
Transpiration	Leaf	Apples, plum, pear, quince, cherry, apricot	Gorin (1963)
Stomatal aperture (microscopy) (impressions) (infiltration)	Leaf	Coffee	Alvim and Havis (1954)
		Apple	Furr and Degman (1932)
		Citrus	Oppenheimer & Mendel (1939)
		Citrus	Oppenheimer & Elze (1941)
		Wheat	Maximov & Zernova (1936)
		Cotton	Ophir & Putter (1959)
		Corn	Ophir & Putter (1959)
		Banana	Schmueli (1953)

(Continued on next page)

need: (i) Select an indicator of plant water deficit which can be observed before growth is checked, (ii) measure plant growth seeking to irrigate just prior to retardation in growth, and (iii) correlate plant growth responses with internal plant water balance using such information as an advanced criterion for irrigation. Indicators of plant water stress often do not appear nor can visible retardation in growth be detected until most plants are damaged with the result that subsequent irrigations can seldom overcome the loss in yield. Measurements of plant water balance provide a more fundamental approach, but the lack of adequate and convenient techniques for plant water measurement has handicapped research and the use of this method as a practical guide to irrigation.

Table 30- 3. Continued.

Plant measurement and technique	Plant organ or constituent measured	Plants investigated	Literature reference
Osmotic potential (cryoscopic)	Cell sap concentration	Cabbages	Lobov (1951, 1957)
		Tomato	Lobov (1951, 1957)
		Potato	Lobov (1951, 1957)
		Alfalfa	Bauman (1955) [†]
		Wheat	Bauman (1955) [†]
		Barley	Bauman (1955) [†]
		Oats	Bauman (1955) [†]
		Sugar beets	Bauman (1955) [†]
		Potatoes	Bauman (1955) [†]
		Bananas	Schmueli (1953)
		Barley	Kreeb (1958)
		Barley	Slavik (1959)
		Cotton	Ophir & Putter (1959)
		Cotton	Filippov (1959a)
		Tomato	Belik (1960)
		Cucumber	Belik (1962)
		Apple	Filippov (1961)
		Corn	Rodionov (1962)
		Tomato	Babushkin (1959)
		Potato	Babushkin (1959)
		Sugar beets	Chunosova (1963)
		Cantaloupe	Davis (1963)
Water potential (Equilibria over solutions known osmotic concentrations, psychometric techniques)	Shoot or leaf	Cotton	Shardakov (1957)
		Cotton	Filippov (1954, 1959b)
		Cotton	Neshina (1955)
		Cotton	Krapivina (1963)
		Alfalfa	Kolesnikova (1957)
		Wheat	Petinov (1959)

* D.W. Henderson, 1955. Unpublished research from University of California, Davis, US Regional Research Project W-29.

† L. Bauman, 1955. Determination of right time for irrigation: Phases, periods, and scales of plant hydrature. Mimeo, Lethbridge, Alberta, Canada. (Cited by Kreeb, 1963).

Hagan and Laborde (1964) give a comprehensive literature review on the use of plants as indicators of irrigation need. Their review, summarized in Table 30–3, includes pertinent literature references for the various methods used to assess plant water stress. The reader is referred to their paper for a more comprehensive discussion. The following section summarizes the usefulness of the various methods now available for assessing plant water stress to schedule irrigations.

A.   Visual Indicators of Water Stress

Plant wilting is the most obvious sign of plant water stress. It has been frequently shown, however, that growth of most plants may be retarded before visible wilting occurs. A visual indicator that would warn the irrigator of plant water deficits soon enough for timely irrigations would serve a useful purpose. Possibilities include plant color, plant movements, and exudation from cut plants as affected by turgor changes. The use of selected or specially managed plants as indicators in cropped fields may warrant greater attention.

1. PLANT COLOR

A practical guide for irrigation of some crops is the distinct color change that occurs in the foliage with the onset of plant water stress. Leaves of beans (*Phaseo-*

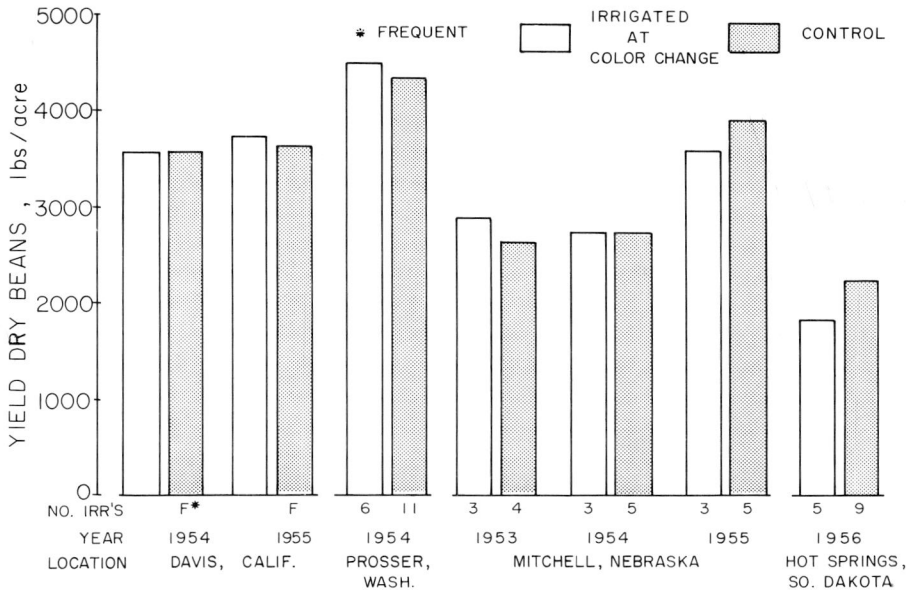

Fig. 30–4. Dry bean yields obtained when irrigated by color change in comparison to more frequent irrigations (control) at five locations in the USA.

lus vulgaris), cotton (*Gossypium hirsutum*), and peanuts (*Arachis hypogea*), for example, become bluish green to dark green as the available soil water is depleted. Burman and Painter (1964) observed that the color change caused by soil-water stress in young bean leaves differed for various stages of growth and was most distinct in the seedling stage. Some bean varieties are more responsive to color change induced by water stress than others. Diseased foliage also makes it more difficult to observe plant water stress symptoms. Field beans have been successfully irrigated by noting leaf color change (light green to dark green) at Prosser, Washington (Robins and Domingo, 1956); at Mitchell, Nebraska (Howe and Rhoades, 1961); at Hot Springs, South Dakota (Fine et al., 1964); and at Davis, California (D. W. Henderson, Irrigation Department, Univ. of California, Davis. *Personal comm.*). At all locations, yields were not significantly reduced even though water stress symptoms were apparent for a 5-day period prior to irrigating the crop (*see* Fig. 30–4). As a result of these studies, farmers in western Nebraska, who previously irrigated beans six or more times during the growing season, now use only three irrigations in most years. Furthermore, the 5-day stress period allows time to order and apply water to a large field without sacrificing yield.

2. PLANT MOVEMENTS

Pronounced diurnal movement of leaves has been repeatedly observed in many plants, notably the *Oxalidaceae*, the *Leguminoseae* and the *Gramineae*. The change in shape is due primarily to variations in turgor pressure of plant cells. Recently, Henderson successfully irrigated sorghum by noting the change in leaf angle of the plant and saved water by irrigating less frequently without yield reductions (D. W. Henderson, 1955. *Unpublished research from Univ. of California, Davis, US Regional Research Project W–29.*). Further study is needed, however, to fully assess the potential of utilizing leaf movement as a means of scheduling irrigations.

3. EXUDATION

Exudation from topped plants depends directly on root pressure. The phenomena was first observed by Litvinov and Gebhardt (1929) and since then has been studied on numerous plants. McDermott (1945) reported that exudation from a cut plant ceased at soil water contents above the wilting point. Similar relations for guttation were found by Gracanin (1963). Filippov recommends (Petinov, 1961) irrigation of cotton before exudation from topped plants ceases permanently. The method offers one of the simplest approaches for determining irrigation need but destroys part of the plant.

4. SELECTION AND MANAGEMENT OF PLANTS AS INDICATORS FOR IRRIGATION

The idea of growing an indicator plant in association with a crop that will exhibit water stress symptoms earlier than the crop itself is not new. Indicator plants could be selected which are naturally more susceptible to soil water deficits, or they could be plants selected from the crop itself but specially managed. One requirement would be a top-to-root ratio exceeding that of the main crop. Growth regulators, pruning of roots, or confining roots within mechanical barriers offer possibilities for controlling the top-root ratio. Indicator plants might also be spaced closer together or placed in a soil mixed with sand to reduce its available water capacity.

B. Plant Growth Indicators of Water Stress

These criteria for irrigation include growth measurements of certain plant organs, leaf angle and elongation. Many problems are involved in using growth as a criterion for irrigation because growth is a slow, complex process strongly affected by environmental factors in addition to water. It is important not only to select a plant organ particularly responsive to water stress but also to establish, for reference, a growth curve for that organ under the local conditions which produce an optimum yield.

1. FRUIT GROWTH

Furr and Taylor (1938) demonstrated a positive correlation between enlargement of lemon (*Citrus limonia*) fruit and availability of soil water. Oppenheim and Elze (1937) and Oppenheimer and Elze (1941) recommend irrigations of orange (*Citrus sinensis spp.*) trees whenever the daily increment in fruit circumference falls below 0.2 to 0.3 mm during the summer (equivalent to an increase in fruit volume of 2 ml/day). Ladin (1959) found that the rate of apple (*Malus sylvestris*) fruit growth decreased gradually as water was extracted from the soil. Uriu et al. (1964) reported similar observations with peaches (*Prunus persica*) and prunes (*Prunus domestica*).

Techniques for measuring fruit growth vary and often lack precision. A tape or caliper is commonly used. An auxanometer is capable of continually measuring enlargement of potato tubers (Dietz and Verner, 1942), daily cycles in the expansion and contraction of well-developed lemon fruits (Bartholomew, 1926), and growth behavior of cotton bolls and elongation of cotton stems and leaves (Anderson and Kerr, 1943). Tukey (1963) adapted a linear displacement voltage transducer for continuously recording the rate of fruit enlargement.

2. LEAF GROWTH

A sensitive apparatus for measuring changes in leaf thickness was devised by Meidner (1952) who showed that leaf thickness changes quickly with variations in leaf water content induced by transpiration, but found no correlation between leaf thickness and soil water tension. The beta ray gauging technique (Mederski, 1961; Whiteman and Wilson, 1963; Nakayama and Ehrler, 1964) yields a value, integrating both leaf thickness and water content. Gale showed that leaf thickness of sugar beets and beans was correlated with relative turgidity (J. Gale, 1962. The influence of transpiration and its depression on plant economy. *Ph.D. Diss., Hebrew Univ., Jerusalem, Israel*). Measurements of leaf thickness appear, however, to be impracticable for scheduling irrigations except possibly for a crop like pineapple (*Ananas sativus*) whose leaves retain a nearly constant basic thickness for a considerable time but are measurably affected by water stress. Aldrich et al. (1946) developed a practical method to measure leaf elongation of the date palm (*Phoenix dactylifera* L.) that is sensitive to soil water deficits. The indicating device, fastened to the trunk of the date palm, is connected by wire to a vertically emerging date leaf. It is time to irrigate when the growth curve begins to flatten.

3. STEM AND TRUNK GROWTH

Stem growth is affected by water stresses, as demonstrated by Gates (1955) on tomatoes, Owen (1958) on sugar beets (*Beta saccharifera*), Vaadia and Kasimatis (1961) on grapes (*Vitis vinifera*), Clements and Kubota (1942), and Clements et al. (1952) on sugarcane; Stockton and Doneen (1957) and Marani and Horwitz (1963) on cotton, and Uriu et al. (1964) on peaches.

Stalk elongation in sugarcane was shown to decrease with increasing soil water suction (Mallick and Venkantaraman, 1957; Leverington, 1960; Cornelison and Humbert, 1960; Robinson, 1963; and Robinson et al., 1963) and to be greatly reduced by soil water suctions > 2 bars (Robinson, 1963). Since stalk elongation is highly correlated with yield, it is possible that stem measurements could be used to indicate irrigation need. Internodal spacing below the growing tip of cotton is quite sensitive to increasing soil water deficits and has been used as a criterion for irrigation (Stockton et al., 1955; Stockton and Doneen, 1957; Stockton, 1961).

A simple dendrometer (*see* Fig. 30–1) for measurement of radial growth of tree trunks (Verner, 1962) may be useful for scheduling orchard irrigations. Measurement of growth rate and diurnal shrinkage of trees under water stress is compared with corresponding values for trees well supplied with water. Whenever growth rates decline below the check, or diurnal shrinkage exceeds the check, the need for irrigation is indicated (R. M. Hagan and P. E. Martin, October 1963. *Unpublished research from Univ. of California, Davis. Report to US Regional Project W–67.*).

C. Leaf Reflectance and Temperature

Leaf reflectance and leaf temperature are other possible approaches to indicate plant water stress.

A clean leaf with turgid mesophyll tissue reflects more infrared light than leaves with flaccid cells. Infrared aerial photography has been used to detect cropped areas affected by disease and salinity (Colwell, 1956, 1961; Myers and Carter, 1965; Marcus, 1963). In California commercial aerial photography service was offered to farmers for several seasons as a guide for scheduling irrigations, particularly for orchards, but it has not been continued.

Tanner (1963) measured leaf temperature with an infrared thermometer. Further study is needed to assess the feasibility of detecting differences in plant temperature imposed by soil water deficits as the basis for scheduling irrigations, but possibilities appear promising.

D. Plant Water Measurements

Many aspects of the plant water balance have been measured. These include relative water content, stomatal opening, transpiration rate, and the osmotic and water potentials. Reviews are available in the UNESCO Arid Zone Research Symposia (UNESCO, 1960, 1961, 1962).

1. WATER CONTENT

Extensive use of absolute water content as the criterion for irrigation has been reported for sugarcane (Clements and Kubota, 1942; Clements et al., 1952; Tanimoto, 1961). Selection of a plant part sensitive to water stress but unaffected by fertility and other environmental variables is difficult. Many studies have been conducted by the sugarcane industry to utilize the plant water content method for scheduling irrigations. Because of greater convenience and better calibration, instruments that measure soil water content and, more recently, evaporation measurements are now preferred by most plantations. Many successful plantations are irrigated according to schedules gained by experience from using plant water measurements and not soil water instruments.

Relative plant water content as a measure of internal water balance of plants or relative turgidity of leaves has been studied by Namken (1965) to schedule irrigations for cotton. The method is simple, but the time of sampling is critical because of diurnal fluctuations in leaf water content. Changes in soil suction and evaporative demand and the influence of these factors on relative turgidity require careful interpretation to schedule irrigations for maximizing yields. Rutter and Sands (1958) working with pine needles (*Pinus sylvestris*) and Halevy (1960) with gladiolus (*Gladiolus* sp.) found that the leaf water deficit appeared best suited as an index to soil water stress. Halevy measured variations in daily transpiration, leaf water content, stomatal aperture, water saturation deficit, osmotic values, and leaf elongation with increasing soil water stress. He considered saturation deficit to be the most sensitive index but that stomatal aperture was almost as good and easier to measure.

To summarize, determinations of relative water content should be a useful indicator of irrigation need because of (i) simplicity of sampling and measurement and (ii) the generally good correlations obtained with plant water potential and presumably plant growth. Diurnal fluctuations in plant water potential and its variation with the plant part selected for measurement present problems of interpretation when scheduling irrigations under variable soil and climatic conditions.

2. TRANSPIRATION AND STOMATAL APERTURE

Gorin (1963) observed that the transpiration rates of leaves of apple (*Pyrus malus*), pear (*Pyrus communis*), quince (*Cydonia japonica Pers.*), plum (*Prunus domestica* L.), cherry (*Prunus cerasus* L.), and apricot (*Prunus armeniaca L.*) trees of different ages could be correlated with soil water content. Transpiration measurements are quite common in ecophysiological studies but are not useful for scheduling irrigations. Measurement of stomatal aperture is more practical and easier to use, but measurements are influenced by age of leaves and exposure to light and wind. Maximov and Zernova (1936) concluded that stomatal aperture was valuable for scheduling irrigation of wheat. Oppenheimer and Mendel (1939) and Oppenheimer and Elze (1941) found infiltration measurements on citrus leaves to be a good indicator of water need. Schmueli (1953) concluded that stomatal aperture of banana (*Musa sapientum*) leaves directly exposed to sunlight at midday was a good indication of soil water deficits. Stomates of plants in soil at 2/3 of field capacity were markedly depressed compared to those in leaves of plants at field capacity. Ophir and Putter (1959) applied the infiltration method using 1:3 mixture by volume of paraffin and turpentine. Halevy (1960) scheduled irrigations for gladiolus using stomatal-aperture measurements made 3 hours after sunrise. In spite of successes indicated above, practical use of the stomatal method to indicate irrigation need requires further study and the development of simple yet reliable methods.

3. OSMOTIC POTENTIAL

The osmotic potential (osmotic pressure) of plants varies considerably with plant species, time of year, and environmental conditions such as water stress, fertility and salinity of the soil. Despite these shortcomings, the ease with which osmotic potential can be determined has led to its use as an indicator of water need for many years (Miller, 1938; Kreeb, 1963). Lobov (1951, 1957, also cited by Chunosova, 1963) reports field experiments with cabbages (*Brassica oleracea* L.), potatoes, and tomatoes. Lobov's method has been modified by Babushkin (1959) who concluded that measurements on exuded sap provide a sound basis for scheduling irrigations especially with tomatoes. Unpublished work by Bauman cited by Kreeb (1963), scheduled irrigations for alfalfa, wheat, barley (*Hordeum*), oats (*Avena*), sugar beets, and potatoes on the basis of osmotic potential determined cryoscopically on expressed leaf cell sap. He found a negative correlation between average osmotic pressure and yield. Other references cited in Table 30–3 describe methods of measurements and, in some instances, give specific recommendations for using osmotic pressure measurements to schedule irrigations. However, these reports often lack descriptions of environmental conditions that prevailed during experiments.

4. WATER POTENTIAL

Measurement of plant water potential is difficult, and hence it has not yet been used extensively to schedule irrigations. However, the use of plant water potential as the criterion for irrigation appears to be theoretically sound.

Rapid progress is being made on techniques for measuring plant water potential. The Shardakov (1957) method has been useful. Lang and Barrs (1965) and Rawlins (1966) have described improved thermocouple psychrometer methods for measuring plant water potentials in leaves attached to plants. Using the wet-loop psychrometer, Gavande and Taylor (1967) have measured plant water potential as influenced by soil water potential and atmospheric environment. Their data clearly indicate that plant water potential is affected by both soil and atmospheric conditions.

Cotton, alfalfa, and wheat have been irrigated on the basis of plant water potential (see Table 30–3 for references). Additional research relating plant growth to plant water potential is needed before the practicality of the latter can be assessed as a criterion for scheduling irrigations.

In summary, many practical limitations now confront the irrigator who may use the plant approach for determining irrigation need. Plant water stress varies markedly with weather, stage of growth, and time of day. Very careful standardization is needed to establish critical levels of plant water stress. Measurements are often time consuming and may require highly refined equipment. Until better instruments are available, plant appearance such as color change and plant growth rates of certain plant parts offer the most practical plant methods available for scheduling irrigations. The plant approach deserves more attention by researchers because of its direct relation to yield and its ability to integrate soil water deficits and the evaporative demand of the atmosphere.

IV. METEOROLOGICAL APPROACHES

Use of meteorological approaches to determine need for irrigation requires a knowledge of: (i) short-term evapotranspiration (ET) rates at various stages of plant development, (ii) soil water retention characteristics, (iii) permissible soil water deficits in relation to evaporative demand, and (iv) the effective rooting depth of the crop grown. It is the purpose of this section to discuss the more practical considerations of estimating and utilizing ET data for developing irrigation guides and on-farm procedures for scheduling irrigations. It is not the intent to repeat unnecessarily the excellent discussions in chapter 29 of the merits and limitations of various empirical methods or procedures available for estimating ET.

Tanner in chapter 29 emphasizes that an empirical equation to estimate ET must be calibrated for the region in which the estimates are made. This is particularly important in arid and semiarid regions where strong advection of energy from dry surroundings adds to incoming solar radiation for evaporation of water. Reasonable estimates of potential ET can be made only over relatively long periods of time, usually for a season and possibly for a month. Information thus obtained is useful in planning the water resource, say for an irrigation project, but generally is inadequate for the development of specific irrigation schedules. One problem that arises concerns the estimate of ET during periods when crop cover is being established and where ET is less than potential (*see* chapter 27). During such periods, estimating formulas for potential ET must be adjusted empirically to reflect more nearly the ET rate during periods of incomplete crop cover. Since calibration procedures to develop prediction equations require that factors such as temperature, humidity, solar radiation, etc. be correlated with field measurements of ET, some argue that the actual ET values determined should be adequate in themselves to develop reasonable irrigation schedules. Experience has shown that, for a crop like cotton, irrigation on a calendar basis in arid and semiarid regions, where evaporative demand is relatively uniform within a given season and from year to year, maintains soil water content within permissible deficits for near optimum crop growth (*see* chapter 33, part IV).

But what about the growing season where evaporative demand is abnormally low or high? Could improvements be made in scheduling irrigations, for example, by extending number of days between irrigations to reduce water application losses if concurrent estimates or measurements of ET were made while the crop is growing? The answer to these questions will require more study but possibilities appear good. For example, use of evaporative devices to be discussed later or energy balance approaches (Tanner, 1960; Pruitt, 1963; Fritschen, 1965) offer possible ways to determine short-term ET rates at a central location for daily broadcast to farmers in an irrigated project. Even a single meteorological measurement such as solar or net radiation may offer improvements in scheduling irrigations. The procedure developed by Jensen and Haise (1963) utilizes solar radiation and provides average short-term estimates of ET for periods ranging from 5 to 10 days. One disadvantage of the method is that a crop characteristic curve is needed for each broad climatic region. The US Bureau of Reclamation modified this procedure to obtain a single crop characteristic curve for each crop. The single crop curve expresses the ratio of ET to potential ET at various stages of crop growth. Estimates of potential ET consider both solar radiation and mean air temperature (Jensen and Haise, 1965).

To use such information requires establishment of a water budget system that balances the amount of water lost through ET against that stored from rainfall or irrigation within the root zone reservoir. Devices like the irrigation scheduling board (*see* Fig. 30–5) simplify the tabulation of data and visually indicate when irrigation water should be applied. Farmers usually require some assistance from an irrigation specialist or other qualified individual to set up a water budget system. Generally speaking, devices like the tensiometer previously described offer a direct means for indicating soil water deficits, are easier to use, and accomplish the same purpose.

Where frequent irrigations are applied, i.e., every 2 or 3 days, to turfgrass and some shallow-rooted, high value vegetable crops using automatic sprinkler systems, irrigation schedules often are based on the capabilities of the irrigation system to just replace the amount of water lost by ET. Under such conditions, soil water content is maintained at a relatively high level, thus eliminating the need for a water budget record. One requirement is a full profile of soil water at the start of the irrigation season.

A. Evaporative Devices

Numerous correlations have been made between ET and evaporation (ET_p) from pans or tanks of various shapes and sizes, and from atmometers (small, wet, porous surfaces). Tanner, in chapter 29, states that "The general belief has been that the evaporation from pans and atmometers and from water bodies, wetted soils, and vegetation (amply supplied with water) should be well correlated since evaporation from both is governed by micrometerological conditions. Over appropriate time periods there is indeed a high correlation of ET with ET_p even though the pans and atmometers are poor analogues of a vegetation surface. This high correlation reinforces the arguments that ET_p is relatively conservative and that any method for estimating departures from the average can be used successfully when calibrated under specific conditions."

Selection of pan site can markedly influence measured evaporation. Work of Pruitt and Angus shows that evaporation from a US Weather Bureau Class "A" pan from June through December amounted to about "135 cm when the pan was sited in a large grass field, 150 cm with the pan surrounded by a 14.6-m fetch of grass in a dryland area, and 175 cm with the pan in a dryland area with no fetch." (W. O. Pruitt and D. E. Angus. 1961. Comparisons of evapotranspiration with solar radiation and evaporation from water surfaces. *First Ann. Rep. USAEPG Contract DA–36–039–SC–80334. Univ. of Calif., Davis. pp. 74–107*). Pruitt (1960) found that even the alternate cutting and regrowth of alfalfa surrounding 10- by 10-m grass sodded weather station in which a 1-m ground pan was installed increased the average ET/E ratio from 0.95 to 1.20. It is clear that transfer of evaporation data through correlations of ET and E from one project to another and even within the same project requires considerable caution. The problem is further complicated by the failure of evaporating devices to estimate water use from soil and plant surfaces when ET is less than potential. Pans generally overestimate ET during early stages of growth and during crop maturation.

Despite the shortcomings of evaporation pans to schedule irrigations, the relative success of several current programs is noteworthy. Wolfe and Evans (1964) report the use of an "oven pan" on sprinkler irrigated pastures designed to catch 1 inch of water for each inch of irrigation water applied and to evaporate 1 inch for each inch evapotranspired by the crop. The amount of water evaporated represents the soil water deficit. Irrigation by sprinkler may commence when the pan is half empty and cease just before the pan overflows. Although designed for pastures, the method can be adapted to other crops by adjusting the ratio of oven space to pan depth so that evaporation rates from the modified pan are nearly the same as the ET rates of the crop grown. The principal advantage as an irrigation scheduling device is freedom from labor. It indicates at a glance when to irrigate, how much water to apply, and when to terminate the irrigation.

Shearer found that oven pan installations on sprinkler irrigated pastures in six areas of the Willamette Valley overestimated seasonal ET by approximately 2 inches where conditions of complete crop cover existed (Marvin Shearer, Oregon State Univ., Corvallis. *Personal communication, March 16, 1965*). Stammers (1963), working with various shapes and sizes of pans to estimate ET at Hermiston, Oregon (including the oven type pan described above), concluded that evaporation measurements offered little promise of scheduling irrigations for alfalfa. Reasons given for lack of success were extreme variability and indications that the pans did not respond to different exposures in the same manner as the crop itself.

Jensen et al. (1961) and Pruitt (1958, 1960) have performed extensive investigations in central Washington to develop procedures for utilizing pan evaporation data in scheduling irrigations for a variety of crops grown in the area. Daily and accumulated evaporation measured at selected locations is supplied to irrigators by daily radio service or through newspapers. The recommended coefficients for a standardized 4-ft diameter pan given in Table 30–4 remain relatively constant after complete crop cover is established. However, coefficients must be adjusted for more accurate predictions of ET during the period of crop development. The irrigation scheduling board shown in Fig. 30–5 is a

Table 30–4. Recommended crop factor values K_c for computing estimated consumptive use from pan evaporation for several crops grown in central Washington (Jensen et al., 1961)*

Crops	K_c†
Corn, grapes, and clean-cultivated peach orchard	0.85
Alfalfa, grains, Ladino-grass pasture, and sugar beets	0.95
Beans, peach orchard with cover crop, and potatoes	1.00
Apple orchard with grass cover	1.05

* Values are for converting evaporation quantities measured with a standardized 4-ft diameter pan into estimated consumptive use quantities: $CU_c = K_c \times E$.
† To provide a safety margin for field use, these recommended values are approximately 0.05 larger than average measured values.

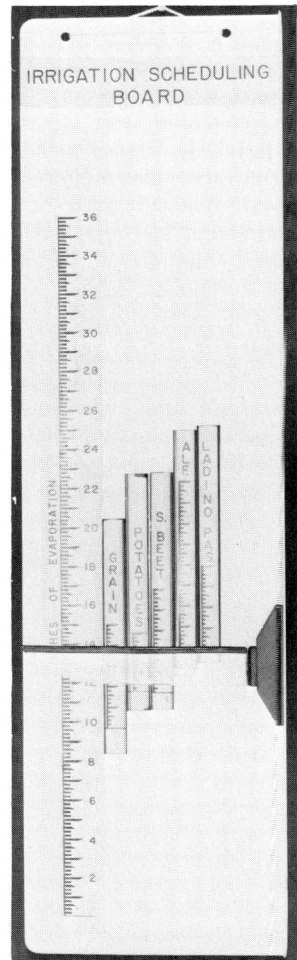

Fig. 30–5. Irrigation scheduling board: Sliding cross arm is moved up according to cumulative evaporation as indicated on left hand scale. Its position in relation to each crop slide at any given time provides an estimate of the amount of water used since the last irrigation. After irrigation of a crop, the bottom of that crop slide is placed under the cross arm. When the cross arm reaches the top of slide, another irrigation is needed in the amount indicated for that crop (from Jensen et al., 1961).

practical innovation developed to systematically utilize ET information obtained from pan evaporation (Pruitt, 1956, 1964; Jensen et al., 1961). The device maintains a visual record of accumulated evaporation and soil water depletion for each

Fig. 30–6. Elevated evaporation pan used by Hawaiian Sugar Planters' Association Experiment Station to estimate ET and schedule irrigations (photo courtesy of Experiment Station, Hawaiian Sugar Planters' Association).

crop grown. Operation is simple but the farmer usually needs assistance from extension specialists to initiate the program. The Washington State Extension Service reports that a number of farmers in the Columbia River Basin Project are using this system to control irrigations.

In a summary of evapotranspiration research being conducted by the Hawaiian Sugar Planters' Association Experiment Station, Chang (1961) emphasized two advantages of pan evaporation compared to soil water sensing devices. First, estimates of drouth probabilities could be determined from evaporation and rainfall data; and second, pan evaporation offers a more economical and convenient procedure upon which to base irrigation schedules. At the present time, 100 evaporation pans in 100,000 acres of sugarcane have been placed on plantations to study variations in climate, and to adapt subsequent evaporation data to various irrigation programs (R. B. Campbell, Senior Agronomist, Experiment Station, Hawaiian Sugar Planters' Association, Honolulu, Hawaii. Personal communication.). A typical pan installation is shown in Fig. 30–6. Various ET/ET_p ratios for estimating evapotranspiration at four stages of sugarcane growth are given by Campbell in Table 33II–2. Campbell states that "Climatic estimation of ET in which evaporation pans are used has a particular appeal in ocean climates for managing irrigation water, because advective heat comprises a small fraction of the total energy budget." Although cumbersome, it is standard practice to periodically elevate evaporation pans to cane top height. Maximum cane height is about 13 to 14 ft. Many of the pans used are on fixed platforms elevated to 5 ft. Evaporation rates at 5 ft are greater than from pans located on the ground over short grass or from pans elevated to the plane of the cane canopy. At present, adjustable towers are being designed to replace the fixed platforms.

Fuchs and Stanhill (1963) found a high correlation between ET from large commercial fields of cotton receiving optimum irrigation and evaporation from Class A pans located in nonirrigated areas upwind of irrigated fields. Measurements were taken during two successive years at 15 sites. The coefficient of variation for mean ET/ET_p ratios was $< 2\%$ for the season and $< 9\%$ for a given irrigation period. They concluded that a single ratio applied to evaporation from the Class A pan could be used in estimating irrigation requirements of cotton.

Lomas (1964) suggested a simple method for assessing relative irrigation requirements from analysis of 10 years of evaporation data in Israel.

Tanner in chapter 29 presents a good summary and literature review on the relative merits of atmometers for estimating ET rates. He cites work of Pruitt, Halkias et al., and Janes (*see* chapter 29) where the difference between black and white spherical atmometers was found to be highly correlated with potential ET and radiation. However, Tanner states that "siting and maintenance of cup condition are severe and because convective heat exchange to atmometers is such a large proportion of the total energy balance relative to that of plants, pans appear preferable to single atmometers. The use of the evaporation from both black and white atmometers offers improved possibility, provided cup condition can be maintained, since more information is available than with black atmometers used singly. If black and white units are used, the geometry of the flat-plate units with one evaporating surface would seem preferable to spherical bulbs or to two-surface units like the Piche atmometer."

Further research is needed to test the practicality of atmometers in predicting ET for scheduling irrigation. At present, there are few atmometer installations being used for this purpose and it would seem that siting and maintenance problems mentioned by Tanner would continue to restrict future use.

Korven and Wilcox (1965) found that a Class A pan gave best correlations with ET for orchards but they favored the black Bellani plate for scheduling irrigations because of convenience. Freezing, weather variation, siting (proximity to dryland areas), and maintenance, however, were among the problems encountered in the use of the Bellani plates. Most growers need continuing help from an extension specialist not only to properly estimate ET but to interpret the results in terms of soil water storage, rooting depth, effective precipitation, and the need for irrigation. Pelton (1964) obtained highly significant relations among daily measurements of evaporation from porous disk and Bellani plate atmometers, Class A evaporation pans, and meteorological factors.

Successful use of the radiation methods and evaporation pans in scheduling irrigations under field conditions requires prediction of water use by crops not only when adequate water is present, but also where potential ET is not realized under conditions influenced by the stage of growth and development, low water availability, or incomplete crop cover. Relatively little work has been directed towards this problem. Pierce (1958) presented data, based on lysimeter studies, on the relative evaporation rates for meadow cover at different stages of growth. He also introduced a factor to account for reduced water use when soil suction was high. Shaw (1964), using a prediction technique based on the work of Pierce, found a close correlation between estimated and observed values of soil water under a meadow crop.

V. PRACTICAL CONSIDERATIONS IN SCHEDULING IRRIGATIONS

Instruments or other approaches for indicating the need for irrigation are most useful in water deficient areas where cost of water is usually high and where high value crops are grown. Crops with established root systems like orchard trees are easier to instrument than annual crops with an expanding root system. In the case of tree crops, installations of tensiometers or resistance blocks can be semipermanent. However, instruments placed in annual crops are normally destroyed or removed and reinstalled each crop season. Where qualified technical personnel are available, as on large or corporate farms, soil water gauging devices or plant water measurements can be more easily managed and the data better interpreted in terms of scheduling irrigations.

It is not always possible for a farmer to schedule irrigations at the precise time water may be needed by the crop. In some projects, the irrigation regime may be limited by the water delivery system under which the farmer operates. The rotation system, for example, allows little choice. For the farmer who elects to pass his turn may indeed need water before the time of the next delivery, generally 1 or 2 weeks later. If he accepts water on schedule, consideration must be given to the crop that should be irrigated. In other projects, the farmer may receive a small stream of water under a continuous delivery system. Here, he must apply water on some part of his farm at all times when evaporative demand is high to cover the irrigated acreage in time for the next irrigation. The continuous delivery system usually is inefficient, and excessive use of water often contributes to drainage problems.

Maximum flexibility and the greatest opportunity to irrigate on a more scientific basis occurs where irrigation water is available on demand. Water is usually ordered 1 or 2 days in advance where project distribution systems have capacities to satisfy crop needs during periods of peak demand. Where groundwater is plentiful, the farmer need only turn on a pump to irrigate. In spite of the advantages, many farmers misuse the demand delivery system by irrigating a crop much more frequently than needed.

Often irrigation schedules and applications must consider other farming operations. For example, a farmer might elect to irrigate alfalfa sooner than necessary because he anticipates harvesting a crop of hay the following week. The maintenance of a favorable salt balance (*see* chapter 51) may require, in some situations, water applications in excess of the ET requirements of the crop. This seldom requires more than a 10% increase. Deep percolation losses occurring in the operation of many surface irrigation systems frequently provide leaching considerably in excess of salt balance requirements. Rains which fall during a portion of the year may also provide much or all of the required leaching.

In areas where water is plentiful and cheap in comparison to other production costs and crop values, there may be little incentive for a farmer to extend irrigation intervals or to improve efficiency of water applications just to conserve water. Irrigation practices are often adopted to minimize labor costs even if, for example, it means changing irrigation sets every 12 hours when a 6-hour set would be adequate. Nor will the farmer show much interest in soil water gauging devices or other means to withhold an irrigation that may cost only $5 to $10/acre when he may have already invested $250 or more in the growing crop. If the farmer is to prosper in the face of rising production costs including labor, machinery, and other factors contributing to the so-called "cost-price squeeze," he must reduce operation costs, avoid any serious risks to achieving favorable yield, and thus maximize net returns wherever possible even if it means wasting water by prolonged irrigation sets.

The extension of irrigation intervals does not always save water. Predicting the possibilities for saving water requires analysis of the prevailing conditions including microclimatic factors affecting evaporation and transpiration, the crop (type, stage of growth, coverage of soil surface, root depth, and effect of plant water stress on yield), the soil (profile characteristics, infiltration and internal drainage rates, and water retention capacity), slope and uniformity of grade, the irrigation system (sprinkler or surface and type of surface system—*see* chapter 61), and management practices—particularly the skill of the irrigator. Extending irrigation intervals by using techniques which allow more accurate assessment of actual need for irrigation can reduce the number of irrigations applied in a season substantially —often to less than one-half as many. The resultant savings in water will depend on its effects on reducing evaporation losses from the soil surface, retarding transpiration losses, reducing nitrogen fertilizer requirements, and improving irrigation efficiencies.

Extending the irrigation interval can reduce evaporation from the soil surface, but such extensions will achieve little saving after the soil is fully covered by the crop. Extending the irrigation interval sufficiently to reduce transpiration will usually lead to lower yields for most crops with possible exceptions during the maturation stages of crops grown for dry weight of a reproductive organ or for some chemical constituent. The extent to which irrigation water can be saved by extending the irrigation interval depends largely on opportunities this provides to reduce application losses. These losses depend on root depth of crop, soil factors as listed above, slope and uniformity of grade, irrigation system, and skill of the irrigator. Losses exceeding 50% may occur with each irrigation, especially where roots are shallow, infiltration is rapid, fields are not graded to a desirable or uniform slope, poorly adapted irrigation methods are used, and the irrigator employs little skill. Application losses will also be higher the wetter the soil at time of irrigation because an already relatively wet soil will retain less of the applied water within the root zone. The minimum depth of water which can be applied uniformly even by a skilled irrigator is limited. Thus, irrigating more frequently than required under the conditions prevailing can substantially raise the seasonal irrigation requirement. However, only minimum water savings can be achieved by extending irrigation intervals if the following conditions prevail: crop covers entire soil surface, roots are deep, water retaining capacity of soil is high, and the surface or sprinkler irrigation system used is capable of high application efficiencies under the existing conditions.

In many situations, extending irrigation intervals to the extent permitted by the crop, the irrigation system employed, and the schedule for water deliveries to the farm constitutes a desirable irrigation program not only in terms of water saving but also in labor saving, accommodation of other cultural operations, and sometimes improvement of root growth and reduction in root diseases. The use of automated self-propelled sprinkler machines, which usually can apply only a shallow depth of water at each irrigation and have an application capacity normally matched closely to the expected evapotranspiration rates, requires special consideration. In such cases, the irrigation interval will depend primarily on the daily ET rate and on the quantities of water which can be applied as the equipment passes over the field.

Wise use of the criteria discussed in this chapter for scheduling irrigations in accordance with actual needs can substantially increase crop yield, save water and labor, minimize deterioration of soil structure and loss of soluble plant nutrients, lessen crop protection problems, and facilitate other cultural operations. These considerations, so important in successful irrigation agriculture, indicate the need to give careful attention to achieving optimum irrigation schedules for the conditions prevailing.

LITERATURE CITED

Aldrich, W. W., and R. A. Work. 1934. Evaporating power of the air and top–root ratio in relation to pear fruit enlargement. Amer. Soc. Hort. Sci., Proc. 32:115–123.

Aldrich, W. W., C. L. Crawford, and D. C. Moore. 1946. Leaf elongation and fruit growth of the Deglet Noor date in relation to soil moisture deficiency. J. Agr. Res. 72:189–199.

Alvim, P. de T. 1960. Moisture stress as a requirement for flowering of coffee. Science 132:354.

Alvim, P. de T., and J. R. Havis. 1954. An improved infiltration series for studying stomatal opening as illustrated with coffee. Plant Physiol. 29:97–98.

Anderson, D. B., and T. Kerr. 1943. A note on the growth behavior of cotton bolls. Plant Physiol. 18:261–269.

Anonymous. Dec. 1954. How much and when to irrigate. Farm Manage. p. 32–34.

Babushkin, L. N. 1959. Diagnostics of the requirements of vegetable plants for irrigation based on sap concentration. Soviet Plant Physiol. 6:493–497.

Bartholomew, E. T. 1926. Internal decline of lemons: III. Water deficit in lemon fruits caused by excessive leaf evaporation. Amer. J. Bot. 13:102–117.

Belik, V. F. 1960. Tomato demands in water based on transpiration and concentration of leaf cell sap. Soviet Plant Physiol. 7:73–75.

Belik, V. F. 1962. Growth and development of cucumbers and the concentration of cell sap in their leaves at different soil moisture. Soviet Plant Physiol. 8:393–396.

Bilbro, J. E., W. Clyma, J. S. Newman, M. E. Jensen, and W. H. Sletten. 1960. A preliminary evaluation of the leaf color change method as an indicator of the optimum time to irrigate cotton on the hardlands of the high plains. Texas Agr. Exp. Sta. Progr. Rep. 2159. 7 p.

Bouyoucos, G. J. 1953. More durable plaster of paris moisture blocks. Soil Sci. 76:447–451.

Bouyoucos, G. J. Aug. 1954. Electrical resistance methods as finally perfected for making continuous measurement of soil moisture content under field conditions. Michigan Agr. Exp. Sta. Quart. Bull. 37:132–149.

Burman, R. D., and L. I. Painter. 1964. Influence of soil moisture on leaf color and foliage volume of beans grown under greenhouse conditions. Agron. J. 56:420–423.

Chang, Jen-hu. 1961. Microclimate of sugar cane. Hawaiian Planters' Record. 56:195–225.

Chang, Jen-hu, R. B. Campbell, and F. E. Robinson. 1963. On the relationship between water and sugar cane yield in Hawaii. Agron. J. 55:450–453.

Chunosova, V. N. 1963. Determination of irrigation regime for sugar beets on the basis of cell sap concentration. Soviet Plant Physiol. 10:189–192.

Clements, H. F., and T. Kubota. 1942. Internal moisture relations of sugar cane: The selection of a moisture index. Hawaiian Planters' Record 46:17–35.

Clements, H. F., C. Shigeura, and E. K. Akamine. 1952. Factors affecting the growth of sugar cane. Univ. Hawaii Agr. Exp. Sta. Tech. Bull. 18. 90 p.

Colwell, R. N. 1956. Determining the prevalence of certain cereal crop diseases by means of aerial photography. Hilgardia 26:223–286.

Colwell, R. N. 1961. Some practical applications of multiband spectral reconnaissance. Amer. Sci. 49:9–36.

Cornelison, A. H., and R. P. Humbert. 1960. Irrigation interval control in the Hawaiian Sugar Industry. Hawaiian Planters' Record. 55:331–343.

Davis, R. M., Jr. 1963. The refractometric reading of muskmelon leaf sap in relation to growing conditions. Amer. Soc. Hort. Sci., Proc. 83:599–604.

Dietz, C. F., and L. Verner. 1942. An auxanometer for continuous recording of potato tuber growth. Amer. Soc. Hort. Sci., Proc. 40:509–512.

Ewart, G. Y. 1951. The mechanics of field irrigation scheduling utilizing Bouyoucos blocks. Agr. Eng. 32:148–151, 154.

Filippov, L. A. 1954. The determination of the time for irrigating cotton by means of the suction force of the leaves. (In Russian) lzv. Timiryazev S.-kh. Akad. No. 3: 99–106.

Filippov, L. A. 1959a. The concentration of the cell sap of the leaves as a physiological index of the moisture regime of cotton. Soviet Plant Physiol. 6:82–85.

Filippov, L. A. 1959b. An appraisal of the effect of several factors in the transpiration pull of cotton leaves. Soviet Plant Physiol. 6:489–491.

Filippov, L. A. 1961. Refractometric method of estimating water content of leaves of apple trees. Soviet Plant Physiol. 8:103–105.

Fine, L. O., N. A. Dimick, R. E. Campbell, and H. M. Vance. 1964. Crop production practices for irrigated land. South Dakota Agr. Exp. Sta. Bull. 517. 18 p.

Fritschen, L. J. 1965. Accuracy of evapotranspiration determinations by the Bowen ratio method. Bull. Int. Ass. Sci. Hydrol. 10:38–48.

Fuchs, M., and G. Stanhill. 1963. The use of class A evaporation pan data to estimate irrigation water requirements of the cotton crop. Israel J. Agr. Res. 13:63–78.

Furr, J. R., and E. S. Degman. 1932. Relation of moisture supply to stomatal behavior of the apple. Amer. Soc. Hort. Sci., Proc. 28:547–551.

Furr, J. R., and C. A. Taylor. 1938. The growth of lemon fruits in relation to the moisture content of the soil. US Dep. Agr. Tech. Bull. 640. p. 1–71.

Gates, C. T. 1955. The response of the young tomato plant to a brief period of water shortage: I. The whole plant and its principal parts. Australian J. Biol. Sci. 8:196–214.

Gavande, S. A., and S. A. Taylor. 1967. The influence of soil water potential and atmospheric evaporative demand on transpiration and the energy status of water in plants. Agron. J. 59:4–7.

Gorin, T. I. 1963. Transpiration of fruit trees in summer. (Russian with English summary) Vestn. sel'sk. Nauki 4:114–117.

Gracanin, M. 1963. Critical soil moisture for guttation: A contribution to the ecological value of active water in soil. (In German) Ber. deut. Bot. Ges. 74:465–473.

Hagan, R. M., and J. F. Laborde. 1964. Plants as indicators of need for irrigation. Int. Congr. Soil Sci., Proc. 8th (Bucharest, Rumania) II:399–422.

Hagan, R. M., and Y. Vaadia. 1960. Principles of irrigated cropping. UNESCO Arid Zone Res., Proc. Madrid Symp. p. 215–225.

Haise, H. R., and O. J. Kelley. 1946. Relation of moisture tension to heat transfer and electrical resistance in plaster of paris blocks. Soil Sci. 61:411–422.

Halevy, A. 1960. Diurnal fluctuations in water balance factors of gladiolus leaves. Bull Res. Counc. Israel 8D:239–246.

Hawthorne, L. R. 1951. Studies of soil moisture and spacing for seed crops of carrots and onions. US Dep. Agr. Circ. 892. p. 26.

Hawthorne, L. R., and L. H. Pollard. 1956. Production of lettuce seed as affected by soil moisture and fertility. Utah Agr. Exp. Sta. Bull. 386. p. 23.

Hoover, M., and L. J. Booher. 1952. Guides in cotton irrigation. Univ. of California Agr. Ext. Serv. unnumbered booklet. 20 p.

Howe, O. W., and H. F. Rhoades. April 1961. Irrigation of Great Northern field beans in western Nebraska. Nebraska Agr. Exp. Sta. Bull. 459. 28 p.

Jensen, M. C., J. E. Middleton, and W. O. Pruitt. May 1961. Scheduling irrigation from pan evaporation. Washington Agr. Exp. Sta. Circ. 386. 14 p.

Jensen, M. E., and H. R. Haise. 1963. Estimating evapotranspiration from solar radiation. Amer. Soc. Civil Eng., Proc. 89(IR4):15–41.

Jensen, M. E., and H. R. Haise. 1965. Closing discussion: Estimating evapotranspiration from solar radiation. Amer. Soc. Civil Eng. Proc. Irrig. Drainage Div. 91(IRI):203–205.

Jensen, M. E., and J. T. Musick. 1960. The effects of irrigation treatments on evapotranspiration and production of sorghum and wheat in the southern Great Plains. Int. Congr. Soil Sci., Trans. 7th (Madison, Wis., USA) I:386–393.

Jensen, M. E., and J. T. Musick. June 1962. Irrigating grain sorghums. US Dep. Agr. Leaf. No. 511. 6 p.

Jones, J. N., G. N. Sparrow, and J. D. Miles. 1960. Principles of tobacco irrigation. US Dep. Agr. Inform. Bull. 228. p. 16.

Kolesnikova, P. D. 1957. Problems of cotton and grass physiology. (In Russian) Izd. Akad. Nauk Uzb. SSR, No. 1. (Cited by Krapivina, 1963)

Korven, H. C., and J. C. Wilcox. 1965. Correlation between evaporation from Bellani plates and evapotranspiration from orchards. Can. J. Plant Sci. 45:132–138.

Krapivina, A. T. 1963. Irrigation of cotton according to leaf suction pressure. Soviet Plant Physiol. 10:87–91.

Kreeb, K. 1958. The relation between osmotic value and yield of barley under sprinkler irrigation and surface flooding irrigation. Iraq Coll. Agr. Sci. Bull. 2. 24 p.

Kreeb, K. 1963. Hydrature and plant production. In A. J. Rutter and F. H. Whitehead [ed.] The water relations of plants. Blackwell Sci. Publ., London. 394 p.

Ladin, C. 1959. An irrigation experiment in an apple orchard in the Upper Galilee. Min. Agr., Tel Aviv, Israel. 16 p.

Lang, A. R. G., and H. D. Barrs. 1965. An apparatus for measuring water potentials in the xylem of intact plants. Australian J. Biol. Sci. 18:487–497.

Leverington, K. C. 1960. The effects of saline conditions on the growth of sugar cane. Bur. Sugar Exp. Sta., Brisbane, Queensland. Australian Tech. Comm. 114:21–25.

Litvinov, L. S., and A. G. Gebhardt. 1929. Bleeding of steppe plants. Bull. Inst. Rech. Biol. Univ. Perm. 6:91–111. (In Russian) Biol. Abstr. (1929) 3:17877.

Lobov, M. F. 1951. Irrigation based on OP schedule. (In Russian) Botan. Zhur. SSSR 36:21–28.

Lobov, M. F. 1957. Collection. The biological bases of irrigation agriculture. (In Russian) Izd. Akad Nauk SSSR. (Cited by Chunosova, 1963)

Lomas, J. 1964. A simple method of assessing relative irrigation requirements. Agr. Meteorol. 1:142–147.

McDermott, J. J. 1945. The effect of the moisture content of the soil upon the rate of exudation. Amer. J. Bot. 32:570–574.

Magness, J. R., E. S. Degman, and J. R. Furr. 1935. Soil moisture and irrigation investigations in eastern apple orchards. US Dep. Agr. Tech. Bull. 491. 36 p.

Mallick, A. K., and S. Venkantaraman. 1957. Elongation of sugar cane in relation to soil moisture. Indian J. Sugar Cane Res. 2:15–18.

Marani, A., and M. Horwitz. 1963. Growth and yield of cotton as affected by the time of a single irrigation. Agron. J. 55:219–222.

Marcus, B. P. 1963. The future of spectral filtration photography in agricultural surveys. Mark Systems, Inc. Report. 2999 San Ysidro Way, Santa Clara, California. 18 p.

Marsh, A. W. 1961. Tensiometers: Key to increased profits. Western Grower and Shipper. 32:15–17, 34.

Maximov, N. A., and T. T. Zernova. 1936. Behavior of stomata of irrigated wheat plants. Plant Physiol. 11:651–654.

Mederski, H. J. 1961. Determination of internal water status of plants by beta ray gauging. Soil Sci. 92:143–146.

Meidner, H. 1952. An instrument for the continuous determination of leaf thickness changes in the field. J. Exp. Bot. 3:319–325.

Miller, E. C. 1938. Plant physiology. 2nd ed. McGraw-Hill, New York. 1201 p.

Myers, V. I., and D. L. Carter. 1965. Photogrammetry and remote temperature sensing for measuring soil salinity. Congr. Int. Comm. Irrig. Drainage, Trans. 6th. p. 19.39–19.49.

Nakayama, F. S., and W. L. Ehrler. 1964. Beta ray gauging technique for measuring leaf water content changes and moisture status of plants. Plant Physiol. 39:95–98.

Namken, L. N. 1965. Relative turgidity technique for scheduling cotton (*Gossypium hirsutum*) irrigation. Agron. J. 57:38–41.

Neshina, A. N. 1955. Transaction of the Ak-Kavak, Central Agrotechnical Station. Tashkent. (Cited by Petinov, 1961)

Ophir, M., and J. Putter. 1959. The response of corn and cotton to different soil moisture regimes. Israel Res. Counc. Bull. 7 D(2):108–109.

Oppenheim, J. D., and D. L. Elze. 1937. Determining the date of irrigation by fruit measurements. Agr. Suppl. 21 to Palestine Gaz. No. 719. p. 168–170.

Oppenheimer, H. R., and D. L. Elze. 1941. Irrigation of citrus trees according to physiological indicators. Palestine J. Bot. Rehovot Ser. 4:20–46.

Oppenheimer, H. R., and K. Mendel. 1939. Orange leaf transpiration under orchard conditions. A bioclimatic study: I. Soil moisture high. Palestine J. Bot. Rehovot. Ser. 2:171–250.

Owen, P. C. 1958. The growth of sugar beets under different water regimes. J. Agr. Sci. (London) 51:133–136.

Pelton, W. L. 1964. Evaporation from atmometers and pans. Can. J. Plant Sci. 44:397–404.

Petinov, N. S. 1959. The physiology of irrigated wheat. (In Russian) Izd. Akad. Nauk SSSR. 554 p.

Petinov, N. S. 1961. Physiological principles of raising plants under irrigation agriculture. UNESCO Arid Zone Res. 16:81–92.

Pew, W. D. 1958. Effects of soil moisture on cantaloupe growth and production. Western Grower and Shipper 29:22–24.

Pierce, L. T. 1958. Estimating seasonal and short-term fluctuations in evapotranspiration from meadow crops. Bull. Amer. Meteorol. Soc. 39:73–78.

Pruitt, W. O. 1956. Irrigation scheduling guide. Agr. Eng. 37:180–181.

Pruitt, W. O. April-May 1958. Irrigation timetable. What's New in Crops and Soils 10:11–13.

Pruitt, W. O. 1960. Relation of consumptive use of water to climate. Amer. Soc. Agr. Eng., Trans. 3:9–13, 17.

Pruitt, W. O. 1963. Application of several energy balance and aerodynamic evaporation equations under a wide range of stability. Ch. 4. *In* Investigation of energy and mass transfers near the ground including the influence of the soil-plant-atmosphere system. Report to US Army Electronic Proving Ground, Task 3A99–27–005–08. p. 107–124.

Pruitt, W. O. 1964. Evapotranspiration–a guide to irrigation. California Turfgrass Cult 14:27–32.

Rawlins, S. L. 1966. Theory for thermocouple psychrometers used to measure water potential in soil and plant samples. Agr. Meteorol. 3:293–310.

Richards, S. J., and A .W. Marsh. 1961. Irrigation based on soil suction measurements. Soil Sci. Soc. Amer. Proc. 25:65–69.

Richards, S. J., J. E. Warneke, and T. F. Bingham. 1962. Avocado tree growth response to irrigation. California Avocado Soc. Yearbook. 46:83–87.

Robins, J. S., and C. E. Domingo. 1956. Moisture deficits in relation to the growth and development of dry beans. Agron. J. 48:67–70.

Robinson, F. E. 1963. Soil moisture tension, sugar cane stalk elongation, and irrigation interval control. Agron. J. 55:481–484.

Robinson, F. E., R. B. Campbell, and J. H. Chang. 1963. Assessing the utility of pan evaporation for controlling irrigation of sugar cane in Hawaii. Agron. J. 55:444–446.

Rodionov, V. S. 1962. Diagnosing the dates, and the possibility of determining the rates of irrigation from an index of the concentration of the cell sap of maize. Soviet Plant Physiol. 9:91–97.

Rutter, A. J., and K. Sands. 1958. The relation of leaf water deficits to soil moisture tension in *Pinus sylvestris*: I. The effects of soil moisture on diurnal changes in water balance. New Phytol. 57:50–65.

Schmueli, E. 1953. Irrigation studies in the Jordan Valley: I. Physiological activity of banana in relation to soil moisture. Bull. Res. Counc. Israel 3:228–247.

Shardakov, V. S. 1957. Principle for determining the watering periods of the cotton plant in relation to the magnitude of the suction force of leaves (from Russian) *In* Physiological Questions in Cotton and Grasses. Akad. Nauk Uzbek SSR (Tashkent) 1:5–32.

Shaw, R. H. 1964. Prediction of soil moisture under meadow. Agron. J. 56:320–324.

Slavik, B. 1959. Gradients of osmotic pressure of cell sap in the area of one leaf blade. Biol. Plant. 1:39–47.

Stammers, W. N. Dec. 1963. Investigations on the improvement of irrigation practices in the Umatilla Irrigation Project. Oregon Agr. Expt. Sta. Spec. Rep. 166. 33 p.

Stanhill, G. 1957. The effect of differences in soil moisture status on plant growth: A review and analysis of soil moisture regime experiments. Soil Sci. 84:205–214.

Stockton, J. R. 1961. Irrigating for profits. Cotton Trade J., 28th Int. Yearbook. p. 64–65.

Stockton, J. R., and L. D. Doneen. 1957. Factors in cotton irrigation. California Agr. 11:16–17, 25.

Stockton, J. R., L. D. Doneen, and V. T. Walhood. 1961. Boll shedding and growth of the cotton plant in relation to irrigation frequency. Agron. J. 53:272–275.

Stockton, J. R., L. D. Doneen, V. T. Walhood, and B. Counts. 1955. Effects of irrigation on the growth and yield of cotton. California Agr. 9:8–11.

Stolzy, L. H., O. C. Taylor, M. J. Garber, and P. B. Lombard. 1963. Previous treatments as factors in subsequent irrigation level studies in orange production. Amer. Hort. Soc., Proc. 82:199–203.

Tanimoto, T. 1961. 4-5 joint as indicators of moisture tension of the sugar cane plant. Hawaiian Sugar Tech. 1961 Rep. p. 265–274.

Tanner, C. B. 1960. Energy balance approach to evapotranspiration from crops. Soil Sci. Soc. Amer. Proc. 24:1–9.

Tanner, C. B. 1963. Plant temperatures. Agron. J. 55:210–211.

Taylor, S. A. 1952. Use of mean soil moisture tension to evaluate the effect of soil moisture on crop yields. Soil Sci. 74:217–226.

Taylor, S. A. 1965. Managing irrigation water on the farm. Amer. Soc. Agr. Eng., Trans. 8:433–436.

Taylor, S. A., D. D. Evans, and W. D. Kemper. June 1961. Evaluating soil water. Utah Agr. Exp. Sta. Bull. 426. 67 p.

Taylor, S. A., J. L. Haddock, and M. W. Petersen. 1959. Alfalfa irrigation for maximum seed production. Agron. J. 51:357–360.

Tukey, L. D. 1963. Electronics record fruit growth. Science for the Farmer 11:3.

UNESCO. 1960. Plant water relations in arid and semiarid conditions. UNESCO Arid Zone Res. 15, Review Res. 225 p.

UNESCO. 1961. Plant water relationships in arid and semiarid conditions. UNESCO Arid Zone Res. 16, Proc. Madrid Symp. 352 p.

UNESCO. 1962. The problems of the arid zone. UNESCO Arid Zone Res. 18, Proc. Paris Symp. 481 p.

Uriu, K., L. Werenfels, G. Post, A. Retan, and D. Fox. 1964. Irrigation of peaches. California Agr. 18:10–11.

Vaadia, Y., and A. N. Kasimatis. 1961. Vineyard irrigation trials. Amer. J. Enol. Viticult 12:88–98.

Verner, L. 1962. Scheduling irrigations by trunk growth. Western Fruit Grower. 16(4):21.

Verner, L., W. J. Kochan, D. O. Ketchie, A. Kamal, R. W. Braun, J. W. Perry, Jr., and M. E. Johnson. 1962. Trunk growth as a guide in orchard irrigation. Idaho Agr. Exp. Sta. Res. Bull. 52. 32 p.

Vissar, W. C. 1959. Crop growth and availability of moisture. Inst. Land Water Manage. Res. (Wageningen) Tech. Bull. 6. p. 16.

Vittum, M. T., R. B. Alderfer, B. E. Janes, C. W. Reynolds, and R. A. Struchtemeyer. 1963. Soil-plant-water relations as a basis for irrigation. Crop response to irrigation in the Northeast Regional Research Publication, New York Agr. Exp. Sta. (Geneva) Bull. 800. p. 66.

Vittum, M. T., W. T. Tapley, and N. H. Peck. 1958. Response of tomato varieties to irrigation and fertility level. New York Agr. Exp. Sta. (Geneva) Bull. 782. p. 78.

Whiteman, P. C., and G. L. Wilson. 1963. Estimation of diffusion pressure deficits by correlation with relative turgidity and beta radiation absorption. Australian J. Biol. Sci. 16:140–146.

Wolfe, J. W., and D. D. Evans. April 1964. Development of a direct reading evaporation pan for scheduling pasture irrigations. Oregon Agr. Exp. Sta. Tech Bull. 75. 31 p.

section **X**

Irrigation of Principal Crops

X

31

Forage Crops

WESLEY KELLER

Agricultural Research Service, USDA
Logan, Utah

CARL W. CARLSON

Agricultural Research Service, USDA
Beltsville, Maryland

I. INTRODUCTION [1]

A. Importance

Forages are the principal as well as cheapest feed for cattle, sheep and big game. In the USA alfalfa is the most important among sown forages. It is also the most important forage grown under irrigation. According to the 1959 Census of Agriculture, there are nearly 5 million acres of irrigated alfalfa in the Western USA. In addition, over 250,000 irrigated acres produce alfalfa seed. In the same area all other irrigated hay, including wild hay (1,318,000 acres), clover (*Trifolium* sp.) and timothy (*Phleum pratense*) hay (875,000), small grains cut for hay (241,000), and 166,000 miscellaneous, totals 2,600,000 acres. Thus, under irrigation, alfalfa for hay occupies nearly twice as many acres as all other hay crops combined. Besides, irrigated alfalfa is usually sown on highly productive cropland and yields average over 3 tons/acre. In contrast, other forages yield an average of only 1.27 tons/acre, because they occupy more land which is unsuited to cultivated crop production. This means that irrigated alfalfa produces nearly 83.3% of all harvested irrigated hay in the Western USA.

In the same area slightly more than 5 million acres are devoted to irrigated pastures. This represents a heterogeneous and poorly defined segment of our agriculture. Irrigated pastures embrace a large number of species and combination of species. These pastures are grown on a wide range of soils and are subject to many different management practices. Irrigated pastures are the only means, on many farms, of utilizing nonproductive land. It is only in comparatively recent years that pastures have begun to be sown on arable lands.

Over 50 years ago Harris (1913) advocated pasture on high-priced irrigated land in Utah, USA, and suggested rotation grazing and other good management practices. Before his concepts could be implemented, nearly 30 years later, it was necessary to demonstrate the effectiveness of rotation grazing and to find mixtures capable of high production when adequately fertilized and correctly managed. This was first accomplished by Bateman and associates in the Intermountain Region, and later in Northwest USA by Heinemann and Van Keuren (1955). As a result, pastures have been sown on some of the most productive irrigated land. Under proper fertilization and management, they have proven their ability to compete with many of the established cash crops.

---

[1] Botanical names will appear in the text only for species not listed in Table 31–1.

Early work of Bateman (Bateman and Keller, 1952) on the Utah Agricultural Experiment Station Dairy Farm, found the generally accepted old pasture mixture (dominated by Kentucky bluegrass (*Poa pratensis* L.) and white clover) capable of yielding slightly less than 3,000 lb total digestible nutrients (TDN)/acre. Fertilized with manure and phosphate, this mixture produced more than 4,000 lb. Newer mixtures, dominated by alfalfa and orchardgrass, and also containing ladino clover and bromegrass, similarly fertilized, produced nearly 5,500 lb/acre. As a direct comparison (Bateman 1958), corn for silage yielded slightly more than 5,500 lb TDN; alfalfa for hay slightly over 4,500 lb; and barley (*Hordeum vulgare*) for grain (85 bushels) a little over 3,000 lb/acre. Since pastures need not be sown every year, and since they are harvested by the grazing animal, and confer a benefit on crops that follow them, their value is not fully expressed in comparative yield of TDN.

In the arid Western USA, irrigated pastures based on new, improved mixtures are clearly in a strong competitive position with the commonly grown cash crops. Harris et al. (1958) reported a net production of over 1,000 lb beef/acre from a 22-acre pasture. Heinemann and Van Keuren (1955, 1956) reported gains of more than 1,000 lb for both beef and sheep from irrigated pastures. Trew and Hoveland (1955) in south Texas, USA, and Staten et al. (1951) in southern New Mexico, USA, both recognized that 1,000 lb or more of beef gain per acre or the equivalent in other animal products was not a difficult goal.

Corn for forage (in 1959) was grown in the Western USA under irrigation on 494,000 acres—Colorado and California together accounting for nearly one-half the total. Sorghum for forage was grown on 185,000 acres, Texas and Arizona together accounting for over one-half the total. Between 1949 and 1959 the corn-for-forage acreage nearly tripled. Sorghum for forage has become increasingly important.

Studies of rainfall patterns in the humid Eastern USA have revealed that summer drouths are not uncommon in areas where annual precipitation may be 40 to 50 inches or more. Irrigation is increasing there, and forage crops have generally given a favorable response, particularly during periods of drouth and when fertilizers are used. The provision to irrigate on any particular farm in the humid Eastern USA is largely an economic problem beyond the scope of this chapter.

B.   The Major Species

Some of the more important forage species grown under irrigation along with reference to areas of adaptation and the more important characteristics are presented in Table 31–1.

II.   IRRIGATING PERENNIAL FORAGE CROPS

A.   Establishment

The perennial grasses and legumes most commonly used in forage production, species with small seeds, require careful seedbed preparation. Weihing et al. (1943) point out that the most common cause of failure to establish perennial forage crops is a deficiency of soil water frequently caused by a loose seedbed.

Table 31–1. The major forage species grown under irrigation and some of their characteristics

Forage crop	Botanical names	Growth habit	Length of life	Salt tolerance*	Rooting habit	Water needs	Other characteristics
				Northern and high altitude			
Alfalfa	Medicago sativa L.	Nonspread.†	Per.	Mod.	V. deep	Mod.	Superior for hay or pasture; danger of bloat
Ladino clover	Trifolium repens L.	Stolon	Per.	V. low	V. shallow	High	Moderate cold hardiness; danger of bloat
Red clover	Trifolium pratense L.	Nonspread.	Sh.Per.	V. low	Mod.	Fairly h.	Hay, mountain meadows; short-lived pasture; danger of bloat
Alsike clover	Trifolium hybridum L.	Nonspread.	Sh.Per.	V. low	Shallow	High	Mountain meadows; tolerates poorly drained land
Strawberry clover	Trifolium fragiferum L.	Stolon	Per.	Mod.	Shallow	High	Pastures on poorly drained valley bottoms
Narrow leaved trefoil	Lotus tenuis Waldst. et Kit.	Nonspread.	Per.	Good	Shallow	High	Slow to establish; does not cause bloat
Birdsfoot trefoil	Lotus corniculatus L.	Nonspread.	Per.	Mod.	Mod.	Mod.	Slow to establish; does not cause bloat
White clover	Trifolium repens L.	Stolon	Per.	Low	V. shallow	High	In old pastures; tolerates continuous close grazing
Smooth brome	Bromus inermis Leyss.	Rhiz.	Per.	Mod-low	Deep	Mod.	Good drouth and cold hardiness; highly palatable
Orchardgrass	Dactylis glomerata L.	Bunch	Per.	Mod-low	Mod.	Fairly h.	Moderate cold hardiness; rapid recovery; moderate palatability
Tall fescue	Festuca arundinacea Schreb.	Bunch	Per.	Good	Deep	Mod.	Low palatability; widely adapted; vigorous
Tall wheat grass	Agropyron elongatum(Host)Beauv.	Bunch	Per.	V. good	Deep	Fairly l.	Low palatability; used in saline land reclamation; vigorous
Perennial ryegrass	Lolium perenne L.	Bunch	Sh.Per.	Mod.	Mod.	Fairly h.	Pastures of central valley of California
Reed canary	Phalaris arundinacea L.	Rhiz.	Per.	Mod-low	Fairly d.	Fairly h.	Difficult and slow to establish; then fairly drouth resistant
Tall oatgrass	Arrhenatherum elatius(L.)Presl.	Bunch	Sh.Per.	Mod-low	Mod.	Mod.	Balances red clover in pasture mixtures.
Sorghum	Sorghum bicolor (L.) Moench.	Nonspread.	Annual	Mod.	Fairly d.	Low	Throughout great plains where too dry for corn
Sudangrass	Sorghum sudanense (Piper) Stapf.	Nonspread.	Annual	Mod.	Fairly d.	Low	Annual or emergency pasture or hay; needs warm weather
Corn	Zea mays L.	Nonspread.	Annual	Mod-low	Fairly d.	Mod.	Best silage crop where water is adequate
				Southern and low altitude			
Alfalfa	Medicago sativa L.	Nonspread.	Per.	Mod.	V. deep	Mod.	Nonhardy forms highly productive
Ladino clover	Trifolium repens L.	Stolon	Per.	V. low	V. shallow	V. high	Irrigation demands very high for general use
Bermuda‡	Cynodon spp.	Rhiz.+Stolon	Per.	V. good	Deep	Mod.	Vegetative reproduction; vigorous summer grower
Blue panic	Panicum antidotale Retz.	Short rhiz.	Per.	Mod.§	Deep	Mod.	Utilize while young
Dallisgrass	Paspalum dilatatum Poir.	Bunch	Per.	Mod.	Deep	High	Slow to establish; resists moderate frost
Pangolagrass	Digitaria decumbens Stent.	Stolon	Per.	--	Deep	Mod.	Subtropical regions only, and low altitudes
Sorghum¶	Sorghum bicolor (L.) Moench.	Nonspread.	Annual	Mod.	Fairly d.	Low	Throughout great plains where too dry for corn
Sudangrass¶	Sorghum sudanense(Piper) Stapf.	Nonspread.	Annual	Mod.	Fairly d.	Low	Annual or emergency pasture or hay
Pearl millet	Pennisetum typhoides (Burm.) Stapf. & Hubb	Nonspread.	Annual	--	Fairly d.	Low	Productive on sandy soils
Corn	Zea mays L.	Nonspread.	Annual	Mod-low	Fairly d.	Mod.	Best silage crop where water is adequate

* Bernstein (1958), except as noted. † Strains or varieties differ considerably. ‡ Improved strains. § Gausman et al. (1954) ¶ Also hybrids between sorghum, sudangrass, and johnsongrass. All may cause prussic acid poisoning.

If the surface soil is dry, early irrigation may be essential to establish a stand. Irrigation is never a substitute for good agronomic practices, however. Bateman and Keller (1956) have listed the following requirements for successful establishment of small-seeded forage crops under irrigation: (i) a firm seedbed to bring the small seeds in close contact with the soil; (ii) a reasonably weed-free seedbed to prevent competition for soil water; (iii) a high level of fertility to promote rapid growth; (iv) accurate coverage of the seed, usually 0.5 inch or less (but deeper on sandy soils) to enable seedlings to emerge; and (v) irrigation in correct amounts to meet the soil water needs of the forage crop. These practices are all interrelated and influence the effectiveness of irrigation.

If surface irrigation methods are to be used, the land should be leveled before seeding. Borders at frequent intervals should be installed to obtain good water control. It is impossible to do any land leveling or border installation after the crop is seeded without destroying the stand.

Generally, the best time to seed a forage crop is in the spring when soil water conditions are good. Tysdal and Westover (1949) advocate irrigation before seeding, if the soil water is not adequate to carry alfalfa to the 3- or 4-leaf stage.

Frequently, it is necessary to irrigate to promote germination or seedling development. If the equipment is available, it is preferable to apply this water with a sprinkler system. Great care must be exercised, if the surface irrigation methods are used. The stream size must be adjusted to prevent erosion. If erosion occurs, some seeds will be covered too deeply and some may be washed away or uncovered.

Land from which grain or peas (*Pisum sativum*) have been harvested makes an ideal seedbed for forages. When provision is made for the irrigation of the initial crop, it also allows for irrigation of the forage crop. Irrigation applications are usually necessary when establishing a pasture in midsummer or later.

When growing a perennial crop, shortcomings in the control of irrigation water are compounded by the number of years the crop is maintained. Except where sprinkler irrigation is used, every effort should be made in advance of planting to prepare the land for excellent water control.

1. COMPANION CROPS

The use of a companion crop, with plantings for either pasture or hay, generally requires particular attention to irrigation needs. It is a highly controversial practice. A vigorous companion crop can aid the emergence of small forage seedlings on soils subject to crusting and can protect seedlings where wind would cause soil blowing. A companion crop may be beneficial where a heavy growth of weeds is sure to develop. The effective use of corn as a companion crop for alfalfa has been reported in Nebraska, USA by Harris (1962) both on sloping land and on level bench terraces. Harris recognized that irrigation management would determine crop success. Madsen (1953) did not object to a companion crop of grain, cut for hay. Norton (1931) recommended use of a companion crop to avoid loss of a crop year, but concluded that the first year's production of alfalfa would be reduced by 0.5 ton. Carlson et al. (1961) reported alfalfa stands to be less vigorous with a companion crop of barley. However, when small amounts of nitrogen (N) fertilizer were applied to the barley, satisfactory stands of alfalfa were established. When as much as 60 lb N/acre were applied, excessive barley growth damaged the alfalfa because of increased competition for light and soil water.

Many farmers understand the needs of both the companion crop and the forage crop and have the ability and resources to meet those needs. A companion crop of barley taken through to grain can then become a highly profitable use of the land while establishing a pasture (Bateman and Keller, 1956). Bateman has also repeatedly used a companion crop to advantage in establishing alfalfa.

B. Irrigating the Established Crop

1. GENERAL RECOMMENDATIONS

Effective practices for irrigation of alfalfa are well known because of the important position the crop has long occupied in the agricultural economy of the Western USA. Alfalfa is a very deep-rooted crop, 8 to 12 feet on deep soils, and irrigation should be adequate to charge the full root zone. The amount of water needed for this depends on the depth of rooting and the water-holding capacity of the soil.

Blaney and Criddle (1962, Table 15) report 14 studies on the "measured seasonal consumptive use of water" by alfalfa and the length of season associated with each. These data provided the information for the construction of Fig. 31–1. A very close relationship is shown between the number of days alfalfa grows and its water need. The relationship is in general agreement with practice in growing alfalfa under irrigation in the arid Western USA as indicated by numerous studies

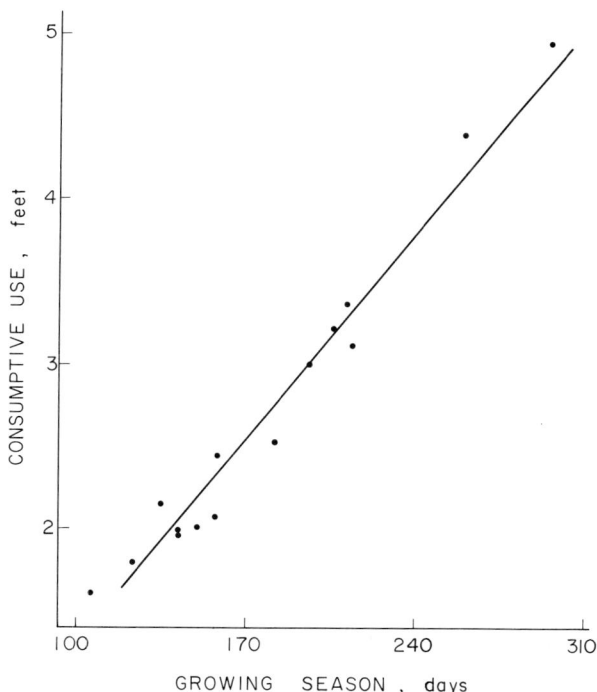

Fig. 31–1. The relationship between length of growing season in days and irrigation water needs of alfalfa at 13 widely scattered locations in the Western USA (taken from Blaney and Criddle, 1962, Table 15). One value was adjusted to compensate for a rest period.

and with calculated requirements based on consumptive use, radiant energy
(Jensen and Haise, 1963), and pan evaporation (Penman, 1948) concepts.
Exceptions have been noted, however, e.g., the data of Tovey (1963), where
consumptive use was much higher than would be indicated by Fig. 31–1. However, advection from drier areas surrounding the experimental site and uncontrolled
water table outside of the lysimeters may have caused higher water use rates
than found in large fields.

Fifty years ago in Arizona, Freeman (1914) recommended two irrigations per
crop during the growing season. In the warmer climates of the West, this is generally considered good practice. The second irrigation should be so timed that at
harvest the ground will be firm and the surface not wet enough to delay drying
of the mown hay. If the soil has become quite dry, irrigating several days before
harvest may get the next crop started more quickly rather than delaying irrigation
until the current crop has been removed. A better job of irrigation can be done,
immediately following removal of the crop, on land difficult to irrigate.

Stanberry (1954) recommended an irrigation schedule that would maintain
between 35% and 85% of readily available soil water in the top 4 ft of soil.

Campbell et al. (1960) found in field studies that alfalfa could draw water

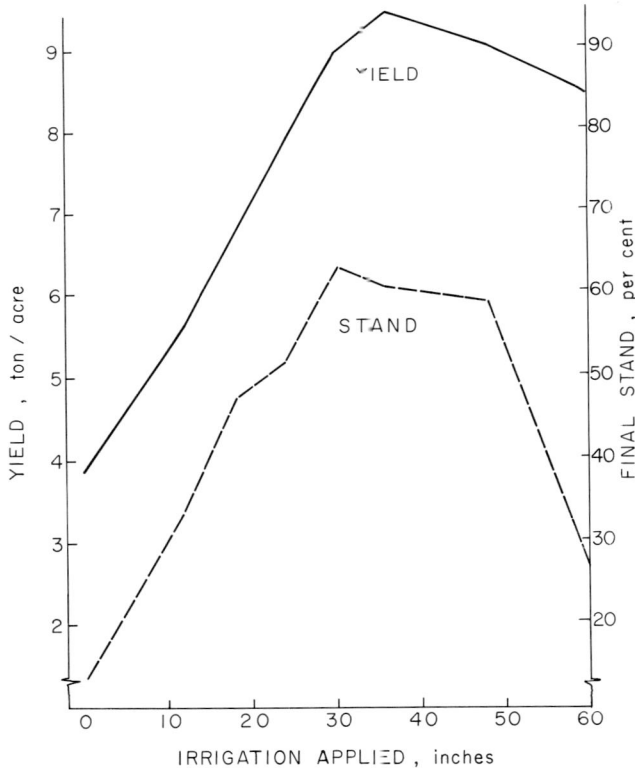

Fig. 31–2. Six-year average yield, and final stand, of alfalfa at Davis, California, in response to different amounts of annual irrigation. (Perfect stand contained 1.5 plants/
ft².) (Taken from Beckett and Robertson, 1917.) The growing season is 242 days.

from a water table at considerable depths. When the water table was 5 to 9 ft below the soil surface, the crop showed little response to irrigation. Carlson et al. (1961) reported that the roots of alfalfa, penetrating into the capillary fringe above a water table, can supply the water needs of the crop even when the top 2 ft of soil are very dry. In their study, the water table was at 7 ft, and the capillary fringe extended to within 3 ft of the surface.

Tovey (1963) from a lysimeter study at Reno, Nevada, USA obtained data indicating the importance of a water table in the production of alfalfa. With a water table at 8 ft and 2/3 of the available moisture extracted, supplemental irrigation increased yield 13% or less than 1 ton/acre. The yield of nonirrigated alfalfa with a water table at 8 ft, was equal to that of irrigated alfalfa in the absence of a water table.

Pasture species are not as deeply rooted as alfalfa and require more frequent irrigation. Peterson et al. (1959) believed a common fault is the failure to irrigate pastures early enough. The irrigation schedule must meet the water needs of the most shallow-rooted component of the mixture (usually clover) and avoid periods where cattle are on the pasture or just prior to being placed thereon. They point out that livestock grazing upon pastures still wet from irrigation compacts the soil and reduces the rate of water intake.

McCormick and Myers (1958) reported that ladino and strawberry clover obtained nearly all their water from the top 1 ft of soil and that frequent light applications of water would be required to keep them productive. They questioned the advisability of planting shallow-rooted species on soils that are capable of growing deep-rooted species.

In the Columbia River Basin, Nelson and Robins (1956) studied an orchard-grass-ladino clover pasture. The forage was harvested each time it had grown to a height of 6 or 12 inches. The highest yields were obtained on the treatment irrigated when 30 to 35% of the available soil water had been extracted from the root zone and when growth had reached 12 inches.

C.   Effects of Irrigation

1.   YIELD

All forage species capable of high production respond to irrigation in the arid Western USA.

Beckett and Robertson (1917) reported the influence of irrigation on yields and ultimate stands of alfalfa at Davis, California. Their data, summarizing 6 years of research, have been utilized to construct Fig. 31–2. On the basis of both yields and plant survival, they recommended applying 30 inches/year to alfalfa. Application of 30 inches of water/season increased yield 5 tons/acre and maintained a good stand. Their data probably adequately represent that part of the arid West having a long growing season and a hot, dry, climate. The frost-free growing season at Davis is 8 months (US Dep. Agr. Yearbook of Agriculture, 1941).

In the mountain valleys of the Western USA, at higher elevations, the response to irrigation is far less striking. Harris and Pittman (1921), summarizing three separate studies covering a period of 12 years, reported that alfalfa yield increased 1.3 tons/acre in response to 15 inches of water. In their studies alfalfa yielded 3.2 tons/acre without irrigation and 4.5 tons/acre when given either 3 irrigations of 5 inches each or 5 irrigations of 3 inches each. The Utah studies were conducted on a very deep loam soil having a high water-holding capacity. Natural precipitation is about 16 inches/year.

Masefield (1961) reported that irrigation of legumes promoted nodulation.

2. WATER USE EFFICIENCY

Schofield (1952) reported a study by Penman in which fertilization of grassland increased yield threefold without any change in the water required. He concluded that the amount of water that could be lost from a growing crop, assuming the soil is kept moist and completely covered by the crop, was almost entirely dependent on the radiant energy available. He showed that there is a close relation between loss of water from a free water surface and from a sward given adequate water. Evaporation from a free water surface is thus recognized as an indicator of crop needs. Tovey (1963) presents data for 28 periods of time in which radiant energy and pan evaporation of water yielded a correlation of +.999. Although wind movement might somewhat lower the relationship, radiant energy and pan evaporation would appear to be about equal in value for estimating water needs.

The above observations point out that after supplying enough water to keep the soil moist, anything that increases yield will increase water use efficiency. Stanberry et al. (1955) found that alfalfa yields were increased 83% on Superstition sand where applications of phosphorus were increased from 44 to 264 lb/acre of P (100 to 600 lb P_2O_5) and water applied increased from 72 inches ("dry" treatment) to 88 inches ("wet" treatment). Fertility alone increased yields 36 and 42% on the "dry" and "wet" treatments, respectively. Water alone gave increases of 29 and 35% for corresponding applications of 44 and 264 lb/acre of P. Averaging three soil water levels, 264 lb/acre of P (600 lb P_2O_5) increased yields 40% (over 44-lb P/acre rate) and decreased water needed per ton of hay produced by 42%. The relationship between yield, fertilizer, and water use efficiency, has been reviewed by Viets (1962).

The texture of the soil apparently has little influence on the water-use efficiency values except in areas where the static water tables are near the ground surface. Briggs and Shantz (1913) concluded from a comprehensive literature review that " . . . there is no indication that the texture of the soil, independently of the plant food it contains, affects the water requirement of plants." Tovey (1963) reported that alfalfa grown in lysimeters yielded 9% more on a loam than on a sandy loam and 14% more than on a clay loam.

3. CROP QUALITY

Numerous studies have reported the chemical composition of forage crops subjected to different fertilizer and cultural treatments. In general they indicate that when soil water is adequate, the more important nutrients in the forage reflect the nutrient providing capacity of the soil. But the natural field situation is very complex because growth and chemical composition are the result of interactions of many factors.

Brezeau and Sommor (1964) studied the nutritive value of alfalfa as influenced by three irrigation treatments where available soil water was depleted 25%, 50%, and 75% before recharging the root zone. The 50% level of depleted moisture yielded 19.8% more protein and 32.8% more digestible cellulose than the wet

level, and 4.8 and 8.3%, respectively, more than the dry level. The same treatment gave a mean nutritive value index 4.2% above the other levels.

Ulrich (1956) has demonstrated that the PO_4 content of alfalfa mid-stems is
a reliable indicator of available phosphorus supplies. In nutrient solutions, with
all other requirements satisfied, the per cent PO_4 in the plant was essentially
unchanged while yield increased 600% (near the maximum of which the plant
was capable) in response to increments of phosphorus. At this point, the PO_4
content was approximately 500 ppm. Further applications of phosphorus had no
further effect on yield, but the PO_4 content of the mid-stems increased to 4,500
ppm. Hylton et al. (1964) have shown that grasses give a similar response to
nitrogen.

4. LONGEVITY OF STANDS

The influence of irrigation on longevity depends not only on species and environment, but on management practices. Frequent close cutting or continuous close
grazing will eliminate nearly all the highly productive forage species, regardless
of irrigation or fertility practices. On deep, well-drained soil, alfalfa responds tc
liberal irrigation, but throughout much of the Western USA, increasing rates of
irrigation may be associated with the increasing invasion of bluegrass (Robertsor
et al. 1958) which hastens deterioration of the stand. Beckett and Robertson
(1917) determined the stand of alfalfa after 6 years of differential irrigation
(Fig. 31–2). Water applications below 30 inches and above 48 inches caused
stand deterioration. Weihing et al. (1943) reported that alfalfa winterkill is
more widespread in dry than in moist soil in Colorado.

The effects of improper irrigation are most pronounced in pasture mixtures
where shallow-rooted clovers may be lost if irrigations are too widely spaced.
Rouse et al. (1955) report that legumes are unable to persist in mountain
meadows where watering is excessive or where water tends to remain in low
places. When water is properly provided, fertilizers play a particularly important
role in the survival of the grass and legume components of pastures. The earliest
investigators learned that legumes were favored by phosphorus, at the expense
of grasses, while grasses were favored by nitrogen, at the expense of legumes.

Several diseases shorten the life of alfalfa stands. Irrigation water is probably
the most important factor in the spread of bacterial wilt. The alfalfa stem nematode is also spread long distances by water as well as in hay. Weimer (1933)
found that frequent irrigation intensified the effects of the dwarf disease of alfalfa,
and Stanford et al. (1954) reported that Rhizoctonia root canker is favored by
a high water table which can be a direct effect of over-irrigation. Richards (1929)
produced white-spot, a physiological disease of alfalfa, at will by applying heavy
irrigations to plants under drouth stress.

5. SEED PRODUCTION

For seed production most forage crops are sown in widely spaced rows. Otherwise, irrigation needs for establishment are the same as for hay or pasture.

Taylor et al. (1959) obtained maximum alfalfa seed yields by maintaining an
annual integrated soil water mean between 2 and 8 bars. Adequate supply of
soil water is needed from initiation of growth until blossoming begins, then water
is withheld. The soil should reach the 15-bar suction value about the time of seed
harvest. Pedersen et al. (1959) found that sprinkler irrigation during blooming

reduced seed yields. Under the conditions of this study, irrigation can be discontinued, if there are 15 inches of available soil water in the root zone when flowering begins. Pedersen and McAllister (1955) advocate keeping adequate soil water during vegetative growth and suggest that one heavy irrigation at bud stage will carry the crop through to harvest. If further irrigation is necessary, only light applications are made.

Miller et al. (1951) recommend frequent light irrigations for ladino clover seed production. The "cupping" together of leaves indicate the need for water. The soil should be kept moist but well drained.

According to Oman and Stark (1951), grass seed production requires two to four irrigations. They stress uniform application of water to promote uniform development and ripening of the seed crop and early irrigations to promote continuous growth.

Sumner et al. (1960) advise that in California, grass seed yields are highest and seed heaviest, if the plants are never subjected to soil water stress. Most of the roots will be in the top 2 ft of soil; this zone should be kept above the wilting point.

Holt and Bashaw (1963) found that supplemental irrigation was desirable when growing seed of dallisgrass, even in areas where natural precipitation was as high as 35 inches/year.

In studies at Woodward, Oklahoma, USA, Kneebone and Greve (1961) considered irrigation for seed production of blue grama grass (*Bouteloua gracilis* (H.B.K.) Lag. ex Stend.) a profitable operation because of the additional ton of hay produced per acre. The field should be kept close cropped until the middle of August when 50 lb N/acre are applied and sprinkler irrigation is begun and continued as needed through September. The resulting seed crop over a 6-year period averaged $76/acre per year. The ton of hay/acre offset irrigation costs.

III.   IRRIGATING ANNUAL FORAGE CROPS

Irrigation is not normally required to establish silage or annual hay or pasture crops. They are large seeded and can be planted deeply. However, annual forage crops are commonly used as emergency crops to occupy the land following the failure of some earlier seeding. The soil water supply may then already be depleted, and irrigation will be necessary for germination and emergence. Sprinkler irrigation reduces the threat of erosion to a minimum and is particularly ideal for this purpose. The farmer may, depending on water supplies, available equipment, soil texture, lay of land, etc., prefer to irrigate the land prior to seeding.

For the coastal areas of southern California, Jones et al. (1957) recommend irrigation of sudangrass every 3 to 4 weeks at 5 to 6 inches/application. On sandy soils or soils with slowly permeable horizons, lighter applications may be required weekly. The development of root systems of annuals is progressive. Annuals should be regarded as shallow-rooted crops until they reach the flowering stage. This means that irrigation should be more frequent than required by an established alfalfa stand on the same soil. Van Keuren and Heinemann (1959) report that sudangrass and pearl or foxtail millet (*Setaria italica* (L.) Beauv.) can be sown as late as August 1 under irrigation in central Washington, USA, although yields decrease as the planting date is delayed. Sudangrass was generally more productive than the millets.

Rumery and Ramig (1962), from studies in Nebraska, reported that iron chlorosis in sudangrass could be avoided by allowing the plants to wilt before irrigating. They found if the soil profile contained 8 or 9 inches of water before planting sudangrass, little water was required thereafter. In dry seasons they applied one 6-inch irrigation in June and one in August. With normal or better precipitation, the irrigation in June was not needed.

Robins and Rhodes (1958) recommend a preplanting irrigation of cornland. This fills the storage capacity of the soil and assures continuous growth of the young crop. They emphasize the importance of a continuous supply of water for high yields of corn. The tasselling-to-silking period is particularly critical. Shank and Kratochvil (1954) reported that irrigation hastened development of corn. This permitted use of hybrids that required more time (and that presumably were more productive) than those adapted to dryland in the same area. McMaster et al. (1962) in Idaho, studied corn growth for silage under 4 moisture and 12 fertilizer treatments. Silage yields on plots irrigated when 70% of the water was depleted were 1 ton/acre lower than on plots irrigated when 35 to 50% of the water was depleted. Yields on treatments irrigated up to the tasselling stage when 60% of the water was depleted and thereafter when 30% was depleted were 1.3 to 2.0 tons/acre more than the treatment irrigated throughout the season when 60% of the water was depleted. Silage yields (76% water) on all treatments were greater than 20 tons/acre. Nitrogen fertilizer was most effective in increasing yield at the higher soil water level employed.

According to Klages (1944) forage sorghums, used as emergency crops in southern Idaho, will produce forage on less water than is required by corn. He advises against overirrigation which cools the soil, and against irrigation after seeds are formed; and he points out that high yields are especially dependent on a uniform soil water supply during flowering.

On the basis of research in Texas, Moore et al. (1962) have indicated the likely production of sorghum silage on hardland soils of the high plains. Assuming four levels of irrigation using 5, 9, 13 or 17 inches of irrigation water and four levels of fertilization (30, 60, and 100 lb N, and 160 lb N + 17.5 lb P/acre, respectively) they predict that yields would be 10, 15, 20 or 27 tons of silage/acre, respectively. The four irrigation levels consisted of one preplanting application of 5 inches and zero, one, two, or three post-planting applications of 4 inches each, respectively.

IV. FORAGE PRODUCTION ON SALINE SOIL

Salinity is a widespread problem in arid regions. Comprehensive treatment has been given this subject. (Magistad, 1945: Hayward and Wadleigh, 1949; Richards, 1954; Bernstein and Hayward, 1958; Allison, 1964.) The problem is particularly acute in areas where irrigation water is obtained by pumping and the water is highly saline (Magee et al. 1959). A number of important forage species possess considerable salt tolerance (Table 31–1). Where salinity is associated with inadequate drainage, forage plants may be the only species capable of utilizing the land, and reclamation of the land may be difficult or impossible. Where drainage is adequate, the salt content of the soil can be lowered in the root zone of plants by periodic leaching. Nielsen et al. (1964) have shown that on Panoche clay loam, chloride in the soil was more effectively leached downward by intermittent than by continuous flooding and with less water. Campbell et al. (1960) pointed out that when alfalfa obtained water from a water table there was danger of salt accumulation within the root zone of the crop. To prevent this, they recommended a minimum of two irrigations annually to flush the salt down.

On moderately saline soil there is a considerable choice of fairly productive crops. As the salinity increases the choice of crops narrows. In the following statement Thorne and Bennett (1952) have identified tall wheatgrass as a highly salt-tolerant forage plant: "Tall wheatgrass is being planted on many highly saline soils in Utah and Nevada. A common procedure is to drill the seed in the bottom of irrigation furrows. Planting is followed by an irrigation. Good emergence and survival of this species have been obtained where most other grasses and legumes have failed." Christensen and Lyerly (1952) recommend flooding or irrigating in closely spaced furrows. They point out also that stands can be obtained by planting in the bottom of furrows under conditions where planting on the ridges between furrows has failed.

Thorne and Bennett (1952) suggest the desirability of a heavy irrigation prior to seeding, to leach the salt deeply into the soil, followed by frequent light irrigations after seeding, to keep the root zone low in salt.

V. IRRIGATION OF FORAGE CROPS IN THE HUMID EASTERN USA

Response of forage crops to irrigation has been the subject of investigation by a number of experiment stations in the Eastern USA with highly variable results (Kottke et al., 1960; Bennett, 1962; Vittum et al., 1963). Shallow-rooted annual forages respond more than deep-rooted perennials. In seasons when rainfall is adequate, supplemental irrigation may actually depress yields. In seasons when irrigation is beneficial, its value is enhanced by the liberal use of fertilizers.

The particular circumstances encountered on any given farm will, in the last analysis, determine whether forage crops can be irrigated profitably in the Eastern USA. On the basis of present data, irrigation in the East will increase primarily because of other crops capable of a greater return per acre. On farms having facilities to irrigate, forage crops will benefit most by irrigation to assure adequate stand establishment, and, in the Southeast, to obtain high late summer and fall production of annual forage crops, particularly in low moisture seasons.

A rather striking illustration of the potential use of irrigation in much of the humid Eastern USA is presented by Crofts et al. (1963) in Australia. In an area with an annual precipitation of 29 inches, they demonstrated that available water could be economically stored and used as needed for irrigation.

VI. THE FUTURE OF IRRIGATED FORAGE CROPS

In 1958, Simmons and Bressler (1958) reported that of the 11 Western States, only Utah, Idaho, and Montana, USA were self-sustaining in dairy products. Yet this area is expected to register a population growth of 67% between 1955 and 1975. If a high standard of living is to be maintained, illustrated not only by such things as automobiles and television sets, but probably better by our protein rich diet which draws heavily on animal products, there will have to be a significant increase in these products from irrigated land. Improvement of land not irrigated in the Western USA is far too slow to keep pace with the population. Hay and pasture crops are highly productive and compete well with the widely grown cash crops. They provide the means in the arid West of maintaining our high standard of living. That there will be a continued expansion of these crops under irrigation is a reasonable expectation.

LITERATURE CITED

Allison, Lowell E. 1964. Salinity in relation to irrigation. Advance. Agron. 16:139–180.

Bateman, George Q. 1958. Irrigated pasture research and management studies. Utah Agr. Exp. Sta. Mimeogr. 36 p.

Bateman, George Q., and Wesley Keller. 1952. Irrigated pastures in arid regions. Int. Grassland Congr., Proc. 6th, (State College, Pa. USA) I:404–410.

Bateman, George Q., and Wesley Keller. 1956. Grass-legume mixtures for irrigated pastures for dairy cows. Utah Agr. Exp. Sta. Bull. 382. 55 p.

Beckett, S. H., and R. D. Robertson. 1917. The economical irrigation of alfalfa in Sacramento Valley. California Agr. Exp. Sta. Bull. 280. p. 273–294.

Bennett, O. L. 1962. Irrigation of forage crops in Eastern United States. US Dep. Agr. Prod. Res. Rep. 59. 25 p.

Bernstein, Leon. 1958. Salt tolerance of grasses and forage legumes. US Dep. Agr. Inform. Bull. 194. 7 p.

Bernstein, L., and H. E. Hayward. 1958. The physiology of salt tolerance. Annu. Rev. Plant Physiol. 9:25–46.

Blaney, Harry F., and Wayne D. Criddle. 1962. Determining consumptive use and irrigation water requirement. US Dep. Agr. Tech. Bull. 1275. 59 p.

Brezeau, L. M., and L. G. Sommor. 1964. The influence of levels of irrigation on the nutritive value of alfalfa. Can. J. Plant Sci. 44(6):505–508.

Briggs, Lyman J., and H. L. Shantz. 1913. The water requirement of plants: II. A review of the literature. US Dep. Agr. BPI Bull. 285. 96 p.

Campbell, R. E., W. E. Larson, T. S. Aasheim, and P. L. Brown. 1960. Alfalfa response to irrigation frequencies in the presence of a water table. Agron. J. 52:437–441.

Carlson, C. W., D. L. Grunes, L. O. Fine, G. A. Reichman, H. R. Haise, J. Alessi, and R. E. Campbell. 1961. Soil, water and crop management on newly irrigated lands in the Dakotas. US Dep. Agr. Prod. Res. Rep. 53. 34 p.

Christensen, Paul D., and Paul J. Lyerly. 1952. Water quality . . . as it influences irrigation practices and crop production. Texas Agr. Exp. Sta. Circ. 132. 19 p.

Crofts, F. C., H. J. Geddes, and O. G. Carter. 1963. Water harvesting and planned pasture production at Badgery's Creek. Univ. Sydney. School of Agr. Rep. 6. 47 p.

Freeman, George F. 1914. Alfalfa in the Southwest. Arizona Agr. Exp. Sta. Bull. 73. p. 233–320.

Gausman, H. W., W. R. Cowley, and J. H. Barton. 1954. Reaction of some grasses to artificial salinization. Agron. J. 46:412–414.

Harris, F. S. 1913. Pastures and pasture grasses for Utah. Utah Agr. Exp. Sta. Circ. 15. p. 33–43.

Harris, F. S., and D. W. Pittman. 1921. The irrigation of alfalfa. Utah Agr. Exp. Sta. Bull. 180. 30 p.

Harris, L. 1962. To establish alfalfa on sloping land—corn good companion crop. Nebraska Agr. Exp. Sta. Quart. 9(1):9–10.

Harris, Lorin E., Milo L. Dew, and George Q. Bateman. 1958. Irrigated pastures for beef cattle. Anim. Husb. Dep. Utah Agr. Exp. Sta. Mimeo. 5 p.

Hayward, H. E, and C. H. Wadleigh. 1949. Plant growth on saline and alkali soils. Advance. Agron. I:1–38.

Heinemann, W. W., and R. W. van Keuren. 1955. Irrigated pasture studies with beef cattle. First year of grazing after establishment of pastures. Washington Agr. Exp. Sta. Circ. 266. 7 p.

Heinemann, W. W., and R. W. van Keuren. 1956. Irrigated pasture studies with sheep, second year of grazing. Washington Agr. Exp. Sta. Circ. 290. 11 p.

Holt, Ethan C., and E. C. Bashaw. 1963. Factors affecting seed production of dallisgrass. Texas Agr. Exp. Sta. MP 662. 8 p.

Hylton, L. O., Jr., D. E. Williams, A. Ulrich, and D. R. Cornelius. 1964. Critical nitrate levels for growth of Italian ryegrass. Crop Sci. 4:16–19.

Jensen, Marvin E., and Howard R. Haise. 1963. Estimating evapotranspiration from solar radiation. J. Irrig. Drainage Div. Amer. Soc. Civil Eng., Proc. 89 (IR 4):15–41.

Jones, L. G., J. R. Goss, M. D. Miller, and M. L. Peterson. 1957. Sudangrass for pasture-hay-seed. California Agr. Exp. Sta. Ext. Serv. Circ. 462. 18 p.

Klages, K. H. 1944. Idaho amber sorgo. Idaho Agr. Exp. Sta. Circ. 97. 8 p.

Kneebone, W. R., and R. W. Greve. 1961. Economic potentials from blue grama seed production under irrigation in northwest Oklahoma. J. Range Manage. 14:138–143.

Kottke, Marvin, Robert McAlexander, Niels Rorholm, and B. F. Stanton. 1960. Evaluating the profitability of irrigation on northeastern dairy farms. Northeast Reg. Publ. New Hampshire Bull. 469. 26 p.

Madsen, B. A. 1933. Alfalfa production. California Agr. Ext. Serv. Circ. 35, 47 p.

Magee, A. C., J. R. Martin, and William F. Hughes. 1959. Production and production requirements of crops—rolling plains and north central prairies. Texas Agr. Exp. Sta. Misc. Publ. 328. 14 p.

Magistad, O. C. 1945. Plant growth relations on saline and alkali soils. Bot. Rev. 11:181–230.

Masefield, G. B. 1961. The effect of irrigation on nodulation of some leguminous crops. Empire J. Exp. Agr. 29(113):51–59.

McCormick, John A., and Victor I. Myers. 1958. Irrigation of certain forage crops. Nevada Agr. Exp. Sta. Circ. 20. 13 p.

McMaster, G., J. G. Walker, and E. W. Owens. 1962. Irrigation and fertilization of field corn grown for silage in southeastern Idaho. Idaho Agr. Exp. Sta. Bull. 392. 15 p.

Miller, Milton D., Victor P. Osterli, Luther G. Jones, and A. D. Reed. 1951. Seed production of ladino clover. California Agr. Ext. Serv. Circ. 182. 30 p.

Moore, D. S., K. R. Tefertiller, W. F. Hughes, and R. H. Rogers. 1962. Production requirements, costs and expected returns for crop enterprises, hardland soils, high plains of Texas. Texas Agri. Exp. Sta. Mimeogr MP–601. 57 p.

Nelson, C. E., and J. S. Robins. 1956. Some effects of moisture, nitrogen fertilizer, and clipping on yield and botanical composition of ladino clover—orchardgrass pasture under irrigation. Agron. J. 48:99–102.

Nielsen, D. R., J. W. Biggar, and R. J. Miller. 1964. Soil profile studies aid water management for salinity control. California Agr. 18(8):4–5.

Norton, J. E. 1931. Irrigated alfalfa in Montana. Montana Agr. Exp. Sta. Bull. 245. 27 p.

Oman, Harry F., and Russell H. Stark. 1951. Grass seed production on irrigated land. US Dep. Agr. Soil Conserv. Serv. Leaflet 300. 8 p.

Pedersen, M. W., and D. R. McAllister. 1955. Growing alfalfa for seed: Section 1. Agronomic practices. Utah Agr. Exp. Sta. Circ. 135. p. 7–23.

Pedersen, M. W., G. E. Bohart, M. D. Levin, W. P. Nye, S. A. Taylor, and J. L. Haddock. 1959. Cultural practices for alfalfa seed production. Utah Agr. Exp. Sta. Bull. 408. 31 p.

Penman, H. L. 1948. Natural evaporation from open water, bare soil, and grass. Proc. Roy. Soc. (London) 193:120–145.

Peterson, M. L., V. P. Osterli, and L. J. Berry. 1959. Managing irrigated pastures. California Agr. Exp. Sta., Ext. Serv. Circ. 476. 31 p.

Richards, B. L. 1929. White-spot of alfalfa and its relation to irrigation. Phytopathology 19:125–141.

Richards, L. A. [ed.] 1954. Diagnosis and improvement of saline and alkali soils. US Dep. Agr. Handbook 60. 160 p.

Robertson, J. H., E. H. Jensen, R. K. Peterson, H. P. Cords, and F. E. Kinsinger. 1958. Forage grass performance under irrigation in Nevada. Nevada Agr. Exp. Sta. Bull. 196. 35 p.

Robins, J. S., and H. F. Rhodes. 1958. Irrigation of field corn in the West. US Dep. Agr. Leaflet 440. 8 p.

Rouse, H. K., F. M. Willhite, and D. E. Miller. 1955. High altitude meadows in Colorado: 1. Effect of irrigation on hay yield and quality. Agron. J. 47:36–40.

Rumery, Myron G. A., and Robert E. Ramig. 1962. Irrigated sudangrass for dairy cows. Nebraska Agr. Exp. Sta. Bull. SB 472. 12 p.

Schofield, R. K. 1952. Control of grassland irrigation based on weather data. Int. Grassland Congr. 6th, Proc. I:757–762.

Shank, D. B., and D. E. Kratochvil. 1954. Corn variety tests. p. 9–11. *In* Irrigation research in the James River area; a 5-year progress rep. South Dakota Agr. Exp. Sta. Circ. 107.

Simmons, R. L., and R. G. Bressler. 1958. California dairy industry—1975. California Agr. 12(11):2.

Stanberry, C. O. 1954. Alfalfa irrigation. Progressive agriculture in Arizona. Univ. Arizona Coll. Agr. Publ. 6(2):9.

Stanberry, C. O., C. D. Converse, H. R. Haise, and O. J. Kelley. 1955. Effect of moisture and phosphate variables on alfalfa hay production on the Yuma Mesa. Soil Sci. Soc. Amer. Proc. 19:303–310.

Stanford, E. H., L. G. Jones, V. P. Osterli, B. R. Houston, R. F. Smith, and A. D. Reed. 1954. Alfalfa production in California. California Agr. Exp. Sta. Ext. Serv. Circ. 442. 44 p.

Staten, Glen, H. B. Pingrey, and Marvin Wilson. 1951. Irrigated pastures in New Mexico. New Mexico Agr. Exp. Sta. Bull. 362. 28 p.

Sumner, D. C., John R. Goss, and Vern L. Marble. 1960. Production of grass seed in California. California Agr. Exp. Sta. Ext. Serv. Circ. 487. 19 p.

Taylor, S. A., J. L. Haddock, and M. W. Pedersen. 1959. Alfalfa irrigation for maximum seed production. Agron. J. 51:357–360.

Thorne, D. W., and W. H. Bennett. 1952. Soil management for grasslands on irrigated salted soils. Int. Grassland Congr., 6th, Proc. I:805–812.

Tovey, Rhys. 1963. Consumptive use and yield of alfalfa grown in the presence of static water tables. Nevada Agr. Exp. Sta. Tech. Bull. 232. 65 p.

Trew, E. M., and C. S. Hoveland. 1955. Irrigated pastures for south Texas. Texas Agr. Exp. Sta. Bull. 819. 20 p.

Tysdal, H. M., and H. L. Westover. 1949. Growing alfalfa. US Dep. Agr. Farmer's Bull. 1722. 33 p.

Ulrich, Albert. 1956. Plant analysis as a guide to the mineral nutrition of alfalfa. Int. Grassland Congr., Proc. 7th (Palmerston, North New Zealand) p. 313–322.

US Dep. of Agriculture. 1941. Climate and Man. US Dep. Agr. Yearbook. US Government Printing Office, Washington, D. C. p. 787.

Van Keuren, R. W., and W. W. Heinemann. 1959. Irrigated sudangrass and millet for forage and seed in central Washington. Washington Agr. Exp. Sta. Bull. 605. 9 p

Viets, Frank G., Jr. 1962. Fertilizers and the efficient use of water. Advance Agron. 14:223–264.

Vittum, M. T., R. B. Alderfer, B. E. Janes, C. W. Reynolds, and R. A. Struchtemeyer. 1963. Crop response to irrigation in the Northeast. Northeast Reg. Res. Publ. New York State Agr. Exp. Sta. (Geneva) Bull. 800. p. 17–22.

Weihing, R. M., D. W. Robertson, O. H. Coleman, and R. Gardner. 1943. Growing alfalfa in Colorado. Colorado Agr. Exp. Sta. Bull. 480. 36 p

Weimer, J. L. 1933. Effect of environmental and cultural factors on the dwarf disease of alfalfa. J. Agr. Res. 47:(6)351–368.

32 | Grain and Field Crops[1]

J. S. ROBINS

Washington State University
Prosser, Washington

J. T. MUSICK

Agricultural Research Service, USDA
Bushland, Texas

D. C. FINFROCK

The Rockefeller Foundation
Bangkok, Thailand

H. F. RHOADES

University of Nebraska (deceased)
Lincoln, Nebraska

I. INTRODUCTION

Chapter 30 in this monograph summarizes and interprets basic soil, plant, water, and atmospheric principles presented in sections V, VI, VII and VIII as they apply generally to determining or predicting irrigation needs of crops grown in a particular physical environment. This chapter presents background information related to and general recommendations for irrigation of corn (*Zea mays*), sorghum (*Sorghum vulgare*), rice (*Oryza sativa*), wheat (*Triticum vulgare*), barley (*Hordeum vulgare*), potatoes (*Solanum tuberosum*), and field beans (*Phaseolus vulgaris*). Reference to the irrigation of oats (*Avena sativa*) will be included also, although specific information on which to base recommendations is largely lacking. Since basic principles are covered elsewhere, the chapter will emphasize specific characteristics and peculiarities of these crops and certain irrigation design and management factors which relate to their irrigation. A concluding section will give summary recommendations for each crop.

Irrigation of field crops is generally practiced solely for the purpose of improving the plants' moisture environment. Secondary effects sometimes are of importance, as with rice where flooding assists in controlling weeds and causes desirable changes in physiochemical soil reactions. Alteration of soil temperature is a side effect which may be beneficial or detrimental depending on the crop, the season, and the water temperature. Microclimate (atmospheric temperature and vapor content) may be favorably affected but little quantitative information exists on this subject. Irrigation for frost control is generally limited to horticultural or high value specialty crops. Fertilizers are sometimes distributed in irrigation water

[1] Contribution from the Soil and Water Conserv. Res. Div. ARS, USDA, the Ford Foundation, and the Nebraska Agr. Exp. Sta. Published with the approval of the Director as paper no. 1543, Journal Ser., Nebraska Agr. Exp. Sta.

but normally water is not applied solely for this purpose. Therefore, aside from rice, the crops considered in this chapter are irrigated specifically to maintain a soil water condition favorable for satisfactory growth and production of the desired plant part or parts.

Except for flooded rice, need for and quantity of water required at irrigation for a given crop grown in a particular environment are determined by (i) the quantity of *usable* water stored in the crop root zone and (ii) the *rate* at which this supply is depleted. Usable soil water storage in this context encompasses water storage and transmission properties of the soil, the crop's effective root depth and proficiency in water absorption, and the ability of the crop to tolerate soil water stress imposed by increasing soil water suction and interacting climatic parameters. These factors interdependently determine the quantity of water that may be extracted safely before irrigation is needed. Rate of depletion of this supply is determined primarily by climatic parameters with certain limitations imposed by soil and plant factors which interact to control the evapotranspiration and deep percolation rates. In the case of submerged rice culture, water is stored above the soil surface and irrigation water must be supplied to keep the depth of water within certain limits.

Considerable research effort has been applied to determining optimum irrigation practices and factors related thereto for corn, sorghum, wheat, and rice, and to a lesser extent for barley. This research has sufficiently clarified crop characteristics and response to climatic, soil and water supply variations to permit outlining rather specific irrigation practice recommendations for near-maximum yield under most conditions. Irrigated oat production is limited; consequently, this crop has received little attention of irrigation researchers. Information presented and guides described for wheat and barley should generally apply for this crop.

Field beans have been studied extensively and, in the absence of diseases, clear-cut recommendations are possible for this crop. Where the root system is partially destroyed by disease, recommendations for efficient irrigation practice become difficult and arbitrary.

Finally, even in the face of considerable research effort, specifying recommendations for potato irrigation is difficult. This is due largely to the apparent sensitivity of the crop during tuber development to variations in soil water, temperature, fertility, and cultural factors as well as meteorologic and other parameters. Even so, it is possible to indicate practices that appear satisfactory under most circumstances.

II.   CROP CHARACTERISTICS INFLUENCING IRRIGATION PRACTICES

A.   Season of Growth

Season of growth influences frequency and quantity of required irrigation applications because of variations in evapotranspiration with seasonal climatic changes (*see* chapter 26). Depletion of a given quantity of water from a particular soil in midspring by winter wheat with full vegetative cover may require two or three times as long as corn or sorghum in midsummer. Early-planted potatoes, on the other hand, will develop full vegetative cover and deplete a given soil water reserve more quickly in late spring or early summer than potatoes planted later which have only partial vegetative cover at that time. Irrigation frequency and quantity for rice, whether produced under upland or lowland environments and whether submerged or not, likewise depend on rate of evapotranspiration. Water depletion is more rapid during midsummer months than during cooler seasons, other conditions being similar. Under submerged conditions, however, rate at which irrigation water must be supplied is often controlled by rate of deep penetration loss and amount of rainfall rather than by evapotranspiration rate.

Winter small grains are normally grown in temperate regions and actively grow for 4 to 12 weeks during the fall months. Combination of low evaporative potential and only partial vegetative canopy results in low rates of evapotranspiration, seldom exceeding 0.2 cm/day. Evapotranspiration during the dormant winter period is generally of the order of 0.02 to 0.1 cm/day. Growth may initiate in midwinter to early spring depending on climatic conditions. Full vegetative cover is soon attained if climatic conditions are favorable, and rate of evapotranspiration generally follows the climatic potential until the crop begins to mature in late spring to midsummer. As the plants mature, evapotranspiration rate gradually decreases and falls below the potential. Peak rates may be as high as 0.8 or 0.9 cm/day in arid, windy areas (Jensen and Musick, 1930; Jensen and Haise, 1963; Musick et al., 1963a).

Spring small grains are seeded in midwinter to early spring and generally require 4 to 8 weeks after emergence to achieve full vegetative canopy. Evapotranspiration rate during this period begins at one-fourth to one-half the climatic potential and gradually increases to approach the potential when full vegetative canopy is achieved. As with winter grains, evapotranspiration, after full cover is present, follows the climatic potential until maturity begins, which is usually 1 or 2 weeks later than winter varieties grown in the same environment (Jensen and Haise, 1963).

Corn, sorghum, potatoes, and field beans grown in temperate climates are seeded in late winter or spring and mature, are harvested, or are killed by frost in the midsummer to midfall season. Full vegetative canopy is generally achieved 6 to 12 weeks after emergence depending on climatic conditions. As with spring grains, evapotranspiration beginning at emergence is generally one-fourth to one-half the evaporative potential and gradually increases to the full potential when good vegetative cover is attained. As with the grains, the rate drops below the potential as the crop matures (Corey and Myers, 1955; Howe and Rhoades, 1961; Jensen and Haise, 1963; Musick and Grimes, 1961; Swanson and Thaxton, 1957).

Rice is grown over a wide range of latitudes and elevations but grows and yields best under warm temperatures. Short and intermediate season varieties (100 to 150 days) are used in cooler climates. Use of long-duration varieties (180 days or more) is limited to tropical latitudes. Using short or intermediate maturing varieties, two and even three crops a year, can be taken from some tropical lands by transplanting seedlings from seedbeds which were sown as the preceding crop neared maturity. Due to the continuous presence of a free water surface under submerged conditions or the very wet condition usually maintained under intermittent irrigation, evapotranspiration by rice generally approaches the climatic potential throughout its growth cycle (Leather, 1910; Vacchani, 1953). This is in contrast to other grain and field crops in which incomplete vegetative cover during a portion of the growth period results in evapotranspiration rates considerably below the climatic potential.

B. Rooting Characteristics

Rate and depth of root growth and degree of root proliferation greatly influence irrigation practice since, except for submerged rice or where a high water table exists, the plant's usable water supply is largely limited to the water storable in that soil volume explored by the crop's root system (*see* chapter 21).

Root systems of corn, sorghum, and the small grains consist of a branching primary system emerging from the seed and a secondary system emerging from growth points above the seed, usually at nodes that are covered with soil or are immediately above the soil surface. The primary system is relatively sparse and extends rapidly to depths of 1 to 1.5 m. Rice is an exception and has a much shallower root system. Initiation of growth of the secondary root system is delayed for several weeks after emergence (until beginning of tillering in the case of small grains and internode elongation in the case of corn and sorghum). The secondary system grows rapidly if soil water conditions are favorable and displays considerable branching. Varietal differences in root systems within these crops probably exist but little research has been devoted to this subject.

Beans normally have a central taproot system with branching secondary roots emerging from the primary taproot. If the central taproot is destroyed near the soil surface by disease or physical injury, a profusion of secondary adventitious roots will emerge from buried nodes above the seed. If the primary system is healthy, development of adventitious roots is limited.

Potatoes generally grow from vegetative buds (eyes) in buried portions of tubers, or whole tubers. One or two such buds normally develop in a given seed piece. The root system develops from these same growth points as many-branched secondary systems. Generally one or two roots from each bud will dominate and appear essentially as branched taproots. Sometimes secondary adventitious roots will develop from buried nodes above the seed piece or even from nodes in the underground stems on which the tubers form.

Winter small grains develop a deep root system (2 to 3 m or more if soil conditions are favorable) by the time spring growth is initiated and will extract significant quantities of water from depths up to 2 or 2.5 m if the surface layers become dry (Kmoch et al., 1957; Musick et al., 1963a). The root system is not only deep but profuse and, at least for wheat, is capable of depleting the soil water content to a level lower than many other crops (Haise et al., 1955). Spring grains also have excellent root systems but differ from winter varieties in that a period of 8 to 12 weeks after emergence in the spring is required for roots to approach full depth development. After full development of the root system, spring wheat is also efficient in water extraction (Robins and Domingo, 1962).

Corn and grain sorghum also have good ultimate root systems. By flowering, 2 m or more of soil will be explored where soil conditions permit. As with spring grains, reasonably full development requires 8 to 12 weeks from seeding (Henderson, 1967a; Musick and Grimes, 1961; Rhoades and Nelson, 1955; Robins and Domingo, 1953). These crops differ from wheat in that they do not extract soil water to levels appreciably below the permanent wilting percentage.

The deep and profuse root systems of winter and spring small grains, corn and grain sorghum permit extending the time between irrigations and increasing the quantity of water applied at each irrigation as compared with crops with shallower and/or sparser root systems. This greatly increases flexibility in irrigation scheduling, simplifies, and generally improves efficiency in irrigation practice where such crops are grown on deep, friable soils.

Potatoes and field beans in contrast have considerably shallower and less proliferous root systems. By maturity, some potato roots may extend to a depth of 2 m or more and bean roots to 1.6 to 1.8 m. Soil layers beyond 1 m are seldom fully explored by either crop, even under most favorable conditions. In temperate regions peak water use occurs 8 to 12 weeks after emergence when often no more than 50 or 60 cm of soil is fully ramified and only a few roots extend beyond the 1-m depth (Corey and Myers, 1955; Howe and Rhoades, 1961; Robins and Domingo, 1956a, 1956b).

Rice roots are capable of extending to depths of 1.25 to 1.5 m when grown under upland conditions with intermittent irrigation or under natural rainfall. Where grown submerged or on upland areas during periods of very high rainfall, most roots are confined to the top 20 or 30 cm with profuse secondary root development on or near the immediate soil surface and in algal growth on the soil surface (Adair et al., 1962; Sethi, 1930).

It must be remembered that numerous soil environmental factors may completely overshadow the plant's genetic capabilities for root development (*see* chapters 20, 21, 23 and 55). Most important of these are soil density and pore size or configuration, soil water status, soil aeration, nutrient status, textural or structural stratification, water tables, soluble salts, and soil-borne organisms that damage or destroy plant roots (Wiersma, 1959). Any one or a combination of these factors may greatly reduce the depth or proliferation of the crop's root system from that obtainable under most favorable conditions. This reduces the usable soil water storage reservoir and thereby may greatly alter the irrigation practices necessary for desired production and water conservation.

C. Reaction to Soil Water Conditions

1. SENSITIVITY TO WATER STRESS

Chapters 19, 20, and 23 cover basic principles governing plant response and sensitivity to deficient soil water. This section will pinpoint the behavior of the crops covered in this chapter including both vegetative and reproductive growth response. Special emphasis will be placed on critical growth periods and on quality of the desired plant product as influenced by inadequate water for maximum plant performance.

"Moisture stress" will be used herein to describe qualitatively the moisture status of the plant. In this context, "stress" denotes a plant condition in which the rate of water supply to the plant is inadequate to maintain it at full turgidity without compensating physiologic or other change. Such compensation may be by wilting, by stomatal closure or by internal adjustment of solute concentration to maintain positive turgor pressure. Moisture stress qualitatively denotes the water potential in the plant.

Most plants are continuously under some moisture stress. As used herein, mild or low stress refers to the condition where leaf wilting is first evidenced during midday. Moderate stress denotes marked loss of turgor during the day but regaining of full turgor at night and no significant leaf desiccation or "firing." Severe or high stress refers to failure of the plant to recover full turgor at night accompanied by killing of leaves by desiccation.

Moisture stress is generally associated with reduction of the soil water content in the plant root zone to a point wherein the rate of evapotranspiration exceeds the rate at which the plant roots can absorb and/or transmit water to the above-ground plant parts. Various facets of this balance between supply and demand are covered in other sections of this monograph. In short, significant stress may occur over an appreciable range of soil water, temperature, plant, and microclimatic conditions. Because of wide variations in plant and water behavior in different soil and climatic situations, no attempt will be made to quantitatively define plant response to soil water energy level at any specific depth or portion of the root zone.

As reviewed elsewhere in this monograph, salts dissolved in the soil solution lower the energy status of soil water. Even in so-called "normal" soils, osmotic potential of the soil solution may reach 2 atm equivalent suction. In moderately saline soils, 5 atm equivalent suction due to dissolved salts is not uncommon. Plants grown on such soils generally behave as if the soil water content were lowered sufficiently to induce the equivalent reduction in potential by capillary forces. Thus, a given degree of moisture stress in the plant will be observed at higher water content in saline than in nonsaline soils. More frequent irrigation with correspondingly less water per irrigation is required where appreciable soluble salts occur in the root zone.

As discussed subsequently, field beans and potatoes are sensitive to depletion of high percentages of the available water from the root zone. Saline soil conditions therefore result in need for considerably more frequent irrigation of these crops. Unless the salt content is quite high, little change in irrigation frequency for wheat, corn, barley, oats or sorghum would be needed. Due to leaching caused by the very wet environment in which rice is grown, soil salinity is seldom a significant problem with this crop.

Both winter and spring wheat when produced for grain or for forage exhibit considerable tolerance to moderate moisture stress throughout the growth cycle (Jensen and Musick, 1960; Musick et al., 1963a; Fernandez-G. and Laird, 1957; Robins and Domingo, 1962). Periods of high stress at any time prior to, during, or for a few days after heading result in reduced vegetative growth—both ultimate height and total yield of vegetative material. Height and vegetative yield reductions are only of limited magnitude unless the stress is permitted to persist until extensive desiccation or "firing" of leaves occurs. High stress at any stage of growth will reduce grain yield appreciably if severe enough to desiccate lower leaves. Grain yield reduction results from a combination of fewer heads, spikelets per spike and kernels per spikelet when stress occurs during and shortly after heading, and from reduced kernel weight, spikelets per spike and kernels per spikelet if stress occurs before boot stage. Reduced kernel weight (shriveling of the grain) is the principal cause of yield reductions caused by late season stress (Robins and Domingo, 1962). Hot, dry winds coincident with high soil water stress during grain development greatly intensifies the shriveling associated with a given soil water deficit. There is evidence for a critical period in wheat of about 2 weeks before pollination during which moderate stress is sufficient to reduce grain yield (Henkel, 1964). Apparently stress during this period has a particularly detrimental effect on pollen viability (Skazkin, 1960).

Moisture stress has increased protein content of winter wheat grain as measured by total nitrogen content. Quality has been reduced, however, as measured by loaf volume per increment of indicated protein in baking tests. Decreased quality offsets the measured increase in protein content of wheat produced under conditions of moderate to high stress (Musick et al., 1963a). Shriveled grain results in increased bran and decreased flour yields during milling.

An extended period of moisture stress during the jointing to heading stage of development followed by a period of adequate water supply may induce growth of additional tillers after the primary tillers have fully developed. This second growth behavior is most pronounced on very fertile soils. Although yield may be improved by the additional heads, this behavior is undesirable since harvest is delayed and dangers of lodging or other loss of the crop increase.

In soils that permit good root development by wheat, moderate stress seldom occurs until 75 to 80% of the available soil water has been extracted from the major root zone. High stress will not be encountered until the moisture content in the root zone approaches the permanent wilting percentage (Jensen and Musick, 1960; Musick et al., 1963a; Robins and Domingo, 1962).

Work with barley indicates a behavior similar to that of wheat with perhaps slightly less tolerance to severe or prolonged stress periods, at least on a coarse-textured soil in western Arizona, USA. On this soil, high stress usually occurred with some residual available soil water (10 to 15%) in the root zone (Bendelow, 1958; Schreiber and Stanberry, 1965; Stanberry and Lowrey, 1965).

Corn generally displays tolerance to appreciable moisture stress except during the growth period from tassel emergence to completion of pollination (Carlson et al., 1961; Howe and Rhoades, 1955; Rhoades et al., 1954; Rhoades and Nelson, 1955; Robins and Domingo, 1953). As with wheat, stress sufficient to induce marked wilting prior to, during and immediately after tasseling, reduces vegetative growth somewhat (*see* Fig. 32–1). Prolonged or severe stress that results in leaf desiccation markedly reduces plant height and yield of vegetation. Significant grain yield reductions are incurred by early season stress only if it is prolonged or severe enough to cause severe wilting. Due to sparseness of the initial root

Fig. 32–1. Inadequate soil water prior to tasseling of corn as shown here reduces vegetative growth. Sorghum and small grains exhibit similar behavior. Grain yield is little affected so long as the water supply is replenished and adequate immediately before, during, and after tasseling in the case of corn and heading in the case of sorghum and small grains. (Photo courtesy of Farm Journal).

system prior to development of the secondary system, however, such wilting in early growth stages may occur when appreciable available soil water remains.

In contrast, even moderate stress that results in significant wilting during the tasseling, silking, and pollination periods may reduce grain yield greatly. Stress adequate to desiccate the lower or the tender top leaves during this interval may cut grain yield in half. Reductions due to stress at this time are largely the result of pollination failure in which grain forms on only part of the ear (Robins and Domingo, 1953). This behavior is probably due to desiccation of either the pollen or the organs that receive and transmit the pollen, and is related to the considerable separation of the sites of pollen formation and pollination.

As with wheat, appreciable stress can be tolerated during grain formation so long as firing of the leaves is avoided. Very high stress with significant firing of the leaves has little effect on grain yield during the maturing (grain drying) period (Robins and Domingo, 1953).

Response of grain sorghum to moisture stress is similar to that of corn (Henderson, 1967a, b; Musick and Grimes, 1961; Musick et al., 1963b; Swanson and Thaxton, 1957). Even though long considered a highly drouth-resistant crop, stress sufficient to cause leaf desiccation results in significant yield reduction, both of vegetative portions and grain. Severe stress during the boot to flowering stage results in pollination failure (head blast) similar to that of corn. Yield reduction is compensated in this case, however, by formation of adventitious heads or by development of additional heads on late-forming tillers, depending on the variety and specific stage of development when blast occurs (D. W. Henderson. *Personal communication. Irrigation Department, Univ. California, Davis;* Musick and Grimes, 1961). As with wheat, development of such second growth delays maturity.

Rice apparently has three critical periods wherein moisture stress reduces grain yield—the seedling establishment or transplanting period, the tillering stage of growth, and a period from about 20 days before to about 5 days after heading (Adair et al., 1962; Erygin, 1936; Matsushima, 1962; Singh et al., 1935). Yield reductions during the latter period are due to unfertilized florets. Marked stress near maturity also may cause yield depressions, presumably by the same mechanism as with other grains. Stress may develop quite rapidly when submerged rice is drained because of the very shallow root system developed under flooded culture. Thus, care is required to avoid sudden stress under these conditions.

Field beans appear to withstand considerable moisture stress without major reproductive yield reductions (Howe and Rhoades, 1961; Robins and Domingo, 1956a). Visible stress symptoms evidenced by foliar color change and wilting for periods up to 5 days or more reduce vegetative growth but do not significantly reduce seed yield (*see* Fig. 32–2). Longer stress periods any time during the flowering period reduce both the number of pods and the number of beans developed per pod. These yield component losses are only partially offset by increased bean weight. Stress periods before blooming delay plant development. Unless stress is prolonged, seed yield is not appreciably reduced if favorable moisture conditions are maintained after the stress period. Stress during the maturing process lowers yield by reducing bean weight but only if the stress is rather severe and prolonged. It should be remembered, however, that the bean root system is sparse compared with many other crops; therefore, irrigation may be necessary at relatively frequent intervals even though the crop is not especially

Fig. 32–2. Darkened appearance of field beans indicates beginning of soil water stress. Such condition can persist for several days without incurring yield reduction, thus is a useful tool in indicating irrigation need.

sensitive. Evidence exists also that too frequent irrigation reduces yield of some varieties such as blackeye and small lima (C. W. Henderson. *Personal communication. Irrigation Department, Univ. California, Davis*).

In contrast to the crops previously discussed, reaction of potatoes to soil water stress is much more complicated and less well understood. Most studies indicate only modest vegetative and tuber yield depressions due to moisture stress unless the stress is severe and prolonged (Box et al., 1963; Carlson et al., 1961; Corey and Myers, 1955; Robins and Domingo, 1956b). Growth rate seems to be reduced, both in aboveground and belowground vegetative organs (tops and tubers) at relatively low stress. Very rapid cell enlargement, especially in the tubers, apparently partially compensates for the slowdown after stress is reduced by irrigation.

Qualitywise, however, even modest stress levels often, but not always, have serious consequences (Box et al., 1963; Corey and Myers, 1955; Robins and Domingo, 1956b). The rapid growth following relief of stress often results in malformed or cracked tubers. The malformation generally is simply enlargement of the "flower" end (the end containing meristematic tissue) of the tuber after stress is relieved by increased water supply as compared to the enlargement of that portion of the tuber developed during the stress period. This results in "spindle"- or "dumbbell"-shaped tubers. Stress early or late in tuber development results in tapering at the stem or flower ends of the tuber, respectively. This type of defect has been reported only on the cylindrically tubered varieties such as Russett and White Rose but probably occurs in less obvious manner on round-tubered varieties. Sometimes cell division and enlargement occurs in the vegetative growth buds (eyes) elsewhere in the tuber. This results in protuberances on the otherwise smooth tuber. The sudden rapid cell enlargement in tubers after stress is reduced by irrigation also may result in "growth cracks" or splitting of tubers. Stress late in the tuber growth period (near harvest) generally decreases dry matter content (specific gravity) slightly (Box et al., 1963; Corey and Myers, 1955).

Some evidence indicates that high temperatures in the soil volume occupied by potato tubers rather than moisture stress is the prime cause of tuber malfor-

Fig. 32–3. Excessive vegetative growth caused by high soil water and nitrogen levels may result in lodging of small grains.

mations (Box et al., 1963; Corey and Myers, 1955). Field or laboratory results to firmly establish specific cause and effect relationships involved in tuber malformation have not been published to date. Wetting of soils by irrigation is generally a cooling process, however, so high temperatures and high moisture stress go hand-in-hand. Thus, whether caused by high temperature or high moisture stress, frequent irrigation does reduce incidence of tuber malformations.

2. SENSITIVITY TO EXCESS SOIL WATER

Remarks in this section will be limited to effects of very high soil water levels in the absence of surface or subsurface drainage problems. Effects of surface water impoundment will be covered in a subsequent section (except as applied to rice irrigation). Subsurface drainage problems and the entire subject of sub-irrigation are covered in chapters 45 and 50, respectively.

Irrigation to maintain a high soil water content during the seedling stage of growth may be detrimental to the growth of any of the crops considered in this chapter except rice. This is due generally to lowering of soil temperature below the optimum but may also relate to leaching of plant nutrients from the very shallow root zone prevalent at this growth stage.

Wet conditions during the jointing and boot stages of development of winter or spring small grains promote luxurious vegetative growth, especially where nitrogen fertility of the soil is high. This often results in increased lodging during grain development. Heavy irrigation of these crops after the grain is well developed also promotes lodging (see Fig. 32–3) because of reduced ability of the wet soil to support the plant in an upright position (Robins and Domingo, 1962).

Very wet soil surrounding potato tubers often increases susceptibility to micro-organisms which may cause tuber damage. Incidence of "pink rot" or "water rot" may be enhanced since wet conditions are favorable to the fungus (*Phytophthora erythroseptica*) and secondary organisms that are associated with this disorder (Corey and Myers, 1955). Prolonged maintenance of soil water contents above field capacity in the soil around the tubers often reduces or eliminates "netting" (sclerotization of the tuber skin) of netted varieties.

Both sorghum and corn are tolerant of soil water contents above the normal field capacity except during seedling establishment when growth may be reduced by lowered soil temperature. Sorghum generally will tolerate flooding or very wet soil conditions for longer periods than corn.

Very wet soil during the seedling stage of growth generally favors the incidence of bean root diseases because of the favorable environment presented the organisms (*see* chapter 55). Similarly, after vegetative growth is well advanced, a wet soil surface may greatly increase incidence of pathogenic organisms such as sclerotinia rot that attack aboveground plant parts. This results in damage to both vegetative and reproductive plant parts (Howe and Rhoades, 1961; Robins and Domingo, 1956a). Wet soil conditions at harvest often seriously delay bean harvesting operations.

III.   IRRIGATION SYSTEMS AND PRACTICES

Sections XI and XII of this monograph are devoted to the several aspects of irrigation system design, methods of water application, and management of water and soils under irrigation. This discussion will point up certain specifics in this general area as related to the crops under consideration.

A.   Water Application Methods and System Design

Water can be successfully applied to all crops discussed in this chapter by any of the three general methods now in extensive use—surface, sprinkler, or sub-irrigation—so long as the systems are adequately designed to meet the crop needs. Rice irrigation, however, is largely limited to surface flooding systems.

With surface systems, the grains, including rice, may be either flooded or irrigated in furrows or rills successfully. None of these crops is especially sensitive to surface water impoundment up to 2 or 3 days so long as ambient temperatures do not raise water temperatures above 40C. Thus, minor topographic irregularities in nearly level fields can be tolerated so long as they do not result in prolonged water impoundment.

Both field beans and potatoes are sensitive to ponded surface water—potatoes because of sensitivities to excess soil water discussed in the previous section and beans because of susceptibility to "drowning" if water impounds the soil for more than 1 or 2 days. Elimination of low areas where water impounds is most desirable if these crops are to be grown. Surface systems for these crops should be designed to avoid water flow depth in excess of about 3 inches in the furrow. In particular, submerging of the tops of potato ridges should be avoided (O. W. Howe. *Personal communication. ARS-USDA, Grand Junction, Colo.*).

Since potatoes are apparently quite sensitive to moisture stress and to marked variations in soil water conditions around the tubers, it is necessary to have a well-designed surface system to irrigate this crop most successfully. Design features such as furrow grade changes which contribute to nonuniformity of water infiltration from one point to another along the furrow should be avoided.

With sprinkler systems, irrigation of corn and sorghum and to a lesser extent the small grains requires extension of the nozzle risers so that these taller crops will not interfere with the sprinkler pattern. Irrigation of the small grains (wheat, barley, oats, and upland rice) late in the season by this method often results in considerable lodging due to the added weight of water collected within the crop canopy. This constitutes a decided disadvantage of this method for these crops in windy areas or on soils where frequent water application is necessary. Sprinkler-irrigated beans and potatoes tend to grow in a more prostrate manner than those surface or subirrigated—thus more of the vegetation contacts wet soil and is susceptible to infection by soil-borne organisms. Finally, the necessity of moving sprinkler laterals generally results in undesirable mechanical damage to the grains and to beans which may reduce yield appreciably. This damage can be reduced by hand moving the sprinkler laterals, but this becomes a most unpleasant task with the wet soil and foliage encountered, especially in the taller crops.

An important aspect of irrigation system design relative to efficient irrigation of all crops considered in this chapter, except the winter grains, is the ability to uniformly apply small increments of irrigation water (2 inches or less) at early irrigations. During seedling establishment and early growth stages these annuals have limited root systems (see previous section of this chapter) and, where early irrigations are required, only a fraction of the total soil water reservoir needs to be replaced by irrigation. Inability to apply water uniformly in small amounts results in waste of water and often loss of soluble nutrients by leaching.

B. Irrigation Practices

Specific time and quantity of irrigation water application is dictated by the rate of removal and the particular quantity of soil water that may be removed from the crop's root zone without adverse crop response. Principles relating to rate of removal, quantity of water stored in the crop's root zone, and physiologic response to different moisture conditions are covered in other sections of this monograph and, specifically for the crops covered, are discussed earlier in this chapter. The discussion to follow relates to effects of inadequate and excessive irrigation water application as they influence these crops.

Application of irrigation water in excess of that required to refill the crop root zone generally results in loss of plant nutrients from the root zone by leaching. Magnitude of such loss depends primarily on the amount of nutrient in the soil solution and subject to displacement by added water. Nitrogen in the nitrate form is the most common nutrient lost in sufficient quantity to influence current crop growth (Robins et al., 1955, 1956). Losses of potassium, calcium, magnesium, phosphorus, and other elements may have serious consequences if repeated over-irrigation is practiced. Yield depressions due to loss of nitrogen are generally of consequence only if heavy water applications are made during early stages of plant growth. Losses later in the growing season are reflected in the growth or fertilizer needs of succeeding crops.

Application of water inadequate to refill the root zone of crops covered in this section advances the time at which a succeeding irrigation will be required, so long as deep soil water reserves are available to the crop. Under-irrigation will result in the onset of moisture stress sooner and more suddenly if the subsoil layers are very dry than if deep reserves are available. Under these conditions, root penetration will also be limited. Thus, the risk of yield-depressing stress periods may be greatly enhanced by too small water applications at any given irrigation. Repeated under-irrigation may result also in salt buildup in the root zone unless periodic leaching with irrigation water or by precipitation occurs.

The practice of alternate-furrow irrigation in graded irrigation systems is employed for some of the crops covered in this chapter—notably on potatoes, corn, and sorghum but sometimes on beans and small grains. In this method, water is applied at a given irrigation in a ternate rather than in all furrows. At the next irrigation, water is applied in furrows that did not receive water at the preceding irrigation. Irrigation applications are generally made a little more often than where every furrow receives water at each irrigation. This system has been applied very successfully in many areas, especially on potatoes (Box et al., 1963). It appears to result in less over-all fluctuation in moisture stress than the every furrow method. Properly applied, it can both simplify the mechanics of irrigation water application and save irrigation water.

C. Irrigation Practice for Rice

Rice generally grows and produces best when submerged during most of the growth period with water depth of 5 to 20 cm. In some cases, rotation irrigation on a schedule sufficient to keep the surface soil layers at or near saturation has produced yields equal to those under submerged culture and with significant saving of water. In other cases, yields comparable to submerged culture were produced by flooding only during seedling establishment and from the early boot stage to maturity (Adair et al., 1962; Aglibut et al., 1960; Chambliss and Jones, 1925; Chow, 1951; Ghose et al., 1960; Matsuo, 1957; Matsushima, 1962; Mitsui, 1954; Sreenivasan and Sadasivan, 1942; Ten Have, 1959).

Need for maintaining flooded conditions for rice production has been attributed to (i) difficulties in plant establishment due to a shallow root system, (ii) weed control, (iii) control of microclimate, (iv) prevention of pollination failure, (v) prevention of high manganese levels which upset the growth regulator balance, and (vi) increase in protein, mineral, and soluble carbohydrate content of the grain.

Detriments from flooding on soils with poor drainage characteristics, low pH and high organic matter include a physiologic disease variously called "Mentak," "Pentek merah," and "browning disease" (Johnston, 1954; Ponnamperuma et al., 1955; Vecht, 1953). Primary visible symptom is damage to the root system. On soils that are low in iron and where sulfur is present, continuous flooding may result in toxic concentrations of H_2S which cause root damage ("Akochi") (Mitsui, 1954). These root-damaging conditions are associated with low soil redox potentials which bring iron, manganese, or H_2S to toxic levels in the soil. Temporary drainage and subsequent aeration as the soil dries help alleviate damage from these "diseases."

IV. RECOMMENDED IRRIGATION PRACTICES

The preceding review of climatic, soil, and plant factors which interdependently determine needed irrigation frequency and quantity for crops covered in this chapter suggests a wide range in irrigation practice. Near-maximum yields of wheat, sorghum, corn, or beans may be achieved with one or two well-timed irrigations when grown in temperate regions on soils that store large amounts of usable water. Appreciable well-timed rainfall may reduce the number of irrigations or quantity of water otherwise required for successful production. In arid semitropical regions, however, up to 15 or more irrigations may be needed to produce high-quality potatoes on shallow or coarse-textured soils. Large amounts of water must be supplied to rice, either by rainfall or through continuous or frequent irrigation.

A. Irrigation Scheduling for Corn, Sorghum, Small Grains, and Field Beans

Generally, yield depressions of corn, sorghum, and field beans will be negligible if no more than 55 to 65% of the available water storage capacity in the soil volume well explored by plant roots is depleted. Near maturity up to 70 or 80% can be safely used. Winter and spring wheat will tolerate withdrawal of 70 to 80% of the supply available in the root zone during most of growth period and up to 90% or more near maturity. Based on limited research, similar depletions for barley and oats would be 65 to 75% until grain is well formed and 80 to 90% near maturity.

With each of the above crops, soil water supply should be at a high level during early growth to assure rapid and complete development of the root system. This can often best be accomplished by a preplanting irrigation, a generally recommended practice except in areas where rainfall has or is expected to fully recharge the soil water storage capacity at or soon after planting. This aids both germination and early growth. A high soil water level is desirable during reproductive and early grain development stages of the grains since this relatively short interval is so critical to successful production. With field beans, short periods of stress are less serious owing to the progressive fruiting habit of this crop. Finally, irrigation timing should be controlled to prevent depletion at maturity beyond the levels given in the preceding paragraph.

Table 32–1 suggests desirable irrigation schedules for the grain crops where the number of irrigations after emergence varies from 1 to 5. The lines in the table denote 10-day intervals over which given irrigations might be applied. It is recommended that the potential root zone of the crop be fully wetted at or soon after planting. Where precipitation is inadequate to assure this, a preplanting irrigation should be applied.

Irrigation scheduling for field beans can best be based on visual stress symptoms. Previously in this chapter, the color change of foliage of this crop from light to dark green was described. It is recommended that irrigation water be applied within 3 to 5 days following this color change on coarse-textured or shallow ($<$ 60 cm or 2 ft) soils and within 5 to 8 days following the color change on deeper, medium- or fine-textured soils. As with the grains, preplanting irrigation is recommended unless the potential root zone is fully wetted by precipitation.

Irrigation of beans after vegetative growth is well advanced may be needed

Table 32–1. Irrigation schedules for corn, sorghum, wheat, barley, and oats

Number of irrigations needed	Growth stage					
	Initial internode elongation	Intermediate internode elongation	Boot or pretassel	Full head or tassel	Grain in milk stage	Grain in dough stage
1			———	———		
2		———	———	———	———	
3	———	———	———	———	———	
4	———	———	———	———	———	———
5	———	———	———	———	———	———

as often as 5 to 7 days on coarse-textured soils, on land where root depth is limited by shallow soil or root disease, or in areas where evapotranspiration is high. On deep, medium- or fine-textured, disease-free soils under lower evaporative demand, irrigation frequency of 15 to 25 days might be expected. Onset of the color change may occur at 5- to 20-day intervals from emergence to development of essentially complete vegetative ground cover depending on soil, climatic, and plant factors.

B. Irrigation Scheduling for Potatoes

In contrast to the grain crops, potatoes are relatively sensitive to moisture stress. Depletion of more than 40 to 50% of the available water supply on coarse-textured soils and 30 to 40% on fine-textured ones from that soil volume fully explored by roots usually results in modest yield reductions but serious quality deterioration due to malformed tubers. In seasons of relatively high evapotranspiration (0.6 to 0.8 cm/day), this may necessitate irrigating every 3 or 4 days on coarse-textured soils and every 4 to 6 days on medium- and fine-textured ones.

All evidence supports the necessity of maintaining a uniform and relatively high soil water level, especially from the time tuber-bearing stolons begin growth until the tubers approach harvestable size. Most successful practice, therefore, entails fairly frequent and more or less uniformly spaced irrigations to avoid large fluctuations in soil water level. This probably can be achieved with minimum irrigation water use in surface irrigation systems by the widely used alternate-furrow system.

Table 32–2 gives general recommendations for scheduling irrigation of potatoes based upon water storage capacity of the surface 90 cm (approximately 3 ft) of soil and evapotranspiration rate. It is assumed that this soil depth contains no root-impeding layers, that the root system is free of depth-inhibiting disease, and that water is applied between each row, either by surface or sprinkler method in sufficient quantity to refill the volume of soil water storage capacity depleted.

During early growth of potatoes, roots generally will not have explored the upper 90-cm soil depth and the time interval between irrigations given in the above table must be reduced proportionately, often by as much as 50%. Actual evapotranspiration rates during this period are generally relatively low, however, so that irrigation at intervals of 7 to 10 days is usually sufficient, even when potential evapotranspiration is high.

Table 32–2. Irrigation frequencies for potatoes

Available soil water storage capacity, 0-90 cm	Rate of evapotranspiration, cm/day			
	0.3	0.5	0.7	0.9
cm	Days between irrigations*			
6.0 - 9.0	10-15	6-8	4-6	3-4
9.0 - 13.0	15-19	8-10	6-8	4-5
13.0 - 17.0	19-23	10-13'	8-10	5-6

* Where alternate-furrow irrigation is used, intervals indicated should be reduced by 1/4 to 1/3.

C. Rice Irrigation Practices

All available evidence confirms that rice produces best when grown under continuous submergence or with frequent water applications of 5 to 20 cm except where continuous saturation of the soil results in chemical reactions that produce physiological or nutritional disorders in the plant. Permitting the soil water content to drop below field capacity for brief periods between seedling establishment and tillering, during the internode elongation growth stage, and during grain formation and ripening, will generally provide oxidation sufficient to eliminate this problem. As a minimum, water sufficient to maintain the soil at or near saturation during seedling establishment, tillering and the 30-day period from 25 days before to 5 days after heading is essential for maximum production.

Where watergrass (*Echinochloa*) or other weeds are a serious problem, continuous flooding with water depth of 15 to 22 cm for 3 weeks after planting generally provides good control.

Use of irrigation water colder than about 15C will generally reduce plant growth and ultimate yield. If use of cold water cannot be avoided, its detrimental effects can be reduced by mixing it with warmer water drained from other fields before introducing it to the crop.

From the above recommendation it is apparent that good water control structures, adequate drainage facilities and land shaping to provide maximum water control during water application and within the field are important if use of irrigation water is to be most efficient. Maximum water control combined with care in scheduling planting can assure meeting the demands of this crop with minimum irrigation water supplies.

LITERATURE CITED

Adair, C. Roy, M. D. Miller, and H. M. Beachell. 1962. Rice improvement and culture in the United States. Advance. Agron. 14:61–108.

Aglibut, A. P., F. L. Valbuena, and A. A. Caoli. 1960. Irrigation and drainage in lowland rice production. Philippine Agr. 44(6):271–278.

Bendelow, V. M. 1958. The effect of irrigation on yield and malting quality of barley in southern Alberta. Can. J. Plant Sci. 38:135–138.

Box, J. E., W. J. Sletten, J. H. Kyle, and Alexander Pope. 1963. Effects of soil moisture, temperature, and fertility on yield and quality of irrigated potatoes in the Southern Plains. Agron. J. 55:492–494.

Carlson, C. W., D. L. Grunes, L. O. Fine, G. A. Reichman, H. R. Haise, J. Alessi, and R. E. Campbell. 1961. Soil, water, and crop management on newly irrigated lands in the Dakotas. US Dep. Agr. Prod. Res. Rep. No. 53. 34 p.

Chambliss, C. E., and J. W. Jones. 1925. Experiments in rice production in southwestern Louisiana. US Dep. Agr. Bull. No. 1356.

Chow, L. 1951. Rotational versus continuous irrigation methods in Taiwan. Joint Comm. Rural Reconstruction Eng. Ser. No. 3.

Corey, G. L., and Victor I. Myers. 1955. Irrigation of Russet Burbank potatoes in Idaho. Idaho Agr. Exp. Sta. Bull. 246. 18 p.

Erygin, P. S. 1936. Change in activity of enzymes soluble carbohydrates and intensity of respiration of rice seed germinating under water. Plant Physiol. 11:821–832.

Fernandez-G., R., and R. J. Laird. 1957. Yield and protein content of wheat in central Mexico as affected by available soil moisture and nitrogen fertilization. Agron. J. 49:20–25.

Ghose, R. L. M., M. B. Ghatge, and V. Subrahmanya. 1960. Rice in India. Indian Counc. Agr. Res., New Delhi. p. 35–39.

Haise, H. R., H. J. Haas, and L. R. Jensen. 1955. Soil moisture studies of some Great Plains soils: II. Field capacity as related to 1/3-atmosphere percentages and "minimum point" as related to 15- and 26-atmosphere percentages. Soil Sci. Soc. Amer. Proc. 19:20–25.

Henderson, D. W. 1967a. Grain sorghum irrigation: I. Root development and yield responses. Agron. J. (in press)

Henderson, D. W. 1967b. Grain sorghum irrigation: II. Plant characteristics and seed yield components. Agron. J. (in press)

Henkel, P. A. 1964. Physiology of plants under drought. Ann. Rev. Plant Physiol. 15: 363–386.

Howe, O. W., and H. F. Rhoades. 1955. Irrigation practice for corn in relation to stage of plant development. Soil Sci. Soc. Amer. Proc. 19:94–98.

Howe, O. W., and H. F. Rhoades. 1961. Irrigation of Great Northern field beans in western Nebraska. Univ. Nebraska Agr. Exp. Sta. Bull. 459. 28 p.

Jensen, Marvin E., and Howard R. Haise. 1963. Estimating evapotranspiration from solar radiation. Amer. Soc. Civ. Eng., Proc. Irrig. Drainage Div. J. 89(IR 4):15–41.

Jensen, Marvin E., and Jack T. Musick. 1960. The effects of irrigation treatments on evapotranspiration and production of sorghum and wheat in the Southern Great Plains. Int. Congr. Soil Sci., Trans. 7th (Madison, Wis., USA). 1:386–393.

Johnston, A. 1954. Preliminary notes on physiological diseases of rice in Malaya. Int. Rice Comm. Newsletter No. 10. p. 16–18.

Kmoch, H. G., R. E. Ramig, R. L. Fox, and F. E. Koehler. 1957. Root development of winter wheat as influenced by soil moisture and nitrogen fertilization. Agron. J. 41:20–25.

Leather, J. W. 1910. Water requirements of crops in India. Mem. Dep. Agr. India, Chem. Ser. 1(8):133–184.

Matsuo, T. 1957. Rice culture in Japan. Yokendo Ltd., Tokyo. p. 104.

Matsushima, S. 1962. Some experiments on soil water plant relationships in rice. Federation of Malaya, Div. Agr. Bull. 112. p. 10–35.

Mitsui, S. 1954. Inorganic nutrition, fertilization, and soil amelioration for lowland rice. Yokendo Ltd., Tokyo.

Musick, J. T., and D. W. Grimes. 1961. Water management and consumptive use by irrigated grain sorghum in western Kansas. Kansas Agr. Exp. Sta. Tech. Bull. 113. 20 p.

Musick, J. T., D. W. Grimes, and G. M. Herron. 1963a. Water management, consumptive use, and nitrogen fertilization of irrigated winter wheat in western Kansas. US Dep. Agr. Prod. Res. Dep. No. 75. 37 p.

Musick, J. T., D. W. Grimes, and G. M. Herron. 1963b Irrigation water management and nitrogen fertilization of grain sorghums. Agron. J. 55:295–298.

Ponnamperuma, F. N., R. Bradfield, and M. Peech. 1955. Physiological disease of rice attributable to iron toxicity. Nature 175:265.

Rhoades, H. F., O. W. Howe, J. A. Bondurant, and Fred B. Hamilton. 1954. Fertilization and irrigation practices for corn production on newly irrigated land in the Republican Valley. Nebraska Agr. Exp. Sta. Bull. 424. 26 p.

Rhoades, H. F., and L. B. Nelson. 1955. Growing 100-bushel corn with irrigation. US Dep. Agr. Yearbook Agr. p. 394–400.

Robins, J. S., and C. E. Domingo. 1953. Some effects of severe soil moisture deficits at specific growth stages in corn. Agron. J. 45:618–521.

Robins, J. S., and C. E. Domingo. 1956a. Moisture deficits in relation to the growth and development of dry beans. Agron. J. 48:67–70.

Robins, J. S., and C. E. Domingo. 1956b. Potato yield and tuber shape as affected by severe soil moisture deficits and plant spacing. Agron. J. 48:488–492.

Robins, J. S., and C. E. Domingo. 1962. Moisture and nitrogen effects on irrigated spring wheat. Agron. J. 54:135–138.

Robins, J. S., C. E. Nelson, and C. E. Domingo. 1955. 1954 irrigation and nitrogen fertilization experiments on sugar beets in the Columbia Basin. Washington Agr. Exp. Sta. Circ. 278. 6 p.

Robins, J. S., C. E. Nelson, and C. E. Domingo. 1956. Some effects of excess water application on utilization of applied nitrogen by sugar beets. J. Amer. Soc. Sugar Beet Tech. IX:181–188.

Schreiber, H. A., and C. O. Stanberry. 1965. Barley production as influenced by timing of soil moisture and timing of nitrogen applications. Agron. J. 57(5):442–445.

Sethi, R. L. 1930. Root development in rice under different conditions of growth. Mem. Dep. Agr. India, Bot. Ser. 18:2–57.

Singh, B. N., R. B. Singh, and K. Singh. 1935. Investigations into the water requirement of crop plants. Indian Acad. Sci., Proc. 1:471–495.

Sreenivasan, A., and V. Sadasivan. 1942. Nutritive value of the proteins and mineral constituents of dry and wet cultivated rices. Cereal Chem. 19:47–55.

Skazkin, F. D. 1960. The effect of surplus moisture on plants of different developmental periods. Soviet Plant Physiol. 7:225–229.

Stanberry, Chauncy O., and Mark Lowrey. 1965. Barley production under various nitrogen and moisture levels. Agron. J. 57(1):31–34.

Swanson, Norris P., and E. L. Thaxton, Jr. 1957. Requirements for grain sorghum irrigation on the High Plains. Texas Agr. Exp. Sta. Bull. 846. 14 p.

Ten Have, H. 1959. Influence of depth of water on the growth of rice. Avergedrukt Uit: De Surinaamse Landbouw. Jaargang 7:13–20.

Vacchani, J. 1953. Water requirements of rice. Int. Rice Comm. Newsletter No. 8. p. 6–10.

Vecht, van der. 1953. The problem of the mentek disease of rice in Java, Pemb. Balai Besar. Penj. Pert. No. 137:1–88. Bogor.

Wiersma, D. 1959. The soil environment and root development. Advance. Agron. XI:43–51.

33 | Sugar, Oil, and Fiber Crops

Part I—Sugar Beets

R. S. LOOMIS

University of California
Davis, California

JAY L. HADDOCK

Agricultural Research Service, USDA
Logan, Utah

The sugar beet (*Beta vulgaris*) is grown for the sucrose which accumulates in its fleshy axis during vegetative growth. The plant is adapted to a wide range of soil and climatic conditions and can be grown with a variety of cultural systems. It is grown widely on irrigated lands in semiarid and arid regions because of its high tolerance to saline conditions and its great productiveness in the long growing seasons common to such regions. Commercial irrigation practices for sugar beets are relatively simple and well developed. Our attention will be given here principally to some of the ecological aspects underlying these irrigation practices. Literature relating to the water and salt relations of the sugar beet under dryland conditions or with only occasional or marginal supplemental irrigation has been ignored in this review. There has been active research on these problems in several areas, particularly the USSR and Israel.

I. CROP DEVELOPMENT

Before examining the water relations of the sugar beet, it will be useful to consider briefly certain aspects of its development as a crop which influence its response to variations in water supply. During its first season of growth, an unvernalized sugar beet plant maintains an indeterminate pattern of vegetative growth. New leaves appear at regular intervals and photosynthates not utilized for growth of leaves and fibrous roots, or in basal metabolism, are available for storage root growth. This pattern dictates a simple set of management principles: Maximum yields will be obtained by maximizing the product of net photosynthesis × the proportion of photosynthate which accumulates in the root as sucrose × time.

The slowness with which young sugar beet seedlings become established limits effective utilization of the growing season and hence yields. Full cover is achieved slowly because of the small size of juvenile leaves and because only three to four new leaves are initiated each week (Loomis and Nevins, 1963; Morton and Watson, 1948). The commercial plant populations which give optimum yields (20,000 to 50,000 plants/acre in 20- to 30-inch rows) also minimize the time required to achieve a full foliage canopy while permitting the necessary tillage operations and the production of roots of a marketable size. A leaf-area index of 3 to 5, which provides full cover and hence maximum evapotranspiration (ET) per unit land area, may not be attained until 60 to 90 days after planting (leaf area index = area of leaves, one side only, per unit area of land surface). The leaf-area index may proceed to higher values of 6 to 12 and crop growth rate (dry matter accumulation per unit land surface) may increase in a diminishing relation with increasing leaf area. However, there is no evidence for an optimum leaf-area index beyond which total crop growth is reduced although maximum storage root yields may be achieved with less than maximum top development (Watson, 1958; and G. F. Worker and R. S. Loomis, 1961–1962. *Unpubl. work. Imperial Valley Field Sta., El Centro, Calif. Exp. no. 1X–8.*)

The root system of the sugar beet is a taproot type with numerous lateral branches, particularly in the first few feet of soil. Except with young stands or wide spacings, the depth and intensity of fibrous root development, as indicated by soil water extraction patterns, extend downward slowly and uniformly. Water is extracted most rapidly from the first few feet of soil but with deep, well-drained soils, appreciable amounts of water will also be obtained from depths of 4 to 5 ft. (Haddock, 1953). About 60% of the total seasonal water used may be acquired commonly from the top 2 feet (Larson and Johnston, 1955; Lawrence, 1953; and L. D. Doneen, *unpubl. work*). This proportion may be higher where root development is poor as on poorly-drained, fine-textured or compact soils or where irrigation is frequent and time does not permit appreciable water depletion in the less densely rooted regions of the profile.

Two components of sucrose yield are commonly considered: root yields and sucrose per cent. Root growth and sucrose storage begin early in the growth of the crop and continue at varying rates throughout its growing period. The storage root growth rate seems to be chiefly dependent on temperature and surplus assimilates from the tops (Ulrich, 1954, 1955). With rapid growth and warm air temperatures, 27 to 38C (80 to 100F), the concentration of sucrose in the roots is low (8 to 14% of the fresh weight). The higher concentrations (about 14 to 18%) desired for processing may be attained when photosynthesis is high but root growth is restricted by cool weather, 10 to 21C (50 to 70F), or by nitrogen deficiency (Loomis and Nevins, 1963; Ulrich, 1954, 1955). Because of this, harvesting is usually confined to the cool fall and spring seasons. These factors have an overriding influence on sucrose concentration and must be considered when interpreting the response to variations in water supply. The sensitivity of the sucrose concentration to environmental variations makes it a very useful index in research and commercial practice.

II.   EVAPOTRANSPIRATION

As with other crops, we do not know enough about the crop and weather factors which determine consumptive use of water by sugar beets. Evaporation from the soil surface is a small but important component of water loss before a full plant cover is attained. With full cover, Pruitt (1960) has shown that ET for the sugar beet will approximate that of a short, green grass sward. Wind causes ET to exceed that of a grass sward apparently because of turbulence induced by the greater roughness of the sugar beet crop. Peak rates at Davis,

California, USA are near 0.3 inches/day with seasonal consumptive use exceeding 40 inches. Evapotranspiration at Logan, Utah, USA also approaches 0.3 inches/day (Taylor and Haddock, 1956); in Montana, Larson and Johnston (1955) reported that seasonal consumptive use varied from 20 to 30 inches. Evapotranspiration values for other environments are not generally available; they can be estimated only approximately by present procedures for calculating potential ET from climate factors. For example, the Blaney and Criddle (1950) formula gives an estimate of only 20 inches for the Davis, California environment. (*See* chapter 29 for discussion of methods for estimating ET values.) Irrigation requirements, varying as they do with soil type, method of irrigation, and other factors in addition to potential ET, range from 20 to 50 inches for various beet-producing areas in the USA.

III.   EFFECTS OF WATER STRESS

Internal water deficits induce a series of morphological and physiological changes in the sugar beet plant. Under stress conditions, lamina area and length of petioles are usually less than with adequate moisture, but the initiation of new leaves is affected little. There are fewer and smaller mesophyll cells in leaf lamina but the number of cell layers (Morton and Watson, 1948; Shah and Loomis, 1965) and leaf color (Burman and Painter, 1964) do not seem to be affected. The senescence of old leaves may be accelerated by stress thus reducing leaf longevity. The resulting reduction in leaf area may represent an increase in xeromorphy (Stocker, 1960), but in dense stands it probably has little effect on total water loss unless cover is reduced below 100%.

Growth of the storage root in size and fresh weight is also retarded by stress but the relative effects of water deficits on cell division and cell enlargement have not been evaluated. Dry matter accumulation in the root and the proportion of that which occurs as sucrose seems unaffected except by prolonged or repeated wilting (Loomis and Worker, 1963; Shah and Loomis, 1965). Sucrose concentration (as per cent of fresh weight) usually increases appreciably with plant-water deficits, due apparently to a slower accumulation of water relative to sucrose rather than to any appreciable net loss of water from the root or more rapid accumulation of sucrose. Experience indicates that the content and composition of molassagenic factors (soluble nonsucrose materials) may increase sometimes.

Under field conditions, considerable attention has been given to whether irrigation should be withheld so that the crop will wilt immediately prior to harvest. Such practices are based in part on a common misconception that the resulting increase in sucrose concentration represents a real increase in quality. An example of the responses observed with such treatment is given in Fig. 33I–1 and 33I–2. While preharvest wilting may offer savings in irrigation and hauling costs it usually increases harvest costs on soils with a high clay content, reduces storage and slicing quality through lower turgidity of the root, and sometimes reduces the extraction of sucrose. Total sucrose yields are not increased and may be reduced significantly by single, brief cycles, and almost invariably are lowered by repeated wilting apparently due to reduction in leaf area and rate of photosynthesis. This is illustrated with field data in Table 33I–1 where, incidentally, as in Fig. 33I–1 and 33I–2, the size of tops is affected relatively more by water deficits than is root

Fig. 33I-1. Yields of fresh tops from a well established sugar beet crop grown on a deep, saline
clay loam soil with high and low nitrogen nutrition at the Imperial Valley Field Station, El
Centro, Calif. The data show the response during a single wilting cycle. The dry plots (D)
received their last irrigation on April 25 and
began to wilt 3 weeks later. The wet plots (W)
continued to receive ample water. The relative
effects of stress at the two nitrogen levels was
similar (Loomis and Worker, 1963).

High N = 600 lb N/acre
Low N = 100 lb N/acre
W = wet plots, irrigated every 10 days
D = dry plots, last irrigated April 25

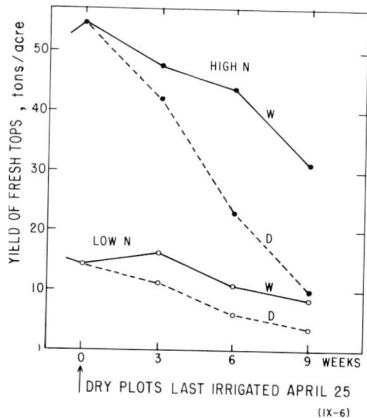

Fig. 33I-2. Root yields and their sucrose percentage (fresh basis) for the experiment
shown in Fig. 33I-1. Restrictions in root
growth due to water deficits were apparent
by the time the plants first wilted, 3 weeks
after the last irrigation, while sucrose per
cent increased above the control only with
continued wilting. Although deficient in nitrogen by tissue analysis and visual symptoms (note top size in Fig. 33I-1), the
low-N plants continued to receive enough
nitrogen for near maximum (but variable)
root growth (1.3 tons/acre per week compared to 1.5 tons/acre per week). Sucrose
yields were similar for all treatments up to 6
weeks but higher for the wet plots at 9
weeks. Apparent purities were higher for
low-N beets but were not affected by stress
(Loomis and Worker, 1963).

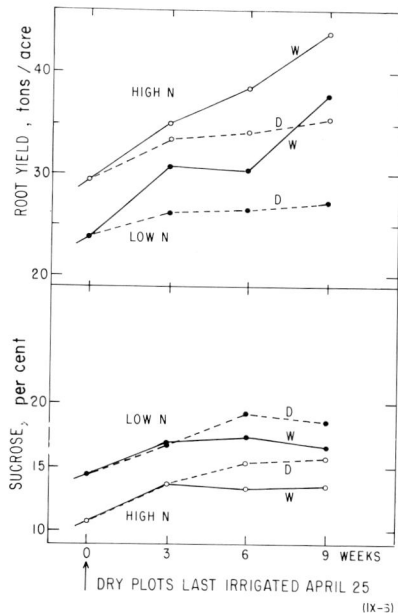

growth. From a physiological standpoint, wilting at any time is undesirable—in
commercial practice, the relative economic advantages and disadvantages will
have to be weighed.

At the molecular level, protein content of various tissues, especially in leaves,
is greatly affected by plant-water deficits even when wilting is not evident (Shah
and Loomis, 1965). Protein hydrolysis apparently increases and at the same time
the ribonucleic acid-dependent apparatus for protein synthesis is disrupted.
Changes in the levels of specific proteins (e.g., enzymes) have not been analyzed,
but this may account for fewer cells being produced in expanding leaves. Rewatering reverses such effects.

IV. RECOVERY FROM STRESS

Restitution of normal growth patterns occurs rapidly when water stress is alleviated. Leaf expansion and longevity and root enlargement rates return to normal (Morton and Watson, 1948; Shah and Loomis, 1965); the regrowth is initiated in some cases by only small amounts of water relative to the soil water deficit (Owen, 1958). If accumulated sucrose were utilized during restitution, growth in size and fresh weight might exceed temporarily the normal rate for plants of similar leaf area and age. While utilization of stored sucrose during regrowth has not been demonstrated, relative leaf growth rates (increases in leaf area per unit leaf area) and net assimilation rates (dry matter accumulation per unit leaf area) may be temporarily higher than with unstressed plants (Orchard, 1963; Owen, 1958). Such effects are expected if stress causes the number and size of leaves and mutual shading to decrease but Orchard (1963) feels that more than changes in leaf area may be involved. Compensatory growth of roots has not been observed. Further studies on the kinetics of regrowth phenomena would be helpful.

V. CORRELATION OF PLANT GROWTH WITH SOIL WATER

Measurement of soil water potentials and their relation to growth and the need for irrigation has led to varying conclusions. Soil water withdrawal patterns cause mean soil water potentials to vary widely with depth during each irrigation cycle. The resulting soil water stress is typified by the tension values shown in Fig. 33I–3 for a sugar beet field in Utah in midseason. From such data and yield information it was concluded that, for maximum root yields, mean integrated soil water tensions in a loam soil should not fall below about 1.0 bars. Doneen's data (Doneen, 1942; and *unpubl. work of L. D. Doneen*) for the same time of season, but at Davis, California, with an earlier planting date, show water extraction from greater soil depths and more complete extraction at comparable depths, presumably reflecting deeper and more completely branched roots. Under these circumstances with moderate ET (0.2 to 0.3 inches/day), water depletion to the wilting point in the first few feet of soil had no influence on root yields and the need of such well rooted crops for irrigation was indicated by integrated soil water potentials

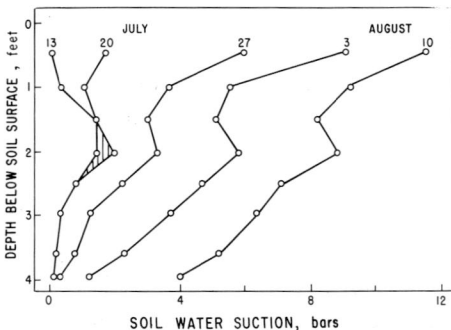

Fig. 33I–3. Soil water suction at various depths in the root zone of sugar beets and at various times during an irrigation cycle. The data were obtained during midseason on a late planted crop grown on Millville silt loam in Utah (after Taylor and Haddock, 1956).

Table 33I-1. Harvest results from a sugar beet experiment which illustrate the effect of repeated irrigations at different levels of available moisture depletion:[*] The "dry" treatment was irrigated at the first sign of wilting; the "dry + stress" treatment was allowed to wilt for several days before irrigation

Irrigation regime	No. of irrigations	Beet root yield	Sucrose	Sucrose	Yield of fresh tops	Thin juice purity[†]
		ton/acre	% fresh wt.	cwt/acre	tons/acre	%
Wet	13	31.6	12.2	76.0	30.4	87.5
Medium	8	29.8	12.9	77.2	24.2	87.8
Dry	5	28.1	13.2	75.1	23.0	88.1
Dry + stress	3	26.4	13.6	72.0	19.3	87.2
LSD (.05)		1.5	0.4	3.9	2.9	N.S.

[*] Unpublished data of L.D. Doneen and R.S. Loomis. The crop was planted at Davis, California on May 14, 1961 on a deep, well-drained, Yolo loam. Means in the table represent four replications of two harvest dates, Sept. 14 and Oct. 27, and two nitrogen levels.

[†] Thin juice purities were performed by the Spreckels Sugar Co. by a modification of Brown and Serro's technique.

diverging much farther from the reference state of pure water. Data from late-planted crops at Davis (Table 33I-1) show an intermediate behavior.

Such results illustrate clearly that the magnitude of the plant-water deficit depends on the relative rates of absorption and transpiration of water and is affected by the extent of root development, leaf area, and evaporative conditions as well as by soil water potential. Correlations of the relative turgidity of various plant parts with growth have not been reported for sugar beet but should provide more satisfactory indices of the need for irrigation than soil water potentials when plant-water deficits develop slowly due to gradual depletion of soil water. However, extreme plant-water deficits may develop quickly within a day under highly evaporative conditions when hot, dry winds supply advective heat for evaporation. The sugar beet appears more susceptible than other field crops to wilting in dry winds even when root development and soil water potentials are near optimum.

Because of the lack of consistent relationships between sugar beet growth and soil water content in the root zone, progress has been slow in adopting tensiometers, gypsum blocks or other means of measuring soil water conditions as guides to irrigation. Within a geographical region where soil, root development, potential ET patterns and other factors are similar, satisfactory empirical relationships can sometimes be established by experiment so that critical water levels in a particular portion of the root zone can be used to indicate the need for irrigation (Richards and Marsh, 1961). In Utah, Haddock recommends that irrigation should begin when a 12-inch tensiometer reads about 0.5 bars in sandy loam, 0.8 bars in loam and 2.0 bars in clay loam, but at Davis, California, these criteria would result in unnecessarily frequent irrigations of beets sown in February on deep soils.

VI. IRRIGATION AND PLANT NUTRITION

Soil water potential may influence availability and uptake of mineral nutrients as well as their assimilation within the plant. Many factors are involved and their interrelationships are so poorly defined that it is difficult or impossible to characterize the variations in sugar beet nutrition which may result.

Nitrogen nutrition is a dominant determinant of sugar beet yield and quality and its relation to irrigation has been studied in some detail. The N-deficient sugar beet plant is quite different in leaf growth and display from the plant high in N and thus might be expected to respond differently to soil water stress. However, as illustrated in Fig. 33I–1 and 33I–2, no interactions between growth habit and soil water stress were observed during single drying cycles (Loomis and Worker, 1963). Low soil water potential (i.e., high stress) may reduce total uptake of N (Mackenzie et al., 1957) while plant nutrient status may be either lowered or improved as evidenced by changes in the concentration of NO_3-N in petioles (Haddock, 1953, 1959; Loomis and Worker, 1963) or of total or soluble-N in various tissues (Haddock, 1954). Apparently reductions in growth and N assimilation due to water stress may be equal to or greater than the reductions in N uptake. However, if root activity in the fertilized soil is limited by low soil water potential, plant N status may be improved by increasing the supply of water (Henderson et al., 1966; Robins et al., 1956. Walker and Hac, 1952).

Such responses may lead to or preclude the occurrence of significant interactions between N and water management. Interactions were not observed in some California studies (Loomis and Worker, 1963; MacKenzie et al., 1957) (e.g., Table 33I–1), but they have been reported (Robins et al., 1956) where excess irrigation led to a low N supply because of leaching. Interactions also may occur if low soil water potential delays N utilization and the development of N deficiency thus preventing early development of a high sucrose per cent (Haddock, 1953, 1959).

The sensitivity of growth and sucrose concentration to N deficiency creates some special problems. With furrow irrigation, soluble N may accumulate in the surface of the beds. Rainfall will move this N into the root zone and cause N-deficient plants to renew growth with a resultant decline in sucrose concentration and perhaps in sucrose yield (Loomis and Worker, 1964; Stout, 1964). Such surface isolation of soluble N may be reduced by employing sprinkler irrigation (Haddock, 1959) or, with furrow irrigation, by deep placement rather than broadcasting (Robins et al., 1956; Stout, 1964).

The uptake and assimilation of other nutrients are generally affected similarly by water stress. In some cases, as with phosphorus, availability of the nutrient is markedly reduced at low soil water potentials leading to low concentrations in the plant (Haddock, 1952). Deep placement of phosphorus (at 16 inches) is effective in improving the phosphorus status of plants on subirrigated lands where available water is depleted from the surface soil (Henderson et al., 1968).

VII.   SOME OTHER ASPECTS OF IRRIGATION MANAGEMENT

Since high sucrose yields are dependent on the extended vegetative growth of the full canopied crop, optimum water management in midseason consists of applying the minimum amount of water required for maintaining evapotranspiration and thus avoiding plant-water deficits which might cause wilting and growth inhibition.

Both sprinkler or furrow irrigations are satisfactory with sugar beets while flood irrigation may contribute to pathogen problems (fungal root rot) and is not suitable. Sprinklers provide the best control of water distribution, prevent surface isolation of nutrients, and, on coarse-textured soils, conserve water and reduce leaching of nutrients. The choice between sprinkler and furrow irrigation is an economic one dependent on the relative costs and advantages of the systems under conditions prevailing in a given situation.

Irrigation management during stand establishment receives too little attention by most growers. Planting on beds permits early furrow irrigation where rainfall is absent or uncertain. Where plantings are made during the dry season, repeated light irrigations are necessary to ensure uniform stands. If successive irrigations are close together, late-germinating seed will emerge early enough to contribute to the final stand. These irrigations and rainfall or preirrigations should be adequate to bring the entire soil profile of 4 to 6 ft to field capacity early in the growing season. Subsequent irrigations need replenish only the depleted water (with allowance for necessary leaching of salts) and may be shallow until evapotranspiration rates approach high levels.

A well-established sugar beet plant easily may tolerate soils with electrical conductivity (EC_e) values near 8 to 12 mmho/cm but germination may be appreciably delayed and reduced at 6 mmho/cm (Richards, 1954). Under saline conditions, furrow irirgations should be conducted with double-row beds, slant beds, alternate furrow irrigation or some other method which avoids salt accumulation near the young seedlings (*see* chapter 51).

VIII. SUMMARY

Sucrose production from sugar beets is dependent on maximizing vegetative growth over a long growing season. Irrigation practices should be directed toward rapid stand establishment and early attainment of a full leaf canopy. Midseason irrigations should be sufficient to avoid periodic water stress which might reduce root growth and the photosynthetic effectiveness of the foliage canopy. Depth and frequency of irrigations are dependent on root distribution and ET, which, with full cover, approximates the ET of a short green crop. Preharvest water stress is of no benefit to sucrose yield or crop quality.

LITERATURE CITED

Blaney, H. F., and W. D. Criddle. 1950. Determining water requirements in irrigated areas from climatological and irrigation data. US Dep. Agr. Soil Conserv. Serv. Bull., Tech. Pap. 96. 48 p.

Burman, R. D., and L. I. Painter. 1964. Influence of soil moisture on leaf color and foliage volume of beans grown under greenhouse conditions. Agron. J. 56:420–423

Doneen, L. D. 1942. Some soil-moisture conditions in relation to growth and nutrition of the sugar beet plant. Amer. Soc. Sugar Beet Technol., Proc. 3:54–62.

Haddock, J. L. 1952. The influence of soil moisture condition on the uptake of phosphorus from calcareous soils by sugar beets. Soil Sci. Soc. Amer. Proc. 16:235–238.

Haddock, J. L. 1953. Sugar beet yield and quality. Utah Agr. Exp. Sta. Bull. 362. 72 p.

Haddock, J. L. 1954. The interrelationships of irrigation method, soil moisture condition, and soil fertility on the yield, quality and nitrogen nutrition of sugar beets p. 121–135. In P. Prevot [ed.] Symp. on Plant Analysis and Fertilizer Problems. Int. Bot. Congr.

Haddock, J. L. 1959. Yield, quality, and nutrient content of sugar beets as affected by irrigation regime and fertilizers. J. Amer. Soc. Sugar Beet Technol. 10:344–355.

Henderson, D. W., F. J. Hills, R. S. Loomis, and E. F. Nourse. 1968. Soil moisture conditions, nutrient uptake, and growth of sugar beets as related to method of irrigation of an organic soil. J. Amer. Soc. Sugar Beet Technol. Vol. 15. (In press)

Larson, W. E., and W. B. Johnston. 1955. The effect of soil moisture level on yield, consumptive use of water and root development by sugar beets. Soil Sci. Soc. Amer. Proc. 19:275–279.

Lawrence, G. A. 1953. Furrow irrigations. US Dep. of Agr. Soil Conserv. Serv. Leafl. 344. 8 p.

Loomis, R. S., and D. J. Nevins. 1963. Interrupted nitrogen nutrition effects on growth, sucrose accumulation and foliar development of the sugar beet plant. J. Amer. Soc. Sugar Beet Technol. 12:309–322.

Loomis, R. S., and G. F. Worker, Jr. 1963. Responses of the sugar beet to low soil moisture at two levels of nitrogen nutrition. Agron. J. 55:509–515.

Loomis, R. S., and G. F. Worker, Jr. 1964. Restitution of growth in nitrogen deficient sugar beet plants. J. Amer. Soc. Sugar Beet Technol. 12:657–665.

MacKenzie, A. J., K. R. Stockinger, and B. A. Krantz. 1957. Growth and nutrient uptake of sugar beets in the Imperial Valley, California. J. Amer. Soc. Sugar Beet Technol. 9:400–407.

Morton, A. G., and D. J. Watson. 1948. A physiological study of leaf growth. Ann. Bot. New Ser. 12:281–310.

Orchard, B. 1963. The growth response of sugar beet to similar irrigation cycles under different weather conditions. p. 340–355. In: The water relations of plants. Blackwell Sci. Publ., Oxford, England.

Owen, P. C. 1958. The growth of sugar beet under different water regimes. J. Agr. Sci. 51:133–136.

Pruitt, W. J. 1960. Correlation of climatological data with water requirement of crops. Dep. Irri. Univ. California, Davis. Mimeogr. Rep. 91 p.

Richards, L. A. [ed.] 1954. Diagnosis and improvement of saline and alkali soils. US Dep. Agr. Handbook No. 60. 160 p.

Richards, S. J., and A. W. Marsh. 1961. Irrigation based on soil suction measurements. Soil Sci. Soc. Amer. Proc. 25:65–69.

Robins, J. S., C. E. Nelson, and C. E. Domingo. 1956. Some effects of excess water application on utilization of applied nitrogen by sugar beets. J. Amer. Soc. Sugar Beet Technol. 9:180–188.

Shah, C. B., and R. S. Loomis. 1965. Ribonucleic acid and protein metabolism in sugar beet during drought. Physiol. Plant. 18:240–254.

Stocker, O. 1960. Physiological and morphological changes in plants due to water deficiency. p. 63–104. In: Plant water relationships in arid and semiarid conditions. Arid Zone Res. Rev. Res. UNESCO., Paris. Vol. 15.

Stout, M. 1964. Redistribution of nitrate in soils and its effects on sugar beet nutrition. J. Amer. Soc. Sugar Beet Technol. 13:68–80.

Taylor, S. A., and J. L. Haddock. 1956. Soil moisture availability related to power required to remove water. Soil. Sci. Soc. Amer. Proc. 20:284–288.

Ulrich, A. 1954. Growth and development of sugar beet plants at two nitrogen levels in a controlled temperature greenhouse. Amer. Soc. Sugar Beet Technol. Proc. 8(2):325–338.

Ulrich, A. 1955. Influence of night temperature and nitrogen nutrition on the growth, sucrose accumulation and leaf minerals of sugar beet plants. Plant Physiol. 30:250–257.

Walker, A. C., and Lucile R. Hac. 1952. Effect of irrigation practices upon the nitrogen metabolism of sugar beets. Amer. Soc. Sugar Beet Technol., Proc. 7:58–66.

Watson, D. J. 1958. The dependence of net assimilation rate on leaf-area index. Ann. Bot. New Ser. 22:37–54.

33

Sugar, Oil, and Fiber Crops

Part II—Sugarcane

ROBERT B. CAMPBELL

Hawaiian Sugar Planters' Association
Honolulu, Hawaii

Sugarcane (*Saccharum officinarum*) is a tropical crop which is grown predominantly between 30°N and 30°S latitude. The climatic boundary for sugarcane in the Northern and Southern Hemispheres is further defined by the coldest mean monthly temperature isotherm of 18.3C (65F). In many of these areas, crop production is often restricted by limited seasonal rainfall. This situation emphasizes the need for irrigation, water development, and effective water management practices.

I. ROOTING HABITS AND MOISTURE EXTRACTION

Sugarcane is a fibrous-rooted crop whose roots are most active in the first 2 or 3 ft of surface soil. Active rootlets have been noted to extend to depths of 8 ft or more. Rooting patterns vary with the physical condition of soils—particularly machine-compacted soil layers, moisture distribution, aeration, and their genetic profile characteristics (Baver et al., 1962; Humbert, 1954; Lee and Weller, 1927; Trouse and Humbert, 1959, 1961; Yamasaki, 1956).

As with most crops after irrigation, the soil water suction increases most rapidly in the surface soil, followed by a gradual rise in successively lower soil layers. The area from which the greatest quantity of water will be withdrawn at any specific time from a uniform soil profile does not depend entirely on the quantitative root distribution or the soil matric potential but rather on a combination of these factors.

This is illustrated in Table 33II–1 where, between two critical stages of cane stalk elongation rate, the greatest water extraction occurred at the 3-ft depth. Under field practice, it is often desirable to irrigate before the cane elongation rate decreases. Robinson (1963a) has shown that the elongation rate of cane stalks declines as the matric potential approaches –200 centibars at the 12-inch depth.

II. EVAPOTRANSPIRATION

Evapotranspiration (ET) represents the quantity of water which must be applied at suitable intervals to maintain a soil water balance favorable to plant growth. The importance of this concept in sugarcane production is evidenced

Table 33II–1. Soil matric potential profile at two stages of growth after irrigation of sugarcane (Robinson, 1963b)

Soil layer	Water suction, cane elongation rate		Water extraction
	Declines	Ceases	
ft	cbars	cbars	inches
1	−400	−600	0.12
2	−230	−490	0.20
3	− 56	−110	0.30
4	− 57	− 66	0.12
5	− 15	− 2.3	0.26

Table 33II–2. Various ET/E_{pan} ratios for estimating evapotranspiration[*] at various stages of growth in the crop cycle (Campbell et al., 1959)

Period in crop cycle	Crop age, months	ET/E_{pan} ratio
Partial canopy		
Planting to 1/4 full canopy	0 - 2.5	0.4
1/4 to 1/2 full canopy	2.5 - 3.5	0.6
1/2 to 3/4 full canopy	3.5 - 4.5	0.8
3/4 to full canopy	4.5 - 6.0	0.9
Peak use	6.0 - 17	1.0
Early senescence	17 - 22	0.7
Ripening	22 - 24	No irrigation

[*] Evapotranspiration (ET) = (ET/E_{pan}) ratio × US Weather Bureau pan evaporation.
[†] ET = evapotranspiration, water evaporation from soil plus water loss to the atmosphere through plants. E_{pan} = water evaporation from standard US Weather Bureau pan.

by several ET studies using field-installed lysimeters (Campbell et al., 1959; Chang, 1961; Cowan and Innes, 1956; Fuhriman and Smith, 1951; Pearson et al., 1961; Thompson et al., 1963). Studies by Blaney (1962) and Penman (1948) on evapotranspiration and particularly the investigations by Chang (1961) and Thompson et al. (1963) on water requirements of sugarcane have suggested the use of evapotranspiration/pan evaporation (ET/E_{pan}) ratios as a basis for estimating potential evapotranspiration for sugarcane irrigation (*also see* chapter 30).

The crop cycle of sugarcane may be conveniently divided into four irrigation periods as shown in Table 33II–2 in which different ET/E_{pan} ratios apply for estimating potential evapotranspiration, i.e. planting to full canopy, peak-use, early senescence, and ripening.

Although it is recognized that these ratios will be modified by frequency of irrigation, capillary conductivity of soil, rate of canopy development, and method of irrigation, Table 33II–2 illustrates a method for estimating the evapotranspiration from pan evaporation data.

The climatological approach for estimating ET of sugarcane has focused attention on the quantitative aspects of the irrigation requirements. The field irrigation requirement for a crop is based upon its ET and the distribution characteristics of the irrigation system used. As the efficiency of these irrigation systems is improved, more exact ET data will be needed for development of better water management practices (*see* chapter 30).

Table 33II–3. Water-to-cane weight ratios derived from sugarcane grown in lysimeters for three important cane production areas in the tropics

Location	Age of crop at harvest	Crop cycle	Water to cane weight ratio	
			(Avg. of 3)	(Avg. of 9)
Puerto Rico, S. Coast	12	Plant	83. 9	–
(Fuhriman and Smith, 1951)	12	Plant	70. 8	–
	12	Plant	72. 6	75. 8
Hawaii, Maui (1957-1958)	20	Plant	73. 3	–
	18. 5	Plant	99. 5	–
	16	Plant	72. 6	81. 8
Hawaii, Maui (1960-1961	13	Plant	76. 5	–
	13	Plant	102. 0	–
	13	Plant	84. 1	87. 5
S. Africa, Chaka Kraal	12	Plant	71. 1	–
(Thompson et al. , 1963)	10. 5	1st ratoon	62. 3	–
	11	2nd ratoon	70. 1	67. 8
S. Africa, Tongaat	12	Plant	73. 7	–
(Thompson et al. , 1963)	11	1st ratoon	58. 5	–
	15. 5	2nd ratoon	102. 2	78. 1
			Grand average	78. 2+

* Standard deviation ± 13. 6 (standard error of mean ± 3. 63). Coefficient of variability = 17. 4%. Difference between countries not significant at $P \leq 0.05$.

III. WATER-YIELD RELATIONSHIPS

For many crops in particular climates, transpiration has been shown to be linearly associated with dry matter production (Arkley, 1963; Briggs and Shantz, 1914; DeWitt, 1958). A transpiration dry matter production study (Hawaiian Sugar Planters' Ass., 1963) in 19.6C (67F) and 23.8C (75F) controlled temperature greenhouses showed that 135 g of water produced 1 g of dry matter.

Cane stalks harvested from this study contained 78.2% water, and they contributed 59% of the total dry matter (*Memorandum from L. G. Nickell, Exp. Sta., Hawaiian Sugar Planters' Ass., Honolulu, Hawaii.*). By taking into account these two percentage values, a calculation shows that it took 50 g of water to produce 1 g of cane. This compares with a value of 56.1 based on 24 replications for Jamaica (Cowan and Innes, 1956). Water-to-cane weight ratios shown in Table 33II–3 for Puerto Rico (Fuhriman and Smith, 1951), Hawaii, and South Africa (Thompson et al., 1963) show a mean value of 78.2 ± 7.8 fiducial limit at 5% probability level. However, these values include evaporation from the soil in addition to transpiration losses. This may indicate that soil evaporation losses may constitute roughly 28% of ET for a 12-month crop. These results may be used as a first approximation for defining ET in relation to cane yield.

A small plot irrigation experiment having controlled uniformity, amount, and frequency of water application provides an insight into the application requirement of sugarcane. For a 2-year crop in Hawaii, 85 to 110 inches of irrigation water, provided by 35 to 44 irrigations, was needed to supplement useful annual rainfall of 19 inches. These data are shown in Table 33II–4.

Under these conditions, approximately 100 lb of water produced 1 lb of cane or about 1.2 tons of cane/acre-inch of water.

Table 33II–4. Yield response to various levels of water application to a 2-year crop of sugarcane*

No. of irrig.	Source of water			Cane yield	Cane weight to water ratio	Water to cane weight ratio
	Net irrig. water	Effective rainfall	Irrig. plus effective rain			
			inches		tons/acre	tons/acre-in.	lb/lb
24	63.3	38.8	102.1	123.4	1.21	94
30	78.5	37.3	115.8	139.4	1.21	94
35	87.9	38.3	126.2	145.3	1.15	99
44	110.6	28.4	139.0	152.9	1.10	102
48	122.1	32.0	154.1	149.1	0.96	118
56	145.0	27.1	172.1	151.7	0.88	129

* Exp. Sta. Hawaiian Sugar Planters' Ass., Waipio Field L, Variety H50-7029, with five replications.

Field irrigation requirements at Ewa Plantation (Renton and Alexander, 1926) were 4.1 inches by irrigation plus 3 inches of rain in January, as compared to 12.5 inches by irrigation and 0.25 inch of rain in July. Water required (Alexander, 1923) at field edge for a level-ditch irrigation system was 230 acre-inches for maximum growth of a 24-month crop.

IV. SCHEDULING IRRIGATION

The rate of stalk elongation in the sugarcane plant is sensitive to soil water changes. This characteristic is fundamental in establishing the opportune time for water application following the drying cycle. Since 1930, when Shaw (1930) initiated the practice of scheduling irrigations on the basis of soil water content, both tensiometer type instruments (Richards, 1942) and resistance blocks (Bouyoucos and Mick, 1940) have been used commercially to schedule irrigation.

Climatic estimation of ET in which evaporation pans are used has a particular appeal in ocean climates (Baver, 1954) for managing irrigation water, because advective heat comprises a relatively small fraction of the total energy budget. It is important to recognize limitations of the climatological approach in irrigation. In areas of high advection, e.g., climatological indices should be based upon climatic parameters for each area and upon the actual ET of the crop grown. As sugarcane ET concepts are used in agriculture, and, in turn, as estimates of ET improve, the water balance method provides a useful technique for irrigation scheduling and for yield prediction work. This method utilizes soil water storage, rainfall, and growth status of the crop and appears to be particularly adaptable to overhead irrigation.

Irrigation scheduling by soil instrumentation has been widely used for timing the application of irrigation water. The following recommendations appear to be satisfactory bases upon which to irrigate with resistance blocks and tensiometers:

1) Place resistance blocks 9 to 12 inches below the cane stool. Irrigate when the block reaches 5,000 ohms resistance (Ewart, 1953).

2) Place the porous tensiometer cup 20 inches below the bottom of the furrow (Waterhouse and Clements, 1953). Irrigate when tensiometer reading reaches 25 to 35 scale units (approximately –25 to –35 centibars).

For a more detailed discussion of irrigation scheduling, see chapter 30.

V. SUMMARY

The sugarcane crop is sensitive to relatively slight changes in soil water suction, and it is also drouth tolerant.

In the tropics, the ET/E_{pan} ratio is potentially a useful method for estimating potential ET for various stages of canopy development. Data indicate that this ratio changes from $ET/E_{pan} = 0.4$ for bare soil to about $ET/E_{pan} = 1.0$ for full canopy.

Evapotranspiration and yield measurements obtained from infield drainage lysimeters show that 78 lb of water produced 1 lb of sugarcane. Depending on the efficiency of irrigation and the amount of effective rainfall, irrigation requirements were found to be 100 lb or more of water/lb of cane.

The gypsum resistance block and tensiometer soil water instruments have been used commercially for scheduling the application of irrigation. Specific recommendations are given.

LITERATURE CITED

Alexander, W. P. 1923. The irrigation of sugarcane in Hawaii. (M.A. Thesis, Univ. Hawaii, 1922) printed by Hawaiian Sugar Planters' Ass. 109 p.

Arkley, R. J. 1963. Relationships between plant growth and transpiration. Univ. California Exp. Sta. Hilgardia. 34:559–584.

Baver, L. D. 1954. The meteorological approach to irrigation control. Hawaiian Planters' Rec. 54:291–298.

Baver, L. D., H. S. Brodie, T. Tantimoto, and A. C. Trouse. 1962. New approaches to the study of cane root systems. Int. Soc. Sugar Cane Technol., Proc. 11th Congr. (ISSCT) p. 248–253.

Blaney, H. F. 1962. Determining consumptive use and irrigation water requirements. US Dep. Agr. Tech. Bull. 1275. 55 p.

Bouyoucos, G. J., and A. H. Mick. 1940. An electrical resistance method for the continuous measurement of soil moisture under field conditions. Michigan Agr. Exp. Sta. Bull. 172. 38 p.

Briggs, L. J., and H. L. Shantz. 1914. Relative water requirements of plants. J. Agr. Res. 3:1–63.

Campbell, R. B., J. H. Chang, and D. C. Cox. 1959. Evapotranspiration of sugarcane in Hawaii measured by in-field lysimeters in relation to climate. Int. Soc. Sugar Cane Technol., Proc. 10th Congr. p. 637–649.

Chang, J. H. 1961. Microclimate of sugarcane. Hawaiian Planters' Rec. 56:195–225.

Cowan, I. R., and R. F. Innes. 1956. Meteorology, evaporation and the water requirements of sugarcane. Int. Soc. Sugar Cane Technol., Proc. 9th Congr. p. 1–20.

DeWit, C. T. 1958. Transpiration and crop yields. Verslag. van Landbouwk. Onderzoek. 64.6. 88 p.

Ewart, G. Y. 1953. Background and results of recent irrigation timing tests in American Factors Plantations. Hawaiian Planters' Rec. 54:257–269.

Fuhriman, D. K., and R. M. Smith. 1951. Conservation and consumptive use of water with sugarcane under irrigation in the south coastal area of Puerto Rico. Univ. Puerto Rico. J. Agr. 35:1–47.

Hawaiian Sugar Planters' Association. 1963. Transpiration related to dry weight production. Ann. Rep. Exp. Sta. p. 4.

Humbert, R. P. 1954. Water distribution studies in the Hawaiian sugar industry. Hawaiian Planters' Rec. 54:211–225.

Lee, H. A., and D. M. Weller. 1927. The progress of sugarcane roots in the soil at different ages. Rep. Hawaiian Sugar Tech. p. 69–72.

Pearson, C. H. O., T. G. Cleasby, and G. D. Thompson. 1961. Attempt to confirm irrigation control factors based on meteorological data in the cane belt of South Africa. So. African Sugar Tech. Ass., Proc. p. 1–6.

Penman, H. L. 1948. Natural evaporation from open water, bare soil and grass. Proc. Roy. Soc. 193:120–145.

Renton, G. F., and W. P. Alexander. 1926. An investigation to determine the relation of irrigation water to maximum sugar yields. Hawaiian Sugar Planters' Ass., 46th Ann. Mtg. p. 36–95.

Richards, L. A. 1942. Soil moisture tensiometer materials and construction. Soil Sci. 53:241–248.

Robinson, F. E. 1963a. Soil moisture tension, sugarcane stalk elongation, and irrigation interval control. Agron. J. 55:481–484.

Robinson, F. E. 1963b. Using neutron meter to establish soil moisture storage and soil moisture withdrawal by sugarcane roots. Rep. Hawaiian Sugar Tech. p. 206–208.

Shaw, H. R. 1930. Studies on the response of cane growth to moisture. Rep. Hawaiian Sugar Tech. p. 127–147.

Thompson, G. D., C. H. O. Pearson, and T. G. Cleasby. 1963. Estimation of the water requirement of sugarcane in Natal. S. African Sugar Tech. Ass., Proc. p. 1–8.

Trouse, A. C., and R. P. Humbert. 1959. Deep tillage in Hawaii: 1. Subsoiling. Soil Sci. 88:150–158.

Trouse, A. C., and R. P. Humbert. 1961. Some effects of soil compaction on the development of cane roots. Soil Sci. 91:208–217.

Waterhouse, A. D., and H. F. Clements. 1953. Irrigation control with tensiometers and irrometers. Hawaiian Planters' Rec. 54:271–285.

Yamasaki, Y. 1956. Root system development under different soil improvement and cultivation practices for main soil groups. Rep. Hawaiian Sugar Tech. 15:10–12.

33

Sugar, Oil, and Fiber Crops

Part III—Oil Crops

D. W. HENDERSON

University of California
Davis, California

With the exception of soybeans (*Glycine* max. L. Merr.), crops grown primarily for oil have become important comparatively recently, and information on irrigation requirements is limited. Furthermore, there is little similarity among the various oil crops. With the possible exception of peak water use rates, characteristics important in irrigation are markedly different.

I. SOYBEANS

Recent reviews of general cultural requirements for soybeans include information on water requirements and on responses to irrigation (Howell, 1960; Carter and Hartwig, 1962). Whitt and Van Bavel (1955) also briefly reviewed soybean irrigation.

A. Water Requirements

Carter and Hartwig (1962) conclude that seasonal water use by soybeans ranges between 20 and 30 inches. Whitt and Van Bavel (1955) indicate a seasonal water requirement of 13 to 23 inches with rates averaging 0.3 inch/day during July and August in Missouri, USA. They feel that these rates may be too high because they were measured on small plots subject to advection. Herpich (1963) lists water requirements of 20 to 24 inches in Kansas, USA with peak use approximately 0.3 inch/day. Somerholder and Schleusener (1960) state that 18 to 25 inches are required in Nebraska, USA. Grissom et al. (1955) give "calculated water use rates for optimum growth" in Mississippi, USA as 6.4, 7.0, and 6.3 inches for June, July and August, respectively.

Water use rates and seasonal requirements depend on many factors. The above data indicate that water use by soybeans is quite similar to that of other crops grown at the same time. Thus local experience or experimental data for other crops may be used for estimates until more specific information is available.

B. Root Development

Depth of rooting and thoroughness of root ramification are important factors in determining the quantity of water which may be depleted before irrigation is required. Howell (1960) gives a general review of root growth, concluding that while the main taproot may penetrate as deep as 5 ft, most of the root system is in the top 2 ft. He cites unpublished data of other workers indicating water extraction from a depth of 51 inches and from a 5-ft depth. D. W. Henderson and L. D. Doneen (*unpublished*), working with a deep, alluvial loam of excellent physical character in California, USA found that growth of Chippewa variety slowed slightly at about 50 days after planting when residual available water remained at all depths below 6 inches and increased with depth. A thorough irrigation was applied, and 4 weeks later when irrigation was again judged desirable, there was essentially no residual available water in the top 3 ft. The treatment from which these data were obtained yielded 2,296 lb/acre with two irrigations while a treatment irrigated weekly during July and August produced 2,264 lb. Plants grown in soil moist to great depth at planting, but without any additional water as rainfall or irrigation, had utilized at maturity nearly all available soil water to the 5-ft depth, one-half that at 7 ft, and none at 8 ft. These plants had undergone moderate prolonged drouth and may not represent the maximum potential in root development. Soil conditions markedly alter rooting characteristics, so it is impossible to generalize. There may be varietal differences as well.

C.   Physiological Responses

The relation of flowering and seed development processes to water supply have not been studied in great detail. Fukui and Ojima (1957) found that drouth and excessively high soil water levels caused excessive shedding of bloom. However, blooming occurs over a period of about 1 month so that short drouth periods have less effect than on a more determinately flowering plant such as corn (*Zea mays* L.) (Morse et al., 1949; Whitt, 1954; Whitt and Van Bavel, 1955). Apparently a common cause of reduced yield through lack of moisture is in reduction of seed weight. Whitt (1954) reported that in unirrigated plots both seed weight and yield were reduced about one-third as compared to irrigated soybeans. This indicates little effect on seed numbers and that yield reduction was largely attributable to smaller seeds. Henderson and Doneen (*unpublished*) obtained substantially the same results in California while comparing unirrigated and frequently irrigated soybeans, whereas other yield components were influenced very little. Whitt's (1954) data indicate a slightly lower oil content in the smaller seeds.

D.   General Irrigation Requirements

In general terms it is difficult to describe irrigation requirements of soybeans. Grissom et al. (1955) obtained essentially equal yields for treatments irrigated when water depletion was 25, 50, and 100% of available soil water at the 2-ft depth. These results are indicative of response expected under comparatively favorable conditions of soil and climate; irrigation practices under less favorable situations should be altered in accord with principles outlined by Hagan et al. (1959). Soybeans develop a moderately deep root system which permeates the soil well at later stages of growth. In comparison with crops whose irrigation requirements are better known Whitt (1954) reported that soybeans did not respond to supplemental irrigation in two seasons in which irrigation increased corn yields 20%. Grissom et al. (1955) concluded that soybeans were less affected by drouth than corn or cotton (*Gossypium herbaceum*).

II. SAFFLOWER

Published information on safflower (*Carthamus tinctorius* L.) irrigation requirements is meager. Knowles (1958) describes water relations briefly. Knowles
and Miller (1965) discuss water requirements, and in a section of this same publication Henderson briefly discusses irrigation recommendations. Erie and French
(1962) report consumptive water use and irrigation requirements in Arizona, USA

A. Water Requirements

Knowles (1958) indicates that annual precipitation averages about 16 inches
in many areas where safflower is adapted. At Davis, California, 20 to 30 inches
are required for maximum yields (Knowles, 1958; D. W. Henderson, *unpublished data*). In the central San Joaquin Valley, California, use increases to 36
inches (P. F. Knowles and C. R. Pomeroy, *unpublished data*), and Erie and
French (1962) report seasonal use of 44 inches near Mesa, Arizona with peak use
rates near one-half inch/day in May.

B. Root Development

Little is known about seedling root development because early growth occurs
during cool weather. However, after the rosette stage under favorable soil conditions, the root system develops rapidly to great depth and thoroughly ramifies
throughout the soil. In deep, open alluvial soil at Davis, California, D. W. Henderson (*unpublished data*) found essentially no residual available moisture at
maturity to a depth of 10 ft and less than 50% available moisture in the 11- to
12-ft depth, the maximum depth of sampling. These data are from unirrigated
plots with no rainfall after midMay but which produced the same seed yield as
plots irrigated several times. Lack of yield increases from irrigation at Davis is
reported for repeated tests (Knowles and Miller, 1965). If the soil at planting
time stores sufficient water in a depth of 10 to 12 ft to supply seasonal requirements, the roots can develop rapidly enough to absorb deep subsoil water at a rate
adequate to produce nearly maximum seed yields. In the author's experience,
safflower is superior in this respect to any other crop.

On the other hand, only moderately dense subsoil will retard root development
and induce drouth. In a test on thoroughly preirrigated soil with dense subsoil
below 18 inches, there was residual available water at maturity at all depths
below 3 ft, the crop suffered severe drouth, and yield was less than one-half that
of unirrigated safflower at Davis, just a few miles distant.

The safflower root system also adapts the plant to obtain its water supply from
a water table within a few feet of the surface. In an area near Sacramento,
California, the soil is fine textured, high in organic matter, with a water table
usually between 3 and 5 ft below the surface. Safflower produces high yields
without supplemental irrigation although the period of rapid water use is one
without rainfall. All other summer grown crops [e.g., sugar beets (*Beta vulgaris*
L.), alfalfa (*Medicago sativa* L.), and tomatoes (*Lycopersicon esculentum* Mill.)]
require supplemental irrigation from the surface. Safflower is also well adapted
to artificial subirrigation.

Paradoxically, safflower is susceptible to root rots associated with wet soil conditions near the surface. The disease is most frequently attributed to Phytophthora and less often to Pythium (Knowles, 1958). In California, susceptibility to root rots restricts the growing of surface irrigated safflower to reasonably permeable soils which permit adequate water infiltration without prolonged irrigation and which drain excess moisture readily from the surface soil. In all but freely draining soils, care in managing irrigation water is necessary to avoid severe injury. The number of irrigations should be minimized by adequate preirrigation and by avoiding application of unnecessary water. However, irrigation should not be excessively delayed because drouth increases susceptibility to root rot injury. Furrow irrigation is often less injurious than flooding, although if beds between furrows wet quickly and thoroughly the advantage is primarily in more complete surface drainage. Irrigation in alternate furrows may reduce waterlogging by providing dry soil between wetted zones beneath furrows to help absorb excess water.

C. Physiological Responses

There is little concrete information on effects of drouth on flowering and seed development. Observations of dryland crops grown with inadequate soil water indicate general reduction in plant size, number of heads, and seed weight. Complete lack of water causes cessation of all plant activity with premature dessication of the plant. A common error in production of irrigated safflower is to apply the final irrigation too soon. In the absence of a reserve of subsoil water the plant runs out of water abruptly much too soon. The result is not readily apparent because the thick hull makes the seed appear normal while it may fail to fill at all or will fill only partially.

D. General Irrigation Requirements

Safflower is considered to be drouth tolerant, but requires an adequate seasonal supply of water for high yields. Water use apparently is not markedly different from other crops grown at the same time. It is not necessary to maintain high levels of soil water in the root zone except possibly during the seedling stage. Its deep, intensive root system allows infrequent irrigation under favorable soil conditions, and this factor can be used to advantage if there is risk of root rot injury. Safflower's adaptation to subirrigation makes it an excellent crop for areas of shallow water tables not well suited to many crop plants.

III. CASTORBEANS

In his general review, Zimmerman (1958) gives seasonal water requirements for castorbeans (*Ricinus communis* L.) as ranging from 2 to 3.5 ft in the irrigated areas of western USA. In general, seasonal water use in any location will probably exceed that of many crops because of the comparatively long growing season (150 to 180 days).

A. Root Development

Castorbeans develop deep root systems, but they are comparatively sparse even in the deep, open soils at Davis, California (L. D. Doneen and P. F. Knowles, *unpublished*). As a result, the plant undergoes water deficit with resultant slowing in growth and yield loss when there is appreciable residual available water throughout much of the root zone. While there may be varietal differences, D. W. Henderson (*unpublished*) found no difference in water extraction patterns of a standard (tall) and a dwarf internode type.

Zimmerman (1958) indicates that castorbeans are subject to root rots, warns against excessive irrigation, and recommends furrow irrigation rather than flooding. He advises irrigating in alternate furrows, especially at the time the first raceme develops. He also indicates that the castorbean root system is well adapted to subirrigation.

B. Physiological Responses

Effects of soil water supply on flowering and seed development are largely unknown. The plant is highly indeterminate in flowering, so it is unlikely that any period is especially critical. However, Zimmerman (1958) reports that if leaves wilt during morning hours it causes blasted or poorly filled seeds. He also warns that very frequent irrigation may maintain sufficiently high atmospheric humidity and that there will be injury from molds attacking seed capsules.

C. General Irrigation Requirements

Because of the comparatively sparse root system, castorbeans require maintenance of high levels of soil water and relatively frequent irrigation. Garton and Barefoot (1959) obtained highest yields for their wettest treatment in Oklahoma, USA (irrigation at 1.3 bars soil suction at the 6- to 12-inch depth). Delaying irrigation to wilting at 4 PM caused about a 10% yield reduction. These data are essentially in agreement with those of Doneen and Knowles (*unpublished*) at Davis, California. The latter workers found 50% to 75% yield increases with seven irrigations as compared to four applications of water. Their unirrigated treatment yielded about one-sixth that obtained with seven irrigations.

IV. FLAX

Flax (*Linum usitatissimum* L.) for oilseed is not widely grown under irrigation. The principal area is the Imperial Valley, California, where seasonal water requirement is 3 to 4 ft for crops planted in December and harvested in June (Knowles et al., 1959). They recommend delaying the first irrigation as long as feasible without severe drouth to minimize root rots, lodging, and weed pests. Subsequently, delaying irrigation excessively during bloom reduces yields. Irrigation should be continued to early June to mature later seeds for best yields. Ten to twelve irrigations are required in all.

Flax is apparently susceptible to root rots associated with excessive soil water levels. Otherwise, there are no special irrigation requirements.

LITERATURE CITED

Carter, J. L., and E. E. Hartwig. 1962. The management of soybeans. Advance. Agron. 14:359–412.

Erie, L. J., and O. F. French. 1962. Safflower—water intelligently. Arizona Farmer-Rancher 41(7)

Fukui, J., and M. Ojima. 1957. Influence of soil moisture content on growth and yield of soybean. Crop Sci. Soc. Japan, Proc. 26:40–42.

Garton, J. E., and A. D. Barefoot. 1959. Irrigation experiments at Altus and El Reno, Oklahoma. Oklahoma Agr. Exp. Sta. Bull. B–534. 19 p.

Grissom, P., W. A. Raney, and P. Hogg. 1955. Crop response to irrigation in the Yazoo-Mississippi Delta. Mississippi Agr. Exp. Sta. Bull. 531.

Hagan, R. M., Y. Vaadia, and M. B. Russell. 1959. Interpretation of plant responses to soil moisture regime. Advance. Agron. 11:77–98.

Herpich, R. L. 1963. Irrigating soybeans. Kansas State Univ. Agr. Ext.: Eng. in Balanced Farming, Land Reclamation no. 12.

Howell, R. W. 1960. Physiology of the soybean. Advance. Agron. 12:265–310.

Knowles, P. F. 1958. Safflower. Advance. Agron. 10:289–323.

Knowles, P. F., W. H. Isom, and G. F. Worker. 1959. Flax production in the Imperial Valley California Agr. Exp. Sta. Circ. 480.

Knowles, P. F., and M. D. Miller. 1965. Safflower. California Agr. Exp. Sta. Circ. 532.

Morse, W. J., J. L. Cartter, and L. F. Williams. 1949. Soybean culture and varieties. US Dep. Agr. Farmer's Bull. 1520.

Somerholder, B. R., and P. E. Schleusener. 1960. Irrigation can increase soybean production. Nebraska Agr. Exp. Sta. Quart. 7(1)16–17.

Whitt, D. M. 1954. Effects of supplemental water on field crops. Missouri Agr. Exp. Sta. Bull. 616.

Whitt, D. M., and C. H. M. van Bavel. 1955. Irrigation of tobacco, peanuts, and soybeans. p. 376–381. In Water. US Dep Agr. Yearbook. US Government Printing Office, Washington.

Zimmerman, L. H. 1958. Castorbeans: A new oil crop for mechanized production. Advance. Agron. 10:275–288.

33

Sugar, Oil, and Fiber Crops

Part IV—Irrigation of Cotton and Other Fiber Crops

J. R. STOCKTON
University of California (deceased)
Shafter, California

JOHN R. CARREKER
Agricultural Research Service, USDA
Athens, Georgia

MARVIN HOOVER
University of California
Shafter, California

I. COTTON

Cotton (*Gossypium* sp.) is sensitive to soil water conditions. Weather conditions, stage of plant development, and type of soil all affect the need for irrigation. In areas where a crop is grown with natural rainfall, irrigation may be quite beneficial when applied at a time of critical need. Under semiarid to arid conditions, irrigation is essential for cotton production. Cotton provides an excellent example of a crop where research has developed opportunities to modify growth processes for optimum production by the use of properly adapted irrigation practices.

A. Rooting Characteristics

The root system of the cotton plant develops rapidly following germination and emergence of the seedling. At the beginning of flowering the taproot may extend to a depth of 1.83 m (6 ft) into a fertile soil with a deep profile and laterally to the adjacent row. In Arizona, USA Erie has found roots 80 inches long growing laterally across adjacent rows from the main stem of the plant (L. J. Erie, *personal communication, Sept. 17, 1965*). Infertile, highly acid, or dense subsoil may limit root penetration (Rios and Pearson, 1964; Camp and Lund, 1964). At the beginning of the flowering period the root system nearly reaches its maximum extent. During the flowering period an intensification of root activity in the shallower depths takes place, reaching maturity near the end of this period. Beckett and Dunshee (1932) obtained soil water extraction patterns that indicate root activity to a depth of 1.52 m (5 ft) by late June in California, USA. Soil water extraction curves for each 30.48-cm (1-ft) increment were identical to a depth of 1.22 m (4 ft) during the latter part of the season. Gerard et al. (1958, 1964) in the lower Rio Grande valley of Texas, USA found that soil water extraction on Harlingen clay was restricted to the surface 61 cm (2 ft) of soil containing 90% or more of the cotton roots. On Willacy loam, water extraction occurred to depths of 0.91 to 1.52 m (3 to 5 ft) even though 80% of the roots were in the surface 61 cm (2 ft). More frequent irrigations (about every 15 days) are required on Harlingen clay for optimal growth of cotton because of restricted rooting depth compared to the same crop produced on medium textures of the Willacy or Hildago soil series. Doss et al. (1964) in Alabama, USA showed that soil water extraction by cotton roots was mainly from the upper 61 cm (2 ft) of the soil profile. Carreker and Cobb (1963) in Georgia, USA found that 82% of the roots under mature cotton plants are in the upper foot of the soil profile and report that irrigation had no effect on the depth of rooting. However, Levin and Shmueli (1964) working on a clay-loam soil in Israel where by time of flowering cotton had extracted most of the water from a 76.2 cm (2.5 ft) depth of soil, report that the irrigation regime had a pronounced effect upon the relative amounts of water extracted at different soil depths. Relatively long intervals between irrigations resulted in greater extraction below the 76.2 cm (2.5 ft) depth.

B.   Growth Stages

Growth stages illustrated in Fig. 33IV–1 show the relation between vegetative growth and fruiting. During the preflowering stage the rate of vegetative growth is low, averaging about 20 mm/day, in terms of plant height. Nutrient levels in the plant are generally high, but growth may be limited by low temperature and soil water availability. The flowering stage generally starts 70 to 80 days after planting. The average main stem growth rate doubles during this period, and rates as high as 25 mm/day are not unusual. Rate of flowering increases rapidly, reaching a peak in about 30 days after first bloom and then continues at a decreasing rate for another 30 days.

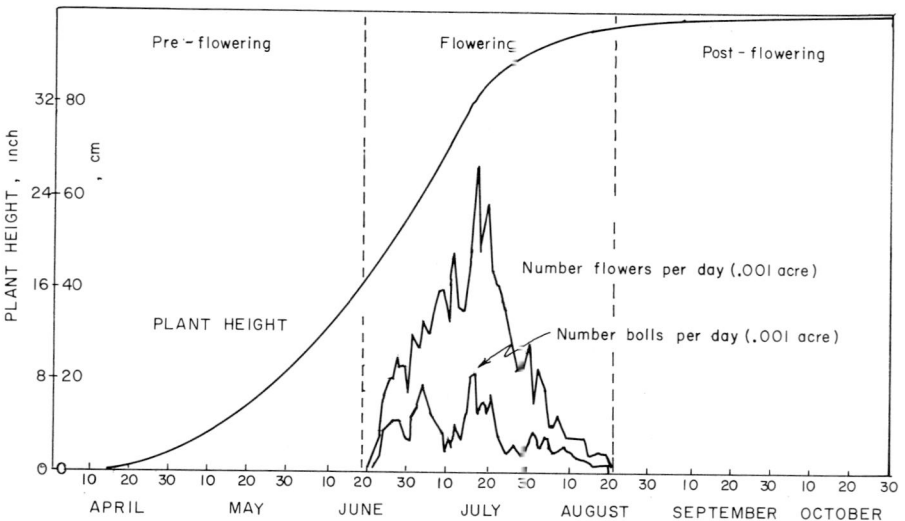

Fig. 33IV–1. Growth stages of cotton in relation to vegetative development and fruiting at Shafter, California, 1954.

At the time of first bloom nearly one-half the flowering curve is already determined since buds are initiated 30 days prior to anthesis. These buds (squares) are in varying stages of development and, barring a catastrophic event, will reach anthesis. Nearly two-thirds of the flowers shed within a few days following anthesis, as shown by the fruiting curves in Fig. 33IV–1. Shedding is influenced by the genetic nature of the plant, irrigation schedules, and fertilizer treatment and by environmental factors such as light, temperature, and soil water availability. As the rate of flowering increases, shedding generally increases. Short periods of intense shedding are generally followed by periods of reduced shedding and in this way, the plant is continually adjusting its fruit load (Stockton et al., 1961).

In Arizona, Erie (1962) reports that certain cotton varieties have two peak flowering periods, one in July and one in late August or September. Long staple cotton 'Pima S–1' (G. barbadense L.), e.g., may produce 70% of its cotton in the last 4 weeks of a 10-week flowering period. The decline in flowering after the July peak is referred to as the "August cutout". Late set for varieties such as 'Acala 44' (G. hirsutum L.) is important for high yields.

After flowering ceases the plant enters a senescent or post-flowering stage during which boll maturation is the principal activity of the plant. The time required for boll maturation and opening ranges from 40 to 60 days, increasing for bolls initiated late in the flowering period. Vegetative growth rate is extremely low, averaging less than 2.5 mm/day for the main stem. Quite often the terminal bud may abort, preventing further increase in plant height or flowering. This has been described as "cutout" (Walhood et al., 1955).

The indeterminate growth pattern of cotton may be influenced by genetic or environmental conditions that cause differing degrees of indeterminateness (Eaton, 1955; McKeever, 1927). Many studies have shown that management practices, including irrigation, that result in maximum vegetative growth, seldom result in maximum reproductive growth. Considerable effort has been expended in measuring relative fruitfulness or the ratio of a yield character to a vegetative growth character, such as number of bolls produced per unit weight of plant or pounds of seed cotton per pound of stalk weight. In most irrigation experiments where measurements are made, an optimum value for this ratio is found. Extremely high or low ratios are generally associated with inferior yields and can generally be related to extremes in determinate or indeterminate tendencies.

C. Scheduling Irrigations

1. PREPLANTING IRRIGATION

Since early root development by cotton is extensive, the soil should be wet to a depth of several feet prior to planting either by rainfall or a preplanting irrigation. Burnett and Fisher (1954) in Texas demonstrated the importance of an adequate preplanting irrigation. Yields were in proportion to the wetted depth of soil at planting. In the lower Rio Grande Valley, Texas, Gerard et al. (1964) found that satisfactory yields on Willacy loam are possible with a preplanting irrigation followed by an irrigation about 30 days after appearance of first bloom. However, the usefulness of this practice depends on soil conditions and climatic factors including rainfall amounts and frequencies. The advantage of preplanting irrigation was shown also by Nagle (1954) in Queensland, Australia. The preplanting irrigation is important to a favorable salt balance since the infiltration rate is generally highest during this irrigation (Marr and Hemphill, 1928; Erie, 1963). The preplanting irrigation consolidates the soil disturbed by deep tillage and land preparation, thereby improving planting conditions. Soil water for seed germination and early plant development usually is supplied by the preplanting irrigation, but sometimes a light irrigation after planting promotes good germination.

2. IRRIGATION DURING THE PREFLOWERING STAGE

The rate of water use is low during emergence and early plant growth. Water used during this period is about 10 cm or about 10% of the total (*see* Fig. 33IV–2). In humid areas, a large portion of the flowering curve may be determined before the next irrigation is needed following emergence. In the San Joaquin Valley of California, two irrigations may be required for normal growth during the same period. Soil water deficits severe enough to restrict vegetative growth during this period consistently result in an increased tendency for indeterminate growth and late maturity (Cowan et al., 1962). In Arizona, Harris and Hawkins (1942) reported average yield reductions of 22% and 9% for Pima and Acala cotton, respectively, when soil water deficits occurred during the preflowering stage of growth. Rank indeterminate plants ranging in height from 5 to 7 ft were used. Beckett and Dunshee (1932) obtained similar results in California where average yield reductions of 39% due to early season water deficits were measured. The stress period extended well into the flowering period. Plant height was 120 cm (48 inches). In recent experiments in the San Joaquin Valley, delaying the first irrigation retarded growth and maturity but increased total growth and yield of cotton (Table 33IV–1).

The date of the first irrigation can vary as much as 3 weeks, depending on climatic conditions. Water should not be applied until soil temperatures are favorable. Under Israeli conditions, Levin and Shmueli (1964) concluded that irrigation could be delayed until flowering, provided the main portion of the root zone (0 to 76.2 cm; 0–2.5 ft depth) did not fall below one-half of the available water. In Arizona, Erie (1961, 1963) recommends irrigation schedules that promote fast vegetative growth until July 1 to obtain good size plants essential for high yields (*see* Fig. 33IV–2). Christidis and Harrison (1955) report studies in Greece that show timing of the first irrigation had little effect on yield. Crowther

Table 33IV–1. Influence of date of first irrigation on growth, maturity, and yield of cotton*

Date of first irrigation	Available soil water†	Plant height		Crop maturity‡ Oct. 2	Seed cotton yield
		June 24	Nov. 4		
	%	cm	cm		lb/acre
May 15	80	58	86	92	2250
June 3	52	56	89	94	2610
June 24	20	41	94	84	2920
LSD, 5%	--	2.5	5	3	290

* For soil preirrigated in early March and cotton emerging in mid-April in San Joaquin Valley, California.
† Available water in soil at time of irrigation.
‡ Based on per cent of total crop harvested on Oct. 2

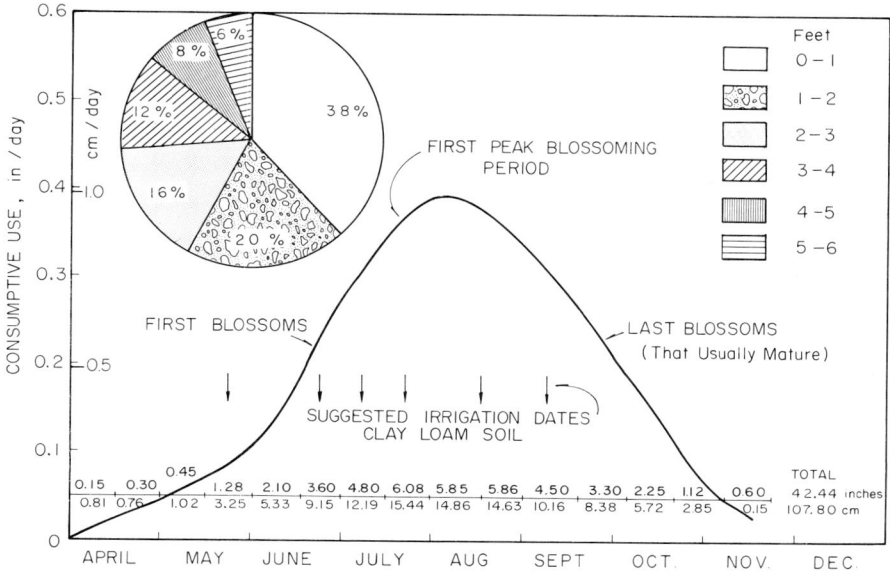

Fig. 33IV–2. Mean consumptive use, per cent soil water extraction by depth, and suggested irrigation dates for Acala 44 cotton grown at Mesa and Tempe, Arizona, 1954–1960, (Erie, 1963).

(1944) determined that differences in early growth of irrigated cotton in the Sudan and in Egypt was due to differences in air temperature rather than differences in water supply during this growth stage. Lower temperatures in Egypt caused slower growth, less nitrogen uptake and less water use in the early stage, but differences did not persist.

Desirable irrigation practices during the preflowering period must take into account the nature of subsequent growth and fruiting patterns. Visual symptoms of plant stress such as darkened color of foliage and transient wilting in the late afternoon generally serve as adequate guides for scheduling early irrigations.

3. IRRIGATION DURING FLOWERING

One-half or more of the total water use occurs during the stage of most rapid plant development and growth. Frequency of irrigation is determined by soil water-holding capacity, rooting depth, degree of plant cover and evaporative demand of the environment. Irrigations should be scheduled to control vegetative growth in relation to reproductive growth. Vegetative growth can be checked by allowing sufficient depletion of available water before each irrigation.

The degree of soil water depletion has been studied by a number of investigators in the USA and remarkable agreement is evident. Beckett and Dunshee (1932) obtained optimum yields by maintaining one-half of the available water to a depth of 1.52 m(5 ft) prior to irrigation. Harris and Hawkins (1942) in Arizona concluded that 55% of the available water can be depleted without reducing yields. Brown et al. (1955) in Arkansas, Grissom et al. (1955) in Mississippi and Bennett et al. (1964) in Alabama favor maintaining available water content near or above 50%. Hamilton et al. (1956) in Arizona and Carreker and

Cobb (1963) in Georgia report no yield difference between 1/3 and 2/3 depletion of available water.

Marr and Hemphill (1928) give a good description of plant stress symptoms related to soil water deficits and claim that withholding irrigation until stress appears does not adversely affect yield. Stockton et al. (1961, 1955) in California and Thurmond et al. (1958) in Texas made similar observations. Curry (1934) in New Mexico, USA and Carreker and Cobb (1963) in Georgia concluded that slight wilting for a few days prior to irrigation is not harmful to yield. Krantz et al. (1955) and Nagle (1954) also report that plant symptoms are a good guide for scheduling irrigation. Erie (1963) obtained maximum yields with six irrigations (Fig. 33IV–2) that allowed 65 percent of available soil water depletion in the top 91.4 cm (3 ft) of a silty clay. He noted that the 65% level of depletion corresponded to visible plant stress symptoms such as " . . . (i) leaves turning dark-bluish green, (ii) very little new growth, (iii) reddening near the terminal growth tips, and (iv) indication of flowering . . ." that could be used to schedule irrigations. Visual plant symptoms of soil water stress develop gradually as soil water is depleted.

Marani and Horwitz (1963) in Israel grew cotton without rainfall on soil brought to field capacity at time of planting and found that growth, flowering, boll set, and yield were affected markedly with the timing of one irrigation application. When soil water was depleted to the wilting point (WP) in the 15– to 30–cm soil layer, and to 30% of available water capacity in the 30– to 60–cm layer, vegetative growth stopped, and flowering ceased 3 weeks later. A single irrigation to wet the soil 120 cm deep resulted in higher yields of lint when applied at the onset of flowering compared to an earlier application at the beginning of budding (squaring) or later at the peak of flowering. Yelsukov (1962) reported that irrigation is needed during the main fruiting period when the soil water content in the 10– to 30–cm soil layer approaches 50% of field capacity. Levin and Shmueli (1964) observed that water stress during the period from onset of flowering to peak flowering reduced yield more drastically than a similar stress occurring after peak flowering.

Shedding of flowers and young bolls, while of major consequence in crop development, is only slightly affected by variations in soil water levels encountered in irrigated cotton. Extreme drouth can result in excessive shedding. Early studies related the "gradual depletion of moisture in the deeper reaches of the soil" to excessive shedding (Lloyd, 1920). These observations were made under conditions of extreme drouth. Hawkins et al. (1933) made similar observations in Arizona, but treatments involved extreme drouth produced by withholding irrigation for an extended period of time. Observations in other irrigation experiments not involving extreme drouth have shown shedding to increase as the number of irrigations increased (Christidis and Harrison, 1955; King, 1922; Stockton et al., 1961). Christidis and Harrison (1955) observed that shedding and flowering are positively correlated. As total flower production increases, shedding increases. Over a moderate range of increasing soil water deficits, both flower production and shedding is reduced. This relationship provides a mechanism whereby, in spite of reductions in flower production, good boll set and better crop maturity can result from soil water deficits intense enough to restrict vegetative growth. An example of this is shown in Table 33IV–2. The first water application in the seven irrigation treatment was given at the first signs of visible water stress. This treat-

Table 33IV–2. Influence of irrigation frequency on a coarse-textured soil on cotton growth and fruiting at Shafter, California, 1954

Irrigations	Plant height	Flower	Shedding	Seed cotton yield
no.	cm	no./row ft	%	lb/acre
21	104	53	70	3640
12	94	55	71	3900
7	89	47	66	3780
3	74	34	58	2790
LSD, 5%	10	6	5	410

Table 33IV–3. Influence of irrigation treatment on fruiting and yield of seed cotton, Watkinsville, Georgia, 1956 to 1957

Irrigation treatment	Number of fruit per 10 ft of row				Yield lb/acre	
	7/23	8/27	7/30	9/3		
	1956		1957		1956	1957
Irrigate at 60% AWC*	456	250	643	280	3257	2759
Irrigate at 30% AWC	410	250	640	261	3463	3282
Irrigate at observed wilting	448	218	635	275	2911	2982
Irrigate 5 days after observed wilting	492	270	532	230	3306	3006
Irrigate at 0% AWC	416	191	633	251	3621	3071
No irrigation	467	181	400	140	1952	1844

* Available water capacity.

ment produced slightly more cotton than the treatment receiving 21 irrigations, although vegetative growth and flowering were reduced approximately 15%. Similar results by Carreker and Cobb (1963) under more humid conditions are shown in Table 33IV–3.

Boll setting may be cyclic or continuous on coarse-textured soils, depending on the frequency of rainfall or irrigation to control the deficit or adequacy of soil water. There seems to be little relationship between soil water deficit and the shedding of bolls on fine-textured soils because soil suction develops less rapidly. The development of an abscission zone is a reversible growth process requiring 4 to 7 days for completion. A soil water deficit severe enough to initiate abscission must persist from 4 to 7 days for shedding to occur. The rate that soil water stress develops and the severity of the stress are important factors to consider in interpreting the influence of soil water levels on shedding of flowers and young bolls. Also, shedding may be due to nutritional relationships or other physiological factors that control the number of bolls on a plant at any one time.

4. IRRIGATION DURING POST-FLOWERING

Fiber and seed development takes place in the post-flowering period as the plant approaches maturity. Water-holding capacity of the soil, root development, and depth of water penetration generally determine the irrigation amount and interval, and, with experience, will also determine the timing of the last irrigation. In the San Joaquin Valley of California, for example, the irrigation interval

Table 33IV–4. Relation of midseason irrigation interval and date of last irrigation for cotton in the San Joaquin Valley, Calif.

Midseason irrigation interval	Date of last irrigation
days	
5	September 10
10	September 1
20	August 20
30	August 5

found to be satisfactory during the midseason can be used to schedule the last irrigation. Under the conditions of this area, midseason irrigation intervals of 5, 10, 20 and 30 days indicate irrigation can normally be terminated on September 10, September 1, August 20 and August 5, respectively. Levin and Shmueli (1964) advocate terminating irrigation 2 weeks earlier than common practice in Israel (by eliminating the last irrigation), thus saving considerable water without appreciable loss of yield. Early cessation of irrigation can be harmful to yield and quality of the fiber because some bolls might not mature properly. Late irrigations in the post-flowering period might cause delay in opening of bolls and greater susceptibility to frost damage, lodging, and boll-rot pathogens. Bennett et al. (1965) showed that topping and pruning rank cotton due to excess water and nitrogen prevented lodging and boll rot.

5. METHODS USED TO SCHEDULE IRRIGATIONS

Soil water measurements by sampling, tensiometers, or resistance blocks (*see* chapters 15 and 30) have been used experimentally to schedule irrigations, but they have not gained much acceptance by cotton farmers.

Most cotton growers prefer to irrigate by the so-called calendar method where prior experience for a given area is available. Irrigation schedules often rely on a certain starting date, followed by irrigations applied weekly, biweekly or at some other interval. To establish efficient schedules, use should be made of information on crop rooting depth, water-retaining capacity of the soil, and the evapotranspiration rates for different portions of the growing season. Such information is given in Fig. 33IV–2 for the Salt River Valley in Arizona (Erie, 1963). In this example, 58% of the soil water extracted by cotton is drawn from the surface 2 ft with some extraction in the sixth foot. The seasonal water use is 107.8 cm (42.4 inches) with peak rates of 1 cm/day (0.4 inches) in July and August during the peak blossoming period. Wherever such detailed information is available to growers, reasonably efficient and practical irrigation schedules can be developed for the given conditions. It may be necessary to provide additional irrigations to plantings in sites exposed to considerable advective energy. Hudson (1964), for example, reported that in the Gezira area of the Sudan, soil water loss was greater in the 50- to 60-m wide strip adjacent to the windward side of an irrigated field due to advective energy and suggested that the timing or amount of water applications should be varied accordingly.

Visible plant water stress symptoms can also be useful in some situations as checks on the irrigation schedule (*see* chapter 30).

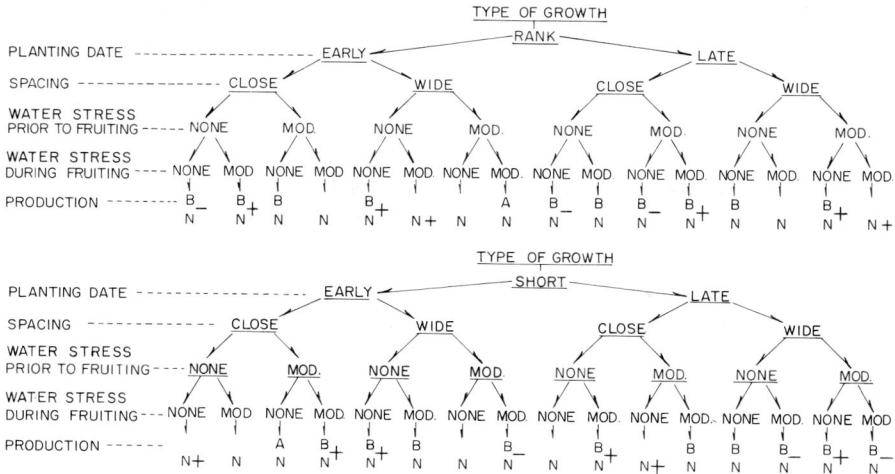

Fig. 33IV–3. Irrigation choice chart for two types of cotton growth showing predicted production as influenced by planting date, spacing, and water stress prior to and during fruiting. The symbol N means normal production (for the area) is predicted. AN above normal production, and BN below normal production. Plus (+) or minus (–) signs show the smaller differences in yield which would be expected. (Adapted from J. R. Stockton and R. M. Hagan, *Unpubl. Rep., Univ. California* 1962.)

D. Interaction of Irrigation and Other Cultural Practices

Soil texture, structure, and depth are examples of edaphic factors that are not easily altered agriculturally and often determine the type of cotton growth for a particular soil. Gussak and Paganyas (1964), for example, reported that yield of irrigated cotton was improved through increased infiltration resulting from greater aggregation of a typical sierozem treated with polymers. However, variations of many cultural practices on a particular soil usually will alter growth and fiber yield only within a narrow range as compared with differences observed among soils varying in physical characteristics. The short, determinate type of cotton growth produced on many soils is not easily altered agronomically. Cultural practices chosen to counteract this determinate tendency often lead to improvements in yield. The irrigation choice chart of Figure 33IV–3 stresses the chain-like dependence of irrigation decisions and predicts their effects on yield of fiber. Rank indeterminate growth, for example, is likely with an early planting date, wide spacing and an irrigation schedule that produces moderate water stress, all of which contribute to better yields and crop maturity. Where very determinate growth is expected there should be less emphasis on earliness of planting and closer spacing between plants. A moderate soil water deficit prior to flowering followed by frequent irrigation has been shown to be desirable under this condition.

E. Influence of Irrigation on Boll and Fiber Properties

Boll and fiber properties of cotton are influenced by irrigation. The properties most generally studied include boll size, lint and seed indices (weight per 100 seeds), lint-to-seed cotton ratio (lint per cent), and fiber properties such as staple length, grade, strength, fineness, and maturity. Spinning tests are frequently employed to determine yarn characteristics such as appearance, strength, nappiness and amount of lint waste in certain spinning operations. Irrigation often has indirect influences on many of these boll and fiber properties. Irrigation practices may alter the fruiting pattern of developing bolls subjected to a different environmental complex. An extremely late crop compared to an early determinate one will develop fruit and fiber under different conditions of temperature and light intensity. Irrigation can affect the ratio of vegetative growth to fruitfulness which in turn can affect the temperature of developing bolls. Increased exposure of a boll to sunlight raises its temperature several degrees, and this in turn increases the rate of cellulose deposition in the fiber (Anderson and Kerr 1938).

In general, as soil water deficits become less severe, boll size, seed index, and lint index increase. The lint per cent or ratio of lint-to-seed cotton tends to decrease since the seed index appears to increase more than the lint index. Relatively low soil water suction during the boll maturation period tends to produce longer and finer fiber. Fiber strength, however, decreases as length and fineness increase. Spooner et al. (1956, 1958) found that irrigation can stabilize and in most cases improve fiber quality.

The direct influence of soil water deficits on fiber properties appears to be slight (Adams et al., 1942; Krantz et al., 1955; Law, 1954; Spooner et al., 1956; Towery et al., 1954). Shedding of young bolls is partly responsible for the lack of response since many bolls shed because of improper fertilization of the embryo or adverse environmental conditions would normally be expected to produce inferior fiber. According to Eaton and Ergle (1952) and Hessler et al. (1955) the cotton plant, when subjected to a soil water deficit, may increase its carbohydrate level and at the same time increase photosynthetic and respiratory activities, all of which tend to offset the disadvantages of reduced soil water supply. This may provide a mechanism whereby the cotton plant can develop normal fiber even when the plant is subjected to drouth.

II. FLAX

Little information is available on the irrigation of flax (*Linum usitatissimum* L.) as a fiber crop. Urzalova (1960) reports a field experiment in Czechoslovakia where fiber yield from a fiber variety was greatest when the soil water supply was high throughout the vegetative period. Fiber and seed yields from an oil-bearing variety were greatest when the soil water supply was low at the beginning of the vegetative period and high thereafter. Unfortunately details on the soil water treatments are not available to the writers.

In parts of southern California and Arizona where flax is grown mostly for seed, the crop is given a pre-planting irrigation to wet the soil to a depth of 1.22 to 1.52 m (4 to 5 ft). The first irrigation may then be delayed 5 to 8 weeks after planting to avoid yellowing of the crop, especially in very cool weather. Additional irrigations are given at 14- to 30-day intervals depending on the stage of growth, evaporative demand, and soil conditions. Recent experiments indicate that highest yields are obtained where a second and even a third blooming period is brought on by adequate soil water and nitrogen during full bloom and later.

Jackson et al. (1963) found that four extra irrigations during May and June and 40 lb of N/acre increased yields by 952 lb/acre (17 bushels). Total water requirement of flax may vary from 3 to 4 acre feet/acre.

III. SISAL

A 4-year study of water use, growth and yield of sisal in various irrigated regions of the Northern Negev (Israel) has shown that applying one irrigation in March and one in August gave optimum results (Archituv et al., 1965). The total irrigation requirement amounted to 200 mm of water plus about 200 mm of rainfall. More frequent irrigations reduced yield. Evapotranspiration in optimum treatments was only 12% of the pan evaporation and most of the water was extracted from the 0- to 50-cm depth.

LITERATURE CITED

Adams, F., F. J. Veihmeyer, and L. N. Brown. 1942. Cotton irrigation investigations in the San Joaquin Valley. California Agr. Exp. Sta. Bull. 668. p. 3–93.

Anderson, D. B., and T. Kerr. 1938. Growth and structure of cotton fiber. Ind. Eng. Chem. 30:48–54.

Archituv, M., J. Rubin, and J. Heller. 1965. The effect of various irrigation regimes on water consumption, growth and yields of sisal in the northern Negev. Israel J. Agr. Res. (In press.)

Beckett, S. H., and C. F. Dunshee. 1932. Water requirements of cotton on sandy loam soils in southern San Joaquin Valley. California Agr. Exp. Sta. Bull. 537. p. 3–48.

Bennett, O. L., D. A. Ashley, B. D. Doss, and C. E. Scarsbrook. 1965. Influence of topping and side pruning on cotton yield and other characteristics. Agron. J. 57:25–27.

Bennett, O. L., B. D. Doss, and D. A. Ashley. 1964. Cotton irrigation in southeastern United States. US Dep. Agr. Inform. Bull. 282. 16 p.

Brown, D. A., R. H. Benedict, and B. B. Bryan. 1955. Irrigation of cotton in Arkansas. Arkansas Agr. Exp. Sta. Bull. 552. p. 3–40.

Burnett, E., and C. E. Fisher. 1954. Correlation of soil moisture and cotton yields. Soil Sci. Soc. Amer. Proc. 18:216–218.

Camp, C. R., and Z. F. Lund. Nov. 1964. Effect of soil compaction on cotton roots. Crops and soils. p. 13–15.

Carreker, J. R., and C. Cobb, Jr. Oct. 1963. Irrigation in the Piedmont. Georgia Agr. Exp. Sta. Bull. New Ser. 29. 65 p.

Christidis, B. G., and G. J. Harrison. 1955. Cotton growing problems. McGraw-Hill, New York. p. 10–12.

Cowan, R., M. Hoover, A. W. Marsh, B. A. Krantz, and S. J. Richards. Nov. 1962. Water, nitrogen and varieties in lower desert cotton production. California Agr. 16:10–12.

Crowther, F. April-July 1944. Studies in growth analysis of the cotton plant under irrigation in the Sudan: III. A comparison of plant development in the Sudan Gezira and in Egypt. Ann. Bot., New Ser. 8:213–257.

Curry, A. S. 1934. Results of irrigation treatments on acala cotton grown in the Mesilla Valley, New Mexico. New Mexico Agr. Exp. Sta. Bull. 220. p. 3–43.

Doss, B. D., D. A. Ashley, and O. L. Bennett. 1964. Effect of moisture regime and stage of plant growth on moisture use by cotton. Soil Sci. 98:156–161.

Eaton, F. M. 1955. Physiology of the cotton plant. Ann. Rev. Plant Physiol. 6:299–328.

Eaton, F. M., and D. R. Ergle. 1952. Fiber properties, carbohydrate and nitrogen levels of cotton plants as influenced by moisture supply and fruitfulness. Plant Physiol. 27:541–562.

Erie, L. J. 1961. Water needs for efficiency. Cotton Gin and Oil Mill Press 62:34–35.

Erie, L. J. 1962. Closing the irrigation gap. Cotton Trade J. and Agr. Reporter. (29th Int. Ed.) p. 82, 234–235.

Erie, L. J. 1963. Irrigation management for optimum cotton production. Cotton Gin and Oil Mill Press 64:30–32.

Gerard, C. J., M. E. Bloodworth, C. A. Burleson, and W. R. Cowley. Sept. 1958. Cotton irrigation in the lower Rio Grande Valley. Texas Agr. Exp. Sta. Bull. 916. 14 p.

Gerard, C. J., C. A. Burleson, W. R. Cowley, L. N. Namken, and M. E. Bloodworth. April 1964. Cotton irrigation in the lower Rio Grande Valley. Texas Agr. Exp. Sta. Bull. B–1014. 15 p.

Grissom, P., W. A. Raney, and P. Hogg. 1955. Crop response to irrigation in the Yazoo-Mississippi Delta. Mississippi Agr. Exp. Sta. Bull. 531. p. 3–24.

Gussak, V. B., and K. P. Paganyas. 1964. Some results of four-year experiments on structural improvement of an irrigated typical sicrozem. Soviet Soil Sci. 5:506–516.

Hamilton, J., C. O. Stanberry, and W. M. Wootton. 1956. Cotton growth and production as affected by moisture, nitrogen, and plant spacing on the Yuma Mesa. Soil Sci. Soc. Amer. Proc. 20:246–252.

Harris, K., and R. S. Hawkins. 1942. Irrigation requirements of cotton on clay loam soils in the Salt River Valley. Arizona Agr. Exp. Sta. Bull. 181. p. 421–459.

Hawkins, R. S., R. L. Matlock, and C. Hobart. 1933. Physiological factors affecting the fruiting of cotton with special reference to boll shedding. Arizona Agr. Exp. Sta. Bull. 46. p. 361–407.

Hessler, L. E., B. K. Power, and J. C. Lowry. 1955. Cotton fiber development study. Cotton Gin and Oil Mill Press. 56:38.

Hudson, J. P. 1964. Evaporation under hot, dry conditions. Empire Cotton Growing Rev. 41:241–254.

Jackson, E. B., D. D. Rubis, and F. Carasso. May-June 1963. More food and drink (nitrogen and water) boost flax yields. Progr. Agr. p. 12–13.

King, C. J. 1922. Water-stress behavior of pima cotton in Arizona. US Dep. Agr. Bull. 1018. p. 1–24.

Krantz, B. A., N. P. Swanson, K. R. Stockinger, and J. R. Carreker. 1955. Irrigating cotton to insure higher yields. p. 381–388. In Water. US Dep. Agr. Yearbook of Agr. US Government Printing Office, Washington.

Law, W. P. July 1954. Effects of irrigation on yield and fiber quality. 8th Annual Cotton Mech. Conf., Proc. p. 13, 15.

Levin, I., and E. Shmueli. 1964. The response of cotton to various irrigation regimes in the Hula Valley. Israel J. Agr. Res. 14:211–225.

Lloyd, F. E. 1920. Environmental changes and their effect upon boll-shedding in cotton. Ann. New York Acad. Sci. 29:1–131.

Marani, A., and M. Horwitz. 1963. Growth and yield of cotton as affected by the time of a single irrigation. Agron. J. 55:219–222.

Marr, J. C., and R. G. Hemphill. 1928. The irrigation of cotton. US Dep. Agr. Tech. Bull. 72. p. 1–37.

McKeever, H. G. 1927. Community production of acala cotton in the Coachella Valley of California. US Dep. Agr. Bull. 1467. p. 1–47.

Nagle, A. 1954. Irrigation practice for cotton production in Queensland. Queensland Agr. J. 79:198.

Rios, M. A., and R. W. Pearson. 1964. The effect of some chemical environmental factors on cotton root behavior. Soil Sci. Soc. Amer. Proc. 28:232–235.

Spooner, A. E., D. A. Brown, and B. A. Waddle. 1956. Irrigation and fiber properties of cotton. Arkansas Farm Res. 5:3.

Spooner, A. E., C. E. Caviness, and W. I. Spurgeon. 1958. Influence of timing of irrigation on yield, quality, and fruiting of upland cotton. Agron. J. 50:74–77.

Stockton, J. R., L. D. Doneen, and V. T. Walhood. 1961. Boll shedding and growth of the cotton plant in relation to irrigation frequency. Agron. J. 53:272–275.

Stockton, J. R., L. D. Doneen, V. T. Walhood, and B. Counts. 1955. Effects of irrigation on the growth and yield of cotton. California Agr. 9:8–11.

Thurmond, R. V., J. Box, and F. C. Elliott. 1958. Texas guide for growing irrigated cotton. Texas Agr. Ext. Serv. Bull. B–896. 12 p.

Towery, J. D., L. E. Hessler, and B. K. Power. March 1954. How are cotton fiber characteristics and spinning performance affected by irrigation. Textile Ind. 118:149–151, 153.

Urzalova, J. 1960. Effect of controlled soil moisture on some properties of flax grown for fiber and oil. (In Czechoslovakian) Sborn. vys. Skol Zemed. Brne A. No. 3–4:425–437.

Walhood, V. T., J. R. Stockton, and B. Counts. Jan. 1955. Defoliation, boll characteristics, and yield of cotton in relation to time of final irrigation and cut-out. 9th Ann. Beltwide Cotton Defoliation Conf., Proc. p. 28–33.

Yelsukov, I. 1962. Minimum permissible soil moisture in pre-irrigation of cotton (applicable thick soils of irrigated valleys). Soviet Soil Sci. 1:72–78.

34 | Vegetable Crops

M. T. VITTUM
New York State Agr. Exp. Sta., Cornell University
Geneva, New York

W. J. FLOCKER
University of California
Davis, California

Vegetables are a very important part of the human diet, supplying major amounts of vitamins A and C, thiamine, riboflavin, niacin, iron, carbohydrates, and roughage; smaller but significant amounts of calcium and phosphorus; and, with the exception of beans (*Phaseolus vulgaris*) and peas (*Pisum sativum*), minor amounts of fat and protein.

Vegetables are grown wherever people live. Geographical distribution encompasses a wide range of soil and climatic conditions. Major commercial production is centered in areas with a particularly favorable climate, accompanied by good soils, transportation, and marketing facilities (Thompson and Kelly, 1957).

Vegetables are high-value crops and require productive, well-drained, level soils. High rates of commercial fertilizer and expensive disease, insect, and weed control measures are required. Harvesting, handling, distributing, and marketing costs are high because the crops are perishable and labor requirements are usually large. Development of mechanical harvesting has been relatively slow because of the indeterminate maturity of vegetables. Mechanization has been adapted more to processing than to fresh market crops. In spite of the difficulties, there has been progress in reducing labor costs for market vegetables, and mechanized operations are being promoted vigorously.

An adequate water supply is one of the prime requirements for growing vegetable crops successfully. Vegetable production demands a large outlay of capital, and, if water is not available at the right time, this investment is no longer safely protected.

In arid and semiarid climates, irrigation has long been basic to vegetable production. However, supplemental irrigation is a relatively new practice in humid areas of the USA such as the Mississippi Valley, the East and Northeast, and parts of Europe, Asia, and Australia. Growers in these areas have found—sometimes through sad experience—that almost invariably, a drouth will occur sometime during the growing season, thereby reducing yield or quality or both. Thus, since World War II, supplemental irrigation has become a common practice in many vegetable fields in humid and semihumid regions.

To further exemplify the benefits of irrigation, consider that good vegetables are tender, crisp, succulent, free from excessive fibers, and usually mildly flavored. High quality requires a rapid and uniform plant growth rate. Without an ample and uniformly available supply of water and plant nutrients in the root zone, this growth would never be realized.

The various factors that determine the need for irrigation have been discussed in sections VI, VII, and VIII and need not be reiterated here. Several good literature reviews covering irrigation of vegetable crops are available (Herner et al., 1961; Janes and Drinkwater, 1959; Stanhill, 1957; Vittum et al., 1963; and Winter, 1957).

I. CROP REQUIREMENTS

Irrigation requirements of certain vegetables are summarized below and arranged according to methods of culture as outlined by Thompson and Kelly (1957).

A. Perennials

Perennials such as asparagus (*Asparagus officinalis*), globe artichoke (*Cynara scolymus*), and rhubarb (*Rheum rhaponticum*) have extensive root systems and require an open, porous, or well-drained soil for optimum growth. Roots occupy such a large volume of soil that irrigation requirements are usually lower than for annual vegetables grown on adjacent fields.

1. ASPARAGUS

Asparagus is a deep-rooted perennial. Several years may be required before nonirrigated plots give any indication of insufficient moisture. Yields on mineral soil in California, USA are increased if asparagus is irrigated with about 20 acre-inches of water in addition to 16 inches of rainfall (Hanna and Doneen, 1958). Older fields in California are commonly flooded with about 12 inches of water for 3 or 4 weeks during December or January. Flooding generally permits harvesting 12 to 14 days earlier and is also an effective method of controlling the garden centipede.

Asparagus is subirrigated on organic soils of the Sacramento-San Joaquin Delta of California (Hanna, 1950). Main ditches about 800 ft apart and 6 to 8 ft deep carry irrigation water into the field. Water is distributed across the field by connecting "spud" ditches about 60 ft apart, 9 inches wide, and 14 to 18 inches deep.

Greatest yield increases of asparagus in Delaware, USA are obtained from sprinkler irrigation applied both during the cutting season and during fern growth the preceding year (Fieldhouse, 1958).

2. GLOBE ARTICHOKE

The globe artichoke requires frequent irrigations throughout the year. Inadequate amounts of water or high temperatures, particularly during the time buds are forming, result in loose buds of unmarketable quality. However, the globe artichoke will not tolerate standing water, and during the rainy season drainage ditches must be provided (Sims et al., 1962).

B. Potherbs or Greens

Potherbs or greens (spinach (*Spinacia oleracea*), kale (*Brassica oleracea* var. *acephala*), mustard (*Brassica juncea*), collards (*Brassica oleracea* var. *acephala*), etc.) are shallow rooted. These crops are usually grown during the cooler part of the growing season when evapotranspiration rates are relatively low.

1. SPINACH

Spinach, the most important potherb or green, requires a supply of available soil water throughout its growth. Because it is shallow rooted and a cool-season crop, a large percentage of the crop is grown for fall, winter, or early spring harvest. This corresponds to the periods of high rainfall in many of the production areas; therefore, much of the crop is grown without supplemental irrigation. Ample moisture is most important near harvest to obtain maximum yield and quality. In arid or semiarid regions, 6 to 12 inches of water are applied to insure ample available water throughout its growth (Beattie and Beattie, 1948; MacGillivray, 1953).

C. Salad Crops

Except for lettuce (*Lactuca sativa*), salad crops such as celery (*Apium grave-olens*), endive (*Cichorium endivia*), and parsley (*Petroselinum hortense*) are relatively long-season crops with limited root systems. Irrigation normally is necessary for maximum yields. Crops with shallow root systems, such as celery and lettuce, are more susceptible to drouth injury than are deep-rooted crops, such as asparagus and tomatoes (*Lycopersicon esculentum*). Proper irrigation increases the quality of the leafy crop which should be crisp and tender at harvest.

1. CELERY

Celery can be direct or field seeded. However, since the seed germinates slowly, soil water must be maintained at about field capacity. In cool, moist areas where seedlings are subject to damping off, the crop is often transplanted to the field from plant-growing beds. The major portion of the celery root system is in the surface 10 to 12 inches; consequently only a limited volume of soil is available to supply water to the crop, and high soil water conditions must be maintained for uniform and rapid growth (Janes, 1959). Most of the celery in California is planted in two rows on beds and furrow irrigated. Recent studies showed that yields increased significantly when soil water tension was maintained between 0 and 0.2 atm (Cannell et al., 1959). Blackheart, a physiological disorder of celery, increased when tension was allowed to reach 0.8 atm before irrigating. Janes (1959) irrigated celery at four different levels of soil water and reported that growth rate was retarded when the soil water was depleted to 40% of that available at field capacity.

2. LETTUCE

Lettuce, the most important of the salad crops, is a shallow-rooted, rapid-growing, cool-season crop requiring a nearly constant water supply from planting to harvest. Some soils upon which lettuce is grown tend to crust badly, a condition which seriously reduces emergence of the tiny seedlings. Wide fluctuation in soil water is undesirable at any time and may cause great damage to the crop in the late stages of development (Thompson, 1951). Fluctuations in soil water frequently cause increased tipburn, a very serious nonparasitic disorder which appears to be most prevalent when bright, warm days follow periods of damp, foggy weather (Grogan et al., 1955). Veihmeyer and Holland (1949) indicate that three irrigations—the first to germinate the seed, the second at time of thinning, and the third 30 days after thinning—will produce a crop of lettuce on loam and silty clay loam soils in the Salinas Valley of California without loss in yield or quality.

With careful management, winter crops of lettuce can be grown successfully in southwestern USA with 6 to 10 acre-inches of applied irrigation water. For a summer crop, 18 to 24 inches may be required in the Imperial Valley of California or Salt River Valley of Arizona and about 18 inches in the Salinas Valley of California. Applications of water during maturation may cause loose, open heads and should be avoided. The wrapper leaves are brittle when completely turgid and damage easily when harvested. Also, a wet soil surface the last week or two before harvest may induce rot and slime on the lower leaves. Good management includes the use of irrigation, either surface or sprinkler, to soften the crust until the crop emerges.

D. Crucifers

Crucifers [cabbage (*Brassica oleracea* var. *capitata*), cauliflower (*Brassica oleracea* var. *botrytis*), Brussels sprouts (*Brassica oleracea* var. *gemmifera*), sprouting broccoli (*Brassica oleracea* var. *Italica*)] thrive best in a comparatively cool, moist climate and well-drained soil. Root systems are shallow to moderately deep. Since relatively high levels of available soil water are necessary, irrigation is essential for top yields whenever precipitation is deficient.

1. CABBAGE

Cabbage should be irrigated to maintain available soil water above 50% in the upper 16 inches of soil. Studies on a silt loam soil in New York indicated very little water was extracted below 16 inches on irrigated plots while on nonirrigated plots practically all of the available water was extracted from the upper 24 inches (Vittum and Peck, 1956). Maximum growth of cabbage occurs when it is grown without interruption from planting to harvest (Janes and Drinkwater, 1959). Any factor which retards growth, such as low soil water, is harmful at any time during the developmental stages.

In humid and semihumid regions, soil water is usually near field capacity when the spring crop is set in the field but is likely to be deficient during the later stages of growth. Since the water requirement of small transplants is low, water in the soil plus rainfall usually is sufficient for normal growth until the crop starts to expand rapidly. Hence, the critical period for irrigation in these areas is usually during the last 3 to 4 weeks of development (Vittum et al., 1963). Irrigation not only increases the acre yield of marketable cabbage but also increases the average weight per head (Nettles et al., 1952; Vittum and Peck, 1956).

In the Western USA, cabbage is grown on slightly raised beds with an irrigation furrow between the rows. Under arid conditions, 12 acre-inches of water is usually needed to raise the crop. It is believed that bursting of heads is caused by the sudden absorption of applied water from irrigating following a period of dryness. Therefore, it is important that water never becomes limiting during the growing period.

2. CAULIFLOWER, BROCCOLI, AND BRUSSELS SPROUTS

Some provision must be made for irrigating cauliflower, broccoli, and Brussels sprouts except in areas where rainfall can be depended on to be adequate and properly distributed during the growing season. Cauliflower plants will not produce marketable heads of good quality if permitted to suffer for water (Salter, 1961). Broccoli can be grown under less favorable moisture conditions than cauliflower (Thompson and Caffrey, 1954). Brussels sprouts must be supplied with plenty of water throughout the season (Sciaroni et al., 1953). In the Western USA, practically all cauliflower, Brussels sprouts, and broccoli are grown under irrigation. Twelve acre-inches of irrigation is generally considered to be ample to maintain soil water above the 50% available level in the upper 24 inches on medium-textured soils.

E. Root Crops

Root crops [beet (*Beta vulgaris*), carrot (*Daucus carota*), radish (*Raphanus sativus*), parsnip (*Pastinaca sativa*), turnip (*Brassica rapa*)] usually are rapid-growing crops with limited root systems that respond readily to irrigation.

1. BEETS

Beets, although not widely grown, are very sensitive to water deficits. Irrigation in Maryland, USA increased the percentage of roots in the largest size, the number and weight of beets 1 inch or more in diameter, and the average weight per root (Reynolds and Rogers, 1949).

2. CARROTS

Carrots grow best in deep sandy or sandy loam soils. Such soils permit good development of marketable roots. Heavy, compacted, or cloddy soils distort the shape of the carrot roots and make them unmarketable. Carrots require an abundant supply of water evenly distributed throughout the growing season (Boswell, 1963; Pew, 1957). In cooler areas along the coast of California where evaporation rates are moderately low, irrigation is applied at intervals by furrow or sprinklers at a rate equivalent to about 1 inch of water/week. This amounts to about 18 to 24 acre-inches of water for the season. In the warmer desert areas of the Imperial Valley, the amount is increased to about 30 to 36 acre-inches of water (Boswell, 1963). In arid regions, the first irrigation is applied to germinate the seed, and, since air temperatures are usually high, the beds are kept constantly wet until the seedlings have emerged. After the plants are well established, the irrigation schedule is regulated to give maximum growth. Care should be exercised to avoid excessive moisture on heavier soils, especially during hot weather, to prevent rotting of the roots. On the other hand, allowing the soil to become too dry tends to cause poorly shaped roots (Pew, 1957). Irrigating after soil water suction has reached 3 atm or more tends to cause cracking.

3. RADISHES

Radishes are a very short-season crop, maturing in 4 to 6 weeks. They are grown in almost every area of the USA in small, market-garden operations, but most of the commercial acreage is found in Florida, Texas, and southern California. In Florida, radishes are planted on flat land in rows 6 to 8 inches apart and are irrigated by sprinklers to maintain soil water near field capacity until shortly before harvest. Essentially the same method is used in Texas and California. In the coastal valleys of California, 8 to 10 rows may be planted on a single bed and furrow irrigated. Since the crop is very shallow-rooted, deep irrigation is not necessary, but the root zone (upper 6 inches) should be maintained at tensions of less than 0.3 atm (very near "field capacity").

F. Bulb Crops

Bulb crops [onion (*Allium cepa*), garlic (*Allium sativum*), leek (*Allium porrum*), shallot (*Allium ascalonicum*), chive (*Allium schoenoprasum*)] have relatively limited root systems and high demands for water.

1. ONIONS

Onions are the most important bulb crop. About one-fourth of the onions grown in the USA are transplanted from seedbeds while the remaining three-fourths are direct seeded in the field. If the soil is dry, transplanted onions should be irrigated as soon as possible after they have been set. Onion roots arise from the stem plate at the base of the plant during most of the time that the plant is growing. New roots do not form unless the stem plate from which they arise is in moist soil; therefore, the soil should be kept at tensions of < 0.3 atm until the crop is nearly mature (Jones et al., 1957). Irrigation increases the yield through increased size of bulb. Early onions at Davis, California, need 5 to 7 sprinkler irrigations, and the late crop, 7 to 9. About 15 to 24 acre-inches of water are required. Onions must continue to grow without interruption to prevent formation of cracks or doubles (Davis, 1943). When plants start to mature, irrigation should be discontinued to allow the tops to desiccate. This also prevents a second flush of root growth that is difficult to stop and complicates the process of curing onions properly (MacGillivray, 1953).

G. Sweet Potatoes

Sweet potatoes (*Ipomoea batatas*) are normally grown in sandy soils and in many parts of the country respond to applications of water. An extensive root system is developed if root penetration is not restricted by impervious soil layers. Roots have been observed at depths up to 5 ft in sandy loam soils. Thus, if the soil profile is at field capacity just prior to planting time, additional water is seldom needed for a month or 6 weeks. Jones (1961), for example, found no significant increase in total yield of US no. 1 sweet potatoes when the soil was permitted to dry to only 20% of available soil water at the 12-inch depth. His rooting depth studies indicated that 51% of the roots were in the upper 9 inches of soil, 30% more were in the upper 18 inches, and the remainder were below. Undoubtedly, the plants were using some of the deeper water as a source necessary for growth. However, this does not mean that sweet potatoes will give maximum yields without ample water. Peterson (1961) reports increases in production with applications of 1 acre-inch of water at weekly intervals during prolonged hot, dry weather. Irrigation increases not only the yield of sweet potatoes but also the average size of each root. With good soil water, yields may double within a period of 2 to 3 weeks during September. To prevent potatoes from becoming oversized, some growers withhold water toward the end of the season. This practice facilitates harvest but may reduce yield. Closer spacing rather than withdrawal of water may be a better practice if it is desirable to hold down size (Minges and Morris, 1953).

H. Legumes

Legumes [snap beans (*Phaseolus vulgaris*), lima beans (*P. limensis* and *P. lunatus*), green peas (*Pisum sativum*), etc.] are rapidly growing vegetables with relatively limited root systems. Warm-season legumes, such as beans, respond much more readily to water than do cool-season legumes, such as peas.

1. SNAP BEANS

Snap beans responded to irrigation in 18 out of 30 different trials in the Northeastern USA. Two additional trials suggested beneficial effects. The maximum increase in yield was 38% in Connecticut, 64% in Maryland, 26% in New Jersey, and 36% in New York (Vittum et al., 1963). The quality of snap beans may be markedly affected by irrigation. Irrigation decreased the percentage of pods that were severely malformed. Irrigated plants produced larger pods containing less fiber and a lower percentage of seed (Reynolds and Rogers, 1949; Vittum et al., 1963).

The most critical water stress period for snap beans is during and immediately following the time of blossoming, although drouth periods at earlier stages of growth obviously will reduce vegetative growth and yield (Gabelman and Williams, 1960; Janes and Drinkwater, 1959; Kattan and Fleming, 1956). However, in many cases a single irrigation at blossomtime may result in substantially improved yields. Higher yields result largely from increased numbers of pods per plant, although pod size is also increased (Vittum et al., 1963).

Although irrigation increases the percentage of blossoms which set and develop into pods (Gabelman and Williams, 1960), it does not always insure a heavy crop of beans. When temperatures are high and other conditions are unfavorable for pod set, poor yields of both lima and snap beans may result even when irrigation is combined with other good cultural practices.

2. PEAS

Peas, a rapidly-growing, cool-season crop, are usually planted early in the spring when soil water reserves are adequate. Thus, irrigation is not so essential as with some of the warm-season crops. Irrigation at blossomtime increases the number of marketable pods and the number of peas per pod, whereas irrigation a few days prior to harvest increases the weight of both pods and peas (Salter, 1962).

I. Solanaceous Crops

1. TOMATOES (*Lycopersicon esculentum*)

Tomatoes are long-season, deep-rooted plants with high requirements for water. Substantial yield increases, averaging 7.8 tons/acre or 67%, were obtained in 13 out of 31 different irrigation experiments conducted in the Northeast USA during 1950–1960 (Vittum et al., 1963). In the remaining 18 tests, differences due to irrigation were not significant. Irrigated plots yielded slightly more than nonirrigated plots in some trials and somewhat less in others. This suggests that irrigation is not essential every year for producing maximum yields of tomatoes in the Northeast. It does indicate, however, that water sometimes limits growth, and irrigation may then be expected to increase yield. When some other factor, such as poor soil aeration and drainage, improper variety, lack of fertilizer, or cool temperatures limits growth or maturation, additional water will not benefit and may even damage the crop. Maximum stress for water develops during the middle or late part of the growing season when most of the fruit is ripening. Irrigation during this critical time appears to be more effective than irrigation before the fruit is set when available soil water is more likely to be adequate.

In California where a large percentage of the canning and fresh market tomatoes for the USA is produced, irrigation is essential throughout the growing season. Being a deep-rooted crop, tomatoes have been observed to withdraw water from depths up to 13 ft when soil conditions permit this depth of rooting. Soil in the entire root zone should be at field capacity prior to planting. Although irrigation is essential, excessive amounts of water are deleterious and should be avoided Plots irrigated when soil water tension reached a mean of 0.7 atm had lower yields than plots irrigated at 2 atm. Excessive irrigation also delayed maturity, made harvesting difficult because of excessive vine growth, and reduced the soluble solids content of the fruits (Flocker and Lingle, 1961).

Tomato fruits from irrigated plots have significantly higher pH and lower titratable acidity than similar fruits from nonirrigated plots. Even though differences are statistically significant, there is little practical significance except when the pH exceeds 4.4 and the flat-sour spoilage organisms develop in the processed tomato product (Vittum et al., 1962; Wight et al., 1962).

Irrigation increases the per cent of flowers that set fruit (Vittum et al., 1963), and the average weight per fruit (Moore et al., 1958; Vittum et al., 1958) but it decreases soluble solids and total solids (Flocker et al., 1961; Lee and Sayre, 1940; Vittum et al., 1962), and blossom-end rot (Vittum et al., 1963). Irrigation practices may have to be altered if ever processors start to purchase tomatoes on the basis of total solids content. High irrigation rates increase the drained weight, the percentage of whole tomatoes, and the percentage of tomatoes that remain firm after canning (Flocker et al., 1961).

As with many other crops, data are not sufficient to prove conclusively that soil water should be maintained above a certain level. Any practice that will delay early establishment of a good root system should be avoided. Moisture levels in the active root zone should not exceed 1 to 2 atm.

The future of the tomato industry is fascinating to anticipate. The advent of mechanical harvesting and the development of smaller, closer-spaced, more determinate varieties will force reevaluation of irrigation and fertilizer techniques.

J. Cucurbits

Cucurbits [cucumber (*Cucumis sativus*), muskmelon (*Cucumis melo*), watermelon (*Citrullus vulgaris*), pumpkin (*Cucurbita pepo*), squash (*Cucurbita maxima*), etc.] are long-season crops with medium or deep root systems that require large amounts of water. As with many other vegetables, irrigation not only increases the yield but also the quality, provided excessive amounts of water are not used.

1. CUCUMBERS

Cucumbers responded to irrigation in 3 out of 4 years on a sandy loam soil in Maryland with an average increase of 265 bu/acre or 82% (Reynolds and Rogers, 1949). In these tests, color of irrigated fruit was a more desirable darker green than that in nonirrigated fruit. In Florida tests, the highest yield of US no. 1 cucumbers was obtained with medium irrigation (Nettles et al., 1952).

2. MUSKMELONS

Muskmelons are able to withdraw water from as deep as 6 ft, if soil conditions permit root penetration to this depth (MacGillivray, 1951). If soil is not at field capacity, preirrigation should be applied (Davis, et al., 1953). In semiarid regions, such as the San Joaquin Valley of California, 12 to 24 acre-inches may be applied, followed by an additional postplanting application of 2 to 4 acre-inches to germinate the seed. This program generally supplies adequate water for the entire season (Doolittle et al., 1961). Soil water suctions of 2 atm do not lower yields of cantaloupes, and suctions of up to 7 atm at a depth of 2 ft reduce yields only slightly, indicating that plants are using water stored at lower depths (Flocker et al., 1962). Melons can be over-irrigated as well as under-irrigated (Pew, 1958). Excessive irrigation, maintaining a suction of < 0.5 atm reduces yields, quality, and shelf life of marketable melons. Excessive irrigation with high nitrogen fertilization produces more cracked, misshapen, and unmarketable cantaloupes than does proper irrigation (Flocker et al., 1962).

3. WATERMELONS

Experiments in central California showed no significant difference in yield of watermelons whether irrigated 7 times or not at all during the growing season (Doneen et al., 1939). Therefore, while it is imperative that plants have adequate available water, very often for the long-season, deep-rooted crops, the source of water is winter rain stored in the soil at depths unavailable to shallow-rooted or short-season vegetable crops.

K. Sweet Corn

Sweet corn (*Zea mays* var. *rugosa*) is harvested at an immature stage so its demand for water and plant nutrients is not as great as that of field corn. On the other hand, sweet corn is very succulent, has a rather shallow root system, and does not yield well if adequate soil water is not readily available.

Irrigation in New York significantly increased the number of marketable ears per plant, the average weight per ear, the gross yield of unhusked ears, and the percentage of usable corn cut from these ears for canning or freezing. Irrigation did not affect quality as measured by percentage of moisture in the cut corn, although it did delay maturity. Irrigated plots were harvested an average of 2.2 days later than nonirrigated plots (Vittum et al., 1959).

The proper timing of irrigation is apparently more important than the total amount of water applied to sweet corn. Likewise, the distribution of rainfall is more important than the total amount. Sweet corn is more sensitive to soil water suction during ear formation than at any other time (Stanberry et al., 1963). Thus, the years that show benefit from irrigation are those when drouth occurs at this crucial period. A general recommendation for irrigation of sweet corn would be to maintain water in the root zone at suction < 2 atm, particularly from pollination to harvest. In the western regions of the USA, the root system becomes more fibrous as the plant approaches maturity, and it will extract water from depths of 2 to 4 ft. Crops grown in the summer in the interior valleys of California may require more water than early spring crops or those grown in the coastal regions (MacGillivray et al., 1962).

II. SUMMARY

Irrigation obviously is needed whenever available water in the soil is depleted to the point where plant growth is retarded. As pointed out in previous chapters, the need for irrigation is influenced by many factors interacting in a complex manner. The amount of water required to grow a crop of vegetables and the frequency of irrigation vary so much with local climate and soil conditions that specific recommendations are futile. However, in general, it may be recommended for most vegetables that available water within the root zone be replenished as soon as one-half has been withdrawn. In most medium-textured soils, this would be at a soil water suction of about 1 atm.

In this chapter an effort was made to bring out two characteristics which could be used as guideposts to irrigation requirements: (i) The depth of the soil profile and the amount of water that can be stored in it, either through rainfall or pre-irrigation, and (ii) the depth to which the roots can penetrate. Deep-rooted crops growing on deep soils filled with water may require very few irrigations or none at all. On the other hand, shallow-rooted crops on the same soils will usually benefit greatly from several irrigations. Shallow soils require frequent irrigations for both deep- or shallow-rooted crops. An attempt to summarize the various conditions is given in Table 34-1.

Table 34-1. Total amounts of water suggested for commercial production of vegetable crops*

Depth of rooting	Amount of water (acre-inches)†			
	12	12 - 17	18 - 24	Over 24
Shallow-rooted, < 2 ft	Lettuce (winter) Spinach	Cabbage, Cauliflower, Onion, Broccoli, Brussels sprouts	Lettuce (summer) Onion (late) Potato (late) Sweet corn	Celery Potato (early)
Moderately deep-rooted, 2-4 ft	Peas (winter)	Beans, pole Beans, snap (spring) Cucumbers	Beans, snap Beets, Carrots (coastal areas) Eggplant, Peas (fall) Pepper, Squash (summer)	Carrots (desert) areas)
Deep-Rooted, > 4 ft		Artichoke Lima beans Tomato (coastal areas) Watermelon	Asparagus Muskmelon (inland areas) Squash (winter) Sweet potato Tomato	Muskmelon (desert areas)

* Adapted from California Agr. Exp. Sta. Lithoprint, 1943.
† Water requirements may be higher in areas where humidity is low and transpiration rates are high. The amounts of water are for medium-textured soils wetted to field capacity to the different depths, either by irrigation or by rainfall.

A major problem is scheduling irrigation so that all of the acreage is covered before the critical level has been exceeded in any part of the field. This requires very careful planning, maintenance, and use of equipment.

In conclusion, irrigation of vegetables must be considered as only a part of the over-all management program. Irrigation is not a cure for all ills, but, when needed, it is of great value. If all factors now known are used in scheduling irrigation, it is possible to use irrigation as a profitable part of farm management in many parts of the world, providing water supplies for irrigation are available. Further improvements in irrigation will come as more basic knowledge of plant physiology and the physics of soil and atmospheric water is obtained.

LITERATURE CITED

Beattie, J. H., and W. R. Beattie. Jan. 1948. Production of spinach. US Dep. Agr. Leafl. no. 128 (Rev.). 8 p.

Boswell, V. R. 1963. Commercial growing of carrots. US Dep. Agr. Leafl. no. 353. 6 p.

Cannell, G. H., K. B. Tyler, and C. W. Asbell. 1959. The effects of irrigation and fertilizer on yield, blackheart, and nutrient uptake of celery. Amer. Soc. Hort. Sci., Proc. 74:539–545.

Davis, G. N. 1943. Onion production in California. California Agr. Exp. Sta. Cir. 357. 8 p.

Davis, G. N., T. W. Whitaker, and G. W. Bohn. 1953. Production of muskmelons in California. California Agr. Exp. Sta. Circ. 429 40 p.

Doneen, L. D., P. R. Porter, and J. H. MacGillivray. 1939. Irrigation studies with watermelons. Amer. Soc. Hort. Sci., Proc. 37:821–824.

Doolittle, S. P., A. L. Taylor, L. L. Danielson, and L. B. Reed. Oct. 1961. Muskmelon culture. US Dep. Agr., Handbook no. 216. 45 p.

Fieldhouse, D. J. 1958. Asparagus irrigation. Peninsula Hort. Soc., Trans. 48:9–13.

Flocker, W. J., and J. C. Lingle. 1961. Field applications of tensiometers and soil moisture blocks as criteria for irrigation of canning tomatoes. Amer. Soc. Hort. Sci., Proc. 78:450–458.

Flocker, W. J., J. C. Lingle, J. H. MacGillivray, and S. Leonard. 1961. Effect of irrigation and nitrogen applications on some quality factors of canning tomatoes. Univ. California, Davis. Veg. Crops Ser. 111. 10 p.

Flocker, W. J., J. R. Wight, J. E. Knott, and R. M. Davis. 1962. Yield and quality of cantaloupes influenced by irrigation and nitrogen fertilization. Univ. California, Dep. Veg. Crops Progr. Rep. 10 p.

Gabelman, W. H., and D. D. F. Williams. 1960. Developmental studies with irrigated snap beans. Wisconsin Agr. Exp. Sta. Res. Bull. 221. 56 p.

Grogan, Raymond G., William C. Snyder, and Roy Bardin. 1955. Diseases of lettuce. California Agr. Ext. Serv. Circ. 448. 28 p.

Hanna, G. C. 1950. Asparagus production in California. California Agr. Ext. Serv. Circ. 91. 20 p.

Hanna, G. C., and L. D. Doneen. 1958. Asparagus irrigation studies. California Agr. 12(9): 8, 14.

Herner, D., C. Tallman, and A. J. Pratt. 1961. A bibliography on irrigation. New York State Coll. Agr., Cornell Univ. Veg. Crops Mimeo. No. 87. 95 p.

Janes, B. E. 1959. Effect of available soil moisture on root distribution, soil moisture extraction and yield of celery. Amer. Soc. Hort. Sci., Proc. 74:526–538.

Janes, B. E., and W. O. Drinkwater. 1959. Irrigation studies on vegetables in Connecticut. Connecticut Agr. Exp. Sta. Bull. 338. 82 p.

Jones, H. A., B. A. Perry, and G. N. Davis. 1957. Growing the transplant onion crop. US Dept. Agr. Farmers' Bull. no. 1956 (Rev.). 2 p.

Jones, S. T. 1961. Effect of irrigation at different levels of soil moisture on yield and evapotranspiration rate of sweet potatoes. Amer. Soc. Hort. Sci., Proc. 77:458–462.

Kattan, A. A., and J. W. Fleming. 1956. Effect of irrigation at specific stages of development on yield, quality, growth and composition of snap beans. Amer. Soc. Hort. Sci., Proc. 68:329–342.

Lee, F. A., and C. B. Sayre. 1940. Effect of soil moisture on acid content of tomatoes. Food Res. 5(1):69–72.

MacGillivray, J. H. 1951. Effect of irrigation on production of cantaloupes. Amer. Soc. Hort. Sci., Proc. 57:266–272.

MacGillivray, J. H. 1953. Vegetable production. The Blakiston Co., New York. p. 177.

MacGillivray, J. H., W. L. Sims, and R. F. Kasmire. 1962. Sweet corn production in California. California Agr. Exp. Sta. Circ. 515. 24 p.

Minges, P. A., and L. L. Morris. 1953. Sweet potato production and handling in California. California Agr. Exp. Sta. Circ. 431. 39 p.

Moore, J. N., A. A. Kattan, and J. W. Fleming. 1958. Effect of supplemental irrigation, spacing and fertility on yield and quality of processing tomatoes. Amer. Soc. Hort. Sci., Proc. 71:356–368.

Nettles, V. F., F. S. Jamison, and B. E. Janes. 1952. Irrigation and other cultural studies with cabbage, sweet corn, snap beans, onions, tomatoes and cucumbers. Florida Agr Exp. Sta. Bull. 495. 26 p.

Peterson, L. E. 1961. The varietal response of sweet potatoes to changing levels of irrigation, fertilizer and plant spacing. Amer. Soc. Hort. Sci., Proc. 77:452–457.

Pew, W. D. 1957. Carrots in Arizona. Arizona Agr. Exp. Sta. Bull. 285. 26 p.

Pew, W. D. May 1958. Effects of soil moisture on cantaloupe growth and production. West. Grower Shipper Mag. 33:6–8.

Reynolds, C. W., and B. L. Rogers. 1949. Irrigation studies with certain fruit and vegetable crops in Maryland. Maryland Agr. Exp. Sta. Bull. 463. 31 p.

Salter, P. J. 1961. The irrigation of early summer cauliflower in relation to stage of growth, plant spacing, and nitrogen level. J. Hort. Sci. 36:241–253.

Salter, P. J. 1962. Some responses of peas to irrigation at different growth stages. J. Hort. Sci. 37:141–149.

Sciaroni, R. H., P. A. Minges, W. H. Lange, and W. C. Snyder. 1953. Brussels sprouts production in California. California Agr. Ext. Serv. Circ. 427. 16 p.

Sims, W. L., R. H. Sciaroni, and H. Lange. 1962. Growing globe artichokes in California. Univ. California Agr. Exp. Publ. AXT–52. 16 p.

Stanberry, C. O., H. A. Schreiber, M. Lowrey, and C. L. Jenson. 1963. Sweet corn production as affected by moisture and nitrogen variables. Agron. J. 55:159–161.

Stanhill, G. 1957. The effect of differences in soil moisture status on plant growth: A review and analysis of soil moisture regime experiments. Soil Sci. 84:205–214.

Thompson, H. C., and W. C. Kelly. 1957. Vegetable crops. McGraw-Hill Book Co., Inc., New York. 611 p.

Thompson, Ross C. 1951. Lettuce varieties and culture. US Dep. Agr. Farmers' Bull. no. 1953 (Rev.). 42 p.

Thompson, Ross C., and D. J. Caffrey. 1954. Cauliflower and broccoli-varieties and culture. US Dept. Agr. Farmers' Bull. no. 1957 (Rev.). 16 p.

Veihmeyer, F. J., and A. H. Holland. 1949. Irrigation and cultivation of lettuce. California Agr. Exp. Sta. Bull. 711. 52 p.

Vittum, M. T., R. B. Alderfer, B. E. Janes, C. W. Reynolds, and R. A. Struchtemeyer. 1963. Crop response to irrigation in the Northeast. New York State Agr. Exp. Sta. Bull. 800. 63 p.

Vittum, M. T., and N. H. Peck. 1956. Response of cabbage to irrigation, fertility level and spacing. New York State Agr. Exp. Sta. Bull. 777. 34 p.

Vittum, M. T., N. H. Peck, and A. F. Carruth. 1959. Response of sweet corn to irrigation, fertility level and spacing. New York State Agr. Exp. Sta. Bull. 786. 45 p.

Vittum, M. T., W. B. Robinson, and G. A. Marx. 1962. Raw-product quality of vine-ripened processing tomatoes as influenced by irrigation, fertility level and variety. Amer. Soc. Hort. Sci., Proc. 80:535–543.

Vittum, M. T., W. T. Tapley, and N. H. Peck. 1958. Response of tomato varieties to irrigation and fertility level. New York State Agr. Exp. Sta. Bull. 782. 78 p.

Wight, J. R., J. C. Lingle, W. J. Flocker, and S. J. Leonard. 1962. The effects of irrigation and nitrogen fertilization treatments on the yield, maturation and quality of canning tomatoes. Amer. Soc. Hort. Sci., Proc. 81:451–457.

Winter, E. J. 1957. The irrigation of vegetables—a review of the present state of our knowledge. Nat. Agr. Adv. Serv. Quart. 36:63–71.

35 | Deciduous Tree Fruits and Nuts

K. URIU

*University of California
Davis, California*

J. R. MAGNESS

*Agricultural Research Service, USDA (retired)
Takoma Park, Maryland*

Deciduous fruits and nuts considered in this chapter are cultivated trees that shed their leaves in the fall. Of these the most important fruit crops are apples (*Malus sylvestris*), apricots (*Prunus armeniaca*), cherries (*Prunus sp.*), figs (*Ficus carica*), peaches (*Prunus persica*), pears (*Pyrus communis*), plums (*Prunus sp.*), and prunes (*Prunus domestica*). Important nut crops are almonds (*Prunus amygdalus*), filberts (*Corylus avellana*), pecans (*Carya illinoensis*), and walnuts (*Juglans regia*).

All of these crops are grown mainly in the temperate zones. All require a period of dormancy to grow and produce well, hence they are not successful in the tropics. Most are grown between 25° and 55° latitude, both north and south of the equator.

Within these temperate zone regions, annual precipitation varies from < 5 inches/year, as in certain areas of north-Africa and Southwest USA, to areas > 100 inches. In general, irrigation of orchards is essential for commercial production of these crops where annual precipitation is < 20 inches. Some supplementary irrigation is needed where precipitation is between 20 and 30 inches while irrigation is rarely used in areas of > 30 to 35 inches precipitation.

Since these deciduous trees are without leaves during the winter months, they use very little water during that time. In the Northern Hemisphere the period without leaves will extend from October to May in northern areas and from late November to April in more southern areas. Precipitation occurring in these periods will be stored in the soil, if the soil is sufficiently deep and retains water. Thus the depth of soil and its water-holding capacity are very important in determining the need for irrigation in areas where total precipitation may be ample but where distribution of rainfall is erratic and uncertain.

I. ROOT SYSTEMS OF DECIDUOUS FRUIT AND NUT TREES

Trees develop very extensive root systems when grown in deep, well-drained soil, free of impervious layers or high water tables. Proebsting (1943) reported apricot roots 16 ft deep in a well-drained soil in California, USA. Veihmeyer and Hendrickson (1938), on the basis of water-extraction studies in well-drained soil, concluded that walnuts had numerous roots at a depth of 12 ft and peaches and prunes at 6 ft, the maximum depths studied. Wiggans (1936) found apple roots 30 to 35 ft deep in the loess soil of Nebraska, USA.

In studies in New York, USA, Oskamp (1932) found cherry and peach roots extended about 5 ft deep in well-drained, uniform, brown soil. Apples and prunes were rooted to 6 ft in such soil. Where a mottled grey, poorly-drained layer occurred, there was no penetration of apple, peach or cherry roots through the layer. Prune roots penetrated it to some extent.

Roots spread laterally to a much greater distance than the branch spread of the trees. A good general estimate is 2 to 3 times the spread of the branches in sandy soils and about 1.5 times on loam and clay soils (Rogers and Vyvyan, 1934). In orchards planted at usual spacing, roots of mature plants occupy all the soil laterally. Where deep rooting occurs, a very large mass of soil is occupied by roots, and this will hold a large amount of water available for plant usage. If such soils are filled to field capacity as a result of winter precipitation, they have a large reservoir of available water for use during periods of inadequate rainfall. In areas where irrigation is necessary, water applications may be less frequent and in larger amounts per application on deeply-rooted trees than where the root system is shallow. Under conditions of natural rainfall, trees with shallow root systems may become deficient in water during short periods of precipitation deficiency.

In general, the concentration of feeder roots becomes less at increasing distances from the trunk, both laterally and vertically. Maximum concentration usually is found in the top 3 ft and within the spread of the branches. Beyond the spread of the branches and at greater depths, the quantity of feeder roots per cubic foot of soil generally diminishes progressively.

A.   Water Absorption by Roots

Water absorption by roots increases significantly with the advent of leaves in the spring. The initial water withdrawal appears to be greater from the surface soils probably because of greater root activity under the more favorable soil temperature near the surface in early spring. With the gradual warming up of the entire soil mass the rate of water withdrawal by the tree becomes more proportional to the feeder root population in any particular soil volume and those of greatest root concentration are depleted of water most rapidly. Thus the soil water may reach the wilting percentage in such volumes—normally the upper 2 to 4 ft of soil within the branch spread—while there is still water available at greater depths and distances from the trunk.

Evidence from many soil water experiments indicates that when a portion of the root system is in soil at the wilting percentage the rate of water uptake will be reduced in that portion. Under conditions of high transpiration this will result in a water deficit within the tree.

II.   EFFECT OF WATER ON CERTAIN PHYSIOLOGICAL PROCESSES

A.   Water Balance of Trees

The water status of fruit trees is continually changing, not only day-to-day, but hour-to-hour. However, changes in soil water supply occur rather slowly. The amount of water used by the tree varies with the air temperature, air movement, relative humidity, and light intensity, all of which affect the transpiration rate. Even when soil water is adequate, temporary deficits will occur in trees during the days when transpiration rate is high. Magness et al. (1933) found that the water content of apple leaves decreased 6 to 7 percentage points between 6 AM and noon on moderately warm, dry days, even when all the root zone had available moisture. Water content of the bark decreased 3 percentage points. Ackley (1954) found minimum water content and maximum water deficits in Bartlett pear leaves occurring around 2 PM each day and the reverse about 2 AM. Hendrickson (1926) found a rapid drop in water content in peach leaves from 6 AM to noon with the minimum about 3 PM. Twig, trunk and root bark and wood tissues behaved similarly. The diurnal change in water content of peach leaves can result in a change in soluble solids content of the expressed sap of leaves. Roberts found the highest soluble solids percentage occurring about 3 PM and the minimum, 6 AM, which corresponds to the minimum and the maximum leaf water contents, respectively. (E. B. Roberts. 1963. Trunk diameter and leaf soluble solids as indicators of moisture stress in peach trees. *Unpubl. M.S. Thesis, Univ. of California, Davis.*).

During the night, with normally lower temperatures, higher humidities and closed stomata in the leaves, water in the tissues is restored, if any appreciable available water is present in the soil.

B. Stomatal Activity and Photosynthesis

Hendrickson (1926) found stomata of peach, prune, and apricot trees to reach their maximum degree of opening between 9 AM and noon, after which they began to close. The stomata on trees in dry soil did not open as widely as those in moist soil and began closing a few hours earlier. Furr and Magness (1931) in a study of stomatal opening in apple leaves found that in irrigated trees, 40% or more of the stomata were open for 4.5 to 7 hours during a period of high temperature and low humidity, while the period was 10 hours or more on cool, humid days. On trees with most of the root zone below the wilting percentage, fewer than 40% of the stomata opened at any time. Furr and Degman (1932) also showed the marked effect of soil water on stomatal behavior.

As water becomes deficient in the leaves, both rate of water loss from the leaves and rate of photosynthesis are reduced. Heinicke and Childers (1935) found CO_2 assimilation by apple leaves on trees beginning to wilt was about 80% of the rate in well-watered trees between 6 and 8 AM, but by 11:00 AM had dropped to about 31%. Reduction in transpiration rate paralleled the reduction in photosynthesis. Closing of the stomata was associated with a reduction both in transpiration and photosynthesis.

Schneider and Childers (1941) found that even before wilting, a reduction in photosynthesis of over 50% occurred in apple trees. When plants showed definite wilting, and the soil water was approximately at the wilting percentage, there was an 87% reduction in both photosynthesis and transpiration. Respiration, on the other hand, was found to increase progressively with the drying out of the soil.

C. Other Processes

In apple trees growing with an inadequate soil water supply, Magness et al. (1933) found sugars in the sap of bark and wood to be higher and the starch to be lower than in trees with ample water. Sugar in the bark of dry trees was 10 to 15% higher and starch 25% lower than in watered trees. Sugar in the wood of dry trees was double that in the watered trees, while starch was 25% lower. Thus the sugar-starch ratio was much higher in the stem tissues of the dry trees.

There is very little known about the activity of auxins in deciduous fruit trees under different soil water stresses. However, Crane and Uriu (1965) applied 2,4,5-T (2,4,5-trichlorophenoxyacetic acid) on apricot trees and found the effectiveness of this material enhanced by irrigation. Fruit size was greater and maturity hastened on sprayed trees that were irrigated compared to nonirrigated trees even though the soil water content in the latter never reached the wilting point (WP).

III. EFFECT OF WATER SUPPLY ON VEGETATIVE GROWTH

A. Shoot Growth

There is no question that vegetative growth suffers when trees are subjected to WP conditions. Growth of shoots in length on bearing deciduous fruit and nut trees occurs mainly during the first 2 months of the growing season while the thickening of such shoots, as well as trunk thickening, continues throughout the growing season. Thus the effect of water stress on vegetative growth depends on the occurrence of stress. Butijn (1961) stressed that shoot growth is very susceptible to dry soil conditions so that ample soil water is needed in the first one-half of the summer to promote good shoot growth. Gamble (1964) found no difference in terminal growth of apple trees under differential irrigation and attributed this to the presence of ample soil water in all plots early in the season during the shoot elongation period. However, circumferences of trees increased with irrigation.

With pears Aldrich and Work (1932) found shoot length to be greater on the trees receiving irrigations throughout the early summer than on those given an irrigation only in August. Trees receiving frequent irrigations early but not late in the summer had comparable shoot lengths but the thickness of the shoots was less than in those frequently irrigated throughout the summer. Furr and Magness (1931) working with apples had also found that the xylem (wood) in apple shoots was 52% thicker on irrigated trees than on similar trees whose soil water content in the top 18 inches of soil was at the WP from mid-July to mid-August.

Goode (1957) stated that the first process that would suffer from water stress is shoot growth. He (1956, 1957) showed that when soil water suction at the 1-ft depth was not allowed to exceed approximately 0.13 bar, new growth of trees was twice as much as that of control trees for which the soil water suction rose to 0.8 bar or over. When the maximum soil water suction was controlled at 0.3 bar, new growth was 1.5 times that of the control trees. The increase in shoot growth was due mainly to an increase in the number of shoots produced (Goode, 1956; Goode and Hyrycz, 1964).

Uriu et al. (1964) took weekly measurements of peach shoots and showed that shoot growth can be affected very early in the season by water stress. An irrigation given when soil water suction reached 0.4 bar at the 2-ft depth maintained the growth rate of the shoots while the trees not irrigated at that time showed a reduced growth rate. As soil water suction rose above 0.8 bar the growth rate was even further reduced. Any subsequent irrigation did not recover the shoot growth lost early in the season due to these water stresses. The total "watersprout" growth was greatest in the plots irrigated when soil water suction reached 0.4 bar, less in the 0.8 bar, and least in the 5.0 bars.

B.   Trunk Growth

Trunk growth increases from irrigation have been reported for many deciduous fruit tree species and by many workers. For apple (Gamble, 1964; Goode, 1956, 1957; Goode and Hyrycz, 1964; Kenworthy, 1949; Taerum, 1964; Verner et al., 1962); apricot (Hendrickson and Veihmeyer, 1950b; Till, 1958); fig (Condit, 1947); pecan (Romberg, 1960); and prune (Hendrickson and Veihmeyer, 1945). In many of these experiments the WP had been reached in the non- or infrequently-irrigated plots.

Trunk growth measurements have served as a very sensitive indicator of the tree's responses to changes in climatic conditions and in soil water content (Goode, 1956; Verner et al., 1962). Even fluctuations in trunk diameter during the day can be observed. On warm days trunks have been observed to contract from early morning until early afternoon, but overnight expansion occurs (Goode, 1956). Goode found that trunk growth was limited before a soil water suction of 0.7 bar was reached. Verner et al. (1962) found trunk growth to be more sensitive than fruit growth in apples and found that the rate of trunk growth started declining well above the WP.

C.   Root Growth

Hendrickson and Veihmeyer (1931) have shown with sunflower plants (*Helianthus annuus*) that roots will not grow into soils at the WP. The soil must be above the WP for root extension to occur. Experimental data on the growth of roots of deciduous fruit trees as related to soil water stress less than the WP are very limited. However, Rogers (1939) found root growth of apples to decrease as soil water suctions reached 0.4 to 0.5 bar.

IV.   EFFECTS OF WATER SUPPLY ON FRUITING

A.   Fruit Bud Initiation and Development

Some limited experimental data indicate that water shortage during the period of fruit bud initiation in deciduous fruits tends to increase the formation of fruit buds. However, other data indicate the opposite. Aldrich (1933) found that water shortage before July 15 tended to increase bloom of pears the following spring. With apples Degman et al. (1932) reported similar results on the Rome Beauty variety but found no difference in the Olderburg variety and actually found a decrease in bloom on York Imperial and Wealthy varieties that had not been subjected to irrigation the previous year. Packer et al. (1963) reported a tendency for increased flower bud size in the dormant period and a greater number of flowers in the spring in Jonathan apples which had been irrigated the previous summer.

Since there is a strong tendency for biennial bearing in most varieties of apples, differences in crop load the previous summer in many of these experiments may have had more influence on subsequent bloom density than differences in irrigation treatments. The biennial bearing habit, however, was reported by Gamble (1962) to be reduced by irrigation on a number of apple varieties in Michigan.

With apricots Hendrickson and Veihmeyer (1950b) reported a nearly complete lack of bloom and reduced yield on trees not irrigated the summer before. Brown (1953) showed that trees kept under prolonged soil water stress during July, August, and September resulted in a reduction in the number of flower buds differentiated, a delay in the time of differentiation of many of the flower buds, and a slower rate of development of these buds. He also showed that bud development was normal on trees irrigated in July immediately after harvest. Uriu (1964) showed that trees subjected to WP conditions for the entire summer beginning the last week in June or earlier will result in no return bloom and a complete crop failure. If during the period of fruit bud initiation and development in July and early August soil water is not depleted, normal bud development can be expected.

With almonds Ross and Meyer (1957) suggested application of water from August to October for proper development of the buds for the following year's crop.

In tung trees (*Aleurites fordii*) Laycock and Foster (1955) report a high correlation between yield and rainfall the previous June-August period which is the time when flower primordia are being formed. It was not clear whether the effect of soil water content was on fruit bud differentiation or development.

B. Fruit Set and Drop

No experimental data are available on the direct effects of soil water deficits on fruit set in deciduous fruits. Since such trees use little water during the winter, in areas of winter rainfall they practically always have water available at time of bloom. However, observations indicate that water deficits during bloom will reduce fruit set. Skepper and Vincent (1962) reported that in apricots and prunes water shortage 4 or 5 weeks before blossoming resulted in a reduction in fruit setting. The degree of reduction was affected by the severity and duration of the shortage.

Some of the early fruit drop of some deciduous fruits might be associated with dry soil conditions. The "June drop" of Conference pears was investigated by Gayner (1941). He found no difference in drop between irrigated and control trees. Modlibowska (1961), however, working with apple trees in pots found "June drop" to be greater with trees grown in dry rather than wet soil conditions.

C. Fruit Growth

In general, when the water supply to the tree is reduced to a critical level, growth rate of fruits is reduced, and in extreme cases, growth will cease. The condition of soil water at which fruit growth is reduced has been extensively studied with different results reported. Some of the effect of soil water stress on fruit growth may depend on the time the stress occurs as the normal fruit growth patterns differ in the various deciduous fruit and nut crops.

1. POME FRUITS

In apples and pears, after the leaf system on the trees is fully expanded, growth of fruit on a volume basis is at an almost uniform rate until maturity, if water is ample (Magness et al., 1935).

On open, well-drained soils in California orchards, it was concluded that growth and final size of pears and apples are not affected as long as the water content of the soil in the root zone is above the wilting percentage (Hendrickson and Veihmeyer, 1937, 1941; Veihmeyer and Hendrickson, 1950a, 1950b, 1952). Fruit growth was reduced when soil water content reached the WP, but irrigation before depletion to the WP had no effect on the rate of growth.

A number of workers (Furr and Magness, 1931; Furr and Degman, 1932; Magness et al., 1935; Boynton and Savage, 1938) studying fruit growth on apples on the more shallow soils of Eastern USA found fruit growth not only decreases when soil water approaches the WP but when soil water content was several percentages above it, especially during dry and hot weather. Forshey (1958) found significantly fewer large (3 inch) apples in a treatment that dropped below 25% available water for 1 week in August than in a treatment kept above 37.5% available water throughout the season.

Taerum (1964) in Idaho, USA, using the dendrometer method developed by Verner et al. (1962) to time irrigations, obtained no difference in growth rates of apples in three differentially irrigated plots. In England, however, fruit size of apples is said to be affected when 70% of available water has been used (Office of Minister for Science, 1962).

2. STONE FRUITS

With stone fruits [peaches, plums (*P. salicina* and *P. domestica*), apricots, cherries (*P. avium* and *P. cerasus*)] there is an initial period of rather rapid fruit enlargement followed by a period during the hardening of the pit when total fruit enlargement is slight. Finally the flesh of the fruit thickens and total enlargement is very rapid prior to maturity. In peaches, about two-thirds of the final volume is attained during the last 30 days before harvest (Batjer and Westwood, 1958; Davis, 1941). In sour cherries (*P. cerasus*), Pollack et al. (1961) found that nearly 80% of the final weight of the fruit developed during the "final swell" of approximately 25 days. Plums (*P. salicina* and *P. domestica*), prunes and apricots have similar double sigmoid growth curves.

Supplemental irrigation on peaches, especially during the final swell, has increased fruit size when compared with plots without irrigation or with summer rainfall only (Feldstein, 1963; Hendrickson and Veihmeyer, 1935; Jones, 1932; Livingston and Hagler, 1957; Morris et al., 1962; Rogers, 1958). When water was added to plots in which soil water content was above the WP, no increase in the rate of fruit growth was found (Hendrickson and Veihmeyer, 1930). However, Cockroft (1963) pointed out that peach fruit size suffers when an average suction of 10 bars is reached in the 0 to 24-inch soil depth. Uriu et al. (1964) presented data showing that the growth rate of cling peaches was reduced when the soil water at the 2-ft depth approached a suction of 5.0 bars, especially during the final swell in the growth of the fruit.

Nasharty and Ibrahim (1961) in Egypt found larger plums (*P. salicina*) in a plot irrigated every 16 days compared to a 24-day interval. Neither treatment reached the WP at any time.

With apricots Crane and Uriu (1965) found fruit size increased as a result of irrigation even though water content in the nonirrigated soil never reached the WP.

3. NUTS

Nuts reach full size by approximately midsummer when the shells harden. There is no "final swell" as found in most stone fruits. In California, almonds reach full size about 50 days after full bloom (Brooks, 1939). There is little development of the kernel at that time. Kernel development, or filling of the nut, occurs mainly during the 2 months prior to harvest.

Early soil water stress can affect the size of nuts while late stress can affect kernel development.

Soil water deficits when pecans are completing their growth can decrease nut size (Alben, 1958; Romberg et al., 1959; Romberg, 1960). Stress late in the season results in poorly filled nuts (Magness, 1931). Veihmeyer and Hendrickson (1949) found this to be true with walnuts also.

D. Fruit Splitting

Splitting or cracking of the fruit is a problem with some species. It is associated, essentially, with rapid fruit enlargement as a result of a marked increase in the water content of the fruit. The uptake of water can occur through the roots or epidermis of the skin. In some cases irrigation has been demonstrated to be a factor in cracking.

In cherries (*P. avium*), fruit cracking results primarily from absorption of water through the skin as a result of rainfall or immersion in water (Verner and Blodgett, 1931).

The splitting or cracking of pecan nuts was noted by Wahlberg (1932) to be most prevalent in orchards that were allowed to dry out and then suddenly saturated with a heavy irrigation. Mathews (1960) also noted the same thing in maturing apples. After a prolonged drouth, heavy rains caused a sudden growth of the fruit and subsequent cracking. Uriu et al. (1962) found that of the two types of cracks in prunes—side and end cracks—end cracks resulted when an irrigation was applied to trees that were under soil water stress, i.e., down to the WP. The end cracking occurred immediately following an irrigation. Prunes were susceptible to this type of cracking during the entire period of fruit growth except at the end of the final swell when the fruits color and the growth rates slow down. At this time end cracking did not occur with irrigation even though the trees were under soil water stress for weeks. Side cracking, on the other hand, was found to occur at a definite time in fruit development, at the beginning of the final swell, no matter what the irrigation treatment. A water relationship in the side cracking phenomena may exist but was not uncovered in this experiment.

With figs, Hendrickson and Veihmeyer (1955b) could find no difference in cracking between irrigated and nonirrigated trees.

E. Preharvest Drop

Some preharvest drop of certain deciduous fruits has been associated with dry soil conditions. Skepper et al. (1961) noted that preharvest drop of prunes can occur under soil water stress. An increase in drop of peaches has been frequently observed on trees under stress following an irrigation close to harvest. However, no published data are available.

Preharvest drop of apples is a widespread problem. Simons (1963) made anatomical studies of apple fruit abscission and found that lack of soil water and presence of high temperatures hastened the abscission zone development in all the varieties observed. Hoffman and Edgerton (1946) reported that in the use of NAA (naphthalene acetic acid) to delay preharvest drop of apples, inadequate soil water may seriously reduce the effectiveness of this material.

Zioni (1963) reduced the preharvest drop of hazel nuts 60% by irrigations every 30 days during the growing season. More frequent irrigations reduced the drop even further. In an extreme drouth, pecan leaves drop before the fruits (Romberg et al., 1959).

F.   Fruit Yield

Meaningful yield data in deciduous fruit trees are difficult to obtain unless differences are large because of the high variability between trees, particularly in the amount of bearing area. Also, differences in amount of flower buds formed, development of the buds, number of fruits that set, amount of "June drop" and preharvest drop, and finally the size of fruit at harvest will all affect yield.

Fruit trees subjected to WP conditions during the fruit growth period can be expected to have reduced yields because of smaller fruit sizes. However, under conditions above the WP there is disagreement. Veihmeyer and Hendrickson (1949) obtained no increase in yield of a number of fruits when soil water was maintained at a high level. However, Goode (1957) reported more apple sizes of 2.25 to 3 inches when soil suction was kept below 0.3 bar compared to similar trees subjected to higher suctions. Uriu et a. (1964) found yield of peaches in plots kept below 0.4 bar suction to be about 8% greater than in the plot allowed to reach 5.0 bars.

Morris et al. (1962) obtained a greater yield response for peaches from irrigation under a heavy crop load compared to a light load. In England it has been recommended that where cropping is heavy soil water content be maintained at a higher level than when cropping is light (Office of the Minister for Science, 1962).

G.   Fruit Quality

Fruit quality is adversely affected when fruit trees are subjected to intense drouth for long periods. Veihmeyer and Hendrickson (1949) reported canned peaches were tough and leathery in texture, pears remained green and hard a week or more after the ripening period, prunes were sunburned, and walnuts were only partly filled as some of the adverse effects that resulted from relatively long periods without water. Rogers (1958) described Elberta peaches from dry plots as "dry and bitter".

Fruits from irrigated plots are generally lower in soluble solids and higher in water content than those from nonirrigated plots: apple (Donoho et al., 1964; Gamble, 1964; Packer et al., 1963); peach (Hendrickson and Veihmeyer, 1927; Morris et al., 1962); and plum (*P. salicina*) (Nasharty and Ibrahim, 1961). Irrigated apples were reported to be less firm (Donoho et al., 1964; Gamble, 1964;

Packer et al., 1963) and less acidic but juicier (Packer et al., 1963). Apples from irrigated plots suffered more storage disorders (Gamble, 1962) and peaches decayed somewhat more rapidly (Hendrickson and Veihmeyer, 1927).

Poorly-filled kernels of nuts that result from soil water stress especially in the few months before harvest reduces nut quality.

In experiments where soil water was kept above the WP, Uriu et al. (1964) found soluble solids in peach fruit to be higher and the water content less when soil water suction approached 5.0 bars. Nasharty and Ibrahim (1961) found higher soluble solids in plums (*P. salicina*) irrigated every 24 days compared to 16 days even though both never reached the WP at any time.

V. EFFECT OF WATER ON DEFOLIATION AND WINTER INJURY

Trees under water stress for extended periods will defoliate prematurely. If such defoliation occurs in late summer and an irrigation is given in the autumn while temperatures are high, the tree can burst into autumn growth and blossoming [prunes: Bowman and Davison, 1941; plums (*P. salicina*) and pears: Skepper and Vincent, 1962], thus reducing the potential crop the following year.

Winter injury can be caused by desiccation of the shoot tissues or by winter freezing. If the soil is too dry when winter begins, the trees may be damaged by the desiccation of tissues from the dry, cold air of winter (Fortier, 1940; Wilcox and Brownlee, 1961). To prevent this type of winter injury irrigation in late fall was recommended in Colorado, USA (Fortier, 1940).

On the other hand, resistance to winter freeze injury was found to be less in apple trees irrigated in the fall than in nonirrigated ones (Way, 1954).

Resistance of apple blossoms to spring frost was also found to be less on well-watered trees (Modlibowska, 1961). However, some resistance to frost injury was obtained only when the trees were under dry condition, i.e., when more than two-thirds of the available water was already removed. Under such stress conditions the water content of blossoms was reduced, the blossoms supercooled to a greater extent, less ice was formed in the tissues, and frost damage was reduced. (*See* chapter 54 for frost protection.)

VI. EFFECT OF IRRIGATION ON DISEASES AND INSECTS

With insufficient irrigation, redmite (*Bruobia praetiosa*) can build up on prune leaves late in the summer (Bowman and Davison, 1941) and contribute to some defoliation. Nonirrigated almonds are also subject to more mite buildup.

Fig trees (English, 1962) under water stress are more subject to sunburn which in turn may invite attack by the branch wilt fungus (*Hendersonula toruloidea*). Maintaining adequate soil water content reduced the infection by this disease. This was also true with walnut (Foott et al., 1955).

With fire-blight of pears, however, Shaw (1935) found that the activity of the inoculated pathogen *Erwinia amylovora* was progressively greater with increasing soil water level from 50% of the water-holding capacity to 70% and finally to 90%. When shoots were wilted, practically no fire-blight appeared. He concluded that there was a positive relation between intercellular relative humidity and fire-blight susceptibility. Till (1958). on the other hand, inoculated the apricot gummosis fungus (*Eutypa armenica*) and found no difference in its growth rate under different irrigation regimes.

The use of over-tree sprinklers (Wilcox, 1953) was observed to have a beneficial effect on mite control in apples, but produced certain harmful effects. It increased crown rot in apples, scattered fire-blight from high to the lower parts of pear trees, and favored the development of fruit rots and twig infections on peaches and apricots due to the organism *Phytophthora cactorum*. Thus, under-tree sprinklers were recommended (Wilcox, 1950). A rise in humidity resulting from sprinkler irrigation and which might favor the spread of fungus diseases was found unimportant. Peikert and Tribbe (1949) showed that the humidity increased only within the area approximately covered by the sprinklers and that it returned to normal in the space of 1 hour (Additional information on effects of irrigation on diseases and insects is given in chapters 55 and 56.)

VII. INTERRELATION BETWEEN IRRIGATION AND PLANT NUTRIENT SUPPLY

Much evidence is available that the nutritional status of deciduous fruit trees can be affected by irrigation.

Hendrickson and Veihmeyer (1950a) found, contrary to expectation, trunk growth and yield of walnuts to be greater in the high water stress plot compared to a low stress plot. They found the leaf nitrogen to be higher in the former and attributed the growth and yield response to a difference in nitrogen nutrition level rather than to soil water differences. In the frequently irrigated plot, N may have been leached out inasmuch as more frequent irrigations have been noted also to decrease the percentage of N content of leaves of other fruits, e.g., apple (Mason, 1958), peach (Morris et al., 1962) and apricot (Branton et al., 1961). Whether the total N uptake, considering the possible increased growth in the frequently irrigated trees, is the same as in less frequently irrigated ones is not clear. The addition of some N fertilizers to irrigated plots has been shown to increase the response to irrigation by increasing yield and trunk growth of peaches (Ballinger et al., 1963) and fruit size and yield of apricots (Till, 1958). The addition of N to nonirrigated plots did not give significant response on peaches in North Carolina, USA (Ballinger et al., 1963). However, in England extra N is said to reduce the effect of a water deficit in fruit trees (Office of the Minister for Science, 1962).

Leaf phosphorus levels have been significantly increased by irrigations in apple and peach (Hibbard and Nour, 1959), and in apricots (Branton et al., 1961). Hibbard and Nour kept soil water content at 70, 40, and 10% of the total available capacity of the soil. There was a progressive decrease in leaf P concentration with a decrease in soil water content. The addition of P to the soil increased P concentration at all water levels, but the effect of soil water suction on leaf P was greater than the effect of the available P supply in the soil. Branton et al. (1961) found P concentration in apricots to rise rapidly in midsummer following an irrigation while P concentration declined in the nonirrigated plot.

Potassium concentration was found to be higher in trees grown under lower soil water suction in apple (Mason, 1958; Hibbard and Nour, 1959), in peach (Hibbard and Nour, 1959; Morris et al., 1962; Feldstein, 1963), and apricot

(Branton et al., 1961). In England, Goode and Hyrycz (1964) working on a sandy soil of limited K content suspected that the plot kept below a soil water suction of 0.13 bar developed K deficiency, indicating leaching of K by frequent watering on this particular soil.

Magnesium deficiency symptoms have been observed in apples in wet seasons (Mason, 1958). In pecans Gammon et al. (1960) reported lower Mg concentration in heavy rainfall years. They attributed this to more feeder roots near the surface under high soil water conditions where the Ca/Mg ratio was high. Under a condition of a high Mg level in the leaves Morris et al. (1962) found no difference in Mg concentration in peach between irrigated and nonirrigated trees in Arkansas, USA. Under a lower but adequate Mg level in the leaves Branton et al. (1961) found the lowest level of Mg in apricot to be in the irrigated plots.

The uptake of boron by tree roots appears to be reduced by a shortage of soil water in the fall and winter. Wilcox and Brownlee (1961) pointed out that a number of cases of boron deficiency at bloom were traced to dry soil the previous fall.

Leaf scorch from excess sodium on apricot has been observed to be worse after a series of dry years. By leaching the soil with excessive amounts of irrigation water the scorch was reduced the year following the leaching (Halsey et al., 1958; Houston and Meyer, 1958). It took about twice the amount of water normally used in irrigation.

The quality of the irrigation water can, in some instances, also affect fruit tree response. Waters high in boron cannot be used because of boron toxicity. High calcium content of the water has induced some chlorosis of fruit trees (Lagache and Pascaud, 1961). Harley and Lindner (1945) found that high bicarbonate waters, when used over the years, also produced chlorosis. In irrigated orchards where incipient chlorosis occurs, sometimes the chlorosis can be reduced by merely reducing the frequency of irrigation.

VIII. METHODS FOR IRRIGATING ORCHARDS

Three methods of applying water to orchards are in wide use, namely: basin or flood, furrow, and sprinkler.

In basin or flood irrigation, small dikes are built along contour lines, and the area between dikes is flooded with several acre-inches of water. This is an economical method of applying water where conditions are suitable. The land as a whole must be fairly level and areas within a flood basin must be level for uniform wetting. Also the soil must be such that 4 to 8 acre-inches or more of applied water can be stored in the soil.

In furrow irrigations, the water is run in small furrows between the tree rows. Furrows are usually spaced about 3 to 4 ft apart with closer spacing on sandy soils. The furrows must have a gentle slope. Generally the whole soil mass is not wet with furrow irrigations, the tree rows remaining dry. Also care is necessary to obtain even distribution of water. With too much slope, water accumulates at the ends of the furrows distant from the source. With too little slope or with coarse-textured soils, areas near the water source may be over-irrigated while areas farther from the source are inadequately watered.

Sprinkler irrigation of orchards has increased greatly since lightweight, portable pipe and small sprinklers on hoses became available. Sprinklers have the advantage over furrows of more uniform and complete wetting, including the surface soil. They can be used on uneven soil and are particularly advantageous on sandy or rocky soils. They are used almost exclusively where some irrigation, supplemental to natural rainfall, is practiced. A disadvantage of sprinklers is the necessity for pumping equipment to maintain pressure. Also the initial cost of equipment may be greater than for furrow or flood irrigation. Provided that over-tree sprinkling is not practiced, under-tree sprinkling has generally not resulted in increased disease problems as originally anticipated.

IX. IRRIGATION PROGRAM

A. Water Usage

The total amount of water needed by fruit trees will vary with the climatic conditions under which they are grown and with the size of trees and density of planting. The amount of water transpired is largely determined by the extent of soil coverage and the total leaf area. Large trees will require 7 to 9 inches of water per month during July and August in hot, dry areas such as the interior valleys of California (Hendrickson and Veihmeyer, 1955a). Usage in the spring months is less, since the leaf system on the trees is not fully expanded. Under hot, dry growing conditions, large trees will require about 40 inches of water per year.

In cooler and more moist climates, water usage is reduced. In the Eastern USA, apples use 4 to 5 acre-inches of water per month in midsummer, peaches somewhat less. In still cooler climates, e.g., northern Europe, water usage is further reduced. (More detail on water use by crops is given in section VIII).

B. Scheduling Irrigation

The timing of irrigation of deciduous fruit trees depends on many factors such as amount of available water present, soil depth, rate of use of water by the tree, climatic conditions, time of year, stage of development of the fruit and tree, and even accessibility of water, water quality itself, and timing of other cultural operations.

Veihmeyer and Hendrickson (1949; 1950a, b; 1952) have repeatedly pointed out that as long as soil water content is kept above the WP no benefit is derived from additional irrigations. In other words, soil water need not be replenished until the moisture approaches the WP. More recent experiments using instruments such as tensiometers, gypsum blocks and dendrometers have resulted in recommendations to begin irrigating at a higher soil water content.

Rogers and Goode (1953) and Goode (1957) recommended irrigation when 50% of the available water in the root zone had been used which was equivalent to about 0.3 to 0.4 bar suction in the soils that were involved. The Ministry of Agriculture in England (1962) recommended the same. Wilcox and Brownlee (1961) in British Columbia recommended irrigation when the average soil water content to root depth was reduced to 50% of the available water or when any part of the soil in the root area approached the WP.

Cripps (1958) recommended irrigation when tensiometers at the 1-ft depth indicated 0.5 to 0.6 bar. Cockroft (1963) in Victoria, Australia, recommended application of water when the soil at the 24-inch level reached a soil water suction of 0.3 bar. Uriu et al. (1964) in California recommended irrigation on peach trees at a tensiometer reading of 0.4 bar at the 2- to 2.5-ft depth, depending on the depth of rooting.

Verner et al. (1962) recommended the use of changes in trunk growth rates as a means to determine the time of irrigation. When the trunk growth rate, as determined by dendrometers, fell below 80% of a very frequently irrigated block of trees used as a standard, irrigation was recommended.

The use of fruit growth measurements to time irrigation was recommended by Cockroft (1963) for peaches. Twenty fruits in a block of trees were tagged and the circumferences measured every 3 days. Irrigation was recommended when, in the flat period (pit-hardening period) of the growth curve, 80% of the fruits stopped growing, in the early part of the final swell when 50% of the fruits stopped growing, and in the last 4 weeks to harvest when 50% of the fruits had less than 0.1 cm increase/day. It would appear that here too a block of trees that is very frequently irrigated should be established as a standard from which to compare the fruit growth rates of the trees to be irrigated.

The use of potential evapotranspiration data to determine the time of irrigation was outlined by Rogers and Goode (1953). Jensen et al. (1962) suggested the use of pan evaporation data and the application of a constant to these data. (For additional discussion of scheduling irrigations see chapter 30.)

LITERATURE CITED

Ackley, W. B. 1954. Seasonal and diurnal changes in the water contents and water deficits of Bartlett pear leaves. Plant Phys. 29:445–448.

Alben, A. O. 1958. Results of an irrigation experiment on Stuart pecan trees in Texas in 1956. Southeastern Pecan Growers' Ass., Proc. 51:61, 63, 65, 67–8.

Aldrich, W. W. 1933. Some response of Anjou pear trees to irrigation. Oregon State Hort. Soc., 25th Annu. Rep. p. 30–35.

Aldrich, W. W., and R. A. Work. 1932. Preliminary report of pear tree responses to variations in available soil moisture in clay adobe soil. Amer. Soc. Hort. Sci., Proc. 29:181–187.

Ballinger, W. E., A. H. Hunter, F. E. Correll, and G. A. Cummings. 1963. Interrelationships of irrigation, nitrogen fertilization and pruning of Redhaven and Elberta peaches in the Sandhills of North Carolina. Amer. Soc. Hort. Sci., Proc. 83:248–258.

Batjer, L. P., and M. N. Westwood. 1958. Size of Elberta and J. H. Hale peaches during the thinning period as related to size at harvest. Amer. Soc. Hort. Sci., Proc. 72:102–105.

Bowman, F. T., and J. R. Davison. 1941. Prunes at Yenda. Results of irrigation and soil management investigations. Agr. Gaz. New South Wales 52:543–4, 585–8.

Boynton, D., and E. F. Savage. 1938. Soils in relation to fruit growing in New York: Part XIII. Seasonal fluctuations of soil moisture in some important New York orchard soil types. New York Agr. Exp. Sta. (Cornell) Bull. 706. 36 p.

Branton, D., O. Lilleland, K. Uriu, and L. Werenfels. 1961. The effect of soil moisture on apricot leaf composition. Amer. Soc. Hort. Sci., Proc. 77:90–96.

Brooks, R. M. 1939. A growth study of the almond fruit. Amer. Soc. Hort. Sci., Proc. 37:193–197.

Brown, D. S. 1953. The effects of irrigation on flower bud development and fruiting in the apricot. Amer. Soc. Hort. Sci., Proc. 61:119–124.

Butijn, J. 1961. Water requirements of pome fruits. (In English and Italian with French and German summaries, 9 lines each.) Reprint from Irrigazione. 8(3):40–51.

Cockroft, B. 1963. Timing of irrigations in Goulburn Valley orchards. J. Agr. Victoria, Australia. 61:492–5, 521.

Condit, I. J. 1947. The fig. Chronica Botanica, Co., Waltham, Mass. p. 119–120.

Crane, J. C., and K. Uriu. 1965. The effect of irrigation on response of apricot fruits to 2,4,5-T application. Amer. Soc. Hort. Sci., Proc. 86:88–94.

Cripps, J. 1958. Orchard irrigation. J. Agr. W Aust. Ser. 3:7:127–128.

Davis, L. D. 1941. Split-pit of peaches. Estimation of time when splitting occurs. Amer. Soc. Hort. Sci., Proc. 39:183–189.

Degman, E. S., J. R. Furr, and J. R. Magness. 1932. Relation of soil moisture to fruit bud formation in apples. Amer. Soc. Hort. Sci., Proc. 29:199–201.

Donoho, C. W., Jr., F. S. Howlett, and R. B. Curry. 1964. Apple irrigation and mulch. Ohio Farm Home Res. 49(3):40–41, 43.

English, H. 1962. Canker and dieback disorders of fig trees. California Fig. Inst., Proc. 16th Annu. Res. Conf. p. 13–15.

Feldstein, J. 1963. Irrigation studies with the peach in Pennsylvania. Diss. Abstr. 23: 4, 4052–4053.

Foott, J. H., A. H. Hendrickson, and E. E. Wilson. 1955. Walnut branch wilt. California Agr. 9(10):11.

Forshey, C. G. 1958. Irrigating New York orchards. New York State Hort. Soc., Proc., 103rd Annu. Mtg. p. 90–93.

Fortier, S. 1940. Orchard irrigation. US Dep. Agr. Farmers' Bull. 1518. 27 p.

Furr, J. R., and J. R. Magness. 1931. Preliminary report on relation of soil moisture to stomatal activity and fruit growth of apples. Amer. Soc. Hort. Sci., Proc. 27:212–218.

Furr, J. S., and E. S. Degman. 1932. Relation of moisture supply to stomatal behavior of the apple. Amer. Soc. Hort. Sci., Proc. 28:547–551.

Gamble, S. J. 1962. Ray Klackle irrigates his apples every year for annual bearing. Michigan State Hort. Soc., Annu. Rep. p. 69–71.

Gamble, S. J. 1964. Will it pay to irrigate your apple orchard. Indiana Hort. Soc., Trans. 103rd Annu. Mtg. p. 32–38.

Gammon, N. Jr., K. D. Butson, and R. H. Sharpe. 1960. Magnesium content of pecan leaves as influenced by seasonal rainfall and soil type. Soil Crop Sci. Soc. Fla., Proc. 20:154–8.

Gayner, F. C. H. 1941. Studies in the non-setting of pears: IV. The effect of irrigation and injection on the June drop of Conference pears. East Malling Res. Sta., Annu. Rep. 1940. A 24:36–41.

Goode, J. E. 1956. Soil moisture relationships in fruit plantations. Ann. Appl. Biol. 44:525–30

Goode, J. E. 1957. Water needs of fruit crops. Sprinkler Irrigation Manual. Wright Rain Ltd. Ringwood, Hampshire, England. p. 125–133.

Goode, J. E., and K. J. Hyrycz. 1964. The response of Laxton's Superb apple trees to different soil moisture conditions. J. Hort. Sci. 39:254–76.

Halsey, D. D., N. L. McFarlane, and R. J. Schmt. 1958. Sodium leaf scorch of apricot. California Agr. 12(9):4–5.

Harley, C. P., and R. C. Lindner. 1945. Observed responses of apple and pear trees to some irrigation waters of North Central Washington. Amer. Soc. Hort. Sci., Proc. 46:35–44.

Heinicke, A. J., and N. F. Childers. 1935. The influence of water deficiency in photosynthesis and transpiration of apple leaves. Amer. Soc. Hort. Sci., Proc. 33:155–159.

Hendrickson, A. H. 1926. Certain water relations of the genus Prunus. Hilgardia 1:479–524.

Hendrickson, A. H., and F. J. Veihmeyer. 1927 Some results of studies on the water relations of clingstone peaches. Amer. Soc. Hort. Sci., Proc. 24:240–244.

Hendrickson, A. H., and F. J. Veihmeyer. 1930. Some facts concerning soil moisture of interest to horticulturists. Amer. Soc. Hort. Sci., Proc. 26:105–108.

Hendrickson, A. H., and F. J. Veihmeyer. 1931. Influence of dry soil on root extension. Plant Physiol. 6:567–576.

Hendrickson, A. H., and F. J. Veihmeyer. 1935. Size of peaches as affected by soil moisture. Amer. Soc. Hort. Sci., Proc. 32:284–6.

Hendrickson, A. H., and F. J. Veihmeyer. 1937. The irrigation of pears on a clay adobe soil. Amer. Soc. Hort. Sci., Proc. 34:224–226.

Hendrickson, A. H., and F. J. Veihmeyer. 1941. Some factors affecting the rate of growth of pears. Amer. Soc. Hort. Sci., Proc. 39:1–7.

Hendrickson, A. H., and F. J. Veihmeyer. 1945. Some effects of irrigation on the interrelations of growth, yields, and drying ratios of French prunes. Amer. Soc. Hort. Sci., Proc. 46:187–190.

Hendrickson, A. H., and F. J. Veihmeyer. 1950a. Growth of walnut trees as affected by irrigation and nitrogen deficiency. Plant Physiol. 25:567–72.

Hendrickson, A. H., and F. J. Veihmeyer. 1950b. Irrigation experiments with apricots. Amer. Soc. Hort. Sci., Proc. 55:1–10.

Hendrickson, A. H., and F. J. Veihmeyer. 1955a. Daily use of water and depth of rooting of almond trees. Amer. Soc. Hort. Sci., Proc. 65:133–8.

Hendrickson, A. H., and F. J. Veihmeyer. 1955b. Results of sprinkling in a fig orchard. California Fig Inst., Proc. 9th Annu. Res. Conf. p. 14–15.

Hibbard, A. D., and Mohsen Nour. 1959. Leaf content of phosphorus and potassium under moisture stress. Amer. Soc. Hort. Sci., Proc. 73:33–39.

Hoffman, M. B., and L. J. Edgerton. 1946. The apparent effect of moisture supply on naphthaleneacetic acid treatments for delaying the drop of McIntosh apples. Amer. Soc. Hort. Sci., Proc. 48:48–50.

Houston, C. E., and J. L. Meyer. 1958. Apricot irrigation studies. California Agr. 12(9):6.

Jensen, M. C., E. S. Degman, and J. E. Middleton. 1962. Apple orchard irrigation. Washington Agr. Exp. Sta. Circ. 402. 11 p.

Jones, I. D. 1932. Preliminary report on relation of soil moisture and leaf area to fruit development of the Georgia Belle peach. Amer. Soc. Hort. Sci., Proc. 28:6–14.

Kenworthy, A. L. 1949. Soil Moisture and growth of apple trees. Amer. Soc. Hort. Sci., Proc. 54:29–39.

Lagache, P., and G. Pascaud. 1961. Nature des eaux d'irrigation et phenomenes de chlorose des arbres fruitiers dans la vallee du Rhone. (Nature of irrigation water and chlorosis effects on fruit trees in the Rhone valley in France.) UNESCO Arid Zone Res. 16:285–294.

Laycock, D. H., and L. J. Foster. 1955. Rainfall and biennial bearing in Tung (*Aleurites montana*). Nature 176:654.

Livingston, R. L., and T. B. Hagler. 1957. Effect of irrigation on peaches. Ass. S. Agr. Workers, Proc. 54:179.

Magness, J. R. 1931. Moisture supply as a factor in pecan production. Nat'l Pecan Ass., Proc. 30:18–23.

Magness, J. R., E. S. Degman, and J. R. Furr. 1935. Soil moisture and irrigation investigations in eastern apple orchards. US Dep. Agr. Tech. Bull. 491. 36 p.

Magness, J. R., L. O. Regeimbal, and E. S. Degman. 1933. Accumulation of carbohydrates in apple foliage, bark and wood as influenced by moisture supply. Amer. Soc. Hort. Sci., Proc. 29:246–252.

Mason, A. C. 1958. The effect of soil moisture on the mineral composition of apple plants grown in pots. J. Hort. Sci. 33(3):202–211.

Mathews, C. D. 1960. Water and the Tasmanian apple grower. Tasmanian J. Agr. 31:346–351.

Ministry of Agriculture, London. 1962. Irrigation. (3rd ed.) Bull. Min. Agr. 138. 88 p.

Modlibowska, I. 1961. Effect of soil moisture on frost resistance of apple blossom, including some observations on "ghost" and "parachute" blossoms. J. Hort. Sci. 36(3):186–196.

Morris, J. R., A. A. Kattan, and E. H. Arrington. 1962. Response of Elberta peaches to the interactive effects of irrigation, pruning and thinning. Amer. Soc. Hort. Sci., Proc. 80:177–189.

Nasharty, A. H., and I. M. Ibrahim. 1961. Progress report on effect of frequency of irrigation on quality and quantity of plum fruits. Agr. Res. Rev. Cairo. 39:100–7.

Office of the Minister for Science. 1962. Irrigation in Great Britain. Natural Resources Tech. Comm. Rep. H. M. Stationery Office, London. 82 p.

Oskamp, J. 1932. The rooting habit of deciduous fruits in different soils. Amer. Soc. Hort. Sci., Proc. 29:213–219.

Packer, W. J., D. J. Chalmers, and P. Baxter. 1963. Supplementary irrigation of Jonathan apple trees. J. Agri. Victoria, Australia. 61:453–60, 475.

Peikert, F. W., and R. T. Tribble. 1949. The effect of irrigation on the humidity in orchards. Michigan Agr. Exp. Sta. Quart. Bull. 31:266–9.

Pollack, R. L., Nancy Hoban, and C. H. Hills. 1961. Respiratory activity of the red tart cherry (*Prunus cerasus*) during growth. Amer. Soc. Hort. Sci., Proc. 78:86–95.

Proebsting, E. L. 1943. Root distribution of some deciduous fruit trees in California. Amer. Soc. Hort. Sci., Proc. 43:1–4.

Rogers, B. L. 1958. Results of peach irrigation in 1957. Maryland State Hort. Soc., Proc. 60:9–12.

Rogers, W. S. 1939. Root studies: VIII. Apple root growth in relation to rootstock, soil, seasonal and climatic factors. J. Pom. Hort. Sci. 17:99–130.

Rogers, W. S., and J. E. Goode. 1953. Irrigation requirements of fruit orchards. East Malling Res. Sta., Annu. Rep. A 36. p. 171–3.

Rogers, W. S., and M. C. Vyvyan. 1934. Root studies: V. Rootstock and soil effect on apple root systems. J. Pom. Hort. Sci. 12:110–150.

Romberg, L. D. 1960. Irrigation of pecan orchards. Southeastern Pecan Growers' Ass., Proc. 53:20, 22–25.

Romberg, L. D., C. L. Smith, and H. L. Crane. 1959. Effects of irrigation and tree re-spacing (thinning) on pecan tree growth and nut production. Texas Pecan Growers' Ass., Proc. 38:60–62, 64–69, 71–75.

Ross, N., and J. L. Meyer. 1957. Almond irrigation timing has important bearing on yields. Almond Facts 22(3):9.

Schneider, G. W., and N. F. Childers. 1941. Influence of soil moisture on photosynthesis, respiration, and transpiration of apple leaves. Plant Physiol. 16:565–583.

Shaw, L. 1935. Intercellular humidity in relation to fire-blight susceptibility in apple and pear. Cornell Agr. Exp. Sta. Mem. 181. 40 p.

Simons, R. K. 1963. Anatomical studies of apple fruit abscission in relation to irrigation. Amer. Soc. Hort. Sci., Proc. 83:77–87.

Skepper, A. H., and A. E. Vincent. 1962. Orchard irrigation. New South Wales Dep. Agr. Publ. 77 p.

Skepper, A. H., J. R. Davison, A. F. Murray, and D. R. Blundell. 1961. Production of prunes under irrigation. Agr. Gaz. New South Wales 72(4):199–202, 215.

Taerum, R. 1964. Effects of moisture stress and climatic conditions on stomatal behavior and growth in Rome Beauty apple trees. Amer. Soc. Hort. Sci., Proc. 85:20–32.

Till, M. R. 1958. Progress report on apricot irrigation and nitrogen trial. S. Australia Dep. Agr. J. 61(6):295–297.

Uriu, K. 1964. Effect of post-harvest soil moisture depletion on subsequent yield of apricots. Amer. Soc. Hort. Sci., Proc. 84:93–97

Uriu, K., C. J. Hansen, and J. J. Smith. 1962. The cracking of prunes in relation to irrigation. Amer. Soc. Hort. Sci., Proc. 80:211–219.

Uriu, K., L. Werenfels, G. Post, A. Retan, and D. Fox. 1964. Cling peach irrigation. California Agr. 18(7):10,11.

Veihmeyer, F. J., and A. H. Hendrickson. 1938. Soil moisture as an indication of root distribution in deciduous orchards. Plant Physiol. 13:169–177.

Veihmeyer, F. J., and A. H. Hendrickson. 1949. The application of some basic concepts of soil moisture to orchard irrigation. Washington State Hort. Ass., Proc. 45:25–41.

Veihmeyer, F. J., and A. H. Hendrickson. 1950a. Responses of fruit trees and vines to soil moisture. Amer. Soc. Hort. Sci., Proc. 55:11–15.

Veihmeyer, F. J., and A. H. Hendrickson. 1950b. Soil moisture in relation to plant growth. Annu. Rev. Plant Physiol. 1:285–304.

Veihmeyer, F. J., and A. H. Hendrickson. 1952. The effects of soil moisture on deciduous fruit trees. Int. Hort. Congr., Rep. 13th, London I:306–319.

Verner, L., and E. C. Blodgett. 1931. Physiological studies of the cracking of sweet cherries. Idaho Agr. Exp. Sta. Bull. 184. 15 p.

Verner, L., W. J. Kochan, D. O. Ketchie, A. Kamal, R. W. Braun, J. W. Berry, Jr., and M. E. Johnson. 1962. Trunk growth as a guide to orchard irrigation. Idaho Agr. Exp. Sta. Res. Bull. 52. 32 p.

Wahlberg, H. E. 1932. Yuma Valley pecan growers prosper. Amer. Fruit Grower. 52(1):8–9.

Way, R. D. 1954. The effect of some cultural practices and of size of crop on the subsequent winter hardiness of apple trees. Amer. Soc. Hort. Sci., Proc. 63:163–6.

Wiggans, C .C. 1936. The effect of orchard plants on subsoil moisture. Amer. Soc. Hort. Sci., Proc. 33:103–107.

Wilcox, J. C. 1950. Sprinkler irrigation experience in British Columbia orchards. Sci. Agr. 30:418–27.

Wilcox, J. C. 1953. Sprinkler irrigation of tree fruits and vegetables in British Columbia. Dep. Agr. Ottawa. Publ. 878. 72 p.

Wilcox, J. C., and C. H. Brownlee. 1961. Sprinkler irrigation requirements for tree fruits in the Okanagan Valley. Can. Dep. Agr. British Columbia Dept. Agr. Publ. 1121. 32 p.

Zioni, E. 1963. The effects of cultivation and irrigation on preharvest drop of hazel nuts in Chiavari. (Italian with English and French summaries, 4 lines each.) Frutticoltura 25:363–7.

36 | Evergreen Tree Fruits

ROBERT H. HILGEMAN

*University of Arizona, Citrus Branch Station
Tempe, Arizona*

WALTER REUTHER

*University of California
Riverside, California*

Evergreen tree fruits are grown commercially in subtropical areas of the world situated in two broad belts above and below the equator. Subtropical zones cannot be delimited by any specific latitude because of the moderating effects of oceans and terrain. Within the subtropical fruit areas three general types of climate occur: humid, subarid and arid. In general, rainfall is sufficient in the humid zones, such as Florida, USA, so that irrigation is required only during drouth periods. In semiarid zones, rainfall is inadequate during dry periods so irrigation is required. In arid zones, such as the Lower Colorado Basin of the USA, irrigation is required throughout the year. A brief summary of certain climatic features is shown in Table 36–1.

Subtropical fruit trees may be classified in two broad categories: (i) Mesophytes that are indigenous to the humid tropics but grow in subtropical areas, and (ii) xerophytes that are apparently indigenous to the subarid and arid subtropical zones.

Citrus is a typical mesophyte that grows in subtropical areas. The trees are characterized by moderately thick, dark green, glossy, evergreen leaves without specialized drouth or heat resisting characteristics. Citrus root systems are comparatively shallow and moderately spreading. Therefore, trees cannot survive long drouth without serious injury.

Xerophytes like the date palm (*Phoenix dactylifera* L.) are apparently indigenous to subarid and arid regions. They survive long drouth periods without serious damage to tissues. Leaves tend to be small containing much sclerenchyma tissue and a thick cuticle which prevents obvious wilting. A wide-ranging, deep root system supplies water to maintain the tree during extreme drouth.

Soil water is removed in close proportion to the feeder root concentration. After an irrigation, rapid water depletion from dense root portions and slow depletion from sparse portions causes a gradual decrease in the rate of water absorbed from the entire root soil mass. As progressively greater portions of the soil approach the wilting range, a gradual reduction in transpiration occurs, water deficits in the plant become greater, and fruit and trunk growth are retarded and finally cease.

The optimum time to replenish soil water for maximum productivity can be established by relating soil water suction to physiological indicators such as fruit growth. Because of variation in root systems, the establishment of a critical level of depletion is specific for each soil and plant type. When transpiration rates are high, maximum growth normally occurs when the entire root system is supplied

Table 36–1. Temperatures and precipitation in subtropical areas

Climate type	Location	Temperature		Mean precipitation	
		Jan.	July	Nov. to Apr.	May to Oct.
		°C		mm	
Humid	Lake Alfred, Fla.	17.2	28.8	360	970
	Harlingen, Texas	16.1	28.9	230	480
	Johannesburg, S. Africa	20.0	10.6	790	150
Subarid	Escondido, Calif.	10.6	22.2	390	47
	Riverside, Calif.	11.1	24.4	250	51
	Sharon Plain, Israel	13.3	25.6	480	25
	Catania, Italy	10.0	26.1	590	150
	Valencia, Spain	10.0	23.9	250	230
Desert	Indio, Calif.	12.2	33.9	51	25
	Cairo, Egypt	12.8	28.3	25	0

with water; when low, a small per cent of the root system may be adequate to supply the water needs of the plant.

I.   IRRIGATION OF MESOPHYTIC TYPE FRUITS

A.   The Citrus Tree

Citrus trees are grown in all types of subtropical and tropical climates but water requirements differ widely. Water may be applied by sprinklers, furrows or flooded basins around the trees. The method of application is dependent on rainfall, water quality, soil infiltration rate, labor and material costs, and other conditions in each area.

1. SEASONAL WATER REQUIREMENTS

Seasonal water requirements for citrus have been reported in California and Arizona, USA by Harris et al. (1936), Beckett et al. (1930) and Pillsbury et al. (1944). Soil water was determined gravimetrically to a depth of 1.85 m (6 ft) below the upper 15-cm (6-inch) disked area. Water usage estimated in this manner does not include water required to wet the upper 15 cm of soil, nor surface evaporation and possibly deep percolation losses.

Water used by Navel or Valencia oranges (*Citrus sinensis* L. Osbeck) and the mean temperatures for each month given in Table 36–2 illustrate typical seasonal changes in the water requirement for the zones indicated. As air temperature increases, the water requirement increases. In the summer, the humidity is higher and day maximum air temperatures lower in Orange and San Diego Counties, California, than in Riverside County, California and the Salt River Valley in Arizona.

In Israel, Oppenheimer and Mendel (1939) estimated transpiration rates from detached leaves of Shamouti oranges. They found that water losses increased from 1.6 g of water/day per g of leaf in January to 7.3 g/day in July. In Arizona, evapotranspiration (ET) loss from the soil surface to a depth of 1.85 m was related by Hilgeman and Rodney (1961) to evaporation (E_{pan}) from a US Weather Bureau pan. The ET/E_{pan} ratio ranged from 0.45 in January to 0.58 in July. Reeve and Furr (1941) obtained a lower ET/E_{pan} ratio of 0.25 to 0.30

Table 36–2. Loss of water from soil in four climatic zones

Month	Maricopa County Arizona* Wash. Navel		Riverside County California† Wash. Navel		Orange County California‡ Val. orange		San Diego County California§ Wash. Navel	
	Surface	Mean temp¶	Surface	Mean temp¶	Surface	Mean temp¶	Surface	Mean temp¶
	cm	° C	cm	° C	cm	° C	cm	° C
April	6. 4	17. 8	5. 3	16 7	3. 3	15. 6	2. 5	13. 9
May	8. 4	21. 7	6. 6	17 8	4. 8	18. 9	2. 5	16. 7
June	9. 9	26. 7	6. 8	20. 0	6. 8	20. 0	3. 0	18. 3
July	11. 2	31. 1	9. 1	23. 9	7. 9	22. 8	3. 3	22. 2
August	10. 7	30. 0	9. 1	23. 3	7. 4	23. 3	4. 6	21. 7
September	9. 6	27. 8	7. 4	20. 6	6. 1	21. 7	4. 3	19. 4
October	6. 6	21. 1	6. 6	18. 9	4. 3	18. 9	3. 6	17. 8
	Total	Mean	Total	Mean	Total	Mean	Total	Mean
	62. 8	25. 2	50. 9	20. 2	40. 6	20. 2	23. 8	18. 6

* Average from two mature Washington Navel orange groves for 3 years (6 values) near Phoenix in Maricopa County, Arizona, 1931-34. Data from Harris et al. (1936).
† Average from four mature Washington Navel orange groves during 5 years (12 values) near Corona in Riverside County, California, 1930-35. Data from Pillsbury et al. (1944).
‡ Average of three mature Valencia orange groves during 2 years (5 values) near Tustin in Orange County, California, 1923-29. Data from Pillsbury et al. (1944).
§ Data from mature Washington Navel orange grove during 1 year (1 value) near Escondido in San Diego County, California. Data from Beckett et al (1930).
¶ Mean temperatures for the periods studied from US Weather Bureau Cooperative Observer records near the area involved.

in southern California during the summer with a shallow black evaporation pan and sampling soil depths from 15 cm to 1.85 m below the surface. The ratio became lower as the irrigation interval was extended and the rate of soil water loss was reduced.

2. VARIETY AND ROOTSTOCK EFFECTS

Varieties differ in transpiration rates. Evapotranspiration losses in large Navel orange and grapefruit groves (*Citrus paradisi* Macfadyen) on sour orange rootstocks, reported by Harris et al. (1936) show average annual water usage of 80 and 104 cm, respectively, in a ratio of 1:1.30. Two studies during dry intervals with Valencia oranges and Marsh grapefruit at Lake Alfred and Lake Wales, Florida by Koo (1961) showed usage ratios of 1.26 and 1.29. It appears that the water requirement of grapefruit is approximately 30% greater than oranges.

The widely different rooting habits of rootstocks observed by Oppenheimer (1936) and Ford (1954) reflect the capacity of the different species to absorb soil water (Table 36–3). The value of greater root proliferation is reflected in the degree of wilting observed by Horanic and Gardner (1959) during a drouth period in Florida. The degree of wilt was closely related to the average root system below 1.5 m in this area as reported by Ford (1954). Deeper root systems on Rough lemon (*Citrus limon* L. Burmann) and Cleopatra mandarin (*Citrus reticulata* Blanco) resulted in no wilting in contrast to excessive wilting of sweet orange and grapefruit with shallow root systems. Sour orange rootstock reduced tree size, and wilt was not serious. Other observations by Mendel (1951) on

Table 36–3. Root development and drouth resistance of common rootstock material

Rootstock	Israel	Florida			
	Root pene-tration*	Per cent of roots†		Wilting index‡	Trunk cir-cumference‡
		Above 0. 8 m	Below 1. 5 m		
	cm				cm
Rough lemon	110	43	32	1. 3	71
Sweet lime	70				
Cleopatra mandarin		52	29	1. 7	70
Sweet orange	76	64	15	2. 3	69
Sour orange	65	52	31	1. 2	57
Grapefruit	45	64	9	2. 6	71

* Root penetration of field grown unbudded trees 19 months after transplanting. Data from Oppenheimer (1936). † Mean of many citrus orchards on each stock located on the deep, well drained fine sand soils of central Florida. Data from Ford (1959). ‡ Observations on wilting and tree size of 17-year-old Valencia trees growing in Lakeland fine sand during a drouth in Florida. Data from Horanic and Gardner (1959). Index 1 - Few exterior leaves curled. Index 3 - All exterior and most interior leaves cupped and rolled.

previously regularly irrigated trees during drouth showed that sour orange wilted more seriously than Rough lemon.

Transpiration can be influenced by rootstocks having different genetic characteristics. Greenhouse tests by Ongun and Wallace (1958) have shown that transpiration rates for Rough lemon were 35% higher and grapefruit 35% lower than sour orange, sweet orange, Cleopatra mandarin and trifoliata (*Poncirus trifoliata* L. Raf.). In field tests with large trees, Mendel (1951) found that transpiration rates for Rough lemon and Palestine lime were 18% higher than for sour orange.

3. ROOT DEVELOPMENT IN RELATION TO IRRIGATION

The primary objective of irrigation is to supply the root system with water in such a manner as to maintain an active, healthy root system which promotes maximum yield of good quality fruit. Limited information shows that citrus root development is largely dependent on rootstock and the profile characteristics of the soil and is modified by soil water.

Under high rainfall with periodic drouth in Florida, Ford (1954) found mandarin rootstock produced larger trees and yields when roots grew into a clay subsoil 75 cm below the surface compared to deep fine sand. However, the size of trees on Rough lemon rootstock were reduced where clay subsoils were less than 1.5 m from the surface. Cleopatra mandarin, growing in deep fine sand, produced larger quantities of feeder roots than Rough lemon, but tree size and yields were less.

At Riverside, California, Cahoon et al. (1961) reported that allowing Ramona sandy loam to approach the wilting point (WP) to a depth of 90 cm between irrigations for many years increased root density of trees on sweet orange root but reduced yields somewhat in comparison with trees irrigated when the upper 30 cm of soil approached the WP. In the same area, Cahoon et al. (1959) found that fertilizers changed water infiltration rates which apparently limited the water available to the tree. Fertilizers that reduced water infiltration rates increased roots and decreased yields; those that increased water infiltration rates produced the opposite effect. In contrast, a survey of 22 groves in California by Cahoon et al. (1956) revealed that greater root density was associated with higher yields.

Near Phoenix, Arizona, Sufi reported that iron chlorosis, dieback, and reduced root density in orange trees on sour orange rootstock growing in calcareous clay loam was associated with high soil water content (S. M. Sufi, 1958. Relationship of root distribution and soil moisture to iron chlorosis in Arizona citrus. *Unpubl. M.S. Thesis. Univ. Arizona, Tucson.*). However, in soil containing 90% fine sand at Yuma, Arizona, maximum root development and top growth were obtained by irrigating at weekly intervals. Water availability was above the wilting percentage at all times (Smith et al., 1931).

The apparent contradictions in the above observations may be explained by the specific adaptation of different rootstocks to soil types and soil water. In coarse-textured soils, ample aeration is provided under all water conditions. Maximum root and top growth is possible only when soil water is adequate. In fine-textured soils, the relationship between soil water, root density and yield appears to be as follows: (i) deficient soil water—large root area and low yield; (ii) adequate soil water—moderate root area and high yield; (iii) excessive soil water—small root area and low yield. The increased root density in the first group illustrates an adaptive response of the tree to obtain water. Within the second group a wide range of root densities occur which is not correlated with yield. Excessive soil water in the third group limits root development possibly by lack of aeration or root decay.

4. FLOWERING AND ABSCISSION OF YOUNG FRUIT

Citrus is mostly grown in subtropical zones where low temperatures in winter induce dormancy and flower differentiation followed by a heavy spring bloom. However, soil water control can be used to induce heavy blossoming in the absence of sufficient chilling to induce dormancy. During a fall drouth in Florida, Horanic and Gardner (1959) observed that the degree of wilting of leaves varied with rootstocks. The following spring, after a warm, wet winter, Gardner observed blossom formation was positively correlated with the degree of wilt the previous fall (F. E. Gardner, 1963. *Personal communication*). In Sicily, stresses induced by soil water depletion in the summer are commonly employed to induce off-season blossoming of lemons.

After blossoming, the abscission of the weaker young fruit, known as "June drop" naturally occurs. Under conditions of severe spring drouth, Sites et al. (1951) noted that drop was increased in Florida and an off-season bloom induced after rainfall resumed. Under moderately high temperature conditions at Riverside, California Huberty (1948) noted that maintaining high soil water reduced drop. However, in desert areas when temperatures of 46C (115F) occurred during the drop period in May, Furr (1955) reported that irrigation did not moderate the physiological drouth which developed and drop was unaffected.

5. EFFECT OF IRRIGATION ON YIELD

Yield of fruit is influenced by many factors such as freezes or high temperatures during bloom and fruit setting, alternate bearing, the presence of the previous crop, and cultural factors. Consequently, the effect of irrigation on yields has been neither uniform nor clear-cut in many experiments.

In a test with newly planted grapefruit in Israel, Samish (1957) found that

Table 36–4. Effect of soil water on tree growth, fruit yield, size, and quality of Valencia oranges°

	Wet	Moderate	Dry	Wet–dry†	LSD
			Treatment		
Number irrigations/year	15	10	5	9	
Soil water, %‡	10. 7	9. 6	7. 6		
Soil water suction, cbar§	32	45	70+		
Trunk growth, cm²/year	20	16	13	15	2. 9
Yield, fruit/tree¶	656	630	486	685	126
Yield, kg/tree¶	104	100	76	107	16
Size fruit 6. 8 cm and larger, %¶	65	61	61	48	11
Total soluble solids, %**	11. 8	12. 2	12. 8	13. 2	0. 6
Ascorbic acid, mg/100 ml**	49	50	53	52	3. 4
Juice by weight, %**	54	53	52	51	NS
Rind thickness, mm**	4. 6	4. 5	4. 9	5. 0	0. 34

* Data of Hilgeman from experiment at Tempe, Arizona, 1949-62.
† Irrigate on wet schedule March to July and on dry schedule Aug. to Feb.
‡ Average per cent soil water in upper 60 cm prior to irrigations. Field capacity = 17%. Wilt range 6. 5% to 8. 5%.
§ Suctions at 75-cm depth prior to irrigations.
¶ Average values for 14 years; fruit harvested in early May.
**Average values for 7 years; fruit quality in Apr.

three irrigation levels receiving 26, 42 and 62 cm of irrigation water plus 350 mm of rainfall had total yields of 272, 379 and 462 kg of fruit/tree, respectively, when 7 years old. Despite these wide differences, tree variation was so great that a statistically significant difference occurred only between the high and low water levels. In a 5-year experiment at Griffith, Australia, Bouna (1954) reported that orange trees, irrigated when available water was one-half depleted from the root zone, did not produce a significantly greater amount of fruit than trees irrigated when soil water approached the WP. Hilgeman (1963) in a 14-year experiment at Tempe, Arizona, reported that trees produced increasing amounts of fruit at successively higher soil water levels, but a significant difference did not exist between the two highest levels (Table 36–4). In all the above tests the trunk size was significantly increased as the amount of water was increased.

During a 14-year test at Riverside, California, Huberty and Richards (1954) and Cahoon et al. (1961) compared orange trees irrigated on a 3-week summer schedule or 6.5 irrigations/year (available water maintained below 30 cm) with trees irrigated on a 6-week schedule or 3.3 irrigations/year (soil at wilting percentage in the upper 60 to 85 cm). The more frequent irrigation significantly increased yields in 11 years with an average 26% gain. At the same location Erickson and Richards (1955) in a 1-year study significantly increased size of Valencia fruit by irrigating after June at 30 centibars suction at a 30-cm depth (about 3-week interval) compared with 70 centibars suction at a 60-cm depth (5- to 6-week interval).

The importance of maintaining ample water in the soil during the blossom and fruit drop period was shown in a 3-year test at Indio, California, when Furr (1955) irrigated grapefruit trees frequently in the spring until the crop was set and subsequently subjected them to drouth. He observed that even though fruit growth was arrested, yields were maintained. Fruit size and trunk growth, however, were slightly decreased. In a similar test with Valencia oranges at Tempe, Arizona, mentioned earlier, severe stress imposed in August and October, following ample water in the spring, maintained production over a 14-year period, although fruit size and trunk growth were always reduced (Table 36–4). In Florida, Koo (1963) reported that irrigating during the spring when one-third of the water was depleted from the upper 5 ft of soil increased production of Marsh grapefruit.

The effect of previous irrigation practices on yield was dramatically illustrated at Riverside, California. Cahoon et al. (1961) report that, after growing trees for 14 years under 3- and 6-week irrigation intervals, a reversal of the treatments for 2 years induced extremely heavy yields of the previously dry trees and reduced yields of the previously wet ones. When treatments were again reversed, no significant differences in yields were noted in the next 4 years. However, Stolzy et al. (1963) did not produce marked changes during a 3-year period when trees irrigated at 6-week intervals were changed to about 3 weeks. Yields gradually increased when the 3-week irrigation interval was changed to a 10-day interval with light irrigations (20 centibars suction at a 50-cm depth).

It is clearly evident that within the limits of common irrigation practices, yield increases induced by irrigation result chiefly from maintaining available water in the entire root system during the sensitive spring fruit set period. Although lavish irrigation to maintain low soil water suctions throughout the year has produced maximum tree and fruit growth, the increase in yield has been small as compared to more moderate regimes. After the fruit has set, soil water depletions that produce moderate physiological stresses between irrigations have had no effect on production and appear to be desirable under the following conditions: (i) In fine-textured soil where continuous low suctions limit root development; (ii) slight reductions in fruit size are not commercially important; (iii) water supply is either limited or water costs are so high that it is economically advantageous.

6. FRUIT QUALITY

The physical and chemical characteristics of fruit are modified by the degree and the time water stress develops in the tree. Sites et al. (1951), Erickson and Richards (1955), and Hilgeman and Sharples (1957) have shown that fruit produced on trees subjected to water stresses during July and August have higher percentages of total soluble solids and acid, and in most cases ascorbic acid. Juice content tends to be decreased by water stress. Peel thickness is more variable but in many instances has been increased by water stress.

Granulation, also known as dry end, or crystallization, is a physiological disorder in which the juice sacs at the stem end become gelatinized. While considerable variation existed between years and different tests, Bartholomew et al. (1941) and Erickson and Richards (1955) reported fruit produced with a drier soil water regime developed less granulation than fruit produced with frequent irrigation.

Cahoon et al. (1964) demonstrated that oleocellosis (rind spot induced by ruptured oil cells during harvest) was reduced as soil water suctions and the vapor pressure deficit (VPD) of the air increased. Injury did not occur when soil water suctions at the 30 cm depth were above 40 centibars accompanied by a VPD above 25 mm Hg.

7. FRUIT GROWTH AS AN INDEX FOR IRRIGATION

Physiological indicators of water deficits in the tree have been studied to determine the proper irrigation interval. Most results have not been correlated with tree growth and yields in long-term experiments. It has been shown that as soil water decreases, the water deficit in the leaves increases (Halma, 1935); stomata in leaves close earlier in the day (Oppenheimer and Mendel, 1939); the daily increment of trunk growth decreases (Hilgeman, 1963); and the rate of enlargement of the fruit decreases (Taylor and Furr, 1937).

The rate of enlargement of immature fruit is one of the guides available to citrus growers for indicating the need for irrigation. Early observations by Compton (1937), noting that the rate of fruit enlargement decreased as soil water was depleted, were further investigated by the extensive work of Furr and Taylor (1939). They noted that growth rates of fruit on deep rooted trees in fine-textured soils changed slowly, whereas in coarse-textured soils, and particularly trees with shallow root systems, rates changed quite abruptly.

Air temperatures, rainfall, and humidity also affect fruit growth rates. Hilgeman et al. (1959) showed rates decreased as daily maximum temperatures increased from 38C to 43C and increased rapidly after rainfall. Lombard et al. (1965) showed that low humidity decreased growth rates. Mendel (1951) and Hilgeman et al. (1959) observed that high temperatures (42–45C) caused growth to stop. Beutel (1964) related maximum air temperatures and soil water suction to lemon fruit growth. With suction between 10 and 40 centibars at 25 cm depth, growth rates increased between 20 and 36C and then decreased slightly at 39C. At higher suctions (50 to 80 centibars), growth rates were lower and a lesser increase occurred between 20 and 34C and a rapid decrease occurred at 39C.

When using fruit growth as an index for irrigation, the above factors must be considered. It is suggested that fruit growth be used to evaluate existing irrigation programs where soil water suction devices described in the following section are used. In order to follow fruit growth, 5 average sized immature fruits on the north one-half of the tree should be tagged when about 10 cm in circumference. The tagged fruits should be measured at their equator with a steel tape within 1.5 hours after sunrise because the fruit shrinks during the day. Measurements should be made just before an irrigation, 3 or 4 days after irrigation, and at intervals of 4 to 7 days until the next irrigation. The circumference values should be converted to a volume measurement and the growth plotted. Experience will show a gradual reduction in growth rate between irrigations followed by a rapid increase after an irrigation.

The amount of reduction which can be permitted before fruit growth and yields are adversely affected has not been precisely established. When growth of grapefruit or Valencia oranges was allowed to stop between summer irrigations, Furr (1955) and Hilgeman and Sharples (1957) showed fruit sizes were reduced but not enough to significantly reduce yields. Oppenheimer and Elze (1941), working with Shamouti orange, showed that a drop in growth rate of 30% to 40% of the seasonal average did not reduce the final size. Later work by Patt (1953) indicated that greater reductions could be allowed. Thus, it is evident that irrigation should be delayed until a marked reduction in fruit growth rate takes place.

8. USE OF TENSIOMETERS AS A GUIDE FOR IRRIGATION

In California, Richards and Marsh (1961) found that the tensiometer which measures the suction forces of water films in the soil is a useful guide for irrigation. It is suggested that soil water suction be correlated with fruit growth by placing tensiometers at the trees where fruit is measured. Two tensiometers should be installed at the drip of the tree: one at the level of maximum root density (possibly 40 to 60 cm) and the other deeper so that about three-fourths of the roots are above it (usually 70 to 100 cm). The number of tree stations required depends on the variability of the soil. Usually a minimum of three stations per irrigation unit are needed. Best correlations between fruit growth and soil water suction usually occur at the deeper depth in deep loam soils and at the shallow depth in sandy soils.

Fruit measurements are only valid on immature fruit, usually during the period of June to November in USA citrus areas. The suction values obtained during this interval may be used as a guide during the winter and spring. In many instances, irrigations at 30 to 40 centibars in the spring and 50 to 70 centibars in the summer have been satisfactory. Irrigations at lower suctions are recommended during extreme hot weather and higher suctions can be allowed to develop when transpiration is low.

Changes in suction after an irrigation show the depth of penetration of the water. This information can be used to adjust the amount of water applied so that only the dry zone is wet at the next irrigation.

Tensiometers have not been satisfactory in the deep, fine sandy soils in Florida. Apparently the soil is not uniformly wet by rainfall and root development is deep and irregular so wide differences in suctions have occurred at similar depths. Measurements of the available water content in the major soil area occupied by roots appears to be the best guide for irrigation in this area.

B.　Irrigation of Other Evergreen Fruit Trees

The results reported with citrus trees apply to other tropical trees in a general way. Trees with larger and more succulent leaves and stems generally are more sensitive to soil water deficiency.

1.　THE AVOCADO

The avocado (*Persea Americana* Mull.) appears to have a more shallow root system and, because its leaves are more subject to heat injury than citrus, it is grown only in subarid or humid zones. Richards et al. (1958) and Moore and Richards (1958) working with young trees at Riverside, California, irrigated avocados at 50, 100 and 1,000 centibars suction at a 30-cm depth. Maximum trunk growth, the largest and deepest colored leaves, and the least tipburn and sunburn were obtained at 50 centibars suction. Trunk growth was extremely sensitive to soil water. Trunks shrank between irrigations in the 1,000-centibar plot and shrank in all plots when hot, dry winds occurred. Combined yields for the sixth and seventh year were 73, 63 and 25 kg for the 50-, 100-, and 1,000-centibar treatments, respectively. These data show that it is necessary to maintain soil water at low suctions throughout the root system for maximum yields and tree growth. However, only slightly less yield and tree growth occurred in the 100-centibar treatment which received about 60% as much water.

Current recommendations in California by Marsh and Gustafson (1962) suggest light irrigations when suctions reach 40 to 50 centibars at 30-cm depth. When similar suctions develop at the 60 cm depth more water is applied to wet this zone. This program requires two or three light irrigations at 8-to 10-day intervals, followed by a heavier one to wet the drier subsoil. Extreme care must be taken to avoid over-irrigation. Huberty and Pillsbury (1943) and Zentmeyer and Richards (1952) report that it is hazardous to maintain subsoil water above field capacity because of root decay by *phytophthera cinnamomi*.

No data is available concerning the effect of irrigation on yields from large, mature fruiting trees. It appears that the most critical period for soil water occurs during blossoming and fruit setting.

2. THE PAPAYA

The papaya (*Carica papaya* L.), which is a rapidly growing, highly succulent upright perennial that bears fruit early, is grown almost exclusively in humid, frost-free, subtropical zones. An irrigation test by Awada and Ikeda (1957) at Oahu, Hawaii, compared two irrigation schedules during a 20-month interval involving two summer dry periods. The frequently irrigated treatment maintained soil water at the 15- to 30-cm depth above field capacity; the dry treatment reduced soil water to two-thirds field capacity on only three occasions. The higher soil water levels increased trunk elongation, fruit size and total weight by inducing a larger per cent of carpellodic fruit but did not change the weight of solo (commercial) fruit.

3. THE BANANA

The banana (*Musa paradisiaca* var. *sapientum* L. Kuntze) is a tropical herb with a broad succulent leaf blade highly sensitive to soil water changes. Normally, it is grown in the humid tropical areas where rainfall ranges from 1,500 to 3,000 mm/year. In certain areas periods of high rainfall are followed by drouth. Experimental data on irrigation programs are not available. However, Ochse et al. (1961) report that soil water stresses seriously retard stem growth and cause premature ripening. Most plantations in tropical America irrigate during drouth with overhead sprinkler systems.

Limited quantities of bananas are grown in semiarid regions. In the Jordan Valley of Israel, Shmueli (1953) found that, as available soil water decreased from 86% to 51%, the water content of leaves was reduced and leaf dry matter decreased from 0.96 to 0.81 g/cm². He associated the reduction with decreased photosynthesis caused by restriction of stomatal opening. Stoler reports a 13 to 17% increase in yield when irrigation intervals were changed from 12 to 6 days (S. Stoler, 1952. *Personal communication. Tel Aviv, Israel*). In the Carnarvan district of Australia, Ticho observed that irrigations are applied at 3- to 7-day intervals, depending on soil texture with an annual application of about 5 m (R. J. Ticho, 1954. Fruit production in Australia. *Mimeo appendix to final report, Food & Agr. Organ., Rome, Italy*).

4. OTHER TROPICAL FRUITS AND NUTS

Other subtropical and tropical fruits and nuts, such as the mango (*Mangifera indica* L.) litchi (*Litchi chinensis* Sonn.), macadamia (*Macadamia ternifolia* F. Muell), and the pineapple, (*Ananas comosus* L. Merr.) which are of major importance, are irrigated in many locations. Extensive irrigation experiments either have not been conducted or not reported so the authors are unable to include information on them.

II. IRRIGATION OF XEROPHYTIC TYPE FRUIT TREES

A. The Date Palm

The date palm (*Phoenix dactylifera* L.) is grown commercially only in hot, dry, arid zones with very low rainfall. Under conditions of extreme drouth, leaf and trunk growth is retarded, finally ceases and older leaves die, but the palm can survive for many years. When the soil is again wet, growth is resumed but a constriction in the trunk that develops during the drouth period remains. In date growing areas, irrigation is required throughout the year, unless a high water table exists.

Aldrich et al. (1942) found that the rate of elongation of the central unexpanded spike leaf reflected soil water depletion. Almost all irrigation tests have since utilized this physiological measurement on a comparative basis for evaluating responses to irrigation. However, specific values have not been established because of interactions with air temperature and differences among varieties. Furr and Armstrong (1955) have shown that root development is deep and wide spreading and that, in stratified alluvial soils, greater root densities occur in silt than in sand layers.

An estimation of the water requirement near Indio, California by Furr and Armstrong (1956), using gravimetric water determinations during intervals when extraction rates were uniform, showed Khadrawy palms used 1.3 m (51 inches) of water/year from the upper 214 cm (7 ft) of soil. Water use ranged from 5.7 cm (2.2 inches) in January to 19 cm (7.5 inches) in July. Since about 20% of the roots were below 214 cm, it was estimated that a total of 1.6 m (63 inches) of water were used. With 75% efficiency in application, this required applying about 2.2 m (7.2 ft) of water. These values are slightly lower than Pillsbury's (1941) findings with the larger Deglet Noor variety.

Applications of excessive amounts of water such as 7.3 m (24 ft) by Reuther and Crawford (1945) and 4.3 m (14 ft) by Furr and Armstrong (1958) did not increase rate of leaf elongation, tree growth, size, yield or grade of fruit in two experiments conducted on silt loam and sandy loam soils. However, with palms growing in a deep fine sand, Reuther (1944) observed leaf growth rates increased by applying a trickle of water continuously. This program applied 6.4 m (21 ft)/year, but growth was not equal to that from 3 m (10 ft) of water applied to a silt soil. Very sandy soils present a particular problem in the supply of water to date roots. Possibly they cannot transmit water to the relatively coarse root system rapidly enough under high transpiration conditions to maintain maximum growth.

Using rate of leaf elongation as an index, Aldrich (1942) and Reuther and Crawford (1945) showed that soil water limited growth of Deglet Noor palms in loam soil during the summer within about 4 weeks after an irrigation. About 50% of the available water had been depleted from the major root zone and suctions near 80 centibars at the 75-cm depth developed.

Aldrich et al. (1942), Reuther and Crawford (1945), Furr et al. (1951) and Hilgeman et al. (1957) found that withholding irrigations to produce a 15 to 20% reduction in rate of leaf elongation during July and August reduced fruit size from 10 to 15%, decreased the moisture content of the fruit and induced earlier ripening. The effect upon fruit quality depended on the variety and weather conditions. In years of low rainfall, the semidry Deglet Noor produced more low grade dry fruit; whereas, with moderate rainfall, checking was reduced. With the soft Maktoom variety under moderate rainfall, limiting irrigation markedly reduced preripening shrivel and blacknose and improved quality. Furr and Armstrong (1958) noted that when water stresses were delayed until the harvest period, no marked effect upon size, grade and yield occurred. The above-mentioned moderate late summer water stresses slightly reduced the growth of the palm but did not affect the total number of leaves and the inflorescences produced.

When establishing a date garden, offshoots are planted directly in the field. Hilgeman and Furr (1948) observed excellent survival and growth of offshoots in loam soils by irrigating at 1-week intervals or at soil water suctions of 15 to 20 centibars.

Research findings to date suggest an irrigation program for fruiting palms which may be summarized as follows. In loam soils apply sufficient water to wet the entire root system in winter. Thereafter, irrigate when about 50% of the available water has been removed from the major rooting area or at suctions of 60 to 80 centibars in the 75- to 100-cm depth. Availability of subsoil water prior to harvest permits withholding irrigations during harvest. More frequent irrigations are required for sandy soils to maintain soil water suction near 20 to 30 centibars at a 75-cm depth. In areas where rainfall and high humidity frequently injure fruit, losses can be reduced by allowing a moderate internal water stress to develop within the palm in late July and August.

B. The Olive

The olive (*Olea europaea* L.) has xerophytic characteristics which enable it to survive and produce fruit during drouth in subarid regions. It is extensively grown in the Mediterranean area under natural winter rainfall, followed by a long dry spring and summer.

In central California, Hendrickson and Veihmeyer (1949) observed that water applied in early June was essentially depleted in the upper 180 cm (6 ft) of soil by mid-August. Fruit grew at a reduced rate thereafter compared to trees irrigated at monthly intervals that required 60 to 75 cm (24 to 30 inches) of water.

In Israel, Samish and Spiegel (1961) found that two or more irrigations, sufficient to wet the 125-cm (4-ft) depth, increased fruit size. However, timing of irrigations was a more important factor. When winter rainfall was subnormal, spring irrigation produced longer fruiting twigs and increased yields the following year. Also, irrigations during the period of rapid fruit growth in August and early September increased fruit size almost as much as continuous summer irrigation. Irrigation delayed ripening and the per cent of fruit oil was decreased; because of increased fruit size, oil yield per unit of land was increased from 23% to 68% by irrigation.

C. The Carob

The carob tree (*Ceratonia siliqua* L.), like the olive, is drouth resistant. In the Mediterranean Basin, it is of commercial importance and grows under natural winter rainfall, usually between 400 and 800 mm. Although summer irrigations increased the water content of pods, susceptibility to insect injury during ripening stages also was increased. Furthermore, the higher water content interfered with the regular harvesting and processing operations (A. Panaretos, *personal communication, Agri. Dep., Nicosia, Cyprus*).

LITERATURE CITED

Aldrich, W. W. 1942. Some effects of soil moisture deficiency upon Deglet Noor fruit. Date Growers Inst. 19:7–10.

Aldrich, W. W., C. L. Crawford, R. W. Nixon, and W. Reuther. 1942. Some factors affecting rate of date leaf elongation. Amer. Soc. Hort. Sci., Proc. 41:77–84.

Awada, M., and W. S. Ikeda. 1957. Effects of water and nitrogen application on composition, growth, sugars in fruits, yield, and sex expression of the papaya plants (*Carica papaya* L.). Hawaii Agr. Exp. Sta. Tech. Bull. 33. 14 p.

Bartholomew, E. T., W. B. Sinclair, and F. M. Turrell. 1941. Granulation of Valencia oranges. California Agr. Exp. Sta. Bull. 647. 63 p.

Beckett, S. H., H. F. Blaney, and C. A. Taylor. 1930. Irrigation water requirement studies of citrus and Avocado trees in San Diego County, California, 1926 and 1927. California Agr. Exp. Sta. Bull. 489. 51 p.

Beutel, J. A. 1964. Soil moisture, weather, and fruit growth. California Citrograph 49:372.

Bouna, D. 1954. A factorial field experiment with citrus at farm 466. Irrig. Res. Sta. (Griffeth, Australia) Int. Rep. 14. 7 p.

Cahoon, G. A., B. L. Grover, and I. L. Eaks. 1964. Cause and control of oleocellosis on lemons. Amer. Soc. Hort. Sci., Proc. 84:188–198.

Cahoon, G. A., R. B. Harding, and D. B. Miller. 1956. Declining root systems. California Agr. 10(9):3, 12.

Cahoon, G A., M. R. Huberty, and M. J. Garber. 1961. Irrigation frequency effects on citrus root distribution and density. Amer. Soc. Hort. Sci., Proc. 77:167–172.

Cahoon, G. A., E. S. Morton, W. W. Jones, and M. J. Garber. 1959. Effects of various types of nitrogen fertilizers on root density and distribution as related to water infiltration and fruit yields of Washington navel oranges in a long term fertilizer experiment. Amer. Soc. Hort. Sci., Proc. 74:289–299.

Compton, C. 1937. Water deficit in citrus. Amer. Soc. Hort. Sci., Proc. 34:91–95.

Erickson, L. C., and S. J. Richards. 1955. Influence of 2,4-D and soil moisture on size and quality of Valencia oranges Amer. Soc. Hort. Sci., Proc. 65:109–112.,

Ford, H. W. 1954. The influence of rootstock and tree age on root distribution of citrus. Amer. Soc. Hort. Sci., Proc. 63:137–143.

Ford, Harry W. 1959. Growth and root distribution of orange trees on two different rootstocks as influenced by depth to subsoil clay. Amer. Soc. Hort. Sci., Proc. 74:313–321.

Furr, J. R. 1955. Responses of citrus and dates to variations in soil-water conditions at different seasons. Int. Hort. Congr., 14th Symp. (Netherlands) 10:400–412.

Furr, J. R., and W. W. Armstrong. 1955. Growth and yield of Khadrawy date palms irrigated at different intervals for two years. Date Growers Inst. 32:3–7.

Furr, J. R., and W. W. Armstrong. 1956. The seasonal use of water by Khadrawy date palms. Date Growers Inst. 23:5–7.

Furr, J. R., and W. W. Armstrong. 1958. The influence of heavy irrigation and fertilization on growth, yield, and fruit quality of Deglet Noor dates. Date Growers Inst. 35:22–24.

Furr, J. R., and C. A. Taylor. 1939. Growth of lemon fruits in relation to moisture content of the soil. US Dep. Agr. Tech. Bull. 640. 71 p.

Furr, J. R., E. C. Currlin, R. H. Hilgeman, and Walter Reuther. 1951. An irrigation and fertilization experiment with Deglet Noor dates. Date Growers Inst. 28:17–20

Halma, F. F. 1935. Trunk growth and water relation in leaves of citrus. Amer. Soc. Hort. Sci., Proc. 32:273–276.

Harris, Karl, A. F. Kinnison, and D. W. Albert. 1936. Use of water by Washington navel orange and Marsh grapefruit trees in Salt River Valley, Arizona. Arizona Agr. Exp. Sta. Bull. 153. p. 441–496.

Hendrickson, A. H., and F. J. Veihmeyer. 1949. Irrigation experiments with olives. California Agr. Exp. Sta. Bull. 715. 7 p.

Hilgeman, R. H. 1963. Trunk growth of the Valencia orange in relation to soil moisture and climate. Amer. Soc. Hort. Sci., Proc. 82:193–198.

Hilgeman, R. H., and J. R. Furr. 1948. How variations in soil moisture affected growth of Deglet Noor date palm offshoots. Date Growers Inst. 25:24–26.

Hilgeman, R. H., and D. R. Rodney. 1961. Commercial citrus production in Arizona. Arizona Agr. Exp. Sta. Spec. Rep. 7. 31 p.

Hilgeman, R. H., and G. C. Sharples. 1957. Irrigation trials with Valencias in Arizona. California Citrograph 42:404–407.

Hilgeman, R. H., G. C. Sharples, and L. H. Howland. 1957. Effect of irrigation and leaf-bunch ratio on shrivel and rain damage of the Maktoom date. Date Growers Inst. 34:1–5.

Hilgeman, R. H., H. Tucker, and T. A. Hales. 1959. The effect of temperature, precipitation, blossom date, and yield upon enlargement of Valencia oranges. Amer. Soc. Hort. Sci., Proc. 74:266–279.

Horanic, G. E., and F. E. Gardner. 1959. Relative wilting of orange trees on various rootstocks. Florida State Hort. Soc., Proc. 72:1–4.

Huberty, M. R. 1948. The citrus industry. Univ. California Press, Berkeley, California. Vol. 1, Ch. 10. p. 445–593.

Huberty, Martin R., and A. F. Pillsbury. 1943. Solid, liquid, gaseous, phase relationships of soils on which avocado trees have declined. Amer. Soc. Hort. Sci., Proc. 42:39–45.

Huberty, M. R., and S. J. Richards. 1954. Irrigation tests with oranges. California Agr. 8(10):8, 15.

Koo, R. C. J. 1961. The distribution and uptake of soil moisture in citrus groves. Florida State Hort. Soc., Proc. 74:86–90.

Koo, R. C. J. 1963. Effects of frequency of irrigations on yield of orange and grapefruit. Florida State Hort. Soc., Proc. 76:1–5.

Lombard, D. B., L. H. Stolzy, M. J. Garber, and T. E. Szuskiewicz. 1965. Effects of climatic factors on fruit volume increases and leaf water deficit of citrus in relation to soil suction. Soil Sci. Soc. Amer. Proc. 29:205–208.

Marsh, A. W., and C. D. Gustafson. 1962. An avocado irrigation program. California Avocado Soc. Yearbook 46. p. 39–41.

Mendel, K. 1951. Orange leaf transpiration under orchard conditions: III. Prolonged soil drought and the influence of stocks. Palestine J. Bot., Rehovot Ser. 8:45–53.

Moore, P. W., and S. J. Richards. 1958. Effects of irrigation treatments and rates of nitrogen fertilization on young Hass avocado trees in relation to leaf tipburn, leaf color, leaf size, tree vigor, and leaf moisture deficits. Amer. Soc. Hort. Sci., Proc. 71:298–303.

Ochse, J. J., M. J. Soule, Jr., N. J. Dijkman, and C. Wehlburg. 1961. Tropical and subtropical agriculture. Macmillan, New York. Vol. 1. p. 384–5.

Ongun, A. R., and A. Wallace. 1958. Transpiration rates of small Washington navel orange trees grown in a glass house with different rootstocks at different root temperatures. Tree Physiology Studies, Univ. California, Los Angeles, No. 1, p. 87–103.

Oppenheimer, H. R. 1936. A citrus rootstock trial on light soil. Hadar 9:35–40.

Oppenheimer, H. R., and D. L. Elze. 1941. Irrigation of citrus trees according to physiological indicators. Agr. Res. Sta. Rehovot, Israel, Bull. 31.

Oppenheimer, H. R., and K. Mendel. 1939. Orange leaf transpiration under orchard conditions. Agr. Res. Sta. (Rehovot, Israel) Bull. 25. 82 p.

Patt, Y. 1953. Experimental studies on irrigation and thinning out of citrus groves. Agr. Res. Sta., Rehovot, Israel. Bull. 60. 45 p.

Pillsbury, A. F. 1941. Observations on use of irrigation water in Coachella Valley, California. California Agr. Exp. Sta. Bull. 649. 48 p.

Pillsbury, A. F., O. C. Compton, and W. E. Picker. 1944. Irrigation water requirements of citrus in South Coastal Basin of California. California Agr. Exp. Sta. Bull. 686. 19 p.

Reeve, J. O., and J. R. Furr. 1941. Evaporation from a shallow black pan evaporimeter as an index of soil moisture extraction by mature citrus trees. Amer. Soc. Hort. Sci., Proc. 39:125–132.

Reuther, W. 1944. Response of Deglet Noor date palms to irrigation on a deep sandy soil. Date Grower's Inst. 21:16–19.

Reuther, W., and C. L. Crawford. 1945. Irrigation experiments with dates. Date Growers Inst. 22:11–14.

Richards, S. J., and A. W. Marsh. 1961. Irrigation based on soil suction measurements. Soil Sci. Soc. Amer. Proc. 25:65–69.

Richards, S. J., L. V. Weeks, and J. C. Johnson. 1958. Effects of irrigation treatments and rates of nitrogen fertilization on young Hass avocado trees: I. Growth response to irrigation. Amer. Soc. Hort. Sci., Proc. 71:292–297.

Samish, R. M. 1957. Irrigation requirements of a young citrus orchard. Ktavim 7:123–139.

Samish, R. M., and P. Spiegel. 1961. The use of irrigation in growing olives for oil production. Israel J. Agr. Res. 11:87–95.

Shmueli, E. 1953. Irrigation studies in the Jordan Valley: I. Physiological activity of the banana in relation to soil moisture. Res. Council Israel Bull. 3:228–247.

Sites, J. W., H. J. Reitz, and E. J. Deszyck. 1951. Some results of irrigation research with Florida citrus. Florida State Hort. Soc, Proc. 64:71–79.

Smith, G. E. P., A. F. Kinnison, and A. G. Carns. 1931. Irrigation investigations in young grapefruit orchards on the Yuma Mesa. Arizona Agr. Exp. Sta. Tech. Bull. 37. p. 413–591.

Stolzy, L. H., O. C. Taylor, M. J. Garber, and P. B. Lombard. 1963. Previous irrigation treatments as factors in subsequent irrigation level studies in orange production. Amer. Soc. Hort. Sci., Proc. 82:199–203.

Taylor, C. A., and J. R. Furr. 1937. Use of soil moisture and fruit growth records for checking irrigation practice in citrus orchards. US Dep. Agr. Circ. 426. 23 p.

Zentmyer, G. A., and S. J. Richards. 1952. Pathogenicity of phytophthera cinnamomi to avocado trees and the effect of irrigation on disease development. Phytopathology 42:35–37.

37

Grapes and Berries

Part I—Grapes

A. N. KASIMATIS

University of California
Davis, California

The culture of grapes (*Vitis vinifera*) had its beginnings in Asia Minor in the dawn of antiquity. World plantings of grapes now include about 25 million acres.

I. CHARACTERISTICS OF THE GRAPEVINE

A. Climatic Requirements

V. vinifera grape varieties, long grown in the countries bordering the Mediterranean, need long, warm-hot, dry summers and cool winters for best development. Their susceptibility to fungus disease limits successful growing in areas subject to humid summers. Neither will they withstand intense winter cold without protection. After growth starts in the spring, frosts may kill the fruitful shoots and drastically reduce the crop.

A long growing season is needed to mature the fruit. While rainfall during the winter is desirable, rains early in the growing season make disease control more difficult but are not otherwise detrimental to growth. Rains or cold, cloudy weather during the bloom period, however, may adversely affect setting of the fruit. When rains occur during the ripening and harvesting season, fruit rotting may result. Where raisins are made by sun drying, a month of warm, clear, rainless weather is essential after the grapes are ripe.

B. Soil Adaptation

Grapes are adapted to a wide range of soil types, even though certain preferences do exist among growers. When a survey is made of the soils which produce the multitudinous varieties of grapes in the many different grape-growing regions of the world, a wide range is encountered—from coarse gravelly sands to heavy clays, shallow to very deep soils, and soils of low to high levels of fertility. They are best adapted to the deep, medium-textured soils; avoid planting them on very shallow soils, on heavy clays or on those that are poorly drained or that contain high concentrations of salts. Since *V. vinifera* varieties grow best in regions of little or no summer rain, enough of the winter rainfall must be stored in the soil to carry the vines through the summer, or irrigation must be supplied. Dry farming of grapes, as practiced in the relatively cool coastal valleys of northern California, USA requires deep soils that can store ample water. Grapes, however, can be successfully grown on soils of 2 ft or less, if irrigation and experienced management are available.

Winkler (1962) notes that most V. *vinifera* varieties are characteristically deep rooted and fully explore the soil mass to depths of 6 to 10 ft or more, if root penetration is not obstructed. In general, there appears to be an inverse relation between depth of rooting and fineness of the soil texture. In deep, coarse sand or gravelly soils, grape roots may penetrate 12 to 25 ft or more. Usually few roots are found deeper than 8 to 12 ft in sandy loam soils, 6 to 8 ft in loams, or 2 to 3 ft in clays. Reports of other workers support this experience (Harmon and Snyder, 1934; Pillsbury, 1941; Hendrickson and Veihmeyer, 1951; Doll, 1955; Kasimatis, 1958; Hemstreet, 1961; Vaadia and Kasimatis, 1961).

C. Annual Cycle of Growth

The grapevine does not begin growth in early spring with the first rise in temperature, but remains dormant until the mean daily temperatures reach about 10C (50F). At first, growth is slow, but as the mean temperature increases, overall growth, including a striking elongation of the shoots, accelerates from day-to-day. After 3 to 4 weeks, the period of most rapid growth is underway. About the time of bloom, the rate of shoot growth declines rapidly, then continues at a slow rate until the end of the season.

D. Aspects of Vine Physiology

In contrast to the growth pattern of deciduous fruit trees which Veihmeyer and Hendrickson (1950b) describe as making one flush of growth during the early part of the season with growth then essentially discontinued, despite availability of soil water, grapevines may continue to grow as long as environmental conditions are favorable. Irrigation after terminal bud formation does not result in additional growth of deciduous fruit trees. Since vines do not form terminal buds, shoots may increase in length at any time, if there is sufficient heat, light, an abundance of soil water, and an adequate level of soil fertility. Young vines, with little or no crop to compete with the vegetative growth for photosynthates, often continue growth into late fall and may be injured by frost. The same condition may exist with table grape varieties harvested early in the growing season. Under the favorable conditions for vine growth that exist in much of California, Winkler (1962) suggests that rigid control of available nitrogen and irrigation water after midsummer is the only means at the disposal of growers to control the growth of such vines.

Since the loss of water occurs almost entirely in the leaves, the amount of water needed by vines is not materially influenced by the load of fruit. The crop, however, through its drain on the carbohydrate nutrition of the vine, may influence the foraging capacity of the roots and hence their ability to obtain water. The developing fruit competes directly with the growing roots and shoots. The roots of heavily cropped vines, consequently, may not be well nourished and thus not capable of exploring the soil, rapidly and thoroughly. A vine with a large crop may show symptoms of water shortage under conditions that a vine with less fruit would not exhibit.

II. RESPONSES TO SOIL WATER CONDITIONS

A. Growth of Vegetative Parts

As noted earlier, the growth of bearing vines is characterized by a very rapid and succulent growth of shoots in the spring and early summer. The rapidly elongating shoot tips have a soft, yellowish-green appearance. Measurements of the rate of growth during this period are quite useful as a sensitive indicator of soil water availability (Vaadia and Kasimatis, 1961). As the soil water content approaches the wilting point (WP), the rate of growth diminishes and the appearance of the shoot tips changes. The internodes become shorter and the yellowish-green color transforms to the darker green of mature leaves (Jacob, 1950). With continued stress during the summer, the leaves on vines with most of their roots in soil at WP do not wilt in the sense of drooping leaves but rather the older leaves become somewhat yellow, dessicated near the margins, and mildly curled (Pruitt and Clore, 1956). Orientation of such leaves in a plane parallel to solar radiation during peak temperatures was reported by Vaadia and Kasimatis (1961). Older leaves at the base of the shoots become brown and dry, die, and drop from the vine.

A severe or sudden reduction in the soil water available to growing vines will result in a wilting of the leaves and succulent shoots followed by a yellowing and shedding of basal leaves (Kobayashi et al., 1963). Such wilting may occur with vines growing in containers when the soil reaches the WP; under field conditions it may happen with sudden rises in temperature with vines growing on shallow soils where the WP is reached in all parts of the root zone at the same time. Wilting occurs infrequently on deep soils since only part of the soil is at the WP at the same time; as readily available soil water is depleted in successive parts of the soil, the vine adjusts to these conditions and makes less shoot growth (Jacob, 1950; Kobayashi et al., 1963). Veihmeyer and Hendrickson (1950a) conclude that although the grapevine has long been considered drouth resistant, this is not due to an inherent ability to extract more water from the soil than other fruit trees which may be injured more easily by dry soil conditions. This characteristic is rather a reflection of the vines' ability to balance the water supplied through the roots and that lost to the atmosphere and still maintain a low level of activity.

The fact that vines with a continuous supply of readily available water during the growing season make more total growth than vines with limited soil water was readily observed but not always measured (Hendrickson and Veihmeyer, 1931; Zineberg and Befani, 1962). A quantitative measure of differences in annual shoot growth from vines under different soil water regimes may be obtained by weighing the prunings removed during the dormant season (Vaadia and Kasimatis, 1961). Since 90 to 95% of the 1-year-old wood is removed during pruning, comparisons of growth can readily be made. Increases in trunk girth can also be used to describe growth responses where differential irrigation programs are maintained for several years (Vaadia and Kasimatis, 1961). Veihmeyer and Hendrickson (1957) regarded the rate of fruit growth as the most sensitive index of soil water availability to the major portion of the root system, though they did not report measures of vegetative response in their studies with grapes. Stanhill (1957) in his review of plant growth responses to different soil water regimes noted a greater ratio of positive results where vegetative growth was measured rather than the character of a reproductive part. The sensitive rate of shoot elongation and the characterization of total shoot growth by the weight of prunings offer important criteria of irrigation response by grapevines in addition to changes in the growth of the fruit.

It should be noted that in most reported research on grape irrigation, the experiments were designed to provide and maintain available soil water during all or part of the growing season for comparison with plots allowed to remain at the WP for a considerable period. Perhaps because suitable instrumentation was lacking at the time, little consideration was given to studies on the effects of soil water conditions within the available range on grapes. Hendrickson and Veihmeyer (1951) stressed the importance of maintaining a supply of available water during the growing season, but they did not measure vine growth per se. More recently tensiometers have been used in vineyards with soil water regimes maintained at soil water suction values in the range of 20 to 70 centibars. Except for clay soils, which are seldom planted to grapes in California, Richards and Marsh (1961) report that tensiometers are capable of indicating soil water conditions over the range of 50% to 75% available water depletion. One study by Kissler et al. (1961) illustrates the use of tensiometers for irrigation scheduling but does not report on vine response. Incomplete tests by the author with soil water suction values maintained at levels from 20 to 50 centibars indicates a marked response in vine growth within this range of comparison. Yields and fruit quality were not similarly affected, however. Much research is needed with different levels of available water to reconcile these observations with the conclusions of earlier workers whose studies largely concerned considerably higher soil water suction values.

B. Changes in the Character of the Fruit

Many workers have noted marked responses in the yield of grapes with applications of irrigation water on table grapes: (Prima, 1960; Spiegel-Roy and Bravdo, 1964; Podoleanu and Alexandrescu, 1964; Hendrickson and Veihmeyer, 1931, 1951); on wine grapes (Vagnoli, 1961; Vaadia and Kasimatis, 1961; Tulloch, 1961; Hendrickson and Veihmeyer, 1951; Zineberg and Befani, 1962); and on raisin grapes (Hendrickson and Veihmeyer, 1951). Few, however, have related these yield increases with soil water extraction (Hendrickson and Veihmeyer, 1931; Vaadia and Kasimatis, 1961; Podoleanu and Alexandrescu, 1964). Yields alone cannot be evaluated without considering corresponding soluble solid contents because of the strong relationship between the two. In this light the yield responses reported above invariably reflected a reduction of soluble solids (Vagnoli, 1961; Spiegel-Roy and Bravdo, 1964; Hendrickson and Veihmeyer, 1931, 1951; Vaadia and Kasimatis, 1961; Podoleanu and Alexandrescu, 1964; Zineberg and Befani, 1962). A delay in the date of harvest of fruit with lower soluble solids would have enhanced the values provided that favorable environmental conditions would persist and that other aspects of quality might not deteriorate, i.e., an increase in fruit rot, a decrease in titratable acidity in the juice, a loss in weight due to shrinkage (Vaadia and Kasimatis, 1961; La Rosa and Nielson, 1956).

Seasonal changes in berry size have been measured for several purposes: (i) to reflect soil water availability (Veihmeyer and Hendrickson, 1957); (ii) to describe growth responses to irrigation (Vaadia and Kasimatis, 1961; Hendrickson and Veihmeyer, 1931, 1951); and (iii) as an index of fruit quality (Safran, 1957). These berry changes have been measured in terms of weight, volume and width, with some preferences for calipering since the same berries could be observed each time. Changes in berry dimensions continue until maturity despite depletion of readily available water (Vaadia and Kasimatis, 1961; Hendrickson and Veihmeyer, 1951). Optimum berry size at harvest, however, does reflect that an adequate supply of soil water was available during the season while the berries continued development. Insufficient water during the early period of rapid berry enlargement prevents the attainment of normal berry size (Veihmeyer and Hendrickson, 1957; Vaadia and Kasimatis, 1961). Applying water after this dynamic growth will not enable undersized berries to become normal. The incidence of bunch rot in wine varieties which have compact clusters has been increased materially with irrigation regimes favorable to the development of large berry size (Vaadia and Kasimatis, 1961).

Water deficits may also be reflected in the production of smaller size bunches as measured in terms of weight and length. Size is further decreased where irrigation programs allow severe water deficits to occur in several successive years (Vagnoli, 1961; Vaadia and Kasimatis, 1961).

The composition of grapes at time of harvest may be influenced by inadequate irrigation practices to varying degrees, depending on the duration of the plant water stress. Where an irrigation response was recorded, lower values of soluble solids and higher values of titratable acidity in the juice were generally obtained with an increasing supply of available water through the growing season (Hendrickson and Veihmeyer, 1951; Spiegel-Roy and Bravdo, 1964; Vaadia and Kasimatis, 1961). It is also true that the degree of crop response is generally directly correlated with improved soil water conditions. Unfortunately, the maturity of grapes is measured mostly by chemical composition, i.e., soluble solids content. Other practical means of relating physical characteristics to maturity have not been devised. It is difficult, therefore, to divorce the separate effects of irrigation on yield and maturity since yield and soluble solids content are so intimately related; further experimental work is needed to define maturity in terms of other characteristics. It seems logical that the improved leaf surface obtained with improved irrigation would result in a higher soluble solids content if the total crop were not also simultaneously increased. The increase of titratable acid in the juice with irrigation is probably directly associated with the improved leaf surface which in turn may lower fruit temperature and light exposure as suggested by the work of Kliewer (1964) and Kliewer and Schultz (1964).

C. Changes in Mineral and Carbohydrate Nutrition

Even though viticulturists have been greatly concerned with the harmful effects of continued growth in the late fall as a result of favorable environmental conditions (Bioletti and Twight, 1921; Winkler, 1962; Osteraas, 1962) which may result in a lack of satisfactory wood maturity, critical studies of the status of carbohydrates or other reserves remain to be undertaken. Vaadia and Kasimatis (1961) in limited studies were unable to show differences in the starch or sugar content of basal cane sections sampled during the dormant season from Chenin blanc vines (*V. vinifera*) which had suffered severely from drouth or from those where adequate soil water levels had been maintained.

It is generally believed that a reduction in the absorption of potassium by the roots of plants may occur under dry soil conditions. Only very limited data are available concerning the effect of drouth for extended periods on the mineral composition of the fruit or leaves of the grapevine. Vaadia and Kasimatis (1961) showed markedly reduced levels of phosphorus and potassium in the analyses of petioles collected just prior to harvest from vines in soil at WP from late June. They attributed these low values to restricted uptake by the roots in dry soil which persisted most of the season. Larsen et al. (1955) suggest that summer drouth is a contributing factor to the degree of expression of potassium deficiency symptoms in the Concord (*V. labrusca*) variety. A deficiency of magnesium was found by Kobayashi et al. (1963) in the leaves of Delaware (*V. labrusca*) vines grown in containers and maintained at very low soil water levels. Low nutrient levels in the leaves of plants might be an expression of low levels of root activity under diminished soil water conditions. The implications of this response in grapes remains to be clarified. The other important area needing investigation is the influence of soil fertility levels on irrigation response.

III. GROWTH PERIODS SENSITIVE TO SOIL WATER CONDITIONS

The grapevine is sensitive to soil water conditions during a number of critical periods in the seasonal growth cycle including dormancy, beginning growth in the spring, following fruit set, approaching fruit maturity, and near the close of the growing season. While such influences have been noted by many workers in the field, an understanding of the vines' reactions during these periods is rather incomplete.

A. During Dormancy and Early Spring

The impact of low winter rainfall situations on the subsequent growth and yields of grapevines grown under nonirrigated conditions has been outlined by several workers (Bioletti and Twight, 1901; Till, 1958; Tulloch, 1961). Winter irrigation is suggested wherever feasible to make up the water deficit; this practice is standard in many California vineyards where rainfall averages less than 12 inches and underground water is available. Flooding of vineyard soils during the dormant season resulting in saturated conditions for several weeks is not harmful to mature vines (Bioletti and Twight, 1921). Hale reported that an 8-week flooding period of dormant potted vines did not affect bud burst nor the subsequent growth of roots or shoots (C. R. Hale. 1959 Response of the grapevine to prolonged flooding of the soil. *Unpubl. M.S. Thesis. Univer. California, Davis*). After budding out, however, such treatment adversely affected both shoots and root development. Soil saturation for prolonged periods during the growing season, especially if hot weather occurs, will result in root death due to anaerobic conditions (Winkler, 1962). If a greater part of the root system is killed in this manner, the vine will likely show symptoms of water stress since the few healthy roots are unable to supply the vine's needs.

The probable influences of irrigation during the period of bloom remains speculative. There is unsubstantiated belief among growers that irrigation applied by surface methods during the period of bloom may be detrimental to set. The possibility that the use of sprinkler irrigation during this period may adversely affect fruit set is suggested by the work of Samish and Lavee (1958) who found that water sprays applied during bloom produced a desirable level of berry thinning with two table grape varieties. In California, grower practices vary, but sprinkling during bloom time is not uncommon.

B. Following Fruit Set

Insufficient water during the period immediately following fruit set when berry enlargement is proceeding at its most rapid rate, prevents the berries from reaching normal berry size (Vaadia and Kasimatis, 1961), even though ample water is applied after this sensitive growth period. Use of this knowledge is applied by growers who withhold irrigation to reduce berry size and bunch compactness as a means of minimizing the incidence of rot resulting from berry splitting during maturation (Winkler, 1962; Osteraas, 1962).

Girdling the trunks of vines at the time of shatter of impotent flowers following bloom, or at the beginning of ripening, is a standard practice in California table grape vineyards as a means of increasing berry size or enhancing the maturity of certain varieties. Girdling, by the removal of a ring of bark about one-fourth inch in width, halts the downward movement of food materials below the wound until after healing (Winkler, 1962). During the healing period of 3 weeks or more, the roots may be undernourished and less capable of growing. Experience dictates the wisdom of irrigating prior to or immediately following girdling, even though the soil mass may be well above the WP, as a protective measure against plant stress brought about by sudden hot weather.

C. Approaching Fruit Maturity

The effects of irrigation during the period of fruit ripening are not clear and the research reports are conflicting. Winkler (1962) states that a severe water deficit in the vine during ripening delays fruit maturity, gives the fruit a dull color, and permits exposure of the fruit. He also indicates that a slight shortage as maturity is approached may actually hasten ripening because of the effect of a water deficit in limiting shoot growth and berry size. Hendrickson and Veihmeyer (1931) showed that no deleterious effects resulted from frequent irrigations during the latter part of the ripening of Thompson Seedless and Emperor vines (V. vinifera). Goosen (1956) found, however, that irrigation of Sultana vines (V. vinifera) just prior to harvest resulted in lower soluble solids. Irrigation at the time of ripening increased the sugar content of berries (Luk' janov, 1961), while the maturation rate of Concords was accelerated by dry conditions and retarded by supplementary irrigation in the findings of Kattan (1963). Early maturity and lower quality resulted from a lack of soil water during the ripening period of several table grape varieties in South Africa (Du Toit and De V. Daniel, 1940). Late irrigations applied in August to Thompson Seedless vines grown for raisin production did not influence berry size but did increase the occurrence of fruit rot (L. P. Christensen. 1959. The effect of late irrigation on the maturation and quality of grapes. Unpubl. M.S. Thesis. Univer. California, Davis). The precise effects of irrigation during the period of fruit maturation are not clear because the reports either indicate the results of different water regimes on a seasonal basis or relate experiences of vine conditions during this period without describing soil water conditions. An adequate test of varying soil water conditions during the ripening period only is needed. This would be of value not only in table grape production where date of maturity is important but also to the raisin industry as well.

Reliable criteria for indicating the timing of the last irrigation prior to harvest of raisin grapes for sun drying have not been developed. Experience suggests that the last irrigation be applied 4 to 6 weeks prior to harvest to enhance the accumulation of sugar in the berries, depending on the ability of the soil to store available water. This situation is unique in grape production since the final product is a dried one (15% water) whose yield is directly correlated with the production of absolute amounts of soluble solids per acre.

In the case of table grapes, berry cracking during maturation has been a serious problem in some years. Where the harvest is extended over a long period so that irrigation becomes necessary on some soils Winkler (1962) reports that a deep irrigation may cause the berries of the Libier and Red Malaga varieties (V. vinifera) to crack across the apical end. He suggests the problem may be avoided by shallow irrigations applied to one side of the vine as a means of minimizing the rapid uptake of water. The variety Opsimo Souflion (V. vinifera), widely grown in Greece, is very susceptible to fruit cracking according to Kelperis (1963) as a result of excess water uptake when the soil water is rapidly increased by heavy rainfall. Safran (1957) describes the Queen of the Vineyard variety (V. vinifera) in South Africa as subject to berry splitting during the ripening period. Safran (1957) and Meynhardt (1957) suggest, however, that the cause is not only the result of irrigation and absorption of water by the vine, but also from restricted transpiration during periods of high humidity. Under these conditions the berries absorb water readily because of a high osmotic value and the excess is not transpired. Safran (1957) and Meynhardt (1957) further recommend that, since grape berries can absorb external water directly, sprinkler applications be avoided during ripening in periods of high humidity and at night. Surface applications or short periods of sprinkling may not cause splitting at this stage of maturity during dry days if soil water conditions dictate an irrigation.

D.   Near the Close of the Growing Season

After the fruit is ripe, and particularly after harvest, vines adjust to a limited supply of soil water. When fall temperatures are relatively cool, the leaves remain functional, the canes ripen, and the vines make little or no shoot growth even though most of the soil in the root zone is at WP. In the hot desert regions of the Coachella Valley, California, for example, where early maturing table varieties— Cardinal, Perlette, Thompson Seedless (V. vinifera)—are harvested in June and July, neglect of irrigation practice may cause damage. The leaves may drop, and where a water table exists, salts may continue building up in the surface soil (Wolfe, 1962). With cooler weather in the fall and reduced water stress, new leaves formed from current season's buds use a portion of the canes' stored reserves and may never develop sufficiently to produce food materials for storage (Winkler, 1962). Such vines may enter dormancy with lower reserves of carbohydrates, and some of the buds may fail to grow next spring or leaf out irregularly, as a result of poor development. Vines grown in the severe conditions of the hot desert require irrigation after harvest, not to continue active growth, but to maintain the established leaf surface.

New shoot growth should not occur late in the season (Winkler, 1962). Vineyard experience indicates that vines of vigorous varieties will continue to grow or start new growth after harvest and fail to ripen the wood (Osteraas, 1962) if supplied with readily available water. Late growth utilizes carbohydrates that should remain as stored reserves until the following year and the shoots, failing to mature properly, are subject to injury by fall frosts or winter cold (Bioletti and Twight, 1921; Winkler, 1962). Such vines are more likely to exhibit bud failure the following year and have less crop than vines which cease growth earlier and mature their canes (Bioletti and Twight, 1921; Winkler, 1962). Nonbearing vines have a greater tendency than bearing ones to grow vigorously until late fall. This late growth can usually be prevented by avoiding unnecessary irrigations after midsummer and by using summer cover crops (Osteraas, 1962). Where irrigations after midsummer are omitted to harden the growth of inherently vigorous varieties, an irrigation during the late fall, beyond the date when temperatures are favorable for the initiation of new growth, is considered necessary to maintain the vines' needs for water, even though it is low at this period (Osteraas, 1962). In contrast to the experience of these viticulturists, Veihmeyer and Hendrickson (1960) did not observe injury due to a lack of maturity of either young growth or buds in any of their irrigation experiments on grapes in northern California by watering late in the season. Unfortunately, however, it is not reported whether temperatures below $-1.1C$ (30F), which could cause injury in the fall, actually occurred during the years of these experiments.

IV. EFFECTS ON QUALITY

Quality is an intangible characteristic subject to varying methods of evaluation mostly by subjective means. It is impossible to correlate appraisals of grape quality from diverse sources since the background of evaluators differ and a common standard or reference is not available. Comparisons within a common experience are valid and do reflect differences resulting from irrigation practice. While Hendrickson and Veihmeyer (1951) conducted an extensive irrigation trial with the raisin variety Muscat of Alexandria (V. vinifera) and reported a positive response in yields and berry size, no appraisal was made of the dried product. Other reports on the influence of irrigation on raisin quality are lacking. Considerably more effort has been made in evaluating table grape quality. Hendrickson and Veihmeyer (1951) concluded that after 2 months of cold storage. the keeping quality of Thompson Seedless grapes was unaffected by high soil water conditions during ripening, and Emperor grapes, from wet and dry treatments following a similar storage test, showed little difference in keeping quality. With the Tokay variety, however, a deeper color was formed when the soil remained at the WP for relatively long periods, and the color of the fruit from nonirrigated plots was better after storage than the fruit from the irrigated vines (Hendrickson and Veihmeyer, 1951). The keeping quality of Emperor grapes irrigated more times than necessary was fully as good as that from plots which received a final irrigation approximately 3 months before picking (Hendrickson and Veihmeyer, 1931). Winkler (1962) emphasizes that the brightness, not the density of color, gives attractiveness to table grapes. He observes that an ample but not excessive supply of water at ripening seems to favor brilliant color; with inadequate water the color tends to develop more abundantly, but is dull and less attractive. Spiegel-Roy and Bravdo (1964) indicate a more favorable sugar-acid ratio resulted by irrigating the Dabouki variety (V. vinifera). Following observations of irrigation treatments with the Regina bianca variety (V. vinifera), Di Prima (1960) found no difference in fruit quality during the first 5 crop years. He concluded that irrigation had only a negligible effect on fruit quality and that quality is conditioned to other factors such as crop size and seasonal climatic conditions. A reduction in the marketable crop and an increase in yield following irrigation with the variety Afuz-Ali (V. vinifera) was experienced by Podoleanu and Alexandrescu (1964). Table grape quality was lowered and the incidence of dry stalk increased in the experience of Du Toit and De V. Daniel (1940) when soil water was deficient.

The quality of the wines made from grapes subject to irrigation treatments was considered in appraising the results of many trials. Hendrickson and Veihmeyer (1951) found no difference in organoleptic quality among the wines made from grapes receiving varying numbers of irrigations in four vineyards of the Carignane and Barbera varieties (V. vinifera). They did report that the intensity of wine color was slightly higher from the dry plots in two of the tests and related this to the greater ratio of skin, where most of the pigments are found, to the volume of the slightly smaller berries. Luden and Sadeh (1963) did not detect an influence on the color of wines produced from Carignane grapes receiving up to five irrigations; Vaadia and Kasimatis (1961) found that irrigation practices with the Chenin blanc variety (V. vinifera) did not influence the quality of wines evaluated by organoleptic means. A more pronounced straw color occurred in the wines from the nonirrigated plots probably resulting from greater fruit exposure and a higher incidence of amber color fruit. In their tests, however, harvest was scheduled by soluble solids content so that picking was delayed by as much as 3 weeks for the wet plots. If the fruit from all treatments were harvested at the same time, a greater difference in chemical composition would have been reflected in differences in wine quality. Ceiko (1955) reports improved organoleptic quality of wine produced from grapes receiving four irrigations.

V.   PROBLEMS INFLUENCED BY IRRIGATION METHOD OR FREQUENCY

Among the problems encountered in grape production that are influenced to some degree by the method or frequency of irrigation are: nematode infestation, salinity build-up, disease and insect control, and sprinkler applications.

California growers widely accept the practice of increasing the irrigation frequency in vineyards planted on light soils where parasitic nematodes are known to be damaging the roots. More frequent applications of water are necessary to partially compensate for the injured root system unable to develop and utilize all of the available soil water in a normal manner. Raski et al. (1965) stress the need for the maintenance of favorable soil water conditions in vineyards subject to nematode infestations.

Vineyards growing in areas subject to a high water table during the growing season have special problems of irrigating in addition to replenishing the soil water. Several reports describing vine conditions and improved vineyard practices cover the salinity-drainage situation in vineyards of the Coachella Valley, California. Halsey (1961); Halsey et al. (1963); and Wolfe (1962) point out that the salinity problem results from the use of a high salt content irrigation water, the presence of impaired drainage and a high water table, and the use of furrow irrigation, all in an arid climate. They suggest that marked improvement will accompany leveling the soil to a flat grade, irrigating frequently by flooding to cover the soil surface, using sufficient water to accomplish leaching, installing tile drains, and irrigating in the fall to reduce salt concentrations even though there may be no need to supply soil water. Conclusions from leaching trials reported by Halsey et al. (1963) point out the sensitivity of grapes to electrical conductivity (EC_e) values greater than 2 to 3 millimho/cm and that grapevines damaged by salinity recover slowly after the salts are leached. Earlier tests of salinity response by Ehlig (1960) outline the sensitivity of vines to chloride, its effects on growth, and the relation of injury to high temperatures. Citrus leaves wetted by sprinkler-applied water can absorb sufficient sodium and chloride to cause leaf burn and defoliation (Harding et al., 1958). Observations by the author in a Thompson Seedless vineyard adjacent to citrus damaged in this manner did not reveal any harm from sprinkling of water from the same source.

Some apprehension about the role of irrigation in vineyards is continually exhibited concerning possible harmful effects on the incidence of disease or insect damage. Considerable observation and widespread experience indicates that insect build-up or control and outbreaks of powdery mildew (*Uncinula necator* Burr.) are not affected by surface applications or by sprinkling (Anonymous, 1956; Meyer, 1963; Winkler, 1962). Vagnoli (1961) reports, however, that the incidence of downy mildew (*Plasmopara viticola* Berl. and De T), was unaffected by irrigation, but powdery mildew and gray mold (*Botrytis cinerea* Pers.) infections increased with an increased water supply. A study of the factors influencing infection of grapes by Nelson (1951) suggests the dangers of sprinkling table grapes near harvest time. He found that infections by *Botrytis cinerea* Pers. require a moisture period and berries high in sugar were more susceptible to invasion.

The influence of sprinkling applications on grape quality have been subject to little study. In a preliminary report Petrucci (1965) suggests that the quality of several table grape varieties and of Thompson Seedless raisins was not different when sprinkling and surface application methods were compared.

VI. SUMMARY OF IRRIGATION RECOMMENDATIONS

A. Methods of Application

No detail of application methods is necessary since this information is generally available from several sources. Comments on the systems used in California vineyards, however, may be useful in summarizing irrigation recommendations.

Where salinity is not a problem, water may be distributed in furrows, either two or three to a middle between rows usually spaced 12 ft apart. Water penetration may be improved in the latter part of the run by cross checking. Two wide-bottom furrows may be used to increase the wetted perimeter and improve an even distribution of water. Table grape vineyards often employ a system of semi-

Table 37I–1. Comparison of annual rainfall, consumptive use, and approximate amounts
of water applied in the commercial production of grapes in various areas
of Calif. and Ariz.

Area	Climatic region*	Range of annual rain-fall†	Amounts of irrigation water‡			Approximate consumptive use
			Wine	Raisin	Table	
		inches	acre-inches/acre			acre-inches/acre
Hot desert	VI	2–8	§	§	48–72	43.6**
Hot	V	6–10	36–42	30–36	42–48	--
Warm	IV	12–18	12–18	§	18–24	18.8–20.9††
Moderately warm	III	12–32	6–12	§	§	20.2††
Moderately cool	II¶	12–32	6–12	§	§	--
Cool	I¶	12–32	4–6	§	§	--

* Climatic regions based on the summation of degree-days above 10C (50 F) for the
period Apr. 1 through Oct. 31:
  Region I   < 2,500 degree-days - cooler parts of coastal valleys of Calif.
  Region II 2,501-3,000 degree-days - middle parts of coastal valleys of Calif.
  Region III 3,001-3,500 degree-days - protected parts of coastal valleys of Calif.
  Region IV 3,501-4,000 degree-days - Southern Calif. interior valleys (except desert);
  intermediate valley area between the San Joaquin and Sacramento Valleys.
  Region V   > 4,000 degree-days - upper San Joaquin and Sacramento Valleys.
  Region VI > 6,000 degree-days - interior desert valleys of Calif., Salt River Valley
  of Ariz.
† Total rainfall, not corrected for evaporation. Rainy season, Sep.-Apr.
‡ After Jacob (1950).   § Grapes not grown for this purpose in this area.   ¶ Many vine-
yards not irrigated.   ** Pillsbury, 1941. †† Hendrickson and Veihmeyer, 1951.

permanent furrows which are renewed annually at the start of the irrigation season.
During the summer the ground is not cultivated and the annual weed growth is
mowed frequently. This practice tends to improve water penetration by reducing
compaction from equipment and tends to minimize dust from settling on the
grapes.

Irrigation by flooding is used on coarse-textured soils where large heads of water
are available and where salinity problems dictate the need for leaching. Basins are
not used because of the high labor requirement and the extensive use of trellising.

Sprinkling is used to good advantage on rolling lands where extensive levelling
would be necessary for surface irrigation, where soils are somewhat shallow and
cannot be leveled, in vineyards where the land was not leveled for irrigation prior
to planting, or where sprinkling is to be used for frost protection.

B. Quantity of Water Needed

The values in Table 37I–1 indicate the amount of water usually applied by
vineyardists in growing grapes for raisins, wine, and table use in California and
Arizona, USA (Jacob, 1950). The annual rainfall figures do not necessarily reflect
the amount of water stored in the soil by winter rains since they may be accumulations of small amounts subject to considerable evaporation. Water use as reported
by Jacob includes the rainfall held in the soil of the root zone and the supplemental irrigation water, but it does not account for surface runoff or deep percolation. For grapes used in wine production he reports a need for 16 to 36 inches/
acre, depending on the climatic region, 24 to 42 inches for raisin production, and
36 to 54 inches for table grapes. For comparison, consumptive use as reported

for similar areas by Hendrickson and Veihmeyer (1951) and by Pillsbury (1941) are included in Table 371-1. That the depths of water used by growers are much greater than the needs as indicated by consumptive use data is a reflection of their desire to make sure of adequate soil water, to provide insurance from damage in case of sudden hot weather, and to meet special conditions imposed with practices such as girdling.

C. Irrigation Water Management

Vineyard irrigation programs must vary widely, depending on the variety grown, the intended use of the crop, and the limitations or influences of the total environment. Certain guides can be given, however, to sound irrigation water management in most vineyards.

1) The soil reservoir should be filled during the dormant season either by rain or supplementary irrigation.

2) Readily available soil water should be maintained during the majority of the growing season for full production. The use of the rate of shoot elongation is suggested as a sensitive indicator of soil water availability.

3) In the production of table fruit, irrigation to minimize the hazards of girdling is strongly recommended.

4) Irrigation practice should be modified in anticipation of ripening of table fruit and raisins as a means of accelerating maturity. This can be done by limiting the length of the irrigation period or by limiting the soil area to which water is applied. Such practice results in some stress in the vine, since not all of the soil in the root zone will be wetted.

5) With wine varieties, irrigations should continue as needed until harvest. If bunch rot is a problem, water may be withheld during the period of rapid berry enlargement and applications restricted during maturation as suggested for table fruit and raisins.

6) In vineyards where vigorous varieties tend to continue growth in the fall irrigations should be withheld after midsummer to promote ripening of the wood.

7) Where salinity is a problem the need for leaching should be satisfied as well as that of replenishing soil water.

LITERATURE CITED

Anonymous. 1956. Sprinklers at the Wente vineyards. Western Fruit Grower 10(4):21–22.

Bioletti, F. T., and E. H. Twight. 1901. Report of conditions in vineyards in portions of the Santa Clara Valley. California Agr. Exp. Sta. Bull. 134. 11 p.

Bioletti, F. T., and E. H. Twight. 1921. Vineyard irrigation in arid climates. California Agr. Exp. Sta. Circ. 228. 4 p.

Ceiko, A. I. 1955. Irrigation of vineyards in the steppe region of the Crimea. (Trans title.) Vinodeli i Vinogradarstvo 4:28–31 (see Hort. Abstr. 24:3663).

Doll, C. C. 1955. Studies of Concord grape roots in loess soil. Amer. Soc. Hort. Sci., Proc. 65:175–182.

Du Toit, M. S., and P. de V. Daniel. 1940. The distribution and nature of Pearl table grape soils. Union S. Africa Dep. Agr. Forest. Sci. Bull. 202.

Ehlig, C. F. 1960. Effects of salinity on four varieties of table grapes grown in sand culture. Amer. Soc. Hort. Sci., Proc. 76:323–331.

Goosen, R. J. 1956. Irrigation of Sultanas alcng the Lower Orange River. Farming S. Africa 32(6):45–48.

Halsey, D. D. 1961. Salinity control in vineyards. Western Fruit Grower 15(5):22.

Halsey, D. D., J. R. Spencer, R. L. Branson, and A. W. Marsh. 1963. Vineyard salinity problems. California Agr. 17(5):2–3.

Harding, R. B., M. P. Miller, and M. Fireman. 1958. Absorption of salts by citrus leaves during sprinkling with water suitable for surface irrigation. Amer. Soc. Hort. Sci., Proc. 71:248–256.

Harmon, F. N., and E. Snyder. 1934. Grape root distribution studies. Amer Soc. Hort. Sci., Proc. 32:370–373.

Hemstreet, C. L. 1961. Irrigating San Bernardino County vineyards. California Agr. Ext. Serv. Pam. p. 1–11.

Hendrickson, A. H., and F. J. Veihmeyer. 1931. Irrigation experiments with grapes. Amer. Soc. Hort. Sci., Proc. 28:151–157.

Hendrickson, A. H., and F. J. Veihmeyer. 1951. Irrigation experiments with grapes. California Agr. Exp. Sta. Bull. 728. 31 p.

Jacob, H. E. 1950. Grape growing in California. California Agr. Ext. Serv. Circ. 116. 80 p.

Kasimatis, A. N. 1958. Irrigation for grapes. Western Fruit Grower 12(4):14–15.

Kattan, A. A. 1963. Seasonal changes in the quality of Concord grapes. Arkansas Farm Res. 12(3): 9.

Kelperis, I. 1963. Skin cracking in the grape variety *Oppsimo souflion* (Trans. title). Athens. Delt. Inst. Ampel. 2:3–6.

Kissler, J. J., C. E. Houston, W. F. Clayton, L. F. Werenfels, and A. N. Kasimatis. 1961. Long-term study on Tokay vineyard irrigation in Lodi area. California Agr. 15(4): 6–7.

Kliewer, W. M. 1964. Influence of environment on metabolism of organic acids and carbohydrates in *Vitis vinifera*: I. Temperature. Plant Physiol. 39:869–879.

Kliewer, W. M., and H. B. Schultz. 1964. Influence of environment on metabolism of organic acids and carbohydrates in *Vitis vinifera*: II. Light. Amer. J. Enol. Vitic. 15:119–129.

Kobayashi, A., M. Kuretani, and H. Oto. 1963. Effects of soil moisture on the growth and nutrient absorption of grapes. (Trans. title). J. Japan. Soc. Hort. Sci. 32:77–84. (See Hort. Abstr. 34:2456.)

LaRosa, W. V., and V. Nielson. 1956. Effect of delay in harvesting on the composition of grapes. Amer. J. Enol. Vitic. 7:105–111.

Larsen, R. P., A. L. Kenworthy, and H. K. Bell. Feb. 1955. Shortages of potash limit grape yields. Better Crops with Plant Foods. 6 p.

Luden, A., and Z. Sadeh. 1963. Studies on the coloration of wines produced from Carignane and Alicante Grenache grapes. Influences of irrigation and influences of temperature conditions. (Abstr.) Israel J. Technol. 1:60. (See Hort. Abstr. 34:4450.)

Luk'janov, A. D. 1961. The efficiency of irrigation of vineyards in the Rostov region. (Trans. title). Vinodelle i Vinogradarstvo 21(4):49–53. (See Hort. Abstr. 32:564.)

Meyer, J. L. 1963. Tractor-tow sprinkler for grapes. Western Fruit Grower 17(4):28–29.

Meynhardt, J. T. 1957. Does spray irrigation cause berry cracking in grapes? Farming S. Africa 33:6–9.

Nelson, K. E. 1951. Factors influencing the infection of table grapes by *Botrytis cinerea* (Pers.). Phytopathology 41(4):319–326.

Osteraas, P. 1962. Vineyard irrigation practices. California Irrig. Inst., Proc. Fresno. 7 p.

Petrucci, V. E. 1965. Storage quality of table grapes grown under overhead and surface irrigation—and quality of raisins from the Thompson Seedless grapes grown under the same conditions. California Irrig. Inst., Proc. Sacramento. 4 p.

Pillsbury, A. F. 1941. Observations on the use of irrigation water in Coachella Valley California. California Agr. Exp. Sta. Bull. 649. 48 p.

Podoleanu, N., and I. Alexandrescu. 1964. Increasing the marketable crop of the variety Afuz-Ali (Trans. title.) Grad. Via Liv. 13(4):28–32. (See Hort. Abstr. 34:6528.)

Prima, S. di. 1960. L'irrigazione del vigneti da uva da tavola nel meridone. Italia Vinicola ed Agraria 50(8):276–280.

Pruitt, W. O., and W. J. Clore. 1956. Irrigating Concord grapes in South Central Washington. Washington State Hort. Ass., Proc. 52nd annual meeting. 174–178.

Raski, D. J., W. H. Hart, and A. N. Kasimatis. 1965. Nematodes and their control in vineyards. California Agr. Exp. Sta. Ext. Serv. Circ. 533:1–23.

Richards, S. J., and A. W. Marsh. 1961. Irrigation based on soil suction measurements. Soil Sci. Soc. America Proc. 25:65–69.

Safran, B. 1957. Grape production in the Union of South Africa with special reference to the export of table grapes. Food Agr. Organ. United Nations (FAO) Rep. 57(7):4368.

Samish, R. M., and S. Lavee. 1958. Spray thinning of grapes with growth regulators. Ktavim 8:273–285.

Spiegel-Roy, P., and B. Bravdo. 1964. Le regime hydrique de la vigne. Bull. Office Int. du Vin (O.I.V.) 37(397):232–248.

Stanhill, G. 1957. The effects of differences in soil moisture status on plant growth: A review and analysis of soil moisture regime experiments. Soil Sci. 84:205–214.

Till, M. R. 1958. What is the value of winter irrigation for Barossa Valley vineyards? J. Dep. Agr. S. Australia 61:489–492.

Tulloch, H. W. 1961. Nurioopta Viticultural Station seeks the answers to problems of dry-land vineyards. J. Agr. S. Australia 64:246–253.

Vaadia, Y., and A. N. Kasimatis. 1961. Vineyard irrigation trials. Amer. J. Enol. Vitic. 12:88–98.

Vagnoli, G. B. T. 1961. Irrigation experiments with wine grapevines in the Marimma area. (Trans. title.) Frutticoltura 23:289–293. (See Hort. Abstr. 31:6049.)

Veihmeyer, F. J., and A. H. Hendrickson. 1950a. Responses of fruit trees and vines to soil moisture. Amer. Soc. Hort. Sci., Proc. 55:11–15.

Veihmeyer, F. J., and A. H. Hendrickson. 1950b. Soil moisture in relation to plant growth. Annu. Rev. Plant Physiol. 1:285–304.

Veihmeyer, F. J., and A. H. Hendrickson. 1957. Grapes and deciduous fruits. California Agr. 11(4):13, 14, 18.

Veihmeyer, F. J., and A. H. Hendrickson. 1960. Essentials of irrigation and cultivation of orchards. California Agr. Exp. Sta. Circ. 486. 28 p.

Winkler, A. J. 1962. General viticulture. Univ. California Press. Berkeley, Los Angeles 612 p.

Wolfe, R. H. 1962. Salt threatens grapes in Coachella Valley. Western Fruit Grower 16(5):39.

Zineberg, M. S., and L. I. Befani. 1962. Irrigation to provide reserve moisture. (Trans. title.) Sadovdstvo 10:34–35. (See Hort. Abstr. 33:2599.)

37

Grapes and Berries

Part II—Strawberries

VICTOR VOTH

University of California, Davis
Santa Ana, California

I. INTRODUCTION

In most areas of the USA, other than California, only supplemental irrigation, if any, is used in commercial strawberry (*Fragaria* sp.) production. More rainfall is encountered in these areas, and the harvest season is short with low production in comparison with California. Strawberries are produced in California and shipped to local and eastern markets for 10 months out of the year. When strawberry plants are required to produce intensively over such a long period, optimum soil water levels are constantly required. Any water stress during the year reduces the total crop and fruit size, and these losses cannot be recovered. This is especially true when an annual planting system is used.

Experiments in areas other than California (Bell and Dawnes, 1961; Fortier, 1961; Simons, 1958, 1961; Newburg, 1960) have consistently shown that supplemental irrigation is advantageous. This is true if the plants are irrigated only during establishment and harvest, as well as when they are irrigated according to need throughout the season. Optimum soil water level during fruit bud formation is shown to increase the number of fruit bud set during fall months (Naumann, 1964; Rom and Dana, 1962). Irrigation during harvest increased total production and fruit size (Simons, 1958, 1961).

II. PLANT CHARACTERISTICS

The strawberry plant is shallow rooted, with 80% to 90% of the roots in the top 12 inches, but some roots penetrate 2 ft or more (Nielsen and Wilhelm, 1957). When daughter plants are set in the field, the first roots produced are laterals from the original roots. As the plant becomes established, the crown enlarges and multiplies, and adventitious roots arise at the base of the crowns directly above the original root system. In order for the newly initiated roots to penetrate the soil, adequate water and satisfactory soil conditions are necessary. Thus, upon enlargement of the root system more crown growth can be subsequently supported. With the matted-row cultural system, moist soil is also necessary in the surface inch to permit runner plants to set and establish themselves as soon as possible so that the daughter plants can make optimum growth.

III. SOIL REQUIREMENTS

Strawberries are grown on all types of soils, from sandy to relatively fine-textured soils. Each soil type has its advantages and disadvantages, depending on cultural system and area. Where culture is very intensive, sandy soils are preferred since they facilitate cultural operations, such as land leveling, soil fumigation, mechanization of planting, runner cutting, laying of plastic mulch, and harvesting soon after irrigation or rain. However, these soils require frequent irrigation. On fine-textured soils, excessive irrigation can reduce plant growth, cause iron chlorosis, and reduce yield.

In all areas of the world, except California, Red Stele, a disease caused by *Phytophthora fragariae*, is a principal limiting factor in strawberry production. The causal organism flourishes in water and wet soils. Any soil that has poor drainage is not desirable for strawberry production. Varieties have been developed which are resistant to this disease. Red Stele is not usually prevalent where irrigation is necessary for strawberry production, since sandy well-drained soils are usually involved and excessive soil water is not a problem.

IV. IRRIGATION METHODS

Furrow or overhead sprinkling systems are used for strawberries. Sprinkling is the more commonly used method in areas where land cannot be leveled and only supplemental irrigation is used. Sprinkling often results in a higher incidence of fungus and bacterial leaf disease (Kennedy and King, 1960). Sprinkler irrigation during harvest causes a much higher rate of fruit rot and also makes fruit softer. Thus, fruit harvested from plantings under sprinkler irrigation cannot be shipped as far and has a shorter storage and shelf life. This problem can be minimized by the use of polyethylene bed mulch, fungicides, and frequent harvest.

Both furrow and overhead sprinkling systems are used on the same plantings in California. Sprinkler irrigation is often employed during the establishment period and winter months. Furrow irrigation is used exclusively during the harvest season. Sprinklers are used exclusively in strawberry nurseries for plant production. Occasionally surface soil compaction and associated crown disease problems cause difficulties in newly set plants under sprinkler irrigation on fine-textured soils. Sprinkler irrigation has been shown (Carolus, 1964) to be beneficial for early spring frost protection and is being practiced by growers, especially in the early areas of California (*see* chapter 54). Sprinklers may also serve to reduce fruit damage when temperatures become too high during harvest.

For furrow irrigation, the strawberry beds must be leveled to zero grade with no slope in the furrows. In order that the irrigation water be raised high enough in the furrow to wet the bed as completely and rapidly as possible, furrows are usually limited to 150 or 200 ft. In most growing areas of California, the double row bed is used, varying in width from 38 to 42 inches from the center of the bed to the center of the adjacent bed. When the water is raised to a high level in the furrow, it leaches the accumulated salt to the center of the bed and below the root zone.

V.   WATER REQUIREMENTS

The amount of irrigation of water required for strawberry production under California conditions varies from 4 to 8 ft/year, depending on soil type and area.

Water quality and soil salinity are also very important factors because strawberry plants are one of the crops least tolerant to salinity (Ehlig, 1961). Ehlig and Bernstein (1958, 1961) have shown that when the electrical conductivity (EC_e) of the soil saturation extract is higher than 2.0 mmho, yield is reduced and plant injury occurs under surface irrigation However, water containing as much as 800 ppm of total salts may be used with very little depression in yield on annual plantings under furrow irrigation. When plantings are to be kept for more than 1 or 2 years, 10 or more inches of rain, or its equivalent, in sprinkler irrigation during the winter months is required to remove the accumulated salt (Brown and Voth, 1955).

VI.   IRRIGATION SCHEDULING

Strawberries, due to their shallow root zone, respond to frequent irrigations. Experiments at Michigan, USA (Carolus, 1964) indicate that the same amount of water applied in 8 applications vs. 2 applications gave an increase in fruit size and yield.

Yield responses of strawberries to soil water may also be affected by the use of polyethylene films as a surface mulch. If a clear polyethylene is used, Cannell et al. (1961) reported that maximum yields were obtained, if the crop were irrigated when tensiometers at the 8-inch depth registered 0.2 bar of suction. Without the clear polyethylene mulch, highest yields occurred under a drier treatment when the tensiometers registered 0.8 bar of suction. There are at least two reasons for this interaction: (i) Under the frequent irrigation required to maintain 0.2 bar suction, the soil surface of the beds without the polyethylene mulch was so wet that incidence of fruit rot was high, but with the mulch present fruit rot was reduced and (ii) soil temperature under the clear polyethylene mulch was much higher than the unmulched soil at a given soil water level (Voth and Bringhurst, 1962) and this promotes growth.

In summary, strawberry plants can tolerate some water stress and survive but should not be subjected to water stress if large fruit size and high total production are to be realized.

LITERATURE CITED

Bell, H. K., and J. D. Dawnes. 1961. Production and size of Robinson strawberries as influenced by plant spacing irrigating and fertilizers. Michigan. Agr. Exp. Sta. Quart. Bull. 44:166–170.

Brown, J. G., and Victor Voth. 1955. Salt damage to strawberries. California. Agr. 9(8):11–12.

Cannell, G. H., Victor Voth, R. S. Bringhurst, and E. L. Proebsting. 1961. The influence of irrigation level and application methods, polyethylene mulch and nitrogen fertilization on strawberry. Amer. Soc. Hort. Sci., Proc. 78:281–291.

Carolus, R. L. 1964. Principles of air conditioning fruit crops with irrigation. Michigan State Hort. Soc., Annu. Rep. 94. p. 39–44.

Ehlig, C. F. 1961. Salt tolerance of strawberries. Under sprinkler irrigation. Amer. Soc. Hort. Sci., Proc. 77:376–379.

Ehlig, C. F., and L. Bernstein. 1958. Salt tolerance of strawberries. Amer. Soc. Hort. Sci., Proc. 72:198–206.

Fortier, F. 1961. L'Irrigation De Fraisier Au Quebec. Rev d'Oka. 35:67–72.

Kennedy, B. W., and T. H. King. 1960. Angular leafspot, a new disease of strawberry. Phytopathology 50:(9):641–42.

Naumann, W. D. 1964. Bewasserungstermin und Blutenknospen differenziering bei Erdbeeren. (Time of irrigation and flower differentiation in strawberries.) Garten-bauwiss 29:21–30.

Newburg, W. B. 1960. Strawberry irrigation research, 3rd progress report. Oregon State Hort. Soc., Annu. Rep. 52. p. 95–97.

Nielson, P. E., and S. Wilhelm. 1957. Some anatomic aspects of the strawberry root. Hilgardia 26:621–641.

Rom, R. C., and M. N. Dana. 1962. Development and nutrition of strawberry plants prior to fruit bud differentiation. Amer. Soc. Hort. Sci., Proc. 81:265–73.

Simons, R. K. 1958. Response of Howard Premier and Vermillion varieties of straw-berries to supplemental irrigation. Amer. Soc. Hort. Sci., Proc. 71:216–223.

Simons, R. K. 1961. The influence of irrigation upon growth and yield of strawberries. Illinois State Acad. Sci., Trans. 63:93–100. (Biol. Abstr. 42:481. 1961.)

Voth, Victor, and R. S. Bringhurst. 1962. Early mulched strawberries. California Agr. 16 (2):14–15.

38

Coffee, Tea, Cacao, and Tobacco[1]

H. C. PEREIRA

Agricultural Research Council of Central Africa
Salisbury, Rhodesia

An important common factor among these four major crops is that each have rather exacting requirements as to climates and soils. As a result the main plantation areas have been developed only in favorable environments so that irrigation does not yet play a significant part in world production of any one crop.

All four are, however, high-priced cash crops in which modest crop increases of the order of 30% to 50% enable them to carry the costs of irrigation. In all of them, progress in the control of pests and diseases and improvements in planting material, processing methods, and soil conservation have left drouth incidence as the most obvious limiting factor in crop production. Irrigation has therefore been attempted with each crop, both to increase yield and to extend the range of climate in which it may be grown.

A substantial amount of irrigation research has been done on both coffee and tobacco, while very little indeed has been done on either tea or cacao.

I. THE IRRIGATION OF COFFEE

Two species, *Coffea arabica* and *Coffea robusta,* provide the bulk of the world's commodity crop: Only the former is irrigated on any scale. *C. robusta* commands a substantially lower price and is usually grown in hotter and wetter areas. *C. arabica,* which provides the critical highly aromatic contribution to most commercial blends of coffee, is indigenous to the high altitude African tropics as a component of the undergrowth of montane forest in the Ethiopian Highlands. Very little coffee is exported from the country of its origin. There is substantial development of coffee plantations in Kenya astride the equator at an altitude range of 4,500 to 6,200 ft above sea level, while the world's main supply of top quality *C. arabica* comes from homoclimes in South America—notably from Colombia where it is grown at similar altitudes. The greatest volume, but mainly of the lower qualities, comes from Brazil. *C. arabica* also flourishes at 2,000 ft above sea level under an island climate in the Blue Mountains of Jamaica.

Climatic limitations are a mean monthly minimum temperature in the meteorological screen of about 10C (50F), and a daily average of more than 7 hours of sun annually. Rainfall requirements are less critical; successful plantations are established in annual rainfalls ranging from 30 to 60 inches or more. This tolerance is due largely to the ability of the vigorously developed root systems of *C. arabica*

[1] Special appreciation is extended to Mr. J. Nick Jones, Jr., Agricultural Engineer, Agricultural Research Service, US Department of Agriculture, Virginia Polytechnic Institute, Blacksburg, Va., for his assistance in preparing the section concerned with irrigation of tobacco.

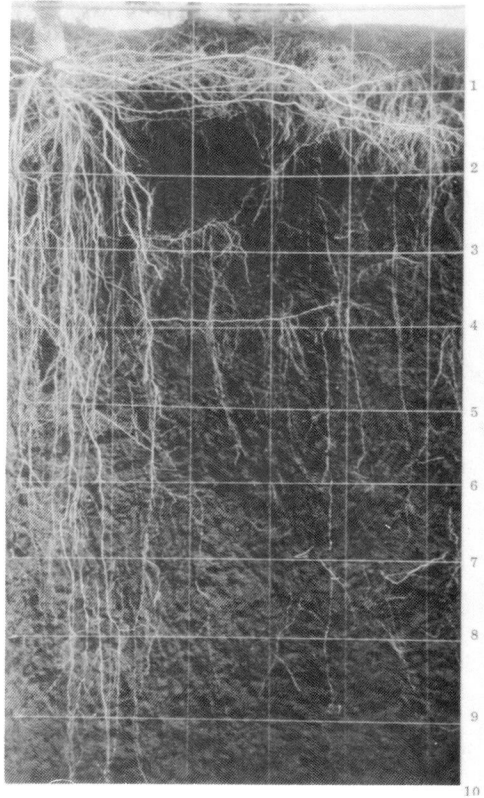

Fig. 38–1. The root system *C. arabica*. Roots of a mature *C. arabica* tree fill the first 10 ft of soil and seasonally reduce it to wilting point. Root transect by R. Rayner at Coffee Research Station, Ruiru, Kenya.

trees to exploit the characteristically deep, well-drained soils on which the crop is grown.

The root systems and water requirements have been studied most intensively in Kenya. Figure 38–1 illustrates a typical root system. Ten years of continuous soil water sampling programs have established the realization that this root system dries the entire 10-ft profile to the wilting point. In this lava-derived Latosol the wilting of coffee in the field agrees well with that of sunflower (*Helianthus annus* L.) seedlings in the laboratory and with the 15-atm pressure plate determination (Pereira, 1957).

Even in a main production area the water balance of coffee without irrigation was shown by Pereira (1957) to be erratic (Fig. 38–2). Biennial bearing in which excessive crops cause leaf shedding, starch deficiency, and root die-back and, thus, depress the yield for the following year, serves to increase the variability of yields. The water balance set out in Fig. 38–2 was continued by Wallis (1963) in a replicated factorial study of irrigation, the results of which are summarized in Table 38–1. The coffee was unshaded and the trees were about 50 years old. The improvement in the size and appearance of coffee beans which resulted from irrigation is of much commercial importance since there is at present in the world an over-production of the smaller sizes. No effects of irrigation on the properties of coffee liquor have been detected in numerous samplings.

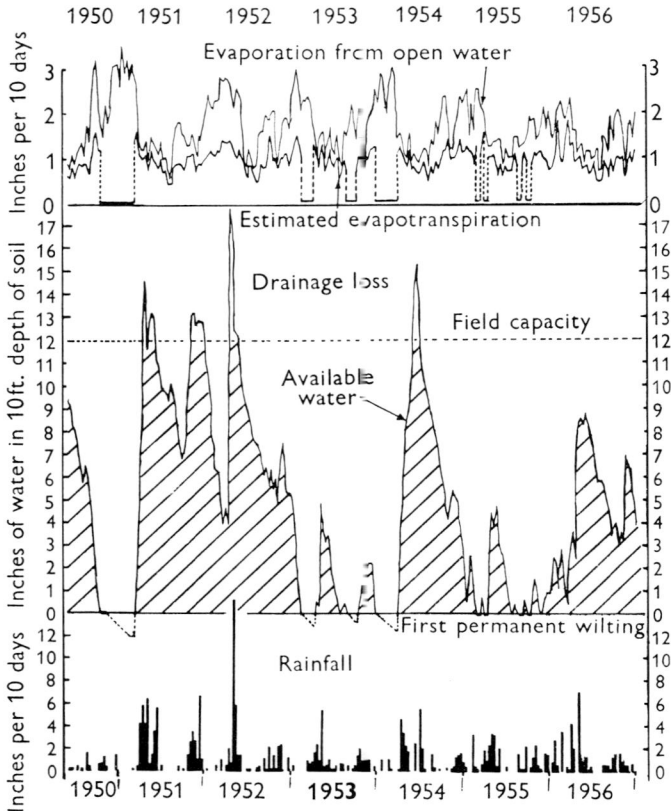

Fig. 38–2. Soil water availability under a plantation of mature *C. arabica*. Water balance computed at 10-day intervals for 6 years and confirmed by soil sampling to a 10-ft depth (Pereira, 1957).

In Brazil, coffee is sometimes irrigated either by furrow or by overhead spray, but only a very small part of the total production is affected. Texeira (1952) reported an observation trial in a 55-inch rainfall area in which 10.5 to 22 inches of supplementary furrow irrigation, given at irregular intervals, effectively doubled the yields. Medcalf et al. (1955) reported from Sao Paulo a 50% increase in yield of mature coffee trees over 3 years on a sandy loam soil. The yield response paid for the entire cost of the irrigation equipment and left a net profit of $60/acre. Similar responses have been reported from commercial estates in East Africa, and a substantial increase in the use of irrigation in coffee is to be expected on a world scale. The late start is partially due to the failure of many earlier irrigation trials to measure irrigation requirements adequately. Trials based on supplementing monthly rainfall totals to arbitrarily fixed amounts have proven quite inadequate to meet water requirements at periods of stress. This is well illustrated by the analysis of the results of a replicated factorial trial in Tanganyika. The coffee showed rather poor responses over 15 years when rainfall was supplemented by irrigation to total 5 or 10 mm/month (2 or 4 inches) (Robinson and Mitchell, 1964). Soil water deficits were found by subsequent calculation by

Table 38–1. Response of C. *arabica* to supplementary irrigation (from Wallis, 1963)

Year	Rainfall	Irrigation	Large (A) beans		Total Clean Coffee	
			Irrigated	Control	Irrigated	Control
	——inches——		————————lb/acre————————			
1957	53.2	14.5	543	548	1,192	1,229
1958	63.7	3.5	125	96	254	196
1959	26.7	20.0	1,015	853	2,054	2,234
1960	38.0	19.0	1,650	688	3,275	1,808
1961	33.5	23.0	673	264	1,894	1,818
Five-year totals			4,006	2,449	8,669	7,285
Total response to irrigation			+63.5%		+11.9%	

Pereira (1963) to have been substantial at the first level while losses by drainage were excessive at the second level (Table 38–2). Blore (1966) has derived improved methods of control.

Useful evidence on the effects of irrigation on the root systems of C. *arabica* trees has recently been reported by Bull (1963). Root systems of 20 trees from the trial summarized in Table 38–2 were excavated, washed free of soil, measured, dried, and weighed. Irrigation appeared to have reduced the depth of penetration of main, vertical roots by 20% and also reduced secondary roots in deep soil. Irrigation had, however, increased the number, length, and weight of both primary and secondary lateral roots. The number of secondary roots on the laterals was decreased by irrigation. Heavy mulching with grass produced an increase in depth of vertical roots and in size and density of the lateral root systems. The combination of both irrigation and grass mulching produced a 30% increase in the dry weights of the total root system with a depth slightly greater than that of untreated trees.

Continuous replenishment of the soil water is not, however, advisable in the monsoon-type climates in which coffee is grown. A deficit of some 6 inches immediately before the main rain season is advisable to permit absorption of heavy storms and to reduce both soil erosion and irrigation costs. Control of flowering and hence better organization of picking is also improved by such a break in water supplies.

The many contradictory reports about coffee flowering appear to be due to the confusion of two effects: (i) the long-term sequence of increase in leaf and hence in starch reserves, resulting in the initiation of additional flowerbuds a year in advance of flowering and (ii) the trigger mechanism of the opening of flowerbuds which are already mature. Coffee flowerbuds, which are formed only when plants are subjected to short days, after differentiation grow slowly for about 2 months to a size of 6 to 8 mm and then stop growing for many weeks and months. Rain or irrigation is known to induce anthesis of many buds within 8 to 12 days, depending on the temperature.

Rayner studied flowering of small groups of individual trees with and without irrigation for 8 years (R. W. Rayner, 1959. *File Rep. Coffee Res. Sta. Ruiru, Kenya*). Very irregular effects were observed in the first 3 years after which the irrigated trees settled to a regime of heavier and more regular flowering. Mes (1957) postulated that it was predominately water stress which keeps flower buds dormant. She submerged branches of coffee trees in water at controlled temperatures. The resultant local reduction of water stress could cause flowering about

Table 38–2. Water-balance of a conventional design of irrigation experiment in *C. arabica* (Pereira, 1963)

	Jan	Feb	March	April	May	June	July	Aug	Sept	Oct	Nov	Dec	Totals
Mean rainfall during experiment, inches	1.4	2.2	2.6	15.1	21.8	4.3	1.4	0.9	1.4	2.0	1.6	2.2	56.9
Mean monthly irrigations, I_1 inches	0.7	0.8	0.7	-	-	0.2	0.5	1.0	1.2	0.8	0.7	0.6	7.2
I_2	2.8	2.4	1.9	-	-	0.8	2.3	3.0	2.8	2.3	2.4	2.2	22.9

5-year mean values in inches of deficits or surpluses at the end of each calendar month summarized from a continuous water budget

	Jan	Feb	March	April	May	June	July	Aug	Sept	Oct	Nov	Dec
					Lost by drainage							
I_0 No irrigation	−14.0	−15.7	−14.5	−6.5	+12.8	+2.0	−3.0	−5.3	−7.1	−10.2	−11.8	−13.0
					Lost by drainage							
I_1 Supplementing of rainfall to total of 2 inches/month	−10.7	−12.3	−13.4	−4.3	+14.1	+2.0	−1.2	−2.5	−3.4	−5.8	−7.3	−8.0
				Lost by drainage								
I_3 Supplementing of rainfall to total of 4 inches/month	−1.7	−1.8	−1.9	+9.8	+18.4	+2.3	+0.9	+0.7	+1.0	+1.0	−0.8	−0.5

* Available water storage in root range at 1/3 atm field capacity is 20 inches.

10 days later, even in trees whose roots were well irrigated. She reported that these experiments confirmed plantation observations that overhead spray has more effect than furrow irrigation on the opening of flowers. However, other investigators (Porteres, 1946; Piringer and Borthwick, 1955) reported flowering of coffee after soil irrigation subsequent to a dry period. This view is supported by more recent work of Alvim (1960a) in Peru who found that the flower buds remained dormant on coffee plants which were irrigated at relatively short intervals to maintain the water content close to field capacity. Rain or irrigation produced flowering only when preceded by a period of water stress. He suggests that water stress leads to the removal of a growth inhibitor responsible for bud dormancy and that a water stress requirement may be just as important for coffee flowering as the chilling requirement is for many species of Temperate Zone plants.

Soil erosion is reduced by the use of overhead, rather than furrow irrigation, since the porous soils in which coffee thrives best tend to deteriorate rather rapidly under furrow applications. Where irrigation water is severely limited, as in Kenya, distribution to basins under trees by plastic pipes perforated at the tree interval has been successfully practiced on a few estates.

Continued world over-production of coffee throws increasing emphasis on size and quality of coffee beans. It is probable that the use of overhead irrigation will therefore increase in this crop.

II. THE IRRIGATION OF TEA

More than one-half of the world's tea exports come from India and Ceylon where tea is grown under conditions of high rainfall ranging from 60 to 200 inches or more per annum. Tea (*Thea sinensis* L.) is also grown on a smaller scale over a wide range of cool subtropics and high altitude tropics, the limitations being, effectively, some 50 inches of rainfall and freedom from frosts in the growing season. Irrigation becomes appropriate only as the environment becomes marginal for tea. In such marginal environments, however, low yield is partially compensated by higher quality. The main tea research station at Tocklai, India has only recently begun irrigation trials.

Plantation textbooks are unanimous that tea is shallow rooted; in a recent research report from India, Barua and Dutta (1961) considered it necessary to study only the first 70-cm depth (28 inches). In East and Central Africa, however, root excavations have shown tea to root very deeply. Laycock and Wood (1963) traced tea roots to an 18-ft depth. Gypsum blocks placed at a 17.5-ft depth showed seasonal drying of soil under unshaded tea (Fig. 38–3). Roots of 7-year-old trees have been shown to reach a 10-ft depth in Kenya (Kerfoot, 1962). (*see* Fig. 38–4.)

Although it is highly probable that irrigation of tea on a peasant scale has been practiced in the Far East, the first record of a tea estate depending entirely on irrigation appears to be from Rhodesia. Here 232 acres of tea have been furrow irrigated since 1929 under an erratic annual rainfall averaging 27 inches and sometimes falling as low as 12 inches/annum. Yields have fluctuated widely in spite of irrigation and have been low but profitable. A nearby estate enjoying 46 inches of rainfall has 700 acres of young tea of which 600 are irrigated. No substantial yield differences have yet been observed as a result of the irrigation.

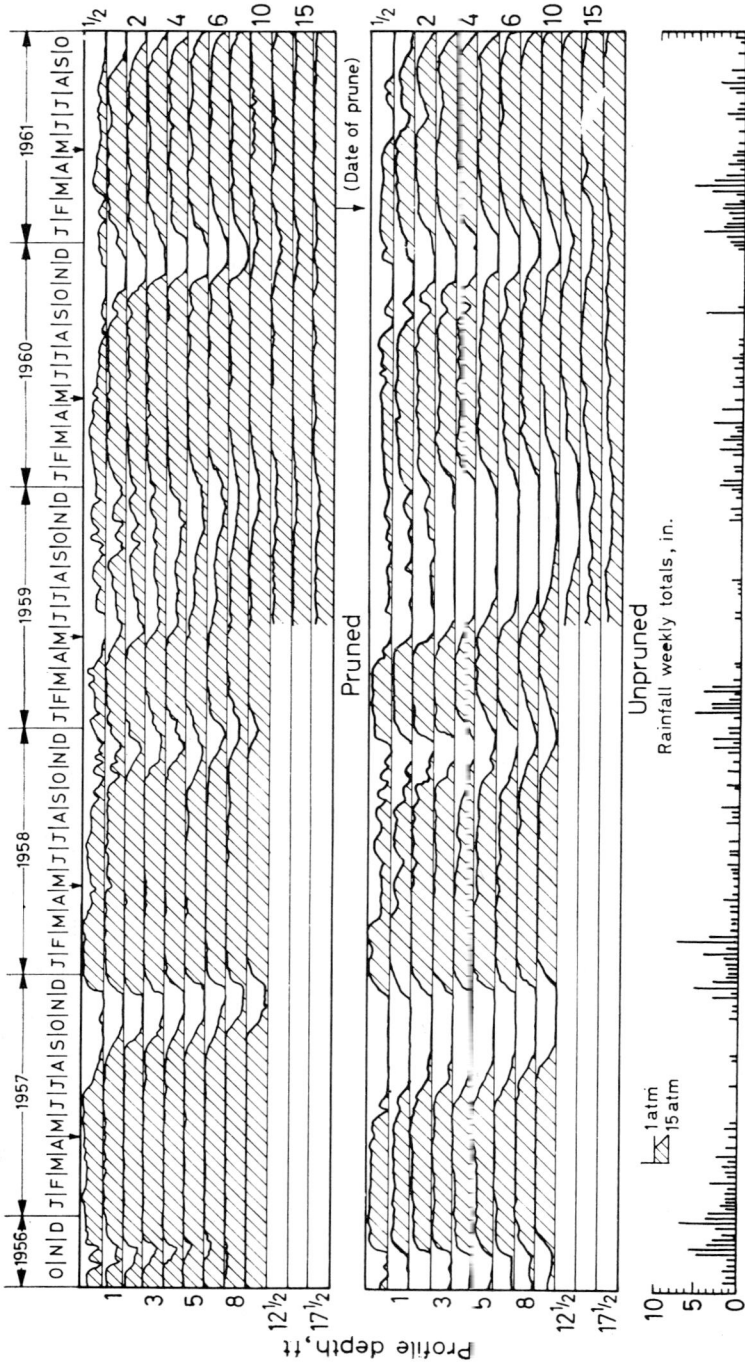

Fig. 38–3. Use by tea bushes of water from deep soils. Seasonal soil water fluctuation under pruned and unpruned tea. The shaded areas indicate the depth and duration of availability of soil water between the suctions of one-third and 15 atm (Laycock and Wood, 1963).

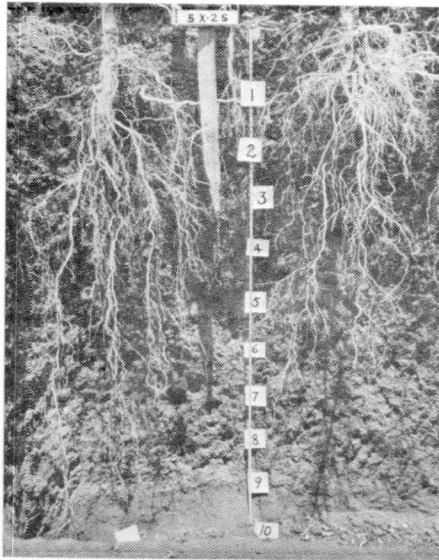

Fig. 38–4. Root systems of young tea bushes. Seven-year-old tea in Kenya rooting freely to a 10-ft depth. Study by O. Kerfoot.

On both estates irrigation is supplied without effective measurement or means of estimating irrigation requirement. Over 1,000 acres of tea in this area are irrigated.

Reports from commercial tea estates in northeast India have described a number of practical exercises in the use of supplementary irrigation. Increases of 20 to 30% are claimed in the annual yield of tea gardens on the basis of comparisons with earlier yields. Evidence from such estate production records is impossible to extract from seasonal effects, but the abundant water supplies and suitable terrain will favor irrigation development if economic returns can be proven. Advisory articles base the discussion of costs on an annual dry season application of 8 to 10 inches of water. Installations for such supplementary irrigation have been established only on a small proportion of estates. A single estate in Assam, India now irrigates 3,000 acres of tea (T. E. Rogers, June 1958; Jan. 1959; Jan. and Nov. 1961; Jan., Feb., April 1962. Articles in the *Tea and Rubber Mail.*)

Opportunities for irrigation appear to be better in high rainfall areas where a dry season occurs annually. In Malawi a small, but successful tea industry is maintained by an annual 70 inches of rain in 6 months. Encouraging responses to dry season irrigation have been obtained from observation plots. Storage dams would, however, be needed. Such irrigation development is probable only on well-established plantations since new tea estates with their elaborate factories require a heavy capital investment of ca. $1,400/acre. It is improbable that a substantial investment in the irrigation of tea can be justified while existing successful plantation districts have large areas of high rainfall and suitable soils remaining to be developed, e.g., in East Africa where a recent hydrological study of a tea estate showed only one 10-day period in 3 years to be without rainfall (Pereira et al., 1962).

III. THE IRRIGATION OF CACAO

The tree, *Theobroma cacao* L., was originally found in Central America where, even in pre-Columbian times, a beverage was made from the seeds. The cocoa, or cacao, crop is now planted in wet, tropical lowlands around the world. Ghana is the largest exporter of cocoa beans, followed by Brazil.

The climatic requirements are high humidity and fairly high temperature. A well-distributed 60 inches of rainfall is a practical minimum, and all main areas of production have rain every month in the year. The Cacao Center of Turrialba, the Inter-American Institute of Agricultural Sciences, receives 104 inches of rainfall per annum.

Cocoa appears to be extremely sensitive to water supply and even in high rainfall areas Murray and Herklots (1955) showed that the annual pattern of crop production reflects previous seasonal rainfall. The cacao tree is reported to have a shallow rooting habit although adequate studies of the root systems of mature trees have not yet been reported. Hurd (1959) made a growth analysis study of the sapling stages and described a network of lateral roots in the top foot of soil with tap roots and other verticals reaching a depth of 18 inches. The water table at this site, however, rose seasonally to within 14 inches of the surface.

The water requirements of cocoa have only been studied in detail in the case of potted seedlings. In the Ivory Coast, where cacao plantations are established in a range of climates, including the marginal areas where rainfall of only 1.25 m leaves 3 months of water stress, Lemée (1955) used a sand, a sandy loam, and a clay soil in a very detailed study of the effects of drouth on the growth of seedlings. He found optimum growth to be near the moisture equivalent of each soil. When only one-sixth of the water available at moisture equivalent had been used up, transpiration, carbon assimilation, and growth were all sharply reduced. Growth became negligible when only two-thirds of the available water storage was exhausted.

Alvim (1960b), also working on seedlings in pots, concluded that yield would be reduced, if soil water fell below 50% to 60% of the total available.

In a study of the effects of cocoa-plantation spacings in Nigeria, Kowal (1959) reported that closer tree spacings (5 by 5 ft to 7.5 by 7.5 ft) which gave closed canopies resulted in higher levels of both water and nitrates as compared with those found on the soils under wider spacing up to 15 by 15 ft. Closer spacing also gave higher yields. Soil water levels remained at or near field capacity for 10 months of the year.

Murray (1960, 1964) found in Trinidad that irrigation could ameliorate but not prevent yield losses from drouth.

At present the only detailed field study of the irrigation of cacao to be published is concerned with the establishment phase of a new plantation. Smith (1964), reporting an irrigation trial on unshaded 6-month-old cocoa, showed that without irrigation the water content of the first foot of soil varied between one-third and two-thirds of the capacity for available water and was near the lower level for 4 months in succession. Irrigation to maintain soil at 65% of available water capacity more than doubled the number of pods and greatly increased growth of leaves, stems, and branches (Table 38–3).

Where cocoa is grown in climates having a dry season which renders production

Table 38–3. Effect of irrigation on unshaded 2-year-old cocoa trees (from Smith, 1964)

	Length of trunk and branches	Leaf area	No. of pods
	cm	cm^2	
No irrigation	612	2600	7
Irrigation to 65% of capacity for available water in the first foot of soil	815	4000	18

marginal, furrow irrigation has been practiced on a commercial scale, but quantitative data are not available. Successful experimental plantings of cocoa dependent on furrow irrigation in hot, dry climates have been made at Bagamoyo in Tanganyika and in the Shire Valley of Malawi, at Occumare de la Costa in Venezuela, and in the coastal districts of Peru, but quantitative water requirements are not yet known. Further large plantation developments of cocoa can be expected in wet, tropical areas, but it appears unlikely that there will be any substantial development of cocoa plantations dependent upon irrigation.

Research within the past decade has, however, initiated a major change in the field practice of growing cocoa under shade trees. Removal of shade trees and substitution of insecticides for the entomological control formerly achieved by shade, has led to increased yields and to the first positive and substantial responses to fertilizers. Such a major change in crop environment requires fresh assessment of all field practices. The inclusion of supplementary irrigation in such studies is already justified in those cocoa-plantation climates in which even a brief dry season occurs.

IV.   THE IRRIGATION OF TOBACCO

The cultivated species of tobacco (*Nicotiana tabacum* L. and *N. rustica* L.) thrive in climates drier than those required for the perennial beverage crops already discussed in this chapter. Although sensitive to frost, tobacco production is worldwide. Garner (1946) indicated that the crop is grown as far north as Sweden at about 60° north latitude and as far south as New Zealand at approximately 40° south latitude; however, 90% of the crop is grown north of the equator.

Tobacco is grown on varying soil types the world over, ranging from coarse sands to heavy clays. The soil is one of the most important factors in determining the type and quality of tobacco produced. However, adequate soil water, proper aeration, and plant nutrients are cardinal requirements for its exacting growth and for chemical composition indicative of quality. As an annual crop, tobacco is grown successfully in regions with an annual rainfall of 20 to 45 inches. In the USA, normal rainfall in the various tobacco areas during the main 3-month growing period averages about 12.5 inches. The drouth tolerance of tobacco is due, in part, to its fairly deep and vigorous root system. Figure 38–5 illustrates a recent study by Robertson of the root development of the flue-cured type of tobacco in Rhodesia. (A. G. Robertson, 1963. *File Rep. Tobacco Res. Board, Rhodesia*)

Drouth during the growing season does, however, result in thickened leaves of

Fig. 38–5. Root system of tobacco in Rhodesia. The depths are indicated at 6-inch vertical intervals. This is a flue-cured variety 'Kutsaga 54' grown in sandy soil without
irrigation. Study by A. G. Robertson, plant physiologist, Tobacco Res. Sta., Kutsaga,
Rhodesia.

a lower quality. Most of the world's tobacco crop is grown under conditions of
moderate drouth hazard which could be expected to result in a widespread use
of supplemental irrigation. Although irrigation has been an increasingly important
feature of the tobacco industry over the past 15 years, water is primarily applied,
in the world's main tobacco-growing areas, to growing seedlings and to plants
at transplanting time; but only a small proportion of the main crop is grown with
the help of irrigation. Australia provides an exception in that the small but
flourishing tobacco production depends entirely on rather heavy supplemental
irrigation.

   In the tobacco-growing areas of the USA, rainfall is often deficient or poorly
distributed and many growers are utilizing a supplemental irrigation system to
apply water at crucial stages of growth to increase yield and maintain quality.

During the past decade, a rapid expansion of irrigated tobacco acreage in the USA occurred in the flue-cured areas with only slight increases for the air-cured types. Difficulties have arisen mainly from overwatering, especially when heavy rain follows an irrigation; but two or three well-controlled irrigation applications usually improve the yield, quality, and uniformity of the crop.

McMurtrey (1961) noted the importance of adequate soil water in producing leaf of high quality. However, he suggested that atmospheric humidity, shading associated with cloudiness, and air movement are all part of the complex that controls plant growth and development.

Harrison and Brothers (1964) in Florida, USA, indicated that, in most years, properly irrigated tobacco will produce better quality and higher yields than nonirrigated tobacco. Anderson in Connecticut, USA (1952) stated that a season with too little or poorly distributed rainfall occurs about every 2 or 3 years and that the defects of a dry weather crop can be overcome by proper irrigation. He reported a 5-year average increase of 37% in crop value through the use of irrigation. Wilson and Van Bavel (1954), in irrigation tests with flue-cured tobacco in North Carolina, USA in 1951 to 1954, obtained increases in crop value from irrigation in each of 4 years ranging from $97/acre to $443/acre. Irrigation demonstrations conducted in South Carolina, USA from 1951 through 1954 by Lynn et al. (1958) gave increases in yield averaging 375 lb/acre from four irrigations. Clark and Myers (1956) in a 3-year study with flue-cured tobacco in Florida, USA found that medium rates of irrigation corresponding to the daily water use (0.06 to 0.25 inch) of the crop at various periods of its growth were more suitable than higher or lower rates. The most critical period for irrigation is from the "knee-high" to the bloom stages, e.g., in two 3-year studies reported in Virginia, USA for Burley and flue-cured tobacco, respectively, Jones et al. (1957) obtained increases in both yield and value of Burley tobacco by irrigation from knee-high through bloom in 1954 to 1956, but found no significant additional benefit from full-season irrigation. Rogers et al. (1961) obtained similar results from flue-cured tobacco in 1958 to 1960, but with small additional advantages from irrigation at transplanting.

Parks et al. (1963) reported that irrigation of Burley tobacco in the Central Basin of Tennessee, USA could be expected to increase yields by 200 to 300 lb/acre in 4 out of 5 years while in the East Tennessee Valley the practice of irrigation would be questionable, since a response to irrigation was obtained in only 2 out of 7 years.

Studies of commercial tobacco irrigation in the USA confirmed that similar responses were obtained. Williams (1957) compared 55 pairs of farms in Virginia, USA on which tobacco was grown with and without irrigation. The average responses to 3 inches of irrigation water was a yield increase of 250 lb of flue-cured tobacco. Miles (1957) reported similar gains in Georgia, USA with improvement in quality.

Jones et al. (1960), in a review of principles of tobacco irrigation, estimate the average yield increase as 10 to 20%, and the rise in quality worth $2 to $5 for 100 lb of leaf. The main limitation is the scarcity of water supplies which have to be obtained, in most cases, by impounding surface runoff. In spite of these difficulties, the general profitability of supplemental irrigation in the USA is well demonstrated by the 10-fold increase in the practice during the past decade.

In Virginia, USA for example, a survey in 1964 of flue-cured tobacco farms indicated that approximately 24% of the farmers are now irrigating; this is 45% of the total allotted acreage. This acreage compares to 12% for fire-cured and < 1% for Burley.

In Canada, Walker and Vickery (1959) compared five different irrigation regimes for 3 years, using evaporation estimates and gypsum blocks to indicate when water was required. Their best treatment showed an average yield response of 20%, which together with a beneficial effect on quality, increased the value of the yield per acre by one-third as compared with nonirrigated controls.

A major part of the skill and science of growing tobacco is in meeting complex and rapidly changing quality specifications. The ability to manipulate water supplies introduces a powerful but complex new form of quality control which calls for scientific study over a wide range of field conditions. Reports on the effects of irrigation on tobacco quality are most numerous from the USA, but detailed studies have been published from Australia, Canada, Czechoslovakia, France, Germany, Hungary, India, Japan, and the Philippines; e.g., Amarell (1957), Australian Tobacco Growers (1963), Hitier et al. (1960), Marseu and Ille (1961), Nishikawa and Ichishima (1954), Peele et al. (1960), and Ramakrishna Kurup and Sastry (1962). These reports have one outstanding result in common. All agree that excessive irrigation reduces quality of chemical content, texture, color, and aroma. Quantitative estimation of irrigation need is therefore important, although the hazard of heavy rain following irrigation remains more serious in tobacco than in other crops.

Hitier et al. (1959), in a very thorough 4-year study in France, found little advantage in irrigation. Amarell (1957), reporting a 3-year experiment at Dresden, Germany claimed a better burning quality in spite of reduced yield from over-irrigation. In drier climates supplemental water will often hasten maturity; a point of some importance in Canada for the avoidance of late summer frosts.

All reports agree that controlled irrigation reduces leaf thickness, nitrogen content, and total alkaloid while several workers report an increase in sugars. Effects on mineral content depend on quality of irrigation water and hence reports are variable. The increase of potassium and chloride content is noted in several cases, particularly from Australia.

Economic factors limiting the use of irrigation are, in the USA and in Europe, the high proportion of crops grown on very small plots of an acre or less for which capital investment in irrigation equipment tends to be excessive. In Africa much larger fields are planted, but a very wide rotation of one tobacco crop in 5 to 7 years of grass or cereals has to be followed to reduce nematode infestation. This raises the cost of applying water to the tobacco. Where tobacco is grown as part of a mixed farming enterprise with other crops or pastures on which irrigation is practiced, the possession of irrigation equipment can be of great advantage for supplemental use during the period of maximum growth. The installation of permanent irrigation equipment is established as normal practice for tobacco only in special circumstances such as the intensive growing of cigar wrapper leaf in Florida and the growing of tobacco in semiarid climates as in Australia.

LITERATURE CITED

Alvim, P. de T. 1960a. Moisture stress as a requirement for flowering in coffee. Science 132:354.

Alvim, P. de T. 1960b. Water requirements of cocoa. Phyton 15:79–89.

Amarell, H. 1957. Effect of supplementary irrigation and shade tests on yield, quality and chemical composition of tobacco (German text). Inst. für. Tabakforchung, Dresden. (1956/8)3–9:203–246.

Anderson, J. B. 1952. Growing tobacco in Connecticut. Conn. Agr. Exp. Sta. Bull. 564.

Australian Tobacco Growers. 1963. Australian Tobacco Growers Bull. Vol. 5, no. 6.

Barua, D. N., and K. N. Dutta. 1961. Root growth of cultivated tea in the presence of shade trees and nitrogenous manure. Empire J. Exp. Agr. 29:287–298.

Blore, T. W. D. 1966. Further studies of water use by irrigated and unirrigated Arabica coffee in Kenya. J. Agr. Sci. 67:145–154.

Bull, R. A. 1963. The effect of mulch and irrigation on root and stem development in Coffea arabica L. Turrialba 13:96–115.

Clark, F., and J. M. Myers. 1956. Effects of rates of irrigation, fertilizers and plant spacing on the yield and quality of flue-cured tobacco in Florida. Soil Crop Sci. Soc. Florida, Proc. 16:249–256.

Garner, W. W. 1946. The production of tobacco. Blakiston Co., Philadelphia. 516 p.

Harrison, D. S., and S. L. Brothers. 1964. Irrigating flue-cured tobacco. Agr. Ext. Serv. Univ. of Florida, Circ. 270.

Hitier, H., J. Chouteau, and A. Mounat, and A. Renier. 1959. Influence of overhead irrigation on the yield and quality of tobacco (French text). Bergerae Inst. Exp. Tabac 3:239–264.

Hitier, H., et al. 1960. Effect of sprinkler irrigation on yield and quality of tobacco (in French). Rev. Int. Tabacs 35:91–92.

Hurd, R. 1959. Annu. Rep. West African Cocao Res. Inst. 1958–1959. p. 51.

Jones, J. N., J. E. Moody, and J. H. Lillard. 1957. Relating irrigation to stage of plant growth. Rep. Virginia Agr. Exp. Sta. Blacksburg.

Jones, J. N., G. N. Sparrow, J. D. Miles. 1960. Principles of tobacco irrigation. US Dep. Agr. Inform. Bull. 228. 16 p.

Kerfoot, O. 1962. Tea root systems. World Crops. 14:140–143.

Kowal, J. M. L. 1959. The effect of spacing on the environment and performance of cocoa under Nigerian conditions. Empire J. Exp. Agr. 27:138–149.

Laycock, D. H. and R. A. Wood. 1963. Some observations on soil moisture use under tea in Nyasaland. Trop. Agr. 40:35–48. Trinidad.

Lemée, G. 1955. Influence de l'alimentation en eau et de l'ombrage sur l'economie hydrique et la photosynthese du cacaoyer. Agron. Trop. 10:592–602.

Lynn, H. P., F. H. Hedden, and J. M. Lewis. 1958. Tobacco irrigation in South Carolina. Ext. Serv. Clemson Agr. Coll. Circ. 438.

Marseu, P., and C. Ille. 1961. The effect of sprinkler irrigation on the quality of flue-cured tobacco (in Hungarian). Lucrarile Inst. Cercet. Aliment. 5:355–361.

McMurtrey, J. E., Jr. 1961. Tobacco production. US Dep. Agr. Inform. Bull. 245.

Medcalf, J. C., W. L. Lett, P. B. Teeter, and L. R. Quin. 1955. Experimental programmes in Brazil (in English) Bull. 6. Int. Basic. Econ. Corp. (IBEC) Res. Inst., New York.

Mes, M. G. 1957. Studies on the flowering of Coffea arabica L. Portugaliae Acta Biologica 4:25–44. (Also IBEC Res. Inst. Pam. 14.)

Miles, J. D. 1957. Influence of irrigation of flue-cured tobacco in Georgia. Georgia Agr Exp. Sta. Circ. New Ser. 8. 20 p.

Murray, D. B. 1960. Soil moisture and cropping cycles in cacao. Rep. Cacao Res. Imperial Coll. Tropical Agr. 1959–60. p. 18–22.

Murray, D. B. 1964. Rep. Cacao Res. ICTA. 1959–60. Trinidad. p. 18.

Murray, D. B., and G. A. C. Herklots. 1955. Rep. Cocoa Conf. London. p. 17.

Nishikawa, G., and K. Ichishima. 1954. Effects of soil moisture upon the leaf quality of bright yellow tobacco. Crop Sci. Soc. Japan, Proc. 22:17–18.

Parks, W. L., B. C. Nichols, R. L. Davis, E. J. Chapman, and J. H. Felts. 1963. Response of Burley tobacco to irrigation and nitrogen. Tenn. Agr. Exp. Sta. Bull. 368.

Peele, T. C., H. J. Webb, and J. F. Bullock. 1960. Chemical composition of irrigation waters in the South Carolina Coastal Plain and effects of chlorides in irrigation water on the quality of flue-cured tobacco. Agron. J. 52:464–467.

Pereira, H. C. 1957. Field measurements of water use for irrigation control in Kenya coffee. J. Agr. Sci. 59:459–466.

Pereira, H. C. 1963. A .five-year water-budget for a coffee irrigation experiment. Turrialba. 13:227–230.

Pereira, H. C., J. S. G. McCulloch, M. Dagg, O. Kerfoot, and P. H. Hosegood. 1962. Hydrological effects of changes in land use. East African Agr. Forest J. 28 (special issue): 16–41.

Piringer, A. A., and H. A. Borthwick. 1955. Photoperiodic responses of coffee. Turrialba. 5(1–2):72–77.

Porteres, R. 1946. Action de l'eau, apres une periode seche, sur le declenchement de la floraison chez Coffea arabica L. Agron. Trop. 1:148–158. Nogent-sur-Marne.

Ramakrishna Kurup, C. K., and A. S. Sastry. 1962. Influence of soil and irrigation water on the chemical composition and quality of cigar tobacco. J. Indian Soc. Soil Sci. 10:99–108.

Robinson, J. B. D., and H. W. Mitchell. 1964. The effects of mulch and irrigation on yield of Coffea arabica L. Turrialba. 14:24–28.

Rogers, M. J., J. N. Jones, and J. E. Moody. 1961. Influence of irrigation during various periods of growth on yield and quality of flue-cured tobacco. Rep. Virginia Agr. Exp. Sta. Blacksburg. 8 p.

Smith, R. W. 1964. The establishment of cocoa under different soil moisture regimes. Empire J. Exp. Agr. 32:249–256.

Texeira, E. F. 1952. Preliminary report of coffee irrigation. Bol. Super. Serv. Cafe. S. Paulo. 27:1038–40.

Walker, E. K., and L. S. Vickery. 1959. Some effects of sprinkler irrigation on flue-cured tobacco. Can. J. Plant. Sci. 39:164.

Wallis, J. A. N. 1963. Water use by irrigated arabica coffee in Kenya. J. Agr. Sci. 60:381.

Williams, F. W. 1957. Evaluating irrigation in flue-cured tobacco production. Virginia J. Sci. 8:264.

Wilson, T. V., and C. H. M. Van Bavel. 1954. Irrigation of flue-cured tobacco. 2nd ed. N.C. St. Coll. Agron. Res. Rep. 3.

39 | Turfgrass, Flowers, and Other Ornamentals

O. R. LUNT

University of California
Los Angeles, California

J. G. SEELEY

Cornell University
Ithaca, New York

I. INTRODUCTION

Commercial turfgrass, flower, and nursery fields are highly specialized segments of horticulture whose irrigation problems are sometimes distinctive because of unusual characteristics of the plant or conditions of culture. Turfgrass and ornamentals are grown also on a large scale for noncommercial use, i.e., recreation or decoration, where irrigation problems may be quite different from those of other plantings.

Many nursery and flower crops are produced in pots, cans, benches, or similar containers in which the soil depth may be typically 10 to 20 cm. Shallower or deeper containers are used also. These cultural conditions have a strong bearing on irrigation practices. Because soil water suction must drop to zero to escape from drainage holes in the bottom of containers, the development of soil water suction after a thorough wetting is limited by the height of the soil column. Under shallow column conditions, soils, sands, or soil mixes retain relatively large amounts of water and may have little free porosity after drainage has ceased. The low free porosity developing in shallow soils following an irrigation has led to the common inclusion of coarse organic additives to increase large pore size porosity. Figure 39–1 illustrates the difference in water retention in the upper 20 cm in two soils after drainage "ceased" following an irrigation. In one case the soil column was 20 cm in depth and in the other it was 125 cm in depth. Note that under shallow conditions the nursery mix composed of equal parts of loamy sand and peat retained more water than did the clay soil under deep conditions. An additional "anomoly" under shallow soil conditions is that a fine-textured soil may require irrigation more frequently when done at moderate suctions, say 50 centibars, than would a sand. This is borne out by appropriate data in Table 39–1 which summarizes some air and water relations in several soils and soil mixes under various suction values. When dealing with plants in shallow containers (and with a small soil volume), the amount of water available to the plant after suctions reach 0.3 bar in typical soil mixes often is surprisingly small (sometimes insufficient to supply the plant for a fraction of a day). Furthermore, capillary conductivity in soil mixes containing large volumes of coarse additives will fall to very low values at 0.1 bar suction (Richards and Wilson, 1936). Therefore, irrigation at low suction values is necessary. Aeration under shallow

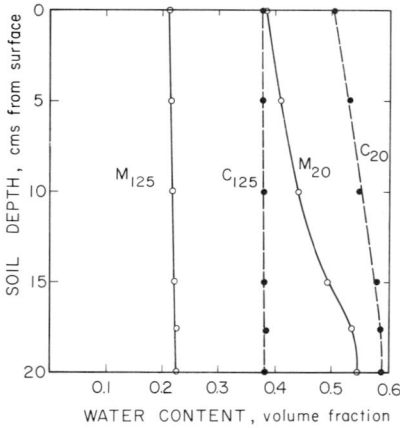

SOIL DEPTH, cms from surface

WATER CONTENT, volume fraction

Fig. 39–1. Water content as a function of depth in the range 0 to 20 cm from the soil surface in two soils and for two soil column lengths after an irrigation. The solid lines M refer to a nursery mix composed of 50% by volume of loamy sand and 50% horticultural grade sphagnum peat. The dashed lines C are for a well aggregated clay soil. The soils were placed in columns 125 cm in length in one case (subscript 125) and 20 cm deep in the other case (subscript 20). All columns were subirrigated and then allowed to drain from an orifice at the bottom of the columns. Evaporation from the soil surface was minimized and the columns were sampled 48 hours after irrigation to yield the data in the figure. Note that the column depth markedly influences water retention in the surface 20 cm and that under shallow soil conditions the nursery mix retains more water than the clay soil under deep conditions.

soil conditions appears adequate according to data of Bunt (1961) and Lunt and Kohl (1957) when free porosities of about 5% or more occur.

In a review by Stanhill (1957) plant response to soil water suction shows the greatest positive trend with annuals, with plants grown in containers, and when measured by vegetative growth. These conditions often apply to ornamentals, and strong responses to water suction usually are reported. Although published research on ornamentals has not always permitted the estimation of a time-weighted average suction value, optimum growth usually requires very low suction—often < 0.5 bar.

II. TURFGRASS

A. Response of Turfgrass to Soil Water Suction

There is a paucity of research data on response of turfgrass to water suction. Furthermore, available data are usually concerned only with yield. Madison (1962) has correctly stressed that the yield of turfgrass clippings is not necessarily an index to high quality. The quality sought in turfgrass culture is a dense, vigorous grass cover left after mowing. He proposed the term "verdure"—the vegetation remaining above the soil after mowing—as a measure of turfgrass quality.

Troughton's (1957) review of literature shows a yield peak for various grasses grown under relatively wet conditions and a rapid decline in yields under moderately dry conditions. Yields appeared greatest at a fraction of a bar suction, and were substantially depressed, in most cases, where soil water conditions approached saturation. Madison and Hagan (1962) doubled the yield of bluegrass (*Poa* sp.) by irrigating at 2-day intervals as compared to 10-day intervals (at which time available water was about one-half depleted). Verdure was un-

Table 39–1. Certain physical properties of soils and typical soil mixes used for container growing of ornamentals

Soil or soil mix*	Bulk density	Total porosity	Free porosity at 1 cbar suction	Volume % water released between 1 and 50 cbars suction	Volume % water released between 1 cbar and 15 bars suction
	g/cc	%	%		
Silt loam	1.52	42	1.5	12	26
Silt loam + fir bark	0.91	59	8.5	26	39
Loamy sand	1.60	39	4.3	23	28
Loamy sand + peat	1.24	54	7.7	37	39
Peat	0.13	73	22	42	46†
Peat + perlite	0.12	74	26	40	43†

* Additives were included at the rates of 50% by volume.   † Water content determined by plant extraction to the wilting point.

affected by the 10-day interval, however. If irrigation was withheld until signs of wilting occurred, both yield and verdure were substantially reduced.

It is generally recognized that very frequent irrigations can detract from the quality of turf, but quantitative data which define the excessively wet conditions have not been developed. In temperate regions, excessively wet conditions in turfgrass favor the development of crabgrass (*Digitaria sanquinalis*). Frequent irrigation appears to encourage shallow rooting (Madison and Hagan, 1962; Roberts and Bredakis, 1960). Madison (1962) found that verdure of two bent-grasses (*Agrostis* sp.) was slightly superior when irrigated at weekly instead of daily intervals. Beach, however, found the quality of bluegrass to be unaffected by irrigation intervals varying between 2 and 8 days provided adequate water was applied (George Beach, 1958. Irrigation of lawns. Results of tests from 1953 to 1957. *Turf Res. litho. Rep. Colorado State Univ.*).

Maintenance of top quality turfgrass may not always be necessary, and if not required, considerable latitude in the irrigation program is possible. Hagan (1955) reports that several creeping fescues (*Festuca* sp.) and bents, bluegrasses, fescue, zoysia (*Zoysia japonica*) and bermuda (*Cynodon dactylon*) which had good vigor and deep roots recovered quickly after irrigation, even though they were nearly brown from lack of water. Healthy zoysia and bermuda could withstand 45 and 75 days, respectively, before wilting at Davis, California, USA. However, prolonged dryness is likely to reduce plant population and quality (Lagenby and Rogers, 1961).

B. Management Problems: Water Application and Distribution on Turfgrass

The use of a turfgrass area often has a considerable bearing on irrigation practices or problems. In general, the use may be described as decorative, general (such as parks or lawns), or that of a sport field. The decorative area typically receives little traffic and may be relatively free of surface compaction which is a common problem in the other types of turfgrass area. Conditions on sports fields are often carefully controlled with respect to water content to provide the proper degree of softness. Soil water control desired or achieved in parks and lawns varies greatly.

1. CONDITIONS AFFECTING IRRIGATION FREQUENCY, RATE, ETC.

Conditions which frequently have an important bearing on turfgrass irrigation include: low infiltration rates, shallow rooting, and nonlevel terrain. Infiltration rates < 0.25 cm/hour are often found where turfgrass is subject to moderate or heavy use. Great variability is also characteristic (R. R. Davis, 1950. The physical condition of putting-green soils and other environmental factors affecting the quality of greens. *Unpubl. Ph.D. thesis, Purdue University, Lafayette, Indiana*). Low infiltration may be due to one or a combination of the following: Compaction due to foot traffic appears to be greatest in the range from 1 to 4 cm below the surface (Lunt, 1956); a thatch layer (a mat of living and relatively undecomposed dead organic matter originating from the grass and which may be difficult to wet); a reduction of capillary conductivity associated with development of a dense root system (Sedgley and Barley, 1958); and the development of low wettability in some soils under turfgrass (Pelishek et al., 1962). Also, in certain situations (i.e., putting or bowling greens) topdressings with soil or soil mixes have caused severe stratification which may impede water movement in turfgrass sod.

Rooting depth has an important bearing on the permissible interval between irrigations. The value of deep grass roots in relation to growth and survival during dry weather was shown by Jacques (1941) and Laude (1953). Troughton's (1957) comprehensive review indicates that most grass species are capable of rooting to depths of 30 to 40 cm and many species to considerably greater depths. However, many studies have shown that often some 60 to 80% of the roots of grass on a weight basis are concentrated in the top 10 or 12 cm of soil.

Management practices have a strong influence on rooting depth. Investigators (Troughton, 1957, p. 94 *et seq.*; Juska and Hanson, 1961; Madison and Hagan, 1962; and Roberts and Bredakis, 1960) are almost unanimous in showing that the lower the cut the lower the root weight, the higher the proportion of shallow roots, and the shorter the life of roots (Weaver and Zink, 1946). Crider (1955) showed that root growth of several species stopped within 24 hours when 40% or more of the top was removed in a given clipping. Rooting depth in putting greens where grass is mowed at about 7 mm is commonly restricted to about 5 cm. High nitrogen rates (Roberts and Bredakis, 1960) and frequent irrigation (Madison and Hagan, 1962) also promote shallow rooting. Inadequate aeration at times also may limit rooting depth. [In California L. H. Stolzy and O. R. Lunt on several typical putting greens have observed oxygen diffusion rates which were apparently well below critical levels (15×10^{-8} g/cm^2 min) for grass at depths of about 6 to 8 cm. Letey et al. (1962) showed an oxygen diffusion rate of 15×10^{-8} g/cm^2 min to be required for root growth in barley (*Hordeum vulgare*)].

It is apparent from the foregoing that many management practices tend to encourage the development of shallow root systems in turfgrass. As a result frequent irrigation has become the *usual* practice with turfgrass. Under favorable conditions, however, bluegrass turf will extract appreciable soil water from depths of 75 cm (Madison and Hagan, 1962). Encouragement of deep rooting and maintenance of reasonable infiltration rates should be major objectives in the management of turfgrass.

2. TECHNIQUES, EQUIPMENT, SPECIAL MANAGEMENT PRACTICES, AND RECOMMENDATIONS

Most turfgrass installations are irrigated with sprinklers because of adaptability to nonlevel terrain, flexibility in regulating quantity and frequency of irrigation, and adaptability to automation techniques. Basin and subirrigation systems are used to a much lesser extent. The general advantages and disadvantages of basin techniques are dealt with elsewhere in this monograph.

With turfgrass a major disadvantage of basin irrigation systems is that they require relatively large quantities of applied water and relatively long periods of submergence compared to sprinkler systems. Subirrigation may be achieved by water table control where feasible. Numerous attempts have been made to develop subirrigation equipment for turfgrass. Success is dependent on adequate capillary conductivity of the soil and proper spacing of the equipment, control of amount of water being applied, and freedom from corrosion and plugging. In arid or subarid regions, accumulation of soluble salts in the soil surface is a major hazard. Subirrigation systems may be useful under ideal conditions where surface obstruction is highly objectionable, such as the infield of baseball diamonds.

Sprinkler irrigation of turfgrass continues to make remarkable advances. Recent developments range from hose attachments which sprinkle in a pattern to complex automatic systems. Space limitations preclude a detailed discussion of equipment advances, but significant developments include: pop-up type sprinkler heads, very low application rate heads (about 3 mm/hour), low trajectory sprays, large radius sprays, "hydrostats" for sensing water suction and controlling valves, new materials for pipelines and equipment (small ditchers) for installation, quick coupling devices for pipelines, and devices for programming irrigation applications. Figures 39–2 and 39–3 illustrate types of pop-up and rotating spray nozzles and an oscillating system which may be adapted for residential use.

Where labor costs are high, automatic irrigation systems recently have gained great favor for large areas of professionally maintained turf such as golf courses (Symp. Golf Course Irrig., 1962). The irrigation sequence is programmed by timing devices which control remote hydraulic valves. Some systems make it possible to deal effectively with the problem of low infiltration rates. For example, a sprinkler may be programmed for 15 min during which time 2.5 mm of water may be applied—insufficient to cause runoff. An "on period" can be repeated as often as needed during the night for adequate irrigation.

As stressed earlier, low infiltration rates are a major problem in turfgrass areas. Most sprinklers for turfgrass apply water in the range from 1 to 5 cm/hour.

Fig. 39–2. Two types of pop-up and rotating spray nozzles for residential use (Photo by J. R. Davis).

Fig. 39–3. An oscillating sprinkler system. The principle may be applied to meet various residential and commercial needs. (Photo by J. R. Davis).

Runoff occurs on many turfgrass areas in 30 min or less with precipitation rates this high. Use of sprinkler heads with low application rates are important in dealing with both of these problems.

Equipment for improving initial infiltration rates in turf includes vertical mowers for thatch removal and control, perforating tools which cut holes about 1 to 2 cm in diameter and 6 to 10 cm deep, and machines which cut a narrow slit into the sod. W. C. Morgan, Los Angeles County Farm Advisor, has shown that the combined effects of perforation and vertical mowing for thatch removal improved infiltration sufficiently that the number of irrigations on turfgrass could be cut in half in summer months. High sand-content putting greens to retain high infiltration rates are advocated (Lunt, 1956). Some wetting agents effectively improve infiltration rates where thatch exists. Tensiometers offer promise for determining when to irrigate if placed near the center of major rooting depth (often about 5 to 8 cm), how long to irrigate if placed below the major rooting depth (often about 10 cm). They may be helpful also in assessing effectiveness of other management practices, i.e. uniformity of irrigation application, variability or change in water infiltration rates, etc. A common complaint in sprinkler irrigation is insufficient overlap of the sprinkling pattern. In designing a system it is better to err on the side of too much overlap than not enough. An overlap of 50% is recommended.

C. Summary

A good irrigation program for high-quality turfgrass should provide for irrigation when suction in the root zone is in the range of about 30 to 50 centibars. Irrigating repeatedly at about 10 centibars or below increases hazards to the development of high-quality turf. Hazards include disease, shallow rooting and reduced infiltration rates. The desirability of maintaining deep rooting is axiomatic. Sprinkler irrigation is the most versatile irrigation technique. Maintenance of infiltration rates is a major management problem.

III. ORNAMENTALS

A. Responses to Soil Water Suction

Soil water suctions favorable to high yields and quality have not been well defined for many ornamental crops. It is apparent, however, that many species grow well in moist or extremely moist soils if soil aeration is adequate. Post and Seeley (1943, 1947a, b) have shown rose (*Rosa* sp.) yields to be unaffected at suctions ranging from 1.5 to 34 centibars. Shanks and Laurie (1949a) observed poor growth when suction was presumably 86 centibars and found (1949b) that soil water had marked effects on types of roots and root systems. Snapdragons (*Antirrhinum majus*) [Cook et al., 1953; Langhans and Hanan, 1962; and Hanan, (J. J. Hanan, 1962. The influence of soil moisture and soil aeration in four root media on the growth and flowering of *Antirrhinum majus* L. Ph.D. Thesis, Cornell University, Ithaca, New York.)] have shown excellent growth at water levels of about 2 to 10 centibars. Hanan and Langhans (1964) observed maximum growth and quality when water equalled about 30% of the total soil volume. Shuel and Shivas (1953) found nectar secretion in snapdragon to be at a maximum "near the moisture equivalent." Some plants including snapdragons are vulnerable to root rot pathogens under relatively moist conditions. The importance of soil water and the incidence of *Pythium* infection and subsequent wilting of snapdragons were stressed by Dimock (1952) and by Hanan et al. (1963). Irrigation techniques which maintain very wet conditions may be made more feasible by promising new fungicides for control of water molds (McCain, 1962; Raabe, 1962).

Sweet peas (*Lathyrus odoratus*), calendula (*Calendula* sp.), larkspur (*Delphinium ajacis*), geranium (*Pelargonium* sp.), and petunias (*Petunia hybrida*) grow well at suctions of a few centibars, but are delayed in flowering at higher soil water suctions (Post, 1943; Reger, 1941; Seeley, 1955). Holley (1953, 1957, 1961) found that carnations (*Dianthus caryophyllus*) grew equally well when suctions of 30 to 50 centibars were allowed to develop before irrigation. Irrigations at 70 centibars impaired quality. Watering at suctions < 30 centibars, especially in tight soils and at low light intensity, led to high water content of plants and soft growth. Caparas (1955) found that extremes of soil water impaired carnation cut flower keeping quality. Van Laan and Cook (1950) found that constant water level subirrigation of carnation soils (37 to 47% water) gave better growth and flower production than surface irrigation of sandy loam and clay soils, but in Oshtemo sand was poor due to exclusion of air.

As soil water treatments decreased from a moist condition near the "maximum water-holding capacity" to a dry situation but well above the wilting percentage Wiggin (1930) observed a consistent decrease in weight of plants, heights, diameter of flowers and stems and number of flowers of chrysanthemums (*Chrysanthemum* sp.). Halevy (1960) found the critical water content (lowest soil water suction which depressed yield) for field gladiolus (*Gladiolus* sp.) depends on the stage of development and according to whether the crop was grown for flowers or for corms. For corm yield, the critical water suction during most stages was 60 to 80 centibars. Wahba (1954) found gladiolus growth and quality were reduced at water levels below field capacity. *Forsythia*, sweet gum (*Liquidambar styraciflua*), and pine seedlings (*Pinus* sp.) (Wenger, 1952; Winkle et al.,

1961) made best growth at a few centibars suction. Lepard observed that by gradually increasing the suction before irrigating, *Hydrangea, Philodendron,* and *Pothos* (*Scindapsus rureus*) could be conditioned not to show wilting at 4 or 5 bars (P. E. Lepard, 1960. Maintaining plants under artificial light. *Unpubl. M.S. Thesis, Rutgers University, New Brunswick, New Jersey*). The treatment not only made plants more adaptable to indoor conditions, but maintained an attractive form for a longer period by restricting growth.

B. General Comments on Irrigation of Ornamental Plantings

1. LANDSCAPED AREAS OTHER THAN TURFGRASS

a. **Trees and Shrubs.** Space limitations permit only brief comments on irrigation in landscaped areas. In general, these areas may be divided into those devoted to deeply rooted plants as trees and shrubs, and those devoted to shallower rooted plants such as annual flowering plants. Deeply rooted plants may require no irrigation or only 2 or 3 irrigations/year to supplement rainfall, depending on soil conditions, plant, and other factors. Normally, substantial quantities of water would be applied in such an irrigation, and basin techniques may be convenient. Where it is desirable to have an area "paved" around a tree, bricks set in coarse sand over a 5-cm layer of gravel provide a means of spreading the water under the brick paving. An injection tube or tubes or other access means are required to transmit the water to the gravel layer. Occasional soil sampling around large trees to determine water content is often of great help in establishing an irrigation program. Over-irrigation is a common cause of tree failure in semiarid regions where an attempt is made to maintain grass under trees. With certain species, particularly those which tend to become diseased when maintained moist, shrubs or vines that require less frequent irrigations may be substituted for grass. In humid regions, poorly-drained soils contribute to tree failure, particularly during the spring.

b. **Annual Plants.** Annual plants, after becoming established, usually root to a depth of 1 or 2 ft and may require less frequent irrigation than turfgrass.

Sprinkler irrigation using fixed, oscillating or rotating nozzles is the most versatile technique for irrigating many of these areas. In the situation where shrubbery is used next to a building and grass beyond, it is usually desirable to provide for independent irrigation of the shrubbery and turfgrass areas since the grass normally requires more frequent irrigation. Low trajectory type spray nozzles are available for shrubbery where it is not desirable to wet the foliage. The same nozzles may also permit better water distribution. A convenient application technique for small landscaped areas is the soaker hose attached to a water hydrant. Apertures which produce a fine spray may either be directed upward or downward in which case water distribution is effected by surface flooding or capillary conduction. Application rates are controlled by regulation of pressure. These devices are inexpensive, conveniently installed, and the fine spray minimizes destruction of surface soil structure. Maximum permissible spacing is 15 ft with better units. Application rates may range from 5 to 12.5 cm/hour at distances of 30 to 60 cm from the hose and may vary with aging (A. F. Pillsbury, *Unpubl. data. Univer. California, Los Angeles*). Soaker hoses may also be conveniently used for flower beds.

Mixed plantings of annuals and perennials, where the rooting depth varies greatly, may or may not be difficult to maintain when growing together. Irrigation programs must be scheduled to meet the needs of the shallower rooted species. Problems arise most frequently when one of the species is susceptible to disease when soil conditions are relatively moist. Many trees and shrubs are less susceptible to disease attacks when the soil is kept relatively dry, and the same is true of many annuals or perennial flowering plants. Unfortunately, there has been very little irrigation research on landscape species and plantings. Judiciously placed tensiometers would probably be very helpful in developing irrigation programs in landscaped areas, particularly where plants of variable rooting depths are grown together.

2. FIELD-GROWN FLOWERS AND NURSERY STOCK

For many species usually grown under field conditions for the nursery or flower industry, there is no specific information on response to soil water. The information available is summarized in part III–A of this chapter. It is quite common, however, to grow at low suctions to achieve fast growth.

In general the techniques for irrigation of field-grown flowers and nursery stock are similar to those used for other crops. Subirrigation, where control of the water table is feasible, is used on various crops in Holland. Furrow irrigation is effective and used for roses, stocks (*Matthiola* sp.), gladiolus, poinsettias (*Ephorbia pulcherrima*), etc. Overhead sprinkling using various techniques has proven effective in certain situations. Portable aluminum pipe with risers several feet in height and supporting rotating sprinklers is used for gladiolus in Florida. The field irrigation techniques are similar to those of other crops and therefore will not be elaborated on here.

C. Application Techniques for Greenhouse and Lathhouse Crops

As summarized in part III, section A, most greenhouse and lathhouse crops grow best when the soil is kept continually moist but not saturated. Irrigations are often made when suctions are in the range of about 10 to 50 centibars for such crops as carnations, roses, chrysanthemums, gardenias (*Gardenia jasminoides*), etc. Traditionally the grower has depended on experience to know when to irrigate. The skill of the grower notwithstanding, tensiometers (Marsh et al., 1962) or measurements of totalized illumination (Furuta et al., 1963; Morris et al., 1956, 1957; Neale, 1956) may offer improved techniques for control of an irrigation program.

In the past, much irrigation has been done by hand, using a hose. High labor costs and new materials have led to the development of automatic or semiautomatic systems which not only reduce labor costs to about one-sixth that of hand watering (Leach, 1959), but also maintain better water uniformity. Tensiometers (Post and Seeley, 1943; Wells, 1961; Wells and Soffe, 1961), gypsum blocks (Bouyoucos, 1952), or float valves in subirrigated capillary systems (Post, 1946; Post and Seeley, 1947b) have permitted complete automation in irrigation.

1. SURFACE WATERING

Numerous variations to surface watering have been developed and may be classified on the basis of those which attempt to: (i) apply water uniformly to the soil surface, and (ii) those which depend on capillary movement for adequate distribution. Both techniques are successfully used, but the latter is dependent on adequate capillary conductivity of the soil which sometimes is poor in greenhouse mixes.

a. Manual Watering. Irrigation with a hose is time consuming, and may not provide good uniformity or control of the amount of water applied. A typical application is 1 to 2 cm of water.

b. Overhead Sprinkling. Sprinkling, by rotating nozzles or nozzles on an oscillating pipe, spreads water over the top of plants and on the soil. The main disadvantage is that the foliage becomes wet, making conditions favorable for disease organisms and walkways become wet. With plants grown in small containers, the system wastes water by irrigating the area between containers. The method is satisfactory for out-of-doors and lathhouse crops, but is seldom used in glasshouses.

A modification of overhead irrigation is used in Europe for watering azaleas (*Rhododendron* sp.) and begonias (*Begonia* sp.) out-of-doors. A boom with sprinklers or nozzles mounted on a cart which is self-propelled by water pressure moves slowly across a field applying water to the plants and soil. Similarly, water "robot" devices may be drawn along aisles in greenhouses with nozzles spraying water on soil in adjacent benches.

c. Low Level Sprinkling. Several variations of this approach have been developed. A widely used method provides a flat spray from the sides of benches from small nylon or brass nozzles in 0.75-inch rigid black plastic pipe. The nozzles, located at 30-inch intervals, are staggered on opposite sides of the bench and may be installed with simple tools (Holley, 1961). A pressure of 15 to 20 lb/inch^2 gives a uniform, flat spray coverage on a bench 4 ft wide. This type of spray on the lower foliage of carnations apparently has not aggravated disease problems.

Another method employs flat spray 360° nozzles in galvanized iron pipe located in the center of the bench (Hasek, 1947; Laurie et al., 1958; Post, 1949). Splashboards at the sides of the bench are needed to prevent water from being wasted in aisles as well as to adequately moisten the soil at the edges of the bench. Considerable pressure and volume are required to water several benches simultaneously. Both methods of low level sprinkling irrigation have been adopted widely.

d. Low Pressure Water Jets or Trickle. Two lines of 0.5-inch copper tubes lengthwise of a bench and drilled with 1-mm orifices 45 to 60 cm apart have been used to distribute water on the soil surface by a series of small (10 to 15 cm) sprinkling fountains (Post and Scripture, 1947). The water moves through the soil to the sides of the bench by capillarity. If capillary movement is too slow, a layer of sand on the soil surface aids in the distribution of water. Flexible or rigid plastic tubing has been used in much the same way (Anon., 1959; Lunt et al., 1956). Oscillating water pressure used by Bean and Wells (1957) minimizes the problem of water distribution by capillarity. Pressure variation by a double float tank or electric float switch varies the distance a jet is thrown by 40 cm.

Water from a low pressure source distributed by flexible tubing laid on the soil with one nozzle per plant (Baumgart, 1959; Bean and Wells, 1957; Blass, 1952) is known as the drip or trickle system. The nozzle consists of a brass screw over which an opaque plastic cap is fitted. By virtue of the thread a long passage is produced resulting in a great resistance to flow, so the output is a slow trickle.

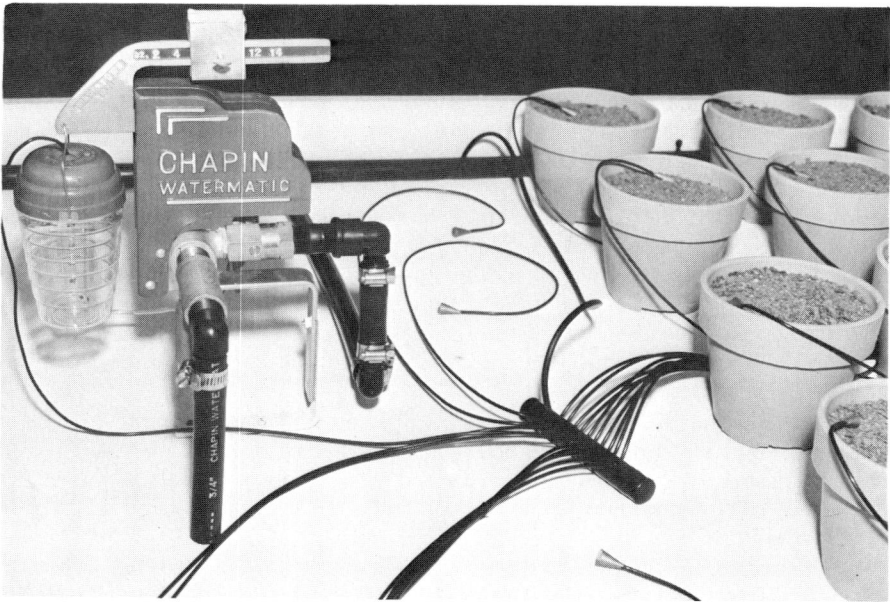

Fig. 39–4. The equipment shown permits automatic irrigation of potted plants. Individual distribution lines lead from a supply source to each pot. Various techniques may be used to make the system automatic. Several hundred pots can be watered simultaneously. Labor savings as compared to hand watering can pay for the cost of the system in a single year.

Each plant is individually watered for 1 or 2 hours at a slow rate so that the soil absorbs water as it is supplied. Water is distributed through the soil by capillarity and gravity. In some soils, however, lateral movement may not be adequate, resulting in unsatisfactory performance of the system.

A similar water distribution method does not require trickle nozzles. Small bore (1.5 to 3 mm) plastic tubing is arranged so that outlets are spaced 45 to 60 cm apart on the soil. A slow flow of water is absorbed by the soil.

A variation of this technique whereby small diameter tubing leads from a supply source to individual pots is proving to be successful for pot irrigation (Hulme, 1961; White, 1961, 1962). Application rates in some systems are such that a typical 5-inch pot can be watered in about 90 sec. Several hundred pots can be watered simultaneously. Pot irrigation is shown in Fig. 39–4. Equipment of this nature is gaining wide acceptance in the USA for the irrigation of potted plants and will probably be used by the nursery trade where plants are grown in containers. In the system shown, excellent uniformity of application may be achieved and careful leveling of the pots is not required.

2. IRRIGATION WITH MIST

Semiautomatic watering of seedlings by 3 to 5 min of mist each morning and evening was found by Salinger (1962) to greatly reduce the time needed for watering. Automatic intermittent mist systems are effective in propagation of woody and herbaceous cuttings (Langhans, 1954; Rowe-Dutton, 1959; Snyder and Hess, 1955).

3. SUBIRRIGATION

Subirrigation from a controlled water table is used in bulb fields and nursery crop areas in Holland. The method is effective in watering glasshouse crops, especially with automatic or semiautomatic systems. The technique has been used occasionally for many years (Green, 1892; Green and Green, 1895; Post, 1939), and currently there is increasing interest in the method for both bench and pot crops [Caparas, 1955; Cook et al., 1953; Post and Seeley, 1943; Seeley, 1948a, b; Wells and Soffe, 1962; (J. D. Triplett, Jr., 1961. The effects of applying water and/or fertilizer through wicks on the growth and flowering of potted plants. *Unpubl. M.S. Thesis. Oklahoma State University, Stillwater.*)]. Three variations of subirrigation have been developed for glasshouse crops: injection subirrigation to saturation, controlled injection subirrigation, and constant water level subirrigation.

a. Subirrigation to Saturation. Water is injected into the tile in the bottom of the bench until the surface of the soil is flooded. Then the outlet plug is removed and the excess water drained. This method is not only wasteful of water, but drastically leaches nutrients from the soil and is not recommended (Post and Seeley, 1947b).

b. Controlled Injection Subirrigation. When the soil water tension or suction reaches the desired level, a predetermined amount of water is injected into the subirrigation tile to moisten the soil by capillarity in about 2 hours. Post and Seeley (1943, 1947a) found that about 13 liters of water per square meter of bench area at each injection was adequate for roses, carnations, and snapdragons in a 12- to 15-cm depth of soil when watered at a maximum suction of 11 or 20 centibars. The technique can be made automatic by use of tensiometer, solenoid valve, and time clock.

The major difficulty is the uneven distribution of nutrients from the lower to the upper levels of the soil as shown by Post and Seeley (1943, 1947a). Recent development of liquid fertilizer applicators (Seeley, 1961) has reduced this difficulty. Caution must be exercised to avoid excessive accumulation of salts at the soil surface. Soil tests may be used to determine the accumulation of salts and are helpful in determining how frequently the soil surface must be flooded to remove the salts.

The injection type of subirrigation also is satisfactory for potted plants (Post, 1949; Seeley, 1948b) and has been used for container grown nursery stock (Harman et al., 1957).

c. Constant Water Level Method. As a result of Post and Seeley's observations that low water suctions favored growth of rose plants, Post (1946) developed a method to maintain a constant water table about 1 inch below the depth of soil mix in a watertight bench (Post and Seeley, 1947b; Seeley, 1948a) by means of a float valve. This is a labor-saving method that may be used satisfactorily (Post, 1946; Stephens and Volz, 1948; Van Laan and Cook, 1950) for production of flower crops if the recommended details of soil preparation and handling are adhered to properly. Difficulties are encountered in accumulation of nutrients at the surface, interpretation of soil tests results, and application of nutrients. The latter is handled best by use of liquid fertilizers (Seeley, 1961).

The principle of supplying water from a constant water table to potted plants works satisfactorily (Seeley, 1948b), and wicks are used occasionally (Post and

Seeley, 1943; Triplett, 1961, *Unpubl. M.S. Thesis, Oklahoma*). Wells and Soffe (1962) have further refined the technique for pot irrigation. Only a small channel of a bench need be watertight, and this is easily accomplished by the use of a plastic sheeting. A constant water level is maintained in the channel which saturates an overlying sand layer by capillarity. Pots are placed on the sand layer and fiber glass plugs conduct the water from the sand into the pots. The technique proved to be highly satisfactory.

A modification of the constant water level method termed the temporary water table was found effective by Caparas (1955). The bottom 0.5 to 1 cm of a level bench was made watertight with plastic liners, providing a reservoir for water at the bottom of the soil column. Water was applied to the soil surface and the excess overflowed just above this watertight liner. The water in the saturated zone was absorbed by the upper soil between irrigations, and the number of irrigations was thus decreased by 35%. The yield and grade of carnations equalled those of surface watered benches.

LITERATURE CITED

Anonymous. 1959. The low-level sprinkler irrigation system for use under glass. Great Brit. Min. Agr., Fish, Food, Nat. Agr. Adv. Ser., Hort. Mach. Leafl. 13. p. 1–4.

Baumgart, M. G. 1959. Automatic control of trickle irrigation. New Zeal. J. Agr. 99:30–31.

Bean, G., and D. A. Wells. 1957. Controlled water applications in glasshouses. J. Agr. Eng. Res. 2:123–134.

Blass, L. Feb. 28, 1952. Trickle irrigation of plants in pots. Fruit-grower, Market Gardener, and Glasshouse Nurseryman. p. 390.

Bouyoucos, G. J. 1952. A new electric automatic irrigation system. Agron. J. 44:448–451.

Bunt, A. C. 1961. Some physical properties of pot plant composts and their effect on plant growth: II. Air capacity of substrates. Plant Soil 15:13–24.

Caparas, J. 1955. Basic methods of irrigating greenhouse carnations. Colorado Flower Growers Ass. Bull. 71:1–3.

Cook, R. L., A. E. Erickson, and P. K. Krone. 1953. Soil factors affecting constant water level subirrigation. Amer. Soc. Hort. Sci., Proc. 62:491–496.

Crider, F. J. 1955. Root growth stoppage resulting from defoliation of grass. US Dep. Agr. Tech. Bull. 1102. p. 1–23.

Dimock, A. W. 1952. Soil aeration and the *Pythium* root rot disease of snapdragons. New York State Flower Growers Bull. 195:1–6.

Furuta, T., W. C. Martin, Jr., and F. Perry. 1963. Precision irrigation with solar energy. Auburn Univ. Agr. Exp. Sta. Circ. 146. p. 1–11.

Green, W. J. 1892. Subirrigation in the greenhouse. Ohio Agr. Exp. Sta. Bull. 43:101–102.

Green, W. J., and E. C. Green. 1895. Subirrigation in the greenhouse. Ohio Agr. Exp. Sta. Bull. 61:47–76.

Hagan, R. M. 1955. Watering lawns and turf and otherwise caring for them. *In* Water. US Dep. Agr. Yearbook Agr. US Government Printing Office, Washington, D.C. p. 462–477.

Halevy, A. H. 1960. The influence of progressive increase in soil moisture tension on the growth and water balance of gladiolus leaves and the development of physiological indicators for irrigation. Amer. Soc. Hort. Sci., Proc. 76:620–630.

Hanan, J. J., and R. W. Langhans. 1964. Soil water content and the growth and flowering of snapdragons. Amer. Soc. Hort. Sci., Proc. 84:613–623.

Hanan, J. J., R. W. Langhans, and A. W. Dimock. 1963. *Pythium* and soil aeration. Amer. Soc. Hort. Sci., Proc. 82:574–582.

Harman, J. S., D. P. Watson, and F. B. Widmoyer. March 1, 1957. Is subirrigation best for your potted stock? South. Florist Nurseryman. p. 66, 79.

Hasek, R. 1947. Irrigating bench crops. Florists' Rev. 100 (2582):31.

Holley, W. D. 1953. Carnations are tolerant to a wide range of soil moistures. Colorado Flower Growers Ass. Bull. 46. p. 2.

Holley, W. D. 1957. Some factors which influence soft growth. Colorado Flower Growers Ass. Bull. 86:1–4.

Holley, W. D. 1961. Watering carnations. Colorado Flower Growers Ass. Bull. 130:1–4.

Hulme, L. 1961. Automatic watering. South. Florist Nurseryman. 74:8–12, 18.

Jacques, W. A. 1941. Root development in some common New Zealand pasture plants: I. Perennial ryegrass (*Lolium perenne*): A. Effect of time of sowing and taking a hay crop in the first harvest year. New Zeal. J. Sci. Tech., Sect. A. 22:237–247.

Juska, F. V., and A. A. Hanson. 1961. Effects of interval and height of mowing on growth of Merion and common Kentucky Bluegrass (*poa pratensis* L.). Agron. J. 53:385–388.

Lagenby, Alec., and H. H. Rogers. 1961. After-effects of irrigation. J. Brit. Grassland Soc. 16:153–155.

Langhans, R. W. 1954. Mist propagation and growing. New York State Flower Growers Bull. 103:1–3.

Langhans, R. W., and J. J. Hanan. 1962. Watering. Ch. 7. p. 35–43. *In* Snapdragon manual. Cornell University, Ithaca, New York.

Laude, H. M. 1953. The nature of summer dormancy in perennial grasses. Bot. Gaz. 114:284–292.

Laurie, A., D. C. Kiplinger, and K. S. Nelson. 1958. Commercial flower forcing. McGraw-Hill, New York. 509 p.

Leach, A. D. 1959. Where does the greenhouse labor go? New York State Flower Growers Bull. 159:4–5.

Letey, J., L. H. Stolzy, N. Valores, and T. E. Szuskiewicz. 1962. Influence of soil oxygen on the growth and mineral concentration of barley. Agron. J. 54:538–540.

Lunt, O. R. 1956. A method for minimizing compaction in putting greens. South. California Turfgrass Cult. 6:(3): 1–4.

Lunt, O. R., and H. C. Kohl, Jr. 1957. Influence of soil physical properties on the production and quality of bench grown carnations. Amer. Soc. Hort. Sci., Proc. 69:535–542.

Lunt, O. R., A. F. Pillsbury, and R. E. Pelishek. Nov. 1956. Irrigation of greenhouse benches. The Bloomin' News. p. 22–24.

McCain, A. H. 1962. Dixon registered for ornamental use. California State Florists Ass. Mag. 12:11.

Madison, J. H. 1962. Turfgrass ecology. Effects of mowing, irrigation, and nitrogen treatments of *Agrostis palustris* Huds., 'Seaside' and *Agrostis Tenuis* Sbith., 'Highland' on population, yield, rooting, and cover Agron. J. 54:407–412.

Madison, J. H., and R. M. Hagan. 1962. Extraction of soil moisture by Merion Bluegrass (*Poa pratensis* L., 'Merion') turf, as affected by irrigation frequency, mowing height, and other cultural operations. Agron. J. 54:157–160.

Marsh, A. W., L. F. Werenfels, and R. H. Sciaroni. 1962. Advantages of tensiometer use in carnation irrigation. Florists' Rev. 130 (3363):13–14, 38–39.

Morris, L. G., F. E. Neale, and J. D. Postlethwaite. 1957. The transpiration of glasshouse crops and its relationship to the incoming solar radiation. J. Agr. Res. 2:111–118.

Morris, L. G., J. D. Postlethwaite, R. L. Edwards, and F. E. Neale. Oct. 1956. The dependence of the water requirements of glasshouse crops upon the total incoming solar radiation. Nat. Inst. Agr. Eng. Tech. Mem. 86 p. 1–14.

Neale, F. E. 1956. Transpiration of glasshouse tomatoes, lettuce, and carnations. Neth. J. Agr. Sci. 4:48–56.

Pelishek, R. E., J. Osborn, and J. Letey. 1962. The effects of wetting agents on infiltration. Soil Sci. Soc. Amer. Proc. 26:595–598.

Post, K. 1939. A comparison of methods of watering greenhouse plants and the distribution of water through the soil. Amer. Soc. Hort. Sci., Proc. 37:1044–1050.

Post, K. 1943. Effect of moisture and nitrate concentrations on growth and bud drop of sweet peas. Amer. Soc. Hort. Sci., Proc. 43:273–280.

Post, K. 1946. Automatic watering. New York State Flower Growers. 7:3–14, 16.

Post, K. 1949. Florist crop production and marketing. Orange Judd Publ. Co., New York. 891 p.

Post, K., and P. Scripture. 1947. Copper tube automatic watering. Amer. Soc. Hort. Sci., Proc. 49:395–404.

Post, K., and J. G. Seeley. 1943. Automatic watering of greenhouse crops. Cornell Univ. Agr. Exp. Sta. Bull. 793. 36 p.

Post, K., and J. G. Seeley. 1947a. Automatic watering of roses, 1943–46. Amer. Soc. Hort. Sci., Proc. 49:433–436.

Post, K., and J. G. Seeley. 1947b. The constant water level method of watering roses. Amer. Soc. Hort. Sci., Proc. 49:441–443.

Raabe, R. D. 1962. Experiments with fungicide drenches to control poinsettia root rot. California State Florists Ass. Mag. 12:9–11.

Reger, M. W. 1941. The relation of soil moisture to the time of bloom of calendulas, larkspurs, and geraniums. Amer. Soc. Hort. Sci., Proc. 39:381–383.

Richards, L. A., and B. D. Wilson. 1936. Capillary conductivity measurements in peat soils. J. Amer. Soc. Agron. 28:427–431.

Roberts, C., and E. J. Bredakis. 1960. Turfgrass root development. Golf Course Rep. 28:12–15, 18–24.

Rowe-Dutton, P. 1959. Mist propagation of cuttings. Commonwealth Bur. Hort. Plantation Crops, Dig. No. 2. p. 1–8.

Salinger, R. H. 1962. Semi-automatic watering of seedlings. New Zeal. J. Agr. 105:69.

Sedgley, R. H., and K. P. Barley. 1958. Effects of root growth and moisture retention. Soil Sci. 86:175–179.

Seeley, J. G. 1948a. Methods of automatic watering of plants. Science 108:65.

Seeley, J. G. 1948b. Automatic watering of potted plants. Amer. Soc. Hort. Sci., Proc. 51:596–604.

Seeley, J. G. 1955. Petunias for sale 8 weeks after potting. Pennsylvania Flower Growers Bull. 51:1, 3–7.

Seeley, J. G. 1961. Soils and fertilizers for florists crops—Liquid fertilizers. Florists' Rev. 128(3307):55, 135, 136, and 128(3308): 27, 99.

Shanks, J., and A. Laurie. 1949a. A progress report on some rose root studies. Amer. Soc. Hort. Sci., Proc. 53:473–488.

Shanks, J., and A. Laurie. 1949b. Rose root studies—some effects of soil moisture content. Amer. Soc. Hort. Sci., Proc. 54:473–476.

Shuel, R. W., and J. A. Shivas. 1953. The influence of soil physical conditions during the flowering period in nectar production in the snapdragon. Plant Physiol. 28:645–651.

Snyder, W. E., and C. E. Hess. 1955. An evaluation of the mist technique for the rooting of cuttings as used experimentally and commercially in America. Int. Hort. Congr. Rep. 14th, Scheveningen. 1957. p. 1125–32.

Stanhill, G. 1957. The effect of differences in soil moisture status on plant growth: A review and analysis of soil moisture regime experiments. Soil Sci. 84:205–214.

Stephens, J. A., and E. C. Volz. 1948. The growth of stocks and China asters on four Iowa soils with constant-level subirrigation. Amer. Soc. Hort. Sci., Proc. 51:605–609.

Symposium on Golf Course Irrigation. 1962. Golf Course Rep. 30:8–50.

Troughton, A. 1957. The underground organs of herbage grasses. Commonwealth Bur. Pastures Field Crops, Hurley, Berkshire, England, Bull. No. 44. 163 p.

Van Laan, G. J., and R. L. Cook. 1950. The effect of three methods of watering on the production of carnations in several soils and soil mixtures. Amer. Soc. Hort. Sci., Proc. 56:415–422.

Wahba, I. J. 1954. The effect of varying minimum soil moisture level on growth and mineral absorption of some young horticultural crops. Hort. Abstr. 25:1644.

Weaver, J. E., and E. Zink. 1946. Length of life of roots of ten species of perennial range and pasture grasses. Plant Physiol. 21:201–217.

Wells, D. A. 1961. A tensiometer for the control of glasshouse irrigation: I. Development of the tensiometer. J. Agr. Eng. Res. 6:16–20.

Wells, D. A., and R. Soffe. 1961. A tensiometer for the control of glasshouse irrigation II. Use of the controlling tensiometer. J. Agr. Eng. Res. 6:20–26.

Wells, D. A., and R. Soffe. 1962. A bench method for the automatic watering by capillarity of plants grown in pots. J. Agr. Eng. Res. 7:42–46.

Wenger, K. F. 1952. Effect of moisture supply and soil texture on the growth of sweet gum and pine seedlings. J. Forest. 50:862–864.

White, J. W. 1961. Handling young plants. Pennsylvania State Univ. Geranium Man. 99 p.

White, J. W. 1962. Untouched by human hands. Pennsylvania Flower Growers Bull. 141:1, 9.

Wiggin, W. W. 1930. The water relations of glasshouse crops. Amer. Soc. Hort. Sci., Proc. 27:323–325.

Winkle, J. S., H. Davidson, and E. A. Erickscn. 1961. Soil moisture studies with container-grown plants. Quart. Bull Michigan Agr. Exp. Sta. 44:125–128.

section XI ━━━━━━

Irrigation Systems

XI

40

Problems and Procedures in Determining Water Supply Requirements for Irrigation Projects

G. G. STAMM

Bureau of Reclamation, US Department of the Interior Washington, D. C.

I. INTRODUCTION

The provision of an adequate water supply for an irrigation project is fundamental to the success of the project. While the water supply for a project may be limited by the initial design and construction of storage, conveyance, and distribution facilities, history reveals that farm sizes, farming methods, farming practices, cropping patterns, crop yields, farm economics, and management efficiency will go through many evolutionary changes over the years of expected project life. Any or all of these inevitable changes can affect the project's water requirement and therefore have an effect on the adequacy of the project's water supply.

All evidence indicates there is much room for improvement in water use. Experience makes it clear, however, that voluntary attainment of theoretically possible efficiencies in water use cannot be expected and, therefore, cannot be fully assumed in water requirement studies. The extent to which the future can be blended with the present in forecasting long-term water requirements demands not only a thorough knowledge of pertinent technical procedures but also the integration of many significant assumptions through the skillful exercise of seasoned judgment.

To compute the water supply requirements for irrigation projects requires a thorough understanding of soil-plant-water relationships at the site of water use, including such factors as the cropping pattern, consumptive use requirements of the crops involved, the water-holding capacity and intake rate of the soil, and other factors. Required assumptions of considerable significance relate to expected cropping patterns, likely intensity of farm cultural practices, and the reasonably attainable level of irrigation efficiency. The validity of basic assumptions will depend on still other variables, not the least of which relate to future farm-market conditions and cost-price relationships.

Experience and judgment are called upon to assess the practicability and appropriateness of assumptions in each case. The best application of technical data is wasted effort if the significant assumptions are unrealistic and therefore unattainable.

In addition, information on soil and topographic features is needed to determine the losses that can be expected in conveying irrigation water from the point of release, from reservoir storage or river diversion to farm turnouts, and also from the turnouts to the irrigated fields. Such losses can be reduced where necessary or desirable by various methods. The extent to which seepage reduction work is provided will depend largely on water supply, harmful effects of seepage, and economic feasibility of canal lining or other indicated seepage control work.

Where reservoir storage is involved, seasonal evaporation loss also must be considered. The extent to which it must be taken into account will vary with physical conditions. Evaporation from deep reservoirs, for example, frequently found in the canyons and valleys of mountainous regions, may be comparatively small and may be fully offset by inflow from minor side streams which is impractical to measure. In contrast, of course, evaporation from shallow storage reservoirs frequently found in the arid plains, prairies, and deserts may be very significant and cannot be ignored. It is in these latter areas that evaporation reduction methods may be most beneficial.

II. WATER SUPPLY

No single factor is of greater importance to the successful operation of an irrigation project than evaluation of the adequacy of the water supply. The following paragraphs will briefly discuss a few of the considerations that enter into this evaluation. A more detailed discussion will be found in chapter 6.

A. Surface Water Supplies

The basic source of water is precipitation, and its occurrence is irregular and, to date, practically uncontrollable. Fortunately, nature provides a basic system of control, and man adds to this his own development. Surface runoff is only one part of the hydrologic cycle which also includes precipitation, evaporation, transpiration, and storage in streams, lakes, and groundwater reservoirs.

Average precipitation of about 76 cm (30 inches), in the form of snow or rain, falls each year on the USA. About 54.61 cm (21.5 inches) or about 72% of the total is evaporated from land and water surfaces or transpired from natural or dryland vegetation. Of the remaining 21.59 cm (8.5 inches), about 6% evaporates while 23% is used by man, and joins the unused portion to make a total of about 20.32 cm (8 inches) discharging into the oceans under present conditions (Langbein et al., 1949).

Most early irrigation developments utilized direct diversion from streams. Although water was plentiful in the early part of the season from snowmelt runoff, the projects frequently sustained deficiencies in the latter part of the season. This situation was corrected by construction of dams for storage of early season runoff for use later in the season. However, practically all streams exhibit not only seasonal but also cyclical fluctuations. Reservoirs, therefore, may be sized to regulate streamflow both on a seasonal and cyclical basis.

For many years, the US Geological Survey has operated gauging stations on all major streams in the USA, and records of discharge are published in a series of water supply papers. At present, about 7,400 gauging stations are in operation which obtain continuous discharge records. Frequently, there will be no records of discharge at the damsite, or, if they do exist, they will encompass only a short time period. There are several different methods that can be used to estimate discharge at a point on a stream where no records or only short-term records are available. These may include either simple or multiple correlation (Searcy, 1960), drainage area ratio, or altitude-area-runoff relationships.

Water supply studies should consider historical streamflow depletions not measured in early years of record, future depletions, water rights, compacts, and treaties. The period of study should encompass years of low, as well as high, flow. Streamflows are usually estimated on a monthly basis and, after adjustment for factors explained above, are used in reservoir operation studies discussed subsequently.

B. Groundwater Supplies

The possibility of developing groundwater supplies should not be overlooked in the study of potential irrigation projects. In many areas of Western USA, groundwater developments preceded those of surface water by many years. Indeed, in some areas of the USA such as Arizona, California, and elsewhere, groundwater has been mined (discharge from the reservoir greater than recharge for long periods of time) for many years, and subsequent surface water developments have been made to supplement the groundwater supply.

Many authoritative publications are available which describe the studies required for a quantitative evaluation of groundwater supplies. Further discussion will be found in chapter 7.

C. Return Flow

Return flow from irrigation comprises the operational waste and seepage from canals, laterals, and farm irrigation operations, as well as increased runoff from precipitation on the project area. Return flow from irrigation permits the available water supply to be used several times as it moves progressively downstream. The number of times water can be diverted for reuse is limited only by its chemical quality and the geography of the area.

The amount of return flow and its distribution throughout the year is dependent on many factors, including the geometry of the drainage system, the irrigation and cropping practices, the aquifer characteristics, and others. A method of predicting the amount and pattern of the groundwater increment of return flow has been developed by Glover (1964) and is applicable to situations where knowledge of the above factors is sufficient to justify its use. This subject is also discussed by Houk (1951).

D. Effective Precipitation

In arid areas of the Western USA, precipitation is so scant that even large errors in estimates would have little effect on diversion requirements, but in semiarid and subhumid areas this effect is of much greater significance. The effectiveness of precipitation depends on several factors such as the amount and intensity of precipitation, character and water-holding capacity of the soil, or plant characteristics.

Table 40–1. Effective precipitation based on increments of monthly rainfall

Precipitation increment		Effective precipitation		
		Per cent	Accumulated	
cm	inches		cm	inches
0.00-2.54	0-1	90-100	2.29-2.54	0.90-1.00
2.54-5.08	1-2	85- 95	4.44-4.95	1.75-1.95
5.08-7.62	2-3	75- 90	6.35-7.24	2.50-2.85
7.62-10.16	3-4	50- 80	7.62-9.27	3.00-3.65
10.16-12.70	4-5	30- 60	8.38-10.79	3.30-4.25
12.70-15.24	5-6	10- 40	8.64-11.81	3.40-4.65
Over 15.24	> 6	0- 10	8.64-12.06	3.40-4.75

One method for estimating effective precipitation has been developed by the US Bureau of Reclamation for use as a guide in arid and semiarid areas. Good judgment, however, must be used in any procedure. Except in rare instances, only the growing season precipitation is used in evaluating effective precipitation. Usually, to be conservative, the average of the growing season precipitation for the five driest, consecutive years is used. Percentage of effectiveness is applied to increments of monthly rainfall ranging from > 90% for the first 2.54 cm (1 inch) or fraction thereof to 0% for precipitation increments above 15.24 cm (6 inches), as indicated in Table 40–1.

The effective precipitation, estimated as described above, should be subtracted from the crop consumptive use as explained in the following paragraphs. Other methods may be used, especially if local data are available.

III.   CONSUMPTIVE USE

A.   Beneficial Consumptive Use

Annual consumptive use may be estimated by a method developed by Lowry and Johnson (1942) who developed a correlation between effective heat and measured consumptive use. This method is adapted to areas of general cropping without a year-long growing season. Temperature records are the only basic data needed to make this estimate. However, if the climatic conditions vary significantly from the areas studied in deriving the relationship, some adjustment may be desirable.

In most areas with cold winters, the winter consumptive use consists of evaporation from snow or frozen soil and is a minor factor in computing water requirements for crops. In many areas, the consumptive use in winter is nearly equal to the winter precipitation, and in other areas the soil water storage is reduced during the winter season and requires replenishment by the application of irrigation water during the following irrigation season. Winter consumptive use is frequently estimated as a percentage of the annual. If winter precipitation exceeds evaporation, the amount retained as available soil water which can be used by crops is subtracted from the seasonal consumptive use requirement.

The Lowry-Johnson method does not provide a means of estimating monthly consumptive use. Such distribution is frequently made on the basis of studies of other areas with similar climatic conditions.

The Blaney-Criddle method (1962) provides estimates of consumptive use by crops and is applicable to year-round growing seasons. The data required are (i) the seasonal crop coefficient, (ii) the mean monthly temperature, and (iii) the monthly percentage of annual daytime hours. From the projected cropping pattern the weighted crop consumptive use may be computed for the entire area to be irrigated. The need for data to determine system capacities and water requirements has led to many investigations of this type, and the US Department of Agriculture has published bulletins giving pertinent information for many areas of the Western USA. Crop coefficients on a monthly basis are available for many crops which enable the hydrologist to compute weighted monthly water requirements for the project if the anticipated cropping pattern is known.

Many other investigators (Amer. Soc. Civ. Eng., 1952) have developed methods of estimating consumptive use of water, and the reader is referred to these for further information.

Neither of the methods described above provides a means of estimating peak requirements for a period of less than 1 month. Some crops, notably corn (*Zea mays*), are sensitive to lack of sufficient water during certain stages of growth. Yields will be reduced if shortages occur during this period of about 10 days to 2 weeks. Most crops have a higher consumptive use rate during certain stages of growth and for a relatively short time period. This is perhaps the most critical factor in the sizing of distribution systems. A method has been developed by Jensen and Haise (1963) for estimating short-period crop consumptive use by using solar radiation data. The main difficulty with this method is that only limited data are available on solar radiation.

J. T. Phelan (1962) of the US Soil Conservation Service, in attempting to correlate the monthly consumptive use coefficient in the Blaney-Criddle formula with mean monthly temperature, noted a loop effect occurred in the plotted points. The computed values of the coefficient were higher in the spring than in the fall for the same temperature. The effects of this loop were corrected by the development of a crop growth stage coefficient. The curves developed for various crops throughout the year are given in a technical release published by the US Department of Agriculture (1964).

B. Nonbeneficial Consumptive Use

Undesirable phreatophytes, including saltcedar (*Tamarix gallica* or *T. pentandra*), cottonwood trees (*Populus sp.*), willows (*Salix sp.*), and saltgrass (*Distichlis sp.*), may grow profusely in some areas and spread rapidly where a high water table prevails along canals, stream and river channels, and lake borders. They consume from 0.003045×10^6 to 0.009251×10^6 m^3/ha (1 to 7.5 acre-ft/acre) of water annually (Thornthwaite, 1952). It has been estimated that in Western USA there are about 6,477,732 ha (16 million acres) of undesirable phreatophytes which annually consume nearly $30,837 \times 10^6$ m^3 (25 million acre-ft) of water, equivalent to about twice the mean annual flow of the Colorado River, USA (Robinson, 1958). When project investigations reveal an anticipated growth of phreatophytes along the waterways, the cost estimates for annual operation and maintenance of the system should provide for control of such plants.

IV.   LEACHING REQUIREMENT

All irrigation waters contain some soluble salts; therefore, unless sufficient leaching occurs to achieve salt balance, excessive salts will accumulate in irrigated soils. The leaching requirement is the fraction of water entering the soil which must be passed through the root zone to control soil salinity at a specified level. This relationship does not directly apply if leaching is partially or wholly taken care of by rainfall. The leaching requirement will depend on the type and quantity of salts in the irrigation water, the type of crop, and the desired yield level.

Excessive salinity is usually associated with restricted drainage and a high water table. Where such conditions exist, the capillary rise of water transports salt to the solum and effective leaching is impossible. Excessive salinity may also develop by encroachment of saline groundwater from adjacent areas or by insufficient irrigation. Highly saline irrigation water requires a substantial portion of the total application to pass through the profile for successful crop production.

Leaching by impoundment of irrigation water on soil surfaces for extended periods is a general practice for reclamation of saline soils. Highly saline water is sometimes more effective than regular irrigation water for initial stages of leaching. Gypsum applications may assist leaching if exchangeable sodium is excessive. Careful attention must be given to the sodium-calcium relationship to assure that the leached soils will have favorable physical properties.

In some instances, leaching of soluble salts from the soil may be necessary before optimum crop production can be attained. Usually, it is not necessary to allow additional water for leaching because this can be done in the early years of project development before the project lands have been fully developed. However, if the water supply contains an unusually high amount of soluble salts, it may be necessary to make allowance for leaching throughout the life of the project in which case this should be considered in computing unit farm delivery requirements.

Further discussion of leaching requirements will be found in chapters 14 and 61.

V.   IRRIGATION EFFICIENCY

Irrigation efficiency is defined as the percentage of total irrigation water supplied to a given area which is made available within the root zone for beneficial consumptive use by crops. There are field efficiencies, farm efficiencies, and project efficiencies, depending on where the water is measured. An aggregate or weighted average of farm efficiencies is used to estimate project irrigation requirements by dividing the estimated potential consumptive use (after deducting the effective precipitation) by the weighted average farm efficiency and then adding to that quotient all transmission and other losses.

The estimation of efficiency in the above calculation is one of the most important and at the same time most difficult decisions facing those who plan irrigation projects today. Irrigation efficiency estimates require a prediction of the cropping pattern, knowledge of the type of farm irrigation layout, an assumption in regard to practices, facts concerning the type and texture of the soils, and probably the most important element of all, the competency and care with which the water will be applied. If an efficiency so established is higher than the project water users can reasonably attain, the project water supply developed on the basis thereof will not provide enough water during years of average precipitation for optimum plant growth. This also will result in a deficient peak period distribution system capacity in most years. An error in estimating irrigation efficiency on the high side, resulting in an inadequate irrigation system capacity, is generally more serious than an error on the low side because it may impair project success.

An example of the consequences which will result from use of too high an irrigation efficiency is illustrated by the US Bureau of Reclamation's (1953) experience on the Eden Project in southwestern Wyoming, USA. This project was constructed jointly by the Bureau and the US Department of Agriculture during the period 1950–58 under provisions of the Case-Wheeler Act. An analysis of the available water supply was made and, using a farm irrigation efficiency of 58%, it was determined that the water supply was adequate to provide full service to 8,097 ha (20,000 acres) of land. Subsequently, while the project was under construction a re-evaluation of the water supply resulted in an agreement between the two agencies to reduce the irrigable area to 7,085 ha (17,500 acres). After irrigation began, it was soon found that the water supply was inadequate even for that reduced acreage. Although a record-breaking, 2-year drouth occurred immediately following the construction period, one of the most important facets in the inadequacy of the water supply was inability of the water users to attain the projected irrigation efficiency. In fact, the project-wide farm irrigation efficiency was not more than 35%. With such low efficiency the available water supply was adequate to provide irrigation service to about 5,263 ha (13,000 acres) of land. Factors contributing to this low irrigation efficiency were a sandy soil with low water-holding capacity, excessive seepage from farm ditches, and inadequate farm-water management by settlers, some of whom lacked the experience necessary to attain higher efficiencies. Inadequate land preparation may also have been an important factor.

The US Bureau of Reclamation (1965) is now restudying the project with a view toward firming up the water supply either by developing additional water, lining canals and laterals, or additional land leveling; by reducing the project's irrigable acreage; or by a combination of these. The new study is being made under the assumption that an average farm irrigation efficiency of 50% can be attained. While this efficiency is considered to be reasonably attainable, it is based on a reduction in farm lateral seepage loss and improved water management which the settlers are expected to achieve. If a substantial reduction of the irrigable area eventually becomes necessary, it will, of course, reduce the repayment ability of the irrigation district to a point below that anticipated when the project was planned.

On the other hand, if the assumed efficiency is too low, a water supply in excess of project needs will result, thus increasing capital investment, reducing the size of the potential area which could be served, and probably contributing to wasteful irrigation practices and drainage problems throughout the life of the project.

Other factors that affect farm irrigation efficiency are intake rate, water-holding capacity, texture of the soil, topography, irrigation practices, and water management.

Coarse-textured soils usually have low water-holding capacities making frequent irrigations necessary. These soils also may have high intake rates and therefore make it difficult to irrigate all portions of a field and at the same time prevent excessive deep percolation losses.

Excessively fine-textured soils may have excellent water-holding capacity but frequently have low intake rates which may cause high farm wastes. Such losses may also be caused by steep slopes.

In circumstances as described above, it would be desirable to consider sprinkler irrigation where amount and rate of application of water can be carefully controlled. A discussion of sprinkler irrigation may be found in chapter 44.

VI. DIVERSION REQUIREMENTS

The losses and wastes inherent in the operation of the conveyance system must be added to the farm delivery requirements to determine the diversion requirements. The losses and wastes incident to the operation of a closed pipe system are very minor, seldom exceeding 5% of the total water conveyed by the system. Losses and wastes from open canal systems may be considerably more, depending on the type of material in which the canal is constructed, whether the canals are

Table 40–2. Factors affecting the estimated average annual diversion requirement

	Weighted average annually	Weighted average annually
	m³/ha	acre-ft/acre
Annual consumptive use (estimated by Lowry-Johnson method)	6,030	1.98
Winter consumptive use (estimated at 15% of annual)	880	0.29
Growing season consumptive use	5,150	1.69
Effective precipitation (estimated at 90% of May-September average for 5 driest years)	365	0.12
Net irrigation water requirement	4,785	1.57
Surface waste and deep percolation losses (estimated at 52% of farm delivery, based on nature of soils and a reasonably efficient irrigation practice)	5,180	1.70
Farm delivery requirement	9,965	3.27
Seepage losses in canals and laterals (estimated at 26% of project diversion requirement)	3,500	1.15
Annual average project diversion requirement at Fontenelle damsite	13,465	4.42
Administrative losses (a 6% increase in project diversion requirement for administrative losses for the project's extensive canals and lateral system)	790	0.26
Annual average gross diversion requirement at Fontenelle damsite	14,255	4.68

lined, the length of canal, and physical control of the water in the system itself. Losses and wastes from lengthy conveyance facilities may be as high as 40%. This subject is discussed in detail in chapter 60.

The following example illustrates the method of estimating diversion requirements for the Seedskadee Project, Wyoming. The estimated diversion requirements average 14,261 m³/ha (4.68 acre-ft/acre) annually measured at the Fontenelle Damsite.

Factors affecting the estimated average annual diversion requirement are summarized in Table 40–2.

To facilitate water requirement studies the project was divided into three divisions; namely, East Side, Big Sandy, and West Side divisions. These divisions were further divided into subareas by the canals and/or laterals required to serve them. Each of the areas was considered individually and irrigation requirements determined for the lands within each area at its head. Seepage losses were then estimated from Fontenelle Dam to the head of each area and a diversion requirement determined for the division. This estimated diversion requirement was then increased 6% for unavoidable administrative losses that would occur in such an extensive conveyance system. The diversion requirements by division and the weighted average diversion requirement for the project are defined in Table 40–3.

There are no data available in the project area by which the monthly distribution of the annual diversion requirements may be estimated. It was necessary, therefore, to rely on data obtained from operating projects in other areas for this purpose. The irrigation diversions for the nearby Eden Project as recorded for

Table 40–3. Summary of estimated irrigation requirements on
Seedskadee Project by division*

Division	Net irrigable	Net irrigation requirement	Farm losses	Average farm delivery requirement	Canal or lateral seepage losses	Diversion requirement at head of division	
	ha	m³/ha	%	m³/ha	%	m³/ha	1,000m³
West Side division	13,290	4,785.38	51	9,814.59	29	13,898.93	184,700
Administrative losses†						792.48	10,500
Total West Side division at West canal outlet works	13,290	4,785.38	51	9,814.59	29	14,691.41	195,200
East Side division	4,083	4,785.38	55	10,546.11	18	12,893.08	52,700
Big Sandy divisions	5,698	4,785.38	52	9,875.55	24	12,923.56	73,600
Total East Side and Big Sandy divisions	9,781	4,785.38	53	10,149.87	21	12,923.56	126,300
Administrative losses†						792.48	7,800
Total East Side and Big Sandy division at East Canal outlet works	9,781	4,785.38	53	10,149.87	21	13,716.04	134,100
Total project area	23,071	4,785.38	52	9,966.99	26	14,264.69	329,300‡

* Weighted averages for the total division or project area. Derived from Appendix Material, Definite Plan Report, Seedskadee Project, US Bureau of Reclamation, Apr. 1959.
† Total diversion requirement at Fontenelle Dam increased ± 6% for unavoidable administrative losses.
‡ Diversion requirement at Fontenelle Dam.

3 years (1954–56) are shown in Table 40–4 as distribution in per cent of the annual requirements.

Eden Project deliveries were curtailed somewhat in May and September each year by construction of new canals and laterals. These deliveries also reflect the practice of excess diversions in June for the purpose of building up the water table for subirrigation in some areas of the Project. In view of the curtailed diversions in May and September and since no subirrigation is anticipated within the Seedskadee Project that would require a buildup of the water table, the above-recorded Eden Project deliveries would not be exactly indicative of the potential Seedskadee Project deliveries.

Comparisons were also made, therefore, with water distribution practice on the Riverton and Shoshone Projects located in northwestern Wyoming. The average deliveries for these two projects during the period 1930–38 are shown in Table 40–5.

Table 40–4. Irrigation diversions for Eden Project for 3 years, 1954–56

Month	1954	1955	1956
	distribution % of annual		
May	10.9	10.9	10.1
June	29.4	28.7	34.6
July	28.6	24.1	27.1
August	24.9	25.7	24.2
September	6.2	10.6	4.0
Total	100.0	100.0	100.0

Table 40–5. Average deliveries for Riverton and Shoshone Projects, 1930–38

Month	Riverton Project	Shoshone Project
	% of annual delivery	
April		0.8
May	8.5	16.4
June	27.5	18.4
July	35.8	26.1
August	18.3	21.2
September	9.9	13.1
October		4.0
Total	100.0	100.0

Table 40–6. Monthly distribution of the Seedskadee Project average monthly diversion requirements—net irrigable 23,031 ha (57,010 acres)

Month	Lowry-Johnson heat unit distribution	Diversion requirement distribution				
	%	%	m³/ha	acre-ft/acre	1×10⁶ m³	acre-ft
May	14	14	1,980	0.65	45.6	37,000
June	20	22	3,138	1.03	72.8	59,000
July	25	26	3,715	1.22	85.1	69,000
August	24	25	3,565	1.17	82.6	67,000
September	17	13	1,857	0.61	43.2	35,000
Total	100	100	14,255	4.68	329.3	267,000

With due consideration given to the difference in the length of growing season, land use, general climatic conditions, etc., the monthly distribution of the Seed-skadee Project average annual diversion requirement is estimated in Table 40–6.

A. Conveyance System Capacity

In determining the capacity of canals to be used for irrigation, the approach should be made from the viewpoint of use; first, determine the requirements for water deliveries at the farm headgate and second, back these to the point of diversion by adding on the losses encountered in transit. To relate these requirements to the design of the canal, it is necessary to consider maximum short time peak demands rather than the average. Numerous studies have shown that virtually all crop plants require more water during periods of above-normal temperatures and below-normal humidity, depending somewhat on the stage of plant

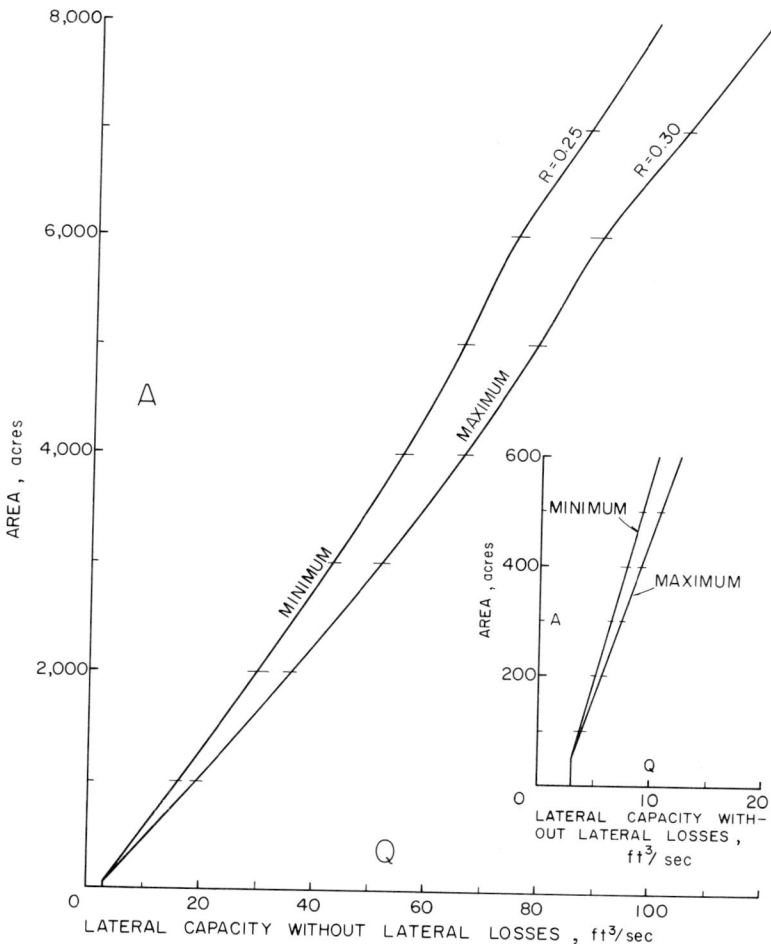

Fig. 40–1. Criteria for lateral capacity for the Great Plains region.

growth, than at any other time. Although this factor has been recognized for many years, only recently data have been available and procedures developed to measure its magnitude (Jensen and Haise, 1963). The peak demands are usually about 10 to 15% greater than the average demands for the maximum month. Consequently, the design capacity should make allowance for peak demands.

Usually in canals serving several thousand hectares, there is sufficient diversity in demands so that the 10 to 15% peaking capacity is adequate to meet them. However, in small laterals which may serve only a few hundred hectares or less, the cropping pattern may be such that a large percentage of the area may be planted to only one crop. In this event, it may be necessary to provide somewhat greater peaking capacity.

There is the practical side that also must be considered. An irrigator cannot afford to spend all his time irrigating his farm. He must have enough flexibility in the system so that he has a sufficient "head" to irrigate his farm in a reasonable period of time and will have time left to do other farm chores. Consequently, the lower end of the lateral is provided with additional capacity. This is frequently referred to as a "flexibility factor." If in order to obtain the water needed for his crops during a 10-day peak demand period a farmer had to irrigate for 10 days, 24 hours a day, he would have a flexibility factor of one. If he had sufficient capacity so that he would only have to irrigate for 5 days, he would have a flexibility factor of two ($10/5 = 2$). Factors of 3 or 4 may be used for the last few farm turnouts on the system.

In some areas, it has been found practical to develop a curve relating canal capacity to the area served. Such a curve for the Great Plains region, USA is shown in Fig. 40–1.

B. Cropping Pattern

In any proposed irrigation project, an estimate must be made of the projected cropping pattern. Consideration must be given to such factors as general climatic conditions, including temperatures and length of growing season, as well as soils, water quality, and economic conditions. A study of cropping patterns in similar irrigated areas may be used as a guide.

Some crops require much more water than others and this fact should be recognized in estimating requirements.

The effect that a different cropping pattern from the one projected can have is illustrated by the experience of the US Bureau of Reclamation on a part of one block of the Columbia Basin Project in Washington, USA. A comparison of the projected and actual cropping patterns is shown in Table 40–7.

The farm units generally contained somewhat less than 40.5 ha (100 acres) of irrigable land each. Water was first available in 1958. In 1961 many of the farmers had leased their holdings and practically the entire acreage served by a particular lateral was planted to potatoes (*Solanum tuberosum*) or beans (*Phaseolus vulgaris*). This concentration of one or two crops having higher than average peak water requirements overtaxed the lateral conveyance capability and it was inadequate to supply the peak period demands in this area. A more diversified cropping pattern, especially on small laterals, along with better land preparation

Table 40–7. Projected and actual cropping patterns, Columbia Basin project, Washington

Crop	Projected per cent in 1952	Actual per cent in 1961
Forage	50	5
Small grain	12.5	14.5
Corn	18.75	6
Sugar beets	18.75	56.5 (including potatoes and truck)
Beans and dried peas	0	18

and improved water management, would largely solve the problem of inadequate capacity.

It is anticipated that as the Columbia Basin Project matures, the improvements described above will be realized and the problem of inadequate capacity will be solved.

VII. RESERVOIR OPERATION STUDIES

The degree of regulation of streamflow which is possible or desirable with a storage reservoir can, in part, be determined through the use of reservoir operation studies. However, these studies are closely related to other factors which are equally important, and are thus only one of the useful methods for examining and testing the factors which constitute the project plan.

Reservoir operation studies are simply an accounting of the water available for supplying the project demands. Historical flows must be modified because of water rights, upstream depletions, etc., to determine inflows to the reservoir which can be stored or diverted for use on the project. Operation studies are usually made by months and include a series of years which incorporate both wet and dry cycles. Allowances must be made for reservoir evaporation, reduction of conservation capacity by sediment deposition, reservoir leakage, etc.

The study is usually set up in tabular form with the years, months or days in the left-hand column, followed by inflow, demands, losses, spills, reservoir content, and shortages. Preliminary studies by mass diagram analysis may be made to determine the approximate size of reservoir for meeting the demands with allowable shortages. Usually, several studies are required using various criteria to determine the most desirable capacity.

A. Evaporation Losses

The most significant loss from a reservoir may be that from evaporation. As an example of its importance, the evaporation losses from Lake Mead on the Colorado River for the period 1953-60, inclusive, averaged about $1,030 \times 10^6$ m^3 (835,000 acre-ft)/year. Usually, evaporation losses are estimated on an average annual amount distributed throughout the year on a per hectare basis. The average monthly reservoir content then can be used to enter the area-capacity table (or curve) for estimating total evaporation from the reservoir.

Although several theoretical methods of estimating evaporation have been developed, direct records of evaporation from large bodies of water in the same general climatological situation are preferable. Frequently, the hydrologist must rely on pan evaporation data with suitable pan-to-lake coefficients. Usually, the monthly distribution of evaporation from a pan is greatly different from a large body of water because of differences of heat storage, and this difference should be estimated if possible by reference to data from large bodies of water, if available. Except for a few areas, pan evaporation records are not obtained in winter months. In such instances, mean monthly temperatures can be plotted against monthly evaporation, using average values for a period of years. This will produce a loop curve which may be entered with mean monthly temperatures to obtain monthly evaporation. For those months of missing record, the curve may be extended by judgment as a basis for estimating winter evaporation.

The evaporation losses chargeable to a reservoir should not include the transpiration and evaporation losses from the area prior to constructing the dam. Plants growing in areas having a high water table such as tules (*Scirpus acutus*), cattails (*Typha sp.*), saltcedar (tamarisk) may consume more water than cultivated crops. These pre-reservoir losses should be subtracted from the evaporation computed by methods described in the preceding paragraph. In areas infested with phreatophytes, it may be desirable to measure plant densities as a means of estimating water use (Horton et al., 1964).

B. Conjunctive Operation—Surface and Groundwater Reservoirs

If groundwater reservoirs are available, consideration should be given to using them in conjunction with surface reservoirs. Frequently, such operation will offer significant advantages. Irrigation will usually provide recharge to the underlying aquifers, and this can be estimated during the planning stage. In addition to reducing the cost of drainage works, the storage capacity of groundwater reservoirs may reduce the capacity of surface storage reservoirs that would otherwise be required for holdover or cyclical storage.

This subject is not discussed in detail here inasmuch as it is thoroughly discussed in the literature (Todd, 1959; Tolman, 1937; Bennison, 1947; Buras, 1963; Amer. Soc. Civ. Eng., 1961; Thomas, 1957).

VIII. WATER SHORTAGES

In any area where water supply is limited, the justifiable annual diversion requirement for irrigation cannot be determined on the basis of providing a full water supply in the year or years of worst drouth. To do so would obligate excessive amounts of water in nondrouth periods and thereby unduly limit irrigation development opportunities. A more practical approach is to recognize that limited shortages can be tolerated occasionally without a severe reduction of farm income. If water shortages can be predicted early in the season, the available water can be used on cash crops, orchards, etc., and application to feed grains and forage crops may be reduced. The economic impact of irrigation water shortages will vary with each project, but generally in an area of diversified cropping, seasonal shortages up to 50% and cumulative shortages for the driest 10 years not to exceed 100% are normally considered tolerable.

In many areas, groundwater pumping can be increased during years of short surface water supplies. Even though the groundwater may be of poorer quality than surface water, little or no significant damage may be experienced if it is used only for brief periods.

It is usually not economical to increase reservoir capacity sufficiently to provide carryover storage for more than a few years, because reservoir evaporation and seepage losses will seriously deplete long time carryover storage.

LITERATURE CITED

American Society Civil Engineers. 1952. Consumptive use of water—A symposium. Amer Soc. Civ. Eng. Trans. Pap. No. 2524. 117:948.

American Society of Civil Engineers. 1961. Manual of engineering practice no. 40. p. 131–145.

Bennison, E. W. 1947. Groundwater, its development, uses, and conservation. H. M. Smyth Printing Co. St. Paul, Minn. p. 444–509.

Blaney, H. F., and W. D. Criddle. 1962. Determining consumptive use and irrigation water requirements. US Dep. Agr. Tech. Bull. 1275. p. 1–59.

Buras, Nathan. Nov. 1963. Conjunctive operation of dams and aquifers. Amer. Soc. Civ. Eng. Hydraul. Div. J. 89 (HY6):111–131.

Glover, R. E. 1964. Groundwater movement. US Dept. Int. Bur. Reclam. Eng. Monog. no. 31. p. 365–393.

Horton, J. S., T. W. Robinson, and H. R. McDonald. 1964. Guide for surveying phrea-tophyte vegetation. US Dep. Agr. Handbook 266. p. 1–35.

Houk, Ivan E. 1951. Irrigation engineering. John Wiley & Sons, Inc. New York. Vol. 1. Ch. 12. p. 365–393.

Jensen, M. E., and H. R. Haise. Dec. 1963. Estimating evapotranspiration from solar radiation. Amer. Soc. Civ. Eng., Proc. 89(IR4):15–41.

Langbein, W .B., B. R. Colby, R. E. Oltman, C. O. Bue, H. C. Troxell, C. C. McDonald, and H. C. Riggs. 1949. Annual runoff in the United States. US Geol. Surv. Circ. 52. 14 p.

Lowry, R. L., Jr., and A. F. Johnson. 1942. Consumptive use of water for agriculture. Amer. Soc. Civ. Eng., Trans. Pap. no. 2158. 107:1243–1265.

Phelan, J. T. 1962. Estimating monthly K values for the Blaney-Criddle formula. Soil Conserv. Serv., US Dep. Agr. Workshop, March 7, 1962.

Robinson, T. W. 1958. Phreatophytes. US Geol. Surv. Water-Supply Pap. no. 1423. p. 1:

Searcy, James K. 1960. Manual of hydrology: Part 1. General surface water techniques. US Geol. Surv. Water-Supply Pap. no. 1541C. p. 67–100.

Thomas, Robert O. 1957. Planned utilization of groundwater. Amer. Soc. Civ. Eng., Trans. Pap. 2869. 122:422–433.

Thornthwaite, C. W. 1952. Evapotranspiration in the hydrologic cycle in the physical and economic foundation of natural resources. Int. Insular Aff. Comm. House of Represent. US Congr. 11:25–35.

Todd, D. K. 1959. Groundwater hydrology. John Wiley & Sons, Inc. New York. p. 200–217.

Tolman, C. F. 1937. Ground Water. McGraw-Hill Book Co., Inc. New York. p. 499–552.

US Bureau of Reclamation. 1953. Eden Project—definite plan report. US Bur. Reclam., Dep. Interior, Washington, D. C.

US Bureau of Reclamation. 1965. Eden Project—water supply improvement report. US Bur. Reclam., Dep. Interior, Washington, D. C.

US Department of Agriculture. 1964. Irrigation water requirements. Soil Conserv. Serv., US Dep. Agr. Technical Release no. 21.

41

Conveyance and Distribution Systems

DARYL B. SIMONS

Colorado State University
Fort Collins, Colorado

I. INTRODUCTION

The design, construction, and maintenance of conveyance and distribution systems is an integral part of most water resource developments, whether the development is small or large, simple or complex, and serves one or several purposes. In most cases, the project will be dual or multipurpose. To optimize the benefits of the development, the entire project must be considered as a unit before the design requirements for the conveyance and distribution systems can be established. The design of the system must consider public health and the nuisance standpoint, and the supply and quality of water must be adequate for successful irrigation. If the distribution system is gravity flow, reservoirs must be located at a sufficient elevation above the irrigated area to provide adequate head for delivery. In most instances in present and future developments the design of conveyance and distribution systems must consider the distribution of water for domestic and municipal purposes, water for industrial use, and stock water. Also, power development and flood control measures may be integrally involved in the design.

Project studies must be conducted to verify project feasibility and must precede design and construction. Basic data required to compile such a study may include maps, aerial photographs, triangulation and bench marks, geology, land classification, climatological data, streamflow data, sediment, quality of water, irrigation and drainage data, etc.

Numerous factors must be considered to derive an adequate conveyance capacity. The annual farm delivery requirement, usually expressed in acre-feet per acre, must be accurately determined by the agronomist working in close cooperation with the irrigation district. It is necessary to evaluate the annual farm delivery demand which is the quantity of water required to bring the crops to maturity exclusive of rainfall. This quantity includes economically unavoidable losses such as percolation, runoff and evaporation and is the base from which conveyance capacity is determined. The annual farm delivery requirement is usually expressed as a monthly demand schedule in acre-feet. This monthly demand varies with crops, climate and soil. In addition, distribution system losses (pipe or open lateral losses) and operational losses must be taken into account. Water losses resulting from seepage, operational losses, and evaporation are of prime importance where water conservation is essential.

With the capacity requirements of the system established, the conveyance and distribution system, including conveyance and control structures, can be designed in accordance with the fundamentals of hydraulics, fluid mechanics, soil mechanics, and structural engineering.

II. PRINCIPLES OF FLOW

The information presented in this part includes some of the basic concepts and fundamental principles of fluid mechanics and hydraulics that are applicable to the design and operation of conveyance systems and hydraulic structures.

A. Physical Properties of Fluids

An understanding of the properties of fluids is based upon considerations of molecular weight, spacing, and activity and mutual attraction of the molecules in the solid, liquid and gaseous states. Water has a molecular spacing that is small compared to the molecular spacing of water vapor. Consequently, the density, ρ, defined as mass per unit volume and the specific weight, γ, defined as weight per unit volume are much larger for the liquid state than for the gaseous state. The activity of the molecules in air and water depends on the heat content. With increasing heat, the molecular activity continually increases until the molecules have sufficient energy to escape the liquid state and become a gas. This phenomena is directly related to evaporation and condensation.

B. Fluid Statics

Fluid statics involve the analysis of pressures that are exerted on boundaries such as gates and dams when there is negligible relative motion in the fluid. Pressure is the force per unit area. Within the fluid it acts in all directions and acts perpendicular to a boundary.

Pressure is usually expressed relative to some reference datum. The two most common data are absolute zero and atmospheric pressure. A pressure less than atmospheric pressure is negative referring to the atmospheric reference datum but is positive when read as an absolute pressure. Pressures and pressure differences are commonly measured with Bourdon gauges, open manometers, differential manometers, mercury barometers, pressure cells or transducers.

Unit hydrostatic pressure within a liquid is composed of the weight of the column of liquid of unit cross section plus the ambient pressure above the column. Hence,

$$P = \gamma h \qquad [41\text{-}1]$$

in which P is the pressure in lb/ft^2, h is the vertical distance in feet from the liquid surface to the area in question, and γ is the specific weight of the liquid in pounds per cubic foot.

Submerged surfaces are subjected to pressures that can be determined from

$$F = \gamma \bar{h} A \qquad [41\text{-}2]$$

where F is the force in pounds, A is the area of the plane submerged surface, and $\bar{h}$ is the vertical height of the liquid column above the centroid of the submerged area. The force acts normal to the plane surface at the center of pressure which is located a distance e below the center of gravity of the plane surface.

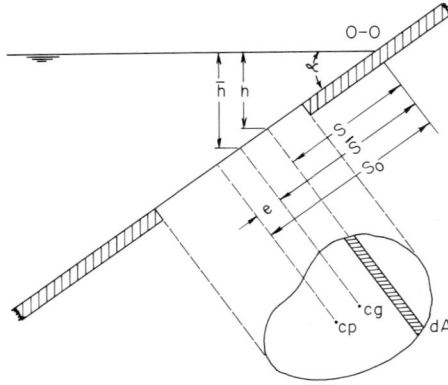

Fig. 41–1. Definition diagram for evaluation of pressure forces on submerged plane surfaces.

The eccentricity e is

$$e = S_o - \bar{S} = \frac{\bar{I}}{\bar{S}A} = \frac{k^2}{\bar{S}} \qquad [41-3]$$

in which S values are moment arms, $\bar{I}$ is the moment of inertia of the area about the centroidal axis parallel to the axis $0 - 0$, and k is the radius of gyration of the surface (see Fig. 41–1).

C.   Types of Flow

With *steady flow* the velocity at a point does not change in magnitude or direction with respect to time. Steady flow is usually easier to analyze than unsteady flow. Many hydraulic and hydrologic phenomena involve changes in velocity with respect to time so that only approximate solutions are possible. These are usually based on statistical averages or a step type analysis for which there is only a small change in velocity over a short increment of time and the steady flow analysis is assumed to be approximately correct.

With *uniform flow* the velocity at a given time does not change with respect to distance. If there is a change (either magnitude or direction) with distance, the flow is nonuniform. In flow around a bend of an open channel the direction changes with distance, and in flow with changing cross section the magnitude changes with distance.

Irrotational flow exists if each fluid element in the flow system has no net rotation about its own mass center. In other words, if any part of a fluid element rotates in one direction, another part of it rotates in another direction so that the net rotation is zero. Most theoretical analyses of fluid flow are based on the assumption of irrotational flow. In most hydraulic phenomena, this assumption is only approximately correct.

Separation is a phenomenon which is encountered in many hydraulic problems. It occurs where there is a boundary configuration which requires a spreading of the stream lines and either the local velocity at the boundary is so great the flow cannot follow the boundary, or the velocity near the boundary is so small that a reduction in velocity causes a negative (reverse) flow. Separation considerations are of great significance in analyzing flow patterns and pressure distributions associated with the design of conveyance systems and hydraulic structures.

The *continuity equation* arises from the basic law of conservation of mass which states that the mass rate of flow is the same at all sections of the flow in a stream tube.

If the average velocity over the section is used and the density is assumed constant from section to section,

$$Q = A_1 V_1 = A_2 V_2 = A_3 V_3 \qquad [41\text{--}4]$$

where Q is the discharge in ft^3/sec, A is the area in square feet, and V is the average velocity in ft/sec.

D. Resistance to Flow

Resistance to flow is encountered wherever there is motion of a fluid adjacent to a moving or stationary boundary. Resistance (or drag) is composed of two parts: (i) shear, which is the force per unit area tangential to the boundary, and (ii) pressure, which is the force per unit area perpendicular to the boundary.

The shear drag can be integrated over the entire boundary to give the total drag due to shear at the boundary. The pressure drag can similarly be integrated over the boundary to obtain the total drag due to pressure on the boundary. The general drag equation is

$$F = C_D A \rho V_o^2/2 \qquad [41\text{--}5]$$

in which C_D is the drag coefficient and is a function of the Reynolds number, the boundary roughness, and the shape of the boundary.

Immediately adjacent to the smooth boundary associated with a turbulent boundary layer, there is a thin laminar flow known as the laminar sublayer. For a turbulent boundary layer, there are two limiting types of boundaries: (i) A smooth boundary for which the roughness elements are covered with the laminar sublayer so that the roughness has no influence on the flow within the boundary layer; and (ii) a rough boundary for which the laminar sublayer is destroyed and the roughness elements are contributing directly to the turbulence.

Boundary roughness is a relative matter. A *hydrodynamically smooth boundary* is one for which the roughness is covered sufficiently by the laminar sublayer so there is no effect of the roughness on the flow. As the height of the roughness e becomes greater relative to the thickness of the laminar sublayer δ', the influence of the roughness becomes increasingly great until the boundary becomes hydrodynamically rough, i.e., the effect of the laminar sublayer is destroyed.

E. Closed Conduit Flow

Closed conduits include all types of pipes and tubes with different shapes and sizes. They are widely used in waterworks, irrigation, drainage, and many aspects of industry.

Flow in closed conduits involves a combination of: (i) steady or unsteady flow; (ii) uniform or nonturbulent flow; and (iii) laminar or turbulent flow.

If flow in a conduit is steady at any given point in the flow, there is no variation in velocity with respect to time, i.e., $\partial v/\partial t = 0$. If there is variation of velocity with time, the flow is unsteady, i.e., $\partial v/\partial t \neq 0$.

When flow is uniform, there is no change at a given time in either the magnitude or the direction of the velocity with respect to distance along a stream line, i.e., $\partial v/\partial s = 0 = \partial v/\partial n$. When the flow is nonuniform, the velocity along a stream line varies either in magnitude, $\partial v/\partial s \neq 0$ or in direction $\partial v/\partial n \neq 0$, or both.

In laminar flow (Reynolds number $\text{Re} = (VD/v) \leq 2,000$), where Re is the ratio of the viscous forces and the inertial forces, the mixing of fluid in one region with that in an adjacent region is accomplished by the extremely slow process of molecular activity or diffusion. Turbulent flow, on the other hand, has finite fluid masses which rotate and move about as eddies, accelerating the mixing process. With turbulent flow, the boundary may be either hydrodynamically smooth or rough. Many experimental equations have been developed.

1. STEADY UNIFORM FLOW

To solve problems involving steady uniform flow, the Bernoulli equation

$$\frac{V_1^2}{2g} + \frac{p_1}{\gamma} + z_1 = \frac{V_2^2}{2g} - \frac{p_2}{\gamma} + z_2 + H_L \qquad [41\text{--}6]$$

is utilized where p is the pressure in lb/ft^2, z is the distance in feet above some arbitrary datum, and g is the acceleration due to gravity in ft/sec^2. The total loss H_L is the sum of the losses caused both by the shear drag h_f and the pressure drag h_L. That is,

$$H_L = h_f + h_L. \qquad [41\text{--}7]$$

The Darcy-Weisbach equation

$$h_f = f \frac{L}{D} \frac{V^2}{2g} \qquad [41\text{--}8]$$

is used to determine the part of the total loss h_f which is caused by the shear resistance at the boundary, L is the distance in feet along which the head loss h_f occurs, and D is a measure of size of conduit. For a pipe flume, D is the pipe diameter in feet.

As in the case of the resistance coefficient C_d, the Darcy-Weisbach resistance coefficient f depends upon the Reynolds number Re and the relative roughness e/D

$$f = \phi\ (\text{Re},\ e/D). \qquad [41\text{--}9]$$

For laminar flow and for turbulent flow with a smooth boundary, the relative roughness e/D is not important and, for a rough boundary, the Reynolds number Re is not important. Hence, for a hydrodynamically smooth boundary,

$$f = \phi_2\ (\text{Re}) \qquad [41\text{--}10]$$

and for a hydrodynamically rough boundary

$$f = \phi_3\ (e/D). \qquad [41\text{--}11]$$

For laminar flow, $\text{Re} < 2,000$

$$f = 64/\text{Re} \qquad [41\text{--}12]$$

In solving steady flow problems, because of the difficulty of determining the values of f from equations, the Moody resistance diagram (Fig. 41–2) can be used. The roughness for various pipe materials and inside coatings is given in Fig. 41–2. The average of the range of e should be utilized unless additional information gives reason to use the smaller or larger values of e. From Fig. 41–2, it may be seen that a large error in the estimate of e results in a smaller error in f.

There are three types of problems which occur most frequently in the design of pipelines and pipe systems. Methods of solving these problems are outlined in Table 41–1. Pipes with noncircular cross section and simple geometric shapes, such as a rectangle, a trapezoid, or an ellipse, that do not differ markedly from circular, can be solved by Fig. 41–2, if the hydraulic radius $R = A/P = D/4$ for a circular pipe is used. The Reynolds number then becomes $\mathrm{Re} = 4RV/v$ and the head loss equation becomes

$$h_f = f \frac{L}{4R} \frac{V^2}{2g}.$$ [41–13]

For turbulent flow, this use of a hydraulic radius gives reasonably accurate results.

2. HAZEN AND WILLIAMS FORMULA

Although the Darcy-Weisbach equation and its accompanying graph (Fig 41–2) is the best rational equation for solving pipe flow problems, various empirical equations for the flow of water in pipes have been developed from data taken in the laboratory and in the field. Perhaps the best known and most extensively used is the Hazen and Williams formula

$$V = 1.32 \, C_1 \, R^{0.63} S^{0.54}$$ [41–14]

in which C_1 is the Hazen and Williams discharge coefficient; R is the hydraulic radius, i.e., the area A divided by the wetted perimeter P (the wetted perimeter for a pipe flowing full is simply the circumference); and S is the slope of the energy line or the hydraulic gradient h_f/L. This equation is widely used in irrigation and waterworks design and is most applicable for pipes 2 inches and larger in diameter and velocities less than 10 ft/sec. The principal advantage of this equation is that the coefficient C_1 does not involve Reynolds number, and all problems have direct solutions. This is also a disadvantage because viscosity variations are ignored. These are at times very important and ignoring them can cause serious error. Common values for the Hazen and Williams coefficient are given in Table 41–2.

3. SCOBEY FORMULA

The Scobey formula is widely used to design concrete pipelines. This equation states that

$$V = C_s \, H^{0.5} \, D^{0.625}$$ [41–15]

where V is the average velocity in ft/sec, H is the head loss in ft/1,000 ft, D is the diameter of pipe in inches, and C_s is a coefficient that varies with smoothness of pipe For modern dry-mix, machine-made concrete pipe, C_s is about 0.31; for wet-mix pipe constructed in short units $C_s = 0.35$. (For a detailed discussion of this relationship see "Flow of water in no-joint concrete pipe lines." No-Joint Concrete Pipe Co., Yuba City, Calif.)

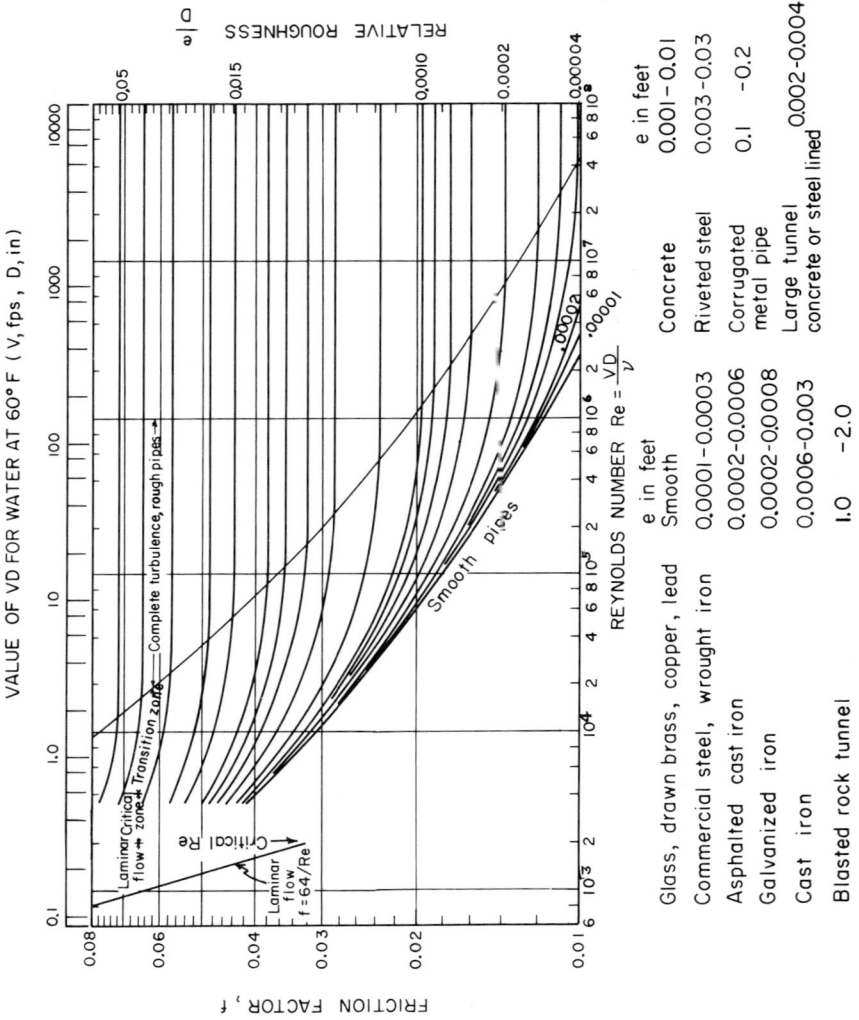

Fig. 41-2. Moody resistance diagram for uniform flow in conduits.

	e in feet
Glass, drawn brass, copper, lead	Smooth
Commercial steel, wrought iron	0.0001-0.0003
Asphalted cast iron	0.0002-0.0006
Galvanized iron	0.0002-0.0008
Cast iron	0.0006-0.003
Blasted rock tunnel	1.0 -2.0

	e in feet
Concrete	0.001-0.01
Riveted steel	0.003-0.03
Corrugated metal pipe	0.1 -0.2
Large tunnel concrete or steel lined	0.002-0.004

Table 41–1. Methods of solving circular pipe problems

Known	Needed	Method of solution
1) V or Q, D, L, e, and υ	h_f	Use Re and e/D scales to obtain f from the Moody Resistance diagram, then use $$h_f = f \frac{L}{D} \frac{V^2}{2g}.$$
2) h_f, D, L, e, and υ Re $(f)^{1/2} = (D^{3/2}/\upsilon)\,(2gh_f/L)^{1/2}$ [1] $$\frac{1}{f^{1/2}} = 1.14 - 2\log\left(\frac{e}{D} + \frac{9.35}{\mathrm{Re}(f)^{1/2}}\right)$$ [2]	V or Q	Compute e/D and Re $f^{1/2}$, eq. [1], determine f directly from eq. [2], or use the e/D-curves of the Moody diagram to determine f by trial and error procedure. Then use $$V = (2gh_fD/fL)^{1/2} \quad\text{or}$$ $$Q = (\pi^2 g h_f D^5/8fL)^{1/2}.$$
3) V or Q, h_f, L, e, and υ	D	Trial and error solution. Estimate f and solve for trial D by $$D = f\frac{L}{h_f}\frac{V^2}{2g} \quad\text{or}$$ $$D^5 = \frac{f\,8L\,Q^2}{\pi^2 g\,h_f}.$$ With trial D, compute Re and e/D as in [1] to find new estimate of f from Moody diagram. Repeat process until calculated f-value agrees with the estimated f-value.

Table 41–2. Hazen and Williams coefficients for flow in pipes

Description of pipe	C_1 value
Polyvinyl chloride pipe	155
Extremely smooth and straight	140
Very smooth	130
Smooth wood and wood stave	120
New riveted steel	110
Vitrified	110
Old riveted steel	100
Old cast iron	95
Old pipes in bad condition	60 to 80
Small pipes badly tuberculated	40 to 50

4. MANNING FORMULA

The Manning formula is widely used for open channels and is often used to design pipelines. For pipelines this formula becomes

$$V = (0.590/n)D^{2/3}\,S^{1/2} \qquad [41\text{–}16]$$

where V is the velocity in feet per second, D is the inside diameter of the pipe in feet, S is the hydraulic gradient, and n is the Manning coefficient of roughness.

Table 41–3. Values of Manning's n for concrete pipe

	Good	Fair
Rough joints	0.016	0.017
Dry mix, rough forms	0.015	0.016
Wet mix, steel forms	0.012	0.014
Very smooth	0.011	0.012

Table 41–4. Values of Scobey's C_s for constant Manning n and varying pipe diameter

Pipe diameter	n = 0.014	n = 0.013	n = 0.012
24	$C_s = 0.292$	$C_s = 0.314$	$C_s = 0.340$
30	$C_s = 0.294$	$C_s = 0.317$	$C_s = 0.342$
36	$C_s = 0.297$	$C_s = 0.319$	$C_s = 0.346$
42	$C_s = 0.298$	$C_s = 0.321$	$C_s = 0.348$
48	$C_s = 0.299$	$C_s = 0.322$	$C_s = 0.350$
54	$C_s = 0.301$	$C_s = 0.325$	$C_s = 0.352$
60	$C_s = 0.303$	$C_s = 0.326$	$C_s = 0.353$
66	$C_s = 0.304$	$C_s = 0.327$	$C_s = 0.355$
72	$C_s = 0.306$	$C_s = 0.329$	$C_s = 0.355$

Suggested values of n for pipes of two different qualities and with different types of joints are given in Table 41–3.

Values of the Scobey coefficient C_s for constant Manning n values and varying pipe diameter are given in Table 41–4.

5. STEADY, NONUNIFORM FLOW

Nonuniform flow exists when either the magnitude or direction of the velocity varies with distance along a stream line. Tangential acceleration occurs if the velocity is changed in magnitude, and normal acceleration occurs if the velocity is changed in direction.

These changes in velocity result in a change in momentum flux (a vector) and the change in momentum flux is accomplished only by pressures against the fluid that are in addition to the pressures associated with uniform flow. When such changes in velocity occur, zones of separation and secondary flow frequently result, and this increases the shear and the turbulence at the expense of the piezometric head. Hence, a head loss h_L results. Since the foregoing changes in velocity and the resulting head losses are caused by nonuniform distribution of pressures on the boundary, the losses are termed form losses, due to pressure resistance and associated increases in shear resistance.

In equation [41–8], the shear resistance loss is expressed as a factor $f(L/D) \times$ the velocity head $V^2/2g$. Thus, $f(L/D)$ may be thought of as the number of velocity heads which are lost due to shear resistance. Form losses h_L can also be expressed as $C_1 \times$ the velocity head $V^2/2g$, where C_1 can be thought of as the number of velocity heads lost due to the form of the conduit—a form loss.

That is

$$h_L = C_1(V^2/2g)$$ [41–17]

in which C_1 is called the form loss coefficient.

Table 41–5. Minor loss coefficients C_1

Sudden expansions, $C_1 = (1 - A_1/A_2)^2$					
A_1/A_2	0	0.2	0.4	0.6	0.8	1.0
C_1	1	0.64	0.36	0.16	0.04	0

Sudden contractions, $C_1 = (1/C_e - 1)^2$					
A_2/A_1	0	0.2	0.4	0.6	0.8	1.0
C_1	0.5	0.45	0.36	0.21	0.07	0

90-degree smooth pipe bends					
r/D	1	2	4	6	8	10
C_1	0.25	0.14	0.10	0.085	0.08	0.08

Gradual expansions ($\theta = 15°$, Re $= 1.5 \times 10^5$)					
A_2/A_1	2	4	6	8	10	20
C_1	0.11	0.21	0.25	0.28	0.29	0.32

Commercial Pipe Fittings

Fitting	C_1
Globe valve, fully open	10
Angle valve, fully open	5
Swing check valve, fully open	2.5
Closed return bend	2.2
Tee, through side outlet	1.8
Short radius elbow	0.9
Medium radius elbow	0.8
Long radius elbow	0.6
45-degree elbow	0.4
Gate valve, fully open	0.2
Gate valve, ¾ open	1
Gate valve, ½ open	5.6
Gate valve, ¼ open	24

These form losses are sometimes called *minor* losses. Such a term represents the true situation when the pipeline is relatively long and the shear loss coefficient $f(L/D)$ is large by comparison with C_1. For shorter pipelines, the form losses caused by pressure resistance may be of major importance instead of minor importance. The so-called minor losses are summarized in Table 41–5 in which A_1 and A_2 are the respective areas upstream and downstream of the source of the minor loss, r is the radius of pipe bend, D is the pipe diameter, $C_c =$ coefficient of contraction $= A_c/A_2$ where $A_c =$ area of contracted jet at vena contracta, and $\theta =$ total angle of expansion. For a detailed discussion of each of these losses, refer to Albertson et al. (1960).

6. COMPOUND PIPELINES

The principles presented in the foregoing sections may be used in combination to solve problems involving compound pipelines. Figure 41–3 is an example of a compound pipeline which involves an entrance, a sudden expansion, a sudden contraction, a valve, a bend, a gradual expansion, an outlet, and flow in pipes of different diameters. Each of these items involves a head loss. The straight pipe involves shear resistance, and each of the others involves both shear and pressure resistance to make up the form losses.

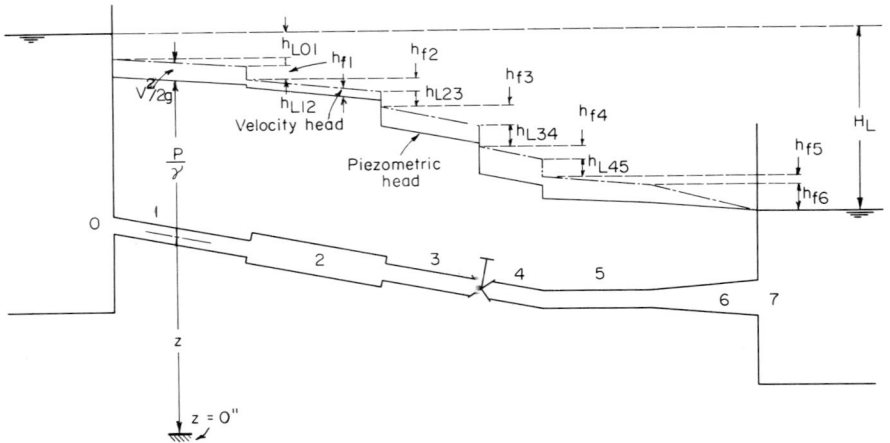

Fig. 41–3. Energy diagram for compound pipelines.

Bernoulli's equation may be written for any reach of pipe as

$$\frac{V_a^2}{2g} + \frac{p_a}{\gamma} + z_a = \frac{V_b^2}{2g} + \frac{p_b}{\gamma} + z_b + H_L \qquad [41\text{–}18]$$

in which H_L includes both shear losses and form losses between sections a and b. If the upstream reservoir is chosen as a and the downstream reservoir as b, then H_L is the sum of all the losses indicated in Fig. 41–3. In other words,

$$
\begin{array}{ccccccccc}
H_L &=& h_{L01} &+& h_{f1} &+& h_{L12} &+& h_{f2} &+& h_{L23} &+& h_{f3} &+& h_{L34} \\
\text{total} && \text{entrance} && \text{pipe} && \text{expansion} && \text{pipe} && \text{contraction} && \text{pipe} && \text{valve} \\
\text{loss} && \text{loss} && \text{loss} && \text{loss} && \text{loss} && \text{loss} && \text{loss} && \text{loss}
\end{array}
$$

$$
\begin{array}{ccccccccc}
&+& h_{f4} &+& h_{L45} &+& h_{f5} &+& h_{L57} &+& h_{L67} \\
&& \text{pipe} && \text{bend} && \text{pipe} && \text{gradual} && \text{exit} \\
&& \text{loss} && \text{loss} && \text{loss} && \text{expansion} && \text{loss}
\end{array}
$$

which are the entrance loss, the sudden expansion loss, the sudden contraction loss, the valve loss, the bend loss, the gradual expansion loss, the exit loss, and the straight pipe losses for sections 1, 2, 3, 4, and 5. Each of these losses must be determined in accordance with the methods already developed and then added to get H_L. These losses have been determined for conditions of uniform upstream and downstream flow for a considerable distance. If one loss is close to another [closer than $50D$ (pipe diameters)], the loss of the two in combination is frequently less than the sum of the two losses individually. Pumps and turbines may also be added to the system. In this case a fourth term H is added to the left-hand side of equation [41–18] which is positive, if the unit is a pump and negative if it is a turbine where H is the theoretical head added to the fluid by the pump or extracted by the turbine.

F. Open Channel Flow

Flow in open channels has been nature's way of conveying water on the surface of the earth through rivers and streams since the beginning of time. Furthermore, these streams have constantly been the subject of study by man since he has been alternately blessed by the life-giving quality of streams under control and plagued by their destructive quality when out of control, such as in time of flood.

Open channels include not only those which are completely open over head, but also closed conduits which are flowing partly full. Examples of such closed conduits are tunnels, storm sewers, sanitary sewers, and various types of pipelines.

Flow in open channels has certain similarities to flow in closed conduits which are flowing full. The boundary of the channel transmits a shearing force to the flow and converts energy into heat. This energy must be supplied by a gradient of piezometric head dh/dx just as in a pipe flowing full. For pipes flowing full, however, the piezometric head gradient includes both the gradient of pressure head dp/dx, and the gradient of elevation head dz/dx, whereas open channels include only the gradient of elevation head dz/dx, since the pressure head is the same everywhere on the surface of the stream, and $dp/dx = 0$.

1. TYPES OF FLOW

Because flow in open channels involves a free surface or interface, it has more degrees of freedom than the flow in closed conduits flowing full. This fact results in additional types of flow which, together with the types already encountered, must be clearly defined and understood. These types include uniform flow and nonuniform (or varied) flow; steady flow and unsteady flow; laminar flow and turbulent flow; and tranquil flow, rapid flow, and ultrarapid flow.

Uniform flow in open channels, like that in pipes, depends on there being no change with distance in either the magnitude or direction of the velocity along a stream line, i.e., both $\partial v/\partial s = 0$ and $\partial v/\partial n = 0$. Nonuniform flow in open channels occurs when either $\partial v/\partial s \neq 0$ or $\partial v/\partial n \neq 0$. Varied flow in open channels is a type of nonuniform flow which occurs when $\partial v/\partial s \neq 0$. Steady flow occurs when the velocity at a point does not change with time, i.e., $\partial v/\partial t = 0$. When the flow is unsteady, $\partial v/\partial t \neq 0$. An example of unsteady flow is a flood wave or a traveling surge. The existence of laminar and turbulent flow depends on the Reynolds number Re of the flow, just as in pipes.

Unlike laminar and turbulent flow, tranquil flow and rapid flow exist only with a free surface or innerface. The criterion for tranquil and rapid flow is the Froude number $Fr = V/(gy)^{1/2}$, which (like the Reynolds number) is the ratio of two types of forces. The Froude number is a ratio of the forces of inertia to the forces of gravity. When $Fr = 1.0$, the flow is critical; when $Fr < 1.0$, the flow is tranquil; and when $Fr > 1.0$, the flow is rapid. Ultrarapid flow involves slugs or waves superposed over the uniform flow pattern which makes the flow both nonuniform and unsteady.

Uniform flow in an open channel occurs with either a mild, a critical, or a steep slope. With a mild slope, the flow is tranquil; with a critical slope, the flow is critical; and with a steep slope, the flow is rapid.

2. FLOW EQUATIONS

One of the common open channel flow equations is the Chezy equation

$$V = (8g/f)^{1/2}(RS)^{1/2} = C\ (RS)^{1/2} = C/g^{1/2}\ (gRS)^{1/2}. \qquad [41\text{--}19]$$

The evaluation of the Chezy C for open channel flow can be made in much the same manner as for pipes. For laminar flow in a wide channel assuming a parabolic velocity distribution

$$C/g^{1/2} = (Re/8)^{1/2} \qquad [41\text{–}20]$$

in which $Re = 4VR/v$.

Referring to equations 1 and 2 of Table 41–1 and replacing D with its equivalent $4R$ for circular pipes

$$\frac{C}{(8g)^{1/2}} = 2 \log \frac{Re}{C/(8g)^{1/2}} - 0.8 \text{ (Smooth Boundary)} \qquad [41\text{–}21]$$

$$\frac{C}{(8g)^{1/2}} = 2 \log \frac{4R}{e} + 1.14 \text{ (Rough Boundary).} \qquad [41\text{–}22]$$

These equations can be used to estimate the Chezy C in equation [41–19]. The magnitude of the Chezy C, however, depends on the form of the boundary roughness in alluvial channels.

Bazin (1897) attempted to shorten the equation for C by producing a new formula in English units

$$C = 157.6/[1 + m/R^{1/2})] \qquad [41\text{–}23]$$

where m is the roughness coefficient varying from 0.11 for very smooth cement or planed wood to 3.17 for earth channels in rough condition. In this equation the upper limit for C is 157.6.

In 1911 Johnston and Goodrich proposed using an exponential formula of the form,

$$V = CR^p s^q \qquad [41\text{–}24]$$

and gave values of C and p, making q uniformly equal to 0.5 for simplicity (Ellis, 1916).This is exactly the same formula as proposed by Chezy where the numerical value of the Chezy's coefficient C is equal to 0.5. Other open channel flow equations that are often used are given in Table 41–6 (N. G. Bhowmik, 1965. The hydraulic design of large concrete-lined canals M.S. Thesis. Colorado State Univ., Ft. Collins, Colo.). (See also Garbrecht, 1961.)

In an effort to correlate and systematize existing data from natural and artificial channels, Manning (1889) proposed an equation which was developed into

$$V = (1.5/n)R^{2/3} S^{1/2} \qquad [41\text{–}25]$$

or

$$Q = AV = A(1.5/n R^{2/3} S^{1/2} \qquad [41\text{–}26]$$

in which n is the Manning roughness coefficient which has the dimensions of $L^{1/6}$. By comparing equation [41–19] with equation [41–25], the Chezy discharge coefficient C can be expressed as follows

$$C = 1.5(R^{1/6}/n) \qquad [41\text{–}27]$$

and is related to the Manning coefficient n and the hydraulic radius R. The Manning n was developed empirically as a coefficient which remained a constant for a given boundary condition, regardless of slope of channel, size of channel, or depth of flow. As a matter of fact, however, each of these factors causes n to vary to some extent. In other words, the Reynolds number, the shape of the channel, and the relative roughness have an influence on the magnitude of Manning's n. Furthermore, for a given alluvial bed of an open channel, the size, pattern, and spacing of the sand waves vary so that n varies. Despite the short-

Table 41–6. Other equations for resistance coefficients

Author	Equation	Definitions
N.N. Pavlovsky (1925)	$C = R^x/n$	C is the Chezy coefficient n is the Manning coefficient R is the hydraulic radius in meters $x = 2.5\,n^{1/2} - 0.13 - 0.75\,R^{1/2}\,(n^{1/2} - 0.10)$
G. H. Keulegan (1938)	$C = 8/f^{1/2}$	C is the Chezy coefficient f is the Darcy-Weisbach coefficient
A. E. Bretting (1948)	$1/f^{1/2} = 2\log\,(K_s/14.83R)$	f is the Darcy-Weisbach coefficient K_s is the equivalent sand grain diameter R is the hydraulic radius
J. H. Thijsse (1949)	$1/f^{1/2} = 2.03$ $[\log 12.2R/(0.282\,\delta + K_s)]$	f is the Darcy-Weisbach coefficient R is the hydraulic radius δ is the thickness of the laminar sublayer K_s is a measure of the roughness
Powell (1950)	$C = 42\log R/e$ (rough channel)	C is the Chezy coefficient e is a measure of channel roughness R is the hydraulic radius
A. D. Alfshul (1952)	$1/f^{1/2} = 1.8\log$ $[\mathrm{Re/Re}\,(\triangle/D) + 7]$	f is the Darcy-Weisbach coefficient Re is the Reynolds number $\triangle$ is a linear measure of roughness height D is depth of flow
P. Ackers (1958)	$C = (32g)^{1/2}\log(14.8R/K_s)$	C is the Chezy coefficient R is the hydraulic radius K_s is a measure of the amplitude of the roughness
W. W. Sayre and M. L. Albertson (1961)	$C/g^{1/2} = 6.06\log\dfrac{(D_n+2.6)}{x}$	C is the Chezy coefficient D_n is normal depth x is a general roughness parameter
H. J. Koloseus and Davidian (1961)	$1/f^{1/2} = 2\log$ $[(0.56\,\lambda^{0.9}\,R)/K_s]$	f is the Darcy-Weisbach coefficient R is the hydraulic radius λ is a measure of the concentration of roughness elements K_s is a measure of the amplitude of the roughness elements

Gunther Garbrecht (1961) lists and evaluates 22 Chezy coefficient formulas that have been used in Europe. The ones that have not been cited follow:

A. Salient formulas from turbulence theory

v. Kármán-Prandtl $\qquad C = (8g)^{1/2}\,[2\log\,(2R/K_s) + 1.74]$

V. Mises $\qquad C = \dfrac{4.43}{0.0024 + (K'/2R)^{1/2}}$

Bazin $\qquad C = 87\dfrac{R^{1/2}}{\mathrm{m}} + R^{1/2}$

(continued on next page)

Table 41–6. (continued)

Agroskin
$$C = \frac{1}{n} + 17.72 \log R$$

Mostkow
$$C = 22 \log \frac{R}{\triangle} + 9.5 \frac{\triangle}{R} + 1.5$$

Kutter
$$C = 100 \frac{R^{1/2}}{n} + R^{1/2}$$

B. Empirical formulas

Eytelwein	$C = 50.9$
Hagen	$C = 43.7 \, R^{1/6}$ (for large canals)
Hessle	$C = 25 + 12.5 \, R^{1/2}$
Lahmeyer	$C = 183.5 \, R^{1/6} \, S^{\,1/6}$
Saint-Vernant	$C \sim 60$
Humphreys and Abbot	$C \sim 5.35 \, S^2$
Beyerhans	$C \sim 27 R^{1/5}$
Hermanek $\quad [V = C(DR)^{1/2}]$	$C = 30.7 \, D^{1/2} \, (D < 1.5 \text{ m})$
	$C = 34 \, (D)^{1/4} \, (1.50 \leqq D \leqq 6.0)$
	$C = 50.2 + 0.5 \, D \, (D > 6.0 \text{ m})$
Strickler	$C = (\text{Constant}/d_s{}^{1/6}) R^{1/6}$
Rinsum $\quad [V = C \, (DS)^{1/2}$	$C \sim K_s + 23$
Eisner	$V = K_1 \, D_r \phi R^{1.5\phi - \xi - 1} S^{0.5\phi}$

Table 41–7. Manning roughness coefficients for various boundaries

Boundary	Manning roughness n in (ft)$^{1/6}$
Very smooth surfaces such as glass, plastic or brass	0.010
Very smooth concrete and planed timber	0.011
Smooth concrete	0.012
Ordinary concrete lining	0.013
Good wood	0.014
Vitrified clay	0.015
Shot concrete, untrowelled, and earth channels in best condition	0.017
Straight unlined earth canals in good condition	0.020
Rivers and earth canals in fair condition--some growth	0.030
Winding natural streams and canals in poor condition-- considerable moss growth	0.035
Mountain streams with rocky beds and rivers with variable sections and some vegetation along banks	0.040-0.050
Alluvial channels, sand bed, no vegetation	
1) Lower flow regime	
ripples	0.017-0.028
dunes	0.018-0.045
2) Transition	0.014-0.030
3) Upper flow regime	
plane bed	0.011-0.015
standing waves	0.012-0.016
antidunes	0.012-0.020

comings of the Manning roughness coefficient it is used extensively in Europe, India, Egypt, and the USA.

The magnitude of Manning roughness is given in Table 41–7 for rigid channels and alluvial channels. Considering alluvial channels, note that as the form of bed roughness changes from dunes through transition to plane bed or standing waves, the magnitude of Manning n decreases by approximately 50%.

The performance of large concrete-lined irrigation canals was investigated by the US Bureau of Reclamation (Tilp and Scrivner, 1964). As a result of its canal capacity test program, a design procedure was presented. The Bureau's design procedure employs Fig. 41–4, 41–5, and 41–6. The procedure it suggested for design is outlined:

1) Determine preliminary canal section using Manning's formula and assuming n from Fig. 41–4.

2) Assume water temperature which will exist at design flow and compute Reynolds number $Re = 4RV/v$ for the preliminary section.

3) Enter Fig. 41–5 with the computed value of Re, project a vertical line upward to the average data curve, and mark the intersection point 1 and read the value of f.

4) Enter Fig. 41–6 and with the value of f locate point 2 at the intersection with the hydraulic radius line from the preliminary section.

5) Project a vertical line from point 2 and read Manning's n from the linear scale. If the n value from Fig. 41–6 does not agree with the assumed value, repeat steps 1 through 5 with the value indicated by Fig. 41–6.

In applying this method, the Reynolds number Re, the slope S, and the Darcy friction factor f are dimensionless, the hydraulic radius R is in feet, the kinematic viscosity v has the units ft^2/sec, the average velocity V is in ft/sec, and the

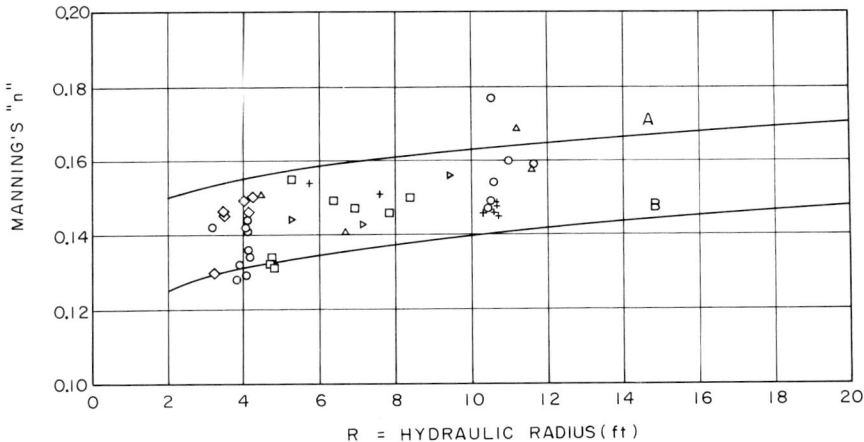

NOTES :

CURVE   A   K = 0.010 ft. PRECAST PIPE WITH MORTAR SQUEEZE AT
            JOINTS. BRICK IN CONCRETE MORTAR (Normal example)

CURVE   B   K = 0.002 ft. MONOLITHIC CONSTRUCTION AGAINST
            STEEL FORMS SMOOTH TROWELED SURFACE
            (Normal example)

Fig. 41–4. Relation between Manning's n and hydraulic radius R for different rigid surfaces [from Tilp and Scrivner (1964), Fig. 7].

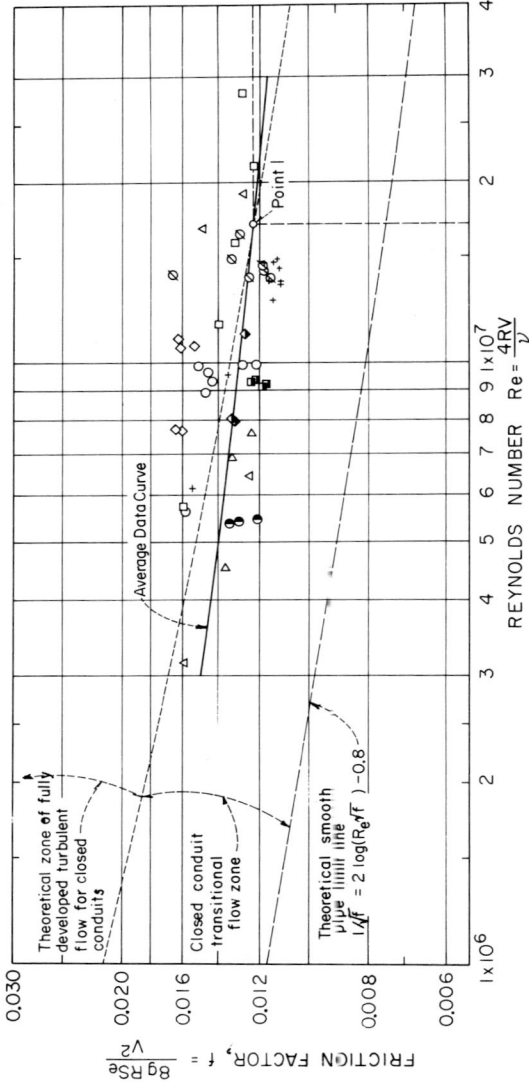

Fig. 41–5. Resistance diagram showing the relation between friction f and Reynolds number Re for straight reaches of all canals [from Tilp and Scrivner (1964), Fig. 8].

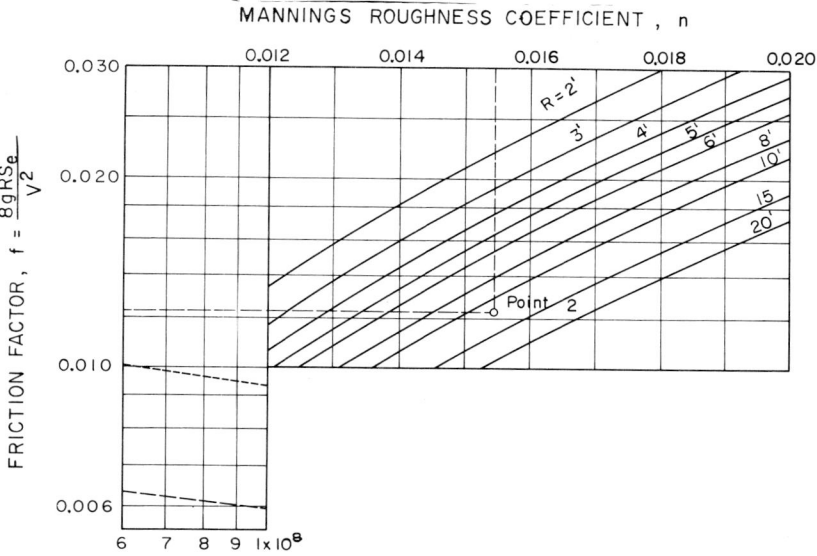

Fig. 41–6. Resistance diagram showing the relation between friction factor f and Manning's roughness coefficient n for straight reaches of all canals [from Tilp and Scrivner (1964), Fig. 8].

Fig. 41–7. Shape of natural channels.

acceleration of gravity g is equal to 32.2 ft/sec^2. For greater detail, refer to Tilp and Scrivner (1964).

3. NATURAL CHANNELS

The natural shape of an open channel may be markedly different from the shapes discussed thus far. However, it is usually possible to break down the complex shape of a natural open channel into simple elementary shapes for analysis. Consider Fig. 41–7, for example, in which flow is occurring not only in the main channel, but also in the overbank or floodplain area. In this case, the hydraulic radius R which would be obtained by using the area and the wetted perimeter for the entire section would not be truly representative of the flow. Furthermore, the roughness coefficient in the overbank area is usually different from the coefficient in the main channel. Therefore, such a section should be divided along AB and treated as two separate sections. The plane AB, however, is not considered as a part of the wetted perimeter, since there is no appreciable shear in this plane.

Along a natural channel, there are frequently pools with a flatter slope and riffles or rapids with a steeper slope than the average slope of the channel over an appreciable distance. Therefore, care must be taken in studies of natural streams to consider the correct slope for the particular discharge and reach of stream in question.

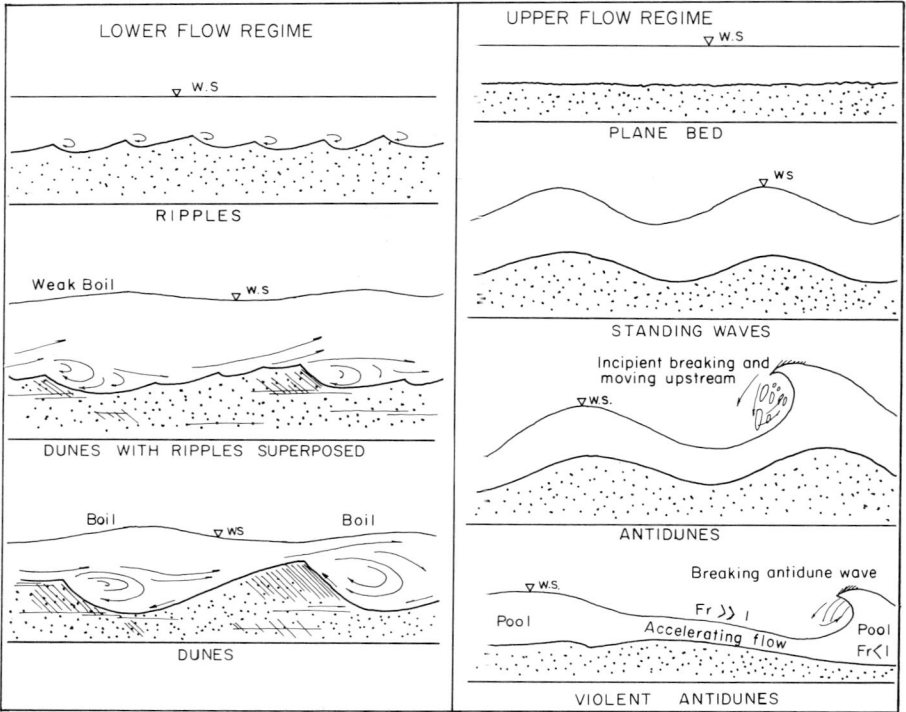

Fig. 41–8. Forms of bed roughness in alluvial channels.

4. FORMS OF BED ROUGHNESS AND RESISTANCE TO FLOW IN ALLUVIAL CHANNELS

The primary variables which affect the form of bed roughness and resistance to flow in alluvial channels (Simons and Richardson, 1962, 1963) include: The slope of the energy grade line, depth, physical size of the bed material as related to grain roughness, and fall velocity or effective median fall diameter as related to form resistance. The fall velocity or effective median fall diameter depends on the viscosity and mass density of the water sediment mixture and the mass density, size, and shape of the bed material. It reflects the principal viscous effect on flow in alluvial channels when Re is large. The effective median fall diameter is defined as the diameter of a sphere having a specific gravity of 2.65 and a fall velocity in distilled water of infinite extent at a temperature of 24C equal to the fall velocity of the particle falling alone in any quiescent stream fluid at stream temperature (Haushild et al., 1961; Simons et al., 1963).

The regimes of flow and various forms of boundary roughness (Simons and Richardson, 1963) which can occur in alluvial channels are illustrated in Fig. 41–8.

In the lower flow regime, flow is tranquil and the water surface undulations are out of phase with the bed undulations. Resistance to flow is large because separation of the flow from the boundary generates large scale turbulence that dissipates considerable energy.

With depths of flow ranging from 0.4 to 1.0 ft, ripple heights range from 0.01

to 0.1 ft and length (crest-to-crest) range from 0.5 to 1.5 ft. When depth is small, the ripples increase in size with depth, but at greater depths, ripple size becomes independent of depth. Therefore, ripples observed in flumes are similar in size and shape to those in natural streams. A decrease in n occurs when depth is increased indicating a relative roughness effect or when effective fall diameter is increased, which causes a decrease in ripple size. The decrease in n with an increase in effective fall diameter is similar to change reported in Leopold and Maddock (1953). Apparently, ripples do not form when the median diameter of the bed material is coarser than 0.7 mm.

With depth of flow ranging from 0.4 to 1.0 ft, dune heights range from 0.15 to 1.0 ft and dune length from 4 to 20 ft, based upon flume studies (Simons and Richardson, 1963). In deep rivers, dunes 30 to 60 ft in height and with lengths of several hundred feet have been observed. In the flume studies n increased with depth because size of the dunes and, hence, scale and intensity of turbulence increased. This may not be the case as larger depths are studied. With an increase in slope, n decreased for the fine sand but increased for the coarse sand because dune length increased appreciably with the fine sand but did not with the coarse sand. An increase in effective fall diameter increased n because dune length decreased and dune angularity increased. The long dunes formed by the finer sands exhibited smaller n values than ripples.

In the transition zone, n varied from the largest value for the lower flow regime to the smallest value for the upper flow regime. In this zone a well-defined relation between n and boundary shear does not exist. The bed form in the transition zone depended, in addition to the other factors, on antecedent conditions. Starting with dunes, slope and/or depth could be increased to relatively large values before plane bed or standing waves occurred. Conversely, with a plane bed and/or standing waves, the slope and/or depth could be decreased to relatively small values before dunes developed.

In the upper flow regime n values are small because surface or grain resistance predominates. However, the energy dissipated by the wave formation with the standing waves and the formation and breaking of the waves with antidunes increases n. Standing waves and antidunes, from the standpoint of wave mechanics, are rapid flow phenomena.

Standing waves are sinusoidal in-phase sand and water waves (Fig. 41–8) that build up in amplitude from a plane bed and water surface and gradually fade away. In the flume studies with depth of flow ranging from 0.2 to 0.6 ft, the water wave height (trough-to-crest) ranged from 0.01 to 0.6 ft and was 1.5 to 2 times the height of corresponding sand waves. Spacing of the waves was from 2 to 5 ft. Both height and spacing of the waves increased with depth. Resistance to flow for standing waves was larger than for a plane bed and, as with a plane bed, increased with an increase in sand size. Standing waves did not occur using the two finer sands because the mobility of the particles (effective fall diameter) allowed the development of antidunes whenever the Froude number equalled one.

Antidunes are similar to standing waves, except they increase in amplitude until they break. Breaking antidunes are similar to the hydraulic jump. The breaking wave dissipates a large amount of energy that is reflected by increased n. The increase of n is in direct proportion to the amount of antidune activity and the portion of the flume or channel occupied by the antidunes. Antidune activity increases with a decrease in effective fall diameter or with an increase in slope.

Table 41–8. Equations to estimate alluvial channel geometry

Engineer	Date	Equations	References
Kennedy	1895	$V = K_c\, D\, Km$ K_c ranges from 0.39-0.84 Km ranges from 0.52-0.73	Lacey, 1958.
Lindley	1919	$V = 0.95\, D^{0.57}$ $V = 0.57\, B^{0.36}$ $B = 3.8\, D^{1.61}$	Lacey, 1958.
Khannaq	1920	$V = 0.0216RS$	Lacey, 1958.
Beleida	1921	$V = 0.02808RS$	Lacey, 1958.
Malakal	1921	$V = 0.046RS$	Lacey, 1958.
Lacey	1929–58	$V = 1.17\, f^{1/2}\, R^{1/2}\,[f = 0.73\, V^2/R]$ $P = 2.67\, Q^{1/2}$ $V = $ Constant $R^{0.619}\, S^{0.357}$ $V = 16\, R^{2/3}\, S^{1/3}$ $A = 1.26Q^{5/6}/f^{1/3}$ $R = 0.47\, Q^{1/3}/f^{1/3}$	Lacey, 1958.
Bose	1936	$V = 1.12\, R^{1/2}$ $S = 2.09 \times 10^3\, d^{0.86}/Q^{.21}$ $A = PR$ $P = 2.8\, Q^{1/2}$ $R = 0.47\, Q^{1/3}$	Amer. Soc. Civil Eng., 1963.
Malhotra	1939–40	$V = 18.18\, R^{0.632}\, S^{0.343}$	Lacey, 1958.
Blench	1939–60	$V = (F_b F_s Q)^{1/6}$ $B = (F_b Q/F_s)^{1/2}$ $D = (F_s Q/F_b^2)^{1/3}$ $S = F_b^{5/6}\, F_s^{1/12}/(1 + C/233)\, KQ^{1/6}$ Bed factor $F_b = V^2/D$ Side factor $F_s = V^3/B$ $K = 3.63\, g\, \nu^{1/4}$	Comrie, 1961.

(continued on next page)

5. DESIGN OF STABLE CHANNELS IN ALLUVIAL MATERIAL

The preceding introduction to flow in alluvial channels indicates the necessity for continued study of alluvial channel flow phenomenon. Until the mechanics of flow in alluvial channels can be more precisely defined in terms of the fundamentals of fluid mechanics, it is essential to utilize such concepts as the regime concept of India as developed and presented by Kennedy, Lindley, Lacey, Bose, Inglis, Blench and others which has been discussed and extended by Simons and Albertson (1963), and the tractive force theory that was developed by the US Bureau of Reclamation under the direction of E. W. Lane (1953).

Although the regime theory is largely empirical, it recognizes that width, depth, and slope are variables and that three independent equations are essential to determine the magnitude of these variables The regime equations can be expressed in many forms. The following equations developed by Lacey (1958) are quite common:

$$V = 1.17\, f^{1/2}\, R^{1/2} \qquad [41\text{--}28]$$

$$P = 2.67\, Q^{1/2} \qquad [41\text{--}29]$$

$$R = 0.47\, Q^{1/3}/f^{1/3} \qquad [41\text{--}30]$$

$$S = 0.000547\, f^{5/3}/Q^{1/6} \qquad [41\text{--}31]$$

in which f is a silt factor equal to $3/4(V^2/R)$ or $1.76(d^{1/2})$, d is the median diameter of the bed material in mm, V is the average velocity in ft/sec, P is the

Table 41–8. (continued)

Engineer	Date	Equations	References
Leliavsky	1955	$V = TR^{0.85} S^{0.72}$ $T = [147 + 3.92 (z - 10)^{0.383}]$	Leliavsky, 1955.
Ning Chien	1955	$V^2/R = C (q_t/q)^{1/2}$ $(R^{1/2}S)^{2/3} = C (q_t/q)^{1/6}$ q_t = Sediment load per unit width q = Discharge per unit width	Chien, 1955.
Inglis-Lacey	1958	$W_s \propto Q^{1/2} I^{1/4}/g^{1/4} m^{1/4}$ $A \propto Q^{5/6} I^{-1/12}$ $S \propto Q^{-1/6} I^{5/12} g^{1/12} m^{5/12}$ $V \propto g^{1/2} D S/Em^{1/2}$ $V^3/W_s \propto g^{3/2} m^{1/2}$ $V^2/gD \propto I^{1/2}$ Inglis no. $I = XV_s/(\upsilon g)^{1/3}$ $E = P/W_s = D/R$	Lacey, 1958.
Liu and Hwang	1959	$V = C_a R_b{}^x S^y$	Liu and Hwang, 1959.
Kansoh	1960	$V = 0.56 D^{0.64}$, ft/sec $V = 0.36 D^{0.64}$, m/sec $B = 2.383 Q^{0.50}$ $D = 0.531 Q^{0.361}$ for sand beds and cohesive banks $D = 0.305 Q^{0.361}$ for coarse noncohesive materials.	Simons and Albertson 1963.
Ghaleb	1960	$V = 284 D^{0.727}$, metric units	Simons and Albertson, 1963.
Jareki	1960	V_b = competent bottom velocity $V_b = 0.645 d^{4/9}$, fine materials $V_b = 0.518 d^{1/2}$, coarse material	US Bur. Reclam., March 1960.
Sethna	1962	$V = \dfrac{66.5}{m^{0.1}} (RS)^{1/2}$, bed material moving $V = \dfrac{2525}{m^{0.2}} RS^{1/2}$, bed material not moving $S = 0.52 f^{0.6}/Q^{0.2}$ $f = (G/L) 8.95m^{0.2}$ G/L = grams/liter of silt charge	Sethna, 1962.
Ahmad and Rehman	1963	$S = K_2 f^{5/3}/Q^{1/6}$ $K_2 = 0.45 \times 10^{-3}$ to 0.7×10^{-3} $S = K_3 f^{5/3}/q^{1/3}$ $K_3 = 0.35 - 0.42$ $f = K_4 d^{1/2}$ $K_4 = 1.1 - 3.0$ $K_{4avg} = 1.9$ $b = K_1 Q^{1/2}$ $K_1 = 2.67 - 3.90$ as $Q/Q_o = 1$ to 0.4 Q_o = design discharge	Ahmad and Rehman, 1963.
Simons and Albertson	1963	Regime type relations modified to include the type of bank material and bed material concentration.	Simons and Albertson, 1963.
		Applicable to Mississippi River cross sections	
Anding	1964	$V = K_v (fR)^{1/2}$ $V = K_r R^{2/3} S^{1/3}$ $S = f^{5/3}/1750 Q^{1/6}$ $f = 8 d^{1/2}$ $P = 2.67 Q^{1/2}$ K_v and K_r = varying empirical constant. d = grain size in inches	Anding, M. G. Potamology studies— Mississippi River— hydraulic analysis of channel characteristics—etc. US Army Eng. Distr., Vicksburg, Mississippi. *Written communication.*

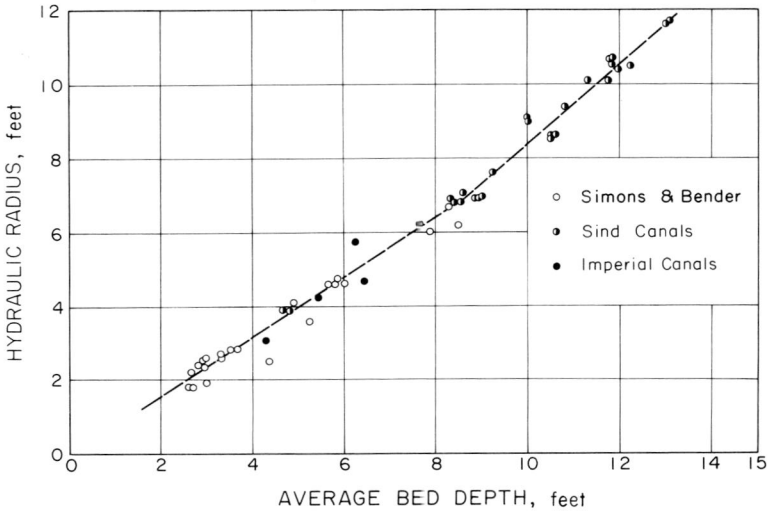

Fig. 41–9. Variation of hydraulic radius R with depth D.

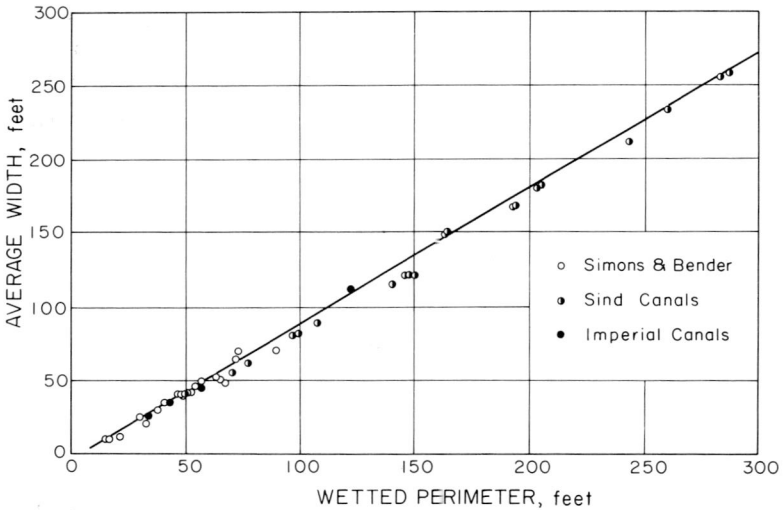

Fig. 41–10. Variation of average width W with wetter perimeter P.

wetted perimeter in feet, R is the hydraulic radius in feet, Q is the discharge in ft^3/sec and S is the slope of channel. Many large canals with sand beds are currently being designed according to the regime concepts. For a more complete list of regime type equations, refer to Table 41–8.

a. **Design Using Regime Type Concepts.** Some of the graphical relations resulting from an investigation of the preceding regime concepts by Simons and Albertson (1963) that are very useful for designing canals constructed in alluvium are presented in Fig. 41–9 through 41–17. These relations are based on field data from the USA, India, and Pakistan.

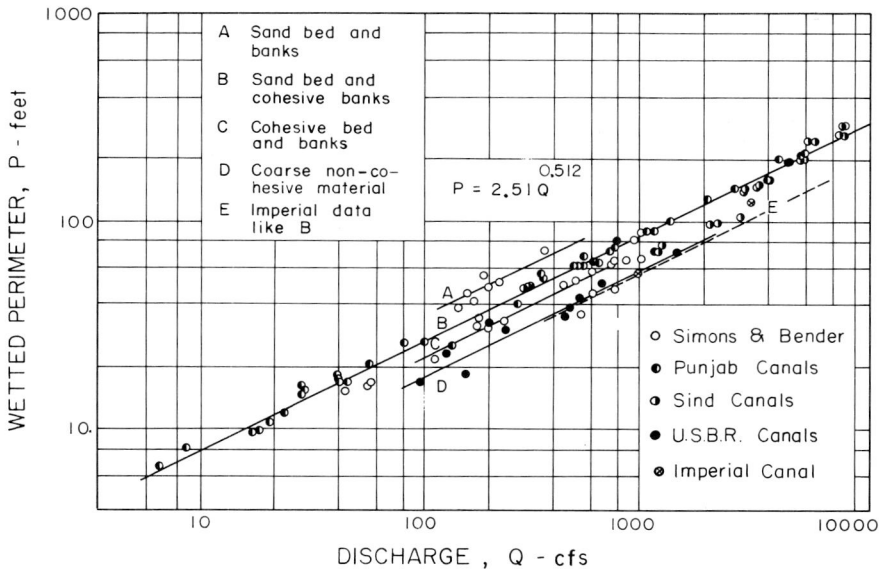

Fig. 41–11. Variation of wetted perimeter P with discharge Q and type of channel.

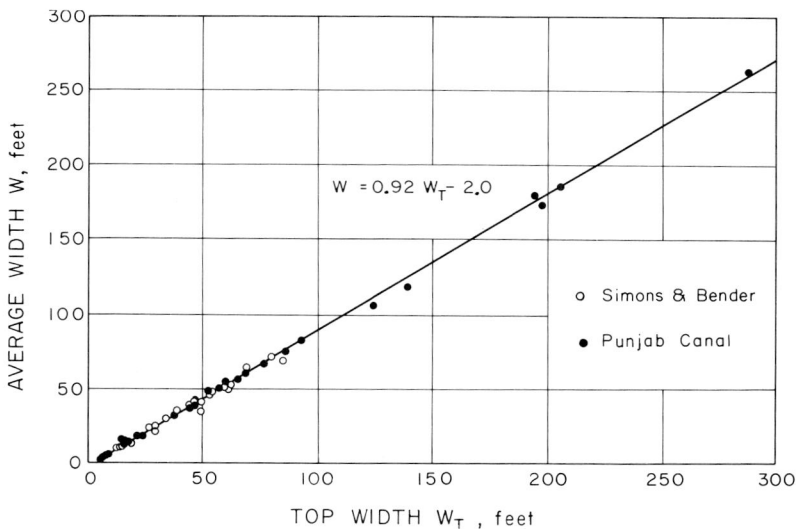

Fig. 41–12. Variation of average width W with top width W_T.

Knowing the design discharge and the type of bank material, one can estimate the hydraulic radius R from Fig. 41–13; the average bed depth from Fig. 41–9; the wetted perimeter from Fig. 41–11; the average width from Fig. 41–10; the top width from Fig. 41–12; the area of water cross section from Fig. 41–14; and knowing Q and A the average velocity can be estimated, and the slope can be estimated from relations such as those of Fig. 41–15, 41–16, and 41–17. One

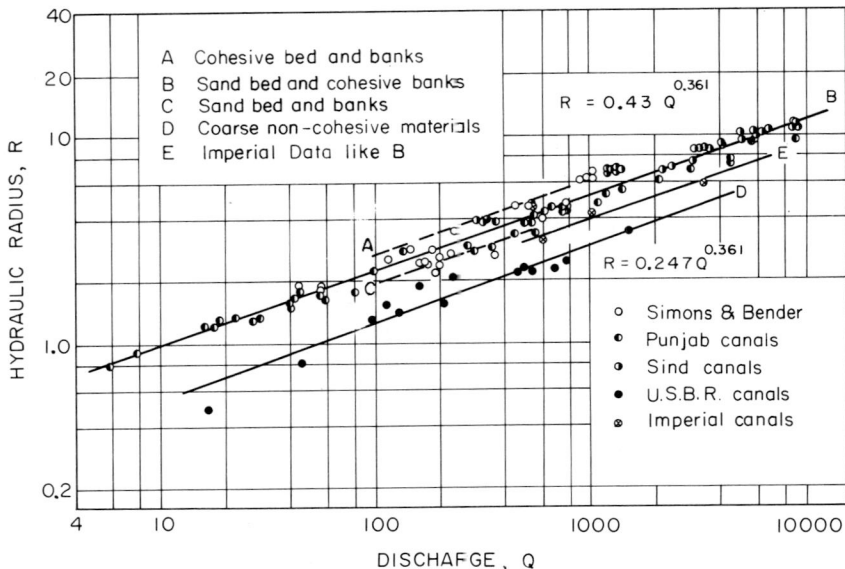

Fig. 41–13. Variation of hydraulic radius R with discharge Q and type of channel, all data.

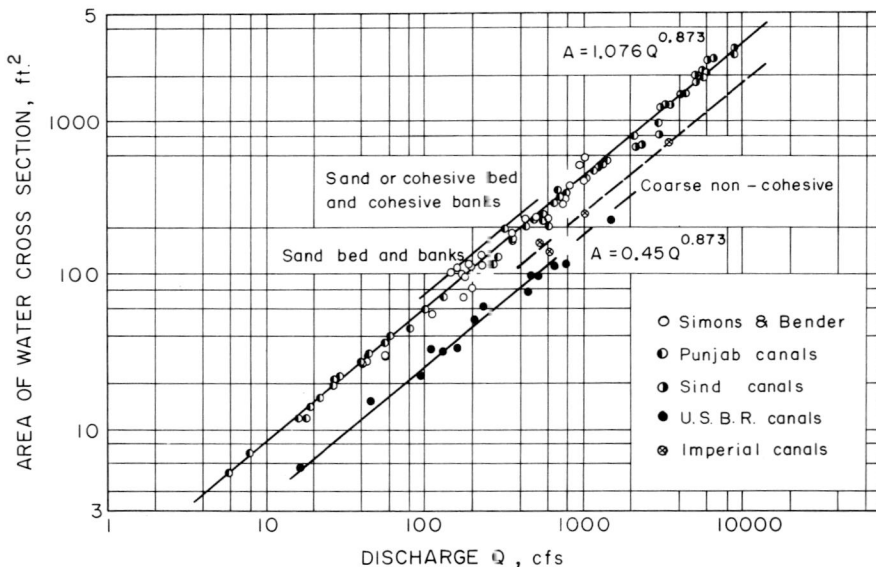

Fig. 41–14. Variation of area of water cross-section A with discharge Q and type of channel.

cannot obtain unique solutions using these relations. The results obtained by one procedure may vary from those obtained by another procedure. More recently, a more basic method of designing stable sand bed canals has been suggested by Simons and Richardson (1966).

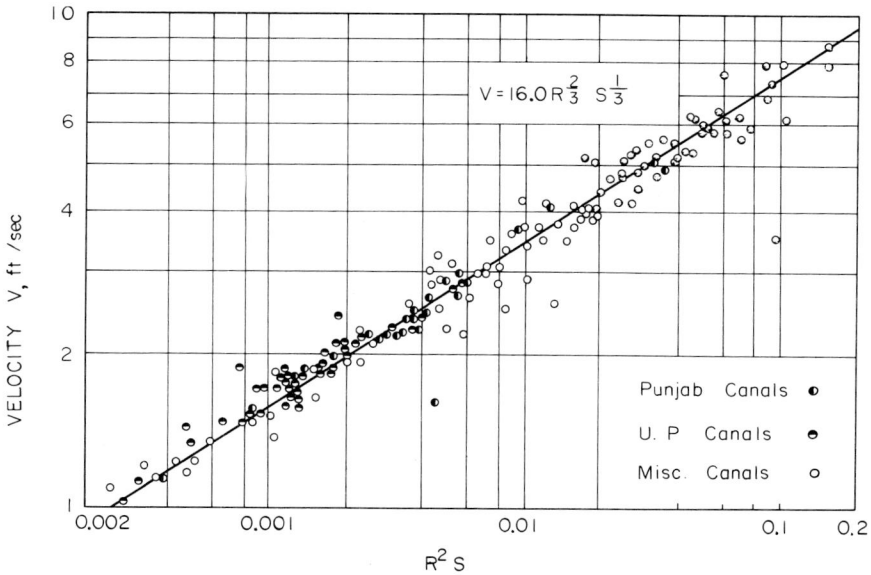

Fig. 41–15. Variation of average velocity V with R^2S, India data.

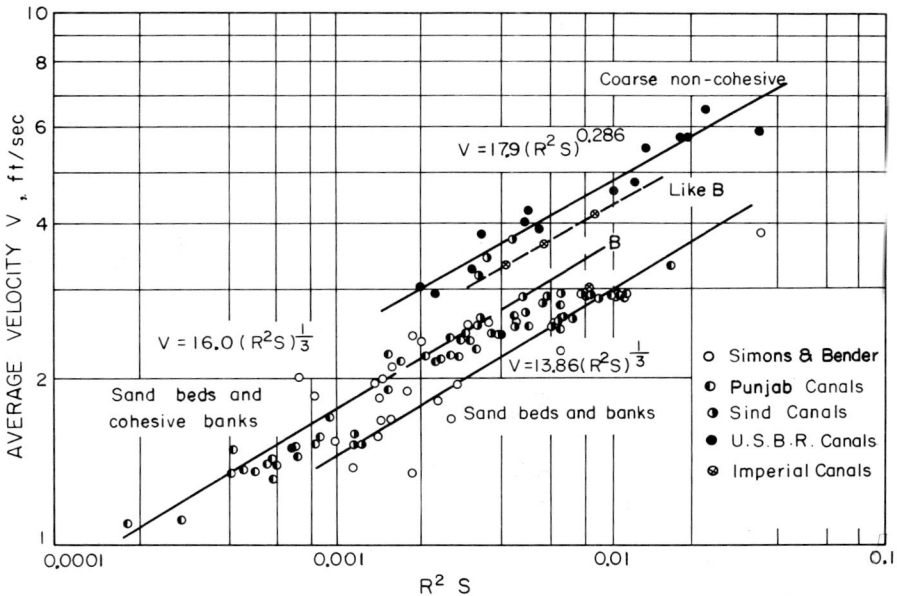

Fig. 41–16. Variation of average velocity V with R^2S and type of channel, all data.

b. Tractive Force. The tractive force theory is formulated on the basis that stability of bank and bed material is a function of the ability of the bank and bed to resist erosion resulting from the tractive force exerted on them by the moving water. The tractive force theory is clearly presented and illustrated by

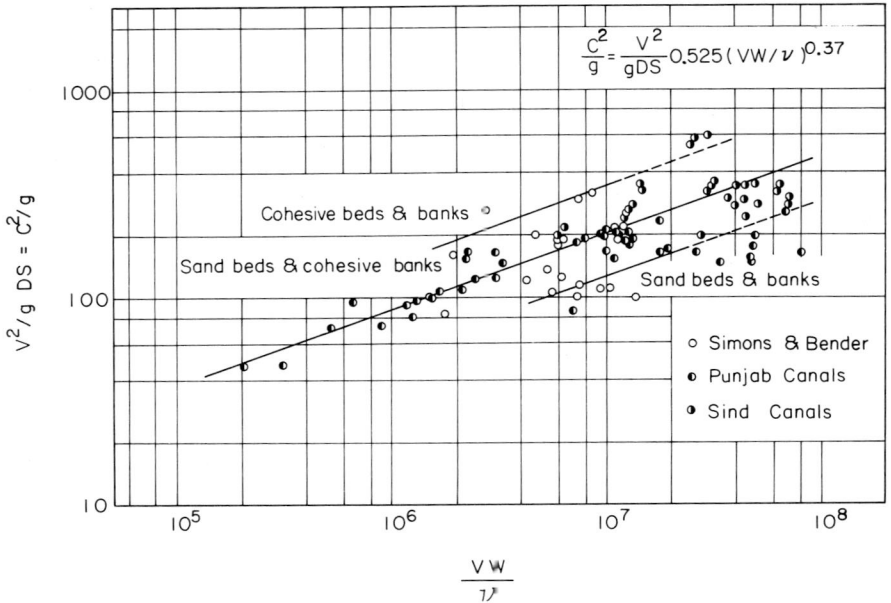

Fig. 41–17. Variation of V^2/gDS with VW/v and type of channel.

Lane (1953) and is applicable when conveying essentially clear water in coarse noncohesive materials and where bank stabilization is to be achieved by armor plating with coarse noncohesive material.

The tractive force design procedure utilizes Fig. 41–18 through 41–21. The procedure is:

1) Arbitrarily select the width W, depth D, and side slopes. In some cases the process can be simplified by using Fig. 41–9 to 41–14, inclusive, to estimate trial dimensions.

2) Obtain the size distribution curve of the natural material in which the canal is to be constructed and from Fig. 41–19 and 41–21, estimate the tractive force τ in pounds per square foot that the material can withstand and the angle of repose of the material.

3) Using Fig. 41–18b, read the value of the maximum tractive force on the bed of the channel divided by γDS where γ is the specific weight of the water in lb/ft³, D is the bed depth in feet, and S is the slope of energy gradient. The only unknown in this relation is the slope S of the energy gradient which can be calculated from this relation.

4) From Fig. 41–20, read the term K corresponding to the angle of repose of the material and the side slopes of the canal. The term K multiplied by the tractive force τ from Fig. 41–19 is the maximum allowable tractive force on the sides.

5) From Fig. 41–18a, read the value of the maximum allowable tractive force on the sides divided by γDS, and from this relation solve for the slope of energy gradient. The smaller value of this slope and that computed in number 4 above is the limiting slope.

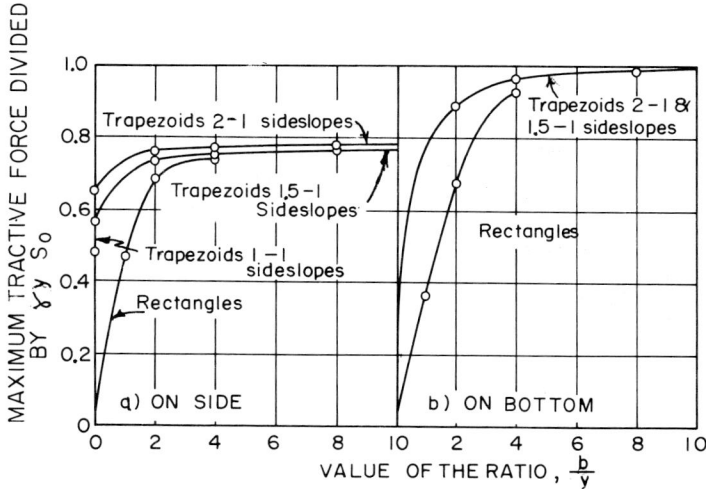

Fig. 41–18. Maximum tractive forces in a channel (after E. W. Lane, 1953).

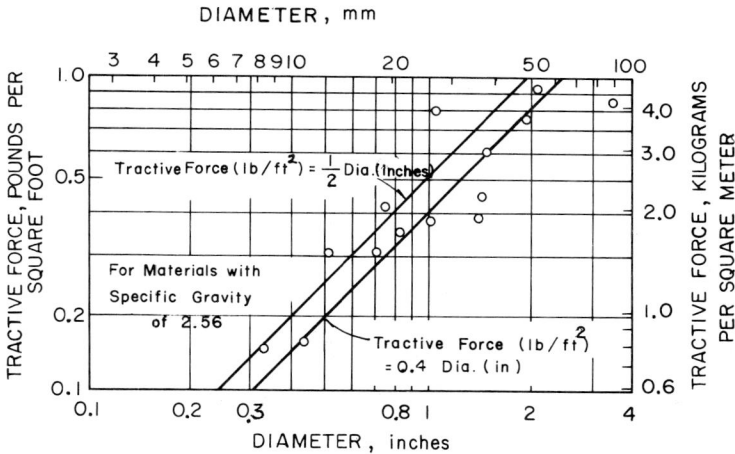

Fig. 41–19. Variation of tractive force with size of alluvial material (after E. W. Lane, 1953).

6) Estimate the value of the Manning n for the size of material using the relation

$$n = (K_{35}/R)^{1/6} \qquad [41\text{–}32]$$

where K_{35} is the sieve size of the material forming the bed of the canals of which 35% of the material is larger and R is the hydraulic radius.

7) Compute the discharge capacity of the canal using the selected value of n, the assumed geometry of the cross section and the limiting slope. Then revise the design as required using the same procedure until a canal is obtained that will just convey the design discharge.

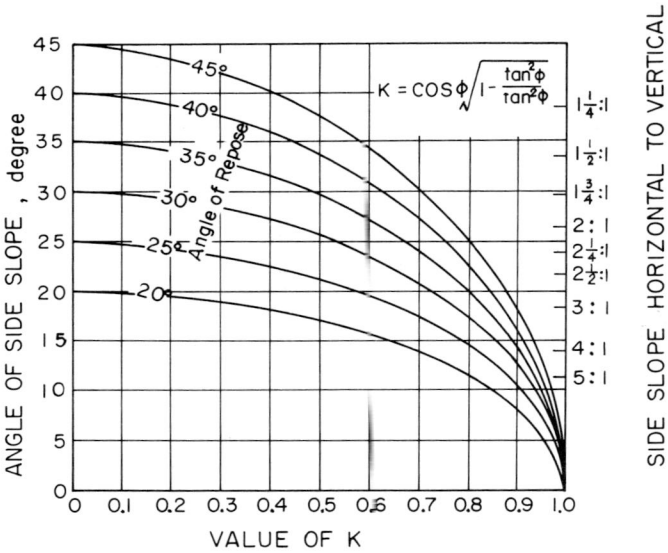

Fig. 41–20. Relationship between side slope and K (after E. W. Lane, 1953).

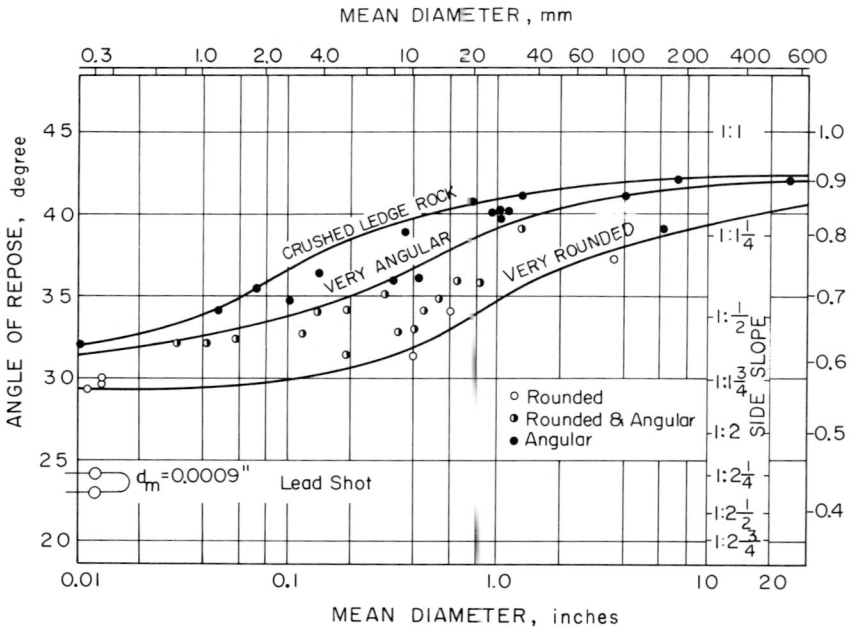

Fig. 41–21. Angle of repose of noncohesive material.

Also, this procedure can be used to design gravel and rock riprap protection for erodible banks of alluvial channels when the average velocity is less than 8 ft/sec.

Fig. 41–22. Characteristic dimensions
of a meander.

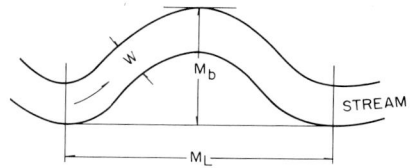

6. MEANDERS

Natural channels can usually be classified as meandering or braided channels. Although the mechanics of meandering are not completely understood, at least four variables, valley slope, sediment discharge, water discharge, and the characteristics of the bed and bank material, control the process of meandering. The process of meandering can be triggered by an excess of sediment load. When the sediment load exceeds that associated with stability, the slope of the stream steepens due to deposition, the boundary shear is increased, and, if the banks of the channel are not sufficiently resistant, the banks begin to erode. The banks erode unevenly and only a slight misalignment of flow is necessary to shift the meander of the stream toward one side or the other which further increases the rate of attack of the water on the bank and may initiate the development of a meander system.

The length and width of meanders, as illustrated in Fig. 41–22, have been studied under both laboratory and field conditions. As a part of the regime theory, the following equations have been prepared by Inglis (1938–1939) for rivers in floodplains.

$$M_L = 29.6 \ (Q)^{1/2} \qquad\qquad [41–33]$$

$$M_b = 84.7 \ (Q)^{1/2} \qquad\qquad [41–34]$$

$$W = 4.88 \ (Q)^{1/2} \qquad\qquad [41–35]$$

Similar equations except for magnitude of coefficients have been prepared for incised rivers.

Experiments conducted by Friedkin (1945) indicate that meandering is caused by local bank erosion, increased sediment load, and deposition of some of the coarser material; the radius of curvature of the bends increases with discharge and/or slope; the meanders move downstream, and many irregularities result due to differences in bank and bed material.

If the slope of a stream is excessive or if the discharge is increased to a relatively large magnitude, the local rate of bank scour and deposition may be of sufficient magnitude to cause the stream to braid. In general, braiding is associated with steeper slopes and larger sediment loads than meandering.

7. SEDIMENT TRANSPORT

Knowledge of sediment transport in alluvial channels is just as important as knowledge of resistance to flow. The ability of a stream to transport bed material is relatively small when the form of bed roughness consists of ripples and/or dunes. In the upper regime of flow, the streams are capable of carrying much larger volumes of sediment per unit volume of water [see qualitative data in Table 41–9 suggested by Simons and Richardson (1963)]. Some of the more useful concepts for estimating bed material discharge have been presented by Einstein (1950), Colby and Hembree (1955), Colby (1964), Simons et al.

Table 41–9. Variation of concentration of tota bed material load with regimes of flow and forms of bec roughness

Regime of flow	Forms of bed roughness	Total bed material load in ppm	
		Median diameter of bed material 0.28 mm	Median diameter of bed material 0.45 mm
Lower flow regime	Ripples	1-150	1-100
	Dunes	150-800	100-1,200
Transition	Zone in which dunes are reducing in amplitude with increasing shear stress.	1,000-2,400	1,400-4,000
Upper flow regime	Plane	1,500-3,100	----
	Standing waves	----	4,000-7,000
	Antidunes	5,000-42,000	6,000-15,000

(1965), and Bishop et al. (1965). For a more detailed treatment of the sediment problems encountered in designing and operating irrigation canals constructed in alluvium, refer to Simons and Miller (1966).

8. SIMPLE WAVES AND SURGES

Various types of waves and surges may occur in open channels and cause a locally unsteady flow. The simplest is the small surface wave which is observed to progress radially outward from the point of a small disturbance. The rate at which this wave progresses is called its celerity c, (Fig. 41–23a). Like velocity, the dimensions of celerity are L/T. The unsteady flow of Fig. 41–23a can be transformed into steady flow simply by superposing a velocity on the system to the right which is equal in magnitude but opposite in direction to the celerity.

Assuming the energy loss in the wave to be negligible, the Bernoulli equation, together with the continuity equation, can be used to develop an analytical solution for rate of propagation of a solitary wave in terms of the depth of flow y and the height of the wave Δy.

$$V = c = (g)^{1/2} \left[\frac{2(y + \Delta y)}{1 + y/(y + \Delta y)} \right]^{1/2} \qquad [41-36]$$

When the wave height Δy is small compared with the depth y, this equation reduces to

$$c = (gy)^{1/2}[1 + (3\Delta y/2y)]^{1/2}. \qquad [41-37]$$

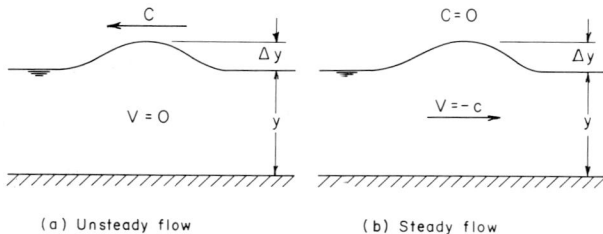

(a) Unsteady flow   (b) Steady flow

Fig. 41–23. Small gravity waves.

Furthermore, as the ratio of wave height Δy to depth y goes to zero, equation [42–37] becomes

$$c = (gy)^{1/2}. \qquad [41\text{--}38]$$

This discussion of waves is based on shallow water waves. However, for the purpose of comparison, the equation for celerity of deep water waves is independent of depth y and is approximately

$$c = (g\lambda/2\pi)^{1/2} \qquad [41\text{--}39]$$

in which λ is the length of the wave and $\lambda < y/2$. As the amplitude or height of the deep water wave increases, this equation must be modified until the limiting condition $2\Delta y/\lambda = 1/7$, where the wave becomes unstable and breaks, as commonly illustrated by ocean or lake waves generated by the wind.

In equations [41–36], [41–37], and [41–38], it can be seen that if velocity is equal to or greater than the celerity of the wave, the wave cannot move upstream, i. e., when $V > c$ the wave moves downstream and when $V = c$ the wave is stationary. For the very small stationary wave

$$V = (gy)^{1/2} \qquad [41\text{--}40]$$

which may be rearranged as

$$V/(gy)^{1/2} = 1.0. \qquad [41\text{--}41]$$

This equation is the principal definition of critical flow. That is, critical flow occurs when the velocity of flow is just equal to the celerity of a small wave in quiet water at the same depth. Equation [41–37] is the Froude number (Fr) for critical flow conditions. The general equation of the Froude number is

$$\text{Fr} = V/(gy)^{1/2} \qquad [41\text{--}42]$$

which shows that the Froude number is the ratio of the velocity of flow to the celerity of a very small gravity wave. When $\text{Fr} < 1.0$ the wave can move upstream, and when $\text{Fr} > 1.0$ the wave is carried downstream.

As the magnitude of the wave height Δy is increased, equation [41–37] (for a small wave) becomes less and less applicable and, at $\Delta y \approx y$, the wave becomes unstable and breaks. This may be understood by considering a large wave or surge as a series of small incremental surges of height Δy (*see* Fig. 41–24a). Each

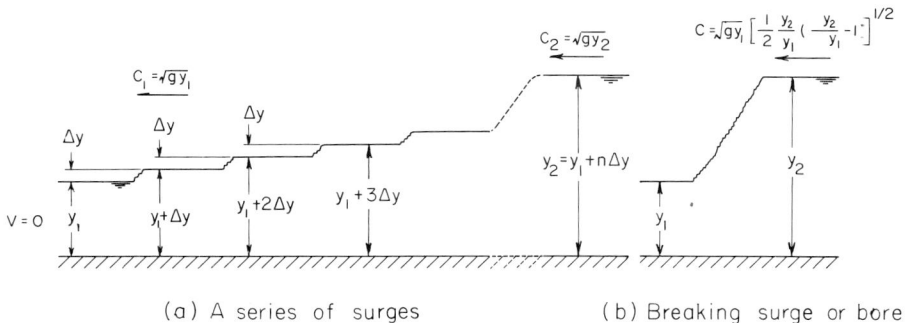

(a) A series of surges   (b) Breaking surge or bore

Fig. 41–24. Analysis of the breaking surge.

of these small individual surges has a celerity corresponding to the depth of flow over which it is moving, i.e., at the left side the depth is y_1 so that the celerity c is

$$c = (gy_1)^{1/2} \qquad\qquad [41\text{–}43]$$

while at the right side the depth is $y_2 = y_1 + n\Delta y$ and the celerity c is

$$c = (gy_2)^{1/2}. \qquad\qquad [41\text{–}44]$$

Since $y_2 > y_1$, the incremental surges on the right have a celerity to the left which is greater than those surges on the left. Hence, the surges on the right will overtake those on the left and a very steep surge will develop. As the height of this surge $n\Delta y$ increases, it becomes undulatory, and as it approaches y_1, the flow at the face of the surge becomes unstable and breaks (*see* Fig. 41–24b).

9. HYDRAULIC JUMP

The breaking surge can be changed to steady flow by superimposing a velocity of flow to the right which is equal in magnitude and opposite in direction to the net celerity of the breaking surge.

One of the most useful ways of expressing the relation for the hydraulic jump is

$$y_2/y_1 = 1/2[(1 + 8\,\mathrm{Fr_1}^2)^{1/2} - 1] \qquad\qquad [41\text{–}45]$$

which shows that when $\mathrm{Fr_1} = 1.0$ (critical depth) the depths y and y_2 are equal.

Figure 41–25 is a dimensionless representation of the characteristics of the hydraulic jump. Because equation [41–45] is not simple to solve, an approximate empirical equation

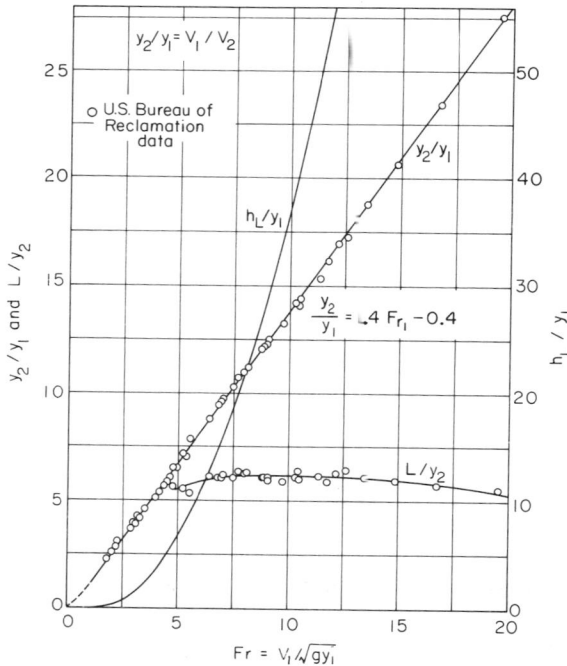

Fig. 41–25. Hydraulic jump, dimensionless form.

$$y_2/y_1 = 1.4 \, \mathrm{Fr}_1 - 0.4 \qquad\qquad [41\text{-}46]$$

may be used for ease of computation.

Certain significant facts may be observed from a study of Fig. 41–25:

1) When $\mathrm{Fr}_1 < 1.0$, no hydraulic jump can exist because the celerity of a wave is greater than the velocity of flow.

2) When $1.0 < \mathrm{Fr}_1 < 1.7$ and $y_2/y_1 < 2.0$, there is very little head loss because the depth ratio $y_2/y_1 < 2$ and hence the surge is undulatory and not breaking.

3) When $\mathrm{Fr}_1 > 2$ and $y_2/y_1 > 2.4$, the surge is breaking and h_L increases very rapidly, showing that the hydraulic jump is an excellent energy dissipator.

4) The relative length of the hydraulic jump L/y_2 is nearly constant at about 5.5 to 6.0 beyond $\mathrm{Fr} = 5.0$.

The hydraulic jump may occur as a moving surge which is traveling either upstream, if $V < c$, or downstream, if $V > c$. To analyze this phenomenon, the flow system is first changed to steady flow by superposing a velocity on the entire system which is equal in magnitude and opposite in direction to the absolute velocity of the surge. Once the flow is steady, the problem may be analyzed by the usual procedure for the hydraulic jump.

A wave caused by a sudden release of additional water upstream in the channel may develop into a moving surge which is either stable or unstable (the moving hydraulic jump), and equation [41–37] is applicable. If the channel is dry, a sudden release of flow may be represented approximately by

$$c \approx 1.53 (gq)^{1/2} \approx 2(g\Delta y)^{1/2} \qquad\qquad [41\text{-}47]$$

in which c is the velocity (celerity) of the wave front, q is the steady discharge introduced into the channel, and Δy is the height of the wave.

10. TRANSITIONS IN OPEN CHANNELS

Flow through relatively short transitions in open channels can be described by the Bernoulli or energy equation, since the resistance forces over these short distances are relatively small. As shown in Fig. 41–26, an increase in velocity caused by a reduction in the area of cross section, either by raising the floor or by contracting the sides, causes an increase in velocity head, and hence a decrease in water surface elevation. The Bernoulli equation

$$(V_1^2/2g) + y_1 + z_1 = (V_2^2/2g) + y_2 + z_2 \qquad\qquad [41\text{-}48]$$

is applicable to this flow pattern and may be used to determine the drop in water-surface elevation.

In connection with a detailed study of Fig. 41–26 and various other transitions,

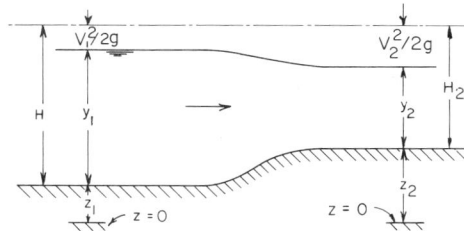

Fig. 41–26. Flow in transitions.

equation [41–48] can be expressed as

$$H_1 + z_1 = H_2 + z_2 \qquad [41\text{–}49]$$

in which

$$H = (V^2/2g) + y. \qquad [41\text{–}50]$$

The new term H is known as the specific head or the total head above the floor of the channel. As will be seen in the following pages, the specific head H is very useful in analyzing the flow in open channels. By use of the continuity equation, equation [41–50] can be expressed in terms of q, the discharge per unit width of a wide rectangular channel, as

$$H = (q^2/2gy^2) + y. \qquad [41\text{–}51]$$

This equation can be analyzed from two viewpoints: (i) Variation of y with H when q is held constant which yields the specific head diagram; and (ii) variation of y and q when H is held constant which yields the discharge diagram.

11. SPECIFIC HEAD DIAGRAM

The specific head equation, equation [41–51], can be plotted as shown in Fig. 41–27a to show how the specific head H varies with the depth of flow y for progressively increasing values of discharge per unit width q_1, q_2, q_3, etc. This diagram shows that two different depths can exist with a given specific head H and discharge q; e.g., see point A in Fig. 41–27a where the depth is small and the velocity is great and point B where the depth is great and the velocity is small. These depths are termed alternate depths because they can occur independently of each other and at the same specific head, depending only on the conditions of the boundaries. Also of significance is the fact that there is a minimum value of specific head. This minimum means that, for a given discharge, there is a minimum specific head below which the discharge will decrease— e.g., from C_2 to C_1 in Fig. 41–27a—until the specific head diagram for the correct discharge q is reached.

The depth of flow for the minimum value of the specific head H can be evaluated by differentiating equation [41–51] with respect to y and setting it equal

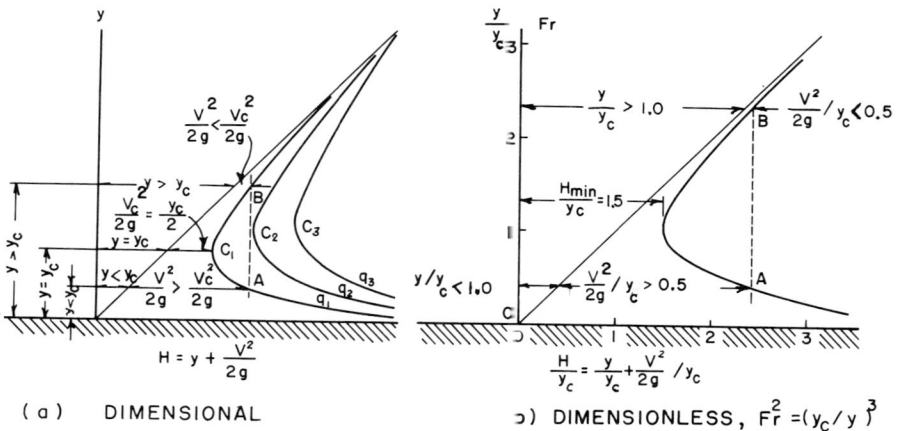

Fig. 41–27. Specific head diagrams.

to zero and rearranging to yield

$$q = (gy_c^3)^{1/2} \qquad\qquad [41\text{--}52]$$

or

$$y_c = (q^2/g)^{1/3} = 2(V_c^2/2g) \qquad\qquad [41\text{--}53]$$

in which y_c and V_c are the critical depth and critical velocity, respectively.

The foregoing discussion of critical depth is confined to flow in rectangular channels. For nonrectangular channels the equation for critical velocity V_c is

$$V_c = (g^A c/K_e B_c)^{1/2} \qquad\qquad [41\text{--}54]$$

where K_e is the kinetic energy correction coefficient.

The critical depth y_c must be determined as the depth which corresponds to the critical area A_c in a plot of A vs. y. The ratio A_c/B_c is an average depth which is less than critical depth, since it is the average over the entire width B_c (*see* Fig. 41–28). The velocity head $V_c^2/2g$ for critical flow

$$K_e(V_c^2/2g) = (A_c/B_c)/2 \qquad\qquad [41\text{--}55]$$

can be seen to be equal to one-half the average depth A_c/B_c when $K_e = 1.0$

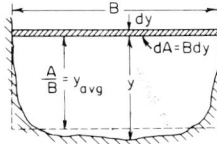

Fig. 41–28. Nonrectangular channels.

a. **Gradually Varied Flow.** The types of flow in open channels which have been considered thus far involve only steady, uniform flow where the depth is called normal depth, or nonuniform flow for which changes in cross section take place in a relatively short distance. For uniform flow, the slope of the water surface, the slope of the bed, and the slope of the total head line are all equal. When the nonuniform flow extends over only a short distance through a streamlined transition, resistance is negligible.

Backwater curves and drawdown curves, which are usually included in the term "backwater curves," describe the water surface in an open channel where changes in cross section take place very gradually with distance along the channels. The assumption is made that the resulting changes in velocity take place so slowly that the accelerative effects are negligible. Because the changes take place gradually, the flow is commonly known as gradually varied flow.

Changes in the cross section of the flow may result either from a change in geometry of the channel—e.g., a change in slope or cross-sectional shape, or an obstruction—or from an unbalance between the forces of resistance tending to retard the flow and the forces of gravity tending to accelerate the flow.

In order to analyze the various types of backwater curves, the total head H_T can be expressed as

$$H_T = K_e(V^2/2g) + y + z \qquad\qquad [41\text{--}56]$$

$$= K_e[(Q^2/A^2)/2g] + y + z \qquad\qquad [41\text{--}57]$$

in which K_e is the kinetic energy correction coefficient, y is the depth of flow, and z is the elevation of the channel bed above some arbitrary datum.

Since the variation of these terms with distance x along the channel is desired, equation [41–57] can be differentiated with respect to x to obtain

$$\frac{dH_T}{dx} = -\frac{Q^2}{gA^3}\frac{dA}{dx} + \frac{dy}{dx} + \frac{dz}{dx} \qquad [41\text{–}58]$$

$$= -\frac{Q^2B}{gA^3}\frac{dy}{dx} + \frac{dy}{dx} + \frac{dz}{dx} \qquad [41\text{–}59]$$

in which K_e is assumed again to be unity and dA is set equal to $B\,dy$. The gradient of total head dH_T/dx can be set equal to the negative of the slope obtained from the Chezy equation $S = (Q/A)^2/C^2R$, and the bed slope is $dz/dx = -(Q/A_o)^2/C_o^2R_o = -S_o$ for normal uniform flow conditions. Making simplifying assumptions (Albertson et al., 1960),

$$dy/dx = S_o\left[\frac{1 - (n/n_o)^2(y_o/y)^{10/3}}{1 - (y_c/y)^3}\right]. \qquad [41\text{–}60]$$

With equation [41–60] it is possible to classify the various surface profiles of backwater curves which may occur in open channels.

12. CLASSIFICATION OF WATER SURFACE PROFILES

The analysis of surface profiles depends first on the sign of dy/dx, the slope of the water surface relative to the bed. If dy/dx is positive, the depth is increasing downstream, and if it is negative, the depth is decreasing downstream. From equation [41–60], it can be seen that the slope dy/dx depends on the slope of the bed S_o, the ratio n/n_o of the Manning n for the actual depth to that for the normal depth which would occur if uniform flow existed, the ratio y_o/y of the normal depth to the actual depth and the ratio y_c/y of the critical depth to the actual depth. In the following analysis, it is assumed that $n/n_o = 1.0$. Although for some conditions this assumption is not justified (*see* the discussion of alluvial channels), it is assumed to be sufficiently exact for the purposes of this analysis. Hence

$$dy/dx = S_o\left[\frac{1 - (y_o/y)^{10/3}}{1 - (y_c/y)^3}\right]. \qquad [41\text{–}61]$$

The slope of the channel serves as the primary means of classification. If the bed slope S_o is negative, the bed rises in the direction of flow. This is called an adverse slope and the curves of the water surface over it are known as A curves. If $S_o = 0$, the bed slope is horizontal and the curves over it are H curves. When $S_o > 0$, the bed slope is either mild, steep, or critical, and the corresponding curves of the water surface are either M curves, S curves, or C curves—depending on the ratio y_o/y_c. When $y_o/y_c > 1.0$, an M curve exists, when $y_o/y_c = 1.0$, a C curve exists, and when $y_o/y_c < 1.0$, an S curve exists.

A further classification of surface curves depends on the ratios y_c/y and y_o/y. If both y_c/y and y_o/y are < 1.0, the curve is designated as type 1—e.g., M_1, S_1, C_1 and H_1. If the depth y is between the normal depth y_o and the critical depth y_c, it is type 2, and if both y_c/y and y_o/y are > 1.0, the curve is type 3.

Each of the foregoing curves is plotted in Fig. 41–29 where the longitudinal distance has been shortened and the slopes have been exaggerated for the sake of clarity.

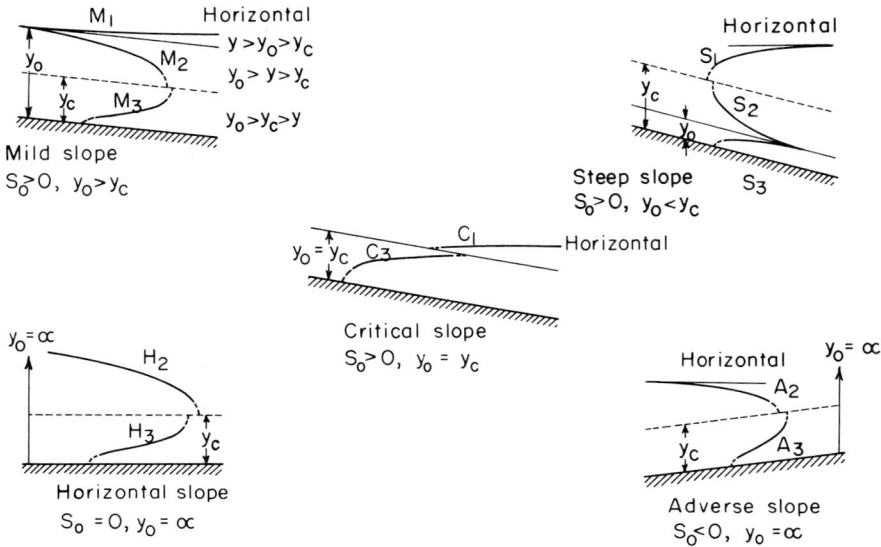

Fig. 41–29. Classification of surface profiles.

The general characteristics of surface profile are summarized in Table 41–10.
Figure 41–30 illustrates actual conditions where backwater curves of the various types occur. Again, the longitudinal distances are very much foreshortened. A sluice gate could be placed where the dam is located in each case and the same curves would exist. It should be observed that a smooth, although relatively sharp, curve is created when the profile crosses the critical depth line from a greater to a lesser depth (see the A_2 curve and the S_2 curve in Fig. 41–30). As the profile crosses from a lesser to a greater depth, however, there is a hydraulic jump which causes a sudden change in elevation.

Table 41–10. Characteristics of surface profiles

Class	Bed slope	$y: y_o: y_c$	Type	Symbol
Mild	$S_o > 0$	$y > y_o > y_c$	1	M_1
Mild	$S_o > 0$	$y_o > y > y_c$	2	M_2
Mild	$S_o > 0$	$y_o > y_c > y$	3	M_3
Critical	$S_o > 0$	$y > y_o = y_c$	1	C_1
Critical	$S_o > 0$	$y < y_o = y_c$	3	C_3
Steep	$S_o > 0$	$y > y_c > y_o$	1	S_1
Steep	$S_o > 0$	$y_c > y > y_o$	2	S_2
Steep	$S_o > 0$	$y_c > y_o > y$	3	S_3
Horizontal	$S_o = 0$	$y > y_c$	2	H_2
Horizontal	$S_o = 0$	$y_c > y$	3	H_3
Adverse	$S_o < 0$	$y > y_c$	2	A_2
Adverse	$S_o < 0$	$y_c > y$	3	A_3

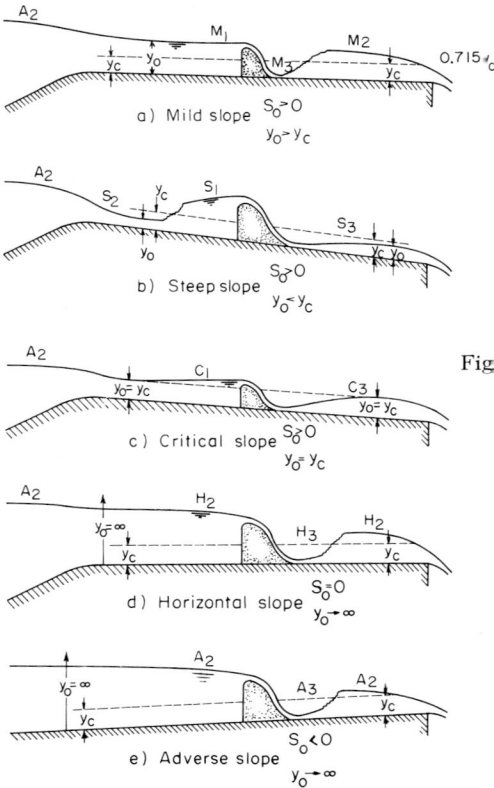

Fig. 41–30. Examples of surface profiles.

The location of the hydraulic jump is a problem which frequently must be solved. This is accomplished by computing the surface profile curve—M_2 and M_3, H_3 and H_2, or A_3 and A_2—in the direction of increasing depth until the computed downstream depth reaches the point where it is the sequent depth for the computed upstream depth. In other words, the surge is propagated upstream or downstream until the approaching velocity of flow is just equal to the celerity of the surge in quiet water at the same depth

In summary, the following statements can be made about the surface profile curves:

1) The type-1 curves are all above both y_c and y_o.

2) The type-2 curves are all between y_c and y_o.

3) The type-3 curves are all below both y_c and y_o.

4) The type-3 curves all approach the bottom perpendicular to it.

5) All curves approaching the y_o line approach it asymptotically, except for the C curves, where $y_o = y_c$.

6) All curves approaching the y_c line approach it perpendicularly, except for the C curves, where $y_o = y_c$.

7) All curves approaching a horizontal line approach it asymptotically, except for the C curves, which are horizontal throughout.

8) All curves where $y_o < y_c$ are unaffected upstream from any disturbance because the wave of the disturbance is swept downstream by the rapid flow. Compute in the downstream direction.

9) All curves where $y_o > y_c$ are influenced by any downstream disturbance because the celerity of the wave of the disturbance is so great it travels upstream against the oncoming flow. Compute in the upstream direction.

The foregoing discussion of the characteristics of surface profiles in open channels is applicable in general not only to wide uniform channels, but also to natural open channels, provided the changes in shape, roughness, and slope are taken into consideration properly. In the following section, the step method is explained for the actual computation of backwater curves.

13. COMPUTATION OF BACKWATER CURVES

The foregoing material explains the various types of backwater curves which can occur in an open channel. The system of computing these curves is explained in this section.

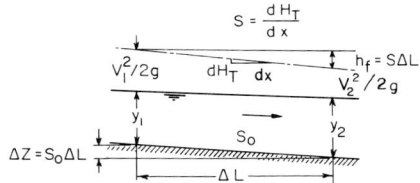

Fig. 41–31. Definition sketch for computation of surface profiles.

Figure 41–31 illustrates a reach of channel sufficiently short so that the water surface can be approximated by a straight line. Equation [41–58] may then be rearranged as

$$\Delta H / \Delta L = S_o - S$$

or

$$\Delta L = \Delta H / (S_o - S) \qquad [41\text{--}62]$$

in which H is the specific head, $S = -\,dH_T/dx$ the average of the total energy gradient, and $S_o = -\,dz/dx$ the slope of the channel bed. By rearranging the Manning equation $S = Q^2 n^2 / (1.5AR^{2/3})^2$ to compute the average of the total energy gradient S for the two ends of the reach, the backwater curve may be computed from equation [41–62] by a step procedure (Chow, 1964).

The basic backwater curve equation can also be derived by writing the Bernoulli equation between points 1 and 2 in Fig. 41–31 and solving for ΔL. That is,

$$(V_1{}^2/2g) + y_1 + \Delta z = (V_2{}^2/2g) + y_2 + h_f$$

or

$$(V_1{}^2/2g) + y_1 + S_o\,\Delta L = (V_2{}^2/2g) + y_2 + S\,\Delta L$$

and

$$\Delta L = (H_2 - H_1)/(S_o - S) = \Delta H/(S_o - S). \qquad [41\text{--}63]$$

The procedure for computation is to assume a depth y and compute the length of reach ΔL required to reach that depth, as follows:

1) Determine or assume the initial conditions of depth, channel characteristics, and discharge. Insert into a table the initial values for the depth y, the area A,

the wetted perimeter P, the hydraulic radius R, the velocity V, the velocity head $V^2/2g$, the specific head $H = y + (V^2/2g)$, and the corresponding slope S for the assumed value of Manning's n.

2) Assume a new depth of flow y for the other end of the reach and insert the corresponding values of A, P, R, V, $V^2/2g$, H and S.

3) Compute the average S values and the change in specific head.

4) Insert the ΔH for the reach value and the average value of S in equation [41–63] and solve for ΔL.

5) Repeat the process for each reach, adding the resulting ΔL values to obtain the total required distance $\Sigma \, \Delta L$ until the depth or distance desired is reached.

The foregoing step procedure for computing backwater curves is very simple and is applicable to channels of variable cross section, roughness and slope, provided this information is known accurately for the entire length of channel under study. Other methods have been developed for computing backwater curves which are more rapid. In most cases, however, they involve assumptions which limit their application to uniform channels, or to first approximation computations.

III.   SUMMARY

The basic concepts of fluid mechanics applicable to the design of conveyance and distribution systems has been presented. For a more detailed treatment of hydraulics and fluid mechanics, see Albertson et al. (1960) and Albertson and Simons (1964) and other fluid mechanics texts. Also, many valuable concepts pertinent to the design of hydraulic structures associated with the conveyance and distribution of water have been presented by the US Bureau of Reclamation (1960, 1963, 1964).

LITERATURE CITED

Ahmad, M., and A. Rehman. Oct. 1963. Appraisal and analysis of new data from alluvial canals of West Pakistan in relation to regime concepts and formulae. West Pakistan Eng. Congr.

Albertson, M. L., J. R. Barton, and D. B. Simons. 1960. Fluid mechanics for engineers. Civil Eng. Eng. Mech. Ser. Prentice Hall, Englewood Cliffs, New Jersey. 568 p.

Albertson, M. L., and D. B. Simons. 1964. Handbook Appl. Hydrol. (V. T. Chow, ed.) McGraw-Hill, New York. Ch. 7.

American Society of Civil Engineers. March 1963. Friction factors in open channels. Amer. Soc. Civil Eng. Proc., J. Hydraul. Div. 89(HY2):97–143.

Bazin, H. 1897. Etude d'une nouvelle formule pour calculer le debit des canaux decouverts. Annales des ponts et chaussees. Memoire No. 41. Vol. 14, Ser. 7, 4 me trimestre. p. 20–70.

Bishop, A. A., D. B. Simons, and E. V. Richardson. 1965. Total bed-material transport. Amer. Soc. Civil Eng., Proc. J. Hydraul. Div. 91(HY2):175–191.

Chien, Ning. 1955. A concept of Lacey's regime theory. Amer. Soc. Civil Eng. Sep. no. 620.

Chow, Ven Te. 1964. Handbook of applied hydrology. McGraw-Hill, New York. 1,418 p.

Colby, B. R. 1964. Discharge of sands and mean-velocity relationships in sandbed streams. US Geol. Surv. Prof. Pap. 462–A. 47 p.

Colby, B. R., and C. H. Hembree. 1955. Computations of total sediment discharge, Niobrara River near Cody, Nebraska. US Geol. Surv Water Supply Pap. 1357. 187 p.

Comrie, J. 1961. Civil engineering reference book. 2nd ed. Butterworth & Co., London.

Einstein, H. A. Sept. 1950. The bed-load function for sediment transportation in open channel flows. US Dep. Agr. Tech. Bull. 1026. 71 p.

Ellis, G. H. 1916. The flow of water in irrigation canals. Amer. Soc. Civil Eng. Trans. Paper no. 1373, p. 1644–1688.

Friedkin, J. F. 1945. A laboratory study of the meandering of alluvial rivers. War Dep. Corps Eng. US Army. Mississippi River Comm. US Waterways Exp. Sta. Vicksburg, Miss. 40 p.

Garbrecht, Gunther. 1961. Flow calculations for rivers and channels. Die Wasser-Wirtschaft. Stuttgart. Parts I & II, p. 40–45 and 72–77. (US Bureau of Reclamation Transl. 402)

Haushild, W. L., D. B. Simons, and E. V. Richardson. 1961. The significance of fall velocity and effective diameter of bed materials. US Geol. Surv. Prof. Pap. 424–D. p. 17–20.

Inglis, C. C. 1938–39. The relationship between meandering belts, distance between meanders on axis of stream, width and discharge of rivers in flood plains and incised rivers. Central Board Irrig. India. Annu. Tech. Rep. p. 49–50.

Lacey, Gerald. 1958. Flow in alluvial channels with sandy mobile beds. Inst. Civil Eng., Proc. (London) 9:145–164.

Lane, E. W. 1953. Progress report on studies on the design of stable channels by the Bureau of Reclamation. Amer. Soc. Civil Eng., Proc. 79:1–31.

Leliavsky, S. 1955. An introduction to fluvial hydraulics. Constable & Co., London.

Leopold, L., and T. Maddock. 1953. The hydraulic geometry of stream channels and some physiographic implications. US Geol. Surv. Prof. Pap. 252. 56 p.

Liu, H. K., and S. Y. Hwang. 1959. A discharge formula for flow in straight alluvial channels. Amer. Soc. Civil Eng. Trans. Paper no. 3276.

Manning, Robert. 1889. On the flow of water in open channels and pipes. Institution Civil Eng. of Ireland, Trans. (Dublin) 20:161–207. (Supplement, 1895. 25:179–207.)

Sethna, T. R. 1962. Uniform flow of water in alluvial channels. Proc. Inst. Civil Eng. Vol. 21, Paper 6524.

Simons, D. B., and M. L. Albertson. 1963. Uniform water conveyance channels in alluvial material. Amer. Soc. Civil Eng., Trans. 128:65–106.

Simons, D. B., and C. R. Miller. Jan. 1966. Sediment discharge in irrigation canals. Int. Comm. Irrig. Drainage, Proc., 6th Congr. (New Delhi, India) Quest. 20, Rep. 12. p. 20, 275–20, 307.

Simons, D. B., and E. V. Richardson. 1962. Resistance to flow in alluvial channels. Amer. Soc. Civil Eng., Trans. 127:927–952.

Simons, D. B., and E. V. Richardson. 1963. Forms of bed roughness in alluvial channels. Amer. Soc. Civil Eng., Trans. 128:284–302.

Simons, D. B., and E. V. Richardson. 1966. Resistance to flow in alluvial channels. US Geol. Surv. Prof. Pap. 422–J (In Press). p. —

Simons, D. B., E. V. Richardson, and W. L. Haushild. 1963. Some effects of fine sediment on flow phenomena. US Geol. Surv. Water Supply Pap. 1498–G. 46 p.

Simons, D. B., E. V. Richardson, and C. F. Nordin. 1965. Bedload equation for ripples and dunes. US Geol. Surv. Prof. Pap. 462–H. 9 p.

Tilp, P. J., and M. W. Scrivner. April 1964. Analyses and descriptions of capacity tests in large concrete-lined canals. US Bur. Rec. Tech. Memo 661. 198 p.

US Bureau of Reclamation. 1960. Design of small dams. US Gov. Printing Office, Washington, D. C. 611 p.

US Bureau of Reclamation. March 1960. Design of stable channels with tractive forces and competent bottom velocity. Sedimentation Sect., Hydrology Branch, Bureau of Reclamation, Denver Federal Center, Denver, Colo.

US Bureau of Reclamation. 1963. Hydraulic design of stilling basins and energy dissipators. Supt. Doc. Washington, D. C. Eng. Monogr. 25. 114 p.

US Bureau of Reclamation. 1964. Design standards no. 3. Canals and related structures. Commissioner's Office. Denver Federal Center, Denver, Colo.

42

Water Control and Measurement on the Farm[1]

A. R. ROBINSON and A. S. HUMPHERYS

Agricultural Research Service, USDA
Kimberly, Idaho

Complete water control and measurement in farm distribution systems are essential requirements for an efficient irrigation system. Problems and hydraulic principles involved in the selection, design and operation of structures for accurate water control and measurement are presented in this chapter. For convenience, the chapter has been divided into two main categories, open channel and closed systems. Control and measurement structures and methods are discussed for each.

1. OPEN-CHANNEL SYSTEMS

An open channel is one in which the streamflow is not completely confined by solid boundaries but has a free surface subject to atmospheric pressure. It may be an open conduit or a pipe flowing partly full. In farm distribution systems, most open channels consist of lined or unlined earth ditches and flumes. In contrast to closed systems, open channels are constructed on a grade corresponding to that which the water surface is expected to assume.

A. Water Control

An efficient irrigation system requires that the operator have complete control of the water with ability to measure it at various points throughout the system. He must be able to apply water to the land in the quantities needed at nonerosive velocities and with a minimum of labor. Open-channel water control on the farm is achieved by using structures to control the water as it is conveyed from the main canal or lateral headgate, natural stream, or other source to its destination on the field. Structures may be required also to control the channel itself when unlined ditches are used. These water-control structures regulate water levels, dissipate excess energy, provide accurate distribution, and deliver water at the desired rate, without erosion, onto the field. Names of the various structures referred to in this chapter will be those in common use in the USA. In other countries, the same structure may be referred to by different names.

1. FUNCTIONAL REQUIREMENTS AND PROBLEMS

Water flowing in a farm ditch is usually below the level required for field diversion. The water surface in the ditch must be raised to allow distribution from the ditch by gravity flow. High water application efficiency also requires

---

[1] Contribution from the Soil and Water Conserv Res. Div., ARS, USDA.

discharge controls that accurately meter and control the flow of water onto the field with a minimum of erosion.

Irrigation water frequently contains sediment or other undesirable materials. Sediment in the water may be deposited in irrigation ditches, pipelines, and measuring structures. This necessitates frequent ditch cleaning and often results in inaccurate flow measurement. Trash in irrigation water may be a source of weed infestation on the farm; it also clogs smaller irrigation structures. Structures are needed to remove trash and excess sediment from the water.

Land slopes on irrigated farms are often greater than the gradient necessary to overcome friction losses in open channels. Thus, problems of grade control, energy dissipation, and maintaining uniform water distribution onto the field are encountered. On most irrigated farms, erosion-control and energy dissipating structures are needed to stabilize unlined ditches.

Structures that remain in place for more than one irrigation season are considered permanent. Those that are moved from place-to-place during each irrigation or installed for one season's use are considered temporary. Some temporary structures are required on most irrigated farms, but permanent structures normally permit better water control with less labor. The cost of farm irrigation structures can often be reduced by combining two or more structures wherever possible. For example, checks, drops, turnouts, divisors and measuring structures can be combined in various combinations as illustrated in Fig. 42–1.

Fig. 42–1. Combination irrigation structures may be built for measuring, checking, dividing, and dropping the stream.

Wooden structures have been used extensively in the USA during the past century but are being replaced more and more by concrete and metal. Structures made of the more durable materials are recommended for permanent installations. Many commercial concerns are now manufacturing precast concrete and modular or component type metal structures. These permanent type structures are very useful, efficient and well adapted for farmer installation. Effective water control on the farm may also be obtained by using improved plastic and rubber devices and structures.

2. WATER LEVEL CONTROL

Water level control structures are perhaps the most common and frequently used. Suggestions and guidelines for the design and installation of farm water

Fig. 42–2. Two types of check structures commonly used on the farm (left photo courtesy of Soil Conserv. Serv., USDA).

level control structures are given in various publications, (Code, 1961; Gilden and Woodward, 1952; Herpich and Manges, 1959; Jensen et al., 1954; and Robinson et al., 1963). Other references are also given by Israelsen and Hansen (1962).

a. Checks. A check is any structure installed in an open channel to raise the water level above its normal flow depth. A variety of checks are used in both lined and unlined ditches. They are usually fitted with grooves to receive checkboards or with metal slide gates which permit flow to bypass while maintaining the desired water level. Some commonly used permanent and portable check structures are shown in Fig. 42–2.

When a constant upstream water level is desired, an overflow type check is normally used. The flow over such a check may be estimated from the general equation

$$Q = C'Lh(2gh)^{1/2} = CLh^n \qquad [42-1]$$

where

$Q =$ discharge, L^3/T,

$C' =$ coefficient of discharge, dimensionless,

$C = C'(2g)^{1/2} =$ coefficient of discharge, $L^{1/2}/T$,

$L =$ overflow crest length, L,

$h =$ head or water depth above the crest measured upstream from the check, L, where L and T denote length and time in convenient dimensions.

The value of the exponent n for most overflow type checks is approximately 1.5. When the crest length L is large, variations in discharge result in relatively small changes in the upstream water level.

When the water level is to be controlled downstream from a structure, an orifice-type check is more desirable because of a more constant discharge. The discharge through an orifice may be determined from the general equation

$$Q = CA(2gh)^{1/2} \qquad [42-2]$$

where C is coefficient of discharge, dimensionless, A is area of opening, L^2, g is acceleration of gravity, L/T^2, and h is head causing flow, L. The coefficient of discharge C ranges from 0.6 to approximately 0.8, depending on the position of

Fig. 42–3. Drop structures used for grade control and energy dissipation in a farm ditch (left photo courtesy of Soil Conserv. Serv., USDA).

the orifice relative to the sides and bottom of the structure and on the roundness of the orifice edge. For free discharge, the head h is the upstream water depth and is measured from the center of the opening. For submerged flow, the effective head is the difference between the upstream and downstream water surface levels. Because of its head–discharge relationship, an orifice-type check is not as well adapted for upstream water level control since fluctuations in quantity of flow result in relatively large upstream water level variations.

Commercial prefabricated checks for unlined ditches can be obtained with or without an apron on the downstream side. The apron is often omitted because of difficulty encountered when backfilling beneath it. If the apron is not part of the check, adequate erosion protection must be provided downstream by a stilling basin, riprap or a rigid apron.

b. Drops. Grade control structures are required to reduce erosive flow velocities in unlined ditches. In a steeply sloping channel, erosion control is accomplished by conveying water from one level to another in a stairstep manner (Fig. 42–3). Grade control structures are called drops when the water surface is lowered in a short horizontal distance. If the drop occurs over longer distances in a channel

Fig. 42–4. Drop structure combined with turnout; the blocks are used to shorten the length of the hydraulic jump.

permitting high velocities, the structure is generally referred to as a chute. Where water is to be diverted from the ditch or lateral to a field or another ditch at a lower level, a drop structure is often combined with a check or turnout (Fig. 42–4).

A primary consideration in the use of drops is to provide adequate downstream protection. The energy of the falling water must be dissipated to prevent undermining of the structure and erosion of the downstream channel. In most field ditches with maximum recommended drops limited to 30 to 60 cm (12 to 24 inches), rock riprap is used to protect the downstream banks from eddy currents and turbulence created in the drop. The apron is set below the downstream grade to form a stilling pool to cushion the falling water. Higher drops or large streams require more elaborate means for dissipating the stream energy and reducing its velocity. Drops in excess of 90 cm (3 ft) require special precautions such as cut-off walls to insure against erosion, uplift, and piping.

A variety of open-type drop structures are used and are constructed from various materials as previously mentioned. Enclosed drops made of concrete or corrugated steel pipe having short right angle elbows are also used (Fig. 42–5). These drops are particularly useful in small ditches and where a combined road crossing is needed. However, this drop is more easily plugged by trash, and riprap protection is needed both at the inlet and outlet.

A recent practice in some areas is that of constructing slipform lined ditches in steps of level sections with a drop at the end of each section. A check, placed just ahead of the drop, raises the water level for diversion to the field. This type of construction results in essentially a horizontal water surface in each level section for uniform distribution to the field. This practice will become more common with increased use of automatic irrigation control structures.

c. Dams. Portable irrigation dams are essentially checks used to raise the water level in the ditch for diversion. They have been used by the irrigation farmer for many years. Dams in common use are made of canvas, plastic, or butyl rubber. They are low in cost but require special care to prevent damage in order to extend their life beyond one season. Many dams are fitted with an adjustable dam stick or weir-type opening. The water level in the ditch is controlled by adjusting the water flow over the top or through the opening of the dam (Fig. 42–6). Portable dams usually have a higher labor requirement than improved permanent structures.

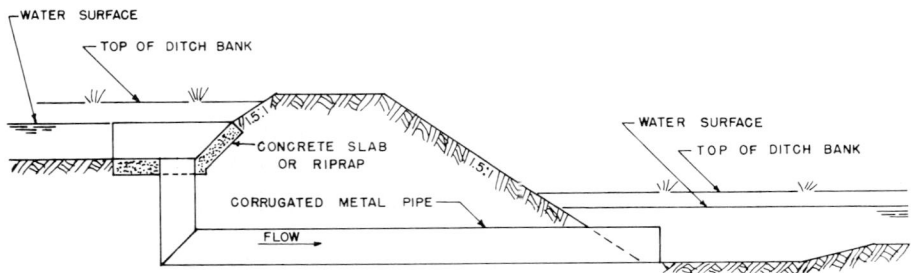

Fig. 42–5. Metal pipe drop structure; in some cases a scour hole may form at the downstream end which may be stabilized by riprap.

Fig. 42–6. Portable dams and siphon tubes for controlling the flow and delivery of water to the crops.

d. Automated Structures. Automatic water level control structures in canals and laterals have been used extensively in Southern Europe and North Africa for many years. Some of the most commonly used structures are described by Thomas (1960). Most of these controls are used to maintain a constant water level in the channel for diversions at turnouts and for water measurement.

Automated water level controls for irrigation on the farm are in the development stage and are currently used only to a very limited extent. Automation of farm water control structures may greatly improve the efficiency of surface irrigation and in particular can reduce the labor requirement considerably. Irrigation systems on many of the larger irrigated farms in the USA are expected to incorporate automatic controls extensively in the near future. At the present time, most automatic structures are of the check type which control the water level in a farm distribution ditch (Fig. 42–7) (Bondurant and Humpherys, 1962). After checking the water level to a raised position for a predetermined time, the automatic gate is released, allowing the water to flow to the next set. Individual furrows or border strips receiving the water must be well graded so they may be irrigated without the farmer's attention.

A timing or sensing device is required to trigger these automatic structures. The energy required to operate the structure itself is usually obtained from the flowing water. Automatic control structures vary from simple alarm-clock-timer

Fig. 42–7. Automatic water control structures for farm ditches.

released checks to elaborate radio and electronically controlled structures with program timers or moisture-sensing devices.

Structures can be classified as fully automatic or semiautomatic. Fully automatic structures usually use sensing devices located in the field or programmed timers to trigger their operation. Other structures are being developed which incorporate the timing means within the structure itself. A fully automatic gate will reset itself after the completion of one irrigation and be ready for the next. Most structures in use at the present time are semiautomatic and require manual resetting between irrigations. They are usually triggered by a mechanical timer.

3. DISCHARGE CONTROL

Discharge control devices are used extensively to control distribution of water from a farm ditch into border strips or furrows. However, discharge control frequently consists of dividing the total flow into two or more specific increments rather than controlling the discharge rate. The hydraulic characteristics of some commonly used devices have been determined in field and laboratory studies (Tovey and Myers, 1959).

a. Turnouts. A turnout structure may be an opening of fixed dimensions in the side of a ditch or one equipped with check boards, gates, or other devices to adjust the area of the opening. Typical examples of commercially available turnouts are shown in Fig. 42–8. If only a portion of the total flow is to be diverted through a given turnout, a more constant discharge is obtained by using an orifice-type device, such as a gated turnout, instead of an overflow or weir-type structure. When the turnout consists only of a fixed opening, flow regulation is achieved by controlling the water level in the ditch.

b. Spiles. Outlets placed in the side of a lined ditch or ditch bank to control the delivery of water to individual furrows or corrugations are called spiles. Individual spiles having adjustable gates are referred to as gated outlets. They are usually placed above the normal water level in a ditch or equalizing bay so the water must be checked before diversion takes place. One of the main advantages in using spiles or gated outlets is that once installed and adjusted, no further adjustment is made during the irrigation season. Thus, after they are installed, the labor requirement is lower than with siphons which must be reset each time or with gated surface pipe which must be moved between irrigation sets.

Fig. 42–8. Commonly used turnouts for farm irrigation ditches.

Irrigation with spiles is similar in principle to using gated surface pipe or lay-flat distribution tubing. Discharge from a pipe or tube is controlled by adjustable gates or outlets uniformly spaced along its length corresponding to the furrow or corrugation spacing.

c. Siphons. Siphon tubes are widely used to distribute water from the ditch onto a field (Fig. 42–2 and 42–6). They are available commercially in several diameters and lengths and are usually made from plastic or aluminum. They may be obtained in sizes that permit control of streams as small as 4 liters (approximately 1 gal)/min or as large as 56 liters (2 ft³)/sec. The larger sizes are usually used to flood border strips or check basins. The use of siphons is normally limited to fields having little cross slope in order to maintain a near-constant operating head on each tube. The discharge depends on tube diameter and length, number and degree of bends, and roughness in addition to the operating head. Siphon-tube discharge is given by the same general equation as that for an orifice, equation [42–2], in which the coefficient of discarge C has different values. The coefficient for siphons may be evaluated, if the entrance coefficient, the inside diameter of the tube, the roughness coefficient, and the tube length are known (US Department of Agriculture, 1962). With the tube size and head known, the discharge may be obtained from charts (Tovey and Myers, 1959; US Department of Agriculture, 1962). Siphons eliminate cutting the ditch bank, thus reducing labor and ditch maintenance, but they must be primed manually.

d. Division Boxes. It is often necessary to divide water from a farm lateral into two or more ditches for distribution to different parts of the farm or to other farms. This may be accomplished with a divisor at the ditch junctions. For accurate flow division, it is best to use measuring structures in each channel. However, if a divisor is used, it should have a long, straight approach channel so that the flow of water approaches it in parallel paths without cross currents. Care must also be taken that downstream conditions do not favor one side or the other. Flow divisions also may be made by dividing the flow as it falls over a control crest (Code, 1961; Thomas, 1960). If the flow is divided by the same user and accurate flow division is not required, two or more regular turnout structures may be used.

e. Automatic Discharge Control. Higher water application efficiency may theoretically be obtained if the flow in a furrow is reduced or cut back after the water has reached the end of the field. This technique is seldom employed in practice because of the time required to readjust the individual furrow streams and the difficulty in managing the surplus water which must be used downstream or wasted. Structures and systems are being developed to automatically reduce the flow of water to the furrow after a prescribed time interval (Garton, 1966).

Thomas (1960) describes some automatic discharge control devices used at farm turnouts where they also function as water measuring structures. Other devices and techniques are being developed to control discharge from the farm ditch or lateral onto the field or into another channel. A self-propelled traveling siphon has been used successfully where a large discharge is needed to flood irrigate borders. Self-operating, float-type turn out structures are being developed to control the discharge from a ditch onto the field.

Pneumatically operated and radio controlled valves are also being developed to control the discharge from turnout structures (Haise et al., 1965). These valves

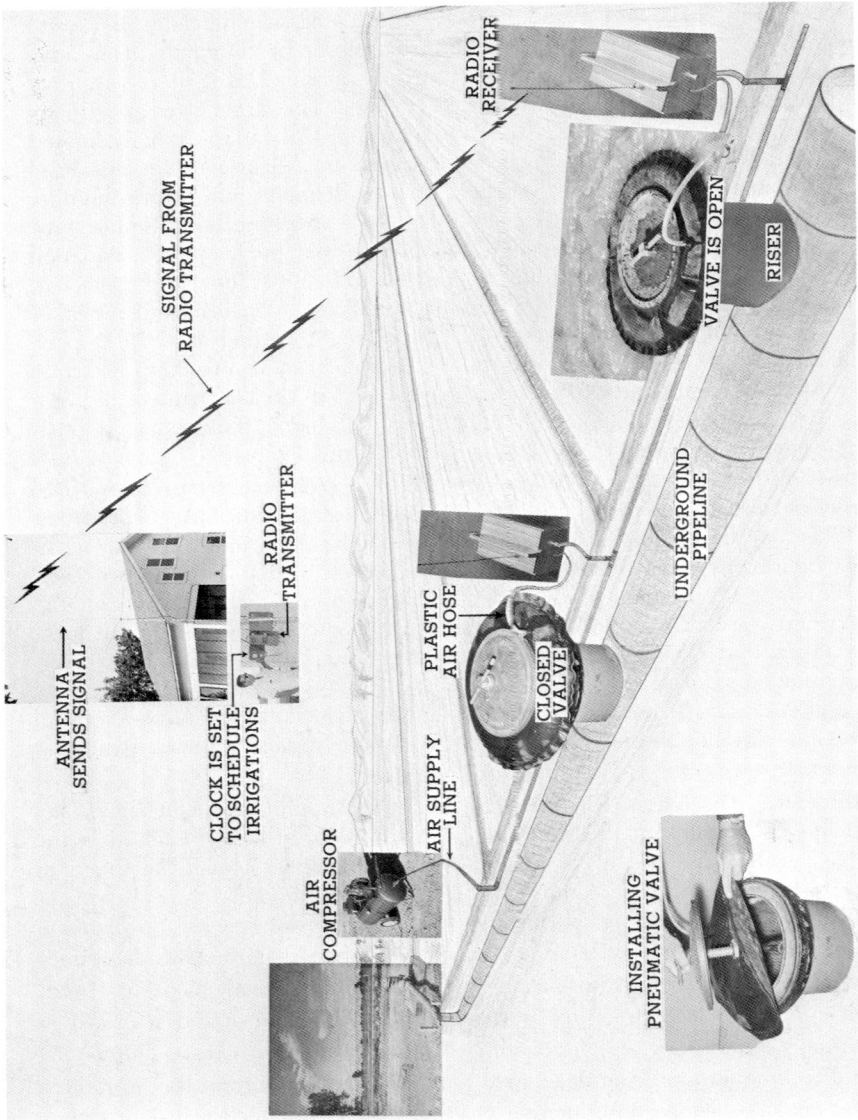

Fig. 42-9a. Radio-controlled pneumatic valves for pipeline distribution.

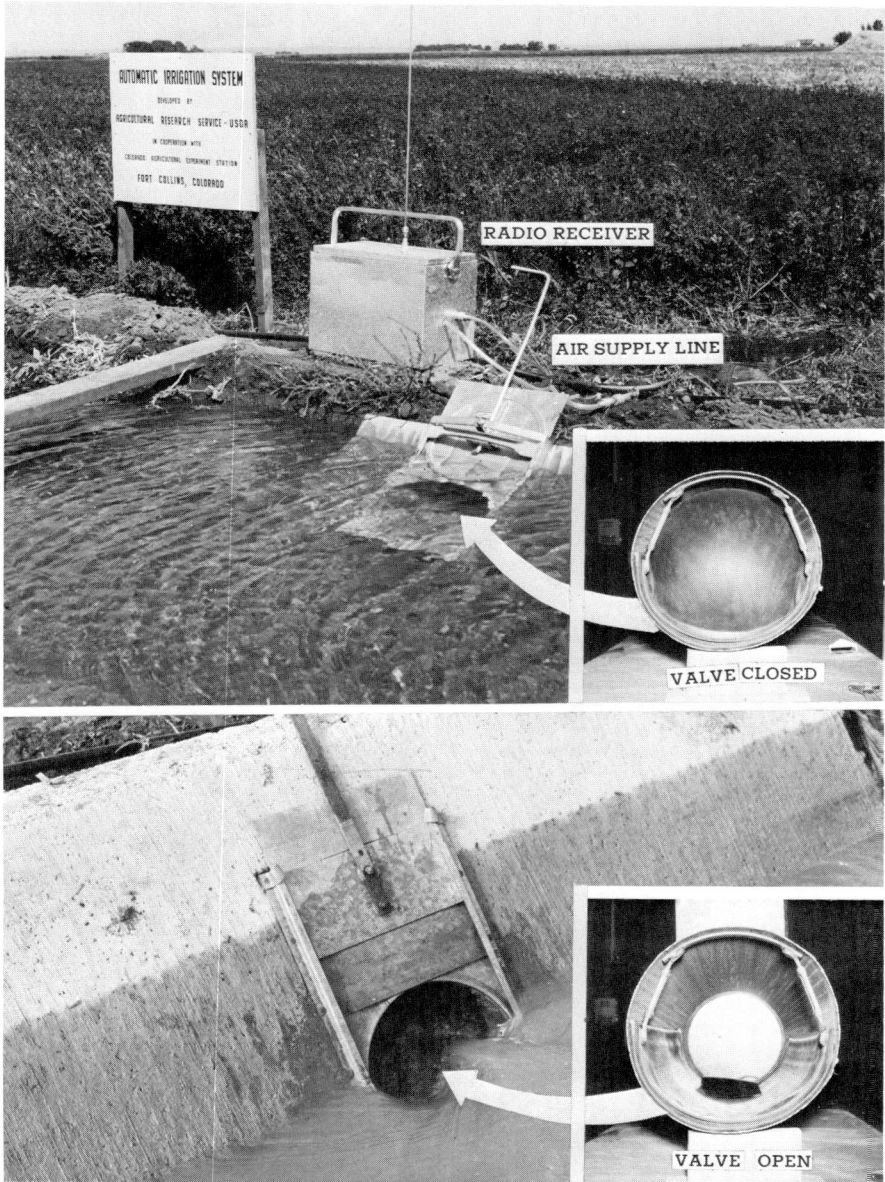

Fig. 42–9b. Radio-controlled pneumatic valves for farm ditch turnouts.

control the discharge from alfalfa-type valves on an underground pipeline system (Fig. 42–9a) or from turnouts in farm ditches (Fig. 42–9b). The pneumatic valve for pipeline distribution systems is essentially an inflatable O-ring which forms an annular seal when inflated between the alfalfa valve seat and valve lid. The lay-flat pneumatic valve for ditch systems is a flat, rectangular tube that inflates to form a closure within the underground portion of the turnout pipe. Inflating and exhausting of air from the valves are remotely controlled by a signal transmitted by wire or by radio from a centrally located timing device.

4. SEDIMENT, DEBRIS AND WEED SEED CONTROL

Trash, weed seed, and sediment in the flowing water present problems in irrigation systems. This is particularly true for sprinkler irrigation systems where nozzles become plugged with debris or become worn because of sediment in the water. The increased use of pipe, both underground and on the surface, demands that the water be relatively free of sediment. Sediment deposits can plug underground pipelines and partially fill surface pipe making it difficult to move. If gated pipe is used, the gates become plugged with debris and the flow is reduced.

Sediment can be troublesome in farm ditches and can necessitate frequent ditch cleaning. Weed seeds are scattered by flowing water and will germinate even after being in the water for long periods. A good weed seed screening device is an essential part of the weed control program on irrigated farms.

Trash racks are necessary at every pipe and underground crossing entrance to prevent entry of floating debris. In general, these are made of spaced steel bars, either round or flat. With this arrangement, however, it is possible to catch and remove only the largest material. The racks are generally slanted for easier cleaning and should be removable.

a. **Screens and Settling-Boxes.** Over the years, a number of devices and designs have been used for screening irrigation water. In some cases a desilting box and trash screen have been combined into one structure. An example of the latter is shown in a US Department of Agriculture Soil Conservation Service design given

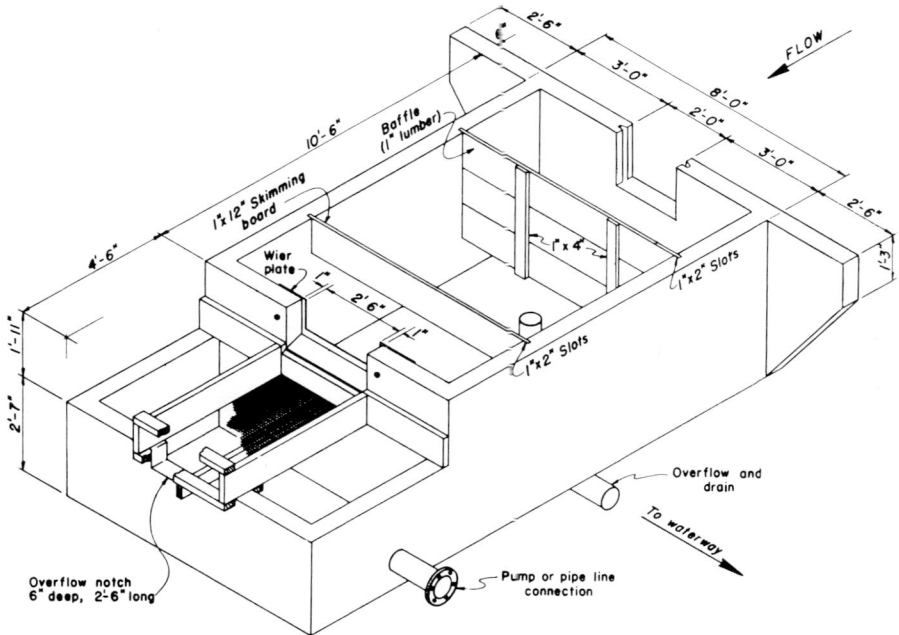

Fig. 42–10. Irrigation water desilting box and trash screen.

Fig. 42–11. Debris screen on an inlet structure to an underground pipeline.

in Fig. 42–10. In this design, provision is made for sediment trapping using a large box which decreases the velocity and, in turn, the transportability of the flow. Provisions are made for flushing this compartment. A second compartment is equipped with a weed screen. An outlet from this compartment discharges into the distribution system. The screen in this design is self cleaning as shown in Fig. 42–11. If desilting is not desired, only the lower section of the structure shown in Fig. 42–10 is needed.

Other types of screens and trash racks for irrigation ditches are described elsewhere (Bergstrom, 1961; Code, 1961; Coulthard et al., 1956). Some of these utilize paddle wheels to power the cleaning mechanism and for screen agitation. When only part of the flow is to be diverted, the screening problem is greatly simplified because a portion of the trash can be floated past the turnout.

b. Sediment Traps. Numerous schemes have been used for trapping sediment and silt in irrigation channels. If the sediment consists of silt or fine sand, the usual trapping method consists of increasing the flow area to reduce the velocity to the point where the material settles. An example of a structure with a settling basin was given in the previous section (Fig. 42–10).

When sediment is moving as a bed load of large sand and small gravel, the trapping problem becomes somewhat different. Settling basins will catch the material but frequency and difficulty of cleaning is increased. Numerous schemes and designs for sand traps which dynamically trap and remove the sediment by flowing water have been proposed (Parshall, 1952; Robinson, 1962; Uppal, 1951). The traps sometimes are constructed depressions or boxes in the bed of a canal covered with a grating and dependent on a transverse sluicing action to move the trapped load.

For small canals or channels, a vortex tube sand trap, as shown in Fig. 42–12, can be very effective in removing material > 0.5 mm. Tests have shown that the following items are optimum for operation of the tube:

1) The Froude number of flow across the tube section should approximate 0.8.

2) The flow usually removed by the tube ranges from 5 to 15% of the total flow.

3) Tube shape is not particularly important, and a pipe with a portion of the circumference removed works very well. The width of opening should be about

Fig. 42–12. Vortex tube sand trap used to remove sand and gravel from canals.

15 cm (0.5 ft), the length-to-opening-width ratio should not exceed 20, and the tube should be set at an angle of about 45°.

5. ENERGY DISSIPATION

The dissipation of kinetic energy in flowing water to prevent excessive erosion is a problem frequently encountered by the engineer and the farmer in irrigated agriculture. In general, kinetic energy, i.e. energy due to the velocity of the flow, is dissipated either in the vertical or the horizontal direction or both. For vertical dissipation, a jet of water is diffused either vertically downward or upward. The energy is dissipated horizontally by channel resistance, form resistance, or by a hydraulic jump and the resulting increase in piezometric head. Figure 42–13, as presented by Smith (1957) and Fiala and Albertson (1963), shows the different methods of energy dissipation classified according to direction.

Dissipation of energy in the horizontal direction may result from surface roughness and drag on the boundary which causes the velocity to be reduced and the depth to be increased (Fig. 42–13a). In many instances, this principle is used to dissipate excess kinetic energy. However, since the roughness and drag are sometimes not great enough, the banks and bed of the channel become badly eroded. The use of boundary resistance alone requires such a long channel that some other means is usually employed.

The hydraulic jump is a very effective energy dissipator and is widely utilized for this purpose. The Froude number of flows in small canals which require energy dissipation normally lies in the range of 2 to 4. The energy loss resulting from a hydraulic jump in this Froude number range is usually only 10 to 20% of the energy involved. In contrast, hydraulic jumps resulting from high Froude numbers (Fr $\geq$ 9.0) may dissipate as much as 85% of the energy. For this reason, a chute, in conjunction with a stilling basin is frequently used to increase the velocity and resulting Froude number of flow.

The distance required to decrease the velocity and dissipate energy in the horizontal can be reduced appreciably, if blocks or sills are used in conjunction with the hydraulic jump as shown in Fig. 42–13b. Since the jump is unstable for variable flows and may move upstream or downstream depending on the dis-

HORIZONTAL ENERGY DISSIPATION

(a) Channel resistance

(b) Hydraulic jump and form resistance

VERTICAL ENERGY DISSIPATION

(c) Downward

(d) Upward

Methods of dissipation of kinetic energy

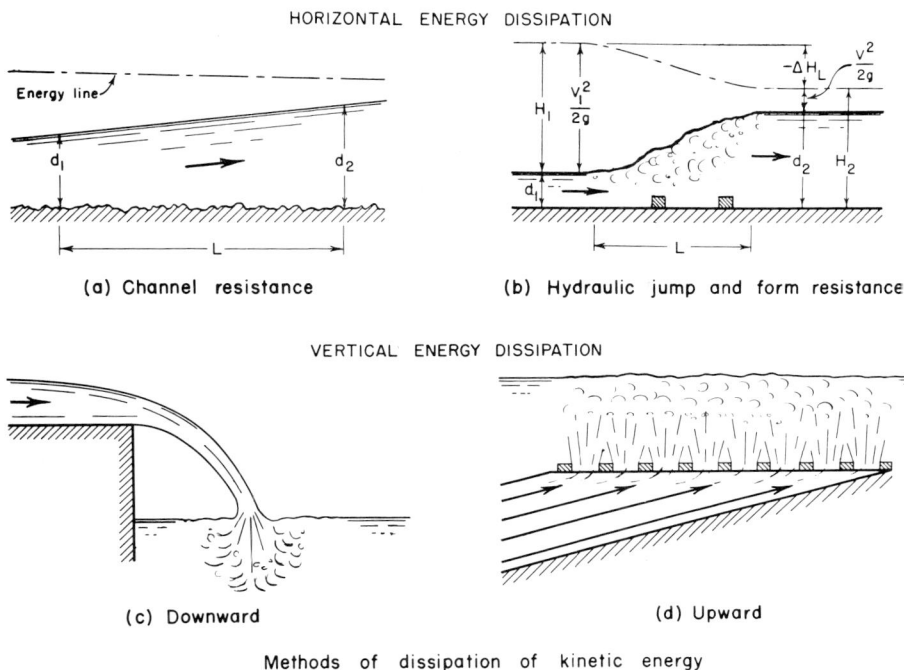

Fig. 42–13. Different methods for dissipation of kinetic energy (from Smith, 1957).

charge, sills or blocks help to stabilize the location. Figure 42–4 shows a standard drop structure where blocks and a sill control the jump.

The dissipation of energy in the vertical direction involves the diffusion of a jet into a pool of water. The jet may be traveling vertically downward as illustrated in Fig. 42–13c and in the common irrigation drop shown in Fig. 42–3. The drop may be equipped with an apron to force the horizontal dissipation of energy (Fig. 42–4) or have a stilling pool as shown in Fig. 42–13c. When a stilling pool is used, special provisions must be made to prevent scour such as maintaining a prescribed depth of water, lining, or armor plating the scour hole with graded riprap material (Smith, 1957). A cantilevered pipe outlet, such as a culvert, is another example of vertical energy dissipation.

A recent development for vertical energy dissipation is the manifold stilling basin (Fiala and Albertson, 1963) illustrated in Fig. 42–13d. Here the flow is upward and the kinetic energy of the jets is dissipated in the depth of tailwater. The manifold device has generally been used on fairly large canals where there is a high velocity inflow to be added to the stream.

6. DESIGN CRITERIA

The type and design of water-control structures depends on the type of irrigation system to which they belong. This in turn is determined by many factors such as soil characteristics, water source, crops grown, and climatic factors. In addition to fitting the irrigation system requirements, a structure must also satisfy hydraulic, structural and operational criteria.

A farm structure must be designed with sufficient capacity to maintain adequate freeboard in the ditch and to handle expected variations in the flow. Because of the head–discharge relationships, variations in discharge from a canal or lateral into the farm ditch or from the ditch onto the field will be minimized by using submerged pipe or orifice-type turnouts to control the discharge and overflow-type checks in the lateral or ditch to maintain a nearly constant water level above the turnout. This is particularly important when siphons are used, since large water level variations cause them to lose their prime.

Sufficient cutoff wall and apron length must be provided to prevent piping, erosion, and undermining of the structure. To avoid concentrating the flow in unlined ditches, the crest and apron width of a structure should be approximately the same as that of the ditch bottom. The apron of an energy dissipating structure should be below grade to provide a pool in which to dissipate energy. Energy dissipation structures such as the drop are frequently not entirely effective and create eddy currents, waves, and zones of high velocity flow. Protection of the earth channel against erosion as a result of these factors is usually accomplished using riprap as an armor-plating material. One of the best forms of bank and bed protection is graded, pit run gravels as opposed to commonly used large rocks or broken concrete. In order to take advantage of large riprap, it is necessary to add a graded material with sizes ranging from that of the bed and bank material up to the largest size of riprap.

Uplift, overturning, and sliding must be considered on larger structures. These factors, however, are usually not important with small structures in farm ditches.

In addition to adequate structural strength, the structures must be resistant to exposure. Concrete structures should contain reinforcing steel to strengthen them against the effects of temperature change and frost action. Soil and water chemical concentrations in certain areas are harmful to concrete and a special resistant cement may be required. Certain metal structures may also be subject to chemical attack.

Good control structures must be easy to operate and provide fingertip control. They must be versatile and able to meet changing irrigation water demands. These demands change throughout the season and also from year-to-year because of different cropping practices.

B. Water Measurement

Many flow measurement devices and methods are in use throughout the world (Thomas, 1960). Progress in developing new methods and devices for measurement of flowing water have not kept pace with available instrumentation using electronic and nuclear techniques. Water measurement in open channels is usually only approximate. The need and demand for good, accurate, and adaptable devices are increasing. The expected accuracy of some methods now in use is no better than ± 5 to 10%, even under the best conditions.

1. METHODS AND PRINCIPLES

Open-channel flow measurement includes all techniques, devices, and methods used to measure flows where a free water surface is involved. Open-channel flow may occur in closed conduits flowing partly full. In general, the flows are turbulent and the boundary surfaces of the conduits are hydraulically rough. Flows heavily laden with sediment and debris occasionally occur. The flow conditions are usually nonsteady and may be nonuniform, resulting in very difficult measurement situations.

Devices and techniques for measurement can be classified as follows: (i) structures which control channel geometry, (ii) instruments which float on, or are immersed in, the flow field, and (iii) techniques which require measurement of the movement or concentration of dispersed material placed in the flow field.

The most universally used and accepted devices are those that control channel geometry. These usually employ the concept of critical depth where flow passes through a point of minimum specific energy within a defined cross section. This method of measurement includes the weirs, both sharp and broad crested, and suppressed or fully contracted. The weir may be the crest of a dam, diversion structure, or ditch check where there is a definite relation between depth and discharge. However, it is generally considered as a specially constructed device having a metal blade with sharp edge and carefully controlled approach conditions.

The measuring flume also employs an open channel constriction. Flumes are used throughout the world and are more commonly called Venturi flumes, standing wave flumes, critical depth flumes, or Parshall measuring flumes. In the USA, Parshall flumes are used in canals almost exclusively. In general, they operate as critical depth devices, but are capable of measurement under conditions where the flow does not go through critical depth. In this case the accuracy is not as great, but a measurement can be made. Rating sections, meter gates, and measuring gates are other examples of structures for flow measurement which control the channel geometry.

There are many devices which float on or are immersed in the flow field for the purpose of measurement such as current meters employing rotating wheels or propellers. The Dethridge meter, widely used in Australia, uses a rotating drum. Surface floats and floating screens are used to estimate the amount of flow. The displacement or drag on a body as a function of the velocity has been used. Examples include deflection vane meters (Robinson, 1963) and deflection wire-strain gauge devices (Sharp, 1964). These devices show promise in the development of new, improved instrumentation. Devices used to determine the velocity by measuring the velocity head include the pitot tube, static tube, and velocity head rods.

The movement, dispersion, and dilution of material in the flowing water can be related to discharge. The amount of dispersion or dilution is measured, using radiation detectors or fluorometric methods. The method using dyes is safe and results are promising from the standpoint of accuracy.

2. DEVICES

a. Weirs. The weir is probably the most common device used for water measurement in ditches. With proper installation and maintenance it gives accurate results (Parshall, 1950; US Bur. Reclam., 1953; US Dep. Agr., 1962). The rectangular weir, as shown in Fig. 42–14 is widely used and can be made of wood, steel or canvas with a steel blade. A weir of this type can also be used as a combination drop for the ditch. In the USA, weirs are usually in widths of 1, 1.5, 2, and 3 ft for which rating tables are available. A prescribed distance must

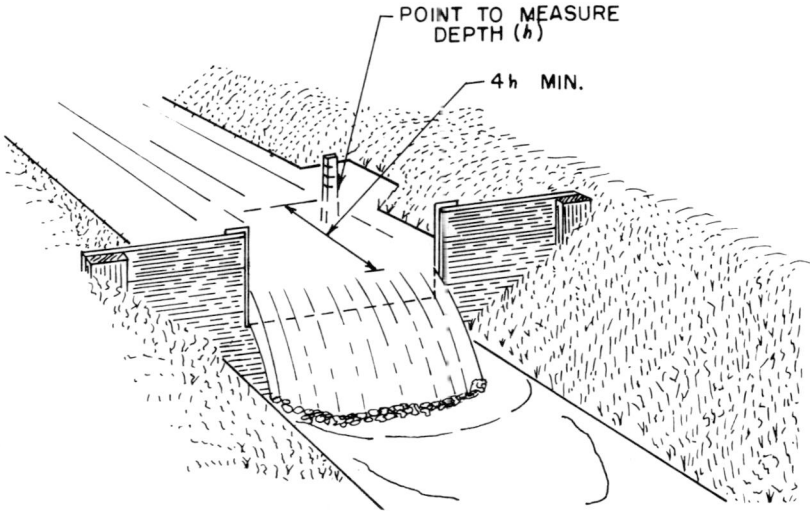

Fig. 42–14. Rectangular weir can be used as a combination measuring device and drop structure.

be maintained between the top of the weir blade and the bottom of the ditch and between the sides of the opening and the ditch banks. Only the depth of water over the weir crest, measured upstream from the weir, and a table are needed to determine the discharge. The depth is usually measured as indicated in Fig. 42–14 or can be made to one side of the weir on the bulkhead but at some distance from the opening. A graduated scale can be fastened at either location with its zero at the crest elevation so that a direct reading of depth can be made.

The 90° V-notch weir is shown in Fig. 42–15. It will measure low flows very accurately and will handle a large range of flows. The weir is easy to construct and install with the aid of a carpenter's square and level.

Another type of weir which has been widely used is the Cipolletti weir shown in Fig. 42–16. This type of weir combines some of the features of the rectangular and V-notch types. A combination headgate and measuring device using the Cipolletti weir is being used successfully in some areas. The height of weir blade in this case is controlled by a hand wheel and the head is measured by a stick held on the crest or a staff gauge mounted on the weir blade.

Any constructed barrier in an open channel over which flow takes place serves as a control and has a fixed relation between head and discharge, if upstream or downstream conditions do not interfere or change. The geometry of the barrier determines both the coefficient of discharge and the exponent in equation [42–1]. Values of the coefficient C and exponent n vary with the type of weir. For the rectangular weir where the width of flow section remains constant, $n = 1.5$. For other weirs where the nappe width and shape changes with depth, the exponent varies. For a triangular weir, n is 2.5, for a parabolic weir, 2.0, and for a Sutro weir, 1.0 (Rouse, 1950).

In all cases when using a weir, the blade should be fairly sharp on the upstream edge. The downstream water surface should be below the level of the blade, if

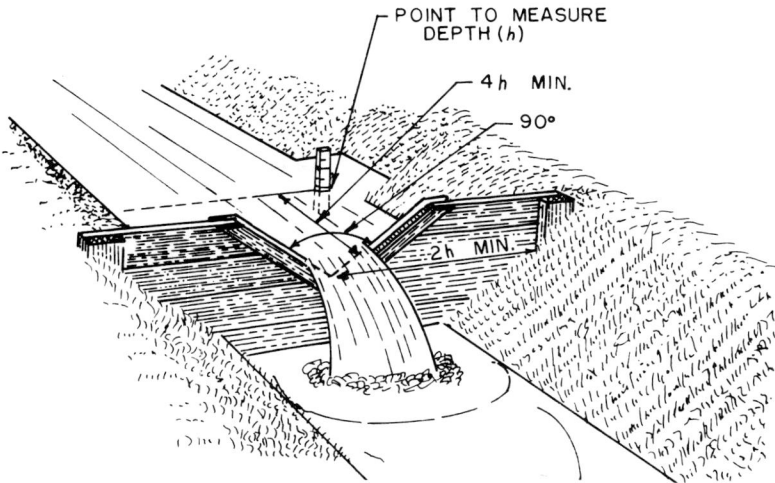

Fig. 42–15. V-notch weir is used for accurate measurement over a large range of flows.

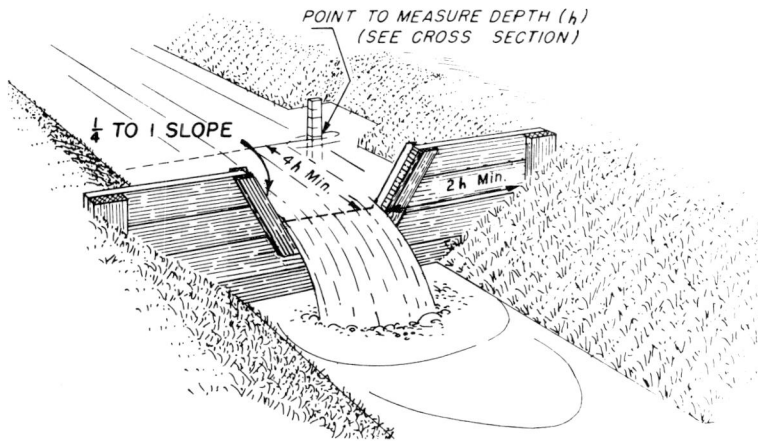

Fig. 42–16. Cipolletti or trapezoidal-notch weir.

a good measurement is to be obtained. It is necessary that a pond of water be formed upstream from the weir. With a properly installed weir, the accuracy of measurement should be ± 2%. However, because of deposition in the upstream pool, misalignment of weir blade, and inaccurate measurement of the head over the weir, discharge errors up to 10 to 15% are possible (Thomas, 1959). In general, this error is positive so that more flow is passing the weir than the measurement indicates.

b. Flumes. The Parshall flume, shown in Fig. 42–17, is widely used for measuring flows in ditches and canals. In contrast to the weir where there must be an appreciable drop between the upstream and downstream water surface, the Parshall flume will measure accurately when there is very little difference in

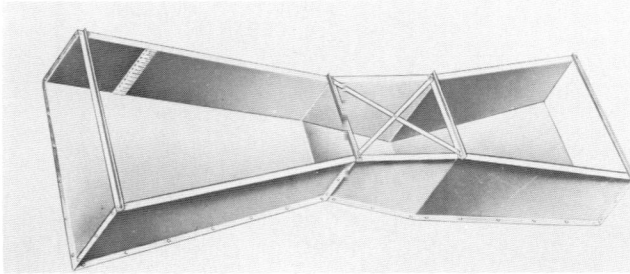

Fig. 42–17. Parshall measuring flumes are commercially available or can be constructed for a large range of discharges.

these levels. One advantage in using a flume is that less head loss is required. This is particularly important for canals on very flat slopes. The device is self cleaning so that it will not silt up as easily as a weir. It can be readily constructed of sheet metal or concrete. Plans and calibrations are available for widths ranging from 1 inch to 50 ft and for flows from 0.01 to 3300 ft³/sec (Parshall, 1950, 1953; Robinson, 1957).

The discharge equation for the Parshall measuring flume is

$$Q = Cbh^n \qquad [42\text{--}3]$$

where b is throat width, L, and h is depth of flow at specified locations, L. Values for C and n given in Table 42–1 have been determined experimentally (Davis, 1963).

The accuracy of measurement when using a Parshall flume should be within ±3%. However, errors in measurement due to faulty construction, installation, or operation can approach ±10%. Probably the most common source of error is in using the free-flow discharge table when the flume is obviously operating under submergence. Errors of –25% or more can be made in this case.

A different type of measuring flume has recently received renewed attention and has some advantages over the rectangular types. This is the trapezoidal flume which has side walls that are sloping rather than vertical. An advantage is that a much larger range of flows can be passed by the structure without backing up the water as much as with a rectangular flume. It will also give an accurate measurement with a smaller difference in the upstream and downstream

Table 42–1. Values of coefficient C and exponent n for Parshall measuring flumes

Flume size		
b	C	n
1 inch	4.06	1.55
2 inches	4.06	1.55
3 inches	3.97	1.547
6 inches	4.12	1.580
9 inches	4.09	1.530
1 ft to 8 ft	4.00	$1.522b^{0.026}$
10 ft to 50 ft	$3.69 + \dfrac{2.50}{b}$	1.60

water surfaces. The general shape of the flume fits the common ditch shape better, particularly when used in a lined ditch with sloping side walls. Designs and ratings are available for flumes with different widths and side wall slopes (Robinson and Chamberlain, 1960; and Robinson, 1966).

Recording instruments are available for use on flumes for a continuous record of water depth. These are simple to install and are needed where the amount of flow in a ditch varies over short periods of time. When using a recorder, the record for all irrigations over a period of years can be maintained by simply filing the charts.

c. Miscellaneous Devices. Numerous other devices are used to measure water flow. Some of these are grouped and briefly described below.

1. Current Meters. Small meters are available which consist of a revolving wheel or propeller to measure the velocity of flowing water (US Bureau of Reclamation, 1953). Besides being used to measure flow in open channels they also can be used to measure the flow discharging from pipe. Current meter measurements are normally made only by those who are trained in this line of work. With practice and care, the accuracy can be within ±3%.

Fig. 42–18. Vane meter for measuring flow where the discharge is given directly on the mounted scale.

2. Vane Meters. A measuring device recently developed and commercially available gives the amount of flow as a direct reading (Robinson, 1963). This device, shown in Fig. 42–18 consists of a vane suspended into the flow and mounted in a section of prescribed size and shape. Flowing water deflects the vane to various degrees depending on the velocity and depth of flow. The amount of flow is read directly from a scale opposite an indicator mounted on the meter. This device is still being improved and at the present time appears to measure within ±5% accuracy.

3. Orifices. There are two general types of orifices which have found limited use in the measurement of irrigation water. These are orifices with fixed dimensions and those with adjustable openings. The discharge from orifices with a fixed opening is determined from the opening area and the depth of water above the center of the opening when the downstream water surface is below the opening. If the downstream water surface is above the opening, the head is the difference in elevation between the upstream and downstream water surfaces. Plans and calibration curves are available for various sizes of fixed orifices (US Bur. Reclam., 1953; US Dep. Agr., 1962). The discharge equation for an orifice is given by equation [42–2].

In certain irrigated sections, a combination headgate and measuring orifice is sometimes used. One design developed by the U. S. Bureau of Reclamation is

Fig. 42–19. The constant-head orifice utilizes two gates and a constant difference in head across the measuring gate.

called the constant-head orifice turnout (Fig. 42–19). This design utilizes two gates; one for amount of opening, and the other for adjustment of head. The operating procedure is to adjust the gates so that a constant difference in head of 0.2 ft is maintained across the upstream gate (US Bureau of Reclamation, 1953). Recently, tests have been made on the constant-head orifice to evaluate some of the problems encountered in the field (Kruse, 1965). The tests considered such factors as: shape of approach section, velocity of the canal flow, clogging by debris, sediment deposits, and accuracy of head difference determination with widely fluctuating water surfaces. The structure provided reasonably accurate measurement of discharge under most operating conditions. However, when the gate opening was obstructed with weeds, discharges were much less than indicated by the calibration curve. For large discharges, the staff gauges furnished erratic indications of the differential head on the orifice gate because of fluctuating water surfaces.

4. *Adjustable Gates.* Headgates are available that control and measure the amount of water being passed. Installation instructions and calibrations are available for a variety of sizes. Two measurements are necessary in finding the discharge: the amount of gate opening and the difference in head across the gate. With these two measurements, one may use tables or curves supplied by the manufacturer of the gate to find the discharge.

Another adjustable gate discussed in the previous section on weirs, utilizes a standard weir, usually of the Cipolletti type, which is mounted to slide up or down with the adjustment of a handwheel. The gate then becomes a combination turnout and measuring device. These have been used to a limited extent by the US Bureau of Reclamation.

5. *Commercial Meters.* A measuring device used with irrigation turnouts is shown in Fig. 42–20. This device utilizes a propeller connected to a dial which records the total amount and rate of flow. The meter can either be used in closed pipe systems or adapted for measurement in canals using a structure such as shown in Fig. 42–20.

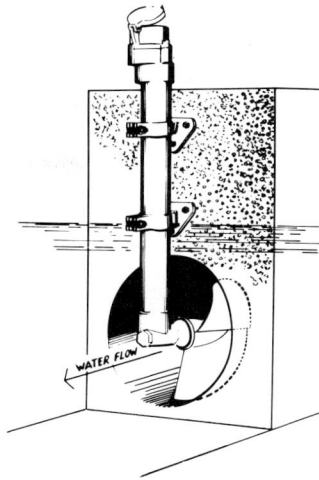

Fig. 42–20. Propeller-type meter which can be installed on the outlet of a farm turnout.

3. SOURCES OF ERROR IN MEASUREMENT

There are a number of sources of error in using weirs, flumes and other devices for flow measurement (Thomas, 1959). The most common sources are:

1) Faulty fabrication or construction
   a) Standard dimensions not maintained
   b) Assembly errors
2) Incorrect setting and improper maintenance
   a) Transverse slope
   b) Weir blade not vertical
   c) Weir blade edge becoming rounded
3) Incorrect head reading
   a) Error in gauge location and setting
   b) Gate marks obliterated
   c) Fluctuation in water surface making reading difficult
   d) Ignoring submerged condition
4) Nonstandard approach conditions
   a) Velocity of approach flow higher than specified
   b) Existence of excessive turbulence and surges
   c) Devices placed immediately below a bend
5) Nonstandard downstream conditions
   a) Inadequate aeration of nappe
   b) Excessive submergence
6) Others
   a) Obstruction of device with debris
   b) Weeds, moss and other vegetation growing in, near or on the device
   c) Poor measurement of gate or orifice opening.

II. CLOSED SYSTEMS

Closed systems are those in which irrigation water is conveyed and distributed with pipelines. They may be operated under high or low pressure. Most high-pressure systems are used for sprinkler irrigation. Special equipment is required for pipelines operating under high pressure. This is available only through commercial sources and is normally provided by the pipeline contractor. Therefore, the following section will be concerned primarily with control structures for low-pressure systems.

A. Water Control

Irrigation pipeline systems provide an efficient means for conveying and distributing water on the farm. In these systems, water is usually conveyed by underground pipe from a well or other source to points on the farm where it is distributed onto the field or into other pipeline laterals. Water distribution onto the field is made from valves, gates, gated surface pipe or lay-flat tubing attached to vertical risers from the pipeline. Because of the high initial cost, at present these systems are generally used with crops having a relatively high economic return or where water costs are high. Underground pipe distribution systems offer many advantages such as minimum seepage and evaporation losses, no loss of productive land occupied by ditches, good control of irrigation water, better weed control by elimination of ditch banks, ease of distribution on rough land, and minimum maintenance.

1. PROBLEMS AND PRINCIPLES INVOLVED

Nonreinforced concrete pipe is used extensively in irrigation pipeline systems. However, steel, asbestos-cement, plastic, and vitrified clays are also used. The latter three types are often better adapted for certain soil and water conditions that are unfavorable to either concrete or steel. Concrete is generally not rcommended for use in saline or alkali soils that have a high water table. Most systems normally operate at heads less than 5 m (approximately 15 ft). If the pressure head exceeds 6.5 m (20 ft), reinforced concrete, steel or other pressure pipe must be used. Because of past failures, concrete pipe 45 cm (18 inches) and larger in diameter on US Bureau of Reclamation systems, is reinforced. Mortared tongue and groove joints on concrete pipe have been extensively used in the past, but improved rubber gasket joints are preferred. These offer a more flexible, leak-proof joint with less flow resistance. Nonreinforced, monolithic, cast-in-place concrete pipe is also used extensively in some areas of the USA. This type is limited to maximum operating heads ranging from 3 to 4.5 m (10 to 15 ft) and is most competitive economically in the 60- to 120-cm diameter (24- to 48-inch) range. Since most low-pressure irrigation pipeline systems use nonreinforced concrete, the principal emphasis in the following discussion will be on this type.

The fundamental principles for the design, installation and operation of underground pipe irrigation systems have been established (Amer. Soc. Agr. Eng., 1964; Pillsbury, 1952; Portland Cement Association, 1952). A number of problems are encountered in the design and operation of the system. Proper installation criteria must be followed to provide satisfactory performance with minimum maintenance. This includes following depth standards to protect pipe from traffic and frost action and adhering to bedding and alignment procedures. Unsteady flow caused by surging is often an operational problem encountered. The hydraulic gradient throughout the system must be controlled and maintained above minimum requirements. This is accomplished by providing stands and other control structures in the system. Concrete pipe is sensitive to stresses created by wetting, drying, differential drying of the shell, temperature changes, and water hammer. The system should be so operated as to minimize these stresses. Thermal contraction and expansion may be caused by soil, water and air temperature variations. Cold water should be turned into the line very slowly to allow the line to adjust gradually to the temperature change. Gates, valves, and covers should be kept closed when not in use to extend their life and to prevent cold air from entering the line. Sulfate fertilizer should not be applied in irrigation water through concrete pipelines. Other fertilizers may be applied, but special precautions are required for some (Pillsbury, 1952).

a. Energy Loss. Energy losses must be accurately determined in designing the system to maintain the hydraulic gradient at the desired field delivery elevation. These losses are classified as either friction or minor losses. In long lengths of pipe the major loss is due to friction; however, the principal loss in short lengths may be due to minor losses.

1. Friction Loss. Laminar flow is rarely encountered in irrigation pipelines. Under normal conditions the flow is turbulent, and the flow resistance depends on conduit roughness, velocity, viscosity and density of water, and on the length and cross section of the conduit. Conduit roughness becomes important in turbulent flow because the laminar boundary layer no longer covers the roughness elements on the inside surface of a rough pipe. When these elements project through the laminar sublayer the flow resistance increases.

A number of formulas have been developed to determine friction losses and velocity in pipes with turbulent flow. These relate loss of head to velocity with empirically determined coefficients. Of those used, the Darcy-Weisbach is the most rational and theoretically correct and includes the effects of temperature and viscosity variations. It is more accurate over a wider range of flows, pipe sizes and types than other formulas. The head loss in this form is expressed as

$$h_f = f(L/D)(V^2/2g) \qquad [42\text{–}4]$$

where

h_f = friction head loss, L,
f = friction factor,
L = length of conduit, L,
D = inside diameter of conduit, L,
V = mean velocity of water in the pipe, L/T, and
g = acceleration of gravity, L/T^2.

The friction factor is determined from graphs relating f to Reynolds number (King, 1954).

A further advantage of this formula is that it is dimensionally homogeneous and appears in the same form in metric units as in English units. The friction coefficient f and Reynolds number are dimensionless and their numerical values are independent of the system used.

The Manning equation is used internationally for solving pipe flow problems. When used for pipe, the conventional form of the equation is modified by substituting $D/4$ for the hydraulic radius. This form for pipes in English units is

$$V = (0.590/n)D^{2/3}\ S^{1/2}. \qquad [42-5]$$

In metric units it is expressed in the form

$$V = (0.397/n)D^{2/3}\ S^{1/2} \qquad [42-6]$$

where n is the roughness coefficient and S is the friction slope. Other terms have the same notation as previously given. The same value of n may be used in both systems of units. Tabulated values of n for various pipes are given by King (1954).

Another widely used formula for solving pipe flow problems is the Hazen-Williams formula

$$V = 1.32\ C_1\ R^{0.63}\ S^{0.54} \qquad [42-7]$$

where C_1 is the Hazen-Williams discharge coefficient (Davis, 1952 and King, 1954) and R is the hydraulic radius which for pipes is $D/4$.

One of the most widely used formulas for computing flow in concrete irrigation pipelines is the Scobey formula

$$V = C_s\ H_f^{0.5}\ d_1^{0.625} \qquad [42-8]$$

where

C_s = Scobey's roughness coefficient which varies from 0.267 to 0.37 (Davis, 1952)
H_f = loss of head/1,000 lineal feet of pipeline, and
d_1 = inside pipe diameter in inches.

The latter two equations are expressed in English units and are not dimensionally homogeneous. Therefore, the coefficients are not dimensionless and commonly used values of C_1 and C_s are valid only when the formulas are used in the English system of units.

2. *Minor Losses.* Minor energy losses are due to entrance, bends, fittings, obstructions, contractions, and enlargements. These are expressed as

$$h = K(V^2/2g) \qquad [42-9]$$

where h is the head loss and K is a loss coefficient determined experimentally for the type of bend or obstruction under consideration. The total minor loss is the sum of the individual losses. Loss coefficients for various conditions are presented in hydraulic handbooks (King, 1954).

In an underground pipe irrigation system losses in risers and distributing hydrants must also be considered. Losses in gated surface pipe and lay-flat tubing must be included when these are used to distribute water from the hydrant.

b. **Flow Capacity.** The capacity of the system is determined by the pipe size and available grade or difference in head between the upper and lower end of the pipe. Assuring adequate capacity is very important, particularly in a gravity feed system. If the system is fed from a pump, the head may be increased to provide sufficient capacity. However, to increase the capacity of a gravity feed system once it is installed is very difficult. If the available grade is insufficient to offset the energy losses in a system having the largest practical size pipe, it may be necessary to use a booster pump in the system design. A common standard is to design the system with the hydraulic gradient 30 to 60 cm (1 to 2 ft) above the ground surface at the discharging hydrant. To assure adequate capacity, care must be taken to maintain proper alignment, clean joints, and avoid any practice which would result in higher loss coefficients than were used in the design.

c. Air Entrainment. Air entrainment is often a problem in the operation of a pipeline system. Air may be entrained in the water at the pump, at a gravity inlet, or in an overflow stand. Air carried into the pipeline tends to collect in pockets at high points and breaks in grade, and reduces the carrying capacity of the pipe. Accumulations of entrained air cause surging and unsteady flow conditions and may contribute to the development of excessive pressures.

d. Hydraulic Transients.

1. Surge. Surge in a pipeline is usually caused by the sudden release of entrapped air from the line. High pressures are not usually encountered in surging. However, the sudden release of large volumes of air may start the process of shock wave generation when the two water surfaces collide. Minor flow fluctuations in the overflow of a pipe stand may initiate surging when amplified in passing through successive stands.

2. Water Hammer. Water hammer results from the sudden stopping of flow in the pipe. This may occur if a valve is closed too rapidly. When this occurs, the kinetic energy of the moving fluid is transformed into pressure energy, generating a pressure wave that oscillates back and forth in the pipe until damped. In actual operation, this is not usually a problem since most systems are fitted with slow-closing, screw-type valves. It is important for this reason, that slow-closing valves be used in the line.

2. STRUCTURES

Special structures are used to provide the necessary water control and to alleviate some of the problems peculiar to this type system. Local conditions may dictate the type of metal used in the structures. Steel fittings may corrode in some waters, requiring the use of cast iron, bronze or brass.

a. Inlet. Water may enter a pipeline by gravity from a ditch or it may be pumped from a well or stream. Inlet structures are needed to protect the system from excessive pressures, to minimize air entrainment, and to develop the full flow capacity. They also may be designed to serve other functions of stands such as trapping trash and sediment or controlling flow into laterals.

1. Pump Stands. Pump stands are installed to receive water from a pump and convey it into the pipeline. They are open topped and usually larger in diameter than the pipeline. This allows the stand to act as a surge chamber and entrained air to escape because of a reduced water velocity. They are built high enough to develop the head needed and to provide sufficient freeboard without overflowing except when unusual pressures occur. A typical stand is shown in Fig. 42–21. If an unusually high stand is required, it is capped and a smaller diameter steel pipe is extended to the necessary height. A flexible coupling is put in the

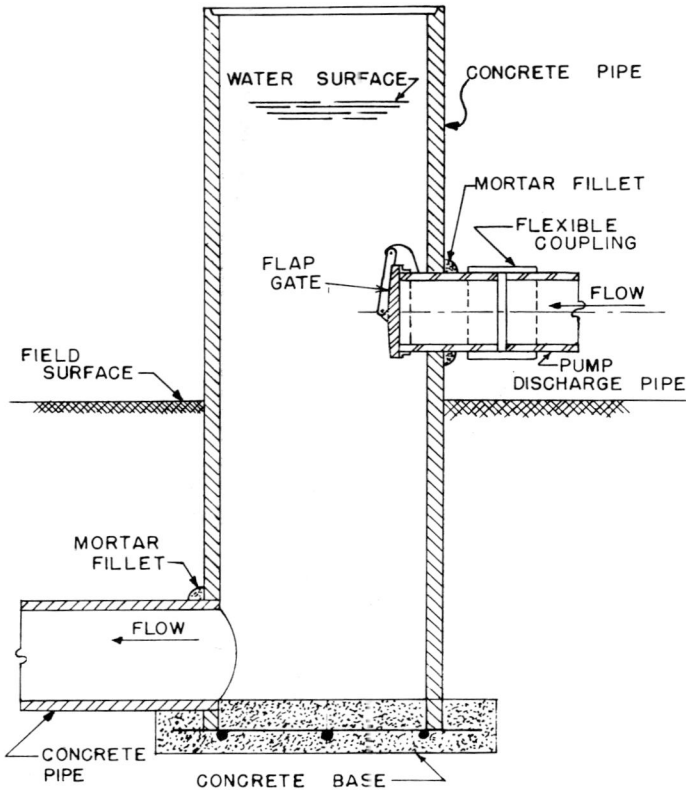

Fig. 42–21. Typical concrete pump stand: The flexible coupling is needed to absorb vibrations from the pump; the flap gate prevents backflow to the pump.

line between the pump and stand to isolate the stand and pipe system from pump vibration.

2. *Gravity Inlet.* When water enters the pipeline from an open ditch, structures such as those shown in Fig. 42–11 and 42–22 are used. They may be concrete block stands or constructed of concrete pipe sections set vertical. They should be equipped with a guard to keep trash out of the line and the top of the stand should be fitted with a cover to prevent accidents and wind-blown trash accumulation.

b. Pressure and Flow Control. Control structures are needed to maintain delivery water levels, to regulate the flow into branching lines, to limit pipe pressures, and to provide for the removal of entrained air.

1. *Gate Stands.* Gate stands are diversion structures to control the flow into laterals. They are also used to increase the pressure upstream, to prevent high pressures, and to act as air vents and surge chambers. The gates are often used to control pressures as required by upstream outlets. A single structure is often built to function as a gate stand and as an overflow stand as shown in Fig. 42–23.

2. *Overflow Stands.* These serve both as check and drop structures in addition to other functions of a stand. As a check, the stand regulates upstream pressures to maintain uniform flow from outlets or into laterals. As a drop, it limits the

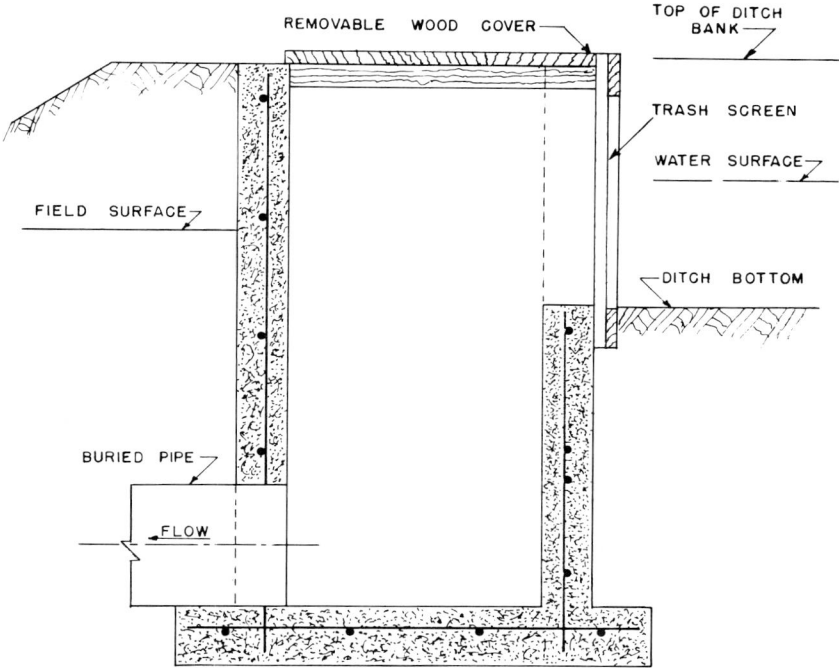

Fig. 42–22. An inlet stand for taking water by gravity from a ditch into an underground pipeline.

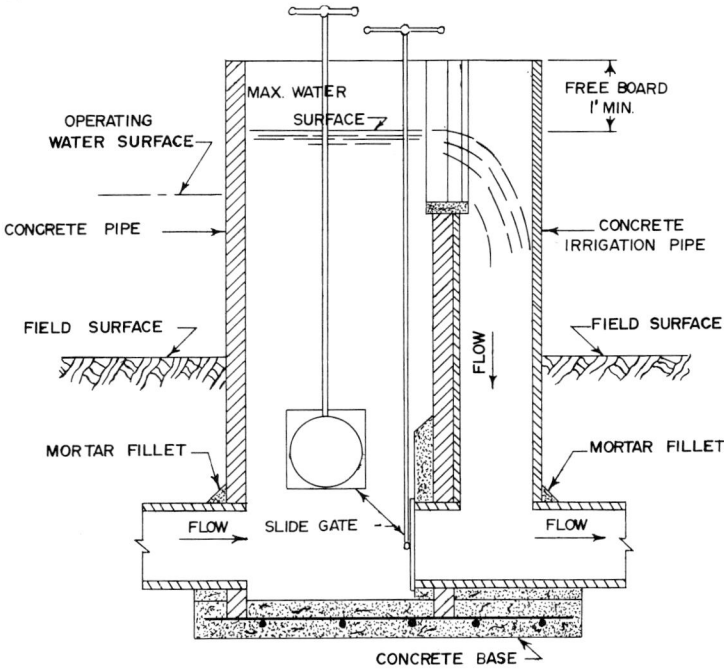

Fig. 42–23. Combination gate and overflow stand used for regulating upstream pressures and diverting water into other pipeline laterals.

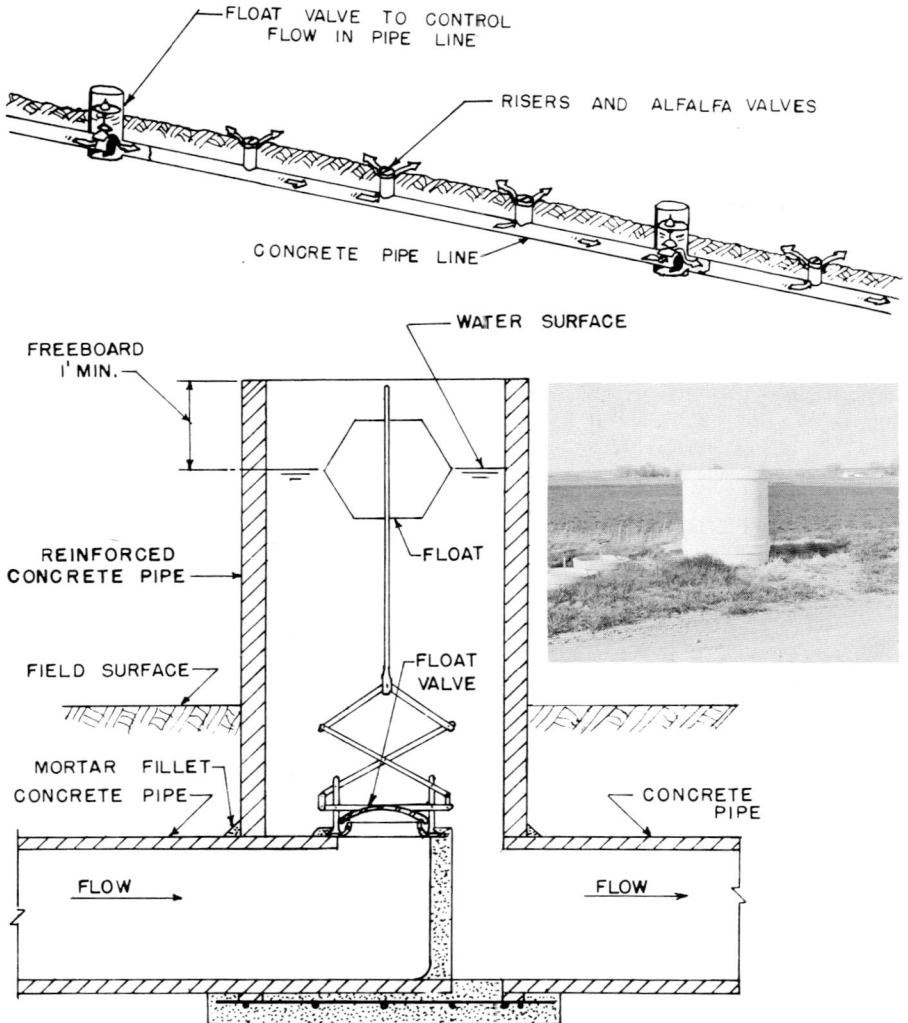

FLOAT VALVE TO CONTROL
FLOW IN PIPE LINE

RISERS AND ALFALFA VALVES

CONCRETE PIPE LINE

WATER SURFACE

FREEBOARD
1' MIN.

REINFORCED
CONCRETE PIPE

FLOAT

FIELD SURFACE

FLOAT
VALVE

MORTAR FILLET
CONCRETE PIPE

CONCRETE
PIPE

FLOW

FLOW

Fig. 42–24. Float valve stand; pressure and flow in the line are automatically regulated by the float valve.

excess head developed by the natural slope. It may be used with or without the side turnout shown in Fig. 42–23. It has the disadvantage that air is often entrained in the water as it spills from the overflow baffle. To minimize this, a gate is installed between the two chambers which is normally open. When pressure is needed for upstream diversion, the gate is closed sufficiently to bring the water level to the crest with only a small overflow. An overflow stand usually is not needed in areas of flat or very slight slopes.

3. *Float Valve Stands.* It is advantageous on steep slopes to install a semiclosed system with float valve stands as shown in Fig. 42–24. The float valves open when the downstream pressure falls to a predetermined level and admit into the

line only as much flow as can be released by the hydrants that are open. Thus, each valve automatically controls pressure in the reach of pipe downstream from it. When a pipeline is served directly from storage, float valves provide full control of the water from the lower end of the line. High overflow stands on steep slopes may be eliminated by using float valves. A semiclosed system is efficient, since surplus water is not wasted at the end of the line as it is sometimes done when overflow stands are used. Tables giving head loss for various size and types of valves at different openings are useful in selecting the proper valve (Pillsbury, 1952).

4. *Line Gate Valves.* Line gates in each lateral are sometimes substituted for gate stands. These valves are regular gate valves with special hubs that are mortared directly into the line. They permit operation from the ground rather than from the stand top. The present trend is toward increased use of line gate valves with adjacent small diameter, capped vent stands instead of large gate stands. Friction losses in wide, open gate valves are low and are often expressed as equivalent lengths of straight pipe (Pillsbury, 1952).

c. Discharge Control. Outlets are necessary to deliver water from the pipeline to the land surface or into some distributing device. They consist of risers built of vertical sections of pipe into which valves or gates are installed to control discharge.

1. *Valves.* Valves are used to distribute water directly into border strips, basins or ditches where relatively large flows are needed. Two general types are used in the USA, alfalfa valves and orchard valves. Alfalfa valves are normally grouted to the top of a pipe riser as shown in Fig. 42–25. This is referred to as an alfalfa valve hydrant. Orchard valves are smaller than alfalfa valves and are used where smaller flows are acceptable. They are usually installed inside the riser as shown in Fig. 42–26. Since water usually flows from an orchard valve with lower velocities, they are commonly used in place of alfalfa valves where erosion is a problem or where the pressure in the riser is extra high.

Portable hydrants and sheet metal stands may fit over the valves for water delivery into surface pipe or ditches. The hydrants are constructed so that the valves may be regulated with the hydrant in place. Gated surface pipe or tubing may be attached to the hydrant from which water is distributed to furrows or corrugations (Fig. 42–27). Sheet metal stands are sometimes fitted with multiple-connections so that one stand may serve several surface pipes individually or simultaneously.

Flow from alfalfa and orchard valves may be determined from equation [42–2] in which the normal value of the discharge coefficient C for alfalfa valves is 0.7 and for orchard valves 0.6. The head h represents the vertical distance between the water surface above the valve and the hydraulic grade line. The recommended height of the hydraulic gradient above the ground for minimum erosion is 30 cm (1 ft). If 15 cm (0.5 ft) of ponding over the valve is assumed, the head loss through the valve would be 15 cm (0.5 ft). Maximum recommended design capacities for different size valves have been tabulated (Pillsbury, 1952).

2. *Pot Hydrants.* There are several types of distributing hydrants, two of which are the alfalfa and orchard type where the water flows from the top of the riser. Another, used for furrow irrigation, consists of a riser pipe extending to the ground level with a larger pipe, called a pot, fitted over it as shown in Fig. 42–28.

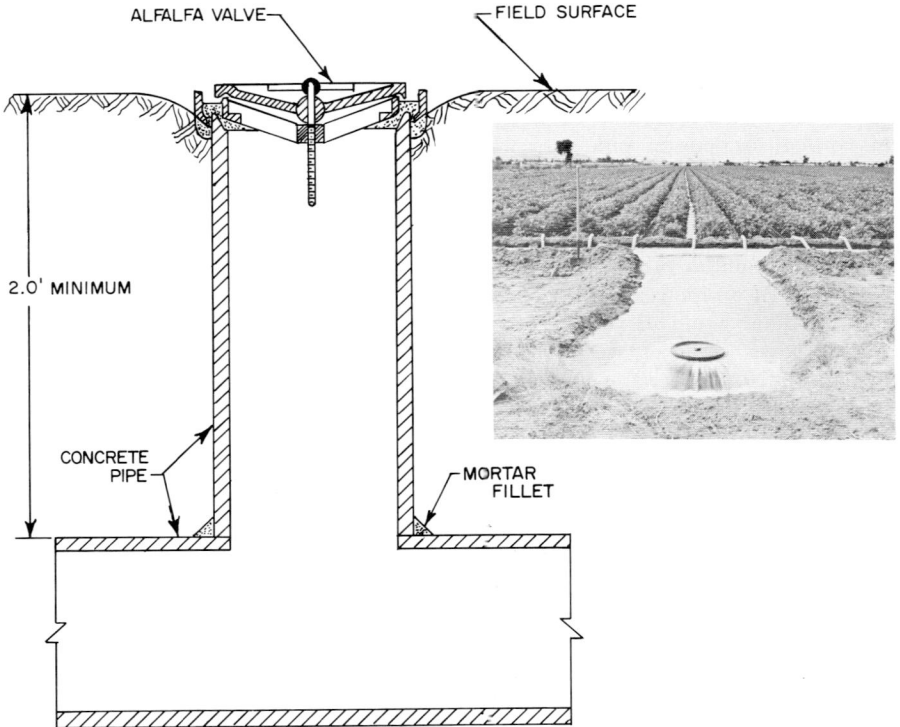

Fig. 42–25. Typical alfalfa-valve hydrant; the riser and valve assembly are sometimes cast to a short section of main pipeline to simplify installation.

The pot has openings fitted with slide gates through which water is distributed to the furrows. The slide gates are placed or the inside of the open pot to minimize erosion. The water level in the hydrant is regulated with an orchard valve. When line pressures are low enough, the valve in the riser may be omitted and the entire control made at the slide gates.

In installations where the hydraulic gradient is not more than 30 to 60 cm (1 to 2 ft) above the ground, the pot may be capped. In this case, the slide gates

at ground surface

Fig. 42–26. Orchard-valve hydrant showing the recommended installation of the valve in the riser.

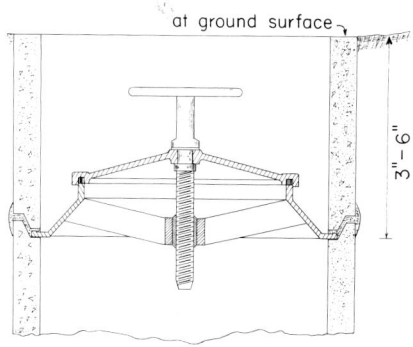

3"–6"

Fig. 42–27. Gated surface pipe and tubing attached to portable hydrants fitted over alfalfa or orchard valves. The flow to the furrows is adjusted from individual outlets in the pipe or tube.

are installed and operated from the outside of the riser. The flow is controlled by adjustment of line pressures and the gate.

Capped pot outlets have the advantage of not allowing leaves or debris to enter the riser to clog the slide gates. However, they provide less control of the flow, and erosion from the water jet is more severe. The slide gates are often replaced by special screw-type valves which cause less erosion. The use of capped pot outlets is usually limited to orchards and permanent crops where small flows are distributed into furrows.

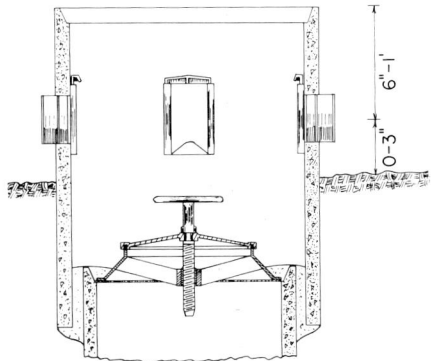

Fig. 42–28. Open-pot hydrant with orchard valve and slide-gate control.

With low line pressures, the pot is sometimes omitted and the slide gate put in the sides of the riser which may be left open or capped. Flow ratings and maximum recommended design capacities for slide gates have been determined (Pillsbury, 1952; Tovey and Myers, 1959).

3. *Surface Pipe Hydrant.* Several different types of hydrants are used to connect the pipelines to surface pipe or tubing. These are essentially variations of those mentioned previously in which the slide gates are replaced by nipples or connections for attachment of the surface pipe. Unless excess pressure is in the pipeline, the riser extends high enough to produce the required pressure in the surface pipe. If the pressure in the pipeline is more than required, the riser may be equipped with an orchard valve to prevent it from overflowing.

The height of an open hydrant should equal or exceed the head loss in the gated pipe or tubing plus freeboard. References are available which are helpful in determining head loss in surface pipe and tubing (Humpherys and Lauritzen, 1964; Pillsbury, 1952; Tovey and Myers, 1959). Discharge into the furrows is controlled by individual outlets along the pipe or tube.

d. Miscellaneous.

1. *Sand Traps.* Sand traps are usually built into the pipe inlet structures. Most of the suspended material may be removed by making the stand extra large in diameter to insure low water velocity and to provide a settling basin. The bottom of the stand is set some distance below the invert of the outlet pipes to provide space for sediment deposition.

Sediment collecting in the pipeline is minimized if a minimum velocity of 60 cm (2 ft)/sec and preferably 90 cm (3 ft)/sec is maintained. Sediment deposits in the pipeline reduce the capacity and eventually may plug the line. It is particularly important that sediment be removed from the water when surface pipe and tubes are used. Sediment deposition in this equipment makes it very difficult to move. The opportunity for sediment to settle out in surface pipe is usually greater since the velocities are lower.

The removal of sediment when water enters the pipeline from a ditch or canal was discussed in the previous section on open channels.

2. *Debris and Weed Screens.* Debris and weed screens should be provided at every gravity inlet. Much of the difficulty caused by this material will be eliminated if provision is made to remove it from the water before entering the pipe. A self-cleaning screen installed on a gravity inlet is shown in Fig. 42–10.

3. *Air Vents.* Vents are required on every pipeline to release air and to prevent high pressures. Vents are needed at all high points of a line, where the pipe slope increases sharply down grade, at sharp turns in the line, at the end of the line, and directly below any structure that entrains air in the flowing water. In addition to releasing air, open vents serve to release surges and prevent damage to the line when gates or valves are opened or closed. They also prevent pipe collapse from vacuum when the line is drained.

The cross-sectional area of the vent riser should be at least one-half the area of the pipeline. A typical installation is shown in Fig. 42–29. It is often recommended that the small vent pipe extend part way down into the larger riser. Air trapped in the space between absorbs pressure waves and the riser thus acts as a surge chamber. The area of the smaller vent pipe should not be less than one-sixtieth of the main line area and in no case less than 5 cm (2 inches) in diameter.

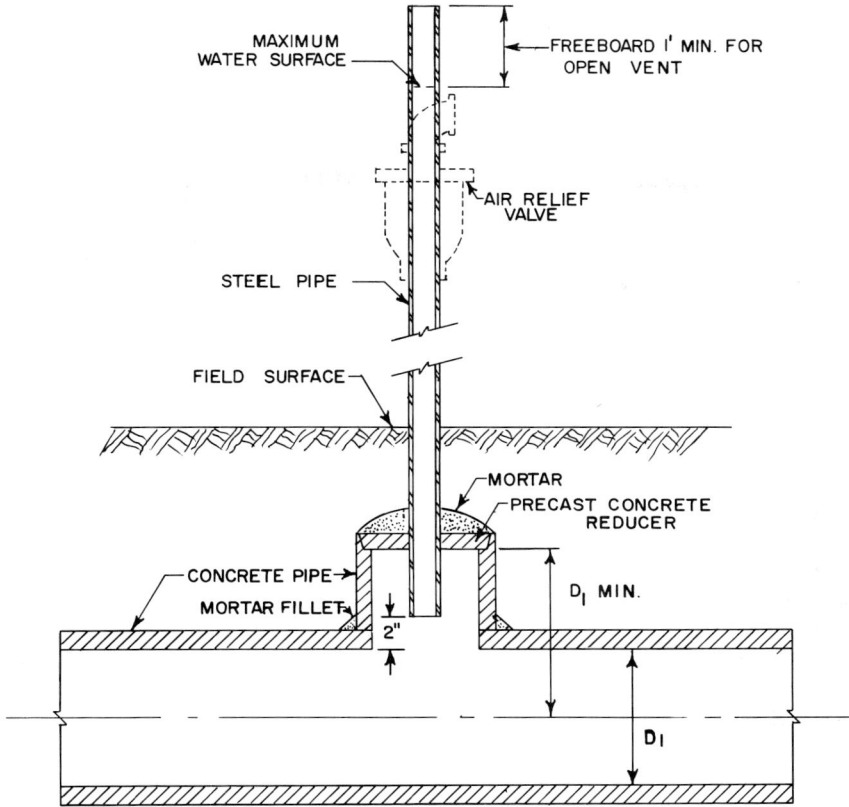

Fig. 42–29. Air vent for underground pipelines. The vent pipe is sometimes allowed to project into the riser to form an air pocket and surge chamber in the top of the riser.

All vents should extend at least 120 cm (4 ft) above the ground or as high as necessary to prevent overflow during normal operation. If an excessively high vent stand is required it may be advisable to install an air-relief valve to reduce the height as indicated in Fig. 42–29. These permit air to escape or enter but do not allow water to pass. They should not be located where it may be necessary to relieve momentary high pressure surges.

B. Measurement

1. CLASSIFICATION AND PRINCIPLES INVOLVED
The measurement of flow in closed conduits can be more complicated than in open channels. It is a common practice to use one of the devices previously discussed for open channels to measure the flow from pipes after the water is discharged into a ditch or canal. One simplifying factor of measurement in closed systems is that the area of flow is generally a constant. There is also no free water surface involved.

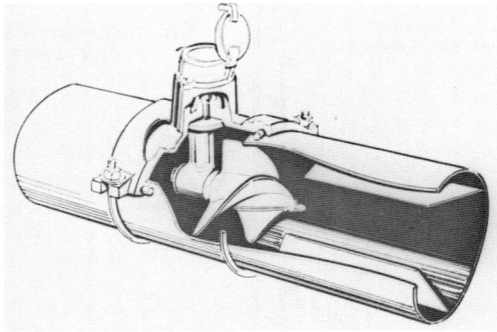

Fig. 42–30. Propeller-type meter for use in pipes.

In general, the devices and techniques available can be classified as those for open channels. The flow is usually contracted through an orifice of some type or a meter using a rotating wheel. There are a number of devices and methods available for pipes which give fairly accurate measurements.

2. DEVICES AND METHODS

a. **Commercial Meters.** Several types of meters, such as the ones shown in Fig. 42–20 and 42–30, give a direct measure of flow by a dial indicator. These vary in design from the disk type frequently used for small diameter lines to propeller types which are used for larger sizes. Some types are relatively easy to install in existing lines. All are subject to clogging if debris is carried in the flow.

b. **Tubes.** Another type of device available commercially for measuring discharge from pipes is the Cox (modified Hall) flow meter (Robinson, 1961). This utilizes a small diameter tube which is inserted across the discharge pipe as shown in Fig. 42–31. The small holes facing the flow are connected to a

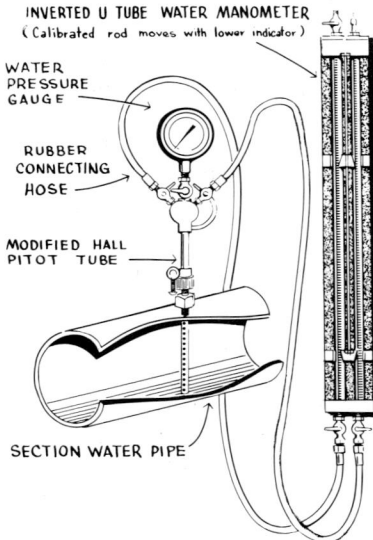

INVERTED U TUBE WATER MANOMETER
(Calibrated rod moves with lower indicator)

WATER PRESSURE GAUGE

RUBBER CONNECTING HOSE

MODIFIED HALL PITOT TUBE

SECTION WATER PIPE

Fig. 42–31. Cox flow meter can be inserted into the pipe for flow measurement.

manifold. In addition, there are three holes, one on each side and one in the rear of the small tube, which are located so that they are at the center of the large pipe diameter. These are connected to a second manifold. Each manifold is then connected by tubing to a differential manometer. With the manometer reading and tables furnished by the manufacturer, the discharge can be determined. With proper care the discharge can be determined with an accuracy of about ±8%.

c. Miscellaneous Devices and Methods.

1. Coordinate Method. A simple method, although not too accurate, is to merely measure the distance out and down from the pipe outlet to some point on the issuing jet (US Dep. Agr., 1962). With these measurements, tables are available to estimate the amount of flowing water. The expected accuracy with this method would not be less than ±10%.

2. Current Meters. Current meters similar to those used in open channels are available for measurement in pipes. The measurement is made by transversing the flow area at the discharge end of the pipe for an integrated measure of the velocity (Rohwer, 1942). Measurements have been made to an accuracy of ±3%, but a wider deviation can generally be expected.

3. End Orifices. In many cases a plate with an orifice hole of smaller diameter than the pipe can be attached to the discharge end of the pipe and used as a measuring device (US Dep. Agr., 1962). If the orifice is installed carefully, an accuracy of about ±5% can be expected.

LITERATURE CITED

American Society of Agricultural Engineers. 1964. ASAE standard S 261.3, design and installation of non-reinforced concrete irrigation pipe systems. Agr. Eng. Yearbook. p. 313–316.

Bergstrom, Walter. 1961. Weed seed screens for irrigation systems. Pacific Northwest Coop. Agr. Ext. Serv. (Washington, Oregon, Idaho). Ext. Publ. Bull. 43. 7 p.

Bondurant, J. A., and A. S. Humpherys. 1962. Surface irrigation through automatic control. Agr. Eng. 43(1):20, 21, 35.

Code, W. E. 1961. Farm irrigation structures. Colorado St. Univ. Agr. Exp. Sta. Bull. 496–S. 60 p.

Couthard, T. L., J. C. Wilcox, and H. O. Lacy. 1956. Screening irrigation water. Univ. British Columbia. Agr. Eng. Div. Bull. A.E. 6. 14 p.

Davis, C. V. 1952. Handbook of applied hydraulics. 2nd ed. McGraw-Hill Book Co., New York. p. 1209–1214.

Davis, Sydney. 1963. Unification of Parshall flume data. Amer. Soc. Civil Eng. Trans. 128(3):399–421.

Fiala, G. R., and M. L. Albertson. 1963. Manifold stilling basins. Amer. Soc. Civil Eng. Trans. 128(1):428–462.

Garton, J. E. 1966. Designing an automatic cut-back furrow irrigation system. Oklahoma Agr. Exp. Sta. Bull. B–651. 20 p.

Gilden, R. O., and G. O. Woodward. 1952. Low cost irrigation structures. Univ. Wyoming. Agr. Ext. Serv. Circ. 122. 16 p.

Haise, H. R., E. G. Kruse, and N. A. Dimick. 1965. Pneumatic valves for automation of irrigation systems. Agr. Res. Serv. US Dep. Agr. ARS 41–104.

Herpich, R L., and H. L. Manges. 1959. Irrigation water control structures. Kansas St. Univ. Dep. Agr. Eng. Agr. Exp. St. Contribution No. 82. 12 p.

Humpherys, A. S., and C. W. Lauritzen. 1964. Hydraulic and geometrical relationships of lay-flat irrigation tubing. US Dep. Agr. Tech. Bull. 1309. 38 p.

Israelsen, O. W., and V. E. Hansen. 1962. Irrigation principles and practices. 3rd ed. John Wiley & Sons, Inc., New York. p. 313–330.

Jensen, M. C., M. A. Hagood, and P. K. Fanning. 1954. Irrigation structures and methods for water control. Washington St. Univ. Agr. Ext. Serv. Bull. 491. 32 p.

King, H. W. 1954. Handbook of hydraulics. 4th ed. McGraw-Hill Book Co., New York. p. 6–8, 11–12, 14–19.

Kruse, E. G. Jan. 1965. The constant-head orifice farm turnout. Agr. Res. Serv. US Dep. Agr. Publ. ARS 41–93.

Parshall, R. L. 1950. Measuring water in irrigation channels with Parshall flumes and small weirs. Soil Conserv. Serv. US Dep. Agr. Circ. 843. 62 p.

Parshall, R. L. 1952. Model and prototype studies of sand traps. Amer. Soc. Civil Eng. Trans. 117:204–217.

Parshall, R. L. 1953. Parshall flumes of large sizes. Colorado St. Univ. Agr. Exp. Sta. Ext. Serv. Bull. 426–A. 39 p.

Pillsbury, A. F. 1952. Concrete pipe for irrigation. Univ. California Agr. Exp. Sta. Ext. Serv. Circ. 418. 51 p.

Portland Cement Association. 1952. Irrigation with concrete pipe. Chicago. 55p.

Robinson, A. R. 1957. Parshall flumes of small sizes. Colorado St. Univ. Agr. Exp. Sta. Tech. Bull. 61. 12 p.

Robinson, A. R. 1961. Study of the Cox flowmeter. Colorado St. Univ. Civil Eng. Dep. Rep. CER 61ARR 5. 15 p.

Robinson, A. R. 1962. Vortex tube sand trap. Amer. Soc. Civil Eng. Trans. 127(3):391–433.

Robinson, A. R. 1963. Evaluation of the vane-type flow meter. Amer. Soc. Agr. Eng. J. 44(C7):374–375.

Robinson, A. R., and A. R. Chamberlain. 1960. Trapezoidal measuring flumes for open channel flow measurement. Amer. Soc. Agr. Eng. Trans. 3(2):120–124, 128.

Robinson, A. R., C. W. Lauritzen, D. C. Mucke, and J. T. Phelan. 1963. Distribution, control and measurement of irrigation water on the farm. US Dep. Agr. Misc. Publ. 929. 27 p.

Robinson, A. R. 1966. Water measurement in small irrigation channels using trapezoidal flumes. Amer. Soc. Agr. Eng. Trans. 9(3):382–385, 388.

Rohwer, Carl. 1942. The use of current meters in measuring pipe discharges. Colorado St. Univ. Agr. Exp. Sta. Tech. Bull. 29. 40 p.

Rouse, H. 1950. Engineering hydraulics. John Wiley & Sons, New York.

Sharp, B. B. 1964. Flow measurement with a suspension wire. Amer. Soc. Civil Eng., Proc. Pap. 3821. Vol. 90. No. HY2. p. 37–54.

Smith, G. L. 1957. Scour and energy dissipation below culvert outlets. Colorado St. Univ. Civil Eng. Dep. Rep. CER 57GLS 16. 96 p.

Thomas, C. W. 1959. Common errors in measurement of irrigation water. Amer. Soc. Civil Eng., Trans. 124:319–340.

Thomas, C. W. 1960. World practices in water measurement at turnouts. Amer. Soc. Civil Eng., Proc. J. Irrig. Drainage Div. Pap. 2530. 86 (IR2):29–52.

Tovey, Rhys, and V. I. Myers. 1959. Evaluation of some irrigation water control devices. Univ. Idaho. Agr. Exp. Sta. Bull. 319. 32 p.

Uppal, H. L. 1951. Sediment excluders and extractors. Int. Ass. Hydraulic Res., Rep. 4th Mg. (Bombay, India). p. 261–316.

US Bureau of Reclamation. 1953. Water measurement manual. US Gov. Printing Office. Washington, D.C. 271 p.

US Dep. of Agriculture. 1962. Measurement of irrigation water. SCS Nat. Eng. Handbook. Sect. 15. Ch. 9. US Government Printing Office. Washington, D. C. 72 p.

43 | Surface Irrigation Systems[1]

A. ALVIN BISHOP

Utah State University
Logan, Utah

MARVIN E. JENSEN

Agricultural Research Service, USDA
Kimberly, Idaho

WARREN A. HALL

University of California
Los Angeles, California

I. SURFACE IRRIGATION

In surface irrigation, water is conveyed to the point of infiltration directly on the soil surface. Thus, the soil surface may be considered as the conveyance channel boundary. Surface irrigation channels vary widely in shape, size, and hydraulic characteristics.

The shape of the channel ranges from the small ditches or corrugations used for furrow irrigation of rowcrops, to a wide shallow channel where the entire land surface is flooded. The hydraulic characteristics of the channel may be extremely variable. It may change with time, with the wetting of the soil during an irrigation, and with the growth of the crop between irrigations. Since infiltration occurs, the stream size decreases along this channel and, since the intake rate is not constant, the flow changes with time at a given point in the channel. The hydraulics of surface irrigation systems therefore must account for nonuniform, unsteady flow.

1. ADAPTABILITY

Surface irrigation can be used on nearly all irrigable soils and most crops. The system can be tailored to accommodate a wide range of stream sizes and still maintain a high water application efficiency.

2. FLEXIBILITY

Surface irrigation systems permit ample latitude to meet emergencies. The capacity of most surface systems is sufficient to permit an entire farm to be irrigated in a small time period as compared to the period between irrigations. The irrigation cycle (period between irrigations), e. g., may be 10 to 14 days whereas the time required to completely irrigate the farm may be only 1 to 3 days. This feature provides an ample factor of safety in case of extreme climatic conditions

[1] Joint contribution from the Coll. Eng., Utah St. Univ.; the Soil and Water Conserv. Res. Div., ARS, USDA; and the Univ. California.

such as hot drying winds and cloudless days that can cause prolonged periods of high water use by crops. The relatively large capacity that can be built into surface irrigation systems without additional cost also provides versatility in meeting changing seasonal requirements. If only small continuous flows are delivered to the farm because of water right or water supply restrictions, on-farm storage ponds may be needed to fully utilize this flexibility.

3. ECONOMY

Surface irrigation systems are usually inexpensive to operate when compared with other methods of application because of low power requirements. Water is usually applied directly to the farmland by gravity flow from the irrigation project's canals and laterals. Where water is pumped from wells, rivers, storage reservoirs, or other sources of supply, only enough power to raise the water slightly above the land surface to be irrigated is needed. Labor requirements and costs may be more or less than other methods of irrigation depending on the systems being compared, the manner in which they are operated, the availability of low cost labor, and whether or not automatic controls are used.

4. DEPENDABILITY

Surface irrigation is as fully dependable as the water supply. The likelihood of having to interrupt the irrigation for repair of mechanical equipment during periods when crops require large amounts of water is small. Therefore, the potential economic loss due to failure of the system is also small.

A.   Types of Systems

Surface irrigation systems may be grouped into two broad classifications, complete flooding of the soil surface and partial flooding or furrow method. Complete flooding which is perhaps the oldest and most widely used method of surface irrigation includes flooding from field ditches, flooding strips between border dikes, and flooding in basins or checks. In this method, the entire land surface in the area being irrigated is covered with water. Water is conveyed to the area in a supply ditch or pipeline, and is distributed over the soil surface in a sheet for the desired time period.

In the partial flooding or furrow method, the entire irrigated area is only partially flooded. Closely spaced furrows (small ditches) contain and distribute the water which moves both laterally and downward from the furrow to moisten the plant root zone.

1. FLOODING FROM FIELD DITCHES

In this method, water from the distribution system is applied directly to the field from ditches without any dikes or levees to control flow (see contour and border ditch irrigation, Fig. 43–1). The advancing sheet of water is controlled primarily by the topography of the field with some guidance from the irrigator's shovel. Additional ditches may be dug to high points or areas difficult to flood. On steep lands, contour ditches generally constitute the distribution system. The spacing of the field ditches varies from 15 to 60 m (50 to 200 ft) or more, depending on the smoothness and slope of the land, texture and depth of soil,

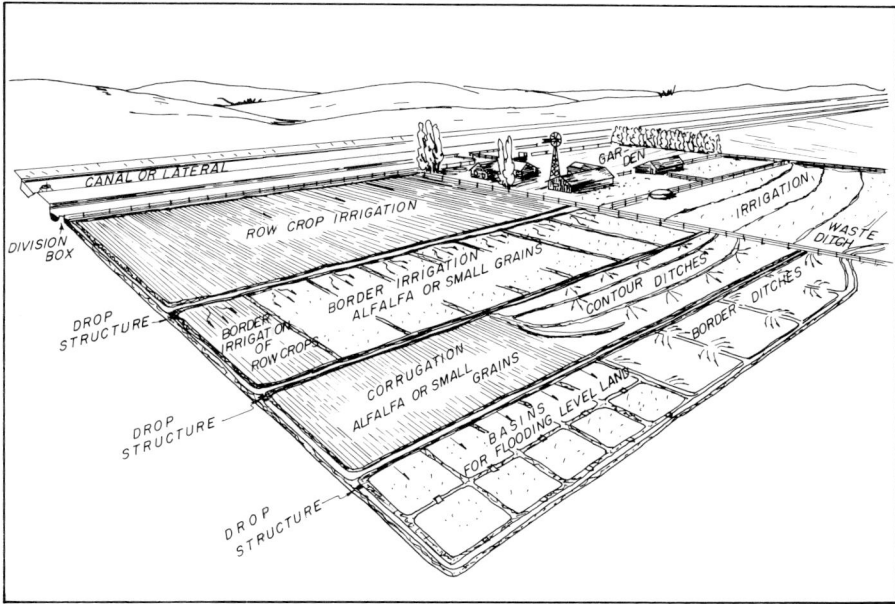

Fig. 43–1. Various methods of applying water to field crops (US Dep. Agr. Farm Security Admin., May 1943).

size of stream, and type and nature of crop. Precise land grading is seldom used to prepare the land for this method of application. Consequently, both the rate of advance and depth of the water sheet may be extremely variable. Uneven distribution of water and low water application efficiencies are common with uncontrolled flooding from farm ditches.

2. BORDER STRIP FLOODING

The border strip method is a controlled flooding process. The area to be irrigated is divided into strips or channels by constructing border dikes or levees (*see* border irrigation, Fig. 43–1). These dikes restrict the lateral movement of water, causing it to flow to the end of the field between the dikes. In reality, the border strips are wide, shallow channels in which the water flows from the head ditch to the end of the border strip in an elongating thin sheet, moistening the soil as it goes. This method of irrigation is commonly used when slopes in the direction of irrigation (parallel to the dikes) range from 0.1% to 1.0% for most crops to as much as 6% for pasturelands. When the field slopes in two directions, most of the slope perpendicular to the direction of irrigation (side fall) is eliminated within the border strip by additional land grading so that the advancing sheet covers the entire width of the strip.

Extensive land grading is usually required for the border strip method of irrigation. On steep slopes with fairly deep soil, border strips with low gradients can be formed by constructing the dikes nearly parallel to the contour. Each border strip then becomes a bench or terrace having the proper grade in the direction of the contour.

On land properly graded, the dikes or levees provide enough control to make this method of irrigation very efficient when properly operated. The dikes should generally be low and rounded on fields with low gradients so that crops can be planted on the dikes as well as on the strip between dikes. In this way no land is taken out of production. Barren dikes may be needed on fields with steeper side slopes and on fine-textured soils to prevent cracking upon drying which could result in lateral movement of the water.

3. BORDER CHECK OR LEVEL BASIN FLOODING

A border check or basin is an area completely surrounded by a dike, Fig. 43–1. The entire desired amount of water is applied quickly and ponded in the area until absorbed by the soil. When properly graded, built to the right dimensions for the soil conditions and size of stream available, and properly operated, checks and basins permit high water application efficiencies and uniform distribution of water.

4. FURROW IRRIGATION

With furrow irrigation small channels or furrows are used to convey the water over the soil surface in small individual, parallel streams, Fig. 43–1. Infiltration occurs through the sides and bottom of the furrow containing water. From the point of infiltration, the water moves both laterally and vertically downward to moisten the plant root zone. The degree of flooding of the land surface depends on the shape, size, and spacing of the furrows, the land slope, and the hydraulic roughness of the furrow.

When crops are grown and cultivated in rows, the construction of furrows between the crop rows can be accomplished as part of the cultivation process. The use of furrows then becomes a natural method for irrigating rowcrops.

Corrugations (small furrows) are often used for irrigating close-growing crops on steep or rolling lands, Fig. 43–1. The corrugations form the major water channels, but some flooding between the corrugations often takes place. This method is especially good for soils that have low intake rates or that disperse when flooded resulting in a hard surface crust upon drying.

Contour furrows enable the irrigator to successfully irrigate steep slopes without erosion, whereas water flowing in furrows directly down the slope would do serious damage. The contour furrows should have just enough slope for water to flow without overtopping (0.1 to 0.5%) but not enough to cause erosion. Deep-furrow rowcrops can be safely irrigated by contour furrows on lands having slopes up to 5% or more. Contour furrows have been successfully used on lands with slopes in excess of 15% when used as permanent deep furrows in orchards.

Different furrow shapes or layouts may be used to achieve special results. A broad bottom, shallow furrow for example, is often used to increase the intake rate or to cool seedbeds and the block-type furrow system is used when irrigating vineyards in California, USA to increase the effective length of the furrow. In this system three furrows are used with water in the middle or second furrow always running opposite the direction of irrigation. When water flows a short distance, approximately 3 m, in the first furrow, it is blocked and diverted to the middle or second furrow and flows in the opposite direction for the same distance. The water is then blocked and diverted to the third furrow and flows in the direction of irrigation to a point about 3 m beyond the block in the middle furrow where it is

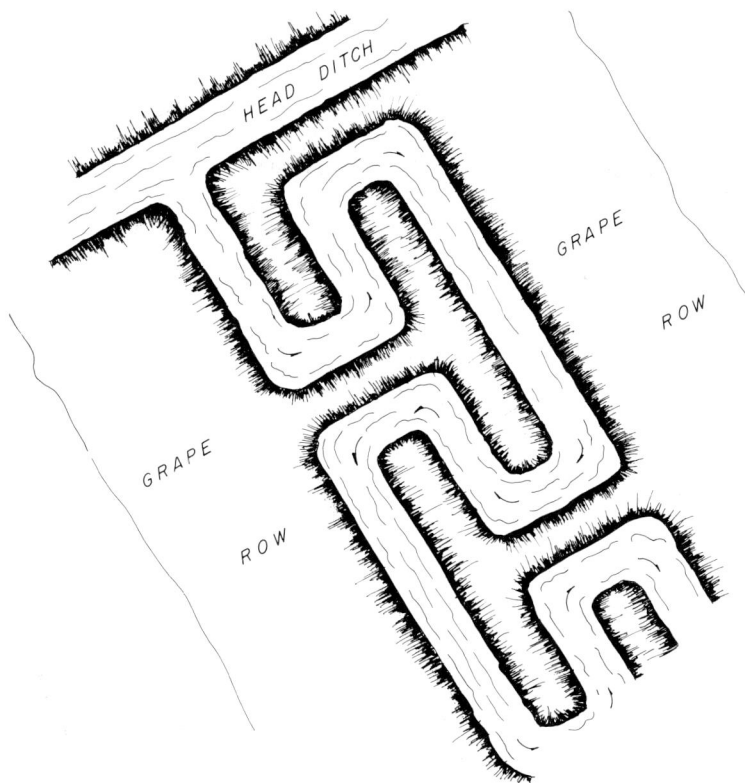

Fig. 43–2. Block system furrow irrigation.

blocked and again diverted to the center furrow and back to the first furrow (*see* Fig. 43–2).

B. General Characteristics of Surface Irrigation Methods

The adaptations, limitations, and advantages of the various methods of surface irrigation are presented in Table 43–1.

II. DESIGN PRINCIPLES AND PRACTICES

A. Design Principles

The design of a surface irrigation system first involves evaluating the general topographic conditions, soils, crops, farming practices anticipated, and farm operator's desires and finances for the field or farm in question. Information collected during the preliminary analysis should be sufficient to permit selecting one or more surface methods that will be most suitable. Then the basic information that will be needed to design the selected system must be secured.

Table 43-1. Adaptations, limitations, and advantages of surface irrigation

Method	Adaptation	Limitations	Advantages
		Flooding	
From field ditches	1) All irrigable soils 2) Close growing crops 3) Slopes up to 10% 4) Rolling lands and shallow soils where land grading is not feasible	1) Subdivides fields 2) High irrigation labor requirements 3) Low water application efficiency 4) Uneven water distribution 5) Possible erosion hazard	1) Low initial cost 2) Adaptable to a wide range of irrigation flows 3) Few permanent structures 4) Runoff from upper areas can be collected and reused
Border strip	1) All irrigable soils 2) Close growing crops 3) Slopes up to 3% for grains and forage crops 4) Slopes up to 7% for pastures	1) Extensive land grading required 2) Engineering designs necessary for high efficiencies 3) Relatively large flows required 4) Shallow soils cannot be economically graded 5) Dikes hinder cultivation and harvesting	1) High water application efficiency possible with good design and operation, regardless of soil type 2) Efficient in use of irrigation labor 3) Applicable on all soil types 4) Low maintenance costs 5) Positive control over irrigation water
Checks or level basins	1) All irrigable soils 2) Orchards and close growing crops 3) Slopes up to 2½% or more when benched or terraced	1) Extensive land grading often required 2) Large flows required 3) Initial cost relatively high 4) Dikes hinder equipment operations 5) Maintenance problems on escarpments on steep slopes 6) May effect crop yields on crops sensitive to inundation	1) Good control of irrigation water 2) High water application efficiency 3) Uniform water applications and leaching 4) Low maintenance costs 5) Erosion control from irrigation and rainfall 6) Large streams can be utilized
		Furrow irrigation	
Corrugations	1) All irrigable soils 2) Slopes up to 10% 3) Close growing crops	1) Moderately high irrigation labor requirements 2) Short runs required on high intake soils 3) Rough on cultivation and harvesting equipment	1) Increase efficiency and uniformity over flooding from field ditches on rolling lands 2) Improves border flooding on new lands
Furrow	1) All row crops 2) All irrigable soils 3) Slopes up to 5% with rowcrops and up to 15% for contour furrows in orchards	1) Moderate irrigation labor requirements 2) Engineering design essential for high efficiencies 3) Some runoff usually necessary for uniform water application 4) Erosion hazard on steep slopes from rainfall	1) Uniform water applications 2) High water application efficiency 3) Good control of irrigation water 4) Control equipment available at low cost such as spiles, siphon tubes and gates

1. DESIGN DATA

The basic data needed to design a system can be grouped into five general categories:

a. Water. Annual allotment, method of delivery (continuous flow, rotation or demand system, pumped, etc.), stream size available at any time and during peak water use period, quality of irrigation water, expected amount and distribution of rainfall, and irrigation water requirement including leaching requirement.

b. Topography. Major land slopes, field sizes and shapes, uniformity of grades, minor topographic undulations, point of water delivery, and surface drainage characteristics.

c. Soils. Feasibility of constructing canals and ditches without excessive seepage losses, structural stability for canals and ditches, maximum root zone depth, available water-holding capacity, effects of surface flooding such as crusting and cracking, cumulative intake as a function of time and expected variability between irrigations, erodibility, salt content, and internal drainage capacity.

d. Crops. Types and proportion of each crop to be grown, rooting depths and allowable soil water deficits at various stages of growth, anticipated germination problems, relative sensitivity to inundation, harvesting procedures required, crop rotation systems, and grazing needs.

e. Other. Availability and cost of labor, financial resources available, local customs, degree of maintenance anticipated and maintenance equipment available, and construction equipment available to the operator or through local contractors.

All of the above items have some bearing on the system selected and its final design. Overlooking or neglecting to consider any one of them can impair the effectiveness of the surface method selected.

2. DESIGN OBJECTIVES

A surface irrigation system should be designed rather than merely built in order to assure satisfactory adaption to the soils, topography and crops, and to guarantee uniform irrigations and high water application efficiencies using the available stream size and water supply. Ideally, the system should be capable of repeatedly replenishing the root zone reservoir uniformly before the soil water has been depleted beyond specified limits. The available stream size, and the length and grade of the land units must be combined to achieve these results without excessive labor, waste of water, erosion, and inconvenience to other farming operations.

Designing a system implies that the behavior of performance of the system can be predicted satisfactorily without a trial and error process in the field. If the intake characteristics are known, the designer then predicts two major occurrences: (i) the advance of the water sheet or furrow flow over the soil surface, and (ii) the recession of this water sheet or furrow flow from the surface.

The water should remain on the surface sufficiently long (required contact time t_{cr}) to allow just the desired amount of water to infiltrate the soil. The required contact time is obtained using the cumulative intake time relationship for the soil in question. For maximum water application efficiency the design objective is to have the actual contact time t_c as nearly equal to the required contact time t_{cr} as practical. The designer accomplishes this by adjusting the size

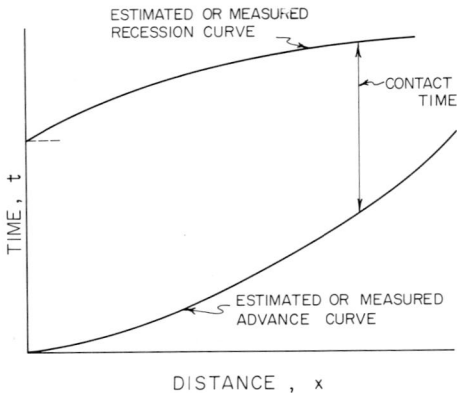

Fig. 43–3. Advance and recession curves (Criddle et al., 1956).

of stream, length of run, and other variables that can be manipulated until a satisfactory agreement is reached.

a. **Advance of the Water.** Predicting the advance of the water sheet is the most critical of the two items mentioned and is done by applying known hydraulic principles to overland flow. Field trials are often made to observe the combined influence of crop and soil roughness, stream size, and cumulative intake on the rate of advance. The results of either the predictions or field trials can be plotted, as shown in Fig. 43–3, to evaluate a given combination of variables.

Most investigators have used the continuity equation or water balance equation to predict rate of advance. Hall (1956) used a water balance equation and presented a numerical method for estimating the advance of the sheet of water in a border strip during equal time increments. This method, illustrated in Fig. 43–4, uses measured cumulative intake as a function of time and assumes a constant depth at the upper end of the border strip based on wide channel flow equations. It also assumes that a ratio or shape factor C_1 of the volume of surface storage to the volume described by $D_o x$ is independent of time, and an additional average depth of water or "puddle factor" ϵ is needed to fill pockets caused by unevenness of the surface of the border strip. The volume of water on the surface of the soil V_i at any time t_i is equal to

$$V_i = w(C_1 D_o + \epsilon) x_i \qquad [43–1]$$

where

V_i = volume of water on the surface at time t_i, L^3,
w = the width of the border check, L,
D_o = depth of water at the upper end, L,
ϵ = depth correction factor, L, and
x_i = distance to leading edge in time t_i, L.

The increment of increased surface storage during any time increment Δt_i is

$$V_i - V_{i-1} = [w(C_1 D_o + \epsilon)][x_i - x_{i-1}] = w(C_1 D_o + \epsilon)\Delta x_i \quad [43–2]$$

The volume of intake by the soil is computed in a similar manner except a shape factor, k, is applied only to the last increment of advance, Δx_i. For other advance increments, the actual intake values based on the measured intake-time relationship are used. When using equal time increments, computation of the average intake depth increment $\overline{\Delta y_i}$ for an advance increment Δx_i during time increment Δt_i reduces to

$$\overline{\Delta y_i} = (y_i - y_{i-2})/2, i \geq 2. \qquad [43–3]$$

The advance of the water during the first time increment is computed using the

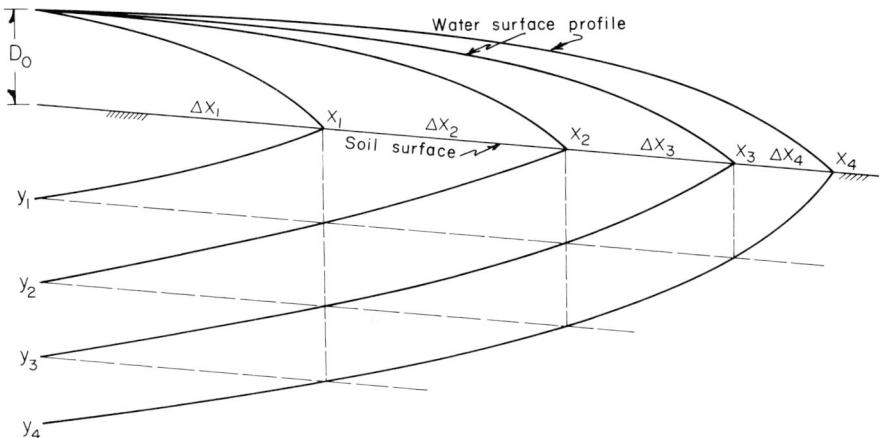

Fig. 43–4. Cumulative infiltration, y_i, advance distance, x_i, and surface storage after equal time increments, Δt_i (Hall, 1956).

equation

$$\Delta x_1 = Q\Delta t / w(C_1 D_o + \epsilon + ky_1) \qquad [43\text{--}4]$$

and for $i \geq 2$ the advance distances are computed as follows

$$\Delta x_i = \frac{\Delta x_1 - (\overline{\Delta y_i} \, \Delta x_1 + \overline{\Delta y_{i-1}} \, \Delta x_2 + \ldots + \overline{\Delta y_2} \, \Delta x_{i-1})}{(C_1 D_o + \epsilon + ky_1)}. \qquad [43\text{--}5]$$

If D_o is computed from the hydraulic characteristics of the border, the value of ϵ will be approximately equal to the tolerance of leveling the field. Severely cracked soils or a loose, porous surface condition may require much larger values of ϵ if such conditions were not present during intake measurements. Tabular forms can be used to simplify the recursive computation of Δx_i.

Less complex approximations of advance distances based on the water balance equation often are justified because hydraulic roughness cannot be predicted accurately and because the intake-time relationship is not constant for different irrigations. These computations are also usually made for a unit width of border strip. One equation used is described below and illustrated graphically in Fig. 43–5.

$$qt = x\overline{D} + x\overline{y} = x(C_1 D_o + C_2 y_o) \qquad [43\text{--}6]$$

where $q = Q/w = $ unit stream size or flow per unit width, $(L^3/T)/L = L^2/T$,

$t = $ total time of flow, T,
$x = $ distance to the leading edge, L,
$\overline{D} = $ average depth of water on the soil surface, L,
$\overline{y} = $ average cumulative intake over distance x, L,
$D_o = $ depth of water at the upper end, L,
$y_o = $ cumulative intake at the upper end, L,
$C_1 = $ surface storage coefficient varying from 2/3 to < 1.0, dimensionless, and
$C_2 = $ intake coefficient varying from 0.5 to < 1.0, dimensionless.

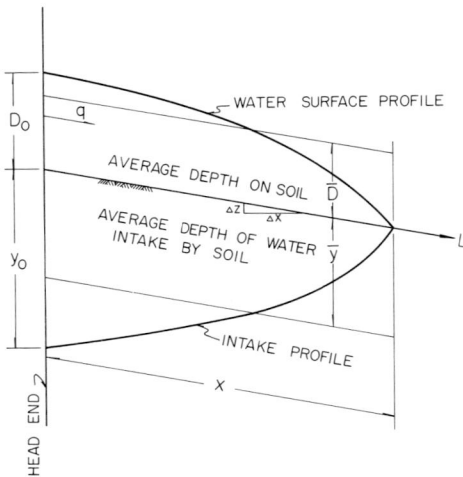

Fig. 43–5. Diagram illustrating the infiltration-advance problem.

The advance distance at any time t will be

$$x = qt/(C_1 D_o + C_2 y_o) \qquad [43\text{–}7]$$

The depth of water at the upper end of sloping fields D_o rapidly approaches a constant (normal depth). This depth can be computed using one of several open channel flow equations. The value of the C_1 will vary somewhat with the advance distance, slope, and hydraulic characteristics of the border strip, but for practical considerations, it can be assumed to be independent of time. For steep slopes, large advance distances, and small intake rates, $C_1 \rightarrow 1.0$. For flat slopes and small advance distances, and for very high intake rates $C_1 \rightarrow 0.67$. Cumulative intake can be the intake for a soil based on actual measurements, or, for design purposes, cumulative intake can usually be represented adequately by the equation $y_o = a t_o{}^b$ where t_o is the time water has been on the upper end. The value of C_2 will approach 1.0 as $b \rightarrow 0$ or when cumulative intake approaches a constant. This condition may occur on fine-textured soils that crack severely. After rapid initial intake the rate becomes very slow when the cracks and voids have filled. C_2 will approach 0.5 with uniform rate of advance as $b \rightarrow 1.0$ or when slopes are steep so that surface storage is small and cumulative intake is nearly linearly dependent on time. C_2 can also be considered independent of time for practical applications.

Analytical solutions for the prediction of advance distance have also been developed. Lewis and Milne (1938) expressed equation [43–7] in differential form essentially as

$$qt = C_1 D_o x + \int_o^t y(t - t_s) x'(t_s)\, dt_s \qquad [43\text{–}8]$$

where

$t_s =$ the value of t at which $x(t) = s$,
$y(t - t_s) =$ the cumulative infiltration at the point $x = s$ at time t,
$x'(t_s) =$ the value of dx/dt at $t = t_s$, and
$t =$ total time irrigation water has been applied.

When cumulative intake can be represented as a function of time, again assuming C_1 to be independent of time, analytical solutions to equation [43–8] can be used. Philip and Farrell (1964), using the Laplace transformation, recently presented a detailed derivation of a general solution to the Lewis-Milne infiltration-advance equation. Particular solutions were also presented for the following forms of the cumulative intake function:

$$y = c[1 - \exp(-rt)], \; y = at + c[1 - \exp(-rt)], \; y = at^b, \; 0 \leq b \leq 1, \text{ and}$$
$$y = at + ct^{1/2}.$$

Some of the particular solutions require the use of real and complex parameters and the use of the error function (or probability integral). A general description of the use of the error function with tabular values can be found in Carslaw and Jaeger (1959).

Several particular solutions were also expressed in simpler forms for either small t or large t. For example, the particular solution for small t and for the case $y = At + Bt^{1/2}$, where A represents the contribution to infiltration caused by gravity and B the contribution caused by capillary pressure gradient is given below:

For small t and $\overline{D} \neq \pi B^2/16A$, where $\overline{D} = C_1 D_o$

$$x = \frac{qt}{\overline{D}} \left[1 - \frac{2B}{3} \left(\frac{t}{\overline{D^2}} \right)^{1/2} + \pi \frac{B^2 - 4A\overline{D}}{8} \left(\frac{t}{\overline{D^2}} \right) - \cdots \right]. \quad [43\text{–}9]$$

For small t and $\overline{D} = \pi B^2/16A$

$$x = \frac{qt}{\overline{D}} \left[1 - \frac{2B}{3} \left(\frac{t}{\overline{D^2}} \right)^{1/2} + \frac{3\pi B^2}{32} \left(\frac{t}{\overline{D^2}} \right) - \cdots \right]. \quad [43\text{–}9a]$$

An evaluation of equation [43–9] is illustrated in Fig. 43–6. In this example the measured average depth $\overline{D}$ was used with the cumulative intake equation in meters and time in minutes, $y = 0.00033t + 0.0066t^{1/2}$. The crop involved was alfalfa (*Medicago sativa* L.), and the border strip was nearly level. Obviously, if the average depth $\overline{D}$ or $C_1 D_o$ can be predicted from hydraulic properties of the soil and crop, and if the cumulative intake function is known, the advance of the water sheet can be readily predicted. Philip and Farrell (1964) also presented a procedure for solving the inverse problem of determining the cumulative intake function using field trial data.

The innumerable variations in soil surface roughness, crop retardance at various stages of growth, and intake rates from one irrigation to the next have resulted in extensive use of field trials to evaluate the combined effects of the variables on rate of advance. Procedures for conducting field trials are given in other publications (Criddle et al., 1956).

b. **Recession of the Water.** Procedures for predicting the recession of water from the soil surface have not been sufficiently developed to allow summarization in this chapter. Approximate methods are being used by the Soil Conservation Service, US Department of Agriculture (Shockley et al., 1964). Field trials should be used to check the predicted advance of the water in the border strip or furrow before major irrigation systems are constructed to evaluate the combined effects of the many variables involved. Such field trials can provide sufficient data on recession for design purposes.

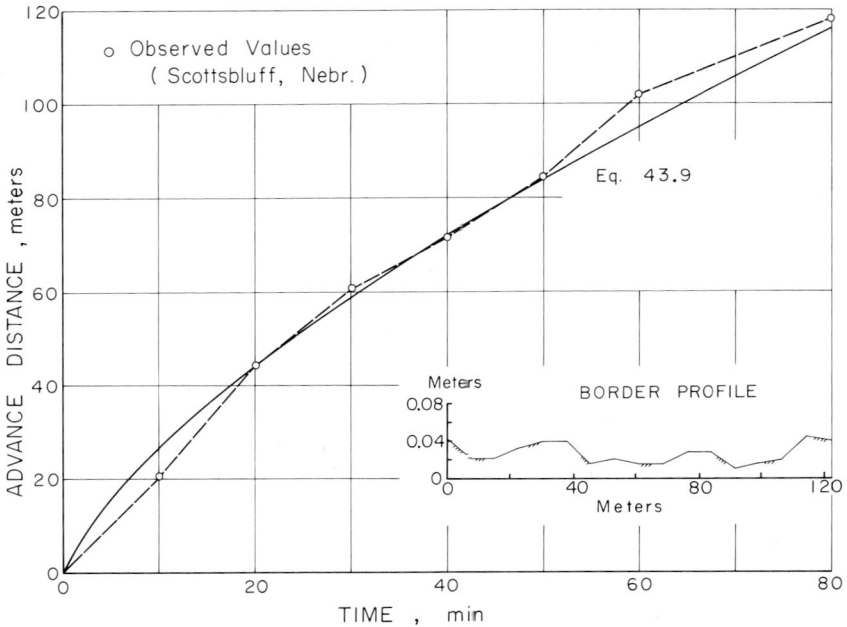

Fig. 43–6. Predicted and observed advance distances, and the soil surface profile of the border strip.

B.   Designing Flood Irrigation Systems

1.   GRADED BORDER STRIPS

Uniform distribution of water, minimum erosion or other crop and soil damage, high water application efficiency, and economical installation, maintenance and operational costs are commonly the broad objectives in the design of graded border strips. The general topographic requirements of border strip irrigation are relatively flat or level land of uniform grade and the assurance of good land preparation. Uniformity of irrigation depends on selecting or modifying the variables involved to provide a nearly constant contact time throughout the border strip, Fig. 43–3.

a. Border Strip Slope and Size. The slope is largely determined by the existing land slope, or by the amount of topsoil that can be economically and safely removed to obtain the desired slope. Economic considerations can be major factors in determining the final field and border slopes.

By properly matching the intake rate of soil with stream size, area to be irrigated, depth of water to be applied, and slope of the land, fairly uniform application can be obtained throughout the border length. Prediction of the rate of advance by one of the methods mentioned previously is a major part of the design of border strips.

Criddle et al. (1956) presented an equation for calculating the contact time necessary using the intake rate equation $dy/dt = At^n$. Integration with respect to time gives the cumulative intake, $y = (At^{n+1})/(n+1)$. The required contact time t_{cr} necessary to apply the desired depth of irrigation Y becomes

$$t_{cr} = \left[\frac{Y(n+1)}{A} \right]^{1/(n+1)} \qquad [43\text{--}11]$$

where

t_{cr} = required contact time, T,
Y = total depth of water to be applied, L, and
n = exponent of t in the intake rate equation.

At the upper end of the border strip, intake begins immediately when water application starts. Intake at the lower end of the field does not begin until some time later depending on the advance time. In order to adequately irrigate the lower end, the total time allotted for applying water must be approximately equal to the contact time required to absorb the desired depth of water plus the advance time.

If the water is in contact with the soil at the lower end of the run just long enough to replenish the soil root zone with the desired quantity of water, deep percolation losses below the root zone can be assumed nil at that point. However, deep percolation losses will occur at all other points in the field, increasing towards the upper end of the border strip, since the actual contact time is greater than the required contact time. The percentage of deep percolation loss will depend on the decrease in the intake rate from $t = 0$ until $t = t_{cr}$ for this soil and on the amount of time by which the required contact time is exceeded. By assuming that the deep percolation loss varies uniformly from a maximum at the upper end of the field to zero at the lower end of the field, Bishop (1962) showed that deep percolation loss P, expressed as a percentage of the total water absorbed, could be obtained from the equation

$$P = \frac{(R+1)^{n+1} - R^{n+1}}{(R+1)^{n+1} + R^{n+1}} (100) \qquad [43\text{--}12]$$

where

P = per cent of water intake which is lost by deep percolation below the root zone,
R = a time ratio = t_{cr}/t_a, where t_{cr} is the required contact time for the desired depth of irrigation water to be absorbed and t_a is the advance time, and
n = the exponent of t in the intake rate equation previously defined.

The percentage of loss is plotted against the values of n for different values of R between $R = 1/2$ to $R = 10$ in Fig. 43–7. By knowing the intake characteristics of the soil and the value of the exponent n in the intake rate equation the designer may select a value of R for the deep percolation loss considered allowable. If the allowable deep percolation loss is 6%, for example, the value of R might be as high as 7 for soils with $n = -0.1$, but a value of R smaller than 0.5 would still be allowable for $n = -0.9$. The smaller the value of R (larger advance time t_a), the longer the allowable length of run for a given soil and stream size. Border strips may be longer with the same percentage of water loss as n approaches -1.0 and shorter as n approaches zero. If the stream can be reduced after the water has advanced to the end of the border strip, thus eliminating any outflow, or when all of the outflow from the border strip is salvaged and used for irrigation on a lower field or recirculated on the same field, deep percolation is the only real

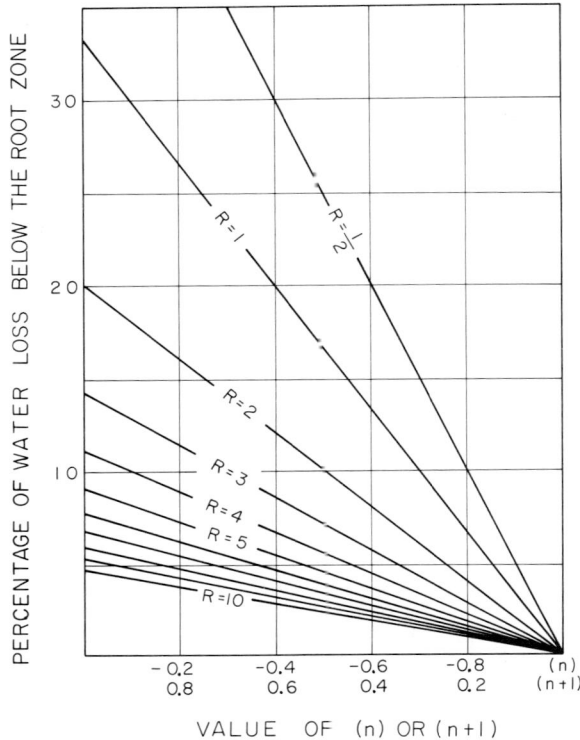

Fig. 43–7. Deep percolation—percentage of water lost below the root zone as a function of the cumulative intake parameter n in $y = at^{n+1}$, and the ratio, required contact time to advance time, t_{cr}/t_a (Bishop, 1962).

loss. Effective irrigation application efficiency then will be related only to the water lost by deep percolation. Under these conditions the water stored in the root zone will be equal to the total quantity applied minus the amount lost through deep percolation. The effective water application efficiency can be estimated using either equation [43–12] or Fig. 43–7 and the equation:

$$E_a = 100 - P \qquad\qquad [43\text{–}13]$$

where

E_a = (water stored/water applied) $\times$ 100 and
P = the percentage lost by deep percolation obtained from equation [43–12] or Fig. 43–7.

b. Stream Size. The most desirable size of stream can be determined by evaluating the contact times throughout the border strip for various combinations of the variables involved. The stream size available to the farm or field may necessitate adjusting the final border strip width to obtain the desired flow per unit width of the border strip.

Empirical procedures have been used extensively to estimate the most efficient stream size for border strip irrigation. Criddle et al. (1956) presented a series of curves to be used in estimating the unit-border stream size as a function of intake rate and depth of water to be applied. A unit-border was defined as 100 ft of border strip 1 ft wide. Shockley (1960) presented a modified procedure for estimating the unit-stream for this unit-border that also considers water application efficiency and the time period before recession begins.

$$Q_u = \frac{1}{E_a} \left(\frac{t_{cr}}{t_{cr} - t_r} \right) \frac{Y}{7.2 t_{cr}}$$ [43–14]

where

Q_u = unit-stream in cubic feet/second,
E_a = water application efficiency expressed as a decimal,
Y = desired depth of water application in inches,
t_{cr} = time in minutes required for infiltration of Y inches of water, and
t_r = recession lag time in minutes (from the time the stream is cut off until recession begins at the upper end).

This equation incorporates increases in the unit-streams to allow for lag in start of recession on small slopes. Usually the correction is not significant for slopes above 0.5%.

The time required for an irrigation is the time it takes to deliver the volume of water that will provide the desired depth of application, adjusted for the expected efficiency level. The total time t in hours can be estimated from equation [43–15] in which the values are as previously described, except t is now time in hours

$$t = Y/432 E_a Q_u$$ [43–15]

where

Y = required net application in inches,
E_a = expected water application efficiency expressed as a fraction,
Q_u = unit-stream in cubic feet/second.

The maximum stream size that can safely be used should also be considered. Criddle et al. (1956) used the following equation to estimate the maximum safe stream in cubic feet per second per foot width of a border strip without sod protection

$$q_{max} = 0.06 S^{0.75}$$ [43–16]

where

q_{max} = maximum stream in cubic feet per second per foot of width of the border strip, and
S = slope in per cent.

Criddle (1961) indicates that on slopes less than 0.3% the maximum stream per unit width will be governed by the height of the border dike. With cover crops on these slopes, streams of 0.15 ft³/sec per foot of width may result in flow depths of 6 to 8 inches and a stream of 0.2 ft³/sec per foot of width may result in flow depths exceeding 8 inches. Because of difficulties involved in maintaining large dikes, designing for streams less than 0.12 to 0.15 ft³/sec per foot of width of the border strip is recommended.

In some cases the minimum flow must also be considered. If the stream size is too small it will not spread laterally across the border strip. The criterion used by Shockley (1960) for the minimum unit stream for graded border irrigation is

$$q_{min} = 0.004S^{0.5} \qquad [43\text{--}17]$$

where

q_{min} = minimum stream size in cubic foot/sec per foot of border strip width, and

S = slope in per cent.

Lawhon (1960) also developed empirical procedures for designing border strip irrigation systems.

2. LEVEL AND LOW GRADIENT BORDER CHECKS

In level or nearly level border checks and basins the flow is unsteady and nonuniform behind the entire advancing stream. Therefore, D_o cannot be assumed independent of time as with graded border strips. Larger unit streams usually are used and the hydraulic gradients generally are smaller. Thus, more accuracy is required in predicting the volume of surface storage because more of the water remains on the surface during the advance of the water sheet as compared to graded border strips.

The solution of equation [43–7] for border checks requires predicting D_o as a function of stream size, soil and crop roughness, gradient, and advance distance. Procedures for predicting D_o as a function of these variables are not generally available although the hydraulic characteristics of this method of irrigation have been observed in field studies. For example, in a field study at Scottsbluff, Nebraska, USA with alfalfa on a fine sandy loam Jensen and Howe (1965) used one stream size, about 4.1 liters/sec per m of width (0.045 ft³/sec per foot of width) and found that the following empirical equation expressed the observed change in depth D_o as a function of advance distance x: For slopes $0 < S < 0.001$ ft/ft and $x < 400$ ft

$$D_o = 0.175x^{0.19} - C_s \qquad [43\text{--}18]$$

where D_o and x are previously defined and C_s = empirical correction for slopes ($C_s = 300 S - 1500 S^2$, $0 < S < 0.001$). Depth and advance distance dimensions in this case are in feet.

When $S = 0$, equation [43–18] gives the depth D_o directly for the one stream size, one soil, and one crop. With small gradients, increasing crop retardance materially reduces rate of advance because surface storage is greatly increased. With a dense growth of sugar beets (*Beta vulgaris*), for example, with $S = 0.00020$, Jensen and Howe (1965) found that the depth D_o could be represented by $D_o = 0.007x^{0.3}$ in contrast to $D_o = 0.0032x^{0.35}$ during the first irrigation with little vegetation on a slope of 0.0015. The value of D_o was nearly doubled as crop retardance increased, thus decreasing the rate of advance of the water sheet. The depth used for the sugar beet data was the average across small furrows and ridges because the water normally overtopped the ridges when retardance was high. The effects of excessive retardance by vegetation can be reduced by maintaining a large open furrow along the border check dikes.

The results of these field studies indicated that maximum efficiency and uniformity of irrigation were obtained when all of the water was applied in 0.2 to

0.33 of the average total intake time. Thus, the width and length of the border check must be related to the stream size available. Also, the width should be some multiple of the normal rowcrop equipment width to be used.

The depth of irrigation water to be applied will have been fixed by the crop and soil factors previously mentioned. The stream size per unit width will be limited by the width selected and flow available. The length will be limited to the existing field length or some fraction thereof such as one-half, one-third, or one-fourth. Thus, the remaining variable that the designer can adjust freely is the total drop Δz or gradient $\Delta z/\Delta x$. Jensen and Howe (1965) derived a prediction equation for estimating the necessary drop to obtain efficient irrigation

$$\Delta z = t_a \overline{y'} \qquad [43\text{--}19]$$

where

Δz = total drop, L,

t_a = advance time or the time for water to reach the end of border check, T, and

$\overline{y'}$ = average intake rate for an irrigation of depth Y, L/T.

This equation also requires predicting the advance of the water sheet. When inadequate data are available for predicting the advance, field trials may be necessary before the design gradient or total drop can be selected. In general when intake rates are extremely small the border checks will be essentially level. When intake rates are large and the contact time is small, the gradient must be increased for the same length of run to compensate for the time required for water to reach the end of the check. More refined surface smoothing to remove low spots may be needed with border checks than with border strips, especially near the lower end of the check.

Other factors to consider in designing border checks or basins are drainage requirements and the effects of inundation on plant growth. In most humid areas, a small gradient and facilities for removing excess water from rainfall are considered essential elements of bench-leveled systems (Phelan, 1960). Some crops are sensitive to inundation only during warm weather. A large percentage of such crops in a rotation may make the use of border checks undesirable.

Procedures for alignment of benches on steeper lands were given by Phelan (1960). Use of border checks is especially advantageous where periodic leaching is required. Large streams can be used where good water control is available. Also, water control structures can be easily automated.

Crops that must be irrigated after planting to assure germination may necessitate combining flat planting beds with deep furrows within the check. This is especially important on soils that develop a dry, hard surface crust after being wetted by flooding.

C. Designing Furrow Irrigation Systems

Furrow irrigation is used for nearly all crops such as corn (*Zea mays*), potatoes (*Solanum tuberosum*), fruit, and vegetables which are grown and cultivated in rows. Corrugations or small furrows are used in close-growing crops such as small grains, hay and pasture when these are grown and irrigated on sloping land or on soils that tend to crust badly after being flooded. Furrow irrigation systems must be designed to meet crop and cultivation equipment requirements. The maximum furrow slope is fixed by the natural slope of the land or the slope to which the land has been graded. Two other primary factors are: (i) the length of the run, and (ii) the size of the furrow stream. Usually these two factors can be adjusted so as to produce the desired water application efficiency (Bishop, 1962).

1. LENGTH OF RUN

From a practical viewpoint, furrows should be as long as possible. The longer the furrows, the greater the economy in handling farm equipment and using the irrigator's time. Long furrows reduce the frequency of turning cultivation equipment and reduce the number of furrow steam settings.

The same general principles of design as discussed for graded border strips apply to furrow irrigation. The advance time can be estimated or determined by field trials using procedures outlined by Criddle et al. (1956). Davis (1961) developed an equation for predicting advance using the same general relationship Hall (1956) used. The equation for furrows assumes an intake function of the form $y_1 = a(\Delta t)^b$, $y_2 = a(2\Delta t)^b$, etc., and is applicable for $i \geq 2$

$$\Delta x_i = \frac{Q\Delta t - \dfrac{Fa(\Delta t)^b}{2}\left[\overline{\Delta y_i}\,\Delta x_1 + \overline{\Delta y_{i-1}}\,\Delta x_2 + \ldots + \overline{\Delta y_2 x_{i-1}}\right]}{[Fa(\Delta t)^b\,k + C_1 D_o^2 + \epsilon]}$$

$$[43\text{–}20]$$

where F = a factor modifying the intake function because of method of measurement. The other variables are as described previously. D_o^2 is used in place of D_o since furrow volume can be described as a function of D_o^2.

Criddle et al. (1956) suggested that the furrow stream should reach the end of the run in one-fourth the required contact time, thus $R = 4$ for average soil conditions (see graded border strip design). However, as previously mentioned, longer runs would be possible with the same percentage of deep percolation loss as n approaches –1.0, but shorter runs would be required as n approaches zero. It is therefore recommended that the value of R used in design should be based on the intake characteristics of the soil to be irrigated (Fig. 43–6).

If runoff from the furrows cannot be salvaged, the size of the runoff stream also plays an important role in the choice of length of run. If the outflow from the furrow for example, amounts to 30% of the inflow stream, then for the average soil conditions assumed by Criddle et al (1956) or $n = -0.5$, the combined deep percolation plus runoff losses would be about the same for all values of $R > 1.0$. Under these conditions, the application efficiency would be about 70% and the combined deep percolation and runoff losses would be about 30%. No advantage would be gained by having short irrigation runs (larger values of R), since the reduction in deep percolation loss would be offset by a longer outflow period and greater runoff losses. When runoff is expected, the size of the runoff stream must be evaluated in relation to soil intake characteristics (values of n) and the contact time-advance ratio R.

2. SIZE OF FURROW STREAM

Once the farm has been prepared for irrigation, i.e. the various fields have been laid out and supply ditches installed, the slope is fixed and the possibilities for altering the spacing and length of furrows becomes limited. Length of furrow can then only be decreased to some fraction of total field length such as one-half, one-third, or one-fourth. The furrow spacing will have been fixed by the farm equipment and crops to be grown. Thus, the furrow stream will be the only variable that can easily be manipulated by the irrigator to achieve adequate and efficient irrigation.

The furrow stream must be large enough to reach the end of the run in the desired time, but small enough to be nonerosive. For most soils, some erosion takes place whenever water flows in the furrow. The larger the stream, the greater the erosion hazard for given conditions. Practical judgment must be used in evaluating the potential erosion problem. What could be considered serious erosion for one farm may be entirely permissible for soil conditions on another. The removal of only 2 to 3 cm ($\sim$1 inch) of topsoil from a very shallow soil may be more damaging than erosion of 25 cm or more (1 foot or more) of a deep soil. Criddle (1961) used the following empirical relationship as a guide for determining the maximum allowable furrow streams for various slopes

$$Q_e = 10/S \qquad\qquad [43\text{--}21]$$

where Q_e is the maximum nonerosive furrow stream, gallons per minute, and S is slope of the land in per cent. The maximum stream size may also be limited by the capacity of the furrow and the erosion potential by rainfall in some areas may further limit the acceptable slope and length of furrows.

The design of a furrow irrigation system must allow for possible variations in the size of furrow stream because intake rates, advance rates, erodibility, and crop requirements change throughout the irrigation season. Thus, the size of furrow stream must be altered occasionally to offset changes in other variables. By modifying the furrow stream, as required, the irrigator can maintain high water application efficiencies. However, this does not eliminate the need for determining the optimum stream for the initial and adverse conditions. Unfortunately with the present status of knowledge, there is no direct method for determining the size of stream. Therefore, considerable judgment in the selection of stream size is necessary.

Field trials are very helpful in providing information about the interrelationships of the variables: length of run, rate of advance, size of stream, and soil intake rates. Details for conducting such trials have been developed by Criddle et al. (1956). In general these instructions suggest measurements of slope, spacing, length of furrow, soil water conditions, and intake rate. Water is then applied to several furrows with different stream sizes whose range is as large as possible to include streams that are too large as well as streams definitely too small. As the water advances down the furrow the rate of advance is measured for each stream size. The extent of erosion under the different conditions is noted as is the flooding or depth of water in the furrow. Analysis of such field trial data provides a basis for selecting the optimum stream size for given conditions.

LITERATURE CITED

Bishop, A. A. 1962. Relation of intake rate to length of run in surface irrigation. Amer. Soc. Civ. Eng., Trans. 127:282–293.

Carslaw, H. S., and J. C. Jaeger. 1959. Conduction of heat in solids. 2nd ed. Oxford Univ. Press, London. 510 p.

Criddle, W. D. 1961. Irrigation. McGraw-Hill Book Co. Inc., New York. Agr. Eng. Handbook, Ch. 44, p. 509–531.

Criddle, W. D., S. Davis, C. H. Pair, and D. G. Shockley. 1956. Methods for evaluating irrigation systems. US Dep. Agr. Handbook No. 82. 24 p.

Davis, J. R. 1961. Estimating rate of advance for irrigation furrows. Amer. Soc. Agr. Eng., Trans. 4(1):52–54, 57.

Hall, W. A. 1956. Estimating irrigation border flow. Agr. Eng. 37(4):263–265.

Jensen, M. E., and O. W. Howe. 1965. Performance and design of border checks on a sandy soil. Amer. Soc. Agr. Eng., Trans. 8(1):141–145.

Lawhon, L. F. 1960. Attempts at improvement of design procedures for border irrigation. Agr. Res. Serv.-Soil Conserv. Serv. Workshop on Hydraul. of Surface Irrig., Proc. US Dep. Agr.-Agr. Res. Serv. 41–43. p. 7–10.

Lewis, M. R., and W. E. Milne. 1938. Analysis of border irrigation. Agr. Eng. 19:267–268.

Phelan, J. T. 1960. Bench leveling for surface irrigation and erosion control. Amer. Soc. Agr. Eng., Trans. 3(1):14–17.

Philip, J. R., and D. A. Farrell. 1964. General solution of the infiltration-advance problem in irrigation hydraulics. J. Geophys. Res. 69(4):621–631.

Shockley, D. G. 1960. Present procedures and major problems in border irrigation design. Agr. Res. Serv.-Soil Conserv. Serv. Workshop on Hydraul. of Surface Irrig., Proc. US Dep. Agr.-Agr. Res. Serv. 41–43, p. 1–6

Shockley, Dell G., Hyrum J. Woodward, and John T. Phelan. 1964. A quasi-rational method of border irrigation design. Amer. Soc. Agr. Eng., Trans. 7(4):420–423, 426.

US Department of Agriculture Farm Security Administration. 1943. First aid for the irrigator. 34 p.

44 | Sprinkler Irrigation Systems

JERALD E. CHRISTIANSEN
Utah State University
Logan, Utah

JOHN R. DAVIS
University of Nebraska
Lincoln, Nebraska

I. INTRODUCTION

Sprinkler irrigation is a versatile means of applying water to the surface of any crop or soil. A sprinkler system can apply water to soils at rates equal to, greater than, or less than the infiltration rate; it can be completely automatic or can be manually operated. In general, a sprinkler system can be employed for most soil and topographic conditions and for those areas where surface irrigation may be inefficient and expensive.

The sprinkler is the most important part of a sprinkler system. Most sprinklers consist of one or more nozzles or orifices that spray water under pressure through the air to irrigate some desired area. The wide range in sprinkler and nozzle sizes and types of sprinklers usually permits the designer to select the proper sprinkler and operating pressure to meet specific soil and crop requirements satisfactorily.

The high degree of water control is the outstanding advantage of sprinklers over other methods of irrigation (Fig. 44–1). Water application rates can be as low as 0.25 cm/hour (0.10 inches/hour), thus permitting the irrigation of steep or undulating lands and shallow soils without the hazards of soil erosion or excessive water losses. Sprinkler systems generally utilize smaller rates of water flow more effectively than surface irrigation methods and are of distinct advantage on soils of high permeability or low water-holding capacity. Under some conditions they can be advantageous on dense soils of low permeability. Where labor costs are high for surface irrigation, sprinkler irrigation may be the most economical method for applying water. Sprinkler systems are especially desirable where the required pressure can be developed by gravity.

Sprinkler systems have uses other than those of meeting the water requirements of a crop. Fertilizer application, frost protection, and temperature control (*see* chapters 52, 53, and 54) are becoming prominent and accepted uses of sprinkler systems all over the world.

Like other economic or physical entities, sprinkler systems also have disadvantages, depending on the conditions involved. Damage to citrus has been observed when poor quality irrigation waters were sprayed on the foliage by sprinklers (Eaton and Harding, 1959). In some cases, poor quality waters will leave undesirable deposits on the leaves or fruit of the crop. Also, in some instances, the influence of free water on crop surfaces as a result of sprinkler irrigation has enhanced the propagation of fungi or foliar bacteria, but the latter

885

Fig. 44–1. A portable, hand-movable sprinkler system irrigating sugar beets on sloping soils. These are medium-sized sprinklers operating at about 45 lb/inch² and spaced 30 ft apart on the lateral pipeline. (Photo courtesy W. R. Ames Co.)

can be avoided by use of disease free seed. Frequently, labor costs of sprinkling may be higher than costs for surface irrigation.

The advantages and disadvantages of sprinkler systems should be assessed economically in comparison with other irrigation methods. Any specific type of sprinkler system should likewise be compared with alternate types.

This chapter describes sprinkler irrigation systems, their design, operational characteristics, and other agricultural uses. Major emphasis is placed on system selection, principles of design, and performance characteristics. Space does not permit a detailed treatise on design which is adequately covered elsewhere (see list of references at end of this chapter).

II. TYPES OF SPRINKLER SYSTEMS—GENERAL DESCRIPTION

There are many types of sprinkler systems. All may be grouped into three general classes according to portability: (i) portable sprinkler systems, (ii) semiportable systems, and (iii) stationary systems.

The portable and semiportable systems are most commonly used in agriculture. A fully portable system consists of either a stationary or a portable pump and portable main lines, laterals, and sprinklers. The semiportable system usually consists of a stationary pump and stationary main pipelines with portable laterals. Stationary systems are those in which the main pipelines, laterals, and sprinklers are fixed permanently throughout the entire area to be irrigated. Combination systems with stationary pipelines and portable sprinklers are frequently used on turfs such as golf courses and parks.

Portable and semiportable sprinkler systems can also be classified according to method of moving the laterals. A common method is the hand moved system

Fig. 44–2 a. Side-roll lateral system uses the pipe as an axle. A small engine at the center of the lateral rotates the pipe, thus moving lateral across field. In some systems, a separate drive shaft is used to rotate the wheels for moving pipe through taller crops. In others, lengths of pipe or hose upon which additional sprinkers are attached, are pulled by the lateral pipe. (Photo courtesy Sprinkler Irrig. Ass.)

Fig. 44–2 b. End-pull system uses a tractor to pull pipeline from one end to a new lateral position. Various types of skids are often used instead of wheels. (Photo courtesy W. R. Ames Co.)

in which the laterals are carried by hand in 6 to 12 m (20 to 40 ft) sections of pipe to the next lateral position, usually a distance of 12 to 24 m (40 to 80 ft). There are also various types of power moveable systems designed to reduce the labor cost of moving pipe. In the side-roll lateral system (Fig. 44–2a), the lateral pipe acts as the axle of large diameter wheels which are spaced 18 to 24 m (30 to 40 ft) apart. A small gasoline engine mounted at the midpoint of the line is used to roll the lateral to the next position. There are also various types of end-pull systems in which the laterals are moved to the next position by pulling one end with a tractor or other power unit (Fig. 44–2b). The side-roll systems are limited in application to crops which will not interfere with movement of the pipe.

The hose-pull system, which is gaining popularity in orchards, employs buried mains and submains and a large number of small diameter flexible hoses upon which one or more sprinklers are mounted. The initial cost is intermediate between a portable and a stationary system; the labor requirement involves pulling the hoses manually to position the sprinkler at its next setting. The skill of the labor required is minimal, and the system need not be shut off during the sprinkler moving.

Boom-type sprinkler systems (Fig. 44–3) employ only one boom sprinkler on each lateral. The boom is a nozzled, slowly rotating pipeline which is suspended from a portable tower. Boom sprinklers are moved by towing the towers to the next position along the laterals with a tractor or winch. This large sprinkler irrigates widths of 75 to 100 m (250 to 350 ft), depending on nozzle sizes and pressures and is particularly useful for tall crops such as corn (*Zea mays*) and

Fig. 44–3. Boom-type sprinkler; boom lengths that vary from 80 to 250 ft permit irrigation of 0.6 to 4.0 acres/setting at application rates of 0.2 to 1.1 inch/hour.

sugarcane (*Saccharum officinarum*), where space is provided at regular intervals for moving the portable towers.

A self-propelled sprinkler system consists of a radial pipeline supported at a height of 1.8 to 2.4 m (6 to 8 ft) at intervals of about 30 m (100 ft) on towers mounted on two wheels or a small track. The radial line is rotated slowly around the pivot point in the center of the field by either water pressure actuators or electric motors at each tower (Fig. 44–4a&b). Conventional sprinklers mounted on the pipe then distribute water to the field as the pipeline is moving. Systems are available to cover square fields ranging in size from 10 to 105 ha (24 to 260 acres). Total capacities range from 1,500 to 4,500 liters/min (400 to 1,200 gal/min). Self-propelled sprinkler systems are often used for crops such as corn where it is difficult to move sprinkler laterals in the conventional manner.

Fig. 44–4 *a*. Aerial photo of self-propelled system in alfalfa; note the increased application of water at the outer end of the pipe to compensate for the increased area to be irrigated per revolution of the pipe. (Photo courtesy Valley Manufacturing Co.)

Fig. 44–4 *b*. Detail of the self-propelled system showing propane engine, gear-driven turbine pump, centrifugal sand separator, and pipeline supported by towers on wheels. (Photo courtesy Valley Manufacturing Co.)

A portable perforated pipe system consists of lightweight tubing with a pattern of small holes in the top side of the pipe (Fig. 44–5). With the pipe laid on the ground, water is distributed up to 7 or 8 m (25 ft) on each side of the pipe at rates from 1.5 to 5.0 cm/hour (0.62 to 2 inch/hour). The pressure requirement is relatively low—from 0.35 to 1.4 kg/cm^2 (5 to 20 lb/inch^2).

Stationary systems are more frequently used for sprinkling lawns and turfs and high value, water-sensitive crops like berries, vegetables, orchards, and nurseries. They are relatively much more expensive in initial cost than other systems, but labor costs for operation are minimized. They can be made completely automatic. Such systems are commonly used for turf where the sprinkling may be done during the night to avoid interference with daytime uses. Stationary systems vary from small lawn and garden systems employing fixed spray heads to large turf systems employing large rotating sprinklers with capacities of 400 liters/min (100 gal/min) or more.

Combination systems using stationary underground pipe systems with portable sprinklers are commonly used for large turf areas, such as golf courses. Special quick coupler valves are used for mounting the sprinklers. The initial cost of such

Fig. 44–5. Perforated pipe system for orchard irrigation. (Photo courtesy W. R. Ames Co.)

systems is lower, but operational costs are higher than those for completely stationary systems.

Nozzle line, or oscillating pipe, systems are usually stationary. The pipelines are spaced 15 to 25 m (50 to 80 ft) and are mounted on rigid supports which are 4 m (12 ft) apart. They are used in nurseries and greenhouses and for special high value crops.

III. TYPES OF SPRINKLERS

Most agricultural sprinklers are of the slowly rotating type with either one or two nozzles that vary in size from about 1.5 mm to 5 cm (1/16 inch to 2 inches) in diameter. They discharge water at rates that vary from less than 4 liters/min (1 gal/min) to more than 4,000 liters/min (1,000 gal/min) and cover circular areas from 6 to over 180 m (20 to 600 ft) in diameter. Thus a large number of sprinkler types are available. The most commonly used types are the small to

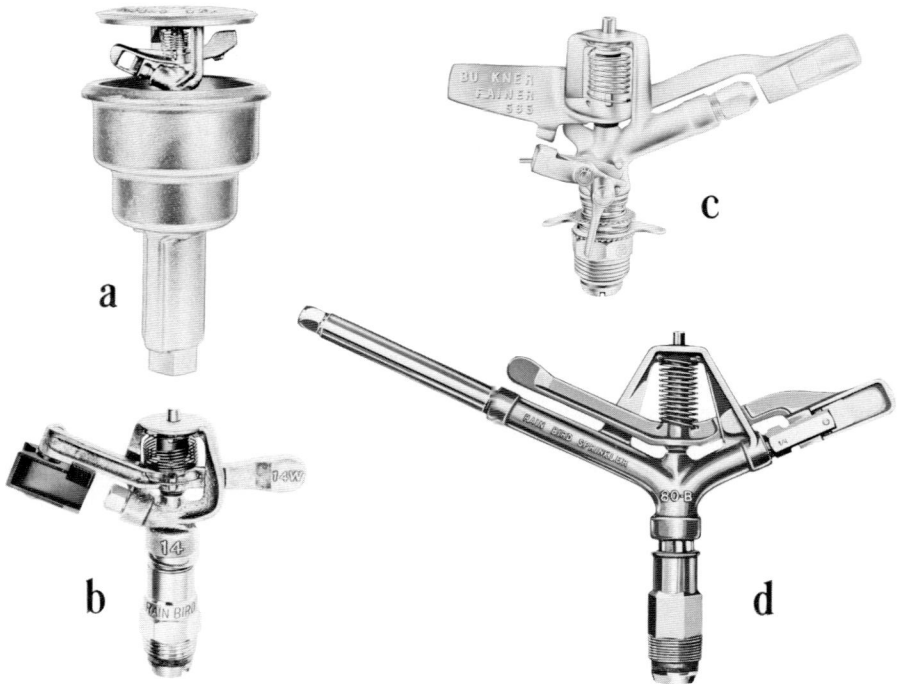

Fig. 44–6.(a) Full-circle, impulse-driven, rotating pop-up sprinkler, used for turf irrigation. (b) Full-circle low-application rate agricultural sprinkler, operates at 35 lb/inch^2, discharges 1.0 to 4 gal/min. (c) Part-circle, medium-sized agricultural sprinkler, operates at 40 to 80 lb/inch^2, discharging 6.5 to 25 gal/min. (d) Full-circle higher-application rate agricultural sprinkler, operates at 70 to 90 lb/inch^2, discharges 40 to 120 gal/min. Larger-sized sprinklers operate at about 100 lb/inch^2 and discharge 200 to 600 gal/mm. [Photos courtesy (a) and (c), Buckner Manufacturing Co., (b) and (d), Rainbird Sprinkler Manufacturing Corp.]

medium sprinklers with capacities ranging from 7.5 to 75 liters/min (2 to 20 gal/min) and which operate at pressures from 1.4 to 4.2 kg/cm² (20 to 60 lb/inch²). These sprinklers cover effective areas of 10 to 40 m (30 to 120 ft) in diameter.

Most sprinklers rotate around a vertical axis (Fig. 44–6). The rotation results from a torque produced by the reaction of the water leaving a nozzle or by the impact of a spring-loaded arm that periodically interrupts the jet from one of the nozzles. Some sprinklers are gear driven by a small water turbine that may be located in the base of the sprinkler or externally on one of the nozzles. Some rotating sprinklers can be adjusted to irrigate any desired segment of a circle.

Fixed spray heads for lawn and garden use depend on diffusing cones or slots to produce the desired fine spray. These cover much smaller areas than rotating sprinklers and have higher rates of application. Half-circle and quarter-circle spray heads are also available.

Small, rapidly rotating sprinklers that depend on jet reaction are sometimes used as garden or undertree orchard sprinklers. These cover areas intermediate between fixed spray heads and slowly rotating sprinklers. They usually have very small nozzles and low rates of application.

Boom-type sprinklers have recently been introduced to cover larger areas than can be covered by the conventional large sprinklers. These sprinklers have long arms 12 m (35 ft) or more in length supported by wires from a central tower. They are used on the boom-type systems described previously. These sprinklers can effectively distribute water over circular areas ranging from about 90 to 120 m (300 to 400 ft) in diameter at application rates of 6 to 25 mm/hour (¼ tc 1 inch/hour).

IV. UNIFORMITY OF APPLICATION OF SPRINKLERS

Basic to the selection and design of sprinkler systems is some knowledge of the performance of rotating sprinklers with respect to the uniformity of application and its relation to the spacing of the sprinklers. Much has been written on this subject, and different methods of characterizing uniformity have been proposed. The most commonly used is the concept of a coefficient of uniformity (Christiansen, 1942) which is defined by the expression:

$$C_u = 100 \ [1.0 - (\Sigma|d|/mn)] \qquad [44\text{–}1]$$

where $|d|$ is the absolute value of the deviation of the individual observations of depth applied from the mean value m, and n is the number of equally spaced observations of depth applied in the sprinkler pattern. Absolute uniformity is represented by a $C_u = 100\%$. A statistical uniformity coefficient SC_u has been proposed by several investigators (Hart, 1961; Korven, 1952; Molenaar, 1954; Wilcox and Swailes, 1947). This coefficient can be expressed by the equation

$$SC_u = 100 \ [1.0 - (S/m)] = 100 \left[1.0 - \sqrt{\frac{\Sigma d^2}{(n-1)m^2}} \right] \qquad [44\text{–}2]$$

where S = standard deviation.

The US Soil Conservation Service (Criddle et al., 1956) proposed the concept of pattern efficiency, PE, which can be expressed by the equation

$$PE = 100 \ (a/m) \qquad\qquad\qquad\qquad [44\text{--}3]$$

where a = average depth for the 25% of the observations having the least water
depth, and m = the mean depth. After studying a large number of actual
sprinkler patterns and comparing the various indices, Dabbous concluded that
the general relationship between these various coefficients could be expressed
approximately by the relations (B. J. Dabbous, 1962. A study of sprinkler uniformity evaluation methods. *Unpubl. M.S. Thesis. Utah St. Univ., Logan*)

$$SC_u = 1.25 \ C_u - 0.25 \qquad\qquad\qquad [44\text{--}4]$$

and

$$PE = 1.45 \ C_u - 0.45. \qquad\qquad\qquad [44\text{--}5]$$

Thus, the reported values of the different indices of uniformity can be compared.

Coefficients of uniformity and pattern efficiency are single-valued indices of
water distribution, however, and do not fully describe the entire distribution
pattern nor are they accurate indicators of crop yields. For more intensive investigation of sprinkler patterns, a cumulative distribution curve E. M. Norum
describes explicitly the distribution of water within the pattern (E. M. Norum,
1961. A method of evaluating the adequacy and efficiency of overhead irrigation
systems. *Paper presented at the Amer. Soc Agr. Eng. 1961 annual meeting at
Ames, Iowa*). This curve is formed by plotting the ratio of depth of application
at a point to the mean application depth, x/m, against the percentage of the area
P that received a depth of water equal to or greater than the point depth. Curves
that approximate the straight line $x/m = 1$ represent uniform distribution of
water. More descriptive expressions for uniformity have been derived by Howell
(1964). These relate the crop yield to the distribution of water, such that the
total yield is expressed in terms of mean depth and moments of the distribution
up to the order of the polynomial in which the yield–depth-of-water relationship is expressed. Use of these expressions thus might require calculation of the
mean depth and the standard deviation, skewness and kurtosis of the
distribution.

The uniformity of application achieved under field conditions depends on (i)
the type of sprinkler pattern, (ii) the spacing of the sprinklers, and (iii) the
effect of such factors as wind, variation in rotation of the sprinklers, tilting of
sprinkler risers, and so forth (Fig. 44–7).

The sprinkler pattern is the shape of the volume of water falling on the area
wetted and can be represented by a vertical section through the sprinkler location or by contours showing the depth of water applied. Under ideal conditions
this pattern is symmetrical around the sprinkler, but under field conditions it is
seldom so. Wind distorts and offsets the pattern. Lack of uniformity of rotation
and tilting of the sprinkler risers also distorts the pattern. Pressures below that
for which the sprinkler is designed produce low applications near the sprinklers
and excessive applications in a ring around the sprinklers.

The triangular-shaped pattern, with maximum application at the sprinkler and
gradual reduction in application to the edges of the area covered produces the
most uniform application when sprinklers are spaced not more than about 55%
of the diameter wetted by the sprinkler. For rectangular sprinkler arrangements
with closer spacing on the laterals, the spacing between laterals can be increased
slightly. For a square arrangement of sprinklers, it is theoretically possible to

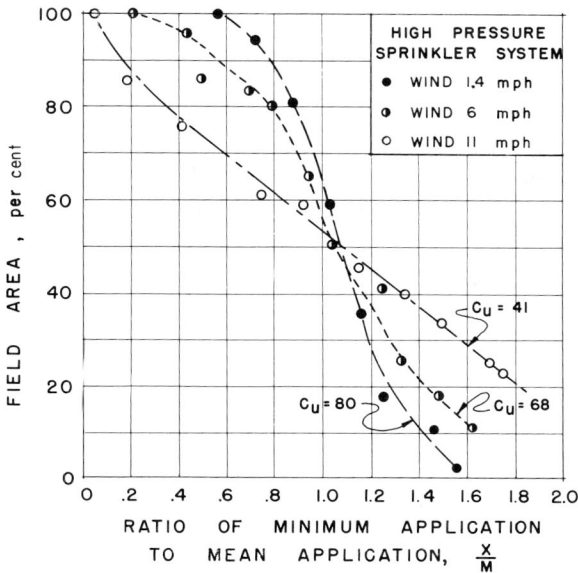

Fig. 44–7. Cumulative distribution curves for sprinkler systems as influenced by wind velocity.

obtain high uniformity coefficients with the spacing between laterals up to about 70% of the diameter covered by the sprinkler; for an equilateral triangle arrangement of sprinklers, up to 75% of the diameter. However, ideal patterns and high uniformity coefficients are seldom obtained under field conditions.

Ideal patterns for each arrangement and spacing are given by Christiansen (1942). The ideal patterns for wide spacings have a uniform application to about 50% of the radius covered; then the application reduces uniformly. Such patterns are very sensitive to correct spacing, and the uniformity obtainable is lower for spacings both less and greater than the optimum because of excessive or insufficient overlap. For triangular patterns, the uniformity remains high for all spacings up to about 55% of the diameter covered.

An effective procedure for securing a greater seasonal uniformity of application is to alternate the position of the lateral for each application; i.e., for successive applications to place the lateral midway between the positions occupied during the previous irrigation. McCulloch and Keller (1962) have estimated that the uniformity achieved by this procedure can be expressed by the relation

$$C_u' = 10(C_u)^{1/2}. \qquad [44\text{–}6]$$

Dabbous (B. J. Dabbous, 1962. A study of sprinkler uniformity evaluation methods. Unpub. M.S. Thesis. Utah St. Univ., Logan) analyzed 40 sprinkler patterns for 19 spacing combinations and found that the relation could more accurately be expressed by the equation

$$C_u' = 27 + 0.73 \, C_u. \qquad [44\text{–}7]$$

Thus, for a normal C_u value of 80%, McCulloch and Keller's C_u' would be 89% and Dabbous' C_u' would be 85%.

A slightly increased seasonal uniformity can also be achieved by altering the position of the sprinklers on the lateral for successive applications. This procedure is known as the double offset method. It is achieved by alternately adding and omitting a length of pipe equal in length to one-half of the sprinkler spacing on the lateral.

V.   SPRINKLER APPLICATION EFFICIENCY

Application efficiency is defined as the ratio of usable water applied to total application. It is dependent on the uniformity of application and on losses from evaporation and deep percolation below the root zone. When sufficient water is applied to furnish most of the area with all its requirements, there will be percolation losses in places receiving excess applications. A small amount of downward leaching is essential, however, to maintain a salt balance within the root zone.

Evaporation losses from the spray depend on temperature, wind, humidity, and on the fineness of the spray which in turn depends on nozzle sizes and pressure (Christiansen, 1942; Frost and Schwalen, 1955). Such losses have been estimated to range from 2% to 8% of the volume of water discharged by the sprinklers.

Numerous tests have shown that the total evapotranspiration from a recently sprinkled crop that fully covers the ground including the direct evaporation from the wet foliage, does not greatly exceed the normal evapotranspiration rate. While free water is available on the foliage, transpiration from the leaves is greatly reduced. Thus, direct evaporation from the wet foliage does not constitute an important loss of water.

Actual application efficiencies obtainable for well-designed and operated sprinkler systems may approach 85%, and for night irrigation and low wind conditions it could be higher. A design value of 70 to 80% is normally used. Full coverage systems with small sprinklers generally have higher efficiencies than the large or boom-type sprinklers (Davis and Hart, 1963).

VI.   HYDRAULICS OF SPRINKLER SYSTEMS

A general knowledge of the hydraulics of sprinkler systems is basic to sprinkler system planning and design. This includes the sprinklers, the pipeline system, and the pumping plant.

The discharge from sprinklers can be expressed by the equation

$$Q = C A (2gH)^{1/2} \qquad\qquad [44\text{--}8]$$

where Q is the discharge, C is the sprinkler coefficient, A is the area of the nozzles, g is the acceleration of gravity, and H is the head, represented by the pressure.

This equation is basic and applies to any consistent set of units, i.e., either the English system or the metric. In the English system, the discharge Q is expressed in cubic feet/second, the area A in square feet, the value of g in feet/second per second, and H in feet. Since the discharge from sprinklers is usually only a small fraction of 1 ft³/sec and the area a very small part of 1 ft², these units are not convenient. The discharge is therefore usually expressed in gallons per minute (gal/min) and the nozzle area in square inches (inches^2), and the head as pressure in pounds/square inch (lb/inch^2). Combining g with the constant and using q to express the discharge in gal/min, equation [44–8] becomes

$$q = 38 \; C \; a \; (P)^{1/2} \tag{44–9}$$

or

$$q = 30 \; C \; (d_1{}^2 + d_2{}^2) \; (P)^{1/2} \tag{44–10}$$

where $P =$ the pressure in pounds/square inch, $a =$ the area of the nozzles in square inches, and d_1 and $d_2 =$ the diameters of the nozzles in inches.

For sprinklers, the values of the coefficient C vary from less than 0.80 to about 0.95, depending on sprinkler and nozzle design. Sprinkler manufacturers provide tables of discharge of sprinklers in gal/min for the range of pressures for which they operate satisfactorily.

Friction losses occur whenever fluids flow through pipelines or other conduits. Many equations have been proposed for estimating friction losses. They can all be expressed by a generalized equation of the form

$$H_f = K L V^m / D^n \tag{44–11}$$

where

$H_f =$ the friction loss in a pipeline of length L,

$K =$ a friction factor that depends on the relative roughness of the walls of the pipe.

$V =$ the mean velocity in the pipeline,

$D =$ the inside diameter of the pipeline,

$m =$ an exponent that varies in the different equations from 1.85 to 2.0 and depends on the roughness, and

$n =$ an exponent that varies in the different equations from 1.0 to 1.25. In some equations n has a value of $3.0 - m$.

Expressed in terms of the flow Q the friction loss is

$$H_f = \frac{K L Q^m}{D^{2m+n}} \; . \tag{44–12}$$

Tables and special slide rules have been prepared for most of the commonly used equations which express the friction loss in feet of head or in pounds/square inch of pressure for a given length of line. Ree (1959) has also summarized many tests of friction losses in pipe and couplers. Any of the equations, tables, or slide rules can be used with reasonable accuracy when the proper friction factor is selected. They can be used directly for the determination of the friction losses in main lines where, for limiting conditions, all of the water flows to the end of the line. For sprinkler laterals, however, the velocity varies along the line as part of the water is discharged through the sprinklers. Christiansen (1942) has shown that, for lines with multiple outlets uniformly spaced, the total friction loss can be estimated from the relation

$$H_f = F(K L Q^m / D^{2m+n}) \tag{44–13}$$

where F depends on the number of sprinklers and the value of m. F can be evaluated from the approximate expression

$$F = \frac{1}{m+1} + \frac{1}{2N} + \frac{(m-1)^{1/2}}{6 N^2} \qquad [44\text{-}14]$$

where N is the number of sprinklers.

When $m = 1.90$, the value of the exponent in Scobey's equation, which is commonly used for sprinkler system design, the value of F is as given below:

Number of sprinklers	Value of F
10	0.396
15	0.379
20	0.370
30	0.362
50	0.350
100	0.345

The power required to pump water for sprinkler systems can be expressed by the equation

$$\text{HP} = \frac{Q\,P_p}{1715E} \qquad [44\text{-}15]$$

where

HP = the horsepower requirement,

Q = the total discharge of the pump in gal/min,

P_p = the total pumping head, including lift and fraction losses, expressed as pressure in lb/inch^2, and

E = the efficiency of the pump expressed decimally.

For any system, the power requirement is proportional to the three halves power of the total pressure requirement

$$\text{HP} = K\,P^{1.5}. \qquad [44\text{-}16]$$

The power unit for sprinkler systems may be an electric motor, a gasoline or diesel engine, or a farm tractor. It should be of ample size to meet the power requirement without overloading.

VII. PRINCIPLES OF SPRINKLER SYSTEM SELECTION

The sprinkler system selected for a field or fields should (i) have a capacity to meet the water requirements of the crop, (ii) apply water at a rate that does not exceed the minimum intake rate of the soil, (iii) apply water with some minimum economic uniformity, (iv) combine capital, labor, and power so as to minimize the total annual cost of irrigation, and (v) produce a crop that economically justifies the use of that system. Each of these points will be discussed in the order given.

The system capacity should be such that the crop is continuously supplied with available soil water during the growing season. The capacity will, therefore, depend on the estimated evapotranspiration rate of the crop, the depth of rooting of the crop, and the water storage characteristics of the soil. It is considered good practice to maintain the soil water above the midpoint between field capacity and wilting percentage for optimum plant growth and production, and to provide an adequate factor of safety.

Sprinkler systems should always apply water at rates below the minimum intake rate of the soil, considering the variation in intake rate with time after the initial wetting of the soil and also variations in intake rate during the season. There should be no runoff from a field with properly designed sprinkler systems. Some investigators believe that application rates well below the intake rate are beneficial, in that the soil is then never completely saturated or puddled and remains in a better physical condition for plant growth.

The amount of water required per irrigation is dependent also on the uniformity and efficiency of the application. In most cases, if water is available and inexpensive, it is more economical to forego achieving good uniformity and application efficiency and to apply more water. On the other hand, if water is precious, the cost of achieving high efficiencies may be less than the cost of the additional water required for a system with a lower application efficiency. Also, if water supplies are limiting, both uniformity and crop yield per unit area may be sacrificed to provide water for irrigating a larger area. For all conditions, an attempt should be made to determine the most economical application efficiency. This will depend largely on the availability of the water, the relative costs of water, equipment, and labor, and the value of the crop produced. The problem is often one of producing the maximum net return with a given amount of water

These same factors affect the choice of the general type of system to be employed. Completely stationary systems have high capital costs and low labor costs. Fully portable systems have minimum capital costs and highest labor costs. Semiportable systems are intermediate and frequently result in the minimum total cost of irrigation. Factors other than cost must be considered such as the ability of the farmer and his family to operate the system without hired help which is sometimes unavailable when needed only on a part-time basis.

Sprinkler systems should be used only for crops that provide sufficient income to justify the added costs of sprinkling as compared with no irrigation or with alternate methods of irrigation. Some low value crops cannot justify sprinkler systems, and if irrigation is required other methods must be employed. Sprinkling, however, is not always more expensive than other methods of irrigation, and added benefits through better uniformity of application and higher efficiency often make it the most economical method for applying water. The individual circumstances must be considered in both the selection of method of irrigation and, if sprinkling is feasible, in the selection of type of system to be used.

VIII.   DESIGN OF SPRINKLER SYSTEMS

Sprinkler system design should preferably be left to specialists who are familiar with the specific problems involved. Space does not permit a full treatment of this subject, hence only a brief description of the procedure involved will be given. Complete discussions of the design of sprinkler systems can be found in several publications (see list of references at end of chapter).

The following procedure is suggested, subject to modifications to meet specific conditions. Although the calculations are shown in English units, the same procedure is followed in those countries using the metric system.

First, the total capacity of the system to be installed is determined from a careful examination of the factors that affect his capacity. These are: (i) the area to be irrigated, (ii) the crops to be grown, (iii) the maximum evapotranspiration rate during an irrigation interval, (iv) the soil water storage capacity of the soil between field capacity and wilting percentage, considering the rooting depth of the crops grown, and (v) source and available flow of water which may sometimes control the design. Whether the system is to be operated continuously day and night during periods of maximum use, only during the night or day, or not on Sundays and holidays, must be taken into account. Considering the application efficiency E_a; the maximum evapotranspiration rate in inches/day ET; the hours of operation per day h; and the area in acres A; the required capacity in cubic feet/second Q, will be

$$Q = (\text{ET} \times A)/(E_a \times h).\qquad\qquad[44\text{--}17]$$

In gallons per minute,

$$\text{Gal/min} = 450\ Q = 450(\text{ET} \times A)/(E_a \times h).\qquad[44\text{--}18]$$

For example, the required capacity of a 40-acre field of alfalfa (*Medicago sativa*) having an estimated maximum evapotranspiration rate of 0.35 inches/day, with an assumed application efficiency of 75%, and 23 hours/day operating time, would be

$$\text{Gal/min} = 450\ (0.35 \times 40)/(0.75 \times 23) = 365\ \text{gal/min}.$$

Consideration is now given to the maximum permissible interval between irrigations I which will depend on the permissible water loss from the soil. Assuming that the soil water is to be maintained above a level midway between field capacity (FC) and wilting percentage (WP) for a root zone depth in inches D and that the soil water reaches field capacity 2 days after the end of the application; the irrigation interval I in days, should not exceed

$$I = \frac{(\text{FC} - \text{WP}) \times D \times a}{2 \times \text{ET}} + 2 + \frac{t}{24}\qquad[44\text{--}19]$$

where t is hours of operation per application, a is apparent specific gravity of the soil, and FC and WP are expressed as a decimal fraction of the weight of dry soil. For very coarse soils and shallow-rooted crops, the value of I in equation [44–19] should be reduced by 1 or 2 days

Thus, for FC = 0.25, WP = 0.10, ET = 0.35, D = 36, a = 1.4 and t = 23 hours.

$$I = \frac{(0.25 - 0.10)\ 36 \times 1.4}{2 \times .35} + 2 + \frac{23}{24} = 10.8 + 2 + 1 = 13.8\ \text{days, say 14.}$$

The amount of water that should be delivered through the system in one irrigation should not exceed

$$d = (\text{ET} \times I)/E_a.\qquad\qquad[44\text{--}20]$$

For the example used

$$d = (0.35 \times 14)/0.75 = 6.5\ \text{inches.}$$

The minimum required period for making such an application will depend on the minimum intake rate i of the soil after several hours of wetting. Thus,

$$t = d/i. \qquad\qquad [44\text{-}21]$$

Assuming a safe value of i to be 0.3 inch/hour, then

$$t = 6.5/0.3 = 21.7 \text{ hours.}$$

The above are then limiting values of I, d, and t, for this example.

IX. PIPELINE DESIGN AND LAYOUT

A. General Principles

Before proceeding with the sprinkler layout and pipeline design, some general principles or rules for good design should be considered. The objective of the overall design is to secure a system that will provide a satisfactory uniformity of distribution with a minimum annual operating cost, including depreciation, power, and labor costs. In the design procedure various alternates might be compared to determine the most economical combinations of capital costs, power costs, and labor costs (Davis and Fry, 1963). Consideration should be given to long-term changes in these costs and in changing cropping patterns.

1) Under most conditions, minimum costs and desirable operating conditions can be assured when the source of the water supply is as near the center of the area to be irrigated as possible. If the source of supply is to be a well, it should preferably be located at the center of the field or farm. If the water is to be pumped from a ditch with a portable pump, the ditch should be placed through the center of the area to be irrigated, if feasible. If a mainline is to be used to supply the sprinkler laterals, it should preferably be placed through the center of the area. This will result in minimum friction losses in laterals, smaller pipe sizes, and generally a more uniform application of water.

2) On sloping fields it is good practice to place the main lines on the slope and to place the laterals along the contour, or slightly downhill, to minimize pressure variations in the pipelines. It is considered good practice to limit the pressure variation in the lateral line to 20% of the design sprinkler pressure. With uniform sprinkler nozzles, the variation in sprinkler discharge will then be limited to 10% of the design discharge. Where the water must be carried up-slope from the source of supply in mainlines, provision should be made to regulate the pressure at the laterals to the design pressure.

3) The design of the system should provide for varied soil types over the area to be irrigated in order to avoid excessive rates of application or excessive intervals between irrigations.

4) When a sprinkler system is being designed for only part of a farm, consideration should be given to the possibility of expanding the system to cover the entire area. Where possible, the design for the initial area should fit into an expanded design.

5) The design should be such that the operation of the system will result in a minimum interference with other farm operations. Portable main lines, or ditches, through the center of the field might seriously interfere with the culti-

vation of the crops grown. Alternatives that might be considered would be to supply the laterals from a mainline or ditch along one side of the field, or to use a buried main line.

6) For irregular fields the design should minimize the variation in the number of sprinklers that would be operating at any given time.

7) Booster pumps and reservoirs should be considered when either might result in a more satisfactory operating schedule or a reduction in over-all costs.

8) Proper safety devices for the power unit and pumping plant which will insure an automatic shutdown of the pump in case of overheating of the motor or engine, or a failure of water supply, should be incorporated into the design. Other auxiliary equipment, such as air relief valves, pressure control valves, pressure gauges, and automatic controls should be considered and used when needed to insure satisfactory operation.

9) Valves on buried lines spaced three times the design spacing of the laterals might result in an appreciable saving in cost of risers and valves and only a nominal increase in the cost of the laterals and added pumping head.

B. Procedure

The design of the sprinkler system includes the most feasible layout of the main lines or ditches, the layout and spacing of the laterals, and the required sizes of the laterals and main lines. A definite plan of operation which will insure irrigating the entire area within the permissible interval between irrigations during the period of maximum evapotranspiration, must be selected.

Under normal conditions, sprinkler systems are designed to be operated continuously night and day. Under some conditions, sprinkler systems are designed to be operated only at night when wind conditions are more favorable or only during the day when nighttime operation is not feasible. The initial cost of the system will be minimum when it is designed to be operated continuously to meet maximum demands. Halftime operation would double the capacity of the system, including the capacity of the well, pumping plant and main lines, and the number of laterals. Very often the intake rate of the soil will not permit a satisfactory application in much less than 24 hours.

Usually the most satisfactory operating schedules involve either one or two moves of the sprinkler laterals per day. For light applications of 60 mm (2.5 inches) or less, and especially on soils with fairly high intake rates, two moves per day with an actual operating time of about 11 hours for each setting is usually satisfactory. For greater applications and for soils with low intake rates, one move per day is more desirable. For any given layout, the number of laterals required to cover the area in the permissible irrigation interval depends on the spacing of the laterals; this can be readily determined. Under most conditions several laterals supplied by central main line will prove most economical.

For the example previously assumed, a 40-acre field, with an application of 6.5 inches, an irrigation interval of 14 days, and a minimum time of application of 21.7 hours, it is obvious that one move per day would be most desirable. Assuming that a main line with valves and portable laterals will be used so the pump can be operated continuously with only the line being moved shut off, the actual operating time can be approximately 23 hours. The 40-acre field will be assumed to have net dimensions 1,300 × 1,300 ft and a net area of 39 acres, approximately. With a lateral spacing of 50 ft assumed, a total of four laterals each 620 ft long, with 16 sprinklers each, will cover the area in 13 days. The revised design capacity of the sprinkler system will be

$$\text{Gal/min} = 450 \ [(0.35 \times 39)/(0.75 \times 23)] = 355,$$

or 5.55 gal/min per sprinkler.

A single nozzle sprinkler with a 11/64-inch nozzle will deliver a flow of 5.4 gal/min under a pressure of 40 lb/inch^2; a dual nozzle sprinkler with 5/32- and 3/32-inch nozzles will deliver a flow of 6 gal/min with a pressure of about 40 lb/inch^2. The effective diameter covered is 90 ft for these sprinklers. The assumed spacing of 40 × 50 ft meets spacing criteria (no greater than 55% of the diameter) and the application rate of 0.29 inch/hour is also satisfactory.

Based on Scobey's K_s value of 0.40, the friction loss in a 2.5-inch diameter lateral would be about 7.5 lb/inch^2, and a 3-inch lateral about 3 lb/inch^2. To limit the over-all friction loss in the laterals and main line to 20% of the design pressure, or about 8 lb/inch^2, the 3-inch lateral would be selected. This limitation would allow for a maximum main line friction loss of about 5 lb/inch^2. With the laterals spaced 650 ft apart and moved in opposite directions on the two sides of the main line, the limiting condition for main line design will be where laterals on opposite sides of the main line are at the one-fourth and three-fourths points along the main. With the source (pumping plant) at the boundary of the field, the maximum friction loss would be about 2.7 lb/inch^2 for a 6-inch main line and 6.5 lb/inch^2 for a 5-inch main line. A main line 6 inches in diameter to the center of the field, and 5 inches for the remaining distance would fully meet requirements with a maximum loss of about 4 lb/inch^2 (Fig. 44–8).

With the source of the water supply and the pump at the center of the field, a 4-inch main line would meet requirements with a total friction loss of 3.5 lb/inch^2. With valves spaced three times the lateral spacing or 150 ft apart, each of the 4-inch main lines would be only 575 ft long, making a total of only 1,150 ft of pipe.

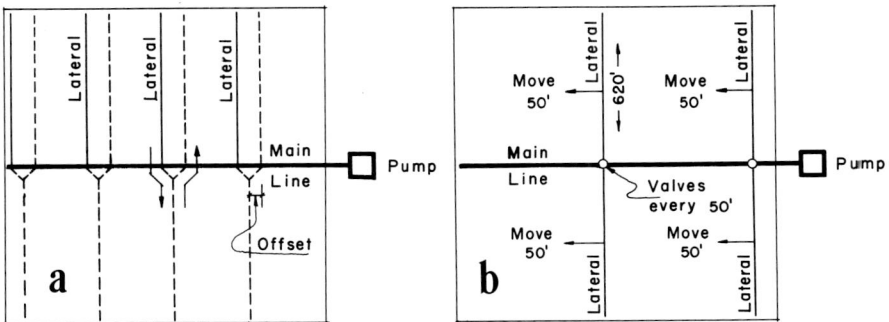

Fig. 44–8. A schematic diagram of the sprinkler system design as discussed in the text: (a) Suitability for end-tow or end pull systems; (b) suitability for hand-move or side roll.

X.   OTHER AGRICULTURAL USES OF SPRINKLER EQUIPMENT

Although all sprinkler systems are designed primarily for meeting the leaching and water requirements of crops, most systems possess uses which, because of the economic benefits, may justify large investments in equipment. In almost any climatic area, these potential benefits should be considered and incorporated whenever possible in the design and management of the system.

A.   Seed Germination

The uniformity of crop stand and time of maturity is of economic importance for many crops—particularly for vegetables. Sprinkler systems often insure adequate seed germination with only one light application of water after seeding. Such applications are more efficient than surface irrigation methods, but care should be taken to avoid soil crusting. Low application rates are thus very desirable.

B.   Application of Fertilizers and Soil Amendments

The distribution of plant nutrients to soils through sprinkler systems is as good as the distribution of water. Such applications are easily adapted to sprinklers and should be considered in the crop management program. Descriptions of methods and equipment are given in chapter 52.

C.   Frost Protection

Most of the conventional sprinkler systems will protect vegetation against damage by radiation frosts to temperatures below $-6C$ (20F). The design and operation of frost protection systems are described in chapter 54.

D.   Cooling Crops and Animals

Many crop yields are seriously depressed by excessively high air temperatures during the fruiting period. Temperatures above 35C (95F) may cause blossom drop of pole beans (*Phaseolus vulgaris*) and fruit drop of citrus, e.g., and temperatures in excess of 38C (100F) for several days can cause losses of grapes (*Vitis* sp.) of up to 50%. Sprinkler systems which apply fine sprays with low application rates will reduce ambient air temperatures and leaf temperatures up to 5C (9F) or more; crop losses can thus be minimized and fruit quality maintained. Cooling crops by sprinkling requires a full-coverage system, but unlike frost protection, water can be applied intermittently (15 min off, 15 min on), thus conserving water (Gray, 1961). This method also requires water of fairly good quality, although tolerance limits have not yet been established.

Cooling livestock and poultry environments by sprinkling has been successful for a number of years in reducing mortality and disease and improving rates of weight gain. Such systems usually employ fine sprays on the roofs of buildings or on feedlots with very low application rates.

E. Other Uses

Sprinkler systems have been adapted for many other agricultural and non-agricultural uses. Some of these include: water distribution for compaction of earth fills, producing snow for ski resorts, settling of dust, log curing, and farm fire protection.

XI. SUMMARY

Sprinkling offers an efficient and economical method of applying irrigation water under a wide variety of conditions. Since sprinkling eliminates soil and topography as important factors in the distribution of water, sprinkling is especially desirable under adverse conditions of soil and topography.

The planning and design of sprinkler systems can best be done by specialists who have the requisite knowledge of irrigation requirements, soil conditions, and the performance of sprinkler equipment.

Many types of sprinklers and sprinkler systems are available from which to choose those best adapted to any specific conditions.

LITERATURE CITED

Allred, E. R., and R. E. Machmeier. 1962. Effect of wind resistance on rotational speed of boom sprinklers. Amer. Soc. Agr. Eng., Trans. 5:218–219, 225.

Anonymous. 1949. Planning sprinkler irrigation systems. US Dep. Agr.-Soil Conserv., Serv., Pacific Reg., Reg. Eng. Handbook. Ch. 6, Part 1, Sect. 6.

Anonymous. 1966. Minimum requirements for the design, installation, and performance of sprinkler irrigation equipment. Amer. Soc. Agr. Eng. Yearbook. p. 371–372.

Christiansen, J. E. 1942. Irrigation by sprinkling. Univ. California. Agr. Exp. Sta. Bull. 670. 124 p.

Criddle, W. D., S. Davis, C. H. Pair, and D. G. Shockley. 1956. Methods for evaluating irrigation systems. US Dep. Agr.-Soil Conserv. Serv. Handbook 82. US Supt. Doc. Washington, D.C. 24 p.

Davis, J. R., and A. W. Fry. May, June-July. 1963. What price sprinkler uniformity? Irrig. Eng. Maintenance. p. 10-11 (May), p. 24 (June-July)

Davis, J. R., and W. E. Hart. 1963. Efficiency factors in sprinkler system design. Sprinkler Irrig. Ass., Open Tech. Conf., Proc. p. 15–30.

Eaton, F. M., and R. B. Harding. 1959. Foliar uptake of salt constituents by citrus plants during intermittent sprinkling and immersion. Plant Physiol. 34:22–26.

Frost, K. R., and H. C. Schwalen. 1955. Sprinkler evaporation losses. Agr. Eng. 36(8): 526–528.

Gray, A. S. 1961. Sprinkler irrigation handbook. Rain Bird Sprinkler Mfg. Corp., Glendora, California. 44 p.

Hansen, V. E. 1960. New concepts in irrigation efficiency. Amer. Soc. Agr. Eng., Trans. 3:55–57.

Hart, W. E. 1961. Overhead irrigation pattern parameters. Agr. Eng. 42(7):354–355.

Howell, D. T. 1964. Sprinkler nonuniformity characteristics and yield. Amer. Soc. Civil Eng. Proc., J. Irrig. Drainage Div. 90(IR 3):55–67.

Israelsen, O. W., and V. E. Hansen. 1962. Irrigation principles and practices. 3rd ed. John Wiley & Sons, Inc., New York. 447 p.

Jensen, M. C., and L. G. King. 1962. Design capacity for irrigation systems. Agr. Eng. 43(9):522–525.

Keller, J. L. S. Willardson, and G. O. Woodward. 1962. Let's take a closer look at sprinkler irrigation interval and system capacity design factors. Sprinkler Irrig. Ass., Open Tech. Conf., Proc. p. 56–66.

Korven, H. C. 1952. The effect of wind on the uniformity of water distribution by some rotary sprinklers. Sci. Agr. 32(4):226–239.

McCulloch, A. W., and J. Keller. 1962. Ames irrigation handbook. Prepared by W. R. Ames Co., Milpitas, California. 196 p.

McDougald, J. M., and J. C. Wilcox. 1955. Water distribution patterns from rotary sprinklers. Can. J. Agr. Sci. 35(3):217–228.

Machmeier, R. E., and E. R. Allred. 1962. Operating performance of a boom sprinkler— a field study. Amer. Soc. Agr. Eng., Trans. 5:220–225.

Molenaar, A. April. 1954. Factors affecting distribution of water from rotating sprinklers. Washington Agr. Exp. Sta. Cir. 248. 18 p.

Ree, W. O. 1959. Head loss in quick-coupled aluminum pipe. US Dep. Agr. Handbook No. 147. US Supt. Doc., Washington, D. C. 20 p.

Strong, W. 1955. Agricultural sprinkler irrigation manual. Prepared by Buckner Mfg. Co., Inc. Fresno, California. 31 p.

US Department of Agriculture. 1955. Water, the Yearbook of Agriculture, 1955. US Supt. Doc., Washington, D. C. 751 p.

Wiersma, J. L. 1955. Effect of wind variation on water distribution from rotating sprinklers. South Dakota Agr. Exp. Sta. Tech. Bull. 16. 18 p.

Wilcox, J. C., and G. E. Swailes. 1947. Uniformity of water distribution of some under-tree orchard sprinklers. Sci. Agr. 27(9):565–583.

Woodward, G. O. 1959. Sprinkler irrigation. (2nd ed.) Sprinkler Irrig. Ass. Santa Monica, California. 377 p.

45 | Subirrigation Systems

WAYNE D. CRIDDLE

Clyde-Criddle-Woodward, Inc.
Salt Lake City, Utah

CORNELIS KALISVAART

Ijsselmeerpolders Development & Settlement Authority
Kampen, Netherlands

I. INTRODUCTION

In arid areas of the world there is only limited practice of subirrigation and this has developed largely through trial and error. In the more humid areas, subirrigation has wider application, often in conjunction with existing drainage systems. The drainflow is sometimes blocked during periods of drouth, thus intentionally holding the water table up to supply water to the crops.

Until recently, principles, possibilities, and limitations of subirrigation have not been studied intensely. Nevertheless, investigations show that subirrigation systems do work satisfactorily in some cases. It is believed that this system of irrigation, if properly designed and operated, might be the best method available for many other areas.

II. SUBIRRIGATION IN HUMID REGIONS

Some of the more extensive subirrigated systems in humid regions of the world are found in the Netherlands. The climate, topography and soils of the Zuiderzee Polders necessitates that drainage facilities be utilized for sustained crop production. However, in the summer when evaporation exceeds rainfall, irrigation is a supplementary expedient on coarser textured soils. Soil water deficiencies during periods of insufficient or too intermittent rainfall often result in wilting of the crop or reduced production.

In the USA, Florida has two extensive areas where conditions are suitable for subirrigation: The Everglades and the Flatwoods of the Coastal Plain. Until recent years, these lands were either too wet or too dry for good crop production. Water table control through subirrigation and drainage has increased production several hundred per cent (Stephens, 1955).

In certain areas of the Great Lakes states of Michigan, Minnesota, Indiana, and Ohio, USA a practice of "controlled drainage" is used which in effect is a subirrigation practice. In this high rainfall area, the proper water table elevation is maintained by control structures in the drainage channels. Ordinarily, additional irrigation water is not used.

In the new Zuiderzee Polders, which may have a surface elevation down to 16 ft (5 m) below sea level, drainage is essential. Drainage is accomplished by pumping the discharge into the surrounding fresh water Yssel Lake during periods of excess subsoil water. In the sandy and peaty areas of the Polder, subirrigation is accomplished by controlling the flow of water back into the drainage system during periods of drouth.

Salinization of soils, an ever present danger where subirrigation is practiced in arid and semiarid regions (to be discussed later) does not occur under climatic and drainage conditions in the Netherlands or in the humid regions as a whole. Any tendency to salinization resulting from the upward movement of the water during the growing season is entirely overcome by the downward movement of water during the winter or noncropped season (Kalisvaart, 1958).

III. PRINCIPLES OF SUBIRRIGATION

Subirrigation might be defined as the act of regulating the elevation of the groundwater table by artificially adding water underground. The water level should be maintained at such heights that the best combination of water and air in the root zone is assured.

The term subirrigation is sometimes applied to a variation of furrow irrigation. Water is supplied through underground shallow tiles or perforated pipes, and the surrounding soil is moistened by gravity and diffusion as from furrow irrigation. This method reduces evaporation and cooling of the soil surface. It may also decrease soil structure decay, and it eliminates the inconveniences of open furrows according to Dorter (1962). Usually, however, the term subirrigation is used to mean irrigating by regulating the position of the groundwater table.

Subirrigation is drainage in reverse. With drainage, the excess water flows through the soil toward and into the drains. Such a drainage system lowers the water table so that the development of the root system and the growth of crops is not limited because of lack of air. With subirrigation, water is diverted into the drains and then infiltrates out into the soil. In this way, the water table is kept at a proper height so crops can derive their water needs from it. The drainage requirements cannot be neglected if the pipe system is used to feed water to the land.

A. The Water Regime in the Root Zone

The various manners of water supply imply that there is a principal difference between the water regime in the root zone of a soil with surface irrigation or sprinkling and those of a soil with subirrigation.

Under surface and sprinkler irrigation systems, soil water in the root zone is continually varying. Immediately following irrigation, the soil is at or near field capacity. Just before irrigation, the soil water content may drop to one-half field capacity. In a homogeneous coarse sandy soil with a pore volume of 35%, this variation may be from approximately 15% down to 7% of the soil volume (see Fig. 45–1). The air content in the root zone will then vary from 25 to 33% of the total volume. However total water plus air content in the root zone, under proper surface irrigation will show only small variations.

Under subirrigation the water content above the groundwater table in the same soil, neglecting evapotranspiration, will vary as shown schematically by

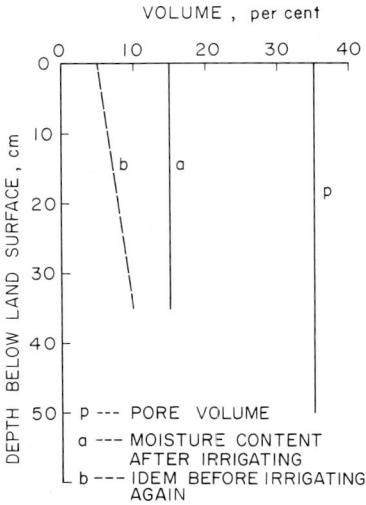

Fig. 45–1. Diagram of water contents in a surface irrigated coarse sandy soil.

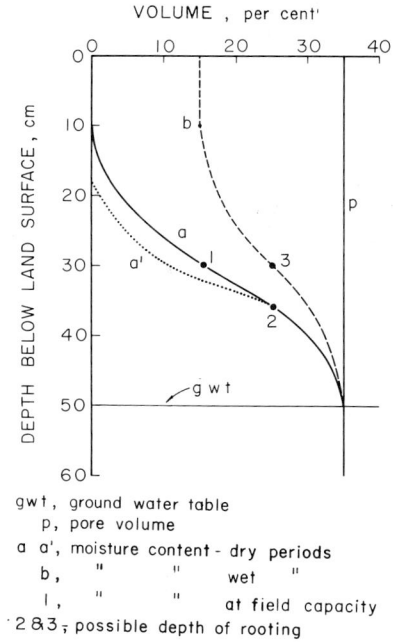

gwt, ground water table
P, pore volume
a a', moisture content - dry periods
b, " " wet "
I, " " at field capacity
2 & 3, possible depth of rooting

Fig. 45–2. Diagrams of water contents above the groundwater table in subirrigated coarse sand.

lines a and b in Fig. 45–2. (To simplify the problem in this and the following schemes unavailable water is ignored.)

Line a represents the water content under dry soil conditions but with the groundwater table maintained at 50 cm. The maximal capillary rise in this soil is assumed to be 40 cm.

Line b indicates the maximal water content possible under these groundwater table and soil conditions. This situation illustrates that the b line, may be reached: (i) when the soil above the water table is kept at or above field capacity to the land surface, (ii) after heavy rainfall has filled storage between a and b, and (iii) when the groundwater table is brought up to within 20 cm of the land surface (point 1 on line a lies then at the surface) and then is allowed to drop again to 50 cm.

When the static water is influenced by evaporation and transpiration, line b will become the same as line a.

When, as in Fig. 45–2, the groundwater table lies below the height of the capillary rise, line a will only be influenced by transpiration. Water for the roots is supplied by capillary rise of the groundwater maintained at a constant height. Transpiration, however, will decrease the water content from line a to line a^1, especially if the rate of capillary movement is slower than the water absorption capacity of the roots. The rate of capillary rise is based on the texture of the soil. In a coarse sandy soil line a may become line a^1.

Figure 45–2 illustrates that more water is available for roots as they grow deeper (or as groundwater table rises). Both, however, are limited because of the need for air by the roots. Assuming plant roots need a minimum of 10% of air in the root zone, then from Fig. 45–2 it seems the roots must have 35 cm depth under situation a (*see* point 2) and 30 cm under situation b (*see* point 3).

If the water is to be most effective in the top layers which have most of the roots, then during dry periods, soil water is not ideal. In the upper 10 cm there is little or no available water. In the second 10 cm only a little usable water exists. Remembering that the rate of capillary rise decreases with height above the water table, it is therefore necessary during dry periods to keep the ground-water table within 40 cm. This leaves a layer with 10% or more air to a depth of 25 cm and limits the rooting depth possibility.

During and immediately after a wet period (situation b) to assure the same 24-cm depth of rooting, the groundwater table must not be higher than about 45 cm if there is to be no injury to the roots. This means that following a heavy rainfall, the groundwater table must be lowered to 45 cm. However, a subsequent dry period would require that the groundwater table be raised to 40 cm. In practice, maintaining the theoretically desired groundwater table level can give difficulties.

B. The Influence of Soil Texture on Water Regime

The influence of soil texture on water regime in the root zone is discussed first by comparing the water content of a coarse sandy soil with that of a homogeneous fine sandy soil. The latter may have a maximum capillary rise of 100 cm, a pore volume of 45%, and a field capacity of 25% (*see* Fig. 45–3).

Using the same reasoning as for the coarse sandy soil, the groundwater table during dry periods in this fine sandy soil must be at least 100 cm beneath the soil surface. This includes a rooting possibility to 60 cm (point 2). But in this fine sandy soil with its much smaller permeability the rate of the capillary rise may not be as rapid as the water removed by evapotranspiration. If line a in Fig. 45–3 can change to line a^1, then in dry periods the groundwater table has to be kept at 70 cm instead of 100 cm. Thus, the rooting possibility will be reduced to a depth of about 40 cm, if during and immediately following a wet period (so that line b occurs), the groundwater table could be kept at 100 cm or lower.

In practice, there often will be situations lying between a^1 and b which require intermediate solutions. A similar rainfall after a dry period, for instance, will only fill the upper part of the storage capacity between b and a^1. This results in a lesser water supply requirement which theoretically would allow for a lowering of the groundwater table.

From the described difference between the water regime above the ground-water table in a coarse sandy soil and in a fine sandy soil, it follows that subirri-gation is easier to control with a coarser and more permeable soil. The extremely fine sandy soils and certainly the silt and clay soils of low permeability and slow capillary rise are less or not at all suitable for subirrigation.

Often the soil profiles above the groundwater table will not be homogeneous. There may be a gradual increase or decrease in the permeability of the soil in a vertical direction. Any restricting layer limits the height of the capillary rise.

Using layers of coarse and fine sand as in Fig. 45–2 and 45–3, and combining

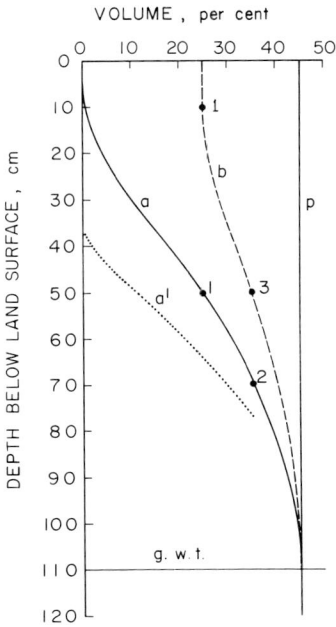

Fig. 45–3. Diagrams of water contents above the groundwater table in subirrigated fine sand.

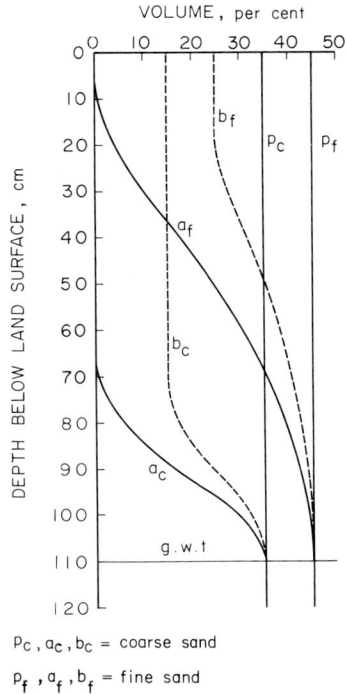

Fig. 45–4. Combination of water diagrams of a coarse and a fine sandy soil (Fig. 45–2 and 3).

these into Fig. 45–4, the influence of different combinations of fine and coarse layers on the desired water table is suggested. During dry periods (line a_c) a layer of fine sand on a coarse sand needs a groundwater table within 40 cm beneath the underside of the fine top layer. If it is deeper than 40 cm, the capillary water will not reach the top layer. The most effective height will depend on the quantity of water needed and the height to which it must rise. These in turn depend on the evapotranspiration rate and the thickness of the top layer of soil.

During wet periods (when the water is represented by line b_c) the groundwater table must be as shown in Fig. 45–3, or, if the top layer is rather thin, something more than 40 cm beneath the underside of this layer. The capillary connection between the top layer and the groundwater should not be broken.

A coarse sand top layer on fine sand requires that the groundwater table must be kept high as shown in Fig. 45–2. Only when the coarse sand top layer is thinner than the accepted rooting depth of 25 cm will the water regime in the fine sand layer be the determining factor. If shallower rooting is accepted, the groundwater may be held higher and require less fluctuation between wet and dry periods.

If layers of low permeability and slow capillary rise occur in the soil profile above the groundwater table, subirrigation is not adaptable. In dry periods capillary rise will be inadequate and in wet periods the drainage will give difficulties.

A profile with layers of silt or clay, even with good permeability, presents two problems: (i) they generally have a slow capillary rise, and (ii) they often lose their permeability under subirrigation practices. Subirrigation on such soils may be satisfactory in the beginning, but under saturated conditions, they often become less permeable or even impermeable.

C. Regulation of Groundwater Table

Controlling the groundwater table by subirrigation requires feeding water underground during periods of drouth and removing excess water during wet periods.

Figure 45–5 illustrates what happens to a cross section of a subirrigated field when the water table is kept constant above the supplying or discharging pipelines or open feeder ditches.

The first conclusions show that it is necessary to accept a certain variation of the groundwater table in the field between the two pipelines. Water will only move into or from the tiles when there is a difference in head. The water table will be a plane surface only in the absence of rainfall, seepage, and evapotranspiration, or if there is a complete equilibrium between these elements (see Fig. 45–5, line O).

When rainfall and/or upward seepage exceed evapotranspiration, the water table will be higher between the pipelines than directly above them. Under these conditions the pipelines would be acting as a drain (see Fig. 45–5, lines d).

When evapotranspiration and/or deep percolation exceeds the rainfall, the water table between the pipelines will be concave and the lines would act as feeders (see Fig. 45–5, lines s).

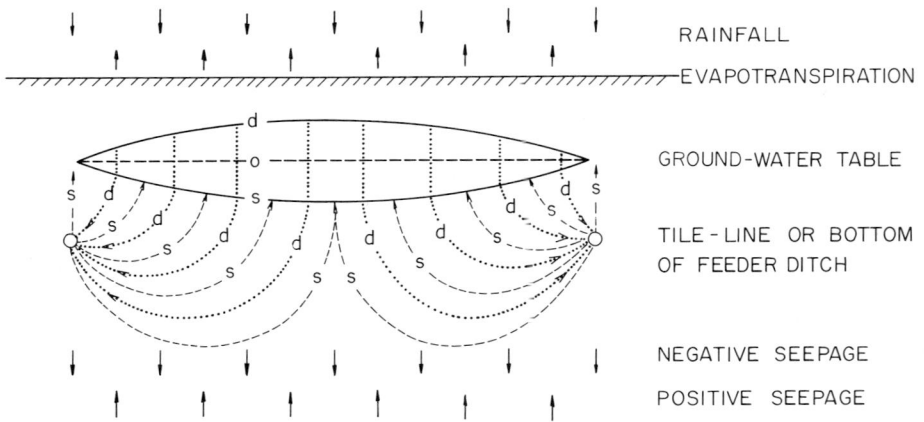

O = No discharge or supply

d = Discharge : Rainfall and /or positive seepage predominate

S = Supply : Evaporation and /or negative seepage predominate

Fig. 45–5. Schematic cross section of a subirrigated field with the possibilities for the groundwater table.

Table 45–1. Characteristics of three sandy soils in the Northeast Polder, Holland and spacing of tile lines (Kalisvaart, 1958)

Characteristic	Blokzijl sand	Ramspol sand	Urk sand
Clay content, %	6	6	2
Humus, %	0.8	0.6	0.3
Calcium carbonate, $CaCO_3$; %	7	7	0.2
Texture of the sand fraction, expressed in U-figure*	275	75	55
Capillary rise, cm	120	50	40
Hydraulic conductivity or permeability; m/24 hr	0.3	3	4
Desirable depth of groundwater table below surface in dry periods, cm	60	40	35
Maximal spacing of the tile lines, m	8†	25	30

* The specific surface, expressed by the U-figure represents the proportion between the total surface of all the sand grains contained in 1 cc of the soil in situ and the surface of a ball with a diameter of 1 cm. The finer the sand the higher the U-figure.
† Because of costs, the spacing of 8 m is considered minimum under existing economic conditions.

The magnitude in the difference of water levels in a subirrigated field depends on (i) the quantity of water to be discharged or supplied, (ii) the permeability of the soil below the groundwater table, (iii) the thickness of the permeable (water conducting) layer, and (iv) the spacing of the pipelines. Table 45–1 gives the characteristics of three sandy soils in the Northeast Polder of the Netherlands and the spacing required as influenced by hydraulic conductivity, capillary rise, etc.

Generally, only the spacing of the conduits can be changed by man. Although changing the spacing will have some minor effect on evapotranspiration, seepage, and thickness of the water conducting layer, the difference in water level will be nearly proportional to the square of the distance between the water conduits.

From the above discussion, it may be concluded that subirrigation will become less possible and finally even totally impossible as: (i) the permeability of the water conducting layer becomes smaller, (ii) the thickness of the permeable layer decreases, and (iii) upward seepage becomes greater. Each of these three characteristics requires a closer spacing of the pipelines. Downward seepage increases the water supply requirements. Seepage upward to the root zone in quantities equal to or surpassing the evapotranspiration requirements makes irrigating superfluous but requires drainage.

Allowable variations in distances to the water table must be decided before designing the system. It will be necessary always to accept some variations which will be a reasonable compromise between the optimum requirements of the crops and the technical and economical limits of shortening the distance between the conduits. It must be noted that the water table above the tile lines is not always the same as the water table as gauged by a piezometer in the tile lines. This phenomenon is particularly noticeable when the tile joints, filter material, and/or soil through which the tile lines run, are less permeable than the general field soils. Under such conditions the groundwater table above the tile lines will be lower when water is being supplied or higher when drainage water is being discharged than the water level in the tile lines or field ditches. Therefore, the feeder or drainage lines must be properly positioned to allow for anticipated variations in water level. In some instances it may be necessary to let the tile lines or the ditches be periodically dry. The use of wider and shorter tiles with the increasing possibility for the water to flow in and out through the joints may be taken into consideration. Use of cover material which will remain permeable is desirable. Cleaning the bottoms and walls of ditches may improve permeability and allow the water to enter or leave more readily.

D.   Control of the Water Table in the Tile Lines or Field Ditches

The purpose of subirrigation is to maintain the water table in the tile lines or feeder ditches at the desirable height so that the water table under the cropped area is maintained at the predetermined elevation.

The question of the maximum quantity of water to be supplied to or discharged from the field is determined by the resultant of evapotranspiration, rainfall, seepage, and length and breadth (the area) of the field.

If the breadth of the field is fixed, the length and the width (diameter) of the pipelines may be varied some to maintain the desired water table. Short and wide pipelines tend to approximate the ideal, although certain variation of the water table above the pipelines must be accepted. Here again, it is necessary to accept a reasonable compromise between the requirements of the crops and the technical and economical possibilities. Especially important is the admissible length and frequency of times when the greatest deviation from the ideal water table will occur. Also the existing topography of the area to be irrigated will help influence the choice.

One must use normal sizes of tiles for economic reasons, and the choice of spacing of pipelines or field ditches and their length is not always left to the

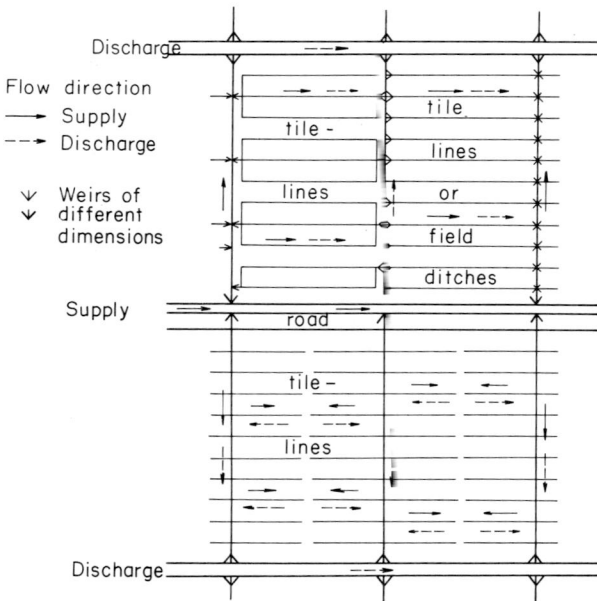

Fig. 45–6. Design for subirrigation in a horizontally laying area.

designer. Instead he must take into consideration the joint influence of all factors on the groundwater table and also on the economical results of farming on the subirrigated soil.

A general plan for water supply and discharge for the irrigated area of which the fields are a part must be considered. This requires a complete system for water supply and drainage. Figure 45–6 shows two possible systems for a flat area; one, the supply and discharge structures totally separated and the other, the same structures partly combined.

If areas of variable slope are to use subirrigation, then adjacent fields must be leveled so that a continuous water table can exist from one field to the next with as little variation in depth as practical.

Slight and regular sloping of the land may be advantageous since the natural slope can be used in designing the total supply and discharge system of the area.

IV. DEVELOPING SUBIRRIGATION IN HUMID REGIONS

Some of the comparative factors in considering the desirability of developing a subirrigation system are listed in Table 45–2. Of particular note is the reduced labor requirement, an important consideration in view of increased costs of operation on irrigated farms.

Basic for design of a good subirrigation system are surveys of soils, hydrology, and topography. It must be possible to decide from the surveys if the soils need irrigation and if they are suitable for it (*see* chapter 10). For this decision and for the system design requires knowledge of: (i) stratification of the soil profile, (ii) texture and the hydraulic conductivity of various strata, (iii) topography of restricting layers, (iv) position of the natural water table and direction of flow, (v) anticipated lateral flow and its consequences after subirrigating, not only on the irrigated project area but on surrounding areas, (vi) suitability of the land topography for construction of the required channels, ditches, tile lines, and other

Table 45–2. Comparative factors in considering subirrigation

Advantages	Disadvantages
1) Effective on drouth soils having low water-holding capacities and high intake rates where other methods may be impracticable due to labor, equipment, and water costs.	1) Requires a more complex combination of physical conditions not readily found in nature.
2) Labor requirements low for well designed system.	2) Generally, adjoining landowner must be following comparable practices.
3) Dispersion of weed seeds reduced, thus reducing weed control costs.	3) More essential that water quality be good to avoid salinity and related problems.
4) Special tillage and frequent land preparation for conveying surface water is eliminated, thus less damage to soil structure.	4) Drainage and leaching practices must be more intensive to assure adequate salinity control.
5) Evaporative loss of water from the land surface is minimal.	5) Germination may be slow or spotty if system does not permit uniform water table control.
6) Crop response is generally good.	6) Choice of crops may be limited to a narrow range of rooting characteristics.

works for supply and discharge of the water, (vii) the probable costs and profits, and (viii) adequacy of suitable water.

Space does not permit a detailed discussion of all possibilities for design of subirrigation systems under the many different site conditions found throughout the world. However, since subirrigation is the reverse of drainage, much of the theory of drainage can be applied directly to subirrigation. Van Deemter (1949) presented formulas for the height of the water table in homogeneously, isotropically permeable land drained or irrigated by pipe drains, under different circumstances of rainfall or evaporation, loss of artesian water, or seepage to great depth. Van Schilfgaarde et al. (1956) have analyzed various physical and mathematical theories on this subject. Other sources of related material can be found

Table 45–3. Subterranean vs. open conduits for subirrigation systems

Advantages of closed conduits	Disadvantages of closed conduits
1) No loss of land	1) Tiles are more costly. (A combination of mole and tile drainage in peat soils might possibly reduce costs)
2) Ease of tillage, weed control, and harvest	
3) Lower cost of maintainence	2) Greater cost of maintainence in cleaning and replacement of clogged pipe-lines
4) Tiles may be placed deeper in soils with high capillary rise (30 cm deeper than capillary rise is desirable)	3) Tile lines are difficult to use in removal of excess surface water
	4) Inspection of water table position more difficult

Table 45–4. Separated vs. nonseparated subirrigation system as illustrated in Fig. 45–6

Advantages of separation	Disadvantages of separation
1) Supply and discharge require a slight slope of the water table along the tile lines. Because of this, it is possible and often advisable to conform to the slope in the field by laying the tile lines in the direction of this surface slope. If supply and discharge are not separated, the desired fall in both cases is opposite and the tile lines and land surface should be completely level.	1) The tile lines are usually longer and therefore must have a greater diameter.
	2) The layout is more expensive because, in general, more controls and greater measuring devices are required. This disadvantage can also lose much of its weight if topography makes more controls necessary.
2) It is possible when a temporary "supply" period is followed by a "discharge" one to maintain the water table in the supply ditch, although naturally the supply must be stopped.	3) The water table in the discharge ditch always must be lower than the lowest water table in the discharge tile lines and opens the possibility of heavy water losses and drouth in the border area of the ditch. Conversely, the water table in the supply ditch, when higher than the water table in the tile lines, can lead to an excessively wet border area.
3) It is possible to give adjacent parcels a somewhat different water table. This advantage can partly or totally lose its worth if topography makes necessary differences in water table.	

in Glover (1964), Todd (1959), Harr (1962), and Donnan (1959) (*see also* chapters 13 and 50 in this monograph.)

The question of open vs. subterranean conduits and separated vs. nonseparated supply and discharge systems is worthy of brief mention here because of practical considerations. Some of the relative merits of each are summarized in Tables 45–3 and 45–4.

There is the need for central control of the water supply and its application. Since the desired water table is dependent on the season and weather conditions, adjustable weirs or controls for passing water at each diversion and division point should be installed. Water passing the end of the supply system into the discharge ditches should also be subject to control.

Using measurements and experience as a guide, it will be possible to regulate the water supply in the whole area with a minimum waste of water and a minimum of labor. This implies that the water supply is centrally controlled. Any local changing of water inlets will automatically influence inlets elsewhere. Therefore, there must be a good understanding between the central controlling agency and the farmers. Usually contact need not be frequent because the weather is the main criterion for the whole subirrigated area, not the individual farmer.

Advantages and/or disadvantages of subsurface irrigation to crops because of variations in the water regimes in the root zone are not yet fully known. Aside from the influences of rainfall on subirrigation, the water content of the soil is always rather constant from the land surface to the groundwater table. With surface irrigation the water content varies periodically. What effect this has on the crops cannot be said. It is, however, generally accepted that when both systems are properly operated, it will not make great differences in crop yields. The main difference between the two systems is the widely different natural conditions required.

V. SUBIRRIGATION IN ARID AND SEMIARID REGIONS

There are a number of locations in the USA where subirrigation is practiced in semiarid and arid regions. Some of the most notable are found in California, Idaho, Colorado, Utah, and Wyoming.

Approximately 160,000 acres (64,750 hectares) of low-lying delta lands and islands are subirrigated successfully in the Sacramento–San Joaquin Delta of California. Before being reclaimed by diking, these tracts were flooded each year by high waters from the rivers. By diking and installing a drainage system, it is possible to maintain the water table at the desired elevation in the peat soils either by pumping from the drains to the river or discharging water by gravity from the river to the feeder ditches as required. The feeder ditches spaced from 150 to 300 ft apart are usually from 2 to 3 ft (0.6 to 0.9 m) deep and 1 ft (0.3 m) wide. Vertical banks are possible in the soils because of the high organic content of the soil.

On the upper Snake River in Idaho there is an area of some 28,000 acres (11,320 ha) known as the Egin Bench which is subirrigated. The land slopes uniformly at about 0.2% and the soils are extremely permeable overlying impervious lava rock. A description of the area is given by Clinton (1948).

The Lewiston area in Cache Valley, Utah, located between the Bear and Cub

Rivers, comprises a flat tableland of deep permeable soils underlain by clay at about 15 ft (4.6 m). Original attempts to irrigate these soils by surface methods resulted in excessive waterlogging and concentration of salts in the surface soils over parts of the Bench. The construction of large open drains, improved land leveling practices, and the development of better control practices for maintaining the water table at the correct level have made crop production profitable. Israelsen and Hansen (1962) gives an account of subirrigation in the Lewiston area.

The San Luis Valley of Colorado is one of the most extensive subirrigated areas in the Western USA containing about 135,000 acres (54,635 ha). This area, part of which is on the headwaters of the Rio Grande River, apparently has surface soils that are slightly less permeable than many other subirrigated tracts. Because of the salts in the water supplies used, special practices have been developed to maintain a proper salt balance in the surface soils and to break up the hard salt lens that forms at the point where the groundwater table is most commonly held. Renfro (1955) reports that subsurface or feeder ditches are normally spaced 50 to 70 ft (15 to 21.3 m) apart and run the lengths of the fields.

Along the Platte River in Nebraska, USA are several areas where the natural water table greatly favors the production of alfalfa (*Medicago sativa*). Many fields in this area are surface irrigated only when the alfalfa needs reseeding.

On the Eden Valley irrigation project in Wyoming, USA an area of sandy land within one segment of the project has been developed for subirrigation. The high intake rate and low water-holding capacity of the soil and the favorable geologic and topographic conditions provided the necessary ingredients for a successful subirrigation system (Fox et al., 1956).

VI.   DEVELOPING SUBIRRIGATION IN ARID AND SEMIARID REGIONS

Most of the principles pertaining to subirrigation discussed in previous sections for humid regions apply to systems developed in semiarid and arid regions. The one drawback in the latter case is the danger of soil salinization where salts occur either in the soil or irrigation water; hence provisions must be made for their removal if the practice of subirrigation is to succeed.

Salinization of a soil profile cropped to alfalfa on the Huntley Reclamation Project of Wyoming is well illustrated in Fig. 45–7 (Campbell et al., 1960). The data show that the "no irrigation treatment" (subirrigation only) resulted in the increased accumulation of excess salts below 2 ft by the upward movement of soluble salts in the groundwater over a 4-year period compared to treatment receiving six irrigations annually, where essentially no accumulation occurred. The salts accumulated at those depths where roots absorbed water for the evapotranspiration process, leaving the salt behind. Maximum concentration of salt as measured by the conductivity of the saturated extract was about 6 to 7 mmhos/cm. In cases like this, it would be necessary to provide drainage and to leach periodically for sustained production of salt-sensitive crops.

As mentioned earlier, subirrigation is limited usually to areas where the soils are relatively permeable for a considerable depth, where surface slopes are gentle, and where natural subdrainage is restricted. It must be practical to hold the

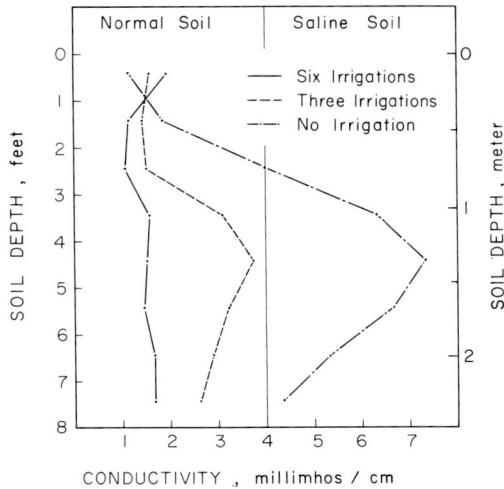

Fig. 45–7. Salinization of a soil profile cropped to alfalfa on the Huntley Reclamation Project of Wyoming, USA (Campbell et al., 1960).

water table at the desired elevations with the water supply available. Thus, a first analysis is to determine the possible lateral flow from the area when the water table is at its desired elevation.

From the Darcy equation of continuity

$$q = p\, h_f\, a/l$$

where

$q =$ flow in cubic feet/second,
$p =$ hydraulic conductivity in cubic feet/second per square foot,
$a =$ the cross-sectional area in square feet through which flow occurs at right angles to the direction of flow, and
$h_f/l =$ slope of the water table in feet/foot.

In the following example of estimating lateral flow, let us assume it is desired to maintain the water table at some predetermined depth below the ground surface in a highly permeable soil underlain by a restricting layer. The water table roughly parallels the ground surface, slopes about 30 ft/mile, and intersects the bottom of a nearby creek which is the only natural drainageway. Therefore, the slope, h_f/l, will be 30/5,280.

From drilling, the average depth to shale or tight clay was found to be about 50 ft. Thus, the cross sectional area from a 1-mile-long section paralleling the creek and through which natural drainage from the irrigated area would have to flow would be 50 × 5,280 feet.

Assume further that p for the soils is 1.0×10^{-4}. Then

$$q = \frac{1}{10,000} \times \frac{30}{5,280} \times \frac{5,280 \times 50}{1} = 0.15 \text{ ft}^3/\text{sec}$$

for the 1-mile long section. If p were as high as 1.0×10^{-3}, outflow would only be 1.5 ft³/sec. If this is not considered adequate water removal, then some drainage system must be installed.

The total water requirement for the area is the drainage plus the consumptive use requirements of the crops grown.

VII.   DESIGN CRITERIA

As a basis for design of a subirrigation system, it is usually necessary to determine the substrata conditions from test borings. These borings should be tied to a common datum and the boring logs and samples analyzed to obtain the following information: (i) existence of any restricting layer and its topography, (ii) contours of the natural water table, and (iii) hydraulic conductivity of the various strata above the restricting layer.

Land mapping and topography should be obtained not only for those areas proposed for irrigation, but also adjacent areas that may be affected by the controlled water table. Location of natural drainage ways or surface outlets must be shown on the map accurately.

Since satisfactory subirrigation depends on being able to control the position of the water table, there must be some provision for getting the water into the soil profile as needed. Experience has shown that under arid conditions, parallel feeder ditches or tile lines which run on the contour, and are spaced sufficiently close to assure proper control of the water table, are often the most practical method. Water is run into these feeders under control and allowed to seep out and feed the groundwater basin. Feeder ditches closed at one end are usually laid out nearly on contour with little or no appreciable slope. The quantity of water allowed into the ditch is held to what the soil will absorb under a given depth of water in the ditch. If either more or less water is desired to raise or lower the water table, the depth of water in the ditches is varied accordingly. If this variation of water depth in the feeders does not give the desired water table control, a change in feeder spacing may be necessary.

The maximum spacing between feeder ditches or drains should be such that the depth from the land surface and the water table will not vary beyond defined limits because of consumptive use or expected rainfall. This limit may be 6 inches or more for such crops as alfalfa or small grain, but may be considerably less for potatoes (*Solanum tuberosum*) or specialized crops. It may also depend on temperatures and consumptive use rates and on the amounts and frequency of summer precipitation.

If the lands were level and no precipitation, lateral flow, or deep seepage occurred, the amount of water necessary to maintain a static water table would equal the consumptive requirement of the crop. However, such a condition is not practical because there must always be some lateral flow and drainage if the groundwater quality is to remain good for crop production; man has little or no control over the weather and must prepare for a certain amount of rainfall except in a few extremely arid areas of the world.

After considering all the influencing factors or water needs for the subirrigated tract, there are other factors that must be considered in the design for the system. Since the slope of the land may affect the water table position, an upslope feeder ditch or pipe will supply water to a greater portion of the area between ditches. Slope tends to make proper control more difficult to maintain.

A. Desirable Slope of Feeders (Ditches and/or Drains)

Feeders may be required to carry considerable water at certain times of the year so that the "sub" or water table may be raised rapidly. Thus, they should be designed to convey several times the amount of water normally required to satisfy the maximum consumptive use rate of the crop. In the arid areas of the USA, this ratio has sometimes been set as high as 20 times the consumptive use rate. Such capacity may not be needed in the more humid areas. However, the size and slope of the ditch should be such that it will convey the maximum flow of water needed.

It is also desirable that the depth of water in the feeder remain reasonably constant throughout its length. Under these conditions, the infiltration opportunity remains constant throughout the ditch length. However, this requirement calls for nearly level feeders.

B. Cross-section of Feeder Ditches

The cross section of the feeder ditch should be so proportioned that adequate infiltration can occur as discussed above. Any open feeder cross section must be designed considering the maximum stable slope the soil permits and the farming requirements. In organic soils, ditches often have vertical sides and are quite deep. In mineral soils planted to pasture, the distributing ditches are sometimes extremely wide and shallow.

Ditches often have a tendency to seal with use and the banks may have to be reworked at intervals to permit adequate infiltration. In some instances, when it is desired to raise the water table quickly, the feeder ditches are permitted to overflow temporarily onto the adjacent cropped area if level.

Because of wide variability in soils and water supplies, field tests of existing ditches in the area under consideration can give the best information as to the expected rate of percolation to the adjacent land mass. From results of these tests and a consideration of other conditions involved, a suitable cross section can be selected.

VIII. GENERAL REQUIREMENTS FOR SUBIRRIGATION

General requirements for subirrigation to be practical and successful under both arid and humid regions are summarized as follows:

1) The soil must be of uniform texture, reasonably deep, and highly permeable.

2) There must be a natural high water table, or a "tight" or restricting layer in the soil profile upon which a perched or temporary water table can be developed beneath the normal root zone of the crops. The restricting layer may be clay, bedrock, or simply natural groundwater.

3) The area to be irrigated must be large so that lateral drainage and return flows will not be excessive in proportion to the water that must be delivered to the area. The land surface must be smooth and level or with only a gentle slope in one direction.

4) Adjacent fields must be in the same general plane. To be practical and successful the difference in elevation of adjacent fields should not exceed about 10 cm (4 inches) and levelling may be required to achieve this limit.

5) The "floor" or restricting layer in the soil profile should be reasonably parallel with the ground surface. An effective natural or artificial drainage system may be necessary to allow for rapid lowering of the water table and for leaching salts. Checks in the drains may be necessary for proper control of the groundwater.

6) Both the soil and the water used for subirrigation must be relatively free of salts, particularly if the lateral movement of the water is limited and if excess water is not available for occasional leaching purposes. They must also be free of suspended silt or clay particles to minimize clogging.

7) When the water table fluctuates regularly in a soil and where the water contains appreciable quantities of certain salts, an impervious layer or lens of salt tends to develop which may retard water movement. This lens usually forms near the phreatophytic surface, and if so, deep chiseling may be necessary to shatter this lens.

8) Where annual precipitation is low, at least one annual surface irrigation may be necessary to leach out the salts. Use of portable sprinklers may be the most practical method of applying this water.

9) During the growing season, the water table must be controlled within limits depending on the crop growth cycle. Few crops will tolerate widely fluctuating water tables. Therefore, water table elevations should not be allowed to change materially, especially during the middle of the growing season.

10) Where independent groundwater control on adjacent farms is not possible, farmers must agree on position of the water table then strive to operate on that schedule. Thus, subirrigation may require special community cooperation.

11) Special provision may be necessary to get a crop germinated and the seedlings started on subirrigated land. This may require raising the water table to wet the surface by capillary action, temporary use of sprinklers or localized surface irrigation. In any case, the surface soil must be kept free of high salt concentrations.

12) Varying water needs and rooting habits of the different crops at various stages of development must be known.

LITERATURE CITED

Campbell, R. E., W. E. Larson, T. S. Aasheim, and P. L. Brown. 1960. Alfalfa response to irrgiation frequencies in the presence of a water table. Agron. J. 52:437–441.

Clinton, F. M. 1948. Invisible irrigation on Elgin Bench. Reclam. Era 34:182.

Donnan, W. W. 1959. Drainage of agricultural lands using interceptor lines. Amer. Soc. Civil Eng. Proc., J. Irrig. Drainage Div. 85(IR1):3–23.

Dorter, K. 1962. Untersuchungen. Zur Losung Bestehender Probleme des Unterflurbewasserung. Kuhn-Archiv. 76½:153–308.

Fox, R. L., J. T. Phelan, and W. D. Criddle. 1956. Design of subirrigation systems. Agr. Eng. 37:103–107.

Glover, R. E. 1964. Groundwater movement. US Bur. Reclam. Eng. Monogr. 31. 67 p.

Harr, M. E. 1962. Groundwater and seepage. McGraw-Hill, New York. 315 p.

Israelsen, O. W., and V. E. Hansen. 1962. Irrigation principles and practices. 3rd ed. John Wiley, New York. 447 p.

Kalisvaart, C. 1958. Subirrigation in the Zuidersee Polders. Int. Inst. Land Reclam. Impr. 53 p.

Renfro, G. M., Jr. 1955. Applying water under the surface of the ground. p. 273–278. *In* Water. US Dep. Agr. Yearbook. US Government Printing Office, Washington, D.C.

Stephens, J. C. 1955. Drainage of peat and muck lands. p. 539–557. *In* Water. US Dep. Agr. Yearbook. US Government Printing Office, Washington, D.C.

Todd, D. K. 1959. Groundwater hydrology. John Wiley, New York. 336 p.

Van Deemter, J. J. 1949. Results of mathematical approach to some flow problems connected with drainage and irrigation. Appl. Sci. Res. A2:33–53.

Van Schilfgaarde, J., D. Kirkham, and R. K. Frevert. 1956. Physical and mathematical theories of tile and ditch drainage and their usefulness in design. Iowa Agr. Exp. Sta. Res. Bull. 436. p. 667–706.

section XII

Irrigation Management

XII

46

Control of Water Intake Rates[1]

D. W. HENDERSON

University of California
Davis, California

HOWARD R. HAISE

Agricultural Research Service, USDA
Fort Collins, Colorado

I. INTRODUCTION

The depth of water applied during irrigation is a basic aspect of all irrigation practice. The total depth applied involves both rate and time factors which are interrelated, but since irrigation time is frequently more subject to control than intake rate, rates have been studied and analyzed in great detail.

While the great effort which has gone into research and investigation on water intake rates is testimony to the complexities of the subject, it is well to reaffirm here that the rate of water infiltration into soils is a universal factor in design and operation of all irrigation systems. Intake rate influences length of run, grade, selection of irrigation method, etc. It also determines irrigation application time in relation to depth of water required to replenish root zone reservoir on systems with grade.

The depth of water applied by the irrigator, in a sense, involves all the physical characteristics of the soil profile which reflect basic mineralogic properties as influenced by natural processes and all the soil management and crop sequence practices of the grower. In addition, intake rates vary with water management practices either through action of the water on soil or through indirect effects of chemicals in the irrigation water.

In qualitative terms, water intake rates range from excessive to negligible. Slow water intake, in particular, may arise from a variety of causes. While it is axiomatic that effective measures will depend on the cause, there is some tendency to consider all problems of slow water intake to be similar and to seek general solutions. Furthermore, even where the source of the problem is apparently similar, many factors influence the response obtained with a given treatment or practice so that results are frequently conflicting.

It is impossible within the scope of this paper to discuss all aspects of water intake in soils without omissions and oversimplification. The reader is referred to the review by Parr and Bertrand (1960), the theoretical series of papers on infiltration by Philip (1957a, 1957b, 1958a, 1958b) and chapters 13 and 15 of this monograph.

---

[1] Contribution from the Dep. Water Sci. and Eng., University of California, Davis. and the Northern Plains Branch, Soil and Water Conserv. Res. Div., ARS, USDA.

II. GENERAL FACTORS AFFECTING INTAKE RATES

In considering various aspects of water intake in relation to soil physical conditions, the entire soil profile throughout the wetted depth must be taken into account. While oversimplification is involved, the concept of some horizon or stratum of the soil profile as the zone limiting water entry is useful. If the surface or near-surface zone is limiting, only the surface will become saturated during intake, and lateral movement is minimized. Practices which alter surface soil characteristics will markedly affect intake. If water entry is limited by the subsoil or substratum, a saturated zone develops above the limiting layer which extends to the surface upon prolonged application of water. Lateral movement is increased, extending for several feet in extreme cases. Only those practices that favorably influence soil physical properties of the limiting layer can have a marked affect on water intake; hence there is little chance that any treatment applied near the surface will be successful.

Infiltration of water into soils is a complex process, but for many practical purposes it is helpful to consider the phenomena involved in terms of hydraulic conductivities and gradients, providing it is understood that the two factors are interrelated.

A. Hydraulic Conductivities

Normally, both saturated and unsaturated conditions exist during water intake, and water conductivity in both states must be considered. However, most water conduction occurs at comparatively low soil water suctions, and water conductivities are well correlated with saturated conductivity or permeability. There are some exceptions to this correlation, but generally intake rate will be high only in soil of high-saturated and near-saturated conductivity.

The principal physical characteristic of soils which determines saturated conductivity is the percentage of the total volume occupied by large pores. Baver (1939), Lutz and Leamer (1940), and Smith et al. (1944) conducted studies which give indications of the relative effectiveness of various increments of pore sizes in conducting water. In near-saturated soils this concept should be amended to exclude those pore sizes that are largely dewatered at the degree of saturation in question.

Pore size distribution undergoes marked change in soils and, since large pores are most readily destroyed, decreases in this size range are most apparent. Water conductivities are very sensitive to changes in soil structure, and intake rates are highly variable from place-to-place and time-to-time. We must be able to characterize the durability of soil structure, or lack of it, as well as describe the pore size distribution at any given time.

While there are exceptions, aggregated structure is less fully developed in arid soils than in those from more humid regions. Aggregates tend to be much smaller, and many irrigated soils approach a single-grain structure, particularly those of coarse texture.

The dynamic nature of the surface soil structure arises because of resultant processes tending to improve structure and those tending to destroy it (Baver, 1956). Beneficial processes include microbiological activity, alternate freezing and thawing, alternate wetting and drying (provided rewetting is slow), proper tillage (frequently to the detriment of soil physical condition just below tillage depth), and possibly the physical incorporation of crop residues; destructive processes include compressive or shear forces due to traffic load or tillage tending to break down soil structural units and the disruptive action of water, or slaking (Yoder, 1936).

Permeable structure is favored by a high proportion of divalent exchangeable cations, a moderate minimum electrolyte concentration, sufficient reactive clay for aggregation but not excessive clay content, and organic matter in a form which contributes to the bonding together of aggregates. Other bonding agencies, particularly lime and the oxides of iron, aluminum, and silicon also contribute to aggregate formation.

While flocculation is usually considered only the first step in aggregation, it is nevertheless very important in water stability because conditions causing flocculation are those which do not lead to peptization. Furthermore, weakly bonded flocs tend to reform after mechanical breakdown in water. Therefore, a favorable proportion of divalent exchangeable cations and a moderate electrolyte concentration is very important in water stability of arid soils. The conditions which tend to cause flocculation also minimize swelling which is both a factor in resistance to slaking and in changes in void dimensions on wetting.

Clay minerals differ in sensitivity to the chemical status. Montmorillinitic clays are highly reactive and when present in small amounts help bind coarser particles to a greater degree than micaceous or kaolinitic clays. However, they are more readily deflocculated and swell more. In both these respects, they are affected to a greater degree by exchangeable sodium and low electrolyte concentration than other types of clays.

Resistance to breakdown by slaking is a complex phenomenon; yet anyone who has observed a freshly tilled soil surface before and after irrigation can have little doubt of its importance. Slaking is caused by the almost explosive escape of entrapped air when aggregate strength is lowered on wetting, and by differential swelling. Slaking action is greatest when dry soil is rapidly inundated—a common condition for surface soil under irrigation. Resistance to slaking depends on the strength of bonding or cementing agents and their resistance to water action. These reactions are not well understood, but a notable example is the marked water stability of soils high in iron oxides (Lutz, 1937).

It is especially difficult to segregate effects of compressive forces and shearing because compression results in shear on a microscopic scale in the zone throughout which the applied pressure is transmitted. Thus, resistance to compression processes should be analyzed in terms of the pattern of transmitted pressures away from the point of stress application. Some soils exhibit marked "bridging" effects, and applied pressures extend only for short distances. The effect of surface loads is then manifest only to shallow depths and the affected zone is easily broken up by shallow tillage.

The principal factor influencing resistance to compression is soil water content with drier soils exhibiting greater resistance. The results have been demonstrated in terms of intake rates (Doneen and Henderson, 1953). Taylor and Henderson (1959) have shown that large applications of organic matter increased resistance to compression in laboratory soil columns as measured by permeability. However, this effect has not been shown conclusively in field tests. Bodman et al. (1958)

found that vinyl acetate-maleic acid (VAMA) increased infiltration until traffic was permitted, after which the beneficial effect largely disappeared. It appears that if aggregate bonding or stabilizing agents can increase resistance to compression and shear, under moist conditions they do not impart sufficient strength to resist heavy loads or other forms of abuse.

Gross aspects of soil physical conditions can be important. In fields where alfalfa (*Medicago sativa* L.) growth is spotty, the vigorous plants often are found where soil physical condition promotes adequate intake and storage of water. Gopher burrows frequently enhance plant growth in the immediate area. A freshly dug drain will exhibit the same effect. In some soils of high clay content, water intake is essentially limited to that entering through cracks, and intake is nil after the cracks close by swelling (Mathews, 1916).

B.   Hydraulic Gradients

Head or depth of water over the soil surface affects intake rate under some conditions. Philip's theoretical analysis (1958a) suggests that in flood irrigation, increased head increases initial intake but that head has negligible effects after prolonged irrigation of homogeneous soil. Field studies usually indicate that differences in head over the surface encountered in normal flood irrigation practice have small effect. On the other hand, increase in depth of water in furrows can markedly increase water intake. Because of furrow shape, however, increased depth likewise increases the wetted area. The two factors have not been segregated, but experience with broad furrows suggests that much of the effect can be attributed to increased area.

Intake rates are decreased with the increase in initial soil water content. This effect is largely caused by reduced hydraulic gradient or a reduction in Philip's "sorptivity" (1957b). Philip's analysis indicates that the reduction in uniform soils is greatest in the early stages of water entry and becomes negligible later. This analysis does not necessarily apply to field soils which are rarely uniform; soil physical conditions may likewise be influenced by antecedent water conditions. As a result, high initial soil water conditions may reduce water intake throughout the irrigation. Mech (1960) and Reitemeyer et al. (1948) provide excellent field data on the magnitude of soil water effects under furrow and basin irrigation, respectively.

Intake rates, with few exceptions (Pelishek et al., 1962), decrease during irrigation. The decrease is rapid initially and the rate tends to approach a constant value. This phenomenon is inherent in the infiltration process in homogeneous soils as indicated by Philip's theoretical analysis (1957a). Breakdown of soil structure by slaking enhances the decrease in field soils, especially in those of low water stability. The coalescense of wetted patterns in furrow irrigation and consequent interference of one furrow with intake from another also affects the intake rate–time relationship.

C.   Other Factors

Wetted area and resultant surface for water entry is a factor in water intake rates which requires consideration. Since two-dimensional flow is involved, the relations are complex, especially after the wetting patterns of adjacent furrows meet. The usual experience is that intake rates per unit of gross area are greater under flooding than in furrow irrigation. This is especially true where intake is limited near the surface or in homogeneous soils. The effect of wetted area in comparing various forms of furrow irrigation is also manifest under these same conditions, especially in early stages of irrigation. No quantitative relation between wetted area and intake rates has been established in general, but in one study by E. L. Jordan, doubling the wetted area increased total depth of infiltration by 1.5 times during the same irrigation period (E. L. Jordan, 1959. Evaluation of irrigation performed with broad and narrow based furrows. *M.S. Thesis, Univ. California, Davis*). It is not known whether the wetted area or the wetted perimeter (i.e., wetted furrow periphery) is most closely related to intake rate.

Where water entry is limited in the subsoil, saturated or near-saturated conditions develop above the limiting layer. Lateral movement above the layer is rapid, and free water soon exists over the entire surface of the limiting stratum. When this condition is reached, the wetted surface area no longer has any effect. In such soils the wetted area may affect the intake rate initially but there would be little influence on the total depth of infiltration except in the case of very widely spaced furrows.

Water and soil temperature affect intake rates although all investigators do not arrive at the same conclusion. Musgrave (1955) reports an excellent correlation between infiltration and water viscosity and utilizes this relation to explain observed diurnal fluctuations in intake rate. Others (Bouyoucos, 1915; Moore, 1941) report an increase with temperature up to about 30C (86F) with a decrease above that temperature. It is generally agreed that fluid viscosity is a factor of direct consequence in permeability, and it seems likely that viscosity is the primary mechanism of temperature effects on infiltration, but it should be remembered that there may be indirect influences as well.

III.   REDUCING EXCESSIVE INTAKE

It may prove desirable to reduce water intake below that attained by current irrigation practice for two reasons. If the soil is permeable throughout, the aim of intake control is to minimize percolation losses below the root zone at least to that required for preventing excessive salt accumulation. On the other hand, if the subsoil is slowly permeable, but overlain by more permeable soil, excessive intake causes waterlogging and injury to susceptible crops. In the latter case, the need for control is dependent on the relative permeabilities of surface and subsoils which determines the degree of waterlogging occurring during irrigation and the permeability of and depth of the limiting layer which determine the duration of waterlogging.

The rate of intake cannot be taken alone as a criterion in judging what constitutes an excessive intake rate. Under many systems of irrigation in current use, an intake rate of 1 inch/hour would be excessive if the desired average depth of application were 0.5 inch, but would be ideal for a 6-inch irrigation. The required irrigation time (i.e., depth of irrigation divided by the average intake rate) is a better criterion. This is especially true in consideration of certain irrigation procedures. In the first instance of the above example, surface irrigation would have to be completed in 30 min. This is manifestly impossible if water must be run for long distances over the soil or if time must be allowed for slow lateral movement from furrows. On the other hand, it could be accomplished by flood irrigation of comparatively small basins provided the water distribution system permitted complete control of the water at all times. Automation or other improvements in control such as increased use of pipe systems under moderate to high pressures (rapid response systems) may very well change what now constitutes excessive intake rates for some situations.

The most obvious means of controlling excessive intake is to irrigate by sprinkling because the application rate is controlled by the system rather than by the soil. In surface irrigation, water application may be speeded up by decreasing the length of irrigation run, increasing the flow rate per unit area, and minimizing factors which retard the flow over the surface or in the furrow.

Where the objective is to minimize waterlogging, irrigating in widely spaced furrows can be effective if lateral movement from furrows is not excessive. To minimize root rot of certain susceptible row crops, irrigation water is turned into alternate furrows or even every third furrow, rotating the furrows used in each subsequent irrigation. Even if there is excessive water applied to the soil below the irrigated furrow, there is drier soil between, permitting redistribution after irrigation. This method obviously fails if water can flow readily from furrow-to-furrow through cracks in the bed and is not effective in preventing deep percolation losses if the soil is highly permeable throughout.

Another approach is to reduce the intake rate of the soil by adoption of practices which are more usually considered abuse. This is more feasible in furrow irrigation where furrow bottoms may be compacted without adversely affecting soil conditions between furrows to the detriment of the crop.

IV.   INCREASING WATER INTAKE

Slow water intake is a widespread problem. The duration of irrigation required for adequate depth of water application may be long enough to cause injury, and runoff waste is excessive. In practice, the actual depth of water applied is frequently inadequate. Drouth results from all but very frequent irrigations, and serious accumulations of salt may arise from water of moderate or higher salt content.

There are two general approaches to the problem. The first is to improve soil structure, which in itself requires a diversity of specific approaches depending on the basic cause of low permeability. This is doubtless the better alternative, since it results in a true solution. A second approach required by failure of the soil to respond or because required soil management practices are not feasible is to alter water management practices to "live with" the condition. Improvement of soil structure is most probable where inherent soil characteristics are favorable. Furthermore, improvement is more lasting in soils which have high structural stability since it is unlikely that all deteriorating forces can be eliminated in an improved management regime. In soils of weak structural stability (including low resistance to compaction by traffic or tillage), improvement is often so temporary that the results are unsatisfactory.

A. Improvement of Soil Structure

Possibilities range from single treatments with immediate response to systems of soil and crop management in which responses are very gradual and, hopefully, accumulative. The latter may include a rather passive approach to minimizing harmful practices.

1. CHEMICAL AMENDMENTS

In general, chemical amendments increase water intake rates only in sodium-affected soils although slow intake rates are sometimes attributed to excessive levels of ammonium and potassium. Very low electrolyte concentration may lead to low permeability even in high calcium soils (Quirk and Schofield, 1955), and such soils respond to applications of chemical amendments which increase concentration of irrigation water or soil solution. In any case, the usual amendment is a source of soluble calcium, either in the form of calcium in solution (most frequently gypsum because of cost factors) or an acidifying agent which releases soluble calcium from lime in the soil.

If the soil is inherently of good structure when calcium saturated and only the surface layer is slowly permeable, response to application of rather small quantities of gypsum or acid can be rapid. Acidifying agents requiring microbial action are slower, and if large quantities of gypsum are required, penetration below the depth of incorporation is slow because of limited solubility.

There is always the question whether it is preferable to apply gypsum in the water or on the soil. If the basic problem arises from high sodium or from very low electrolyte content in irrigation water, best results may be obtained with application in the water (Doneen, 1949). However, the amount of gypsum which can be quickly dissolved is only a few millequivalents per liter, and it is not possible to alter the chemical composition of the irrigation water unless it is quite low in electrolyte content. If the source of the cation causing low permeability is in the soil or in fertilizer, soil applications are usually performed.

It has been shown that under certain conditions ammonium accumulates from applications of ammonium sulfate, causing reduced water intake (Huberty and Pillsbury, 1941). This fertilizer is residually acidic, and if the soil becomes sufficiently acid, nitrification is seriously impaired. The key factors are the buffer capacity of the soil and alkalinity of irrigation water. In soils of low buffer capacity such as the soil at Riverside, California, USA, acidification occurred within a few years. On the other hand, if the soil is highly buffered or sufficient alkalinity occurs in irrigation water to neutralize the acidic fertilizer, ammonium will not accumulate to harmful degrees.

A variety of chemicals have been considered as possible soil conditioners. Most have not been effective under any wide range of field conditions. Polymeric conditioners and polyionic flocculants are very strongly absorbed at the point of contact with soil and are therefore difficult to distribute properly except in surface soil where they can be mixed by tillage. At present, they are prohibitively expensive. Detergent compounds have proven largely ineffective (Lunt and Huberty, 1954) except in cases where poor wettability is a factor (Pelishek et al., 1962).

2. TILLAGE

Too few safe generalizations can be made on influences of tillage, and effects of tillage on infiltration must be considered relative to the basic factors already discussed. Proper tillage temporarily improves the physical condition of the tilled layer. Long-term tillage is frequently considered to be detrimental, especially from the standpoint of structural stability, but there are instances where tillage results in marked general improvement if structure of the native soil is poor. The most universal deleterious effect occurs in the soil below the tillage depth.

The effectiveness of a tillage operation in increasing water intake is determined by whether or not it can produce improvement in the layer limiting water entry, and maximum effect can be attained only if the limiting layer can be completely penetrated. Furthermore, the efficacy of a tillage operation depends on whether or not mixing and reorientation of strata of differing chemical or physical characteristics is essential or whether mere displacement or shattering is the only action required.

If normal tillage proves harmful, deleterious effects can be minimized by decreasing numbers of operations, reducing traffic loads, and by proper timing. Minimum tillage for irrigated field and vegetable crops has not been evaluated from the standpoint of effects on water intake but should prove desirable in many instances. Nontillage of orchard soils has been shown to be beneficial (Jones et al., 1961; Stolzy et al., 1960). However some soils slake drastically, sealing the surface. Such soils must be cultivated for maximum water intake. Pillsbury (1947) and Reitemeyer et al. (1948) obtained marked increases in water intake of basins following cultivation. Werenfels et al. (1963) evaluated nontillage treatments in four orchards with ring infiltrometers and found intake rates lower under nontillage than with clean cultivation in all orchards.

Doneen and Henderson (1953) found less reduction in infiltration rate after tractor passage in drier than in moist soils. Again, the authors have observed exceptions in that weakly cohesive soils are more adversely affected by cultivation when dry than when moderately moist.

Rough tillage in the Salt River Valley of Arizona, USA, permits greater depth of water penetration and storage during prep ant irrigation (Harris et al., 1954). Soil plowed when relatively dry results in an extremely cloddy surface. Applied water moves and infiltrates beneath the clods on nearly level basins. Minimum tillage to break up the moistened clods several days later allows preparation of adequate seedbed and a surface receptive to intake of subsequent water applications.

Deep tillage may be performed once only or repeated at extended intervals. One of two procedures may be selected: (i) Subsoiling, which results in shattering and heaving if performed when the soil is dry, and (ii) deep plowing which produces at least partial mixing and soil inversion as well. Since subsoiling is cheaper, it is preferred when mixing and inversion are not required.

To be at least temporarily effective, subsoiling should be performed when the soil is nonplastic and hard, which for all but very coarse-textured soils means when the soil is dry. Even then a ridge several inches in height is left between subsoiler prints (Trouse and Humbert, 1959). Subsoiling in two directions may help to break out these ridges and increase the effective depth.

Soil shattering by subsoiling may produce dramatic permanent results, but

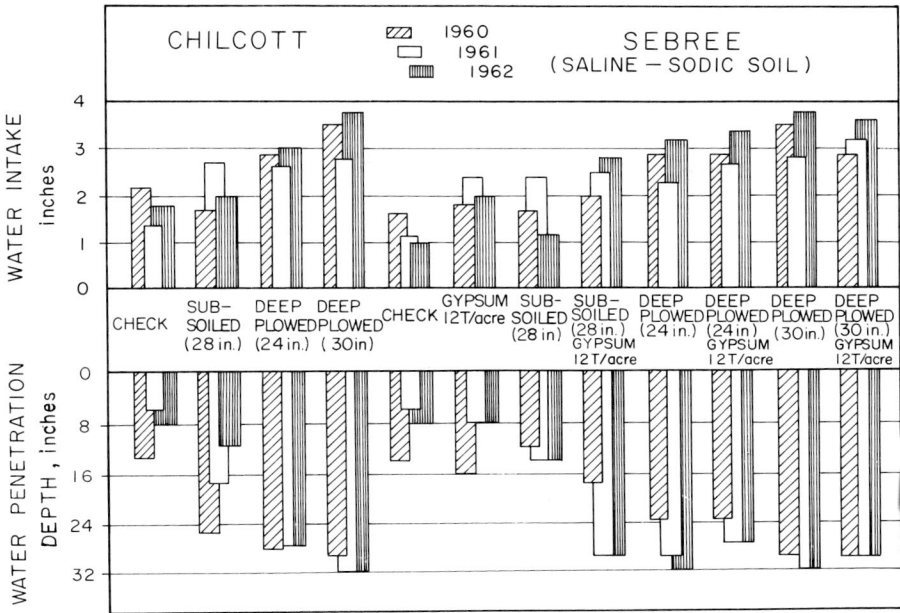

Fig. 46–1. Average quantity of water absorbed and depth of water penetration during 24-hour irrigations. Infiltration was measured in small furrows—Black Canyon Irrigation Project, near Caldwell, Idaho, 1960–1962 (Rasmussen et al., 1964).

frequently the effect is disappointingly temporary. Marsh et al. (1952) and Stolzy et al. (1960) report no increase in intake after subsoiling. Even if the tillage penetrates the limiting layer, structurally unstable soils tend to break down under action of water so that duration of improvement basically depends on structural stability. In serious cases of low water intake, however, even if subsoiling effects last no more than through a single irrigation, it may materially aid in deep preplanting or early season irrigation which can mean the difference between drouth and a crop more adequately supplied with water.

Deep plowing is sufficiently expensive that improvement must last several years. Soil breakup is more complete than in subsoiling; there is at least partial mixing of layers of differing physical or chemical character and partial inversion. Inversion is less complete than in shallower plowing, and in very deep plowing the principal effect is to move the soil layers from horizontal to more nearly vertical orientation. It may thus be very effective in anisotropic soils of high horizontal but low vertical permeability.

Rasmussen et al. (1964) compared deep plowing, subsoiling, and additions of gypsum with and without tillage on a Chilcott (nonsaline, nonsodic) and Sebree (saline, sodic) soil series overlying a calcareous silt loam soil material beneath a clayey B horizon. They found that both intake rate and depth of water penetration were greatly increased by deep plowing on both soil series (see Fig. 46–1). Yields of wheat (*Triticum vulgare*) and alfalfa were increased from three to five times compared to nontreated areas. Subsoiling alone did not influence total intake on the Sebree after the first year but subsoiling with gypsum resulted in marked improvement of intake and penetration depth. Lime-cemented layers at about the 16-inch depth restricted root growth. Deep plowing to the 24- and 30-inch depths on both soil series essentially doubled depth of root penetration and available soil water.

It should be emphasized again that soil conditions were "right" for improvements brought about by the tillage operations noted. However, without the right conditions, deep plowing is often ineffective or of such temporary benefit that it cannot be justified economically.

Vertical mulching is the term applied to filling subsoiler slots with organic matter, usually chopped crop residue (Spain and McCune, 1956). The objective is to keep the slots open for water passage in unstable soils or more open in soils of higher stability. The mulched slots must communicate directly with free water at the surface (Swartzendruber, 1960), so they are largely ineffective in crops requiring surface tillage after vertical mulching, and the practice is limited to more permanent crops such as alfalfa and untilled orchards. The large bulk of organic matter required prohibits close spacing. If the mulched slots are spaced several feet apart they must penetrate the limiting layer to be effective, and the soil below the limiting layer must be moderately permeable. A variation of vertical mulching is to fill augered holes with organic matter. The principal advantage is the lower volume of organic matter required which is of especial importance where it cannot be produced on the area to be treated.

3. CROP SEQUENCE

As in discussion of tillage effects, it is very difficult to generalize influences of crop sequence. Cropping sequences which require a minimum of tillage and traffic may be beneficial. Such sequences often include perennial close-growing crops such as alfalfa, clovers (*Trifolium* sp.) and pastures which involve favorable factors other than nontillage. Mech (1960) reported an approximate threefold increase in intake rate of furrow-irrigated alfalfa on a fine sandy loam Sierozem between the first and third year after seeding, although the infiltration rate was approximately the same in 1-year alfalfa as in preceding rowcrops. Mech's data are reproduced in Fig. 46–2 as an example of crop sequence effects.

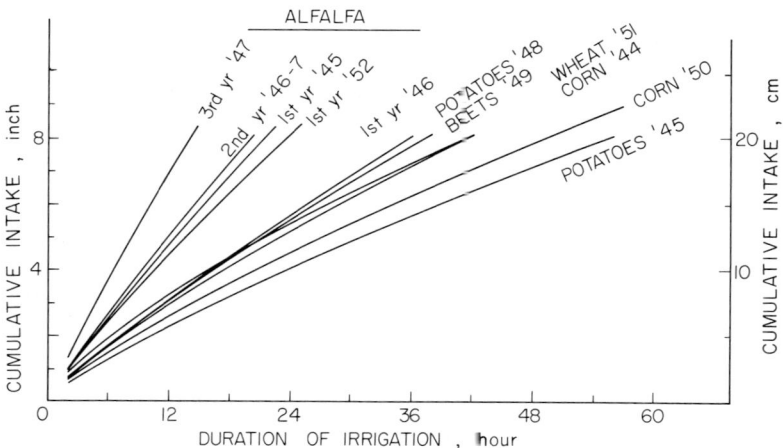

Fig. 46–2. Effect of crop and crop sequence on water intake (Mech, 1960).

Marsh et al. (1952) reported a nearly twofold increase in intake rate in 4-year-old, furrow-irrigated grass over that of other cropping treatments. However, the absolute increase was small. Jensen and Sletten (1965) found increases of 43, 37, 25, and 11% in intake rates of Pullman silty clay loam (a hardlands soil with dense subsoil) in the second, third, fourth, and fifth seasons after plowing down alfalfa, indicating considerable persistence of beneficial effect relative to continuous cropping. However, reported rates were very low, and absolute increases were small.

Crops which permit or cause extreme drying of fine-textured wet subsoils have been observed to be beneficial, indicating that shrinkage of subsoils is sometimes slowly reversible.

4. ORGANIC MATTER, GREEN MANURING, AND COVER CROPPING

Manure or other imported organic matter is probably not effective in improving soil structure much below depth of incorporation and can be expected to have rapid or pronounced effects on water intake only in soils where water entry is limited near the surface. This is especially true in short-term treatments. Effects of roots of green manure and sod cover crops may extend to greater depths, but such crops appear to improve deeper subsoils slowly if at all (except by drying effects).

Marsh et al. (1952) measured only a small increase in infiltration from two annual applications of manure. Responses of a recent alluvial sandy loam to applications of manure, alfalfa meal, and cotton hulls (*Gossypium herbaceum*) were studied by Huberty and Pillsbury (1941). Several-fold increases were measured within a few weeks, and the differences persisted through several months in the relatively undisturbed small test basins. Gypsum also increased infiltration rate of the low-salt irrigation water to a lesser degree. Chopped alfalfa and gypsum applications both essentially doubled the intake rate of a low-salt, high-sodium irrigation water into a stratified alluvial fine sandy loam in basin irrigation tests with date palms (*Phoenix dactylifera* L.) by Reitemeyer et al. (1948). Pillsbury (1947) reported a several-fold increase in water intake rate of an alluvial loam from cornstalk (*Zea mays*) mulch added to miniature undisturbed test basins.

In longer term organic matter studies, Parker and Jenny (1945) found after 11 years of differential treatment that the time for a given furrow stream to reach a point 100 ft away from the inlet was about one-third greater where 1 lb nitrogen/orange tree (*Citrus sinensis*) was supplied as manure than where the nitrogen source was urea. Manure added to supply 3 lb N/tree doubled the 100-ft advance time when compared to 1 lb of N as manure. These data are difficult to interpret in terms of intake rates. Jones et al. (1961), reporting on the same experiment after 15 years of additional treatment, showed that the average seasonal depth of water infiltered from furrows in 48-hour irrigations was 2.4 inches in the urea treatment and 3.7 inches in the manure treatment under clean cultivation. After conversion of the same plots to nontillage comparable values are 3.3 and 5.1 inches for urea and manure treatments, respectively. Pillsbury and Richards (1954), studying effects of cultural practices on infiltration in basin-irrigated oranges on the same soil series, found that during 7 years of treatment, additions of organic matter increased infiltration rate substantially under nontillage. Mazurak et al. (1955) report large increases in intake rates of a Chestnut very fine sandy loam measured by ring infiltrometers following 8 years of annual manure application under continuous cropping and fewer applications in crop rotations.

Williams et al. (1957) and Williams and Doneen (1960) evaluated influences of green manures on furrow infiltration during the following season in several tests. They conclude that gramineous green manures increased furrow intake in five out of six tests on loam and clay loam soils. In some tests green manures approximately doubled the rate late in the irrigation as compared to fallow, although absolute increases were small. Legumes did not increase infiltration significantly. Williams and Doneen (1960) attributed the more favorable effect of grass green manures to lower nitrogen content and slower decomposition, indicating that increased infiltration was associated with the physical presence of largely undecomposed organic matter. In two tests on sandy loam soils they report no increase in infiltration, although one test involved winter green manure crops for 4 consecutive years between cotton crops.

Marsh et al. (1952) found no appreciable increase in intake following Hubam sweet clover (*Melilotus alba*) plowed under as green manure. Flocker et al. (1958) reported no differences in infiltration in studies of irrigated crop rotations which included green manures.

Annual cover crops in orchards are essentially green manures, except that if they reseed without tillage they may be mowed or chopped and replanted only frequently. This type becomes a cover crop with minimum or nontillage. Proebsting (1952) reports that annual cover crops increased water intake in a basin-irrigated orchard on alluvial loam which had a plow sole when the trees were planted and treatments begun. Maximum difference between cover cropped and clean cultivated plots occurred after 11 treatment years, since intake rate of the cultivated treatment increased as well. Werenfels et al. (1963) measured ring infiltrometer rates after 37 years of treatment. A summer legume cover which required reseeding every third year doubled intake rate; winter covers which were disked under and reseeded annually produced smaller increases.

Long-term winter cover cropping of furrow irrigated citrus (Jones et al., 1961) has resulted in only small infiltration rate increase (in 1953 after 26 years of treatment, 2.7 inches water/48-hour irrigation in cover cropped; 2.4 inches in clean cultivated treatments). The effects of long-term cover cropping were completely overshadowed by the increase due to nontillage in the first year after the change from clean cultivation to nontillage in their tests.

Effects of perennial cover crops on water intake of orchard soils have received little attention. In Proebsting's experiment, continuous alfalfa reseeded every 5 or 6 years more than doubled infiltration (Proebsting, 1952; Werenfels et al., 1963), and proved the best of all treatments. Myer and Henderson (1965) found that grass sod increased infiltration several-fold within 4 years after establishment as compared to clean cultivation. They found that soil permeability increased with depth to 12 inches under clean cultivation and decreased under sod; permeability was the same under both treatments at the 12-inch depth. Under clean cultivation the layer limiting water entry was near the surface in this case and therefore in a position for cover crop roots to exert a rapid and pronounced effect.

B. Water Management

Soil improvement sometimes is a completely satisfactory solution to slow water intake problems. On the other hand, results are often disappointing, and soil management practices must be supplemented by changes in management of irrigation water. Water management may involve the irrigation methods, various aspects of the system, timing, and duration of irrigation.

1. UTILIZING SOIL WATER STORAGE

Slow intake problems can be minimized by providing a maximum of stored water at the beginning of the period of rapid water use by plants. If the storage capacity of the soil of the root zone is high and appreciable soil water depletion can be permitted, water can be replenished by prolonged irrigation when the land is bare, or plants dormant, or during cool weather when many diseases are less severe or slower in appearing. The subsoil water reserve produces marked yield increases where slow water intake tends to cause drouth.

2. INCREASING FREQUENCY OF IRRIGATION

Of all management changes, irrigating more often is the simplest because the irrigation system and mode of operation need not be altered. In judging irrigation frequency, however, the tendency for decreased intake with higher soil water levels must be considered (Mech, 1960). The principal disadvantage is that more irrigation labor is required.

3. PROLONGING IRRIGATION

If the intake rate does not drop to a negligible value during irrigation, prolonging infiltration time increases the depth of water applied although not in direct proportion. It is most readily accomplished by using basin irrigation where water is ponded until it infiltrates. In other methods water application time and infiltration time are equivalent, so the irrigation stream must be run for a longer period. The principal limitations are increase in runoff water and possible injury of susceptible crops.

Except where standing water causes or increases severity of crown rots or injuries such as drowning of alfalfa, prolonged irrigation can be practiced with a minimum of risk if surface soil limits water entry. On the other hand, if the subsoil is limiting, prolonged irrigation combined with slow internal drainage may injure all but the most tolerant crops. For these soils, shallow, frequent irrigations are the only feasible management practices.

4. INCREASING WETTED AREA

The basic aspects of wetted area effects have been discussed earlier. Gross water intake is generally higher in flood irrigation than in furrows, and there are special furrow modifications designed to increase wetted area. In orchards, vineyards, and widely spaced rowcrops, wide flat bottom furrows may be used. On even moderate slopes, however, flow is shallow and tends to concentrate in a small portion of the furrow bottom unless it is roughened or made undulating. Yamada and Doneen found that two small furrows between 40-inch spaced croprows effectively increased wetted area and water intake (H. Yamada and L. D. Doneen, 1957 *Unpublished report Dep. Irrig., Univ. Calif., Davis*). In orchards and vineyards various patterns of cross-blocked furrows have been designed for the same purpose. Furrows can be spaced more closely if crop spacing permits (*see* Marsh et al., 1952). For example, sugar beets (*Beta vulgaris* L.) are commonly planted on beds with two rows per bed. Where water intake is slow, they are grown on single-row beds with closer furrow spacing.

5. ALTERATION OF FLOW CHARACTERISTICS

Depth of water applied during irrigation is affected by the flow regimen in a variety of ways. In flood irrigation, the principal effects are on irrigation time rather than on infiltration rate. Factors which retard the flow and increase flow depth only prolong the irrigation. In furrow irrigation, however, depth of flowing water influences intake rate as well, not only through effect of hydraulic head but by changes in wetted area.

The flow depth is determined by stream size, slope, retardance or roughness factors, and furrow shape. Larger streams increase depth and intake rate but reduce irrigation time or increase runoff so that it is difficult to predict total intake from different flow rates. Holmen et al. (1964) showed that increasing furrow flow from 6 to 14 gal/min essentially doubled the intake rate while increasing runoff from 18% to 33% of the applied water. Mech (1960) reports that doubling the flow rate increased intake rate < 15%. Mech's results are averages for 2% and 7% furrow slope whereas those of Holmen et al. are for 0.5% slope, and these data are in qualitative agreement with the fact that increasing flow rate on steeper slopes will have less effect on flow depth than the same change on lesser grades. However, the different effects on intake could also be attributed in part to soil differences as well.

Mech (1960) presents the only data for effect of slope (furrow grade) on intake rates. He showed that intake rate averaged approximately 30% higher on 2% grades than on 7% grades. The effect can be attributed partly to increased flow depth, but the small change in intake with increased stream size suggests other causes as well. Effects of changes in furrow grade on flatter slopes have not been evaluated quantitatively.

Factors affecting flow retardance vary greatly, influence flow depth and irrigation time, and presumably have important effects on intake rates, although direct information is lacking. Loose soil and clods in freshly formed or cultivated furrows retard flows markedly, as does organic litter and plant foliage. Yamada et al. (1963) grew grasses in furrows between cotton rows to heights of 12 to 18 inches. The grasses were then killed with oil spray. Sudangrass (*Sorghum sudanese*) increased intake rates from 0.17 inch/hour in a control to 0.30 and nearly doubled the time required for the irrigation stream to reach the end of the 450-ft test furrows. Yamada et al. (1963) attribute the effects principally to flow retardation although there may be effects on the soil as well.

There has been a marked increase in research interest in recent years in hydraulics of surface flows during irrigation. This work will aid greatly in proper design and operation of surface irrigation systems and will provide impetus for studying the effects of various flow factors on water intake because intake must be evaluated for complete description of the flow regimen. Irrigators will then have still more bases for control of water intake.

LITERATURE CITED

Baver, L. D. 1939. Soil permeability in relation to noncapillary porosity. Soil Sci. Soc. Amer. Proc. (1938) 3:70–76.

Baver, L. D. 1956. Soil physics. 3rd ed. John Wiley, New York.

Bodman, G. B., D. E. Johnson, and W. H. Kruskal. 1958. Influence of VAMA and depth of rotary hoeing upon infiltration of irrigation water. Soil Sci. Soc. Amer. Proc. 22:463–468.

Bouyoucos, G. J. 1915. Effect of temperature on some of the most important physical processes in soils. Michigan Agr. Exp. Sta. Tech. Bull. 22.

Corey, G. L., and D. W Fitzsimmons. 1962. Infiltration patterns from irrigation furrows. Idaho Agr. Exp. Sta. Res. Bull. 59.

Doneen, L. D. 1949. The quality of irrigation water and soil permeability. Soil Sci. Soc. Amer. Proc. (1948) 13:523–526.

Doneen, L. D., and D. W. Henderson. 1953. Compaction of irrigated soils by tractors. Agr. Eng. 34:94–95, 102.

Flocker, W. J., J. A. Vomocil, M. T. Vittum, and L. D. Doneen. 1958. Effect of rotation with green manuring and irrigation on physical characteristics of Hesperia sandy loam. Agron. J. 50:251–254.

Harris, K., D. C. Aepli, and W. D. Pew. 1954. Tillage practices for irrigated soils. Arizona Agr. Exp. Sta. Bull. 257.

Holmen, H., J. M. Schaak, and R. I. Strand. 1964. Infiltration of water into potentially irrigable soils. North Dakota Agr. Exp. Sta. Reprint No. 613.

Huberty, M. R., and A. F. Pillsbury. 1941. Factors influencing infiltration rates into some California soils. Amer. Geophys. Union, Trans. 22:686–697.

Jensen, M. E., and W. H. Sletten. 1965. Effects of alfalfa, crop sequence, and tillage practice on intake rates of a Pullman silty clay loam and grain yields. US Dep. Agr. Conserv. Res. Rep. No. 1.

Jones, W. W., C. B. Cree, and T. W. Embleton. 1961. Some effects of nitrogen sources and cultural practices on water intake by soil in a Washington navel orange orchard and on fruit production, size, and quality. Amer. Soc. Hort. Sci., Proc. 77:146–154.

Lunt, O. R., and M. R. Huberty. 1954. Effect of wetting agents in water on infiltration rates into soils. California Agr. 8(1):12.

Lutz, J. F. 1937. The relation of free iron in the soil to aggregation. Soil Sci. Soc. Amer. Proc. (1936) 1:43–45.

Lutz, J. F., and R. W. Leamer. 1940. Pore-size distribution as related to permeability of soils. Soil Sci. Soc. Amer. Proc. (1939) 4:28–31.

Marsh, A. W., L. R. Swarner, F. M. Tileston, C. A. Bower, and E. N. Hoffman. 1952. Irrigation management investigations on nonsaline soils. Oregon Agr. Exp. Sta. Tech. Bull. 23.

Mathews, O. R. 1916. Water penetration in the gumbo soils of the Belle Fourche Reclamation Project. US Dep. Agr. Bull. No. 447.

Mazurak, A. P., H. R. Cosper, and H. F. Rhoades. 1955. Rate of water entry into an irrigated Chestnut soil as affected by 39 years of cropping and manurial practices. Agron. J. 47:490–493.

Mech, S. J. 1960. Soil management as related to irrigation practice and irrigation design. Int. Congr. Soil Sci., Trans. 7th (Madison, Wis., USA). I:645–650.

Moore, R. E. 1941. The relation of soil temperature to soil moisture: pressure potential, retention, and infiltration rate. Soil Sci. Soc. Amer. Proc. (1940) 5:61–64.

Musgrave, G. W. 1955. How much of the rain enters the soil? p. 151–159. In Water. US Dep. Agr. Yearbook. US Government Printing Office, Washington, D.C.

Myer, J. L., and D. W. Henderson. 1967. Physical characteristics of a sandy loam orchard soil under clean cultivation and permanent sod. Soil Sci. Soc. Amer. Proc. (in press)

Parker, E. R., and H. Jenny. 1945. Water infiltration and related soil properties as affected by cultivation and organic fertilization. Soil Sci. 60:353–376.

Parr, J. F., and A. R. Bertrand. 1960. Water infiltration into soils. Advance Agron. 12:311–363.

Pelishek, R. E., J. Osborn, and J. Leyte. 1962. The effect of wetting agents on infiltration. Soil Sci. Soc. Amer. Proc. 26:595-597.

Philip, J. R. 1957a. The theory of infiltration: 1. The infiltration equation and its solution. Soil Sci. 83:345–357.

Philip, J. R. 1957b. The theory of infiltration 5. The influence of the initial moisture content. Soil Sci. 84:329–346.

Philip, J. R. 1958a. The theory of infiltration: 6. Effect of water depth over soil. Soil Sci. 85:278–286.

Philip, J. R. 1958b. The theory of infiltration: 7. Soil Sci. 85:333–337.

Pillsbury, A. F. 1947. Factors influencing infiltration rates into Yolo loam. Soil Sci. 64: 171–181.

Pillsbury, A. F., and S. J. Richards. 1954. Some factors affecting rates of irrigation water entry into Ramona sandy loam soil. Soil Sci. 78:211–217.

Proebsting, E. L. 1952. Some effects of long continued cover cropping in a California orchard. Amer. Soc. Hort. Sci., Proc. 60:87–90.

Quirk, J. P., and R. K. Schofield. 1955. The effect of electrolyte concentration on soil permeability. J. Soil Sci. 6:163–178.

Rasmussen, W. W., G. C. Lewis, and M. A. Fosberg. 1964. Improvement of the Chilcott-Sebree (solodized-solonetz) slick spot soils in southwestern Idaho. US Dep. Agr., Agr. Res. Serv. 41–91. 39 p.

Reitemeyer, R. F., J. E. Christiansen, R. E. Moore, and W. W. Aldrich. 1948. Effect of gypsum, organic matter, and drying on infiltration of a sodium water into a fine sandy loam. US Dep. Agr. Tech. Bull. 937.

Smith, R. M., D. R. Browning, and G. G. Pohlman. 1944. Laboratory percolation through undisturbed soil samples in relation to pore size distribution. Soil Sci. 57:197–213.

Spain, J. M., and D. L. McCune. 1956. Something new in subsoiling. Agron. J. 48:192–193.

Stolzy, L. H., T. E. Szuszkiewicz, M. J. Garber, and R. B. Harding. 1960. Effects of soil management practices on infiltration rates. Soil Sci. 89:338–341.

Swartzendruber, D. 1960. Water movement into soil from idealized vertical mulch channels. Soil Sci. Soc. Amer. Proc. 24:152–156.

Taylor, H. M., and D. W. Henderson. 1959. Some effects of organic additives on compressibility of Yolo silt loam soil. Soil Sci. 88:101–106.

Trouse, A. C., Jr., and R. P. Humbert. 1959. Deep tillage in Hawaii: 1. Subsoiling. Soil Sci. 88:150–158.

Werenfels, L., E. L. Proebsting, R. M. Warner, and R. Tate. 1963. Cover crops improve infiltration rates. California Agr. 17(5):4–5.

Williams, W. A., and L. D. Doneen. 1960. Field infiltration studies with green manures and crop residues on irrigated soils. Soil Sci. Soc. Amer. Proc. 24:58–61.

Williams, W. A., L. D. Doneen, and D. Ririe. 1957. Production of sugar beets following winter green manure cropping in California: II. Soil physical conditions and associated crop response. Soil Sci. Soc. Amer. Proc. 21:92–94.

Yamada, H., J. Miller, and J. Stockton. 1963. Dessicated grass mulch increases irrigation efficiency for cotton. California Agr. 17(11):12–13.

Yoder, R. E. 1936. A direct method of aggregate analysis and a study of the physical nature of soil erosion losses. J. Amer. Soc. Agron. 28:337–351.

47 | Soil Aeration

JOHN LETEY, JR. and L. H. STOLZY

University of California
Riverside, California

W. D. KEMPER

Agricultural Research Service, USDA
and Colorado State University
Fort Collins, Colorado

Russell (1952) presented an excellent review of the literature on soil aeration and plant growth. Research conducted prior to that review had provided much information on the need for adequate aeration and the effect of poor aeration on various plant life processes. Van't Woudt and Hagan (1957) reviewed the literature on crop response at excessively high soil water levels. One aspect of excessive soil water is inadequate soil aeration. Effects of many soil management practices such as irrigation on crop production had been attributed to soil aeration by various authors. Direct measurements of soil aeration to support the assumption were made in only a few cases. Russell (1952) concluded: "This situation undoubtedly is a consequence of the lack of satisfactory methods of characterizing soil aeration in terms of parameters that are of significance in plant growth." Development of satisfactory measuring techniques was therefore considered to be a key factor in the progress of soil aeration research. Progress in soil aeration research since these reviews will be discussed in this chapter. Emphasis will be placed on methods of measuring and characterizing soil aeration and on information which would be helpful in interpreting the effect of field conditions on crop production from a soil aeration point-of-view. This chapter will not be an all-inclusive literature review on the general subject of soil aeration.

I. MECHANISMS FOR GAS EXCHANGE

To determine the soil aeration parameters which have the greatest significance to plant growth, it is necessary to understand the mechanisms involved in the gas exchange between the respiration sites and the aerial atmosphere.

Research has revealed that diffusion is the most important single mechanism in soil aeration and therefore requires the greatest amount of attention. Oxygen diffuses in response to a partial pressure gradient from the aerial atmosphere through the gas-filled pore spaces. Oxygen then dissolves in the soil solution and diffuses to the respiration site. Carbon dioxide diffuses in the opposite direction from oxygen. The rate of diffusion in the gas phase depends on the diffusion coefficient through gas and the fraction of the soil which is occupied by gas. Penman (1940) reported that $D/D_o = 0.66$ S for the diffusion of vapors through porous solids whereas Van Bavel (1952) found $D/D_o = 0.6$ S where D is the effective diffusion coefficient through the soil mass, D_o is the diffusion coefficient in air, and S is the air filled porosity. Millington (1959) and Marshall (1959) suggested $D/D_o = S^{4/3}$ and $D/D_o = S^{3/2}$, respectively. The oxygen diffusion rate through the liquid surrounding a root depends on the diffusion coefficient in water, fraction of the area not occupied by solids, tortuosity of the diffusion path, and concentration gradient which will, in turn, depend on the thickness of the water layer around the root and rate of use. Kristensen and Lemon (1964) presented a good discussion on oxygen diffusion in the liquid phase surrounding the root. The diffusion coefficient of oxygen in water is about 2.56×10^{-5} and about 1.89×10^{-1} cm^2/sec in air. The relatively low diffusion coefficient in water indicates that diffusion from the gas phase to the respiration site cannot be completely ignored as a limiting factor in the gas exchange.

Although it has been traditionally considered that the pathway for gas exchange is through the soil, there is evidence that plant roots may obtain oxygen by internal diffusion through the plant and root. Jensen et al. (1964), e.g., demonstrated that O^{18} which was supplied around corn leaves (*Zea mays*) moved through the plant and out into the soil environment. The extent that plants have an internal gas exchange would influence the amount of exchange required through the soil and therefore the interpretation of any measurement on soil aeration made in the soil. An obvious example is the rice plant (*Oryza sativa*) which can get its entire oxygen supply internally. A measurement in the soil which would indicate an aeration deficiency for one plant species would not be deficient for rice. Very likely plant species differ in the amount of "internal aeration." More research is required on internal aeration of various plant species before it can adequately be evaluated with respect to soil conditions. Of particular interest would be the extent of movement through the root from a zone of adequate aeration to a zone of inadequate aeration.

II.   MEASURING AND CHARACTERIZING SOIL AERATION

The earliest method of measuring soil aeration was based on determining the concentration of O_2 and CO_2 in soil pores. Considerable difficulty was encountered in extracting gas samples without contamination from the aerial atmosphere and in running the analysis. The rather recent development of the gas chromatograph has simplified this procedure. Only small samples (0.5 ml) need be extracted and these can be quickly analyzed. Membrane-covered electrodes such as described by Willey and Tanner (1963) allow O_2 measurement in pores without extraction of samples. These electrodes are particularly useful for periodic monitoring of O_2 at a given location. Because the electrode must be kept dry to measure gas concentration, it is usually sealed in an access tube for insertion into the soil. Equilibration of gas between the volume surrounding the electrode and the soil pores must be achieved before making a measurement. Equilibration will be extremely slow if water blocks the passage between the electrode area and the remainder of the soil.

Oxygen diffusion in the gas phase of the soil system was proposed as a method of characterizing soil aeration by Raney (1950).

Results of measurements made on the gas phase concentration or diffusion are useful for some purposes. However, upon reviewing the soil aeration literature,

Russell (1952) stated: ". . . the use of oxygen and carbon dioxide concentrations as a means of characterizing soil aeration for plant growth has not been entirely satisfactory." None of the methods include the impedance of the barrier between the gas phase and respiration site. It is possible to have high O_2 concentration in the open pore and yet have an inadequate supply in a root if the root is removed far enough from the pore (Kristensen and Lemon, 1964).

Wiegand and Lemon (1958, 1963) calculated the concentrations at the root surface and found them to be less than optimum at the 8- and 12-inch depths of Miller clay. Concentrations in the pore were above the so-called critical values.

Lemon and Erickson (1952) introduced the method of measuring O_2 diffusion rate to a platinum wire electrode which is inserted in the soil. The factors which generally influence the O_2 diffusion rate to the electrode also affect diffusion to a root. This method therefore showed promise of characterizing soil aeration in terms of parameters that are of significance in plant growth. It has been used with varying degrees of success by investigators. Some of the difficulties encountered have been overcome by a better recognition of the factors involved in making the measurement and how these may affect the results. Fortunately, most of the difficulties are associated with making measurements where the soil is relatively dry or where high O_2 diffusion rates (ODR) occur. Accurate determination under these conditions is less important than under conditions where the ODR is low and can be limiting to plant response. A description of the equipment, method of making the measurement, and factors which can affect the results have been presented (Birkle et al., 1964; Letey and Stolzy, 1964).

Plant tissue analysis has become a very useful diagnostic tool for mineral nutrients. Fulton and Erickson (1964) found that ethanol in the xylem exudates of tomato (*Lycopersicon esculentum* Mill.) increases as the ODR decreases. Presently, it is very speculative whether analysis for ethanol in tissue sap will serve as a useful diagnosis for poor aeration but such possibility stimulates the imagination. Research along these lines is warranted.

III.   SOIL AERATION AND PLANT GROWTH

The two reviews mentioned in the introduction (Russell, 1952; Van't Woudt and Hagan, 1957) presented much information on the effect of soil aeration on plant growth based on analysis of the gas phase as an index of soil aeration. Because of these reviews, less attention will be given in this chapter on results associated with gas analysis except for results obtained with the membrane-covered electrode which is a rather recent development.

The measurements of ODR with the platinum electrode provided a new parameter to characterize soil aeration. The question remained whether this parameter would be of significance with respect to plant response.

Stolzy and Letey (1964a, b) reviewed the literature on plant response to the measured ODR values. There was generally good agreement between various investigators on the relationship of ODR and root growth. The roots of many plant species will not grow in an environment where the ODR is less than about 0.20 μg cm^{-2} min^{-1}. Oxygen diffusion rate measurements can therefore be made in the soil and the volume available for root growth at any given time can be determined. An understanding of the relationship between the rooting system and top growth and reproductive cycles is then required to assess the resultant effect on the entire plant. It is conceivable that a relatively shallow root system could produce as large a plant as a deeper root system if sufficient nutrients and water were maintained in the root zone. Plant species will differ in this respect. Good correlation between ODR and top growth is not generally possible without an understanding of the root-top relationship.

The platinum electrode does not measure CO_2 and raises the question whether under certain circumstances the CO_2 may reach detrimental concentrations even though ODR values are sufficiently high. Grable and Danielson (1965a, b) conducted an extensive study on the effect of CO_2 on the germination and growth of corn and soybean (*Glycine max*) seedlings. Very high concentrations (20% or more) were necessary to decrease seedling growth. Low CO_2 concentrations stimulated growth over the regular air treatment. Factors which influence O_2 diffusion also influence CO_2 diffusion. It is very doubtful that CO_2 could accumulate to toxic levels under conditions which would allow adequate O_2 diffusion rates.

A difficulty in interpreting any soil aeration data with respect to plant response under field conditions is that the soil aeration condition is not static. The conditions vary both with respect to time and with respect to position. The plant somehow integrates all of these factors. Most controlled experiments conducted on soil aeration establish a given soil aeration level and keep it fairly constant during the course of the experiment.

Letey et al. (1962) found that when the O_2 supply was very low, root growth stopped and did not recover immediately after adequate O_2 was supplied. The effect on the vegetative growth was greatest when low O_2 treatments were applied before the plant had established a good root system. Irrigation practices which would cause O_2 deficiency when a plant is becoming established could stunt plant growth and because of the slow recovery after adequate aeration, decrease yield or increase the growth period required for good production.

Three hours of very low O_2 altered some of the physiological processes of the plant. The effect of these low O_2 periods was still evident 24 hours after the plant was subjected to adequate O_2 (Stolzy et al., 1964).

Erickson and Van Doren (1960) found that 1 day of O_2 deficiency reduced the production of peas (*Pisum sativum*). Reduction was greatest when the deficiency occurred just before or during bloom. The dry weight of tomato plants was reduced by 1 day of O_2 deficiency when it occurred shortly after planting. Short periods of poor aeration caused by irrigation may have more effect on production than has generally been recognized.

Barley (*Hordeum vulgare*) yields were decreased by short periods of flooding at all growth stages during a hot and dry summer (Shazkin, 1960). Unfortunately, the report does not specify the length of flooding period or specific weather data. These results are in agreement with others (Letey et al., 1962) that show that low O_2 is more detrimental to plant growth during high soil or air temperatures than lower temperatures. Irrigation must be done at a lower soil water suction when temperature is high than when the temperature is lower to avoid a decrease in vegetative growth due to insufficient water (Letey and Blank, 1961). Irrigation practice is therefore quite critical in hot arid regions if maximum production is to be achieved. The soil cannot be allowed to become too dry and yet excess water cannot be allowed to accumulate in any part of the soil profile. An extended period of poor aeration will cause reduced production. Ideal soil conditions would allow for fairly rapid water intake so that prolonged flooding can be avoided. Application of excess water for leaching purposes would best be done in the coolest parts of the season except when needed to prevent salt damage.

IV.   IRRIGATION PRACTICES AND SOIL AERATION

How often is soil aeration reduced to a damaging level by irrigation and does the type of irrigation such as sprinkler or flooding have a great influence on the resultant soil aeration, are practical questions which unfortunately, cannot adequately be answered at this time. The difficulties are that there have been relatively few fairly continuous measurements of soil aeration under field conditions and only a small amount of information available on the effect of short periods of O_2 deficiency on crop production.

Results of Kemper and Amemiya (1957) and Willey and Tanner (1963) show that when the soil surface is sealed either by prolonged flooding or sprinkling at a relatively high rate, the O_2 concentration in the pores decreases. As soon as the pores have an open channel to the aerial atmosphere, the O_2 concentration in the pores increases. Surface sealing occurred for about 4 and 2 days respectively for the two cases. Alfalfa (*Medicago sativa*) production did not appear to be limited by the 4 days of surface sealing; however, a lower soil water stress and decrease in salts may have over-shadowed the low aeration effect.

Williamson (1964) conducted a 4-year study with field lysimeters in which the water table was established at various depths below the soil surface. Oxygen diffusion rate measurements were made at different depths for the various treatments. The ODR was low under the water table and for a few inches above the water table but increased with increasing distance above the water table. The average ODR increased as the depth to water table increased. Crop yield increased with increased average ODR. In about one-half of the cases, however, the yield reached a maximum at an average ODR less than the greatest ODR. This decrease was attributed to a water stress on the plant because the water table was too low and insufficient water was added to the surface to maintain optimum water conditions for plant growth.

Williamson and Willey (1964) studied the effect of the depth of water table on the growth of tall fescue (*Festuca elatior*) in greenhouse and growth chamber experiments. The soil aeration was determined by monitoring the O_2 partial pressure at various soil depths with the membrane-covered electrode. After an application of water at the surface, the O_2 partial pressure decreased to lower values and stayed at these low values longer at positions nearest the water table. Tall fescue, being a shallow-rooted plant, grew best when the water table was 16 inches deep if supplemental irrigation was applied at the top. When surface water was not applied, yield was about the same for water tables at 9 and 17 inches. These water tables which were relatively near the surface were beneficial in supplying water for the plants. The decrease in O_2 was not detrimental because most of the roots grow near the surface anyway and very little could be gained by providing an environment which would be conducive to deeper root growth. This would not apply to deep rooting crops.

Rate of recovery of adequate soil aeration following an irrigation depends on the soil physical conditions. Erickson and Van Doren (1960) found differences

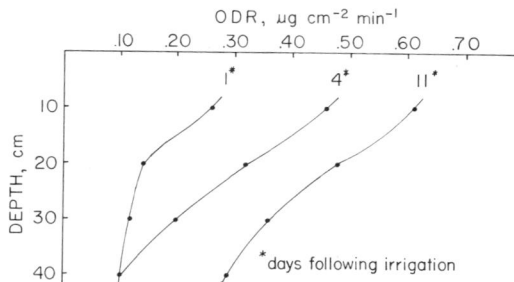

Fig. 47–1. Oxygen diffusion rate (ODR) as a function of soil depth on various days following irrigation in a cotton field.

in the number of days required to reach optimum ODR after irrigation on soils which had been subjected to different cropping rotations.

Oxygen diffusion rates were measured by the platinum electrode technique in cotton (*Gossypium herbaceum*) fields of the San Joaquin Valley of California, USA. (The authors acknowledge the assistance of T. E. Szuszkiewicz and Melvin Jeffers, Department of Soils & Plant Nutrition, University of California, Riverside, in making these measurements.) The cotton was furrow-irrigated and the measurements were made in the plant row about equal distance between the furrows. The results of one case are presented in Fig. 47–1. The ODR was very low throughout the profile 1 day after irrigation and increased gradually with time. The ODR was adequate for good root growth to a depth less than 40 cm on the fourth day after irrigation.

In one case, water did not get through a furrow. The ODR listed in Table 47–1 were measured at various distances from the furrow which did have water. These measurements were made 4 days after irrigation. The diffusion rates increase with increased distance from the furrow as should be expected. These data illustrate that if the wetted zones of adjacent furrows do not overlap, the root system is subjected to zones of various ODR. The effect of having parts of the root system under low ODR and the rest under high ODR on plant production has not been conclusively determined. The effect will probably depend upon the type of plant. All of the results reported in the figure and table were measured with .64-mm diameter electrode with an applied potential of −.65 v. The soil is a Panoche clay loam.

Sprinkling at a rate less than the infiltration rate of the soil would theoretically

Table 47–1. Oxygen diffusion rates at various depths and distances from the wetted furrow

Soil depth, cm	Distance from furrow, cm		
	20	40	60
	Oxygen diffusion rate, $\mu g \ cm^{-2} \ min^{-1}$		
10	0.67	0.88	dry
20	0.18	0.42	1.00
30	0.07	0.32	0.66
40	0.06	0.28	0.58
50	0.04	0.18	0.33

provide a situation where soil aeration is better than flooding. Rubin et al. (1964) pointed out that the lower application rate causes a lower water content in the wetted profile and would therefore allow greater O_2 diffusion. The ODR measured in the San Joaquin Valley under sprinkler irrigation were slightly higher than furrow irrigation, but not significantly higher. More studies are required to determine the value of sprinkling vs. other forms of irrigation from a soil aeration point-of-view.

Irrigation may lead to soil aeration problems if the subsoil is kept very wet for deep-rooted perennial crops. The time to irrigate is often based on water content in the surface soil; water content at greater depths is not checked. Tensiometers were installed in a chlorotic citrus orchard in southern California to check the existing irrigation practices. It was found that water was generally applied when the surface was sufficiently dry to warrant irrigation. The duration of water application was always long enough, however, to wet the subsoil which did not need water. As a result the subsoil was always wet. Based upon this information, the time of water application was decreased so that only the upper part of the profile was wet except when the lower tensiometer indicated that the subsoil required water. About a year after the revised irrigation practice, the chlorosis disappeared. Soil aeration affects nutrient uptake (Stolzy and Letey, 1964a). Low soil O_2 has been demonstrated to decrease the iron concentration in leaves of orange seedlings (*Citrus sinensis*) (Wallihan et al., 1961). It is logical to conclude that the chlorosis was caused by inadequate O_2 in part of the root zone. This case also points out the value of using water-sensing devices in the soil, such as tensiometers, for guidance in an irrigation program. They are useful not only to prevent the soil from becoming too dry, but also in avoiding the application of too much water to parts of the soil profile.

The effect of excessively wet subsoil may not be so great for some of the annual crops. The root system would be restricted to relatively shallow depths, but some crops if properly managed otherwise may still have good production. Again, the requirement for knowledge of the relationship of root development to production is required for the various plant species for proper interpretation.

Both water and soil particle arrangement and packing influence soil aeration. Emphasis has been placed on water in the chapter because the monograph is on irrigation. It is very likely, however, that inherent differences between soils will have greater effect on soil aeration than will irrigation practices.

In summary, the fact that plants require an O_2 supply to the root for growth has long been recognized. Many experiments have been conducted to determine what effect low O_2 supply has on plants. Most of these experiments were conducted under conditions such that the results could not be quantitatively translated to the field because O_2 conditions as measured in the field were not comparable to those measured under controlled conditions. The rather recent introduction of the platinum electrode by Lemon and Erickson (1952) to measure ODR provides a method to bridge the results of greenhouse experiments to field data. Very few measurements have been made under various irrigation practices so it is presently impossible to evaluate the possible effect of these practices on soil aeration. Results indicate that O_2 is low for a day or more following irrigation on some soils. The overall effect of short periods of low soil aeration on crop production has not been thoroughly investigated. There are reports, however, that short periods of O_2 stress at critical stages of growth can reduce production.

LITERATURE CITED

Birkle, D. E., J. Letey, L. H. Stolzy, and T. E. Szuszkiewicz. 1964. Measurement of oxygen diffusion rates with the platinum microelectrode: II. Factors influencing the measurement. Hilgardia 35:555–566.

Erickson, A. E., and D. M. van Doren. 1960. The relation of plant growth and yield to soil oxygen availability. Int. Congr. Soil Sci., Trans. 7th (Madison, Wis., USA) 4:428–434.

Fulton, J. M., and A. E. Erickson. 1964. Relation between soil aeration and ethyl alcohol accumulation in xylem exudates of tomatoes. Soil Sci. Soc. Amer. Proc. 28:610–614.

Grable, A. R., and R. E. Danielson. 1965a. Effect of carbon dioxide, oxygen, and soil moisture suction on germination of corn and soybeans. Soil Sci. Soc. Amer. Proc. 29:12–18.

Grable, A. R., and R. E. Danielson. 1965b. Influence of CO_2 on growth of corn and soybean seedlings. Soil Sci. Soc. Amer. Proc. 29:233–238.

Jensen, C. R., J. Letey, and L. H. Stolzy. 1964. Labeled oxygen: Transport through growing corn roots. Science 144:550–552.

Kemper, W. D., and M. Amemiya. 1957. Alfalfa growth as affected by aeration and soil moisture stress under flood irrigation. Soil Sci. Soc. Amer. Proc. 21:657–660.

Kristensen, K. J., and E. R. Lemon. 1964. Soil aeration and plant root relations: III. Physical aspects of oxygen diffusion in the liquid phase of the soil. Agron. J. 56:295–301.

Lemon, E. R., and A. E. Erickson. 1952. The measurement of oxygen diffusion in the soil with a platinum microelectrode. Soil Sci. Soc. Amer. Proc. 16:160–163.

Letey, J., and G. B. Blank. 1961. Influence of environment on the vegetative growth of plants watered at various soil moisture suctions. Agron. J. 53:151–153.

Letey, J., and L. H. Stolzy. 1964. Measurement of oxygen diffusion rates with the platinum microelectrode: I. Theory and equipment. Hilgardia 35:545–554.

Letey, J., L. H. Stolzy, and G. B. Blank. 1962. Effect of duration and timing of low soil oxygen content on shoot and root growth. Agron. J. 54:34–37.

Letey, J., L. H. Stolzy, N. Valoras, and T. E. Szuszkiewicz. 1962. Influence of oxygen diffusion rate on sunflower growth at various soil and air temperatures. Agron. J. 54:316–319.

Marshall, T. J. 1959. The diffusion of gases through porous media. J. Soil Sci. 10:79–82.

Millington, R. J. 1959. Gas diffusion in porous media. Science 130:100–102.

Penman, H. L. 1940. Gas and vapor movements in the soil: I. The diffusion of vapors through porous solids. J. Agr. Sci. 30:437–461.

Raney, W. A. 1950. Field measurement of oxygen diffusion through soil. Soil Sci. Soc. Amer. Proc. (1949) 14:61–65.

Rubin, J., R. Steinhardt, and P. Reiniger. 1964. Soil water relations during rain infiltration: II. Moisture content profiles during rains of low intensities. Soil Sci. Soc. Amer. Proc. 28:1–5.

Russell, M. B. 1952. Soil aeration and plant growth. p. 253–301. In B. T. Shaw. [ed.] Soil physical conditions and plant growth. Academic Press, Inc., New York.

Shazkin, F. D. 1960. The effect of surplus moisture on plants at different developmental periods. Plant Physiol. (USSR) [English Transl.] (Fiziologiya Rastanii) 7:225–229.

Stolzy, L. H., and J. Letey. 1964a. Characterizing soil oxygen conditions with a platinum microelectrode. Advance. Agron. 16:249–279.

Stolzy, L. H., and J. Letey. 1964b. Measurement of oxygen diffusion rates with the platinum microelectrode: III. Correlation of plant response to soil oxygen diffusion rates. Hilgardia 35:567–576.

Stolzy, L. H., O. C. Taylor, W. M. Dugger, Jr., and J. D. Mercereau. 1964. Physiological changes in and ozone susceptibility of the tomato plant after short periods of inadequate oxygen diffusion to the roots. Soil Sci. Soc. Amer. Proc. 28:305–308.

Van Bavel, C. H. M. 1952. Gaseous diffusion and porosity in porous media. Soil Sci. 73:91–104.

Van't Woudt, Bessel D., and Robert M. Hagan. 1957. Crop responses at excessively high soil moisture levels. In J. N. Luthin (ed.) Drainage of agricultural lands. Agronomy 7:514–578.

Wallihan, E. F., M. J. Garber, R. G. Sharpless, and W. L. Printy. 1961. Effect of soil oxygen deficit on iron nutrition of orange seedlings. Plant Physiol. 36:425–428.

Wiegand, C. L., and E. R. Lemon. 1958. A field study of some plant-soil relations in aeration. Soil Sci. Soc. Amer. Proc. 22:216–221.

Wiegand, C. L., and E. R. Lemon. 1963. Correction in paper "A field study of some plant soil relations in aeration." Soil Sci. Soc. Amer. Proc. 27:714–715.

Willey, C. R., and C. B. Tanner. 1963. Membrane-covered electrode for measurement of oxygen concentration in soil. Soil Sci. Soc. Amer. Proc. 27:511–515.

Williamson, R. E. 1964. The effect of root aeration on plant growth. Soil Sci. Soc. Amer. Proc. 28:86–90

Williamson, R. E., and C. R. Willey. 1964. Effect of depth of water table on yield of tall fescue. Agron. J. 56:585–588.

48

Water Erosion Under Irrigation[1]

STEPHEN J. MECH *(deceased, July 1968)*
Agricultural Research Service, USDA
Prosser, Washington

DWIGHT D. SMITH
Agricultural Research Service, USDA
Beltsville, Maryland

I. INTRODUCTION

Soil erosion has long been considered a threat to agriculture in humid regions. Recognition of erosion as a problem on irrigated lands is relatively recent, yet many who have intimate knowledge of irrigation are deeply concerned. Israelsen et al. (1946) reporting on a 6-year study of soil erosion in irrigation furrows stated that ". . . excessive soil erosion on irrigated lands is adverse to the perpetuation of permanent agriculture in arid regions. The authors have seen, on many occasions, sugar beet (*Beta vulgaris* L.) lands in Utah, USA in which, after the first irrigation season, the furrow depths near the head ditches have been eroded from 2.5 to 10.2 cm (1 to 4 inches). On the other hand, the lower ends of these furrows have been completely filled with soil eroded from the upper ends."

It is ironical that the water brought in to give life to the desert lands could indirectly also be the means of destroying them. Though irrigation implies that man has control of the water, there are limits below which one cannot reduce the flow rate or the length of run or slope. Also, the soil over which this water flows must be kept friable and cultivated.

Robins and Neff (1963) stated although many of the problems associated with land and water resources are accompanied by a cyclic opportunity for improvement or renewal, soil erosion is a conspicuous exception to this "renewable" concept. Once a portion of the soil mantle is removed by erosion, it can be restored only by extreme measures that are generally too slow or too expensive for consideration in the foreseeable future.

Early irrigation literature dealt almost exclusively with water and its relation to crop production. As irrigation became more widespread, references to erosion and to erosion damage became more frequent. Taylor (1935) reported results of a study on the influence of tillage on infiltration and erosion under furrow irrigation. Taylor (1940) reported on the relative transporting power of furrow streams.

In 1937 the Soil Conservation Service measured irrigation furrow flow and resulting soil losses from potato (*Solanum tuberosum*) lands near Ellensburg,

---

[1] Contribution from the Soil and Water Conserv. Res. Div., ARS-USDA, and the Washington Agr. Exp. Sta.

Washington, USA. Unpublished progress reports covering this work state ". . . that 20 to 75% of the water applied . . . was lost as waste water with soil losses ranging from 67 to 134 metric tons/ha (30 to 60 tons/acre). Serious erosion has occurred in furrows with deposition near the end . . ."

A. Erosion Types and Causes

Erosion may occur whenever water flows over cultivated land. Where irrigation water is applied on level or nearly level land or where properly managed flood irrigation of close-growing crops is practiced, erosion usually is slight. Likewise, no serious erosion occurs with properly designed sprinkler systems, since the water is generally applied no faster than the soil will absorb it and no surface flow occurs. However, when water is applied on steeper land by furrows, erosion is a constant threat. These irrigated lands, because of their steeper slopes, are generally well drained, free from excess salts, and highly productive when properly irrigated.

Erosion by irrigation water may be due to (i) uncontrolled concentration of runoff water, (ii) excessive flow of water in furrows, or (iii) normal flow of water in graded furrows—when the soil is loose and transportable.

Uncontrolled concentration of runoff water generally produces a spectacular and devastating type of gully erosion. The control of such erosion is relatively simple from the technical standpoint. It can be controlled by reducing the amount of water leaving the field or by protecting the waterway with resistant vegetation or mechanical structures.

Erosion caused by applying water faster than the furrow can absorb it is shown in Fig. 48–1. This excessive flow increases erosion along the entire length of the furrow. Here, too, control is relatively uncomplicated. Erosion along furrows can be reduced by positive control of the irrigating stream so that the water is applied only fast enough to irrigate the row, with little or no runoff.

The erosion associated with normal furrow flows is much more difficult to reduce. Some erosion will always occur when flowing water comes in contact with soil in a friable condition and even normal flow in an irrigation furrow will often produce unavoidable erosion.

Fig. 48–1. Erosion is severe on potato field at left. The field on the right is ready to be irrigated. Erosion similar to that on the left can be prevented if extreme care is used in applying the water.

Fig. 48–2. Severe erosion near water distribution pipeline (foreground) and deposition on lower end of 122-m (400-ft) run on a 2% slope.

This paper will be confined to soil erosion associated with the normal flow of irrigation water in irrigation furrows as distinguished from erosion caused by rain on irrigated land or erosion under spinkler irrigation.

Soil deposition at the lower end of the field in Fig. 48–2 indicates that very little water was wasted as runoff. The irrigating streams were not much above the intake rate of the furrows and could not be reduced much more without decreasing irrigation efficiency and effectiveness.

Severe erosion near the upper end of the furrow will, in a few years, produce a dip in the surface a short distance downstream from the pipeline. Such stairstep development generally occurs in varying degrees where cultivated land is furrow irrigated. This erosion cannot be completely eliminated. The best that can be done is to reduce it to a minimum by observing a few basic precautions.

II. PRINCIPLES OF EROSION UNDER IRRIGATION

A. Nature of Erosion

Because of the limited amount of available information on erosion under furrow irrigation, much of the thinking on this problem was necessarily patterned after that under rainfall conditions. Subsequent research pointed out that many of the erosion concepts and techniques developed and used so successfully under rainfall conditions do not apply to furrow irrigation.

Under furrow irrigation the problem is essentially one of streambed and streambank erosion in miniature. The amount of erosion depends not on the over-all condition of the field but on the condition in the narrow channel where the irrigating stream and the soil surface are in intimate contact. It matters very little whether the space between the furrows is vegetated or bare. There is no raindrop impact. Borst and Woodburn (1942) reported on the interrelation of overland flow and rain impact.

Serious erosion may be found on the upper end of irrigated fields even when neither soil nor water is lost from the lower end. Under irrigation the soil loss as expressed in tons per unit of area removed from the field can be reduced and almost eliminated by simply increasing the length of the run until the water is absorbed by the furrow and all the sediment deposited. That erosion *per se* can be so reduced is an obvious fallacy.

Unfortunately, no other unit for describing soil movement along the furrow length has the general acceptance of tons per acre at the present. Mech and Free (1942) suggested the term "foot-pounds" regardless if the soil movement is caused by tillage, flowing water, or any other cause. Ellison (1947) uses "foot-pounds" to describe soil transportation in splash erosion.

B.   Soil Movement in a Furrow

To more fully understand this erosion process it is necessary to know what is occurring along the entire length of the furrow instead of merely obtaining a measurement at the lower end of the field. The end of the run is, incidentally, the point where both the runoff and soil loss are less than anywhere else along the furrow. This is contrasted to rainfall conditions where the soil loss and runoff are usually greatest at the bottom of the plot and decrease as the measuring point is moved uphill.

Table 48–1 shows some of the characteristics of the flow and silt loads along an irrigation furrow 274 m (900 ft) long as reported by Mech (1949).

Because serious difficulties are encountered in measuring the silt content of water at different points along the stream without disturbing the flow pattern and thus affecting the rate of erosion, measurements in this study were made at points 91, 183, and 274 (300, 600, and 900 ft) from the top of the separate plots. Each of the three plots in a test received identical application rates in furrows with a grade of 2% in the channel. For the first test 26.6 liters (7.03 gal)/min were applied to each furrow. Since this application took over 11 hours to reach the lower end of the furrow, the flow for the second test in a separate set of furrows was increased to 30.6 liters (8.08 gal)/min. This increased flow required 7 hours and 16 min to travel the 274 m (900 ft). This too was longer than that usually recommended for efficient irrigation.

Table 48–1. Characteristics of flow and silt load along an irrigation furrow

Distance from upper end		Flow per furrow per minute		Soil loss per furrow		Runoff	Travel time	
							From point of application	For 91-m (300-ft) distance
m	ft	liters	gal	kg	lb	%	hr-min	hr-min
				Test no. 1				
0	0	26.6	7.03	0	0		0	
91	300	17.0	4.49	43.3	116	61	0-48	0-48
183	600	7.3	1.94	4.8	13	21	3-31	2-43
274	900	2.5	0.67	0.4	1	2	11-22	7-51
				Test no. 2				
0	0	30.6	8.08	0	0		0	
91	300	20.7	5.46	51.1	137	66	0-24	0-24
183	600	11.9	3.14	14.2	38	35	1-38	1-14
274	900	5.4	1.42	0.7	2	8	7-16	5-38

The results support the maxim that the greatest erosion takes place where the flow is greatest. In furrow irrigation, however, the greatest flow is at the upper end of the run, and it is there that the greatest erosion takes place.

The results presented in Table 48–1 are for irrigation flows adjusted much more precisely than practicable; and considering the 274-m (900-ft) run as a whole, the runoff and soil losses as measured at the end are negligible in both tests. In actual field practice, the flow and the resulting erosion would be considerably greater. Nevertheless, these data show that the upper portions of a run are subjected to severe runoff and erosion. The first 91-m (300 ft) test section, e.g., had an average runoff of 61% and 66% and a soil loss of 43.3 and 51.1 kg (116 and 137 lb)/furrow, respectively.

The middle third receives a milder treatment, but here too the sediment load and runoff are considerably higher than those at the end of the 274-m (900-ft) length. The runoff from the first 91-m (300-ft) increment became the application to the rest of the run. The sediment load in this flow, 43.3 and 51.1 kg (116 and 137 lb)/furrow, was delivered to the upper end of the 91- and 183-m (300- to 600-ft) increment while that leaving the lower end was 4.8 and 14.2 kg, respectively. This means that 38.5 and 36.9 kg (103 and 99 lb) of sediment were deposited in the middle third of furrow length during the tests.

The 183- to 274-m (600- to 900-ft) increment was subjected to a very small flow resulting in very little soil movement. This decreased soil movement was obtained at considerable reduction in irrigation efficiency because the low rate of advance of the irrigating stream resulted in a wide difference in the duration of wetting at the top and bottom ends of these increments.

It is evident from the above data that erosion measurements at the bottom end of the plot furnish practically no information on what occurs up the slope. These data also suggest that soil movement along a furrow consists of removal of soil, i.e., "picking up a sediment load," near the upper end followed by gradual deposition as the magnitude of the furrow flow is decreased by infiltration into the soil.

The above tests furnish additional information. Separately, the three length increments represent the erosion and runoff picture for a 91-m (300-ft) run irrigation with (i) an excessively large stream, (ii) a medium stream, and (iii) a stream adjusted very close to the intake rate of the furrow. In addition, combining the first two increments provides information on a 183-m (600-ft) run irrigated with a runoff of 21% and 35%. Combining the two lower sections shows what is happening on the same furrow length irrigated with negligible runoff.

C.   The Role of Infiltration

One of the significant differences between erosion by rainfall and that by furrow irrigation is the influence of infiltration on surface flow. Under rainfall increased infiltration is always accompanied by a corresponding decrease in rate of runoff. For effective and efficient irrigation, however, water must be delivered to the furrow at a rate high enough to satisfy its total intake demand, and increased infiltration will require a correspondingly greater demand along the entire furrow.

Each increment of furrow functions not only as an absorbing surface for adding water to the soil, but also as a channel for conducting the water required to irrigate the remainder of the run. For example, the first meter of a 100-m run not only absorbs the water required to irrigate this increment of furrow length, but also acts as a channel for conducting the flow required to irrigate the remaining 99 m. Similarly, the ninety-first meter must conduct enough water to supply the remaining 9 m of furrow in addition to absorbing its share. Infiltration demand, therefore, determines the minimum flow needed to irrigate the entire furrow.

Practices such as rotation and incorporation of organic matter generally increase infiltration. Because under rainfall conditions they decrease runoff, they decrease erosion. Under furrow irrigation, increased infiltration necessitates a greater furrow flow, and thus increases the erosion hazard. If decreased flow reduces erosion, conversely, increasing the flow will increase erosion regardless of whether the flow is increased by accident or design. It may be concluded therefore, that, under furrow irrigation, practices which increase infiltration increase the erosion hazard.

Whether the other changes that accompany changes in infiltration have any effect on erodibility is a question for which no satisfactory answer is currently available. Even if the accompanying changes decrease erodibility, there will still be the question of whether this decrease is enough to offset the increased erosion caused by the greater flow demand. Much has been reported showing that rotations, contouring, organic matter, etc., decrease erosion, but most of such measurements were comparisons under equal precipitation or water applications and were generally associated with a decrease in the rate of runoff. It is true that such treatments cause less erosion under the same rainstorm or rate of rainfall, but these tests provide no information on what the erosion would have been under identical runoff flow—a much truer test of erodibility per se.

D. The Influence of Flow Rate

It seems reasonable to assume that a given flow will produce a certain amount of erosion at a given point regardless if this flow is collected in a tank at the point of measurement or is used to irrigate additional land farther down the slope. It is assumed also that erosion caused by a given flow is independent of the position along the furrow, whether the 30.3-liter application, previously discussed, is used to irrigate a 274-m run or is all absorbed in a shorter run with greater infiltration, the stream would still carry 51.1 kg of sediment at the point where the flow is 20.7 liters. The influence of length-of-run on erosion and silt load can, therefore, be resolved into the influence of different stream sizes— assuming other factors remain constant.

Figure 48–3 shows the relationship between the season's average flow and total erosion for rowcrops and alfalfa (*Medicago sativa* L.) as reported by Mech (1949). It shows only the ability of different rates of flow to pick up and carry soil under the existing crop and furrow conditions. It is not a comparison of erosion on rowcrops and that on alfalfa. Nor does it compare erosion from a field on a 2% slope with one on a 7% slope. Gardner et al. (1946) present additional data on the rate that soil is worn down with slope as the independent variable and stream size the parameter.

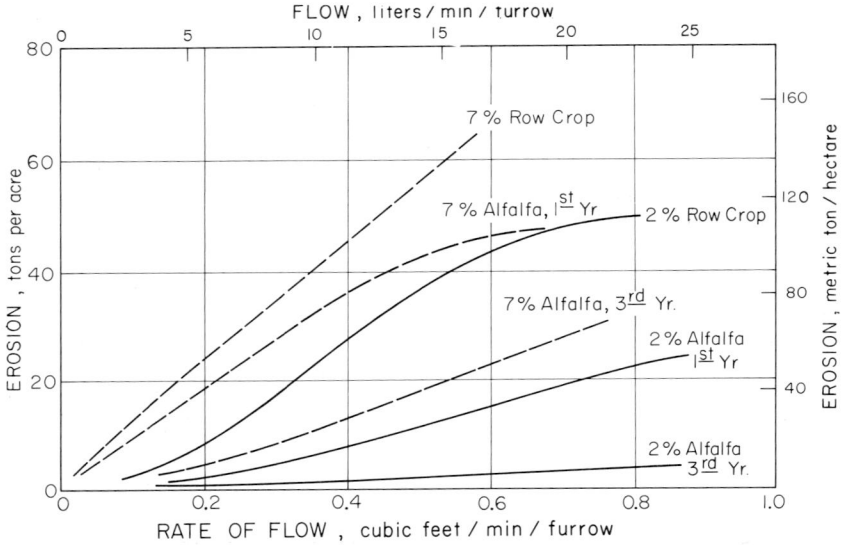

Fig. 48–3. Relationship of runoff rate and erosion. Erosion is the total; flow is the average for the irrigation season (from Mech, 1949).

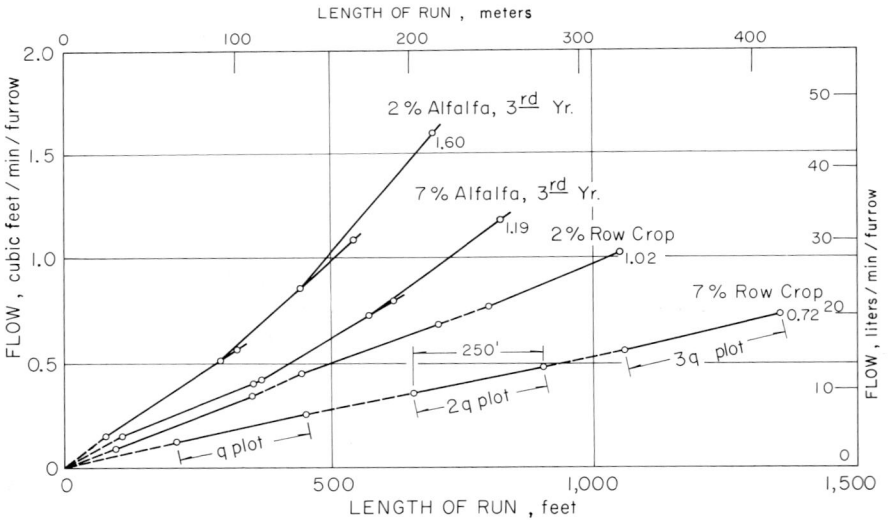

Fig. 48–4. Influence of crop and furrow grade on size of stream along the irrigation channel (from Mech, 1949).

To compare erosion between fields or crops it is necessary to find out first what the stream flows are under the conditions in question and then apply the erosion rates applicable to these flows and conditions.

The determination of stream sizes and subsequent erosion comparison must be made for similar runoff flow rates, not similar per cent runoff. In practice, the irrigator tries to set the water to each furrow to produce a certain flow of runoff

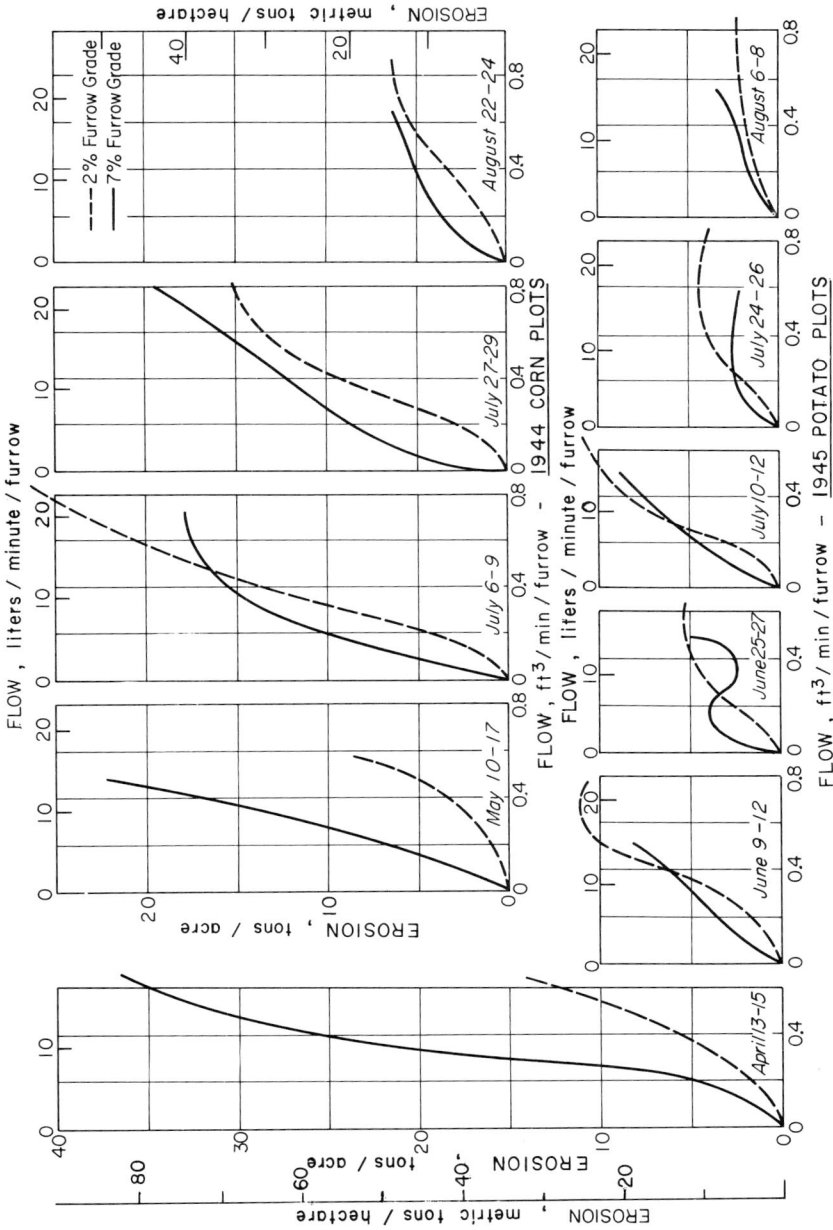

Fig. 48–5. Relationship between erosion per 250-ft plot and rate of flow. August 22-24 irrigation on corn and those on potatoes on June 25-27, July 24-26, and August 6-8 were made in furrows not disturbed after the previous irrigation (from Mech, 1949).

or tailwater from the lower end. He strives to keep this tailwater uniform in all furrows. In general, this tailwater flow will be about the same regardless of whether the field has a 2% slope or a 20% slope; whether the infiltration is 3 cm/hour or 0.3 cm/hour. An evaluation of the above is presented by Mech (1949).

Figure 48–4 shows the flow for longer slopes synthesized from data from 76-m (250-ft) experimental plots. It shows, e.g., that an application of 20 liters (0.72 ft^3)/min per furrow was enough to irrigate 411 m (1,350 ft) of rowcrops on the 7% slope but that, on a 2% slope, an application of 29 liters (1.02 ft^3)/min would irrigate only 320 m (1,050 ft). Looking at it another way, an application of 14 liters (0.5 ft^3)/min per furrow on a 7% grade will irrigate 290 m (950 ft) of rowcrops or about 91 m (300 ft) of well-established alfalfa on a 2% grade.

Figure 48–5 shows the relationship between erosion and rate of flow for individual irrigations during the season. Irrigations preceded by cultivation produced heavy soil losses, whereas similar flows in furrows not disturbed since the previous irrigation produced only a small amount of erosion. In some instances the erosion was greater from the 2% furrow grade. This is attributed not to any change in inherent transporting power of the streams on the two grades, but to the differences in the availability of soil in a condition susceptible to erosion. The low erosion rates occurring in the uncultivated furrows point out the importance of keeping cultivation to a minimum.

E. Erosion vs. Duration of Irrigation

The fact that erosion takes place early in irrigation has been reported by many. Tovey et al. (1962) and Evans and Jensen (1952) presented time rate of erosion for different flows and different furrow slopes. Mech (1949) presented data showing that most of the erosion takes place within 3 to 4 hours after the flow past a given point begins. One corn (*Zea mays* L.) plot which had a total soil loss of 50.9 metric tons/ha (22.7 tons/acre) during a 24-hour irrigation lost 39.9 metric tons/ha (17.8 tons/acre) during the first 30 min of flow. The entire loss occurred within 4 hours, and irrigation after the fourth hour added nothing to the total erosion, though the furrow flow continued at a slowly increasing rate because of decreasing intake.

F. The Influence of Furrow Shape

The influence of furrow shape has been the subject of considerable interest. Israelsen et al. (1946) reported that rectangular channels with flat furrow beds on slopes of 3% or less permitted less erosion than V-shaped ones. But on steeper slopes the stream developed its own special channel, effectively eliminating any influence of the original furrow shape.

G. Influence of Soil Water

The amount of water in the soil at the time it is exposed to erosion-producing stresses had a decided effect on the soil strength or erosion potential. Trask and

Close (1958) have demonstrated that a soil mass offers decreasing resistance to erosion as the intergranular bond is weakened by increasing soil water content. Hough (1957) has shown the relationship between water content and shearing strength.

Flaxman (1962) states, "Because most soils have less strength when saturated . . . erosion may be resisted by soils of low cohesion if low permeability retards water intake. Soils with high intake rate and higher permeability are least resistant to overland flow . . ."

Flaxman further states, ". . . Saturation must in general be preliminary to loss of erosion resistance. A positive correlation exists between unconfined compressive strength and resistance of soils to erosion in a channel flow."

The influence of flow depth, weight of overlaying water, and slope of the energy gradient, generally called "tractive force," is widely used as a measure of the hydraulic force required to initiate transport of discrete particles. This was evaluated by Smerdon and Beasley (1959) and Dunn (1959).

H. Other Factors Affecting Erosion

Erosion is the result of many interrelated forces. Some are natural and cannot be modified by man; some are man-induced. Because infiltration is a basic consideration in determining the size of the irrigating stream, it is a very important component in the erosion problem. Mech (1960) found that management of crops, soil, and irrigation has a pronounced effect on the rate at which soil absorbs water. Factors that affect furrow intake rate including crop, tillage, stream size, furrow grade, and soil water are discussed in chapter 46. Tovey et al. (1962) reported, "Erosion is extremely critical on irrigated lands in excess of 1% slope. Influencing erosion are the degree of slope, size of furrow stream, infiltration rate, water content of the soil, furrow shape, furrow roughness, size of soil particles, and some other factors."

III. AN EROSION EQUATION

It is often desirable to reduce the variables in a problem to an equation form. The factors are thus identified and the magnitude of their influence is specified. The empirical erosion equation reported by Smith and Wischmeier (1962) and Wischmeier and Smith (1962) for predicting rainfall erosion losses has proved very effective for rainfall conditions.

The rainfall erosion equation is as follows:

$$A = RKLSCP$$

where A is the computed average annual soil loss in tons per acre; R is the average annual erosion potential of the rainstorm; K is the soil erodibility factor; L and S are functions of length and steepness of slope; C is the function of cover, crop sequence, productivity, tillage, residue management and stage of crop growth; and P modifies the soil loss estimate for effects of erosion control practices such as contouring, strip cropping, or terracing.

But as discussed in the preceding pages, the difference between rainfall erosion and that under irrigation precludes the application of this equation in its present form to irrigated conditions.

A suitable equation for furrow erosion might be developed by (i) Modifying the factors and factor values in the rainfall equation to fit irrigation conditions, (ii) adapting channel stability and sediment transport concepts, and (iii) developing a new furrow irrigation erosion equation. Gardner and Lauritzen (1946) suggested a framework for such an equation.

Under irrigation, the rainfall factor R would not apply since there is no rainfall. The erosive force is the furrow flow. The development of indices for the erosion potential of different streams will be necessary. Considerable information on this phenomenon already exists in related fields.

Considerable information on scour and sedimentation with large flows is also available. This is similar to the soil factor K. Establishing K values would involve relating soil erodibility to flow and soil water conditions associated with furrow irrigation. Flaxman's (1962) suggestion of soil strength as a measure of erodibility may be useful here.

The length of slope factor L would be resolved into stream flow. The steepness of slope factor S should be determined as slope, not on the field but in the irrigation furrow. This would essentially eliminate the land slope factor and replace it with furrow grade. Whether this grade is the natural slope of the land or is made by contouring would be immaterial.

The crop management factor C is a most important one. Management often determines the compaction, detachment, intake rate, permeability, and other characteristics existing in the furrow at the time the water is applied.

The P factor would not apply because terracing and contouring effects would be reflected in the furrow grade, factor S.

Channel stability equations used in sedimentation studies provide a means of determining conditions under which scour occurs. Two common approaches to this that have possibilities for application to furrow erosion are: (i) permissible velocity concept, which includes work by Fortier and Scobey (1926) and Ree and Palmer (1949), and (ii) the tractive force concept which has been discussed by Lane (1955), Schroeder and Hansen (1953), Chow (1959), Moore and Masch (1962), and Smerdon and Beasley (1959).

Another phase of sedimentation that could have application is that of sediment transport. Articles in the US Department of Agriculture Proceedings Inter-Agency Sedimentation Conference (1965) by Vanoni, Sheppard, Brooks, and others are especially applicable.

In the development of an equation for furrow erosion, consideration must be given to (i) The capacity of a stream of clear water to initiate erosion under different channel conditions, (ii) the transporting capacity of streams, and (iii) the rate of silt load decrease associated with the decreased stream flow caused by infiltration—as affected by the availability of soil particles and the magnitude of forces involved at different time increments during the stream flow period.

Admittedly there is a need for an irrigation erosion equation, but as Ihde (1948) stated ". . . The primary factor in bringing about scientific discovery is not necessity or individual genius, but the relentless pressure of accumulating knowledge. Seldom is it possible to foresee all those factors that are essential to the solution of a certain problem until they have been accumulated and become a part of man's scientific heritage. Once the accumulation is complete, the next step becomes inevitable."

IV. PRACTICAL EROSION CONTROL

Mech (1959) presented a number of practical measures for reducing erosion on irrigated land. They are as follows:

1) A field 200 m long will have less erosion when irrigated in two 100-m runs than when irrigated as one 200-m run. In addition, shortening the length of run requires smaller flows which makes it easier to control water application. However, economical use of labor and machinery must also be considered in designing the length of runs.

2) Erosion is greater on steep land than on flat. Under furrow irrigation, steep land can be irrigated more like flat land if the irrigation furrows are constructed across the slope or on the contour. Actually, the irrigation furrow on the contour must have some slope to irrigate effectively. This slope may range from almost nothing to as much as 2 or 3%, depending on the texture of the soil and the steepness of the land. Steeper land may require a steeper grade in the furrow to minimize overtopping of the ridges.

3) Irrigating on the contour not only decreases the amount of erosion but usually improves irrigation. The water infiltrates faster and tends to move farther out from the furrows.

4) Contour irrigation is well suited for crops that require considerable ridging or hilling such as corn, potatoes, and many perennials. The ridge confines the water and reduces the danger of overtopping.

5) Applying water in excess of what the furrow can absorb is wasteful of both water and soil. Water running from the end of the furrow is of no benefit to the crop. It merely carries soil from the field and aggravates silting and drainage problems.

6) There is an optimum stream for a given furrow, soil, and crop condition. For effective erosion control and efficient irrigation, positive control devices that meter the desired amount of water from the pipeline, flume, or ditch into each furrow are practical necessities. Valves, gates, spiles, and other similar devices not only permit repeating desirable settings but permit small adjustments in flows. These devices also assure that the setting made for the desirable flow will remain unchanged until it is reset.

7) Sometimes, in starting irrigation, a greater initial flow is desirable to get the water through to the end of the furrow. Because erosion occurs early in irrigation, it is equally important that this initial greater flow is not excessive and that it be cut back as soon as possible.

8) On most cultivated land, the smallest stream that will irrigate to the end of a furrow will add just about as much water to the soil as will a stream many times as large. If more water is needed, the number of furrows or the duration of irrigation can be increased. Either method is better than increasing the size of the stream.

9) Some erosion is inevitable during the first irrigation after seedbed preparation because tillage operations loosen the soil. Small flows to settle and consolidate

loose soil are used where erosion is especially serious. Subsequent irrigations can be applied with reduced erosion.

10) It is more efficient to irrigate thoroughly and less often. Each irrigation adds to the total soil loss. Fewer irrigations will reduce erosion as well as irrigation labor.

11) Alternate furrow irrigation is another means of reducing labor and minimizing erosion. Odd-numbered furrows are irrigated one time and even-numbered ones are irrigated the next time. Thus, in a field that receives eight irrigations, each furrow will receive water four times. This method requires that each irrigation be about twice as long as in the case where every furrow is irrigated. This method is limited to soils that permit adequate lateral movement of irrigation water.

12) Tilling the soil contributes to erosion. For example, first-year alfalfa on Sagemoor fine sandy loam soil on a 7% slope irrigated immediately after reditching lost 178 kg of soil/furrow. The next irrigation, made with no intervening cultivation but using approximately the same flow, lost only 19 kg. Third-year alfalfa, on the same slope, lost 93 and 9 kg, respectively, for disturbed and undisturbed furrows. Similar results were obtained for rowcrops. When the field was in corn, each furrow lost 168 kg of soil during the irrigation after cultivation. But during the next irrigation with no intervening tillage, the loss was only 29 kg.

LITERATURE CITED

Borst, H. L., and R. Woodburn. 1942. The effect of mulching and methods of cultivation on runoff and erosion from Muskingum silt loam. Agr. Eng. 23:12–22, 24.

Chow, Ven Te. 1959. Open channel hydraulics. McGraw-Hill Eng. Ser., New York. 680 p.

Dunn, I. S. 1959. Tractive resistance of cohesive channels. Amer. Soc. Civ. Eng. Proc. J. Soil Mech. Found. Div. 85(SM3):1–24.

Ellison, W. D. 1947. Soil erosion studies: Part V. Soil transportation and the splash process. Agr. Eng. 28:349–353, 355.

Evans, N. A., and M. E. Jensen. 1952. Erosion under furrow irrigation. North Dakota Agr. Exp. Sta. Bimonthly Bull. 15:7–13.

Flaxman, E. H. 1962. A method of determining the erosion potential of cohesive soils. Symp. Soil Erosion. (Bari, Italy) Int. Ass. Sci. Hydrol., Comm. of Land Erosion 59:114–123.

Fortier, S. and F. C. Scobey. 1926. Permissible canal velocities. Amer. Soc. Civ. Eng., Trans. 89:940–956.

Gardner, W., and C. W. Lauritzen. 1946. Erosion as a function of the size of the irrigating stream in the slope of the eroding surface. Soil Sci. 62:233–242.

Gardner, W., J. H. Gardner, and C. W. Lauritzen. 1946. Rainfall and irrigation as related to rainfall erosion. Utah Agr. Exp. Sta. Bull. 326. 12 p.

Hough, B. K. 1957. Basic soils engineering. Roland Press Co., New York. 158 p.

Ihde, A. J. Dec. 1948. The inevitability of scientific discovery. Sci. Mon. (Address presented at Annual Meeting Wisconsin Academy of Sciences, Arts, and Letters.) p. 427–429.

Israelsen, O. W., G. D. Clyde, and C. W. Lauritzen. 1946. Soil erosion in small irrigation furrows. Utah Agr. Exp. Sta. Bull. 320. 39 p.

Lane, B. W. 1955. Design of stable channels. Amer. Soc. Civ. Eng., Trans. 120:1234–1279.

Mech, S. J. 1949. Effect of slope and length of run of erosion under irrigation. Agr. Eng. 30:379–383, 389.

Mech, S. J. 1959. Soil erosion and its control under furrow irrigation in arid west. Agr. Res. Serv. US Dep. Agr. Inform. Bull. 184. 6 p.

Mech, S. J., 1960. Soil management as related to irrigation practices and irrigation design. Int. Congr. Soil Sci., Trans. 7th (Madison, Wis., USA) 1:645–650.

Mech, S. J., and G. R. Free. 1942. Movement of soil during tillage operations. Agr. Eng. 30:379–383, 389.

Moore, W. L., and F. D. Masch, Jr. 1962. Experiments on scour resistance of cohesive sediments. J. Geophys. Res. 67:1437–1447.

Ree, W. O., and V. J. Palmer. 1949. Flow of water in channels protected by vegetative linings. US Dep. Agr. Soil Conserv. Serv. Tech. Bull. 967. 115 p.

Robins, J. S., and E. L. Neff. 1963. Principles of the erosion process. Proc. Forest Watershed Manage. Symp. Amer. Soc. Forest. and Oregon State Univ. 13 p.

Schroeder, K. B., and O. C. Hansen. 1953. Interim report on channel stability of natural and artificial drainageways in Republican, Loup, and Little Sioux River areas, Nebraska and Iowa. US Bur. Reclam. Hydrol. Br., Denver, Colorado. 32 p.

Smerdon, E. T., and R. P. Beasley. 1959. The tractive force theory applied to stability of open channels in cohesive soils. Missouri Agr. Exp. Sta. Bull. 715. 36 p.

Smith, D. D., and W. H. Wischmeier. 1962. Rainfall erosion. Advance. Agron. 14:109–148.

Taylor, C. A. 1935. Orchard tillage under straight furrow irrigation. Agr. Eng. 16:99–102.

Taylor, C. A. 1940. Transportation of soil in irrigation furrows. Agr. Eng. 21:307–309.

Tovey, R., V. I. Meyers, and J. W. Martin. 1962. Furrow erosion on steep irrigated lands Idaho Agr. Exp. Sta. Bull. 53. 20 p.

Trask, P. D., and J. E. H. Close. 1958. Effect of clay content on strength of soils. Coastal Eng., Proc. 6th Conf. p. 827–843.

US Department of Agriculture. 1965. Proc. Inter-agency Sedimentation Conf., Jackson, Mississippi. US Dep. Agr. Misc. Publ. 970. p. 8–13, 272–287, 229–237, 320–330.

Wischmeier, W. H., and D. D. Smith. 1962. Soil-loss estimation as a tool in soil and water management planning. Symp. Soil Erosion (Bari, Italy) Int. Ass. Sci. Hydrol. Comm. Land Erosion. 59:148–159.

49

Wind Erosion on Irrigated Lands[1]

STEPHEN J. MECH *(deceased, July 1968)*
Agricultural Research Service, USDA
Prosser, Washington

NEIL P. WOODRUFF
Agricultural Research Service, USDA
Manhattan, Kansas

I. GENERAL

It is ironical that wind erosion is a hazard on irrigated land. Wind erosion and dust storms are usually associated with arid or semiarid farming areas suffering from drouth. Yet, soil erosion by wind is a problem in many irrigated areas in spite of an abundance of water and generally high soil water content. In fact, occurrences such as that shown in Fig. 49–1 prompted one of the USA national farm magazines to call the Columbia Basin Irrigation Project in central Washington, USA the "Little Dust Bowl." Woodruff (1964) noted that the Pacific Southwest, the Colorado Basin, the muck and sandy soils around the Great Lakes, the Gulf of Mexico, and the Atlantic Seaboard are other regions in the USA where wind erosion is a hazard. Wind erosion is also a problem on irrigated lands in other parts of the world.

An understanding of the mechanics of wind erosion processes and the factors influencing the rate of soil erosion by wind is a prerequisite to the development of suitable methods for wind erosion control on irrigated as well as dryland areas.

The wind erosion problem under irrigation differs from that on nonirrigated lands, and few references on the subject occur in the literature. Fortunately, there is sufficient similarity so much of the basic information obtained for non-irrigated conditions can be applied directly to irrigated lands.

Erosion by wind may occur whenever a loose, dry, and pulverized soil has a surface that is smooth and unprotected by vegetative cover, and the field is sufficiently large and the wind sufficiently strong. Woodruff (1964) and Zingg (1953a) concluded that wind erosion is generally most serious in those areas where the climate is characterized by low and variable precipitation, high frequency of drouth, high temperatures and evaporation rates, and variable high wind velocities. Irrigated lands usually have the above climatic characteristics. In addition, soils are often sandy and low in organic matter, with prolonged periods when the surface is not protected by vegetation. All these combine to make wind erosion an inherent hazard on many irrigated lands. Chepil et al.

[1] Contribution from the Soil and Water Conserv. Res. Div., ARS, USDA, in cooperation with the Washington and Kansas Agri. Exp. Sta. Grateful acknowledgment is made to the late W. S. Chepil for his participation in the early planning of this chapter.

(1962) have evaluated the wind and some other climate factors associated with erosion.

Irrigated lands produce a large variety of crops. Many are tender and succulent, often harvested in their entirety before full maturity, leaving very little or no vegetative residue. The result is that the fields may be bare from harvest until the next crop is established.

Though the erosion hazard is greatest in the spring when the tillage is most intense and the plants are most vulnerable, and in late winter when vegetation is absent or dormant, serious erosion may occur at any time of the year if conditions are right. The wide variety of crops, a large percentage of which are row-crops, can create vulnerable conditions at any time of the year.

II. NATURE OF PARTICLE MOVEMENT

Soil movement is initiated when the forces of lift and drag exerted by the wind against the surface of the ground overcome the force of gravity of the individual soil particles (Chepil, 1960). If the particles are sufficiently large or are attached to others to form sufficiently large aggregates or clods, they can effectively resist the force of the wind. However, if they are not heavy enough, the wind may lift them from or roll them along the surface and thus initiate soil movement.

Soil movement is therefore dependent on the size and weight of detachable soil particles and the turbulent forces exerted on them. The threshold or eroding velocity for soils varies from 20.9 km (13 miles)/hour to an indefinite limit depending on previous history of the soil surface; but a bare, previously eroded soil surface usually starts eroding when the wind reaches 20.9 to 24.1 km (13 to 15 miles)/hour at the 30.5-cm (1-ft) height (Chepil and Milne, 1939; Chepil, 1945a).

A velocity gradient exists whenever wind blows over a surface. The velocity is lowest near the ground surface and increases as the logarithm of the height above the surface. It is this turbulent flow which produces the forces that cause soil movement (Food Agr. Organ., 1960).

Fig. 49–1. Eroded fields, drifted roads, and filled irrigation ditches such as this occur when irrigated land is left unprotected.

The movement of the soil particles by wind has been described by Bagnold (1943) as taking place in three forms: surface creep, saltation, and suspension. Saltation, the dominant type of particle movement, is the bouncing or jumping of the soil particles. Surface creep is the rolling or sliding of larger particles along the surface. Suspension represents the floating of small-sized particles in the airstream. The size of the particles involved in these processes is greatest for surface creep, and only the finest are moved in suspension. Bisal and Nielsen (1962) conclude that the particles in saltation range from 0.1 to 0.55 mm. Udden (1894) suggests that surface creep involves particles 0.5 to 1.0 mm in diameter.

Once soil grains are loosened and movement is initiated, the impact of particles in saltation breaks down the clods, destroys soil crusts, wears down vegetative residue, destroys living plant tissues—and otherwise abrades the surface. Thus saltating grains not only accelerate movement of other erodible particles, but also break clods into erodible sizes and reduce the effectiveness of protective vegetative cover.

Chepil (1960) said that over 90% of the soil movement in saltation occurs below the height of 30.5 cm (1 ft) and that the average height of a jump is about 10.16 cm (4 inches). Finer and lighter particles tend to move faster and farther than the coarser and denser ones.

Bisal and Nielsen (1962) suggest that movement of particles in saltation is initiated by the impulsive forces of instantaneous differences in air pressure near the ground and that particles need not roll on the surface to gather energy sufficient to translate them into an upward direction. This is contrary to the commonly accepted view expressed by Chepil (1945b) and Zingg (1953b) that particles roll along the ground and gather sufficient momentum which causes them to bounce and become airborne.

III. THE ROLE OF VEGETATION AND ORGANIC MATTER

A good cover of growing plants or vegetative residue will protect land from wind erosion better than any other practice. Such cover acts as a barrier between the soil and the wind and effectively reduces the force of the wind on the soil surface.

Destruction of vegetative cover is the most common cause of wind erosion. Drouths have at times reduced or stopped vegetative growth, but drouth alone is seldom the cause of severe wind erosion. Even dead roots in an unbroken surface tend to bind the surface particles together and provide a measure of protection. A rupture in an otherwise adequate protective layer often is the location of incipient erosion (Fig. 49–2).

High organic matter content in soil is usually associated with high fertility and good tilth. Literature reveals that numerous cementing substances produced by soil microorganisms as they attack vegetative matter bind the soil particles to form aggregates. However, observations by Hopkins (1935) and Hopkins et al. (1946) indicate that organic matter may facilitate erosion by wind. Later experiments by the Canada Department of Agriculture (1949) verify these observations and show that while decomposing wheatstraw (*Triticum vulgare*) increases soil cloddiness and decreases erodibility, the reverse is true after the straw is decomposed. Chepil (1955) concludes that the aggregating effect during decomposition

Fig. 49–2. Wind erosion often starts as a rupture in an otherwise protected surface. The erosion started in the gashes made during the building of border dikes.

is due to the products of the decomposition and not particularly to any binding action of the vegetative fibers.

Chepil (1955) reports that the soil aggregation produced during the decomposition of vegetative matter persists from 2 to 5 years. He further states that, from the wind erosion control standpoint, the benefits obtained from the primary products of decomposition are small compared to the detrimental secondary effects of decomposition. It is concluded, therefore, that far greater protection from wind erosion will be obtained by keeping the vegetative matter anchored on top of the ground instead of mixing it into the soil (Fig. 49–3).

Fig. 49–3. Organic matter plowed under provides almost no protection against wind erosion. Note the loose sandy surfaces following the plowing under of a 91.44-cm (3-ft) high winter cover crop of rye and vetch (*Vicia villosa*, Roth). Irrigation and subsequent planting converted the surface to that shown in Fig. 49–5.

The favorable growing conditions under irrigation stimulates growth of both the desirable crop and the volunteer growth associated with crop residues. Such volunteer growth often creates conditions that are incompatible with the main crop. Some low-growing crops such as beans (*Phaseolus vulgaris*) have been abandoned because the volunteer growth within the row made harvesting impossible. This is less of a problem in taller crops such as corn (*Zea mays* L.). Residues are very effective and are recommended for erosion control where wind erosion is critical and measures less objectionable than residues are inadequate. Mech (1962) concludes that under irrigation such special residue treatment should be confined to the generally more severe but smaller problem areas.

Where the use of residues is not objectionable, a tough, fine-stemmed residue securely anchored to the soil surface will provide excellent control. Residue should be evaluated on the basis of dry weight. Because of its fineness, a unit weight of straw provides considerably more protection than an equal weight of coarser residue such as sorghum (*Sorghum vulgare*) or corn stubble.

IV. THE EFFECT OF WATER

Timely applications of irrigations for wind erosion control make possible the growth of vegetation and create a moist condition amenable to the development of a cloddy surface by tillage. Chepil (1956) and Belly (1964) point out moist soil has a threshold or eroding wind velocity considerably higher than that for dry soil. All natural soil materials, even dune sands, exhibit some degree of cementation after they are wetted and dried (Fig. 49–4). The cohesive force of water films enveloping the soil particles holds them in fragile larger units. Under a slight amount of compaction as from ordinary tillage at optimum water, this cohesion is further increased. Compaction is a form of "puddling" and is, in general, undesirable. It is, however, a much smaller hazard on the coarser tex-

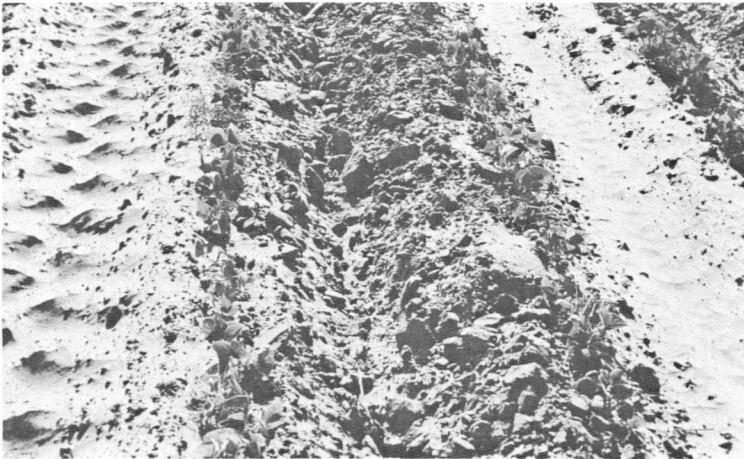

Fig. 49–4. This rough, cloddy surface was developed by wet tillage of loose sandy soil, and conversely, this rough, cloddy dry surface was pulverized by the action of implement wheels and tires.

tured soils usually associated with wind erosion—and a lesser evil than severe wind erosion.

The method of applying irrigation water to the soil has considerable influence on the susceptibility of the soil to wind erosion. Mech (1955) reports that the impact of falling water, as from rainfall or sprinklers, tends to break the clods into particles that erode easily (Fig. 49–5). Most clods tend to slake down when wet. In general, wetting cloddy surfaces by furrow irrigation, even without drop impact, tends to produce a smoother, slightly crusted surface over the wetted area which can blow more readily. Fortunately this wetted area is only a small portion of the field surface and even this is confined to the furrow depression.

Mech (1955) further states that under certain conditions, it is impractical and wasteful of water to irrigate often enough to prevent a finely pulverized surface soil from blowing. The depth of drying may be only a fraction of an inch and the soil below this may be wet, but if the immediate surface is dry and the wind is strong enough, the top layer can erode unless the soil particles are consolidated into clods or protected by vegetation. Under certain conditions it would require almost continuous irrigation to keep the immediate surface layer wet enough to prevent erosion.

V. WIND EROSION CONTROL BY TILLAGE

Tillage is probably the most common and deliberate erosion control practice. It can be very effective if done properly, but its protection is only temporary. Freezing and thawing, wetting and drying, and other weather elements all tend to break the aggregates into more erodible particles.

The basic element in erosion control by tillage is the creation of a rough, cloddy surface which will resist the force of the wind, decrease its velocity at the ground

Fig. 49–5. The same field shown in Fig. 49–3—after corn (*Zea mays* L.) planting. The area on the right received a 1.6 cm (0.63 inch) rain the day after planting. That on the left was planted after the rain and represents what the surface on the right looked like before the rain. Impact of raindrops has reduced the surface roughness.

Fig. 49–6. Ridge planting on the left provides a rougher and more protective surface than the flat planting on the right. The stand of beans on both plots is the same, but the plants on the left are hidden and sheltered by the rougher surface produced by "ditching out" ahead of the planter.

level, and trap moving soil. Tillage may be used as an emergency measure when severely erosive conditions are encountered and vegetative cover cannot be established. Once the erosion is stilled or reduced, plant cover must be established for longer-lived protection. Sandy soils usually found in irrigated areas are far more difficult to protect by emergency measures than fine-textured ones.

Since, under irrigation, the production of vegetative growth is practically assured, tillage is used most often during that part of the season before vegetation is well established. Basic control, however, should be obtained through the wide use of vegetation and compatible crop residues.

A few clods will be formed with most tillage tools, but some tools are better than others. Lyles and Woodruff (1962) report that plows which invert the soil layer produce a rougher, cloddier, and therefore more resistant surface than do sweeps or disks. Chepil et al. (1961) reports that listers and narrow-point chisels are generally very effective in emergency tillage operations.

Some of the recommendations for controlling wind erosion on irrigated land may be at variance with principles of good farming where wind erosion is not a problem. For example, ridge planting accompanied by furrowing out ahead of the planter may look like a very poor seedbed, but in wind erosion areas it is excellent. Mech (1962) points out that ridge planting provides furrows, should it be desirable to apply water to the germinating seeds or seedling plants (Fig. 49–6). In addition, the furrows and ridges, together with a rough cloddy surface, combine to effectively protect the young seedlings from the abrasive blowing sand. Armbrust et al. (1964) present an excellent discussion of the mechanisms by which ridging provides wind erosion protection.

Although cultipackers and rollers are used extensively in different regions for smoothing the surface, any rolling implement or wheel will tend to break down existing clods, decrease surface roughness and thus increase the erosion hazard. This is especially true if the clods are below the optimum water content for compaction (Fig. 49–4).

Farming operations that may rupture any existing protective surface should be avoided during the nongrowing season. Land preparation should wait until the growing season is near and the irrigation water is available or in the ditch. The interval between land preparation and irrigation should be as short as possible.

The direction of the rows and most of the tillage on irrigated farms is dictated by the irrigation systems adapted to the topography of the land. The field and direction of tillage should be oriented at right angles to the prevailing wind wherever possible, but only after full consideration of the problem. In the Columbia Basin of central Washington, e.g., the prevailing spring winds come from the west or southwest. These cause considerable erosion. However, during the late winter, strong, dry winds from the north frequently cause an equal amount of damage. Any orientation of tillage should consider the prevailing wind direction during the period that protection is desired.

VI. WIND EROSION CONTROL BY WINDBREAKS

Tree windbreaks or wind barriers function as do any surface roughness element in providing wind erosion control; i.e., they absorb or deflect some of the wind force and thus lower the wind velocities to the leeward. The effectiveness of any barrier depends on factors such as wind velocity and direction; shape, width, height, and porosity of the barrier; and threshold velocities of soils.

Wind tunnel tests by Woodruff and Zingg (1952) indicated that full protection from a 64.4-km (40-mile)/hour wind was provided only for a distance equal to 9 times the height of model tree barriers. Chepil (1949) reported that willow barriers form a protective influence extending only 6 or 7 heights in some of the highly erodible sandy regions of China. Field studies of various narrow tree windbreaks by Woodruff et al. (1963) showed the effective zone of protection to range from 8 to 18 times the height of the barrier for winds of 64.4 km (40 miles)/hour measured at the 15.2-m (50-ft) elevation. Iizuka (1950) has observed that a windbreak which reduced wind velocities to 61%, 69%, and 77% of that in the open at leeward distances of 10, 20, and 30 times the barrier height, respectively, decreased soil blowing at those distances to 14%, 18% and 50% of that in the open, respectively.

The relatively limited areal extent of windbreak protection means that closely spaced windbreaks would be required to provide extended protection across fields. Irrigated land is expensive and the application of irrigation water is an additional cost. For this reason, windbreaks on irrigation farms are generally limited to the vicinity of the homestead to protect the garden and buildings. In general, these windbreaks consist of 1 or 2 rows of conifers reinforced by an auxiliary row of short-lived deciduous shrubs. On lands where truck crops are grown, high picket fences or single-row deciduous tree windbreaks are sometimes used.

VII. WINTER COVER CROPS

A good winter cover crop in strips or solid seeding is almost a necessity for the protection of an otherwise bare soil during the winter and early spring. Since winter and spring are periods of highest wind velocities and lowest vegetative growth and vigor, only an adapted, well-established winter crop should be seeded. This will not only remain alive but may even grow a little during the warmer winter days and protect an otherwise susceptible soil surface under the severe conditions experienced during the "nongrowing" period. Winter wheat (*Triticum aestivum* spp. *vulgare*, Vill.) or rye (*Secale cereale*, L.) are usually satisfactory.

The winter cover crop must be winter hardy and it must be planted early enough to make a satisfactory cover before the end of the growing season. Neither oats (*Avena sativa* L.) nor sweetclover (*Melilotus alba* Desr.) is an effective winter cover crop in the Columbia Basin. Sweetclover freezes down to the crown early in the fall and provides no protection. Oats make a lush growth in the fall but winterkills, leaving practically no residue.

VIII.   THE EROSION EQUATION

It is often desirable to reduce the variables in any problem to an equation form. The factors involved are thus identified and the magnitude of their influence is specified.

An equation to indicate relationships between the amount of wind erosion and the various field and climatic factors that influence erosion has been developed by Chepil and Woodruff (1963). The equation is:

$$E = f(IKCLV).$$

It states that the average annual erosion potential, or soil loss E, expressed in tons per acre is a function of soil erodibility index I, soil ridge roughness factor K, climatic factor C, field length along the prevailing wind erosion direction L, and equivalent quantity of vegetative cover V. These five variables are obtained by grouping and converting the 11 primary variables now known to govern wind erodibility.

The equation is designed as a tool to: (i) determine the potential erosion from a particular field, and (ii) determine what field conditions are necessary to reduce potential erosion to a tolerable amount. Techniques for estimating some of the factors in the field are presented by Chepil (1959), and Chepil and Woodruff (1963).

LITERATURE CITED

Armbrust, D. V., W. S. Chepil, and F. H. Siddoway. 1964. Effect of ridges on erosion by wind. Soil Sci. Soc. Amer. Proc. 28:559–560.

Bagnold, R. A. 1943. The physics of blown sand and desert dunes. William Morrow & Co., New York. 265 p.

Belly, Pierre-Yves. 1964. Sand movement by wind. Dep. Army Corps Eng. Tech. Memo. 1. 38 p.

Bisal, F., and K. F. Nielsen. 1962. Movement of soil particles in saltation. Can. J. Sci. 42:81–86.

Canada Department of Agriculture. 1949. Soil moisture, wind erosion, and fertility of some Canadian prairie soils. Soil Res. Lab. Tech. Bull. 71. 78 p.

Chepil, W. S. 1945a. Dynamics of wind erosion: I. Nature of movement of soil by wind Soil Sci. 60:305–320.

Chepil, W. S. 1945b. Dynamics of wind erosion: II. The initiation of soil movement. Soil Sci. 60:397–411.

Chepil, W. S. 1949. Wind erosion control with shelterbelts in North China. Agron. J. 41:127–129.

Chepil, W. S. 1955. Factors that influence clod structure and erodibility of soil by wind: V. Organic matter at various stages of decomposition. Soil Sci. 80:413–421.

Chepil, W. S. 1956. Influence of moisture on erodibility of soil by wind. Soil Sci. Soc. Amer. Proc. 20:288–292.

Chepil, W. S. 1959. Wind erodibility of farm fields. J. Soil Water Conserv. 14:214–219.

Chepil, W. S. 1960. The cycle of wind erosion. Int. Congr. Soil Sci., Proc. 7th (Madison, Wis., USA) 1:225–231.

Chepil, W. S., and R. A. Milne. 1939. Comparative study of soil drifting in the field in a wind tunnel. Sci. Agr. 19:249–257.

Chepil, W. S., F. H. Siddoway, and D. V. Armbrust. 1962. Climatic factors for estimating wind erodibility on farm fields. J. Soil Water Conserv. 17:162–165.

Chepil, W. S., and N. P. Woodruff. 1963. The physics of wind erosion and its control. Advance. Agron. 15:211–302.

Chepil, W. S., N. P. Woodruff, and F. H. Siddoway. 1961. How to control soil blowing. US Dep. Agr. Farmers' Bull. 2169. 16 p.

Food and Agricultural Organization of the United Nations. 1960. Soil erosion by wind and measures for its control on agricultural lands. Food Agr. Organ. Agr. Develop. Pap. 71. 78 p.

Hopkins, E. S. 1935. Soil drifting in Canada. Int. Congr. Soil Sci., Trans. 3rd (Oxford, Great Brit.) 1:403–405.

Hopkins, E. S., A. E. Palmer, and W. S. Chepil. 1946. Soil drifting control in the prairie provinces. Canada Dep. Agr. Farmers' Bull. 32. 8 p.

Iizuka, H. 1950. Wind erosion prevention by windbreaks. Meguro Forest, Exp. Sta. (Tokyo) Bull. No. 45. p. 95–129.

Lyles, Leon, and N. P. Woodruff. 1962. How moisture and tillage affect soil cloddiness for wind erosion control. Agr. Eng. 43:150–153, 159.

Mech, Stephen J. 1955. Wind erosion control in the Columbia Basin. Washington Agr. Exp. Sta. Circ. No. 268. 5 p.

Mech, Stephen J. 1962. Wind erosion control on irrigated lands. US Dep. Agr. Leafl. No. 506. 8 p.

Udden, J. A. 1894. Erosion, transportation, and sedimentation performed by the atmosphere. J. Geol. 2:318–331.

Woodruff, N. P. 1964. Wind erosion laboratory finds answers. Soil Conserv. 39:152–154.

Woodruff, N. P., D. W. Fryrear, and Leon Lyles. 1963. Reducing wind velocity with field shelterbelts. Kansas Agr. Exp. Sta. Tech. Bull. No. 131. 26 p.

Woodruff, N P., and A. W. Zingg. 1952. Wind tunnel studies on fundamental problems related to windbreaks. US Dep. Agr.-Soil Conserv. Serv. Tech. Pap. No. 112. 25 p.

Zingg, A. W. 1953a. Speculation of climate as a factor in wind erosion problems of the Great Plains. Kansas Acad. Sci., Trans. 56:371–377.

Zingg, A. W. 1953b. Some characteristics of aeolian sand movement by saltation process. Ed. Centre Nat. Rechereche Sci., 13 Quai Anatole France, Paris (7e). p. 197–208.

50 Drainage Related to Irrigation Management

WILLIAM W. DONNAN

Agricultural Research Service, USDA
Riverside, California

CLYDE E. HOUSTON

University of California
Davis, California

The basic principles related to drainage of agricultural lands have been covered and discussed in great detail in *Drainage of Agricultural Lands* edited by Dr. James N. Luthin (no. 7 in the monograph series *AGRONOMY*) and published in 1957 by the American Society of Agronomy. The purpose of this chapter is to discuss practical consideration in the design and operation of drainage systems, particularly as related to drainage of irrigated lands in arid and semiarid regions.

I. MAGNITUDE OF THE PROBLEM

Drainage problems usually develop as a consequence of irrigation. Historic evidence of this fact can be found on every continent of the world. A major contribution to the decline and disappearance of some ancient civilizations can be attributed to their failure to heed the drainage hazard.

Even during the present century, some seemingly excellent irrigation schemes have failed or been weakened by the subsequent development of drainage problems. The threat of waterlogging and salt accumulation hangs over nearly every irrigated acre. It is this circumstance that has prompted socio-agriculturists to raise the question of whether irrigated agriculture is a permanent enterprise. The answer is that it can be made relatively permanent if proper drainage works are provided and operated properly. Israelsen and Ayazi (1957) state that irrigation and drainage are inseparable. The recognition and application of this basic fact is essential to the continued productivity of arid land soils.

A. Existing Drainage Problems

It is difficult to gauge the magnitude of existing drainage problems in irrigated areas because of the dynamic nature of their development. In some areas where irrigation water is obtained from pumped wells, a generally low groundwater table is maintained and drainage problems are minimized. A good example of this is the Salt River Valley of Arizona, USA (Marr, 1926), where 40 years ago much of the land in the valley was waterlogged. Development of pump irrigation has eliminated most of the waterlogging problem. The same situation existed in parts of the San Joaquin Valley of California, USA (Weir, 1925), where today over 600 drainage wells are used for water table control and as an irrigation supply.

In other areas where water has been imported, the natural regime is disrupted and waterlogging usually develops rapidly. An example of this is the Indus Basin of West Pakistan. Modern-day irrigation development began about 1870. A comprehensive survey of this area (Hamid, 1961) has revealed that about 100,000 acres/year are being rendered unfit for agriculture as a result of waterlogging or salinity, or both. Recent developments indicate that pump drainage may offer a partial solution to the problem.

The Imperial Valley of California, USA was developed for irrigation beginning in about 1905 when water was imported from the Colorado River. By 1920 acute drainage problems had developed. Beginning in 1928 an accelerated program of tile drainage was inaugurated to supplement a system of open drains (Donnan et al., 1954). Today, over 250,000 acres have been tile-drained and it is estimated that an additional 200,000 acres will require drainage to safeguard and maintain optimum crop production.

What is the magnitude of existing drainage problems in irrigated areas of the world? A conservative estimate based on observations would be that 150 to 200 million acres of cropland are affected to some degree. This figure is obtained by taking the world acreage of irrigated land, 300 million acres (Gulhati, 1955), and assuming that about one-half to two-thirds of this land has developed drainage problems.

B. Potential Drainage Problems

Waterlogging and salinity pose a threat to irrigated areas. Future developments tend to maximize these threats. Many irrigation enterprises have plans to expand and develop additional areas. Any expansion upslope from existing irrigated lands becomes a direct threat to the waterlogging of downslope areas. For example, the fertile and productive irrigated lands in the trough of the San Joaquin Valley of California are now threatened by upslope irrigation projects (Berry and Stetson, 1959). Some productive areas in the Yuma Valley of Arizona have been rendered unfit for agriculture by irrigation developments on the Yuma Mesa (Jacob, 1960).

An even more dangerous threat is propounded when the salt balance problem is considered. Continuous recirculation of water by pumpage from underground basins or reuse of return flows sometimes results in a gradual buildup of mineral elements in the irrigation water. The mineral content of the water supply of an entire river basin system can be materially altered by upstream development and use. This problem has already become apparent in the lower reaches of the Colorado, Rio Grande, Gila, San Joaquin, and other rivers of southwestern USA (Hill, 1961). With each increase in mineral content comes an implied increase in the leaching requirement and a corresponding increase in the drainage need.

The development of an irrigation enterprise, including impoundment of the water, its conveyance and its application, upsets the natural hydrologic cycle of an area. The recognition and solution of the resulting drainage problems requires an intensive application of scientific knowledge.

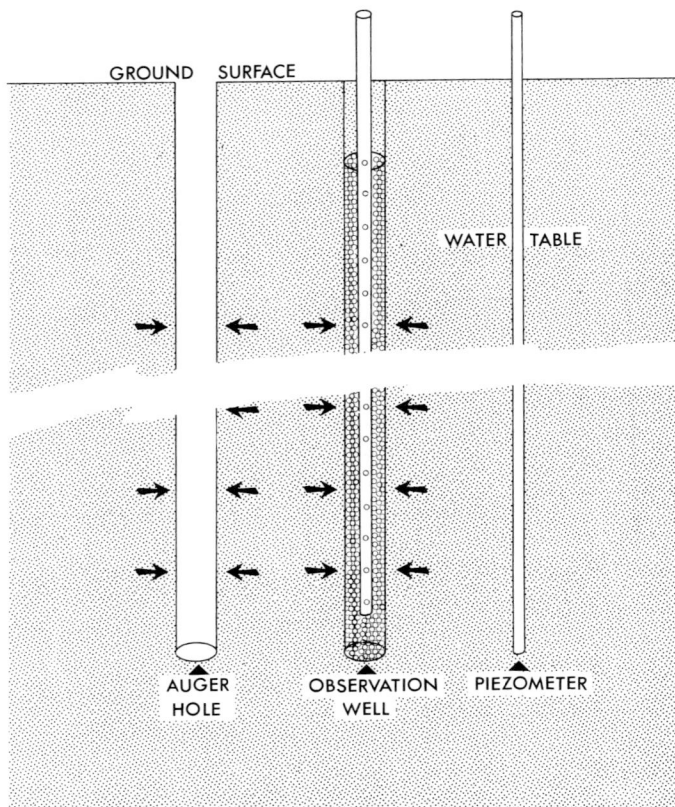

Fig. 50–1. The auger hole, observation well, and piezometer are necessary tools for studying drainage problems.

II.   DRAINAGE REQUIREMENTS

A.   Relationships to Water Supply

Why do drainage problems occur when man can regulate the application of water for irrigation? The answer is that man does not have complete control of the water used. Appreciable losses occur in bringing the water to the point of use (Houston, 1961). Also, it is extremely difficult to apply the exact amount of water needed to sustain plant growth. The natural tendency is to apply too much water rather than risk applying not enough. A third factor inherent in irrigated areas is the necessity for maintaining a desired salt balance in the root zone of the plant. This requires applications of excess water to leach harmful mineral elements (Hill, 1961).

A survey of the water table is one of the most important parts of any drainage investigation (Donnan and Bradshaw, 1952). Information obtained by soil borings and water table measurements is needed on the source of the water, its movement, quality and quantity, and the cyclic trend of the water table (Fig.

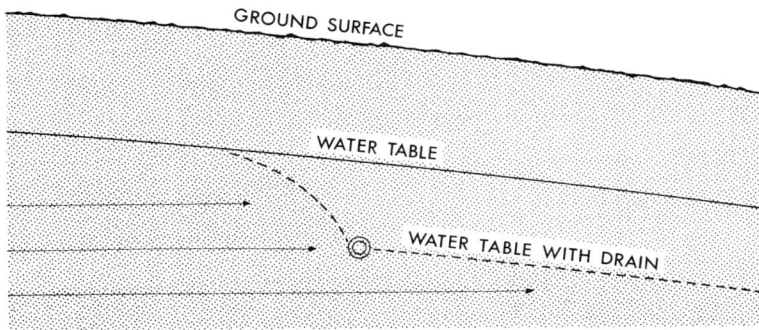

Fig. 50–2. As subsurface water moves downslope, it may be intercepted and removed by a tile drain.

50–1). If artesian pressure is discovered from deep water-bearing strata, relief wells may be needed. Where rainfall is a factor influencing the rise of the water table, surface drains may be indicated to remove excess water. If seepage from an adjacent canal or reservoir can be detected, an interceptor drain may solve the problem (Fig. 50–2). If the source of excess water is over irrigation, a tile grid system is needed. If the drainage water is of good quality, plans can be made for its reuse for irrigation purposes.

B.   Relationships to Soils

Some soils drain easily, others are extremely difficult to drain. Generally speaking, coarse-textured soils drain better than fine-textured soils. In most irrigated areas, the soils consist of stratified sands, silts, and clays. Fine-textured clay layers are often underlain or overlain by coarse-textured sands. Thus it is important to define the soil profile to locate the drainable layers. A discussion of methods for making soil profile surveys is found in Luthin (1957) beginning on p. 448.

The sequence of permeable and impermeable soils and their ability to transmit water or impede its flow determine both the type of drainage system that should be installed and its design (Fly, 1961). For example, open drains at intervals of 1 mile may be adequate drainage for coarse gravel subsoils, whereas a fine-textured clay soil to a depth of 10 feet might require drains spaced at 40 feet. Lack of drainable aquifers at the 3- to 8-foot level may make drainage by tile lines infeasible. A discussion of the various methods of measuring the permeability of soil strata is found in Luthin (1957) beginning on p. 395.

C.   Relationship to Topography

The natural topography of the land will influence the type of drain system required. Irrigation schemes are usually developed in areas characterized by broad, flat expanses of land that are devoid of natural channels and streamways. Thus most development plans must include the construction of a trunk drainage outlet system. A basin-type topography underlain by suitable aquifer material often lends itself to pumping for drainage. In general, broad, flat fields are ideal for the grid systems while benches and swales may call for interceptor or meandering lines.

D.   Relationship to Crops

Drainage requirements for shallow-rooted crops are usually different from that of deep-rooted ones. Some plants require well-drained soils; others are classed as "water-loving." Also, some crops are extremely sensitive to saline conditions while others are relatively salt tolerant (Bernstein, 1961). The type of crops to be grown and their drainage requirements are important factors influencing the prescribed drainage plan for a given area.

III.   DRAINAGE DESIGN

It is evident that there are many factors that contribute to the cause of a drainage problem and many others that will affect the ultimate solution. This is particularly applicable to drainage problems in irrigated areas. The best drainage system for a given area thus becomes a blend, or compromise, of individual dictates. The success of any prescribed drainage system depends on the quality and intensity of investigation made of causitive factors.

A.   Location Criteria

If the source of the excess water has been determined, the location of the drain device can be fixed with confidence. A leaking canal suggests an adjacent interceptor drain. Seepage down a slope would call for interception, probably above the toe of the slope. In general, tile grid systems should be oriented to the prevailing slope of the land. The main outlets are usually located at the lowest corner of the field. Pump drainage wells must tap the aquifer. If the pump wells are placed in a grid they should be located so as to maximize the drawdown (Peterson, 1961). If the water being pumped is usable, pump locations should take advantage of existing irrigation canals.

B.   Depth Criteria

Where excess mineral elements are present in the waterlogged areas, the depth to which the water table is lowered is important. In most cases the water table must be lowered to a depth of 4 to 5 feet to prevent salt injury and to provide an aerated root zone for growing plants. In tile systems, the depth of the drain is often dictated by the stratification of the soil and the location of permeable layers of soil. It may be expedient to install the drain at a shallower depth in order to place it within the most drainable soil. In an interceptor drain the depth would be dictated by the depth of the path of the migrating water (Donnan, 1959). Depth of the system is sometimes determined by the depth of outlet.

Where pump wells are used for water table control, they should be operated to maintain water table conditions similar to other types of drainage.

C. Spacing Criteria

Spacing will influence both the performance of the drain device and the economic feasibility of the system. Drain devices should be spaced to individually or collectively drain the waterlogged area and maintain the groundwater level at the desired depth. Parallel open drains or tile drains are spaced so the minimum drawdown of the groundwater at the midpoint between drains is about 4 to 5 feet from the ground surface (Donnan, 1946). This provides a root zone for plant growth. Pumped wells are spaced so their drawdown curves provide sufficient interception and the desired lowering of the water table.

Spacings are relatively wide in coarse-textured soils and narrow in fine-textured soils. Thus the hydraulic conductivity of the soil being drained becomes an important function of drain spacing. In addition, the depth at which a drain device functions becomes an important factor in spacing.

1. FORMULAS

A number of formulas have been developed for the calculation of drain-device spacing. For theories on the spacing of pumped wells Wensel (1942) has presented perhaps the best treatise on the problem. Van Schilfgaarde (1957) makes a rather complete summary of formulas used to calculate the spacing of tile lines and open drains. Many of these formulas were developed for humid area conditions but they can be adapted and applied to arid irrigated land problems.

In using these formulas, one must first determine the optimum depth at which the drains are to be installed. Then the spacing between drains will depend on: (i) the permissible water table depth between drains; (ii) the hydraulic conductivity of the soil to be drained; and (iii) the quantity of water to be drained. This last factor is influenced by quantity and distribution of precipitation or irrigation, water-holding capacity of the soil, and evapotranspiration.

IV. DRAINAGE SYSTEMS

The solution of a drainage problem may require the installation of one or several types of drain devices. These are: (i) open drains, (ii) covered drains, (iii) wells, and (iv) sumps.

A. Open Drains

The greatest advantages of open drains are the low initial cost and their ability to convey large quantities of water. The cost factor is partially offset by high maintenance costs. Open drains are used to solve many different drainage problems, but their primary use is to intercept lateral underground or surface flow and to function as a conveyance device (Fig. 50–3). In irrigated areas they should be constructed to a minimum depth of 5 feet. A shallower depth would

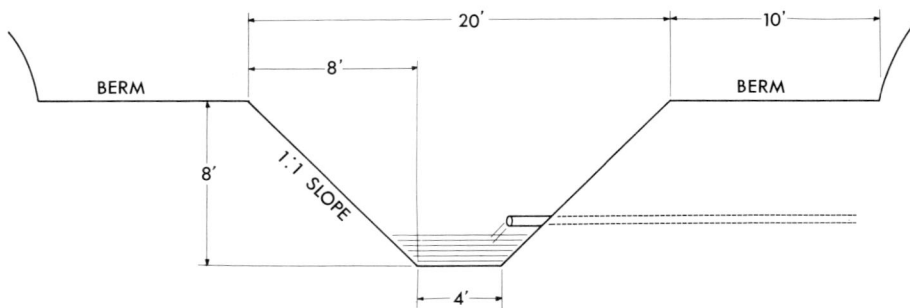

Fig. 50–3. Dimensions of open drains depend on soil characteristics, capacity desired, and depth to water to be removed.

nullify their performance as an effective groundwater drain and would preclude their use as a tile outlet. Depth of open drains in irrigated areas ranges from 5 to 10 feet for laterals and 8 to 15 feet for main drains. The size of open drain necessary to carry a given quantity of water depends on the slope or grade of the channel and, to some extent, on the shape of its cross section (Houston, 1961).

Open drains usually have a 4-foot-wide bottom width and sloping sides. The slope of the sides depends on the soil through which the drain traverses. In clays and fine-textured soils, the side slopes can be quite steep.. In coarse-textured soil, the side slopes must be moderate. The excavated soil is usually deposited on each side of the drain as a berm, but may be distributed over the field. Open drains may require a strip of land from 50- to 100-feet wide. This is a disadvantage since it removes a considerable amount of land from production. Other disadvantages of open drains are: (i) They encourage the growth of weeds, and (ii) they constitute a barrier to farm machinery and must be bridged at crossing points.

B.   Covered Drains

Covered drains take many forms and are built with various types of materials. The most common type is the tile line (Fig. 50–4). It consists of short 1- or 2-foot sections of pipe butted together to form a continuous line and laid in a narrow trench. Filter material is laid over or around the pipe and the trench is back-filled (Fig. 50–5). This forms a buried conduit. Water enters the line at the pipe joints or other openings. The entire line or system of lines is laid on a fixed predetermined grade so that the drain water flows toward the outlet. Tile lines have several advantages such as: (i) low maintenance costs, (ii) no interference with farming operations, and (iii) no land taken out of production.

Concrete and clay are the commonest types of tile used for drainage in irrigated areas although perforated pipe made from bituminized fiber is being used. Acceptable quality tile should conform to American Society for Testing and Materials Standards. Standards have also been prescribed for filter materials such as sand and fiber glass.

Fig. 50–4. Common types of drain tile, 4 inches in diameter. Left, concrete; upper center, polyethylene; lower center, bituminized fiber; right, clay.

Fig. 50–5. Common types of envelope material. Bottom, sand and gravel; upper left, thin fiberglass; upper right, 1-inch thick fiberglass.

The size of the tile to be used can be computed when the quantity of water to be drained is known. This factor, however, is usually so nebulous that rules of thumb must be resorted to for prescribing the size. These rules are as follows: (i) Lateral lines should be at least 4 inches in diameter; (ii) the downstream reach of any lateral line exceeding 1,300 feet should be at least 5 inches in diameter; (iii) collecting lines should be at least 6 inches in diameter; and (iv) collecting lines that serve more than 15,000 lineal feet of tile should be at least 8 inches in diameter.

In designing large main drains there is a need to compute aggregate flows

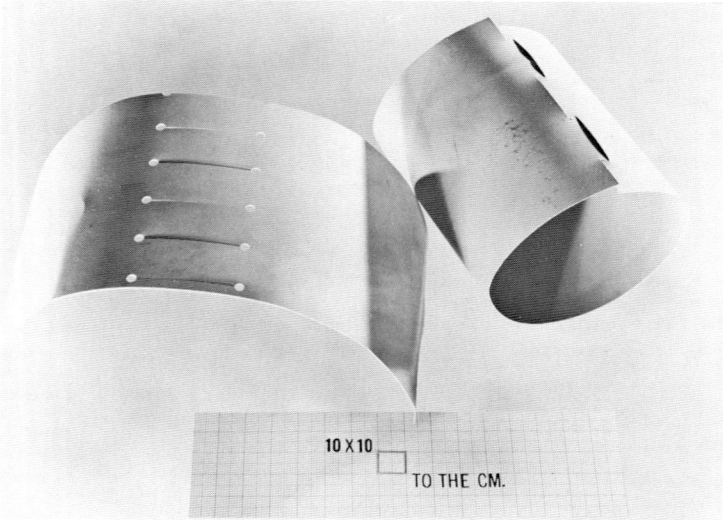

Fig. 50–6. Zipper type PVC mole drain lining material.

Fig. 50–7. Equipment for installing PVC mole drain lining material.

from a diversity of drainage areas. Weeks (1959) has proposed a set of design criteria based on many years' measurements of flows from tile systems in irrigated areas.

Other types of covered drains include French drains and mole drains. French drains consist of a narrow trench partially back-filled with rock, brush, or straw. These are seldom used in irrigated areas. Mole drains are constructed by pulling a torpedo-shaped object through the subsoil. This leaves a cylindrical opening in the subsoil that acts as a drain. This drain device has not found wide acceptance under irrigated conditions due to the temporary nature of the practice and due to the fact that it is difficult to pull the mole device at depths exceeding 30 inches. Since depths of at least 5 feet are desirable in irrigated areas, the mole drain seems to have limited adaptation. Recent research (Fouss and Donnan, 1962) to develop thin-walled, plastic-lined mole drains, and the machines to install them, holds some promise of success (Fig. 50–6 and 50–7).

C. Wells and Pumps

Drainage wells and pumps vary in size and capacity according to the nature of the drainage problem involved. The depth of a drainage well and the effective

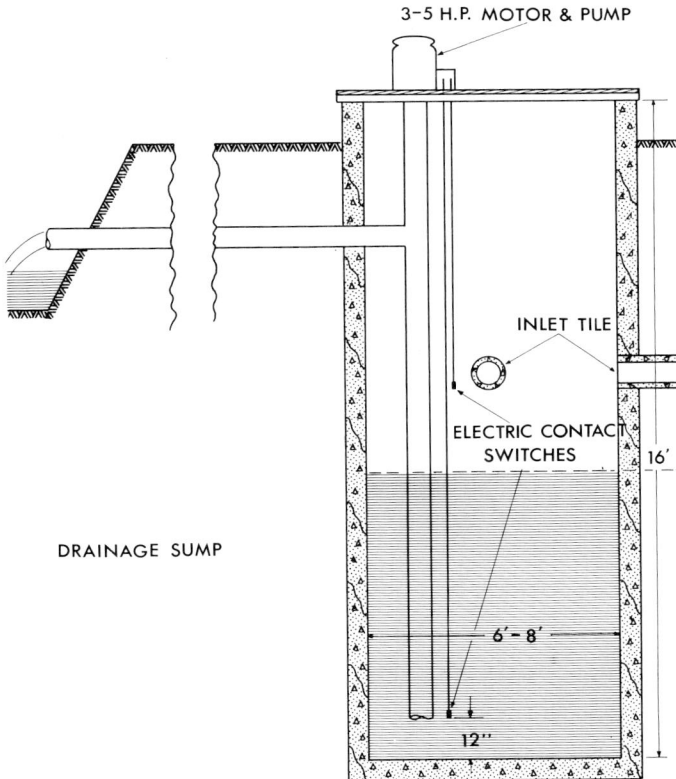

Fig. 50–8. Typical sump for use where main drain is at a higher elevation than tile outlets.

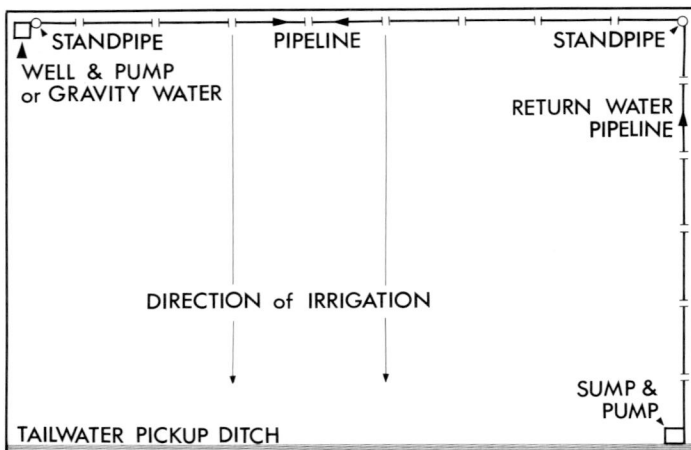

Fig. 50–9. Irrigation water return system for reuse of irrigation tailwater.

radius of drainage depends mainly on the water-bearing materials encountered. A good discussion of the theory of drainage by pumping from wells can be found in Luthin (1957) beginning on p. 181. Even under favorable conditions, drainage wells seldom are economical unless the drainage water can be used for irrigation or other purposes. The value of the water for irrigation nearly always offsets the pumping costs. Even when quality of the water is poor, it may be mixed with better quality water and used.

In artesian areas, wells may be installed in or alongside open drains. The artesian water rises in the wells and flows into the drains, thus alleviating the upward pressure of the groundwater in the aquifer.

D. Sumps

Small pumps are used to dewater sumps. Sumps can be built to depths of about 20 feet to drain small waterlogged areas. Also, they are often used as outlets for tile systems (Fig. 50–8). Another type of sump is the shallow collector type constructed at the lower end of an irrigated field. The excess surface irrigation water collects in these sumps and is pumped back upslope in a pipeline to the irrigation head ditch for reuse (Fig. 50–9). This technique is a good conservation practice since it saves water, saves nutrients in the water, and eliminates a potential drainage hazard.

V.   INSTALLATION AND MAINTENANCE

Proper installation and maintenance of the drain device is mandatory. Engineering survey techniques should be used to locate the site of the installation. Each line of the drain should be suitably marked with stakes indicating the alignment of the drain, grade, slope, depth of cut, and location of junctions and outlets. Installation always begins at the outlet and proceeds upslope.

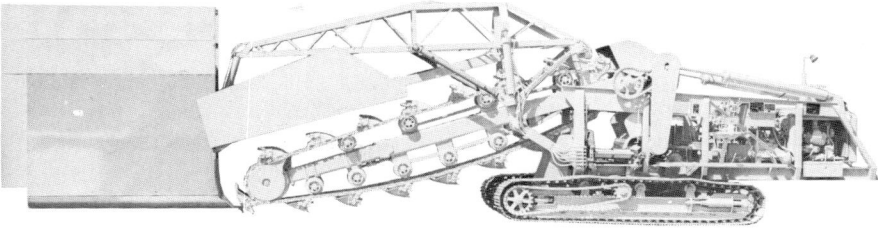

Fig. 50–10. Modern ladder type trenching machine capable of tile installations to 13 feet depth. Man placing tile rides in metal box at extreme left and is protected from cave-in by trench.

A. Open Drains

Open drains may be constructed by hand or with machinery. Hand-dug drains are time-consuming and require tremendous amounts of labor but they are the only method used in many parts of the world. Open drains can be dug with machines using the tractor-scraper, bulldozer, tractor-loader, trencher, machine shovel, backhoe, dragline, clamshell, or grader (Houston, 1961).

The material excavated from an open drain should be either spread over adjacent areas or placed far enough from the edge so that it will not slip back into the drain.

Open drains can be kept in efficient working condition only by careful maintenance. They quickly become clogged by brush, weeds, and silt and require frequent cleaning. Water-loving plants and weeds grow in the bottom and sides of open ditches. This vegetation must be removed or the drain will cease to function as designed. Machines have been developed for cleaning drains but, in general, the cleaning is accomplished by the same types of machines or methods used to construct the drains. Open drains should be cleaned to the same depth as their original design. Burning and chemical control of vegetative growth are becoming more popular and in many instances more economical than hand- or machine-cleaning techniques.

B. Covered Drains

Covered drains are installed by hand or with machines. Great care must be exercised to see that the drain is installed to the proper grade since the line is usually covered over immediately and mistakes are hard to locate. Hand-dug trenches are difficult to accomplish because of the depth requirements. Thus most covered drains in irrigated areas are installed using machines.

Machines have been developed which dig the trench to the required depth, lay the filter material, position the tile or fiber pipe, cover the conduit with additional filter material, and then backfill the trench, all in one continuous operation (Fig. 50–10). After the covered drain has been installed, careful management must be exercised to insure proper consolidation of the backfill material. The initial application of irrigation water after installation is monitored to check for excessive leaks into the drain. Heavy farm machinery will crush and destroy mole drains where the soil has not been allowed to come into complete repose around the drain.

Maintenance of covered drains is not difficult but frequent inspections are desirable. Deep-rooted trees and brush may clog covered drains, requiring a cleanout with special reaming tools. Where silt clogging is a problem, manhole silt collectors are sometimes provided at tile drain junctions. The outlet of a covered drain is perhaps the most critical point from a maintenance standpoint. Covered drains often outlet to an open drain. Care should be taken to maintain a free overfall to the outlet. Provisions should also be made to prevent cave-ins and the eventual erosion upslope from the tile outlet.

C. Wells and Pumps

Wells and pumps for drainage purposes require the same criteria for installation and maintenance as for irrigation wells and pumps.

LITERATURE CITED

Bernstein, Leon. 1961. Tolerance of plants to salinity. Amer. Soc. Civ. Eng. Irrig. Drainage Div. J. 87(IR4):1–12.

Berry, W. L., and E. D. Stetson. 1959. Drainage problems in the San Joaquin Valley, Amer. Soc. Civ. Eng. Irrig. Drainage Div. J. 85(IR3):97–106.

Donnan, William W. 1946. Model tests of a tile-spacing formula. Soil Sci. Soc. Amer. Proc. (1947) 11:131–136.

Donnan, William W. 1959. Drainage of agricultural land using interceptor lines. Amer. Soc. Civ. Eng. Irrig. Drainage Div. J. 85(IR1):13–25.

Donnan, William W., and G. B. Bradshaw. Sept. 1952. Drainage investigation methods for irrigated areas in western United States. US Dep. Agr. Tech. Bull. 1065. 45 p.

Donnan, William W., G. B. Bradshaw, and H. F. Blaney. Sept. 1954. Drainage investigation in Imperial Valley, Calif. (A 10-year summary). US Dep. Agr.-Soil Conserv. Serv. Tech. Publ. No. 120. 71 p.

Fly, Claude L. 1961. Soil drainability factor in land classification. Amer. Soc. Civ. Eng. Irrig. Drainage Div. J. 87(IR3):47–63.

Fouss, James E., and William W. Donnan. 1962. Plastic-lined mole drains. Agr. Eng. 43:512–515.

Gulhati, N. D. 1955. Irrigation in the world, a global review. Int. Comm. Irrig. Drainage. 130 p.

Hamid, Sayyid. May 1961. Programs for waterlogging and salinity control in the irrigated areas of West Pakistan. W. Pakistan Water Power Develop. Authority. 29 p.

Hill, Raymond A. 1961. Leaching requirements in irrigation. Amer. Soc. Civ. Eng. Irrig. Drainage Div. J. 87(IR1):1–5.

Houston, Clyde E. Nov. 1961. Drainage of irrigated land. California Agr. Exp. Sta. Circ. No. 504. 40 p.

Israelson, O. W., and Manouchehr Ayazi. 1957. Interrelation between irrigation and drainage. Int. Comm. Irrig. Drainage, Congr. Irrig. Drainage, Proc. 3rd. p. 10.221–10.242.

Jacob, C. E. Oct. 1960. Groundwater and drainage of Yuma Valley and contiguous area. Yuma Valley Water Users Ass., Yuma, Arizona. 50 p.

Luthin, J. N. (editor) 1957. Drainage of agricultural lands. Agronomy no. 7. American Society of Agronomy, Madison, Wis. 620 p.

Marr, J. C. Dec. 1926. Drainage by means of pumping from wells in the Salt River Valley, Arizona. US Dep. Agr. Dep. Bull. No. 1456. 22 p.

Peterson, D. E. 1961. Intercepting drainage wells in artesian aquifer. Amer. Soc. Civ. Eng. Irrig. Drainage Div. J. 87(IR1):7–14.

Van Schilfgaarde, Jan. 1957. Approximate solutions to drainage-flow problems. *In* J. N. Luthin (ed.) Drainage of agricultural lands. Agronomy 7:79–112.

Weeks, Lowell O. 1959. Drainage in the Coachella Valley of California. Amer. Soc. Civ. Eng. Irrig. Drainage Div. J. 85(IR3):83–89.

Weir, Walter W. Jan. 1925. Pumping for drainage in the San Joaquin Valley, California. California Agr. Exp. Sta. Bull. 382. 28 p.

Wensel, L. K. 1942. Methods for determining permeability of water-bearing materials. US Geol. Surv. Water-Supply Pap. 887. 192 p.

51

Salt Problems in Relation to Irrigation

RONALD C. REEVE

Agricultural Research Service, USDA
Columbus, Ohio

MILTON FIREMAN

Tipton and Kalmbach, Inc.
Denver, Colorado

I. INTRODUCTION

The water soluble mineral constituents of weathered rocks present in soils and soil solutions constitute an important part of the environment in which plants grow. The extent to which salts in the soil solution or sodium on the exchangeable fraction of the soil are in excess are measures of the salt problem. Inasmuch as the problem involves water-soluble materials, the occurrence of salt problems, whether created during a geologic era or recently, is directly related to the transport and disposition of salts by water (Kelley, 1951; US Salinity Lab. Staff, 1954). Infiltration, drainage, evaporation, and transpiration are the major processes involved. Important aspects of the salt problem in irrigation agriculture are its extent and nature prior to irrigation, the contribution of the irrigation process *per se*, and the management practices that can be applied to ameliorate problem soils and/or satisfactorily control the salinity environment of crop plants. This chapter summarizes current knowledge on the subject with respect to definition, diagnosis, amelioration, and/or control. References to the literature are intended to bring out the salient points pertaining to the subject. A literature review is not intended. Hayward (1954) reviewed the literature pertaining to plant growth under saline conditions and Bernstein (1962) reviewed the literature relative to salt-affected soils and plants.

A. Extent of the Problem

Salt problems commonly occur in regions with arid and semiarid climates and lower the productivity of extensive areas of agricultural land throughout the world—an estimated one-third or more of the 300 million acres of land presently irrigated as well as untold millions of acres of potentially irrigable land. The widespread nature of the problem is demonstrated by the fact that at least 25 nations throughout the world have a million or more acres each (up to 50 million in the case of China) presently under irrigation, and serious salinity problems exist in every one of these countries!

The total area of the earth's land surface is approximately 37 billion acres, and about one-third of this has been classified, on the basis of both climate *and* vegetation, as arid and semiarid. Of these 12 billion acres of arid and semiarid lands,

$< 3\%$ are irrigated at present, although the potentially irrigable land greatly exceeds 1 billion and may exceed 2 billion acres. A greater proportion of the irrigable lands undoubtedly would have been under production by now if it were not for the uncertainties resulting from the development of salt-affected soils in the older irrigated areas.

In the past, saline and alkali (sodic) soils were most often formed as a result of salt accumulation due to natural causes, such as floods, impaired drainage, and the evaporation of salty ground waters. In recent centuries, vast areas of salt-affected soils have developed from man-made causes, such as irrigation without provision for adequate drainage, application of insufficient amounts of irrigation water, use of poor quality irrigation waters, or from a combination of these.

B. Historical Implications

For various reasons, most, if not all, of the early civilizations arose in arid and semiarid regions where irrigation was required for the production of crops. All of these early civilizations, with the exception of Egypt, sooner or later declined in importance and finally disappeared. From rather sketchy evidence available to us, it can be inferred that neglect of, or inability to cope with, salinity and waterlogging problems contributed significantly, if not predominantly, to the decline of these nations. In fact, recent consideration of the agricultural history of these irrigated countries has led a number of authorities to seriously question the permanence of *economically feasible* irrigation agriculture in the Western USA and elsewhere. We think that such reasoning is unduly pessimistic; however, it is clear that the best possible use must be made of present day knowledge and resources if we are to avoid repeating the mistakes of the past and thus limiting the "life" of irrigation projects.

II. NATURE OF THE SALT PROBLEM

Salt-affected soils have excessive concentrations of soluble salts (saline soils) or adsorbed sodium (alkali or sodic soils), or both (saline-alkali soils). The soluble salts that occur in soils consist mainly of the ions: sodium (Na^+), calcium (Ca^{2+}), magnesium (Mg^{2+}), potassium (K^+), chloride (Cl^-), sulfate (SO_4^{2-}), bicarbonate (HCO_3^-), carbonate (CO_3^{2-}), borate (BO_3^{3-}), and nitrate (NO_3^-).

The original sources of these salts are the exposed rocks and minerals of the earth's crust. As a result of chemical decomposition and physical weathering, the soluble constituents (salts) are gradually released. In humid areas these soluble salts are carried downward through the soil profile by rain, and ultimately are transported by streams to the oceans. In arid regions, however, leaching (washing salts out of the soil) may be local and the soluble salts may not be transported far because of the relative scarcity of rainfall. Also, the high evaporation and transpiration rates characteristic of arid climates tend to decrease the limited amount of water available for leaching and transporting salts.

Salts also may be imported into an area via irrigation water. All irrigation waters, whether derived from springs, streams, or pumped from wells, contain appreciable quantities of soluble salts. Therefore, wherever irrigation water is used, it may be the major source of the soluble salts that give rise to salinity problems.

Inadequate natural drainage is intimately associated with, and contributes to, the severity of saline and alkali (sodic) conditions. Because of the low rainfall characteristic of arid regions, surface drainageways may be poorly developed, and consequently large drainage basins may have no satisfactory outlets to permanent streams.

Salt problems may also result in some areas, even with good irrigation water and good irrigation practices, if the soils, for either economic or physical reasons, *cannot* be drained adequately.

Man-made saline and alkali soils most often develop from the irrigation of lands that may have been adequately drained under "natural" conditions (no irrigation) but are not adequately drained when irrigated. As a result of irrigation, large quantities of water percolate into the subsoil; consequently, the groundwater level may be raised from a considerable depth to within a few feet of the surface in a relatively short time. When this occurs, water, more or less saline, moves upward by capillarity and salts are deposited in the surface soil as a result of both evaporation and transpiration.

Equally important, but less well recognized, is the fact that saline and alkali problems arise even when drainage facilities are adequate unless sufficient irrigation water is applied to provide for both crop needs (consumptive use) and the necessary leaching of excess salts out of the soil (leaching requirement).

A.   Salt Effects on Soils

Saline soils contain enough soluble salts so distributed in the soil profile that they decrease the growth of most plants. For purposes of definition, the electrical conductivity (EC) of the saturation extract is at least 4 mmho/cm at 25C, and the exchangeable sodium percentage (ESP) is < 15. Ordinarily the soil is only slightly alkaline in reaction (pH = 7.1 to 8.5). Saline soils are often recognized by the presence of white surface crusts, by damp oily-looking surfaces devoid of vegetation, by stunted growth of crop plants with considerable variability in size, and sometimes by tipburn and firing of the margin of the leaves, particularly in trees and vines. Soil analyses rather than visual observations, however, are needed to properly assess salinity.

The adsorbed (exchangeable) ions in saline soils are principally Ca^{2+} and Mg^{2+}. Calcium- and magnesium-saturated soils are stable in water, are flocculated, and are easily worked into granules and crumbs; their soil permeabilities are equal to, or higher than, those of similar nonsaline soils, and they usually provide a desirable environment for seed germination and plant growth aside from the effect of salinity *per se*.

Alkali (sodic) soils contain sufficient adsorbed (exchangeable) sodium to interfere with the growth of most crop plants. Alkali soils (ESP > 15) may be highly alkaline in reaction (pH = 8.5 or above), but do *not* contain excessive amounts of soluble salts.

As a consequence of the electrical charges on their surface, soil particles adsorb and retain certain amounts of cations such as Na^+, Ca^{2+}, and Mg^{2+}. The adsorbed ions are bound or fixed firmly to the soil particle when it is dry, but these adsorbed ions can interchange freely with other soluble cations in the soil solution. This reaction is called cation exchange. The proportion of the various cations held by the exchange complex is related mainly to their concentration in the soil solution and to their valence.

As the proportion of exchangeable sodium increases, the soil tends to become dispersed and impermeable to water and air. Because of decreased permeability, it becomes increasingly difficult to replenish the water supply in the root zone by irrigation. Also because of the dispersed condition of the soil, it becomes more difficult to establish a condition of surface tilth favorable for seed germination and plant growth.

The sodium adsorption ratio (SAR) is an expression pertaining to the cation makeup of waters and soil solutions and is useful for predicting the exchangeable cation status of a soil that is in equilibrium with a water or solution of known composition (US Salinity Lab. Staff, 1954). SAR is defined by the equation

$$\text{SAR} = \text{Na}^+ / [(\text{Ca}^{2+} + \text{Mg}^{2+})/2]^{1/2} \qquad [51\text{--}1]$$

where the cation concentrations of Na^+, Ca^{2+}, and Mg^{2+} are given in milliequivalents/liter. Although it is an empirical relationship, it has a theoretical basis. It is particularly useful in characterizing the sodium hazard of irrigation waters, and has application in both diagnosis and reclamation of alkali soils (US Salinity Lab. Staff, 1954). For example, an irrigation water having Na^+ and $\text{Ca}^{2+} + \text{Mg}^{2+}$ concentrations of 10 and 2 meq/liter, respectively, would have an $\text{SAR} = 10$. A soil equilibrated with this water would have an exchangeable sodium percentage of about 12.

Saline-alkali soils generally are more common than either saline or alkali soils. A saline-alkali soil contains excessive quantities of both soluble salts (> 4 millimhos) and adsorbed sodium ($> 15\%$). The soil is seldom highly alkaline (pH is generally < 8.5). As long as a considerable excess of salts is present, the appearance and properties of these soils usually are similar to those of saline soils. However, if the excess soluble salts are leached out, the soil properties may change markedly and become similar to those of alkali soils. They then become strongly alkaline, the particles disperse, and conditions become unfavorable for the entry and movement of water and air, and for the preparation of a good seedbed.

B. Effects of Salinity on Plants

Excess salinity delays or prevents seed germination, and reduces the amount and rate of plant growth. These effects primarily are associated with high osmotic pressures of the soil solution which impair the plant's ability to absorb water but some of these effects may be due to nutritional imbalance or toxicity caused by specific ions (Bernstein, 1961).

1. OSMOTIC EFFECTS

Growth retardation due to salinity has been shown (Wadleigh et al., 1951) to be related directly to the osmotic pressure of the soil solution, and to be largely independent of the kinds of salts present. Because the electrical conductivities of soil solutions are highly correlated with their osmotic pressures, the very simple

Table 51–1. Salt concentrations expressed in meq/liter and in ppm in solution, and as per cent salt on a dry soil basis (assumed saturation percentage of soil = 50) for three common salts at osmotic pressure = 2 atm

Salt	Salt concentration		
	Saturation extract		Dry soil basis
	meq/liter	ppm	%
NaCl	48	2,806	0.14
CaCl$_2$	64	3,552	0.18
Na$_2$SO$_4$	72	5,114	0.26

conductivity measurement can be substituted for the much more tedious osmotic pressure determination in assessing the salinity of soils. Another simplification is the substitution of the saturation extract for the more difficultly obtainable soil solution (US Salinity Lab. Staff, 1954). Soil salinity is sometimes expressed on a soil weight basis (weight of salts per unit of soil or as a percentage of soil weight) or in terms of salt concentration of the saturation extract (meq/liter or ppm). Table 51–1 shows the differences in these various ways of expressing salinity for three common salts at an equal osmotic pressure (OP) level (OP = 2 atm). Although these methods of expressing salinity may have some application in reclamation operations, they are of little value in relating soil salinity to plant growth response.

The relationship between osmotic pressure in atmospheres (OP) and electrical conductivity in millimhos per centimeter at 25C ($EC_e \times 10^3$) is given approximately by

$$OP = 0.36 \ (EC_e \times 10^3). \qquad [51\text{–}2]$$

2. SPECIFIC ION EFFECTS

Although osmotic pressure accounts for the major effect of salinity on plant growth, some ions have specific nutritional or toxic effects that also depress growth and yield. Because plants vary widely in their nutrient requirements and in their ability to absorb specific nutrients, the effects of salinity on nutrition differ markedly among species. For example, high soluble calcium in a soil reduces potassium uptake by carrots (*Daucus carota* L.) and beans (*Phaseolus vulgaris* L.); high magnesium or sodium in the soil may result in calcium or potassium deficiencies in some crops; and high bicarbonate may cause iron chlorosis, particularly in fruit trees and ornamentals. Also, high nitrate may promote undesirable vegetative growth in grapes (*Vitis vinifera* L.) or sugar beets (*Beta vulgaris* L.) at the expense of sugar content.

3. TOXIC ION EFFECTS

An ion is said to be toxic when its presence causes direct damage to the plant, and many common soluble constituents of soils which are essential or harmless to plant growth at low or moderate concentrations are harmful at high concentrations. In this respect, chloride and sodium are toxic to many fruit crops. Boron, which is essential in small quantities for plant growth, is injurious to many crop plants at solution concentrations of more than 1 or 2 ppm. Moderate concentrations of chloride (700 to 1,500 ppm) in the saturation extract usually cause chloride to accumulate in leaves of fruit crops to the extent of 1 to 2% of the dry weight.

As a result, marginal leaf burn develops which leads to leaf drop, twig dieback, and sometimes death of the plant. Sodium accumulations of just a few tenths of a per cent (dry weight) produce similar leaf burn symptoms and extensive tissue injury. It must be emphasized, however, that most crops, including vegetable, grain, forage, and fiber crops, are *not* specifically sensitive to chloride or sodium and may accumulate high concentrations of these ions without developing any symptoms of injury.

4. MODIFICATIONS OF RESPONSES TO SALINITY

From the foregoing it is evident that tolerance of most plants to salinity can be adequately described by reference to the electrical conductivity of the saturation extract of the soil. Except for the very tolerant crops, most plants show a progressive decline in growth and yield with increasing salinity. Tolerance of crops to salinity is expressed in terms of the electrical conductivity of the saturation extract of the root-zone soil at which a specified reduction in crop yield is experienced. Salt tolerance values have been expressed both in terms of a 50% (US Salinity Lab. Staff, 1954) and a 10% to 15% (Bernstein, 1958, 1959, 1960) yield reduction. The former values are most useful in research for comparing the response of one crop with that of another, whereas the latter values are most useful to farmers and nursery men who are primarily interested in maintaining salinity levels in the soil that will give an economic return. Figure 51–1 shows curves relating yield to electrical conductivity of the saturation extract for crops of different salt tolerance. It should be noted that the information on salt tolerance of crops has been obtained under optimum conditions with respect to seed variety; insect, disease, and weed control; soil and water management; and, in particular, fertilization. Very little reliable information is available on the relationship between plant growth and salinity levels where other factors (fertility, water, etc.) may be limiting.

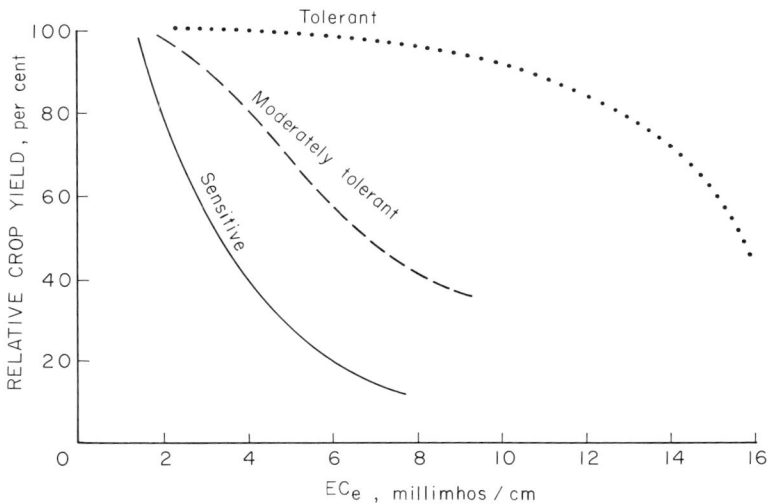

Fig. 51–1. Salt tolerance curves for tolerant crops (barley, cotton, etc.), moderately tolerant (forage, maize, etc.), and sensitive crops (beans, trees, etc.) (from US Salinity Lab. Staff, 1954).

Plants may be more sensitive to salinity at one stage of growth than at another. For example, the germination of some crops may be affected by a conductivity of 4 to 5 mmho/cm, but once the plants are established they may tolerate conductivities up to 12 or 15 mmho/cm (e.g., sugar beets). Also, many crops are more sensitive during the seedling stage than at later growth stages. Rice (*Oryza sativa* L.), for example, can germinate at salinities up to 10 or 15 mmho/cm, but the plants usually die if the salinity during the seedling stage exceeds 5 or 6 mmho/cm. Rice is also quite sensitive during fruiting; a moderate salinity level at this stage of growth may reduce grain production by more than 50% while hardly affecting straw yield. In contrast, barley (*Hordeum vulgare* L.) and cotton (*Gossypium hirsutum* Cav.) can be quite stunted, vegetatively, by salinity but still produce normal yields of grain and fiber.

Climatic factors also may modify the salt tolerance of plants. Some crops are much more sensitive to given levels of salinity during hot, dry periods than during cooler, more humid weather.

The soil water regime may also modify plant response to a given salinity level. Immediately following an irrigation, soil water is at a maximum and soluble salt concentration is at a minimum. As water is lost from the soil by evaporation and plant use, almost all of the salts are left behind in a decreasing volume of water. The salt concentration, therefore, increases progressively; the drier the soil is allowed to become before irrigating again, the higher the average concentration of the soil solution and the greater its effect on crop growth. More frequent irrigations, therefore, tend to minimize the harmful effects of a given salinity level. Because all irrigation waters contain salt, each irrigation adds an increment of salinity to the soil. If only enough water is added to meet the crop requirements, there will be an increase in the salinity of the root zone. Saline soils, therefore, not only require more frequent irrigations than do nonsaline soils, but also require the application of extra water for leaching.

5. FOLIAR ABSORPTION OF SALT CONSTITUENTS

It has long been known that some plants can absorb nutrients and other ions through their leaves. Sprinkling has been used widely with satisfactory results for the application of some fertilizer elements, particularly to citrus and ornamental crops. In hot, dry climates, sprinkling with water that is not sufficiently saline to affect plant growth if applied directly to the soil can result in excess foliar absorption, particularly sodium and chloride, with consequent tip and marginal burn and defoliation of plants. Crops known to be sensitive to damage by foliar absorption are stone fruits [peaches (*Prunus persica* Stokes), apricots (*Prunus armeniaca* L.), almonds (*Prunus amygdalus* Stokes)], citrus (*Citrus sp.* L.), walnut (*Juglans regia* L.) and woody ornamentals. A few crops such as avocado (*Persea americana* Marsh) and strawberries (*Fragaria* sp. L.) seem to absorb little, if any, salts. Excessive absorption of harmful ions occurs when the concentration of these ions in the water that remains on the plant leaf becomes high. Harmful accumulations can occur by use of irrigation waters containing as little as 5 meq/liter of chloride or sodium. Moreover, irrigation waters with even lower concentrations of these ions can result in harmful accumulations if applied so that the sodium or chloride concentrations in the water adhering to the leaves are increased excessively. Such may be the case with intermittent sprinkling or where atmospheric conditions favor rapid evaporation (low relative humidity, high temperatures, windiness, etc.).

Keeping the foliage continuously wet during sprinkler irrigation, sprinkling when it is cool and cloudy, or at night, and using low head sprinklers for tall crops such as trees so that a minmum of foliage is wetted, will reduce foliar absorption of salts and, therefore, damage from sprinkling.

III. CRITERIA AND METHODS OF DIAGNOSIS

Saline, alkali (sodic), and saline-alkali soils are defined solely on the basis of (i) salinity and (ii) percentage of exchangeable sodium.

Whether saline, alkali (sodic), or specific and toxic ion conditions have affected plant growth or soil characteristics may be deduced by observation of plant or soil abnormalities, or can be determined more accurately by appropriate *significant* measurements of (i) plant growth or crop yield under controlled conditions, (ii) chemical composition of plant tissue, or (iii) soil permeability rates or other physical parameters, or a combination of these measurements.

A. Soil Tests

Accurate diagnosis for purposes of soil reclamation or management generally requires detailed information on soils in addition to the salinity and exchangeable sodium status. Such information almost always includes location and thickness of the affected soil layer, some measure of texture (saturation percentage or particle size distribution), reaction (pH), and amount of soluble calcium required for reclamation, e.g., gypsum requirement. For special purposes, other tests may be required. Among these are content of alkaline-earth carbonates ("lime"); gypsum content; exchangeable calcium, magnesium, and potassium; cation-exchange capacity; clay type; surface area; aggregate size distribution; water retention characteristics; soluble chloride, sulfate, carbonate, bicarbonate, nitrate, and boron in the saturation extract; etc. Details and discussions of these methods are given in the US Department of Agriculture Handbook 60 (US Salinity Lab. Staff, 1954).

Water from many irrigation wells, a few surplus supplies, and some soils contain micronutrients such as boron and lithium in amounts that are toxic to plants. The plant symptoms are unique in a few cases, but generally are similar to those caused by excess salinity. In most cases, therefore, soil and plant analyses must be used for diagnosis (Chapman and Pratt, 1961; US Salinity Lab. Staff, 1954).

B. Plant Symptoms and Tests

Crops growing on saline soils usually have barren spots and may be stunted with considerable variability in size and sometimes with a deep blue-green foliage. Since most crop plants are more sensitive to salinity during germination than in later stages of growth, barren spots may be more indicative of salinity near the seed during germination than they are of a general condition of salinity in the soil profile. The vigor of plants adjacent to bare spots is probably a better indication of the general level of soil salinity. Size and distribution of plants often indicate the concentration and distribution of salt in the soil. However, caution

Table 51–2. Modified Scofield scale relating electrical conductivity of the saturation extract to plant growth (US. Salinity Lab. Staff, 1954)

Plant Response				
Salinity effects mostly neg- ligible	Yields of very sensitive crops may be restricted	Yields of many crops restricted	Only tolerant crops yield satisfactorily	Only a few very tolerant crops yield satisfac- torily
Electrical Conductivity of the Saturation Extract (EC$_e$), mmho/cm				
0	2	4	8	16

should be exercised in using crop distribution and growth indications for diagnosis because of the similarity between the effects of salinity and of other factors such as excess water (ponding), soil infertility, disease, etc.

Some species of plants develop characteristic necrotic areas, tip burn, and firing of the margin of the leaves when grown on saline soils. For some crops cupping or rolling of leaves, a common manifestation of water deficiency in plants, also may be indicative of salinity when it occurs in the presence of adequate water and in the absence of root diseases and high water table. It is apparent, therefore, that whereas the appearance of the crop may be *indicative* of saline conditions, a reliable diagnosis of salinity usually requires additional evidence derived from appropriate soil and plant tests.

The Scofield Scale, shown in Table 51–2, gives crop response to salinity under average conditions in terms of the conductivity of the saturation extract in millimhos/centimeter at 25C (US Salinity Lab. Staff, 1954). It should be empha- sized that the plant response given in this scale is referenced to the salt status of the soil solution in the active root zone.

Plant species vary greatly in the amounts of the various ions that they may accumulate. Furthermore, many species may exclude certain ions from their leaves while accumulating them in their stems or roots, and vice versa. Therefore, since the normal mineral composition of plant parts is frequently altered by saline or sodic conditions, appropriate plant analyses may serve for diagnosing such problems, even though the offending salts may have been eliminated previ- ously from the soil by leaching or other means.

Caution must be exercised in relating the malfunctioning of plants to specific ions or salts. Frequently, excessive accumulation of an ion or ions in a plant may be a result of conditions other than the high concentration of the ion or ions in the soil solution; this is so because any of the factors, such as water stress and plant disease, that inhibit plant growth may cause abnormal accumulation of some ions in the plant tissues. Therefore, the chemical composition of plant parts usually should be considered only one line of evidence in the diagnosis of crop injury on saline or alkali soils, and corroborative evidence should be sought by appropriate soil tests.

The relationships between foliar composition and the principal ions that occur in excess in saline soils can be summarized, as follows:

1) Chloride concentration in leaves usually bears a close relationship to the chloride concentration in the soil solution.

2) Excess soluble sodium may or may not be reflected in the sodium content of leaf tissues.

3) Increases of calcium in the soil solution usually result in increases in the calcium content of leaves.

4) Excess sulfate in the soil solution causes relatively small increase in total sulfur of the leaf tissue.

Other factors that may affect the level of accumulated ions include age of leaf, season, aeration, climatic conditions, and diseases, especially of the root.

Foliar analyses are useful in the diagnosis of boron injury of many plants. The boron content of normal mature leaves of such plants as citrus, avocado, walnut, fig (*Ficus carica* L.), grape, cotton, and alfalfa (*Medicago sativa* L.) tops is about 50 ppm. Boron contents of 20 ppm or less indicate deficiency, and values > 250 ppm usually are associated with boron toxicity. Stone fruits, apples (*Pyrus malus* L.), and pears (*Pyrus communis* L.) do not accumulate high concentrations of boron in their leaves, although these species are sensitive to excess boron. If due allowance is made for varietal specificity in boron accumulation, foliar analysis may provide a better basis for diagnosis than analysis of soil or water.

In addition to the determinations for chloride, sulfate, calcium, sodium, and boron, it may be useful to obtain other information such as content of magnesium, potassium, nitrogen, phosphorus, and other ions. Methods of analysis and discussion of known critical limits are given by Chapman and Pratt (1961).

IV. SALINITY CONTROL

Salts move and accumulate in soils largely as a result of the movement of water. Controlling or reducing salinity levels in the soil, therefore, depends largely on the management practices that have to do with water movement and its control. Among these practices are: amount, frequency, and method of applying water; leveling, forming, and preparing land; drainage; and other farm operations that influence soil structure or tilth.

Salinity control is important not only to the farmer who is concerned with crop production on his farm, but also to the irrigation project planner and to organizations that are concerned with development, storage, diversion, and distribution of irrigation water within a drainage basin or valley area. *Salt balance* and *leaching requirement* are two basic concepts relating to salinity control that have useful applications at all stages of the water development-distribution-use regime.

A. Salt Balance

The salt balance concept relates the quantity of dissolved salts carried into an area in the irrigation water to the quantity of dissolved salts removed by the drainage water. This concept was introduced and used by Scofield (1940) in the statement: "If the mass of the salt input exceeds the mass of the salt output, the salt balance is regarded as adverse, because this trend is in the direction of the accumulation of salt in the area and such a trend is manifestly undesirable."

Wilcox and Resch (1963) illustrated the usefulness of this concept and summarized the salt balance data of a 15-year study on the Rio Grande project. They concluded that salt balance is a reliable and useful indicator of year-to-year trends in salinity on irrigation projects. Short-term records can be misleading, but where salt balance data are available for a number of years, unfavorable salinity trends can be identified and appropriate action can be taken to remedy the situation. Salt balance analyses thus serve as a useful tool in the control of salinity. Although such analyses have been used mostly for irrigation projects, the concept is applicable to any size unit—a valley, an irrigation project, a farm, or a garden tract—provided the appropriate measurements of salt input and salt output are obtainable. Irrigation districts, and other agencies concerned with irrigation water development and use, are becoming more aware of the usefulness of salt balance data and are providing for their collection.

B. Leaching Requirement

The leaching requirement (LR) is defined as the fraction (or percentage) of the irrigation water that must be leached through the root zone to keep the salinity of the soil below a specified value (US Salinity Lab. Staff, 1954).[1] The leaching concept embodies the concept of salt balance for the root zone, but in addition requires that the salinity at the bottom of the root zone be kept below a specified value.

The defining equation for leaching requirement given by the US Salinity Laboratory (1954) is

$$\mathrm{LR} = D_d/D_i = \mathrm{EC}_i/\mathrm{EC}_d \qquad\qquad [51\text{--}3]$$

in which LR is the leaching requirement, D_d/D_i is the ratio of the depth of drainage water that passes beyond the bottom of the root zone to the depth of irrigation water entering the soil, and $\mathrm{EC}_i/\mathrm{EC}_d$ is the ratio of the electrical conductivity of the irrigation water to the electrical conductivity of the drainage water. This equation takes into account only the salt that is contained in the irrigation water, and the salt removed from the root zone by the drainage water. The amounts of other salts added (fertilizers, amendments, etc.) and the salts that are lost from the soil solution (salts that are precipitated in the soil and removed by the crop) are usually small compared with the amounts brought in with the irrigation water or removed by drainage and are, therefore, assumed to be negligible. In some instances, however, these other salts are not negligible. When such is the case, equation [51–3] must be corrected for such effects. Equation [51–3], although derived for the steady state case, is applicable to the longtime average or quasisteady state. Where rainfall is involved, an adjusted

---

[1] Several different terms have been used to describe the process whereby irrigation waters applied to the soil are concentrated by evapotranspiration, and different and distinct defining equations have been applied to each term. Hill and Scofield (1938) used the term "service equivalence." L. A. Richards introduced the term "leaching requirement" (US Salinity Lab. Staff, 1954). Eaton (1954) used the terms "leaching percentage" and "leaching requirement."

depth $(D_i)_{adj}$ and an adjusted electrical conductivity of the applied water $(EC_i)_{adj}{}^2$ should be substituted in equation [51–3].

The depth of irrigation water required to maintain a given level of salinity in the soil in terms of the evapotranspiration or consumptive use of the crop is given by the equation

$$D_i = D_c[1/(1 - \text{LR})] \qquad [51\text{–}4]$$

where D_c is the equivalent depth of water representing evapotranspiration or consumptive use. In terms of electrical conductivity of the irrigation and drainage waters, the depth of irrigation water required is

$$D_i = [EC_d/(EC_d - EC_i)] \, D_c. \qquad [51\text{–}5]$$

The depth of irrigation water D_i is thus expressed in terms of the electrical conductivity of the irrigation water and the other conditions determined by crop and climate; namely, salt tolerance of the crop and consumptive use. The salt tolerance of the crop is taken into account in the selection of permissible values of EC_d. Although the relationship between EC_d and the salt tolerance of the crop has not been well defined, especially for field conditions where salinity in the root zone may not be uniform, the published salt tolerance values, which were obtained for uniform salinity in the soil, serve as a guide for estimating EC_d. The electrical conductivity of the saturation extract at a 50% reduction in crop yield has been used as an estimate of EC_d (US Salinity Lab. Staff, 1954). For a linear increase of salts with depth, it can be shown that with this estimate of EC_d the average salinity of the root zone is approximately equal to the salt tolerance value for a 90% to 95% yield for many crops.

C. Irrigation Methods

Salinity levels in the root zone are influenced by the methods of irrigation. For example, the distribution of salt within the root zone will be vastly different under furrow irrigation than with flooding or sprinkling. In general, the salt content of the soil solution varies over a greater range and varies much more from one place to another with furrow irrigation than with either flooding or sprinkling. Salts move with the water by both saturated and capillary flow into ridges and borders where, upon evaporation of the water, the salts are concentrated. The spatial distribution of salts in the soil as a function of time has been experimentally evaluated for furrow and other surface configurations (Wadleigh and Fireman, 1949; Bernstein and Fireman, 1957).

The distribution of salts resulting from wetting of the ridges between the furrows is of particular importance in seed germination and seedling growth. Some crops, such as sugar beets and cotton have a high tolerance to salinity in

---

[2]

$$(D_i)_{adj} = D_i + D_r$$
$$(EC_i)_{adj} = [D_r (EC)_r + D_i (EC)_i]/(D_r + D_i)$$

where the subscripts adj and r stand for "adjusted value" and "effective rainwater" (that which percolates into the soil), respectively.

the later growth stages but are sensitive to salinity during germination and in the seedling stage. The sloping-bed method of irrigation (Bernstein and Ayers, 1955; Bernstein et al., 1955) provides low salinity levels in the vicinity of the seed, even though the initial soil salinity may be high. With an initial soil salinity of 10 mmho/cm in the saturation extract, the emergence of sugar beet seedlings (*Beta vulgaris* L.) was 85% for the *sloping bed* (seed planted 2/3 distance up the sloping furrow), compared with 60% for the *lettuce bed* (seeds planted near the shoulder of a flat ridge), and < 10% for the standard beet bed (seed planted near the center of a flat ridge). Similar results were obtained with cotton (*Gossypium hirsutum* L.), lettuce (*Lactuca sativa capitata* L.), broccoli (*Brassica oleracea italica* L.), cantaloupe (*Cucumis melo reticulatus* L.), and sorghum (*Sorghum vulgare* Pers.). Satisfactory stands (> 60% emergence) were obtained with all crops planted on the sloping bed, even when the initial EC_e value of the saturation extract of the soil was as high as 43 mmho/cm; but emergence was reduced to < 50% when the EC_e of the saturation extract of the soil was 12 mmho/cm for the lettuce bed and 5 mmho/cm for the standard bed.

D.   Drainage

Salinity control depends on the drainability of the soil. The leaching requirement can be met only if the required amount of water can pass beyond the root zone. An estimate of the minimum amount of water that is required to be drained, $D_{d(\mathrm{min})}$, is obtained by solving the leaching requirement equation [51–3] for the depth of drainage water (Reeve, 1957).

$$D_{d(\mathrm{min})} = D_i(EC_i/EC_d) \qquad [51\text{–}6]$$

or in terms of consumptive use

$$D_{d(\mathrm{min})} = D_c[EC_i/(EC_d - EC_i)] \qquad [51\text{–}7]$$

where the terms are as previously defined. The term $D_{d(\mathrm{min})}$ of equations [51–6] and [51–7] does not include water that moves laterally from adjacent areas or that is lost by seepage from canals or from other sources. To obtain the total quantity of water that must be drained, the depth of water from all other sources must be added to the value $D_{d(\mathrm{min})}$. From these equations it can be seen that the quantity of water to be drained increases as the salt content of the irrigation water EC_i increases, and as the permissible salt content of the drainage water EC_d decreases. Inadequate drainage may result from low permeability of the soil or from inadequate outlet facilities (tile lines, open drains or natural channels). If the leaching requirement of a given area or field is not being met because of impaired drainage, with the result that salts are accumulating in the root zone, the problem may be met in one of the following ways, or by a combination of these ways:

1) Improved drainage outlet facilities (more tile, more ditches, etc., see chapter 50).

2) Improved soil permeability [improvement of soil structure by cropping and management practices; or by other means, such as reducing exchangeable Na content in the case of alkali (sodic) soils].

3) Selection of a more salt-tolerant crop with a corresponding reduction in

the amount of water to be drained. By changing to a more tolerant crop, a higher salt content of the drainage water (EC_d) can be tolerated, thereby decreasing the quantity of water required to be drained (see equation [51–7]). If the amount of drainage is decreased, the corresponding decrease in water application must be accompanied by improved water management practices. The water must be applied more uniformly to control salinity and to meet the water needs of the crop.

Control of the depth of the water table is necessary in addition to passing and removing the required amount of water from the root zone. The water table must be at a depth sufficient to permit adequate aeration in the crop root zone and to allow the required farming operations to be carried out on the land surface. Control of the water table at a suitable depth also minimizes the salinization of the soil that occurs as a consequence of upward movement of water and salts from a water table. Between irrigations, or during periods when the land is idle, groundwaters move upward by capillarity, and the salts they carry are deposited in the surface soil as a result of evaporation. The rate of upward movement of water has been expressed as an inverse exponential function of depth to water table (Gardner, 1958). The maximum evaporation rate from a water table (for values that are less than those imposed by external evaporative conditions) is given by the equation

$$E = C \, a \, d^{-n} \qquad\qquad [51\text{–}8]$$

where E is the limiting evaporation rate, d is the depth to water table below the soil surface, and C, a, and n are constants. Values for a must be determined experimentally. Table 51–3 gives values of n, a, and C for several soils with textures from sand to clay. E is given in centimeters/day when units indicated in Table 51–3 are used in the equation. Values of b and K which are also included in Table 51–3 relate to the capillary conductivity equation given by Gardner (1958).

$$k = a/(S^n + b) \qquad\qquad [51\text{–}9]$$

where k is the capillary conductivity of the soil when the matric suction of the soil water is S; a, n, and b are soil constants. When $S = 0$, the capillary conductivity k becomes the hydraulic conductivity K and from equation [51–9] $k = K = a/b$.

Equation [51–8] can be used to estimate evaporation rates from fallow soils. An upper limit for the salinization rate can also be obtained if the electrolyte concentration of the groundwater is known. The evaporation rate times the electrolyte concentration gives the rate of salt accumulation in the soil. Doering et al.

Table 51–3. Experimental values of the soil parameters of equations 51–8 and 51–9 that are useful for estimating evaporation from water tables and evaluating soil flow characteristics [*]

Soil	Texture	n	a, $cm^{(n+1)}$/day	b, cm^n	C	K, cm/day
River sand	sand	4	1.7×10^8	2.5×10^6	1.52	68
Pachappa	fine sandy loam	3	3.2×10^5	2.6×10^4	1.76	12.3
Chino	clay	2	1.1×10^3	5.65×10^2	2.46	1.95
Diablo	loam	2	7.0×10^2	1.45×10^3	2.46	0.48
Yolo	light clay	2	4.0×10^2	4.0×10^2	2.46	1.0

[*] Data are from Gardner (1958), Gardner and Fireman (1958), and Willis (1960).

(1964) found a differential between the net rates of salt movement and water movement upward from a water table because of downward diffusion of salts in response to a concentration gradient. In a field trial involving a silty clay soil, a correction of 15 to 20%, added to the measured salinization rate, was required to account for the counter diffusion of salts. The salts generally accumulate in a relatively shallow depth of the surface soil. This depth has been shown to be < 15 cm for a fine sandy loam (Richards et al., 1956), and < 30 cm in a silty clay (Doering et al., 1964). Both counter diffusion and openness of structure (depth of cracking, etc.) among other things influence the distribution of the accumulated salts in the soil profile.

E. Soil Management

In general, soil management practices that improve permeability or improve irrigation also aid in the control of salinity. Applying water uniformly is perhaps the most important way in which salinity can be controlled. Where water is uniformly applied in quantities sufficient to meet both the consumptive use and the leaching requirement, excess salts are leached downward through the soil and discharged with the drainage waters. Where land surfaces are uneven, such as when leveling is uneven or where ridges or banks are formed, excess leaching occurs in the low spots and salts accumulate in the high spots. On sloping runs, such as are used with border irrigation, it is essential to have uniform grades and very little, if any, cross slope to achieve uniform water distribution. Salts are leached from areas of excess water penetration but accumulate where penetration is deficient. In areas where water infiltration rates are low, level-border irrigation has been used successfully for uniform distribution of water to the soil, and for effective control of salinity.

Where land is irrigated, attention should be given to areas of possible salt accumulation (i.e., parts of a valley or project; an entire farm or areas within a farm) and to the cultural practices that when combined with irrigation tend to move the salts away from or beyond the active crop root zone. If salinity control cannot be maintained during the regular cropping and irrigation regime with normal practices, consideration should be given to the possibility of off-season leaching. Excess salts can often be removed by leaching during the winter or other noncropping seasons.

Mulching has been used (Fanning and Carter, 1963) to enhance the removal of salts from localized saline soil areas. Mulching tends to improve salinity control by (i) reducing evaporation from the soil surface and thereby reducing the rate of salt accumulation, and (ii) increasing infiltration rates so that runoff is reduced and leaching by irrigation or rainfall is accomplished more effectively. This practice may be particularly useful in dryland farming areas where salinity is a problem.

Crop selection is an important factor that affects the control of salinity. The necessary quantities and frequency of water application and the method of distributing the water vary with crops. If salts tend to increase in the soil under a given cropping regime, leaching often can be accomplished and salinity controlled by changing to a crop that requires the application of more water, or changing to a crop that will allow water to be flooded or more evenly spread over the soil surface. For example, on soils of low permeability in the Imperial Valley, California, USA, salts accumulated when alfalfa was grown (Kelley et al., 1949). By adopting a crop rotation system including vegetable crops, which require greater quantities of water and which are not subject to scald, the salinity of the soil was reduced in one or two irrigation seasons.

V. RECLAMATION OF SALT-AFFECTED SOILS

Leaching with water is the only way known for effectively reclaiming salt-affected soils. Salts can be removed from saline soils by leaching alone, but for alkali (sodic) soils, a source of calcium or other divalent cation generally is required.

A. Reclaiming Saline Soils

Attempts have been made to reclaim saline soils by growing a crop that takes up relatively large amounts of salt, but this method has been neither effective nor practical. Flushing salts from the soil surface by passing water over the land surface also has been tried (Huberty et al., 1948). Theoretically, this method could be effective on soils of very low permeability where high concentrations of salt have accumulated on the soil surface. It was shown (Reeve et al., 1955), however, that flushing was ineffective in removing salts from a silty clay loam soil of relatively low permeability. In this case, flushing removed an insignificant amount ($< 1/10$ of 1%) of salt from the surface foot of soil.

The effectiveness of leaching varies from one soil to another, but for most soils, the quantity of water that passes through the soil is the principal factor governing the rate of salt removal. Figure 51–2 gives the results of experimental leaching studies (Reeve et al., 1955) showing the percentage of salt remaining in the soil profile as a function of depth of water applied per unit depth of soil, D_w/D_s. In general, about 50% of the salt is removed from the soil when the depth of water applied per unit depth of soil, $D_w/D_s = 0.5$; and about 80% of the salt is removed (20% remained) when $D_w/D_s = 1$. Theoretical analyses of the leaching problem and experimental studies with laboratory soil columns (Gardner and Brooks, 1957) showed for several soils that from 1.5 to 2.0 pore-volume replacements were required to reduce soil salinity to $< 20\%$ of its initial value. This is in general agreement with the results given in Fig. 51–2.

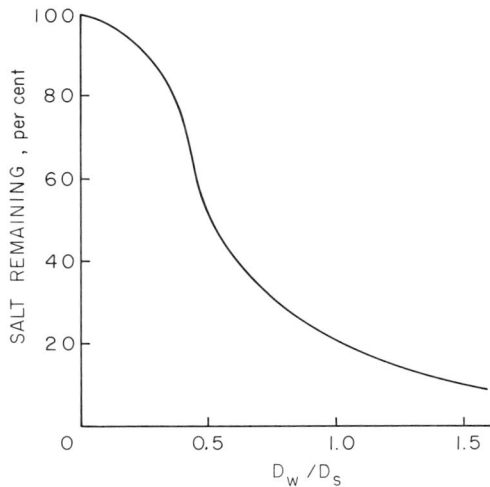

Fig. 51–2. Salt removal by leaching (expressed as percentage salt remaining in the soil) as a function of the ratio of depth of water to depth of soil, D_w/D_s (from Reeve et al., 1955).

The efficiency of leaching is influenced by variations in water flow velocities within a soil (Biggar and Nielsen, 1962). Soils having both very large interconnected pores and very small essentially isolated or blind pores, tend to transmit large quantities of water relative to the amount of salts removed. Leaching intermittently to allow more time for movement of water through the small pores, or applying the water at a rate less than the maximum ponded intake rate (i.e., allowing water entry under a slight suction) to minimize water movement through the large pores, will improve the leaching efficiency in such soils. For further details, *see* chapter 14.

B. Reclaiming Alkali (Sodic) Soils

Alkali soils are difficult and costly to reclaim for two reasons: First, because soluble calcium must be supplied to replace the excess exchangeable sodium, and secondly, because cation exchange usually proceeds slowly during leaching, owing to the characteristic low permeability of alkali soils to waters of relatively low electrolyte content. This results in a slow rate of replacement of the exchangeable sodium within the soil mass by the soluble calcium (gypsum, etc.). The usual practice is to apply a chemical amendment either directly to the soil or indirectly in the water applied for leaching. If the alkali problem is not severe, the necessary leaching may be accomplished by excessive irrigation during cropping. Where the exchangeable sodium content of the soil is high and the soil is leached with the usual irrigation waters, the time required for reclamation may be so great as to be impractical (because of the extremely low permeability of the soil), even though ample amounts of chemical amendment may have been applied.

The need for an amendment and the amount and kind of amendment to apply is best ascertained by soil tests. A common reclamation practice is to apply sufficient amendment to replace the adsorbed sodium from the top 6 to 12 inches of soil. In this way, the physical condition of the surface soil may be improved sufficiently to permit shallow-rooted crops to be grown. Additional reclamation will result from the passage of irrigation water through the soil; from the growth and decomposition of organic matter, and from the application of additional amounts of amendments.

The most common amendments used for alkali soil reclamation are gypsum, calcium chloride, sulfur, sulfuric acid, iron sulfate, aluminum sulfate, and lime-sulfur (Bower, 1958; US Salinity Lab. Staff, 1954). The soil tests that are needed to adequately diagnose an alkali soil problem and provide information on the amount and kind of amendment needed include: exchangeable sodium, gypsum, and lime contents and permeability of the soil (US Salinity Lab. Staff, 1954). Sometimes alkali soils naturally contain enough gypsum to replace part or all of the exchangeable sodium. In such cases, a relatively small amount or no amendment is needed.

Soluble salts of calcium (gypsum and calcium chloride) supply calcium directly to the soil, whereas acids act directly, and acid-forming amendments (sulfur, iron sulfate, aluminum sulfate, and lime-sulfur) are converted into acids that react with lime ($CaCO_3$) in the soil to form a soluble calcium salt (gypsum) that is available for exchange. Lime must be present or must be added to the soil if acids or acid-forming amendments are to be used successfully. The application of limestone alone for reclamation of an alkali lime-free soil will be of little benefit unless the pH of the soil is 6 or less.

The less costly amendments generally are slower to react than the more expensive ones. Calcium chloride, iron sulfate, and aluminum sulfate are quick-acting amendments but usually are too costly to be used for reclamation of most agricultural soils. Sulfuric acid is a quick-acting, but dangerous, amendment sometimes cheap enough for field use. Gypsum and sulfur, because of their relatively low cost, are the most commonly used amendments. The rate of reaction of gypsum is limited by its relatively low solubility (much less than 0.2% under field conditions), and that of sulfur, because it is slowly oxidized to a reactive compound by soil bacteria. *Under field conditions, the application of approximately 1 acre-foot of irrigation water is required to dissolve 1 ton of high-grade gypsum.*

In some instances, plowing to a depth of 36 to 48 inches may accelerate the reclamation of alkali soils. Sometimes a subsurface soil layer may be high in gypsum and, if brought to an overlying position or mixed with the soil, will provide a ready source of Ca. If a subsurface layer of highly permeable material such as sand or decomposed granite can be turned up and/or mixed with the soil, the water transmitting properties of the soil may be improved sufficiently to speed up reclamation. However, deep plowing of a relatively uniform soil high in exchangeable sodium seldom improves permeability or aids reclamation.

Plowing to a depth of 30 inches was shown to be effective in reclaiming sodic "slick spot" soils of the Sebree series in Idaho (Rasmussen et al., 1964; Rasmussen, 1965). In this instance, the subsoil contained sufficient amount of lime and gypsum to replace the exchangeable sodium in the profile. When plowing was followed by ordinary irrigation and cropping practices, the excessive exchangeable sodium was reduced to a safe level in the plant root zone within from 2 to 3 years.

Deep plowing of nonsodic saline soils (Chilcott series) also improved crop yields and rate and depth of both water and root penetration. In this case, the soil physical condition was improved by disruption of the hardpan and cemented subsoil layers and by thorough mixing of the soil horizons. With improved water intake and increased penetration soluble salts were effectively leached from the crop root zone in from 2 to 3 years by conventional irrigation methods.

It was further demonstrated that a 4-foot moleboard plow was satisfactory for plowing these soils, and that the cost varied from $35 to $45/acre. In most instances, the cost of plowing was repaid by increased yields in one or two cropping seasons.

While deep plowing is beneficial in improving tilth, the removal of excess salts depends on subsequent passage of excess water either by leaching or by over irrigation, and the exchangeable Na is replaced only if soluble Ca is supplied by bringing it up from a deeper depth or as an amendment. In many instances, the fertility of the soil may be altered considerably by this treatment. Practices should be instituted that will provide the needed fertility.

High-salt waters have been used as a flocculant and as a source of divalent cations for reclaiming alkali soils (Reeve and Bower, 1960). A distinct advantage of this method is that a relatively high soil permeability is maintained during leaching. Because of the pronounced effect of electrolyte concentration on the permeability of soils high in exchangeable sodium, the rate of water movement through the soil may be increased one, two, or more orders of magnitude. Thus, the time required for reclamation may be reduced to only a small fraction of that required when low-electrolyte waters are used.

Also involved in reclamation by this method is the "valence dilution" principle whereby divalent cations (Ca^{2+} and Mg^{2+}) contained in the applied water tend to exchange for the monovalent cations (Na^+) on the soil exchange complex simply as a result of adding water to the system, and thereby diluting the soil solution. The dilution may be accomplished continuously or stepwise.

In either case, on dilution some of the divalent cations in the applied water become available for exchange with the Na^+ held by the soil. As the concentration of the soil solution is reduced by the leaching water, the cation-exchange equilibrium between the solution and the soil is unbalanced. This causes the exchange reaction to proceed in the direction of the divalent cations in the soil solution replacing the monovalent cations held by the soil until a new equilibrium has been established between the leaching water and the soil. As the dilution of the soil solution by the applied water proceeds, the net result is for the Ca^{2+} and Mg^{2+} in the applied water to replace adsorbed Na^+ on the soil.

The possible magnitude of such changes between the soil solution and solid phases is shown by the change that occurs in the sodium adsorption ratio (SAR) of a water simply upon dilution and the fact that the SAR of an equilibrium solution is directly related to the ESP of the soil. If $(SAR)_o$ is the initial SAR of the water and d is the dilution factor [3] the initial SAR after dilution, $(SAR)_{dil}$ is given by the equation (Reeve and Bower, 1960)

$$(SAR)_{dil} = (SAR)_o d^{-1/2}. \qquad\qquad [51-10]$$

The change in SAR on dilution as given by equation [51–10] may be illustrated by considering sea water for which $(SAR)_o \approx 60$. The SAR values for successive dilutions of sea water, where $d = 2, 4, 8, 16$, and 32, are 42, 30, 21, 15, and 11, respectively. The corresponding estimated exchangeable sodium percentages for soil at equilibrium with waters of these SAR values are 39, 30, 23, 17, and 13, respectively.

It has been established experimentally that alkali (sodic) soils can be reclaimed by this method. However, the effect of electrolyte concentration on permeability of an alkali soil is known only qualitatively. For this method to be used successfully the electrolyte concentration of the leaching water must be regulated such that water continues to penetrate the soil at a rate that will allow for the required exchange reactions to occur in a reasonable period of time. Present knowledge of the relative proportions of divalent to monovalent cations in the water required for the method to be feasible is limited. However, it is known that the quantity of water required for leaching by this method increases very markedly as the percentage of divalent cations in the water decreases. The following are results of a field trial in which a soil high in exchangeable sodium was reclaimed by leaching with successive dilutions of a high-salt water: A high sodium soil (ESP = 79%, CEC = 0.25 meq/g) near Hemet, Calif. was reclaimed (Reeve & Doering, 1966) by the use of simulated sea water [concentration = 600 meq/liter, SAR = $58(meq/liter)^{1/2}$]. A total of 10 m of water was used to reduce the exchangeable

---

[3] $d = V_w/V_s + 1$, where V_w = volume of salt-free water added to a volume of salty water V_s.

Table 51–4. Time required to reclaim a highly sodic soil (ESP = 79) near Hemet, California by several different methods

	Leaching solution			
	Colorado River water	Saturated gypsum solution	Successive dilutions of sea water	Calcium chloride solution
Reclamation method	A	B	C	D
Time required	42 years*	7 years*	$\frac{1}{2}$ year†	3 days†

A-Colorado River Water (C = 11 meq/liter)
B-Saturated Gypsum solution (C = 35 meq/liter)
C-Successive dilutions of synthesized sea water (C = 600, 300, 150 and 75 meq/liter R = 0.217)
D-Successive dilutions of calcium chloride solutions (C = 600, 300, 150, 75, and 38 meq/liter) R = 1.00

* Calculated reclamation time required to give a reduction in exchangeable sodium equal to that of method D, based on the divalent cation content of the applied solution and the average intake rate during 228 days of continuous ponding.
† Observed reclamation time.

sodium percentage of the soil in the 0- to 90-cm depth from an initial value of 79 to a value of 24. Exchangeable sodium was reduced further by a light application of gypsum and continued cropping under irrigation with Colorado River water.

The time required for reclamation by the high-salt-water dilution method as compared with other methods is shown in Table 51–4.

The simulated sea water was very high in sodium—the ratio R of the divalent cation concentration to the total salt concentration was only 0.217. If R in their experiment had been 0.30, 0.40, and 0.50 instead of 0.217 with the total concentration remaining constant (C = 600 meq/liter), the calculated depth of water required to reclaim the soil would have been reduced by 50%, 75%, and 85%, respectively. Thus, the feasibility of the high-salt-water dilution method depends markedly on the availability of high-salt waters having R values of at least 0.30 or greater.

LITERATURE CITED

Bernstein, Leon. 1958. Salt tolerance of grasses and forage legumes. US Dep. Agr. Agr. Inform. Bull. 194. 7 p.

Bernstein, Leon. 1959. Salt tolerance of vegetable crops in the West. US Dep. Agr. Agr. Inform. Bull. 205. 5 p.

Bernstein, Leon. 1960. Salt tolerance of field crops. US Dep. Agr. Agr. Inform. Bull. 217. 6 p.

Bernstein, Leon. 1961. Tolerance of plants to salinity. Amer. Soc. Civ. Eng. Proc. J. Irrig. Drainage Div. 87:(IR4):1–12.

Bernstein, Leon. 1962. Salt-affected soils and plants. UNESCO Arid Zone Research 18:139–174.

Bernstein, Leon, and Robert S. Ayers. 1955. Sloping seedbeds. California Agr. 9(1_):8.

Bernstein, Leon, and Milton Fireman. 1957. Laboratory studies on salt distribution in furrow-irrigated soil with special reference to the pre-emergence period. Soil Sci. 83: 249–263.

Bernstein, Leon, A. J. Mackenzie, and B. A. Krantz 1955. The interaction of salinity and planting practice on the germination of irrigated row crops. Soil Sci. Soc. Amer. Proc. 19:240–243.

Biggar, J. W., and D. R. Nielsen. 1962. Miscible displacement: II. Behavior of tracers. Soil Sci. Soc. Amer. Proc. 26:125–132.

Bower, C. A. 1958. Chemical amendments for improving sodium soils. US Dep. Agr. Agr. Inform. Bull. 195. 9 p.

Chapman, Homer D., and Parker F. Pratt. 1961. Methods of analysis for soils, plants, and waters. Univ. California Div. Agr. Sci. 309 p.

Doering, E. J., R. C. Reeve, and K. C. Stockinger. 1964. Salt accumulation and salt distribution as an indicator of evaporation from fallow soils. Soil Sci. 97:312–319.

Eaton, F. M. 1954. Formulas for estimating leaching and gypsum requirements of irrigation waters. Texas Agr. Exp. Sta. Misc. Publ. 111. 18 p.

Fanning, Carl D., and David L. Carter. 1963. The effectiveness of a cotton bur mulch and a ridge-furrow system in reclaiming sodic soils by rainfall. Soil Sci. Soc. Amer. Proc. 27:703–706.

Gardner, W. R. 1958. Some steady-state solutions of the unsaturated moisture flow equation with application to evaporation from a water table. Soil Sci. 85:228–232.

Gardner, W. R., and R. H. Brooks. 1957. A descriptive theory of leaching. Soil Sci. 83:295–304.

Gardner, W. R., and Milton Fireman. 1958. Laboratory studies of evaporation from soil columns in the presence of a water table. Soil Sci. 85:244–249.

Hayward, H. E. 1954. Plant growth under saline conditions. UNESCO Arid Zone Research 4:37–71.

Hill, R. A., and C. S. Scofield. 1938. Salt concentration and service equivalence. The Rio Grande Joint Investigation in the Upper Rio Grande Basin in Colorado, New Mexico, and Texas. US Nat. Resources Comm. US Government Printing Office. Washington, D.C. I(7). 566 p.

Huberty, M. R., A. F. Pillsbury, and V. P. Sokoloff. 1948. Hydrologic studies in Coachella Valley, California. Univ. California, Berkeley, Spec. Publ. 31 p.

Kelley, W. P. 1951. Alkali soils, their formation properties and reclamation. Reinhold Publ. Co., New York. 176 p. + illus.

Kelley, W. P., B. M. Laurance, and H. D. Chapman. 1949. Soil salinity in relation to irrigation. Hilgardia 18:635–665.

Rasmussen, W. W. 1965. Deep plowing as a management practice for improving solonetzic (slick spot) soils. US Dep. Agr. ARS 41–91. 36 p.

Rasmussen, W. W., G. C. Lewis, and M. A. Fosberg. 1964. Improvement of the Chilcott-Sebree (solodized-solonetz) slick spot soils in southwestern Idaho. US Dep. Agr. ARS 41–91. 36 p.

Reeve, R. C. 1957. The relation of salinity to irrigation and drainage requirements. Int. Comm. Irrig. Drainage, 3rd Congr. (San Francisco, California.) Question 10, Report 10, p. 10.175–10.187.

Reeve, R. C., and C. A. Bower. 1960. Use of high-salt waters as a flocculant and a source of divalent cations for reclaiming sodic soils. Soil Sci. 90:139–144.

Reeve, R. C., and E. J. Doering. 1966. Field comparison of the high-salt-water dilution method and conventional methods for reclaiming sodic soils. Int. Comm. Irrig. Drainage. 6th Congr. (New Delhi, India) Question 19, Report 1, p. 19.1–19.14.

Reeve, R. C., A. F. Pillsbury, and L. V. Wilcox. 1955. Reclamation of a saline and high boron soil in the Coachella Valley of California. Hilgardia 24:69–91.

Richards, L. A., W. R. Gardner, and Gen. Ogata. 1956. Physical process determining water loss from soil. Soil Sci. Soc. Amer. Proc. 20:310–314.

Scofield, C. S. 1940. Salt balance in irrigated areas. J. Agr. Res. 61:17–39.

United States Salinity Laboratory Staff. 1954. Diagnosis and improvement of saline and alkali soils. US Dep. Agr. Handbook 60. 158 p.

Wadleigh, C. H., and Milton Fireman. 1949. Salt distribution under furrow and basin irrigated cotton and its effect on water removal. Soil Sci. Soc. Amer. Proc. (1948) 13:527–530.

Wadleigh, C. H., H. G. Gauch, and Maria Kolisch. 1951. Mineral composition of orchard grass grown on Pachappa loam salinized with various salts. Soil Sci. 72:275–282.

Wilcox, L. V., and W. F. Resch. 1963. Salt balance and leaching requirement in irrigated lands. US Dep. Agr. Tech. Bull. 1290. 23 p.

Willis, W. O. 1960. Evaporation from layered soils in the presence of a water table. Soil Sci. Soc. Amer. Proc. 24:239–242.

52 | Fertilizers in Relation to Irrigation Practice[1]

FRANK G. VIETS, JR.

Agricultural Research Service, USDA
Fort Collins, Colorado

R. P. HUMBERT

American Potash Institute, Inc.
Los Gatos, California

C. E. NELSON

Washington State University
Prosser, Washington

I. INTRODUCTION

Irrigation poses some special problems in the use of fertilizers and also provides some unique ways to supply nutrients not encountered in nonirrigated agriculture. First, irrigation imposes a greater demand for fertilizer nutrients. For irrigation to be profitable, yields must be high. Higher yields mean greater nutrient uptake by crops with the nutrient uptake being roughly proportional to the crop yield (*see* chapter 24). The increased needs for N were discussed in a recent review (Viets, 1965). Therefore nutrient needs for irrigated crops must be met by an adequate fertilizer program if irrigation water is to be used efficiently (Viets, 1962).

The movement of soluble nutrients in the soil and their availability to plant roots are highly dependent on the method and frequency of irrigation. Some basic aspects of nutrient availability in relation to soil water suction were discussed in chapter 24. On the other hand, irrigation itself often provides a convenient and economical method of fertilizing the crop.

The purpose of this chapter is to discuss some of the special problems of nutrient movement arising from irrigation, some of the fertilizer problems posed by the choice of irrigation method, and the application of fertilizers in irrigation systems. No further mention is made of fertilizer requirements as affected by crop yield, but it should be emphasized that the failure to meet increased fertilizer requirements is one of the more common faults of new irrigation projects. Some nutrients are applied to crops as foliar sprays dissolved in water, injected into soils in water, or applied in water to the roots of transplants; the use of water in this sense can scarcely be called irrigation and is beyond the scope of this chapter.

[1] Contribution of the Soil and Water Conserv. Res. Div., ARS, USDA; the American Potash Institute; and Washington Agr. Exp. Sta.

II. MOVEMENT OF FERTILIZER NUTRIENTS IN SOIL WATER

All of the 14 mineral nutrients required by crop plants can be purchased in a water-soluble form in fertilizers and occur in the soil solution. Regardless of their water solubility, most of the fertilizer nutrients added to soils either as dry fertilizers or in solution in irrigation water quickly undergo chemical reactions with constituents of the soil so that the concentration remaining in solution is vastly changed. The extent of these reactions depends on the kind of ion and the kind and amount of adsorption complex in the soil. Knowledge of these chemical reactions is essential to understanding how irrigation method affects nutrient distribution in soil and why some nutrients can be effectively applied in irrigation water and others cannot. This discussion of movement in soil water is the gross movement over considerable distance and is distinct from the movement over microdistances of great nutritional significance discussed in chapter 24.

The nutrient anions, NO_3^-, Cl^-, and SO_4^{2-}, are completely mobile in the soil water because of the absence of strong adsorption forces in the neutral and alkaline soils, characteristic of most irrigated regions, that hold most other ions to the clay and humus.

The NO_3^- form of N remains exclusively in the soil water and is subject to appreciable movement downward, laterally, or upward with the soil water. Gardner (1965) has presented a theoretical treatment of NO_3^- and water movement. Experiments and theory on miscible displacement in soils and its possible significance in the movement of NO_3^- and other soluble ions are also discussed in chapter 14 of this monograph. Burns and Dean (1964) discussed another phenomenon called "drop out" that may be of considerable significance in the movement downward of soluble materials like NO_3^- placed in bands. According to their explanation and experiments with $NaNO_3$, water moves to the fertilizer solution of high osmotic pressure in the band. Water content of the soil is increased in the band and if this water movement is sustained, the soluble nutrients move downward because of the high density of the solution compared to the density of the rest of the soil solution.

The amount of NO_3^- in the soil solution at a given time depends on the recency of adding NO_3^--containing fertilizers, the nitrification rate of NH_4^+ or NH_3 from fertilizers and the decomposition of soil organic matter, and the amount removed by plants and soil microbes decomposing low-N materials like straw and corn stalks. Some knowledge of the amount of NO_3^- in the soil is extremely important in the management of irrigation water. For example, soils should not be leached to get rid of salts soon after N fertilization. The best time is in the fall after the crop has been removed and NO_3^- has been depleted by crop growth. Also, Robins et al. (1956) have shown that excessive irrigation of sugar beets (*Beta vulgaris* L.) early in the season after N fertilization markedly reduced yields, but that the same amount of excess water applied late in the season has no such effect. Nitrate in significant amounts often occurs in drainage from tile lines, demonstrating the mobility of this ion in soil water.

Gross movement of the other nutrient anions in irrigation water is much less significant in irrigation practice than the movement of NO_3^- simply because soils are generally better supplied with them and areas of deficiency are much less common. Plants get their sulfur from SO_4^{2-} in fertilizers and irrigation water and by the slow release of S from organic matter. Sulfate is a common constituent of saline soils and its movement is important in reclaming such soils. No areas of chloride deficiency on irrigated soils are known and are not likely to be found. Chloride, like SO_4^{2-}, is important in saline soils. Some of the boron in soils is soluble and moves with the irrigation water. Soluble B can accumulate with other salts where drainage is poor. Irrigated areas with B deficiency and toxicity are known, but are not very common. Molybdenum deficiency is virtually unknown in irrigated areas because the neutral to alkaline soil pH favors Mo availability.

Urea-N is weakly held by the soil and most of it is free to move with the soil water as long as it remains as urea. Loomis et al. (1960) found that sidedressing sugar beets with urea, then immediately irrigating, corrected late season N deficiency more quickly than NH_4^+-N sources because the urea was carried deeper into the root zone. However, urea is so quickly hydrolyzed to NH_4^+ in most soils that its movement is of little practical importance and generally should be considered relatively immobile like NH_4^+.

Ammonia- or ammonium-N occurs in soils in two forms: the NH_4^+ ion and NH_3 gas or NH_4OH. The equilibrium between the two forms in solution depends on the solution pH. The form in which they are added to the soils has a marked effect on pH near the point of application, NH_3 or aqua NH_3(NH_4OH) producing a very alkaline solution. In solution at pH 6, there is 1,800 times as much NH_4^+ as NH_4OH plus NH_3, but at pH 8 there is only 18 times as much, and at pH 9 there is only 1.8 times as much. The important point for this discussion is that NH_4^+ is relatively immobile in the soil, but NH_4OH and NH_3 move much more freely.

Ammonium-N (the NH_4^+ ion) is held rather tightly by the exchange complex of soil clays and organic matter, but the extent of this adsorption depends on the amount of clay and organic matter in the soil, the composition of the percolating water, and the degree of base saturation. Ammonium will move much farther with the soil water in a sandy soil than it will in one of clay. Movement is greater in alkaline soils than in slightly acidic ones. Movement also depends on how much is applied per unit of soil volume, movement being greater when the NH_4^+ adsorption capacity of an exchange site is satisfied. The amount of total NH_4^+ in the soil depends largely on the recency of fertilization with NH_4^+ salts, anhydrous or aqua NH_3, and the rate of nitrification. For practical purposes, the movement of NH_4^- in soil water can almost be disregarded on all soils except sands and loamy sands. For example, Nelson (1953, 1961) found that NH_4^+ moved only 3 or 4 inches when applied as NH_4NO_3 to two fine sandy loams that were furrow irrigated. Ammonium moves further in a soil when injected as anhydrous NH_3 or applied as aqua NH_3 than when applied as $(NH_4)_2SO_4$ or $NH_4H_2PO_4$ (Dow et al., 1953). They found that NH_4^+ applied as a salt in irrigation water did not move far and displaced exchangeable Ca^{2+}, but that NH_4OH (aqua NH_3) moved much farther and did not displace Ca^{2+}. Ammonium also moves up more readily than K^+ and other cations, particularly in alkaline soils. Part of these behaviors can be explained by movement of NH_3 as gas. Ammonium seldom occurs in drainage waters from soils.

Phosphorus as phosphate is so strongly adsorbed by the absorption complex in the soil that there is little in the soil solution except in the very acidic solution immediately surrounding a superphosphate granule or near a H_3PO_4 application. Most irrigated soils are Ca-dominated systems in which the solubility of P is controlled by the solubility product of a calcium phosphate system. The concentration of P in solution is in the range of 0.05 ppm for soils low in available P to about 0.5 ppm for soils liberally fertilized with superphosphate (Olsen and Watanabe, 1963). The solubility of P is so low that it can be considered immobile with respect to appreciable movement in irrigated soils, except on sand. For example, after irrigation Converse (1948) noted deep penetration of P from superphosphate applied to Superstition sand in Arizona. In spite of the immobility of P in soil water, topdressings of superphosphate applied to established alfalfa (*Medicago sativa*) fields on alkaline soils have been effective. Phosphoric acid applied in irrigation water or to the soil surface moves deeper into the soil than other soluble forms of phosphate. Ulrich et al. (1947) e.g., applied H_3PO_4 at a rate of 1,000 lb of P/acre to a soil with high phosphate fixing capacity and found that P had penetrated 11 inches in 43 days. Water had penetrated at least 20 inches. Olsen et al. (1950) found that P in H_3PO_4 moved deeper than P from concentrated superphosphate and that amount of water applied had little effect on depth of movement. At least 85% of the applied H_3PO_4 remained in the 4-inch surface soil. Significant amounts of phosphate never escape in drainage waters from fields.

Potassium, Ca^{2+}, and Mg^{2+}, like NH_4^+, are adsorbed by the exchange complex of the soil in an equilibrium with other ions in the soil solution. Ions in excess of adsorption demands, remain in solution and are free to move with the soil water. Movement depends on how much and where the fertilizer is applied, the salt status of the soil, the salt content and amount of water applied, and the exchange capacity as determined by kind of clay, soil texture, and the humus content. Potassium, because of its movement away from concentrated bands, should not be placed with seeds or in bands close to them. Nelson and Wheeting (1941) found that K^+ moved 4 to 8 inches laterally from a band into the ridge by furrow irrigation of a fine sandy loam. Appreciable amounts of K^+, Ca^{2+}, and Mg^{2+}, along with Na^+, can be removed from saline soils by leaching. In nonsaline irrigated soils, there is little movement of these cations out of the profile unless N fertilization greatly exceeds crop demands. Then they move out in association with NO_3^-.

The micronutrient cations Zn, Cu, Co, Mn, and Fe are so tightly held in well aerated soils, except quartz sands, that there is practically no movement in the soil by water. In summary, the mobile nutrients are NO_3^-, Cl^-, B, and SO_4^{2-}, and urea-N for 1 or 2 days after application. The ions with limited mobility are K^+, Ca^{2+}, Mg^{2+}, and NH_4^+, depending on conditions, and the immobile ones are P and the micronutrient cations.

III. MOVEMENT OF FERTILIZER NUTRIENTS AS AFFECTED BY METHOD OF IRRIGATION

Irrigation methods can be divided into three general classes: furrow, flood, and sprinkler. Redistribution of nutrients by movement of soil water has been studied more under furrow irrigation systems than under the others. Movement from bands has been studied more than movement from broadcast applications or from those in irrigation water simply because furrow irrigation and banded applications lead to the greatest variation in distribution of available nutrients both laterally and vertically in the soil.

Furrow irrigation leads to marked redistribution of NO_3^- and other mobile ions like SO_4^{2-} and Cl^- but has little influence on the movement of the immobile ions. Some of the effects of NO_3^- movement are desirable and some are not. Among the desirable effects are the movement of NO_3^- from a band of fertilizer, established at planting or by later sidedressing between the furrow and the seed or plant, over to the root zone of the seedling or deeper into the profile where roots are established. In fact, often there is no benefit from a sidedressing until the crop is irrigated or rain washes the NO_3^- into the root zone. Nelson and Wheeting (1941) studied the movement of nutrients from a band of mixed fertilizer containing $NaNO_3$, superphosphate, and KCl placed 8 inches from an irrigation furrow on a fine sandy loam. Although irrigation infiltrated water to a depth of almost 5 ft, NO_3^- moved laterally about 14 inches and down only 6 inches. Greatest downward movement was to the lee side of the band. The movement of K^+ was only about 10 inches laterally and 4 inches deep. No measurable movement of P occurred. Nelson (1953) showed that NO_3^- from NH_4NO_3 broadcast and plowed down was distributed much more extensively in the profile after furrow irrigation than when the same amount of N was banded midway between furrows. In the latter situation, NO_3^- concentrated between the band and the soil surface and some downward movement occurred directly below the band. Nelson (1961) later studied the movement of NH_4^+ and NO_3^- from bands of N from five different N sources placed on each side of potato (*Solanum tuberosum*) rows in two different arrangements of row widths which changed the elevation of the furrow with respect to the fertilizer band. With potatoes in 34-inch rows and ridged, the rill was level with the fertilizer band. In this case, the predominant movement of NO_3^- was into the row of potatoes and upward into the ridge with the capillary movement of water. With potatoes in 26-inch rows and unridged, the fertilizers were 3 inches below the irrigation furrow. In this situation, NO_3^- moved into the row and downward. Movement of NO_3^- downward was much greater in this system. Movement of NH_4^+ was much less than NO_3^- and was mostly lateral movement into the row in both systems of culture.

The mentioned experiments point up the fact, previously investigated by Haise (1950), that the flow lines of water control the direction and magnitude of NO_3^- movement. If the NO_3^- is above the water level, NO_3^- will concentrate at the surface and in a ridge, if there is one. If the NO_3^- is placed below the water level, the movement will be lateral and downward. The closer the furrows, the greater the downward movement of NO_3^- and the less movement into the ridges and middles between them.

The upward and lateral movement of NO_3^- and soluble salts into ridges has some very practical applications. Seeds placed on the margins of furrows are more likely to escape salinity since salts move away from the roots. The concentration of NO_3^- in ridges can lead to poor N recovery by the crop because the ridge is dry much of the time and there is little root activity in it (Nielson and Banks, 1960). Stout (1961) discussed NO_3^- accumulation in the ridges of beet rows in the low summer rainfall, intermountain area of the western USA and its implications as to lower sugar content and purity of sugar beets if fall rains wash the accumulated NO_3^- into the active root zone.

In flood irrigation that covers all of the surface with water in level systems of irrigation and with drilled crops on gently sloping systems, water and soluble ions move down. Flooding is more effective than furrow irrigation in removing soluble salts from saline soils. The same considerations and theory apply to movement of NO_3^- and other mobile ions of nutritional significance. Tyler et al. (1958) found that NO_3^- moved down with the wetting front on four flood-irrigated soils in laboratory boxes.

Miscible displacement principles discussed in chapter 14 may have real significance in the relative movement of soluble nutrient ions in comparisons of flood vs. sprinkler irrigation systems. However, these principles have not been extensively tested except for ions involved in salinity. This theory predicts that soluble ion movement on most soils will be deeper in the profile with a given amount of water when the water is applied slowly or intermittently by sprinklers than when the water is applied by flood irrigation.

IV. FIELD COMPARISONS OF IRRIGATION METHOD ON FERTILIZER REQUIREMENTS

Few rigid comparisons have been made on the effects of method of irrigation on crop yields and fertilizer requirements. The most extensive studies have been conducted at Logan, Utah, USA, on Millville loam, a black soil containing > 2.4% organic matter and > 50% $CaCO_3$ equivalent. Dr. J. L. Haddock and associates have compared crop yields and composition of canning peas (*Pisum sativum*) 2 years of alfalfa, potatoes, and sugar beets grown in rotation as affected by soil water regime, fertilizers and furrow vs. sprinkler irrigation. Data for 1950 (Haddock, 1954) showed that N and P fertilizers did not increase the yields of sprinkled beets but did increase the yields of furrow-irrigated beets (Table 52–1). Without fertilizer, sprinkled beets produced as much roots and sugar as beets that got 80 lb of N/acre as $(NH_4)_2SO_4$ plowed down before planting. Sprinkled beets had more tops and a greater top-to-root ratio than furrow-irrigated beets. Furrow-irrigated beets had higher purity and sucrose percentage than sprinkled beets. Under these soil conditions, sprinkler irrigation favored N uptake (presumably due to greater decomposition of soil N) which increased root yields and sugar yields, but decreased the sugar content and purity. Similar results were reported for the 1956 crop of beets (Haddock, 1959), except the favorable effects of sprinkler irrigation as compared to furrow irrigation on roots and sugar yields were less. Petioles of sprinkled beets were much higher in NO_3^- than the petioles of furrow-irrigated beets, in line with the lower root purity and greater top-root ratio noted with sprinkled beets. For 'Russett

Table 52–1. Effect of furrow (F) vs. sprinkle (S) irrigation and fertilizer on sugar beets at Logan, Utah, USA (from Haddock, 1954)

Fertilizer		Yield of roots		Sugar content		Sugar yield		Top-root ratio	
N	P	F	S	F	S	F	S	F	S
lb/acre		tons/acre		%		tons/acre			
0	0	15.3	17.7	17.1	16.7	2.6	3.0	0.67	0.93
0	44	15.8	17.9	17.2	16.8	2.7	3.0	0.66	0.83
80	0	18.4	17.8	16.7	16.0	3.0	2.9	0.83	1.00
80	44	19.1	18.3	16.8	16.0	3.2	2.9	0.80	1.01

Burbank' potatoes grown in these rotations, "No advantage in yield or quality of potatoes was observed for either sprinkler or furrow irrigation when soil water suction in the root zone was comparable" (Haddock, 1961).

These studies have emphasized the complexities of comparing furrow with sprinkler irrigation. Attention must be given in such studies to the soil water suction and its uniformity in the root zone, the rate of fertilizer application and its movement in the soil water, and the effect of the soil water regime on mineralization of nutrients from the soil, to mention the principal factors. Certainly, the relative performance of the two irrigation systems is going to depend on the crop and the kind of soil.

V. APPLICATION OF FERTILIZERS IN IRRIGATION WATER

When men of early civilizations developed crude sanitary systems in their cities and used the canal water to irrigate crops, they started fertilizer application in water. In fact, water used for irrigation often contains enough K, SO_4, B, Cl, Ca, and Mg from mineral sources to meet all crop requirements on a sustained basis. Hence, water analysis at the start of a project is just as basic to predicting the fertilizer program as it is to assessing the salinity hazards and drainage requirements.

Irrigation water is often the cheapest and most convenient means of applying fertilizers. When an irrigated crop runs out of N in mid- or late season, application in water or by air is the only way fertilizer can be applied. In California, USA, the fertilizer industry surveys show that about 5% of the N and 1% of the total fertilizer sold in 1964 were applied in irrigation water. Irrigation systems for new golf courses, athletic fields, greenhouses, and specialty crops are now being designed to permit application of fertilizers in the irrigation water.

A. Kinds of Fertilizer Materials Applied

Fertilizers suitable for use in irrigation water must be quickly and completely soluble. They must not react with each other or with constituents of the water to form screen-, pump-, and nozzle-clogging precipitates. They must not react with dissolved oxygen and precipitate, as ferrous iron will react and settle to the bottom of the canal or pipeline as rust.

Nitrogen is, by far, the nutrient used in greatest quantities in irrigation water. Anhydrous NH_3, first marketed in the Western USA over 20 years ago, accelerated interest in fertilizer application in irrigation water. Research indicated it was possible to use this cheaper method of application, unless poor water distribution and other factors restricted its use. Tanks of anhydrous and aqua NH_3 are commonly seen along irrigation ditches today (Fig. 52–1). The gas or liquid N solutions are metered into the irrigation water and flow with the water to its point of infiltration. Ammoniated solutions of NH_4NO_3, anhydrous or aqua NH_3, and urea are the most commonly used liquid N fertilizers. Their low relative cost and ease of handling and application have resulted in their increased use recently. Certain dry N fertilizers are favored over others because of differences in water solubility in making simple solutions. The common dry N carriers in order of

Fig. 52–1. Application of anhydrous ammonia to potatoes in California (Courtesy Shell Chemical Company).

decreasing solubility are: NH_4NO_3, $Ca(NO_3)_2$, urea, $NaNO_3$, and $(NH_4)_2SO_4$; the range in solubilities being 118 to 71 parts of material in 100 parts of water. Ammoniated solutions of these compounds are now commonly used as liquid N fertilizers.

Liquid H_3PO_4 and solutions of K fertilizers are also being applied in irrigation water (Humbert, 1963; Jones and Green, 1946). The low solubilities of the commonly used muriate and sulphate of potash make these materials less suitable than KNO_3 which is now on the market at reasonably competitive prices. Mixed solid and liquid fertilizers are used in both surface and sprinkler irrigation systems. Special care is taken to have compatible mixtures of the most soluble macronutrients and micronutrients. The rapidly increasing use of liquid fertilizers in the western USA, however, will probably mean smaller quantities of solids being applied directly into irrigation water. Gypsum is added to irrigation water in parts of the San Joaquin Valley, California to counteract the deleterious effects of high Na in the water on infiltration rates.

B.   Application Equipment and Methods

Irrigation and fertilizer equipment companies now make the needed tanks, pumps, and injection systems for mixing either soluble dry or liquid fertilizers into the irrigation water for distribution in ditches, pipelines, or sprinkler systems (Fig. 52–2). Anhydrous NH_3 is injected directly into the irrigation stream by its own pressure, the amount being known from gauge pressure and orifice size. Names of equipment manufacturers are available in irrigation and fertilizer trade journals. Harrison (1965) provides a partial list. A satisfactory system must provide a constant flow of the fertilizer solution per unit of water applied, but this constant rate must be variable so that different rates of fertilizer application can be fit to the requirements of different crops and to the depth of water applied. The system must provide turbulence in the water flow after the point of fertilizer injection to ensure adequate mixing. All components of the system must be resistant to chemical corrosion. Ammonium nitrate, H_3PO_4, and $(NH_4)_2SO_4$ are very corrosive. Bronze and brass bearings and sprinkler heads are attacked by phosphate, particularly in the presence of NH_4^+. All equipment and lines

Fig. 52–2. Application of fertilizer in sprinkler irrigation system (Courtesy Dragon Engineering Company).

should be thoroughly flushed with water toward the end of the irrigation period. Nutrients that will not move much in the soil should be applied toward the beginning of the irrigation period so that water will have maximum opportunity to carry them into the root zone. Nitrate (NO_3^-) should be applied late in the irrigation period to ensure against too deep penetration. If irrigation water is hard, NH_3 or ammoniated solutions will precipitate $CaCO_3$ which clogs valves and screens. The addition of a polyphosphate water conditioner will prevent this precipitation.

C. NH₃ Losses by Volatilization

In application of aqua NH_3 or NH_4^+ salts in water, losses of NH_3 by vaporization from the water as well as the soil must be considered. In studies of losses between the sprinkler jet and the ground, Henderson et al. (1955) report that losses as great as 60% may be experienced when anhydrous NH_3 is used, and that NH_3 loss from $(NH_4)_2SO_4$, NH_4NO_3, and $(NH_4)H_2PO_4$ may be as high as 10%. Their results are shown in Figs. 52–3 and 52–4. Losses from aqua NH_3

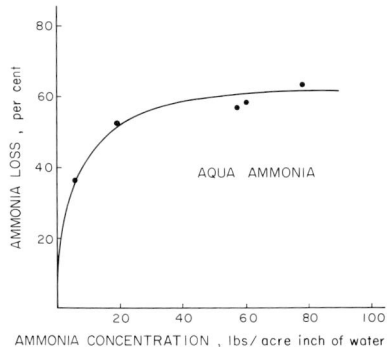

Fig. 52–3. Losses of aqua ammonia in relation to concentration of ammonia in the irrigation water (Henderson et al., 1955).

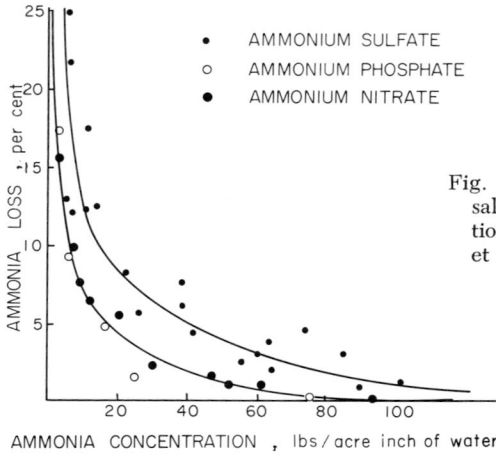

Fig. 52–4. Loss of ammonia from fertilizer salts in relation to ammonia concentration in the irrigation water (Henderson et al., 1955).

solutions are very high, and increase with initial concentration. Percentage losses from solutions of NH_4^+ salts were appreciable at low initial concentration, but decreased rapidly as the initial concentration was increased. Losses from NH_4NO_3 and $(NH_4)H_2PO_4$ were essentially the same at equal NH_4^+ concentrations and are represented by a single curve in Fig. 52–4. Losses from $(NH_4)_2SO_4$ were appreciably higher.

The losses of NH_3 in relation to the pH of the fertilizer solution are shown in Fig. 52–5. These data show that if NH_3 losses from the sprinkler jets are to be kept below 10%, the pH of the fertilizer solution should be 8.0 or less; approximately 5% loss can be expected at pH 7.5, and loss is negligible if the pH is 7.0 or below. The pH depends on the kind and amount of fertilizer and the initial pH and buffer capacity of the water. High temperature of irrigation water

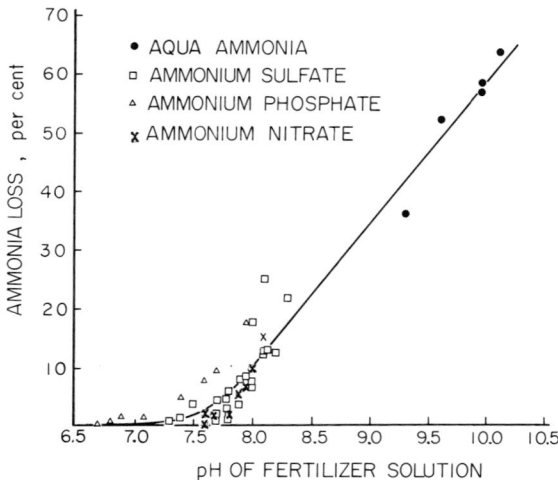

Fig. 52–5. Losses of ammonia from aqua ammonia and ammonium salts in relation to the pH of the fertilizer solution (Henderson et al., 1955).

increases NH_3 losses. Henderson et al. (1955) in a series of tests with $(NH_4)_2SO_4$ applied at 14 lb/acre-inch at temperatures of 20, 25, and 32C (68, 77, and 90F) got losses of 5.2%, 6.6%, and 7.6%, respectively.

Field studies show that there may be large losses of NH_3 from aqua NH_3 in an irrigation furrow. Humbert and Ayres (1956) got a 21% average decrease in the NH_3 content of irrigation water as it flowed the length of an irrigation furrow. Even in furrows lined with plastic, the loss was 20%. Losses are generally small if water is not too alkaline and NH_3 concentration is kept below 100 ppm (20 lb of N/acre-inch).

Another potential source of N loss is the volatilization of NH_3 from the surface after irrigation ceases and the soil begins to dry. This volatilization loss is dependent on the temperature, the rate of drying, and the soil pH. Ammonia in the irrigation water increases the pH of surface soil until subsequent nitrification of the NH_3 lowers the pH. Many studies have shown that there are appreciable volatilization losses of NH_3 from surface applications of aqua NH_3 or NH_4 salts to soils even at pH 6.5, and that losses are greater from calcareous soils. Few studies of NH_3 loss from soils after irrigation have been conducted, however. Humbert and Ayres (1957) applied aqua NH_3 and $(NH_4)_2SO_4$ to the surface of a neutral Gray Hydromorphic soil of pH 7.3 in Hawaii and found the NH_3-N losses over 70 days to be about 16% of the N applied. The initial losses were considerably higher from aqua NH_3 than from $(NH_4)_2SO_4$, but after several drying cycles, the cumulative losses were comparable. Loomis et al. (1960) found that aqua NH_3 injected into irrigation water was unsatisfactory for the late season correction of N deficiency of sugar beets grown on a calcareous soil in the Imperial Valley, California because of volatilization losses during and after irrigation.

In view of the losses of NH_3 from irrigation water and from soil surfaces and the widespread use of NH_3 and NH_4^+ salts in water, it is surprising that so few quantitative data exist on the efficiency of this method of fertilization.

D. Application Efficiency

Poor distribution of water in either sprinkler or furrow irrigation results in uneven application of fertilizer dissolved in the irrigation water. Humbert (1954) and co-workers used radioactive materials and perfected tracer techniques to study water distribution under various systems of surface irrigation. Their field studies on sugarcane (*Saccharum officinarum*) resulted in modifications in field layouts, lengths and slopes of lines, volume of water discharge, and time of irrigation to improve irrigation performance so that smaller quantities of water would be more evenly distributed, thus permitting fertilization in irrigation water. Meek et al. (1964) used similar techniques in studying water distribution in California. The concentrations of NH_3 usually used in water are 75 to 100 ppm. The 100-ppm concentration results in the application of 82 lb of N/acre in a 4-inch irrigation. At concentrations of 75 to 100 ppm, distribution of NH_3 along the irrigation furrow is reasonably uniform. Leavitt (1957) reports concentrations within 5% variation between the head and the end of the furrows when irrigation water is properly controlled. Bryan and Thomas (1958) reported that concentrations of Na^+ from $NaNO_3$, K^+ from a mixed fertilizer, and NO_3^- from NH_4NO_3 were

Table 52–2. Onion yields in hundred weights/acre from N and P fertilizers applied in irrigation water compared with yields from placed dry fertilizers (from Lorenz et al., 1955)

Source of material and application*	Year of test			
	1950	1951	1952	1953
No fertilizer	289	336	132	161
Dry fertilizers placed under row				
Ammonium sulfate	314	432	222	---
Treble superphosphate	396	542	145	---
Ammonium sulfate and treble superphosphate	571	618	412	736
Nitrogen in irrigation water (treble superphosphate under row)				
Aqua ammonia	413	516	285	537
Ammonium sulfate	---	530	264	550
Calcium nitrate	407	578	---	---
Nitric acid	---	---	179	---
Urea	444	590	196	---
Phosphoric acid in irrigation water (Ammonium sulfate under row)				
Phosphoric acid	459	546	315	---
Nitrogen and phosphoric acid in irrigation water				
Aqua ammonia and phosphoric acid	395	546	---	---
LSD (.05)	83	85	45	73

* Fertilizer application was 120 lb of N and 52 lb of P/acre.

satisfactorily uniform as measured by uniformity coefficients in tests with a sprinkler system. They did not report on the uniformity of water distribution which is just as important for uniform distribution of fertilizers.

E. Yield Results

Lorenz et al. (1955) compared the yields of onions (*Allium cepa* 'San Joaquin') grown on calcareous Hesperia sandy loam (pH 7.8) and fertilized with liquid materials applied both by injection into the bed and by furrow irrigation with yields obtained from dry fertilizers placed under the plant row. The results are shown in Table 52–2. Highest yields were obtained from $(NH_4)_2SO_4$ and treble (conc.) superphosphate placed under the plant row. Response was obtained to both N and P. $Ca(NO_3)_2$, $(NH_4)_2SO_4$, aqua NH_3, and urea applied in irrigation water produced nearly equal yields, but these were considerably less than those from $(NH_4)_2SO_4$ placed in the bed. In one test HNO_3 applied in the water produced yields much below those obtained with aqua NH_3 or $(NH_4)_2SO_4$. Yields from H_3PO_4 applied in irrigation water were much lower than those from concentrated superphosphate placed under the plant row.

F. General Comments

The pitfalls and hazards of obtaining efficient fertilizer application in water are greater than in other systems of fertilizer application unless the operator has a properly engineered system and has competent counsel from fertilizer dealers and soil scientists as to the special problems that may exist with a particular soil and crop. If direct fertilizer loss is to be prevented, there must be no loss of tail water from surface irrigation systems. The application rate from sprinklers must not exceed the percolation rate of the soils. The uniformity of applications of fertilizers can be no more uniform than the application of water.

From consideration of nutrient movement into the soil, fast crop response to nutrients like P and K that are held near the soil surface cannot be expected on deep-rooted crops, but may be expected on shallow-rooted crops such as turf-grasses. Although immobile nutrients like P and K may have limited immediate value, they may be used by future crops when mixed into the soil by later tillage, because these elements remain available for many years in most commonly irrigated calcareous soils. Some acidic soils, sometimes supplementally irrigated, may have high fixation capacities.

In spite of the large amount of N materials being added in irrigation water, the published information points up the high hazards of NH_3 volatilization by this method unless all conditions are optimum. Definitely, more information is needed on this subject. Of course, lower efficiency of crop use can be tolerated to the extent that this loss is offset by lower fertilizer and application costs.

VI. FERTILITY CONSIDERATIONS IN THE SELECTION OF AN IRRIGATION SYSTEM

Topography of some land is suitable with some "touch-up" planning for almost any type of irrigation system, but slope and undulations of other parcels of land require considerable earth moving to make them suitable for gravity irrigation. The alternative is the use of a sprinkler system. When extensive earth moving is required and results in extensive deep cuts and fills, the physical and chemical properties of the resulting surface soil should be carefully evaluated in choosing an irrigation system. If leveling will uncover material with undesirable physical properties, experience has shown that man can do little to improve them. However, the chemical properties affecting the nutrition of plants can usually be improved. The costs of this improvement with additional fertilizers should be considered in assessing comparative costs of systems.

A detailed soil survey and appropriate soil tests for fertilizer requirements should be made on all horizons of the soil profile. Soils developed in arid regions usually show little difference in chemical properties with depth, and leveling has little effect on future fertilizer requirements. However, soils developed in regions of greater rainfall show greater differences in profile development, and leveling may produce serious problems. At the extreme are the dark Chestnut and Chernozemic soils that have marked differences among horizons. Most of the available N, P, Fe, and Zn are contained in the surface soil. Its removal may induce serious deficiencies of any of these elements. A Chernozem may contain about 0.3% N in the surface 6 inches. This amounts to about 6,000 lb of N/acre which is worth about $500 at current USA prices. However, in any one year only 2 to 3% of the N becomes available so only 120 to 180 lb might have to be replaced. Conversely, depressions filled with surface soil may have too much available N. Extreme soil heterogeneity and yield of crops may result. Phosphorus and zinc deficiencies resulting from land leveling can be corrected at a lower cost with appropriate fertilizers than can N deficiency. The problem of iron deficiency or lime-induced chlorosis is much more difficult to handle. Fertilizer treatments for the soil are either ineffective or too expensive. Foliar applications of iron compounds or selection of chlorosis-resistant crops are the alternatives. When considering land leveling, farm managers should think twice before heavily cutting soils subject to serious chlorosis.

Some field experiments on restoring the productivity of cut areas under irrigation have been conducted. Whitney et al. (1950) reported that P and K deficiencies induced by leveling a Fort Collins clay loam could be corrected with application of manure or commercial fertilizers. Carlson et al. (1961) found that manure, N, P, and Zn were essential to produce as much corn on subsoils of Gardena fine sandy loam as on intact soils in North Dakota, USA. Reuss and Campbell (1961) found that extra N and P, or manure, were needed to restore productivity to two irrigated Brown soils of fine texture in Montana, USA.

Special methods of leveling have been developed for keeping most of the surface soil on cut areas by stockpiling the surface soil or strip cutting the area. Such methods are more costly than conventional land leveling.

LITERATURE CITED

Bryan, B. B., and E. L. Thomas, Jr. 1958. Distribution of fertilizer materials applied through sprinkler irrigation systems. Univ. Arkansas Agr. Exp. Sta. Bull. 598. 12 p.

Burns, G. R., and L. A. Dean. 1964. The movement of water and nitrate around bands of sodium nitrate in soils and glass beads. Soil Sci. Soc. Amer. Proc. 28:470–474.

Carlson, C. W., D. L. Grunes, J. Alessi, and G. A. Reichman. 1961. Corn growth on Gardena surface and subsoil as affected by applications of fertilizer and manure. Soil Sci. Soc. Amer. Proc. 25:44–47.

Converse, C. D. 1948. Phosphorus fertility and movement studies. Soil Sci. Soc. Amer. Proc. (1947) 13:423–427.

Dow, A. I., C. D. Moodie, and C. O. Stanberry. 1953. Movement of ammonia nitrogen and phosphorus in alkaline irrigated soil. Agron. J. 45:353–356.

Gardner, W. R. 1965. Movement of nitrogen in soil. In W. V. Bartholomew and F. E. Clark [ed.] Soil nitrogen. Agronomy 10:555–577.

Haddock, J. L. 1954. The interrelationships of irrigation method, soil moisture condition, and soil fertility on the yield quality and nitrogen nutrition of sugar beets. Congr. Int. de Bot., Proc. 8th (Paris) 1954: 1–15.

Haddock, J. L. 1959. Yield, quality, and nutrient content of sugar beets as affected by irrigation regime and fertilizers. J. Amer. Soc. Sugar Beet Technol. 10:344–355.

Haddock, J. L. 1961. The influence of irrigation regime on yield and quality of potato tubers and nutritional status of plants. Amer. Potato J. 38:423–434.

Haise, Howard R. 1950. Flow pattern studies in irrigated coarse textured soils. Abstr. of Doctoral Diss. no. 59. Ohio State Univ. Press, Columbus, Ohio.

Harrison, D. S. 1965. Injection of liquid fertilizer materials into irrigations systems. Univ. Florida Agr. Ext. Serv. Circ. 276. 10 p.

Henderson, D. W., W. C. Bianchi, and L. D. Doneen. 1955. Ammonia loss from sprinkler jets. Agr. Eng. 36:398–399.

Humbert, R. P. 1954. Water distribution studies in the Hawaiian sugar industry. Hawaiian Planter's Rec. 54:211–225.

Humbert, R. P. 1963. The growing of sugar cane. Elsevier Publ. Co., Amsterdam. 710 p.

Humbert, R. P., and A. S. Ayres. 1956. The use of aqua ammonia in the Hawaiian sugar industry. Int. Soc. Sugar Technol. Proc. 9th Congr. 1:524–538.

Humbert, R. P., and A. S. Ayres. 1957. The use of aqua ammonia in the Hawaiian sugar industry: II. Injection studies. Soil Sci. Soc. Amer. Proc. 21:312–319.

Jones, R. A., and J. Green. 1946. Liquid H_3PO_4 as a fertilizer. Amer. Soc. Sugar Beet Technol., Proc. 4:36–39.

Leavitt, H. 1957. The application of liquid gaseous fertilizer. Nat. Joint Comm. on Fert. Appl., Proc. p. 85–95.

Loomis, R. S., J. H. Brickey, F. E. Broadbent, and G. F. Worker. 1960. Comparisons of nitrogen source materials for midseason fertilization of sugar beets Agron. J. 52:97–101.

Lorenz, O. A., J. C. Bishop, and D. N. Wright. 1955. Liquid, dry, and gaseous fertilizers for onions on sandy loam soils. Amer. Soc. Hort. Sci., Proc. 65:296–306.

Meek, B. D., A. J. MacKenzie, and K. R. Stockinger. 1964. Evaluation of a radioactive tracer method for measuring water intake of soils. Soil Sci. Soc. Amer. Proc. 28:153–155.

Nelson, C. E. 1953. Methods of applying ammonium nitrate fertilizer on field corn, and a study of the movement of NH_4^+ and NO_3^- nitrogen in the soil under irrigation. Agron. J. 45:154–157.

Nelson, C. E. 1961. Movement of NH_4^+ and NO_3^- nitrogen from five nitrogen carriers banded in two row-treatments under irrigation. Washington Agr. Exp. Sta. Circ. 380. 14 p.

Nelson, C. E., and L. C. Wheeting. 1941. Fertilizer placement under irrigation in Washington. J. Amer. Soc. Agron. 33:105–114.

Nielson, R. F., and L. A. Banks. 1960. A new look at nitrate movement in soils. Utah Farm Home Sci. 21:2–3, 19.

Olsen, S. R., W. R. Schmel, F. S. Watanabe, C. O. Scott, W. H. Fuller, J. V. Jordan, and R. Kunkel. 1950. Utilization of phosphorus by various crops as affected by source of material and placement. Colorado Agr. Exp. Sta. Tech. Bull. 42. 43 p.

Olsen, S. R., and F. S. Watanabe. 1963. Diffusion of phosphorus as related to soil texture and plant uptake. Soil Sci. Soc. Amer. Proc. 27:648–653.

Reuss, J. O., and R. E. Campbell. 1961. Restoring productivity to leveled land. Soil Sci. Soc. Amer. Proc. 25:302–304.

Robins, J. S., C. E. Nelson, and C. E. Domingo. 1956. Some effects of excess water application on utilization of applied nitrogen by sugar beets. J. Amer. Soc. Sugar Beet Technol. 9:180–188.

Stout, Myron. 1961. A new look at some nitrogen relationships affecting the quality of sugar beets. J. Amer. Soc. Sugar Beet Technol. 11:388–398.

Tyler, K. B., F. E. Broadbent, and V. Kondo. 1958. Nitrogen movement in simulated cross sections of field soil. Agron. J. 50:626–628.

Ulrich, A., L. Jacobson, and R. Overstreet. 1947. Use of radioactive phosphorus in a study of the availability of phosphorus to grape vines under field conditions. Soil Sci. 64:17–28.

Viets, Frank G., Jr. 1962. Fertilizers and the efficient use of water. Advance Agron. 14:223–264.

Viets, Frank G., Jr. 1965. The plant's need for and use of nitrogen. In W. V. Bartholomew and F. E. Clark [ed.] Soil Nitrogen. Agronomy 10:508–544.

Whitney, R. S., R. Gardner, and D. W. Robertson. 1950. The effectiveness of manure and commercial fertilizer in restoring the productivity of subsoils exposed by leveling. Agron. J. 42:239–245.

53 | Water and Soil Temperature

F. C. RANEY

Western Washington State College
Bellingham, Washington

YOSHIAKI MIHARA

National Institute of Agricultural Sciences
Tokyo, Japan

I. INTRODUCTION

Today as competitive agriculture in technically advanced nations becomes operationally more like other business enterprises the well-known economic cost-price squeeze forces us to examine more closely the effect of each factor in the production equation. Certainly, the results of plant selection, pest control, soil moisture management, and fertility manipulation have been spectacular.

Soil temperature is another one of the dependent environmental factors which the land user is able to manage to some extent (Pessi, 1958), although probably only a small part of the irrigable land of the earth will ever suffer a great reduction in plant production because of avoidable extreme root temperatures.

In 1952 Richards et al. published a comprehensive review of soil temperature literature, in 1964 Willis prepared a bibliography on this subject, and a review and annotated bibliography of soil temperature and plant growth by Nielsen and Humphries appeared in *Soils and Fertilizers* in 1966. Therefore, the present discussion will be confined to soil and root temperatures in irrigated areas.

II. TEMPERATURES OF IRRIGATION WATER SOURCES

The primary "source regions" for low temperature water are at high altitude and high latitude. Snow-covered mountain ranges, regions of low latent heat storage, yielding water to basal plains through short, steep drainways are typical conditions. When the distance is short from the melting snow to storage reservoir, irrigation headgate, or point of well recharge, little thermal gain can occur. These conditions are reported in irrigated areas of Japan (Takatuki et al., 1955a, b, c; 1956; 1957); Italy (Piacco, 1953; 1954); and California, USA (Raney, 1963; Raney et al., 1957). Similar geographic conditions are found near the highest mountains of Africa, Argentina, Canada, Chile, China, India, Iran, Russian Central Asia, Turkey, and Tibet. Therefore, we may expect to hear of concern with low water temperatures in these areas as agriculture becomes more intensively irrigated.

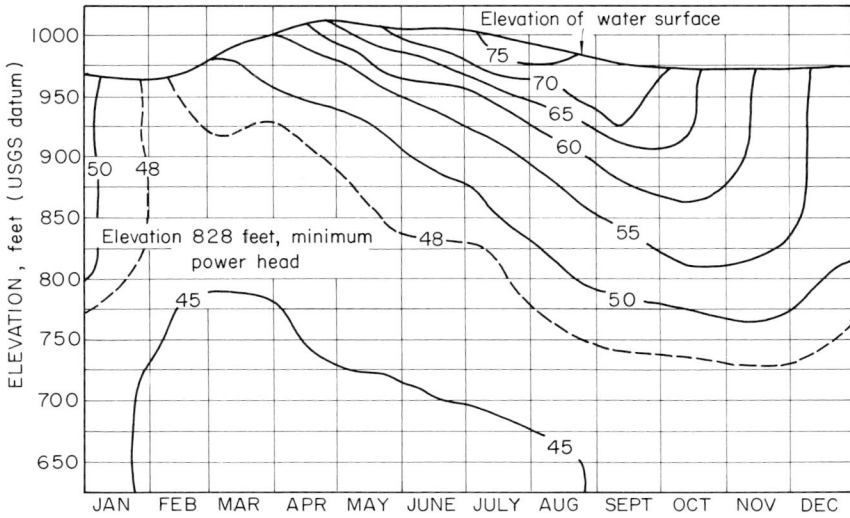

Fig. 53–1. Isothermobaths in Shasta Reservoir for 1947 (Anon., 1959).

A.   Effects of Reservoirs on Water Temperature

Large reservoirs impound such great quantities of water that thermal stratification often produces a cold isothermal bottom layer and a warmer surface layer
(Ruttner, 1953). The usual procedure is to discharge bottom colder water into
the river below. Often large quantities of this cold water are then taken rapidly
into irrigation canals for immediate application to cropped land. In California,
when the Shasta Dam, impounding 4.5 million acre-ft of water was completed
in 1946, the temperature of the Sacramento River below the Dam dropped suddenly from 16C (61F) to 7C (45F), a fall of 9C (16F). During the same year
river water temperatures fell 2C (5F) at the city of Sacramento, 260 river miles
below the Dam (Fig. 53–1 and 53–2). Similar cold discharges of water from
reservoirs are reported in Tennessee, USA (Dendy and Stroud, 1949), Japan
(Takatuki et al., 1955a, b, c; 1956; 1957), and the USSR (Krivosheyeva, 1962).

B.   Effects of Ponding on Rice Field Water Temperatures

Regardless of its initial temperature, water diverted for crop irrigation very
rapidly acquires a temperature corresponding to the net balance of energy fluxes
in its new environment. Research is currently vigorous in this area.

The study of energy budget components involved in warming of cool irrigation
water is most advanced in rice (*Oryza sativa*) growing areas where effects would
be expected because the rice crop is flood-irrigated for prolonged periods. Depth
of submergence depends on the irrigation method used, phase of growth of the
rice plant, topographical situation, and methods used for weed control. Sometimes
in Asia the water is several meters deep where streams overflow during the growing season. Under controlled irrigation practice, water is generally held much

shallower. Three to 10 cm of water are common in areas like Japan and in some countries of southeast Asia where the fields are small and carefully irrigated by hand. In the USA, 10 to 20 cm of water are commonly used since fields are much larger and submergence is important for weed control.

After warming rapidly to some terminal temperature independent of the original temperature, standing water in the field will maintain a relatively steady temperature pattern of variation. Figure 53–3 shows examples of seasonal variation in daily maximum temperature of air and shallow water (5 cm) in a paddy field of Japan and Malaya. The temperature of the standing water was 4C to 6C (7F to 11F) higher by day than air, although it gradually approached air temperature as the rice plants grew larger. Analogous diurnal patterns in a California rice field show how much the water temperature may differ in flowing (Fig. 53–4) compared with stagnant water (Fig. 53–5).

The temperature of underground water supplies is usually within a few degrees of the annual mean air temperature but ordinarily lower than the dew-point temperature of air in summer. However, Mihara et al. (1959a, b) have shown that the daily or monthly mean terminal temperature of a shallow water body $< 1m$ deep can be expressed as follows:

$$\Theta_{eq} = \Theta_a + \frac{S/h - 2D}{1 + 2a} \qquad [53\text{--}1]$$

where Θ_a is the mean temperature of the air (°C), S is the net radiation at the water surface (cal/m² per sec), h is the sensible heat transfer coefficient (cal/m² per °C per sec), D is the saturation deficit (mm Hg), and a is the value of Θ_a at the tangent to the curve of saturation vapor pressure curve plotted against air temperature (mm Hg/°C). In a shallow pond without an overflowing outlet the irrigation water is usually warmed to the terminal temperature while canals, being relatively narrow and having large discharge rates, are not very effective in modifying the over-all temperature of the water conveyed.

C. Temperature Gains in Canals and Warming Basins

Many investigators have used the following equation for estimating temperature gains by water flowing steadily through a pond or canal without stagnant water present (Nakanish and Yamada, 1957):

$$\Delta\Theta = \Theta_t - \Theta_o = (\Theta_{eq} - \Theta_o)(1 - e^{-Bt}). \qquad [53\text{--}2]$$

Mihara et al. (1959a, 1959b) define the following terms for this equation:

$$B = h(1 + 2a)/c\rho H; \; t = HA/q$$

where H is water depth, A is area of the water surface, q is discharge, c and ρ are specific heat and density of water respectively, and h, a, and Θ_{eq} are the same as in the previous equation. Water temperature at the intake is Θ_o. The value of h was empirically selected to be:

$$h = 1.05 + 0.75U^2 \; (\text{cal/m}^2 \text{ per °C per sec})$$

where U is the wind speed (m/sec) at a height of 2 m. In these equations the

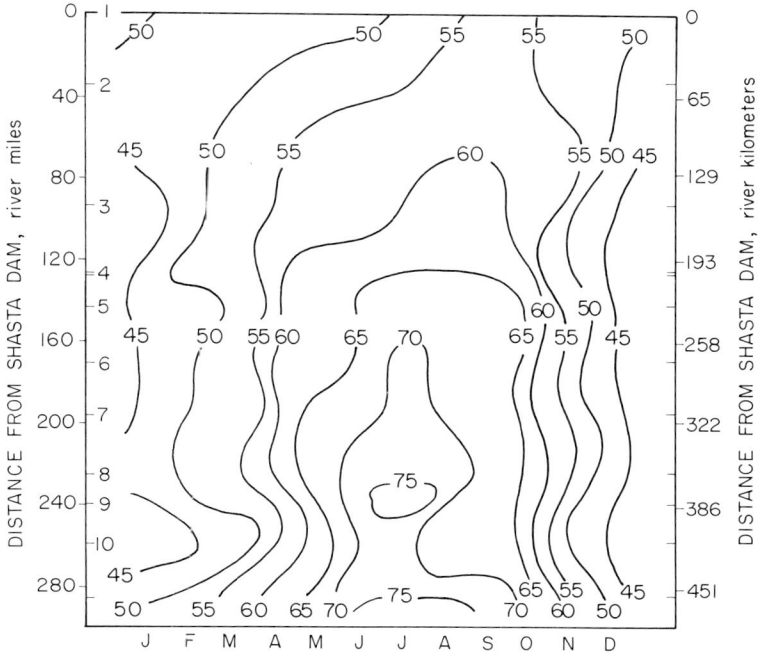

Fig. 53–2. Temperatures in the Sacramento River from Shasta Dam to Sacramento fcr 1947 (Anon., 1959).

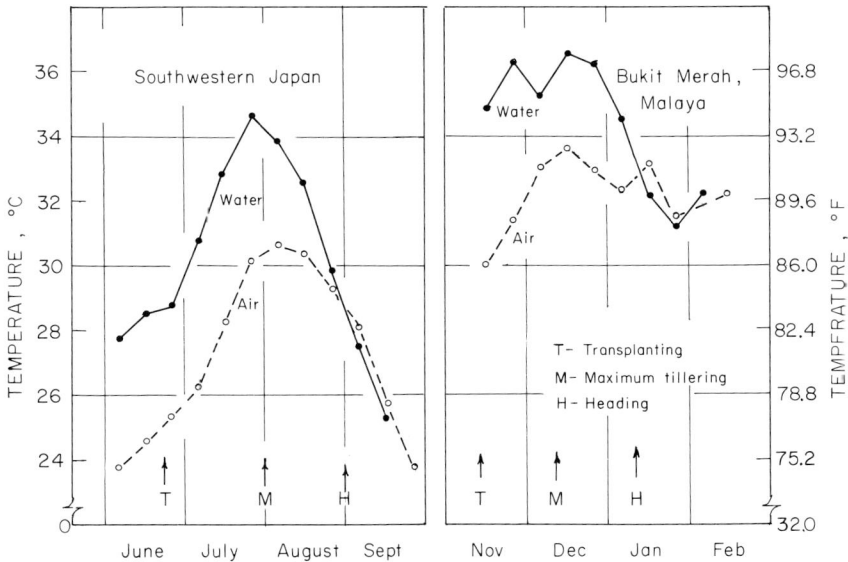

Fig. 53–3. Seasonal variation in air and water temperatures in paddy fields of Japan and Malaya. Water depth about 5 cm. T = transplanting time, M = maximum tillering time, H = heading time (Mihara, 1961).

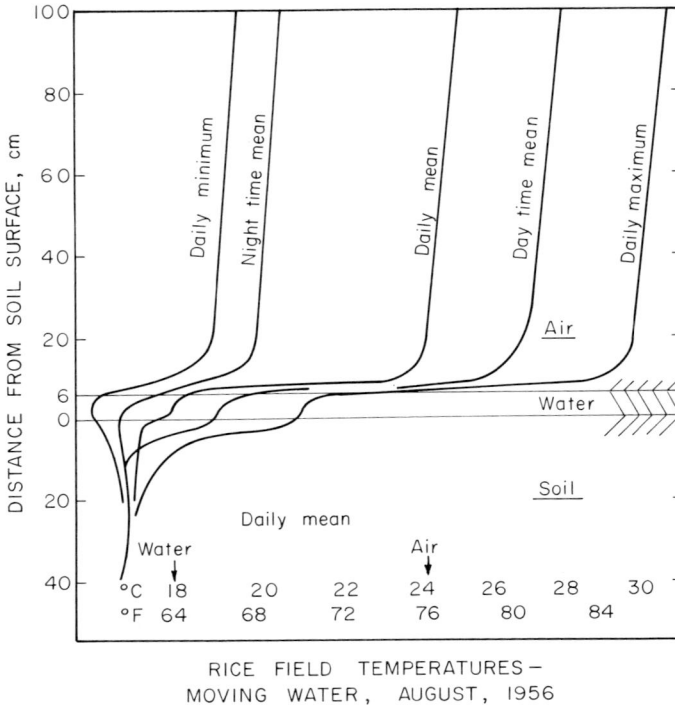

RICE FIELD TEMPERATURES —
MOVING WATER, AUGUST, 1956

Fig. 53–4. Temperatures of moving water in rice fields during August 1956, Glenn County, California.

latent heat transfer coefficient is considered implicitly to be twice the value of h as obtained by Ferguson (1952).

It will be noted that the mean gain in water temperature for a given time interval is exponentially proportional to the water surface area. Thermal gain is not much affected by water depth although water depth affects diurnal variation of water temperature inversely. From a physiological standpoint it is worthy of note that the minimum temperature of the root zone under deep water is reached during the daytime when vegetative activity of the leaves is greatest.

D. Warm Water of Possible Agricultural Importance

Sometimes nonagricultural activities in a region tend to increase rather than decrease the temperature quality of water supplies. Use of rivers or reservoirs for cooling atomic reactors or thermoelectric plants (Kumarina, 1962; Harbeck, 1953), steel mills (Robertson and Horton, 1961; Arnold, 1962; Anon., 1964), as drainage sumps (Moore, 1958), and programmed discharge from impoundments for evaporation control (Vaughn, 1963) can raise water temperatures and may possibly affect agriculture. Water quality characteristics other than temperature may be downgraded. At this time there seems to be no data relating crop growth to water temperatures of such water supplies.

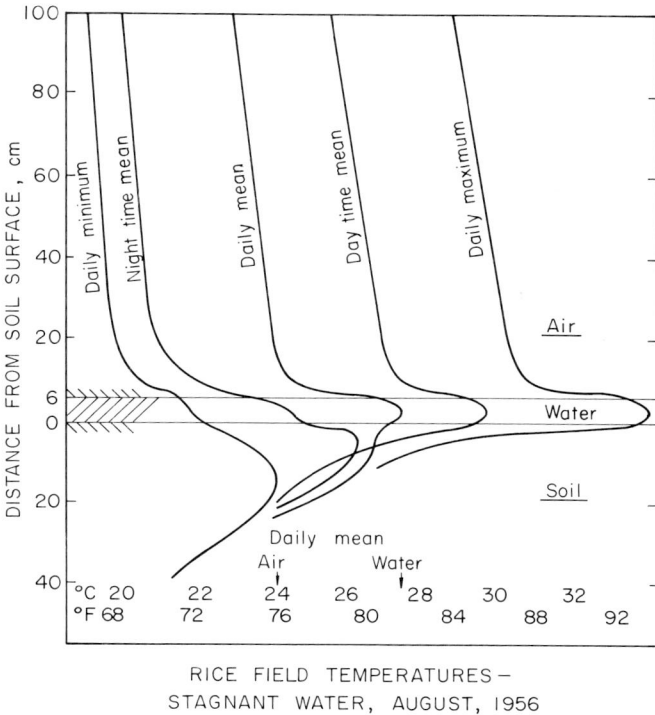

RICE FIELD TEMPERATURES –
STAGNANT WATER, AUGUST, 1956

Fig. 53–5. Temperatures of stagnant water in rice fields during August 1956, Glenn County, California.

III. CROP GROWTH RELATIVE TO ROOT TEMPERATURE

The idea that soil, root, and water temperatures affect plant growth and yield has been held for a long time (Kikkawa, 1929). The subject has been reviewed by Koeppen (1870), Belehradek (1935), Richards et al. (1952) and recently in a biochemical frame of reference by Langridge (1963). Shul'gin (1957) and Korovin (1961) have reviewed much of the Russian literature on soil temperature and plant growth.

Although soil temperature is of practical importance (Pessi, 1958; Post, 1959), field studies alone have been a disappointing source for understanding relationships for at least three reasons: (i) There is inherent variability in field conditions; (ii) environmental factors that produce the observed soil temperatures also directly influence the growth of the plant itself; and (iii) at any given time under field conditions a root system is exposed to a considerable range of temperature between its various parts and to differing temperature fluctuations (Richards et al., 1952).

Ultimately, the effect of soil temperature on plant growth and development can be determined in detail only in controlled experiments maintaining specified shoot conditions while providing the desired variation in soil conditions (Hansel, 1951; Kramer, 1958). Most of the information we now have was obtained in studies where the relation between soil temperature and plant performance was not a primary objective. However, the present literature tends to support the generalization that vegetative growth is largely correlated with root temperature and reproductional events with shoot temperature.

Since recent literature relating soil temperature to seed germination and growth, plant water relationships, soil microorganisms, plant mineral absorption, translocation, and possible mechanisms affected is reviewed by Nielsen and Humphries (1966), we will here confine discussion to effects of soil temperature on production of certain crops reported to be specifically affected.

A. Effects of Cold Water on Crop Production

In some instances cool irrigation water appears to favor growth of certain plant species which may have soil temperature optima lower than commonly encountered in cultivation where environment conditions may be quite different from those in the indigenous habitats from whence the plant came. For these species the importance of soil temperature as a growth factor may be greatly altered, even critical (Smith et al., 1931; Kezer and Robertson, 1927; Matsubayashi et al., 1957; Box et al., 1963). Although in a routinely irrigated field root growth of many plants appears to be greatest in cool spring soils, to decrease in warm summer soils, and resume in cooling fall soils, coincidence of environmental and management factors must be recognized.

Lorenz (1950) has reported that midsummer alternate-row irrigation of potatoes (*Solanum tuberosum*) in the Central Valley of California, USA on sandy loam depressed soil temperatures about 3C (5.4F) for a period of somewhat less than 48 hours while Box et al. (1963) reported that minimum soil temperatures and highest potato yields in Texas, USA occurred in his every-row irrigated, 80–0–0 fertility treatment (80 lb N).

On the other hand, shoot and root growth may both be reduced if cold irrigation water, frequently applied, keeps the soil cool beneath plant species favored by higher soil temperatures or a narrow range of temperatures. This is recognized by greenhouse operators (Bodine, 1917; Post, 1959). Irrigation water at temperatures below 20C (68F) has resulted in damage to greenhouse cucumbers (*Cucumis sativus*) (Schroeder, 1941; Raleigh, 1940). This may well occur in the field with other crops of tropical origin. Damage to the cucumbers was greatest when (i) plants were growing vigorously, (ii) sunny days were followed by a cloudy or cold period, and (iii) soil and water temperatures differed greatly. Whenever irrigation forced the soil temperature below 15C (59F) the plants ceased to grow and damage was economically important.

Low water temperature is considered an important limiting factor in Japanese rice production (Miyamoto, 1958; Kondo and Okamura, 1932). In Italy also, rice yields are reported to be depressed markedly by the cold water drawn from short, swift streams draining the Alps southward toward the Po River Valley (Piacco, 1953, 1954). When the gates of the Shasta Dam in northern California were closed in 1946, the temperature of the irrigation water dropped appreciably. Immediately, the rice growers found that often as much as 5% of acreage planted did not mature in time for harvesting at the end of the available cropping season of 160 days. Irrigation water taken from the river more than 100 miles below the Dam had become too cold for satisfactory rice growth. Growers farthest from

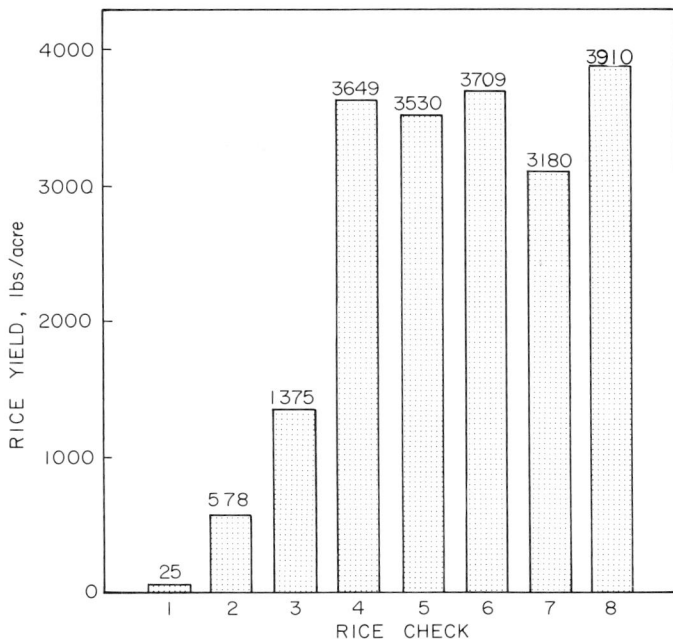

Fig. 53–6. Rice yields downfield from the cold water intake (Raney et al., 1957).

the river diversion point and those using well waters or warmer drain waters were little affected. Experience from this area concurs with experience elsewhere in demonstrating that when the mean temperature of irrigation water lies below 20C (68F) (i) Rice seed germination and seedling emergence from the water is severely retarded; (ii) heading is prevented or delayed; (iii) grains fill poorly; and (iv) maturity date is delayed.

Figure 53–6 indicates how severely rice yields in northern California are reduced in the first four checks of a contour-leveed field on Stockton Clay. Only after the water coming into the field at a temperature of 10C (50F) had traversed 500 m (1,700 ft) at a depth of about 20 cm (outfall of check number four) did the rice yields cease climbing. At this point the mean water temperature increased nearly 10C (18F). Since inflow rate, the geometry of the pathway and percolation heat losses will influence the pattern of thermal gain and rice yield, the example given will be useful only as an analog for prediction.

B. Effects of Warm Water on Plants

In general, agricultural use of the large quantities of warm water available in certain areas of the world has not been economic because (i) The cold soils involved were too far from the warm water, (ii) other crop production costs were so high that root temperature was not a consideration, (iii) the unreliability of the supply of warm water did not merit considering its use, or (iv) the salt content of the warm water would be detrimental to plants or animals. Although hot water derived from geothermal steam condensed during power production at certain locations should have a low salt content and be favorable for agricultural use, quantities so far reported appear small.

Except for the knowledge that a warmer root zone may alter yields of rice (Raney, 1963; Ehrler and Bernstein, 1958), Kentucky bluegrass (*Poa pratensis*) (Pellett and Roberts, 1963), and greenhouse crops (Nelson, 1944), we know little about the effect of elevated root temperatures in commercial crop production. Crop quality would not necessarily be increased, since plant production of certain materials or organs is not to be assumed as a single-valued function of root temperature over the whole range of plant growth temperatures (Khlebnikova, 1937; Al'tergot, 1937).

Temporary increases in root temperature for control of nematodes or viruses (Maggenti, 1962) and other soil-borne plant pests (Baker, 1962) might be one benefit from field use of high temperature water provided that enduring spore-formers did not become a problem (Evans, 1958). A number of workers have expressed interest in seeing warm water relationships with plants investigated (Gusev, 1959; Sycheva and Bystrova, 1959; Went and Hull, 1949; Hasagawa and Yahiro, 1957; Molotkovskii, 1960).

IV. METHODS FOR MODIFYING WATER TEMPERATURE

At the farm level several methods have been used to circumvent the detrimental rice cold water situation: (i) water warming in a basin or in the conveyance canal (Piacco, 1954; Mihara and Onuma, 1955; Mihara et al., 1955; Raney et al., 1957; Yakuwa, 1955a, 1955b, 1956a, 1956b; Yakuwa and Yamabuki, 1955; and Yakuwa and Maeda, 1956); (ii) reducing water losses and thus energy losses by evaporation suppression (Mihara, 1961); (iii) compaction of the deeper soil layers to reduce percolation of warmed water (Mitsui, 1954); and (iv) providing several inlets to a field, flooding the field quickly, and holding the water with a minimum of inflow.

Proposed partial solutions for the problem of maintaining irrigation water temperatures above the minimum threshold before the water reaches the user include:

1) Skimming warmer water from the reservoir surface for release into stream channels as the US Bureau of Reclamation is doing on the Sacramento River drainage near Redding and at Folsom Dam on the American River principally for the benefit of wildlife downstream. Japanese engineers are using this technique to benefit rice growing areas, and California is incorporating a surface water skimming tower into the Oroville Dam on the Feather River to benefit both wildlife and agriculture downstream through the blending of cold bottom water and warmer surface water in order to permit temperature control of the stream below the dam.

2) Providing an afterbay with sufficient area to permit water to warm up. The planned Oroville Dam afterbay of 23,000 acres will warm the water 2C to 5C (4F to 9F) at the proposed flow rates.

3) Use of broad, shallow canals for joint water warming and conveyance and/or water warming basins located at the point of use as being done in Japan, California, and Italy.

4) In certain localities in the future, surplus heat energy such as from deep-earth hydrogen fusion reactors may be used for water warming. In each case economic and engineering considerations will dictate choice.

V.   USE OF IRRIGATION TO MODIFY EFFECTS OF EXTREME SOIL AND AIR TEMPERATURES

Hilgeman and Howland (1955), Hilgeman et al. (1964), and Turrell et al. (1961) have shown that water in furrows in a citrus orchard during a freeze provides cold protection because of the large specific heat of water and the release of latent heat during freezing. Sprinkler irrigation is more widely used for frost protection (*see* chapter 54). However, if application of irrigation water in either furrows or by sprinklers promotes growth of the plants after a freeze, damage from a subsequent freeze may be enhanced.

Irrigation has also been used for the purpose of reducing high soil and plant temperatures. Smith et al. (1931) devoted a chapter to the effect of irrigation on soil temperature in citrus orchards. On the deep calcareous sands of arid sub-tropical Yuma Mesa, Arizona, USA water diverted from surface canals ranged in temperature from 19C (64F) in April to 34C (93F) in December. During this period an irrigated orchard would reduce soil temperature as much as 0.5C (.9F) at a 1-ft depth, 0.3C (.54F) at 2 ft, and 0.1C (.18F) at 3 ft in April or December, as compared with an unirrigated orchard area. In midsummer an irrigation reduced soil temperatures as much as 1.1C (1.98F) at a 1-ft depth, 0.5C (.9F) at 2 ft, and 0.3C (.54F) at 3 ft. The soil temperature was evidently lower on the irrigated areas compared to the dry areas even on the fifth day and was only completely undetectable on the 14th day. On the average the root systems of citrus irrigated at a 4- to 6-week interval were about 5C cooler (9.0F) in the top 2 ft than in the case of unirrigated trees. Since studies indicated that root growth ceased between 33.9C (93F) and 36.7C (98F), this small cooling because of irrigation may be critically important.

Poor establishment of seedling stands of head lettuce (*Lactuca sativa capitata*) in the Salt River Valley of Arizona has been attributed to a variety of causes including high soil temperatures 32C (90F). Wharton and Hobart (1931) have found that the common practice of continual irrigation with 21C to 26.7C (70F to 80F) water during hot weather did not directly cool the soil in the root zone of the seedling beds. So long as the soil surface was moist with minimum irrigation, evaporation would be the major factor reducing soil temperatures and lettuce stands would be good.

It is apparent that the soil energy budget, soil texture, water temperature, irrigation frequency, and plant rooting depth must be included in any estimate of the pattern to be expected in root temperature resulting from irrigation.

Published data suggest that wide differences exist between species in their response to root temperature. A crop plant under stress because of other environmental factors may suffer economic yield reductions when something like root temperature is altered. If the change is sufficient, the crop may be "pushed over" the economic brink.

If we can prevent or circumvent undesirable extremes in irrigation water temperatures, we will place ourselves in a position (i) to keep a maximum area in irrigated crops, (ii) to optimize plant growing conditions and broaden the spectrum of crops that can be grown profitably in a given locality, and (iii) to optimize the water temperature needs of agriculture, wildlife, and recreation. To satisfy these objectives further research into root temperature relationships is necessary.

LITERATURE CITED

Al'tergot, V. G. 1937. Self-purging of plant cells at high temperatures as a result of irreversible biochemical processes. Akad. Nauk. Inst. Fiziol. Rast. Timiryazeva Trudy 1:5–79.

Anonymous. 1959. United States Department of Interior, Bureau of Reclamation, Division of Irrigation Operation. Annu. Rep. p. 10–17.

Anonymous. 1964. Cooling ponds. Radioisotopes—Physical Sciences. Sec. 5:604–605.

Arnold, G. E. 1962. Heated discharges, their effect on streams. Div. Sanitary Eng., Penn. Dep. Health Pub. 3:1–107.

Baker, K. F. 1962. Principles of heat treatment of soil and planting material. Australian Inst. Agr. Sci. J. 28:118–126.

Belehradek, J. 1935. Temperature and living matter. Protoplasma-Monogr. 8:1–277.

Bodine, W. G. 1917. The forcing of plants by means of warm water immersions. Univ. Vermont Agr. Exp. Sta. Bull. 203. p. 9–10.

Box, J. E., W. H. Sletten, J. H. Kyle, and A. Pope. 1963. Effects of soil moisture, temperature, and fertility on yield and quality of irrigated potatoes in the Southern Plains. Agron. J. 55:492–494.

Dendy, J. S., and R. H. Stroud. 1949. The dominating influence of Fontana reservoirs on temperature and dissolved oxygen in the Little Tennessee River and its impoundments. Tennessee Acad. Sci. J. 24:41–51.

Ehrler, W. L., and L. Bernstein. 1958. Effects of root temperature, mineral nutrition and salinity on the growth and composition of rice. Bot. Gaz. 120:67–74.

Evans, C. 1958. Heating pad for asparagus. California Farmer. 208:671.

Ferguson, J. 1952. The rate of natural evaporation from shallow ponds. Australian J. Sci. Res. A:315–330.

Gusev, N. A. 1959. Effect of increased temperature on the plant water regime. Akad. Nauk SSSR Izvest., Ser. Biol. 1959. p. 79–86.

Hansel, H. 1951. The effect of varying temperature upon the growth and development of certain plants, a review. Wetter und Leben. 3:161–166.

Harbeck, G. E. 1953. The use of reservoirs and lakes for the dissipation of heat. US Geol. Sur. Circ. 282:1–16.

Hasegawa, H., and T. Yahiro. 1957. Effects of high soil temperatures on the growth of the sweetpotato plants. Crop Sci. Soc. Jap. Proc. 26:37–39.

Hilgeman, R. H., and L. H. Howland. 1955. Report on the frost situation and the effect of a wind on temperatures during 1954–55. Arizona Citrus Inst. Proc. 1:1–3.

Hilgeman, R. H., C. E. Everling, and J. A. Dunlap. 1964. Effect of wind machines, orchard heaters, and irrigation water on moderating temperatures in a citrus grove during severe freezes. Amer. Soc. Hort. Sci., Proc. 85:232–244.

Kezer, A., and D. W. Robertson. 1927. The critical period of applying irrigation water to wheat. J. Amer. Soc. Agron. 19:80–116.

Khlebnikova, N. A. 1937. The chemical nature of plant resistance to the effect of the high temperature factor. Akad. Nauk. Inst. Fiziol. Rast. Timirayazev Trudy 1:93–110.

Kikkawa, S. 1929. The influence of temperature of irrigation water on the growth and yield of rice. Imperial Acad., Japan, Proc. 5:303–305.

Koeppen, W. 1870. Waerme and pflanzenwachsthum. Soc. Imper. Nat. Moscou Bull. 63. Pt. 2, p. 41–110.

Kondo, M., and T. Okamura. 1932. The relation between water temperature and growth of the rice plants. Ohara Inst. Agr. Res. Rep. 5:347–374.

Korovin, A. I. 1961. Soil temperature and plant growth in the north. Karel'skii Filial Akademii Nauk SSSR. Gozizdat, Petrozavodsk. p. 191.

Kramer, P. J. 1958. Thermoperiodism in trees. p. 573–580. In K. V. Thimann [ed.] The physiology of forest trees. Ronald Press, New York.

Krivosheyeva, I. T. 1962. Concerning the thermal conditions of the V. I. Lenin Lake. Sov. Hydrol. (Transl.) 1962:217–223.

Kumarina, M. H. 1962. Selection of a method for dumping warm water into cooling ponds. Leningrad Univ. Vestnik Ser. Geol. i Geograf. No. 2:55–64.

Langridge, J. 1963. Biochemical aspects of temperature response. Ann Rev. Plant Physiol. 14:441–462.

Lorenz, G. A. 1950. Air and soil temperatures in potato fields, Kern County, California, during spring and early summer. Amer. Potato J. 27:396–407.

Maggenti, A. R. 1962. Hot water treatment of hop rhizomes for nematode control. California Agr. 16:11–12.

Matsubayashi, M., S. Hatta, A. Shimoda, and S. Sekimura. 1957. Studies on the flood irrigation of upland crops: 1. Effect of irrigation on wheat and soil temperature. J. Agr. Eng. Soc. Japan 24:313–316.

Mihara, Y. 1961. The microclimate of paddy rice culture and artificial improvement of the temperature factor. Pacific Sci. Congr. Proc. 10th. p. 181–210.

Mihara, Y., and K. Onuma. 1955. Thermal efficiency of warming ponds for cold irrigation water: 1. The temperature rise and efficiency of heat acquisition in water warming ponds. Nat. Inst. Agr. Sci. Japan Bull. Ser. A. 4:45–66.

Mihara, Y., Z. Uchijima, and S. Nakamura. 1959a. A study of the heat balance of water warming channels. Nat. Inst. Agr. Sci. Japan Bull. Ser. A. 7:45–67.

Mihara, Y., Z. Uchijima, S. Nakamura, and K. Onuma. 1959b. A study of the heat balance and temperature increase in a warming pond. Nat. Inst. Agr. Sci. Japan Bull. Ser. A. 7:1–43.

Mihara, Y., Z. Uchijima, K. Onuma, S. Yamamoto, and M. Hagiwara. 1955. Thermal efficiency of warming ponds for cold irrigation water: 3. Estimation of the efficiency of water warming ponds from model experiments. Nat. Inst. Agr. Sci. Japan Bull. Ser. A. 4:79–92.

Mitsui, S. 1954. Inorganic nutrition, fertilization, and soil amelioration for lowland rice. Yekendo, Ltd. Tokyo.

Miyamoto, K. 1958. Of the relation between temperature of irrigation water and growth and yield of rice plants. Nagyo Kisho. J. Agr. Meteorol. (Nippon Mogyo Kisho Gakkai, Tokyo) 13(4):147–152.

Molotkovskii, Yu. G. 1960. Changes in adenosine triphosphate activity of subcellular units in heat-treated plants. Fiziol. Rast. 8:669–672.

Moore, E. W. 1958. Thermal "pollution" of streams. Ind. Eng. Chem. 50(4):87A–83A.

Nakanish, K., and M. Yamada. 1957. Temperatures of water in paddy fields in Japan. Water Temperature J. 1:173–185.

Nelson, C. H. 1944. Growth responses of hemp to differential soil and air temperatures. Plant Physiol. 19:294–309.

Nielsen, K. F., and E. C. Humphries. 1966. Soil temperature and plant growth. Soils and Fert. 29:1–7.

Pellett, H. M., and E. C. Roberts. 1963. Effects of mineral nutrition on high temperature induced growth retardation of Kentucky bluegrass. Agron. J. 55:473–476.

Pessi, Y. 1958. On the significance of soil temperature in plant cultivation. Finnish State Agr. Res. Board Pub. (Maatalouskoetoiminnan Julkaisuja Valtion). No. 167:1–20.

Piacco, R. 1953. Irrigation of rice. I. Soil and water temperature during the irrigation period. L'Italia Agr. 90:493–496.

Piacco, R. 1954. Irrigation of rice. II. Distribution of maximum water temperature. L'Italia Agr. 91:455–456.

Post, K. 1959. Florist crop production and marketing. Judd, New York. p. 32–55.

Raleigh, G. J. 1940. The effect of culture solution temperature on water intake and wilting of the muskmelon. Amer. Soc. Hort. Sci., Proc. 38:487–488.

Raney, F. C. 1963. Rice water temperature. Rice J. 66:19–22.

Raney, F. C., R. M. Hagan, and D. C. Finfrock. 1957. Water temperature in irrigation of rice. California Agr. 11(4):19–20, 37.

Richards, S. J., R. M. Hagan, and T. M. McCalla. 1952. Soil temperature and plant growth. In B. T. Shaw [ed.] Soil physical conditions and plant growth Agronomy 2:303–480.

Robertson, D. A., and R. K. Horton. 1961. River-quality conditions during a 16-week shutdown of upper Ohio Valley steel mills. Ohio River Valley Water Sanitation Comm. 1961:1–39.

Ruttner, R. 1953. Fundamentals of limnology. Univ. Toronto Press. p. 32–39.

Schroeder, R. A. 1941. Root temperature as it affects growth of greenhouse cucumbers. Market Growers J. 69:451–454.

Shul'gin, A. M. 1957. Temperature regime of the soil. Gidrometizdat, Leningrad, p. 241.

Smith, G. E., A. F. Kinnon, and A. G. Cairns. 1931. Irrigation investigations in young grapefruit orchards on the Yuma Mesa. Arizona Agr. Exp. Sta. Tech. Bull. 37. p. 554–589.

Sycheva, Z. F., and Z. A. Bystrova. 1959. Effect of reduced soil temperatures on uptake of minerals and nitrogen by plants. Karel'skogo i Kol'skogo Filiala Akad. Nauk SSR Izvest. 4:68–75.

Takatuki, T., I. Takahasi, and S. Tezima. 1955a. Study of the effect of water power works on temperature of irrigation water: 1. Change of water temperature through the penstocks. J. Agr. Eng. Soc. Japan 22:551–566.

Takatuki, T., I. Takahasi, and S. Tezima. 1955b. Study of the effect of water power works on temperature of irrigation water: 2. Temperature of water in a regulating reservoir and its intakes. J. Agr. Eng. Soc. Japan 23:35–44.

Takatuki, T., I. Takahasi, and S. Tezima. 1955c. Study of the effect of water works on temperature of irrigation water: 3. On entraining warm water. J. Agr. Engr. Soc. Japan 23:233–240.

Takatuki, T., I Takahasi, and S. Tezima. 1956. Study of the effect of water power works on temperature of irrigation water: 4. Temperature of river water before the construction of water power works. J. Agr. Eng. Soc. Japan 24:277–284.

Takatuki, T., I Takahasi, and S. Tezima. 1957. Study of the effect of water power works on temperature of irrigation water: 5. General considerations. J. Agr. Eng. Soc. Japan 24:336–344.

Turrell, F. M., S. W. Austin, and R. L. Perry. 1961. Water, heaters, and wind machines. Calif. Citrogr. 46(5):154–160.

Vaughn, C. S. 1963. Selective withdrawal as a means of reducing evaporation from reservoirs. Univ. Texas Hydraul. Eng. Lab. Tech. Rep. HYD 02–6301:1–72.

Went, F. W., and H. M. Hull. 1949. The effect of temperature upon translocation of carbohydrates in the tomato root. Plant Physiol. 24:505–526.

Wharton, M. F., and C. Hobart. 1931. Studies in lettuce seedbed irrigation under high temperature conditions. Arizona Agr. Exp. Sta. Tech. Bull. 33. p. 283–303.

Willis, W. D. 1964. Bibliography on soil temperature. US Dep. Agr. Agr. Res. Serv. Ser. 41-94. p. 1–82.

Yakuwa, R. 1955a. Studies on raising irrigation water temperature: 2. Temperature in a warming pond. Hokkaido Univ. Fac. Agr. Memoirs. 1:37–42.

Yakuwa, R. 1955b. Studies on raising irrigation water temperature: 3. Microclimate of a warming pond. Hokkaido Univ. Fac. Agr. Memoirs. 1:43–51.

Yakuwa, R. 1956a. Studies on raising irrigation water temperatures: 4. Effect of irrigating with warm water on the temperature of water and soil in a paddy field and on air temperature near the water surface. Hokkaido Univ. Fac. Agr. Memoirs. 2:1–10.

Yakuwa, R. 1956b. Studies on raising irrigation water temperature: 5. Effect of a winding channel and warming pond upon water warming efficiency Hokkaido Univ. Fac. Agr. Memoirs. 2:11–22.

Yakuwa, R., and T. Maeda. 1956. Studies on raising irrigation water temperature: 6. Effect of vinyl covering on water temperature in a warming basin. Hokkaido Univ. Fac. Agr. Memoirs. 2:15–22.

Yakuwa, R., and F. Yamabuki. 1955. Studies on raising irrigation water temperature: 1. Water temperature in shallow water tanks. Hokkaido Univ. Fac. Agr. Memoirs. 1:28–36.

54 | Irrigation for Frost Protection

JOHN N. LANDERS
IRI Research Institute, Inc., USAID
Recife, Brazil

K. WITTE
University of Bonn, Germany

Frost damage to crops can be financially devastating for farmers the world over In California, USA alone, estimated losses to fruit and vegetable crops sometimes run into millions of dollars. This has resulted in the adoption of various methods of frost protection. Oil for the traditional heaters is expensive, and their use demands much labor. Wind machines represent a large investment. Surface methods of irrigation have been utilized in frost protection for many years, but their application is limited. Perhaps for these reasons, in the last decade sprinklers have come into quite widespread use for frost protection in parts of the USA, Europe, and elsewhere. Protection is normally profitable only where frosts in the growing season are infrequent—winter frosts in Mediterranean climates, spring and autumn frosts in more temperate regions.

Ice formation in the intercellular spaces is lethal to plant tissues. This is the normal cause of frost damage in plants; the affected parts take on a wilted appearance and often become black in color. Less frequently, intracellular ice formation may also have the same effect.

I. FROST TYPES AND FORMATION

Radiation and advection frosts are distinguished by the circumstances which produce them. A common feature of both is the local influx of a relatively cold and dry polar air mass. The difference is in the degree of coldness and the speed with which the new air is entering the local area.

A. Advection Frost

This type of frost is accompanied by a wind below 0C which may be in excess of 15 miles/hour. Consequently, it is not limited to nighttime as radiation frosts generally are. Since protection from windborne frost has generally proved to be of little value, the physical conditions of its microclimate will not be discussed in detail.

B. Radiation Frost

The ideal meteorological condition for a radiation frost is a clear, cold, dry, and calm atmosphere with the temperature of the invading air mass above freezing. Under these conditions the net loss of radiant energy to outer space by the solid objects (plants, soil, etc.) of the earth's surface is high for the percentage of return radiation from the water vapor in the sky is small. Other things being equal, a high rate of radiant energy loss results in rapidly falling temperatures. Objects with a large heat capacity in relation to their surface area, e.g. mature citrus fruit, will cool more slowly than those with a small heat capacity per unit surface area, e.g., young deciduous friut and especially leaves. For this reason, soil cools more slowly than plants.

At night, part of the net energy radiated to space by the earth's surface is drawn from the air layers in contact with it. Thus, the coldest air temperatures during a radiation frost are found close to the surface, a condition known as "temperature inversion" (Fig. 54–1). The turbulent mixing associated with wind effectively increases the depth of air from which the surface withdraws energy, producing comparatively higher air temperatures near the ground and a correspondingly weaker inversion. Thus, the occurrence of a slight wind (over 2 miles/hour) during radiation frost conditions often prevents frost damage. Thick, low clouds return a large percentage of the earth's outgoing long wave radiation, reducing the rate of radiative cooling of the surface (Fig. 54–2). Thin, high clouds have little effect in reducing radiative cooling (Sutton, 1953).

If the crop surfaces cool to temperatures below the dewpoint of the atmosphere, dew will condense on the crop. The heat of condensation released in this change of state slows the rate of cooling, although the amount of water involved is not large enough to retard cooling for long. When this dew freezes it forms the familiar white coating of tiny ice crystals loosely referred to as frost. Frost is better defined as the condition when atmospheric temperatures fall below the

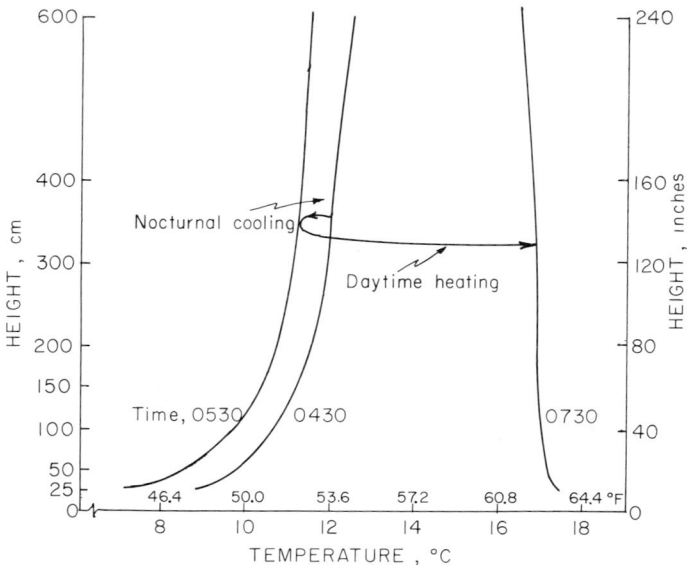

Fig. 54–1. The nocturnal temperature inversion near the ground produced by radiative cooling (Aug. 14, 1963). The pattern is reversed after sunrise. Data supplied by T. V. Crawford, Agr. Eng. Dep., Univ. California, Davis.

freezing point of water; if the dewpoint of the air mass is below this point, freezing conditions can occur without dew formation—a so-called "black frost."

The above discussion has concerned the rate of nocturnal cooling. The actual temperature attained also depends on the temperatures of the crop and ground at sunset when the cooling process begins. Under radiation frost conditions Brooks (1959) states that the temperature of the soil at sundown practically governs the night minimum temperature. If crop and soil are comparatively warm at sunset, cooling must be rapid in order to reach freezing before sunrise. Also, longer nights allow more time for cooling to take place.

Frost pockets tend to form in hollows in the topography because on an otherwise calm night, cold, heavy air flows downhill under the influence of gravity and collects in these low spots in a manner similar to the surface drainage of water.

II. ENVIRONMENTAL AND CROP FACTORS IN FROST PROTECTION

A. The Atmosphere

During a frost, plants exchange heat with their environment by radiation, evaporation, and convection processes.

The net radiation loss of exposed outer crop surfaces is chiefly dependent cn the cloudiness of the sky (Fig. 54–2) and may be considered constant for the same atmospheric conditions during a single night. On a clear, frost night Niemann (1957) suggests an average value of 60 kcal/m² per hour for vegetation surfaces, and Brooks (1959) measured the same value for the net nocturnal radiation lcss of a California citrus orchard. Using a correlation between air temperature in a standard weather shelter and the effective sky temperature proposed by Businger (1965), the maximum value of the earth's net outgoing radiation during a frost night would not exceed 76 kcal/m² per hour. Raschke (1960) states that plants have long wave emissivities $\geqq 0.95$ and thus can be considered to approximate a black body.

The rate of evaporative and convective energy exchanges for a particular crop are determined by three primary variables: (i) The temperature gradient be-

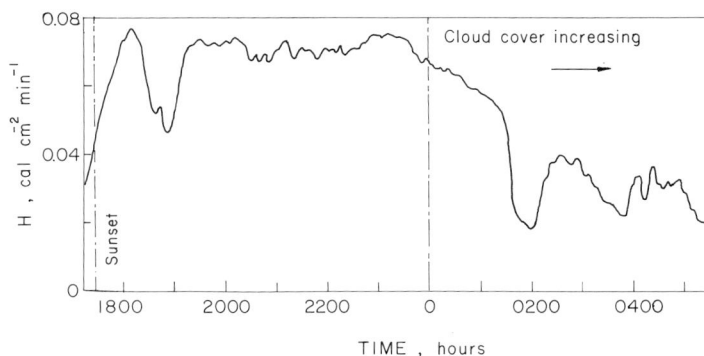

Fig. 54–2. Net radiation H during a night with variable cloud cover. Observation at Wageningen, the Netherlands (adapted from Wijk, 1963).

tween the plant and its environment, (ii) the corresponding vapor pressure gradient, and (iii) the windspeed over the plant surfaces.

Loss of heat by evaporation occurs any time that the vapor pressure at the crop surfaces exceeds that of the surrounding air. Evaporation can thus take place in a saturated atmosphere, providing it is cooler than the crop. The evaporative heat flux is directly proportional to the vapor pressure gradient. This proportionality constant is called the evaporative heat transfer coefficient. The coefficient increases with windspeed and decreases with the size of the body being cooled, and it also depends on the body's shape. The vapor pressure gradient is also affected by windspeed, but it is the product of this gradient and the evaporative heat transfer coefficient which determines evaporative heat flux. The flux may be negative as in evaporation or positive when dew condenses. In an analogous fashion the convective heat flux is determined by the product of the temperature gradient and the convective heat transfer coefficient. Attempts at theoretically describing the energy balance of a plant require some knowledge of heat transfer coefficients of plant parts. Empirical values have been obtained by Raschke (1960) and Linacre (1964).

B. The Soil

The surface temperature at night is greatly influenced by the volumetric heat capacity and thermal conductivity of the surface layers of soil and by the nature of the vegetative cover. The effects of soil water on frost incidence are discussed in a later section.

Low thermal conductivities are exhibited by mulches and organic soils, when compared with mineral soils. The probability of frosts over the former is greater, because the nocturnal release of soil heat to the air is slow. Compaction improves conductivity and hence reduces the frost risk. Cultivation has the reverse effect on a homogeneous soil but may be beneficial when it mixes an accumulated surface layer of organic matter with the lower soil layers (Georg, 1960). Diurnal fluctuations in soil temperature only penetrate to a limited depth (Table 54–1) which is determined by the soil heat capacity and conductivity. It is this soil layer which is therefore to be considered in frost protection.

A cover crop results in lower surface temperatures during radiation frosts. This is partly a result of the reduction in daytime solar radiation which reaches

Table 54–1. Average thermal properties of soils (Wijk, 1963)

Soil type	Volumetric water content χ_w*	Thermal conductivity λ	Volumetric heat capacity C	Damping depth D_d*
		10^{-3} cal cm^{-1}°C^{-1}	cal cm^{-3} °C^{-1}	cm
Sand	0. 0	0. 7	0. 3	8. 0
	0. 2	4. 2	0. 5	15. 2
	0. 4	5. 2	0. 7	14. 3
Clay	0. 0	0. 6	0. 3	7. 4
	0. 2	2. 8	0. 5	12. 4
	0. 4	3. 8	0. 7	12. 2

* χ_w is the volume of water per unit volume of soil and D_d is the damping depth for the diurnal variation. The porosity in both cases was 0. 4.

the soil and partly an effect of the low thermal conductivity of the calm air layer within the vegetation at night which impedes the release of soil heat to the overlying air. The net depression of night minimum temperatures caused by a cover crop may be from 0.6C to 1.7C (1 to 3 F) in an orchard (Brooks, 1959).

C. The Crop

The critical temperature below which frost damage occurs is of paramount importance. Usually this temperature refers to a weather shelter observation and is normally quoted in conjunction with a time period after which damage occurs (Phillips et al., 1962). The phase of growth of the crop, degree of cold hardening, specific and varietal variations, mineral nutrition and moisture stress all affect the critical temperature. A detailed discussion of these factors is to be found in volume 6 of this series (Levitt, 1956). Proebsting (1964) shows the effects of hardening and stage of growth on the temperature (T_{50}) which killed 50% of the buds on peach trees (*Prunus persica*) (Fig. 54–3). He also reports less frost damage in this crop when the winter irrigation was withheld; this is in agreement with the field observations of Young (1940) on citrus.

The ability of plant tissues to supercool is a widespread phenomenon which explains why the critical temperature may be appreciably below 0C. Pogrell and Kidder (1960) showed that the effect of the presence of ice or mechanical disturbances, such as the impact of sprinkler drops, diminished the ability to supercool, while a combined treatment eliminated it completely.

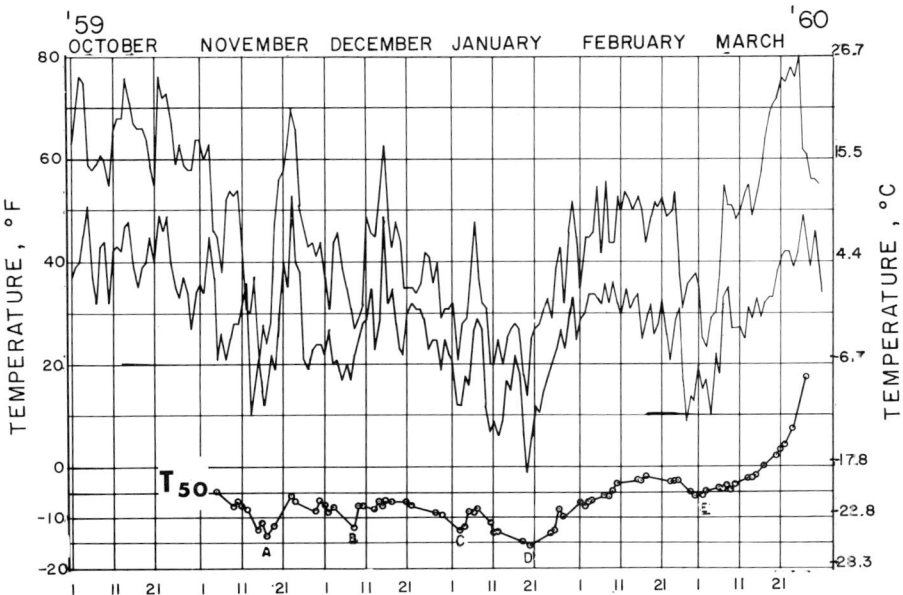

Fig. 54–3. Seasonal changes in T_{50} of Elberta peach buds as related to maximum (upper line) and minimum temperatures (middle line) during the winter of 1959–60 (Proebsting, 1964).

In overhead sprinkling, the structural strength of the plant and its height above the ground are important, since the ice thickness can exceed 1.3 cm. Young (1940) reported extensive damage to mature citrus trees, while young trees in a citrus nursery withstood heavy ice loads with negligible breakage because of their suppleness (Dean, 1963). Gladiolus (*Gladiolus* Sp.) and other tall thin plants can withstand only a small ice load (Kidder and Davis, 1956).

D. The Heat Balance of the Soil-Plant-Atmosphere System at Night

Analyses of the nocturnal heat loss from a crop during frost with respect to overhead sprinkling for frost protection have been made by Niemann (1957), Zeeuw (1960), Gerber and Harrison (1963), Businger (1965), and Beahm (R. B. Beahm, 1959. Experimental and theoretical study of frost protection by water application under simulated radiation frost conditions. *Unpubl. M.S. Thesis. Michigan State Univ., East Lansing, Mich.*). Use of geometrical analogies and heat transfer theory enables a heat balance to be set up for a single plant part. There appears to be general agreement that spheres and cylinders (as models for buds and branches) represent more severe cooling conditions than flat plates (as models for leaves). This has not been adequately verified in practice.

Several difficulties arise in translating these results to the more complex case of an entire plant or crop. For example, temperatures vary within the crop, depending on height and exposure of the surfaces. Also the critical temperature of such plant parts as buds may be quite distinct from that of leaves. These difficulties may be partly overcome by considering all parts of the crop to be cooling at the same rate and taking a representative critical temperature. Perhaps the greatest problem in applying this approach to sprinkler frost protection is that the total surface area of the crop may be several times greater than that of the surfaces which directly intercept precipitation. However, Zeeuw (1960) points out that the coldest surfaces during a radiation frost are also those most exposed to the sprinkler drops from an overhead system. During a severe frost when the more sheltered surfaces require protection, the "precipitation receipt" of the intercepting surfaces must be high enough to satisfy the "precipitation demands" of less exposed surfaces by a process of redistribution through splashing and runoff. (Precipitation demand may be defined as the precipitation rate, with no runoff, required to maintain a plant part just above its critical temperature.) To arrive at a final precipitation rate, Zeeuw (1960), Gerber and Harrison (1963), and Businger (1965) apply a correction factor to the precipitation demand of the intercepting surfaces. It is a variable multiplier (≥ 1.0) which is called here the "distribution factor" and is an expression of water in excess of precipitation de-

Table 54–2. Some values for the distribution factor

Crop	Distribution factor	Researcher
Strawberries and other low-growing crops	1.5	Zeeuw (1960)
Deciduous fruit trees	3.0 – 4.0	Zeeuw (1960)
Tomatoes, potatoes, and beans	1.5	Businger (1955)*
Citrus trees	1.0 – 1.5	Gerber and Harrison (1963)*

* Calculated from experimental results.

mand which must be redistributed onto nonintercepting surfaces by splashing or runoff. Ideally analogous to the leaf area index, the distribution factor actually has to absorb uncertainties in the heat transfer coefficients and the effective body temperatures used in the above calculations.

Table 54–2 shows some values for the distribution factor.

III. IRRIGATION AS A FROST CONTROL TECHNIQUE

Of the various methods of utilizing irrigation for frost protection, overhead sprinkling during the frost has achieved, by far, the most widespread success; consequently it will be treated in more detail than the other methods.

A. Irrigation During Frost

1. OVERHEAD SPRINKLING

The release of 80 cal/g of latent heat as water freezes is utilized to offset the heat lost by the crop to its cooler surroundings. Overhead sprinkling, during frost, forms ice on the crop, but a mere coating of ice does not act as an insulating layer and prevent frost damage. In fact, ice is a comparatively good conductor of heat. Thus a film of continuously freezing water must be maintained by the sprinklers on the surface of the ice. This local release of heat holds the plant near 0C and above its critical temperature. Additional amelioration of the microclimate may be gained by the processes described later for ground sprinkling. (*See* Fig. 54–4.)

Fig. 54–4. Ice on apple blossoms caused by overhead sprinkling during frost (Rogers and Modlibowska, 1962).

The first question which arises in considering overhead sprinkler irrigation for frost protection is how much water to apply? Plant and environmental factors determine the maximum amount of protection required. The rate of water application necessary to achieve this is dictated to some extent by the characteristics of the sprinkler system employed.

As temperatures in the environment around the crop being protected decrease, precipitation rates must increase. This effect has been demonstrated in laboratory experiments by Rogers et al. (1954); Pogrell and Kidder (1959); and Perraudin (1961). Their work has been corroborated in the field by Rogers et al. (1954); Businger (1955); Pogrell (1958); Perraudin (1961); and Rogers and Modlibowska (1962). Table 54–3 gives some typical results. Note also the effect that the stage of growth has on damage.

Wind increases cooling by evaporation and convection. Pogrell (1958) and Witte and Pogrell (1958) demonstrated, on low growing crops, that a corresponding increase is required in the amount of water applied (Table 54–4). Wheaton and Kidder (1964) showed that the temperature of a wet leaf subjected to moderate wind, but no radiative cooling, declined rapidly and approached the wet bulb temperature of the air.

The precipitation demand of a crop becomes higher as the dewpoint of the surrounding air decreases. With the air at –5C (23F) and the crop maintained at 0C by sprinkling, Niemann (1957) calculated that the heat lost by evaporation would increase by 50%, if the dewpoint fell from –5C (100% relative humidity) to –8C (18F) (80% relative humidity).

As discussed in the previous section, the morphology of the crop influences precipitation rates required for frost protection. Wettability, too, varies from crop-to-crop and with the stage of growth. Where the sprinkled plant surfaces are distinctly hydrophobic, as in the case of blackcurrant (*Ribes nigrum*) blossoms,

Table 54–3. Frost damaged blossoms in relation to precipitation, 1957
(Rogers and Modlibowska, 1962)

Apple variety	Stage of development	Precipitation, mm/hr				
		0 - 1.0	1.1 - 1.5	1.6 - 2.0	2.1 - 2.5	2.6 - 3.6
James Grieve	open	77.7	32.7	7.2	0.5	0.0
	swelling	99.2	96.9	31.3	15.7	0.0
Cox's Orange	open	95.1	72.6	37.8	9.9	0.0
Pippin	petal fall	99.9	90.4	58.1	28.5	0.0

Table 54–4. Influence of windspeed on required precipitation rates (Pogrell, 1959)[*]

Air temperature		Wind speed		Precipitation rate	
°F	°C	meters/sec	miles/hour	mm/hr	inches/hour
-3.5 to -2.5	25.7 to 27.5	0.5	1.1	1.1 - 1.5	0.04 - 0.06
		1.4 - 2.5	3.1 - 5.5	1.5 - 2.5	0.06 - 0.10
-5.1 to -4.7	22.9 to 23.5	0.5	1.1	2.5 - 3.6	0.10 - 0.14
		1.4 - 2.5	3.1 - 5.5	3.6 - 4.6	0.14 - 0.18
-8.8 to -7.2	16.2 to 19.0	0.5	1.1	4.6 - 5.6	0.18 - 0.22
		1.4 - 2.5	3.1 - 5.5	5.6 - 6.6	0.22 - 0.26

Applicable to low growing crops and rotation rates of 1 min or less.

higher precipitation rates are necessary to produce effective wetting in comparison with otherwise similar crops. Modlibowska et al. (1962) mention that when vineyards in the Rhineland are at the "wool" stage of growth, it is necessary to ensure a thorough wetting of the crop before freezing starts.

Witte (1962) points out that failures of overhead sprinkler frost protection in German vineyards occurred because the precipitation rates used were those recommended for low growing crops. Higher rates are necessary on grapes (*Vitis* Sp.) than on low growing crops, because of the growth habit of the vine and the difficulty of wetting the crop at some stages of growth.

A further plant factor concerned is the critical temperature for damage. This varies with the stage of development. Table 54–3 exhibits the effects of the variation on the results of overhead sprinkling for frost protection. If the critical temperature, even when the crop is wet, is appreciably below 0C throughout the frost season it may be possible to apply less water.

Sprinkler system characteristics also have an effect. In a freezing environment, higher frequencies of rewetting reduce the temperature fluctuations in ice-covered plant parts (Fig. 54–5). As a result, lower precipitation rates for frost protection may be possible when sprinklers rotate faster (Rogers et al., 1954; Pogrell, 1958; Pogrell and Kidder, 1959). In the field the former found no difference between a rotation rate of 0.5 sec/revolution and another of 48 sec/revolution. Businger (1955) observed more damage to potatoes (*Solanum tuberosum*) with a 3-minute rotation, as opposed to one of 25 to 30 sec/revolution. At variance with the weight of this evidence, Perraudin (1961) observed no significant effect on precipitation demand with rewetting frequencies between 29 and 119 sec. A 4-sec rotation time

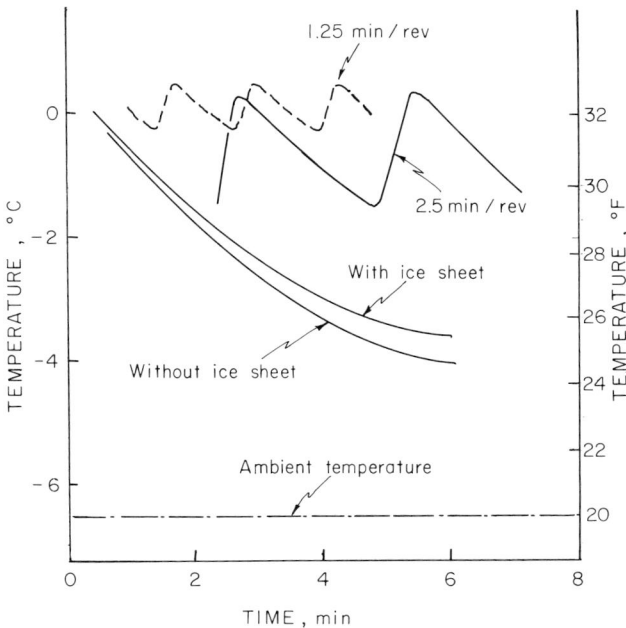

Fig. 54–5. The effect of (a) rewetting frequency and (b) stopping the sprinkling on the temperature of buds sprinkled during frost (redrawn from Jenny, 1961).

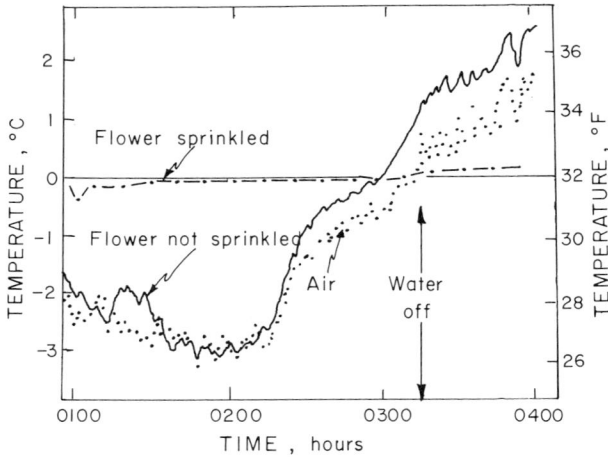

Fig. 54–6. Temperatures of sprinkled and unsprinkled flowers, and of air traced from recorder chart (Rogers and Modlibowska, 1962).

(compared with 60 sec) reduced the wetted diameter of a small, impact type sprinkler by 10 ft (W. P. Annable, 1960. An investigation of frost protection by sprinkler irrigation. *Unpubl. M.S. Thesis. Univ. Massachusetts, Amherst, Mass.*). Rogers and Modlibowska (1962) and Witte (1962) recommend a maximum rotation time of 60 sec/revolution; in view of the practical limitation demonstrated by Annable, a lower limit of about 30 sec might also be imposed.

Measurements of precipitation rate made in laboratory tests and some field trials are related to a very small area. If protection is to be 100% effective in practice, such values should be regarded as the absolute minimum which must be achieved in every part of the sprinkled area. Distribution of water within a sprinkler coverage pattern is never completely uniform. It follows that a high degree of uniformity over the sprinkled area (coefficient of uniformity over 80%) allows a lower overall precipitation rate (sprinkler discharge divided by effective area covered per sprinkler) to be used and still achieve the minimum rate over the whole area. For a complete discussion of uniformity within a sprinkler pattern see chapter 44.

Apart from the limitations of water supply, low total applications are desirable since waterlogging, leaching of nutrients, and crop breakage from heavy ice loads are possible when large quantities of water are applied for frost protection. The water economy, attendant upon high uniformities, may be small when compared with the possibilities of reducing safe operation time. Rogers and Modlibowska (1962) discuss the possibility of utilizing the initial supercooling period of fruit blossoms to start sprinkling at approximately –1 to –2C (28 to 30F), but this would only be possible if no difficulties are encountered in wetting the crop when sprinkling starts. In this respect Annable (1960) observed a tendency for small sprinkler nozzles to clog with ice when the water was turned on at temperatures below 0C. The practice of starting sprinkling below 0C is not yet proven and should be approached with extreme caution (Fig. 54–6). If sprinkling is started too late (or stopped too early), damage will occur which may exceed that in an

Fig. 54–7. Minimum thermometer at proper location in cranberries—bulb at elevation of vine tips (Norton, 1964).

unsprinkled crop. Detailed research is required on this important aspect of overhead sprinkler frost protection.

Meteorologists always quote temperatures from a shielded thermometer located at a standard height (4.5 to 5 ft) above ground in a weather shelter. Night temperatures measured in this way will always be a degree or so higher than what the vegetation is experiencing at the same time. An exposed thermometer at the plant level may be regarded as a reasonable approximation to the actual plant temperature (Fig. 54–7). It should be located in the coldest part of the protected area. If wind or low humidity conditions are likely to occur, a wet bulb thermometer should be employed in order to include the evaporative cooling effect in the temperature measured, e.g., Wheaton observed ice formation under wind conditions at air temperatures above freezing (R. Z. Wheaton, 1959. An experimental study on the effect of wind and water application factors on frost protection by sprinkling. Unpubl. M.S. Thesis. Michigan State Univ., East Lansing, Michigan).

In Germany it is recommended that sprinkling start when the wet bulb temperature reaches 0C. Kidder and Davis (1956) recommend starting irrigation when the temperature of an exposed dry bulb thermometer is 0.6 to 1.2C (33 to 34F), to allow time for starting the sprinklers. Both recommendations are sound, the latter giving a safety margin which might be eliminated with experience. Rogers and Modlibowska (1962) have developed a reliable thermostat switch for automatic operation; in this case no safety margin is required. Some reports suggest that it is necessary to continue sprinkling until all the ice is washed off the crop. Sprinkling may be stopped when the ice is melting on its own, and the air temperature remains above 0C (W. S. Rogers, 1964. Personal communication). On grounds of physical reasoning, the latter view is more tenable. Sprinkling may be stopped with absolute certainty when the wet bulb temperature has risen above 0C.

Droplet size has not been demonstrated to influence sprinkler frost protection except that extremely small droplets are vulnerable to maldistribution by wind (Businger, 1955).

Jenny (1961) noted a marginal reduction in the rate of cooling because of the presence of an ice sheath (Fig. 54–5), and Annable found that an ice thickness of .25 inch reduced the temperature fluctuations in a sprinkled leaf (W. P. Annable, 1960. An investigation of frost protection by sprinkler irrigation. Unpubl. M.S. Thesis. Univ. Massachusetts, Amherst, Massachusetts.). These effects are

Table 54-5. Some results of overhead sprinkler frost protection

Crop	Damage		Temperature		Rotation time	Precipitation rate	Windspeed	Researcher
	Unsprinkled	Sprinkled	Unsprinkled	Sprinkled				
	per cent		°C		sec	inches/hr	miles/hr	
Strawberries (Green fruit & blossoms)	96.2	5.5	-6.2* -2.4 (air 3 ft)	0.6	--	0.10†	--	Braud & Hawthorne (1963)
Potatoes in pots	--	nil	-3.5*	--	25-30	0.12	0-3.4	Businger (1955)
Tomatoes & beans in pots	--	nil	-3.5*	--	25-30	0.12	1.3	Businger (1955)
Citrus nursery	killed	negligible	-7.2 (air 4.5 ft)	-5.6 (air 4.5 ft)	--	0.40†	10-20	Dean (1963)
Mature citrus	--	killed	-9.4 (air)	-10.6 (air)	--	0.10†	5-10	Gerber & Harrison (1963)
Apples (variety "Cox" at petal fall)	99.9	90.4	-3.0 (blossom)	-0.5 (blossom)	--	0.04-0.06	0.7-0.9	Rogers & Modlibowska (1962)
	99.9	0.0	-3.0 (blossom)	-0.5 (blossom)	--	0.11-0.15	0.7-0.9	Rogers & Modlibowska (1962)
Apples	100	0.0	-4.5	--	--	0.13†	--	Perraudin (1961)
Bell Peppers§	killed	nil	-6.1	--	12-20	0.10	--	Kidder & Davis (1956)
Potatoes & tomatoes in pots	100	0.0	-9.0*	--	30	0.28	--	Pogrell (1958)
Cranberries (not at susceptible stage)			-9.2*	-0.2*		0.125†	--	Norton (1964)
Grapes	--	serious	-4.9 to -4.5	--	--	<0.08	--	Eisel (1958)
Grapes	--	negligible	-4.9 to -4.5	--	--	>0.08	--	Eisel (1958)

* Temperature at top of crop.   † Overall precipitation rates.   § (Capsicum frutescens)

short lived. Records of pump failures during frost show that damage occurred in spite of the presence of an ice sheath, proving that the ice has no useful insulating properties.

The effect of deficient sprinkling rates (*see* Table 54–5) is increased damage (Rogers and Modlibowska, 1962; Businger, 1955; Pogrell, 1958). Many growers have experienced this phenomenon. The presence of water or ice has the dual effect of introducing an evaporative cooling factor and preventing the natural supercooling of plant tissue. Pogrell (1958) observed the supercooling of un-sprinkled lettuce plants (*Lactuca sativa*) to –5C (23F), but the crop was completely damaged at the periphery of a sprinkled area where insufficient water fell. Inadequate precipitation rates may result because the designed overall rate was initially too low, from poor design uniformity (due to over-extended sprinkler spacings, incorrect nozzle discharge pressures, or poor sprinkler characteristics), and from distortion of the sprinkler patterns by wind.

The importance of high level equipment maintenance is obvious. Power plant or water supply failure during a frost period could result in a total crop loss; blocked sprinkler nozzles can also cause damage. The equipment should be installed and tested several days in advance of the expected frost season.

2. GROUND (UNDERTREE) SPRINKLING

Ground sprinkling during frost has been employed in deciduous fruit and nut orchards in northern California for several years. A small rise in air temperature in the sprinkled area is responsible for the successes claimed for this method. However, insufficient experimental evidence precludes an authoritative explanation of the processes involved. Heat could be made available to the air in three ways: Enhanced soil heat conduction, heat lost by cooling sprinkler drops, and the release of latent heat by water freezing on the ground. Evaporation would reduce these effects. In a large, sprinkled area, a local rise in humidity would reduce the heat lost by evaporation and possibly reduce the outgoing radiation a little. All these effects are likely to be small. Where water wets low foliage, protection would be analogous to the overhead method.

Brooks and Leonard (1956) report a maximum temperature rise of 0.6 to 1.1C (1 to 2 F) in an almond orchard (*Prunus amygdalus*), when the outside temperature was –3.3C (26F). The undertree system, in this test, wet about 25% of the ground area and discharged approximately 0.1 inches/hour. Had a larger percentage of the ground area been wetted, the effect of sprinkling could have been somewhat greater. Hansen (1964) reports several instances of good protection gained in almond orchards when the minimum temperature (shielded thermometer) was –4C (25F) and the critical level at that stage of growth was –2.8 to –2.2C (27 to 28F) (shielded thermometer). Schultz (1964) found no effect on air temperature with a precipitation rate of 6.4 mm/hr (0.25 inches/hr) on a 2-acre sprinkled area in a California deciduous orchard. In this experiment the sprinklers were not turned on until the air temperature had reached –4.5C (24F). In Michigan, USA, a 2.2C (4F) rise in air temperature was measured on a small overhead sprinkled plot of vegetables with a precipitation rate of 2.5 mm/hr (0.10 inches/hr) (W. K. Bilanski, 1954. Protection of garden crops against frost damage by use of overhead irrigation. *Unpubl. M.S. Thesis. Michigan State Univ., East Lansing, Michigan*). A lack of windspeed and humidity measurements in the above tests makes comparisons difficult.

As the size of the sprinkled area increases, heat losses at the edges will become less important, and a slight increase in the temperature rise produced could be expected (cf. with orchard heaters, Crawford, 1964).

The sprinkling is normally started at about 1.1C (34F) (shielded thermometer at 4 to 5 ft) and continued until outside air temperatures rise above 0C. Precipitation rates on installations specifically for frost protection are as low as 1.3 mm/hr (0.05 inches/hr). They may be somewhat higher when a normal irrigation system is utilized for frost protection (0.10 to 0.25 inches /hr or 2.5 to 6.4 mm/hr).

This method has the advantage of not being likely to elevate the critical temperature by interfering with the natural supercooling of the crop. The possibility of using very low application rates and widely extended spacings is an additional asset. The small temperature rise which can be achieved limits effective undertree sprinkling to use in areas where temperatures, in most years, do not fall below the critical values for frost damage. High humidity and calm atmosphere conditions are also important in achieving success. Under these circumstances, the low initial cost of the undertree system allows a measure of frost protection to be provided where an overhead system would prove too expensive.

3. SURFACE IRRIGATION METHODS

The practice of flooding or furrow irrigation to combat frost is used in many areas where large supplies of water are readily available.

Complete immersion of a crop gives greatest protection. The heat capacity of water plus crop is far greater than that of the crop alone, and as a result, cooling is slowed considerably. Warm water in some areas provides a substantial amount of sensible heat as it cools (1 acre-ft of water loses approximately 100 million BTU in cooling from 21.1 to 0C (70 to 32F). In addition, latent heat of freezing is also available in the case of a severe frost. As opposed to other methods described here, total immersion would be equally effective against both wind and radiation frosts.

The flooding of cranberry (*Vaccinium macrocarpon*) bogs for frost protection is a common practice (Franklin, 1940; Norton, 1964). The depth of water used depends on the severity of the frost. Complete immersion is often the rule in winter while only 2 to 3 inches may be used in May and June (Franklin, 1940). During the time that partial flooding is practiced, the rise in temperature necessary to effect protection is small. In the Imperial Valley of California, complete immersion has successfully protected vegetable rowcrops from frost damage.

Most crops cannot be immersed, but the heat available for warming the air from surface irrigation is often sufficient to give useful frost protection. Furrow irrigation of vegetables and other rowcrops is likely to give a small amount of protection in light frosts. Flooding and broad furrow irrigation have been used for frost protection in citrus groves for many years; the latter is less effective because of the smaller contact area between water and air. Young (1940) observed a temperature rise of 0.8C (1.5F) in an orange (*Citrus sinensis*) grove when it was furrow irrigated with water at 22C (72F). Based on similar field experience, Simpson (R. R. Simpson, 1965. *Personal communication*) suggests that the air temperature in a citrus orchard might be raised by 2.0C (3.5F) by flooding with adequate water at 21.1–26.7C (70–80F). This represents rather ideal conditions. With running water at 18.3C (65F) Hilgeman et al. (1964) obtained variable temperature gains of 1.4 to 2.2C (2.6 to 4F), but with this water stand-ing only 0.5C (0.9F) gain was observed at a minimum temperature of −4.2C (24.5F) under calm conditions. The practice is employed only during severe frost (below −3.9C to −3.3C or 25 to 26F approximately) because citrus leaves and fruit can supercool to a considerable degree, approximate figures being −6C (21F) and −3.3C (26F), respectively; Hendershott (1962). Over-irrigation in winter may cause new growth, reducing frost hardiness greatly; root damage is also possible when the soil is waterlogged for extended periods. Although flood and furrow irrigation are not capable of a substantial degree of frost protection, they are well suited to citrus and other crops in areas where severe frosts are rare.

B. Irrigation Prior to Frost (Advance or Preirrigation)

On a comparatively dry soil, any form of irrigation prior to a frost increases the heat capacity of the soil and improves its ability to absorb and release heat Given high daytime solar radiation and little chilling of the ground by evaporation or cold irrigation water, the amount of solar heat stored in the soil will be raised. Thus, at night, more heat is released from the soil to counteract radiational cooling. This practice might prove hazardous, since it is dependent on favorable weather conditions in the days preceding a frost.

Studies by Brooks and Rhoades (1954) in a pear (*Pyrus communis*) orchard showed a doubling of diurnal soil heat flows as a result of a light flood irrigation. They attributed an initial fall in the 10 cm soil temperature to "evaporation chill." The soil required nearly 2 weeks to show a comparative advantage in temperature over an unirrigated control. The high soil temperature and other features of their data (Fig. 54–8) indicate that the major effect could possibly have been due to the cooling of the soil by comparatively colder irrigation water. If this were the case, the initial disadvantage of irrigation would be avoided by using water which was at least as warm as the average temperature of the soil layers affected by diurnal fluctuations. Earlier in the frost season, or in areas where warm well water is available, irrigation water may in fact be as warm as the soil, in which case the practice of advance surface irrigation would be useful.

Using sprinkler applications of 8.4 mm (0.33 inches) every 4 days, Harvey (1963) reports a 1.1C (2F) rise in the temperature of air, 1 ft above the irrigated soil, sufficient to raise the yield of early potatoes in England by 3 tons/acre. According to field experience of the US Weather Bureau, Simpson (1965) con-cludes that advance irrigation would always be beneficial when a general dry ground situation coincides with a frost period.

Using surface or sprinkler irrigation to provide frost protection in advance, the water should be applied at least 1 day, preferably 2 or 3 days, before the expected frost. Use of water which is cooler than the soil should be avoided. Radiation frost forecasts are not reliable this far in advance, and a policy of maintaining a moist soil by light applications at fairly short intervals is the best approach. The temperature rise obtained by Harvey is probably near maximum for this method and would decrease with height above the ground. Hence this method is best suited to low growing crops in light radiation frosts, although some advantage could be realized in taller crops.

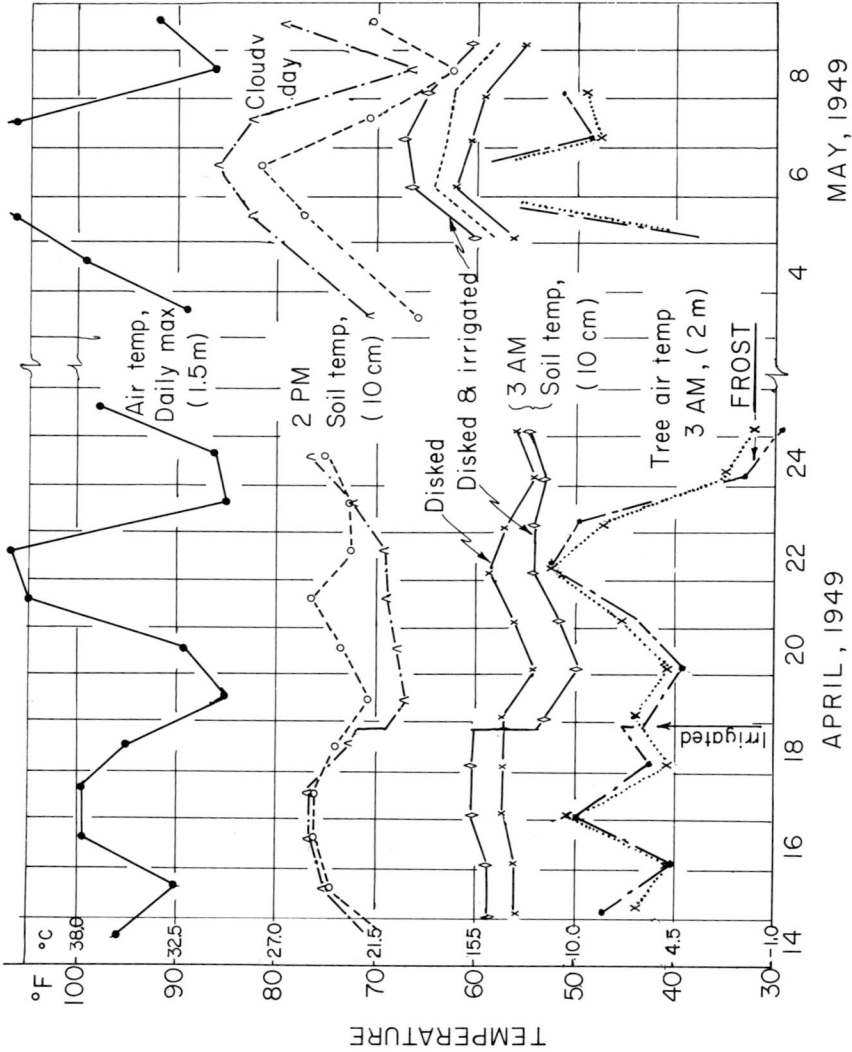

Fig. 54–8. Effects of advance irrigation on pear orchard temperatures (Brooks and Rhoades, 1954).

IV. SELECTION AND DESIGN OF SYSTEMS

The selection of the most suitable method of irrigation depends on: (i) the degree of protection required; (ii) the economic value of the expected increase in yield; (iii) the costs involved in providing the irrigation system, or in adapting the present system to frost protection; (iv) the water supply available; and (v) the effects of the additional irrigation on crop and soil. To choose the most appropriate method and system design, the designer should have a clear picture of the worst condition against which protection is required. A good example of presenting frost probabilities useful for design purposes may be found in Phillips et al. (1962).

Using irrigation during frost, the whole area to be protected must be irrigated at one time. This represents a large increase in water supply, compared with the staggered applications of normal irrigation. Water supply, drainage, and, in overhead sprinkling, ice load problems may limit application rates or the total amount of water applied.

A. Overhead Sprinkling

Assuming none of the above limitations, uniformity of application is not so important when sprinkler irrigation equipment is used for frost protection provided that all parts of the sprinkled area receive the required minimum precipitation rate. In Table 54–6 an attempt has been made to set up guidelines for over-all precipitation rates for design purposes based on the findings of several authors.

For slower rotating sprinklers, higher windspeeds, or lower uniformities, precipitation rates should be increased. Since these rates take no account of the effects of humidity on precipitation demands, an increase of 10% for every deg C that the dewpoint is below 0C might be prudent. An increase of 25% over the rate for deciduous trees should be adopted for grapes, blackcurrants, and other crops which are not easily wetted during the frost season. The design of sprinkler systems for frost protection on cranberry bogs is discussed in detail by Norton (1964).

Table 54–6. Overall precipitation rates for overhead sprinkler frost protection [*]

Deciduous fruit trees (critical temperature −2.2 C)				
Precipitation rate, inches/hr	0.10	0.12	0.18	0.25
Minimum shelter temp.,°C	−2.2	−3.3	−4.7	−5.8
Approx. unshielded thermometer temperature,°C	−3.9 to −3.3	−5.0 to −4.4	−6.7 to −5.8	−7.8 to −6.9
Low growing crops (critical temperature −0.6 to 0 C)				
Precipitation rate inches/hr	0.08	0.10	0.14	0.13
Minimum shelter temp.,°C	−2.2	−3.3	−5.0	−6.7
Approx. unshielded thermometer temperature,°C	−3.9 to −3.3	−5.0 to −4.4	−6.7 to −6.1	−8.3 to −7.8

[*] These values are for winds of < 2 miles/hour, sprinkler rotation times of 30 to 60 sec/revolution and a minimum uniformity coefficient of 90%.

In some areas there is a trend toward permanent sprinkler installations which are used for frost protection and heat control, as well as normal irrigation. The latter use will normally dictate the spacings used. In orchards, high risers are necessary and in trellised grapes, laterals are sometimes placed along the tops of the stakes. The latter systems cost $550 to $625/acre, installed. Systems with underground plastic mains and risers cost about the same. Approximate cost involved for cranberry bogs is $400 to $500/acre, without installation.

The irrigator with a portable system usually has to buy additional pipe and laterals to make a solid set system, covering the whole area to be protected. Although a high uniformity of application is desirable, it also requires closer spacings and therefore demands a higher investment in equipment. As long as sufficiently high application rates ensure that no areas receive too little water, extended spacings may be employed. Using a triangular sprinkler arrangement with 60 ft between laterals and 60-ft sprinkler spacing on the lateral, a 9/64 inch nozzle sprinkler would achieve a 0.10 inch/hr minimum precipitation rate throughout the area and protect low growing crops down to about −3.9C (25F) shelter temperature. Sprinkler rotation times should be 30 to 60 sec/revolution. It is important to use optimum operating pressures recommended by the manufacturer for even distribution of water.

B. Ground Sprinkling

More latitude is allowed in the design of undertree systems. Precipitation rates are not so critical as for the overhead method and, while complete coverage of the area would improve protection, it is not essential. In areas where waterlogging is likely, precipitation rates as low as 1.3 mm (0.05 inches)/hour are used. No information is available on the relative effectiveness of different precipitation rates, but higher precipitation rates can be expected to increase protection.

Sprinklers employed are low angle, undertree types as used for irrigation, but on more extended spacings. Again, if the complete area is to be covered, extra laterals and sprinklers must be purchased. The cost for an undertree sprinkler system using portable aluminum pipe would be approximately $250/acre.

C. Surface Irrigation

There must be few, if any, surface irrigation systems designed primarily for frost protection. Consequently, the latter use, although involving larger supply and delivery capacities than for normal irrigation, is most often a subordinate consideration in design. Ideally, sufficient delivery capacity should be provided to irrigate the whole crop rapidly. Complete immersion of low growing crops requires a great deal of water, and requirements are increased, if the surface is uneven. The amount of water required for flooding cranberry bogs ranges between 6 inches and several feet (Norton, 1964). Availability of water may limit the number of irrigations possible during a frost season.

Since frost protection with surface irrigation is essentially an auxiliary use of existing facilities, costs involved are minimal.

D.   Advance Irrigation

The sole design consideration for frost protection by this method is to maintain a moist soil. Costs involved are minimal as with surface irrigation; if sprinkling is the normal irrigation method, little, if any, extra equipment is needed.

V.   SUMMARY AND CONCLUSION

Overhead sprinkling generally provides the most effective frost protection, but if extra equipment must be purchased, the investment cost is high. Crop damage from ice loads and waterlogging are potential problems. Detailed studies are required on the effects of starting and stopping the sprinklers at various temperatures close to, and below, the freezing point, for here lies the greatest potential for water economy.

Ground sprinkling involves a smaller investment than the overhead method A great deal of work must be done to investigate the means by which this practice produces its effect. Relationships have yet to be established between temperature rise, precipitation rate, thermal properties of the soil, size of the sprinkled area, windspeed, and humidity of environment.

Both sprinkler methods offer considerable water economy, when compared with surface irrigation during frost. Flooding the surface, but not covering the crop, offers a protection potential similar to undertree sprinkling with a slight advantage, if abundant supplies of warm water are available. Key factors in surface irrigation are water supply, delivery capacity, soil drainage, and effects on plant growth.

Advance irrigation, in common with surface methods used during frost, offers a considerable cost advantage offset by limited effectiveness. More research is required to establish limits of effectiveness in this relatively unproven field.

All methods of irrigation for frost protection are only effective under radiation frost conditions (with the exception of total immersion of the crop in water). If sufficiently high precipitation rates were possible, overhead sprinkling would be effective under advective frost conditions, provided the accompanying wind is not strong enough to distort sprinkler patterns severely. In the Florida, USA freeze of 1962–63 sprinkler irrigation failures on citrus trees were due to high winds and insufficient precipitation rates.

A greater appreciation of the limitations of the various methods of irrigation for frost protection will enable them to be applied more beneficially in the future.

ACKNOWLEDGEMENT

The authors wish to recognize the numerous valuable criticisms and suggestions of Dr. T. V. Crawford (formerly of the Agricultural Engineering Dept., University of California, Davis) in the preparation of the manuscript.

LITERATURE CITED

Braud, H. J., and P. L. Hawthorne. 1963. Frost control in strawberries. Louisiana Agr. 6:6–7.

Brooks, F. A. 1959. An introduction to physical microclimatology. Associated Students Store. Univ. California, Davis. 264 p.

Brooks, F. A., and A. S. Leonard. 1956. Temperatures and frost damage. California Agr. 10(8):7–8, 13.

Brooks, F. A., and D. G. Rhoades. 1954. Daytime partition of irradiation and the evaporative chilling of the ground. Amer. Geophys. Union, Trans. 35:145.

Businger, J. A. 1955. Nachtvorstbestrijding Door Middel van Besproeiing. Tuinbouw Techniek Inst. 18:21–34.

Businger, J. A. 1965. Frost protection with irrigation. Amer. Meteorol. Soc. Agr. Meteorol. Monogr. 6. Boston.

Crawford, T. V. 1964. Computing heating requirements for frost protection. J. Appl. Meteorol. 3:750–760.

Dean, R. H. 1963. Use of water sprinklers for protection of nursery trees against freeze damage. Citrus Ind. 44:14–16, 22.

Eisel, H. 1958. Untersuchungen uber die frostschutzberegnung. Weinberg u. Keller 5:120.

Franklin, H. J. 1940. Cranberry growing in Massachusetts. Massachusetts Agr. Exp. Sta. Bull. 371. 44 p.

Georg, J. G. 1960–61. The effects of soil moisture and tillage on the nocturnal minimum air temperature over sandy soils. US Weather Bur. Lakeland, Florida Weather Forecasting Mimeo no. 1. 2 p.

Gerber, J. F., and D. S. Harrison. 1963. Research with sprinkler irrigation for cold protection of citrus. Amer. Soc. Agr. Eng. (Winter Meeting, 1963). Pap. 63. p. 734.

Hansen, H. B. April 1964. US Weather Bur. Mimeo, Chico, California. 3 p.

Harvey, P. N. 1963. Protecting early potatoes against frost. J. Min. Agr. 70:214–215.

Hendershott, C. H. 1962. Responses of citrus to freezing temperatures. Amer. Soc. Hort. Sci., Proc. 80:247–254.

Hilgeman, R. H., C. E. Everling, and J. A. Dunlap. 1964. Effect of wind machines, orchard heaters, and irrigation water on moderating temperatures in a citrus grove during severe freezes. Amer. Soc. Hort. Sci., Proc. 85:232–244.

Jenny, J. 1961. Probleme der frostbekämpfung: Der einfluss der dicke der eisschicht bei der Beregnung. Traktor U. die Landmaschinen 23:258.

Kidder, E. H., and J. R. Davis. 1956. Frost protection with sprinkler irrigation. Coop. Ext. Serv. Michigan State Univ. Ext. Bull. 327 (rev.) 12 p.

Levitt, J. [editor] 1956. The hardiness of plants. Agronomy Vol. 6. 278 p.

Linacre, E. T. 1964. Determinations of the heat transfer coefficient of a leaf. Plant Physiol. 39:687–690.

Modlibowska, I., J. MacCormick, and C. H. W. Slater. 1962. Fruit growing and frost protection in France and the Rhineland. Annual Rep. E. Malling Research Sta., Kent, England. (1961)A45:128–132.

Niemann, A. 1957. Neue untersuchungen zur physik der frostberegnung und deren bedeutung fur die bemessung und bedienung der anlagen. Akten der International Tagung fur Frostberegnung, Bolzano Sept., 1957. Camera di Commercio, Industria ed Agricoltura Bolzano, Italy. p. 25–36.

Norton, J. S. 1964. Design of minimum gallonage sprinkler systems for cranberry bogs. Massachusetts Agr. Exp. Sta. Bull. 532. 20 p.

Perraudin, G. 1961. Lalutte contre le gel. Imprimerie Rhodanique S. A. Prom. No. 3121., Valais, Switzerland. 20 p.

Phillips, E. L., M. D. Magnuson, A. H. Jones, A. van Doren, E. L. Proebsting, and P. C. Crandall. 1962. Washington State freeze circular. Washington Agr. Exp. Sta. Circ. 400. 28 p.

Pogrell, H. von. 1958. Grundlegende Fragen der Direckten Frostschutz beregnung. Buch und Verlagsdruckerei Ludw. Leopold. Bonn. 48 p.

Pogrell, H. von. 1959. Experiences in frost protection work by sprinkling in Germany. Michigan State Univ. Agr. Eng. Dep. Mimeo. 5 p.

Pogrell, H. von, and E. H. Kidder. 1959. The effectiveness of water use in sprinkler irrigation for frost protection. Michigan Agr. Exp. Sta. Quart. Bull.. 42:323–330.

Pogrell, H. von, and E. H. Kidder. 1960. Experimental study of critical leaf temperature in regard to frost protection by sprinkling. Michigan Agr. Exp. Sta. Quart. Bull. 42:615–621.

Proebsting, E. L. 1964. Recent advances in cold hardiness research. Washington State Hort. Ass., Proc. 60th Ann. Meeting. p. 37–40.

Raschke, K. 1960. Heat transfer between plant and the environment. Annual Rev. Plant Physiol. 11:111–126.

Rogers, W. S., and I. Modlibowska. 1962. Automatic protection from frosts by water sprinkling. Advance Hort. Sci., and their Applications. 3:416–425.

Rogers, W. S., I. Modlibowska, J. P. Ruxton, and C. H. W. Slater. 1954. Low temperature injury to fruit blossom: IV. Further experiments on water-sprinkling as an anti-frost measure. J. Hort. Sci. 29:126–141.

Schultz, H. B. 1964. Ground sprinkling limitations in orchard frost protection. California Agr. 18(4):14–15.

Sutton, O. G. 1953. Micrometeorology. McGraw-Hill, New York. 333 p.

Wheaton, R. Z., and E. H. Kidder. 1964. The effect of evaporation on frost protection by sprinkling. Michigan Agr. Exp. Sta. Quart. Bull. 46:431–437.

Wijk, W. R. van. 1963. Physics of the plant environment. North Holland Publ. Co.. Amsterdam. 382 p.

Witte, K. April-May 1962. Frostschadenverthütung durch direkte Beregunug. L'Irriga-zione. p. 23–29.

Witte, K., and Pogrell, H. von. 1958. Untersuchungen uber der einfluss des windes auf die pflanzen temperatur bei der direkte frostschultzberegnung. Rhein. Monatsschr. f. Gemuse Obst. p. Gartenbau. 2:121–128.

Young, F. D. 1940. Frost and the prevention of frost damage. US Dep. Agr. Farmer's Bull. 1588 (rev.) 62 p.

Zeeuw, J. de. 1960. Sproeien als Middel Tegen Nachtvorstschade. Meded. Dir. Van de Tuinbouw 23:43–51.

55

Plant Diseases Related to Irrigation[1]

J. D. MENZIES

US Soils Laboratory, ARS, USDA
Beltsville, Maryland

A farming practice like irrigation, which not only changes the ecological relations of a crop or region but also requires major adjustments in crop management, can be expected to have important effects on plant disease and insect problems. Modification of the natural water supply changes the biological equilibria between crops and their pests in numerous and sometimes complicated ways. Even the changes in farm operations that are required when irrigation is practiced may, in turn, make it necessary to develop new pest-control methods or schedules. The implications of irrigation for plant disease control are discussed in this chapter and insect disease problems in chapter 56.

Irrigation, especially when it involves the sprinkler application of water over a growing crop, might be expected to increase the danger of losses from fungus and bacterial foliage diseases where high humidity, and often free water, is necessary for infection to take place. Many years of farming experience and observation, however, have shown that this danger is relatively slight; but only recently have these observations been supported by experimental studies that help explain the factors involved.

For a parasitic disease to develop to damaging proportions in a crop the climate must not only be suitable for spore germination and infection, but must also permit sufficient sporulation for secondary infections. It is these secondary infections that lead to widespread damage in the crop. Considerable detailed knowledge is available on the minimum requirements for high humidity or free water for these processes with most of the important disease producing organisms (Gottlieb, 1950). In general, free water must be present on the plant surface from 6 to 12 hours or more to permit infection, and a much longer time is necessary for secondary sporulation. In addition, many of these disease organisms depend on splashing water to loosen the spores from the infected plant surface and to spread them to other plants. Usually, the optimum temperatures for spores production, germination, and infection are lower than those normally prevailing during clear summer weather.

The normal processes of fungus disease development in plants are thus uniquely adapted to the natural humid climate. It is obvious that artificial irrigation, even by sprinklers, may provide far more water to a crop than does natural rain, without the prolonged periods of high humidity and lowered temperatures conducive to disease.

The general absence of humid climate diseases on irrigated farms in arid regions is probably not due to the lack of introduction of inoculum of the disease produc-

---

[1] Contribution from the US Soils Laboratory, Soil & Water Conserv. Res. Div., ARS, USDA, Beltsville, Md.

ing organisms. These are readily introduced on seed or planting stock and may even persist from year-to-year at an undetectable level (Menzies, 1952). The occasional outbreaks can usually be related to abnormally high precipitation or cool weather during the growing season.

In arid or semiarid regions, overhead irrigation may raise the relative humidity within a crop only for 2 or 3 hours beyond the period of actual water application (Schnathorst, 1960). In a normally arid climate this may not be sufficient to bring about a disease situation. For example, sprinkler irrigation does not cause significant increase in bacterial diseases of beans (*Phaseolus vulgaris* L.) in the Columbia Basin region of Washington, USA until late in the season when cooler temperatures and higher rainfall are normally encountered (Menzies, 1954). It has also been shown that while an overhead irrigation, even of short duration, is sufficient to spread bacterial inoculum and lead to infection, the hot, dry conditions between irrigations may be so unfavorable to bacterial multiplication in the infected leaves that little or no inoculum survives between irrigations.

Though sprinkler irrigation may not encourage enough bacterial disease in beans to cause yield losses, it may still cause a low level of infection. In the seed bean areas of Idaho and Washington, USA, where complete freedom from bacterial disease is necessary for the production of certified seed, sprinkler irrigation thus introduces an additional hazard (Menzies, 1952). In such special cases, the safer practice of furrow irrigation is recommended.

Late blight is a destructive disease of potatoes (*Solanum tuberosum* L.) in cool, humid climates (Cox and Large, 1960), but is virtually unknown in the intensively irrigated potato areas of the Western USA, even when this crop is irrigated by sprinklers. Experiments in Israel, however, showed that in the northern Negev, blight developed on plots irrigated by sprinklers while being almost absent on furrow-irrigated plots (Rotem et al., 1962). Apparently, when the conditions were such that a trace of blight occurred under furrow irrigation, the disease spread was accentuated by sprinkling. It was further found that long periods of low-rate application had about the same effect as short periods of heavy application. The explanation offered was that the former schedule favored more successful infection, whereas the latter was more effective in splashing inoculum among the plants. The net effect was the same.

Forages and pastures are particularly well suited to sprinkler irrigation and this method is widely used in the arid areas of the Western USA. Even though the low-growing, dense foliage tends to maintain high humidity within the crop, there is little evidence that either surface or overhead irrigation has induced economic losses in this type of crop. An unusual exception to this is damage associated with flood irrigation of alfalfa (*Medicago sativa* L.) during hot summer weather on sandy soil in the irrigated desert areas of southern California (Erwin et al., 1959). Under these conditions the high soil and water temperatures apparently intensify anaerobic damage to root systems. In the Southeastern USA the heavy use of supplemental sprinkler irrigation was reported to cause a slight increase in forage crop diseases compared to that occurring on comparable plots protected from natural rainfall and irrigated as infrequently as possible (Bennett, 1962; Curl and Weaver, 1958).

Sprinkler irrigation may create special disease problems in orchards, not so much because of high humidity, but because of inoculum spread by the driving action of the water through the lower limbs of the tree. Phytophthora fruit rot has become an orchard disease rather than a storage disease, when sprinkler irrigation is used (Kienholz, 1946). The fungus is probably splashed from the soil and introduced into small cuts or abrasions on ripening fruit through the action of the sprinklers. The Phytophthora rot problem is aggravated if a high cover crop is grown under the trees or if the sprinklers distribute the water high into the foliage (Luce, 1953).

Probably the severest test of the danger of sprinkler irrigation in spreading disease is on low growing fruits like tomatoes (*Lycopersicon esculentum* Mill.) or strawberries (*Fragaria* sp.), where field rotting of the ripening fruits is a constant problem. In one detailed study (Cannel et al., 1961) in the Sacramento Valley of California, it was found that excessive overhead irrigation (32 inches of water applied in 17 irrigations) produced 3.6 tons of rotted strawberries/acre, compared to 2.7 tons/acre where furrow irrigation was used. A more moderate irrigation of 18 inches in 8 applications, however, proved to be sufficient for maximum yield while fruit rot was the same as in furrow irrigation (1.7 tons/acre). This loss, incidentally, could be further reduced to less than 1/2 ton/acre by using plastic ground cover under the plants. The results show that over-irrigation with sprinklers can lead to disease damage and that extra care may be required in irrigating tender crops by this method.

Where irrigation is used in normally humid regions to supplement the natural rainfall there is more likelihood of an accompanying increase in disease. The diseases in question are normal for the area and occur, perhaps at a low level, without irrigation. In such cases the extra irrigation may have sufficient effect on humidity to increase disease significantly. In Delaware, USA, for example, overhead supplemental irrigation significantly increased tomato anthracnose, and fruit rot when 1-inch applications of water were made at weekly intervals during the crop ripening period (Crossan and Lloyd, 1956). In this experiment it was also noted that the supplemental irrigation reduced the physiological disorder, blossom end rot, so that the net result was a highly significant increase in the percentage of sound fruit. This advantage was in addition to the increase in total yield as a result of the extra water.

In the above experiments sprinkler irrigation resulted in higher relative humidities and longer periods of dew formation within the crop foliosphere, that persisted as long as 5 days after the water application (Raniere and Crossan, 1959). This favored the incidence of tomato anthracnose, presumably by prolonging spore viability and permitting germination of the pathogen. These humidity effects, extending over several days, must have been due to the higher soil water content rather than to the wetting of foliage and may have been accentuated by heavier foliage growth on the irrigated plots.

Supplemental irrigation of tobacco (*Nicotiana tabacum* L.) in the Southeastern USA may lead to increased disease losses, not only by lengthening the periods of high humidity around the plants, but also by serving as a means of spreading the disease organisms (Jones et al., 1960; McMurtrey, 1961). The black shank disease, e.g., is caused by a species of Phytophthora that readily contaminates ponds and streams receiving drainage from infested fields. These sources of water are unsafe to use for tobacco irrigation.

There is very little published information on the effect of supplemental irrigation on the spread of diseases of cotton (*Gossypium hirsutum* L.). Apparently, the danger is not great if the irrigation is used with moderation (Presley, 1954).

Boll rot may be increased indirectly because of the rank growth resulting from a combination of increased water and higher nitrogen fertilization (Scarsbrook et al., 1961). Sometimes this leads to increased lodging, which tends to favor boll rot.

Whether supplemental irrigation induces serious disease outbreaks is closely related to weather conditions. If an irrigation is applied just prior to, or following, a rainy period, the slight prolonging of the time in which conditions are favorable for spread of disease may be important. Growers can avoid irrigation just after a rain (when the irrigation is least necessary) but cannot always predict the weather several days ahead. Nevertheless, with a proper awareness of the hazards and reasonable judgment of the weather, a grower can use supplemental irrigation with little risk.

In dense-growing row crops, irrigation may favor diseases that attack vines or fruits in contact with the ground. Examples are sclerotinia rot of beans; fruit rots of tomatoes, strawberries, and melons (*Cucumis melo* L.); and bottom rot of lettuce (*Lactuca sativa* L.). Whether the irrigation is applied by furrows or sprinklers, the disease-producing environment is the high humidity trapped at the soil surface by the low, dense foliage. Such damage can be minimized by wider row spacing or plant training to permit wind movement between the plants.

The root diseases of irrigated crops differ in some respects from those of dryland crops, but it is difficult to show that this is directly related to the increased soil water. Irrigated crops are managed differently, fertilized differently, and grown more luxuriantly. If any generalization is to be made, it would be that as water supply is increased up to the optimum for the crop in question any increase in disease is overshadowed by a greater crop response. But as excessive water is applied, root rots can be expected to increase without any compensating benefits to the crop.

Root rots of irrigated vegetable crops have been carefully observed by the author over a period of 20 years in irrigated areas of the Northwestern USA. Rhizoctonia, verticillium wilt, and fusarium wilts of potatoes will usually do more harm to a crop under water stress than when moisture is optimum. If the crop is over-irrigated, however, there is likely to be more damage from tuber rots. Excessive irrigation appears to favor the root rot complex of peas (*Pisum sativum* L.) and may induce various root rots on sugar beets (*Beta vulgaris* L.). With fusarium root rot of beans, the damage can be minimized by encouraging the development of adventitious roots from the hypocotyl above the infection. This is accomplished by hilling soil around the plants and irrigating heavily to keep the soil moist to the surface. In this case over-irrigation is more apparent than real, because the functioning roots are confined to a restricted zone that is difficult to wet by furrow irrigation. The same effect can be achieved by over-head application with lesser amounts of water.

The powdery mildews constitute a class of foliage diseases that do not require high humidity (Erwin et al., 1959; Gottlieb, 1950; Schnathorst, 1960). The general impression that irrigation, especially with sprinklers, favors mildew now seems to be somewhat erroneous. Studies in California showed that the development of mildew in peach (*Prunus persica* L. Batoch.) orchards was not related to relative humidity (Weinhold, 1961) and similar results were obtained in a careful study of mildew development in lettuce (Schnathorst, 1960). Since it has been shown that mildew spores may germinate and infect plants at very low

humidities, it is now believed that the important factor is plant tissue susceptibility. Succulent, fast-growing tissues are generally more susceptible to mildew,
and insofar as irrigation induces this type of growth, it can be considered an
indirect cause (Schnathorst, 1960). Maximum growth is usually desired, however,
for high yield, so the best solution may be to combine irrigation, if needed, with
chemical control of mildew.

Certain physiological disorders of plants associated with chronic or periodic
water stress can be prevented by careful use of irrigation. At least some of the
internal damage of potato tubers, known as heat necrosis or drouth spot, is
reduced by adequate irrigation. As mentioned above, in one Delaware study,
blossom end rot of tomato was markedly reduced by supplemental sprinkler
irrigation (Crossan and Lloyd, 1956). In other studies the black heart disease of
celery (*Apium graveolens* L. var. *dulce* DC) was controlled by preventing
water stress (Cannell, 1959). The drouth spots of fruit caused by boron deficiency
are so called because of the common observation that boron deficiency becomes
more pronounced when the crop is allowed to suffer water stress.

It should also be mentioned that other physiologic disorders of crops may be
aggravated or caused directly by irrigation. Lime-induced chlorosis, e.g., is generally more severe under high soil water conditions (Reuther and Crawford, 1946).
Water quality may also be a factor since irrigation waters high in bicarbonates
have been shown to lead to chlorosis (Brown et al., 1959). One unusual case of
direct foliage toxicity from saline water used in overhead irrigation has been
noted in California citrus groves (Hardy, 1957). It was found that a foliage tip
burn was being caused by toxic levels of Na and Cl absorbed by the sprinkled
leaves. The water in question contained 131 ppm Cl^- and 190 ppm Na^+, and
damage occurred when leaf tissue levels of these ions exceeded 0.25%.

Irrigation water unquestionably can disseminate disease-producing agents not
only within a field, but for much longer distances in distribution and drainage
systems. The rapidity with which these pathogens become distributed over new
irrigation projects furnishes presumptive evidence that the interconnecting water
distribution systems also carry pathogens. In this respect, an irrigation project
differs from the natural rainfall and drainage situation. Rainfall is relatively uncontaminated with disease-producing organisms, and except for heavy storms,
there is no surface runoff. There is no reuse of rainfall water comparable to the
storage, diversion, runoff, and rediversion that occurs in an irrigation system. Even
the water in reservoirs and streams has drained from vegetated watersheds and
has had an opportunity to become contaminated with pathogenic organisms. This
possible danger is well illustrated by the report that Phytophthora rot of fruits
in orchards can be traced to reuse of drainage water for orchard sprinkling (Luce,
1953). The presence of this particular plant disease organism could be demonstrated even in the main canal waters. Similar problems attend the use of drainage
water for irrigation in tobacco culture (McMurtrey, 1961).

Irrigation practice usually will require modification in spraying or dusting
operations. Sprinkling, of course, will tend to wash fungicide residues from the
foliage, necessitating more frequent applications of fungicide or use of formulations more resistant to washing. Also, the use of wheeled spray equipment must
be suspended during irrigation. Where careful timing of spray applications is
important, it will be necessary to modify irrigation schedules accordingly.

Sometimes fungicides are included in irrigation water. For example, copper sulphate has been used in sprinkler systems to reduce the spread of Phytophthora rot in orchards (Luce, 1953), and the nematocide 1-2-dibromo-3-chloropropane has been successfully applied in irrigation water for root-knot nematode control (Morton, 1959). Fungicides in general, however, are not applied with sprinklers because, if the fungicide is included in the entire application, excessive amounts must be used; and if the fungicide is injected into the system only near the end of an irrigation, there are problems of obtaining uniform distribution and coverage. It is usually more practical to make a conventional fungicide application after the irrigation.

Experience has demonstrated that careful use of irrigation will rarely lead to significant increase in crop disease. It is also of interest that the greater the requirement for irrigation (due to lower natural rainfall), the less will be the danger of increasing disease loss. The borderline situation, where a small amount of supplemental water is needed, has the greatest hazard, especially if this water is applied by overhead methods. A general rule for safe sprinkler irrigation is to apply it as rapidly as feasible, considering the crop and the soil intake rate and to avoid, where possible, irrigating immediately before or after a natural rain.

LITERATURE CITED

Bennett, O. L. 1962. Irrigation of forage crops in eastern United States. US Dep. Agr. Prod. Res. Rep. 59. 25 p.

Brown, J. C., O. R. Lunt, R. S. Holmes, and L. O. Tiffin. 1959. The bicarbonate ion as an indirect cause of iron chlorosis. Soil Sci. 88:260–266.

Cannell, G. H. 1959. Effect of irrigation and fertilizer on yield, black heart and nutrient uptake of celery. Amer. Soc. Hort. Sci., Proc. 74:539–545.

Cannell, G. H., Victor Voth, R. S. Bringhurst, and E. L. Proebsting. 1961. The influence of irrigation levels and application methods, polyethylene mulch, and nitrogen fertilization on strawberry production in southern California. Amer. Soc. Hort. Sci., Proc. 78:281–291.

Cox, A. E., and E. C. Large. 1960. Potato blight epidemics throughout the world. US Dep. Agr. Handbook 174. 230 p.

Crossan, D. F., and P. J. Lloyd. 1956. The influence of overhead irrigation on the incidence and control of certain tomato diseases. Plant Disease Rep. 40:314–317.

Curl, E. A., and H. A. Weaver. 1958. Diseases of forage crops under sprinkler irrigation in the southeast. Plant Disease Rep. 42:637–644.

Erwin, D. C., B. W. Kennedy, and W. F. Lehman. 1959. Xylem necrosis and root rot of alfalfa associated with excessive irrigation and high temperatures. Phytopathology 49:572–578.

Gottlieb, D. 1950. The physiology of spore germination in fungi. Bot. Rev. 16:229–257.

Harding, R. B., M. P. Miller, and M. Fireman. 1957. Leafburn on sprinkled citrus. California Agr. 11(1):9–10.

Jones, J. N., Jr., G. N. Sparrow, and J. D. Miles, 1960. Principles of tobacco irrigation. US Dep. Agr. Inform. Bull. 228. 16 p.

Kienholz, J. R. 1946. Phytophthora rot of pears under sprinkler irrigation at Hood River, Oregon. Plant Disease Rep. 39:31.

Luce, W. A. 1953. Observations of phytophthora rot of fruits in the Wenatchee and Yakima valleys of Washington and the Hood River district of Oregon. Washington State Hort. Ass., Proc. 49:147–152.

McMurtrey, J. E., Jr. 1961. Tobacco production. US Dep. Agr. Inform. Bull. 245. 58 p.

Menzies, J. D. 1952. Observations on the introduction and spread of bean diseases into newly irrigated areas of the Columbia Basin. Plant Disease Rep. 36:44–47.

Menzies, J. D. 1954. Effect of sprinkler irrigation in an arid climate on the spread of bacterial diseases of beans. Phytopathology 44:553–556.

Morton, D. J. 1959. The control of cotton root rot by the addition of 1-2-dibromo-3-chloropropane to irrigation water. Plant Disease Rep. 43:243–247.

Presley, J. T. 1954. Cotton diseases and methods of control. US Dep. Agr. Farmer's Bull. 1745. 19 p.

Raniere, L. C., and D. F. Crossan. 1959. The influence of overhead irrigation and microclimate on *Colletotrichum phomoides*. Phytopathology 49:72–74.

Reuther, W., and C. L. Crawford. 1946. Effect of certain soil and irrigation treatments on citrus chlorosis in calcareous soils: I. Plant responses. Soil Sci. 62:477–491.

Rotem, J., J. Palti, and E. Rawitz. 1962. Effect of irrigation method on development of *Phytopthora infestans* on potatoes under arid conditions. Plant Disease Rep. 46:145–149.

Scarsbrook, C. E., O. L. Bennett, L. J. Chapman, R. W. Pearson, and D. G. Sturkie. 1961. Management of irrigated cotton. Alabama Agr. Exp. Sta. Bull. 332. 23 p.

Schnathorst, W. C. 1960. Relations of microclimate to the development of powdery mildew of lettuce. Phytopathology 50:450–454.

Weinhold, A. R. 1961. The orchard development of peach powdery mildew. Phytopathology 51:478–481.

56 | Insect Problems of Irrigated Lands

E. C. KLOSTERMEYER

Washington State University
Irrigated Agriculture Research & Extension Center
Prosser, Washington

The insect problems of irrigated agriculture are essentially the same as those on crops grown under natural rainfall, but in some ways they are unique and require a different approach for their solution. Irrigation projects in arid lands are often "islands" of lush vegetation surrounded by or interspersed with uncultivated nonirrigated land. This dry land is a harsh environment, usually supporting a limited fauna of highly adapted insect species. Some of these may, however, occur in large numbers. When arid land is removed from its native state through irrigation a change in the existing insect fauna takes place. Insects which feed on the foliage retreat to the nonirrigated land, or, if they have a sufficiently wide host range, may transfer to irrigated crops. Many existing sucking insects such a lygus bugs, leafhoppers, and mites feed readily on certain irrigated crops and may become serious pests. Some chewing forms such as cutworms, grasshoppers, and crickets may also find irrigated crops suitable food.

In the soil, directly affected by a changed water relationship, some species cannot survive. Thus, the Great Basin wireworm, [*Ctenicera pruinina noxia* (Hyslop)] often present in dry land soils in Western USA is gradually replaced by the Pacific Coast wireworm (*Limonius canus* LeC.), and the sugar beet wireworm [*L. californicus* (Mann.)], which normally inhabit damp soils near streams and lakes (Lane, 1941).

Irrigation may provide conditions enabling a species to develop populations not possible under normal circumstances. Thus, the alkali bee (*Nomia melanderi* Ckll.) originally nested in moist alkaline soils along streams and ponds. Such locations, rare, in the arid countries, did not supply sufficient nesting areas nor did the native plants provide sufficient food for development of large populations of this species. Seepage areas along irrigation canals provided an initial increase of the species. Irrigation on shallow soils resulted in extensive moist alkaline areas suitable for nesting of this bee. Large acreages of alfalfa (*Medicago sativa* L.) and certain other crops afford ample forage and permit this species to develop populations numbering a million or more per acre. Alfalfa seed yields in such areas may exceed 1,000 lb/acre (Frick et al., 1960). So valuable has this bee become as a pollinator of alfalfa that many growers now provide artificial nesting sites where alkalinity and soil water are controlled. Under such conditions, farmers may resist attempts to reclaim such alkaline land for crop production, since to do so would destroy the bees (Bohart and Knowlton, 1960).

The effects of irrigation on insects usually is brought about by irrigation's influence on plant growth. More clover root borers [*Hylastinus obscurus* (Marsham)] were found at lower soil water levels but there was no significant effect of water on root borer damage (Pruess and Weaver, 1959). Borer numbers increased as the root size increased but the root size has a more direct influence on borer numbers than soil water (Koehler and Gyrisco, 1959). Water indirectly affects borer numbers owing to its influence on root size. In India, increased irrigation increased the infestation of the rice stem borer (*Sesamia inferens* Wlk.) through changed ecological and nutritional conditions. With inadequate irrigation the plants were hardened, sparse, and of poor growth whereas the wet plots produced plants of succulent and luxuriant growth more attractive and probably more nutritious to the stem borer (Chowdhry and Sharma, 1960).

A similar situation occurred in Russia with the Hessian fly [*Mayetiola destructor* (Say)] and the frit-fly (*Oscinella pusilla* Mg.) on wheat (*Triticum vulgare*). Irrigation provided better conditions for multiplication of the Hessian fly so that more eggs were laid (1.2 per stem) on irrigated wheat than on nonirrigated wheat (0.05 eggs per stem). Irrigation also induced greater tillering of the wheat with a resultant infestation of 46% of the stems infested with frit-flies compared with 29% on nonirrigated wheat (Pavlov, 1959).

Two-spotted spider mites [*Tetranychus telarius* (L.)] developed higher populations on flood-irrigated alfalfa watered at 10-day intervals (1,024 mites/100 leaves) than on alfalfa watered at 3 week intervals (304 mites/100 leaves) or with no irrigation after March (no mites per 100 leaves). This was attributed to the lush growth of the plants under irrigation and to the sparse open growth of the plants in the dry plots resulting in higher temperatures detrimental to the mites (Butler, 1955).

Irrigated alfalfa plants were better able to sustain attacks of the potato leafhopper [*Empoasca fabae* (Harris)] in Ohio. On dry plots the leafhoppers reduced the plant growth by 28% that of sprinkler irrigated alfalfa. Insecticidal control of the leafhoppers resulted in greater growth on the dry plots than the irrigated plots (Wilson et al., 1955). In Canada, controlling pea aphids [*Acyrthosiphon pisi* (Harris)] with insecticides did not increase yields if the infested alfalfa plants had adequate irrigation (Hobbs et al., 1961). On cotton (*Gossypium herbaceum*), there were no significant differences in mirid bug populations due to irrigation until drouth symptoms became evident. Then the highly fertilized, irrigated plots were infested with the greatest number of insects (Adkisson, 1957).

The uncultivated areas interspersed among irrigation projects prevents the direct spread of many insects. For example, the spotted alfalfa aphid [*Therioaphis maculata* (Buckton)], which spread so rapidly through the USA after its introduction is unable to survive the winter in the northern states. In the Midwest it migrates northward each year and may achieve damaging populations by early summer, its northern spread evidently aided by contiguous cultivation. In the Western USA the aphid does not appear in northern Oregon and in Washington until September or October, perhaps brought in by wind currents at high altitudes. Its northern spread in this area is evidently impeded by intervening areas of uncultivated, alfalfa-free land.

On the other hand, continuous development of irrigated lands results in "bridges" over deserts across which pests can travel. The banded cucumber beetle (*Diabrotica balteata* Le Conte) and Egyptian alfalfa weevil (*Hypera brunneipennis* Boheman) migrated northward across such bridges (Smith, 1957).

The dry lands act as a reservoir for only a few pest species. Of greatest importance are locusts or grasshoppers (*Acrididae*) which develop on rangelands or uncultivated land in many parts of the world and then migrate when native vegetation dries up or is consumed (Gunn, 1960). Irrigated lands in the path of such migrations suffer heavy losses. Dry lands may also harbor vectors of plant diseases, as well as host plants of such diseases. Probably the most important example is the beet leafhopper [*Circulifer tenellus* (Baker)] (Carter, 1961). Now known to be a native of the Middle East where the curly-top virus disease it transmits also occurs, the beet leafhopper achieved its greatest importance in the Western USA where the combination of favorable over-wintering conditions on nonirrigated rangeland and adequate host plants in irrigated areas for summer development proved an ideal combination. Long distance migration of this species across desert areas is common. The virus disease it transmits, curly-top, has a wide host range and is destructive to beets (*Beta vulgaris* L.), tomatoes (*Lycopersicon esculentum* Mill.), and beans (*Phaseolus vulgaris* L.).

Although soil inhabiting insects can survive relatively long submersion, they will eventually succumb. Prior to the development of effective synthetic insecticides, flooding, as well as drying out the soil were tested and used to control certain soil pests. Wireworms [*Limonius californicus* (Mannerheim) and *L. canus* LeConte] can be killed by flood irrigation with 95 to 100% control if done when soil temperatures under the water layer can be kept at 21C (70F) or higher for 1 week (Lane, 1941). This may require 4 acre-inches of water and is possible only on level land. Wireworms may also be killed by the reverse procedure—drying the soil. However, water must be withheld for a full season while growing a crop of fall grain or alfalfa to dry out the soil. This will result in 80% reduction of the wireworms but with some crop loss due to the lack of water (Lane, 1941).

Flood irrigation will kill enough of the lesser corn stalk borer [*Elasmopalpus lignosellus* (Zell)] attacking corn (*Zea mays*) and sorghum (*Sorghum vulgare*) in the seedling stage to permit a satisfactory stand. The treatment could also be used in furrows if the seeds were planted in the bottom of the furrow (Reynolds et al., 1959). Very likely many other soil-inhabiting insects are washed out of their burrows or drowned during normal irrigation practices but the effects on the total pest population is likely minor.

Irrigation ditches may serve as a barrier and control for certain species, e.g. migrating Mormon crickets (*Anabrus simplex* Hald.). This wingless species migrates in large numbers from rangeland. When they encounter an irrigation ditch they enter it and float or swim. Some cross the canals while others may be carried downstream for several miles. If oil is added to the surface by drip applicators the crickets are killed. The floating oil is removed by baffles before it is turned onto cropland (Cowan et al., 1943).

Irrigation may also be used as a means of applying insecticides. In Puerto Rico irrigation was effective for applying soil insecticides but was not satisfactory for controlling moths of the sugarcane borer [*Diatraea saccharalis* (Fabricius)] (Mortorell and Burleigh, 1954). Dieldrin and aldrin were applied to soils in Canada with sprinklers to control tuber flea beetles (*Epitrix tuberis* Gent.). The insecticides were applied in the first hour of irrigation followed by a 6-hour irrigation. The materials were introduced into the system with a pressurized injector on the intake side of the pump. Mechanical incorporation into the soil was needed since the irrigation did not leach the insecticide into the soil deeper than 1 inch (Banham, 1960).

A nomogram is available for computing rates of application in applying insecticides in irrigation water. Although developed for soil fumigants, the procedure is suitable for any chemical application (Warren, 1961). Systemic insecticides and some contact insecticides may be applied through sprinkler systems, if added at the end of the set so the residue will not be washed away. However, the method has not been widely used for foliage applications since the uniformity of application is low.

Irrigation complicates insecticide application to some extent, particularly where furrow or flood irrigation is used, and the time of application is critical as in corn earworm [*Heliothis zea* (Boddie)] control (Eden, 1955). Irrigation schedules may prevent movement of ground application equipment into the field at such critical times. For this reason aerial application is a common practice in irrigated areas.

Most insecticides are not water soluble and hence not readily leached from the soil but they may be transported by mechanical action and move downslope to a considerable extent. Lindane moved into the lower levels of the soil when applied to the upper area but could not be recovered from water leached through such soils. Parathion apparently moved in all directions by a diffusion phenomenon but it too could not be detected in the leachate (Lichtenstein, 1958). Granular insecticides applied to the soil usually do not become available until dissolved by water. Systemic insecticides of varying water solubilities were not absorbed by the plant until the plants were furrow irrigated (Reynolds and Metcalf, 1962).

Sprinkler irrigation following insecticide application by any method has the same effect on insecticide residues that rain has. Five days after spraying, when 70% of the Guthion applied still remained on apple (*Malus sylvestris*) tree foliage, sprinkler irrigating equivalent to 1 inch of rain reduced the residue to 15% of the original deposit. When overhead sprinklers are used the irrigation should be delayed as long as possible after spraying (Williams, 1961).

Biological control may also be affected by irrigation practices. Fungus diseases of spotted alfalfa aphids may be increased by increasing the number of irrigations between cuttings, thus aiding in the control of the spotted aphid as well as washing off the honeydew and aphids (Smith, 1957).

The intensive production of fruit and seed crops under irrigation has resulted in the development of an extensive agricultural industry. Honey bees are needed to supplement the usually inadequate populations of native pollinators, and rental of bee colonies for this purpose constitutes an extra source of income for the beekeeper. Irrigated areas also provide an abundant and varied supply of nectar and pollen plants resulting in good honey crops. However, the usually greater need for insect control on irrigated crops often results in a serious bee poisoning problem.

Irrigation permits growing crops where it is not possible to grow them otherwise, but at the same time it provides conditions conducive to maximum population development of pest species. Insect control must not be neglected if the maximum yield potential is to be attained.

LITERATURE CITED

Adkisson, P. L. 1957. Influence of irrigation and fertilizer on populations of three species of mirids attacking cotton. U. N. Food Agr. Organ. Plant Protect. Bull. 6 p. 33–36.

Banham, F. L. 1960. Sprinkler irrigation as a means of applying insecticides for controlling the tuber flea beetle, *Epitrix tuberis* (Gent.) in the interior of British Columbia. Can. J. Plant Sci. 40:172–177.

Bohart, G. E., and G. F. Knowlton. 1960. Managing alkali bees for alfalfa polination. Utah Agr. Ext. Leafl. 78. 8 p.

Butler, G. D., Jr. 1955. The effect of alfalfa irrigation treatments on the two-spotted mite in alfalfa. J. Econ. Entomol. 48:221–222.

Carter, Walter. 1961. Ecological aspects of plant virus transmission. Ann. Rev. Entomol. 6:347–370.

Chowdhry, S., and R. G. Sharma. June 1960. Effect of methods and frequency of irrigation on the incidence of stem borer of wheat (*Sesamia inferens* Wlk.) at irrigation research farm, Madhipura. Indian J. Agron. 4:264–268.

Cowan, F. T., H. J. Shipman, and C. Wakeland. 1943. Mormon crickets and their control. US Dep. Agr. Farmers Bull. 1928. 17 p.

Eden, W. G. Aug. 1955. Effects of irrigation on insect problems. Alabama Polytech. Inst. Agr. Ext. Circ. 485:32–34.

Frick, K. E., H. Potter, and H. Weaver. 1960. Development and maintenance of alkali bee nesting sites. Washington Agr. Exp. Sta. Circ. 366. 10 p.

Gunn, D. L. 1960. The biological background of locust control. Ann. Rev. Entomol. 5:279–299.

Hobbs, G. A., N. D. Holmes, G. E. Swailes, and N. S. Church. 1961. Effect of the pea aphid, *Acyrthosiphon pisum* (Harr.) (Homoptera Aphididae), on yields of alfalfa hay on irrigated land. Can. Entomol. 43:801–804.

Koehler, C. S., and G. G. Gyrisco. 1959. Effect of root size and soil moisture on the number of clover root borers present in red clover roots. J. Econ. Entomol. 52:658–660.

Lane, M. C. 1941. Wireworms and their control on irrigated lands. US Dep. Agr. Farmers Bull. 1866. 21 p.

Lichtenstein, E. P. 1958. Movement of insecticides in soils under leaching and non-leaching conditions. J. Econ. Entomol. 51:380–383.

Mortorell, L. F., and C. H. Burleigh. 1954. Ineffectiveness of the overhead irrigation method for the application of insecticides to control the sugar cane stalk-borer, *Diatracea saccharalis* (Fabricius). Univ. Puerto Rico J. Agr. 38:38–60.

Pavlov, I. F. 1959. The influence of shelter-belts, irrigation and perennial grasses on the frequency of the Hessian fly and frit-fly (*Diptera*) *Cecidomycidae* and *Chloropidae* (In Russian). Entomol. Obozr. 38:326–340. (Entomol. Rev. Amer. Inst. Biol. Sci. Engl. transl. p. 297–308.)

Pruess, K. P., and C. R. Weaver. 1959. Effects of moisture on the clover root borer and red clover yields. J. Econ. Entomol. 52:1166–1167.

Reynolds, H. T., L. D. Anderson, and L. A. Andres 1959. Cultural and chemical control of the lesser cornstalk borer in southern California. J. Econ. Entomol. 52:63–66.

Reynolds, H. T., and R. L. Metcalf. 1962. Effect of water solubility and soil moisture upon plant uptake of granulated systemic insecticides. J. Econ. Entomol. 55:2–5.

Smith, R. F. July 1957. The effects of irrigation on insects and other arthropods. Irrig. Eng. Maintenance 7:20–21.

Warren, L. E. Winter 1961. Computation chart for applying chemicals to irrigation water. Down to Earth 17:2, 16.

Williams, K. 1961. Note on the effect of rain, and sprinkler irrigation, on the persistence of spray residues of Guthion and Sevin on apple leaves. Can. J. Plant Sci. 41:449–451.

Wilson, M. C., R. L. Davis, and G. G. Williams. 1955. Multiple effects of leaf-hopper infestation on irrigated and nonirrigated alfalfa. J. Econ. Entomol. 48:323–326.

57 Public Health Problems Related to Irrigation

M. B. RAINEY

Federal Water Pollution Control Administration,
US Department of the Interior
Denver, Colorado

A. D. HESS

Public Health Service, US Department of HEW
Greeley, Colorado

I. INTRODUCTION

Environmental changes resulting from the development of irrigation agriculture often create serious public health problems. Malaria, encephalitis, and annoyance by blood-sucking insects are among the more important. Vector-borne diseases such as bilharziasis, as well as typhoid fever, and other diseases related to polluted water (Henderson, 1952; World Health Organ., 1955, 1962a; Stead, 1957; Russell et al., 1963) are also troublesome but not necessarily limited to irrigated areas. Ecological conditions associated with irrigation are often responsible for their occurrence or intensification. Public health problems have frequently restricted agricultural, industrial, and economic development in irrigated areas, particularly in underdeveloped countries (May, 1954; Russell, 1952, 1956). Malaria alone has imposed a great burden of death, illness, and economic loss on the human population of irrigated areas in many parts of the world, and the prevention or control of this disease removes a tremendous barrier to economic and social progress (Russell, 1951).

Irrigation agriculture will play a major role in providing the additional food supplies needed to keep pace with the increasing world population. If maximum benefits are to be derived from existing and future irrigation developments, adequate provision must be made for the prevention and control of public health problems. This requires a mutual understanding of these problems and close cooperation and coordination between the various agencies and groups concerned with irrigation agriculture and public health.

II. VECTOR-BORNE DISEASES AND RELATED PROBLEMS

Many insects of public health importance as well as the snail hosts of bilharziasis are produced in aquatic habitats associated with irrigation. Mosquitoes are by far the most important of these insects. Several species serve as vectors of human diseases such as malaria and encephalitis, and some create public health problems because of their vicious biting habits. Other insects of public health importance that may be produced in habitats associated with irrigation include horse flies and deer flies (*Tabanidae*), black flies (*Simuliidae*), and several species of small gnats (*Heleidae* and *Ceratopogonidae*). Improper preparation and maintenance of reservoirs and conveyance canals; inadequate drainage systems for removal of excess water; excessive seepage losses; poor land preparation; use of farm layouts and irrigation methods that do not fit the land, crops, or water supply; and the application of water in excess of crop requirements are important factors involved in the development of man-made aquatic habitats suitable for snails, mosquitoes, and other insects of public health importance.

In a number of countries malaria mosquitoes are produced in reservoirs and other aquatic habitats associated with irrigation. The magnitude of the malaria problem in many countries has been greatly reduced by extensive residual spray programs and the use of improved drugs for treatment of human cases. These programs have proved to be an effective means of rapidly controlling malaria, but must be carried out on a repetitive basis and are expensive. In addition, experience in a number of countries has shown that malaria mosquitoes may develop high levels of resistance to the insecticides used for residual spraying (World Health Organ., 1959a, 1962b). Attention should, therefore, be given to preventing and eliminating mosquito production wherever possible.

Conditions associated with impoundments that are conducive to malaria mosquito production include emergent or floating vegetation, or both in shallow water areas, accumulations of floatage and debris in shallow water areas or embayments protected from wave action (Fig. 57–1), and undrained depressions within the seasonal fluctuation zone. The basic principles and practices of malaria

Fig. 57–1. In irrigation reservoirs, conditions favorable for mosquito production are created by emergent and/or floating vegetation in shallow water areas and embayments that are protected from wave action. (Photograph—Irrigation reservoir at Whitney, Nebraska, USA).

control on impounded water have been presented in a manual prepared jointly by personnel of the US Public Health Service and the Tennessee Valley Authority (1947). This manual deals primarily with impoundments in the USA, but much of the information pertaining to reservoir preparation and maintenance, as well as water level management, can be adapted for use in other countries. Excessive production of malaria mosquitoes can be foreseen and, in almost all cases prevented, if proper precautions are taken from the time of initial planning of impoundments (Russell et al., 1963).

In some countries the major sources of malaria mosquitoes are related to faulty irrigation and drainage practices, e.g., in Saudi Arabia and India (Daggy, 1959; Russell, 1938). These sources include seepage areas (Fig. 57–2), obstructed canals and ditches, and collections of irrigation waste water in undrained areas. Many of the conditions that result in such mosquito sources can usually be foreseen and minimized by the application of good soil and water conservation practices as discussed in other chapters of this monograph. Provision of irrigation water without adequate drainage systems for the removal of excess water destroys the productivity of the land and endangers the welfare of its occupants (Russell et al., 1963). Rice (*Oryza sativa*) fields often produce large numbers of malaria mosquitoes (Freeborn, 1917; Barber et al., 1926, 1929; Russell et al., 1942a). It has been shown that many sources of malaria mosquitoes associated with irrigation can be prevented or eliminated by modified agricultural methods and improved irrigation and drainage practices (Knipe and Russell, 1942; Russell et al., 1942b; Ejercito, 1951; Henderson, 1955).

Fig. 57–2. Borrow pits and numerous other depressions are often flooded by seepage from unlined irrigation canals and laterals. These aquatic habitats are major sources of mosquitoes in many irrigated areas. (Photograph—Pathfinder irrigation canal, North Platte Valley, Nebraska).

Encephalitis is now the most important arthropod-borne disease in North America (Hess, 1956–1957) and related arbovirus diseases are widespread in other parts of the world. There are no known effective chemotherapeutic measures for preventing or treating human cases of these viral diseases; and some individuals, particularly children, suffer permanent mental disability after a severe attack of encephalitis. Production of mosquito vectors of encephalitis is frequently associated with irrigation, particularly in the USA. *Culex tarsalis*, the primary vector of western and St. Louis encephalitis in the Western USA (Hess and Holden, 1958), is often called an irrigation mosquito. It is produced in vegetated margins of impoundments, obstructed canals, seepage areas, and numerous collections of irrigation waste water. House mosquitoes (*C. pipiens* and *C. quinquefasciatus*) are primary vectors of St. Louis encephalitis in the Central and Eastern USA and transmit Bancroftian filariasis in other parts of the world (Foote and Cook, 1959). These mosquitoes often occur in artificial containers as well as ground pools and sometimes in habitats associated with irrigation, especially where sewage effluent is involved. The primary vector of Japanese B encephalitis in Japan and possibly other Asiatic countries, *C. tritaeniorhynchus*, is produced in a variety of habitats including artificial impoundments, rice fields, field drains, and irrigation ditches (Foote and Cook, 1959). In Australia several suspected mosquito vectors of Murray Valley encephalitis are produced in habitats associated with irrigation (Reeves et al., 1954). Production of encephalitis mosquitoes in habitats associated with irrigation may be minimized through proper reservoir preparation and management techniques and by the application of the same irrigation and drainage practices previously discussed for the prevention of malaria mosquito production (US Public Health Serv., 1951; Rainey, 1955; Hess, 1958; Rainey and Hess, 1957).

Several species of pest mosquitoes (*Aedes* and *Psorophora*) are produced in large numbers in both on-field and off-field habitats (Fig. 57–3 and 57–4) in irrigated areas of the USA (US Public Health Serv., 1951; Husbands, 1955; Rainey, 1955; Harmston et al., 1956; Mezger, 1963). These mosquitoes often create public health problems because of their vicious biting habits, and some of them may be involved in basic transmission cycles of encephalitis viruses. The types of problems involved are illustrated by a survey conducted in irrigated areas in northern Montana by personnel of the US Public Health Service (Hess and Quinby, 1956). In three-fourths of the families surveyed, mosquitoes caused severe annoyance to both adults and children and interfered with their normal outdoor activities during summer months. Mosquito bites caused some degree of injurious reaction in 8 out of 10 people interviewed; and in one section, 40% of the individuals examined by the physician showed evidence of secondary infection of mosquito bites. As mosquito-borne diseases such as malaria are brought under control in other countries, public health problems related to pest mosquitoes will undoubtedly be considered more important.

Production of *Aedes* or *Psorophora* mosquitoes is associated with natural or artificial fluctuations of water levels, whereas production of encephalitis mosquitoes (*Culex* sp.) and malaria mosquitoes (*Anopheles* sp.) is favored by permanent or semipermanent water levels. Rice fields constitute a special source for both types of mosquitoes. The intermittent flooding of rice fields is favorable for the production of *Aedes* and *Psorophora*, while constant flooding results in production of *Anopheles* and *Culex* mosquitoes.

Fig. 57–3. Prolific mosquito production occurs in ponded areas on irrigated fields used for pastures, hay meadows, and other close-growing crops. When irrigation water is ponded on such fields long enough to produce mosquitoes, the forage grasses and legumes are usually killed and replaced by wet land plants. (Photograph—Tomichi Valley, Gunnison, Colorado, USA).

Fig. 57–4. Improperly maintained drainage ditches with sluggish flows and ponding are favorable for mosquito production. (Photograph—Milk River Valley, Montana, USA).

Excessive production of malaria, encephalitis, and pest mosquitoes is often associated with poor irrigation, drainage, and management practices that also result in serious soil and water problems such as excessive water losses, waterlogging, salt and alkali accumulations, damage to soil structure, leaching of plant nutrients, and reduced crop yields. In the USA most of the mosquito control agencies in irrigated areas are now carrying out source reduction programs which are aimed at preventing, eliminating, or reducing man-made aquatic habitats through improved irrigation and drainage practices which are of mutual benefit to mosquito prevention and irrigation agriculture. The basic principles and practices involved in the prevention and elimination of mosquito sources associated with irrigation have been outlined by a special committee of the Soil and Water Division of the American Society of Agricultural Engineers (1964).

In addition to mosquito-borne diseases, bilharziasis, a parasitic disease, also known as schistosomiasis, is a major public health problem in many tropical and subtropical regions. The disease is widely distributed throughout the Eastern Mediterranean, Africa, China, Japan, the Philippines, and some countries in Latin America (World Health Organ., 1959b). Several species of aquatic and amphibious snails serve as intermediate hosts of the bilharzia parasites. These snails become infected when their aquatic habitats are contaminated with human or animal excreta containing schistosome eggs, and humans are infected when they come in contact with water containing cercariae which have emerged from infected snails (Mackie et al., 1958).

Bilharziasis is becoming increasingly important in irrigated areas of many countries (Lanoix, 1958). The snail hosts of this disease may be produced in storage reservoirs, rice fields, irrigation conveyance systems, drainage ditches, and seepage areas. The first report of the Joint Office Int. d'Hygiène Publique/World Health Organ. Study Group on Bilharziasis in Africa (World Health Organ., 1950) stated that: "The introduction or development of irrigation schemes, as well as the change from basin to perennial irrigation, has always resulted in a considerable increase in the incidence and intensity of bilharziasis wherever that infection existed or was introduced by outside laborers. The severity of the infection may be such as to cause the abandonment of an irrigation scheme created at considerable expense." In the Middle East and Africa the construction or extension of irrigation systems has resulted in the creation of habitats which, within a short period of time, produce enormous populations of the snail vectors of bilharziasis (Watson, 1958). The cultivation of rice in endemic areas also provides favorable habitats for these snails, and snail control is more difficult in paddies than in channels, drains, and culverts because of the greater extent of infested water which has to be treated with chemicals.

Extension of agriculture by means of irrigation leads to increased human pollution of water, increased production and infection of snail vectors, and more contact between man and the infective cercariae. The use of irrigation water for washing, bathing, or other household purposes is common where residents in the infected areas are not fortunate enough to have a piped supply of filtered chlorinated water. Even when this is the case, children are tempted to bathe in infested waters unless cercaria-free swimming pools are provided for them (Watson, 1958).

The various methods of bilharziasis control involve attacking the schistosome parasite at one or more points in its complex life cycle. Control methods related to the human host include environmental sanitation, treatment of infected persons with drugs, and preventing human contact with water infested with infected snails. The snail host may be attacked directly through the use of molluscicides or other means of destruction such as trapping, mechanical barriers, and the use of fish and other snail predators (McMullen et al., 1962). Indirect attack may be made upon the snail host by preventing, reducing, eliminating, or modifying its aquatic habitats (World Health Organ., 1957; Pesigan et al., 1958; McMullen et al., 1962). This includes such measures as removal of vegetation from reservoirs or other habitats, lining of irrigation canals and laterals, use of pipe in irrigation and drainage channels, and proper placement of culverts. Permanent control of schistosomiasis in many endemic areas would require major improvements in environmental sanitation and water supplies, but in most areas this goal is impossible to attain, at least within the next few decades (Olivier, 1955). However, the source reduction approach should be the ultimate goal, and in many situations it will be of mutual benefit to irrigation agriculture and bilharziasis control. For immediate results, the World Health Organization Expert Committee on Bilharziasis recommends control of the intermediate snail hosts (World Health Organ., 1953). For new irrigation developments in endemic areas, full consideration should be given to certain basic environmental sanitation measures such as location and layout of villages, safe water supplies, proper disposal of human and animal excreta, snail control, and health education to minimize the transmission of bilharziasis (McMullen et al., 1962). These basic environmental sanitation measures should be carefully planned and built into each new irrigation project.

III. WATER POLLUTION PROBLEMS

Water pollution is becoming increasingly important in many parts of the world (World Health Organ., 1963). In some areas limited water supplies must meet the needs of a growing population as well as the expansion in agriculture and industry, and at the same time, handle an ever increasing pollution load. Water pollution may be defined as adding to water any substance which changes the physical, chemical, or biological properties of the water thus making it unfit for use. In addition to domestic sewage and related organic wastes which deplete the dissolved oxygen in water as they decompose, there are many other polluting substances which may affect water use or be detrimental to public health. These include infectious agents, minerals that serve as nutrients for aquatic plants, organic chemicals such as insecticides and detergents, numerous other mineral and chemical substances, sediments from land erosion, radioactive substances, and heat (US Senate, 1960).

In arid and semiarid regions throughout the world, the same water supplies that are used for irrigation of crops must also be used for domestic, industrial, recreational, and various other purposes. The prevention and control of water pollution is of vital importance to social and economic progress in areas with a limited water supply. It is, therefore, a problem of mutual concern to irrigation, agricultural, conservation, and public health agencies.

The streams and canals which convey irrigation water are sometimes polluted by municipal and industrial wastes that are discharged into them. These effluents may include raw or partially treated sewage as well as wastes from slaughtering houses and food processing plants. In addition to decreasing the dissolved oxygen in the water, these organic wastes often contain disease organisms. On the other hand, in a number of countries, a large proportion of the people in irrigated areas live in compact, rural villages. Irrigation canals are frequently the only source of water for these villages, and the canal water is used for drinking, religious ablutions, bathing, swimming, washing of clothes and dishes, and all other domestic purposes. In many situations, facilities for the disposal of human and other wastes are not provided or are inadequate; the canal banks and adjacent areas are often used for these purposes. Such environmental sanitation practices inevitably result in gross pollution of the canal water and serious health hazards. The problem of bilharziasis and its relation to polluted water has already been discussed. Outbreaks of typhoid fever, cholera, and other gastrointestinal diseases are also caused by polluted water and each year these water-borne diseases cause a tremendous amount of illness and death in underdeveloped countries. The importance of safe water supplies as well as proper disposal and treatment of human wastes in preventing and controlling outbreaks of water-borne diseases has been well documented.

Municipal sewage often serves as a valuable supplemental source of irrigation water, particularly in arid regions. Sewage farming is practiced in some areas to keep sewage effluents out of surface waters, thereby avoiding pollution. Outbreaks of typhoid and several other diseases have been caused by the consumption of raw vegetables that were irrigated with sewage-polluted water (Sepp, 1963). In order to avoid this disease hazard, the use of raw, settled, or untreated sewage on vegetable crops for human consumption is now prohibited in many areas of the USA. Similar rules and regulations have also been adopted by the governments of several other countries.

Certain water pollution problems are directly related to agricultural activities that are carried out in irrigated areas. Soil erosion has long been recognized as a serious land problem associated with irrigation (McCulloch and Criddle, 1950). This problem is often intensified by the use of improper irrigation methods and practices as well as poor soil management and cultural practices. Erosion of irrigated land not only causes economic loss to farmers, but it is a major source of sediment pollution in water courses and reservoirs. Suspended sediment degrades the quality of domestic and industrial water supplies, increases the cost of treatment, and damages equipment.

Irrigation return flows also contain various dissolved minerals and chemical substances that degrade water quality (Sylvester and Seabloom, 1962; Wilcox, 1962). The concentration of dissolved constituents in return flows may be considerably greater than that of the applied irrigation water due to the leaching of salts from the soil and the loss of water through evapotranspiration (see chapters 9 and 51). These salts reduce water potability, damage equipment, and increase irrigation requirements. Irrigation return flows also increase the hardness of the receiving waters (Eldridge, 1960). When this occurs, complex and costly treatment processes may be required to remove compounds from the water to make it suitable for industrial and domestic uses. When the same water is used over and over for irrigation and numerous return flows are discharged into a given stream, the water quality may be degraded to the point where it is unsuitable for irrigation and other uses unless adequate flows of satisfactory quality are maintained for dilution of the dissolved minerals.

In recent years there has been a tremendous increase in the use of commercial fertilizers, insecticides, herbicides, and other toxic substances in irrigated areas (Webb, 1962). Some of these chemicals are often discharged into surface and groundwater supplies by means of return flows. Present methods of water treatment are not effective in removing these constituents from domestic supplies. Health officials are concerned about the human hazard that may result from long-term ingestion of these chemicals that may be present in drinking water. These chemicals are highly important in the control of disease vectors, the production of food and fiber, as well as the control of nuisance insects and plants; however, they must be used wisely to avoid water pollution and other problems that may seriously affect human health and welfare.

Excessive applications of fertilizer and excessive irrigation may result in high concentrations of certain nutrients in drainage water. These nutrients, especially nitrogen and phosphorus, encourage aquatic growths of bacteria, algae, and vegetation in reservoirs. Some of these aquatic growths cause undesirable tastes and odors in domestic water supplies that are costly and difficult to remove. It has been indicated that nitrates in water supplies may cause methemoglobinemia in infants, if the concentration exceeds about 10 ppm as nitrogen (Walton, 1951). Wastes from the processing of agricultural products and animal feeding lots also create serious water pollution problems if they are not handled and treated in a satisfactory manner (Webb, 1962; Schleusener, 1964). The water pollution problem is being intensified in some parts of the USA by the practice of raising and feeding cattle, hogs, and chickens in small areas.

IV. SUMMARY

Important public health problems associated with irrigation in various parts of the world include vector-borne diseases such as malaria, encephalitis, and bilharziasis; severe annoyance by blood-sucking insects; as well as typhoid fever, cholera, and other gastrointestinal diseases resulting from polluted water. Adequate provision must be made for the prevention and control of these problems in irrigated areas in order to realize maximum benefits from water, land, and human resources.

The same faulty irrigation, drainage, and water management practices that contribute to serious soil and water problems often create or intensify conditions that result in vector-borne diseases and water pollution problems. Proper irrigation and drainage practices that insure high crop yields without excessive water losses or damaged soils are usually beneficial to both agriculture and public health.

The solution to many public health problems associated with irrigation will require a joint attack by those who are concerned with public health, irrigation, and agriculture. These groups should have considerable knowledge and understanding of each other's problems and be willing to cooperate in working toward the common goal of developing and utilizing water and land resources in a manner that minimizes public health problems.

LITERATURE CITED

American Society of Agricultural Engineers. 1964. Principles and practices for prevention and elimination of mosquito sources associated with irrigation. Agr. Eng. Yearbook. 325 p.

Barber, M. A., M. H. W. Komp, and T. B. Hayne. 1926. Malaria in the prairie rice regions of Louisiana and Arkansas. Public Health Rep., Washington, D.C. 41:2527–2549.

Barber, M. A., M. H. W. Komp, and C. H. King. 1929. Malaria and the malaria danger in certain irrigated regions of southwestern United States. Public Health Rep., Washington, D.C. 44:1300–1315.

Daggy, R. H. 1959. Malaria in oasis of Eastern Saudi Arabia. Amer. J. Trop. Med. Hyg. 8:223–291.

Ejercito, A. 1951. Agricultural control of malaria. J. Philippine Med. Ass. 27:591–607.

Eldridge, E. F. 1960. Return irrigation water—characteristics and effects. US Dep. Health, Educ., Welfare-Public Health Serv., Region IX, Portland, Oregon. p. 28–36.

Foote, R. H., and D. R. Cook. 1959. Mosquitoes of medical importance. US Dep. Agr., Agr. Handbook 152. 158 p.

Freeborn, S. B. 1917. The malaria problem in the rice fields. California State J. Med. 15:412–414.

Harmston, F. C., G. R. Schultz, R. B. Eads, and G. C. Menzies. 1956. Mosquitoes and encephalitis in the irrigated high plains of Texas. Public Health Rep. 71:759–766.

Henderson, J. M. 1952. Irrigation and mosquitoes in the United States of America. Indian J. Mal. 6:72–116.

Henderson, J. M. 1955. Water management planning for malaria prevention in the Damodar Valley, India. Amer. J. Trop. Med. Hyg. 4:1091–1102.

Hess, A. D. Dec. 1956-Jan. 1957. Public health importance of insect problems. Soap Chem. Spec. p. 191, 193, 195, 197, 219–220 (Dec. 1956) and p. 82–83 (Jan. 1957).

Hess, A. D. 1958. Vector problems associated with the development and utilization of water resources in the United States. Proc. Int. Congr. Entomol., 10th (1956) 3:595–601.

Hess, A. D., and P. Holden. 1958. The natural history of the arthropod-borne encephalitides in the United States. Ann. New York Acad. Sci. 70:294–311.

Hess, A. D., and M. D. Quinby. 1956. A survey of the public health importance of pest mosquitoes in the Milk River Valley, Montana. Mosquito News. 16:266–268.

Husbands, R. C. 1955. Irrigated pasture study—a review of factors influencing mosquito production—weather, ponding and irrigation, and soil moisture. Conf. California Mosquito Contr. Ass., Proc. 23rd, and Ann. Meeting Amer. Mosquito Contr. Ass., 11th. p. 112–117.

Knipe, F. W., and P. F. Russell. 1942. A demonstration project in the control of rural irrigation malaria by antilarval measures. J. Mal. Inst. India 4:615–631.

Lanoix, Joseph N. 1958. Relation between irrigation engineering and bilharziasis. World Health Organ. Bull. 18:1011–1035.

McCulloch, A. W., and W. D. Criddle. 1950. Conservation irrigation. US Dep. Agr. Soil Conserv. Serv., Agr. Inform. Bull. 8. 15 p.

McMullen, B. D., Z. J. Buzo, M. B. Rainey, and J. Francotte. 1962. Bilharziasis control in relation to water resources development in Africa and the Middle East. World Health Organ. Bull. 27:25–40.

Mackie, T. T., G. W. Hunter, and C. B. Worth. 1958. A manual of tropical medicine. W. B. Saunders Co., Philadelphia. p. 461–485.

May, S. 1954. Economic interest in tropical medicine. Amer. J. Trop. Med. Hyg. 3:412–421.

Mezger, E. G. 1963. Entomological evaluation of a proposed mosquito source reduction pilot operation. Mosquito News 23:85–88.

Olivier, L. 1955. The natural history and control of the snails that transmit the schistosomes of man. Amer. J. Trop. Med. Hyg. 4:415–423.

Pesigan, T. P., M. Farooq, N. G. Hairston, J. J. Jauregui, E. G. Garcia, A. T. Santos, B. C. Santos, and A. A. Besa. 1958. Studies on *schistosoma japonicum* infection in the Philippines. World Health Organ. Bull. 19:223–261.

Rainey, M. B. 1955. Good irrigation vital in mosquito control. Agr. Eng. 35:185–187, 191.

Rainey, M. B., and A. D. Hess. 1957. Integration of vector control into federal water resource developments. California Vector Views. 4:55–59.

Reeves, W. C., E. L. French, E. N. Marks, and N. E. Kent. 1954. Murray Valley encephalitis: A survey of suspected mosquito vectors. Amer. J. Trop. Med. Hyg. 3:147–159.

Russell, P. F. 1938. Malaria due to defective and untidy irrigation. J. Mal. Inst. India. 1:339–349.

Russell, P. F. 1951. Malaria and society. J. Nat. Mal. Soc. 10:1–7.

Russell, P. F. 1952. The present status of malaria in the world. Amer. J. Trop. Med. Hyg. 1:111–123.

Russell, P. F. 1956. World-wide malaria distribution, prevalence, and control. Amer. J. Trop. Med. Hyg. 5:937–965.

Russell, P. F., F. W. Knipe, and H. R. Rao. 1942a. On the intermittent irrigation of rice fields to control malaria in South India. J. Mal. Inst. India. 4:321–340.

Russell, P. F., R. W. Knipe, and H. R. Rao. 1942b. On agricultural malaria and its control with special reference to India. Indian Med. Gaz. 77:744–756.

Russell, P F., L. S. West, R. D. Manwell, and G. MacDonald. 1963. Practical Malariology. 2nd ed. Oxford Univ. Press, Oxford, England. 750 p.

Schleusener, P. E. 1964. Research needs in rural waste utilization. Agr. Eng. 45:490–495.

Sepp, E. 1963. The use of sewage for irrigation. Bur. Sanit. Eng., California State Dep. Public Health. 33 p.

Stead, F. M. 1957. Health problems as affected by irrigation agriculture. California's Health. 14:217–224.

Sylvester, R. O., and R. W. Seabloom. 1962. A study on the character and significance of irrigation return flows in the Yakima River Basin. Univ. Washington, Dep. Civ. Eng., rev. 104 p.

US Public Health Service and Tennessee Valley Authority. 1947. Malaria control on impounded water. US Government Printing Office, Washington, D.C. 422 p.

US Public Health Service, Communicable Disease Center. 1951. Mosquito problems in irrigated areas and their prevention. Communicable Disease Center. Atlanta, Georgia. 18 p.

US Senate, Select Committee on National Water Resources. 1960. Water resources activities in the United States—Pollution Abatement. US Government Printing Office, Washington, D.C. 38 p.

Walton, Graham. 1951. Survey of literature relating to infant methemoglobinemia due to nitrate-contaminated water. J. Amer. Public Health Ass. 41:986.

Watson, J. M. 1958. Ecology of *Bulinus truncatus* in the Middle East. World Health Organ. Bull. 18:888–889.

Webb, H. J. 1962. Water pollution resulting from agricultural activities. J. Amer. Water Works Ass. 54:83–87.

Wilcox, L. V. 1962. Salinity caused by irrigation. J. Amer. Water Works Ass. 54:217–222.

World Health Organization. 1950. Joint Study-Group on Bilharziasis in Africa. Rep. 1st Sess. Tech. Rep. 17:16.

World Health Organization. 1953. Expert committee on bilharziasis. 1st. Rep. Tech. Rep. Ser. 65. 45 p.

World Health Organization. 1955. Malaria: A world problem. World Health Organ. Chron. 9:31–100.

World Health Organization. 1957. Study group on the ecology of intermediate snail hosts of bilharziasis. Tech. Rep. Ser. 120. 38 p.

World Health Organization. 1959a. Insecticide resistance in *Anophelines*. World Health Organ. Chron. 13:357–358.

World Health Organization. 1959b. Nature and extent of the problem of bilharziasis. World Health Organ. Chron. 13:3–19.

World Health Organization. 1962a. Occupational health problems in agriculture. Tech. Rep. Ser. 246. 61 p.

World Health Organization. 1962b. Expert Committee on Malaria. The operational implications of the development of insecticide resistance and irritability. 9th Rep. World Health Organ. Tech. Rep. 243:30–32.

World Health Organization. 1963. Water pollution. World Health Organ. Chron. 17:62.

58 Irrigation Problems of Humid-Temperate and Tropical Regions

W. A. RANEY
Agricultural Research Service, USDA
Beltsville, Maryland

I. INTRODUCTION

Humid region agriculture embraces farming in areas where rainfall is generally sufficient to permit sustained production of crops and pasture. The level of production may be quite low in years of deficient rainfall but irrigation or arid region practices, such as alternate years of crops and fallow, are not essential. The rainfall seldom coincides completely with crop needs, however, so water excesses and drouths are common occurrences.

Most of the incoming solar energy that is received in the humid region of the Northern Hemisphere comes between March 21 and September 21. During the first 3 months of that period, nearly all of the energy is used to evaporate the excess water that has accumulated during the winter months. This usually amounts to 5 to 6 mm/day. Unless rains replenish the diminished soil water, late summer energy cannot be expended in evaporating water so most of it is used to heat the soil and the air. Late summer is always hot for this reason and not because solar radiation has increased. We may never be able to markedly reduce the amount of water required to produce an acre of crop but we most certainly can increase the yield and quality of the crop. This will require, in addition to good germ plasm and protection from insects and disease, a major increase in efficiency of management of soil and water resources (Quackenbush et al., 1957; Quackenbush and Thorne, 1957).

II. DROUTH CRITERIA

Procedures have been developed for evaluation of recurrence probabilities of drouths and water excesses (Van Bavel, 1953). The amount of water that the soil can store in a form available to plants is treated as a bank account, rainfall as input, and evapotranspiration as withdrawal. When more rain falls than the soil can store, it is treated as excess, and when evapotranspiration exceeds rainfall for sufficient time to exhaust soil water, it is treated as a deficit or drouth. Such studies in Eastern USA have shown that water excesses occur with sufficient frequency to make drainage absolutely essential and that drouths of sufficient duration to reduce crop yields are likely to occur every year (Van Bavel, 1959; Blake et al., 1960; Palmer, 1958). Crop yields are reduced by these drouths; yet, a prominent role for irrigation is not assured. In some cases, limitations on use of irrigation result from inadequate water supplies; in other cases, costs of irrigation exceed economic benefits.

III. DROUTH EVASION

Although we cannot yet control the weather, the problems of erratic rainfall may be tackled in at least two ways. First, water received during periods of surplus can be stored in the soil profile for use during deficient periods. Even then, the efficiency of stored soil water must be extended, for profile storage alone is inadequate to meet demands in extremely dry years. In some cases, water deficiencies may be evaded by using crop varieties that mature before the time of year when drouths are most likely to occur, or by using varieties that are tolerant to dry conditions. A higher water intake can be maintained with mulches which increase effectiveness of rainfall for profile storage of water. Such mulches have very little effect on evaporation.

The amount of water that the profile can supply may be increased by providing conditions conducive to increasing root penetration and proliferation. Mechanical measures to increase rooting by improving drainage and aeration and by reducing soil impedance have been used for a long time. More recently, attention has been given to nutrient deficiencies and to situations where toxic levels of elements, like aluminum or manganese, restrict rooting (Foy and Brown, 1963, 1964).

The water storage per unit volume of soil may also be increased by stratifying the profile with layers that impede drainage (Miller, 1964). Placing a coarse-textured layer under a finer textured layer increases storage in the finer layer because of the lower water flux rate through the coarse-textured layer when the soil is not saturated with water.

Chemicals which control transpiration from plants or evaporation from soil, among other things, may be promising in the future (*see* chapters 27, 28, and 62).

In most situations, however, the only sure approach to rectifying the problem is irrigation. Irrigation, far from a general panacea, is an expensive production practice. It usually should not be considered until every effort has been made to conserve and efficiently use precipitation.

IV. IRRIGATION WATER SOURCE

In the humid part of the USA, irrigation waters are drawn from streams, lakes, ponds, or groundwater, but groundwater is the major source of irrigation water (US Dep. Com., 1959). Groundwater rights are not clearly defined but the rule of reasonable use applies in many states. That means that it may be used for irrigation if other users are not adversely affected. There seems little likelihood, however, that one would ever be allowed to develop more resource than that which falls on his own land. An aquifer would thus be treated as a reservoir to be replenished during periods of water excesses. There are fluctuations in groundwater supplies associated with both weather and pumping, and in coastal areas, water tables must not be pumped down below sea level or salt intrusion into the aquifer may cause permanent damage.

In humid regions, more water may be needed than that which falls on the land if both crop production and crop processing are involved. The use of brackish waters for irrigation has resulted in saving crops that otherwise would have been severely damaged by drouth. Winter rains usually flush out the excess salts that are added in this manner (Lunin et al., 1964). The possibilities for using waste waters from crop processing plants for irrigation have not yet been adequately explored.

V. WATER MANAGEMENT REQUIREMENTS

Water management programs in humid areas should not be limited merely to rectification of water deficiencies for favorable crop yield and quality. Consideration must also be given to water control in relation to machinery requirements.

In arid lands, water management may be orderly programmed, but in humid regions randomly occurring rains make regular irrigation schedules impractical. Examination of soil and plants has not been entirely satisfactory for controlling irrigation timing. The use of a weather station for precipitation information that is correct for a particular field is unreliable because of point variability, but such measurements can be made by the irrigator himself. Weather stations can provide solar energy data for estimates of evapotranspiration. For most efficient irrigation, soil and plant observations should be supplemented by meteorological observations (see also chapter 30).

Since water surpluses occur every winter in the humid temperate regions, soils are invariably wet just prior to planting. Where drainage is inadequate, machinery inventory has to be expanded to get in the number of machine hours necessary for seedbed preparation within a reasonable time. When soils do dry out at this time of year, there is also the hazard of losing excessive soil water during seedbed preparation. This delays or prevents establishment of a crop stand. Furthermore, if drainage is not adequate, the subsequent use of irrigation will aggravate a bad situation and keep a severe drainage problem in the field all year. Machinery requirements dictate that drainage must have precedence over irrigation.

Machinery requirements also place limitations on field size and topography. Even though adequately drained, small uneven fields must be consolidated for efficient use of machinery. This is not always feasible, not only for economic reasons, but because fields must contain soils that are fairly similar in water retention and drainage characteristics or timeliness of operations will be seriously restricted. Dissimilar soils may also differ in management requirements other than water.

VI. ECONOMICS

The cost of installation of an irrigation system is about the same per unit of land area in both humid and arid regions. There is no difference in systems required for humid regions than those described for arid regions in chapters 43, 44, and 45. In humid regions, however, good crop yields can be obtained in many years without irrigation so the margin of profit from the practice is lower. One could then rarely economically use irrigation except to assure high quality as well as quantity yields of high cash value crops.

But in addition to considerations of value of crop product, there are also increased erosion hazards and drainage problems in humid areas to be handled. In several studies in the humid area, greater runoff has been measured with irrigation than without.

VII. PRESENT USE OF IRRIGATION IN HUMID AREAS

Less than 1% of the cropland in the humid part of the USA is now irrigated, except in the Lower Mississippi River Valley where planters irrigate about 10% of the cropland acreage (US Dep. Com., 1959). The valley contains large areas of very productive soils that are consolidated into operational units that can be adequately drained and efficiently mechanized. Even more important, there are subsurface aquifers from which irrigation waters may be obtained.

Since water management is essential for maintaining high yields and for control of quality, why has irrigation not been generally adopted as a practice? First of all, soil water cannot be controlled by just adding water—excesses must also be removed and drainage must be effected before irrigation can be considered. Timing of managerial operations is very critical for many practices, such as control of insects and weeds, and irrigation has generally accentuated these problems. This suggests that management skills required are much greater when irrigation is used.

Secondly, irrigation generally extends the production season into the fall. This is all right when there is a dry harvest season. Even though hurricanes do not always hit each state, hurricane season in the fall sometimes brings extended periods of wet weather. If the production season is extended into this period by irrigation, lowered quality and yield often offsets any advantage from irrigation.

VIII. THE FUTURE OF IRRIGATION IN THE HUMID REGION

Even though current acceptance of irrigation is modest in humid areas, the demand for controlled crop quality will make irrigation in future years a necessity. This will tend to shift production to locations where water resources are available, where good cropland is available, and where the management system is extremely efficient. Irrigation will never be a useful practice until these conditions are met.

Increased research emphasis is now being placed on practices to assure efficient utilization of water resources in production of high quality crops and on measures that will offer protection to the crop until it is harvested. Research has thus been extended from just the chemical and physical characteristics of the environment to include the meteorological conditions in and above the crop canopy. There are presently no methods available for crop canopy control of humidity during extended periods of wet weather, but certainly crop canopy geometry can be altered to reduce humidity as quickly as possible after rain ceases. Furthermore, much more attention to drainage design offers possibilities for more quickly removing excess soil water.

There is every reason to believe that irrigation will be an essential practice for the required system of production in future years.

LITERATURE CITED

Blake, G. R., E. R. Allred, C. H. M. van Bavel, and F. D. Whisler. 1960. Agricultural drought and moisture excesses in Minnesota. Minnesota Agr. Exp. Sta. Tech. Bull. 235. 36 p.

Foy, C. D., and J. C. Brown. 1963. Toxic factors in acid soils: I. Characterization of aluminum toxicity in cotton. Soil Sci. Soc. Amer. Proc. 27:403–407.

Foy, C. D., and J. C. Brown. 1964. Toxic factors in acid soils: II. Differential aluminum tolerance of plant species. Soil Sci. Soc. Amer. Proc. 28:27–32.

Lunin, J., M. H. Gallatin, and A. R. Batchelder. 1964. Effects of supplemental irrigation with saline water on soil composition and on yield and cation content of forage crops. Soil Sci. Soc. Amer. Proc. 28:551–554.

Miller, D. E. 1964. Estimating moisture retained by layered soils. J. Soil Water Conserv. 19:235–237.

Palmer, R. S. 1958. Agricultural drought in New England. New Hampshire Agr. Exp. Sta. Tech. Bull. 97. 51 p.

Quackenbush, T. H., G. M. Renfro, K. H. Beauchamp, L. F. Lawhon, and G. W. Eley. 1957. Conservation irrigation in humid areas. US Dep. Agr. Handbook 107. 52 p.

Quackenbush, T. H., and M. D. Thorne. 1957. Irrigation in the East. In: Soil. Yearbook of Agriculture. US Government Printing Office, Washington, D.C. p. 368–378.

US Dep. of Commerce. 1959. Irrigation in humid areas. Census of Agriculture. US Government Printing Office, Washington, D.C. p. 6–59.

Van Bavel, C. H. M. 1953. A drought criterion and its application in evaluating drought incidence and hazard. Agron. J. 45:167–172.

Van Bavel, C. H. M. 1959. Drought and water surplus in agricultural soils of the Lower Mississippi Valley area. US Dep. Agr. Tech. Bull. 1209. 93 p.

section XIII

Water Conservation
Related to Irrigation

XIII

59 | Watershed Management[1]

R. H. BURGY

*University of California
Davis, California*

J. E. FLETCHER

*Utah State University
Logan, Utah*

A. L. SHARP

*Agricultural Research Service, USDA (retired)
West Linn, Oregon*

I. INTRODUCTION

An understanding of the factors that make water valuable for irrigation is paramount to any discussion of watershed management in relation to irrigation water supply. Briefly, these factors are: quantity of water available, time of availability, quality of water, and reliability and location of water supply. Any management system is aimed at improving one or more of these factors.

It is also important to understand the factors involved in the precipitation-water-yield-time relationship (PY_wT). Estimates indicate that some 30 parameters are involved. The most important, according to Chow (1962), are: precipitation form, intensity, duration, time, and areal distribution, and frequency of occurrence; evapotranspiration; size, slope, shape, and drainage density of the catchment; soil type, cover, and condition; geology; topography; vegetation, channel size, shape, slope, and roughness; and storage capacity of the watershed. The parameters of major importance are discussed; those of lesser importance are mentioned; none will be fully treated.

II. PRECIPITATION

Some form of precipitation is the ultimate source of all soil water, groundwater, or streamflow, whether from direct surface runoff, immediate storm runoff, snowmelt, return flow, or groundwater storage. In general, groundwater recharge and streamflow are residuals after soil water deficits are replenished. In addition to the kinds, amounts, and characteristics of precipitation, this residual is intimately related to many factors such as soil, geology, topography, climate, and vegetation.

In areas of humid to perhumid climate, precipitation is sufficient and well enough distributed in time to satisfy soil water deficits. Water yields are residual after evapotranspiration. In the absence of soil water stress, evapotranspiration is near its potential but may be limited by the available solar energy. Under such climatic conditions, land treatment will have little long-term effect on the PY_wT relationship.

Table 59–1. Major climatic factors affecting runoff

Precipitation		Relative disposition of runoff		
		Surface	Subsurface	Groundwater
Form:				
Rain		High	Moderate	Low
Snow		Low	High	High
Sleet		Moderate	Moderate	Moderate
Rime frost		Low	Low	High
Hail		Moderate	Moderate	Low
Intensity:				
> 3 inches/hr	High	High	High	Low
0.5 - 3	Medium	Medium	Medium	Medium
< 0.5	Low	Low	Low	High
Amount:				
> 2 inches	Large	High	High	High
1 - 2	Medium	Medium	Medium	Medium
< 1	Small	Low	Low	Low
Duration:				
< 1 hr	Short	High	High	Low
1 - 12	Medium	Medium	Medium	Medium
> 12	Long	Low	Low	High
Time distribution:				
Close together		High	High	High
Hours, medium		Medium	Medium	Medium
Days, long separation		Low	Low	High
Time of year:				
Winter		Low	Moderate	High
Summer		High	Low	Low
Areal distribution:				
Uniform		Low	Low	Low
Cellular		High	High	High

Arid or semiarid climates produce chronic soil water deficits, and when precipitation occurs on soils which have water storage capacity available, the characteristics, amounts, and kinds of precipitation are more closely related to water yields than under humid or perhumid climates. Table 59–1 summarizes the effects of different properties, kinds, and times of precipitation on the type of runoff occurring. Precipitation as rain causes more surface runoff than precipitation as snow, sleet, rime frost, hail, or any other solid form. Snowfall goes into storage for later melt. The same is true of rime frost. Sleet or hail, which form at warmer temperatures, may produce immediate surface runoff. If other surrounding conditions are suitable, however, sleet and hail also may go into storage but are subject to rapid melt.

A. Intensity

Precipitation from high intensity storms in arid areas usually falls at rates greatly exceeding the infiltration capacity. Thus, its chief distribution is to surface runoff. Low intensity rains in equal volume may all move into the soil profile and return again as groundwater flow.

B. Amount

High volume storms produce more water than medium or low volume storms in all categories of runoff—surface, subsurface, or groundwater.

C. Duration

Frequently, short duration storms are merely storms of low volume. If the volumes are not small, the duration is an inverse measurement of intensity. Under this condition, short durations produce high surface runoff, medium storms produce medium runoff, and long duration storms produce low surface runoff and high groundwater return flow.

D. Time Distribution

When storms are spaced very close together, the succeeding storms produce much higher runoff. Time distribution can also make storms appear to "move" down the watershed, the resulting runoff yielding higher peaks than if the time lapses cause the storms to appear to be moving up the watershed. The interval between storms, in relation to the rate of travel of the runoff wave down the watershed produces the apparent movement.

E. Time of Year

Storms which occur in cool weather yield greater quantities of water than those in hot weather, particularly when cold enough to reduce precipitation to a solid form since evaporation losses are reduced.

F. Areal Distribution

Precipitation spread uniformly over a watershed is much more subject to losses than when concentrated in comparatively small cells. Runoff is inversely proportional to the areal distribution of the precipitation. Losses from extremely deep snowdrifts are much lower, and water yields much higher, than when uniformly distributed over the surface.

In arid areas the bulk of groundwater recharge takes place from channels. The volume of recharge is roughly proportional to the length of time that water is in contact with the channel. The authors have observed that 5% of storms produce events which yield over one-half of the streamflow, and approximately 15% of the flood events produce nearly 90% of the water yield. Thus small rains, and even large rains with very low intensities, may yield no water.

To summarize, the total amount of precipitation in humid areas bears a close relationship to water yield; whereas in arid areas, the character of the precipitation becomes the dominant factor in water yield.

G. Snow and Water Yield

Precipitation as snow may accumulate to great depths and appear as stream-flow at some later date. Precipitation in this form becomes the source of water for a large portion of the irrigated land in the USA.

The precipitation-streamflow-time relation of snow may be separated into two divisions, namely, accumulation and melt. The factors influencing snow accumulation are shown in Table 59–2. Other factors being equal, the greater the amount of snow accumulation, the greater the water yield, and the more uneven the distribution of the snow, the later and greater is the water yield. Uneven distribution caused by wind may affect water yield by increasing sublimation losses from the snowpack and by increasing the runoff from drifted snow. Again, wind is the causal factor of snow depth in undershrubs as well as the improvement in snowpack because of topography. Major vegetation intercepts snow similarly to rain except in larger quantities. Connaughton (1935) and Haupt (1951) have estimated interception losses between 4% and 36% of the snowfall. On the other hand, snow that filters down through major vegetation is protected from melt and evaporation, giving a greater water yield per unit of snow accumulation than snow in the open. Gartska et al. (1958) have shown that snow persists in these glades for many weeks later than it does in the open places at equal elevation. They also showed that evapotranspiration losses from a snow area can be computed by Light's (1941) equation. Rosa (1956) and Gartska et al. (1958) studied the hydrology of snowmelt and noted that essentially all snowmelt comes from direct solar radiation. Light's equation is a good index of snowmelt. Among the data readily available, however, the best index of radiation is the degree days of heating above 0C (32F). They point out that while maximum water yields are obtained with the most rapid snowmelt, flood and erosion damage is best controlled by retardation of snowmelt. Time of availability of runoff for optimum beneficial use is also better served by retardation of the snowmelt.

H. Physiography

The physical features of the area on which precipitation falls also influence the PY_wT relationship. Chow (1962) has tabulated these factors as shown in Table 59–3. The effects of the factors whose relationships are more or less obvious are shown in the table; however, those which are more complex are discussed in the text.

I. Vegetation Cover

Vegetative cover exerts its influence on runoff and water yields through its infiltration capacity, water storage and detention, and evapotranspiration and interception. Some of the precautions to be observed in dealing with vegetation in various parts of the USA, and some of the effects of vegetative materials on the PY_wT relationship were outlined by Fletcher (1960), Dortignac (1960), Free et al. (1940), and Mannering and Meyer (1963). They point out that soils, geology, topography, and climate largely determine the vegetation encountered on

Table 59–2. Factors influencing snow accumulation

Factor	How related
1) Amount falling	Direct
2) Wind	Increases uneven distribution
3) Under shrubs	Depth proportional to height of vegetation
4) Topography	Snow accumulates in low places
5) Aspect	Snow accumulation greater on north, east, west, and south in order
6) Forests	Snow accumulates proportional to number of openings 2-10 times height of trees
	Interception directly proportional to density may be 4-36% of snowfall; melt delayed proportional to density; maxiumum efficiency as water producers when openings are 2-5 times as wide as tree heights; late season yields are improved if openings are 1-1.5 times tree heights.
7) Elevation (temperature)	Precipitation equal, more snow accumulates at low temperatures or high elevations
8) Potential evapotranspiration	Snow accumulation inversely proportional

Table 59–3. Physiographic factors affecting runoff

Factor	Kind of relationship
1) Size of catchment	Direct
2) Shape	Affects lag time, time to peak, etc. --length is proportional
3) Slope	Direct proportion to peak yield
4) Drainage density	Inverse relation
5) Vegetation or use	See text
6) Infiltration	Direct proportion to abstraction
7) Soils	See text
8) Geology	See text
9) Topography	See text
10) Storage capacity	Duration proportional to storage
11) Channel factors	See text

a watershed and the effects to be expected from it on the PY_wT relationship. It is generally assumed that infiltration is roughly proportional to the amount of vegetation; exceptions to this general rule are shown. In addition to its effect on the soil surface, dense vegetation extracts water from the lower layers at a much higher rate than sparse vegetation. Thus, a much greater water deficit in the subsoil allows a greater infiltration capacity.

Vegetation also influences the quality of the water through its effect on erosion and sedimentation. It is well known that denuded areas produce greater quantities of sediment than well vegetated ones. Rich (1960) states that sediment yields were decreased by revegetation. He attributed a good portion of this decrease in sediment yields to the increase in the infiltration capacity. Vegetative material on the surface of the soil in the form of mulch also affects the PY_wT relationship. While mulches greatly increase the infiltration capacity, they also increase the surface storage component. Thus, they decrease runoff and increase the duration. Mulches also decrease the amount of sediment in the water.

J. Soils

Soils play a significant role in the PY_wT relationship. Vegetative cover, evapo-transpiration, soil water storage capacity, infiltration rate, and other factors may be intimately related to soil properties. Sandy or gravelly soils may have enormous infiltration capacities but low storage capacities, and finer textured soils may have low infiltration capacities with greater storage for comparable soil depths. Coarse-textured soils may have low direct runoff and high groundwater recharge rates. Such soils may have high return flow rates, if they lie over relatively impermeable strata. Fine-textured soils tend to yield high direct runoff and low return flow or groundwater recharge under similar conditions. Total water storage capacity available for evapotranspiration is high in deep, fine-textured soils, contrasted with coarse-textured soils in which water storage capacity is smaller. Free et al. (1940) indicate that chemical composition of soils may also affect infiltration rates. The depth, texture, aggregation and chemical composition of the soil are therefore important in the PY_wT relationships of watersheds.

K. Geology

The geology of a watershed affects soils, topography, land aspect and slopes, vegetation, permeability, and the presence or absence of groundwater aquifers. These factors in turn affect water yields and the relation of precipitation to them.

L. Topography

The topography of watersheds affects the efficiency of watershed channel systems as conveyers of water from the watershed. Topography also affects overland flow velocity and stage of flow in channels, and the amount of temporary and depression storage on the watershed affect the PY_wT relationship.

M. Channel Factors

Channel size, shape, roughness, slope, and densities affect the PY_wT relationship. Any factor in the channel which tends to slow the movement of water acts as a storage component and increases the duration of runoff. Any factor in the channel which accelerates the velocity of flow in turn decreases the duration of the runoff period. Thus, channels as conveyance systems greatly influence the runoff and water yield from watersheds. For example, Keppel (1960) indicates that channels in Arizona, USA, which are choked with fine gravels and sands, may store as much as 5 acre-ft/mile of channel, and flows that traverse these channels soon become depleted and disappear.

N. Other Factors

Many other factors affect the PY_wT relationship: wind, humidity, fog, altitude, and latitude affect evaporation and transpiration; fires, both natural and man caused, affect vegetation, soil, and litter; erosion and deposition affect runoff and streamflow regimes; loessial deposits affect soils; and volcanic action and earthquakes affect soils, geology, and topography. These and many other factors affect PY_wT relationships in various ways and magnitudes.

III. EFFECTS OF WATERSHED TREATMENT ON WATER YIELD

It seems desirable to explore the potentials of selected land and watershed treatment practices that may affect water yields.

A. Vegetation Changes

Love (1960) discussed the effects of vegetation management in the Rocky Mountains, USA. The cutting of pine-fir to commercially "clear cut" gave responses close to 50%, and strip cutting increased water yield about 30%. He points out that, "The rate of melt in alpine snowfields can be regulated to a limited degree at least." Treatments tried, which looked promising, were dark materials to accelerate melt and sawdust or drift fences to slow the melting. Dortignac (1960) indicated that increasing water yields from piñon juniper (*Pinus* sp.—*Juniperus* sp.) lands are not promising. If brush which uses water the year around is replaced by grass which uses water only part of the year, water yields may be increased, provided there is enough precipitation during the dormant season of the grass to develop a precipitation excess.

The removal of riparian vegetation, particularly along stream channels, may increase water yield if the streams are perennial. If streams are ephemeral, removal of riparian vegetation may or may not enhance water yields. Blaney (1961) estimated that a saving of about 32,600 acre-ft of water could be obtained by converting *tamarix* sp. to bermudagrass (*Cynodon dactylon* L. Pers.) from a 56-mile stretch of the Pecos River channel in New Mexico, USA.

It is important to consider the entire watershed system when evaluating the influence of vegetation on water yield. In some situations, complex hydrologic changes, which are not immediately apparent, result from management of the surface vegetation. Watersheds in mountainous areas and foothills may support extensive growth of scrub timber and brush. Such regions may be situated geologically with a groundwater table beneath conforming to the land form of the local stream network.

Recent studies on California, USA, experimental watersheds by Burgy (1958) and Lewis and Burgy (1964) have demonstrated that oak trees (*Quercus* sp.) and other species extend root systems to depths as great as 80 ft through fractured rock and transpire water directly from the groundwater system. Removal of this transpirational loss, which occurs under conditions of high water stress and high evapotranspiration potential, reduces the loss and allows substantial quantities of water to be supplied as groundwater outflow and ultimately as water yield to downstream points (Burgy, 1958; Lewis and Burgy, 1963).

Removal of heavy canopied vegetation and replacement with grasses and low growing species changes the interception loss which occurs from the rainfall (Burgy and Pomeroy, 1958; McMillan and Burgy, 1960). Greater quantities of rain thus reach the soil surface for infiltration and runoff. Combined with reduction of rooting depth and consequent lower transpirational loss, smaller amounts of water are required to recharge soil water deficits under such grassy vegetation, generally resulting in greater water yields.

The effectiveness of vegetative treatment for increasing water yields is a function of soil depth, vegetative type and characteristics, and amounts of precipitation. Precipitation must be adequate to satisfy the losses and recharge the profile. Lesser quantities of precipitation will affect the water yield little if soil surface conditions are unchanged.

Costs of treatment of vegetation must be considered. Hundreds, even thousands, of square miles may be involved in some watersheds. It may not be easy to eradicate brush and trees on areas this large, and it may be even more difficult to prevent re-invasion. Research workers at Spur, Texas reported privately to one of the authors that in one pasture they counted 7,000 mesquite (*Prosopis* sp.) per acre, 2 years after the mesquite trees had been cleared. Certain types of vegetation persist in parts of watersheds because they are better adapted to that particular environment. It is not only difficult to eradicate one type of vegetation and substitute another, but maintenance may be difficult or expensive.

A notable example of an applied program of vegetation management designed to increase available water supplies is reported by Barr (1956) in Arizona. A long-range research and action program was developed and implemented to evaluate the potential for greater water yields through manipulation of the vegetation on the Salt River watershed, USA.

B.　Phreatophytes

Any plant growing where its roots are in free water, such as a high water table, and extracting water directly is known as a phreatophyte. In the Western USA, according to Robinson (1958), about 80 species of plants fall into this category and occupy nearly 17 million acres. They consume an estimated 25 million acre-ft of water annually in the Western USA without benefit to man. In fact, some phreatophytes actually pose a flood hazard and induce sedimentation in unwanted places.

Blaney (1961) studied a section of the Pecos River and estimated that the 36-mile stretch of river between Artesia and Carlsbad, New Mexico would produce 32,600 acre-ft more water if the *tamarix* sp. were converted to bermudagrass.

Cremer (1958) summarizes the data on control of phreatophytes. He points out the meager nature of information on conversion of phreatophytes to more xerophytic type vegetation and the lack of economic studies of the conversion.

C.　Timber Stands and Snow

Thick stands of large coniferous timber intercept great quantities of snow, much of which sublimates or melts and evaporates directly into the atmosphere. It has also been well demonstrated that judicious logging and thinning of such timber may increase water yield where most of the water yield comes from snowmelt (Goodell, 1958). Obtaining and maintaining the optimum timber stands (densities, age groups, and cutting patterns) to enhance water yields can be expensive but rewarding (Love, 1960; and Anderson, 1963).

D. Range Development

The effect of improving range vegetation on water yields is not commonly understood. Fletcher (1960) showed that soils under improved range conditions may acquire increased infiltration capacities and result in reduced runoff. Range improvement may be obtained by substituting desirable species for undesirable rather than revegetating barren areas, but such treatment usually increases grass canopy. Under semiarid to arid conditions, any vegetation (and no vegetation in very shallow soils) will use all the soil water available; hence, water use is limited by supply rather than by type of vegetation. In arid and semiarid areas essentially all water yield comes from surface runoff. Any change in the infiltration capacity or interception directly influences water yield, even though the soil has a water deficit. Considering the highly intense rains characteristic of arid and semiarid zones, the effect of changing the infiltration capacity of most soils is well within experimental errors found in runoff measurement. Thus, for practical purposes, improving range condition may have little effect on the water yield of large acreages except to decrease their sediment content.

E. Stockponds

Criddle et al. (1962) have reviewed the whole picture of livestock water facilities in the Pacific Southwest, USA. They point out that range livestock water facilities contribute to economic stability and growth. Even though it is possible to construct a sufficient number of ponds on a single drainage to allow minimum runoff, the likelihood is small, if there are vested interests downstream. If there

Table 59–4. Estimated effects of stock ponds on downstream water yield

| | Unit | Water resource basin, USA | | | |
		Colorado River	Great Basin	Central Pacific	South Pacific
Area of basin	Sq mile	243,318	174,648	103,564	41,023
Natural runoff at	Acre-ft/				
downstream points*	sq mile	160.08	50.01	640.08	35.01
(no ponds)					
Downstream yield	Acre-ft/	160.00	50.00	640.00	35.00
in 1959†	sq mile				
Downstream yield	Acre-ft/	159.91	49.98	639.92	34.99
in 1980 †	sq mile				
Downstream yield	Acre-ft/	159.83	49.97	639.85	34 98
in year 2000 †	sq mile				
Per cent effect in	%	0.15	0.08	0.04	0.10
year 2000†					

* Runoff estimated for drainage areas of 1,000 sq miles assuming no regulation due to stockponds.
† Downstream effects due to stockponds were estimated for a drainage area of 1,000 sq miles.

are no downstream vested interests, perhaps such development is desirable. Most states require recognition of prior use even for stockponds.

Criddle et al. (1962) show the downstream effects of stockponds as given in Table 59–4 for areas of 1,000 square miles. For areas of any other size the values must be recomputed.

F. Flood-Prevention Reservoirs

Many small floodwater-retarding reservoirs are being constructed in the USA under the provisions of Public Law 566, the Small Watershed Protection and Flood Prevention Act of 1953. Most structures have a small pool below the elevation of the principal spillway (a sediment storage pool), flood storage capacity between the principal and emergency spillways, and surcharge storage capacity above the emergency spillway. The sedimentation pools of such structures perform much as stockponds in reducing runoff, except that such structures are generally closer to groundwater; hence, deep percolation may cause groundwater mounds below them that become an integral part of the area groundwater.

The area around, and above, the sediment pool is only periodically inundated. There are evaporation losses from this body of water and losses by irrigation of the temporarily flooded area. Such irrigation water is lost largely by evapotranspiration. This loss may be sufficiently small on a watershed basis to be ignored. The total water cost of this type of structure is dependent on soils, geology, climate, numbers and sizes of structures, water stages, frequency of flooding, and other factors.

In arid and semiarid areas of all of the recharge to regional groundwater tables is either through channels or ponds. The amount of recharge is directly proportional to the length of time the water is in a channel or a pond, rather than the size of flow or the depth of storage. Thus, since stockponds and flood-prevention channels hold water for longer periods of time than would be normal under flashy floodflows, they serve to increase the water yields by storage of additional water in regional groundwater basins. This is particularly true in California, Arizona, and New Mexico, where considerable quantities of water are being artificially recharged for storage.

G. Large Flood-Control and Conservation Reservoirs

In dry countries, the total water yield of a watershed is decreased when large reservoirs are constructed. Indeed, there is a point-of-no-return in watersheds in arid climates, a point at which it is unprofitable to build more reservoirs because evaporation and other losses become greater than the usable water gained, thus resulting in net deficits, rather than gains in terms of usable water. It has been reported that this point is now being approached on the Colorado River of Southwestern USA.

Total water yield will be reduced by the presence of large reservoirs. The amount of usable water, however, will be increased. If not, the reservoirs must have been built for other benefits. Evaporation from water surfaces of such reservoirs in hot dry climates is enormous. It is generally greater than evapotranspiration from the original site. Such reservoirs sometimes cause channels upstream to aggrade, thus providing deposits that may become infested with phreatophytes that waste water. Phreatophytes and riparian vegetation may also cause great water losses downstream, if the discharge from the reservoir provides more water for phreatophytes than was available before reservoir construction.

H.   Cultivated Land Treatment

In many watersheds where water conservation works are planned, much of the land is cultivated. Such use of land, and the concomitant treatment of it. may affect water yields.

Nearly all surface water works, whether diversion dams, storage dams and reservoirs, canals, or city pipelines, are affected, to a greater or lesser degree, by sedimentation. Two great sources of sediment in most surface waters are sheet erosion and channel erosion. Sheet erosion is intimately related to use and treatment of cultivated land; hence, the extent of the sediment hazard to which water control works are subjected is partially due to the manner in which cultivated land is used, or misused and treated. This aspect of the relations of land use and treatment to water conservation programs should not be overlooked.

1.   CROP ROTATIONS

Crop alternatives possible on watersheds may affect water yields. A watershed in continuous alfalfa (*Medicago sativa* L.) might, for instance, yield less water than one in mixed-row-crops. Crops grown are generally commensurate with economic and social demands and are not determined by their relative effects on water yields; hence, alteration of cropping patterns by individuals influences water yields from river basins only to a minor extent.

2.   TERRACES

Level terraces and concomitant contour tillage probably affect on-site water yield. Graded terraces *per se* have little or no effect on water yield, although the near contour tillage practiced in conjunction with graded terraces probably reduces water yield over conventional straight row farming.

3.   LAND-USE CONVERSIONS

Converting clean tilled land to perennial vegetation, such as grass pastures or meadows, probably reduces on-site water yield. The effect of this practice will be smaller on thin, steep, eroded land than on better land.

4.   IRRIGATION

Irrigation of formerly dry-farmed land probably increases on-site water yield, but reduces total watershed water yield, if the irrigation water comes from surface waters within the watershed. If such irrigation water comes from deep wells drilled into aquifers that do not contribute to streamflow within the basin, or if such water is imported to the basin, basin water yield may be increased.

5. DRAINAGE

The opening up of drains from sink holes or pot holes that, in effect, increase the tributary drainage areas of rivers probably increases water yield in direct proportion to the formerly nontributary areas made tributary. Drainage of swamps will probably increase water yield, since evapotranspiration from the crops grown on the drained swamps will generally be less than evapotranspiration from the original hydrophytes and phreatophytes and the direct evaporation from open water surfaces.

6. WATER-SPREADING

Water-spreading is one of a variety of management practices that prevent channelization of surface flow and force it to remain a diffuse, laminar type flow. This practice, if it affects substantial areas, probably reduces water yield, since water flowing in upstream channels is diverted and spread onto land rather than going downstream.

I. Evaluation of Land Treatment

Research data available in the USA indicates that land treatment affects water yield as indicated by the indices shown in Table 59–5 and in the manner shown in Fig. 59–1. Table 59–5 lists land treatment practices thought to be most important in their effects on water yield and shows indices of the relative magnitudes of each. Figure 59–1 shows the relation of percentage surface runoff de-

Table 59–5. Relative effects of land use and treatment measures in depleting or increasing water yields by surface runoff *

1	2	3
Practice	Index to convert from base curve†	Effect on runoff
All level closed-end terraces	1.0	Depleting
Row crops, straight-row‡	0	Base
Row crops, contour tillage with or without graded terraces	0.5	Depleting
Row crops, level open-end terraced with contour tillage	0.7	Depleting
Small grain, straight-row	0.3	Depleting
Small grain, contour tillage with or without graded terraces	0.6	Depleting
Small grain, level open-end terraces with contour tillage	0.7	Depleting
Land conversions, cultivated to range, pasture, and brush -- noncultivated, deep, permeable soils	0.7	Depleting
Ditto, shallow, eroded, slowly permeable soils	0.4	Depleting
Irrigation (as compared to former dryland farming)	−0.4	Increasing

* To be used in conjunction with curve in Fig. 59–1.
† These are, in effect, percentages of the maximum depleting effect of closed-end level terraces, as compared to straight rowcrops, shown by the curve in Fig. 59–1.
‡ This is the base from which effects of all other practices are referenced.

pletion to the ratio of annual watershed precipitation to average annual potential watershed evapotranspiration. A weighted annual practice index may be computed to apply to the portion a watershed affects. Annual observed surface streamflow, minus base flow, multiplied by this weighted index, becomes an estimate of the effect of a given practice on water yield.

The effect of ponds and reservoirs on water yields can be determined by estimating annual net evaporation, seepage, and percolation losses and adjusting on-site losses for transmission losses that would have occurred had the water not been stored upstream. Transmission losses also vary inversely with the ratio of annual precipitation to average annual potential evapotranspiration (*see* Fig. 59–2).

The effect of upstream treatments on downstream water yield was intensively studied by Sharp et al. (1966). They devised a method for the evaluation of the effects of conservation use and treatment of land on streamflow, water yield, and data analysis.

Sharp and Saxton (1962) studied 57 flood events on 18 Great Plains rivers, USA, whose channel reaches averaged 53 valley miles and found an average transmission loss equal to 40% of the flow. Culler (1961) working on the Cheyenne River near Hot Springs, South Dakota, USA, found similar differences between upstream runoff to stockponds and downstream streamflow (Allis, 1961). Cornish (1961) found transmission losses around 42% in a 173-mile stretch of the South Canadian River. Keppel and Renard (1962) measured transmission losses on Walnut Gulch, Arizona at 25-acre-ft/mile and estimated that losses might be as high as 80 acre-ft/mile on some of the ephemeral streams in the Southwest.

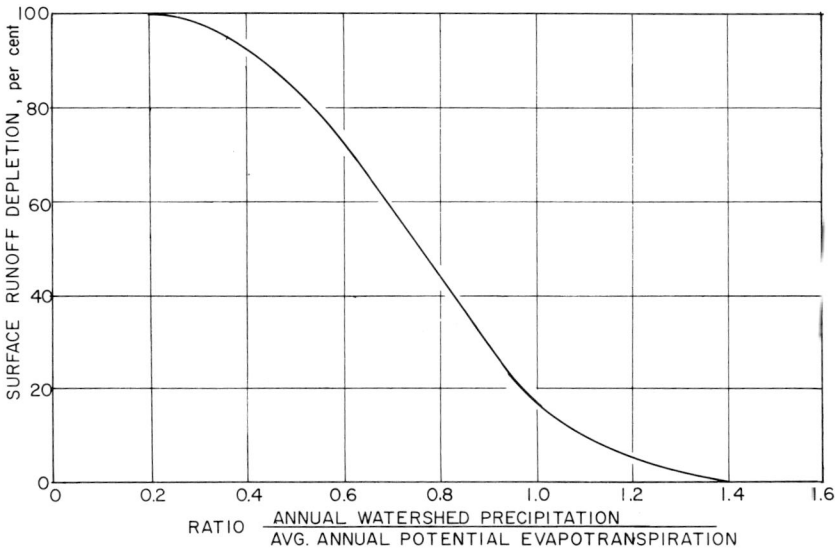

Fig. 59–1. Base curve of surface runoff depletion, row crops in straight row farming to level closed end terraces with contour tillage, as related to relative aridity of climate.

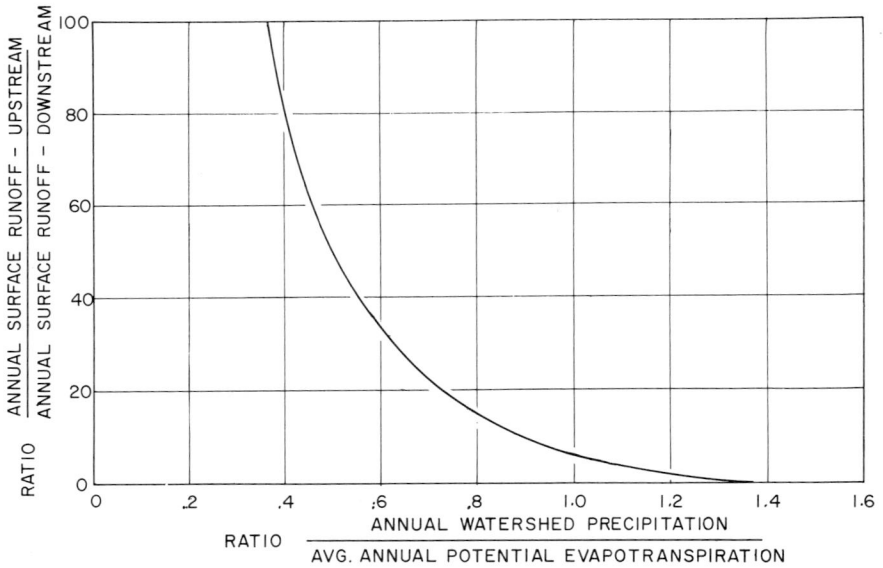

Fig. 59–2. Effects of valley transmission losses on upstream unit-source-area surface runoff enroute to downstream gauging stations of large creeks and rivers.

Reduction of transmission losses through collection and differentiation of low flows, lining of stream channels, and the use of pipelines to convey water or direct recharge from the channel to groundwater aquifers may be profitable watershed management practices.

J. Water Harvesting

Meyers (1962) outlined the feasibility and distribution of water harvesting for special purposes. He estimated that water presently available for water harvesting would be equal to about 3.5 times the streamflow in the USA. Many systems have been used to increase the amount of water available, ranging from denuding small areas to paving areas with concrete or collecting runoff from roofs.

Development of efficient and low cost methods for water harvesting are underway in many locations and are in actual operation in many parts of the Southwest.

IV. SUMMARY

Large-scale manipulation of vegetative cover and watershed characteristics to increase water yields has not yet been demonstrated. Such manipulation has increased water yield from small areas. Before any large-scale manipulation is attempted, a thorough investigation of all of the surrounding parameters should be made. Precipitation on a watershed is divided between soil water deficits, direct surface runoff, subsurface runoff, transmission losses, evapotranspiration, and recharge. Watershed management may be applied to any one of these processes. Those largely neglected are transmission losses and evapotranspiration reduction.

Watershed management in its several forms is amenable to economic analysis and should be evaluated as an alternative means of obtaining increased water yields. Alternative methods of achieving a desired increased water supply need to be evaluated as economic alternatives, as with any engineering design, and the relative benefits and costs compared to determine the optimum method.

LITERATURE CITED

Allis, John A. 1961. Transmission losses in loessial watersheds. Presented to joint ARS-SCS Hydrology Workshop. Reno, Nevada. Amer. Soc. Agr. Eng. p. 4.1–4.32.

Anderson, Henry W. 1963. Managing California snow zone lands for water. US Forest. Serv. Res. Pap. PSW–6. 28 p.

Barr, George W. 1956. Recovering rainfall. The Arizona Watershed Program Univ. Arizona. Vol. 1, 33 p., Vol. II, 218 p.

Blaney, Harry F. 1961. Consumptive use and water waste by phreatophytes. J. Irrig. Drainage Div. Amer. Soc. Civil Eng., Proc. Pap. 2929. 87(IR 3):37–46.

Burgy, Robert H. April 1958. Water yields as influenced by watershed management. J. Irrig. Drainage Div. Amer. Soc. Civil Eng., Proc. Pap. 1590. 84 (IR 2). 10 p.

Burgy, Robert H., and C. R. Pomeroy. 1958. Interception losses in grassy vegetation. Amer. Geophys. Union, Trans. 39:1095–1100.

Chow, Ven Te. 1962. Hydrologic determination of waterway areas for the design of drainage structures in small drainage basins. Univ. Illinois. Eng. Exp. Sta. Bull. 462. 104 p.

Connaughton, C. A. 1935. The accumulation and rate of melting of snow as influenced by vegetation. J. Forest. 35:564–69.

Cornish, John H. 1961. Flow losses in dry sandy channels. J. Geophys. Res. 66:1845–1853.

Cremer, Henry J. 1958. Eradication and control of phreatophytes in reservoir delta areas and replacement of ground cover. p. 42–45. In Symposium on phreatophytes. Pacific Southwest Inter-Agency Comm. US Dep. Interior P.O. Box 360, Salt Lake City 10, Utah.

Criddle, Wayne D., Henry W. Anderson, Virgil T. Heath, Loren D. Morrell, Harold V. Peterson, and Robert E. Rallison. 1962. Stock water facilities guide. Pacific Southwest Inter-Agency Comm. US Dep. Interior P.O. Box 360, Salt Lake City 10, Utah. 68 p.

Culler, R. C. 1961. Hydrology of the upper Cheyenne River Basin. US Geol. Surv. Water Supply Pap. 1531. 198 p.

Dortignac, E. J. 1960. Water yield from pinyon-juniper woodlands. p. 16–27. In Barton H. Warnock and J. L. Gardner [ed.] Symposium on water yield in relation to environment in the southwestern United States. Amer. Ass. Advance. Sci. and Sul Ross Coll., Alpine, Texas.

Fletcher, Joel E. 1960. Some effects of plant growth on infiltration in the southwest. p. 51–63. In Barton H. Warnock and J. L. Gardner [ed.] Symposium on water yield in relation to environment in the southwestern United States. Amer. Ass. Advance. Sci. and Sul Ross Coll., Alpine, Texas.

Free, G. R., G. M. Browning, and G. W. Musgrave. 1940. Relative infiltration and related characteristics of certain soils. US Dep. Agr. Tech. Bull. 729. 51 p.

Gartska, W. U., L. D. Love, B. C. Goodell, and F. A. Bertle. 1958. Factors affecting snowmelt and streamflow. US Forest Serv. and Bur. Reclamation. 189 p.

Goodell, Bertram C. 1958. Watershed studies at Frazier, Colorado. Amer. Soc. Forest., Proc. p. 42–45.

Haupt, H. F. 1951. Snow accumulation and retention on ponderosa pine lands of Idaho. J. Forest. 49:869–71.

Keppel, R. V. 1960. Water yield from grassland. p. 39–46. In Barton H. Warnock and J. L. Gardner [ed.] Symposium on water yield in relation to environment in the southwestern United States. Amer. Ass. Advance. Sci. and Sul Ross Coll., Alpine, Texas.

Keppel, R. V., and K. G. Renard, Jr., Aug. 1962. Transmission losses in ephemeral stream beds. J. Hydraulics Div. Amer. Soc. Civil Eng. 89 (HY–3):59–68.

Lewis, D. C., and Robert H. Burgy. 1963. Water use by native vegetation and hydrologic studies. Dep. Irrig. Univ. California, Davis, Annual Rep. 4. 108 p.

Lewis, D. C., and R. H. Burgy. 1964. The relationship between oak tree roots and groundwater in fractured rock as determined by tritium tracing. J. Geophys. Res. 69:2579–2588.

Light, P. 1941. Analysis of high rates of snowmelting. Amer. Geophys. Union Trans. Part 1. p. 195–205.

Love, L. D. 1960. Water yield from mountain areas. p. 3–15. In Barton H. Warnock and J. L. Gardner [ed.] Symposium on water yield in relation to environment in the southwestern United States. Amer. Ass. Advance. Sci. and Sul Ross Coll., Alpine, Texas.

Mannering, J. V., and L. D. Meyer. 1963. The effects of various rates of surface mulch on infiltration and erosion. Soil Sci. Soc. Amer. Proc. 27:84–86.

McMillan, W. D., and R. H. Burgy. 1960. Interception loss from grass. J. Geophys. Res. 85:2389–2394.

Meyers, L. E. 1962. Water harvesting. Presented at Nevada Water Conf., 17th (Carson City, Nevada), 14 p.

Rich, L. R. 1960. Water yield from oak woodland. p. 28–38. In Barton H. Warnock and J. L. Gardner [ed.] Symposium on water yield in relation to environment in the southwestern United States. Amer. Ass. Advance Sci. and Sul Ross Coll., Alpine, Texas.

Robinson, T. W. 1958. The phreatophyte problem. p. 1–11. In Symposium on phreatophytes. Pacific Southwest Inter-Agency Committee.

Rosa, J. M. 1956. Forests and snowmelt floods. J. Forest 54:231–235.

Sharp, A. L., A. E. Gibbs, and W. J. Owen. 1966. Development of a procedure for estimating the effects of land and watershed treatments on streamflow. US Dep. Agr. Tech. Bull. 1352. 57 p.

Sharp, A. L., and K. E. Saxton. 1962. Transmission losses in natural stream valleys. J. Hydraulics Div. Amer. Soc. Civil Eng. 88(HY–5):121–142.

60

Reducing Water Losses in Conveyance and Storage [1]

C. W. LAURITZEN

Agricultural Research Service, USDA
Logan, Utah

P. W. TERRELL

Bureau of Reclamation, US Department of the Interior
Denver, Colorado

I. INTRODUCTION

A. Source and Extent of Losses

It has been estimated that one-fourth to one-third of all the water diverted for irrigation purposes is lost in conveyance. US Bureau of Reclamation records from 46 irrigation projects show that losses range from 3% to 86%. If we assume one-fourth of the 15,650,000 acre-ft of water reported as diverted by these projects is lost, this represents 3,900,000 acre-ft (US Senate, 1962a). Considering the USA as a whole, if only one-fifth of the total water diverted for irrigation purposes is lost to the user, the quantity seeping from canals would be 22 million acre-ft/year. If the intended users could retain this water, they would be able to irrigate 5.5 million additional acres, using 4 acre-ft/acre.

The question of whether seepage losses are actual losses is discussed fully in other articles (US Senate, 1960). For our purpose, seepage will be called a loss; and it is an actual loss to its intended user.

Other losses in a system consist of evaporation from canals and reservoirs, evapotranspiration from areas waterlogged by seepage or from backwaters of reservoirs, and tail water associated with scheduling and delivery.

Evaporation losses from canals are usually 1.0 to 1.5% of the diverted water. Based on present research, it appears that about 14 million acre-ft/year are lost from large lakes and reservoirs in the western USA, and the total evaporation from all water surfaces in this area is almost 25 million acre-ft. In addition to losses by evaporation from free water surfaces, appreciable quantities of water are lost through evapotranspiration from plants and land surfaces along the canals (Robinson, 1958). It appears that more than one million acre-ft/year of water lost to evapotranspiration could be salvaged by channeling natural waterways, diking flood areas, and controlling phreatophytes. Assuming again a water requirement of 4 acre-ft/acre, the water saved would irrigate 250,000 acres. It is possible that some of this projected saving may be duplication in part, since control of seepage during conveyance and storage will in itself limit evapotranspiration by restricting the supply of water to land area and vegetation responsible

[1] Contribution from the Soil and Water Conservation Research Division, ARS, USDA and Bur. of Reclamation, USDI.

for these losses. Actually, the most effective way of controlling evapotranspiration losses is to cut off the water supply to phreatophytes.

II.   WATER LOSS MEASUREMENTS

A.   General

Seepage rates depend on the permeability of the soil, the existence or absence of natural or artificial linings, the depth of the water and its velocity, the elevation of the water table, and other variables.

In most canals, and in many reservoirs, the greatest part of seepage occurs through relatively short reaches. Since there are seldom sufficient funds to line a canal in its entirety, the problem evolves into one of locating and delineating the reaches where the greatest seepage is taking place and concentrating on the control of seepage in these reaches.

B.   Methods of Measuring Seepage

1. INFLOW-OUTFLOW

The over-all loss in a canal system can be determined by inflow-outflow measurements, using current meters or available measuring structures. Recently, fluorescent dyes and isotopes have been used to measure flow with improved accuracy, but these methods are not fully developed. The problem is to isolate those sections of canal where the greatest loss is taking place. In a short reach of canal the quantitative accuracy of inflow-outflow measurements with existing methods is not adequate to establish the extent of the loss, or even to determine whether or not there is a loss. This is an inherent weakness of the method and a limitation that confronts us whenever we attempt to establish losses based on small differences involving the measurement of large volumes (Kaufman and Orlob, 1956; Robinson and Rohwer, 1959).

2. PONDING

Ponding measurements are obtained by sealing off a section of canal with dikes, or watertight structures, and determining the rate at which the water elevation in the ponded section lowers with time. This is probably the most accurate method that we have for evaluating seepage losses in short reaches of canal. Ponding, however, interrupts water deliveries and, in larger canals, involves considerable expense. Inaccuracies arise from the fact that seepage varies from year-to-year and during the irrigation year within any given reach of canal (US Dep. Interior, 1963).

3. SEEPAGE METERS

Since the first meter was developed at the US Salinity Laboratory, several meters of this general design, but with modifications, have been used (Rohwer and Stout, 1948; US Dep. Interior, 1963; Warnick, 1963). Figure 60–1 shows one type used. These devices, while differing in design, have many things in common. They all provide for confining a small area of the perimeter and measur-

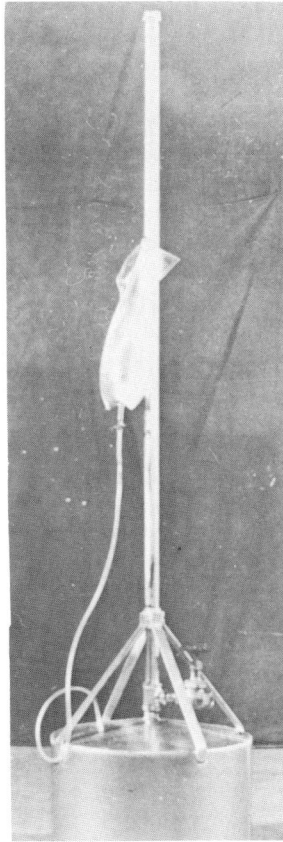

Fig. 60–1. Seepage meter
with plastic bag for use
in unlined operating channels.

ing the quantity of water that passes through it. Falling head seepage meters
have been developed which reportedly measure the hydraulic conductivity and
gradient in the material below the seepage cup (Bouwer, 1961).

All seepage meters have a serious disadvantage in that the seepage meter
location may not be representative of the canal section being measured. Added
to this is the problem of operating the meter throughout the full cross section of
the canal. Since the seepage through the bottom and the sides of a canal is frequently different, the meter measurements may not be indicative unless measurements are made in both bottom and sides. Even when a large number of measurements are made, there is no assurance that seepage meter measurements will
provide data that are indicative of losses (Rasmussen and Lauritzen, 1953).
Under some conditions, however, they do give some indications of losses and the
those reaches where high losses can be expected (Bouwer, 1963; Beers, 1958;
Bouwer, 1963).

4. GROUNDWATER ELEVATIONS

Inflow-outflow, ponding, and seepage meter measurements are quantitative.
There are other methods which provide data indicative of losses but which are
not quantitative within themselves. For example, an indication that seepage is

occurring can frequently be established by observing the position of the ground-water table in the vicinity of a canal. If it slopes away from the canal, it can be assumed that seepage is occurring.

5. IN-PLACE MEASUREMENT OF PERMEABILITY

Indicative also of the seepage which is taking place, or could be expected to take place, is the rate at which water will enter the ground as measured by the auger-hole or double-tube methods. While the data assembled by this method have little quantitative significance, they are indicative of the permeability of the material and, therefore, of the seepage potential. If measurements are made along a reach of canal, it is sometimes possible to get an indication of the places where the greatest seepage is occurring or, in the case of unconstructed canals, those reaches where high losses can be expected (Bouwer, 1963; Beers, 1958; Warnick, 1963).

6. LABORATORY PERMEABILITY TESTS

Information similar to the above can be obtained by sampling the soil profile along the course of the canal and making laboratory measurements of permeability. It is hazardous to assume, however, that similar values from different canals obtained by the auger hole or double-tube methods, or that soil of like permeability as determined in the laboratory from soils of different canals, will be associated with corresponding canal losses, if tests are on dissimilar mineralogical materials. If field and laboratory permeability measurements are supplemented by a few ponding measurements in the area being tested, and the results correlated, more reliability can be expected (US Dep. Interior, 1963).

7. ELECTRIC LOGGING

A recent development which is being watched with much interest is the use of electric logging for estimating or locating those areas in which seepage losses are heavy. This method, like some of the others, is indicative only. Electric logging has several practical advantages over other methods. The measurement is continuous and gives point-to-point information, assuming, of course, the method is reliable and enough is known about the earth profile to make proper interpretation. The resistance measurement indicates relative saturation of the bed material, and the self-induced soil electrical potential indicates the movement of soil water. The measurements can be made with water in the canal at any stage without interfering with deliveries of water and at any season of the year when the canal is free from ice. The equipment consists of two lead electrodes connected with wires to a source of current, an extremely sensitive microvolt meter, and truck-mounted recorders for both readings. The depth to which the resistivity of formations is measured is controlled by the spacing of the electrodes (US Dep. Interior, 1963).

C.   Methods of Measuring Evaporation

Evaporation is usually determined by measuring the loss of water from an evaporation pan. The water-filled pans are located so that the wind, sun, and exposure condition is closely representative of the water surface under consideration. By keeping pans filled to a constant level, the amount of water lost through evaporation can be measured. (Amer. Soc. Civil Eng., 1934). *See also* chapter 29.

D.   Method of Measuring Evapotranspiration

Evapotranspiration is difficult to measure and evaluate. It is sometimes estimated by the water budget method. This entails measuring the water into and out of a natural watershed. "As a rule, normal annual evapotranspiration can be reliably computed as the difference between long-time averages of precipitation and streamflow, since the change in storage over a long period of years is inconsequential" (Linsley et al., 1958). Estimates for shorter periods require the evaluation of soil water, groundwater, and surface storage at the beginning and end of the period considered. As it applies to specific plants, it can be estimated by measuring soil water depletion through field sampling or by growing plants in a block of soil confined in a metal or plastic tank, commonly referred to as a lysimeter. The amount of water that must be added to the lysimeter to sustain plant growth and maintain soil and surface water gives a measure of the evapotranspiration rate and quantity in the lysimeter (Robinson and Johnson, 1961). For further detail see chapter 29.

III.   CANAL AND RESERVOIR LININGS

A.   General

Canals and reservoirs are lined for a great number of reasons—seepage control, erosion control, safety, reduction of maintenance, reduction in right-of-way, appearance, public acceptance, and maintenance of better water quality. The justification may include any or all of these, but seepage and erosion control are usually present on any list of benefits from linings. Because even the least costly lining is expensive, the selection of type must be made with care and its design and construction undertaken only by trained personnel (Chowdhry, 1957; US Senate, 1960; US Dep. Interior, 1963).

Linings used in canals can be classified into five general types according to their hydraulic properties, function, and other characteristics. These types are: paved or hard-surfaced linings, exposed membrane linings, conditioned earth linings, buried-membrane linings, and earth and chemical sealants.

The types differ in their adaptation and the function they perform. The selection for a particular job must, therefore, be governed by site conditions and functional requirements. Although not necessarily related to performance, comparative costs are generally a determining factor in selection. In order to be comparable, costs should be in terms of annual costs including, in addition to the initial cost amortized over the expected service life, maintenance and repair and a reasonable return on the original investment. Where such factors as weed control and removal of sediment differ significantly between types of lining, the advantages of one type of lining over another in this regard should be considered when making a comparison. Costs, likewise, will vary from area-to-area (Lauritzen, 1957, 1960; South, 1955).

All effective linings have certain characteristics in common. Since seepage control is of primary importance, either directly or indirectly, all linings must be reasonably watertight. A lining should be resistant to damage, reasonably long lasting, and easy to maintain. The degree to which the different types of linings fulfill these requirements depends on the materials used and the construction practices.

B. Hard-Surface Linings

Common materials for these linings are portland cement concrete, asphaltic concrete, brick, and rock. Where properly constructed and maintained, these linings embody more desirable features than any other type. In addition to controlling seepage, they prevent scour, afford maximum weed control, and can be designed for velocities which will be conducive to minimum sedimentation in canals. Furthermore, they are resistant to mechanical damage. Seepage through hard-surface linings in good condition can be expected to range from 0.1 to 0.2 ft^3/ft^2 per day, depending on the extent of cracking, the nature of the subgrade material, and other factors. Mechanization offers the best opportunity for reducing the cost of concrete linings. Normally, costs are lower when the concrete is placed with slipforms and pavers, but concrete linings can be constructed in panels with a minimum of equipment and semiskilled labor, if supervised by trained personnel. Linings constructed in panels are normally used on small jobs where mechanized equipment is not justified. A typical section for concrete is shown in Fig. 60–2.

Linings of brick, tile, or rock are usually too expensive if the cost of labor is high with respect to materials. Where the reverse is true, they may be used to advantage and will give excellent service if carefully constructed.

Asphaltic concrete linings when properly constructed are comparable to portland cement concrete linings in many respects, but their expected service life may be shorter. It is advisable to include subgrade sterilization as a part of this type construction to prevent plants from penetrating the lining. The use of asphaltic concrete for lining is justified only when asphaltic concrete is less costly than portland cement concrete. Usually, this depends on the adaptability of local aggregates to asphalt construction rather than to portland cement (US Dep. Interior, 1963; Asbeck, 1957).

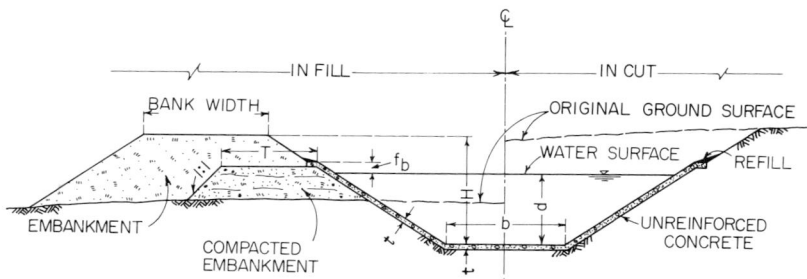

Fig. 60–2. Typical section of concrete canal lining.

C.   Exposed Membrane Linings

The ideal lining would be slightly flexible; this would allow it to adjust to small settlements in the subgrade without damage to the lining itself. At the same time, the lining must be watertight and should be reasonably resistant to deterioration from weathering and mechanical damage. In an attempt to provide such a lining, many materials have been tested. Asphalt structures and butyl rubber have shown the most promise. Asphalt is subject to rapid deterioration from exposure to sunlight, and it is this property that has presented the greatest difficulty in its use for exposed lining.

Two types of asphalt structures recently developed are showing promise. The first is a prefabricated liner consisting of heavily filled asphalt sandwiched between two thin layers of asphalt-saturated felt (US Dep. Interior, 1963). The installed cost is about the same as concrete, but this varies considerably with distances from the place of manufacture because of freight charges. While this liner has been used to some extent for lining irrigation canals and reservoirs, it is being used more extensively for lining industrial and municipal reservoirs. Another lining of this general type is a prefabricated asphalt-coated jute liner. Data on performance are extremely limited, but this lining looks promising and will be lower in cost than the thicker plank-type liner.

Butyl rubber sheeting, either with or without fabric reinforcing, looks promising. Sheeting, either 32- or 64-mil, is satisfactory. Installation requirements are the same as for buried membrane linings. The material is watertight and ages very slowly. Hydraulically, it appears excellent (US Dep. Interior, 1963).

D.   Compacted Earth Linings

Where suitable earth material is available near the site of construction, a lining of thick, compacted earth is an economical and effective means of controlling seepage. Gravelly and sandy clays, with plasticity indexes ranging from 12 to 24, are the best soils for this use. Materials having plasticity indexes higher than this are difficult to work and tend to be unstable, and those having lower values have less resistance to scour and are susceptible to frost damage. Thick, compacted earth lining is constructed by rolling selected earth materials in 6-inch layers while at a water content which will result in maximum compaction. Linings of this type, shown in Fig. 60–3, are commonly constructed with side slopes of 2 horizontal to 1 vertical and with a horizontal width of 8 ft in larger canals. This width is reduced so that a minimum lining thickness of 1 ft is obtained in very small canals. To be competitive costwise, it is necessary to use heavy equipment for excavating, conveying, and compacting the earth material, since the volume of material is large.

Stream velocities in canals lined with any type of earth lining must not exceed the permissible nonerosive velocity characteristic of the material used in lining. This normally restricts velocities to 3 ft/sec maximum. Even then, scour below structures and on bends is often serious, and a topping of rock or gravel must be used in these critical areas. Unless the material is gravelly clay, a more serviceable lining, requiring less maintenance, results if the compacted earth is topped with rock or gravel.

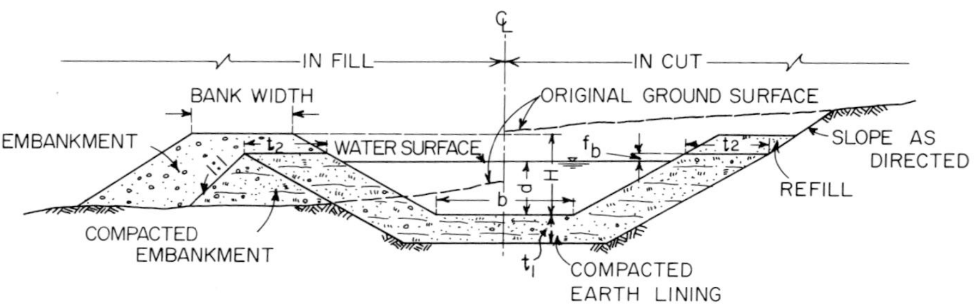

Fig. 60–3. Typical section of thick compacted earth lining.

Since seepage losses are governed primarily by the permeability of the material, the seepage can be reduced only to a degree by compaction alone. The probable effectiveness of compaction alone, or other treatments, can be estimated from laboratory permeability measurements. A more positive seepage control is likely to result if substandard materials are treated to overcome their natural deficiencies. This treatment may be the addition of fine-grained soils, as in blended linings, or the addition of portland cement or asphalt to the soil which reduces permeability and increases scour resistance. Some testing is being done on the addition of chemicals, but current results are inconclusive (US Dep. Interior, 1963).

E. Buried Membrane Linings

Buried-membrane linings can control seepage effectively and, because they are cheaper than other linings in some areas and under some conditions, they have been used extensively in recent years. The membrane should consist of any watertight material that has a long life in soil.

Where an earth membrane such as bentonite or other impervious soil is used, a minimum of subgrade preparation is required. It is essential only that the canal be shaped to proper line and grade. If the subgrade material is extremely coarse textured, a topping of some type should be applied as a filter bed to prevent piping the membrane into the subgrade. The membrane thickness varies from 1.05 inches for a high grade sodium bentonite to 1 ft for marginal clays or silts. The thinner membranes should be topped by a layer of fine- to medium-textured earth, and this—as well as the thicker membranes—covered by a layer of gravel to prevent erosion. To obtain better initial stability, it is desirable to roll earth membrane linings before topping with gravel. A smooth, or pneumatic roller, is well adapted to obtaining the necessary compaction.

Buried asphaltic membrane (BAM) linings are a development dating back only 10 to 15 years. This lining consists of a thin layer of asphalt sprayed on the prepared invert of the canal and topped first with a 6-inch layer of fine-textured earth and then with a 6-inch layer of gravel or with a 12-inch layer of graded sand and gravel that is erosion resistant and stable on the slope (see Fig. 60–4). The canal must be over-excavated to allow for the thickness of the lining. It is extremely important that the subgrade on which the asphalt is applied be firm, fine textured, and smooth and rolled with a smooth roller. Where the subgrade

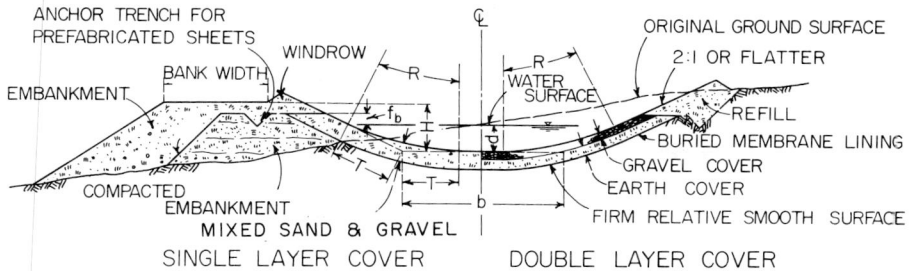

Fig. 60–4. Typical section of buried-membrane lining.

material is coarse, a fine-textured material must be brought in, spread, and rolled to provide a satisfactory surface for the asphalt. For details of the application, see reference US Dep. Interior (1963). The chief shortcomings of this and other buried-membrane linings are tendencies for the cover to slide downslope—particularly when the canal is being emptied—and the extremely slow drying of the invert after the canal is emptied. This latter characteristic results in prolific weed growth for a month after the canal is empty.

The BAM lining is particularly adapted to the lining of old canals where seepage control is the major consideration. It can be installed quickly and may be constructed in cold weather. If the subgrades are ready, or can be prepared, the asphalt can be applied on frozen ground and even over a thin snow cover without jeopardizing the quality of the installation.

Plastic film and rubber sheeting can be substituted very satisfactorily for the asphalt in the buried-membrane types of lining. These materials are frequently more nearly watertight than other types of lining. Either vinyl (polyvinyl chloride) or polyethylene film may be used. Both the film and the sheeting are much more resistant to penetration by vegetation than the asphalt membrane, but sterilization of the subgrades is still considered advisable. Much work has been done on adapting plastic film, particularly the vinyl film, to use as linings. Through the use of special formulations, the addition of pigments, and a selection of plasticizers, vinyl and polyethylene films available today are superior to earlier products. The minimum recommended film thickness is tentatively 8 mils (0.008 inch).

Vinyl film is available in widths up to 61 ft. The length is controlled only by the limitations of weight which should not be in excess of 500 lb because of handling problems. Vinyl film, 8 mils thick, weighs about 0.05 lb/ft² and polyethylene film about 0.04 lb/ft². Film linings are quick to install and require a minimum of equipment, but wind creates a problem during installation, and a calm day should be chosen for the job. The cover materials are the same as those for asphalt membranes. Damage to the lining when covering can be minimized by placing the cover from a crane-operated materials bucket and working from the bottom upward.

Like the polyethylene and vinyl film, butyl sheeting can be used as a buried-membrane lining but is usually more expensive than either polyethylene or vinyl film. In the few installations that have been made, 32-mil minimum material was used. It is expected that these will give the best performance of all the buried membranes because of the thickness, strength, and durability of the material.

Thinner butyl may become commercially available in the future. Installation procedures and general precautions recommended are similar to those outlined for vinyl and polyethylene film (Lauritzen, 1960; US Dep. Interior, 1963).

Plastic film and butyl sheeting may also be used in repairing breaks in canals. If the new earth fills are covered with a sheet of plastic film and the film carried well past the break and topped with a covering of earth, it is very effective in stabilizing the fill material and preventing washouts. Film and sheeting have also been used as waterproof barriers in cutoff trenches to stabilize banks and cuts.

F. Earth and Chemical Sealants

The fact that muddy water tends to seal canals has long been known, and from this the practice of sluicing fine earth material into canals to stop leakage has developed. Muddy water also tends to reduce erosion and inhibit the growth of waterweeds. One method currently having limited use is that of introducing colloidal material, such as bentonite, into the flowing water (Dirmeyer, 1962). However, in many instances, the degree of control obtained was too slight to justify a continuation of the practice. In spite of these failures, the idea of a waterborne sealant and channel stabilizer is too attractive to discard, so many materials and methods have been proposed and tested. So far, success has been only partial, but progress is being made.

A number of chemical sealants have been developed which, when added to the water flowing down the channel, are ultimately filtered out in the soil or settle on the perimeter. These include waxes, asphalt, resins, lignin, and polymers. All are prepared in the form of emulsions. A problem at present is that many trials result in only a surface seal which is easily damaged. It is desirable to have the sealant penetrate the soil, yet stop within 6 to 12 inches of the surface. Several methods are being developed to obtain this result, but none are completely successful as yet. Nevertheless, substantial reduction in seepage losses has been reported, and at least one supplier guarantees an immediate reduction of 60%. The greatest uncertainty is how lasting is this effect. Many tests show a great loss in effectiveness in two seasons. Other problems are the influence which canal and reservoir management and site conditions will have on both the initial and continued effectiveness. If the treatment is inexpensive enough, periodic applications may compensate for short-term effectiveness. At the present time, the cost ranges from 10 to 25 cents/yard^2 per application (US Dep. Interior, 1963).

G. Reservoir Linings

The location of a reservoir can greatly influence losses. Evaporation is less at higher elevations and in protected sites than at lower elevations and in areas subject to heavy winds. The characteristics of site material will govern seepage losses; where formations are extremely permeable, an alternate site may be mandatory. In selecting sites, consideration should be given to both evaporation and seepage potentials of the area. Many sites can be improved by diking off shallow areas, thus reducing both evapotranspiration and seepage losses. The possibility of restricting losses by this method should always be considered.

At the present time it does not seem practical to line the larger reservoirs. It is estimated from the best data available that seepage losses in the reservoirs constructed by the US Bureau of Reclamation only range from 1 to 5% of the inflow.

The lining of equalizing and small storage reservoirs is entirely feasible, however. Frequently, unless these reservoirs are lined, the increased efficiency of the system justifying their construction is nullified by the losses which they create. All the various types of linings used in canals can and have been used in such reservoirs. Of these, buried-membrane linings are particularly well adapted, since a velocity factor is not involved.

Generally, the higher grade linings—such as concrete and shotcrete—cannot be justified for irrigation reservoirs, considering the good performance which can be expected from earth and membrane linings. If buried-membrane linings are objectionable, exposed membrane linings—such as prefabricated asphaltic structures and butyl sheeting—can be used. In some locations, bentonite or sodium salts tend to seal soils, but calcium water will frequently destroy the effectiveness of sodium bentonite or salts. Where satisfactory earth material is available adjacent to the job, the use of earth as a lining material should be given serious consideration. With all types of buried linings, side slopes should be flat enough to insure stability of the cover material, with slopes of 3 horizontal to 1 vertical being the steepest usually recommended.

H. Cost of Linings

The cost of lining, like other construction, varies widely from location-to-location. Table 60–1, which gives average comparative costs, shows that portland cement concrete and asphaltic concrete are the most costly to construct and that thin compacted earth lining is the least costly. These figures, however, apply only to the cost of construction and do not take into consideration service life or maintenance requirements. Neither do they take into consideration differences

Table 60–1. Appropriate cost of canal linings

Lining	Material	Other*	Total
		Cost per square yard	
		dollars	
Portland cement concrete, 3-inch unreinforced	1. 20	1. 80	3. 00
Portland cement concrete, 2-inch, unreinforced	. 80	1. 70	2. 50
Asphaltic concrete, 2-inch	1. 00	1. 50	2. 50
Asphaltic membrane, buried (BAM)	. 33†	. 77	1. 10
Prefabricated plank type, exposed	1. 66	. 50	2. 16
Asphalt-coated jute, exposed	. 85	. 30	1. 15
Asphalt-coated jute, buried	. 85	. 90	1. 75
Polyethylene film, 10-mil, buried	. 23	. 62	. 85
Vinyl film, 10-mil, buried	. 45	. 62	1. 07
Butyl sheeting, 32-mil, buried	1. 25	. 62	1. 87
Butyl sheeting, 60-mil, exposed	2. 40	. 10	2. 50
Earth, compacted, 3 feet			1. 10
Earth, compacted, 1 foot			. 50

* Subgrade preparation, installation, and covering, when applicable. † Asphalt in place.

in the area of lining required per unit length of canal caused by the differences in permissible velocities applicable to the linings nor the differences in required perimeters on the same longitudinal slope. As an example of the way retardation coefficients influence lining perimeter and hence costs, let us compare concrete lining with BAM lining. In Manning's equation $V = (1.486/n)r^{2/3}s^{1/2}$ the value of n for concrete lining is 0.014 and for gravel-protected BAM 0.025. Thus, the cross sectional area of the BAM-lined channel will be 1.7 times larger than the concrete-lined and the perimeter about 1.5 times greater. Therefore, on average costs, the ratio per foot would be $1.10 $\times$ 1.5: or $1.65 to $3.00, instead of $1.10:$3.00. Other factors affecting cost are discussed by Lauritzen (1960) and US Dep. Interior (1963).

IV.   IRRIGATION PIPE

A.   Reasons for Use

Pipe in a project system costs from 2 to 10 times as much as a lined canal and, for this reason, is used much less extensively. In spite of this cost, there are valid reasons for using pipe systems. One of these is that seepage and evaporation are reduced to almost zero. Other justifying reasons are: reduced maintenance, reduced or eliminated right-of-way use, reduced or eliminated weed seed contamination, topography that renders an open channel unfeasible, requirement for high turnout pressures for on-the-farm sprinkler systems, elimination of severances, and other reasons having to do with land use and public acceptance.

B.   Pipe Types

Pipe types are extremely varied, and each has its strong and weak points. The most widely used material is concrete, because of its extremely long life and maintenance-free characteristics. A simple list of pipes in more-or-less common use are: (i) cast-in-place unreinforced concrete pipe; (ii) unreinforced rubber-jointed, precast-concrete pipe; (iii) reinforced-concrete pipe with either mortar or rubber gasketed joints; (iv) prestressed, precast-concrete noncylinder pipe; (v) pretensioned, precast-concrete cylinder pipe; (vi) steel pipe with mortar or coal-tar linings and coatings; (vii) asbestos-cement pipe; and (viii) plastic pipes. From the above, it can be seen that space does not permit an evaluation of each in this chapter (Amer. Concrete Pipe Ass., 1959; Amer. Water Works Ass., 1964; Shipley, 1957; Rippon, 1962).

C.   Pipe Systems

Pipe systems, in spite of cost, are being used much more generally because of the extreme pressure being made on the available land and water. It seems probable that much more extensive systems will be justified in the near future. For further information, see reference US Dep. Interior (1961). See chapter 42 for additional discussion.

V. EVAPORATION SUPPRESSION

A. Monomolecular Layers

The first step in undertaking evaporation control should be to reduce the area of water surface to the minimum within practical limits. This has always received some attention, but it is entitled to even more. At the present time, the use of monomolecular layers is receiving considerable testing. A film of hexadecanol is placed on the water surface by spraying, spreading from boats or airplanes, or dispersing from anchored containers. As long as the film remains intact, evaporation is considerably reduced in the area covered. However, winds of more than 15 miles/hr force the layer ashore where it is lost and the cover must be reestablished (Crow, 1964; US Dep. Interior, 1959, 1962; Anonymous, 1960; Cruse and Harbeck, 1960; Myers, 1965).

B. Floating Plastic Film Covers

Evaporation on small reservoirs has been controlled by as much as 70% with floating polyethylene covers. Difficulty was experienced, however, with keeping the covers in place. In laboratory tests, foamed polyethylene, with an apparent specific gravity of less than 0.6, or a two-layer quilted structure, reduced evaporation approximately in proportion to the area covered (US Senate, 1962b).

C. Storage Bags

Complete control of storage losses from small volumes of water is readily accomplished by the use of collapsible bags, such as those shown in Fig. 60–5 (US Senate, 1962b). The bag is fabricated from light-gauge material not designed to withstand pressure. Because of this, it can be employed only in an invert of the same approximate dimension. Little is known about the life and maintenance requirements of these bags. Assuming that damage can be controlled, a bag is a relatively economical method of controlling both seepage and evaporation.

Fig. 60–5. Water storage bag, 50,000-gal capacity, controls both seepage and evaporation.

LITERATURE CITED

American Concrete Pipe Association. 1959. Concrete pipe handbook. Chicago. 442 p.

American Society of Civil Engineers, 1934. Subcommittee on Evaporation, Special Committee on Irrigation Hydraulics. Standard equipment for evaporation stations. In Symposium on evaporation from water surfaces. Amer. Soc. Civil Eng. Trans. 99:671–747.

American Water Works Association. 1964. Steel pipe design and installation. Steel Pipe Manual, M–11. New York. 260 p.

Anonymous. 1960. World-wide survey of experiments and results on the prevention of evaporation losses from reservoirs. Int. Comm. Irrig. Drainage. (New Delhi, India). 74 p.

Asbeck, Baron W. F. van. 1957. Asphalt irrigation canal linings. Int. Comm. Irrig. Drainage, 3rd Congr., Proc. 7:1–16.

Beers, W. F. T. van. 1958. The auger-hole method, a field measurement of the hydraulic conductivity of soil below the water table. Int. Inst. Land Reclamation and Improvement. (Wageningen, Netherlands) Bull. 1. 32 p.

Bouwer, H. 1961. A variable head technique for seepage meters. J. Irrig. Drainage Div., Amer. Soc. Civil Eng. 87(IR 1):31–44.

Bouwer, Herman. 1963. Application of seepage meters. Seepage Symp., Proc. (Phoenix, Arizona) US Dep. Agr., Agr. Res. Serv. 41–90. p. 77–81.

Chowdhry, Abdul Hamid. 1957. Canal linings as practices in Pakistan. Int. Comm. Irrig. Drainage, 3rd Congr., Proc. (San Francisco, California) 7.191–7.204:191–204.

Crow, F. R. May 1964. Wind effects on chemicals for reducing evaporation from small reservoirs. Dep. Agr. Eng., Oklahoma State Univ., Stillwater. 37 p.

Cruse, R. R., and G. E. Harbeck, Jr. 1960. Evaporation control research, 1955–1958. US Geol. Surv. Water Supply Pap. 1480. 45 p.

Dirmeyer, R. D., Jr. January 1962. Progress report of clay-sealing investigation during 1961. Civil Eng. Sect., Colorado State Univ. Exp. Sta. (Fort Collins, Colorado) 8 p.

Kaufman, Warren J., and Gerald T. Orlob. May 1956. Measuring ground water movement with radioactive and chemical tracers. J. Amer. Water Works Ass. 48:559–572.

Lauritzen, C. W. May 1957. Canal and reservoir lining materials. Int. Comm. Irrig. Drainage, 3rd Congr., Trans. (San Francisco, California) Rep. 16, Quest. 7. p. 7.245–7.257.

Lauritzen, C. W. 1960. Linings for irrigation canals. Irrig. Eng. Maintenance. (2 parts). Dec. 1959. p. 10–11. Jan. 1960. p. 12–13, 21.

Linsley, Ray K., Max A. Kohler, and Joseph L. H. Paulhus. 1958. Hydrology for engineers. McGraw-Hill, New York. p. 113.

Myers, L. E. 1965. Evaporation retardants: Application by means of a water soluble matrix. Science 148:70–71.

Rasmussen, W. W., and C. W. Lauritzen. May 1953. Measuring seepage from irrigation canals. Agr. Eng. 34:326–329, 331.

Rippon, F. E. December 1962. General design considerations for canals and canal structures. US Dep. Interior. Bur. of Reclamation. (Denver, Colorado) 57 p.

Robinson, A. R., and Carl Rohwer. September 1959. Measuring seepage from irrigation channels. US Dep. Agr. Tech. Bull. 1203. 82 p.

Robinson, T. W. 1958. Phreatophytes. US Geol. Surv. Water Supply Pap. 1423. 82 p.

Robinson, T. W., and A. I. Johnson. 1961. Selected bibliography on evaporation and transpiration. US Geol. Surv. Water Supply Pap. 1539–R. Washington, D.C. 25 p.

Rohwer, Carl, and Oscar Van Pelt Stout. 1948. Seepage losses from irrigation channels. Colorado Agr. Exp. Sta. Colorado A & M Coll. Tech. Bull. 38. 100 p.

Shipley, H. Nov. 1957. Cast-in-place pipe for irrigation. Western Const. 32:48, 50, 52.

South, G. P. Oct. 1955. Canal linings. Asphalt in Hydraulics, 1st Western Conf. Utah Eng. Exp. Sta. Bull. 78. p. 103–112

US Dep. Interior. Bur. of Reclamation. June 1959. Water loss investigations: Lake Hefner, 1958, evaporation reduction investigations. (Denver, Colorado) 131 p.

US Dep. Interior. Bur. of Reclamation. 1961. Design standards No. 3, canals and related structures. (Denver, Colorado) Ch. 1, 30 p. Ch. 6. 10 p.

US Dep. Interior. Bur. of Reclamation. 1962. Water loss investigations: Lake Cachuma, 1961, evaporation reduction investigations. Chem. Eng. Lab. Rep. S1–33. (Denver, Colorado) 81 p.

US Dep. Interior. Bur. of Reclamation. 1963. Linings for irrigation canals. (1st ed.) Sup. of Doc., Washington, D.C. 134 p.

US Senate Commission on Public Works. 1962a. Study and investigations of use of materials and new designs and methods in public works. 87th Congr. Comm. Print 5. 61 p.

US Senate Commission on Public Works. 1962b. New materials and methods for water resource management. 87th Congr. Comm. Print 6. 71 p.

US Senate Select Commission on National Water Resources. 1960. Water resources activities in the United States—evaporation reduction and seepage control. S. Res. 48. 86th Congr. 1st Sess. Comm. Print 23. 18 p.

Warnick, C. C. Feb. 1963. Problems in seepage evaluation and control. Seepage Symposium, Proc. (Phoenix, Arizona) US Dep. Agr. ARS 41–90. p. 132–136.

61

Improving Irrigation Efficiencies[1]

MARVIN E. JENSEN

Agricultural Research Service, USDA
Kimberly, Idaho

LAWRENCE R. SWARNER

Bureau of Reclamation, US Department of the Interior
Boise, Idaho

JOHN T. PHELAN

Soil Conservation Service, USDA
Washington, D.C.

I. EVALUATING IRRIGATION EFFICIENCY[2]

The term "irrigation efficiency" has been used extensively during the past 30 years to express the performance of a complete irrigation system or components of a system. Though specifically defined by its users, on occasion, irrigation efficiency is not rigidly defined and has many interpretations. The term is frequently modified to assure specific interpretation such as "water application efficiency," but consistency in the use of modified terms is also lacking. Before considering techniques for improving irrigation efficiencies, terminology involved and the factors affecting irrigation efficiency will be discussed.

A. Terminology

Irrigation was defined by Israelsen (1950) as the artificial application of water to soil for the purpose of supplying water essential to plant growth. He also stated that irrigation is essentially a practice of supplementing natural precipitation for the production of crops. This definition generally has been accepted with minor modifications. However, a quantitative definition of essential water for plant growth is lacking. Irrigation, with the exception of subirrigation, is usually not a continuous process, but the application of water to soil after the soil water has been depleted to some level. Numerous studies have shown that soil water can be depleted to specific energy levels, depending on the crop and root zone depth, before the yield or quality of the crop or both are materially affected. The allowable energy level is an additive function of mechanical energy (soil water suction) and chemical energy (osmotic pressure). Deliberately permitting depletion of

---

[1] Joint contribution from the Soil and Water Conserv. Res. Div., ARS; the Eng. Div., SCS, USDA; and the Div. Irrig., Bur. of Reclamation, USDI.
[2] Part I of this chapter was written by Marvin E. Jensen.

soil water to prescribed energy levels is practiced to stimulate or retard vegetative plant growth to obtain the most desired marketable product from a crop.

The effect of salts in the soil solution on plant growth is an important factor in arid areas. Thorne and Peterson (1954) indicated that any attempt to control soil water without recognizing osmotic pressure of the soil solution might fail entirely. The soil water energy level is the major controlling factor in the availability of soil water to plants and is a function of both soil water suction and osmotic pressure. Therefore, if irrigation is for the purpose of supplying water essential for plant growth, it must maintain not only favorable soil water levels, but also a favorable salt concentration in the soil solution.

The term "efficiency" is used in many ways, e.g., (i) as an index of performing a task with a minimum of waste effort, and (ii) as a ratio of the results actually obtained from an operation compared to results that theoretically could be obtained. The latter definition is used extensively in engineering such as the ratio of energy output from a machine to energy input. This definition provides a numerical value that has direct and useable applications in engineering design and will be used in defining the following terms.

Irrigation efficiency (E_i) is the ratio, usually expressed as per cent, of the volume of the irrigation water transpired by plants, plus that evaporated from the soil, plus that necessary to regulate the salt concentration in the soil solution, and that used by the plant in building plant tissue to the total volume of water diverted, stored, or pumped for irrigation.

Today, water returned to the atmosphere by transpiration from plants, evaporation from the soil, and water used in building plant tissue is accepted as the basic water requirement (consumptive use or essentially evapotranspiration). Irrigation water requirement generally used in computing irrigation efficiency is evapotranspiration minus effective rainfall. It should, however, also include the leaching requirement. The current concept of effective rainfall for most field crops includes all light showers. However, a light rain shower may increase evapotranspiration from a crop having a partial canopy of vegetation, or a crop showing signs of water stress, for 1 to 2 days because of increased evaporation. Under these conditions, light showers may not reduce the irrigation requirement. Effective rainfall is total rainfall minus deep percolation that may occur during heavy rains or when rain follows a thorough irrigation. In practice, rainfall runoff is often overlooked. When evaporation from the soil can be economically prevented, evaporation may some day be eliminated as part of the basic water requirement used in computing irrigation efficiency.

The definition of irrigation efficiency given above is affected by all losses of water that occur after the water in a natural stream or aquifer is controlled or removed specifically for irrigation purposes. As with all efficiency terms, the theoretical maximum efficiency is 100%. Sustained operation of an irrigation project in an arid area maintaining high crop yields with an irrigation efficiency of 100% would not be theoretically possible unless water necessary to control salts in the soil solution is included in the numerator of the efficiency term. Water necessary to control salts cannot logically be considered waste when computing irrigation efficiencies because this water is beneficially used. Israelsen (1950) stated that irrigation must provide water for growth of crops and at the same time allow enough water to pass through the soil to leach out excess salts.

The definition of irrigation efficiency as presented differs from that given by

Israelsen (1932, 1950), Myers (1955), US Dep. Agr. (1954), and Thorne and Peterson (1954), by including in the numerator the water necessary to maintain a favorable salt concentration in the soil solution, and including in the denominator the water losses that occur when water is stored in a reservoir for irrigation. Dividing the volume of irrigation water used in evapotranspiration and building of tissue plus that amount necessary for salt control per unit area of land by the irrigation efficiency, expressed as a decimal, gives the volume of water per unit area that must be available for storage or for direct diversion for irrigation. This definition is applicable to any size project for any specified period of time, and would theoretically sustain permanent irrigation agriculture, if continuous operation at 100% efficiency were possible.

Major irrigation projects generally store water for later distribution to tracts of irrigated land. However, many smaller projects may divert directly from a natural stream, or pump from an aquifer directly onto the land. Separation of the various components of irrigation efficiency is necessary to evaluate the efficiency of segments of the entire irrigation system. Efficiency terms for segments of the system are defined below beginning with the reservoir.

Reservoir storage efficiency (E_s) is the ratio, usually expressed as per cent, of the volume of water delivered from the reservoir for irrigation to the volume of water delivered to the storage reservoir, surface or underground, for irrigation.

Water conveyance efficiency (E_c) is the ratio, usually expressed as per cent, of the volume of water delivered by an open or closed conveyance system to the volume of water delivered to the conveyance system at the supply source or sources.

Water application efficiency (E_a) is the ratio, usually expressed as per cent, of the volume of irrigation water used in evapotranspiration in a specified irrigated area, plus that necessary to maintain a favorable salt content in the soil solution, to the volume of water delivered to this area.

Reservoir storage efficiency and water conveyance efficiency terms have been in general use for many years (Israelsen, 1932). Water application efficiency, pertaining only to the water actually stored in the soil, was defined earlier by Israelsen (1932, 1939, 1950) and has been used extensively, although occasionally interchangeably with irrigation efficiency. The major reason for including the leaching requirement in E_a is because, as defined by Israelsen, water application efficiency of 100% could not be theoretically obtained, if salt control by leaching were necessary for permanent agriculture. Other authors have expressed a need to define E_a as given. Reeve (1957) indicated the desirability to redefine the term so as to represent operational procedures that are essential in providing a soil environment favorable to a crop in respect to both water and salinity. Hall (1960) also defined water application efficiency to include the volume applied for intentional leaching.

The preceding efficiency terms may be applied to any size project, or segment thereof, for any specified period of time. For clarity and comparative purposes, all efficiency values reported or used should be identified as to the size of the unit, the period of time or number of irrigations involved, the adequacy of irrigations, and the computational procedure used in obtaining the efficiency value. Willardson (1960) presented several terms that, in essence, limit the defined efficiency terms to specific components of an irrigation project. Hall (1960) also presented several parameters that may be used for evaluating irrigation system performance.

Uniformity in definition and measurement of efficiencies still makes rigid comparisons of the capabilities or potential efficiencies of similar systems difficult, because of the human element involved in the operation of the system. Variation in operational procedures can cause marked differences in irrigation efficiencies of identical systems. Reliable evaluation of basic differences in system performances, such as between surface and sprinkler systems, can be made only when systems are operated to give the same adequacy of irrigation over the same percentage of the field, and when operated as designed.

B. Factors Affecting Irrigation Efficiencies

1. COMPONENTS OF EFFICIENCY TERMS

The efficiency of individual components of an irrigation system should be so defined and computed so that the product of the component efficiency terms, expressed as ratios, gives the over-all irrigation efficiency for the area considered. Terms describing uniformity and adequacy of an irrigation should not be labeled as efficiency terms, if the product of all such terms does not give over-all irrigation efficiency. Irrigation efficiency should be the product of E_s, E_c, and E_a when expressed as ratios. The relationships of component efficiency terms, expressed as per cent, are described below:

$$E_i = \frac{E_s}{100} \frac{E_c}{100} \frac{E_a}{100} \times 100 = \frac{W_{et} + W_l - R_e}{W_i} \times 100 \qquad [61\text{--}1]$$

where

$E_i =$ irrigation efficiency, per cent,

$W_{et} + W_l - R_e =$ the volume of irrigation water per unit area of land transpired by plants and evaporated from the soil under favorable soil water levels, plus that necessary to regulate the salt content of the soil solution minus effective rainfall (Irrigation Water Requirement), and

$W_i =$ the volume of water per unit area of land that is stored in a reservoir or diverted for irrigation.

$$E_s = (W_s/W_{so}) \times 100 \qquad [61\text{--}2]$$

where

$E_s =$ reservoir storage efficiency, per cent,

$W_s =$ volume of water delivered from the reservoir for irrigation, and

$W_{so} =$ volume of water delivered to the reservoir to be stored for irrigation.

$$E_c = (W_c/W_{co}) \times 100 \qquad [61\text{--}3]$$

where

$E_c =$ water conveyance efficiency, per cent,

$W_c =$ volume of water delivered by the conveyance system, and

$W_{co} =$ volume of water delivered to the conveyance system, at the source of supply (where reservoir storage is used, $W_{co} = W_s$).

$$E_a = [(W_{et} + W_l - R_e)/W_a] \times 100 \qquad [61\text{--}4]$$

where

E_a = water application efficiency, per cent,

W_{et} = volume of irrigation water in a specified area transpired by plants and evaporated from the soil under desirable soil water levels,

W_l = volume of water necessary for leaching (salt control) in the given area,

R_e = volume of effective rainfall in the given area, and

W_a = volume of water applied to the given area (where a main conveyance system is used, $W_a = W_c$).

Differences in opinion arise as to whether computed water application efficiency for an irrigation should be less if only a portion of the total root zone water storage capacity were filled during the irrigation. For example, assuming no leaching requirement, if a soil could hold a 15-cm irrigation but only 5 cm of water were applied and all of it stored, should the computed water application efficiency be 100% or 33%? Obviously if only 5 cm were necessary to mature a crop, and heavy rainfall was anticipated before the next season, the 5-cm application would be adequate and the water application efficiency should be 100%, if the water was uniformly applied. One attempt to circumvent this difference in opinion was presented by Myers and Haise (1960). Water application efficiency was defined as $(W_n \, W_{et})\ 100/W_a^2$ or the product of the ratios: water needed to water applied, and water stored in the soil for evapotranspiration to water applied. Water needed may, or may not, be the amount required to fill the root zone storage capacity. Hansen (1960) proposed a term called water storage efficiency, referring to storage in the soil. It was defined as the ratio of water stored in the root zone during an irrigation to water needed to fill the root zone prior to the irrigation; it is useful in evaluating the operation of an irrigation system. However, it is possible to irrigate more frequently and produce an excellent crop, but never filling the potential root zone more than one-half its capacity each time. Therefore, this term should not be labeled efficiency because a calculated average water storage efficiency, in this example of 50%, would not necessarily mean that twice as much irrigation water as necessary is being delivered. Likewise, a light irrigation may be given to cover a field in a short time with more water applied during the next irrigation to bring the soil to its full capacity. This would not necessarily increase the gross water requirement, though this would be implied if the soil water storage efficiency term were used. The degree of actual storage obtained is an indication of the adequacy of an irrigation and does not necessarily affect the efficiency of the system.

Similarly, terms have been devised to describe the uniformity of irrigation. A term similar to the uniformity coefficient proposed by Christiansen (1942) has been used: $C_u = 100\ [1.0 - (\Sigma x/nM)]$, where x is the deviation of individual observations (absolute values) from the mean M, and n is the number of observations.

Hansen (1960) proposed a similar term and called it water distribution efficiency. However, this term also should be labeled uniformity coefficient as originally intended because the additional water to be delivered to assure adequate irrigation over 75% of the area is not necessarily obtained by dividing the amount of irrigation water desired by this uniformity coefficient as proposed. For example, if 20% of the area received 80% of the average water applied and the other 20% increments received 90, 100, 110, and 120% of the average, the uniformity coefficient would be:

$$100 \left[1.0 - \left(\frac{0.20 + 0.10 + 0 + 0.10 + 0.20}{(1.0)\ 5} \right) \right] = 85\%.$$

Dividing the amount of irrigation water to be applied by this coefficient, expressed as a fraction, would mean that 18% more water would be applied, but if the same distribution pattern persisted, 20% of the area would receive 98% of the original mean depth intended to be applied, and 80% would receive 108 to 138% of the original mean depth. In an example presented by Hansen (1960) all areas received adequate irrigation but some areas received excessive irrigation. The distribution efficiency of 80% could not be used as proposed, since adequate water was already being applied to all areas.

When the distribution pattern of an irrigation system is known with either a sprinkler system or a surface system, the distribution pattern can be used to evaluate the adequacy of an irrigation or provide a numerical value that can be used to adjust the duration of an irrigation to obtain the desired adequacy of irrigation. This procedure is illustrated in Fig. 61–1. For example, first assume that under-irrigation can be tolerated on some arbitrary percentage of the irrigated area, but the remaining area must be adequately irrigated for economic reasons, and the same distribution pattern will exist for all depths applied. The distribution pattern is represented by the solid curve in Fig. 61–1. The distribution coefficient is the ratio of the depth applied at the percentage of the area where inadequate irrigation is barely tolerable to the mean depth applied. Dividing the

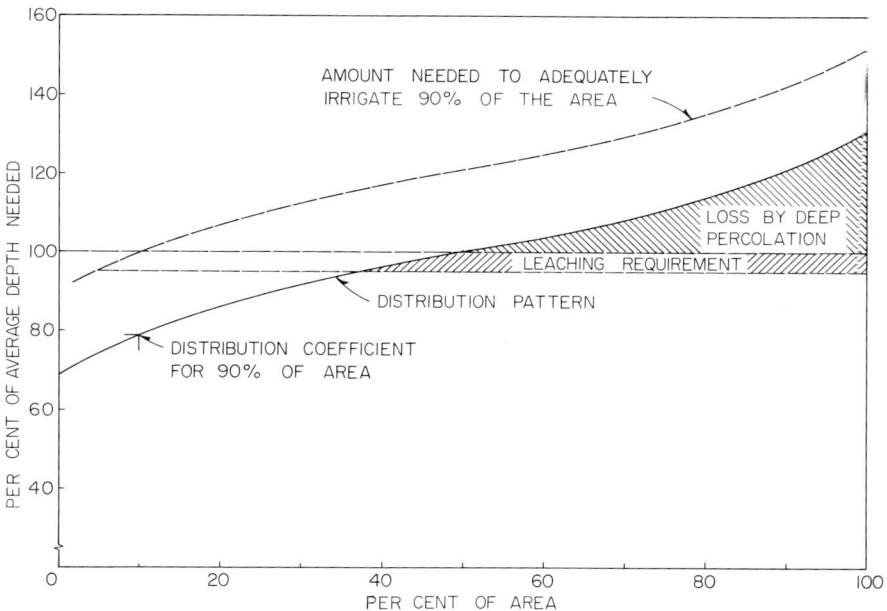

Fig. 61–1. Illustration using the distribution pattern in evaluating irrigation adequacy and adjusting the amount of water applied to obtain the desired adequacy.

desired depth to be applied at this point by the distribution coefficient gives the mean depth actually needed to assure that all but the percentage of area selected would be adequately irrigated. The percentage of the area on which under-irrigation is tolerable would depend on the economic value and sensitivity of the crop grown to both under-irrigation and over-irrigation. An economic evaluation of the net returns obtained in terms of crop response with the distribution pattern of the system should be the deciding factor in selecting the percentage of the area on which under-irrigation can be tolerated. In areas where leaching is important, maintenance of soil productivity must also be considered.

Distribution patterns for sprinkler systems tend to follow a normal distribution (Hart, 1961). Thus distribution patterns for sprinkler systems can be reproduced knowing only the mean application rate and the standard deviation or variance.

The distribution coefficient would also be applicable, even if all areas were being over-irrigated and the amount being applied needed to be reduced. For example, if the depth of water being applied during an irrigation is indicated by the dashed line in Fig. 61–1 and inadequate irrigation on 20% of the area would give the greatest net returns, then the adjusted amount or duration of irrigation needed would be 1/1.07 or 93.5% of the original amount. The distribution pattern can be assumed to remain the same for a sprinkler system, but may change with depth applied with a surface system.

2. WATER LOSSES AFFECTING IRRIGATION EFFICIENCIES

Irrigation efficiency is affected by evaporation from water surfaces in reservoirs and conveyance channels, transpiration by nonbeneficial riparian vegetation along reservoirs and channels, seepage losses in reservoirs and conveyance channels, deep percolation losses in fields, and unavoidable operational waste. The magnitude of these losses varies widely among irrigation projects because of the different physiographic features, water control and conveyance structures, and management practices.

3. MEASUREMENT AND CALCULATION PROCEDURES

a. **Water Application Efficiency and Farm Delivery Requirement.** When evaluating water application efficiencies on existing projects, one of two methods is generally used. For individual irrigations, the soil water content is measured at representative points throughout the field prior to, and 2 or more days after, an irrigation. If the leaching requirement is zero, the water application efficiency of the irrigation is calculated from the volume of water stored per unit area, or the mean depth stored adjusted for normal evapotranspiration between sampling dates, and the mean depth applied:

$$E_a = [(W_2 - W_1 + nE_t)/W_a] \times 100 \qquad [61\text{--}5]$$

where

W_2 and $W_1 =$ mean depths of water in the soil after and before irrigating,

$\qquad n =$ number of days between sampling dates,

$\qquad E_t =$ consumptive use or evapotranspiration rate between sampling dates expressed as depth of water used per day, and

$\qquad W_a =$ mean depth of water applied.

The evapotranspiration rate should be either an estimate for the period between sampling dates or the average of the rate measured before and after irrigating.

Water application efficiencies computed in this manner only approximate water application efficiency because deep percolation losses do not cease in 2 to 3
days after an adequate irrigation. Consequently, water application efficiencies will
vary, depending on the time of sampling after the irrigation. For example, on a
bare, sandy loam soil at Prosser, Washington, USA, the mean total depth of water
in the soil profile 5, 10, and 20 days after an irrigation with evaporation prevented was 31.5, 30.2, and 28.2 cm, respectively (D. E. Miller. Moisture retention in synthetic soil profiles. SWC, ARS, USDA, Irrigation Exp. Sta., Prosser,
Washington. *Personal correspondence*). If there were 21.3 cm of water in the
profile before irrigating and 12.7 cm of water was applied, the computed water
application efficiency would have been 80%, 70%, and 54%, respectively, depending on when the after-irrigation samples were taken. The decrease in soil
water with time was caused by unsaturated flow from the profile due to gravity.
When evapotranspiration is occurring, the rate of unsaturated flow from the soil
profile decreases more rapidly because of more rapid withdrawal of water from
the upper layers.

Suitable techniques are not available to easily evaluate the actual amount of
water that drains through the soil profile, as compared with the amount required
for leaching purposes. This is one reason why leaching requirement generally
has not been considered in water application efficiency studies. Reported water
application efficiencies with present measurement procedures, therefore, are only
approximations because of the difficulty in evaluating water needed and used for
leaching, and because slow drainage from the profile is difficult to evaluate. More
comparable water application efficiencies could be determined for individual irrigations, if uniformity in time of sampling after irrigation was used, such as when
gravity drainage decreased to less than a fixed rate of 0.5 or 1 mm/day or 0.1
of the mean evapotranspiration rate. The most accurate comparisons of water
application efficiencies for various systems can be made only when total deep
percolation losses between irrigations are known. Thus, water used in evapotranspiration, instead of water stored in the soil as presented in equation [61–3], is
essential for accurate water application efficiencies.

Measurement of water application efficiencies for individual irrigations is time
consuming and expensive. The second method of evaluating water application
efficiencies involves estimating field or farm irrigation efficiency which includes
conveyance and application losses on the farm or field. Estimated farm irrigation
efficiency is the ratio, expressed as per cent, of estimated evapotranspiration and
leaching requirement on the farm minus effective rainfall to irrigation water
delivered to the farm:

$$\text{Estimated farm or field irrigation efficiency} = \frac{(E_t + W_l - R_e)\,100}{W_d} \qquad [61\text{--}6]$$

where
E_t = estimated volume of evapotranspiration,
W_l = volume of water required for leaching,
R_e = volume of effective rainfall, and
W_d = volume of irrigation water delivered.

These estimates are reasonable and useable when good water management is practiced. However, when poor irrigation practices are involved, such as inadequate irrigations on part of the farm or field, the estimated evapotranspiration may be too high, resulting in apparent farm or field irrigation efficiencies that also are too high. Farm or field irrigation efficiency should be based on measured evapotranspiration whenever possible, or the estimated consumptive use should be adjusted for actual soil water levels maintained, actual duration of growing seasons, and adequacy of the irrigations.

Most studies evaluating irrigation efficiencies by estimating evapotranspiration have not included the leaching requirement in the numerator of equation [61–6]. If leaching requirement were included, farm irrigation efficiency values reported would be higher in many areas.

Efficiency studies of this type should be interpreted cautiously. For example, if an annual water allotment is involved, the farmers tend to take their full allotment, thus predetermining estimated farm irrigation efficiency.

b. Water Conveyance Efficiency and Diversion Requirement. Water conveyance efficiency is easily determined, if water delivered to farms, known operational waste, and the water diverted from a stream or released from a reservoir to the channel is measured. Reliable measurements are more difficult with short reaches of conveyance channels because of the difficulties encountered in accurately measuring a small difference in large flows between two points. Diversion requirement is the farm delivery requirement plus necessary operational waste divided by the conveyance efficiency expressed as a fraction.

c. Irrigation Efficiency and Storage Requirement. The over-all irrigation efficiency is the product of reservoir storage efficiency, water conveyance efficiency, and farm irrigation efficiency. Evapotranspiration plus leaching requirement, minus effective rainfall on the irrigated area of the project, plus operational waste, divided by the over-all project irrigation efficiency gives the total storage requirement. Over-all irrigation efficiency is often quite low. For example, assuming no leaching requirement, farm irrigation efficiency with gravity systems often averages 45 to 55% and conveyance efficiency 70 to 75%, depending on the type of channels. Reservoir storage efficiency can be quite high, if underground storage or deep, tight surface reservoirs are involved. Assuming a 90 to 95% range in reservoir storage efficiency, the over-all project irrigation efficiency would range from 28 to 39%. Often canal seepage and runoff from farms can be collected and brought back to the conveyance system, thereby increasing the over-all irrigation efficiency of a project.

4. RIVER BASIN CONSIDERATIONS

Several projects are frequently supplied by a single reservoir or diversion structure. If canal and reservoir seepage, deep percolation losses, and runoff from farms can be collected and rediverted to another project in the river basin, the over-all river basin irrigation efficiency can be greater than in any single project. Likewise under these conditions, the need for extremely high efficiencies on a given project may not be necessary. However, low project irrigation efficiency means that the irrigation conveyance system must have a larger capacity, which increases construction costs. Also, only a portion of the runoff, deep percolation, and canal and reservoir seepage can be collected and reused in a river basin because of evapotranspiration losses on waste or non-cropland and transpiration from vegetation bordering channels.

The quality of the water collected for reuse is often impaired. The evapotranspiration process on both crop and wasteland sends salt-free water into the atmosphere. Dissolved solids in irrigation water become more concentrated in the seepage collected and reused in projects further downstream.

II. IMPROVING PROJECT IRRIGATION EFFICIENCIES[3]

Project irrigation efficiency encompasses water storage efficiency (when storage is involved), water conveyance efficiency, and farm irrigation efficiency. Although project irrigation efficiency may vary considerably during the irrigation season it is commonly considered as a seasonal value.

The water supply source affects project irrigation efficiency. The water supply for many irrigation projects is stored in a reservoir during periods of excess flows for later release and distribution to the irrigable lands of the projects. Other projects secure their water from natural flow by direct diversion from a stream or by pumping from an underground aquifer. Storage and conveyance losses may not be involved in the latter project efficiencies.

A. Current Project Irrigation Efficiencies

Project irrigation efficiencies vary widely from area to area. A recent study, using information secured on 21 selected Bureau of Reclamation projects in the 17 Western States, USA, indicated that the average project water conveyance efficiency for the years 1949 to 1960, was 63.1%, but ranged from 47.5 to 82.7%, US Dep. Interior (1962). Reservoir storage efficiencies for these projects were not available. Average farm irrigation efficiencies for these projects for the same years ranged from 32.3 to 78.2%, with an average of 59.3%.

There is almost universally a correlation between abundance and/or cost of water and project irrigation efficiency. Where water is scarce or high in cost, the efficiencies are higher. Conversely, where water is abundant and/or low in cost, the efficiencies are lower. Thus, in a sense, economics play a major role in existing project irrigation efficiencies. Project management, as well as farm management, involves balancing the immediate cost of water against the higher labor and investment costs required to use it more efficiently. In many cases, the true costs of using excess water are not recognized immediately but may be reflected in reduced yields due to leaching of plant nutrients, reduced yields caused by accumulation of soluble salts or exchangeable sodium, or in extra drainage installations which will be required later to control rising water table levels.

---

[3] Part II of this chapter was written by Lawrence R. Swarner.

B.   Factors Affecting Project Efficiencies

1.   NATURE OF LOSSES

The losses and wastes in storage reservoirs and in the conveyance and distribution system to the individual farm occur as seepage, evaporation, consumptive use, and operational losses and wastes. These losses and wastes vary with the type and design of the irrigation project. Many irrigation projects store water in a mountainous area during the winter. During the irrigation season, water is released at the reservoir, either directly into a distribution system or into the natural stream channel from which it may be diverted into a distribution system downstream.

Some projects merely divert or pump natural flows from a stream into a distribution system. On these projects the distribution channels may be excavated through natural material and left unlined, or they may be earth, concrete, asphalt, or plastic lined. Other systems deliver water in concrete, metal, or transite pipe for gravity irrigation, or adequate pressure may be provided for sprinkler irrigation. Because opportunity for seepage, evaporation, and operational wastes and losses are different in each distribution system, project irrigation efficiencies vary widely.

Some reservoirs or dams serve multipurpose functions and, in addition to irrigation, regulation for flood control, power generation, recreation, and fish and wildlife requirements, must be considered and met, insofar as possible. All water required for multipurpose operations should not be charged to irrigation. For this reason, care must be exercised in delineating and explaining project irrigation efficiencies, where multipurpose reservoirs are involved.

2.   RESERVOIR STORAGE LOSSES AND WASTES

Seepage from reservoirs reduces reservoir storage efficiencies. In the selection of a reservoir site, the permeability of the soil, or earthen mantle, covering the reservoir area is evaluated. In some instances where high permeabilities are found, either compacting the earth or applying a compacted earthen blanket may effectively decrease the seepage losses. Polyethylene or vinyl film has been used to line smaller reservoirs, but is currently considered to be too expensive for large reservoirs.

When a reservoir is filled, some of the water is absorbed by the bank of the reservoir. When the reservoir water level is lowered, water drains from the bank into the reservoir. This water is referred to as bank storage and may amount to a sizeable volume in a large reservoir. For example, inflow-outflow measurements for Lake Mead on the Colorado River, USA, (capacity 3.85×10^{10} m^3 (3.125×10^7 acre-ft)) show the bank storage to be about 4.1×10^9 m^3 (3.3×10^6 acre-ft) when the lake is filled to capacity (Langbein, 1960). If a portion of the bank storage does not return to the reservoir because of use by nonbeneficial riparian vegetation, reservoir storage efficiency will be reduced.

A small amount of leakage may occur through nearly every dam, especially if it is an earthfill dam. This water generally finds its way back into the channel below and, although it may lower the reservoir efficiency insofar as that reservoir is concerned, it may be recovered at a lower elevation for irrigation or be put to some other beneficial use.

3. LOSSES IN PROJECT SYSTEMS

One of the greatest causes of water losses from a reservoir area, or a canal distribution system, is the phreatophytes which in general are nonbeneficial water-using plants that transpire large volumes of water annually. Salt cedar (*Tamarix gallica*) is the number one offender in the Southwest, USA. It is not uncommon for its water consumption to reach 1.5×10^4 m^3/ha (5 acre-ft/acre). Under ideal conditions of growth and density it may exceed 2.7×10^4 m^3/ha (9 acre-ft/acre). In the 17 Western States, USA, it is estimated that 6.1×10^6 ha (1.5×10^7 acres) are infested with phreatophytes, consuming 3.1×10^{10} m^3 (2.5×10^7 acre-ft) annually.

Studies made in the Rio Grande Basin, USA, indicated that, prior to channelization, there were approximately 3,000 ha (75,000 acres) of salt cedar, cottonwoods (*Populus* sp.), willows (*Salix* sp.), Russian-olive trees (*Eloeagnus augustifolia* L.), and other water-consuming vegetation in, and adjacent to, the river channel and flood plain. About 3.7×10^8 m^3 (3.0×10^5 acre-ft) of water, or about one-third of the normal annual flow of the river at the head of the irrigated lands, was used nonbeneficially by phreatophytes (Hill, 1963). The eradication of the phreatophytes is an obvious remedy, although a difficult one to apply. Improved chemical sprays, as well as improved mechanical devices, have aided in the eradication, especially when used in conjunction with the improved channelization of water courses. Deliberate lowering of the groundwater table beyond the reach of the phreatophyte roots has been used successfully in Utah for shallow-rooted phreatophytes.

The US Geological Survey estimates the average annual evaporation from freshwater bodies in the 17 Western States, USA, at over 2.8×10^{10} m^3 (2.3×10^7 acre-ft). To meet the growing demand for fresh water, all known losses are being reevaluated to increase the effectiveness of water resource management. One of the most promising developments in the reduction of evaporation losses from reservoir or lake surfaces, which does not interfere with multipurpose usage of a body of water, is the use of hexadecanol and octadeconal formulations. These chemicals form a monomolecular film over the water surface which provides an invisible, but effective, barrier to evaporation from the water surface, if heavy weed or algae growth does not exist. The application and distribution problems have been difficult on large bodies of water, or where the water surface is commonly subjected to violent wave action. Work is continuing on the application and distribution methods (US Dep. Interior, 1963). It appears possible to cut evaporation losses one-third or more by this method of evaporation control.

Seepage losses from canal and distribution systems, and resulting damage from waterlogging and sodium or salt accumulation, may be reduced by canal lining. In the USA, the extensiveness of any canal lining program depends on economic feasibility after considering many factors, including value of land protected from seepage, value of water, increased capacity available in lined canals, and lower maintenance costs. Considerable work is being done on the development of low-cost canal linings and is fully discussed in chapter 60.

Other losses occurring in conveyance and distribution systems are caused by leakage at canal gates. Even though high maintenance standards are adhered to, some leakage through the control structures is inevitable and the distribution of water throughout a project area will necessitate some unavoidable operational wastes. Operational wastes of 5% are considered reasonable.

As a general rule, low project irrigation efficiencies are considered to be undesirable; however, there are instances where the losses and wastes, which occur in one system, are not harmful and are recovered and utilized as return flows providing irrigation water on another project. In such cases, it would be false economy on a basin basis to expend funds to reduce these harmless wastes. However, if operational costs for the upper project can be materially reduced and the two projects are operated independently, then reduction of wastes in the upper project may be economically feasible.

4. METHODS OF WATER DELIVERY

The method of water delivery has a pronounced effect on project irrigation efficiency. Application of irrigation water must be closely adjusted, both to the requirements of the crop and to the available water-holding capacity of the soil root zone, if satisfactory production is to be obtained, losses of both soil and water on the farm are to be held to a minimum, and the needs of the farmer are to be served. Attainment of these conditions calls for flexibility, rather than rigidity, of water deliveries through the project distribution system. Flexibility is especially important for satisfactory operation of an irrigation system in areas such as Nebraska and Kansas, USA, where intense precipitation storms occur. The occurrence of these storms requires sudden and frequent changes in water deliveries. The system must be constructed to permit flexible operation because alternate operational procedures must frequently be used.

Three distinct methods of delivery of irrigation water are commonly recognized in the USA: demand, rotation, and continuous flow. Seldom, if ever, is all of the irrigation water delivered on a project strictly according to any one of these methods. Rather, modifications or combinations of two or all three are used at various times or in various locations as needed.

a. The Demand System. The demand system involves the delivery of water to the farms at times, and in quantities, as requested by the water user. It is ideal from the water user's point-of-view, as it enables him to irrigate each crop when irrigation is needed and to use a stream size that he finds to be most economical and efficient. This system of delivery offers many opportunities for a project to encourage wise use of water and generally results in higher project irrigation efficiencies. Demand deliveries require an alert, ingenious and flexible operational organization capable of matching daily supply with demands. Because it is not economically feasible to design unlimited capacity in the canal and laterals, the water user's demand may exceed project capacity during the peak of the irrigation season. If so, a change to one or both of the other systems of water delivery may be necessary.

b. The Rotation System. The rotation system of water delivery is probably the most flexible of all methods, since it can be varied greatly. Rotations may be made between two water users, two or more groups of water users under a single lateral, two or more different laterals, or between definite divisions of the whole project canal system. Although local conditions will determine which kind of rotation is most applicable, in all probability the divisions of the canal system or the size of the groups of the water users, between which rotation is practiced, will be varied throughout the irrigation season to secure the most economical water distribution. Under the rotation method, water is delivered to each user in sufficient quantity for a fixed period of time under a prearranged schedule.

Under careful management good project irrigation efficiencies can be secured under the rotation method, but the fixed schedule makes it impossible for a water user to delay his irrigation even a few days. This would be possible, if he received his water under the demand system. To forego an irrigation when his rotation period is due would subject his crop to severe water stress before the next rotation period was due.

c. The Continuous Flow System. Under this method each water user receives his share of water as a continuous flow. It reportedly had its start in the USA under the early miner's inch appropriations where users demanded their legal allotment constantly, whether needed or not. Generally speaking, on small tracts or on sandy soils, and when the irrigation stream is small, this method wastes water and time and contributes to waterlogging of the soil. Because of the resulting low farm irrigation efficiencies it should not be used except when extreme conditions render rotation or demand systems impractical.

5. WATER CHARGE SCHEDULE

A project water charge schedule affects project irrigation efficiencies. There are two schedules in general use in the USA: The flat rate charge and the graduated or excess water charge.

a. Flat Rate Charge. Under the flat rate charge schedule each water user pays the same rate either on a hectare (or acre) basis or on a cubic meter (acre-ft) basis. Where the rate is based on the irrigable land owned or operated by the water user, each operator pays the same amount per hectare (or acre) irrespective of how much water he uses per hectare (or acre) during the irrigation season. In many areas water storage supplies may limit the amount of water per hectare (or acre). However, where supplies are adequate and the seasonal water charge is based on irrigable land, it is possible for one water user to secure more water per hectare (or acre) than needed at no extra cost. Under this condition there is no financial incentive for the water user to conserve or make efficient use of his water.

b. Graduated or Excess Water Charge. Under this system an allotment is established for each hectare (or acre) of irrigable land. Unless limited by storage capacity an allotment is the amount of water normally required to produce crops under reasonably efficient irrigation practices. Each water user is required to pay a minimum amount for the water allotment. On some projects the allotment varies, depending on the characteristics of the soils. Coarser textured soils, because of the associated problems of distributing water efficiently, receive more water than finer textured soils. In a few cases all water users are required to pay for a "base amount" which is usually 1,500 m³/ha (0.5 acre-ft/acre) less than the allotment. This provides a monetary incentive to apply water efficiently. The additional 1,500 m³/ha (0.5 acre-ft/acre) which makes up the allotment, may be purchased, if needed, at the same rate as the base amount.

When additional water is needed or desired above the allotment, it may be purchased at a higher rate. This is known as an excess water charge and is generally graduated upward with each additional 3,000 m³ (acre-ft) increment made available. On many projects the first 3,000 m³ (acre-ft) of excess water costs 120% of the allotment rate, the second increment costs 140%, and the third increment or more mosts 160% of the allotment rate. The excess water charge has been found to be very effective in encouraging good farm irrigation practices. In addition to creating an awareness among water users that using water carelessly costs more money, there is the additional incentive in the pride a water user takes "in staying within his allotment."

c. **Methods of Improving Project Irrigation Efficiencies.** Since water for irrigation is a natural resource available to those making application for it and beneficially utilizing it in the USA, charges greater than those incurred in the construction of facilities to bring the water to the land and to operate and maintain these facilities cannot be levied without the consent of the water users. For this reason in an area of abundant and readily available water, the charges will be low and will provide little incentive to use water efficiently. In such cases it is difficult to increase existing project irrigation efficiencies. Project water requirements are generally based on the consumptive irrigation requirement for an anticipated cropping pattern for the area and a reasonable water application efficiency. However, there are some older projects in existence where present water use is unreasonably high. In some of these areas, particularly in Utah, USA, new water allotments have been imposed and upheld by the courts. These were based on estimated consumptive irrigation requirement and a reasonable, but firm, water application efficiency. This action has forced water users to improve their water application efficiencies (Bagley, 1965). There are a number of practices that can be used within the project, or imposed by the district, to improve present efficiencies. These practices are briefly described below.

1. *Water Measurement.* Experience has shown that where water measurement is not practiced throughout a project, irrigation efficiencies are generally very low. On one irrigation project in eastern Oregon, USA where water is measured only at the diversion to the main canal, it is estimated that project irrigation efficiencies range from 20 to 30% (Stammers, 1963). Measuring devices are being installed at individual farm turnouts to provide better water control and management throughout the irrigation system and to insure each water user of his equitable supply. Preliminary observations indicate that the project irrigation efficiency will be increased substantially as a result of these installations. Although there may be a few projects where the abundance of water makes it unnecessary to measure water, more efficient project and farm operations would result from measurement of water. In the future, as water supplies are more fully utilized, water measurement will become more important to farmers and irrigation districts alike. Water measurement will permit equitable water distribution to the farmers of a project area and will allow the project management to properly regulate and control the water throughout the irrigation system.

2. *Modification of Delivery Schedule.* On many projects it is possible to modify present delivery schedules, especially where the continuous flow method is found to be inefficient. In some cases modification of the system will be necessary, but the greatest obstacle to overcome is the long established local custom of water delivery and use. Often it is possible to change from a rotation system of delivery to a demand system, except during peak delivery period, with considerable savings in water and substantial increases in project irrigation efficiency.

3. *Water Charge Schedule.* Experience with irrigation projects has shown that when excess water charges have been levied, water use has remained reasonably

low without reductions in crop yields. Records also disclose that when excess water charges have been removed, or the allotment has been raised, there has been a definite increase in water use without a noticeable increase in crop production. For example, on one large division of a project in southwest Idaho. USA where the allotment was 9,150 m³/ha (3.00 acre-ft/acre) with excess charges being made for additional water, the water use averaged 13,600 m³/ha (4.47 acre-ft/acre) for the period 1951 to 1955, inclusive. At the start of the 1956 season the allotment was changed to 15,200 m³/ha (5.00 acre-ft/acre). An immediate increase in water use occurred with the average water use being 16,000 m³/ha (5.21 acre-ft/acre) for the period 1956 to 1963, inclusive. On another project in eastern Oregon where the allotment was increased in 1945 from 10,700 m³/ha (3.50 acre-ft/acre) to 12,200 (4.00), the average water use was 13,200 m³/ha (4.33 acre-ft/acre) for the years 1941 to 1945, inclusive. For the years 1946 to 1963, inclusive, the average water use was 14,300 (4.83). In many areas efficiencies could be increased by the irrigation districts levying an excess water charge, thus making the farmer cognizant of the need for good irrigation practices.

4. *Improvement in Operational Practices.* On some projects, particularly small ones, the operational practices, insofar as water deliveries are concerned, are not conducive to either high farm or project irrigation efficiencies. If a farmer is to make efficient use of his irrigation water, he must have reasonable assurance that he can obtain water when he needs it or is entitled to receive it. Operational procedures that control and regulate the water throughout the project distribution system will increase irrigation efficiencies on many projects. In some instances operational wastes and losses are considerably higher than necessary. These losses and wastes may be held to a minimum with improved operational practices.

5. *Consolidation of Irrigation Districts.* Although it is difficult to accomplish, the consolidation of several irrigation districts into one operational unit will allow the employment of a better qualified manager than could be employed by a small district. The better management would generally be reflected in lower operating costs per unit area and more efficient regulation and control of the water supply, thus increasing project irrigation efficiencies. The combination of small irrigation districts often allows exchanges or regulation of water between districts which improve the project efficiency. In many cases overlapping portions of distribution systems or duplicate operational structures and equipment may be eliminated.

6. *Modernization of Project Facilities.* Many of the control and regulating structures on projects constructed in the USA during the early 1900's are obsolete or nearly worn out. This is particularly true of checks, turnouts, and some canal linings. A planned program of replacement and modernization of the delivery system to keep the system abreast of the changes being made in other farming practices will increase the operational efficiency of the project both from a water use and economic standpoint. The use of automatic control structures will reduce the waste and losses necessary to operate the system.

7. *Education Programs.* Undoubtedly the greatest factor in the increase of project irrigation efficiencies is a strong educational program geared to reach those responsible for the operation of the project distribution system that they may have complete regulation and control of the water throughout the entire system. Likewise, an educational program should be directed to the water users, impressing them with their responsibility for efficient water use. The potential

limits of project irrigation efficiencies are unknown, but it is reasonable to expect considerable increases as the demand for water becomes greater.

III. IMPROVING FARM WATER CONVEYANCE AND APPLICATION EFFICIENCIES[4]

A. Irrigation Water Losses on the Farm

Only a portion of the irrigation water delivered to a farm fulfills its intended purpose, that of providing essential water for the crops grown. Some of the water will be lost by evaporation or seepage from farm ditches, more will be lost from runoff or percolation below the root zone in the field.

Conveyance losses from the headgate or farm water source to the individual furrow, border, or sprinkler head may vary from almost zero when the water is conveyed through a watertight pipe to as much as one-half of the initial supply in sandy ditches. In open ditches, some of the loss is attributed to evaporation, but in most instances this portion is relatively small. Weeds and aquatic plants along or in ditches can use significant amounts of water, but generally the most important loss occurs as seepage through the ditch sides and bottom. The magnitude of the seepage loss is dependent primarily upon the permeability of the soil material in which the ditch is built. It may also be affected by such factors as nearness to a water table, silt transported by the irrigation system, and chemical composition of the water.

On the field itself waste will occur by runoff from the surface of the land and uneven distribution of the water, or excessive duration of irrigations, may cause excessive amounts to percolate below the rooting depth of the crop. The relative magnitudes of these two types of losses are greatly affected by the intake characteristics of the soil, the slope of the land, the method of irrigation, coupled with the size of the irrigation stream and the field, and the management ability of the irrigator. On slowly permeable soils, runoff is usually greater than deep percolation, whereas on sandy, open soils, runoff may be practically nil and deep percolation losses may be great. Likewise, the use of large irrigation streams tends to increase runoff and decrease deep percolation losses. The ability of the irrigator to adjust the size of his irrigation stream to the soil and the field conditions is probably the most important element in reducing losses from these causes.

There are also some losses by evaporation in distributing the water over the field. These losses are largely a function of climatic conditions, the method of irrigation used, the type and quantity of ground cover, and the duration of the irrigation set. With surface irrigation and an actively growing, dense crop, evaporation may be no greater than normal evapotranspiration.

The US Dep. Agr. (1960) has estimated that on the average about 47% of the irrigation water available on the farm enters the soil and is held in the root zone where it is available to crops. It also points out that it is not unusual for farmers to attain irrigation efficiencies of 70 to 75% by proper selection, design, and operation of their system.

---

[4] Part III of this chapter was written by John T. Phelan.

B. Factors Affecting Farm Irrigation Efficiency

1. AVAILABILITY OF WATER

For most efficient use, the irrigation water must be delivered at a rate to: (i) Satisfy the water needs of the crop during the peak use period and (ii) permit uniform distribution of water over the land. The volume available over the season must be sufficient to maintain the desired water level throughout the growing season and to provide leaching water that may be required.

These requirements for stream size and volume cannot always be met. As a result of competition for water, there has developed a governmental, social, and economic structure which greatly affects the flexibility of the irrigator in managing his water.

In the USA, individual states have developed water codes that fix limits as to the stream size and volume diverted. These, together with regulations on priority of use of the available supply, often have resulted in attitudes conducive to overuse of water when it is available. Seldom, if ever, is the natural supply of water in phase with the crop demands and there are periods of surplus and deficiencies. Under such conditions, it is human nature to use water lavishly when it is available, knowing that drouth periods will probably follow.

2. ECONOMIC FACTORS

The manager of an irrigated farm must balance all the production costs to derive the greatest profit from his operation. Efficient use of water will require a greater capital investment in physical facilities or will require greater amounts of labor. When water is cheap, there is little incentive to invest more capital or to make the effort to use water judiciously. On the other hand, when the cost of water is great, there is great concern and much justification for care in its application.

Efficient use of water is also affected by the intensity of the farm enterprise and the value of the crop produced. In many instances, even when water is cheap, the operator will economically benefit from proper application of his water through increased production per unit area and improved quality of his product.

3. INDIVIDUAL AND COMMUNITY HABITS

Unfortunately, the irrigation systems on many farms have not been maintained or improved and irrigation methods have not kept abreast with new crops, changed economic conditions, and technological advances. Irrigation habits develop and the modernization of facilities and procedures nearly always fails to keep pace with changing conditions. The habits the community develops for operating irrigation group organizations follow the same pattern as on individual farms. However, when a breakthrough in the adoption of a new practice or technique is made, it is not uncommon to see its use spread rapidly throughout the community.

C. Improving Farm Distribution and Conveyance Efficiency

1. FUNCTIONAL REQUIREMENTS

The farm distribution system conveys water from the headgate, or source, to the individual furrow, border, basin, or sprinkler head. It must do this without excessive conveyance losses and must deliver the water in the quantity needed at the point of application with a minimum labor requirement.

2. MINIMIZING DISTRIBUTION LOSSES

Much can be done to assure that the water will be distributed uniformly. Of primary importance is the arrangement of the irrigated fields so they conform to topographic and farm boundaries, yet keep the required length of ditches or pipelines to a minimum.

Weeds, brush, and trees growing in and along ditch banks can use considerable quantities of water. They are also undesirable because they reduce ditch capacity, increase maintenance costs, and may provide a source of weed seed that irrigation water carries onto the fields. They may be controlled by chemicals, by burning, or by mowing or pasturing.

Canal linings on the farm are very effective in reducing seepage losses and minimizing undesirable plant growth. They sometimes permit the use of smaller ditches, stabilize erosive grades, and prevent breaks and washouts.

Pipelines have these same advantages and in addition reduce evaporation losses to a minimum. They are also capable of conveying water under pressure. Pipelines are essential with sprinkler irrigation and are very useful when water is to be conveyed across a swale or pumped to a higher point. Permanent pipelines are usually buried so as not to obstruct farm operations. Portable pipelines are used to carry the water to the individual furrow, border, basin, or sprinkler head.

3. MINIMIZING OPERATION LOSSES

Efficient use of irrigation water requires adequate controls to measure, check, divide, or divert the irrigation stream. Inadequate controls in the farm distribution system result in increased labor for irrigating and in water losses because of inaccurate methods of apportioning the stream to the individual outlet.

In many instances, the volume of water delivered to a farm over a 24-hour period may be adequate to satisfy crop requirements, but the stream size may be so small that it does not permit efficient irrigation. In these instances, overnight storage reservoirs may be helpful. The small flow for the 24-hour period may be stored and a large stream withdrawn from the reservoir for irrigation. Overnight reservoirs are helpful only if seepage and evaporation losses in the reservoir can be held to a practical minimum. This may require treating the storage area chemically or mechanically by lining the sides and bottom of the pond.

4. ON-FARM REUSE OF SURFACE RUNOFF

Efficient use of irrigation water often results in some runoff or tail water with surface irrigation systems. Too often, irrigators are under the impression that tail water is a sign of poor water management and their attempts to reduce runoff to zero results in considerably greater losses by deep percolation. Often prudent irrigation requires that the principal attention be given to minimizing the deep percolation losses by allowing considerable surface runoff which is then collected and reused. On-farm reuse of surface runoff is particularly well adapted to fields that have fine textured soils or considerable slope.

Tail water recovery systems consist of pickup ditches, which convey the water to a sump or storage reservoir, and often a pump and pipeline to deliver this water to a point where it can be used again. When the volume of the reservoir is small in relation to the rate of flow in the pickup ditches, automatic controls may be needed on the pump.

Tail water recovery systems should not be considered as a substitute for good irrigation water management. The stream sizes used for irrigation must be non-erosive and proportioned to suit the needs of the crop and the soil. The systems often permit a saving in the labor for irrigating by eliminating "cutback" streams used with furrow irrigation and by compensating for small errors in stream adjustment with other surface methods. Design, operation, and general practices being used are described by Davis (1964) and Kasmire et al. (1955).

D. Improving Farm Irrigation Efficiencies

Proper use of irrigation water is only partially accomplished when it has been efficiently delivered to the field. Frequently insufficient attention has been given to the design and management of the facilities for distributing the water over the field. Improvements in water distribution probably offer the greatest opportunity to conserve irrigation water.

1. IMPROVEMENT OF PHYSICAL FACILITIES

The land surface often needs modification to make it better adapted to the method of irrigation used. Land leveling is a popular improvement on fields that are irrigated by surface or subsurface methods. In some instances this intensive practice has been adopted, not to save water but to provide greater ease in irrigating. Land leveling can only make management of irrigation water easier; in itself, it does not improve efficiency. It provides good surface drainage and permits more precise cultural and harvesting operations. When land leveling reduces slope, as is common with bench leveling, it greatly reduces runoff of natural precipitation, reduces soil erosion, and permits efficient water application with lower management skills.

Land smoothing or grading for surface drainage is often helpful, even when the irrigation water is applied by sprinklers. Fine-textured, nearly level soils often benefit from smoothing when sprinkler irrigated because it eliminates small over-irrigated spots caused by ponding.

It is essential that the irrigation methods used be adapted to the crops, soil, and slope of the land. No one method is superior to others for all conditions. Usually, several methods are adapted to a particular site and the irrigator may choose one over another for personal preference or convenience. For example, a field may be adapted to borders, corrugations, or sprinklers. If the irrigation stream is small, corrugations or sprinklers may be the best choice. In another instance, the irrigator may choose to use the sprinkler method to irrigate his crop up and then use borders for later irrigations. The most efficient use of water will, result if the farm irrigation system is designed with sufficient flexibility to permit the irrigator to respond to changed physical and economic conditions.

The irrigator needs good controls and measuring devices. It is just as important for the manager to know how much water has been applied as to know the quantities of seed or fertilizer used. The system should allow him to put the water where he wants it in the quantity that will permit efficient use.

Facilities for the disposal or reuse of surface waste are essential with many methods of irrigation, yet their importance frequently is not recognized. The design and layout of the water disposal or reuse system should be coordinated with the design and layout of the water distribution system.

2. MANAGEMENT IMPROVEMENT

While an adequate water supply and facilities for its distribution are essential, efficient use of irrigation water requires good management. Irrigation water management is defined by the US Soil Conservation Service (US Dep. Agr., 1965) as "the use and management of irrigation water, where the quantity of water used for each irrigation is determined by the water-holding capacity of the soil, and the need of the crop, where the water is applied at a rate and in such a manner that the crop can use it efficiently and significant erosion does not occur."

Conditions on an irrigated field never remain static. As the crop roots develop, the amount of available soil water that can be stored increases. The soil intake characteristics and susceptibility to erosion change with tillage practices, climatic conditions, and crop influences. Thus, the irrigator is faced with an ever-changing set of conditions requiring adjustments in his irrigation techniques. To keep abreast of these needs, he must continually evaluate his irrigation operation to insure himself that he is making best use of his water.

The irrigator can detect erosive streams by simple observation, but he must rely on moisture meters, soil examinations, or time estimates to evaluate the amount of water remaining in his field. He must be able to measure the amount of water applied and compare this volume with the volume needed to fill the root zone. He needs to estimate how uniformly the water is being applied over the field and how much is being lost as surface waste and by deep percolation.

Techniques for estimating surface waste and deep percolation vary with different methods of irrigation. With sprinklers, for example, it is quite simple for the irrigator to know how much water has been applied for the system delivers a predetermined amount each hour. With subirrigation systems, the water condition can be observed by the depth to the water table as measured in a shallow well. With surface methods, an experienced irrigator can judge the uniformity of his application by noting the times of advance and recession of the irrigation water over the border, furrow, or field.

Level borders or basins must be filled quickly with the correct volume of water to make the desired application. With graded borders, the irrigator must have his stream large enough to spread across the border strip and should have the correct volume applied by the time the water has approached the lower end of the border. If the water does not get there soon enough, he should increase the size of his stream or shorten his length of border and irrigate for a shorter period of time. More details on operating surface systems are given in chapter 43.

The irrigator can best judge the uniformity of application in furrows or corrugations by observing the time water is running in the upper and lower ends of the furrow. Irrigation should be continued until adequate water is applied at the lower end. If the furrow length and stream size are in proper proportion, the upper end will not have received an excessive application.

The irrigator can compensate somewhat for distortion of the sprinkler pattern by wind. In severe instances, irrigations may be scheduled during the part of day when the normal wind velocity is lowest.

With all systems, the irrigator must schedule his operations so irrigation water will be applied to all parts of the field as it is needed. He must be equally aware of the magnitude of the deep percolation losses as he is of the water lost as surface runoff. Only by keeping the two in balance can he attain the highest over-all farm irrigation efficiency.

3. REQUIREMENTS TO REALIZE POTENTIAL EFFICIENCIES

Knowledge obtained through research and development of new equipment and materials is constantly improving the prospects of attaining truly high irrigation efficiencies. Almost any device or procedure available at a reasonable cost that simplifies the management problem will prove valuable.

The precision of irrigation attained through management cannot exceed the precision of knowledge of the physical conditions in the field. One of the most fertile areas for improvement of irrigation efficiency through management is in the development of better instruments to measure the amount of available soil water in a field. Simple devices to measure volumetrically the amount of water applied to small areas or the amount that runs off an area are needed. Instruments which will reflect a buildup of saline or alkali conditions would be helpful in maintaining adequate control without waste of leaching water or soil amendments.

As noted, the irrigator often must balance the cost of water against the cost of labor to apply it more efficiently. Automation of irrigation potentially can reduce labor costs. The future is almost sure to bring more and better sensing devices to control the starting and stopping of water application, automatically controlled gates, valves, and other devices to change the point of application within a field.

Attainment of the highest possible irrigation efficiencies requires that all variables that affect the application of water be controlled. At the present time, one of the factors most variable in nature is the intake characteristic of the soil. Intake rates on most soils vary from very low to very high because of changes in density or structure. If the intake characteristics could be controlled by mechanical or chemical means, the most desirable conditions for the irrigation method used could be achieved and maintained.

While the potential farm irrigation efficiency is very high—probably in the range of 85 or 90%—we must not lose sight of the benefits that can be attained by even small increases. It is well within our capabilities at the present time to raise the average farm irrigation efficiency from the present 47% to 65%. This would mean that an irrigated farm could increase its irrigated acreage more than one-third with the available water supply. Of course, more efficient use would mean that less return flow was available for reuse downstream, but if it is assumed that only 55% of the waste is recovered and reused, the improvement in farm efficiency would still permit irrigating about 20% more land within the basin.

LITERATURE CITED

Bagley, J. M. 1965. Effects of competition on efficiency of water use. Amer. Soc. Civil Eng., Proc., J. Irrig. Drainage Div. 91 (IR 1):69–77.

Christiansen, J. E. 1942. Irrigation by sprinkling. California Agr. Exp. Sta. Bull. 670. 124 p.

Davis, J. R. 1964. Design of irrigation tailwater systems. Amer. Soc. Agr. Eng., Trans. 7:336–338.

Hall, Warren A. 1960. Performance parameters of irrigation systems. Amer. Soc. Agr. Eng., Trans. 3:75–76, 81.

Hansen, Vaughn E. 1960. New concepts in irrigation efficiency. Amer. Soc. Agr. Eng., Trans. 3:55–57, 61, 64.

Hart, W. E. 1961. Overhead irrigation pattern parameters. Agr. Eng. 42:354–355.

Hill, Leon W. 1963. Improving irrigation efficiencies in a river basin. Int. Comm. Irrig. Drainage, Trans. 5th Congr. (Tokyo) Rep. 13, Quest. 16, p. 16.197–16.223.

Israelsen, O. W. 1932. Irrigation principles and practices. 1st ed. John Wiley, New York. 422 p.

Israelsen, O. W. 1939. Water-application efficiencies in irrigation and soil conservation. Agr. Eng. 20:423–425.

Israelsen, O. W. 1950. Irrigation principles and practices. 2nd ed. John Wiley, New York. 405 p.

Kasmire, N. L., N. L. McFarlane, and O. A. Harvey. 1955. Beneficial use of waste water in the Coachella Valley. California Agr. Ext. Serv. Riverside. 8 p.

Langbein, W. B. 1960. Water budget. Comprehensive survey of sedimentation in Lake Mead, 1948–49, US Dep. Interior. Geol. Surv. Prof. Pap. 295. p. 95–102.

Myers, L. E., Jr. 1955. Glossary of terms used in irrigation practice. California Mosquito Contr. Ass., Proc., 23rd Annual Conf., and Amer. Mosquito Contr. Ass., 11th Meeting, Los Angeles, California. p. 64–70.

Myers, L. E., Jr., and H. R. Haise. 1960. Concerning water application efficiency of surface and sprinkler methods of irrigating. Int. Comm. Irrig. Drainage, 4th Congr. (Madrid, Spain) Rep. 1, Quest. 12, p. 12.1–12.14.

Reeve, R. C. 1957. The relation of salinity to irrigation and drainage requirements. Int. Comm. Irrig. Drainage, 3rd Congr. Rep. 10, Quest. 10, p. 10.175–10.187.

Stammers, W. N. 1963. Investigations on the improvement of irrigation practices in the Umatilla Irrigation Project. Oregon State Univ. Agr. Exp. Sta. Spec. Rep. 166. 33 p.

Thorne, D. W., and H. B. Peterson. 1954. Irrigated soils. 2nd ed. The Blakiston Co., Inc., New York. p. 1–15.

US Department of Agriculture. 1954. Diagnosis and improvement of saline and alkali soils. US Dep. Agr. Handbook 60. 160 p.

US Department of Agriculture. 1960. Estimated water requirements for agricultural purposes and their effects on water supplies. In Water resource activities in the United States. US Senate Select Comm. Nat. Water Resources, 86th Congr. Comm. Print 13. 24 p.

US Department of Agriculture. June 1965. National catalogue of practices used in soil and water conservation. Soil Conserv. Serv. US Dep. Agr. 20 p.

US Department of Interior. 1962. Use of irrigation water, a report on selected federal reclamation projects, 1949–1960. Bur. of Reclamation. 116 p.

US Department of Interior. 1963. Research-engineering methods and materials. Bur. of Reclamation. Res. Rep. 1. 137 p.

Willardson, L. S. 1960. What is irrigation efficiency. Irrig. Eng. Maintenance. 10:13–14, 18.

62

Reducing Irrigation Requirements[1]

J. S. ROBINS

Washington State University
Irrigated Agriculture Research & Extension Center
Prosser, Washington

I. INTRODUCTION

The use of irrigation requirement in this chapter will refer to the quantity of irrigation water per unit area of land required at the farmer's headgate or pump discharge, annually, to adequately irrigate the crops. It includes seepage, evaporation, and transpiration from the farm distribution system; deep percolation; leaching requirement; surface runoff from the farm; and requirements for evapotranspiration in excess of usable precipitation.

Irrigation requirement thus defined is greatly influenced by many factors elaborated in other sections of this monograph. Many of these are covered in sections II, IV, VII, VIII, XI, and XII, along with chapters 60 and 61. In essence, any climatic, soil, plant, irrigation system, or management factor that affects a water consumption mechanism alters irrigation requirement. Many of the factors will be merely mentioned in this chapter, and the reader will be referred to other chapters for more detail. Emphasis in this chapter will be placed on fundamentals of the new, the possible, and the unproven mechanisms generally not elaborated elsewhere.

Although not specifically embraced in irrigation requirement as defined, the concept of water use efficiency (crop production per unit of water used) has notable implications when considering irrigation requirement. This applies both from an on-farm and national or regional standpoint as related to total production possible with present and potential irrigation water supplies (Haise et al., 1960; Viets, 1962). Frequent reference to this concept will be made in this discussion relative to the efficiency of water utilization on a national, regional, and individual farm basis.

II. IRRIGATION SYSTEM DESIGN, WATER CONTROL, AND MANAGEMENT

Positive water control and management on the farm can greatly reduce irrigation requirement. Reducing seepage loss from farm ditches, preventing farm runoff, improving water distribution over the field, and reducing unnecessary deep percolation are probably the most significant areas for improvement (*see* chapters 41, 42, 43, 44, 45, 60 and 61).

Related to distribution system losses is the use of water by weeds growing in, on, or adjacent to farm ditches or canals. Such plants not only transpire water

---

[1] Contribution from the Soil and Water Conserv. Res. Div., ARS, US Department of Agriculture, with which the author was formerly associated.

directly from the supply in transit but also extract water from the soil under and adjacent to the ditch when flowing water is not present. This increases seepage when water reenters the distribution system. Additionally, retardance of flow by such vegetation often increases seepage and evaporation from the ditch and, in severe cases, causes water waste by overflowing or breaking of the ditchbank.

Appropriate frequency of irrigation water application, in relation to crop needs and soil water conditions, is an additional management area subject to significant improvement. Pertinent to system design and management is irrigation scheduling to reduce runoff and deep percolation losses and to improve uniformity of distribution over a field. Due to marked changes in infiltration rates from one crop to the next in a crop sequence and from one irrigation to the next in a given crop, the design and management of irrigation systems must permit flexible operation and management if these losses are to be kept at a minimum (Mech, 1960). Other aspects of this factor will be noted in subsequent sections of this chapter.

Paramount to reducing irrigation requirements through the above mechanisms are (i) knowledge of the magnitude of seepage losses and development of technology essential to economic control methods; (ii) adequate water control structures and water measurement to permit application of a known quantity of water to a field or portion thereof; (iii) use of a suitable, properly designed system which will distribute the required quantity of water uniformly across the field without unnecessary runoff; and (iv) workable and economic methods, instruments, and criteria on which to base the decision to irrigate a given crop and the quantity of water needed. Significant strides have been made in the development of knowledge and in its application for each of the above areas. However, both the science and the practice of irrigation are presently far short of the ultimate in terms of knowledge, instrumentation, methods, and materials and their application to effectively and economically minimize irrigation water losses.

Finally, it is essential to recognize that losses promoting high irrigation requirements on a given farm may not constitute a net water loss to the irrigated area or river basin in which the farm is located. Much of the water lost to the individual farm by seepage, deep percolation, or runoff re-enters the supply and is used elsewhere. Therefore, from a regional or national water supply standpoint, these losses, so far as volume of water is concerned, usually are not of serious consequence. However, the recovery of such water for reuse is never complete, and such losses often are the cause of drainage and saline and sodic soils problems. Furthermore, chemical, physical, and biological quality generally deteriorates as the water moves through loss and recovery cycles. Finally, the farmer must manage and usually pay for the added volume of water and any drainage facilities required to remove the water. For these reasons, it is generally in the interest of both the farmer and the public to control such losses whenever possible.

III. CROP ADAPTATION, VARIETY SELECTION AND IMPROVEMENT, AND CULTURAL PRACTICES

Numerous possibilities for reducing irrigation requirements exist through better crop adaptation, variety selection and improvement, and improved cultural practices. Some of these factors relate to increasing water use efficiency rather than reducing irrigation requirement but they will be mentioned here, since efficient use of irrigation water is a matter of utmost concern.

Probably the most significant crop factor for reducing irrigation requirement is the selection of properly adapted crops. Since evapotranspiration rate is chiefly dependent on climate (*see* chapter 26), it is apparent that short season crops require less irrigation water than long season ones, other conditions being equal. This is especially true for such crops as small grains or forage producers that mature or are harvested during midsummer when evapotranspiration potential is high. Shortening the season by 1 week at this time of year may reduce evapotranspiration by 2 inches or more and irrigation requirement correspondingly. Use of cool season species in lieu of warm season ones (grasses, for example), where adapted, can greatly reduce irrigation water needs again by avoiding the season of highest evapotranspiration. These conditions apply to selection of both plant species and varieties.

A second crop factor related to irrigation requirement is the relative sensitivity to soil water depletion which governs irrigation frequency and thus irrigation requirement (*see* chapters 21 and 23). Such sensitivity differences may be due to inherent physiologic behavior, nature of the root system, or other factors. In any event, a crop such as alfalfa (*Medicago sativa* L.) which can deplete deep stored soil water without adverse yield effects will generally have lower irrigation requirements during an equal time period than one such as Ladino clover (*Trifolium repens* L.) which cannot. This is due primarily to reductions in deep percolation, runoff, and related losses with less frequent irrigation and, secondarily, to a probable reduction in evaporation from the soil surface due to less frequent moistening of the surface soil layers.

Related to the matter of root proliferation is the problem of diseases or insects that limit rooting depth which in turn necessitates more frequent irrigations. An example is root rot in beans (*Phaseolus vulgaris*) which may limit effective root development to a depth of 1 ft or less, thus making essential very frequent irrigation with attendant water losses, even on deep soils.

A third area in the crops field where irrigation requirement may be reduced is in use of species or varieties that inherently have a low transpiration rate (*see* chapter 17). Although past studies indicate appreciable differences in control of transpiration rate due to leaf structure, number or location of stomates, or composition of the cuticular layers of different plant species, systematic exploration of varietal differences or differences between related species of agricultural crops has been limited. Even though evapotranspiration is primarily governed by climate, significant reductions in transpiration might be effected through plant breeding or selection to incorporate a reduction in number or location of stomates or a change in other leaf characteristics which influence transpiration. Such a development may have great significance when effective and economic methods for control of evaporation from the soil are developed, a matter to be discussed in a subsequent portion of this chapter.

Possibilities also exist for crop varietal or species selection for improved energy consumption in photosynthesis. Recent results indicate that only 5 to 10% of the net solar radiation is consumed in this process (Lemon, 1960; Moss et al., 1961). Any increase in this percentage would correspondingly lessen energy available for evapotranspiration but, more important, would increase productive potential per unit of evapotranspiration. Again, this approach has received little direct attention in crop improvement studies.

Further related to transpiration control by crop selection or development is the possibility of increasing plant leaf reflectance. An increase in reflectance would reduce incident radiant energy available for evapotranspiration. Limited studies to date indicate only minor plant differences in this regard but the approach should be further explored.

Crop cultural practices warrant consideration in assessing possibilities for reducing irrigation requirement. The adjustment of planting dates to shorten the growing season, to avoid high evapotranspiration in midsummer, or to permit plant establishment and root development during cool weather represents one of the likely possibilities. Row direction, plant spacing, and other geometric factors, though incompletely understood at present, offer other means of controlling evapotranspiration by altering interception and disposition of incident radiant energy (Tanner and Lemon, 1962). These latter methods are related to microclimate control which will be discussed later. Such geometric control, in addition to altering microclimate, could (i) reduce evapotranspiration by lessening leaf surface or density of the crop canopy without a corresponding yield depression, (ii) increase the proportion of energy consumed in photosynthesis, or (iii) increase reflectance of the cropped surface and thus reduce radiant energy available for evapotranspiration.

A final significant cultural factor is weed control in crops. Although not proportional to relative weed population, evapotranspiration and thus irrigation requirement is generally increased by weeds growing in the crop, especially in row crops with wide row spacing. Because such growth also competes with the crop for light energy and nutrients, it has implications considerably beyond that of irrigation requirement. Related to this competition problem is the use of cover crops in orchards and vineyards. Although often beneficial for nutrition and erosion control, to improve water infiltration or to reduce cultivation costs, such cover crops do increase evapotranspiration appreciably (Jensen and Haise, 1963). In water short areas, this may prohibit their use.

As indicated in the chapter introduction, most crop factors influencing irrigation requirement also inherently relate to water use efficiency. For example, species or varieties selected for their seasonal characteristics, rooting habits, transpirational properties, or other factors elaborated above may use less water but produce so much less than the alternate crop or variety that the production per unit of water actually decreases. Similarly, cultural factors to minimize irrigation requirement may not be rational from a water use efficiency standpoint. In applying any principles related to either irrigation requirement or water use efficiency, the economics of such application places added limitations on the practices ultimately found to be feasible.

In line with the above, one final matter related to crop selection and culture is worthy of note. From a regional or national standpoint, minimum evapotranspiration for the food and fiber needs demanded of irrigated agriculture is achieved by use of plant species; crop varieties; cultural and management practices which produce the maximum forage, fiber, grain, produce, or other product per unit of time. Exceptions may occur in areas such as the Great Plains of the USA where rainfall contributes significantly to evapotranspiration needs, and where land area is large compared to irrigation water supplies. This presumes that efficiency of delivery and application of irrigation water is constant and all other factors remain equal. The above is true of any management factor that influences crop growth, and results from the fact that evapotranspiration is dominantly controlled by climate.

IV. IMPROVING USE OF PRECIPITATION

Stored soil water from precipitation contributes significantly to crop evapotranspiration in most irrigated areas. In regions such as the Southwest USA, the contribution is largely through winter precipitation with little addition during the growing season. In other regions such as the Great Plains, significant contributions more often come during a part of the growing season. In either event, or where a combination of the two exist, effective use of this water can appreciably reduce irrigation requirements.

Basically, effective use of precipitation resolves to assuring (i) that storage capacity in the soil exists when precipitation occurs; (ii) that, with such capacity, received precipitation actually enters and is stored in the crop root zone; and (iii) that crops and cultural practices are properly selected for effective utilization of precipitation stored as soil water.

Available storage capacity in the soil can be most effectively assured by proper timing and control of the amount of water added in relation to anticipated precipitation. For example, in the southern Great Plains, periods of high precipitation are normally expected in May and early June and in August (Jensen and Hildreth, 1962). Thus, wherever possible, complete recharge of the profile by irrigations before and during these periods should be avoided unless a rapidly growing crop is on the land. In such case, the crop will deplete 2 inches or more of soil water within 1 week. Provision should be made also for use of runoff that usually occurs from those parts of the field where irrigation water has recently been applied. Where winter precipitation dominates, the soil should be allowed to dry as much as feasible prior to crop maturity or harvest, both to make available maximum storage capacity and to lessen winter evaporation losses.

Turning now to assuring storage of precipitation, several factors are important. Probably most significant is providing time to permit complete infiltration. A most effective means of accomplishing this is land leveling to impound the water and hold it in place until infiltration is accomplished. Such systems have been highly successful in Texas, Kansas, Nebraska, and in other areas of the USA where soil conditions are suitable (Ross and Swanson, 1957). A variation on this method is leveling of contour strips so that runoff from sloping parts of a field is impounded on leveled strips (Zingg and Hauser, 1959). A second and less effective method is to decrease the furrow grade on sloping lands by running furrows across the slope of the land (nearly on the contour). This increases concentration time and infiltration (see chapter 43).

A second important factor in assuring on-site infiltration is maintaining an infiltration rate as high as possible (see chapter 46). This can be done in some cases by maintaining crop residues on the soil surface, by avoiding soil compaction, by certain tillage operations (chiseling in furrow bottoms, for example), by incorporating green manure or other crop residues in the soil, and by using certain crops (alfalfa or grass) in the crop sequence. In some cases, deep plowing (2 to 4 ft deep) to break up impeding layers or to change the surface soil texture may improve infiltration significantly. Finally, water content of the surface soil layers has a major effect on infiltration rate. In addition to providing added storage capacity as indicated above, permitting significant water depletion before and during periods when rainfall is expected also helps to assure infiltration and storage of precipitation when it occurs.

An area of special interest related to precipitation storage is that of snow catchment and snowmelt infiltration. Two problems are encountered here: control of snow deposition and infiltration when soils are frozen or when snow melts rapidly.

Control of snow deposition generally is possible in windy areas such as central and northern Great Plains. Since the amount of water in snow is generally small, it may be advisable to concentrate the snow on portions of the field rather than to distribute it uniformly over the whole field. This can be done by using barriers such as snowfences, very narrow strips of tall stubble, or alternating strips of tall and short stubble. The resulting drifts contain an amount of water sufficient to wet the soil to appreciable depth, thus eliminating at least a part of the evaporation loss that would occur were the water spread uniformly over the field (Greb and Black, 1961).

Where snowfall is heavy and/or storage capacity small, it may be best to uniformly spread the snow over the whole field. Standing crop stubble is a most effective means of accomplishing this. Where stubble is absent, ridging the soil or the initial snow deposit will generally aid in snow catchment.

In regard to snowmelt infiltration, the primary problem is existence of frozen soil under the snow which greatly reduces infiltration (Willis et al., 1961). Deep snow deposition eliminates or reduces soil freezing. Inducing deeper snow deposits offers one possibility for increasing storage of snowmelt water. Secondly, the rate of thawing is appreciably faster for dry than wet soil, owing to the much lower heat capacity of dry soil. This indicates another advantage for entering the winter season with the profile relatively dry. Finally, a rough, cloddy, open soil surface will help trap and hold snowmelt until infiltration is accomplished. Practices such as contour chiseling, rotary subsoiling, or plowing when the soil is dry assist in creating the desired open surface condition.

Control of time and rate of snowmelt is another possible means for controlling storage of snow water. Changes in exposure to wind and incident radiation as well as radiation absorption due to color control offer possibilities for either hastening or slowing snowmelt. Although some studies of this type have been made in mountain snowpack areas, little attention has been given the subject on agricultural lands. Thus, the matter remains only a possibility for future study.

Before leaving the subject of effective precipitation storage, the matter of precipitation waste on nonproductive areas should be considered. This applies both to snow deposition and runoff accumulation in fence rows, borrow pits, unfarmable draws, sloughs, and streambeds, and to precipitation infiltrated in areas adjacent to productive lands where soil conditions, topography, or other factors prevent its use for desirable plant growth. Definite possibilities exist for special treatment of these waste areas to limit infiltration and for collection and use of resulting runoff on productive lands or for other purposes. Some use of this concept is now being made for livestock water and other needs (Lauritzen, 1960; Myers, 1961). However, to date only limited exploration of this possibility for irrigation purposes has been made aside from spreading of naturally occurring runoff. As water needs and competition for existing supplies increase, such measures will surely deserve more attention and use.

In regard to effective use of stored precipitation, the general concepts outlined in the previous section on crop selection are most applicable. Evaporation, a subject to be discussed later, and transpiration by weeds are most important factors in limiting retention of soil water stored during the fallow season. Where winter precipitation storage is significant, its utilization can be greatly enhanced by spring- rather than summer-planted crops. Advancing the planting date insofar as climate will permit is also beneficial. Where late summer and early fall precipitation is significant, selection of fall- or spring-sown crops which will grow and make use of water stored therefrom will result in reduced irrigation requirement. Attempts to store and retain such water for use by a subsequent summer crop will result in a high percentage loss. Other considerations in relation to evaporation and its control will be treated in the following part.

V. EVAPORATION AND TRANSPIRATION CONTROL

Irrigation water is generally applied for the specific purpose of meeting crop evapotranspiration needs. Therefore, any reduction in evapotranspiration reduces the quantity of irrigation water required for successful crop production. This section deals with the subject of evapotranspiration control as related to irrigation requirement.

Climatic, plant, soil, and cultural factors related to evapotranspiration and aspects of transpiration are covered in sections VIII and VI of this monograph, respectively. This part will be limited mainly to the discussion of control methods. The reader should consult other chapters for discussion of the fundamentals of these processes, fundamentals on which the possible controls are based.

As indicated previously and elaborated in chapter 26, evapotranspiration is an evaporative process largely controlled by climatic factors, notably incident solar radiation and advected heat which provide energy for the process, and atmospheric vapor content and wind movement which control vapor and heat flow phenomena. The plant exercises control over transpiration and radiation utilization in photosynthesis; soil factors, especially water content and transmission properties, may limit evaporation; and both soil and plant factors, especially surface albedo, radiation interception, heat storage, and conduction influence the partitioning of radiation and advected energy. These factors alter, within limits, the proportion of total energy consumed in the evapotranspiration process (Tanner and Lemon, 1962).

This complex of interrelated and as yet incompletely understood factors influencing evapotranspiration suggests a number of possible control measures. These factors may be broadly classed in three categories—climatic, soil, and plant—that may be controlled to alter evapotranspiration.

The three climatic factors subject to control are incident radiation, the turbulent wind layer at and immediately above the surface, and the atmospheric vapor content and distribution in the boundary layer.

The primary means of controlling incident radiation is through shading. Limited studies using mechanical shading devices such as wood or metal strips have demonstrated, as expected, appreciable reduction of evapotranspiration and sensible heat production in the crop. Light energy during midday hours is generally far in excess of the 2,000 to 3,000 ft-c and covers a broader spectrum than is required for normal photosynthesis. Also, especially in midsummer and at other seasons in semitropical and tropical latitudes, high total radiation may cause temperatures in excess of the optimum for some crops. Selective shading at appropriate seasons or times of day may prove beneficial both in reducing evapotranspiration and in lowering soil, plant, and air temperatures. Complete or partial shading also influences reproductive processes of certain plant species.

Commercial application of shading techniques except for control of temperature or plant reproduction is seriously limited by economic factors. In addition, the degree and timing of shading must be such that sufficient light of proper wavelength reaches and penetrates the crop foliage to assure maximum photosynthesis. In early morning and late evening hours, and in the spring and fall months in northern latitudes, intensity and/or duration of incident radiation may become limiting for maximum CO_2 assimilation (Lemon, 1960; Moss et al., 1961).

A second possible radiant energy control mechanism is selective absorption, reflection, filtering, or otherwise excluding those light wavelengths not essential to normal plant functions. Light wavelengths from about 3,900 to 4,800A and 6,200 to 6,800A are largely responsible for the chlorophyll-activated photochemical reactions that combine CO_2 and H_2O to form initial carbohydrates. These wavelengths constitute only a small fraction of the total solar radiation. Exclusion of the ultraviolet, the intermediate visible, the far red, and the infrared portions could greatly reduce evapotranspiration and heating effects in the surface boundary layers. Glass, for example, does not efficiently transmit long (infrared) wavelengths; thus incident radiation in glass houses is appreciably reduced, yet plant functions are unimpaired unless shading is excessive. However, this same mechanism prevents back radiation to the atmosphere of similar wavelengths by reradiation from soil, plant, or other materials within the glass structure. This radiation is trapped in the house and causes heating which must be removed by ventilation. If, however, the covering were treated or impregnated with material to either absorb or reflect a large percentage of all except the effective wavelengths, a much higher percentage of transmitted radiation would be used in photosynthesis, thus reducing energy for evapotranspiration and heating.

Gas molecules, including water vapor, water droplets in the form of clouds or mist, and dust or smoke particles in the atmosphere reflect, refract, and/or absorb light energy. Generally speaking, long wavelengths are less affected by very small dust particles or gas molecules than short ones. However, individual atoms and molecules as well as crystalline and amorphous particles have specific absorption bands at which light energy is absorbed and is reradiated, used in photochemical reaction, or lost by conduction to adjacent particles as sensible heat.

Considerable attention has been directed toward developing an understanding of the characteristics and magnitude of absorption by various materials naturally occurring in the atmosphere. However, aside from use of fluid layers between the light source and interior of plant growth chambers, almost no attention has been given the possibility of controlling radiation quality or quantity by this mechanism. It would appear possible, although probably not economically feasible at present, to impose a fluid boundary or add molecular substances to the atmosphere between the radiation source and a cropped surface to significantly reduce light intensity in certain wavelengths, and thus reduce evapotranspiration and heating of the soil, plants, and adjacent air layers.

A second climatic factor subject to alteration is the flow characteristics of the lower atmospheric boundary. This generally turbulent air layer is largely responsible for transmission of water vapor and sensible heat from the earth's surface to the upper air layers and for carrying CO_2 from the outer atmosphere to the crop leaves. So long as air movement parallel to the surface is sufficient to develop a turbulent layer near the earth (wind velocity about 2 miles/hour at a 1-ft height), the zero-plane displacement is fixed and the velocity distribution increases logarithmically with height above the surface. With a growing crop, however, the boundary is not rigid; thus, the zero-plane displacement changes with wind speed. Where wind speed is < 2 miles/hour, the importance of turbulent transfer becomes less and diffusion becomes the dominant transfer mechanism.

Increased turbulence accelerates the interchange of water vapor, heat, and CO_2 between the soil-plant environment and the upper atmosphere; therefore, reduction of wind velocity near the surface should, within limits, reduce evapotranspiration. It should be remembered, however, that at certain times of day, reduced CO_2 delivery due to lowered turbulence may become a limiting factor, thus reducing assimilation and water use efficiency. In addition, reduced heat removal might result in plant injury as well as an offsetting increase in evapotranspiration due to increased temperature.

The chief means of altering wind velocity near the surface is through wind barriers, both artificial and vegetative. Recent estimates indicate reductions in evaporative potential downwind from shelterbelts in Kansas to vary from 5% at 25 times the height of the barrier to 10 to 30% in the zone up to 10 times the barrier height (Woodruff et al., 1959). It should be mentioned that vegetative windbreaks themselves use water and, therefore, partially offset any moisture saving unless their needs are satisfied from a water table at significant depth. If irrigation water is required to maintain such a windbreak, the net irrigation requirement may be more than if the windbreak were eliminated.

Finally, evapotranspiration might be reduced by increasing the vapor content of the air, thus reducing the vapor gradient near the earth's surface. A possible practical mechanism to accomplish this would be control of the season at which evapotranspiration occurs in upwind areas to conform to the crop growth season on irrigated lands. This could be accomplished by changing vegetation on such areas to summer rather than cool-season types or by eliminating vegetation during certain seasons. In either case, more soil water stored in surrounding areas would be liberated to the atmosphere during periods of high evapotranspiration on irrigated lands. A second method to increase vapor content would be artificial injection of vapor upwind from the field. Again, this method would likely increase net irrigation requirement if this water were considered a part of the irrigation requirement as it properly should.

Turning now to soil properties affecting evapotranspiration, these can generally be classed as (i) storage properties, (ii) transmission characteristics, (iii) factors affecting uptake of water by plants, and (iv) surface properties related to evaporation or radiation absorption.

In general, the deeper water is stored in the soil, the more slowly it will be removed by evapotranspiration. Thus, depth of water storage is a most important consideration in potential reduction of evapotranspiration. Soil texture and profile stratification are the principal properties that control distribution of water storage in the soil. Coarse-textured soils generally hold less water per unit volume than fine-textured ones, and thus pass more of a given amount of water to greater depth. This beneficial effect is at least partially offset by the fact that coarse soils hold less water in the crop root zone, thus require more frequent irrigation than fine-textured ones. Although no precise data are available on this point, it is likely that the two effects for any given crop are approximately offsetting. Ideally from the water storage standpoint, the surface layers should be coarse and subsurface layers fine textured to combine both attributes in a single profile.

A stratified profile usually has a higher water content above the discontinuity for some time following water addition than does an isotropic profile (Miller and Bunger, 1963; Miller and Gardner, 1962; Robins, 1957). This is true for coarse-textured layers underlying fine-textured surface soil and where underlying layers have lower conductivity than the overlying mantle. This retardance of downward flow generally increases evapotranspiration, because of the higher water content near the soil surface. These profile factors can sometimes be altered by deep plowing or mixing. It must be remembered, however, that the higher surface water condition greatly benefits crop establishment and often simplifies irrigation practice. Therefore, effects of such alterations must be carefully weighed from standpoints other than evapotranspiration reduction.

Artificial inducement of deep water storage can reduce wetting of the immediate surface layers and thereby reduce evaporation. Deeper storage of irrigation water can be induced to some degree by increasing the distance between or by deepening irrigation furrows. Benefit can sometimes be derived from chiseling beneath the furrow bottom. Such chisel trenches could be filled with crop residue or coarse inorganic material to increase the proportion of water stored at greater depth. Increasing aggregation (cloddiness) of the surface soil between furrows restricts the upward unsaturated flow of water into this layer, thus also offers a possibility for controlling storage depth.

Evaporation of soil water is intimately related to soil water transmission properties, since water flow up to or near the immediate surface layer must occur before evaporation occurs (Hanks, 1958). Soils generally drain rather rapidly to soil water suctions or tensions in the range of 100 to 400 millibars. Evaporation is slower from soils with low conductivity than from soils with high conductivity in this tension range. Medium- and fine-textured soils generally have higher conductivity in this range than coarse-textured ones. This, combined with the greater volume of stored water in fine-textured soils, often results in high evaporation losses from them as compared to coarse-textured soils.

As in the case of water storage, profile discontinuities also influence water transmission upward to the surface (Eagleman and Jamison, 1962; Lemon, 1956; Willis, 1960). Thus, providing a textural or other discontinuity near the surface can materially reduce evaporation. Again, a thin layer of coarse material on or near the surface seems a promising technique.

Certain chemicals, particularly organics, alter the surface tension, viscosity, or wetting angle of soil water (Bowers and Hanks, 1961; Lemon, 1956). Reduced surface tension will reduce unsaturated flow rate and, in the absence of evaporation, will increase storage per unit volume shortly after watering but reduce storage after extended free drainage. Reduced viscosity increases water flow rate with attendant reductions in storage. Materials that increase the wetting angle greatly reduce flow rate and, in the extreme, may completely prevent film flow.

Any of these mechanisms could be effective in altering evaporation by limiting transmission and/or water storage. The most promising appears to be use of materials that limit film flow, yet permit flow through large pores. Such materials applied as a complete or partial layer at some small distance below the soil surface could alter water transmission sufficiently to greatly reduce evaporation.

Factors affecting water uptake by plants include soil aeration, density, biological or chemical properties which limit plant root growth or metabolic functions, and soil chemical or physical properties which limit water transmission to roots or increase the total soil water potential. In general, any factor that limits transpiration by reducing water uptake will adversely affect plant water balance and result in growth reductions (*see* Sections VI and VII). Possible exceptions are where plants insensitive to water stress are grown and water adequate for stomatal transpiration can be maintained, yet cuticular transpiration and exudation can be reduced. Such delicate balance, in view of the constantly changing water environment, seems unlikely.

Finally, the important area of soil surface property alteration to control evaporation or radiation partitioning deserves consideration. Certain aspects related to water transmission and storage are elaborated in previous paragraphs. In considering the immediate surface, however, the matter of vapor transfer and liquid flow is important. Since vapor flow is proportional to porosity, a most effective mechanism to reduce it is to reduce total porosity in the immediate surface layer through which diffusion must occur. Although a sound mechanism for evaporation control, this method is limited practically since such a layer also restricts infiltration (Hanks, 1958). In the extreme, however, use of an essentially impervious covering such as plastic, paper, or foil is highly effective and irrigation water can be applied beneath or through perforations or slits in the covering (Willis, 1962). Such materials are presently used commercially for certain crops for weed and produce quality control and for evaporation reduction. As materials become cheaper and water more costly, such methods will warrant increased study and use to control moisture losses.

Since vapor diffusion is relatively slow compared to film flow in moist soil, rapid drying of the surface few millimeters greatly reduces evaporation (Hanks, 1958). Limiting water storage in this thin layer by textural or structural control has been mentioned previously. Reduced storage in this zone will also result in more rapid drying. This introduces the question of evaporation control with porous mulches such as vegetative residues, gravel, and clods. The effectiveness of vegetative mulches is limited, unless they are rather thick, because their high porosity permits rapid diffusion (Hanks and Woodruff, 1958; McCalla and Army, 1961). Evaporation rate for short periods is usually appreciably reduced. For extended periods, however, such mulches may keep the soil surface more moist and thus effect no net saving or even result in a net loss because of increased evaporation. A secondary effect of such mulches is on radiation which will be discussed later in this section. Gravel and clod mulches, on the other hand, effectively reduce evaporation, both in the short and long run, because the immediate surface dries rapidly and such materials act as a diffusion barrier. In windy areas, however, the diffusion barrier effect of any porous mulch may be largely offset by pressure fluctuations (extension of the turbulent boundary) into the mulch (Hanks and Woodruff, 1958).

Finally, a most interesting area of consideration is control of incident radiation partitioning. Possibilities for controlling incident radiation *per se* and for increasing CO_2 assimilation have been mentioned previously.

Under normal conditions, 40 to 50% of the incident radiation reaching the land surface is back radiated, reflected, or refracted to outer space. Any increase in this proportion would correspondingly lessen energy available for heating and evaporation processes. Increasing plant reflectance has been previously considered but is automatically included here along with soil properties in consideration of land surface properties.

Appreciable differences in net radiation exist over bare soil surfaces owing to color differences. Lightening the surface color will reduce absorption of incident radiation during daylight hours 10 to 15%. However, it should be stressed that such reduction lessens soil heating which reduces back radiation at night and loss of sensible heat to the atmosphere. Thus, the net reduction in energy use for evaporation is considerably less than the daytime reduction in net radiation. This mechanism along with alteration of vegetative reflectance properties is not fully understood and deserves further study.

Soil surface geometry alteration offers a second possibility for controlling radiation absorption and consumption in evapotranspiration. Ridging techniques to concentrate absorption in localized steep south-facing slopes with intervening less steep north-facing slopes is one possibility. This would localize heating and rapid evaporation, and therefore enhance back radiation, sensible heat production, and possibly reflectance on these south-facing slopes. Such effects would be lessened on the north-facing slopes and, through proper shaping and geometrical dimensioning, a net reduction in evaporation may accrue. Other possibilities surely exist and need to be fully explored experimentally. It must be remembered that such geometric changes also influence the nature and degree of turbulence in the lower atmosphere which can also affect evaporation. Thus, the influence of such alteration of wind flow must be considered along with radiation partitioning for their combined effect on evapotranspiration.

Several plant factors relating to evapotranspiration reduction have been previously discussed; namely, genetic development or selection for transpiration control; increased radiation reflectance or energy utilization in photosynthesis; or crop selection and cultural factors to favorably alter evapotranspiration, utilization of rainfall, or irrigation need. The plant factors to be considered here are artificial transpiration control by foliage treatment and certain plant characteristics related to sensible heat exchange.

Nurserymen have practiced effective transpiration control during shipment and transplanting operations by using a variety of organic materials, usually of fatty or waxy nature. Such materials form a vapor barrier over the plant surface and sometimes even cover the stomates. General use of such a mechanism in the field is limited by (i) economics, (ii) difficulties in coverage during initial treatment and retreatment of new growth, (iii) problems in maintaining a cover of such materials, and (iv) the fact that reduction in transpiration results in availability of more energy for plant and soil heating and evaporation. The fourth problem cited is perhaps the greatest long-run deterrent, since major transpiration reduction under field conditions without control of soil water evaporation seems unlikely. The same applies for chemical treatment, either applied externally or through the plant for reducing evaporation in or on the leaf surface. Chemical control of stomatal opening and closure might be possible and certainly warrants further study.

Reducing leaf area is another possibility for reducing transpiration. Since, within limits, both transpiration and photosynthesis are related to leaf surface, there may be some cases where eliminating leaf surface, especially older or senescent leaves, might reduce evapotranspiration without appreciably affecting growth. To the author's knowledge, attempts to improve this balance have met with failure, especially where water was adequate. Such mechanism may have limited application under conditions of deficient water. Application in irrigated agriculture may be limited to certain crops where defoliation after a certain stage of growth has no adverse effect on product quantity or quality, as with potatoes (*Solanum tuberosum*) or cotton (*Gossypium herbaceum*).

Sensible heat exchange between plants and the atmosphere has a bearing on evapotranspiration both where advected energy is appreciable and where sensible heat is developed on site. Such exchange alters plant temperature and the corresponding vapor pressure at the evaporative surface. In addition to exposed leaf surface, the heat exchange rate is greatly conditioned by leaf and stem flexibility. Greater motion (thus greater opportunity for convective heat transfer) occurs with flexible leaves and stems than with stiff ones. Thus, where sensible heat dissipation is desired, flexible plants may be best, whereas in the presence of advected heat, the reverse may be true.

Concern has been expressed that widespread use of evaporation or transpiration control methods would produce adverse changes in climate including higher temperatures, lower humidity, and less precipitation. It should be stressed that widespread use of any such method on presently irrigated lands in semiarid or arid areas would have little influence on the amount of water annually returned to the atmosphere within a given region. If appreciable reductions in evapotranspiration were possible, most of the water saved would be used to irrigate additional land or for other consumptive use needs. On nonirrigated semiarid and arid lands, the amount of water consumed would be practically unchanged since the prime limiting factor is the annual water supply. Thus, the energy consumed annually in this process as well as the over-all hydrology would be little affected in such areas. Seasonal distribution of energy consumption might be shifted appreciably, however, and result in warmer temperature and lower humidity during the spring and early summer, and cooler temperature and higher humidity in late summer and fall months.

In subhumid and humid areas, more effect on local climate may result since present annual water supply is more nearly adequate to consume available radiant energy. Any reduction in evapotranspiration under such conditions would increase correspondingly the surface or subsurface water supply as well as make the climate slightly more arid. Neither of these effects should be considered detrimental.

In regard to influences on precipitation caused by reduced evapotranspiration, it should be remembered that mass air movement from the oceans to land areas is the major source of significant regional precipitation. Water derived from the land surface generally has only limited influence either on the airborne moisture supply or on precipitation occurrence or magnitude. This, along with previously described limitations on the annual control of evapotranspiration, assures that such controls would have a negligible effect on precipitation.

VI. SOIL AND PLANT MANAGEMENT PRACTICES

Numerous soil, climatic, plant, and other factors relating to irrigation requirement have been discussed in previous parts of this chapter. We will now examine
a few of the most common management practices as they relate to irrigation
requirement.

Tillage for seedbed preparation, for weed control, or other purposes in growing
crops, is perhaps the most ubiquitous of soil management practices. Soil disturbances by tillage implements generally result in loosening of the tilled layer, which
generally increases evaporation from the disturbed soil. It also breaks the continuity
of soil pores and reduces film flow from lower depths into the tilled layer. These
mechanisms produce offsetting results and the net effect on irrigation requirement
depends on (i) depth, degree, and frequency of soil disturbance; (ii) initial water
distribution in the profile; and (iii) the time lapse between tillage and irrigation
or rainfall that reconsolidates the disturbed layer.

In seedbed preparation, tillage depth is usually appreciable. Generally, therefore, if the tilled layer is relatively moist, efforts should be made to artificially
reconsolidate the surface layer as soon as possible, either by compactive tillage
or by water addition to reduce net evaporative losses. Depth of primary tillage
operations should be as shallow as feasible consistent with preparation of the
desired surface condition and disturbance of soil layers that restrict water and/or
root penetration. If the surface soil is dry to the tillage depth, irrigation requirement will be little affected.

Primary tillage should always be performed before preplanting irrigation if
water saving is an important consideration. Where appreciable rainfall is expected
prior to planting, the need for preplanting irrigation with its attendant water
waste often can be eliminated by performing most seedbed preparation operations
well ahead of planting.

Tillage for weed control should be as shallow as possible. Deep cultivation not
only increases evaporation but often adversely affects the crop because of root
pruning. Cultivation to break up soil crusts in furrows is generally ineffective in
enhancing intake rate, thus does not improve irrigation efficiency. As mentioned
earlier, however, chiseling or otherwise disturbing dense or highly compacted
layers may enhance infiltration and therefore reduce runoff and improve distribution of applied water.

Use of irrigation water merely to refill the soil reservoir between crops is usually
less efficient than irrigation during the period of crop growth. Higher evaporation
losses and inefficient precipitation storage generally are the results of this practice.
Only where the irrigation system is inadequate to cover the land area served or
where excess water is available during the noncrop season should this practice
be employed.

Related to off-season irrigation is the practice of irrigating green manure or
cover crops. Such crops, of course, use appreciable water and thus add to the
annual irrigation requirement unless the water used would otherwise be lost to
evaporation, deep percolation, or runoff. Even where extra water is required, use
of such crops may be justified for land protection or for production of nitrogen
where the crop is leguminous and fixes appreciable atmospheric nitrogen. Production of such crops merely for organic matter addition in lieu of high residue

producers in the crop sequence may be justified from a water conservation standpoint when such crops grow during seasons of low evapotranspiration where highly intensive cropping is practiced. Organic material supply and turnover may be effected with less water by rapid-growing, cool-season crops than by crops such as alfalfa or clovers grown in regular rotation.

The subject of fertilizer use in relation to irrigation requirement has recently had an excellent review (Viets, 1962). It was concluded that use of fertilizer to promote maximum plant growth increases evapotranspiration only slightly, if at all. Thus, irrigation requirement of an adequately nourished crop is, at most, only slightly increased over one whose growth is adversely affected by inadequate nutrition. This indicates a reduction in irrigation water required per unit of production and production of the quantity of the commodity needed with less total water.

LITERATURE CITED

Bowers, S. A., and R. J. Hanks. 1961. Effect of DDAC on evaporation and infiltration of soil moisture. Soil Sci. 92:340–346.

Eagleman, J. R., and V. C. Jamison. 1962. Soil layering and compaction effects on unsaturated moisture movement. Soil Sci. Soc. Amer. Proc. 26:519–522.

Greb, B. W., and A. L. Black. Feb. 1961. New stripcropping pattern saves moisture for dryland. Crops and Soils. 13:23.

Haise, H. R., F. G. Viets, Jr., and J. S. Robins. 1960. Efficiency of water use related to nutrient supply. Int. Congr. Soil Sci., Trans. 7th (Madison, Wis., USA) I:663–671.

Hanks, R. J. 1958. Water vapor transfer in dry soil. Soil Sci. Soc. Amer. Proc. 22:372–374.

Hanks, R. J., and N. P. Woodruff. 1958. Influence of wind on water vapor transfer through soil, gravel, and straw mulches. Soil Sci. 86:160–164.

Jensen, M. E., and H. R. Haise. Dec. 1963. Estimating evapotranspiration from solar radiation Amer. Soc. Civ. Eng. Irrig. Drainage Div. J. 89 (IR 4): 15–41.

Jensen, M. E., and R. J. Hildreth. May 1962. Rainfall at Amarillo, Texas. Texas Agr. Exp. Sta. Misc. Publ. No. 583. 9 p.

Lauritzen, C. W. 1960. Ground covers for collecting precipitation. Utah Farm Home Sci. 21:66–67.

Lemon, E. R. 1956. The potentialities for decreasing soil moisture evaporation loss. Soil Sci. Soc. Amer. Proc. 20:120–125.

Lemon, E. R. 1960. Photosynthesis under field conditions: II. An aerodynamic method for determining the turbulent carbon dioxide exchange between the atmosphere and a corn field. Agron. J. 52:697–703.

McCalla, T. M., and T. J. Army. 1961. Stubble mulch farming. Advance. Agron. 13:125–196.

Mech, S. J. 1960. Soil management as related to irrigation practices and irrigation design. Int. Congr. Soil Sci., Trans. 7th (Madison, Wis., USA) I:645–650.

Miller, D. E., and W. C. Bunger. 1963. Moisture retention by soils with coarse layers in the profile. Soil Sci. Soc. Amer. Proc. 27:586–589.

Miller, D. E., and W. H. Gardner. 1962. Water infiltration into stratified soil. Soil Sci. Soc. Amer. Proc. 26:115–119.

Moss, D., R. B. Musgrave, and E. R. Lemon. 1961. Photosynthesis under field conditions: III. Some effects of light, carbon dioxide, temperature, and soil moisture on photosynthesis, respiration, and transpiration of corn. Crop Sci. I:83–87.

Myers, L. E. 1961. Waterproofing soil to collect precipitation. J. Soil Water Conserv. 16:281–282.

Robins, J. S. 1957. Moisture movement and profile characteristics in relation to field capacity. Int. Comm. Irrig. Drainage, Trans. 3rd Congr., San Francisco, California. Questions for Discuss., Quest. 8, Report 26, p. 8.509–8.521.

Ross, P. Earl, and Norris P. Swanson. 1957. Level irrigation. J. Soil Water Conserv. 12:209–214.

Tanner, C. B., and E. R. Lemon. 1962. Radiant energy utilized in evapotranspiration. Agron. J. 54:207–212.

Viets, F. G., Jr. 1962. Fertilizers and the efficient use of water. Advance. Agron. 14:223–264.

Willis, W. O. 1960. Evaporation from layered soils in the presence of a water table. Soil Sci. Soc. Amer. Proc. 24:239–242.

Willis, W. O. 1962. Effect of partial surface covers on evaporation from soil. Soil Sci. Soc. Amer. Proc. 26:598–601.

Willis, W. O., C. W. Carlson, J. Alessi, and H. J. Haas. 1961. Depth of freezing and spring runoff as related to fall soil moisture level. Can. J. Soil Sci. 41:115–123.

Woodruff, N. P., R. A. Read, and W. S. Chepil. 1959. Influence of a field windbreak on summer wind movement and air temperature. Kansas Agr. Exp. Sta. Tech. Bull. 100. 24 p.

Zingg, A. W., and V. L. Hauser. 1959. Terrace benching to save potential runoff for semi-arid land. Agron. J. 51:289–292.

SUBJECT INDEX